REVIEWS IN MINERALOGY
AND GEOCHEMISTRY

VOLUME 58 2005

LOW-TEMPERATURE THERMOCHRONOLOGY: TECHNIQUES, INTERPRETATIONS, AND APPLICATIONS

EDITORS:

Peter W. Reiners — *Yale University, New Haven, Conneticut*

Todd A. Ehlers — *University of Michigan, Ann Arbor, Michigan*

COVER: ***Upper left***: Apatite crystals from the Bighorn Mountains, Wyoming (scale bar is 300 μm). ***Upper right***: Arrhenius plot for step-heating helium diffusion experiment on titanite crystal fragments; after Reiners PW, Farley KA (1999) He diffusion and (U-Th)/He thermochronometry of titanite. *Geochim Cosmochim Acta* 63:3845-3859. ***Lower left***: 3D thermo-kinematic model of the Himalayan front and major structures, Central Nepal; courtesy of D. Whipp and T. Ehlers. ***Lower right***: View of the Washington Cascades, from Sahale Arm; photo by Drew Stolar.

Series Editor: ***Jodi J. Rosso***

MINERALOGICAL SOCIETY OF AMERICA
GEOCHEMICAL SOCIETY

REVIEWS IN MINERALOGY AND GEOCHEMISTRY

(Formerly: REVIEWS IN MINERALOGY)

ISSN 1529-6466

Volume 58

Low-Temperature Thermochronology: Techniques, Interpretations, and Applications

ISBN 093995070-7

Additional copies of this volume as well as others in this series may be obtained at moderate cost from:

THE MINERALOGICAL SOCIETY OF AMERICA
3635 CONCORDE PARKWAY, SUITE 500
CHANTILLY, VIRGINIA, 20151-1125, U.S.A.
WWW.MINSOCAM.ORG

DEDICATION

Dr. William C. Luth has had a long and distinguished career in research, education and in the government. He was a leader in experimental petrology and in training graduate students at Stanford University. His efforts at Sandia National Laboratory and at the Department of Energy's headquarters resulted in the initiation and long-term support of many of the cutting edge research projects whose results form the foundations of these short courses. Bill's broad interest in understanding fundamental geochemical processes and their applications to national problems is a continuous thread through both his university and government career. He retired in 1996, but his efforts to foster excellent basic research, and to promote the development of advanced analytical capabilities gave a unique focus to the basic research portfolio in Geosciences at the Department of Energy. He has been, and continues to be, a friend and mentor to many of us. It is appropriate to celebrate his career in education and government service with this series of courses in cutting-edge geochemistry that have particular focus on Department of Energy-related science, at a time when he can still enjoy the recognition of his contributions.

LOW-TEMPERATURE THERMOCHRONOLOGY:

TECHNIQUES, INTERPRETATIONS, AND APPLICATIONS

58 *Reviews in Mineralogy and Geochemistry* **58**

FROM THE SERIES EDITOR

This volume, *Low-Temperature Thermochronology: Techniques, Interpretations, and Applications*, was prepared in advance of a short course of the same title, sponsored by MSA and presented at Snowbird, Utah, October 13-15, 2005 prior to the fall GSA meeting in Salt Lake City, Utah. The editors, Peter Reiners (Yale University) and Todd Ehlers (University of Michigan), carefully selected a diverse group of authors in order to produce this volume that assesses the current state-of-the-art in low-temperature thermochronology and provides a convienent context for evaluating advances in analytical and interpretation techniques, future potential, and outstanding issues in the field that have emerged in recent years. The editors and contributing authors have done an excellent job in generating this volume that should find broad use by researchers seeking to incorporate low-temperature thermochronologic constraints in their research.

Readers are encouraged to visit the MSA website (*http://www.minsocam.org/MSA/RIM/*) in order to access the computer programs outlined in Chapter 22. Errata (if any) can be found at the MSA website *www.minsocam.org*.

Jodi J. Rosso, Series Editor
West Richland, Washington
August 2005

PREFACE

The publication of this volume occurs at the one-hundredth anniversary of 1905, which has been called the *annus mirabilus* because it was the year of a number of enormous scientific advances. Among them are four papers by Albert Einstein explaining (among other things) Brownian motion, the photoelectric effect, the special theory of relativity, and the equation $E = mc^2$. Also of significance in 1905 was the first application of another major advance in physics, which dramatically changed the fields of Earth and planetary science. In March of 1905 (and published the following year), Ernest Rutherford presented the following in the Silliman Lectures at Yale:

> *"The helium observed in the radioactive minerals is almost certainly due to its production from the radium and other radioactive substances contained therein. If the rate of production of helium from known weights of the different radioelements were experimentally known, it should thus be possible to determine the interval required for the production of the amount of helium observed in radioactive minerals, or, in other words, to determine the age of the mineral."*
>
> *Rutherford E (1906) Radioactive Transformations. Charles Scriber's Sons, NY*

Thus radioisotopic geochronology was born, almost immediately shattering centuries of speculative conjectures and estimates and laying the foundation for establishment of the

1529-6466/05/0058-0000$05.00 DOI: 10.2138/rmg.2005.58.0

geologic timescale, the age of the Earth and meteorites, and a quantitative understanding of the rates of processes ranging from nebular condensation to Quaternary glaciations.

There is an important subplot to the historical development of radioisotopic dating over the last hundred years, which, ironically, arises directly from the subsequent history of the U-He dating method Rutherford described in 1905. Almost as soon as radioisotopic dating was invented, it was recognized that the U-He [or later the (U-Th)/He method], provided ages that were often far younger than those allowed by stratigraphic correlations or other techniques such as U/Pb dating. Clearly, as R.J. Strutt noted in 1910, He ages only provided "minimum values, because helium leaks out from the mineral, to what extent it is impossible to say" (Strutt, 1910, Proc Roy Soc Lond, Ser A 84:379-388). For several decades most attention was diverted to U/Pb and other techniques better suited to measurement of crystallization ages and establishment of the geologic timescale.

Gradually it became clear that other radioisotopic systems such as K/Ar and later fission-track also provided ages that were clearly younger than formation ages. In 1910 it may have been impossible to say the extent to which He (or most other elements) leaked out of minerals, but eventually a growing understanding of thermally-activated diffusion and annealing began to shed light on the significance of such ages. The recognition that some systems can provide cooling, rather than formation, ages, was gradual and diachronous across radioisotopic systems. Most of the heavy lifting in this regard was accomplished by researchers working on the interpretation of K/Ar and fission-track ages.

Ironically, Rutherford's He-based radioisotopic system was one of the last to be quantitatively interpreted as a thermochronometer, and has been added to K/Ar (including $^{40}Ar/^{39}Ar$) and fission-track methods as important for constraining the medium- to low-temperature thermal histories of rocks and minerals.

Thermochronology has had a slow and sometimes fitful maturation from what were once troubling age discrepancies and poorly-understood open-system behaviors, into a powerful branch of geochronology applied by Earth scientists from diverse fields. Cooling ages, coupled with quantitative understanding of crystal-scale kinetic phenomena and crustal- or landscape-scale interpretational models now provide an enormous range of insights into tectonics, geomorphology, and subjects of other fields. At the same time, blossoming of lower temperature thermochronometric approaches has inspired new perspectives into the detailed behavior of higher temperature systems that previously may have been primarily used for establishing formation ages. Increased recognition of the importance of thermal histories, combined with improved analytical precision, has motivated progress in understanding the thermochronologic behavior of U/Pb, Sm/Nd, Lu/Hf, and other systems in a wide range of minerals, filling out the temperature range accessible by thermochronologic approaches. Thus the maturation of low- and medium-temperature thermochronology has led to a fuller understanding of the significance of radioisotopic ages in general, and to one degree or another has permeated most of geochronology.

Except in rare cases, the goal of thermochronology is not thermal histories themselves, but rather the geologic processes responsible for them. Thermochronometers are now routinely used for quantifying exhumation histories (tectonic or erosional), magmatism, or landscape evolution. As thermochronology has matured, so have model and interpretational approaches used to convert thermal histories into these more useful geologic histories. Low-temperature thermochronology has been especially important in this regard, as knowledge of thermal processes in the uppermost few kilometers of the crust require consideration of coupled interactions of tectonic, geodynamic, and surface processes. Exciting new developments in these fields in turn drive improved thermochronologic methods and innovative sampling approaches.

The chapters

This volume presents 22 chapters covering many of the important modern aspects of thermochronology. The coverage of the chapters ranges widely, including historical perspective, analytical techniques, kinetics and calibrations, modeling approaches, and interpretational methods. In general, the chapters focus on intermediate- to low-temperature thermochronometry, though some chapters cover higher temperature methods such as monazite U/Pb closure profiles, and the same theory and approaches used in low-temperature thermochronometry are generally applicable to higher temperature systems. The widely used low- to medium-temperature thermochronometric systems are reviewed in detail in these chapters, but while there are numerous chapters reviewing various aspects of the apatite (U-Th)/He system, there is no chapter singularly devoted to it, partly because of several previous reviews recently published on this topic.

Chapter 1 by Reiners, Ehlers, and Zeitler provides a perspective on the history of thermochronology, comments on modern work in this field and general lessons on the potential for noise to be turned into signal. This chapter also provides a summary of the current challenges, unresolved issues, and most exciting prospects in the field.

Much of the modern understanding of kinetic controls on apparent ages, thermal histories, and sampling approaches comes from decades of progress in fission-track dating, a method that remains as essential as ever, partly because of the power of track-length measurements and the depth of (at least empirical) understanding of the kinetics of track annealing. Tagami, Donelick and O'Sullivan review the fundamentals of modern fission-track dating (Chapter 2). Two of the most commonly dated, well-understood, and powerful minerals dated by fission-track methods are apatite and zircon, and the specifics of modern methods for these systems and their kinetics are reviewed by Donelick, O'Sullivan, and Ketcham (Chapter 3), and Tagami (Chapter 4).

Although $^{40}Ar/^{39}Ar$ and (U-Th)/He dating methods followed somewhat different paths to their modern thermochronologic incarnations, they have many features in common, especially in the kinetics of diffusion and closure. Zeitler and Harrison review the concepts underlying both $^{40}Ar/^{39}Ar$ and (U-Th)/He methods (Chapter 5). Zircon was one of the first minerals dated by the (U-Th)/He method, but has only just begun to be used for thermochronometry of both bedrock and detrital samples, as reviewed by Reiners (Chapter 6). Continuous time-temperature paths from intracrystalline variations of radiogenic Ar proven perhaps the most powerful of all thermochronologic approaches, and an innovative analogous approach in He dating ($^{4}He/^{3}He$ thermochronometry) is revealing remarkably powerful constraints on the extreme low temperature end of thermal histories, as reviewed by Shuster and Farley (Chapter 7).

Thermochronology of detrital minerals provides unique constraints on the long-term evolution of orogens, sediment provenance, and depositional age constraints, to name a few. Bernet and Garver (Chapter 8) review the essentials of detrital zircon fission-track dating, one of the most venerable and robust of detrital thermochronometers, and in Chapter 9, Hodges, Ruhl, Wobus, and Pringle review the use of $^{40}Ar/^{39}Ar$ dating of detrital minerals, demonstrating the power of detrital muscovite ages in illuminating variations in exhumation rates in catchments over broad landscapes.

(U-Th)/He thermochronometry presents several unique interpretational challenges besides new kinetics and low temperature sensitivity. One of these is long-alpha stopping distances, and its coupling with diffusion and U-Th zonation in age corrections. Dunai reviews modeling approaches to deal with these issues in interpreting low-temperature thermal histories (Chapter 10). Ketcham (Chapter 11) reviews the theory and calibration of both forward and inverse models of thermal histories from fission-track and (U-Th)/He data, and makes some important points about the interpretations of such models.

Translating thermal histories into exhumational histories and their tectonic or geomorphic significance across a landscape requires quantitative understanding of the thermal structure of the crust and how it is perturbed, a review of which is presented by Ehlers (Chapter 12). Braun (Chapter 13) illustrates the power of low-temperature thermochronometry to constrain topographic evolution of landscapes over time, using PECUBE. Gallagher, Stephenson, Brown, Holmes, and Ballester present a novel method of inverse modeling of fission-track and (U-Th)/He data for thermal histories over landscapes (Chapter 14).

Continuous time-temperature paths from closure profiles or their step-heating-derived equivalents are, to some degree, the holy grail of thermochronology. Harrison, Zeitler, Grove, and Lovera (Chapter 15) provide a review of the theory, measurement, and interpretation of continuous thermal histories at both intermediate and high temperatures, derived from both K-feldspar $^{40}Ar/^{39}Ar$ and monazite U/Pb dating.

Extensional orogens provide a special challenge and opportunity for thermochronometry because tectonic exhumation by footwall unroofing often outstrips erosional exhumation, and often occurs at high rates. As Stockli shows (Chapter 16) thermochronology in these setting provides opportunities to measure rates of a number of important processes, as well as obtain a snapshot of crustal thermal structure and its imprint on thermochronometers with varying closure temperatures. Spotila (Chapter 17) reviews the use of thermochronology applied to tectonic geomorphology in a wide range of orogenic settings, introducing the concept of denudational maturity.

Thermochronology has found great utility in economic geology, and newly developed approaches pose great potential in this area, and shown by McInnes, Evans, Fu, and Garwin in their review of the use and modeling of thermochronology of hydrothermal ore deposits (Chapter 18). The thermal histories of sedimentary basins are also critical to understanding thermal maturation of hydrocarbons, but are also critical for understanding basin formation, erosional histories of source regions, fluid flow, and climate change and other temporal signals preserved in sedimentary rocks. Armstrong (Chapter 19) reviews these issues and the use of thermochronology in deducing the thermal histories of sedimentary basins. Drawing on large datasets of bedrock apatite fission-track dates, Kohn, Gleadow, Brown, Gallagher, Lorencak, and Noble demonstrate the power of modeling, and, importantly, effectively visualizing, integrated thermotectonic and denudational histories over large regions (Chapter 20).

Thermal histories of meteorites provide constraints on a wide range of fundamentally important processes, including nebular condensation and early solar-system metamorphic histories, and the dynamics of interplanetary collisions and shock metamorphism. Min reviews thermochronologic approaches to understanding meteorite thermal histories (Chapter 21), including new methods and approaches.

Finally, the importance of robust models with which to interpret thermochronologic data is underscored by the review of the Software for Interpretation and Analysis of Thermochronologic Data (Chapter 22), summarized and compiled by Ehlers, for programs associated with the work of authors in this volume and others. The programs outlined in this chapter are available for download through the RiMG Series website: *http://www.minsocam.org/MSA/RIM/*.

Acknowledgments

There are numerous individuals and groups deserving of thanks for this *Thermochronology* RiMG volume for their efforts from the point of initial conception through final publication the convening of the shortcourse.

We gratefully acknowledge the generous financial support of the U.S. Department of Energy, and National Science Foundation in sponsoring the short course, as well as contributions from Apatite-to-Zircon, Inc., the Yale Department of Geology and Geophysics, and the University of Michigan Department of Geological Sciences.

We appreciate helpful chapter reviews from all the reviewers, whose efforts helped make this a real community project. They include: Bruce Idleman, Barry Kohn, Diane Seward, Kerry Gallagher, Richard Stewart, John Garver, Ethan Baxter, Danny Stockli, Raphael Pik, Andy Carter, Max Zattin, Mike Cosca, Fin Stuart, Cristina Persano, Marlies ter Voorde, Dale Issler, Ed Sobel, Kevin Furlong, Phil Armstrong, Jean Braun, Mark Brandon, Peter Van der Beek, Terry Spell, Kip Hodges, Uwe Ring, Annia Fayon, Laura Webb, Anthony Harris, Ken Hickey, Peter Kamp, Paul O'Sullivan, Geoff Batt, Paul Andriessen, Mario Trieloff, Francis Albarède, Phaedra Upton, Stuart Thomson, Andrew Meigs, Barbara Carrapa, Tibor Dunai, Ann Blythe, and the anonymous reviewers.

We appreciate the patient and wise counsel of Alex Speer who helped us navigate the logistics of the volume and shortcourse; John Valley also provided helpful encouragement and advice in the early stages. We also thank Jodi Rosso for thoroughly professional and efficient editorial handling of the production phase of the volume. We appreciate the perspective and advice of Peter Zeitler has provided to us in putting together this volume and shortcourse, and are grateful to him and all the other veterans of the field who have encouraged and inspired us and many other researchers to enter into the great thermochronologic conversation. We hope that this volume will likewise encourage continued innovation and discoveries in this field that synergizes the work and imaginations of geochemists, geodynamicists, tectonicists, geomorphologists, and others.

August 2005

Peter W. Reiners	***Todd A. Ehlers***
New Haven, CT	Ann Arbor, MI

TABLE OF CONTENTS

1 Past, Present, and Future of Thermochronology

Peter W. Reiners, Todd A. Ehlers, Peter K. Zeitler

2 Fundamentals of Fission-Track Thermochronology

Takahiro Tagami, Paul B. O'Sullivan

3 Apatite Fission-Track Analysis

Raymond A. Donelick, Paul B. O'Sullivan
Richard A. Ketcham

4 Zircon Fission-Track Thermochronology and Applications to Fault Studies

Takahiro Tagami

5 Fundamentals of Noble Gas Thermochronometry

T. Mark Harrison, Peter K. Zeitler

6 Zircon (U-Th)/He Thermochronometry

Peter W. Reiners

7 $^{4}He/^{3}He$ Thermochronometry: Theory, Practice, and Potential Complications

David L. Shuster, Kenneth A. Farley

8 Fission-track Analysis of Detrital Zircon

Matthias Bernet, John I. Garver

9 $^{40}Ar/^{39}Ar$ Thermochronology of Detrital Minerals

K.V. Hodges, K.W. Ruhl, C.W. Wobus, M.S. Pringle

10 Forward Modeling and Interpretation of (U-Th)/He Ages

Tibor J. Dunai

11 Forward and Inverse Modeling of Low-Temperature Thermochronometry Data

Richard A. Ketcham

12 Crustal Thermal Processes and the Interpretation of Thermochronometer Data

Todd A. Ehlers

13 Quantitative Constraints on the Rate of Landform Evolution Derived from Low-Temperature Thermochronology

Jean Braun

14 Exploiting 3D Spatial Sampling in Inverse Modeling of Thermochronological Data

Kerry Gallagher, John Stephenson, Roderick Brown, Chris Holmes, Pedro Ballester

15 Continuous Thermal Histories from Inversion of Closure Profiles

T. Mark Harrison, Marty Grove, Oscar M. Lovera, Peter K. Zeitler

16 Application of Low-Temperature Thermochronometry to Extensional Tectonic Settings

Daniel F. Stockli

17 Applications of Low-Temperature Thermochronometry to Quantification of Recent Exhumation in Mountain Belts

James Spotila

18

Application of Thermochronology to Hydrothermal Ore Deposits

Brent I. A. McInnes, Noreen J. Evans, Frank Q. Fu, Steve Garwin

19 Thermochronometers in Sedimentary Basins

Phillip A. Armstrong

20 Visualizing Thermotectonic and Denudation Histories Using Apatite Fission Track Thermochronology

Barry P. Kohn, Andrew J.W. Gleadow, Roderick W. Brown, Kerry Gallagher, Matevz Lorencak, Wayne P. Noble

21 Low-Temperature Thermochronometry of Meteorites

Kyoungwon Min

22 Computational Tools for Low-Temperature Thermochronometer Interpretation

Todd A. Ehlers, Tehmasp Chaudhri, Santosh Kumar, Chris W. Fuller, Sean D. Willett, Richard A. Ketcham, Mark T. Brandon, David X. Belton, Barry P. Kohn, Andrew J.W. Gleadow, Tibor J. Dunai, Frank Q. Fu

Reviews in Mineralogy & Geochemistry
Vol. 58, pp. 1-18, 2005

Past, Present, and Future of Thermochronology

Peter W. Reiners

Dept. of Geology and Geophysics
Yale University
New Haven, Connecticut, 06520, U.S.A.
peter.reiners@yale.edu

Todd A. Ehlers

Dept. of Geological Sciences
University of Michigan
Ann Arbor, Michigan, 48109, U.S.A.
tehlers@umich.edu

Peter K. Zeitler

Dept. of Earth and Environmental Science
Lehigh University
Bethlehem, Pennsylvania, 18015, U.S.A.
peter.zeitler@lehigh.edu

INTRODUCTION

In one form or another, geochronologists have been practicing thermochronology[1], the use of radioisotopic dating to constrain thermal histories of rocks and minerals, for over 40 years. Building from lessons learned over these four decades, thermochronology continues to evolve due to technical developments, increasingly sophisticated theoretical models, and an expanding range of applications in geologic and planetary science. Most recently, interest in earth-surface processes and interactions between tectonics, erosion, and climate has drawn attention to techniques that can address the timing and rates of processes operating at temperatures below about 300 °C.

The purpose of this RiMG volume is to assess the current state of thermochronology, as of *circa* 2005, which is, coincidentally, the 100th anniversary of the first radioisotopic date (Rutherford 1905; 1906). Excellent review papers and books on specific topics within this field have been published, but no single volume has yet provided a comprehensive review of current practices, basic theory, and illustrative examples. The motivation for this volume stems from these considerations. Knowing that in a fast-developing field a book like this can

[1] Several nomenclative conventions have evolved around variations of the term "thermochronology." We consider the following most appropriate: 1) *Thermochronometer*: a radioisotopic system consisting of radioactive parent, radiogenic daughter or crystallographic feature, and the mineral in which they are found. 2) *Thermochronometry*: the analysis, practice, or application of a thermochronometer to understand thermal histories of rocks or minerals. 3) *Thermochronology*: The thermal history of a rock, mineral, or geologic terrane. In practice, however, *thermochronology* is often also used to denote the study of *thermochronologies*, in which case it is synonymous with *thermochronometry*. The use of the term *thermochronology* in the latter sense is too common and deeply ingrained in the community (and parallel with conventional usage of geochronology), to attempt any corrective usage recommendation (e.g., the title of this book is *Thermochronology*, referring to the study of, not specific, *thermochronologies*).

1529-6466/05/0058-0001$05.00 DOI: 10.2138/rmg.2005.58.1

quickly become dated, we tried to include sufficient review of fundamentals and the literature to offer students and new users a useful introduction to thermochronology that may have some staying power.

In this chapter, we first review the salient points of thermochronology's history before assessing our current capabilities and challenges and then taking the risk of suggesting where the field is headed. We do not provide a comprehensive history of the method that does full justice to the work of the large and growing cohort of thermochronologists. In this short space, we instead opted to give our perspectives on where the intellectual and technological roots of the discipline lie, which run deeper and go back farther than is sometimes appreciated.

Geochronology vs. thermochronology

Given that all radioisotopic systems are subject to disturbance and resetting at sufficiently high temperatures, it might be surmised that all radioisotopic dating is essentially thermochronology. Distinctions between geo- and thermochronology are indeed often fuzzy, but thermochronology is largely different in several ways. Some phases form or are stable only at temperatures much lower than the closure temperature for the isotopic system of interest (see Harrison and Zeitler 2005), effectively disqualifying them as thermochronometers. Zircon can form or dissolve in felsic magmas at temperatures where Pb diffusion rates are negligibly slow, for example, and in such cases it provides no useful thermal history information. Authigenic phosphates in supergene deposits provide an analogous example at low temperatures.

The nature of thermochronologic and geochronologic questions are also often fundamentally different. The formation age of minerals is typically irrelevant in thermochronology, whereas rates of processes are often of paramount interest. The numerical values of thermochronologic ages across an orogen, for example, may have little to no geologic significance aside from their inverted value in estimating steady-state cooling (and inferred exhumation) rates and their spatial variations. Geochronologic applications, on the other hand, aim exclusively to determine a singular absolute stratigraphic or magmatic formation age, with little direct concern for durations or rates of processes. There are exceptions to these generalizations. Thermochronology can estimate the absolute timing of events, such as bolide impacts or magmatic processes that may only reset low-temperature systems, and geochronology may estimate rates of processes such as evolutionary change or landscape fluvial incision by bracketing formation ages around paleontologic or geomorphic features. Despite this overlap, the bottom line is that thermochronology is distinguished from geochronology by its ability to resolve both temporal and thermal aspects of geologic processes, and thus both timing and rates of processes.

HISTORY

1950s and 1960s – development of fundamentals

The field of radioisotopic geochronology is now a century old (e.g., Rutherford 1905, 1906), and the understanding that diffusion is a means of resetting or perturbing ages is not a new idea either (e.g., Hurley 1954 and references therein). In the scientific boom years following the second world war, many ages were measured by an expanding range of techniques. One immediate observation was that many of the ages obtained by different techniques on the same samples did not agree and many were too young to represent formation ages. In general, geochronologists and petrologists were acutely aware that the measurements they were making might be dating processes other than rock formation, with diffusion and thermal resetting being major suspects, although radiation damage was also considered a potential culprit. This can be seen in the work of Patrick Hurley and his interpretations of previous studies of He retention in a wide variety of minerals (Hurley 1954), in the work of Paul Damon and colleagues on He retentivity in zircon (Damon and Kulp 1957), in examinations of Ar diffusion in various rock-forming minerals (Evernden et al. 1960; Fechtig and Kalbitzer 1966; Musset 1969), and in Richard Armstrong's notion of a "metamorphic veil" (Armstrong 1966). Thermal resetting was also directly or indirectly invoked in a number of early field studies including Mason (1961) and Hurley et al. (1965) on mineral ages near the Alpine Fault, Hart (1964) and Hanson and Gast (1967) in examining suites of mineral ages near contact aureoles, and Westcott (1966), Harper (1967), Jäger et al. (1967), and Dewey and Pankhurst (1970) in interpreting regional suites of mineral ages from orogens.

Although thermal effects on radioisotopic ages were generally recognized, they received little quantitative attention in the early part of the 20th century and throughout the 1950s, as technological developments allowed both the U-Pb and K-Ar method to develop through advances in chemical methods and static gas mass spectrometry. In hindsight, given geochronologists' focus at that time in determining reliable formation ages rather than thermal histories, and given what we now know to be the greater retentivity of minerals for Pb and Ar compared to He, it is clear why these methods became dominant at a time when U-He workers were struggling with ages that were often too young. In any event, this period saw the rapid development of U-Pb and K-Ar dating and the abandonment of the U-He method. Other important technical developments towards the end of this period include the development of fission-track dating of geological materials by Chuck Naeser and Gunther Wagner (e.g., Naeser 1967; Wagner 1968; Naeser and Faul 1969), development of the Ar-Ar method [see McDougall and Harrison (1999) for a full account], and the impetus given to geochemistry by the lunar program.

By the late 1960s, all of the basic techniques that we use today were in fact in existence and most were under active development, with the exception of (U-Th)/He dating, which was seeing only sporadic use. Geochronologists were aware that high temperatures and diffusion could reset ages, they were conducting laboratory and field studies to study this phenomenon, and they were beginning to use mineral ages to constrain orogenic processes. The stage was clearly set for the development of modern thermochronology.

1970s – a decade of closure

Three developments in the 1970s were to prove essential to the birth of modern thermochronology. First, E. Jager and colleagues including Gunther Wagner, having accumulated considerable numbers of mineral ages from the Alps, concluded that they were recording the thermal history of the region, and published papers in support of this conclusion (e.g., Purdy and Jäger 1976; Wagner et al. 1977). Further, they used petrological data and petrogenetic grids to assign temperature values to specific isotopic systems. This directly leads to the notion of dating suites of minerals to establish thermal histories. Over the same interval, Chuck Naeser and colleagues were applying the fission-track method to a variety of geological settings of known thermal history such as boreholes and contact aureoles and relating their results to laboratory annealing experiments (Calk and Naeser 1973; Naser and Forbes 1976), and Berger (1975) and Hanson et al. (1975) revisited the studies of Hart (1964) and Hanson and Gast (1965) for K-Ar and Ar-Ar data. Finally and most significantly, in 1973, Martin Dodson published his landmark paper introducing the concept of closure temperature (defined as the temperature of a system at the time given by its apparent age), thereby providing a clear theoretical basis for understanding many mineral ages as cooling ages owing to the interplay between the kinetics of diffusion (or annealing) and accumulation rates in cooling radioisotopic systems (Dodson 1973, 1979; see discussion in Harrison and Zeitler 2005).

1980s – modern thermochronology is born

The first literature use of the word "thermochronology" appears in a paper by Berger and York (1981). By modern thermochronology, we mean the explicit use of kinetic data to interpret suites of isotopic ages in terms of thermal histories. During this period, Andy Gleadow and the Melbourne group pursued improvements in the interpretation of fission-track data, incorporating the use of confined track length data into their kinetic models for apatite data, and developing the notion of the partial annealing zone (PAZ) (Gleadow et al. 1993), which has been adapted to the diffusive context with the notion of the partial retention zone (PRZ) (Baldwin and Lister 1998; Wolf et al. 1998). Concurrently Mark Harrison put Ar-Ar thermochronology on a firm theoretical footing in a series of papers starting with a precocious undergraduate thesis devoted to the cooling history of the Quottoon pluton and extending through a series of contributions spanning his PhD work with Ian McDougall at the Australian National University (e.g., Harrison and Clarke 1979; Harrison et al. 1979, 1985; Harrison and McDougall 1980a,b, 1981, 1982). Martin Dodson extended his ideas about closure to include the notion of closure profiles (Dodson 1986). The end of the decade saw publication of papers which set the stage for the development of the multidomain model for K-feldspar age spectra (Gillespie et al. 1982; Zeitler 1987; Lovera et al. 1989, 1991; Richter et al. 1991), and also a paper on He diffusion in, and (U-Th)/He dating of, apatite (Zeitler et al. 1987) that engendered an appreciation for the potential utility of the method in many geologic applications. Finally, the interval spanning the 1980's some of the first applications of "detrital thermochronology" were attempted (Wagner et al. 1979; Zeitler et al. 1986, Cerveny et al. 1988).

During the 1980s some backlash and confusion developed over the significance of ages from thermally sensitive systems. Dodsonian closure temperatures, which are strictly only relevant for samples that have cooled montonically from high temperature and that have known kinetic properties and diffusion length scales, were often portrayed more like magnetic blocking temperatures; in fact many people at the time (and some still do) used "blocking temperature" to refer to isotopic closure. While the original choice of terms might have been arbitrary, at this point it is important to distinguish between systems with such high activation energies that in effect have a single "off-on" blocking temperature (like remagnetization of single-domain magnetite), and isotopic systems with more modest activation energies which allow both time and temperature to play significant roles in daughter isotope retention or fission-track annealing. The term closure properly serves to remind us that the conceptualization only applies to systems that have closed due to cooling; closure temperature has little to no relevance to interpretations of samples whose ages primarily reflect thermal histories involving reheating or prolonged isothermal stagnation.

Finally, the 1980s saw an increasing number of applied papers using mineral ages to constrain and solve tectonic problems. Most applications focused on quantifying the timing of geologic events such as when faults became active within an orogen, or the timing of exhumation. Many studies also used thermochronometers to estimate rates of tectonic processes (e.g., rates of exhumation) by assuming steady-state 1D thermal gradients and a closure temperatures. The later third of this volume discusses applications of thermochronometers to different settings and summarizes work by many different groups over the last 2 decades.

1990s and 2000s

One of the most important developments in the 1990s was the theoretical maturation of the multidomain model for K-feldspar $^{40}Ar/^{39}Ar$ age spectra, largely by the group at UCLA, and its application in a large number of regional studies (Harrison et al. 1991, 1993; Fitzgerald and Harrison 1993; Lovera et al. 1993, 1997, 2002). Time-temperature models provided by K-feldspar $^{40}Ar/^{39}Ar$ methods are in some sense the holy grail of thermochronology in that they provide continuous histories, approximations to which are often the focus of laborious efforts of multiple thermochronometers providing single ages corresponding (theoretically) to a single temperature. K-feldspar $^{40}Ar/^{39}Ar$ dating continues to attract wide use, as well as occasional theoretical criticism (e.g., Villa 1994; Parsons et al. 1999), the merits of which are debated. Provided certain sample and analytical criteria are met (Lovera et al. 2002), there is as yet no convincing empirical evidence suggesting problems with the theory or application of multidomain K-feldspar $^{40}Ar/^{39}Ar$ thermochronometry (Harrison et al. 2005).

The 1990s also saw the pioneering work of Ken Farley and coworkers in bringing (U-Th)/He geochronology of apatite and other minerals to become routine measurements capable of exciting new applications (e.g., Farley et al. 1996; Wolf et al. 1996, 1998). Over the same time period, widespread adoption of laser heating for gas extraction (e.g., Kelley et al. 1994; House et al. 2000) and automation in noble-gas laboratories greatly reduced the effort required to obtain data, providing increased throughput, greater quality control, and lower system blanks. From a phenomenological standpoint, it is not clear whether lasers have increased insight into the workings of thermochronometers, as the nature of diffusion boundaries within mineral grains remains cryptic and often smaller and more complex than can be internally sampled using a laser on a routine basis (although there are exceptions in which crystal and domain sizes scale similarly and can be used to model thermal histories; Hess et al. 1993; Hawkins and Bowring 1999; Reiners and Farley 2001). Nevertheless, this area remains a frontier in which there is a great deal of interest and ongoing work.

A potentially promising development in radiogenic He chronometry is reminiscent of the shift from K/Ar to Ar/Ar dating in the 1960s and 70s: the bombardment of samples by high-energy protons to form abundant and homogeneously distributed ^{3}He provides the opportunity to simultaneously degas ^{3}He and ^{4}He and examine, with high precision, the internal distribution of ^{4}He within crystals (Shuster et al. 2003; Shuster and Farley 2003, 2005). Modeling intradomainal ^{4}He profiles allows detailed constraints on subtle but critical features of cooling in temperature ranges near and significantly below those of the closure temperature of the system of interest. The method has also been used to date formation of weathering horizons (Shuster et al. 2005).

Another development in this period has been the building of more complex numerical models used for interpretation of thermochronometer data. For example, forward and inverse models of thermochronologic data became available and widely used (e.g., Laslett et al. 1987; Gallagher 1995; Ketcham 1999, 2005; Willett 1997). Furthermore, forward modeling of crustal thermal fields to aid in interpretation of thermochronometer data also became more common. It is worth noting that the influence of topography, erosion, and faulting has long been appreciated in the geothermics community, going back to early publications by Lees (1910), Bullard (1938) and Benfield (1949). Unfortunately, much of the geothermics literature has been under utilized by thermochronologists until work by Parrish (1985) provided a fairly complete analysis of the implications of erosion on geotherms and thermochronometer interpretation.

In the last decade modeling approaches have moved beyond 1D applications to consider the influence of 2D and 3D heat transfer on thermochronometer age interpretation. Recognition in the thermochronology community that thermal gradients in the upper 1–5 km are spatially variable within orogens due to topography, faulting, and other processes has sparked interest in using thermochronometer data to quantify the rates of tectonics and erosional processes. As a results of this interest, thermochronometer data are now increasingly used to constrain thermal-kinematic and geodynamic models of tectonic (e.g., Rahn and Grasemann 1999; Batt et al. 2001; Beaumont et al. 2001; Ehlers 2005), topographic, and erosional processes (e.g., Stuwe et al. 1994; Mancktelow and Grasemann 1997; Ehlers and Farley 2003; Braun 2005). A more thorough discussion and overview of modeling procedures for thermochronometer grain scale and crustal processes are presented in this volume.

CURRENT PRACTICE

At present, by far the most commonly used thermochronometers are $^{40}Ar/^{39}Ar$ in micas and amphiboles (e.g., McDougall and Harrison 1999; Kelley 2002), and in K-feldspar (Harrison et al. 2005), fission-tracks in apatite (e.g., Donelick et al. 2005) and zircon (e.g., Tagami et al. 2005), and (U-Th)/He in apatite (e.g., Farley 2002) and zircon (e.g., Reiners 2005). These techniques have typical closure temperatures ranging from as high as 400–600 °C to as low as 60–70 °C (Table 1). Other thermochronometers used less commonly include fission-tracks in titanite (e.g., Coyle and Wagner 1998), (U-Th)/He in titanite (Reiners and Farley 1999; Stockli and Farley 2004), and monazite (Farley and Stockli 2002), and U/Pb or Th/Pb in monazite (Harrison et al. 2005), apatite (Chamberlain and Bowring 2001), titanite (Schmitz and Bowring 2003), and other phases. Thermochronologic constraints from U/Pb and Th/Pb work on accessory phases is seeing increasing use owing to the greater number of ion probes now available for radioisotopic dating, as well as advances in understanding of Pb diffusion (e.g., Cherniak 1993; Cherniak and Watson 2001) and closure profiles within minerals (Harrison et al. 2005).

Although not a focus of this volume, it is worth noting that the Rb/Sr (e.g., Jenkin et al. 2001), Sm/Nd, and Lu/Hf systems have also seen increasing use as thermochronometers, with the latter two finding increasing use in garnet (Scherer et al. 2001; Ducea et al. 2003) and apatite (Barfod et al. 2002). In general, the Sm/Nd, Lu/Hf, and (U-Th)/Pb systems provide relatively high temperature thermochronologic constraints (>450 °C). Their utility as thermochronometers has been facilitated in part by increased temporal resolution of high-temperature portions of time-temperature paths that have come from both increased precision and numbers of phases dated in the same rocks. The recognition that distinct systems provide ages that are consistently resolvable has helped extend the reach of thermochronology to higher temperatures (e.g., Hawkins and Bowring 1999).

In practice, given the sorts of minerals that yield consistent results and their fairly low closure temperatures, almost all thermochronological studies have been directed at the more felsic rocks of the continental crust. Minerals suitable for dating are also found in meteorites (e.g., Min 2005) and, perhaps more commonly than realized, in oceanic crust (e.g., John et al. 2004), and thermochronometry poses potential for understanding a range of processes in these settings as well.

The various papers in this volume discuss the systematics and kinetics of different mineral systems in more detail, but to provide an overview, Table 1 summarizes the approximate temperature ranges and time-temperature responses of thermochronometer systems commonly used today [also see Hodges (2003) for a more complete summary].

It is beyond the scope of this chapter to explore how well various thermochronometers perform when compared. There has been little if any community effort devoted to controlled comparisons of diffusion measurements or in developing standards to facilitate this. Although laboratories tend to use similar technologies, this is not universally the case, and most thermochronologists know of the pitfalls that can afflict, for example, the accuracy of temperature measurements made using thermocouples or pyrometers. Fortunately, the results from the numerous applied studies that have now been done suggest that to at least first order, we know the relative performance of mineral systems fairly well, as expressed by the closure temperatures listed in Table 1. It is extremely important to understand that the closure temperatures listed in this table serve as a useful shorthand for representing the "retentivity" of a system, but that in the actual case of reheating, these temperatures have little to no significance, the response of the system being a function of both the duration and magnitude of the heating event.

Table 1. Summary of commonly used thermochronometers and features.

Decay System	*Mineral*		*Approximate precision* (%, 1σ)	*Closure Temperature* (°C)	*Activation Energy* (kJ/mol)	*References*
(U-Th)/Pb	zircon		1–2	>900	550	Cherniak and Watson (2001); Cherniak (2001)
	titanite		1–2	550–650	330	Cherniak (1993)
	monazite		1–2	~700	590	Cherniak et al. (2004)
	apatite		1–2	425–500	230	Chamberlain and Bowring (2001); Cherniak et al. (1991)
$^{40}Ar/^{39}Ar$	hornblende		1	400–600	270	Harrison (1981); Dahl (1996)
	biotite		1	350–400	210	Grove and Harrison (1996); Harrison et al. (1985)
	muscovite		1	300–350	180	Robbins (1972); Hames and Bowring (1994)
	K-feldspar		1	150–350	170–210	Foland (1994); Lovera et al. (1991; 1997)
Fission-track	titanite		6	(a) 240–300 (b) 380–420	440–480	(a) Coyle and Wagner (1998); (b) Watt and Durrani (1985); Naeser and Faul (1969)
	zircon	(a) zero-damage (b) "natural"	6	(a) 330–350 (b) 230	(a) 300–350 (b) 210	(a) Tagami et al. (1998); Rahn et al. (2004) (b) Brandon and Vance (1992); Brandon et al. (1998)
	apatite		8	90–120	190	Laslett et al. (1987); Ketcham et al. (1999)
(U-Th)/He	titanite		3–4	160–220	190	Reiners and Farley (1999)
	zircon		3–4	160–200	170	Reiners et al. (2004)
	apatite		3–4	55–80	140	Farley (2000)

Note: Approximate precisions are estimated values for age determinations; for TIMS U/Pb measurements precisions can be considerably better than cited here. Closure temperatures calculated using Dodson (1973) [or, for fission-track, Dodson (1979) using the 50% annealing isopleth (fanning models); also see Brandon et al. (1998)] using typical ranges of grain sizes and cooling rates (1–100 °C/m.y.) (small grains/low cooling rate and large grains fast/cooling rate). Also see Hodges (2003) for a similar and more complete compilation.

At the simplest level, thermochronometric data are used to determine the time-temperature history of a sample (e.g., Ketcham et al. 2005). After a thermal history is determined, numerical or analytical thermal models or simply geologic constraints can be used to interpret the processes responsible. Thermal models can be used to simulate the exhumation, and/or burial, history of the thermochronometric record as a function of geologic processes that are free parameters in the model. In practice, the range of geologic processes simulated in this type of model can be large and encompass geomorphic, faulting, magmatic, thermo-physical (e.g., basal heat flow) and basin evolution processes. Model thermal histories can be compared to observed thermal histories to determine the combinations of model parameters that provide a good fit to the data.

Unfortunately, interpretations of thermochronometric data are seldom unique because different combinations of model parameters (e.g., basal heat flow and erosion rate) can produce an equally good fit to the data. Thus, a rigorous interpretation of a data set usually results in the identification of *the range* of solutions that satisfy the observation rather than a single solution. Unfortunately, all too often studies that attempt to quantify, for example, the kinematic history of a fault fail to report the range of solutions that satisfy a thermochronometric data set.

Most recently, attempts have been made to couple crustal scale thermal models with increasingly complex process based models to better understand rates of landscape evolution, or the dynamics of orogenesis. The coupling of thermal and landform evolution and/or geodynamics models with thermochronometry pushes the utility of thermochronometric data to its limit. Coupled models introduce many more free parameters and require a careful evaluation of these parameters as well as large, carefully sampled data sets to ascertain meaningful results. The next decade will undoubtedly reveal the limit of thermochronometric data to quantify processes such as the evolution of drainage basins and/or couplings between climate and tectonics.

PROSPECTS

Existing and emerging techniques and approaches

It is likely that the recent trend towards highly automated sample processing and smaller sample sizes will increase sample throughput from laboratories, and this will be aided by lower costs, (e.g., if inexpensive quadrupole mass spectrometers can be shown to serve adequately for Ar-Ar as well as He work), because if the capital cost of new laboratories were lessened more facilities could be established. Should it reach a sufficient threshold, higher throughput opens the possibility of constructing synoptic data sets for critical portions of orogens, or even of whole orogens themselves. In addition, as interest in detrital thermochronology grows, there will be a great need for increased throughput in order to meet demand.

Work on laser microsampling in $^{40}Ar/^{39}Ar$ dating has revealed the potential of intracrystalline ^{40}Ar profiles to reveal details of cooling histories and other powerful constraints (e.g., Kelley and Wartho 2000). The prospect of measuring in-situ ^{4}He diffusion profiles by laser ablation may also be realized (Hodges and Boyce 2003), though several additional complications will need to overcome including complications posed by combined effects of U-Th heterogeneity and long alpha-stopping distances, as well as attaining sufficient resolution relative to the features in minerals that serve as diffusion pathways. If these hurdles can be overcome, continuous time-temperature histories at the low temperatures accessible by closure profiles in the (U-Th)/He system will be a particularly valuable tool for understanding near-surface processes. Laser profiling in single crystals has also found application in characterization of parent distributions, which can be vital for accurate alpha-ejection corrections in zoned crystals (Hourigan et al. 2005).

The application of standard radioisotopic decay schemes to "new" phases has been a standard source of progress in thermochronometry for some time and may continue to be in the future, because of the unique thermal sensitivity and natural occurrence patterns of each mineral. Specific systems that have shown particular promise recently include $^{40}Ar/^{39}Ar$ and $^{4}He/^{3}He$ thermochronometry of supergene weathering deposits (Vasconcelos 1999; Shuster et al. 2005). Development of thermochronometers with temperature sensitivities lower than that of apatite (U-Th)/He may prove powerful in some circumstances such as subsurface weathering profiles and submarine samples, but such sensitivities may also make the systems susceptible to diurnal heating or other surficial temperature fluctuations, restricting their application to certain settings. On the other hand, the sensitivity of low-temperature thermochronometers, and especially the contrasting kinetic responses of different systems, to surficial (or very nearly surficial) thermal processes, may prove valuable in understanding such phenomena if strategically applied (Mitchell and Reiners 2003; Shuster et al. 2005).

Combining multiple thermochronometers in the same samples to increase the temperature range of thermal histories is fairly common, but few examples exist of combinations of thermochronometers with electron-spin-resonance, thermoluminescence, or cosmogenic nuclide analyses that constrain rates and timing of exposure. Comparisons of erosion rates from steady-state interpretations of bedrock thermochronometric ages and both in situ and basin-scale (stream sediment sample) cosmogenic nuclide abundances have been made in several cases, often with intriguingly different erosion rate estimates over the contrasting timescales of each system (e.g., Kirchner et al. 2001; Vance et al. 2003; Stock et al. 2004). But another approach that may hold potential for understanding surface processes is measuring cosmogenic and radiogenic (and in some cases nucleogenic) abundances in the same crystals. Possible examples include ^{4}He, ^{3}He, and Ne isotopes in detrital zircons or supergene weathering deposits, to determine relationships between formation, thermochronologic, and exposure ages.

Thermochronology of detrital minerals has been used since the mid-late 1980s to constrain provenance and thermal histories of source terrains (e.g., Cerveny et al. 1988; Brandon and Vance 1992; Garver and Brandon 1994). Several new approaches have emerged recently that hold promise. One is modeling of observed probability density functions of many detrital $^{40}Ar/^{39}Ar$ grains ages from alluvium from a drainage basin, using predictions of various tectonic models combined with the basin's hypsometry (Hodges et al. 2005). While this or similar approaches have been used with zircon fission-track dating for some time, its application in detrital $^{40}Ar/^{39}Ar$ methods has been one of the fruits of vast improvements in automation and sample throughput, and once again demonstrates that in as is often the case in thermochronology, quantity has a quality all its own. Another advance in detrital thermochronology is measurement of both formation (U/Pb) and cooling ages [(U-Th)/He and/or FT] in single zircons, providing improved resolution of provenance, depositional ages, and long-term orogenic histories of source terrains (e.g., Rahl et al. 2003; Reiners et al. 2005).

Kinetics, partitioning, and other fundamentals

There are still many important unresolved issues associated with the fundamental kinetics and systematics of diffusion and annealing that to some degree limit the robustness of interpretations from thermochronology. Fission-track annealing models remain vigorously debated, especially how realistically various models capture the dynamics of long-time, low-temperature annealing that is characteristic of many slowly-cooled terrains (e.g., Ketcham et al. 1999). The precise causes of annealing kinetic variations caused by composition, radiation damage, and other effects is still somewhat primitive. The potential effects of pressure on track stability have also been recently debated (Wendt et al. 2001; Kohn et al. 2002; Vidal et al. 2002). With some exceptions, fission-track annealing is generally regarded as sufficiently complex that calibrations are largely empirical and do not seek quantitative modeling of the mechanistic atomic scale processes that lead to annealing. Better understanding of the atomic-scale processes controlling annealing may be an area of fruitful progress.

Helium diffusion in commonly dated phases also bears several poorly understood and enigmatic phenomenon such as the >280 °C "rollover" in apatite diffusion experiments (Farley 2000, 2002), and the anomalously high (compared with later stages) rates of He diffusion observed in early stages of step heating experiments in zircon (Reiners et al. 2004), titanite (Reiners and Farley 1999), and possibly other minerals (Stockli and Farley 2002). Changes in He diffusion properties at high temperature are generally considered irrelevant to retention during cooling through lower temperatures where ages actually evolve (Farley 2000), and the anomalously high He diffusivity observed at small gas fractions during experiments can be modeled as minor amounts of gas residing in low-retentivity domains which do not significantly affect the bulk crystal's thermochronometric properties (e.g., Reiners et al. 2004). Nevertheless, it is possible that these and other poorly understood non-Arrhenius phenomena may turn out to be more important than currently realized.

Other potentially important but as yet poorly understood aspects of He diffusion that may have broader implications are crystallographic anisotropy in diffusion characteristics (Farley 2000), and the fact that ^{3}He and ^{4}He appear to diffuse from apatite at essentially the same rate (Shuster et al. 2003), rather than with the inverse-root-mass dependence expected from kinetic theory. Shuster et al. (2003) suggested that a possible reason may be that movement of He is actually limited by diffusion of crystallographic defects, not the intrinsic diffusion properties of He atoms. If this is true, it raises questions about what other features of noble gas diffusion phenomena may actually be proxies for migration of crystallographic or impurity defects in minerals and how this may affect thermochronologic interpretations.

Radiation damage has long been recognized as affecting diffusion and annealing properties that control thermochronometric systems. The effects of radiation damage on annealing and diffusion are generally most evident in zircon (e.g., Hurley 1954; Nasdala et al. 2004; Rahn et al. 2004; Reiners et al. 2004; Garver et al. 2005), partly because of the high activation barrier to annealing and high U-Th concentrations. Although quantitative understanding of these effects is relatively primitive, there is potential that in certain cases, the effects of radiation damage could be used to an advantage, by essentially providing a range of thermal sensitivities similar to multi-domain behavior in a single sample (Garver et al. 2005).

Finally, there is growing recognition of the potential importance of equilibrium partitioning of noble gases into minerals, and the relationship between this and assumptions of zero-concentration boundaries and infinite reservoirs surrounding minerals of thermochronologic interest. Baxter (2003) has reformulated some of the basic theoretical constructs of noble gas chronometers to effectively explore the relationships between "excess" Ar or He, and the efficiency with which these gases can be transported away from minerals in which they were produced during cooling and closure. Examples in recent literature of ^{40}Ar/^{39}Ar ages that make little to no sense in terms of closure temperatures and classic Dodsonian theory (Kelley and Wartho 2000; Baxter et al. 2002), show that "excess" Ar can be quantitatively interpreted to provide important insights into Ar mobility, partitioning, and geologic processes. A firmer understanding of the significance and potential utility of "excess" Ar or He awaits clever experimentation, theoretical investigations, and natural examples consistent with predictions.

Quantitative interpretations of data with numerical models

In the last decade significant advances have been made in coupling thermochronometric data with numerical models to interpret topographic, erosional, and tectonic histories of orogens. The future will likely follow this trend as computing power continues to be faster, better, and cheaper and the source code for simulating different geologic processes matures.

The development and dissemination of more powerful computer models will most likely require larger data sets and carefully planned sampling strategies to optimize the signal of interest.

A variety of computer programs are currently available for quantifying different aspects of the thermochronometric record of exhumation processes. For example, forward and inverse programs for predicting apatite fission track ages and track lengths and (U-Th)/He ages a function of temperature histories are freely available (e.g., Ehlers et al. 2003; Ketcham 2005; Dunai 2005). 2D and 3D thermal-kinematic, and 2D dynamic models of orogenesis have been successfully used to interpret thermochronometer data (e.g., Batt and Braun 1997; Batt et al. 2001; Beaumont et al. 2001; Ehlers et al. 2003; Braun 2005; Braun and Robert 2005). Several different landform evolution models are also in use to study long-term landscape evolution as a function of hillslope and fluvial processes (e.g., Braun and Sambridge 1997; Ellis et al. 1999). However, with the exception of recent work by Braun (2005), few attempts have been made so far to couple all of the previous types of models into one comprehensive tool for thermochronometric analysis. Future prospects for model intensive interpretations of thermochronometric data clearly include the development and dissemination of refined coupled landform evolution and thermal models, as well as creative applications of these models to multiple thermochronometric systems. The development of more complex numerical models will also allow additional rigor in data analysis because non-uniqueness in interpretations and more complete propagation of uncertainties can be more easily explored.

Inevitably, the increased complexity of numerical and geodynamic models for predicting and interpreting thermochronometric datasets make their routine application somewhat difficult. Historically, thermal and geodynamic modeling have often been fields of study to themselves because of the extensive time and training required to learn and practice modeling that is both geologically useful and sufficiently sophisticated to advance the field. Similarly, collection of thermochronometric data requires unique skills and time investments as well. Much of the future in creative applications of thermochronology will likely rely on either the training of students with interdisciplinary skills (e.g., data collection and programming and modeling) and/or expansion of symbiotic collaborations between modelers and analysts.

General comments on the future of thermochronology

There are challenges facing thermochronology, some of which will undoubtedly lead in surprising directions. But we suggest that many if not most of these apparent unresolved issues or outstanding problems will eventually bear fruits that expand and strengthen the field. In some ways, thermochronology is the inevitable outgrowth of the empirical and theoretical maturation of geochronology in general, as it also dealt with challenging issues. As datasets from various radioisotopic systems grew in quality and quantity, complications that once confounded explanation (such as intra- and inter-method age discrepancies or inconsistent experimental diffusion results) have, through hard-won quantitative understanding of kinetic properties, been transformed into powerful tools for piecing together detailed thermal histories. It is in this lemons-to-lemonade context that we mention several challenges we see facing thermochronology today, which could be important in the next forty years of the fields' evolution.

In general one of the greatest needs in thermochronology is better quantitative understanding of the kinetics of diffusion (and annealing), especially from experimental approaches. There are many important limitations to our understanding of Ar and He diffusion in many phases, and not nearly as many attempts to resolve these issues by direct experiments as there are applications with heuristic assumptions and attempted empirical "calibrations". Particularly important in this regard would be development of routine ways of measuring kinetic data directly on unknowns, rather than assuming all samples are the same, as is done for many types of minerals. Kinetic variations among specimens of the same mineral and strategic exploitation of these properties in sampling and analyses may in fact lead to great advances in understanding kinetic mechanisms and controls in general, not to mention more detailed and accurate thermal histories. A related challenge that is more specific to (U-Th)/He dating is the need for agreed-upon, cost-effective, and reliable protocols for treating alpha-ejection and U-Th zonation on grain-by-grain bases.

Much thermochronologic interpretation assumes that daughter nuclides diffuse simply across zero-concentration grain boundaries into an infinite-sink reservoir. Whereas this idealized model has allowed a great deal of progress in interpreting noble gas thermochronometry, there are some natural examples that could be interpreted as evidence for violations of this behavior. Better theoretical and experimental understanding of the phenomena of excess Ar and He like those of Baxter (2003) may improve the robustness of thermochronologic interpretations in some cases. A better quantitative understanding of these issues may in fact provide important thermochronologic interpretations in unexpected areas (e.g., Kelley and Wartho 2001).

Another issue is the inherent uncertainty involved in inferring exhumational histories from thermal histories. One of the reasons for increasing interest in thermochronology in the last decade is the prospect of bringing radioisotopic dating techniques to bear on near-surface crustal processes, especially those constraining erosional exhumation. Even if the thermal histories themselves bore no uncertainty, inferring exhumation histories from them requires assumptions about geothermal gradients and their variations in space and time. These assumptions and their uncertainty range widely in complexity depending on the problem being addressed, but often involve spatial and temporal transients in exhumation rates, topography, fluid circulation, and deep-derived (basal) heat flow. Some of the currently most exciting issues such as evolution of paleotopography and erosion rates on 10^5-10^6 yr scales may rely on assumptions that are difficult to test or require extensive integration of other data sets such as heat flow or hydrologic data to constrain models. In many cases, convincing arguments can be made that the essential aspects of interpretations are insensitive to some of these assumptions. But in others, uncertainties such as how groundwater circulation patterns affect the thermal field at depths less than 2–3 km (especially in regions of high topographic relief, e.g., Ehlers 2005), or how magmatic events that may produce little surface expression affect thermal fields to greater depths, are difficult to constrain. At least some of these issues may be addressed by focused high-density sampling of currently active orogens, structures, or topographic features. If groundwater flow significantly deflects isotherms in the uppermost 2–3 km of high relief areas, for example, a careful heat-flow, hydrologic, and thermochronologic study could illuminate the details. Non-uniqueness in interpretations is not new to the Earth sciences, and thermochronologists and modelers can address uncertainties in interpretations by reporting the range of processes and solutions that satisfy a set of observations rather than looking for a single solution.

A general issue facing thermochronology is the problem of nonmonotonic thermal histories. It is generally acknowledged that while most datasets do not uniquely constrain model thermal histories, when data from enough different systems or high-quality multi-domain or closure profile data are available, a range of thermal histories emerges that is commonly sufficiently small to be geologically useful. It may be generally underappreciated, however, that many models often focus on solutions assuming monotonic cooling. When nonmonotonic cooling histories are allowed, it is often more difficult to find a family of thermal histories with a restricted enough range to be useful (e.g., Quidelleur et al. 1997; McDougall and Harrison 1999; Lovera et al. 2002). This is explicitly recognized in most formal inversions of fission-track age and length data, as well as multi-domain $^{40}Ar/^{39}Ar$ K-feldspar cooling models, but many casual users fail to recognize the importance or limitations of this assumption. Thermal histories from multiple thermochronometers with certain shapes (e.g., concave down, followed by concave up) may be consistent with simplified expectations of reheating, but they are not diagnostic (e.g., Harrison et al. 1979). In this respect, fission-track dating bears a distinct advantage over noble gas methods, because track-length analysis allows resolution of distinct thermal histories for tracks for different ages. Nonetheless, realistic modeling of age and track-length data for thermal histories involving reheating is often subject to considerable uncertainty (e.g., Ketcham 2003, 2005). Ultimately, geologic considerations may provide critically important information in considering non-monotonic thermal histories, but development of techniques for diagnosing reheating from thermochronologic data alone may be an important goal for future studies. One potential tack may be to use contrasting responses of thermochronometers with strongly varying activation energies. Extremely short duration (1–100 yr) and relatively high temperature reheating events, for example, can produce diagnostic age inversions in fission-track and (U-Th)/He ages in the same minerals, because of their distinct kinetics.

Finally, compared with some disciplines in geophysics and geodesy (e.g., IRIS or UNAVCO), the thermochronological community is not particularly cohesive at the moment and there has been little formal effort to improve shared resources or improve the access to thermochronological data. There are considerable historical reasons for the current culture of each thermochronologist having their own facility and their own protocols. However, we suggest that the time may have come to consider development of a community-wide vision for sharing facilities to help support and encourage comprehensive regional studies. As we outlined earlier and as is discussed elsewhere in this volume (e.g., Ehlers 2005), the demand for large datasets will increase, given intellectual developments driven by modeling, the growing focus on low-temperature systems much affected by complex high-frequency boundary conditions (Braun 2005), intense interest in detrital thermochronology, and analytical demands stemming from community initiatives like Earthscope's USArrray and the Plate Boundary Observatory. These sorts of demands may completely overwhelm the capabilities of a single laboratory, and both cooperative work among laboratories and development of new high-throughput facilities will be required. We suggest that thermochronologists might benefit from pursuing a grander vision. There are important problems in geodynamics that could be solvable given high-enough sampling densities that, while a far stretch for current analytical capacity, are not economically out of reach, even today—investigators are often successful in procuring funds for seismic lines costing millions of dollars, a sum that could support tens of thousands of mineral ages, even using current systems not optimized for analytical throughput.

The future of thermochronology is bright. Intense interest in understanding near-surface processes and links between tectonics, erosion, and climate will continue to motivate advances in a number of areas that should increase the resolution and accuracy of thermochronologic models. New thermochronometeric systems are being developed that will expand accessible temperature ranges, better models of crystal-scale kinetic processes are clarifying age and thermal history interpretations, and analytical innovations will likely soon permit generation of large datasets that can invert for thermal histories of entire drainage basins or orogens, or provide routine closure profiles in single crystals. Improved geodynamic and surface process models will also undoubtedly allow increasingly sophisticated interpretations of tectonogeomorphic evolution. Motivated by exciting problems linking geologic processes such as tectonics, erosion, and climate, the next few decades will undoubtedly witness ingenious innovations and development of powerful analytical, interpretational, and modeling approaches that rival the progress and changes in radioisotopic dating in the last hundred years.

REFERENCES

Armstrong RL (1966) K-Ar dating of plutonic and volcanic rocks in orogenic belts. *In*: Potassium-Argon Dating. Schaeffer OA, Zähringer J (eds) Springer–Verlag, Berlin, p 117-133

Barfod GH, Frei R, Krogstad EJ (2002) The closure temperature of the Lu-Hf isotopic system in apatite. Geochim Cosmochim Acta 66:15A:51

Batt GM, Brandon M, Farley K, Roden-Tice M (2001) Tectonic synthesis of the Olympic Mountains segment of the Cascadia wedge, using two-dimensional thermal and kinematic modeling of thermochronological ages. J Geophys Res 106:26731-26746
Batt G, Braun J (1997) On the thermomechanical evolution of compressional orogens, Geophys J Int 128: 364-382
Baxter EF (2003) Quantification of the factors controlling the presence of excess ^{40}Ar or ^{4}He. Earth Planet Sci Lett 216:619-634
Baxter EF, DePaolo DJ, Renne PR (2002) Spatially correlated anomalous ^{40}Ar/^{39}Ar 'Age' variations about a lithologic contact near Simplon Pass, Switzerland: a mechanistic explanation for excess Ar. Geochim Cosmochim Acta 66:1067-1083
Beaumont CR, Jamieson RA, Nguyen MH, Lee B (2001) Himalayan tectonics explained by extrusion of a low-viscosity crustal channel coupled to focused surface denudation. Nature 414:738-742
Benfield A (1949) The effect of uplift and denudation on underground temperatures. J Appl Phys 20:66-70
Berger GW (1975) ^{40}Ar^{39}Ar step heating of thermally overprinted biotites, hornblendes and potassium feldspars from Eldora, Colorado. Earth Planet Sci Lett 26:387-408
Berger GW, York D (1981) Geothermometry from ^{40}Ar^{39}Ar dating experiments. Geochim Cosmochim Acta 45:795-811
Brandon MT, Roden-Tice MK, Garver JI (1998) Late Cenozoic exhumation of the Cascadia accretionary wedge in the Olympic Mountains, northwest Washington State. Geol Soc Am Bull 110:985-1009
Brandon MT, Vance JA (1992) Tectonic evolution of the Cenozoic Olympic Subduction Complex, Washington State, as deduced from fission-track ages for detrital zircons. Am J Sci 292:565-636
Braun J (2005) Quantitative constraints on the rate of landform evolution derived from low-temperature thermochrohonology. Rev Mineral Geochem 58:351-374
Braun J, Robert X (2005) Constraints on the rate of post-orogenic erosional decay from low-temperature thermochronological data: application to the Dabie Shan, China. Earth Surf Proc Land, in press
Braun J, Sambridge M (1997) Modelling landscape evolution on geological time scales; a new method based on irregular spatial discretization. Basin Res 9:27-52
Bullard EC (1938) The disturbance of the temperature gradient in the Earth's crust by inequalities of height. Monthly Notices Roy Astro Soc, Geophys Suppl 4:360-362
Calk LC, Naeser CW (1973) The thermal effect of a basalt intrusion on fission-tracks in quartz monzonite. J Geol 81:189-198
Cerveny PF, Naeser ND, Zeitler PK, Naeser CW, Johnson NM (1988) History of uplift and relief of the Himalaya over the past 18 Ma–Evidence from fission-track ages of detrital zircons from sandstones of the Siwalik Group. *In*: New Perspectives in Basin Analysis, Kleinspehn K, Paola C (ed) Univ. Minnesota Press, p. 43-61
Chamberlain KR, Bowring SA (2001) Apatite–feldspar U–Pb thermochronometer: a reliable, mid-range (~450 °C), diffusion-controlled system. Chem Geol 172:173-200
Cherniak DJ (1993) Lead diffusion in titanite and preliminary results on the effects of radiation damage on Pb transport. Chem Geol 110:177-194
Cherniak DJ, Watson EB (2001) Pb diffusion in zircon. Chem Geol 172:5-24
Cherniak DJ, Lanford WA, Ryerson FJ (1991) Lead diffusion in apatite and zircon using ion implantation and Rutherford Backscattering techniques. Geochim Cosmochim Acta 55:1663-1673
Cherniak DJ, Watson EB, Grove M, Harrison TM (2004) Pb diffusion in monazite: a combined RBS/SIMS study. Geochim Cosmochim Acta 68:829-840
Clark SP, Jäger E (1969) Denudation rate in the Alps from geochronologic and heat flow data. Am J Sci 267: 1143–1160
Coyle DA, Wagner GA (1998) Positioning the titanite fission-track partial annealing zone. Chem Geol 149: 117-125
Dahl PS (1996) The effects of composition on retentivity of argon and oxygen in hornblende and related amphiboles: A field-tested empirical model. Geochim Cosmochim Acta 60:3687-3700
Damon PE, Kulp JL (1957) Determination of radiogenic helium in zircon by stable isotope dilution technique. Transaction, Am Geophys Union 38:945-953
Dewey JF, Panhurst RP (1970) The evolution of the Scottish Caledonides in relation to their radiometric age patterns. Trans Royal Soc Edin 69:361-389
Dodson MH (1973) Closure temperature in cooling geochronological and petrological systems Contrib Mineral Petrol 40:259-274
Dodson MH (1979) Theory of cooling ages. *In:* Lectures in Isotope Geology. Jager E, Hunziker JC (eds.) Springer-Verlag, Berlin, p. 194
Dodson MH (1986) Closure profiles in cooling systems. *In:* Materials Science Forum, Vol 7, Trans Tech Publications, Aedermannsdorf, Switzerland, p 145-153

Ducea MN, Ganguly J, Rosenberg EJ, Patchett PJ, Cheng W, Isachsen C (2003) Sm-Nd dating of spatially controlled domains of garnet single crystals; a new method of high-temperature thermochronology. Earth Planet Sci Lett 213:31-42

Ehlers TA (2005) Crustal thermal processes and the interpretation of thermochronometer data. Rev Mineral Geochem 58:315-350

Ehlers TA, Farley KA (2003) Apatite (U-Th)/He thermochronometry: methods and applications to problems in tectonic and surface processes. Earth Planet Sci Lett 206:1-14

Ehlers TA, Willett SD, Armstrong PA, Chapman DS (2003) Exhumation of the central Wasatch Mountains, Utah: 2. Thermokinematic model of exhumation, erosion, and thermochronometry interpretation. J Geophys Res 108:2173, doi:10.1029/2001JB001723.

Ellis MA, Densmore AL, Anderson RS (1999) Development of mountainous topography in the Basin and Ranges, USA. Basin Res 11: 21-41

Evernden JF, Curtis GH, Kistler RW, Obradovich J (1960) Argon diffusion in glauconite, microcline, sanidine, leucite and phlogopite. Am J Sci 258:583-604

Farley KA (2000) Helium diffusion from apatite: General behavior as illustrated by Durango fluorapatite. J Geophys Res 105:2903-2914

Farley KA (2002) (U-Th)/He dating: techniques, calibrations, and applications. Rev Mineral Geochem 47: 819-844

Farley KA, Stockli DF (2002) (U-Th)/He dating of phosphates: apatite, monazite, and xenotime. Rev Mineral Geochem 48:559-577

Farley KA, Wolf RA, Silver LT (1996) The effects of long alpha-stopping distances on (U-Th)/He ages. Geochim Cosmochim Acta 60:4223-4229

Fechtig H, Kalbitzer S (1966) The diffusion of argon in potassium-bearing solids. *In:* Potassium-Argon Dating. Schaeffer OA, Zähringer J (eds) Springer, New York, p 68-107

Fitzgerald JD, Harrison TM (1993) Argon diffusion domains in K-feldspars I: microstructures in MH-10. Contrib Mineral Petrol 113:367-380

Foland KA (1994) Argon diffusion in feldspars. *In:* Feldspars and Their Reactions. Parsons I (ed.), Kluwer, Dordrecht, p. 415-447

Garver JI, Brandon MT (1994) Erosional exhumation of the British Columbia coast ranges as determined from fission-track ages of detrital zircon from the Tofino basin, Olympic Peninsula, Washington. Geol Soc Am Bull 106:1398–1412

Gleadow AJW, Duddy IR, Lovering JF (1983) Fission track analysis: a new tool for the evaluation of thermal histories and hydrocarbon potential. Aust Petrol Explor Assoc J 23:93-102

Grove M, Harrison TM (1996) $^{40}Ar^*$ diffusion in Fe-rich biotite. Am Mineral 81:940-951

Hames WE, Bowring SA (1994) An empirical evaluation of the argon diffusion geometry in muscovite. Earth Planet Sci Lett 124:161-167

Hanson GN Gast PW (1967) Kinetic studies in contact metamorphic zones. Geochim Cosmochim Acta 31: 1119-1153

Hanson GN, Simmons KR, Bence AE (1975) $^{40}Ar^{39}Ar$ spectrum ages for biotite, hornblende and muscovite in a contact metamorphic zone. Geochim Cosmochim Acta 39:1269-1277

Harrison TM (1981) Diffusion of ^{40}Ar in hornblende. Contrib Mineral Petrol 78:324-331

Harrison TM, Armstrong RL, Naeser, CW, Harakal JE (1979) Geochronology and thermal history of the Coast Plutonic Complex, near Prince Rupert, British Columbia. Can J Earth Sci 16:400-410

Harrison TM, Clarke GKC (1979) A model of the thermal effects of igneous intrusion and uplift as applied to Quottoon pluton, British Columbia. Can J Earth Sci 16: 411-420

Harrison TM, Duncan I, McDougall I (1985) Diffusion of ^{40}Ar in biotite: temperature, pressure and compositional effects. Geochim Cosmochim Acta 49:2461-2468

Harrison TM, Heizler MT, Lovera OM (1993) In vacuo crushing experiments and K-feldspar thermochronometry. Earth Planet Sci Lett 117:169-180

Harrison TM, Lovera OM, Heizler MT (1991) $^{40}Ar/^{39}Ar$ results for alkali feldspars containing diffusion domains with differing activation energy. Geochim Cosmochim Acta 55:1435-1448

Harrison TM, McDougall I (1980a) Investigations of an intrusive contact, northwest Nelson, New Zealand-I. Thermal, chronological, and isotopic constraints. Geochim Cosmochim Acta 44:1985-2003

Harrison TM, McDougall I (1980b) Investigations of an intrusive contact, northwest Nelson, New Zealand-II. Diffusion of radiogenic and excess ^{40}Ar in hornblende revealed by $^{40}Ar/^{39}Ar$ age spectrum analysis. Geochim Cosmochim Acta 44:2005-2020

Harrison TM, McDougall I (1981) Excess ^{40}Ar in metamorphic rocks from Broken hill, New South Wales: Implications for $^{40}Ar/^{39}Ar$ age spectra and the thermal history of the region. Earth Planet Sci Lett 55: 123-149

Harrison TM, McDougall I (1982) The thermal significance of potassium feldspar K-Ar ages inferred from $^{40}Ar/^{39}Ar$ age spectrum results. Geochim Cosmochim Acta 46:1811-1820

Harrison TM, Grove M, Lovera OM, Zeitler PK (2005) Continuous thermal histories from inversion of closure profiles. Rev Mineral Geochem 58:389-409
Harrison TM, Zeitler PK (2005) Fundamentals of noble gas thermochronometry. Rev Mineral Geochem 58: 123-149
Harper CT (1967) The geological interpretation of potassium-argon ages of metamorphic rocks from the Scottish Caledonides. Scottish J Geol 3:46-66
Hart SR (1964) The petrology and isotopic mineral age relations of a contact zone in the Front Range, Colorado. J Geol 72:493-525
Hess JC, Lippolt HJ, Gurbanov AG, Michalski I (1993) The cooling history of the late Pliocene Eldzhurtinskiy granite (Caucasus, Russia) and the thermochronological potential of grain-size/age relationships. Earth Planet Sci Lett 117:393-406
Hawkins DP, Bowring SA (1999) U-Pb monazite, xenotime and titanite geochronological constraints on the prograde to post-peak metamorphic thermal history of Paleoproterozoic migmatites from the Grand Canyon, Arizona. Contrib Mineral Petrol 134:150-169
Hodges K (2003) Geochronology and thermochronology in orogenic systems. *In:* Treatise on Geochemistry. Turekian KK, Holland HD (eds) Elsevier, p 263-292
Hodges K, Boyce J (2003) Laser-ablation (U-Th)/He geochronology. Eos, Trans, Am Geophys U 84(46), Fall Meeting Supplement, Abstract V22G-05
Hourigan JK, Reiners PW, Brandon MT (2005) U-Th zonation-dependent alpha-ejection in (U-Th)/He chronometry. Geochim Cosmochim Acta 69:3349-3365
House MA, Farley KA, Stockli D (2000) Helium chronometry of apatite and titanite using Nd-YAG laser heating. Earth Planet Sci Lett 183:365-368
Hurley PM (1954) The helium age method and the distribution and migration of helium in rocks. *In:* Nuclear Geology. Faul H (ed), Wiley & Sons, 301-329
Hurley PM, Hughes J, Pinsoons WH, Fairbairn HW (1962) Radiogenic argon and strontium diffusion parameters in biotite at low temperatures obtained from Alpine fault uplift in New Zealand. Geochim Cosmochim Acta 26:67-80
Jäger E, Niggli E, Wenk E (1967) Rb-Sr Alterbestimmungen an Glimmern der Zentralalpen. Beitrage Geol. Karte Schweiz 134:1-67
Jenkin GRT, Ellam RM, Rogers G, Stuart FM (2001) An investigation of closure temperature of the biotite Rb-Sr system: The importance of cation exchange. Geochim Cosmochim Acta 65:1141-1160
John BE, Foster DA, Murphy JM, Cheadle MJ, Baines AG, Fanning CM, Copeland P (2004) Determining the cooling history of in situ lower oceanic crust—Atlantis Bank, SW Indian Ridge. Earth Planet Sci Lett 222:145-160
Kelly SP (2002) K-Ar and Ar-Ar dating. Rev Mineral Geochem 47:785-818
Kelley SP, Arnaud NO, Turner SP (1994) High spatial resolution $^{40}Ar/^{39}Ar$ investigations using an ultra-violet laser probe extraction technique. Geochim Cosmochim Acta 58:3519-3525
Ketcham RA (2005) Forward and inverse modeling of low-temperature thermochronometry data. Rev Mineral Geochem 58:275-314
Kelley SP, Wartho J-A, (2000) Rapid Kimberlite Ascent and the Significance of Ar-Ar Ages in Xenolith Phlogopites. Science 289:609-611
Ketcham RA (2003) Effects of allowable complexity and multiple chronometers on thermal history inversion. Geochim Cosmochim Acta (Abs), 67(18), Supp. 1, A213.
Ketcham RA, Donelick RA, Carlson WD (1999) Variability of apatite fission-track annealing kinetics: III. extrapolation to geological time scales. Am Mineral 84:1235-1255.
Kirchner JR, Finkel RC, Riebe CS, Granger DE, Clayton JL, King JG, Megahan WF (2001) Mountain erosion over 10 yr, 10 k.y., and 10 m.y. time scales. Geology 29:591-594
Kohn BP, Belton D, Brown RW, Gleadow AJW, Green PF, Lovering JF (2003) Comment on: "Experimental evidence for the pressure dependence of fission track annealing in apatite" by A.S. Wendt et al. [Earth Planet. Sci. Lett. 201 (2002) 593–607]. Earth Planet Sci Lett 215:299-306
Laslett GM, Green PF, Duddy IR, Gleadow AJW (1987) Thermal annealing of fission tracks in apatite. 2. A quantitative analysis. Chem Geol 65:1-13
Lees CH (1910) On the isogeotherms under mountain ranges in radioactive districts. Proc Roy Soc 83:339-346
Lovera OM, Richter FM, Harrison TM (1991) Diffusion domains determined by ^{39}Ar release during step heating. J Geophys Res 96:2057-2069
Lovera OM, Richter FM, Harrison TM (1989) The $^{40}Ar/^{39}Ar$ geothermometry for slowly cooled samples having a distribution of diffusion domain sizes. J Geophys Res 94:17917-17935
Lovera OM, Grove M, Harrison TM, Mahon KI (1997) Systematic analysis of K-feldspar $^{40}Ar/^{39}Ar$ step heating results: I. Significance of activation energy determinations. Geochim Cosmochim Acta 61:3171-3192

Lovera OM, Grove M, Harrison TM (2002) Systematic analysis of K-feldspar $^{40}Ar/^{39}Ar$ step heating results II: Relevance of laboratory argon diffusion properties to nature. Geochim Cosmochim Acta 66:1237-1255
Lovera OM, Heizler MT, Harrison TM (1993) Argon diffusion domains in K-feldspar II: Kinetic properties of MH-10. Contrib Mineral Petrol 113:381-393
McDougall I, Harrison TM (1999) Geochronology and Thermochronology by the $^{40}Ar/^{39}Ar$ Method. 2nd ed., Oxford University Press, New York
Mason B (1961) Potassium-argon ages of metamorphic rocks and granites from Westland, New Zealand. New Zealand J Geol Geophys 4:352-356
Mancktelow N, Grasemann B (1997) Time-dependent effects of heat advection and topography on cooling histories during erosion. Tectonophys 270:167-195
Mitchell SG, Reiners PW (2003) Influence of wildfires on apatite and zircon (U-Th)/He ages. Geology 31: 1025-1028
Musset AE (1960) Diffusion measurements and the potassium-argon method of dating. Geophys J Royal Astronom Soc 18:257-303
Naeser CW (1967) The use of apatite and sphene for fission track age determinations. Bull Geol Soc Am 78: 1523-1526
Naeser CW, Faul H (1969) Fission track annealing in apatite and sphene. J Geophys Res 74:705-710
Naeser CW, Forbes RB (1976) Variation of fission-track ages with depth in two deep drill holes. EOS Trans Am Geophys Union 57:353
Nasdala L, Reiners PW, Garver JI, Kennedy AK, Stern RA, Balan E, Wirth R (2004) Incomplete retention of radiation damage in zircon from Sri Lanka. Am Mineral 89:219-231
Parrish RR (1985) Some cautions which should be exercised when interpreting fission-track and other dates with regard to uplift rate calculations. Nucl Tracks 10:425
Parsons I, Brown WL, Smith JV (1999) $^{40}Ar/^{39}Ar$ thermochronology using alkali feldspars: real thermal history or mathematical mirage of microtexture?. Contrib Mineral Petrol 136:92-110
Purdy JW, Jäger E (1976) K-Ar ages on rock-forming minerals from the Central Alps. Mem 1st Geol Min Univ Padova 30, 31 p.
Quidelleur Z, Grove M, Lovera OM, Harrison TM, Yin A, Ryerson FJ (1997) The thermal evolution and slip history of the Renbu Zedong Thrust, southeastern Tibet. J Geophys Res 102:2659-2679
Rahl JM, Reiners PW, Campbell IH, Nicolescu S, Allen CM (2003) Combined single-grain (U-Th)/He and U/Pb dating of detrital zircons from the Navajo Sandstone, Utah. Geology 31:761-764
Rahn MK, Grasemann B (1999) Fission track and numerical thermal modeling of differential exhumation of the Glarus thrust plane (Switzerland). Earth Planet Sci Lett 169:245-259
Rahn MK, Brandon MT, Batt GE, Garver JI (2004) A zero-damage model for fission-track annealing in zircon. Am Mineral 89:473-484
Reiners PW (2005) Zircon (U-Th)/He thermochronometry. Rev Mineral Geochem 58:151-179
Reiners PW, Campbell IH, Nicolescu S, Allen CM, Hourigan JK, Garver JI, Mattinson JM, Cowan DS (2005) (U-Th)/(He-Pb) double dating of detrital zircons. Am J Sci 305:259-311
Reiners PW, Campbell IH, Nicolescu S, Allen CA, Hourigan JK, Garver JI, Mattinson JM, Cowan DS (2005) (U-Th)/(He-Pb) "double-dating" of detrital zircons. Am J Sci 305:259-311
Reiners PW, Farley KA (1999) He diffusion and (U-Th)/He thermochronometry of titanite. Geochim Cosmochim Acta 63:3845-3859
Reiners PW, Farley KA (2001) Influence of crystal size on apatite (U-Th)/He thermochronology: an example from the Bighorn Mountains, Wyoming. Earth Planet Sci Lett 188:413-420
Reiners PW, Spell TL, Nicolescu S, Zanetti KA (2004) Zircon (U-Th)/He thermochronometry: He diffusion and comparisons with $^{40}Ar/^{39}Ar$ dating. Geochim Cosmochim Acta 68:1857-1887
Richter FM, Lovera OM, Harrison TM, Copeland P (1991) Tibetan tectonics from $^{40}Ar/^{39}Ar$ analysis of a single K-feldspar sample Earth Planet Sci Lett 105:266-278
Robbins GA (1972) Radiogenic argon diffusion in muscovite under hydrothermal conditions. M.S. Thesis, Brown University. Providence, Rhode Island
Rutherford E (1905) Present problems in radioactivity. Pop Sci Monthly (May):1-34
Rutherford E (1906) Radioactive Transformations. Charles Scribner's Sons, NY
Scherer EE, Cameron KL, Blichert-Toft J (2000) Lu-Hf garnet geochronology; closure temperature relative to the Sm-Nd system and the effects of trace mineral inclusions. Geochim Cosmochim Acta 64:3413-3432
Shuster DL, Farley KA (2003) $^{4}He/^{3}He$ thermochronometry. Earth Planet Sci Lett 217:1-17
Shuster DL, Farley KA (2005) $^{4}He/^{3}He$ thermochronometry: theory, practice, and potential complications. Rev Mineral Geochem 58:181-203
Shuster DL, Farley KA, Sisterson JM, Burnett DS (2003) Quantifying the diffusion kinetics and spatial distributions of radiogenic ^{4}He in minerals containing proton-induced ^{3}He. Earth Planet Sci Lett 217: 19-32

Shuster DL, Vasconcelos PM, Heim JA, Farley KA (2005) Weathering geochronology by (U-Th)/He dating of goethite. Geochim Cosmochim Acta 69:659-673

Stock GM, Anderson RS, Finkel RC (2004) Pace of landscape evolution in the Sierra Nevada, California, revealed by cosmogenic dating of cave sediments. Geology 32:193-196

Stockli DF (2005) Application of low-temperature thermochronometry to extensional tectonic settings. Rev Mineral Geochem 58:411-448

Stuwe KL, White L, Brown R (1994) The influence of eroding topography on steady-state isotherms; application to fission track analysis. Earth Planet Sci Lett 124:63-7

Tagami T, Galbraith RF, Yamada R, Laslett GM (1998) Revised annealing kinetcs of fission tracks in zircon and geological implications In Van den haute P, De Corte F (eds) Advances in fission-track geochronology. Kluwer academic publishers, Dordrecht, The Netherlands, p 99-112

Vance D, Bickle M, Ivy-Ochs S, Kubik PW (2003) Erosion and exhumation in the Himalaya from cosmogenic isotope inventories of river sediments. Earth Planet Sci Lett 206:273-288

Vasconcelos PM (1999) K-Ar and $^{40}Ar/^{39}Ar$ geochronology of weathering processes. Ann Rev Earth Planet Sci 27:183-229

Vidal O, Wendt AS, Chadderton LT (2003) Further discussion on the pressure dependence of fission track annealing in apatite: reply to the critical comment of Kohn et al. Earth Planet Sci Lett 215:307-316

Villa IM (1994) Multipath Ar transport in K-feldspar deduced from isothermal heating experiments. Earth Planet Sci Lett 122:393-401

Wagner GA (1968) Fission track dating of apatites. Earth Planet Sci Lett 4:411–415

Wagner GA, Miller DS, Jäger E (1979) Fission-track ages on apatite of Bergell rocks from Central Alps and Bergell boulders in Oligocene sediments. Earth Planet Sci Lett 45:355-360

Wagner GA, Reimer GM, Jäger E (1977) Cooling ages derived from apatite fission track, mica Rb-Sr and K-Ar dating: The uplift and cooling history of the Central Alps. Mem Instit Geol Mn Univ Padova 30:1-27

Watt S, Durrani SA (1985) Thermal stability of fission tracks in apatite and sphene: Using confined-track-length measurements. Nuclear Tracks 10:349-357

Wendt AS, Vidal O, Chadderton LT (2002) Experimental evidence for the pressure dependence of fission track annealing in apatite. Earth Planet Sci Lett 201:593-607

Westcott MR (1966) Loss of argon from biotite in a thermal metamorphism. Nature 210:84-84

Wolf RA, Farley KA, Silver LT (1996) Helium diffusion and low-temperature thermochronometry of apatite. Geochim Cosmochim Acta 60:4231-4240

Wolf RA, Farley KA, Kass DM (1998) Modeling of the temperature sensitivity of the apatite (U-Th)/He thermochronometer. Chem Geol 148:105-114

Zeitler PK, Herczeg AL, McDougall I, Honda M (1987) U-Th-He dating of apatite: a potential thermochronometer. Geochim Cosmochim Acta 51:2865-2868

Zeitler PK (1987) Argon diffusion in partially outgassed alkali-feldspars: insights from $^{40}Ar/^{39}Ar$ analysis. Chem Geol (Isotope Geoscience Section) 65:167-181

Zeitler PK, Johnson NM, Briggs ND, Naeser CW (1986) Uplift history of the NW Himalaya as recorded by fission-track ages on detrital Siwalik zircons. *In:* Proceedings of the Symposium on Mesozoic and Cenozoic Geology in Connection of the 60th Anniversary of the Geological Society of China. Jiqing H. (ed) Geological Publishing House, Beijing, p. 481-494

Reviews in Mineralogy & Geochemistry
Vol. 58, pp. 19-47, 2005

2

Fundamentals of Fission-Track Thermochronology

Takahiro Tagami

Division of Earth and Planetary Sciences
Graduate School of Science
Kyoto University
Kyoto 606-8502, Japan
tagami@kueps.kyoto-u.ac.jp

Paul B. O'Sullivan

Apatite to Zircon, Inc.
1075 Matson Road
Viola, Idaho 83872-9709, U.S.A.
osullivan@apatite.com

INTRODUCTION

Fission-track (FT) analysis has developed into one of the most useful techniques used throughout the geologic community to reconstruct the low-temperature thermal history of rocks over geological time. The FT method is based on the accumulation of narrow damage trails (i.e., fission tracks) in uranium-rich mineral grains (e.g., apatite, zircon, titanite) and natural glasses, which form as a result of spontaneous nuclear fission decay of ^{238}U in nature (Price and Walker 1963; Fleischer et al. 1975). The time elapsed since fission tracks began to accumulate is estimated by determining the density of accumulated tracks in a particular material in relation to the uranium content of that material. Chemical etching can be used to enlarge fission tracks that have formed within a mineral in order to make them readily observable under an ordinary optical microscope (Price and Walker 1962).

If a host rock is subjected to elevated temperatures, fission tracks that have formed up to that point in time are shortened progressively and eventually erased by the thermal recovery (i.e., annealing) of the damage (Fleischer et al. 1975). Because thermal diffusion basically governs the annealing process, the reduction in FT length is a function of heating time and temperature. Importantly, fission tracks are partially annealed over different temperature intervals within different minerals. This characteristic allows for the construction of time-temperature paths of many different rock types by (a) plotting FT (and other isotopic) ages from different minerals versus their closure temperatures, which is applicable in the case of a monotonous cooling history (e.g., Wagner et al. 1977; Zeilter et al. 1982), and/or by (b) the inverse modeling of observed FT age and confined track length data (e.g., Corrigan 1991; Lutz and Omar 1991; Gallagher 1995; Ketcham et al. 2000; see also Ketcham 2005).

Fleischer et al. (1975) summarized the early studies of the solid-state nuclear track detection and its geological applications. Subsequently, a comprehensive overview of FT dating and thermochronology was provided by Wagner and Van den haute (1992), followed by more recent advances of the FT method in a number of review articles by Ravenhurst and Donelick (1992), Gallagher et al. (1998), Dumitru (2000), and Gleadow et al. (2002).

The aim of this particular chapter is to present a simplified and historical overview of some of the basic fundamentals of FT thermochronology, and is not written in an attempt to supersede

1529-6466/05/0058-0002$05.00 DOI: 10.2138/rmg.2005.58.2

all or any of the detailed review articles. This overview is designed to assist a reader of this volume in gaining a better understanding of some of the fundamentals of FT thermochronology prior to reading in-depth discussions of the apatite fission track (AFT) and zircon fission track (ZFT) methods, which are the focus of the following two chapters. Furthermore, there is bound to be some overlap between the content presented within this chapter and that within the following chapters discussing the AFT and ZFT methods. The authors have attempted to minimize this overlap wherever possible, however, readers should once again view this chapter as a historical overview of FT fundamentals, whereas the following chapters discuss some of the same fundamental ideas, but in much greater detail applicable to a particular technique.

FORMATION AND REGISTRATION OF NUCLEAR FISSION TRACKS

Spontaneous and induced nuclear fission decay

Nuclear fission is a process during which a heavy, unstable nucleus splits into a pair of fragments of similar size. This reaction takes place both spontaneously in nature and artificially during bombardment by neutrons and other high-energy particles or γ-rays. Each reaction is accompanied by the release of a few neutrons and ~210 MeV of energy, of which the majority (~170 MeV on average) is the kinetic energy of the fission fragments. As a result of this energetic disintegration, the fission fragments with massive positive charges are propelled from the reaction site in opposite directions. If the nucleus is located within a dielectric solid, the reaction will create a damage trail, called a fission track, along the trajectories of the two fragments.

Spontaneous fission occurs in very heavy nuclides that belong to the actinide series of elements. Of those nuclides, ^{232}Th and three U isotopes (^{234}U, ^{235}U and ^{238}U) are the typical candidates that produce a significant number of spontaneous fission tracks in solids. However, in regards to the relative abundances and spontaneous fission half-lives, ^{238}U is the only source of spontaneous tracks in terrestrial materials including natural apatite and zircon, except for those anomalously enriched in Th (Wagner and Van den haute 1992). Note that ^{238}U also decays through a chain comprised of eight α (^{4}He) and six β^- emissions. This decay, which occurs 2×10^6 times more frequently than the spontaneous fission events, needs to be taken into account in calculating an FT age. This will be discussed in more detail later.

Track formation process in solids

When a heavy ionized particle travels at a high velocity through a solid, it interacts with the host lattice, gradually loses its kinetic energy and slows down until it eventually stops. During this process, a couple of interactions are predominantly responsible for the deceleration of the particle: (1) electric interaction, more effective at high velocities, in which the particle surrenders its energy by stripping electrons from target atoms, by having its own electrons stripped away, and by raising the excitation level of the lattice electrons; and (2) nuclear interaction, more important as the particle slows down, in which the particle loses its energy by elastic collisions with lattice atoms. In addition, the energy is also partitioned to radiation (e.g., Chadderton 2003).

Although there is good agreement on the deceleration process of the charged particle and induced interactions, the debate has raged as to which mechanism causes the resultant motion of the lattice atoms that forms and registers the damage trails (e.g. Fleischer et al. 1975; Durrani and Bull 1987; Wagner and Van den haute 1992; Gleadow et al. 2002; Chadderton 2003). One of the widely accepted theories is the "ion explosion spike" model (Fleischer et al. 1965a, 1975) that treats electrostatic displacements as the primary process. In this model, track formation occurs during three stages (Fig. 1): (1) the rapidly moving positively-charged particle strips lattice electrons along its trajectory, leaving an array of positively-ionized lattice atoms; (2) the resulting clusters of positive ions are displaced from their original lattice sites as a result of Coulomb repulsion, creating interstitials and vacancies; and (3) the stressed region relaxes elastically, straining the surrounding undamaged lattice. The creation of the long-ranged strains in the third stage makes possible the direct observation of unetched (or latent) tracks by transmission electron microscopy (TEM). This model explains the primary observation that particle tracks are observed only in dielectric solids having electric resistivity >2000 Ωm (Fleischer et al. 1975).

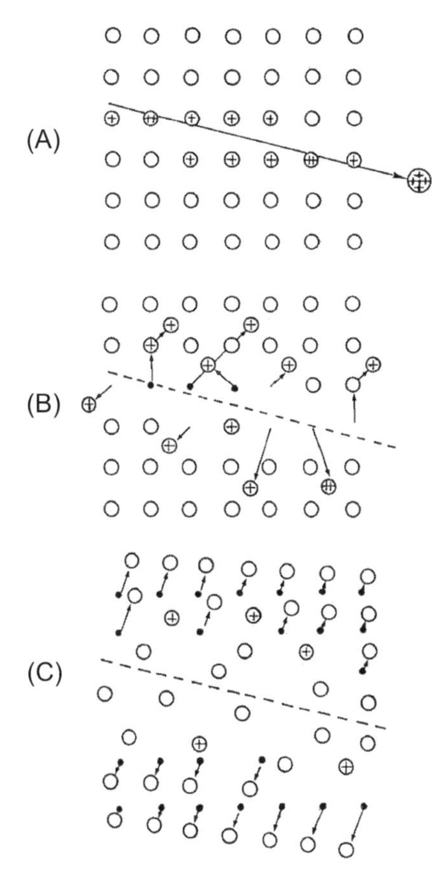

Figure 1. Formation and registration process of charged particle tracks in a dielectric solid by the ion explosion spike model (Fleischer et al. 1965a), which involves three stages: (A) the rapidly moving positively-charged particle strips lattice electrons along its trajectory, leaving an array of positively-ionized lattice atoms; (B) the resulting clusters of positive ions are displaced from their original lattice sites as a result of Coulomb repulsion, creating interstitials and vacancies; and (C) the stressed region relaxes elastically, straining the surrounding undamaged lattice.

An alternative is the "thermal spike" model (Seitz 1949; Bonfiglioli et al. 1961; Chadderton and Montagu-Pollock 1963), in which the passage of an energetic particle is assumed to produce instantaneous, intense heating of the lattice along the trajectory. The track core is rapidly heated to a high temperature and subsequently quenched by thermal conduction into the surrounding lattice. As a result, lattice defects are created by the thermal activation and the track core is left disordered. Chadderton (2003) argued that both ion explosion and thermal spikes could be present to different degrees for track formation and registration in a variety of solids.

Structure of the latent track

Knowledge on the structure of fission tracks in their latent state is critical not only to unravel their formation process but also to serve as a physical basis for their thermal annealing behaviors, which have conventionally been observed and quantified after chemical etching. Since the pioneering work by Silk and Barnes (1959), the atomic-scale characterization of latent tracks has been conducted using TEM and other analytical techniques on apatite (Paul and Fitzgerald 1992; Paul 1993), zircon (Yada et al. 1981, 1987; Bursill and Braunshausen 1990), muscovite mica (Thibaudau et al. 1991; Vetter et al. 1998) and other dielectric materials (see overviews by Fleischer et al. 1975; Wagner and Van den haute 1992; Neumann 2000; Chadderton 2003) (Fig. 2). In general, latent tracks represent a cylindrical shape of amorphous material in a crystalline matrix, with a sharp amorphous-crystalline transition. The elastic strain field around the disordered core extends a short distance into the matrix, likely with $1/R^2$ dependence (Bursill and Braunshausen 1990; R, radial distance), with no evidence for structural defects in its vicinity. The cross section of the track has a nearly circular shape with widths of 6–10 nm in apatite (Paul 1993), ~8 nm in zircon (Bursill and Braunshausen 1990) and ~4–10 nm in muscovite mica, depending on the ion energy loss (Vetter et al. 1998). The track is a linear, continuous feature and has approximately

(A)

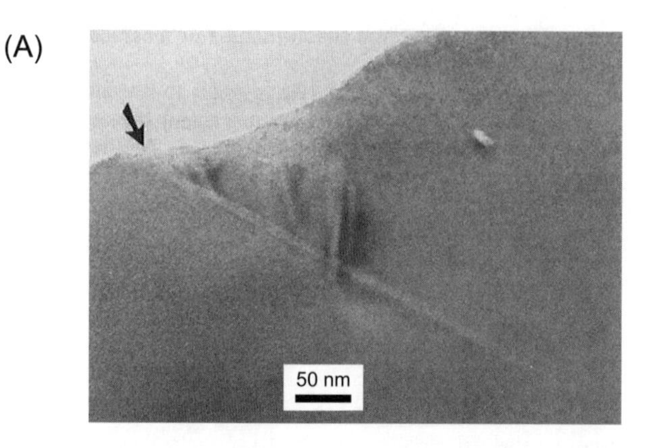

(B)

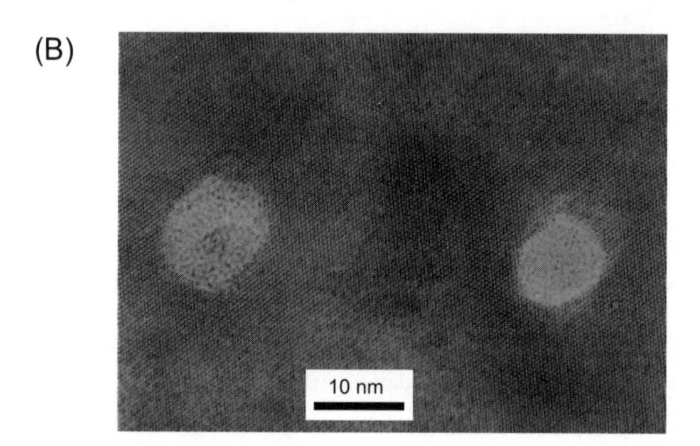

Figure 2. Atomic-scale images of latent (unetched) tracks: (A) an induced track in Durango fluorapatite observed subparallel to its length by transmission electron microscopy (Paul and Fitzgerald 1992) and (B) intersections of Pb ion tracks in muscovite mica imaged by 400 kV high-resolution transmission electron microscopy (Vetter et al. 1998).

a uniform diameter along most of its length. For its entire range, however, the track is a cylinder over a certain length and has a tapering-down in diameter near its end (Dunlap et al. 1997; Chadderton 2003), which may be the case of both terminals of a fission track (Carlson 1990).

The axial variation in damage density is deduced from a theoretical consideration on the rate of kinetic energy loss along a trajectory of nuclear fission fragments (Green et al. 1988; Wagner and Van den haute 1992). Since the kinetic energies of a pair of fission fragments are both below the Bragg peak in the energy loss curve, both fragments are most intensely ionizing, and therefore create most intensive damages, in the early part of their passage through the crystal lattice, i.e., around the site of the nuclear fission. As they slow down and lose energy, the damage intensity falls off away from this point, with the result that the damage density along the track peaks around the center and falls to zero at each end (Green et al. 1988).

CHEMICAL ETCHING AND OPTICAL MICROSCOPE OBSERVATION

Latent fission tracks can be viewed only by TEM and other high-resolution microscopic techniques because of their limited widths of only several nm. They, however, are conveniently scanned only when samples have very high track densities (i.e., > 10^{10}cm^{-2}), which is generally not the case of terrestrial minerals such as apatite and zircon. Another potential problem with viewing latent fission tracks is that the entire range of each track cannot be observed at such a high magnification, despite the necessity to analyze track lengths for the modern FT thermochronometry method (Paul 1993). In this regard, a technique is needed to enlarge fission tracks in order to be visible at a lower magnification under an optical microscope. Of the visualization techniques available, by far the most general and widely used has been chemical etching, which utilizes the preferential chemical attack along the entire length of the fission tracks.

Basic process of track etching

Fission tracks can be enlarged by chemical etching because the disordered region of the track core is more rapidly dissolved than the surrounding undamaged bulk material, due to its lowered binding energy. Chemical etching is conducted by immersing a dielectric material into a particular reagent under strictly controlled temperature and time conditions. Thus, only tracks that intersect the material's surface are etched and enlarged for observation. The reagent and etching conditions are selected on an empirical basis and have been established for a variety of minerals and other dielectric solids (Fleischer et al. 1975; Durrani and Bull 1987; Wagner and Van den haute 1992).

The geometry of an etched track is controlled by the simultaneous action of two etching processes: chemical dissolution along the track at a rate V_T and general attack on the etched surface and on the internal surface of the etched track at a lesser rate V_G (Fleischer and Price 1963a,b) (Fig. 3). Under given etching conditions, V_T in general increases with ionization rate and thus varies along a track, whereas V_G is generally constant for a given material but can depend on crystallographic orientations. In natural glasses, which are typically amorphous, V_T is only a few times higher than V_G and hence the etched tracks appear as rounded cones. In crystals, however, V_T is generally much higher than V_G by more than a factor of 10 and this results in a needle-like track shape (Fig. 4). In the course of such etching, dissolution and removal of the latent track core is completed before the track is sufficiently enlarged and becomes visible under the optical microscope. Therefore, the visible enlargement process is predominantly governed by the progressive widening of the track channel to ~1 μm at a rate V_G. See previous literature for more details (Fleischer et al. 1975; Durrani and Bull 1987; Wagner and Van den haute 1992).

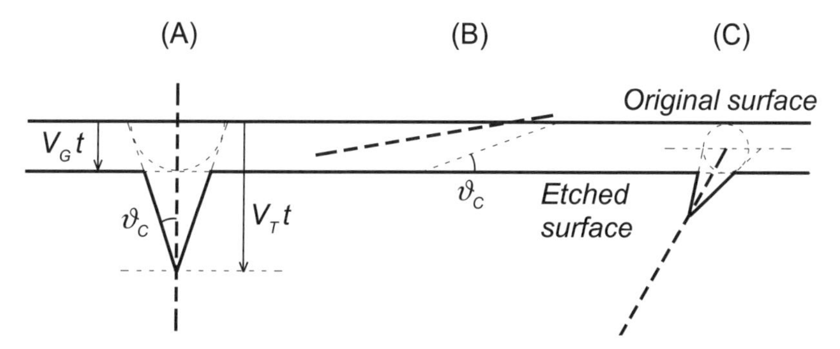

Figure 3. (A) The geometry of an etched track is controlled by two etching processes, i.e., chemical dissolution along the track at a rate V_T and general attack on the etched surface and on the internal surface of the etched track at a lesser rate V_G (Fleischer and Price 1963a,b): (B) Tracks inclined at relatively low angles to a surface (i.e., less than the critical angle, ϑ_C) are not etched to be observable under optical microscope: (C) New tracks that began and ended beneath the original surface are revealed by progressive removal of the surface itself. Here V_G/V_T is given as 1/3 throughout.

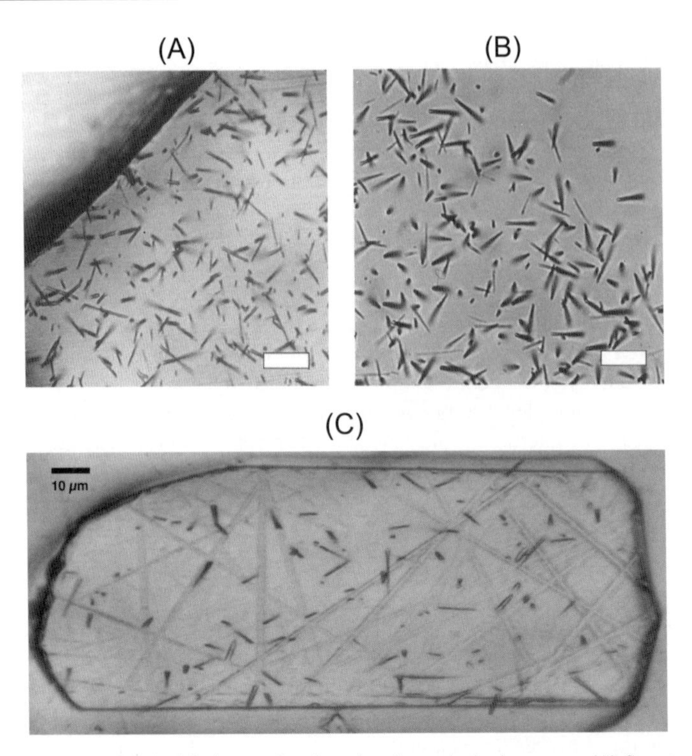

Figure 4. Photographs of etched fission tracks viewed under optical microscope. (A) Spontaneous tracks revealed on a polished internal surface of ~27.8 Ma Fish Canyon Tuff zircon. The crystallographic *c*-axis lies approximately vertical. (B) Induced tracks implanted on a muscovite detector (Brazilian Ruby clear) that were derived from the region of the photograph (A). (Photos by TT) (C) Spontaneous tracks on a polished internal surface of ~33 Ma apatite crystal. The *c*-axis lies approximately horizontal. (Photo by POS) Scale bars are 10 μm.

Etching efficiency and prolonged-etching factor

Because the surface of the mineral grain or glass is attacked and progressively removed during etching, tracks inclined at relatively low angles to the original surface are etched away. Such low-angle tracks are therefore not observable under the optical microscope (Fig. 3), which results in the number of latent tracks intersecting a given surface no longer being equal to that of etched ones on the surface. The minimum angle to the surface above which tracks are etched is called the critical angle ϑ_C, which is equal to arcsine (V_G/V_T) (Fleischer and Price 1964; Fleischer et al. 1975). An etching efficiency η is defined by the fraction of tracks intersecting a surface that are etched on the surface, and is given by $\cos^2\vartheta_C$ for the case of internal "thick" sources, where particle tracks originate throughout the volume of the detector itself. Thus:

$$\rho_E \equiv \eta\rho_L = \rho_L \cos^2 \vartheta_C = \rho_L \left(1 - \frac{V_G^{\,2}}{V_T^{\,2}}\right) \tag{1}$$

where ρ_E and ρ_L are the areal density of etched and latent tracks, respectively. ϑ_C can be experimentally determined by the etch-test of particle tracks that are implanted by the bombardment of collimated heavy ions at a certain angle to the sample's surface. Such experiments yielded ϑ_C values of ~25–35° for natural glasses and < 10° for crystals (Khan and Durrani 1972).

When tracks originating from the internal source are etched, new tracks that began and ended beneath the original surface are revealed by progressive removal of the surface itself (Fig. 3). As a result, the total number of tracks will increase monotonically as etching proceeds because the tracks etched at an early stage continue to be visible even after long etch times, although they grow and become less distinct. This effect is called the "prolonged-etching factor" (Kahn and Durrani 1972) and is quantified by:

$$\Delta\rho = N_F h(1-\sin\vartheta_C) \tag{2}$$

where $\Delta\rho$ is the areal density of new tracks added, N_F is a number of tracks per unit volume and h is the thickness of the layer removed. Because $\rho_L = gN_FR_L$, where g is a geometry factor of the detector (0.5 and 1 for external and internal surfaces, respectively) and R_L is an etchable range (i.e., length) of the latent track, thus:

$$\Delta\rho = \rho_L \frac{h(1-\sin\vartheta_C)}{gR_L} = \rho_E \frac{h}{gR_L(1+\sin\vartheta_C)} \tag{3}$$

Since $h = V_G t$, where t is a time duration of etching:

$$\Delta\rho = \rho_L \frac{V_G t\left(1-{V_G}/{V_T}\right)}{gR_L} \tag{4}$$

The total observed density of etched tracks ρ_O is thus given by:

$$\rho_O = \rho_E + \Delta\rho = \rho_L\left\{1-\frac{V_G^{\,2}}{V_T^{\,2}} + V_G t\frac{\left(1-{V_G}/{V_T}\right)}{gR_L}\right\} \tag{5}$$

Etching criteria and their influences on the observed track density and length

For reliable track density determination, the observed track density ρ_O needs to be as close to the latent track density ρ_L as possible. It is therefore desirable to analyze material that has lower V_G, so that ρ_E and $\Delta\rho$ come close to ρ_L and 0, respectively. Because V_G is anisotropic in crystals, it is necessary to employ a crystal surface having the lowest V_G. This is particularly important with highly anisotropic minerals, such as zircon and sphene (Fig. 5). A conventional way to identify surfaces of low V_G is to check whether the etched surface has clear and sharp polishing scratches that indicate relatively low $V_G t$ (Naeser et al. 1980;

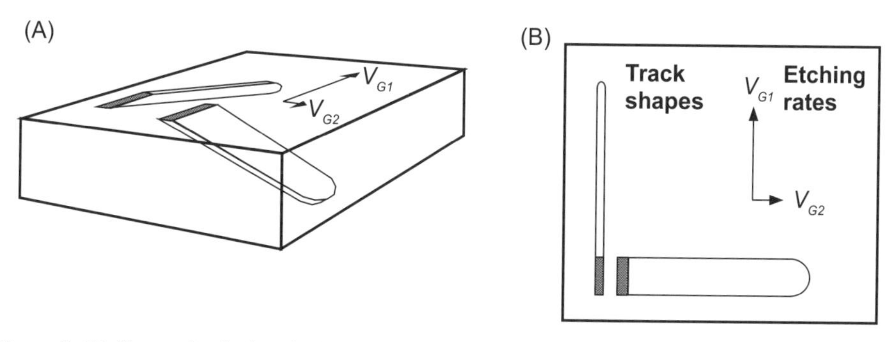

Figure 5. (A) Shape of etched tracks in minerals having highly anisotropic V_G (i.e., a big contrast between V_{G1} and V_{G2}). (B) Top view of the etched mineral surface, on which track revelation is also highly anisotropic. After Gleadow (1981).

Gleadow 1981). Secondly, etch time t needs to be as short as possible to minimize $\Delta\rho$. Thus, in theory, etching should be stopped when all tracks that intersect the original surface become clearly visible under the optical microscope. Because V_G changes between individual samples, particularly zircon and sphene, the optimum etch time must be determined by the step-etch and observation procedure (Gleadow 1981; Hasebe et al. 1994).

Care should be taken when etching minerals having highly anisotropic V_G, on which track revelation is also anisotropic. That is, tracks lying subparallel to a certain crystallographic orientation (e.g., c-axis in case of zircon) are more slowly enlarged than others because they have lower rates of widening of the track channel due to lower V_G perpendicular to the track orientation (Gleadow 1981). For such samples, track etching should be continued until the tracks that are etched most slowly and weakly become visible under the microscope, so that the number of etched tracks is approximately equal for all crystallographic orientations (Gleadow 1981; Sumii et al. 1987). Otherwise, ρ_O will be grossly underestimated relative to ρ_L. Further complexity and difficulty is due to the fact that the V_G anisotropy is not constant between samples but shows a systematic decrease as a result of accumulation of radiation damages. Hence, special care is needed to etch young zircons and sphenes because they have higher anisotropy due to the low level of radiation damage and because for such samples, ρ_L is in general so low that it is often difficult to judge the optimum etching condition using a small number of tracks therein (Gleadow 1980; Watanabe 1988).

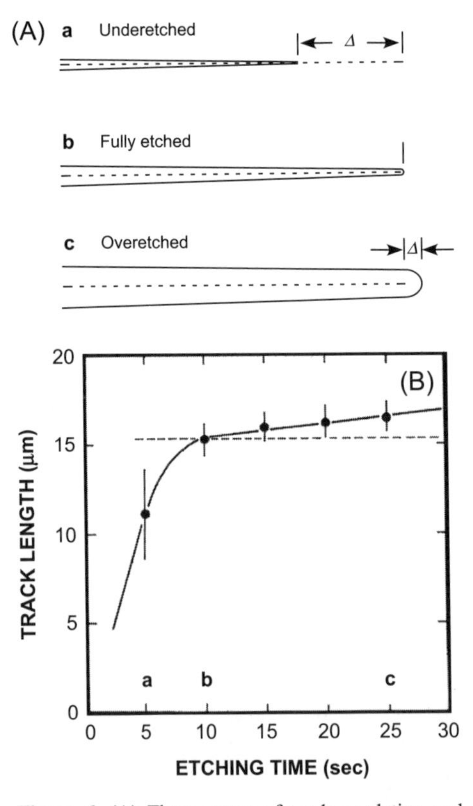

Figure 6. (A) Three stages of track revelation and associated errors (Δ) of the observed track length in apatite. The dot line represents an unetched (latent) track. (B) Observed track length vs. total etching time. Note the steady increase in the observed length after being fully etched, due primarily to the outer growth of the track channel as the etching proceeds. After Laslett et al. (1984).

The observed range (i.e., length) of etched track, R_E, is also different from R_L, the etchable range of the latent track, due primarily to the outer growth of the track channel as the etching proceeds (Fig. 6). As the overestimation of the range, ΔR, is approximately given by $\Delta R = 2V_G t$ (Laslett et al. 1984), thus:

$$R_E \equiv R_L + \Delta R = R_L + 2V_G t \qquad (6)$$

If V_G is isotropic and $V_T >> V_G$, the width of etched track, W_E, is approximately equal to $2V_G t$ and is accordingly a good measure of ΔR. In the case of crystals, however, V_G is generally anisotropic and W_E merely offers a rough estimate of ΔR. For an ordinary etching condition, $\Delta R \approx W_E = \sim 1$ μm and therefore is not negligible for confined track length analysis that will be mentioned later. Since ΔR is a function of t and increases as the etching proceeds, standardization of etching condition is required for reliable track length measurement as well as for inter-laboratory comparison of length data.

DERIVATION OF AGE CALCULATION EQUATION

In principle, a radiometric age is given by three parameters, i.e., the numbers of parent and daughter nuclides in a material, and the decay constant for the parent nuclide. In the FT method, they are respectively the number of ^{238}U per unit volume, ^{238}N, the number of spontaneous fission tracks per unit volume, N_S, and the decay constant for spontaneous nuclear fission, λ_F. Because ^{238}U also decays by α-emission with a much greater decay constant (1.55125×10^{-10}y^{-1}), λ_D, the basic formula for a FT age, t, is given by:

$$N_S = \frac{\lambda_F}{\lambda_D}\, {}^{238}N\left\{\exp(\lambda_D t) - 1\right\} \tag{7}$$

To measure ^{238}N, the nuclear fission reaction of ^{235}U that is artificially induced by thermal neutron irradiation is utilized. The number of induced fission tracks per unit volume, N_I, is given by:

$$N_I = {}^{235}N\sigma_F\Phi \tag{8}$$

where ^{235}N is the number of ^{235}U per unit volume, σ_F is the cross section for induced nuclear fission of ^{235}U by thermal neutrons (580.2×10^{-24}cm^2), and Φ is the thermal neutron fluence. From these equations,

$$t = \frac{1}{\lambda_D}\ln\left\{1 + \left(\frac{\lambda_D}{\lambda_F}\right)\left(\frac{N_S}{N_I}\right) I\,\sigma_F\,\Phi\right\} \tag{9}$$

where I is the isotopic abundance of U, $^{235}N/^{238}N$ (7.2527×10^{-3}). Only tracks intersecting the etched surface are observable under an optical microscope and thus:

$$t = \frac{1}{\lambda_D}\ln\left\{1 + \left(\frac{\lambda_D}{\lambda_F}\right)\left(\frac{\rho_S}{\rho_I}\right) Q\,G\,I\,\sigma_F\,\Phi\right\} \tag{10}$$

where ρ_S is the surface density of etched spontaneous fission tracks, ρ_I is the surface density of etched induced fission tracks, Q is the integrated factor of registration and observation efficiency of fission tracks, and G is the integrated geometry factor of etched surface. In this equation, λ_F is not well determined (Bigazzi 1981; Van den haute et al. 1998), whereas Q and Φ are generally difficult to measure accurately (Van den haute et al. 1998). Although, in principle, it is possible to determine these constants and values individually and calculate t absolutely (Wagner and Van den haute 1992; Van den haute et al. 1998), the empirical zeta calibration based on the analysis of age standards (Hurford and Green 1982, 1983; see also Fleischer et al. 1975) was recommended by the IUGS subcommission on Geochronology (Hurford 1990) and has since been used universally. Φ is conventionally measured by the induced fission-track density on a U-doped standard glass, ρ_D, irradiated together with the sample, and is thus given by:

$$\Phi = B\rho_D \tag{11}$$

where B is a calibration constant empirically determined. Thus, the zeta age calibration factor, ζ, is defined accordingly by:

$$\zeta \equiv B\frac{I\sigma_F}{\lambda_F} \tag{12}$$

Hence,

$$t = \frac{1}{\lambda_D}\ln\left\{1 + \lambda_D\zeta\rho_D\left(\frac{\rho_S}{\rho_I}\right) QG\right\} \tag{13}$$

ζ is determined empirically by analyzing a set of standard materials of known ages (Hurford 1990). Q is the same between the age standards and age-unknown samples as long as they are analyzed using identical experimental procedures and criteria. In the routine analysis, therefore, Q is ignorable and ζ, is given by inputting the measured data of standard (i.e., ρ_D, ρ_S and ρ_I) and the reference age t into the equation above.

STABILITY AND FADING OF TRACKS

Basic process of track fading

Knowledge of the stability and fading of fission tracks is critically important in order to decode the geological history recorded by the host mineral. Of several possible environmental effects that could alter fission tracks, the most pervasive and recognized one is thermal annealing (Fleischer et al. 1975). The thermal annealing of fission tracks at increased temperature (or time) is a process during which the amorphous, disordered core is gradually restored to the ordered structure of crystalline matrix, as confirmed by TEM observation (Paul and Fitzgerald 1992; Paul 1993) (Fig. 7). The process is characterized by (1) the development of irregular morphology at the track-matrix boundary, (2) segmentation of the track, i.e., appearance of gaps, (3) an increase in the spacing between segments, the shape of which approaches a sphere, and (4) instantaneous healing of spheres, with an observable minimum diameter of ~3 nm (Paul 1993). The annealing likely occurs through the diffusion of interstitial atoms and lattice vacancies to recombine as a result of their thermal activation due to heating.

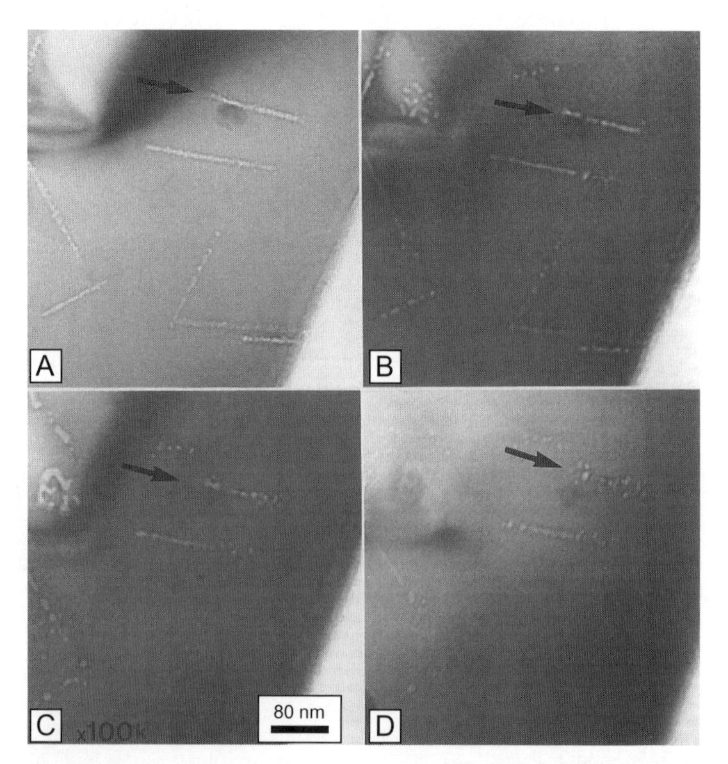

Figure 7. (A) – (D): Radiolytic-thermal annealing processes of induced tracks in Durango fluorapatite (after Paul and Fitzgerald 1992). The annealing proceeds from (A) to (D), as a consequence of electron-beam exposure and heating on the sample.

The thermal annealing behavior of fission tracks has been characterized and quantified using etched tracks, upon which routine thermochronologic studies are based. At the optical-microscopic scale, fission tracks progressively shrink from each end during early stages of annealing, as suggested by confined track-length analyses on apatite (Green et al. 1986; Donelick 1991; Donelick et al. 1999) and on zircon (Yamada et al. 1995a) (Fig. 8). At later stages of annealing, fission tracks continue to shrink, and appear to become broken into separate segments, with a result of a gross increase in standard deviation of track lengths for a specific crystallographic orientation (Green et al. 1986; Hejl 1995; Yamada et al. 1995a).

Track annealing at geological timescales: procedures and findings

Understanding the long-term annealing behaviors of the FT system over geological timescales is essential to unravel thermal histories of terrestrial and planetary materials. The conventional approach towards this is to: (1) make a systematic, quantitative description of FT annealing behaviors in laboratory under well-controlled conditions of temperature, time,

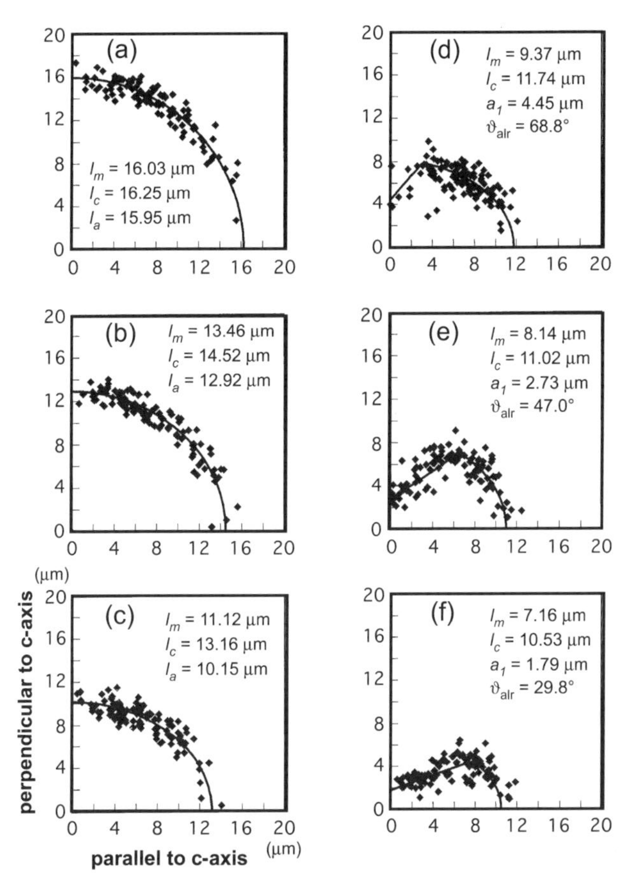

Figure 8. Polar coordinate plots of the induced fission-track lengths measured for Durango apatite at various degrees of annealing (Donelick et al. 1999). At relatively low degrees of annealing (a-c), fission-track lengths are approximately uniformly distributed about an ellipse with respect to track angle to the crystallographic *c*-axis. At higher degrees of annealing (d-f), the ellipse collapses, with fission tracks at relatively high angles to the crystallographic *c*-axis experiencing systematic, accelerated length reduction. As the degree of annealing increases, the initiation of the collapse of the ellipses rotates toward the *c*-axis.

pressure, etc., (2) predict FT annealing under plausible geological conditions by extrapolating a numerical annealing model constructed using the laboratory annealing data, (3) test the prediction by analyzing fission tracks in natural materials for which the thermal history and other environmental conditions are known or inferred with some confidence, and (4) modify and recalibrate the numerical annealing model, if necessary.

The long-term annealing characteristics of the FT system have been studied at a variety of natural settings; such as deep boreholes (e.g., Naeser and Forbes 1976; Gleadow and Duddy 1981; Green et al. 1989a; Corrigan 1993; Wagner et al. 1994; Tagami et al. 1996; Coyle and Wagner 1998; Hasebe et al. 2003) (Fig. 9), exposure of deeply buried sections (e.g., Zaun and Wagner 1985; Stockli et al. 2002), contact metamorphic aureoles (e.g., Naser and Faul 1969; Calk and Naeser 1973; Tagami and Shimada 1996), regional diagenetic and metamorphic zones (e.g., Green et al. 1996; Brix et al. 2002), monotonous cooling of plutonic and metamorphic rocks (e.g., Harrison et al. 1979; Hurford 1986), etc.

Some important results discussed in the aforementioned natural annealing studies are briefly summarized below (details will be discussed in following chapters):

- In geologic studies, even those involving thermally undisturbed volcanic rocks, the measured lengths of spontaneous tracks are not always equal to the measured lengths of induced tracks. Of note: length measurements are made on "confined" tracks (Laslett et al. 1982); i.e., tracks that are located within a crystal and are etched through a host track or cleavage, so that they exhibit their entire lengths (Fig. 10). For instance, confined track length data from apatite indicate that spontaneous tracks are always shorter than induced tracks (Wagner and Storzer 1970; Bertel et al. 1977; Green 1980; Gleadow et al. 1986), whereas this is not the case for zircon (Hasebe et al. 1994) and sphene (Gleadow 1978).
- FT annealing is indistinguishable between laboratory and nature in terms of the appearance and patterns of track shortening (Green et al. 1989a; Tagami and Shimada 1996), implying that the same physical processes consistently govern track annealing. This is also supported by the fact that the existence of fossil and modern

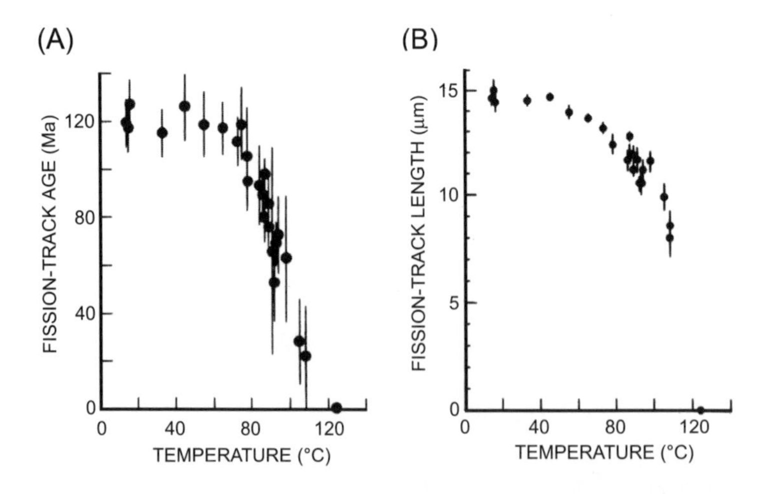

Figure 9. Long-term natural annealing of fission tracks in apatite observed in borehole samples from the Otway Basin, Australia, for which the geological evolution was well constrained (after Green et al. 1989a). Both fission-track age (A) and length (B) are reduced progressively down to zero in the temperature range of ~60–120 °C due to the increase in geothermal temperature with depth. Error bars are ± 2σ.

(A)

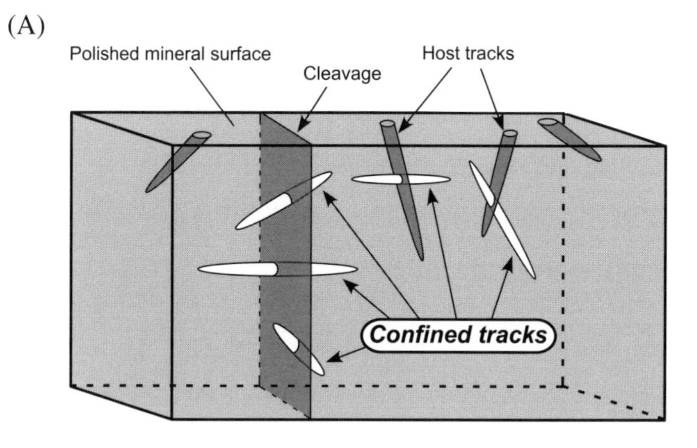

(B)

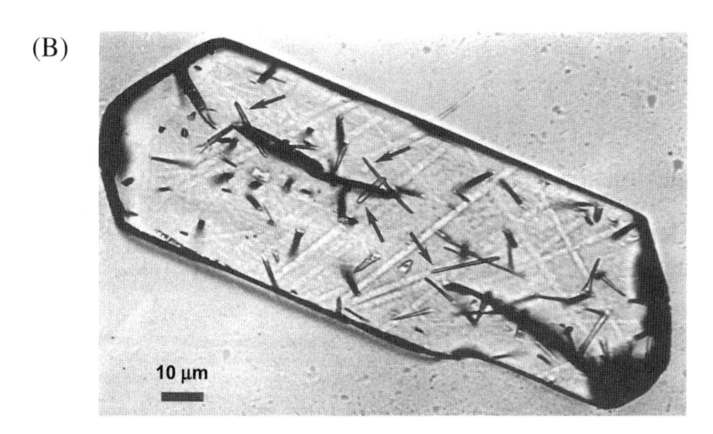

Figure 10. (A) A schematic illustration of an etched mineral that reveals confined tracks of different dimensions, i.e., tracks-in-cleavage (TINCLEs) or tracks-in-track (TINTs). (B) A top-view photograph of etched spontaneous tracks on a polished internal surface of apatite crystal (after Gleadow et al. 1986). Most of the visible tracks are surface-intersecting spontaneous tracks, which are used for age determination. Arrows point to four individual confined tracks (revealed as TINCLEs) exhibiting original entire track lengths, which are useful for estimating the true length distribution.

partial annealing zones (PAZ's; e.g., Fitzgerald and Gleadow 1990) was confirmed within a variety of geological settings (e.g., Calk and Naeser 1973; Naeser and Forbes 1976; Gleadow and Duddy 1981; Zaun and Wagner 1985; Green et al. 1989a; Corrigan 1993; Wagner et al. 1994; Tagami and Shimada 1996; Coyle and Wagner 1998; Stockli et al. 2002).

- Thermal sensitivities of fission tracks and other thermochronometers at geological timescales were inferred by relative positions of the PAZs (or partial retention zones), showing a reasonable consistency with the sensitivities observed in the laboratory (Calk and Naeser 1973; Coyle and Wagner 1998; Stockli et al. 2002; Reiners 2003; Tagami et al. 2003).
- The FT annealing behaviors predicted by extrapolating numerical annealing models have been tested on apatite (Green et al. 1989b; Corrigan 1993; Ketcham et al. 1999) and on zircon (Tagami et al. 1998), with a result that the predictions are generally

consistent with observations. However, at least in the case of apatite, this really only holds true when the mineral species being modeled is identical in composition to the mineral species for which the model was generated. This accordingly requires the use of a "multi-kinetic" model that theoretically incorporates both compositional and crystallographic effects (e.g., Ketcham et al. 1999).

Laboratory heating experiments: procedures and findings

Where laboratory-heating experiments are conducted on pieces of a material containing fission tracks at a series of temperatures (T), the time (t) to produce a given degree of annealing of the fission tracks was classically described by:

$$t = A\exp\left(\frac{E_A}{kT}\right) \tag{14}$$

where k is Boltzmann's constant, A is a constant and E_A is the activation energy (e.g., Naeser and Faul 1969; Fleischer et al. 1975). The degree of annealing is represented by iso-density or iso-length contours, when the data are displayed on a variation of an Arrhenius plot, i.e., the "$1/T$ vs. log t" diagram (Fig. 11). In earlier studies, the rate equation that describes FT annealing was given by first-order kinetics (e.g., Mark et al. 1973, 1981; Zimmermann and Gaines 1978; Bertagnolli et al. 1981; Huntsberger and Lerche 1987), in forms equivalent to:

$$\frac{dr}{dt} = -\alpha(T)r, \text{ with } \alpha(T) = \alpha_0 \exp\left(-\frac{E_A}{kT}\right) \tag{15}$$

where r is the degree of annealing, which is measured by the reduced FT density (or length) normalized to the unannealed value, and α_0 is a constant. This formulation is valid if the annealing results from diffusion of lattice defects having a unique E_A.

Later studies demonstrated, however, that the first-order kinetics gave a poor description of track annealing of minerals, such as apatite and zircon (e.g., Green et al. 1988; Green and Duddy 1989). Higher-order processes using a rate equation gave the more preferred description:

$$\frac{dr}{dt} = -\alpha(T)(1-r)^n \tag{16}$$

where n is approximately −3 to −4 for the Durango apatite (Laslett et al. 1987). Green et al. (1988) argued that, whereas the diffusive motion of individual disordered atoms may fundamentally be controlled by first-order kinetics, the annealing actually observed on etched fission tracks is a result of a number of complicated processes: i.e., the reduction in r is a manifestation of the reduction in atomic disorder that consists of a number of defects in an anisotropic, multi-component mineral.

Originally, laboratory heating experiments were aimed at determining the reduction in etched track density, upon which the FT dating method was based (e.g., Wagner 1968; Naeser and Faul 1969; Nagpaul et al. 1974; Bertel and Mark 1983). More recent studies, however, focused primarily on: 1) the reduction in length of etched tracks with which FT annealing is more precisely quantified, and 2) the distribution of etched confined track lengths, the shape of which is indicative of a rock's thermal history (e.g., Green et al. 1986; Carlson et al. 1989; Crowley et al. 1991; Donelick 1991; Yamada et al. 1995ab; Donelick et al. 1999; Barbarand et al. 2003ab). Here it is noted that the observed mean confined track length of spontaneous fission tracks in apatite and zircon grains separated from rapidly cooled and thermally undisturbed volcanic rocks is approximately constant (~14.5–15.0 μm for apatite; Gleadow et al. 1986; ~10.5 μm for zircon; Hasebe et al. 1994). Thus, these values offer good controls to measure length reduction of unknown samples.

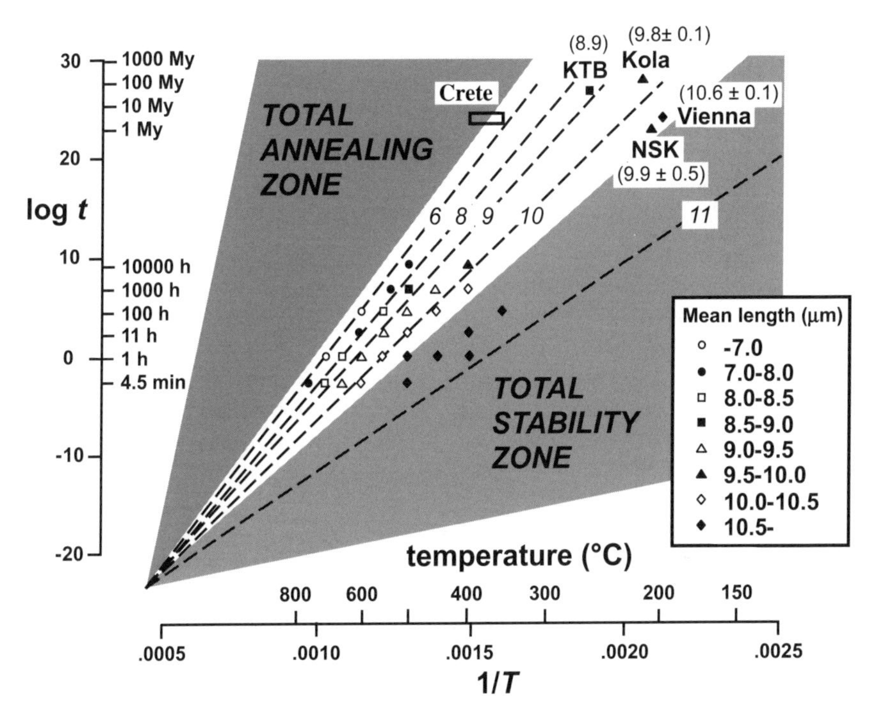

Figure 11. Arrhenius plot showing the design points of the laboratory annealing experiments of spontaneous fission tracks in zircon as well as contour lines for the fitted fanning model extrapolated to geological time scale (Yamada et al. 1995b; Tagami et al. 1998; model after Galbraith and Laslett 1997). Also shown are data points of four deep borehole samples subjected to long-term natural annealing, which were used to test the extrapolation of the annealing model (Tagami et al. 1998; Hasebe et al. 2003), and a box that indicates the time-temperature condition estimated for the higher temperature limit of the zircon PAZ on Crete (Brix et al. 2002). Three temperature zones can be defined as first-order approximation of fission-track annealing: the total stability zone (TSZ) where tracks are thermally stable and hence accumulated as time elapsed; the partial annealing zone (PAZ) where tracks are partially stable and slowly annealed and shortened; and the total annealing zone (TAZ) where tracks are unstable and faded soon after their formation. Note that the zircon PAZ is defined here as a zone having mean lengths of ~4 to 10.5 μm (Yamada et al. 1995b) and shown as a white region intervened by the TSZ and TAZ.

Heating experiments have typically been conducted under muffle-type or other electric furnaces, with samples being heated to varying temperatures under dry conditions, and pressures equal to one atmosphere. However, to assess the influence of various geological environmental factors, such as pressure, deformation, etc., a variety of heating systems and settings were employed during some tests, including the belt apparatus for dry and high-pressure conditions (Fleischer et al. 1965b), and the hydrothermal synthetic apparatus to synthesize hydrothermal-pressurized conditions (Wendt et al. 2002; Yamada et al. 2003). These experiments were usually carried out isochronally, i.e., at a set temperature with a variable time duration, or isothermally, for a controlled time duration with variable temperature. Once etched, fission tracks are no longer useful for further annealing runs because the chemical etching completely dissolves the damage core of latent tracks thus permanently preserving the etched track length. Therefore, each aliquot of a sample is annealed and analyzed for each step of heating experiments.

The results from numerous annealing experiments have previously been summarized by Fleischer et al. (1975), Durrani and Bull (1987), and Wagner and Van den haute (1992), and a similar scale of summarization is beyond the scope of this chapter. However, some of the important results reported from these experiments, other than those mentioned in the last section, are briefly summarized below (some details will be discussed in the following chapters):

- The track annealing rate, *dr/dt*, depends upon the crystallographic orientation of the tracks. For instance, fission tracks parallel to the *c*-axis anneal at a slower rate than do those perpendicular to it for both apatite (Green et al. 1986; Donelick 1991; Donelick et al. 1999), and zircon (Tagami et al. 1990).
- The annealing rate also depends upon the chemical composition of a mineral, e.g., fission tracks in Cl-rich apatites anneal more slowly than do those in OH- and F-rich apatites (Green et al. 1986; Crowley et al. 1991; Carlson et al. 1999; Barbarand et al. 2003a).
- The annealing rate may depend upon the accumulation of radiation damage in crystals, such as zircon and sphene. Experimental data suggest that spontaneous fission tracks in older zircons (i.e., having greater damage accumulation) anneal slower than do those in young zircons (Kasuya and Naeser 1988; Carter 1990; Yamada et al. 1998; Rahn et al. 2004).
- The annealing rate possibly depends to some degree on pressure. While three studies using different experimental systems show no such dependence on the annealing rate in zircon (Fleischer et al. 1965b; Brix et al. 2002; Yamada et al. 2003), similar studies report conflicting results in terms of a pressure influence on apatite annealing (Wendt et al. 2002; Kohn et al. 2003; Donelick et al. 2003). See Donelick et al. (2005) for further discussion of this topic as it pertains to apatite.
- The use of different etchants can give rise to different track annealing rates: an example is HCl vs. NaOH etchants for sphene (Fleischer et al. 1975).
- Fission tracks in some materials undergo slow annealing at low ambient temperatures. For instance, extremely fresh fission tracks induced by thermal neutron bombardment in apatite show appreciable reduction in etchable length at temperatures < 23 °C (Donelick et al. 1990). This is consistent with the observation that unannealed spontaneous fission tracks are always shorter than unannealed induced fission tracks in apatite (Gleadow et al. 1986).
- The effects of various experimental biases are assessed in track length measurement of both unannealed and annealed fission tracks, which is indispensable for reliable interpretation of FT data (Green 1988; Galbraith et al. 1990; Yamada et al. 1995a; Donelick et al. 1999; Barbarand et al. 2003b).
- Numerical FT annealing models (i.e., kinetics) have been investigated and improved on apatite (Laslett et al. 1987; Carlson 1990; Crowley et al. 1991; Laslett and Galbraith 1996; Ketcham et al. 1999) and on zircon (Yamada et al. 1995b; Galbraith and Laslett 1997; Tagami et al. 1998), opening a way to the inverse modeling of geological thermal history (Crowley 1985, 1993; Green et al. 1989; Corrigan 1991; Lutz and Omar 1991; Gallagher 1995; Issler 1996; Willet 1997; Ketcham et al. 2000). See following chapters for more details (Donelick et al. 2005 on apatite kinetics; Tagami 2005 on zircon kinetics; Ketcham 2005 for inverse modeling).

EXPERIMENTAL PROCEDURES

Brief descriptions of individual experimental steps are given in this section, with a particular focus on the external detector method (EDM), the most widely accepted method of dating mineral grains at this time (Fig. 12). See previous discussions by Naeser (1976), Gleadow

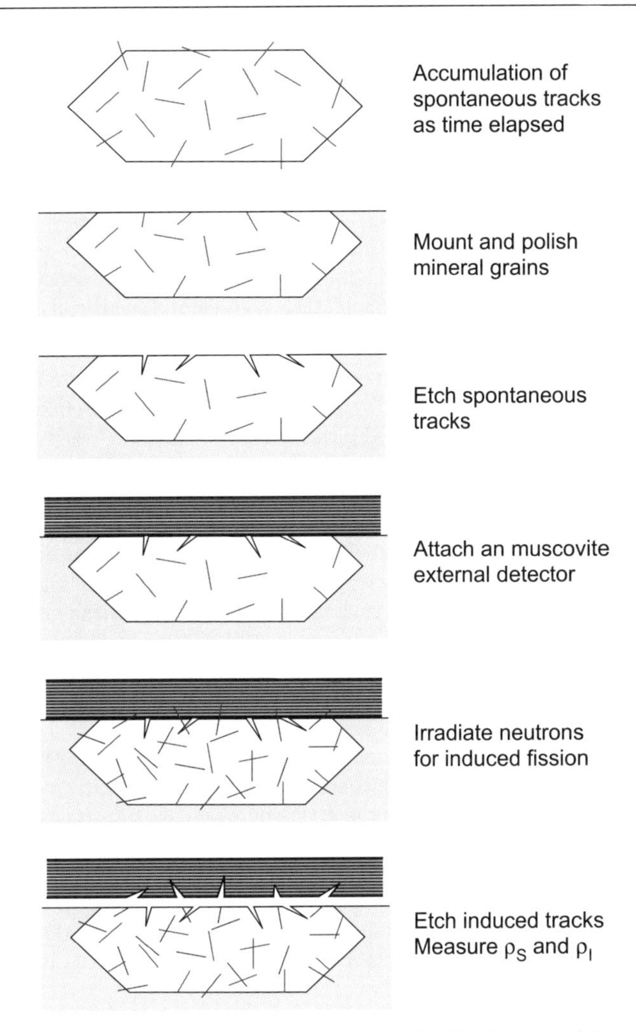

Figure 12. Schematic illustration of the experimental steps involved in the external detector method. The technique is by far the most popular dating procedure for mineral samples collected from a variety of geologic settings. ρ_S = surface density of etched spontaneous fission tracks; ρ_I = surface density of etched induced fission tracks.

(1984), Tagami et al. (1988), Ravenhurst and Donelick (1992), Wagner and Van den haute (1992) and Dumitru (2000) for further details as well as for other procedures and materials. See Donelick et al. (2005) for further discussion and suggestions of proposed experimental procedures as they pertain to apatite.

Methods of analysis

The FT age determination of an unknown sample is based upon the measurement of three different track densities, i.e., ρ_S, ρ_I and ρ_D. The first two densities are measured directly from the sample. At one time or another, a variety of experimental procedures have been employed in order to make these measurements, depending to some extent on the nature of the material to date (i.e., Naeser 1979; Gleadow 1981; Hurford and Green 1982; Ravenhurst and Donelick 1992; Wagner and Van den haute 1992).

Originally, the "population method" was the most popular for FT analysis of many different minerals, but is now recommended only for dating of glasses. In this procedure, ρ_S and ρ_I are both measured on the internal mineral surface using separate aliquots of the same sample. Induced fission tracks are etched and revealed on surfaces of mineral grains from which spontaneous fission tracks were removed by laboratory heating prior to the neutron irradiation. This procedure has the advantage that ρ_S and ρ_I are measured on surfaces of the same material that have identical track registration efficiencies. However, the disadvantage of this procedure is that FT ages cannot be measured on individual mineral grains. In addition, this procedure assumes that the two aliquots possess equal uranium concentrations and no zoning within individual grains, assumptions that need careful statistical and experimental assessments for individual samples.

At the time of publication of this chapter, by far the most widely used procedure is the external detector method (EDM) (Fig. 12). In this procedure, spontaneous fission tracks (ρ_S) are etched and revealed on an internal polished surface of a mineral grain, whereas induced fission tracks (ρ_I) are recorded on an external detector (e.g., thin sheet of a low-U muscovite mica) that is firmly attached onto the etched grain surface during neutron irradiation. Thus, ρ_S and ρ_I are measured respectively on the internal mineral surface and on the external detector surface. The advantage of this procedure is that FT ages can be determined on individual mineral grains, which is particularly important when attempting to date grains with varying ages and compositions within individual samples. Previous studies of both sedimentary rocks (e.g., Green et al. 1989a; Hurford and Carter 1991; Burtner et al. 1994; Barbarand et al. 2003a) as well as some basement rocks (O'Sullivan and Parrish 1995), have shown that significant variations in grain ages can occur within individual samples, and thus having the ability to date individual grains instead of simply determining a bulk sample age is imperative. The disadvantage is that because ρ_S and ρ_I are measured on different materials (i.e., original grain vs. mica detector), great care is needed to standardize etching conditions for each, particularly in the case of zircon and sphene that have variable and anisotropic etching characteristics (Gleadow 1978, 1981; Sumii et al. 1987). It is also noted that the sensitivity for track etching can be different between the mineral and external detector, as experimentally shown for apatite and zircon and muscovite detector (Iwano et al. 1992, 1993; Iwano and Danhara 1998). These points should be taken into account when applying an absolute age calibration (Wagner and Van den haute 1992; Van den haute et al. 1998), or when comparing observed zeta factors between different minerals or laboratories (Green 1985; Tagami 1987; Iwano and Danhara 1998).

Of special note, new procedure that utilizes a laser ablation ICP-MS to measure uranium values from individual grains, has recently been proposed (e.g., Cox et al. 2000; Hasebe et al. 2002, 2004; Svojtka and Kosler 2002; Kosler and Sylvester 2003), and may soon develop into the most widely recommended and utilized method. See Donelick et al. (2005) for further discussion on the use of laser ablation ICP-MS in FT age calculation and analysis of apatite.

Sample preparation and track etching

Minerals suitable for FT dating, such as apatite, zircon and sphene, are concentrated and separated from their host rock samples by conventional mineral separation techniques. The mineral grains are mounted into grain-holding materials, such as epoxy resin or a Teflon sheet (Gleadow et al. 1976; Naeser 1976; Ravenhurst and Donelick 1992). The sample mount is ground and polished to expose internal mineral surfaces (which should have at least $R/2$ distance to the original mineral surface for having 4π geometry). Spontaneous fission tracks that intersect the polished surface are revealed by placing the sample mount into an appropriate chemical reagent, such as acid or eutectic alkali (e.g., Fleischer et al. 1975), under particular time and temperature conditions. After etching, the sample is rinsed thoroughly to remove the reagent from the sample.

Neutron irradiation

An external detector, such as a thin low-uranium (<5 ppb) muscovite mica sheet, is firmly attached to each etched mineral mount. Individual sample-detector pairs are stacked between standard glass dosimeters, such as NIST-SRM 612 (or 962a) or Corning CN1 through CN6, on each of which the detector is also attached. The sample-standard stacks are then bombarded by neutrons at a well-thermalized irradiation facility in a nuclear reactor. Note that a substantial error is introduced in the age calculation if the sample is irradiated at a poorly-thermalized facility because induced fission tracks of the same size are also formed by epithermal- and fast-neutron induced fissions of ^{235}U as well as by fast-neutron induced fissions of ^{238}U and ^{232}Th (Crowley 1985; Tagami and Nishimura 1989, 1992). After irradiation and radioactive cooling, the muscovite detectors are detached from the samples, and are then etched to reveal tracks that were induced into the mica following decay of uranium within mineral grains during irradiation. See Green and Hurford (1984) and Wagner and Van den haute (1992) for more details of the neutron irradiation issue.

Track density determination

To determine ρ_S, ρ_I and ρ_D, the number of etched tracks that intersect the surface within a known area are counted using an optical microscope at magnifications of at least ~1000×. The following points need to be taken into account for reliable track density determination (e.g., Green 1981):

(1) Only mineral grains (or areas in individual grains) that satisfy the following criteria should be measured to determine ρ_S.

  (a) Grains should be free of spurious, non-track features, such as dislocations or tiny inclusions, and grains having many non-track features are best avoided. In many cases, however, non-track features are readily distinguishable from spontaneous fission tracks, which are straight, of limited length and randomly oriented.

  (b) Fission tracks on the grain surface should be well etched and clearly visible. This is particularly critical to dating zircon and sphene as they have, as mentioned above, variable and anisotropic etching characteristics (Gleadow 1978, 1981; Sumii et al. 1987).

  (c) Uranium distribution should be approximately homogeneous within the grain. This can generally be judged by variation in ρ_S on the grain surface. For samples of low ρ_S, variation in ρ_I on the muscovite detector will help.

(2) When using the EDM, ρ_S is counted on the etched internal surface of each mineral grain, whereas ρ_I is counted on the external detector surface within the corresponding area. Thus, for precise FT dating a reliable correlation is needed in order to move between the two areas.

Track length measurement

Analysis of horizontal confined tracks. As defined earlier, and shown in Figure 10, confined tracks are those contained entirely within the boundaries of the crystal being analyzed. For the purpose of measurement, these are etched to be observable with help of two primary types of host features having different geometries, known as Track-in-Track (TINT) and Track-in-Cleavage (TINCLE) (Lal et al. 1969). Laslett et al. (1982) documented that, of various length parameters, the greatest information about the true length distribution can be obtained from the measurement of "horizontal" confined tracks. This technique is by far the most popular approach to estimate the true length distribution of spontaneous fission tracks in minerals that is indispensable for quantitative modeling of rocks' thermal histories as well as for determining FT thermal annealing kinetics. The following points need to be taken into account for reliable determination of confined track lengths (for further details, see Donelick et al. 2005 and Tagami 2005):

(1) Not only the length but also the crystallographic orientation should be measured for individual fission tracks because the track-annealing rate depends upon the crystallographic orientation.

(2) Only tracks having appropriate etched width should be measured because, as mentioned above, the observed FT length is a function of t and increases significantly as the etching proceeds. This is particularly enhanced when we analyze partially annealed samples that contain segmented tracks because the observed length of such tracks is very sensitive to the degree of etching (Green et al. 1986; Hejl 1995; Yamada et al. 1995a).

(3) While many FT analysts still measure TINCLE's in apatite in order to boost the number of confined track lengths measured for individual samples, it should be noted that natural etching of tracks does occur along pre-existing cracks or cleavage planes; thus, measuring TINCLE's can result in length distributions that are unrepresentative of the sample's true thermal history (Jonckheere and Wagner 2000).

Techniques to increase the number of measurable confined tracks. It is often difficult to measure sufficient number of confined tracks for reliable estimation of track length distributions when analyzing samples with low spontaneous track densities. To resolve this, three techniques have been developed to artificially increase the number of host tracks or cleavages that raises the probability of observing more confined tracks: 1) irradiation of nuclear fission fragments using the ^{252}Cf source (Donelick and Miller 1991), 2) irradiation of heavy nuclides using the accelerator (Watanabe et al. 1991) and 3) artificial fracturing using a well-controlled grinding machine (Yoshioka et al. 1994). See Yamada et al. (1998) for comparison of the three techniques in analyzing young zircons. See Donelick et al. (2005) for details and procedures for using a ^{252}Cf source for apatite and Tagami (2005) for procedures of artificial fracturing for zircon.

DATA ANALYSIS AND GRAPHICAL DISPLAYS

An analytical scheme for single-grain (or grain-by-grain) data obtained by the EDM is briefly presented below. See Wagner and Van den haute (1992) for data analysis of the population method. See Donelick et al. (2005) for a discussion of data analysis using the laser ablation ICP-MS.

Statistical test of single-grain data and error calculation of sample mean age

Routinely 10–30 single-grain ages are determined for a single FT analysis. If the grains within the sample have a common age, the variation in single grain ages is governed only by the Poissonian statistics concerned with the determination of ρ_S, ρ_I and ρ_D. In this case, ρ_S and ρ_I are obtained from:

$$\rho_S = \frac{\sum N_{Sj}}{\sum A_j} \qquad \rho_I = \frac{\sum N_{Ij}}{\sum A_j} \tag{17}$$

where N_{Sj} is the number of spontaneous fission tracks in area A_j of the jth crystal and N_{Ij} is the number of induced fission tracks in the same area of the corresponding grain print on the muscovite detector (e.g., Green 1981). Then, the uncertainty in the mean age t is given by the formula:

$$\frac{\sigma(t)}{t}=\left(\frac{1}{N_S}+\frac{1}{N_I}+\frac{1}{N_D}\right)^{1/2} \tag{18}$$

where N_S, N_I and N_D are the total numbers of counted tracks for ρ_S, ρ_I and ρ_D, respectively. In reality, however, the assumption mentioned above is not necessarily valid for a number of factors (see Green 1981) and thus a statistical procedure, called χ^2-test (Galbraith 1981; see also Green 1981), was developed to assess the validity. It was shown that results from the conventional formula give the best estimate of (ρ_S/ρ_I) and $\sigma(\rho_S/\rho_I)$, as long as the observed track counts are acceptable under a χ^2-criterion:

$$\chi^2=\sum\left\{\frac{\left(N_{Sj}-P_{Sj}\right)^2}{P_{Sj}}\right\}+\sum\left\{\frac{\left(N_{Ij}-P_{Ij}\right)^2}{P_{Ij}}\right\} \tag{19}$$

where $P_{Sj} = N_S(N_{Sj}+N_{Ij})/(N_S+N_I)$; $P_{Ij} = N_I(N_{Sj}+N_{Ij})/(N_S+N_I)$. Here the χ^2-value is tested at a desirable critical level, say 5%, with $(n-1)$ degree of freedom where n is the number of grains counted.

If the χ^2-value is unacceptable, it suggests that the data suffer from extra- Poissonian variation(s) due to a variety of experimental and geological factors (Burchart 1981; Green 1981), such as:

(a) incomplete revelation of spontaneous fission tracks, particularly in the case of zircon and sphene, which have more complex etching characteristics as mentioned above,

(b) imprecise track counting, particularly in the case of samples having high track densities or those having a great proportion of non-track features,

(c) incomplete contact between crystals and muscovite detectors,

(d) inhomogeneity of uranium within the measured grain,

(e) spatial variation of the thermal neutron flux on a scale of the areal distance of mounted grains (i.e., ~0.1 to 10 mm),

(f) inherent variability of single-grain ages in the sample dated, such as the case of reworked volcanic ash beds, sedimentary rocks having multiple detrital age components, and partially annealed samples with a significant variation in the thermal sensitivities among grains dated (e.g., apatites with a spread of Cl contents).

In the above cases, usage of conventional formula for the sample mean age and its uncertainty is no longer valid. Where the values of log (ρ_S/ρ_I) can be approximated by a Normal distribution, the central age and age dispersion will yield a better approximation of the population ages (Galbraith and Laslett 1993).

Graphical displays of single-grain age distribution

If the extra-Poissonian variation(s) are statistically detected in the single-grain age data, the structure of the spread in measured single-grain ages should be assessed in order to constrain the source of errors. For this, graphical methods have been employed using (a) the one-dimensional plot, or histogram, (b) the interval plot using, e.g., 95% confidence intervals (Seward and Rhoades 1986), (c) the age spectra, or weighted histogram (Hurford et al. 1984), and (d) the radial plot (Galbraith 1988, 1990) (Fig. 13). Of these, the most widely used today is the radial plot, in which the uncertainty in a single age estimate is isolated so that it is easier to judge the variation in ages between crystals. When multiple age populations are deduced in the radial plot of the sample data, statistical models can be applied to estimate the component ages, particularly the youngest age population (Galbraith and Green 1990; Brandon 1992; Galbraith

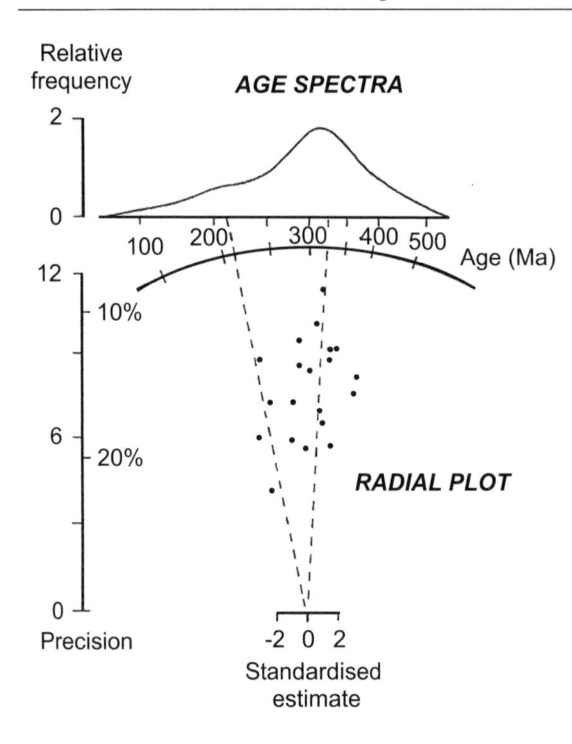

Figure 13. The age spectra (top) and radial plot (bottom) of twenty apatite grains from a sample artificially composed of two age groups (i.e., ~240 and ~340 Ma) (After Galbraith and Green 1990; Wagner and Van den haute 1992). The two dashed lines show the component age estimates. In the radial plot, the horizontal and vertical axes represent the standardized age estimate and reciprocal error, respectively.

and Laslett 1993) (Fig. 13). This technique is critical to the study of provenance ages using the detrital FT thermochronology (see Hurford and Carter 1991; Carter and Moss 1999; Garver et al. 1999; Ruiz et al. 2004) as well as to identify and extract the "essential age" in dating volcanic ash layers (e.g., Naeser et al. 1973; Gleadow 1980; Kowallis et al. 1986; Kohn et al. 1992; Andriessen et al. 1993; O'Sullivan et al. 2001), and even granitic rocks containing apatite grains with varying compositions (O'Sullivan and Parrish 1995).

Graphical displays of track length distribution

Approximately 50–100 confined tracks are routinely measured for each sample, even though even this number should be considered an absolute minimum. The distribution of track lengths is classically shown using a histogram with 1 μm bins since the general overall uncertainty of measured lengths is around 0.5 μm, due primarily to the outer growth of etched track channels as mentioned above. Because the track annealing is anisotropic, the angular distribution diagram is also used widely to rigorously judge the degree of partial annealing of tracks (Green et al. 1986; Donelick 1991; Ravenhurst and Donelick 1992; Yamada et al. 1995ab; Donelick et al. 1999; Barbarand et al. 2003b) (Fig. 8). In the case of apatite, mean track lengths either parallel or perpendicular to the *c*-axis can be used, instead of the usual mean track length, as fitted parameters of track annealing models (Carlson et al. 1999; Donelick et al. 1999; Ketcham et al. 1999). Mean track lengths for zircon are calculated using tracks ≥60° to *c*-axis so that the effects of anisotropic annealing and etching are minimized (Yamada et al. 1993, 1995ab).

The key issue in interpreting the observed length distribution is that longer tracks tend to be over-represented because they have a greater probability of being etched and observed as TINTs or TINCLEs than shorter tracks (Laslett et al. 1982) (Fig. 10). This sampling bias needs to be appropriately corrected to obtain the real length distribution, which is necessary for the quantitative thermal history modeling of the observed FT data. In theory, the bias is proportional to the track length (Laslett et al. 1982) and this was empirically confirmed by the "two-component mixtures" experiment (Galbraith et al. 1990). In practice, however, other geometrical complexities will further be introduced into the track length analysis; i.e., the anisotropy in track etching and annealing for TINTs as well as the effective thickness of host fractures for TINCLEs (Galbraith et al. 1990). Furthermore, these factors can vary according to different experimental and analytical conditions/criteria used between laboratories as well as between observers (Barbarand et al. 2003; see also Yamada et al. 1995a). A proposed methodology to overcome much of this bias is described in further detail by Donelick et al. (2005).

CONCLUDING REMARKS

More than four decades have passed since Price and Walker (1963) first documented the possibility of FT dating, and more than two decades since the introduction of quantitative modeling of a rock's thermal history based upon the confined track length analysis of apatite (e.g., Laslett et al. 1982; Green et al. 1989a). Fission track thermochronology is now considered a mature method. Even so, it has recently been the subject of calibration and tuning towards more reliable reconstruction of thermal histories, as represented by the "multi-kinetic" thermal modeling (Carlson et al. 1999; Donelick et al. 1999; Ketcham et al. 1999). Furthermore, new lines of research are now emerging to potentially refine the technique further: (a) FT analysis of monazite (Gleadow et al. 2002, 2004; Fayon 2004), (b) FT analysis using laser ablation ICP-MS, with which one can potentially obtain both fission track and U/Pb ages on the same single crystal (Cox et al. 2000; Hasebe et al. 2002, 2004; Svojtka and Kosler 2002; Kosler and Sylvester 2003; Donelick et al. 2005), and (c) experimental characterization of very short-term track annealing using the graphite furnace, which simulates frictional heating of fault motions and thus may offer an innovative tool in the geo- and thermochronology of fault systems (Murakami et al. 2002; Murakami and Tagami 2004; Tagami 2005).

ACKNOWLEDGMENTS

We thank Barry Kohn and Diane Seward for their constructive reviews on the manuscript. TT has been supported by a Grant-in Aid (no. 12440137) as well as by a Grant-in-Aid for the 21st Century COE Program (Kyoto University, G3) from the Japanese Ministry of Education, Culture, Sports, Science and Technology.

REFERENCES

Andriessen PA, Helmes M, Hooghiemstra H, Riezebos PA, Van der Hammen T (1993) Absolute chronology of the Pliocene-Quaternary sediment sequence of the Bogota area, Colombia. Quatern Sci Rev 12:483-501

Barbarand J, Carter A, Wood I, Hurford AJ (2003a) Compositional and structural control of fission-track annealing in apatite. Chem Geol 198:107-137

Barbarand J, Hurford AJ, Carter A (2003b) Variation in apatite fission-track length measurement: implications for thermal history modeling. Chem Geol 198:77-106

Bertagnolli E, Mark E, Bertel E, Pahl M, Mark TD (1981) Determination of palaeotemperatures of apatite with the fission-track method. Nucl Tracks 5:175-180

Bertel E, Mark TD (1983) Fission tracks in minerals: annealing kinetics, track structure and age correction. Phys Chem Mineral 9:197-204

Bertel E, Mark TD, Pahl M (1977) A new method for the measurement of the mean etchable fission track length and of extremely high fission track densities in minerals. Nucl Track Detection 1:123-126

Bigazzi G (1981) The problem of the decay constant λ_f of ^{238}U. Nucl Tracks 5:35-44

Bonfiglioli G, Ferro A, Mojoni A (1961) Electron microscope investigation on the nature of tracks of fission products in mica. J Appl Phys 32:2499-2503

Brandon MT (1992) Decomposition of fission-track grain-age distributions. Am J Sci 292:535-564
Brix MR, Stockhert B, Seidel E, Theye T, Thomson SN, Kuster M (2002) Thermobarometric data from a fossil zircon partial annealing zone in high pressure-low temperature rocks of eastern and central Crete, Greece. Tectonophys 349:309-326
Burchart J (1981) Evaluation of uncertainties in fission-track dating: some statistical and geochemical problems. Nucl Tracks 5:87-92
Bursill LA, Braunshausen G (1990) Heavy-ion irradiation tracks in zircon. Phil Mag 62:395-420
Burtner RL, Nigrini A, Donelick RA (1994) Thermochronology of Lower Cretaceous source rocks in the Idaho-Wyoming Thrust Belt. Bull Am Assoc Petrol Geol 78:1613-1636
Calk LC, Naeser CW (1973) The thermal effect of a basalt intrusion on fission tracks in quartz monzonite. J Geol 81:189-198
Carlson WD (1990) Mechanisms and kinetics of apatite fission-track annealing. Am Mineral 75:1120-1139
Carlson WD, Donelick RA, Ketcham RA (1999) Variability of apatite fission-track annealing kinetics: I. Experimental results. Am Mineral 84:1213-1223
Carter A (1990) The thermal history and annealing effects in zircons from the Ordovician of North Wales. Nucl Tracks Radiat Meas 17:309-313
Carter A, Moss SJ (1999) Combined detrital-zircon fission-track and U-Pb dating: a new approach to understanding hinterland evolution. Geology 27:235-238
Chadderton LT (2003) Nuclear tracks in solids: registration physics and the compound spike. Radiat Meas 36: 13-34
Chadderton LT, Montagu-Pollock HM (1963) Fission fragment damage to crystal lattices: heat sensitive crystals. Proc Roy Soc A274:239-252
Corrigan JD (1991) Inversion of apatite fission track data for thermal history information. J Geophys Res 96: 10347-10360
Corrigan JD (1993) Apatite fission track analysis of Oligocene strata in South Texas, U.S.A.: Testing annealing models. Chem Geol 104:227-249
Cox R, Kosler J, Sylvester P, Hodych J (2000) Apatite fission-track (FT) dating by LAM-ICO-MS. Abstr. Goldschmidt 2000, Oxford, p 322
Coyle DA, Wagner GA (1998) Positioning the titanite fission-track partial annealing zone. Chem Geol 149: 117-125
Crowley KD (1985) Thermal significance of fission-track length distributions. Nucl Tracks 10:311-322
Crowley KD (1986) Neutron dosimetry in fission-track analysis. Nucl Tracks Radiat Meas 11:237-243
Crowley KD (1993) Lenmodel—a forward model for calculating length distributions and fission-track ages in apatite. Computer Geosci 19:619-626
Crowley KD, Cameron M, Schaefer RL (1991) Experimental studies of annealing of etched fission tracks in fluorapatite. Geochim Cosmochim Acta 55:1449-1465
Donelick RA (1991) Crystallographic orientation dependence of mean etchable fission track length in apatite: An empirical model and experimental observations. Am Mineral 76:83-91
Donelick RA, Miller DS (1991) Enhanced TINT fission track densities in low spontaneous track density apatites using ^{252}Cf-derived fission fragment tracks: a model and experimental observations. Nucl Tracks Radiat Meas 18:301-307
Donelick RA, Farley KA, Asimow P, O'Sullivan PB (2003) Pressure dependence of He diffusion and fission-track annealing kinetics in apatite?: Experimental results. Geochim Cosmochim Acta 67:A82
Donelick RA, Ketcham RA, Carlson WD (1999) Variability of apatite fission-track annealing kinetics: II. Crystallographic orientation effects. Am Mineral 84:1224-1234
Donelick RA, Roden MK, Mooers JD, Carpenter BS, Miller DS (1990) Etchable length reduction of induced fission tracks in apatite at room temperature (~23 °C): Crystallographic orientation effects and initial mean lengths. Nucl Tracks Radiat Meas 17:261-265
Donelick RA, O'Sullivan PB, RA Ketcham (2005) Apatite fission-track analysis. Rev Mineral Geochem 58:49-94
Dumitru TA (2000) Fission-Track Geochronology. In: Quaternary Geochronology: Methods and Applications. Noller JS, Sowers JM, Lettis WR (eds) Am Geophys Union Ref Shelf 4, Washington, DC, American Geophysical Union, p 131-155
Dunlap A, Jaskierowicz G, Jensen J, Della-Negra S (1997) Track separation due to dissociation of MeV C_{60} inside a solid. Nucl Instr Methods B 132:93-98
Durrani IR, Bull RK (1987) Solid State Nuclear Track Detection (Principles, Methods and Application). Pergamon Press, Oxford
Fayon AK (2004) U electron microprobe analyses and monazite fission-track thermochronology. Abstr. 10th International Conference on Fission Track Dating and Thermochronology, Amsterdam, p 36
Fitzgerald P, Gleadow AJW (1990) New approaches in fission track geochronology as a tectonic tool: Examples from the Transantarctic Mountains. Nucl Tracks Radiat Meas 17:351-357
Fleischer RL, Price PB (1963a) Charged particle tracks in glass. J Appl Phys 34:2903-2904

Fleischer RL, Price PB (1963b) Tracks of charged particles in high polymers. Science 140:1221-1222
Fleischer RL, Price PB (1964) Techniques for geological dating of minerals by chemical etching of fission fragment tracks. Geochim Cosmochim Acta 28:1705-1714
Fleischer RL, Price PB, Walker RM (1965a) Ion explosion spike mechanism for formation of charged particle tracks in solids. J Appl Phys 36:3645-3652
Fleischer RL, Price PB, Walker RM (1965b) Effects of temperature, pressure, and ionization on the formation and stability of fission tracks in minerals and glasses. J Geophys Res 70:1497-1502
Fleischer RL, Price PB, Walker RM (1975) Nuclear Tracks in Solids: Principles and Applications. University of California Press, Berkeley
Galbraith RF (1981) On statistical models for fission track counts. Mathem Geol 13:471-488
Galbraith RF (1988) Graphical display of estimates having differing standard errors. Technometrics 30:271-281
Galbraith RF (1990) The radial plot: Graphical assessment of spread in ages. Nucl Tracks Radiat Meas 17: 207-214
Galbraith RF, Green PF (1990) Estimating the component ages in a finite mixture. Nucl Tracks Radiat Meas 17:197-206
Galbraith RF, Laslett GM (1993) Statistical models for mixed fission track ages. Nucl Tracks Radiat Meas 21: 459-470
Galbraith RF, Laslett GM (1997) Statistical modeling of thermal annealing of fission tracks in zircon. Chem Geol 140:123-135
Galbraith RF, Laslett GM, Green PF, Duddy IR (1990) Apatite fission track analysis: geological thermal history analysis based on a three-dimensional random process of linear radiation damage. Phil Trans R Soc Lond A 332:419-438
Gallagher K (1995) Evolving temperature histories from apatite fission-track data. Earth Planet Sci Lett 136: 421-435
Gallagher K, Brown R, Johnson C (1998) Fission track analysis and its applications to geological problems. Ann Rev Earth Planet Sci 26:519-572
Garver JI, Brandon MT, Roden-Tice M, Kamp PJJ (1999) Exhumation history of orogenic highlands determined by detrital fission-track thermochronology. In: Exhumation Processes: Normal Faulting, Ductile Flow and Erosion. Ring U, Brandon MT, Lister GS, Willett SD (eds) Geol Soc Lond Spec Pub 154:283-304
Gleadow AJW (1978) Anisotropic and variable track etching characteristics in natural sphenes. Nucl Track Detection 2:105-117
Gleadow AJW (1980) Fission track age of the KBS Tuff and associated hominid remains in northern Kenya. Nature 284:225-230
Gleadow AJW (1981) Fission-track dating methods: What are the real alternatives? Nucl Tracks 5:3-14
Gleadow AJW (1984) Fission track dating methods II. A manual of principles and techniques. Workshop on fission track analysis: principles and applications, James Cook University, Townsville
Gleadow AJW, Duddy IR (1981) A natural long-term track annealing experiment for apatite. Nucl Tracks 5: 169-174
Gleadow AJW, Belton DX, Kohn BP, Brown RW (2002) Fission track dating of phosphate minerals and the thermochronology of apatite. Rev Mineral Geochem 48:579-630
Gleadow AJW, Duddy IR, Green PF, Lovering JF (1986) Confined fission track lengths in apatite: a diagnostic tool for thermal history analysis. Contrib Mineral Petrol 94:405-415
Gleadow AJW, Hurford AJ, Quaife RD (1976) Fission track dating of zircon: improved etching techniques. Earth Planet Sci Lett 33:273-276
Gleadow AJW, Raza A, Kohn BP, Spencer SAS (2004) The potential of monazite as a new low-temperature fission-track thermochronometer. Abstr. 10th International Conference on Fission Track Dating and Thermochronology, Amsterdam, p 3
Green PF (1980) On the cause of the shortening of spontaneous fission tracks in certain minerals. Nucl Tracks 4:91-100
Green PF (1981) A new look at statistics in fission-track dating. Nucl Tracks 5:77-86
Green PF (1985) Comparison of zeta calibration baselines for fission-track dating of apatite, zircon and sphene. Chem Geol 58:1-22
Green PF (1988) The relationship between track shortening and fission track age reduction in apatite: combined influences of inherent instability, annealing anisotropy, length bias and system calibration. Earth Planet Sci Lett 89:335-352
Green PF, Duddy IR (1989) Some comments on paleotemperature estimation from apatite fission track analysis. J Petrol Geol 12:111-114
Green PF, Hurford AJ (1984) Thermal neutron dosimetry for fission track dating. Nucl Tracks 9:232-241

Green PF, Duddy IR, Gleadow AJW, Tingate PR, Laslett GM (1986) Thermal annealing of fission tracks in apatite 1. A qualitative description. Chem Geol 59:237-253
Green PF, Duddy IR, Laslett GM (1988) Can fission track annealing in apatite be described by first-order kinetics? Earth Planet Sci Lett 87:216-228
Green PF, Duddy IR, Gleadow AJW, Lovering JF (1989a) Apatite fission-track analysis as a paleotemperature indicator for hydrocarbon exploration. In: Thermal History of Sedimentary Basins: Methods and Case Histories. Naeser ND, McCulloh TH (eds) Springer-Verlag, New York, p 181-195
Green PF, Duddy IR, Laslett GM, Hegarty KA, Gleadow AJW, Lovering JF (1989b) Thermal annealing of fission tracks in apatite 4. Quantitative modeling techniques and extension to geological timescales. Chem Geol 79:155-182
Green PF, Hegarty KA, Duddy IR, Foland SS, Gorbachev V (1996) Geological constraints on fission track annealing in zircon. Abstr. International Workshop on Fission Track Dating, Gent 1996, Gent, p 44
Harrison TM, Armstrong RL, Naeser CW, Harakal JE (1979) Geochronology and thermal history of the Coast Plutonic Complex, near Prince Rupert, British Columbia. Can J Earth Sci 16:400-410
Hasebe N, Barbarand J, Jarvis K, Carter A, Hurford AJ (2002) Can FT ages be derived using LA-ICP-MS? Geochim Cosmochim Acta 66: A314
Hasebe N, Barbarand J, Jarvis K, Carter A, Hurford AJ (2004) Apatite fission-track chronometry using laser ablation ICP-MS. Chem Geol 207: 135-145
Hasebe N, Mori S, Tagami T, Matsui R (2003) Geological partial annealing zone of zircon fission-track system: additional constraints from the deep drilling MITI-Nishikubiki and MITI-Mishima. Chem Geol 199: 45-52
Hasebe N, Tagami T, Nishimura S (1994) Towards zircon fission-track thermochronology: Reference framework for confined track length measurements. Chem Geol 112:169-178
Hejl E (1995) Evidence for unetchable gaps in apatite fission tracks. Chem Geol 122:259-269
Huntsberger TL, Lerche I (1987) Determination of paleo heat-flux from fission tracks in apatite. J Petrol Geol 10:365-394
Hurford AJ (1986) Cooling and uplift patterns in the Lepontine Alps South Central Switzerland and an age of vertical movement on the Insubric fault line. Contrib Mineral Petrol 92:413-427
Hurford AJ (1990) Standardization of fission track dating calibration: Recommendation by the Fission Track Working Group of the I. U. G. S. Subcommission on Geochronology. Chem Geol 80:171-178
Hurford AJ, Carter A (1991) The role of fission track dating in discrimination of provenance. *In:* Developments in Sedimentary Provenance Studies. Morton AC, Todd SP, Haughton PDW (eds) Geol Soc Spec Pub 57: 67-78
Hurford AJ, Green PF (1982) A user's guide to fission track dating calibration. Earth Planet Sci Lett 59:343-354
Hurford AJ, Green PF (1983) The zeta age calibration of fission-track dating. Isot Geosci 1:285-317
Hurford AJ, Fitch FJ, Clarke A (1984) Resolution of the age structure of the detrital zircon populations of two Lower Cretaceous sandstones from the Weald of England by fission track dating. Geol Mag 121: 269-277
Issler DR (1996) Optimizing time-step size for apatite fission-track annealing models. Computer Geosci 22: 67-74
Iwano H, Danhara T (1998) A re-investigation of the geometry factors for fission-track dating of apatite, sphene and zircon. *In:* Advances in Fission-Track Geochronology. Van den haute P, De Corte F (eds) Kluwer Academic Publishers, Dordrecht, p 47-66
Iwano H, Kasuya M, Yamashita T, Danhara T (1992) One-to-one correlation of fission tracks between zircon and mica detectors. Nucl Tracks Radiat Meas 20:341-347
Iwano H, Kasuya M, Danhara T, Yamashita T, Tagami T (1993) Track counting efficiency and unetchable track range in apatite. Nucl Tracks Radiat Meas 21:513-517
Jonckheere RC, Wagner GA (2000) On the occurrence of anomalous fission tracks in apatite and titanite. Am Mineral 85:1744-1753
Kasuya M, Naeser CW (1988) The effect of α-damage on fission-track annealing in zircon. Nucl Tracks Radiat Meas 14:477-480
Ketcham RA (2005) Forward and inverse modeling of low-temperature thermochronometry data. Rev Mineral Geochem 58:275-314
Ketcham RA, Donelick RA, Carlson WD (1999) Variability of apatite fission-track annealing kinetics: III. Extrapolation to geological time scales. Am Mineral 84:1235-1255
Ketcham RA, Donelick RA, Donelick MB (2000) AFTSolve: A program for multi-kinetic modeling of apatite fission-track data. Geol Mater Res 2:1-32
Khan HA, Durrani SA (1972) Prolonged etching factor in solid state track detection and its applications. Radiat Effects 13:257-266

Kohn BP, Pillans B, McGlone MS (1992) Zircon fission track ages for middle Pleistocene Rangitawa tephra, New Zealand: Stratigraphic and paleoclimatic significance. Palaeogeog Palaleoclim Palaeoecol 95:73-94

Kohn BP, Belton DX, Brown RW, Gleadow AJW, Green PF, Lovering JF (2003) Comment on: "Experimental evidence for the pressure dependence of fission track annealing in apatite" by A.S. Wendt et al. [Earth Planet Sci Lett 201 (2002) 593-607]. Earth Planet Sci Lett 215:299-306

Kosler J, Sylvester PJ (2003) Present trends and the future of zircon in geochronology: Laser ablation ICPMS. Rev Mineral Geochem 53:243-275

Kowallis BJ, Heaton JS, Bringhurst K (1986) Fission-track dating of volcanically derived sedimentary rocks. Geology 14:19-22

Lal D, Rajan RS, Tamhane AS (1969) Chemical composition of nuclei of Z > 22 in cosmic rays using meteoritic minerals as detectors. Nature 221:33-37

Laslett GM, Galbraith RF (1996) Statistical modeling of thermal annealing of fission tracks in apatite. Geochim Cosmochim Acta 60:5117-5131

Laslett GM, Kendall WS, Gleadow AJW, Duddy IR (1982) Bias in measurement of fission-track length distributions. Nucl Tracks 6:79-85

Laslett GM, Gleadow AJW, Duddy IR (1984) The relationship between fission track length and track density in apatite. Nucl Tracks 9:29-38

Laslett GM, Green PF, Duddy IR, Gleadow AJW (1987) Thermal annealing of fission tracks in apatite 2. A quantitative analysis. Chem Geol 65:1-13

Lutz TM, Omar G (1991) An inverse method of modeling thermal histories from apatite fission-track data. Earth Planet Sci Lett 104:181-195

Mark E, Pahl M, Purtscheller F, Mark TD (1973) Thermische Ausheilung von Uran-Spaltspuren in Apatiten, Alterskorrekturen und Beitrage zur Geothermochronologie. Tschermaks Min Petr Mitt 20:131-154

Mark TD, Vartanian R, Purtscheller F, Pahl M (1981) Fission track annealing and application to the dating of Austrian sphene. Acta Phys Austriaca 53:45-59

Murakami M, Yamada R, Tagami T (2002) Detection of frictional heating of fault motion by zircon fission track thermochronology. Geochim Cosmochim Acta 66:A537

Murakami M, Tagami T (2004) Dating pseudotachylyte of the Nojima fault using the zircon fission-track method. Geophys Res Lett 31: 10.1029/2004GL020211

Naeser CW (1976) Fission track dating. US Geol Surv Open-File Rep 76-190 65P

Naeser CW (1979) Fission-track dating and geologic annealing of fission tracks. *In:* Lectures in Isotope Geology. Jager E, Hunziker JC (eds) Springer-Verlag, Berlin, p 154-169

Naeser CW, Faul H (1969) Fission track annealing in apatite and sphene. J Geophys Res 74:705-710

Naeser CW, Forbes RB (1976) Variation of fission track ages with depth in two deep drill holes. EOS Trans Am Geophys Union 57:353

Naeser CW, Izett GA, Obradovich JD (1980) Fission-track and K-Ar ages of natural glasses. US Geol Surv Bull 1489:1-31

Naeser CW, Izett GA, Wilcox RE (1973) Zircon fission-track ages of Pearlette family ash beds in Meade County, Kansas. Geology 1:187-189

Nagpaul KK, Metha PP, Gupta ML (1974) Annealing studies on radiation damages in biotite, apatite and sphene and corrections to fission track ages. Pure Appl Geophys 112:131-139

Neumann R (2000) Characterization of ion tracks in solids by near-field microscopy and TEM. Abstr Internat Conf Fission Track Dating Thermochr, Lorne, p 241-243

O'Sullivan PB, Parrish RR (1995) The importance of apatite composition and single grain ages when interpreting fission track data from plutonic rocks: a case study from the Coast Ranges, British Columbia. Earth Planet Sci Lett 132:213-224

O'Sullivan PB, Morwood M, Hobbs D, Aziz F, Suminto, Situmorang M, Raza A, Maas R (2001) Archeological implications of the geology and chronology of the Soa Basin, Flores, Indonesia. Geology 29:607-610

Paul TA (1993) Transmission electron microscopy investigation of unetched fission tracks in fluorapatite —physical process of annealing. Nucl Tracks Radiat Meas 21:507-511

Paul TA, Fitzgerald PG (1992) Transmission electron microscopic investigation of fission tracks in fluorapatite. Am Mineral 77:336-344

Price PB, Walker RM (1962) Chemical etching of charged-particle tracks in solids. J Appl Phys 33:3407-3412

Price PB, Walker RM (1963) Fossil tracks of charged particles in mica and the age of minerals. J Geophys Res 68:4847-4862

Rahn MK, Brandon MT, Batt GE, Garver JI (2004) A zero-damage model for fission-track annealing in zircon. Am Mineral 89:473-484

Ravenhurst CE, Donelick RA (1992) Fission track thermochronology. *In:* Short Course Handbook on Low Temperature Thermochronology. Zentilli M, Reynolds PM (eds) Mineral Assoc Can, Ottawa, p 21-42

Reiners PW (2003) (U-Th)/He dating and calibration of low-T thermochronometry. Geochim Cosmochim Acta 67:A395

Ruiz GMH, Seward D and Winkler W (2004) Detrital thermochronology – a new perspective on hinterland tectonics, an example from the Andean Amazon Basin, Ecuador. Basin Res 10.1111/j.1365-2117.2004.00239.x

Seitz F (1949) The disordering of solids by the action of fast massive particles. Disc Faraday Soc 5:271-282

Seward D, Rhoades DA (1986) A clustering technique for fission track dating of fully to partially annealed minerals and other non-unique populations. Nucl Tracks Radiat Meas 11:259-268

Silk ECH, Barnes RS (1959) Examination of fission fragment tracks with an electron microscope. Phil Mag 4:970-972

Stockli DF, Benjamin ES, Dumitru TA (2002) Thermochronological constraints on the timing and magnitude of Miocene and Pliocene extension in the central Wassuk Range, western Nevada. Tectonics 21:10.1029/2001TC001295

Sumii T, Tagami T, Nishimura S (1987) Anisotropic etching character of spontaneous tracks in zircon. Nucl Tracks Radiat Meas 13:275-277

Svojtka M, Kosler M (2002) Fission-track dating of zircon by laser ablation ICPMS. Geochim Cosmochim Acta 66:A756

Tagami T (1987) Determination of zeta calibration constant for fission track dating. Nucl Tracks Radiat Meas 13:127-130

Tagami T (2005) Zircon fission-track thermochronology and applications to fault studies. Rev Mineral Geochem 58:95-122

Tagami T, Nishimura S (1989) Intercalibration of thermal neutron dosimeter glasses NBS-SRM612 and Cornig 1 in some irradiation facilities: a comparison. Nucl Tracks Radiat Meas 16:11-14

Tagami T, Nishimura S (1992) Neutron dosimetry and fission-track age calibration: insights from intercalibration of uranium and thorium glass dosimeters. Chem Geol (Isot Geosci Sect) 102:277-296

Tagami T, Shimada C (1996) Natural long-term annealing of the zircon fission track system around a granitic pluton. J Geophys Res 101:8245-8255

Tagami T, Carter A, Hurford AJ (1996) Natural long-term annealing of the zircon fission-track system in Vienna Basin deep borehole samples: constraints upon the partial annealing zone and closure temperature. Chem Geol 130:147-157

Tagami T, Farley KA, Stockli DF (2003) Thermal sensitivities of zircon (U-Th)/He and fission-track systems. Geochim Cosmochim Acta 67:A466

Tagami T, Galbraith RF, Yamada R, Laslett GM (1998) Revised annealing kinetics of fission tracks in zircon and geological implications In: Advances in fission-track geochronology. Van den haute P, De Corte F (eds) Kluwer academic publishers, Dordrecht, The Netherlands, p 99-112

Tagami T, Ito H, Nishimura S (1990) Thermal annealing characteristics of spontaneous fission tracks in zircon. Chem Geol (Isot Geosci Sect) 80:159-169

Tagami T, Lal N, Sorkhabi RB, Ito H, Nishimura S (1988) Fission track dating using external detector method: A laboratory procedure. Mem Fac Sci Kyoto Univ 53:14-30

Thibaudau F, Cousty J, Balanzat E, Bouffard S (1991) Atomic-force-microscopy observations of tracks induced by swift Kr ions in mica. Phys Rev Lett 67:1582-1585

Van den haute P, De Corte F, Jonckheere R, Bellemans F (1998) The parameters that govern the accuracy of fission-track age determinations: a re-appraisal. *In:* Advances in Fission-Track Geochronology. Van den haute P, De Corte F (eds) Kluwer Academic Publishers, Dordrecht, p 33-46

Vetter J, Scholz R, Dobrev D, Nistor L (1998) HREM investigation of latent tracks in GeS and mica induced by high energy ions. Nucl Instr Methods B 141:747-752

Wagner GA (1968) Fission track dating of apatites. Earth Planet Sci Lett 4:411-415

Wagner GA, Storzer D (1970) Die Interpretation von Spaltspurenaltern am Beispiel von naturlichen Glasern, Apatiten und Zirkonen. Eclogae Geol Helv 63:335-344

Wagner GA, Van den haute P (1992) Fission-Track Dating. Kulwer Academy, Norwell, Massachusetts

Wagner GA, Hejl E, Van den haute P (1994) The KTB fission-track project: methodical aspects and geological implications. Radiat Meas 23:95-101

Wagner GA, Reimer GM, Jager E (1977) Cooling ages derived by apatite fission-track, mica Rb-Sr and K-Ar dating: the uplift and cooling history of the Central Alps. Mem Inst Geol Mineral Univ Padova 30:1-27

Watanabe K (1988) Etching characteristics of fission tracks in Plio-Pleistocene zircon. Nucl Tracks Radiat Meas 15:171-174

Watanabe K, Izawa E, Kuroki K, Honda T, Nakamura H (1991) Detection of confined ^{238}U fission tracks in minerals and its application to geothermal geology. Annual Rep Tandem Accel Lab (Kyushu Univ) 3: 151-155

Wendt AS, Vidal O, Chadderton LT (2002) Experimental evidence for the pressure dependence of fission track annealing in apatite. Earth Planet Sci Lett 201:593-607

Willet SD (1997) Inverse modeling of annealing of fission tracks in apatite 1: A controlled random search method. Am J Sci 297:939-969

Yada K, Tanji T, Sunagawa I (1981) Application of lattice imagery to radiation damage investigation in natural zircon. Phys Chem Mineral 7:47-52

Yada K, Tanji T, Sunagawa I (1987) Radiation induced lattice defects in natural zircon ($ZrSiO_4$) observed at atomic resolution. Phys Chem Mineral 14:197-204

Yamada K, Tagami T, Shimobayashi N (2003) Experimental study on hydrothermal annealing of fission tracks in zircon. Chem Geol 201:351-357

Yamada R, Tagami T, Nishimura S (1993) Assessment of overetching factor for confined fission track length measurement in zircon. Chem Geol (Isot Geosci Sect) 104:251-259

Yamada R, Tagami T, Nishimura S (1995a) Confined fission-track length measurement of zircon: assessment of factors affecting the paleotemperature estimate. Chem Geol (Isot Geosci Sect) 119:293-306.

Yamada R, Tagami T, Nishimura S, Ito H (1995b) Annealing kinetics of fission tracks in zircon: an experimental study. Chem Geol (Isot Geosci Sect) 122:249-258

Yamada R, Yoshioka T, Watanabe K, Tagami T, Nakamura H, Hashimoto T, Nishimura S (1998) Comparison of experimental techniques to increase the number of measurable confined fission tracks in zircon. Chem Geol (Isot Geosci Sect) 149:99-107

Yoshioka T, Tagami T, Nishimura S (1994) Experimental technique to increase the TINCLEs in zircon. Fission Track News Lett 7:50

Zaun PE, Wagner GA (1985) Fission-track stability in zircons under geological conditions. Nucl Tracks 10: 303-307

Zeitler PK, Tahirkheli RAK, Naeser CW, Johnson NM (1982) Unroofing history of a suture zone in the Himalaya of Pakistan by means of fission-track annealing ages. Earth Planet Sci Lett 57:227-240

Zimmermann RA, Gaines AM (1978) A new approach to the study of fission track fading. US Geol Surv Open-file Rep 78-701:467-468

Reviews in Mineralogy & Geochemistry
Vol. 58, pp. 49-94, 2005

3

Apatite Fission-Track Analysis

Raymond A. Donelick

Apatite to Zircon, Inc.
1075 Matson Road, Viola, Idaho 83872-9709, U.S.A.
Department of Geological Sciences
University of Nevada-Reno, Reno, Nevada 89557-0138, U.S.A.
donelick@apatite.com

Paul B. O'Sullivan

Apatite to Zircon, Inc.
1075 Matson Road, Viola, Idaho 83872-9709, U.S.A.
Department of Geological Sciences
University of Idaho, Moscow, Idaho 83844-3022, U.S.A.

Richard A. Ketcham

Jackson School of Geosciences,
University of Texas at Austin
Austin, Texas, 78712-0254, U.S.A.

INTRODUCTION

Apatite fission-track (AFT) analysis has been widely used during the past 30+ years to constrain the low-temperature thermal histories of many igneous, metamorphic and sedimentary rocks in a wide range of geological settings. Applicable geological settings include orogenic belts, rifted margins, faults, sedimentary basins, cratons, and mineral deposits. The types of geologic problems that can be addressed include the timing and rates of tectonic events, sedimentary basin evolution, the timing of hydrocarbon generation and ore mineralization, the absolute age of volcanic deposits, the effects of major climatic changes on the near-surface geothermal gradient, and long-term landscape evolution. Early work by Naeser (1967) and Wagner (1968, 1969) first established the basic procedures that enabled the fission-track dating method to be applied routinely to these geologic problems. Fleischer et al. (1975) summarized the early studies of the broader discipline of nuclear-track detection in different solid-state materials. More recent comprehensive overviews of fission-track applications have been provided by Naeser and McCulloh (1989), Wagner and Van den haute (1992), Gallagher et al. (1998), Van den haute and De Corte (1998), Dumitru (2000), and Gleadow et al. (2002).

Etched, natural fission tracks in several apatite grains are shown in Figure 1. Successful AFT analysis is limited by the following: 1) the availability of apatite from which useful AFT data can be obtained, often due to a lack of apatite of sufficient grain size and quality within available rock types, or, alternatively, the lack of available rock samples due to minimal outcrop exposure or other reasons, 2) economic considerations in terms of the time and money required to obtain sufficient AFT data, 3) the inherent limitations of AFT data to resolve geological thermal history information, often related to limited numbers of accumulated spontaneous fission tracks due to low uranium concentration within the available apatite and/or young apatite grain fission-track ages, 4) the limitations of the model(s) upon which AFT interpretations are based, and 5) self-imposed limitations by various laboratories, often in the form of the measurement and interpretative schemes employed.

1529-6466/05/0058-0003$05.00 DOI: 10.2138/rmg.2005.58.3

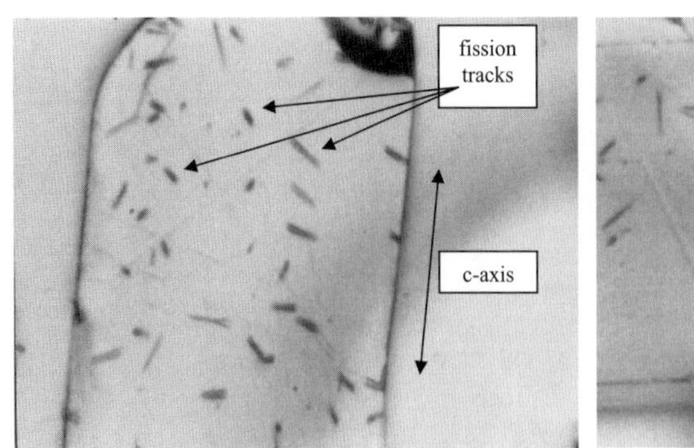

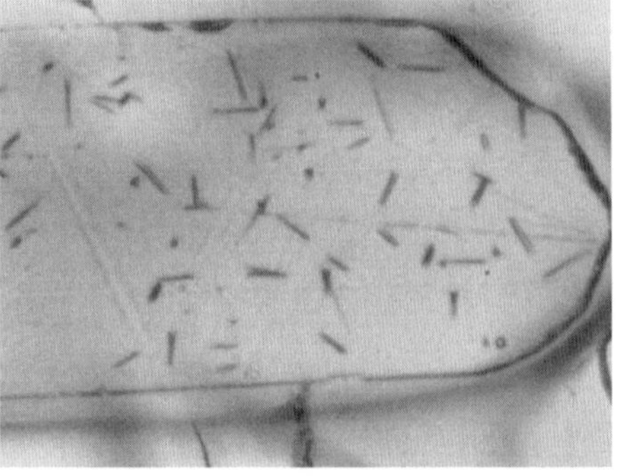

Figure 1. Fission tracks on a polished and etched surface of apatite.

The recent resurgence of (U-Th)/He dating, in particular for the minerals apatite, zircon, and titanite (e.g, Reiners 2003 and references therein), complements and challenges AFT methods. The complementary aspect resides in the low-temperature nature of the various (U-Th)/He systems. The challenge resides in the periodically-encountered, apparent incompatibility of the data derived from the various techniques (e.g., apatite (U-Th)/He ages > AFT ages). Incompatible (U-Th)/He and fission-track data demonstrate that the systematics of one or both methods are incompletely understood. Therefore, considerable study and debate in the AFT community now centers on the type of data to measure, which methods are used to collect those data, and which models are developed and employed to interpret those data in terms of geological thermal histories.

The body of literature regarding AFT methods and procedures is immense and a thorough review of that literature is beyond the scope of this paper, particularly because many older approaches have been superceded by superior methods. The intent of this chapter is to highlight some of these recent methodological developments and demonstrate their justification, efficacy, and ease of implementation. This is followed by several appendices describing sample preparation tips and data collection schemes.

APATITE AS A FISSION-TRACK ANALYSIS MATERIAL

General

Several fundamental characteristics of apatite make it an excellent mineral for fission-track analysis and provide the basis for the AFT methods presented here. These are: 1) its nearly ubiquitous natural occurrence in many common crustal rock types, 2) its physical properties, 3) its major and minor element chemistries, 4) its trace element chemistry, in particular the presence of uranium and thorium in most natural crystals, 5) its ability to retain fission tracks in the geological environment, and 6) the ability of an experimentalist to mimic important features of the behavior of spontaneous fission tracks in the geological environment using laboratory analogues.

Natural occurrence

Apatite is a nearly ubiquitous accessory mineral, found in many types of igneous, metamorphic, and sedimentary rocks. As used throughout this discussion, an apatite grain or simply a grain may be a whole apatite crystal, a fragment of an apatite crystal, or an assemblage of multiple whole crystals and/or crystal fragments that may or may not have experienced a sedimentary transport history. Apatite grains composed of whole crystals or crystal fragments occur as: 1) anhedral to euhedral accessory grains in igneous rocks, 2) detrital grains in clastic sedimentary rocks, 3) primary and secondary grains in shales, limestones, and coals, and 4) primary mineral grains or porphyroblasts in metamorphic rocks including marbles. Apatite grains composed of aggregate crystals occur as: 1) biologically-derived apatite or phosphatic material including teeth, bones, and conodonts, 2) chemical precipitates or nodules in phosphate-rich sedimentary rocks, and 3) diagenetic or hydrothermal grains in clastic and non-clastic sedimentary rocks. Apatite usually occurs as small sand-sized grains, typically < 300 μm (<medium sand or <phi scale value of +1 to +2) across the short diameter, in amounts less than 1% of the total rock volume. AFT analysis methods generally require apatite grains to have a minimum diameter of approximately 50 μm (≥very fine sand or ≥phi scale value of approximately +4), although smaller grains can be used.

Physical properties

Most naturally occurring apatite is a member of the 6/*m*-hexagonal dipyramidal crystal class (Deer et al. 1969); the known exception being the uncommon, near-end-member chorapatite which is monoclinic (e.g., Hughes et al. 1989). The typical crystal habit of apatite derived from igneous rocks is prismatic. In a sedimentary transport system, such primary accessory apatite grains usually experience some degree of rounding of the crystal edges. In sandstones, any degree of rounding may be evidenced from undetectable to complete, but it is somewhat uncommon to find apatite grains that are completely rounded into a nearly spherical shape. Apatite possesses a weak parting perpendicular to the crystallographic *c*-axis, but possesses no strong cleavage. Furthermore, apatite is soluble in acidic aqueous solution and except for the rounding effects mentioned above and possible interaction with natural acidic solutions (e.g., humic acid in tropical weathering regimes) apatite commonly survives surface and near-surface geological processes.

Apatite has a specific gravity of between 3.15–3.20 g/cm^3, making it a heavy mineral relative to quartz, feldspar, calcite, and other rock-forming minerals that tend to have specific gravities less than 2.90 g/cm^3. In a standard 30 μm thick thin section, apatite appears as first-order gray or white (perpendicular to the *c*-axis) and can be distinguished from quartz based on its greater relief due to its higher refractive index (~1.63) compared to that of quartz (1.54). In hand specimen, apatite is typically difficult to impossible to identify, even with a hand lens, due to its generally low abundance level.

Major and minor element chemistries

Apatite is the most common phosphorous-bearing mineral in the Earth's crust. The chemical formula for apatite is generally written as $Ca_5(PO_4)_3$[F,Cl,OH] (Deer et al. 1969). The three anions F^{-1}, Cl^{-1}, and OH^{-1} are believed to substitute for one another between the various end-member compositions fluorapatite, chlorapatite, and hydroxyapatite, respectively. Appreciable CO_2 is known to occur in some natural apatites. Additional substitutions include Mn, Sr, Fe, Na, and rare-earth elements (particularly Ce) in the Ca-site and Si, S, and C in the P-site (Deer et al. 1969; Young et al. 1969; Roeder et al. 1987). Near-end-member calcian-fluorapatite is the dominant apatite variety in most crustal rocks. However, it is quite common for relatively non-fluorapatite compositions to be found in crustal rocks, especially in sandstones that contain apatite grains from multiple provenance sources.

Uranium and thorium as trace elements

Uranium and thorium are important trace elements in apatite, the former being important for AFT analysis and both being important for apatite (U-Th)/He analysis. The concentration of natural uranium in apatite typically ranges from 1–200 ppm. Under oxidizing conditions, uranium forms the uranly ion (UO_2^{+2}), which in turn forms compounds that are soluble in aqueous solution. Because uranium in apatite is protected from the environment by its host crystal lattice, it is not lost from the host grain when the apatite is immersed in water, but it can be lost by a process of dissolution and re-precipitation. The combination of the high charge and relatively small ionic radius of the uranium cation (U^{+4} = 1.05 Å) makes uranium incompatible with other cations in the common silicate minerals. Consequently, uranium tends to be concentrated in accessory mineral phases such as apatite, zircon, titanite, and monazite.

Natural uranium is composed of the three isotopes ^{234}U, ^{235}U, and ^{238}U. The ratio $^{238}U/^{235}U$ of natural uranium is adopted by convention (Steiger and Jäger 1977) to be exactly 137.88. Natural uranium is distinguished from enriched uranium (enriched in ^{235}U; depleted of ^{238}U) or depleted uranium (depleted of ^{235}U; enriched in ^{238}U), enriched and depleted uranium being dominantly the result of human processing of natural uranium for fuel or weapons. A summary of the natural uranium and thorium isotopes, including their relative abundances and nuclear properties relevant to AFT analysis, is given in Table 1. Nuclear fission of the isotope ^{238}U produces the spontaneous fission tracks in apatite. The isotope ^{235}U, on the other hand, is important because it is often used to measure the amount of ^{238}U present in an apatite grain. Additionally, ^{235}U is used extensively in laboratory-based, experimental studies of fission tracks in apatite because it can be easily induced to form fission tracks under carefully controlled conditions (see below). ^{238}U experiences spontaneous nuclear decay by two processes: α-decay (emission of a ^4He nucleus), the dominant process, and spontaneous fission (splitting of the nucleus into two fragment nuclei, rarely three, plus 2 or 3 high-energy neutrons; Friedlander et al. 1981). The ^{238}U α-decay process is composed of a complicated series of individual α-particle-emitting nuclear reactions that ultimately lead to a stable ^{206}Pb nucleus. For every approximately two million ^{238}U nuclei that undergo α-decay, a single ^{238}U nucleus will experience spontaneous nuclear fission. For purposes of AFT analysis, the large thermal-neutron capture cross-section of ^{235}U relative to those for ^{234}U, ^{238}U, and ^{232}Th is of great importance (see Table 1). Thermal neutrons are neutrons traveling at speeds similar to the speeds traveled by gas molecules at room temperature (approximately 2200 m s^{-1}). Thermal neutrons are available from the core of a nuclear fission reactor that is surrounded by a medium (such as light or heavy water with a high spatial density of hydrogen or deuterium nuclei, respectively) that moderates the velocities of the neutrons emanating from the core. The thermal-neutron capture cross-section is the effective target area that a nucleus presents to a thermal neutron. By exposing a natural mixture of uranium and thorium to a flux of thermal neutrons, it is possible to induce fission of ^{235}U with negligible induced fission of the other uranium and thorium isotopes. The process of inducing fission of ^{235}U in this manner permits ^{235}U to be used as a means of determining the amount of ^{238}U present in an apatite grain because the ratio of ^{238}U to ^{235}U is a fixed number in most natural materials (Steiger and Jäger 1977).

Table 1. The abundances, half-lives, and decay constants of naturally occurring uranium isotopes (Lederer et al. 1967; Steiger and Jäger 1977; Friedlander et al. 1981).

Isotope	*Abundance* (%)	*Half-Life* (yr)	*Decay Constant* (yr^{-1})	*Thermal-Neutron Capture Cross-Section* (σ) (10^{-24} cm^2)
^{232}Th	100.0000	1.41×10^{10} (α)	4.916×10^{-11} (α)	7.4
^{234}U	0.0057	2.47×10^{5} (α)	2.806×10^{-6} (α)	100
^{235}U	0.7200	0.7038×10^{9} (α)	9.8485×10^{-10} (α)	580
^{238}U	99.2743	4.468×10^{9} (α) $\sim 1.3 \times 10^{16}$ (s.f.)	1.55125×10^{-10} (α) $\sim 7.5 \times 10^{-17}$ (s.f.)	2.7

NOTES: (α) indicates alpha-decay series, (s.f.) indicates spontaneous fission-decay

Fission-track retention in the geological environment

Spontaneous fission tracks in apatite remain relatively stable if exposed to temperatures below approximately 100 °C over geological time-scales (Fig. 2). Naeser and Forbes (1976) found that AFT ages decrease from ~100 Ma at the surface to ~12 Ma at 3 km depth (~95 °C present-day temperature) in the Eielson Air Force Base, Alaska, deep drill hole. Naeser (1981) reported AFT ages decreasing to 0 Ma at 2 km depth (~135 °C present-day temperature) from the Los Alamos, New Mexico, geothermal test wells 1 and 2. Gleadow and Duddy (1981) reported AFT ages from a Cretaceous stratigraphic section in the Otway Basin of southeastern Australia that decrease to 0 Ma age at ~125 °C.

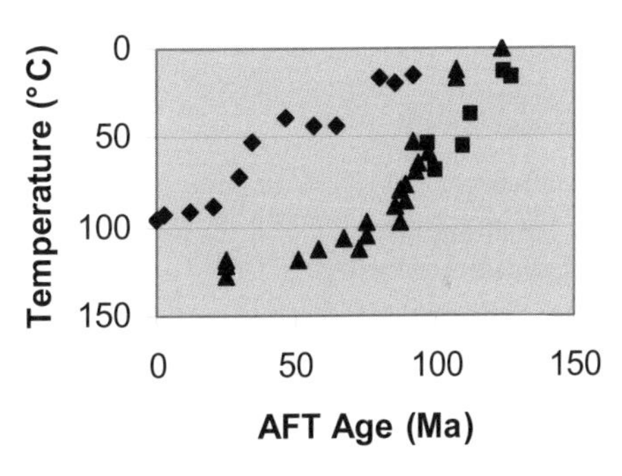

Figure 2. Preservation of spontaneous fission tracks in apatite in the geological environment. ◆ = Eielson (Naeser and Forbes 1976; Naeser, written communication), ■ = Los Alamos (Naeser 1981), and ▲ = Otway (Gleadow and Duddy 1981).

Laboratory analogues to spontaneous fission-track behavior

Ideally, laboratory studies and resultant interpretation schemes should use and be based upon, respectively, spontaneous fission tracks in apatite. However, the very nature of the spontaneous fission-track system complicates this. Because fission tracks in a given apatite grain or population of grains span a continuous range of ages from recent to the age of the oldest fission track present, the net effect is that different tracks have different initial conditions going into any laboratory study. Therefore, extreme caution must be exercised to account for this range of initial conditions when utilizing spontaneous fission tracks in laboratory studies for ultimate use in interpretive schemes. To avoid this pitfall of spontaneous fission tracks, fission tracks resulting from the thermal-neutron-induced fission of ^{235}U are most often used in laboratory experiments because single populations of these tracks can be formed under controlled conditions and thus all of these fission tracks have essentially the same initial condition. One must, of course, recognize the inherent noise in fission-track length data due to the variation of fission-decay energies and other effects such as crystallographic orientation that impart different lengths to different tracks having a single history. Partly on this basis, it is widely accepted that fission tracks resulting from the thermal-neutron induced fission of ^{235}U in apatite are good proxies for spontaneous fission tracks resulting from the spontaneous fission of ^{238}U over geological time (Green et al. 1988 offer a fairly comprehensive description of this problem).

Numerous studies have been undertaken using induced fission-track behavior as a proxy for spontaneous fission-track behavior as a function of time and temperature (e.g., Green et al. 1986; Donelick et al. 1990; Crowley et al. 1991; Carlson et al. 1999; Barbarand et al. 2003a), crystallographic orientation (e.g., Donelick et al. 1990, 1999; Donelick 1991; Ketcham 2003; Barbarand et al. 2003a), apatite chemistry and etching characteristics (Carlson et al. 1999; Barbarand et al. 2003a,b), and pressure (Donelick et al. 2003; Kohn et al. 2004); it is noteworthy that the pressure study of Wendt et al. (2002) utilized spontaneous fission tracks so it does not apply here. These studies hold as their fundamental premise that a population of thermal-neutron-induced fission tracks, formed at low ambient temperatures and then subsequently partially annealed, has as its natural equivalent a population of spontaneous fission tracks formed during a brief but significant time interval. By dividing the time over which spontaneous fission tracks accumulated in any given apatite into discrete intervals, this premise holds that the tracks formed during any given time interval represent a single population that has experienced, in effect, a single integrated geological history characterized by the same time, temperature, pressure, and other environmental conditions.

AFT SAMPLE PREPARATION

Over the years a multitude of innovative methods have been conceived of and implemented for AFT sample preparation and data collection and many of these methods have been reviewed elsewhere (e.g., Naeser 1976, 1979; Gleadow 1984; Crowley et al. 1989; Ravenhurst and Donelick 1993; Burtner et al. 1994). Considering the number of different methods presently or potentially in use, it is well beyond the scope of this chapter to discuss and compare all of these methods in detail. Several of the approaches to AFT sample preparation preferred by the authors are offered in Appendix 1.

APATITE FISSION-TRACK AGE EQUATIONS

The AFT age of a single apatite grain determined using the external detector method (EDM) is given by:

$$t_i = \frac{1}{\lambda_d} \ln\left(1 + \lambda_d \zeta g \rho_d \frac{\rho_{s,i}}{\rho_{i,i}}\right) \tag{1a}$$

where subscript i refers to grain i, t_i = fission-track age of grain i, λ_d = total decay constant of ^{238}U, ζ = ζ-calibration factor based on EDM of fission-track age standards, g = geometry factor for spontaneous fission-track registration (see below), ρ_d = induced fission-track density for a uranium standard corresponding to the sample position during neutron irradiation. $\rho_{s,i}$ = spontaneous fission-track density for grain i which equals $(N_{s,i}/\Omega_i)$ where $N_{s,i}$ is the number spontaneous fission tracks counted over area Ω_i, $\rho_{i,i}$ = induced fission-track density for grain i which equals $(N_{i,i}/\Omega_i)$ where $N_{i,i}$ is the number induced fission tracks counted over area Ω_i.

Similarly, the AFT age of a single apatite grain determined using laser-ablation inductively coupled plasma-mass spectrometry (LA-ICP-MS) is given by (Donelick et al. 2005):

$$t_i = \frac{1}{\lambda_d} \ln\left(1 + \lambda_d \zeta_{MS} g \frac{\rho_{s,i}}{\mathcal{P}_i}\right) \tag{1b}$$

where $\mathcal{P}_i = (^{238}U/^{43}Ca)$ for apatite grain i, ζ_{MS} = ζ-calibration factor based on LA-ICP-MS of fission-track age standards adjusted for the sample position during the LA-ICP-MS session and ^{238}U = background-corrected ^{238}U cps, ^{43}Ca = background-corrected ^{43}Ca cps.

The symmetrical error of the AFT single-grain age determined using the EDM is given by:

$$\sigma_i = \left[\frac{1}{N_{s,i}} + \frac{1}{N_{i,i}} + \frac{1}{N_d} + \left(\frac{\sigma_\zeta}{\zeta}\right)^2\right]^{1/2} \tag{2a}$$

where N_d = number of induced fission tracks counted to determine ρ_d, σ_ζ = error of ζ.

Similarly, the symmetrical error of the AFT single-grain age determined using LA-ICP-MS is given by:

$$\sigma_i = \left[\frac{1}{N_{s,i}} + \left(\frac{\sigma_{\mathcal{P}_i}}{\mathcal{P}_i}\right)^2 + \left(\frac{\sigma_{\zeta_{MS}}}{\zeta_{MS}}\right)^2\right]^{1/2} \tag{2b}$$

where $\sigma_{\mathcal{P}_i}$ = error of $\mathcal{P}_i$, $\sigma_{\zeta MS}$ = error of ζ_{MS}. Equations (2a) and (2b) can be modified to provide asymmetrical errors as described by Donelick et al. (2005).

The pooled AFT age of a population of apatite grains determined using the EDM is given by:

$$t_{pooled} = \frac{1}{\lambda_d} \ln\left(1 + \lambda_d \zeta g \rho_d \frac{\sum N_{s,i}}{\sum N_{i,i}}\right) \tag{3a}$$

where t_{pooled} = the pooled AFT age. Similarly, the pooled AFT age of a population of apatite grains determined using LA-ICP-MS is given by:

$$t_{pooled} = \frac{1}{\lambda_d} \ln\left(1 + \lambda_d \zeta_{MS} g \frac{\sum N_{s,i}}{\sum \mathcal{P}_i \Omega_i}\right) \tag{3b}$$

The symmetrical error of the pooled AFT age determined using the EDM is given by:

$$\sigma_{pooled} = \left[\frac{1}{\sum N_{s,i}} + \frac{1}{\sum N_{i,i}} + \frac{1}{N_d} + \left(\frac{\sigma_\zeta}{\zeta}\right)^2\right]^{1/2} \tag{4a}$$

Similarly, the symmetrical error of the pooled AFT age determined using LA-ICP-MS is given by:

$$\sigma_{pooled} = \left[\frac{1}{\sum N_{s,i}} + \frac{\sum \sigma_{\mathcal{P}_i}^2 \Omega_i^2}{\left(\sum \mathcal{P}_i \Omega_i\right)^2} + \left(\frac{\sigma_{\zeta MS}}{\zeta_{MS}}\right)^2\right]^{1/2} \tag{4b}$$

Equations (4a) and (4b) can be modified to provide asymmetrical errors of the pooled AFT age as described by Donelick et al. (2005).

Further discussion of these and related equations can be found in Tagami and O'Sullivan (2005) and Donelick et al. (2005).

AFT DATA AND DATA COLLECTION

General

There are four measurable parameter types associated with the collection of apatite fission-track data:

- Spontaneous fission-track densities are used to calculate fission-track ages of either individual grains or populations of grains.
- Relative uranium concentrations are used to calculate fission-track ages of either individual grains or populations of grains.
- Horizontal, confined fission-track lengths are used to constrain the style of cooling histories of cooling-only samples (e.g., exhumed orogenic belts) or other low-temperature phenomena including peak burial temperatures of sedimentary rocks or mineralization temperatures in ore bodies.
- Apatite fission-track annealing kinetic parameters are used to provide an estimation of the annealing behavior of individual grains or populations of grains, and allow one to group together apatite grains having similar annealing-kinetic response during later quantitative thermal-history modeling of AFT data.

These data types are described below. This is preceded, however, by a rare, but extremely important discussion of analyst bias accompanied by suggestions for minimizing this bias. The method of data collection most commonly used by the authors is compared to another common approach in Appendix 2.

Analyst bias

AFT analysis is presently highly dependent upon decisions made by the analyst, although computer-driven image analysis systems hold promise for the future provided intelligent software can be developed and made economical and dependable. Bias on the part of the analyst concerning these decisions, conscious or not, can significantly impact AFT analyses and it must be minimized.

For spontaneous fission-track density measurements (described below), a grain to be dated is selected, an area of that grain to be counted is selected, and fission tracks within that area are viewed and deemed countable or not using criteria learned and practiced by the analyst. A similar process is employed for counting induced fission tracks on external detectors. Provided the criteria for these choices remain invariant for calibration purposes (i.e., standards and experiments) and for analyses of geological unknown samples, bias is minimized. Questions facing the analyst from which bias may arise include (Fig. 3a): Which etched features are fission tracks? Which features deemed to be fission tracks lie within the area being counted or meet the analyst's criteria for selecting tracks that are centered on the border of the area being counted? For external detectors these questions apply as well as the following (Fig. 3b): Precisely which area of the external detector corresponds to the counted area of the apatite grain? What orientation of the counting grid is appropriate?

For fission-track length measurements (described below), sources of potential bias similarly derive from decisions made by the analyst. Questions confronting the analyst include (Fig. 3c): Which etched features meet the criteria learned and practiced by the analyst as candidates for fission-track length measurement? Where precisely is the measurement device (e.g., a LED image projected onto the microscope field of view via a projection tube; the LED being affixed to a cursor for a digitizing tablet) to be placed at both ends of the fission track? As with fission-track density measurements, bias is minimized provided the criteria used to make these decisions are invariant for calibration purposes and for analyses of geological unknown samples.

Many laboratories use commercially obtained or in-house developed analytical software that permits the analyst to view results as they are being generated. Results available for viewing may include a running fission-track age and associated statistics for a sample (for the latter, e.g., a chi-squared value for the age data) and/or a running mean fission-track length and associated statistical parameters (e.g., moments of the distribution of the length data). Viewing these running values during an analysis combined with an expected analytical result in the mind of the analyst can severely bias the analytical results themselves. This is not often admitted by fission-track analysts but it is a prevalent issue. Although the ability to view running values during an analysis may appear to give analytical software increased "power" or "utility", this paradigm is misplaced in fission-track analysis. The human tendency to be influenced and ultimately biased by expectations and/or preliminary results makes it critical that all possible sources of such contamination be removed from the analytical process. Put another way, no apparent advantage is conferred by being able to see running values, but the potential pitfall is undeniable. Therefore, these values should not be viewed and options to view them should be eliminated from analytical software. For fission-track density measurements, the age data can be biased if, for example, the analyst biases borderline decisions toward either young or old ages. Similarly, for fission-track length measurements, the length data can be significantly biased toward either a low or high mean length based on an expected final result.

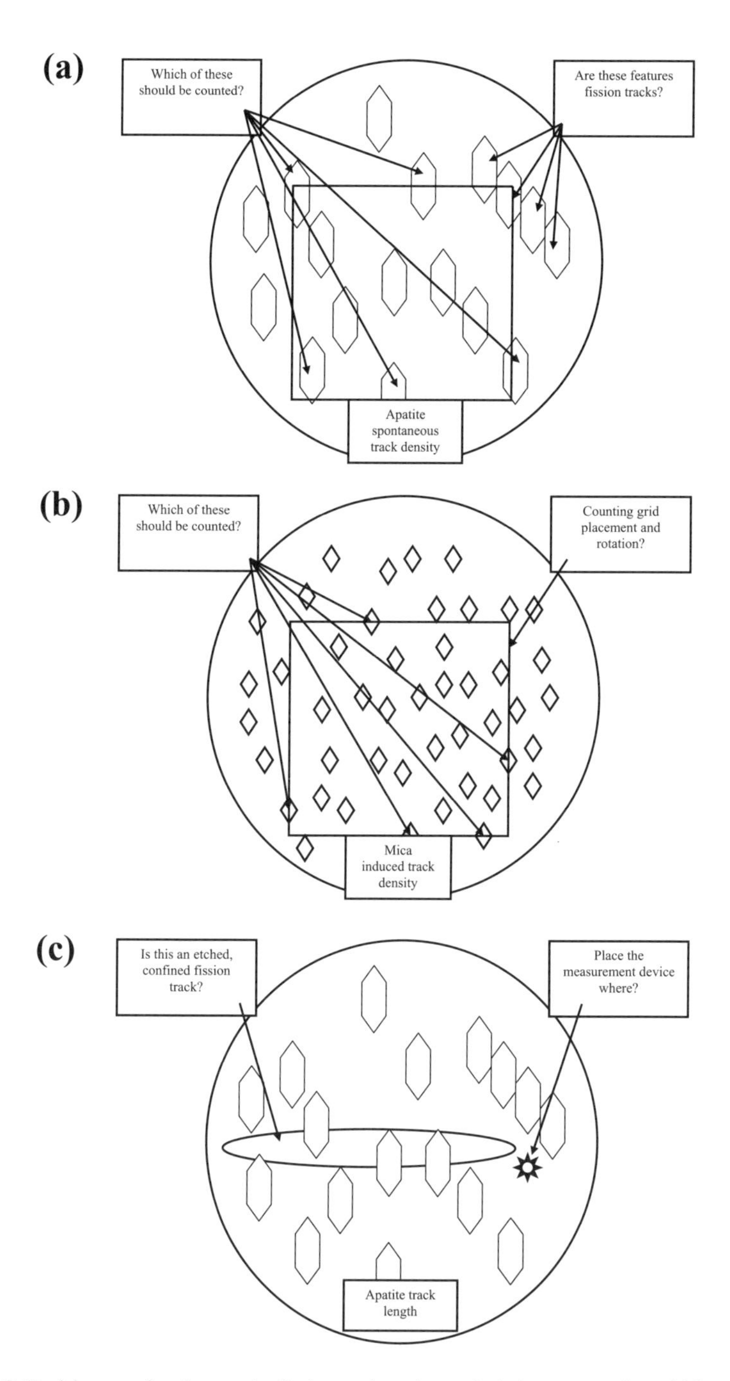

Figure 3. Decisions confronting apatite fission track analysts. a) choices concerning which spontaneous fission tracks to count, b) choices concerning where to count induced fission tracks on an external detector, and c) choices concerning confined fission-track length measurement.

To minimize bias, the analyst should do the following:

- Prior to analysis, mask as effectively as possible all sample identity of geological unknowns, standards, and experiments.
- Use identical decision criteria for all samples including geological unknowns, standards, and experiments.
- If possible, disable the viewing of running results of measurements when performing an analysis. If this is not possible, mask those values on the computer screen or simply resist, totally, the urge to view those results.

For many apatite samples, analyst bias can be useful but this bias derives from the choice of which apatite grains are selected for measurement, not from how the data obtained from those grains are actually measured. For example, consider a sample with apatite grains exhibiting a range of etching characteristics (i.e., a range of D_{par} values; parameter D_{par} is discussed below), with most grains having small D_{par} values but a few grains having large D_{par} values. The analyst may choose to bias the analysis by seeking out and measuring a disproportionate amount of data from large D_{par} grains. Overall, the data are biased. However, within a given sub-population of apatite grains (defined by a subset of the D_{par} values), the data should be unbiased. This is achieved by minimizing bias once the analyst selects a grain for measurement. The point here is that intelligent, human analysts can be both biased (regarding which grains to select for providing data) and unbiased (by minimizing bias when collecting data from those grains).

Spontaneous fission-track densities

Spontaneous fission tracks form continuously as the result of the spontaneous fission, over geological time, of ^{238}U nuclei within the host apatite crystal lattice. Fission tracks are randomly oriented and approximately linear, composed of zones of damaged crystal lattice typically less than 20 μm in length and 3 to 14 nm wide (e.g., Paul and Fitzgerald 1992). Fission tracks in this state are termed latent fission tracks and they are not visible using standard optical microscopy. To calculate a fission track age it is necessary to obtain a measure of the spatial density of these fission tracks within an apatite grain. Unlike other radiometric dating systems (e.g., U-Pb, K-Ar, Rb-Sr) for which the spatial density of radiogenic products can be determined chemically, the spatial density of fission tracks cannot be measured chemically due to their being composed of damaged apatite. However, the damaged apatite material forming the core of the fission tracks, being at a higher free-energy state relative to the surrounding undamaged crystal and thus more chemically reactive, can be chemically etched and the resultant fission-track etch pits enlarged sufficiently for viewing using an optical microscope (Fig. 1; e.g., Fleischer et al. 1975 and references therein). For apatite, this is usually accomplished by polishing an apatite crystal to an internal surface that, with the exception of edge effects, intersects fission tracks derived equally from both above (from the material polished away) and below (from the material that remains after polishing). This geometry of fission-track registration on an apatite surface is referred to as a 4π geometry (for 4π geometry $g = 0.5$ in Eqns. 1a,b and 3a,b) versus the 2π geometry ($g = 1.0$) of natural apatite grain surfaces that intersect fission tracks derived only from the apatite side. The authors prefer to etch apatite fission tracks in 5.5 M HNO_3 for 20 s at 21 °C and to view the tracks using un-polarized, transmitted and/or reflected light at 1562.5× or 2000× magnification (100× dry objective, 1.25× projection tube, 12.5× or 16× oculars, respectively). The result after chemical etching is a flat surface of apatite from which emanate etched semi-tracks (partial fission tracks because a portion of each has been removed by polishing). Laslett et al. (1984) offer a detailed argument regarding the relationship between the spatial density of fission tracks in a crystal, the fission-track density on a polished and etched 4π geometry surface, and the etchable length distribution of full-length, horizontal, confined fission tracks.

The number of spontaneous fission tracks ($N_{s,i}$ in Eqns. 3a and 3b) counted over a selected apatite grain area divided by the area gives the spontaneous fission-track density ($\rho_{s,i}$ in Eqns. 1a and 1b). For large apatite grains, the area over which the spontaneous fission-track density is measured should not be closer than approximately 10 μm from the grain boundary. This is because there is a transition from 2π geometry to 4π geometry from the grain boundary inward to about 10 μm in from the grain boundary, respectively. The area counted should be free of large surface imperfections such as cracks and large etch pits and possess a minimum of spurious etch features such as inclusions and crystallographic defects. If a grain is too small to count over a meaningful area located at least 10 μm from the grain boundaries, the whole grain can be counted provided it is reasonable to assume that the polished and etched surface is approximately halfway through the original grain and that no significant surface area is removed by polishing and etching including along the grain boundaries.

Relative uranium concentrations

Two methods of relative uranium concentration determination are described here: 1) induced fission-track density measurements obtained using external detectors for both apatite grains and standard materials of known uranium concentration, commonly referred to as the external detector method (EDM), and 2) laser-ablation inductively coupled plasma-mass spectrometry (LA-ICP-MS) measurements.

Induced fission-track densities. The EDM of AFT age analysis is a form of thermal-neutron activation, a detailed description of which is provided by Tagami and O'Sullivan (2005). To obtain relative measurements of the uranium concentration in apatite grains selected for age determination, an external detector commonly composed of low-uranium, fission-track-free muscovite mica is placed in intimate contact with the grains and the grain mount-mica pair is irradiated with thermal neutrons in the vicinity of the core of a nuclear reactor. A high thermal neutron to fast neutron ratio is preferred because thermal-neutron-induced fission of ^{235}U is highly favored over thermal-neutron-induced fission of ^{238}U and ^{232}Th; the reverse is true for fast neutrons (e.g., Fleischer et al. 1975). It is extremely difficult to confidently determine the integrated thermal-neutron flux resulting from an irradiation session. To overcome this problem, it is common practice to irradiate silicate glass standards of known and/or constant uranium concentration affixed with their own external detectors at either end of a stack of apatite grain mount samples being irradiated. For each of these uranium glasses, an induced fission-track density is measured and from these a corresponding induced fission-track density is calculated for each sample position in the stack (parameter ρ_d in Eqns. 1a and 3a) by linear interpolation. Examples of uranium-doped silicate glasses commonly used for EDM age analysis include Corning Glasses CN-1 through CN-6 or NIST glasses 612 or 962a. Induced fission-track densities are determined for the uranium-doped glasses irradiated along with the apatite grain mounts. The number of induced fission tracks (N_d in Eqns. 2a and 4a) counted over an area on an external detector in contact with a uranium-doped glass divided by the area gives the induced fission-track density corresponding to the glass (ρ_d in Eqns. 1a and 3a). These induced fission-track densities need to be measured using randomly selected areas of the external detector that was in contact with the glass. A total of 4000 induced fission tracks for ρ_d determinations is a reasonable number as this total is likely to exceed the number of spontaneous fission tracks counted in most samples and hence will not significantly limit the precision of the age measurement.

Induced fission-track densities are measured on the external detectors from the apatite grain mounts and the uranium-doped glasses. Prior to removal of the external detectors from the apatite grain mounts, three widely-spaced pin-holes are punched through each external detector and into its respective apatite grain mount to serve as easily found reference points. A combination of a) the coordinates of the reference points on the apatite grain mount, b) the coordinates, in the same coordinate system as the reference points, of the apatite grains analyzed, and c) the coordinates of the reference points on the external detector permits the coordinates of the locations on the external detector corresponding to the grains analyzed to be calculated. Most laboratories use only two reference points, which permit solving the in-plane translation and rotation components of the coordinate transformation problem (two variables, two unknowns). However, a third reference point is important for a) establishing if there is a reflection component to the transformation, and b) accounting for any tilting that would otherwise degrade focus when automated analyses are arranged on other pieces of equipment such an electron microprobe or laser ablation stage.

The authors prefer etching mica detectors in 49% HF for 15 min at 23 °C to reveal fission tracks induced within the apatite grains by this irradiation and registered in the detector. The number of induced fission tracks ($N_{i,i}$ in Eqns. 2a, 3a, and 4a) counted on an external detector for an apatite grain divided by the area counted gives the induced fission-track density ($\rho_{i,i}$ in Eqn. 1a). For a large apatite grain, induced fission tracks are counted over the area of the external detector that was in intimate contact with the area of the apatite grain over which the spontaneous fission-track density was determined. For a small apatite grain for which every spontaneous fission track was counted up to the grain boundaries, every induced fission track on the external detector from that grain needs to be counted.

LA-ICP-MS measurement of uranium concentrations. Hasebe et al. (2004) discussed two approaches to fission-track age determinations using laser ablation combined with inductively coupled plasma-mass spectrometry (LA-ICP-MS). First they developed an absolute calibration based on the ^{238}U fission-decay constant, a model factor describing the registration efficiency of fission tracks, and a calibration factor for etching and observation conditions, each of which involves some degree of uncertainty. Alternatively, they followed the zeta calibration approach (Hurford and Green 1983) of subsuming all of these factors into a single term that is calibrated by reference to an age standard. Donelick et al. (2005) followed the latter approach, combining LA-ICP-MS analysis of $^{238}U/^{43}Ca$ ratios for apatite with a modified zeta-calibration approach to permit routine measurement and calculation of apatite individual grain and pooled fission-track ages. For apatite the primary assumption is that the ^{43}Ca, at least during a given LA-ICP-MS session, is directly related to the volume of apatite ablated. Under this assumption, the ratio of $^{238}U/^{43}Ca$ gives a relative measure of the uranium concentration in apatite. A key difficulty is dealing with uranium zoning in the target mineral grains. Only uranium within ~10 μm of the polished and etched target mineral surface contributes spontaneous fission tracks to that surface. The approach taken by Donelick et al. (2005) used spot laser analyses (the laser beam remained centered at a fixed point on the polished and etched target mineral surface) and the fitting of 3rd-order polynomials to $^{238}U/^{43}Ca$ ratio versus analysis point at each spot to obtain the best-fit estimate of relative uranium concentration at the surface. Fission track ages were then calculated using a modified zeta-calibration approach. This approach requires determination of an initial, primary zeta factor of the desired precision during a single, extensive LA-ICP-MS session and, for each subsequent LA-ICP-MS session, determination of a secondary zeta calibration factor, related directly to the primary factor for each sample in the session.

Confined fission-track lengths

Lengths of etched, horizontal, confined fission tracks were first reported in the literature by Bhandari et al. (1971). Confined fission tracks are tracks for which both ends are confined within the volume of the polished and etched crystal and are reached by the chemical etchant via other etched features that intersect the polished and etched surface. Wagner and Storzer (1972) first reported projected semi-track length measurements (spontaneous fission tracks that intersect the polished and etched surface and for which part of the track has been removed by polishing) and proposed that different distributions of lengths were indicative of different geological settings and histories. Laslett et al. (1982) argued that the lengths of horizontal, confined fission tracks in apatite provide superior resolution of the true fission-track length distribution relative to the lengths of projected semi-tracks. As used in this chapter, fission-track length refers to the length of an etched, horizontal, confined fission track. Lal et al. (1969) described two types of confined fission tracks: TINT (Track-IN-Track) fission tracks and TINCLE (Track-IN-CLEavage) fission tracks. A third type of confined fission tracks, TINDEF (Track-IN-DEFect or fluid or soluble mineral inclusion), is recognized by the authors (Fig. 4).

Fission tracks form continuously through time at a rate determined solely by the concentration of ^{238}U in the host apatite grain. After their formation, fission-tracks in apatite progressively anneal at a rate that depends primarily on temperature (Fleischer et al. 1975; Gleadow et al. 1986). Mean fission-track length and spontaneous fission-track density of single populations of thermal-neutron-induced fission tracks in apatite are positively correlated (e.g., Green 1988). In nature, younger fission tracks, by virtue of their age, experience only a portion of the integrated time-temperature history experienced by older fission tracks in the same apatite crystal or population of crystals. Hence, older fission tracks in natural apatite grains have experienced a greater degree of partial annealing than younger tracks. If the time-temperature history is characterized by overall cooling of the apatite crystals, the older fission tracks may be significantly shorter than their younger counterparts. The overall fission-track length distribution exhibits properties indicative of the nature of the cooling history experienced. For the case of rapid initial cooling followed by a prolonged period of slow cooling to the present-day temperature (Fig. 5a), the fission-track length distribution is dominated by mostly long fission tracks. For a prolonged period of slow cooling followed by recent rapid cooling to the present-day temperature (Fig. 5b), a length distribution dominated by relatively short fission tracks is expected. This concept represents the greatest strength of apatite fission-track analysis and it has received intense study since the early 1980s.

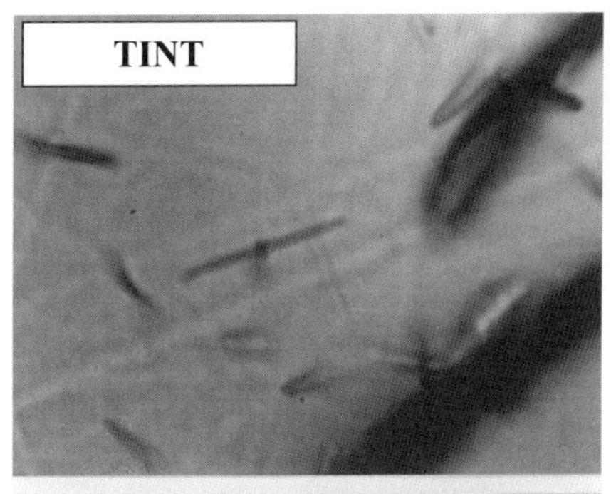

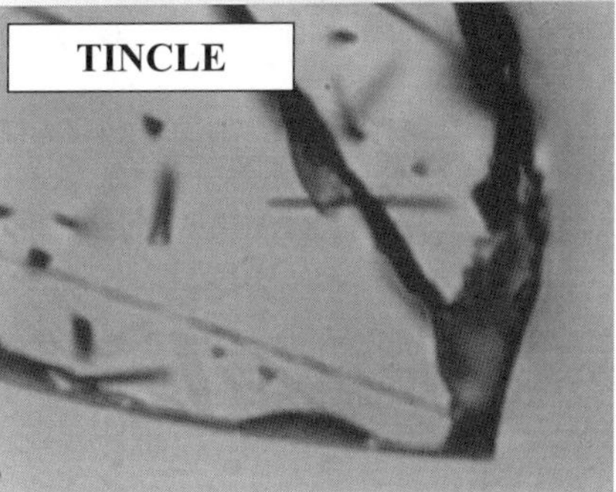

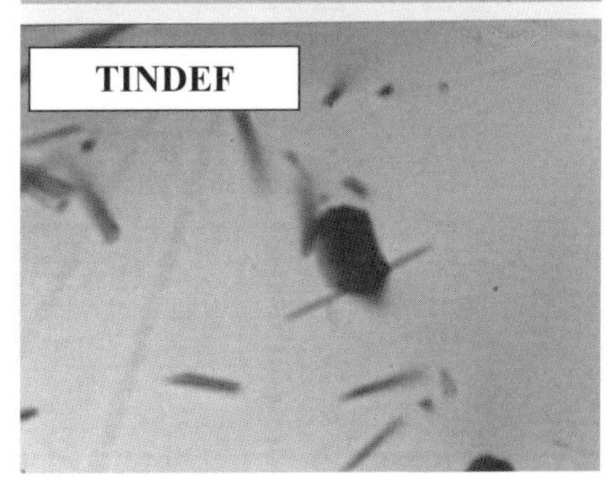

Figure 4. Types of etched, confined fission tracks in apatite.

The measurable parameters regarding fission track lengths include:

- The complete length of each etched, horizontal, confined fission track.
- The angle of each confined fission-track relative to the crystallographic *c*-axis.
- The type of fission-track length measured in terms of the pathway via which the chemical etchant reached each fission track measured (i.e., TINT, TINCLE, or TINDEF).

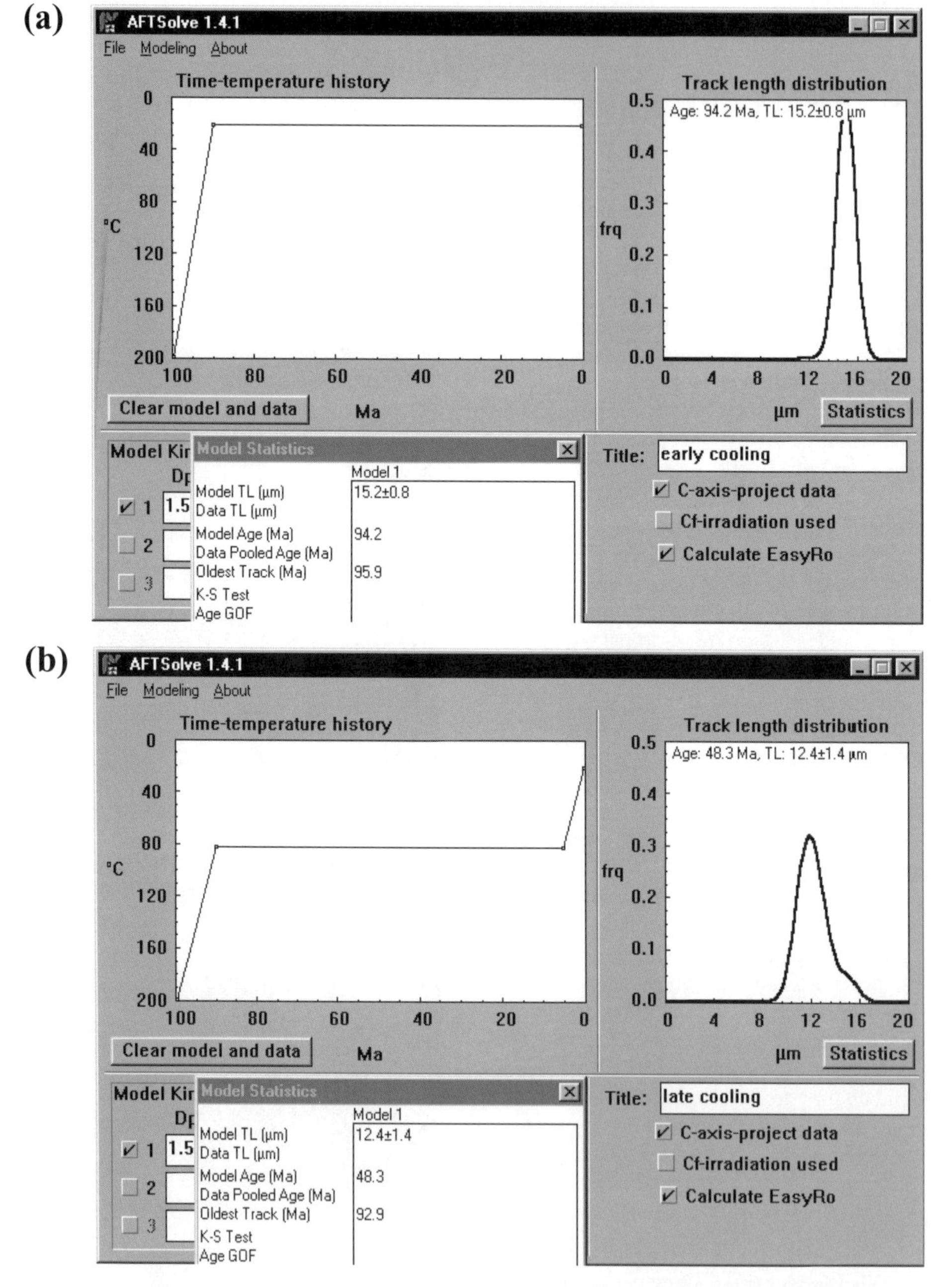

Figure 5. Model fission-track age and length data for: a) rapid cooling to surface temperature in the distant past and b) recent rapid cooling to surface temperature. The thermal histories were calculated using AFTSolve (Ketcham et al. 2000; Ketcham et al. 1999 multi-kinetic model, D_{par}=1.50 µm, fission-track lengths projected onto the crystallographic *c*-axis using Donelick et al. 1999).

Total confined fission-track lengths in apatite grains. It is most common to measure the lengths of etched, nominally horizontal confined fission tracks in *c*-axis-parallel crystallographic planes. Confined fission tracks within approximately ±10° from horizontal are acceptable candidates for length measurement because the length apparent to the analyst equals the true length multiplied by cosine of the track angle of inclination to the horizontal plane (cos[10°] equals 0.985 and this deviation from one is near the level of precision for measuring a single fission-track length; Donelick 1991). When selecting confined fission tracks for length measurement, it is essential that only confined TINT fission tracks with well-etched and clearly visible ends be considered (Donelick et al. 1990; Carlson et al. 1999; Jonckheere and Wagner 2000; restricting measurements to TINT fission tracks is discussed further below). Analyst-derived measurement bias can be minimized by adopting and using consistent criteria for selection of fission tracks for length measurement and for placement of the measuring device at the ends of each track measured. It is also very important to determine whether or not any residual fluid may be trapped in the tracks, thus lowering the relief of the track ends and potentially masking them from view (Fig. 6). If trapped fluid is apparent or suspected, the grain mount should be washed thoroughly with acetone and then dried with a bead of acetone on the surface in order to withdraw by surface tension any acetone remaining in the tracks. Acetone effectively dissolves aqueous fluid that may be entrained in the etched fission tracks. It is essential that etched apatite grain mount surfaces not be touched directly as oil from a human hand can enter the fission tracks, obscure the etched track ends, and this oil is very difficult to effectively remove with solvents. In extreme cases, immersing the grain mounts in soapy water and agitating them vigorously, followed by cleaning with acetone or ethanol can result in the successful removal of any residual fluid from within the tracks.

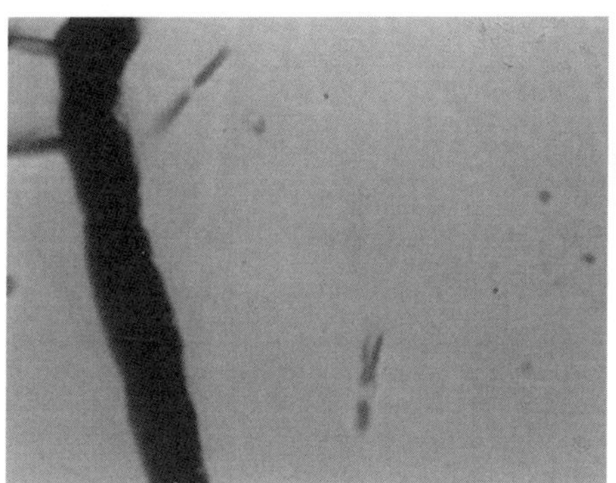
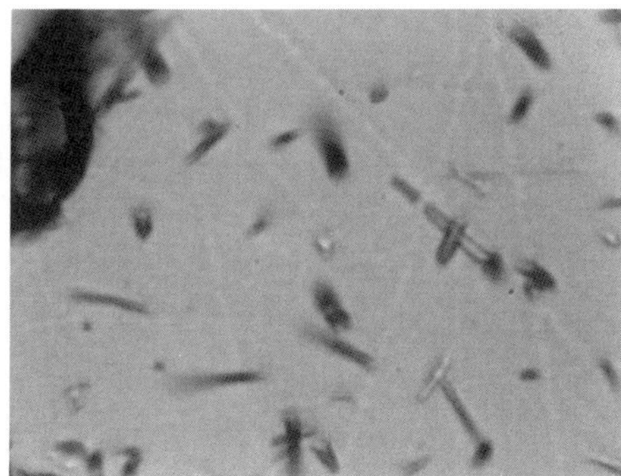

Figure 6. Fluid trapped in etched fission tracks in apatite. Note the discontinuous nature of the several of the fission tracks emanating from the polished and etched surface. For confined fission tracks, fluid near the track tips can obscure the ends from view and render accurate measurement impossible.

Angle to the crystallographic c-axis of confined fission tracks in apatite grains. Fission tracks in apatite exhibit a well-documented variation in mean length with crystallographic orientation (e.g., Donelick 1991 and references therein; Donelick et al. 1999; Ketcham 2003), even for minimally annealed laboratory-induced (Donelick et al. 1990) and natural (Donelick and Miller 1991; Vrolijk et al. 1992) fission tracks (Fig. 7). Donelick (1991) showed that mean fission track length decreases faster for tracks orientated perpendicular to the crystallographic *c*-axis relative to their *c*-axis parallel counterparts and this behavior is effectively modeled using ellipses when the data are plotted in polar coordinates (coordinates [length, angle to *c*-axis]). Donelick et al. (1999) extended these observations to show that, beginning for track populations having an overall mean track length equal to approximately

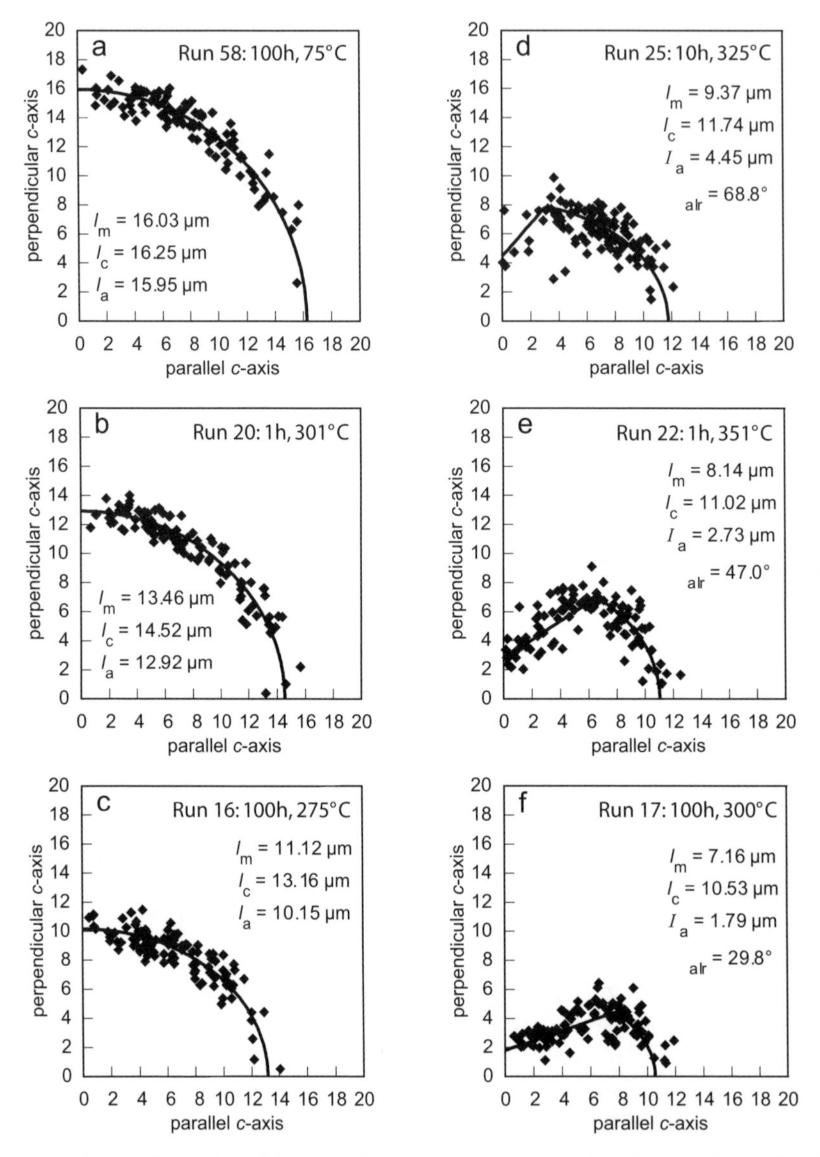

Figure 7. Polar coordinate plots of fission-track lengths for Durango apatite (after Donelick et al. 1999). Continuous solid curves in parts a, b, and c are ellipses fitted to lengths at all orientations. Discontinuous solid curves in parts d, e, and f are ellipses fitted to lengths at angles less than ϕ_{alr} to the *c*-axis combined with lines fitted to lengths at angles greater than ϕ_{alr}.

0.65 of the initial overall mean track length (e.g., Green 1988), fission tracks at high angles to the crystallographic *c*-axis undergo a systematic (non-random), accelerated length reduction. As the degree of annealing increases systematically beyond the onset of accelerated length reduction, the accelerated length reduction involves tracks at systematically lower angles to the crystallographic *c*-axis. In polar coordinates, this effect can be described as a collapse of the ellipse at high-angles to the *c*-axis, with the angle of initiation of the collapse (parameter ϕ_{alr} in Fig. 7) rotating toward the *c*-axis as the degree of annealing increases (Donelick et al.

1999; Ketcham 2003). At some point, the angle of initiation of the collapse appears to closely approach the *c*-axis and the track population simply becomes unetchable as a whole (i.e., the track population attains a state of total annealing, at least as far as the ability to image the tracks using chemical etching is concerned).

Accounting for fission-track orientation to the crystallographic *c*-axis in apatite permits the removal of significant scatter from the track length data. The models of Donelick et al. (1999) and Ketcham (2003) offer a means to convert any fission track length at any angle to the crystallographic *c*-axis into an equivalent track length parallel to the *c*-axis (or any other orientation but *c*-axis-parallel is preferred because tracks in this orientation are most resistant to thermal annealing). Ketcham et al. (2004) noted that the details of the projection model may depend upon the etching protocol used. Such converted tracks are considered projected onto the *c*-axis (distinguished from projected semi-track lengths that are no longer routinely used). To date, no studies quantifying the advantages of this projection have been published but because significant scatter is removed from the data, this projection should be a matter of routine practice for purposes of quantitative thermal history modeling of AFT data.

It has been questioned whether an ellipse is the optimal model for depicting length variation with respect to angle at low to modest levels of annealing (e.g., Galbraith 2002, and references therein). However, a comparison by Ketcham (2003) demonstrated that the elliptical model characterized track length populations with no evidence of accelerated length reduction at least as well, and usually better, than the proposed alternatives. In fact, the experience of the authors strongly suggests that non-elliptical results are likely to be indicative of experimental error, examples of which may include non-adherence to strict measurement criteria for track ends at low and high *c*-axis angles, and inclusion of TINCLE tracks. This may partially explain the incompatibility of the Durango annealing data of Green et al. (1986) and Carlson et al. (1999) because the latter were measured using these strict criteria.

Fission tracks that have experienced accelerated length reduction (see above) are often referred to as segmented fission tracks possessing unetchable gaps. In support of this notion Paul and Fitzgerald (1992) used transmission electron microscopy to observe that latent apatite fission tracks are at times discontinuous, with intensely damaged apatite material becoming increasingly separated by zones of material that are indistinguishable from the host crystal with increasing degree of track annealing. This discontinuity of latent fission tracks has been proposed for other minerals (e.g., Dartyge et al. 1981). The so-called gaps in the fission tracks range in length from 4–100 nm (0.04 to 0.10 μm; Paul and Fitzgerald 1992). The data of Carlson et al. (1999) and their modeling by Donelick et al. (1999) and Ketcham (2003) strongly suggest that these gaps alter the etching efficiency of the fission tracks in a systematic fashion. There exists no published evidence to support a model of random track segmentation such as proposed by Green et al. (1986).

Type of confined fission tracks in apatite. Only TINT fission tracks (Fig. 4) should be measured (e.g., Carlson et al. 1999). While many or even most TINCLE fission tracks probably anneal similarly to TINTs, there is well-documented evidence that some TINCLEs are abnormally resistant to annealing (Barbarand et al. 2003b), and some may not anneal at all (Jonckheere and Wagner 2000). Proposed mechanisms for such radically different behavior include chemical infiltration and natural etching; whether such processes occur probably varies on a sample-by-sample, or even a track-by-track basis. Utilization of TINCLE fission tracks in an analysis thus amounts to nothing less than irresponsible inclusion of unreliable data. Similarly, some TINDEF fission-track lengths may also be abnormally long; Jonckheere and Wagner (2000) argue for example that fission particles may pass through large fluid inclusions without significant energy loss, leading to a corresponding lengthening. The number of TINT fission tracks available for length measurement can be greatly increased using ^{252}Cf-fission-fragment irradiation in a nominal vacuum (Fig. 8). Donelick and Miller (1991) demonstrated that irradiating apatite grains with ^{252}Cf-derived fission fragments can yield a 20-fold increase in the number of available TINT fission tracks available for length measurement. Although Ketcham (in press) provides evidence that ^{252}Cf-fission-fragment irradiation may bias length measurements in some cases, this effect is largely eliminated if *c*-axis projection is employed. Furthermore, irradiation with ^{252}Cf-derived fission fragments eliminates the justification used by many analysts that TINCLE fission tracks must be measured to obtain sufficient data for interpretation purposes. The notion that one can measure TINCLE fission tracks in geological unknown samples and quantitatively interpret them using a calibration based in part upon TINCLE fission tracks is untenable. TINCLE fission tracks too often offer spurious length results in experiments. This is especially true for heavily annealed experiments where only a few TINT fission tracks may be available for measurement and especially for experiments conducted without the use ^{252}Cf-derived fission-fragment irradiation to enhance TINT availability. The presence of even a few spurious TINCLE fission-track lengths mixed with sparse TINT fission-track lengths in an experimental dataset renders that data set useless and repeating this measurement scheme for geological unknowns simply extends an untenable method. Therefore, TINCLE and TINDEF fission tracks should be avoided as a matter of routine practice for experiments and geological unknowns.

AFT annealing kinetic parameters

General. Once formed, the crystallographic damage that constitutes a latent fission track immediately begins to spontaneously heal or anneal (e.g., Donelick et al. 1990). The annealing process is not well understood, but it is known to be highly temperature dependent

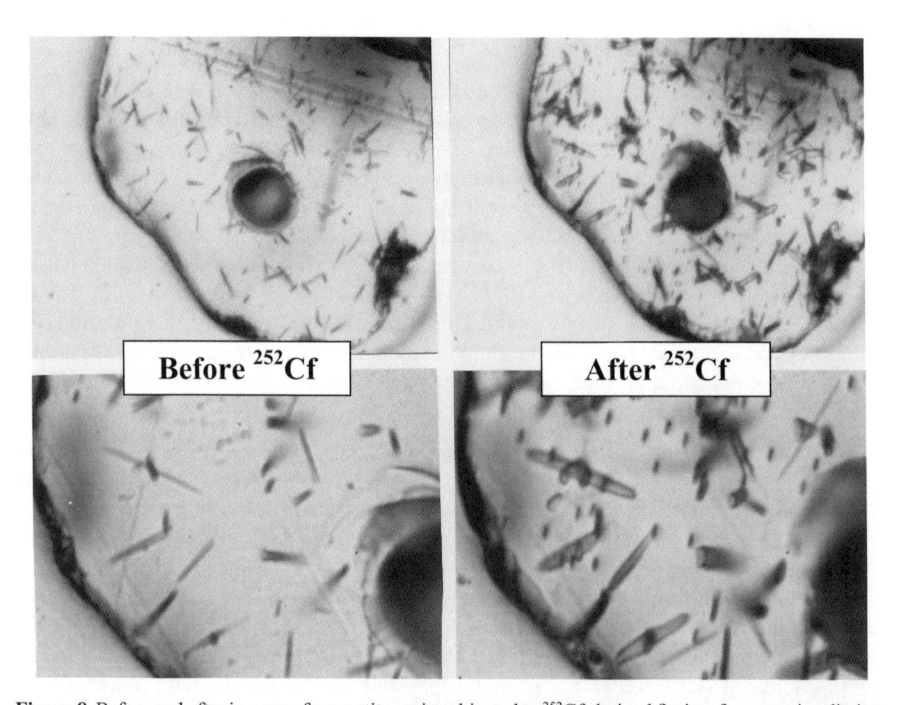

Figure 8. Before and after images of an apatite grain subjected to ^{252}Cf-derived fission-fragment irradiation. Upper images are at 625× magnification; lower images are of the same grain at 1562.5× magnification. Note the obvious increase in TINTs available for length measurement. The large pit represents a laser ablation pit created during LA-ICP-MS analysis for AFT age dating.

and moderately dependent upon crystallographic orientation. It is also correlated with the chemical composition of the host apatite (e.g. Green et al. 1985; Crowley et al. 1991; Carlson et al. 1999; Barbarand et al. 2003a), the etching characteristics of the host apatite (Donelick 1993; Burtner et al. 1994; Donelick 1995; Carlson et al. 1999; Barbarand et al. 2003b), and possibly confining pressure (Wendt et al. 2002). This section is concerned with the correlation between fission-track annealing kinetics and the following measurable parameters, commonly referred to as kinetic parameters:

- D_{par}, the arithmetic mean fission-track etch figure diameter parallel to the crystallographic *c*-axis
- Chlorine content, either Cl wt% or Cl apfu (atoms per formula unit)
- Hydroxyl content, as OH apfu (ions per formula unit)
- Infra-red (IR) microspectroscopy, where the IR absorption characteristics are a function of the F and Cl contents of the target apatite
- α-particle damage potentially accumulated in apatite due to the α-decay of trace U and Th

It is very common for apatite grain populations from a single rock sample, particularly from sedimentary rocks but also from some igneous and metamorphic rocks, to exhibit a significant range in one or more of the above listed parameters. An example is shown of several grains from a Triassic-aged sandstone in Figure 9 that is discussed in more detail later in this chapter. It is desirable to identify an easily and reliably measured parameter that permits grouping apatite grains from such rocks according to their kinetic response and by which the fission-track age and length data provided by those grains can be plotted, grouped, and modeled. Burtner et al. (1994) utilized parameters D_{par} and Cl wt% to discretely group both apatite fission-track age and fission-track length data from individual samples for purposes of thermal history modeling. Unfortunately, these two apatite fission-track annealing kinetic parameters, as well as Cl apfu, OH apfu (Carlson et al. 1999), IR microspectroscopy (Siddall and Hurford 1998), and possibly accumulated α-particle damage (Hendriks and Redfield 2004; Garver et al. 2004), are imperfect. The discussion below concentrates on D_{par} and Cl wt%, the two most commonly applied and better studied kinetic parameters. The other parameters are likely to receive attention in the future as they offer fertile ground for further understanding of how fission tracks anneal in apatite.

Kinetic parameter D_{par}. As used here, an etch figure represents the geometrical figure formed by the intersection of an etch pit (e.g., a fission track or other crystallographic imperfection) and the *c*-axis parallel polished and etched apatite surface. Kinetic parameter D_{par} is the mean, maximum etch figure diameter parallel to the crystallographic *c*-axis. Fission track etch pits in apatite, etched in 5.5 M HNO_3 for 20 s at 21 °C are elongate parallel to the *c*-axis and tend to exhibit hexagonal shaped etch figures (Fig. 10). The maximum diameter of one of these hexagonal etch figures is the distance between the two tips. A minimum total of four individual D_{par} values are recommended from which to calculate the mean D_{par} value although this number is not always attainable. Also, other etched features such as dislocations that intersect the surface, polishing scratches, and minute fluid inclusions can often be used for D_{par} measurement. Experience is necessary to make the distinction between an etched feature with a representative D_{par} value and one which gives a value that is too large (e.g., for a large fluid inclusion) or too small (for a feature that only began etching part-way through the etching process).

A frequent misconception on the part of fission-track workers is that parameter D_{par} is a proxy for Cl wt%. This is not the case, although it is true that D_{par} is positively correlated with Cl wt% and OH wt% and negatively correlated with F wt% in apatite (Donelick 1993; Donelick 1995; Burtner et al. 1994; Fig. 11). D_{par} is a stand-alone parameter and should be

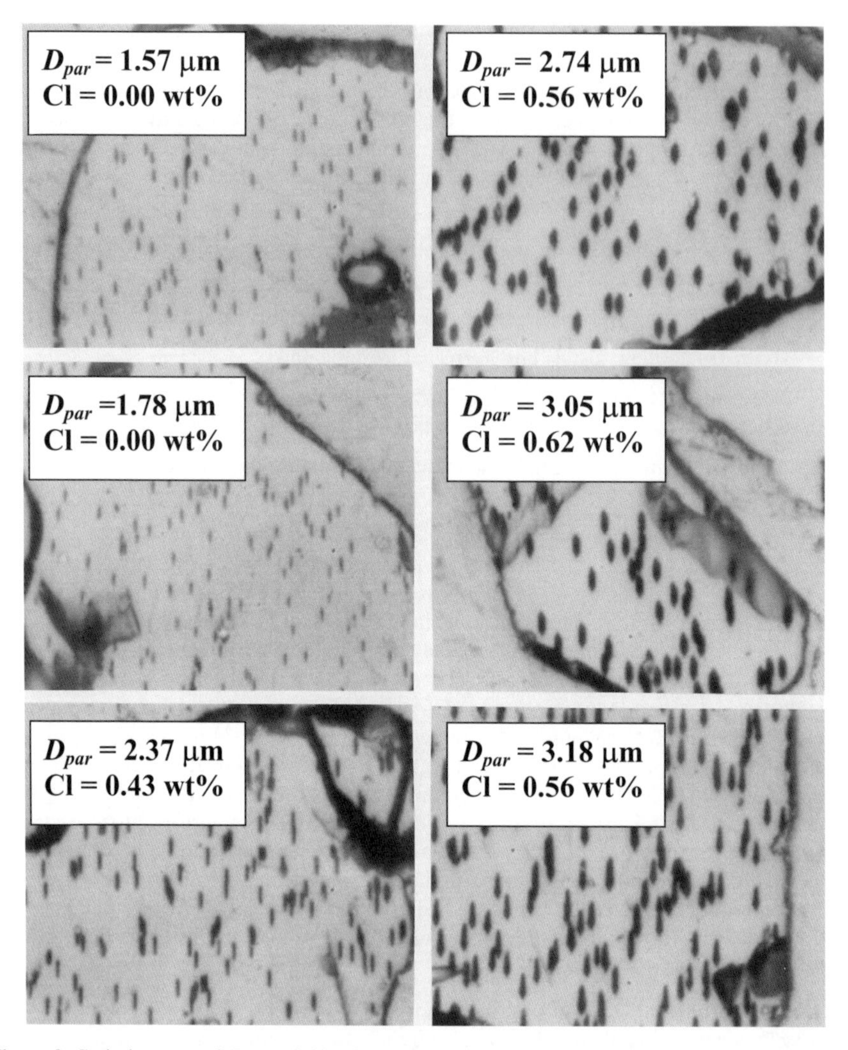

Figure 9. Grain images and D_{par} and Cl wt% values for six track length grains for a sandstone sample. Most of the etch figures visible in these grains represent fission tracks from ^{252}Cf-derived fission-fragment irradiation.

considered independent of any single chemical composition variable because a number of chemical composition variables as well as variables as yet to be fully explored or envisioned control it (Carlson et al. 1999). Some properties of kinetic parameter D_{par} follow:

- Apatite grains with relatively low values of D_{par} (≤1.75 μm for apatite grains etched for 20 s in 5.5 M HNO_3 at 21 °C) anneal rapidly and can generally be considered fast-annealing, typical near-end-member calcian-fluorapatites (Carlson et al. 1999). No reliable exceptions to this statement have been observed by the authors. Apatite grains of this type are quite common.
- Apatite grains with relatively high values of D_{par} (>1.75 μm for apatite grains etched for 20 s in 5.5 M HNO_3 at 21 °C) usually, but not always, anneal more slowly than their low D_{par} counterparts (Carlson et al. 1999). Apatite grains of this type are also

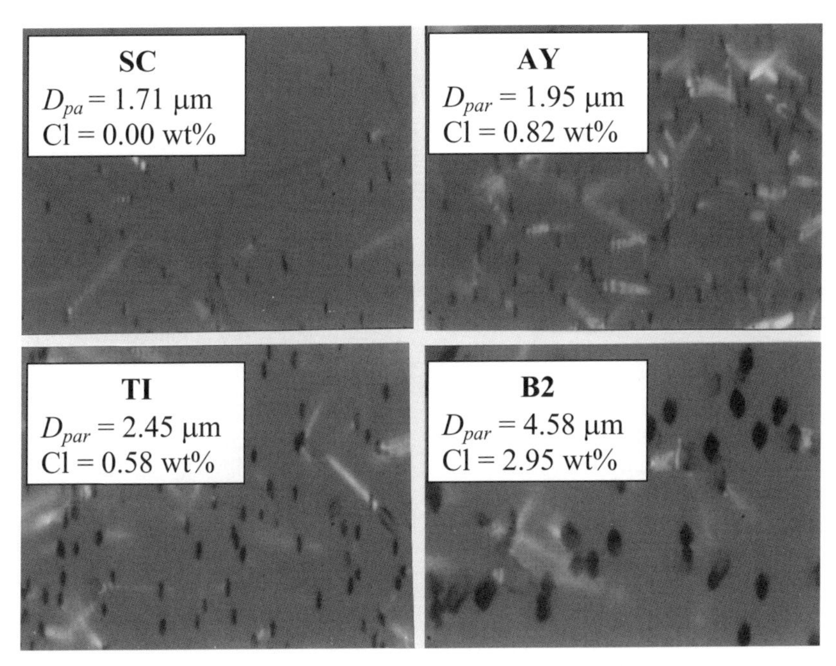

Figure 10. Images showing etch figures for several apatite species studied by Carlson et al. (1999).

quite common, but they are not as common as apatite grains with relatively low values of D_{par}.

- Failure of kinetic parameter D_{par} is most pronounced for near-end-member hydroxyapatite grains that exhibit high D_{par} values and anneal faster than typical calcian-fluorapatites. D_{par} appears to work well for near-end-member fluorapatite grains with unusual cation substitutions such as Mn, probably Fe (apparently, even in small amounts), possibly rare earth elements, and possibly some combination of OH and Cl (Carlson et al. 1999).
- Measured values of D_{par} need not be converted to some equivalent Cl wt% value (or other chemical compositional parameter) for interpretation purposes. Quite simply, apatite fission-track age and length data are best related directly to D_{par}, without reference to or need of Cl wt% or other chemical composition information.
- Because of the extremely fine-scale measurements involved with D_{par}, it is of critical importance that great analytical care be taken when preparing samples for measurement, especially in terms of precisely controlling etchant strength, duration, and temperature. Lack of attention to any of these parameters can easily result in poor and/or inconsistent data quality. Furthermore, it is strongly encouraged that labs interested in using D_{par} conduct careful cross-calibrations (for example, using Durango and Fish Canyon apatite; e.g., Sobel et al. 2004) to compensate for subtle but inevitable variations in measurement equipment and procedures.
- It is not known exactly what controls D_{par} in apatite. Certainly, chemical composition plays a role, Cl wt% being of particular apparent importance, but other parameters related to the concentration and type of crystallographic imperfections in apatite (e.g., accumulated α-particle damage, crystallization age, temperature of formation, deformation history) may also be very important.

When using ^{252}Cf-derived fission-fragment irradiation, it is important to consider how the ^{252}Cf-derived fission-fragment tracks mask the orientation of the crystallographic *c*-axis and hence give rise to potentially spurious fission-track orientation measurements and estimates of D_{par}. When an apatite grain mount is irradiated with ^{252}Cf-derived fission-fragments some distance from the ^{252}Cf source (>5 cm for a 0.5 cm diameter source), the resultant ^{252}Cf-derived fission-fragment tracks are nearly parallel to each other and their etch figures are nearly parallel to each another as well, regardless of the true orientation of the *c*-axis within the target apatite grain. In such cases, it is necessary to seek out spontaneous fission tracks (from the spontaneous fission of ^{238}U) that are randomly orientated and verify that their etch figures are parallel to one another and parallel to the ^{252}Cf-derived etch figures. Other features such as mineral and/or fluid inclusions and the presence of a prismatic face along the apatite grain boundary can aid the identification of properly orientated grains.

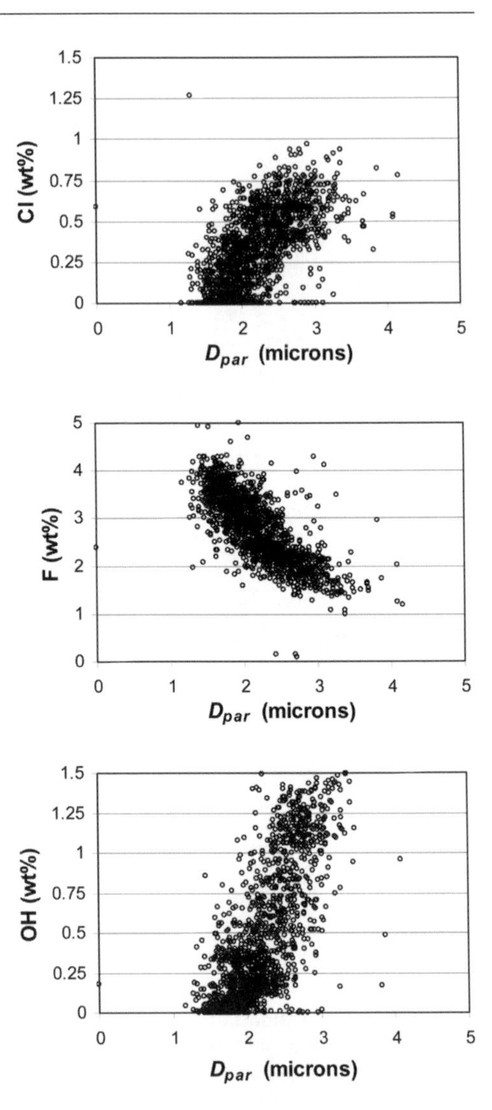

Figure 11. Correlation of kinetic parameter D_{par} with a) Cl wt%, b) F wt%, and c) OH wt% (Donelick, unpublished data).

Kinetic parameter Cl wt%. To obtain useful chemical composition data for apatite, particularly for the halogen site, it is common to use electron probe microanalysis (EPMA). A variety of analytical schemes may be employed but all should include the following basic criteria: 1) for useful halogen determinations in apatite, particularly F and to a lesser extent Cl, it is necessary to perform EPMA analyses on *c*-axis parallel crystallographic planes due to induced diffusion of the halogens under the influence of the electron beam (Stormer et al. 1993) and 2) at least Ca or P should also be measured so that a signal of appropriate strength can be used to assess the quality of the analysis and the true presence of apatite under the electron beam. Finally, it is suggested that one measure at least Ca, Cl, F, and Fe and assume stoichiometry for P or measure P directly.

Some properties of kinetic parameter Cl wt% follow:

- Apatite grains with relatively high values of Cl wt% (>1–2 wt%) anneal relatively slowly. However, there exists significant kinetic variability among apatite grains of this type (Carlson et al. 1999; Barbarand et al. 2003a).

- Apatite grains with relatively low values of Cl wt% (<1–2 wt%) usually, but not always, anneal rapidly relative to their high Cl wt% counterparts.
- Failure of the kinetic parameter Cl wt% occurs for apatite grains having unusually high concentrations of Mn, probably Fe (apparently, even in small amounts), possibly rare earth elements, and possibly some combination of OH and Cl (Carlson et al. 1999).
- Kinetic parameter Cl wt% alone does not completely account for the variation of fission track annealing kinetics in apatite. Other elements play a role and there is evidence that apatite solubility (related directly to D_{par}) plays an important role (Carlson et al. 1999).
- Kinetic parameter Cl wt% is typically measured using an electron microprobe but this fact alone does not make it better than kinetic parameter D_{par}, which is measured using an optical microscope. Green (1995) stated that "...it is essential to know the chlorine content of every grain analyzed..." Green's statement blatantly ignores the reality that parameter D_{par} (which does not require EPMA of Cl content) is at least as effective as Cl wt% as an apatite fission-track annealing kinetic parameter.

The necessity of kinetic parameter measurement in AFT studies. Both D_{par} and Cl wt% are very useful as indicators of fission-track annealing kinetics in apatite and they appear to be approximately equal in their effectiveness (Carlson et al. 1999). In a general sense, for 90% of all apatite grains likely to be encountered, these parameters appear to work as desired, although they work differently; for 10% of the grains, these parameters appear to fail, although they fail differently (Carlson et al. 1999). Considering the risk of interpretation failure when neither parameter is used (or a parameter as yet, undiscovered, untested, or not presented in the public domain), the 10% risk of failure when either is used outweighs the greater risk of failure if a kinetic parameter is not utilized at all for interpreting AFT data. Recognizing the existence and nature of the failures of D_{par} and Cl wt% is an essential step in the understanding of how fission tracks anneal in apatite. Any 100% successful model of fission-track annealing in apatite must account for the failures as well as the successes of both of these parameters.

Whereas the least resistant to annealing apatite grains (i.e., D_{par} = 1.50 μm; Cl wt% = 0 wt%) appear to experience total fission-track annealing at around 100–110 °C in the geological environment, the most resistant to annealing apatite grains (i.e., $D_{par} \geq 3.00$ μm; Cl wt% > 3 wt%) experience total fission-track annealing at temperatures greater than 160 °C (Ketcham et al. 1999). When the least resistant to annealing apatite grains are entering the so-called total annealing zone, the most resistant apatite grains are just entering the so-called partial annealing zone. As an example of the importance of this effect, consider a hypothetical series of sedimentary samples analyzed over a range of temperatures in a drill hole (Table 2). Suppose all of the samples exhibit a typical distribution of apatite annealing kinetics, say 75% of the apatite grains of the least resistant to annealing kinetic type (D_{par} = 1.50 μm in Table 2) and 25% of the grains of the most resistant type (D_{par} = 3.00 μm in Table 2; actually, as a rule, a continuum of D_{par} or Cl wt% values is usually observed in sedimentary rocks but the grouping assumed here helps to simplify the following argument). Suppose further that all apatite grains would give the same fission-track age if the effects of heating could be removed and that all exhibit the same uranium concentration. In the relatively unannealed samples at low temperature, the least-resistant-to-annealing apatite grains dominate both the fission-track age and length data. However, with increasing temperature, the fission-track age data continue to be dominated by the least resistant-to-annealing apatite grains whereas the fission-track length data become increasingly dominated by the most resistant-to-annealing apatite grains because the probability of observing TINT fission tracks is proportional to the product of the spontaneous fission-track density squared, the mean fission-track length, and D_{par} (Donelick and Miller 1991). Near the bottom of the drill hole, if temperatures are sufficient

Table 2. Grain age and mean track length characteristics of apatite grains exhibiting different annealing kinetics and their relative contributions to whole sample age and length data in a hypothetical drill hole.

	Apatite with D_{par} 1.50 μm			Apatite with D_{par} 3.00 μm				
Sample Temp. (°C)	Grain Fission Track Age (Ma)	Grain Mean Fission Track Length (μm)	TINT Density Weighting Factor (per area)	Grain Fission Track Age (Ma)	Grain Mean Fission Track Length (μm)	TINT Density Weighting Factor (per area)	Age Data Contribution Ratio for 3 grains D_{par} =1.50 per 1 grain D_{par} =3.00 (1.50:3.00)	Length Data Contribution Ratio for 3 grains D_{par} =1.50 per 1 grain D_{par} =3.00 (1.50:3.00)
20	10.1	14.7	2250	10.3	15.3	4870	3:1	6750:4870
40	9.7	14.2	2004	10.0	14.9	4470	3:1	6012:4470
60	9.2	13.5	1714	9.7	14.5	4093	3:1	5142:4093
80	8.3	12.3	1271	9.3	13.9	3607	3:1	3813:3607
100	4.3	9.3	258	8.7	13.1	2975	3:1	774:2975
110	0.7	9.2	7	8.3	12.6	2604	3:1	21:2604
120	0.1	9.0	<1	7.8	11.9	2172	3:1	<1:2172
140	0	0	0	5.1	9.9	772	3:1	0:772
150	0	0	0	2.6	8.5	172	3:1	0:172
160	0	0	0	0.4	8.4	4	3:1	0:4

NOTES: TINT Density Weighting Factor = (Grain Fission Track Age)^2 % Mean Track Length % D_{par} (after Donelick and Miller 1991; their Equation 1). Ages and mean lengths calculated using AFTSolve (Ketcham et al. 2000) assuming isothermal temperature conditions for a period of 10 Ma.

to totally anneal fission tracks in the least resistant apatite grains but not sufficient to totally anneal fission tracks in the most resistant apatite grains, the track length data become totally represented by the most resistant kinetic type while the fission-track grain age data maintain their initial 3:1 ratio. Throughout the drill hole, the combination of the fission track age and length data is a mixture in terms of the annealing kinetics of their host apatite grains. Without a useful kinetic parameter by which to classify the fission-track age and length data into kinetic groups, an interpretation that treats this mixture as a single kinetic system is, quite obviously, erroneous. Importantly, this data mixing effect predominates at geological temperatures above approximately 90 °C, even for apatite mixtures with a fractional component of the most resistant to annealing apatite grains significantly less than 25%. There is no doubt that a large quantity of data has been and continues to be generated and reported in the literature that suffers from this problem, owing to the rather common occurrence of apatite grains that are resistant to annealing and the dearth of kinetic data reported. This is especially true for studies that utilize samples from drill holes. Simply assuming a particular kinetic behavior for a mixture of apatite grains under study (e.g., many workers assume Durango apatite is representative of their grain mixtures) does not adequately address this problem.

How many AFT grain ages and lengths should be measured?

Commonly a fission-track analyst selects the first 20 randomly encountered, acceptable grains for age measurement and the first 100 or so randomly encountered, acceptable confined fission-tracks for length measurement (too often including unacceptable TINCLE fission tracks). This approach to AFT analysis has a basis in both statistics and economy of measurement. Suppose 20 apatite grains yield, on average, 5 spontaneous fission tracks per grain for a total of 100 spontaneous fission tracks. Assuming other measurements contributing to the fission-track age calculations have greater counts, such as induced fission tracks for the same grains, the relative error on the pooled fission-track age would be approximately 1/sqrt(100) = 10%. This level of precision for a pooled fission-track age is generally acceptable. For fission-track length data, 100 track lengths might give a mean of 12.00 μm and a standard deviation of 1.50 μm, for example. The precision of the mean is given by its standard error and would be 1.50 μm/sqrt(99) = 0.15 μm. For an estimate of the mean track length, this precision is generally acceptable and it is similar to the precision for measuring a single fission-track length (Donelick 1991).

Considerably better constraints on the geological problem at hand can be obtained by being flexible with both the number of ages and lengths measured and the choice of apatite grains selected for data collection. Consider an example where 1000 small D_{par} grains co-exist on a grain mount with only 10 large D_{par} grains. It is far better to obtain age data from 30 small D_{par} grains and all 10 large D_{par} grains, and obtain length data from those same grains, than to obtain data randomly from, say, 39 small D_{par} grains and only 1 large D_{par} grains. The approach of 20 grain ages and 100 track lengths works well for a sample exhibiting a narrow range of D_{par} values among the available apatite grains. It does not work well for the more common situation of apatite grains exhibiting a significant (and often wide) range of D_{par} values. Suppose that an apatite grain population exhibits a wide range of D_{par} values and that this range could easily be divided into two end-members encompassing, say, 1/3 each of the data (with the 1/3 of the data from the transitional etch pit population being disregarded). Suppose that the analyst measures 21 grain ages and the 7 small D_{par} grains yield 1 spontaneous fission-track each on average for a total of 7 tracks; for this sub-population of grains the relative error of the AFT age is approximately 1/sqrt(7) = 38%. Suppose the 7 large D_{par} grains yield 9 spontaneous tracks each on average for a total of 63 tracks; for this sub-population of grains the relative error of the AFT age is approximately 1/sqrt(63) = 13%. If the analyst doubles the number of analyzed grains to 42, all else being equal, the relative error of the small D_{par} grains would be approximately 1/sqrt(14) = 27% and of the large D_{par} grains approximately 1/sqrt(126) = 9%. Depending on the distribution of grains analyzed in terms of their D_{par} values and their respective track counts, the analyst can high-grade the analysis to obtain optimal AFT age precision for each sub-population of apatite. According to this approach, there is no single preferred number of grains to be analyzed for age measurement. A similar argument can be made regarding fission-track length data. The analyst should worry less about filling lines in data tables and more about obtaining the most useful data possible within reasonable time constraints. Considering the effort made to collect and prepare a sample for analysis, doubling the data measured, if this is possible, is almost trivial.

LABORATORY CALIBRATION OF THE APATITE FISSION-TRACK SYSTEM

General

Fission-track annealing behavior in apatite has been the subject of numerous studies since the late 1960s. The discussion here is not intended to be a comprehensive review of all of this work but is intended to illustrate the current level of understanding of this important process.

Setting up a calibration procedure

The following considerations are made when setting up an AFT calibration procedure:

- Choice of apatite species or range of species for study.
- Choice of fission-track parameter(s) to be calibrated and the measurement method(s) used.

- Choice of experimental conditions to be controlled and the control methods used.
- Method or assumptions relating laboratory-derived data to data from the geological environment.

Choice of apatite species for study. The choice of apatite species for several major studies of AFT annealing is given below:

- Durango fluorapatite (Green et al. 1986; Carlson et al. 1999; Barbarand et al. 2003a).
- Near-end-member fluorapatite and Sr-rich fluorapatite (Crowley et al 1991; Carlson et al. 1999).
- A range of Cl-contents for otherwise dominantly calcian-rich apatites (Carlson et al. 1999; Barbarand et al. 2003a).
- A range of cation substitutions with significant halogen-site substitutions (Carlson et al. 1999).

The goal or goals of the experiments should dictate the apatite species to be included in an experimental study. If the goal is to show the general nature of AFT annealing, then a study of a single apatite suffices (e.g., Green et al. 1986). If the goal is to show how Cl-content affects AFT annealing kinetics, then varying Cl with minimal variation in other parameters is desirable (e.g., Barbarand et al. 2003a). If the goal is to explore wider variations based on cation and/or anion substitutions then a range of apatite species spanning both types of substitutions can begin to address this problem (e.g., Carlson et al. 1999). Most importantly, within the limitations of the data themselves (quality, accuracy, precision) and the laboratory calibrations (accuracy, precision), all data from all sources should ultimately be reconciled in any resultant model. For a model that does not address one or another aspect of the overall data, then that model is limited, and the data excluded and reasons for excluding them should be disclosed.

Fission tracks in apatite are considered to be easily studied in the laboratory. This is because apatite, unlike zircon and titanite, accumulates minimal radiation damage from the α-decay of trace amounts of uranium and thorium in its structure. As such, it is common practice to study the behavior of fission tracks formed by the thermal-neutron-induced fission of ^{235}U as a proxy for the behavior of spontaneous fission tracks formed by the fission of ^{238}U. This is accomplished using the following procedures:

- Select an apatite for study.
- Heat the apatite for sufficient time at sufficient temperature to completely anneal the spontaneous fission tracks present.
- Place the apatite near the core of a nuclear reactor so that thermal neutrons can react with ^{235}U in the apatite and induce fission of the ^{235}U creating fresh, induced fission tracks.
- Subject the apatite containing thermal-neutron-induced fission tracks to a range of laboratory conditions necessary to study the fission-track annealing characteristic of interest.

Choice of AFT parameter and method of measurement. Experimental design can greatly affect the ability of the experimenter to successfully calibrate a parameter of interest. Several parameters are typically measured and reported in AFT calibration studies:

- The overall mean length of confined fission tracks, the standard error of the mean, and moments of the distribution about the mean (assuming Guassian distributions) for either a) relatively high temperatures in the laboratory (generally 75–400 °C; Green et al. 1986; Crowley et al. 1991; Carlson et al. 1999; Barbarand et al. 2003a) or b) relatively low temperatures (essentially room temperature following induced fission track creation; Donelick et al. 1990).

- The mean lengths of confined fission tracks parallel and perpendicular to the crystallographic *c*-axis obtained by fitting whole or partial ellipses, the standard error of the means, and the moments of the track length distribution about the whole or partial fitted ellipses (Donelick et al. 1990; Donelick 1991; Crowley et al. 1991; Carlson et al. 1999).
- The frequency distribution of fission-track length measurements at different angles to the crystallographic *c*-axis (e.g., Ketcham 2003).
- The fission track density (e.g., Green 1988).

It is has been common practice in these experiments to utilize a digitizing tablet interfaced with a personal computer and use as the measurement device a projected LED image from the tablet cursor onto the microscope field-of-view via a projection tube (the exception is that only a fraction of the Green et al. 1986 data were measured this way; the remaining data were measured using a calibrated eye-piece graticule with significantly poorer resolution compared to the digitizing tablet). The precisions of individual fission-track length and angle to the *c*-axis measurements are approximately 0.15 μm (1σ) and 2° (1σ), respectively (Donelick 1991; other researchers have not reported similar estimates but these likely apply).

Choice of controlled parameters and methods of control. Most studies vary time and temperature of annealing of induced fission tracks. However, pressure has recently been shown to be of concern (Wendt et al. 2002), but this study was performed using spontaneous fission tracks in apatite. Donelick et al. (2003) performed a limited range of variable pressure experiments on spontaneous and induced fission tracks in Durango apatite but did not observe a significant effect up to 1 kbar. It is noteworthy that there may, in fact, be a significant effect beyond 1 kbar as noted by Wendt et al. (2002) and these high pressures may be necessary to study this problem at laboratory timescales in a manner analogous to the need for high laboratory temperatures to study geological thermal effects.

Temperature is a difficult, but not impossible, parameter to control accurately and precisely in the laboratory. Carlson et al. (1999) offer a detailed discussion of their temperature calibration scheme and conclude overall temperatures are most likely within ±2 °C of the reported values (taking into account the calibration of the thermocouples relative to NIST standards and anomalous excursions due to power surges, etc.). No other published AFT annealing study reports a temperature calibration scheme in such detail, although this level of detail should be the norm. It is common practice for some workers to assume all variation among experimenters is due to variation derived from the analysts. This assumption holds further that temperature and time are both well controlled to within the reported errors. This assumption almost certainly fails in reality and the lack of information on temperature calibrations in the published literature precludes thoughtful inter-comparison of temperature calibration approaches and potential sources of incompatibility. Therefore, inter-comparison of different datasets must consider both analyst-derived and experimental design differences.

Method or assumptions extrapolating experimental data to geological data. In pseudo-Arrhenius-space ([$1/T$], ln[t]), laboratory experiments can routinely be performed over 1 hour (ln[t]=8.2) to 1 year (ln[t]=17.3) timescales. Geological timescales of 1 Ma (ln[t]=31.1) to 500 Ma (ln[t]=37.3) are of greatest interest for AFT interpretation purposes. Effectively, 1/3 of pseudo-Arrhenius space is sampled in the laboratory and a calibration is obtained. That calibration skips over 1/3 of pseudo-Arrhenius space to reach the other 1/3 of the space of interest to geological studies. Such an extrapolation requires the best control possible on temperature for the laboratory data (e.g., the abscissa for each calibration point) and a mathematical model that is believable and properly predicts known behavior at geological time. This requires the following for confident extrapolation:

- A mathematical model (empirical or physical) with sufficient independent variables that, once optimized, provide an adequate calibration of the observed data.
- Sufficient data in the laboratory portion of the pseudo-Arrhenius space to permit the form of the mathematical model used for extrapolation to be obtained.
- One or more good control points in the geological portion of pseudo-Arrhenius space through which the extrapolation based on the laboratory data can be passed and against which the mathematical model tested.

Importantly, the empirical mathematical model of Laslett et al. (1987) and the physical model of Carlson (1990) offer a vantage point from which to consider the form of the mathematical model. Both of these studies yielded similar contours of fission-track length in pseudo-Arrhenius space. However, the model of Carlson (1990) was predicated upon a track shape. If that track shape is allowed to vary by including an inflection point, then this would cause an inflection to occur among the track length contours. Many, including Laslett and co-workers, have argued vigorously that the mathematical model in Laslett et al. (1987) is the best choice. However, the natural behavior of fission tracks in apatite does not preclude a length-contour inflection. In fact, such an inflection may be warranted if one combines the data of Donelick et al. (1990) and Carlson et al. (1999) for Durango apatite. Hence, more work on developing mathematical models for empirical fitting may produce better models of AFT length in pseudo-Arrhenius space.

The Ketcham et al. (1999) study uniquely satisfies the requirements listed above in comparison to other published mathematical models (Laslett et al. 1987; Crowley et al. 1991). These authors used the AFT annealing data of Carlson et al. (1999) as the basis for their calibration of the laboratory portion of pseudo-Arrhenius space and used geological control points at both low temperature (Vrolijk et al. 1992) and high temperature (Gleadow and Duddy 1981) as tests of the extrapolation of their calibration. Ketcham et al. (2000) and Ketcham (2005) provide further discussion of laboratory calibration and modeling in general.

DISCUSSION AND FUTURE WORK

General

Many approaches are used to obtain and interpret AFT data within the fission-track community today. New approaches continue to be developed and this is a healthy sign of a science that continues to move forward. With the resurgence of apatite (U-Th)/He age dating, many fission-track workers are experiencing pressures from internal and external sources to both justify their science and integrate it with the broader thermochronology community. The authors note the following questions that deserve attention and further study on the part of fission-track workers:

- What type of data should be measured for fission-track ages (EDM versus LA-ICP-MS)? What type of data should be measured for fission-track lengths (TINTs-only, track crystallographic orientation)?
- What kinetic parameter for apatite should be measured, if any, for each age and length grain? For parameter D_{par}, how does one analyst inter-calibrate to the data of another analyst?
- What process should be applied to laboratory annealing data to obtain a useful and geologically believable extrapolation?

- Can AFT annealing models be improved? If so, how (new experiments, new analogues, inter-calibration with (U-Th)/He methods)?

There are no 100% correct answers to any of these questions but there are certainly wrong answers. The rapid and insightful development of the (U-Th)/He age dating field during the past decade can serve as a useful model to the fission-track community for how to approach these questions. After all, the (U-Th)/He community has used the experience of the fission-track community, at least in part, to guide its development.

Type of data to measure for AFT ages and lengths

Regarding single-grain age data obtained using either the EDM or LA-ICP-MS approaches, the key issues regarding choice should be quality of data obtained and economy of obtaining those data. A secondary consideration could be the detrimental health and societal effects of neutron irradiation for EDM. The EDM, at least presently, poses an interesting advantage over LA-ICP-MS because it leads to minimal physical damage to the apatite grain being dated. This has the key advantage of permitting one to pluck an apatite grain from a grain mount and confidently perform (U-Th)/He age dating of it. This advantage may disappear, however, as the authors are vigorously trying to determine whether or not the actual damage and peripheral effects of laser ablation ultimately affect (U-Th)/He ages. A key advantage of LA-ICP-MS analysis of apatite grains is that one is directly viewing the grains themselves instead of fission-track maps of the grains on external detectors. As fission-trackers universally know, it is often difficult to identify the image of a grain on a fission-track map, regardless of the precision of the microscope stage used. So, at least for low-uranium grains being dated, LA-ICP-MS analysis is the superior choice. Another advantage of LA-ICP-MS analysis is the elimination of analyst bias during age measurement.

Regarding data measured during AFT length analysis, the work of Donelick et al. (1990), Carlson et al. (1999) and especially Jonckheere and Wagner (2000) clearly demonstrate the importance of measuring only TINT fission tracks and the risks of measuring TINCLE and TINDEF fission tracks. Accounting for fission-track orientation to the crystallographic *c*-axis removes significant noise from the track length data. In a manner analogous to avoiding mineral inclusions during (U-Th)/He age dating, fission-track analysts should measure only TINT fission tracks and measure and correct for the track orientation to the *c*-axis in apatite.

Measurement of kinetic parameter for AFT analysis

Kinetic variations among apatite grains, either revealed by parameter D_{par} or Cl wt%, are significant and very common in the geological environment. It is common practice today to assume that all grains in a sample behave like Durango apatite in terms of their apatite fission-track annealing kinetics (this assumption was the state-of-the-art in the late 1980s). However, published data and calibrations provide a reliable means to move beyond this assumption. Parameter D_{par} is readily available for measurement (it can be done when the analyst is measuring age or length information for an apatite grain at minimal time and money costs), Cl wt% is also easily obtained (at somewhat greater time and money costs relative to D_{par}), and both have been calibrated and published on (Carlson et al. 1999; Ketcham et al. 1999). It is inexcusable for a laboratory to ignore this important parameter and it should be measured for each and every age and length grain that provides AFT data. Ignoring this parameter is analogous to not measuring fission-track lengths at all because this parameter has the potential to significantly shift interpretations of AFT data. An example is shown in Figure 12 for a Devonian sandstone sample from the Andean Foothills belt of Bolivia where the data have been classified according to parameter D_{par} and Cl wt% (Figs. 12a and 12b, respectively). Note the small D_{par} value grains give a pooled age of 4.3 ± 1.6 Ma (26 grains) and mean fission-track length of 14.7 ± 1.1 μm (68 track lengths) whereas the large D_{par} grains give a pooled age of

221 ± 35 Ma (5 grains) and a mean fission-track length of 12.3 ± 1.0 μm (72 track lengths; the track lengths were projected on the *c*-axis using Donelick et al. 1999; data grouped according to Cl wt% show similar values). Qualitatively, these data indicate that the sample was exhumed at ca. 4.3 ± 1.6 Ma (age given by the small D_{par} grains) and was not heated above ca. 130 °C after burial in the Devonian (due to the survival of fission tracks indicated by the age and shortened fission tracks of the large D_{par} grains). Ignoring D_{par} and simply combining these data into a single population gives a meaningless age of 18.0 ± 5.7 Ma (40 grains) and meaningless mean fission-track length of 13.0 ± 1.8 μm (210 track lengths). The erroneous approach of combining the complete dataset into a single age and mean track length continues to appear often in the published literature. There exists no credible reason for not measuring a kinetic parameter for each apatite grain that yields fission-track age and length data given the ease of measuring parameter D_{par}. The calibrations of Ketcham et al. (1999) for D_{par}, Cl apfu (similar to Cl wt%), and OH apfu give similar results in the several hundred samples to which the authors have subjected tests. An example of this is shown in Figures 13, 14, and 15 for sample 171-1, a Triassic sandstone from the Paradox Basin in Utah that was collected in close proximity to a ca. 25 Ma intrusive body. Here, it is shown that a single acceptable thermal history predicts the measured AFT parameters (age, length distribution) for all three kinetic parameters within 95% confidence limits. For sample 171-1, simply grouping all AFT data into a single population gives a meaningless age of 42.5 ± 3.0 Ma (40 grains) and meaningless mean fission-track length of 13.3 ± 1.5 μm (187 track lengths). Finally, it is noteworthy that the range of variation in Cl wt% values for apatite grains from sample 171-1 (shown in Fig. 9) does not extend very far beyond the value of 0.43 for Durango apatite (Carlson et al. 1999 data for Durango apatite).

Inter-calibrating parameter D_{par} between analysts was the subject of several recent presentations (Sobel et al. 2004; O'Sullivan et al. 2004). To date, a comprehensive study of this important topic has yet to be published. The Ketcham et al. (1999) multi-kinetic annealing model based on D_{par} is based on data presented by Carlson et al. (1999). To use the Ketcham et al. model, the analyst must obtain a relationship between his/her D_{par} values and those published by Carlson et al. It is recommended that the Durango and Fish Canyon Tuff apatite species be used for this purpose.

Extrapolation of calibrations to geological time

To obtain a useful mathematical model of laboratory-derived data for AFT annealing, it is only necessary that the laboratory data being fitted are defensible and that sound mathematical principles are involved with the calibration of those data. The modeler should only be concerned with whether or not the calibration predicts reasonable behavior at geological timescales and not be concerned with whether the calibration follows previously used methods. The works of Vrolijk et al. (1992) and Gleadow and Duddy (1981) provide reasonable geological benchmarks and were used by Ketcham et al. (1999) for selection of their preferred model; other benchmarks may be available now or in the future. There is no single equation form that one must be limited to and there is no published data set or calibration that one must seek to replicate.

Can AFT models be improved?

Where should experimentalists go from here? Given the complexities encountered trying to account for kinetic variability among apatite species, the authors believe it is time to seek novel solutions to potential problems. Below are some parting thoughts.

What is known of the initial condition (as measured) of fission tracks? Carlson et al. (1999) studied 15 apatite species, of which only four were studied in great detail. Ketcham et al. (1999) demonstrated that adequate apatite-apatite inter-comparisons can be made with 10+ track-length measurements instead of the traditional 50+ to characterize a single apatite. Thus, it would be useful to conduct detailed experiments on a single highly resistant-to-annealing

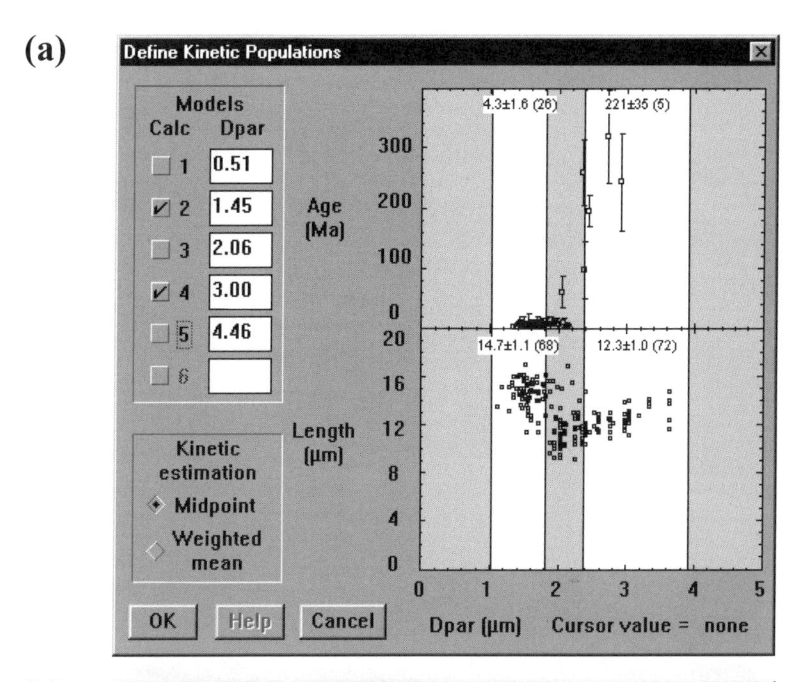

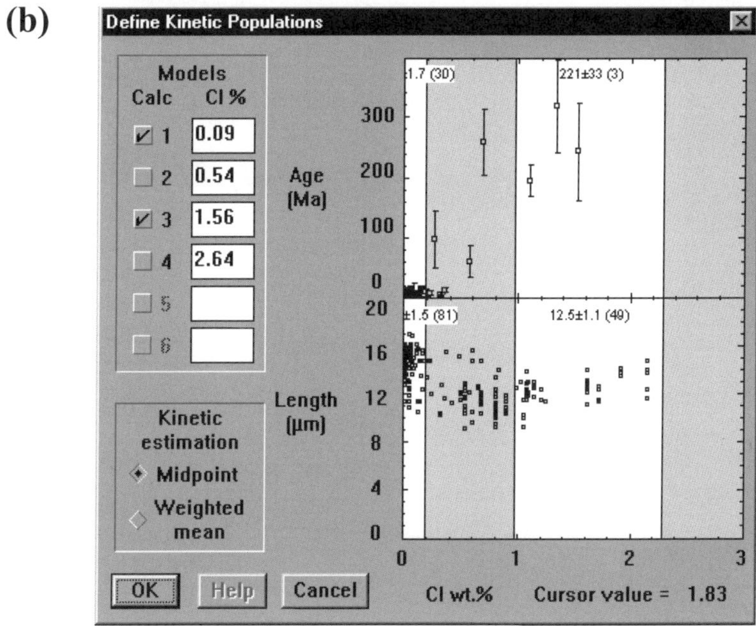

Figure 12. AFT data from a Devonian-aged sample from the Andean Foothills in Bolivia classified into groups according to kinetic parameter a) D_{par} and b) Cl wt% (Donelick unpublished data).

(a)

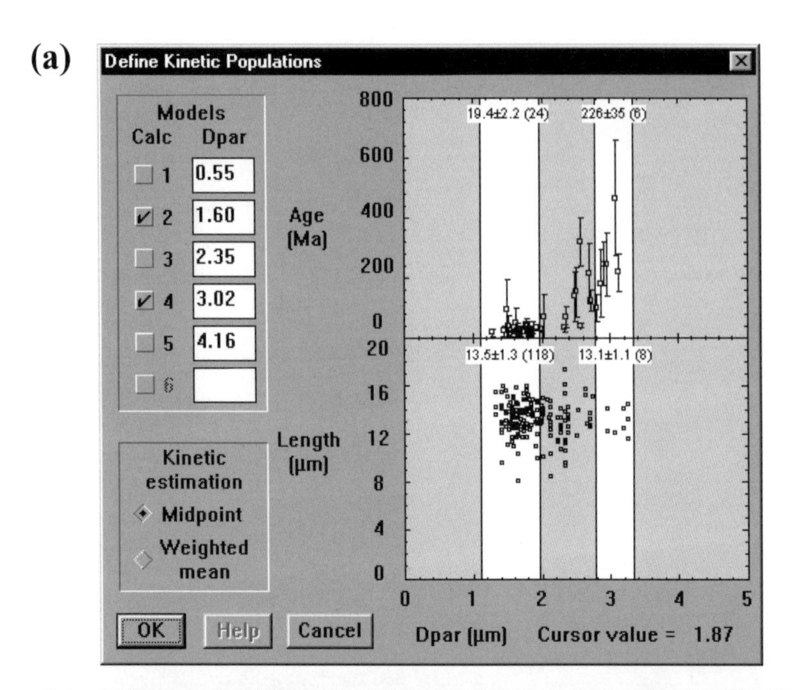

(b)

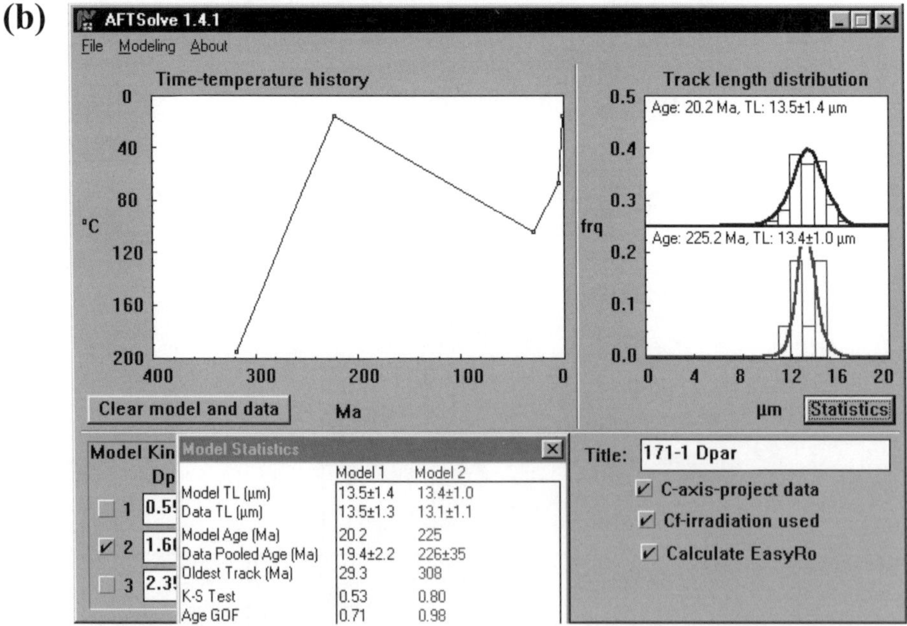

Figure 13. a) AFT data from a Triassic-aged sample from the Paradox Basin in Utah classified into groups according to kinetic parameter D_{par}. b) Acceptable model thermal history for the data in a) calculated using AFTSolve (Donelick, unpublished data; Ketcham et al. 2000; Ketcham et al. 1999 multi-kinetic model, fission-track lengths projected onto the crystallographic *c*-axis using Donelick et al. 1999).

(a)

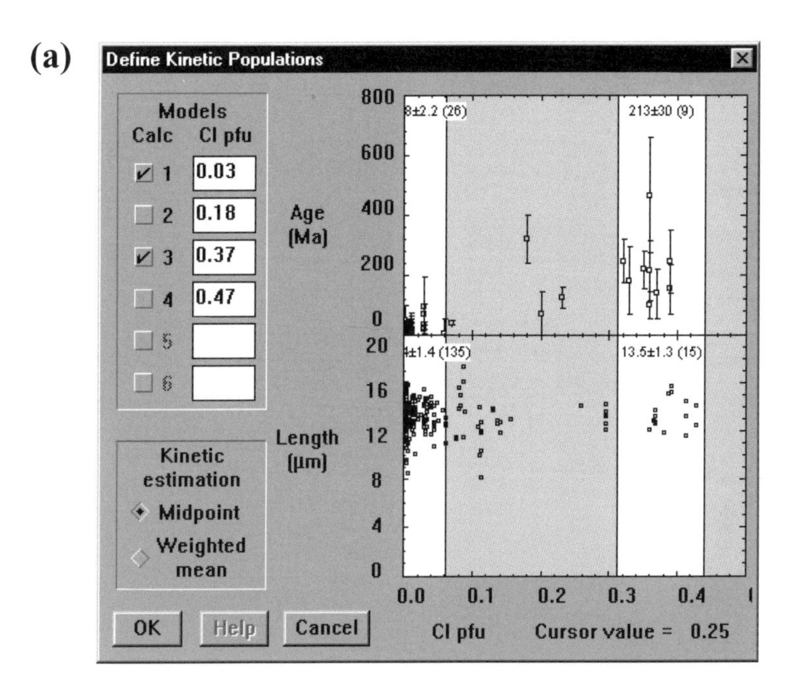

(b)

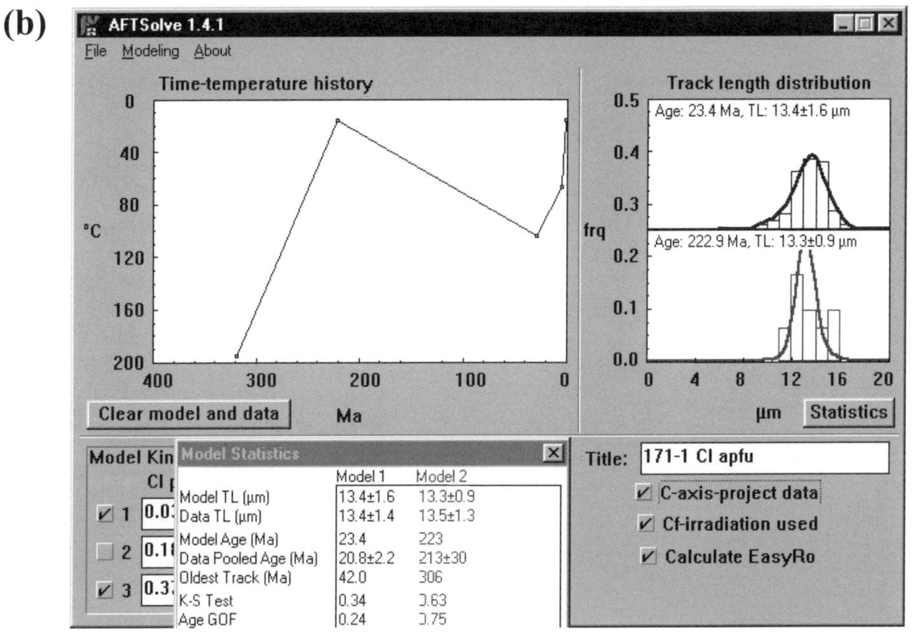

Figure 14. a) The same AFT data as shown in Figure 13a classified into groups according to kinetic parameter Cl apfu. b) The same thermal history as shown in Figure 13b gives an acceptable fit to the data based on Cl apfu.

(a)

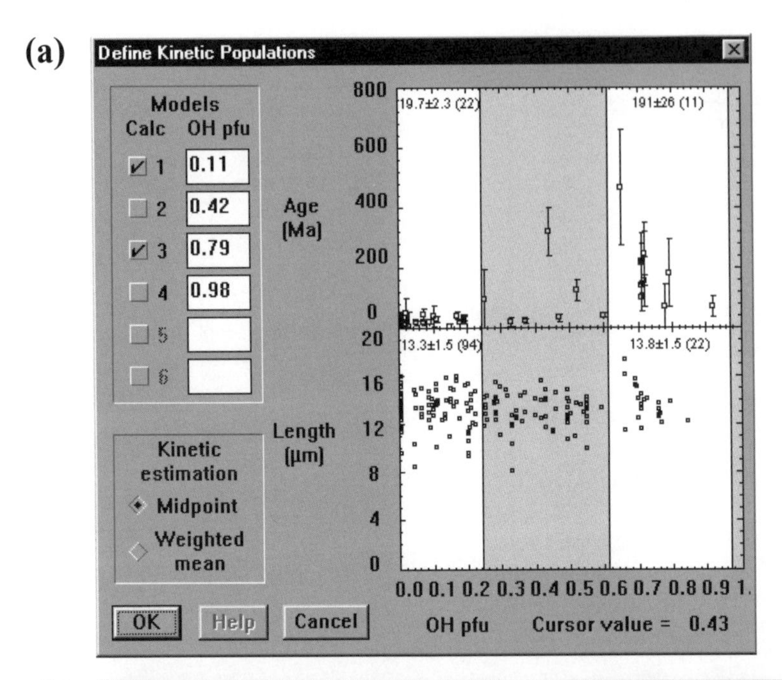

(b)

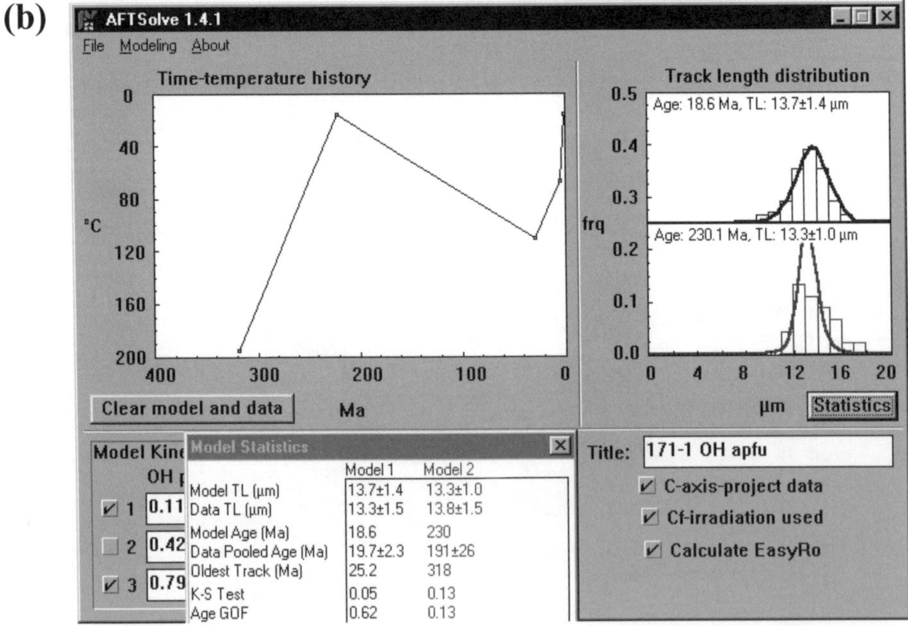

Figure 15. a) The same AFT data as shown in Figure 13a classified into groups according to kinetic parameter OH apfu. b) The same thermal history as shown in Figure 13b gives an acceptable fit to the data based on OH apfu.

apatite and lesser-detailed experiments on many more apatite species than the 15 of Carlson et al. (1999). For all of these apatite species, a detailed careful study of the initial condition should be pursued. The intuition of the authors suggests that 50 or more apatite species spanning a wide range of cation and anion substitutions will yield significant additional insight into the controls on the initial condition of fission-tracks and fission-track annealing rates.

Are laboratory calibrations a direct analogue to the behavior of spontaneous fission tracks? Attention should focus on whether or not spontaneous fission tracks (resulting from the spontaneous fission of ^{238}U) exhibit annealing rates inherently different from those of their laboratory-produced analogues (fission tracks resulting from the thermal-neutron-induced fission of ^{235}U). In particular, attention should be given to the possible effects on AFT annealing kinetics of pre-heating apatite at temperatures of ca. 350–550 °C to remove spontaneous fission tracks for experimental studies. Should these conditions be normalized? What physical processes occur during this treatment and are they significant? Does alpha-damage matter? Does gas loss (e.g., helium, other?) matter? These questions, and no doubt others, remain unanswered.

Novel analogue to the AFT and apatite (U-Th)/He systems. An alternative and intriguing laboratory analogue to spontaneous fission-track behavior and α-emission would be to synthesize apatite doped with an isotope that undergoes spontaneous fission and an isotope that emits α particles, both at laboratory timescales. The isotope the undergoes spontaneous fission, for example ^{252}Cf, would serve as an internal source of continuous, spontaneous fission track production (similar to ^{238}U over geological time in natural apatite grains; Table 3). Similarly, the apatite could be doped with an appropriate α-emitter such as ^{248}Cf (α-decay half-life = 333 days) to facilitate creating a laboratory analogue to the apatite (U-Th)/He system to simultaneously accompany the fission-track analogue. The apatite grains obtained could then be subjected to heating schedules, mimicking geological thermal histories, and the AFT parameters and (U-Th)/He ages measured and compared to laboratory-derived kinetic models.

Table 3. Analogous apatites containing spontaneously fissioning isotopes.

^{238}U natural spontaneous fission
40 ppm U (which is essentially all ^{238}U)
(200 Ma = geological age)/(4468 Ma = half-life) = 0.0448
($\sim 7.5 \times 10^{-17}$ = spontaneous fission decay constant)/(1.55125×10^{-10} = α-decay decay constant) = 4.83×10^{-7}
$(40) \times (0.0448) - (4.83 - 10^{-7}) = 8.66 \times 10^{-7}$
A 40 ppm U natural apatite with a 200 Ma fission-track age contains a workable density of fission tracks.
^{252}Cf-doped spontaneous fission
? ppm ^{252}Cf
(43 days = laboratory age)/(964 days = half-life) = 0.0446
(3.09% = spontaneous fission probability)/(96.91% = α-decay probability) = 0.032
$8.66 \times 10^{-7} / [(0.0446) \times (0.032)] = 6 \times 10^{-4}$ ppm ^{252}Cf
6×10^{-4} ppm ^{252}Cf over 43 laboratory days mimics 40 ppm U over 200 Ma of geological time.
^{250}Cm-doped spontaneous fission
? ppm ^{250}Cm
(43 days = laboratory age)/(4017640 days = half-life) = 1.07×10^{-5}
100% = spontaneous fission probability = 1
$8.66 \times 10^{-7} / [(1.07 \times 10^{-5}) \times 1)] = 8 \times 10^{-2}$ ppm ^{250}Cm
8×10^{-2} ppm ^{250}Cm over 43 laboratory days mimics 40 ppm U over 200 Ma of geological time.

ACKNOWLEDGMENTS

The authors wish to thank their spouses Margaret Donelick, Andrea O'Sullivan, and Denise Ketcham for putting up with us during the writing of this chapter. Also greatly appreciated are thorough reviews by Kerry Gallagher and Stuart Thompson, and fruitful discussions with the editors Peter Reiners and Todd Ehlers.

REFERENCES

Barbarand J, Carter A, Wood I, Hurford T (2003a) Compositional and structural control of fission-track annealing in apatite. Chem Geol 198:107-137

Barbarand J, Carter A, Hurford T (2003b) Variation in apatite fission-track length measurement: implications for thermal history modelling. Chem Geol 198:77-106

Bhandari N, Bhat SC, Lal D, Rajagoplan G, Tamhane AS, Venkatavaradan VS (1971) Fission fragment tracks in apatite: recordable track lengths. Earth Planet Sci Lett 13:191-199

Burtner RL, Nigrini A, Donelick RA (1994) Thermochronology of Lower Cretaceous source rocks in the Idaho-Wyoming Thrust Belt. Bull Am Assoc Petrol Geol 78:1613-1636

Carlson WD (1990) Mechanisms and kinetics of apatite fission-track annealing. Am Mineral 75:1120-1139

Carlson WD, Donelick RA, Ketcham RA (1999) Variability of apatite fission-track annealing kinetics: I. Experimental results. Am Mineral 84:1213-1223

Crowley KD, Cameron M, Schaefer RL (1991) Experimental studies of annealing of etched fission tracks in fluorapatite. Geochim Cosmochim Acta 55:1449-1465

Crowley KD, Naeser CW, Naeser ND (1989) Fission Track Analysis: Theory and Applications. Geological Society of America, Boulder, CO

Dartyge E, Duraud JP, Langevin Y, Maurette M (1981) New model of nuclear particle tracks in dielectric materials. Phys Rev B 23:5213-5228

Deer WA, Howie RA, Zussman J (1969) An Introduction to the Rock-forming Minerals. John Wiley and Sons, New York

Donelick RA (1991) Crystallographic orientation dependence of mean etchable fission track length in apatite: an empirical model and experimental observations. Am Mineral 76:83-91

Donelick RA (1993) A method of fission track analysis utilizing bulk chemical etching of apatite. Patent 5267274, U.S.A.

Donelick RA (1995) A method of fission track analysis utilizing bulk chemical etching of apatite. Patent 658800, Australia

Donelick R, Farley K, Asimow P, O'Sullivan P (2003) Pressure dependence of He diffusion and fission-track annealing kinetics in apatite?: Experimental results. Abst. Geochim Cosmochim Acta 67:A82

Donelick RA, Ketcham RA, Carlson WD (1999) Variability of apatite fission-track annealing kinetics: II. Crystallographic orientation effects. Am Mineral 84:1224-1234

Donelick RA, Miller DS (1991) Enhanced TINT fission track densities in low spontaneous track density apatites using ^{252}Cf-derived fission fragments tracks: a model and experimental observations. Nucl Tracks Radiat Meas 18:301-307

Donelick RA, O'Sullivan PB, Ketcham RA, Knaack C, Donelick MB, Kirstein L (2005) A practical approach to apatite and zircon fission-track dating using laser ablation ICP-MS. Chemical Geology, accepted

Donelick RA, Roden MK, Mooers JD, Carpenter BS, Miller DS (1990) Etchable length reduction of induced fission tracks in apatite at room temperature (~23°C): crystallographic orientation effects and "initial" mean lengths. Nucl Tracks Radiat Meas 17:261-265

Dumitru TA (2000) Fission-Track Geochronology. In: Quaternary Geochronology: Methods and Applications. Noller JS, Sowers JM, Lettis WR (eds) Am Geophys Union Ref Shelf 4, Washington, DC, American Geophysical Union, 131-155

Fleischer RL, Price PB, Walker RM (1975) Nuclear Tracks in Solids: Principles and Techniques. University of California Press, Berkeley

Friedlander G, Kennedy JW, Macias ES, Miller JM (1981) Nuclear and Radiochemisty. John Wiley and Sons, New York

Galbraith, RF (2002) Some remarks on fission-track observational biases and crystallographic orientation effects; discussion Am Mineral 87:991-995

Gallagher K, Brown R, Johnson C (1998) Fission track analysis and its applications to geological problems. Ann Rev Earth Planet Sci 26:519-572

Garver JI, Soloviev AV, Reiners PW (2004) Field observations of the stability of fission tracks in radiation-damaged zircon. Abstr. 10th International Conference on Fission Track Dating and Thermochronology, Amsterdam, p 56

Gleadow AJW (1984) Fission track dating methods - II: a manual of principles and techniques. Workshop on fission track analysis: principles and applications: James Cook University, Townsville, Australia, 4-6 September 1984

Gleadow AJW, Duddy IR, (1981) A natural long-term track annealing experiment for apatite. Nucl Tracks Radiat Meas 5:169-174

Gleadow AJW, Duddy IR, Green PF, Lovering JF (1986) Confined fission track lengths in apatite: a diagnostic tool for thermal history analysis. Contrib Mineral Petrol 94:405-415

Gleadow AJW, Belton DX, Kohn BP, Brown RW (2002) Fission track dating of phosphate minerals and the thermochronology of apatite. Rev Mineral Geochem 48:579-630

Green PF (1988) The relationship between track shortening and fission track age reduction in apatite: combined influences of inherent instability, annealing anisotropy, length bias and systems calibration. Earth Planet Sci Lett 89:335-352

Green PF (1995) AFTA today. On Track (Newsletter of the International Fission-Track Community (not peer-reviewed), v.5, n.2:8-10

Green PF, Duddy IR, Gleadow AJW, Tingate PR, Laslett GM (1985) Fission track annealing in apatite: track length measurements and the form of the Arrhenius plot. Nucl Tracks 10:323-328

Green PF, Duddy IR, Gleadow AJW, Tingate PR, Laslett GM (1986) Thermal annealing of fission tracks in apatite: A qualitative description. Chem Geol 59:237-253

Green PF, Duddy IR, Laslett GM (1988) Can fission track annealing in apatite be described by first order kinetics? Earth Planet Sci Lett 87:216-228

Hasebe N, Barbarand J, Jarvis K, Carter A, Hurford AJ (2004) Apatite fission-track chronometry using laser ablation ICP-MS. Chem Geol 207:135-145

Hendriks BWH, Redfield TF (2004) AFT and (U-Th)/He data from cratonic interiors: evidence of denudation or low-temperature annealing. Abstr. 10th International Conference on Fission Track Dating and Thermochronology, Amsterdam, p 51

Hughes JM, Cameron M, Crowley KD (1989) Structural variation in natural F, OH, and Cl apatites. Am Mineral 74:870-876

Hurford AJ, Green PF (1983) The zeta age calibration of fission-track dating. Isotope Geoscience 1:285-317

Jonckheere RC, Wagner GA (2000) On the occurrence of anomalous fission tracks in apatite and titanite. Am Mineral 85:1744-1753

Ketcham RA (2003) Observations on the relationship between crystallographic orientation and biasing in apatite fission-track measurements. Am Mineral 88:817-829

Ketcham RA (2005) Forward and inverse modeling of low-temperature thermochronometry data. Rev Mineral Geochem 58:275-314

Ketcham RA (in press) The role of crystallographic angle in characterizing and modeling apatite fission-track length data. Rad Meas

Ketcham RA, Carter A, Barbarand J, Hurford AJ (2004) Analysis of the UCL apatite annealing data set. Abstr. 10th International Conference on Fission Track Dating and Thermochronology, Amsterdam, p 6

Ketcham RA, Donelick RA, Carlson WD (1999) Variability of apatite fission-track annealing kinetics: III. Extrapolation to geological time scales. Am Mineral 84:1235-1255

Ketcham RA, Donelick RA, Donelick MB (2000) AFTSolve: A program for multi-kinetic modeling of apatite fission-track data. Geological Materials Research, v.2, n.1.

Kohn BP, Gleadow AJW, Raza A, Mavrogenes J, Raab MJ, Belton DX, Kukkonen IT (2004) Revisiting apatite under pressure: further experimental data on fission track annealing and deep borehole measurements. Abstr. 10th International Conference on Fission Track Dating and Thermochronology, Amsterdam, p 54

Lal D, Rajan RS, Tamhane AS (1969) Chemical composition of nuclei of $Z > 22$ in cosmic rays using meteoric minerals as detectors. Nature 221:33-37

Laslett GM, Kendall WS, Gleadow AJW, Duddy IR (1982) Bias in measurement of fission-track length distribution. Nucl Tracks 6:79-85

Laslett GM, Gleadow AJW, Duddy IR (1984) The relationship between fission track length and track density distributions. Nucl Tracks 9:29-38

Laslett GM, Green PF, Duddy IR, Gleadow AJW (1987). Thermal annealing of fission tracks in apatite, 2. A quantitative analysis. Isotope Geoscience 65:1-13

Lederer CM, Hollander JM, Perlman I (1967) Table of isotopes, 6th Edition. John Wiley, New York

Naeser CW (1967) The use of apatite and sphene for fission track age determinations. Bull Geol Soc Am 78: 1523-1526

Naeser CW (1976) Fission track dating. US Geol Surv Open-File Report 76-190

Naeser CW (1979) Fission-track dating and geologic annealing of fission tracks. *In:* Lectures in Isotope Geology. Jäger E, Hunziker JC (eds) Springer-Verlag, Berlin, p 154-169

Naeser CW (1981) The fading of fission tracks in the geological environment – data from deep drill holes. Nucl Tracks Rad Meas 5:248-250

Naeser CW, Forbes RB (1976) Variation of fission track ages with depth in two deep drill holes. EOS, Trans Am Geophys Union 57:353
Naeser ND, McCulloh TH (eds) (1989) Thermal History of Sedimentary Basins: Methods and Case Histories, Springer-Verlag, Berlin
O'Sullivan PB, Donelick RA, Ketcham RA (2004) Etching conditions and fitting ellipses: What constitutes a proper apatite fission-track annealing calibration measurement. Abstr. 10th International Conference on Fission Track Dating and Thermochronology, Amsterdam, p 5
Paul TA, Fitzgerald PG (1992) Transmission electron microscopic investigation of fission tracks in fluorapatite. Am Mineral 77:336-344
Ravenhurst CE, Donelick RA (1993) Fission track thermochronology. *In:* Short Course Handbook on Low Temperature Thermochronology. Zentilli M, Reynolds PM (eds) Mineral Assoc Can, Ottawa, p 21-42
Reiners PW (2003) (U-Th)/He chronometry experiences a renaissance. EOS, Trans Am Geophys Union 83: 21,26-27
Roeder PL, MacArthur D, Ma X-P, Palmer GR, Mariano AN (1987) Cathodoluminescence and microprobe study of rare-earth elements in apatite. Am Mineral 72:801-811
Siddall R, Hurford AJ (1998) Semi-quantitative determination of apatite anion composition for fission-track analysis using infrared microspectroscopy. Chem Geol 150:181-190
Sobel ER, Seward D, Ruiz G, Kuonov A, Ege H, Wipf M, Krugh C (2004) Influence of etching conditions on Dpar measurements: Implications for thermal modeling. Abstr. 10th International Conference on Fission Track Dating and Thermochronology, Amsterdam, p 7
Steiger RH, Jäger E (1977) Subcommission on geochronology: Convention on the use of decay constants in geo- and cosmochronology. Earth Planet Sci Lett 36:359-362
Stormer JC, Pierson ML, Tacker RC (1993) Variation of F and Cl X-ray intensity due to anisotropic diffusion in apatite during electron microprobe analysis. Am Mineral 78:641-648
Tagami T, O'Sullivan PB (2005) Fundamentals of fission-track thermochronology. Rev Mineral Geochem 58: 19-47
Van den haute P, De Corte F (eds) (1998) Advances in Fission-Track Geochronology. Kulwer Academic Publishers, Dordrecht
Vrolijk P, Donelick RA, Queng J, Cloos M (1992) Testing models of fission track annealing in apatite in a simple thermal setting: Site 800, ODP Proceedings 129:169-176
Wagner GA (1968) Fission track dating of apatites. Earth Planet Sci Lett 4:411-415
Wagner GA (1969) Spuren der spontanen Kernspaltung des ^{238}Urans als Mittel zur Datierung von Apatiten und ein Beitrag zur Geochronologie des Odenwaldes. N Jahrb Mineral Abh 110:252-286
Wagner GA, Storzer D (1972) Fission track length reductions in minerals and the thermal history of rocks. Trans Am Nucl Soc 15:127-128
Wagner GA, Van den haute P (1992) Fission-track dating. Kulwer Academic Publishers, Dordrecht
Wendt AS, Olivier V, Chadderton LT (2002) Experimental evidence for the pressure dependence of fission track annealing in apatite. Earth Planet Sci Lett 201:593-607
Young EJ, Myers AT, Munson EL, Conklin NM (1969) Mineralogy and geochemistry of fluorapatite from Cerro de Mercado, Durango, Mexico. US Geol Surv Prof Paper 650-D, p D84-D93

APPENDIX 1: AFT SAMPLE PREPARATION TIPS

General

The following steps (not necessarily in this order) can be taken to prepare AFT samples:

Preparation Step 1. Separate apatite from host rock and then mount, polish, and etch an apatite grain mount.

Preparation Step 2. Irradiate an apatite grain mount with ^{252}Cf-derived fission fragments and then etch the grain mount to reveal TINT fission tracks.

Preparation Step 3. Prepare an apatite grain mount for external detector (EDM) age analysis.

Preparation Step 4. Affix the apatite grain mount and its external detector to a microscope slide for analytical and archiving purposes.

Preparation Step 5. Prepare an apatite grain mount for bulk chemistry measurements using EPMA.

Preparation Step 6. Prepare an apatite grain mount for relative uranium concentration (or relative concentration) measurements using LA-ICP-MS.

Tips for apatite mineral separation

Determining the presence of apatite in a sample. There are three ways to determine whether or not apatite is present in a rock sample of interest: 1) positively identify apatite in hand-sample, 2) prepare and inspect a thin section of the rock sample (or more than one thin-section, if necessary), or 3) perform mineral separation procedures on the rock sample using a combination of density and magnetic mineral separation techniques. With rare exceptions, hand-sample inspection is not a viable means of determining the presence of apatite in a sample because of the scarcity (<1%) and small grain size of apatite in most rocks. Thin-section analysis may be useful, but only if other petrographic information is required from the inspection process. This is because apatite is often so sparse in some rocks that it is unlikely that one would find more than a grain or two in a standard thin-section. Mineral separation procedures provide the most definitive result regarding apatite availability. In terms of economy of time and money, it is recommended that thin-section analysis be avoided if its sole purpose is to determine the presence of apatite in a rock sample for AFT analysis.

Crushing samples. In order to obtain apatite from a rock, it is necessary to disaggregate the rock into constituent grains. Ideally, this is done with minimal damage to each grain and maximum separation of mineral phases. The use of a disk mill and/or percussion pulverizer should be avoided, if sufficient rock is available, or minimized for small samples. Such an approach has the key advantage of preserving a maximum number of whole apatite and other accessory mineral grains that can be potentially used for other analytical techniques (e.g., whole grains are desirable for (U-Th)/He analysis due to the need to correct for He loss along grain boundaries). For sufficiently large samples, this can be accomplished by running samples through a jaw-crusher multiple times in order to produce sufficient sand-sized material for mineral separation.

Of the thousands of mineral separations performed by the authors (notably, rocks collected usually for the purpose of obtaining apatite), the success rate for obtaining sufficient apatite for AFT analysis is about 90% when approximately 500 g of 50–300 μm washed and dried sand is processed. Sufficient material is obtained from another 5% of samples if a total of 1000 g of such sand is obtained and processed. The remaining samples usually yield only a few apatite grains (<10) or none at all (for the latter, processing three times the rock would yield three times zero grains or nothing). Therefore, unless data from a particular poor-yielding sample are absolutely necessary, 1000 g of sand should be the upper limit of the amount of sample processed.

Sieving and washing samples. Once samples have been run through a jaw-crusher multiple times in order to produce sufficient sand-sized material for mineral separation, the crushed material is then sieved using 300 μm sieve cloth and the <300 μm fraction is processed further. Prior to heavy liquid separation, it is advisable to avoid use of a washing table containing riffles; the authors prefer to simply wash the sieved material directly with tap water and decanting until the water attains clarity within 10 seconds or so. Washing directly with tap water requires significantly less time than using a table with riffles, it requires minimal cleaning between samples, and small accessory mineral grains (apatite among them) are not washed away. The latter point is particularly an issue when separating low-yield samples, small grain-sized samples, or samples collected for detrital studies where all grain sizes (large and small) need to be represented in the final analysis.

Heavy liquid separation. Two types of heavy liquids are routinely used for mineral separation purposes: organic-based (tetrabromoethane, bromoform, and diiodomethane) and inorganic solutions (Na-polytungstate and Li-metatungstate). Organic-based fluids exhibit low viscosities and low reactivities to rocks but high toxicity (the toxicity requires a fumehood, provisions for air and sewerage discharge, and special handling procedures for protection of personnel). Inorganic-based fluids exhibit high viscosities and relatively high reactivities to many rocks (in particular Ca-rich rocks) and low toxicity (no fumehood is required, there are minimal discharge problems, and personnel protection is simplified). To overcome the high viscosity and reactivity of an inorganic-based fluid, use of a high-volume centrifuge is preferred. The combined use of inorganic-based fluids and a centrifuge, even for Ca-rich rocks such as limestones and marbles, is preferable due to the low toxicity, easy handling procedures, and the high rate of successful mineral separations possible (the numbers quoted above for the authors on success rate are based on inorganic-based fluids and centrifuge usage).

Zircon and other quite heavy, relatively non-magnetic minerals such as sulfides or sulfide composite grains, tourmaline, etc. are easily separated from the mineral fraction containing apatite by sinking it in 3.3 g/cm^3 diiodomethane. An extremely simple (but efficient) way to accomplish this step is to place the fraction containing apatite and zircon into approximately 10 ml of diiodomethane within a 50 ml beaker. Swirl the mineral fraction+diiodomethane mixture in the beaker and let heavy minerals settle to the bottom of the beaker. Then, swirl mildly or create a mild circular wave within the beaker causing the material that has settled to the bottom to collect at the center of the beaker. If a large quantity of grains has settled to the bottom of the beaker, tilt the beaker at a ca. 45° angle, tap the bottom, and the material that has settled will collect along the bottom edge of the beaker. This material can then be extracted using an eyedropper and then be placed into a second beaker. If a small amount of material has collected in the center of the beaker, use an eyedropper to collect it, doing so with care taken to not stir up the material that has floated to the surface of the diiodomethane. This process can be repeated until virtually no material settles to the bottom of the beaker. When extracting mineral grains that have sunk to the bottom of the beaker (Sp.g. >3.3 g/cm^3), a small amount of float material will adhere to the outside of the eyedropper and inadvertently be transferred to the second beaker. Much of this transferred float material can be transferred back to the original beaker containing the bulk of the float material by decanting the excess fluid from the second beaker into the original beaker. A combination of eyedropper transfer and decanting of excess fluid can result in a relatively pure mineral separate of material having density greater than that of diiodomethane. Once the settled material is in the second beaker and the float material in original beaker, disassemble the eyedropper and squirt a small amount (no more than several ml) of acetone or methanol onto the outside of the eyedropper glass to wash the adhered float material into the original beaker. Then wash the grains adhering to the inside the eyedropper into the second beaker by squirting a small amount of acetone or methanol into the interior of the eyedropper. Vigorously swirl the beaker containing the float material+diiodomethane+several ml acetone or methanol and repeat the whole process. This repeat process is now aimed at sinking apatite from other relatively light, non-magnetic material that might be present in the float material (such as quartz that was dragged down into the initial heavy mineral separation or otherwise included inadvertently). Practice at this, with care taken to adjust the density of the fluid to sink apatite but float other mineral grains, can result in successful apatite purification.

Magnetic separation. Apatite is often slightly magnetic. Several examples include Tioga (Bed B, near Old Port, Pennsylvania), and Fish Canyon Tuff (San Juan Mountains, Colorado); both of these apatite species are relatively rich in Fe (0.07 ± 0.06 wt% and 0.82 ± 0.18 wt% FeO, respectively; Carlson et al 1999). Use a Frantz Isodynamic magnetic separator to split the heavy mineral fractions according to magnetic susceptibility set at a slope of 25° and a tilt of 10° (the old model Frantz works quite well for apatite separation). Obtain a non-magnetic fraction at 1.0 A current and from this a magnetic and non-magnetic fraction at 1.8 A current. If sufficient apatite is not observed in the 1.8 A non-magnetic fraction, look for apatite in the magnetic fraction from the 1.8A run; for about 50% of samples processed by the authors, sufficient apatite for apatite fission-track analysis can be found in this fraction.

Tips for mounting and polishing apatite grain mounts

The goal here is to create a grain mount containing a sufficient number of apatite grains that are individually distinguishable (i.e., sufficient space exists between them) and properly polished to reveal clear and flat internal grain surfaces across which spontaneous fission tracks have accumulated from both above and below (4π geometry).

Concentrating the apatite fraction (panning method). To prepare grain mounts for apatite fission-track analysis, one should avoid simply pouring mineral grains from the apatite fraction onto a mounting medium unless it is known in advance that the separate is relatively rich in apatite (say, 50% apatite or greater). For relatively apatite-poor mineral fractions, a panning method is desirable in order to concentrate the apatite grains prior to mounting. Place the apatite separate into a petri dish and add a small quantity of ethanol to the petri dish to thoroughly wet the apatite grains and permit swirling of the grains along the bottom edges of the dish. While tilting the petri dish at a slight angle, swirl the grains. The swirling motion required depends on the nature and amount of the mineral separate present, percentage of apatite present and relative grain sizes of the apatite and non-apatite species, size of the petri dish, and amount of ethanol used. Practice at swirling in this manner will permit one to obtain zones of large apatite grains (typically on the leeward side of the swirling direction) and small grains (typically on the windward side). Zones of relatively pure apatite can be located and identified using a binocular microscope. Swirling or creating unusual wave patterns in the petri dish can permit grouping of apatite grains relative to other, non-apatite grains. When panning relatively impure apatite separates, one should not take the first result as an indication of potential success unless that result yields a zone of sufficiently pure apatite. When no zone of relative apatite purity is identified, one should try panning the separate again, either by repeating the same swirling motion for a longer time or by trying a different swirling motion or wave pattern. For apatite-poor mineral fractions, one should try the following: identify a zone exhibiting high zircon concentration (yes, some zircon may make it through the zircon separation step) in contact with obvious quartz of similar grain size. Apatite has a density between that of zircon and quartz and if apatite is present, it will be concentrated between the zircon-rich and quartz-rich zones. Using an eyedropper, isolate a sufficient number of grains from the apatite rich zone and place them on the mounting medium.

Mounting medium. Two general approaches have been taken to create apatite grain mounts: epoxy-on-glass and epoxy-only. While the epoxy-on-glass method still remains the standard choice of many researchers, epoxy-only is preferred herein for several reasons:

- The epoxy-on-glass method requires that when neutron irradiation is used during analysis, the excess glass material is also irradiated in a nuclear reactor. This commonly results in the creation of abundant radioactive byproducts associated with the chemical components of the glass itself. This is particularly a problem with Na-glass, but not much of a problem with Si-glass.
- Epoxy-only grain mounts permit one to create mounts of a given size and shape that does not require subsequent post-mount-formation size-minimization. When using the epoxy-on-glass method, a standard petrographic glass slide (~1 x 2 in) or equivalent is commonly used. Mounts of this size however, are much too large to be loaded into standard irradiation tubes and must therefore be broken down to a smaller size prior to irradiation. This size-minimization for the epoxy-on-glass approach requires the scoring and breaking of glass slides that all too often leads to undesirable breakage of the glass, injury to the technician, and usually partial or sometimes total loss of the grain mount.
- Once mounted using the epoxy-only method, the individual apatite grains (independent of grain size) are all located at the surface of the epoxy mount. Polishing epoxy-only grain requires less time and results in a greater fraction of the apatite grains polished to reveal internal surfaces (4π geometry).

Frankly, the epoxy-only approach is easier, less likely to result in partial to total loss of sample, and less dangerous concerning the resultant radioactive nuclides (the latter, if applicable).

Polishing apatite grain mounts. Polishing apatite grain mounts can be done manually, with or without a sample holder, or by using any number of available mechanical polishing apparatus. The hierarchy of abrasive materials used can also vary by type and configuration. The method preferred by the authors is to polish each grain mount manually while held fast to the polisher's finger using double-stick tape, and using a 12-inch (30 cm) lap wheel rotating at several hundred rpm. First, a slurry of 3.0 μm Al_2O_3 on paper is used; the sample is rotated in the opposite direction of the direction of wheel rotation 25 to 35 times (about 1 second per sample rotation) or for a sufficient time to expose internal grain surfaces (as noted above, this works well for epoxy-only grain mounts as the grains are already situated at the epoxy surface prior to the initiation of polishing). The grain mount is then inspected using a binocular microscope and any areas requiring further polishing are identified. The process is repeated, adjusting the number of sample rotations and the pressure point(s) on the grain mount until the whole mount exhibits grains with well-exposed internal surfaces. A second lap wheel is then used containing a slurry of 0.3 μm Al_2O_3 on paper; again, 25 to 35 rotations of the sample are used to impart a fine (near glass-like) polish to the whole grain mount surface. This process avoids the use of Al_2O_3 wet-dry paper, which is usually required for epoxy-on-glass grain mounts, and results in minimal fracturing of the apatite grains during polishing, which is desirable to avoid unwanted TINCLE fission tracks (see ^{252}Cf-fission-fragment irradiation discussion below).

Tips for etching apatite grain mounts

After polishing, the exposed grains on each apatite mount are easily etched by immersing the grain mount in dilute HNO_3 to reveal any spontaneous fission track that intersects its respective polished apatite grain surface. The authors prefer etching apatite fission tracks in 5.5 N HNO_3 for 20.0 s (± 0.5 s) at 21 °C (±1 °C), the same recipe used by Carlson et al. (1999) and upon which the calibrations of Ketcham et al. (1999) and Ketcham (2003) are based. This etching recipe offers superior results relative to 5.0 N HNO_3 for 20.0 s (± 0.5 s) at 20 °C (± 1 °C) because fission tracks at low angles to the crystallographic *c*-axis are insufficiently etched using the 5.0 N HNO_3 recipe, especially for the common calcian-fluorapatite species. The 5.5 N HNO_3 etching recipe also offers superior results relative to 1.6 N HNO_3 for 40.0 s (± 0.5 s) at 24 °C (± 1 °C) because fast-etching, often Cl- or OH-rich apatite species become over-etched using the 1.6 N HNO_3 recipe. Regardless of etching recipe choice, it is essential that the three etching condition parameters be tightly controlled during etching (HNO_3 strength, duration of etching, temperature of etching), especially if one chooses to measure and use kinetic parameter D_{par}.

During etching, two apatite grains mounts are placed back-to-back (polished surfaces directed outward) and held together tightly at one corner using tweezers. A stopwatch is used to time the duration of etching which begins counting etch time at the moment the grain mounts contact the etchant and stops counting at the moment the grain mounts are removed from contact with the etchant. While immersed in the etchant the grain mounts are moved back and forth sufficiently vigorously to ensure continuous movement of fresh etchant over the grain mount surfaces. Once removed from the etchant, the grain mounts are dipped as quickly as possible into a bath of room-temperature distilled water, stirred vigorously for several seconds, removed from the distilled water and dipped into a second bath of room-temperature distilled water and stirred vigorously again for several additional seconds. The grain mounts are then removed from the second water bath, dried quickly (all surfaces) with a clean, dry paper towel, and then placed in open air. This last step is extremely important, as minute quantities of etchant may be trapped within etched fission tracks even after the grain mounts have been thoroughly washed in distilled water and dried. Allowing the grain mounts to remain in open air for several hours, as opposed to placing them into a closed sample holder, permits any residual etchant to dissipate into the atmosphere without causing additional, uncontrolled etching of the grain mounts.

Once etched, the feasibility of measurement of the AFT parameters (i.e., grain ages and confined track lengths) is then assessed by quickly scanning the polished and etched grain mount to determine the amount and quality of any apatite present. If adequate apatite is observed during this scan measurement of the apatite fission-track parameters for that sample can begin.

Tips for ^{252}Cf-derived fission-fragment irradiation of apatite grain mounts

The use of ^{252}Cf-fission-fragment irradiation has three major advantages: 1) it provides superior data quality since TINT are highly desired relative to TINCLE and TINDEF fission tracks and they have simple etchant pathway geometries versus the more variable geometries of TINCLE and TINDEF fission tracks, 2) it provides enhanced data quantity permitting more precise constraints to be placed on the geological problem under study, and 3) it permits an operator to measure a desired number of TINT fission tracks (say 200) in a shorter period of time. In some cases, the time required to measure AFT length data using ^{252}Cf-fission-fragment irradiation is actually increased. This is especially true for low-spontaneous fission-track density apatite grains. However, it is the view of the authors that the abundant data usually obtained from this method are desirable over insufficient data or no data at all.

Epoxy-only apatite grain mounts facilitate preparation for ^{252}Cf-fission-fragment irradiation because the grains mounts can be made to the desired dimensions and hence do not need to be broken down to the desired size, which often results in partial or complete loss of sample material. Square or rectangular apatite grain mounts are preferred over circular grain mounts because this geometry permits the closest packing of the grain mounts during ^{252}Cf-fission-fragment irradiation. Typically, once the spontaneous fission-track densities have been counted in apatite grains for which grain ages are sought and the grain locations digitally recorded, the grain mounts are irradiated with approximately 10^7 tracks/cm^2 fission fragments from a 50 μCi ^{252}Cf source in a vacuum chamber (the actual activity of the ^{252}Cf source is not important as the exposure time in the vacuum chamber can be adjusted to the source activity). Irradiated grain mounts are then re-etched to reveal any horizontal, confined fission tracks, and the track lengths can be measured.

Tip for preparing apatite grain mounts for EDM age dating

Epoxy-only apatite grain mounts facilitate preparation for EDM because the grains mounts can be made to the desired dimensions. No other special sample preparation procedures are required beyond collecting digital information concerning the coordinates of three reference points and the apatite grains for which ages are to be determined (the former permits a three-dimensional transformation from the microscope coordinate system relative to the counted apatite grains and the microscope coordinate system relative to the external detector).

Tips for preparing apatite mounts for the LA-ICP-MS age dating

Preparing apatite grain mounts for LA-ICP-MS is facilitated by the use of epoxy-only grain mounts. No other special sample preparation procedures are required beyond collecting digital information concerning the coordinates of three reference points and the apatite grains for which ages are to be determined (the former permits a three-dimensional transformation from the microscope coordinate system to the laser ablation stage coordinate system).

Preparing apatite grain mounts for electron probe microanalysis (EPMA)

Preparing apatite grain mounts for EPMA is facilitated by the use of epoxy-only grain mounts. No other special sample preparation procedures are required beyond collecting digital information concerning the coordinates of three reference points and the apatite grains from which data have been collected (the former permits a three-dimensional transformation from the microscope coordinate system to the EPMA stage coordinate system).

APPENDIX 2: DATA COLLECTION SCHEMES FOR APATITE FISSION-TRACK ANALYSIS

Measurement of grain ages

The following steps (not necessarily in this order) can be taken measure AFT grain ages:

Age Measurement Step 1. Select an apatite grain for fission-track age analysis.
Age Measurement Step 2. Measure parameter D_{par} for the selected apatite grain for age analysis.
Age Measurement Step 3. Count spontaneous fission tracks over some area of the selected apatite grain for age analysis.
Age Measurement Step 4. Count induced fission tracks in an external detector over the area corresponding to the area counted in Measurement Step 3.
Age Measurement Step 5. Measure uranium concentration (or relative concentration) for the selected apatite grain for age analysis using LA-ICP-MS.
Age Measurement Step 6. Measure bulk element and selected minor element concentrations for the selected apatite grain for age analysis using EPMA.

Measurement of lengths

The following steps (not necessarily in this order) can be taken to measure confined AFT lengths:

Length Measurement Step 1. Select an apatite grain for fission-track length measurement.
Length Measurement Step 2. Measure kinetic parameter D_{par} for the selected apatite grain for track length measurement.
Length Measurement Step 3. Measure the length of the first confined fission track observed.
Length Measurement Step 4. Measure the inclination angle of the first confined fission track to the c-axis of the host apatite grain.
Length Measurement Step 5. Note whether the first confined track is a TINT, TINCLE, or TINDEF fission track.
Length Measurement Step 6. Repeat Length Measurement Steps 4, 5 and 6 until all measurable confined fission tracks are measured for the selected apatite grain.
Length Measurement Step 7. Measure bulk element and selected minor element concentrations for the selected apatite grain for track length measurement using EPMA.

AFT analysis

AFT analysis involves some combination of sample preparation steps (see **Appendix 1**) and age and length measurement steps. A common scheme, practiced by many laboratories and largely unchanged since the 1980s, is to perform the following combination of steps in the following order (referring to the lists above):

Common scheme

Preparation Step 1
Preparation Step 3
Preparation Step 4 (permanent bonding requiring storage of bulky material)
Age Measurement Step 1
Age Measurement Step 3
Age Measurement Step 4
repeat Age Measurement Steps 1, 3, and 4 until 20 grains are measured

Length Measurement Step 1
Length Measurement Step 3
Length Measurement Step 6
repeat Length Measurement Steps 1, 3, and 6 until 100 track lengths are measured

This is the state-of-the-art in many laboratories today: obtain 20 grain ages and 100 track lengths (if possible, but most importantly generate sufficient data to fill a line in a data table). There is no kinetic information for the apatite grains from which ages and lengths are measured and no use of fission-track orientation to the *c*-axis in the length data. ^{252}Cf-fission-fragment irradiation is not used so the track length data are usually dominated by TINCLE fission tracks of questionable quality and quantity.

The authors prefer not to fill lines in data tables but to constrain the geological problem at hand. The emphasis is placed partly upon different apatite populations as exhibited by the apparent spread of D_{par} values on a grain mount. Where significant spread in D_{par} is apparent, 40 grain ages and 200 track lengths are sought. The following combination of steps is the scheme preferred by the authors:

Preferred scheme

Preparation Step 1
Age Measurement Step 1
Age Measurement Step 2
Age Measurement Step 3
Preparation Step 2
Length Measurement Step 1
Length Measurement Step 2
Length Measurement Step 3
Length Measurement Step 4
Length Measurement Step 5 (TINTs only are selected)
Length Measurement Step 6
Preparation Step 3
Preparation Step 4 (non-permanent bonding; samples removed from slide after analysis)

-or-

Preparation Step 6
Age Measurement Step 4 (preferred if the authors desire to pluck apatite grains from the grain mount for (U-Th)/He age dating)

-or-

Age Measurement Step 5 (preferred if no grains are to be plucked for (U-Th)/He age dating)

Several hardware and software systems sold commercially and used by many laboratories offer analytical solutions to the *Common scheme* above but offer little variation regarding the type of data that can be collected or the order in which those data can be collected. The *Preferred scheme* utilizes a broadly general hardware and software system that permits the analyst to determine which type of data to collect and in which order. For the latter system, other schemes can be accommodated as the AFT sample and geological problem warrant.

Reviews in Mineralogy & Geochemistry
Vol. 58, pp. 95-122, 2005

Zircon Fission-Track Thermochronology and Applications to Fault Studies

Takahiro Tagami

Division of Earth and Planetary Sciences
Graduate School of Science
Kyoto University
Kyoto 606-8502, Japan
tagami@kueps.kyoto-u.ac.jp

INTRODUCTION

Zircon is one of the most useful minerals used to unravel the Earth's history recorded in rocks. After the success of etching fission tracks in zircon (Fleischer et al. 1964; Naeser 1969; Krishnaswami et al. 1974), early studies focused primarily on dating young volcanic horizons for stratigraphic purposes (e.g., Naeser et al. 1973; Hurford et al. 1976; Seward 1979; Gleadow 1980; Kowallis et al. 1986; Kohn et al. 1992). Subsequently, fission-track analysis of zircon has been extensively employed, along with other radiometric dating methods, such as U/Pb and (U-Th)/He techniques (e.g., Davis et al. 2003; Reiners 2005), to understand the thermochronology of rocks in a variety of geological settings: i.e., thermal history analysis of basement rocks for orogenic studies (e.g., Zeilter et al. 1982; Hurford 1986; Kamp et al. 1989; Sorkhabi 1993; Seward and Mancktelow 1994; Hasebe and Tagami 2001; Spikings et al. 2001), thermochronology using detrital zircon grains in sedimentary rocks for provenance analysis and thermal history analysis in or near faults, which will be highlighted below.

In this present chapter, the thermal sensitivity of the zircon fission-track thermochronometry is reviewed in the context of laboratory and field studies. Then, laboratory procedures of the zircon fission-track analysis using the external detector method are outlined, as practiced in the fission-track laboratory of Kyoto University. For basic fundamentals of these fission-track analyses, including data analysis and graphical displays, see the chapter by Tagami and O'Sullivan (2005). Finally, as an example of recent geological applications of the technique, the results of zircon fission-track analysis of the Nojima fault zone, Japan, are presented. These rocks were the target of systematic drilling of an active fault system using five boreholes and a trench. Applications to other geological settings and the inverse modeling of FT data will be discussed in detail in the following chapters.

THERMAL SENSITIVITY OF ZIRCON FISSION-TRACK THERMOCHRONOMETRY

Laboratory heating data and annealing models

Qualitative description of track annealing. Early studies of track stability in zircon were aimed primarily at determining the reduction in etched track density using different etching conditions; i.e., boiling H_3PO_4 at ~375–500 °C (Fleischer et al. 1965), equal volume mixture of 48% HF and 98% H_2SO_4 at 150–180 °C under pressure (Krishnaswami et al. 1974), and 100 N NaOH solution at 220 °C (Nishida and Takashima 1975). Later studies focused on reduction of etched track lengths, with which fission-track annealing is more precisely quantified, in the

1529-6466/05/0058-0004$05.00 DOI: 10.2138/rmg.2005.58.4

particular case of zircons having inhomogeneous uranium distribution. Kasuya and Naeser (1988) conducted 1 hour isochronal annealing experiments of four zircon samples having different spontaneous track densities of ~9×10^5, ~1.5×10^6, ~4×10^6 and ~1×10^7 cm^{-2}. For each sample, they etched both spontaneous and induced tracks at the same condition, regardless of annealing temperatures, using an eutectic mixture of NaOH, KOH, and LiOH (6:14:1 mol%; Zaun and Wagner 1985) at 215 °C. They determined that thermal stability of spontaneous tracks is independent of initial spontaneous track density, and is significantly lower than that of induced tracks in pre-annealed zircon. Consequently, they suggested that the presence of α-damage lowers the thermal stability of fission tracks in zircon (for this issue, see also Carter 1990; Rahn et al. 2004).

A series of laboratory studies of temperature-dependent track retention in zircon has been conducted over the last decade, using the eutectic NaOH:KOH (1:1) etchant throughout (Tagami et al. 1990, 1998; Yamada et al. 1993, 1995a,b, 1998; Hasebe et al. 1994; Yamada et al. 2003; Murakami et al. 2005.; Matsuura and Tagami 2005). The etching was performed at 225 °C in early studies (Tagami et al. 1990; Yamada et al. 1993; Hasebe et al. 1994) and at 248 °C in the subsequent studies in order to attain shorter etching durations. The annealing experiments were mainly carried out using the zircon crystals separated from the Nisatai Dacite (a.k.a. NST zircons), which has a K-Ar biotite age of 20.6 ± 0.5 (1σ) Ma (Tagami et al. 1995). Zircons are euhedral and clear, and have a spontaneous track density of ~4×10^6 cm^{-2} with relatively small variation in uranium content within as well as between individual grains. The spontaneous tracks are not geologically annealed after their formation, as confirmed by the concordance of mean horizontal confined track lengths between the spontaneous and induced tracks (Yamada et al. 1995b). Hence, spontaneous tracks are primarily used for horizontal confined track analysis of annealed NST zircons (Fig. 1). Important findings on the qualitative nature of track annealing are summarized below:

- Observed lengths of unannealed spontaneous tracks in zircons from rapidly cooled volcanics are slightly anisotropic, with shorter tracks having greater angles to *c*-axis (Hasebe et al. 1994). Zircons show large disparity in the number of observed horizontal confined tracks due to the etching anisotropy, particularly for induced tracks in pre-annealed zircons (Hasebe et al. 1994).
- The prolonged etching factor for horizontal confined spontaneous track lengths is anisotropic, with slower growth rate for tracks having greater angles to *c*-axis (Yamada et al. 1993). The growth rate of track length is greater than the general attack rate of the surrounding bulk material for all crystallographic directions, implying that the etching rate of a latent track decreases rather continuously towards both ends (Yamada et al. 1993).

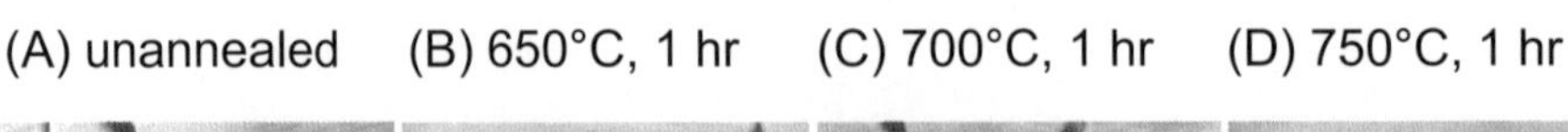

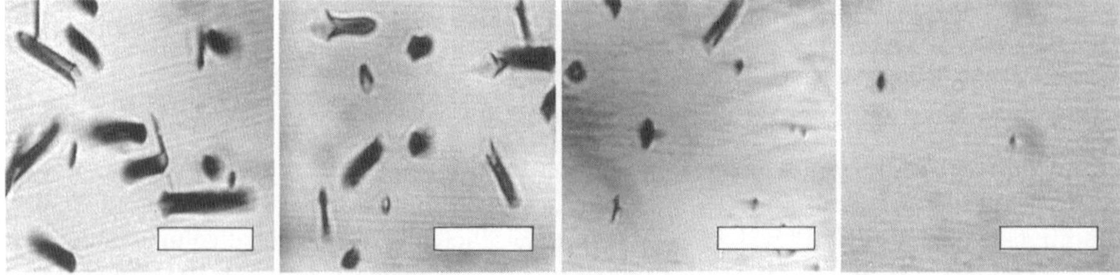

Figure 1. Photographs of thermal annealing of spontaneous fission tracks in zircons from the Nisatai Dacite (i.e., NST zircons): (A) unannealed, (B) – (D) annealed at 650°, 700° and 750 °C for 1 hr, respectively. Scale bar is 10 μm.

- As annealing proceeds, the length of spontaneous tracks decreases gradually and becomes slightly more anisotropic, with shorter tracks having greater angles to *c*-axis (Tagami et al. 1990; Yamada et al. 1995a) (Fig. 2). The standard deviation of track length distribution shows a marked increase in highly annealed samples (i.e., ~650 to 700 °C; Fig. 2), probably as a result of track segmentation.

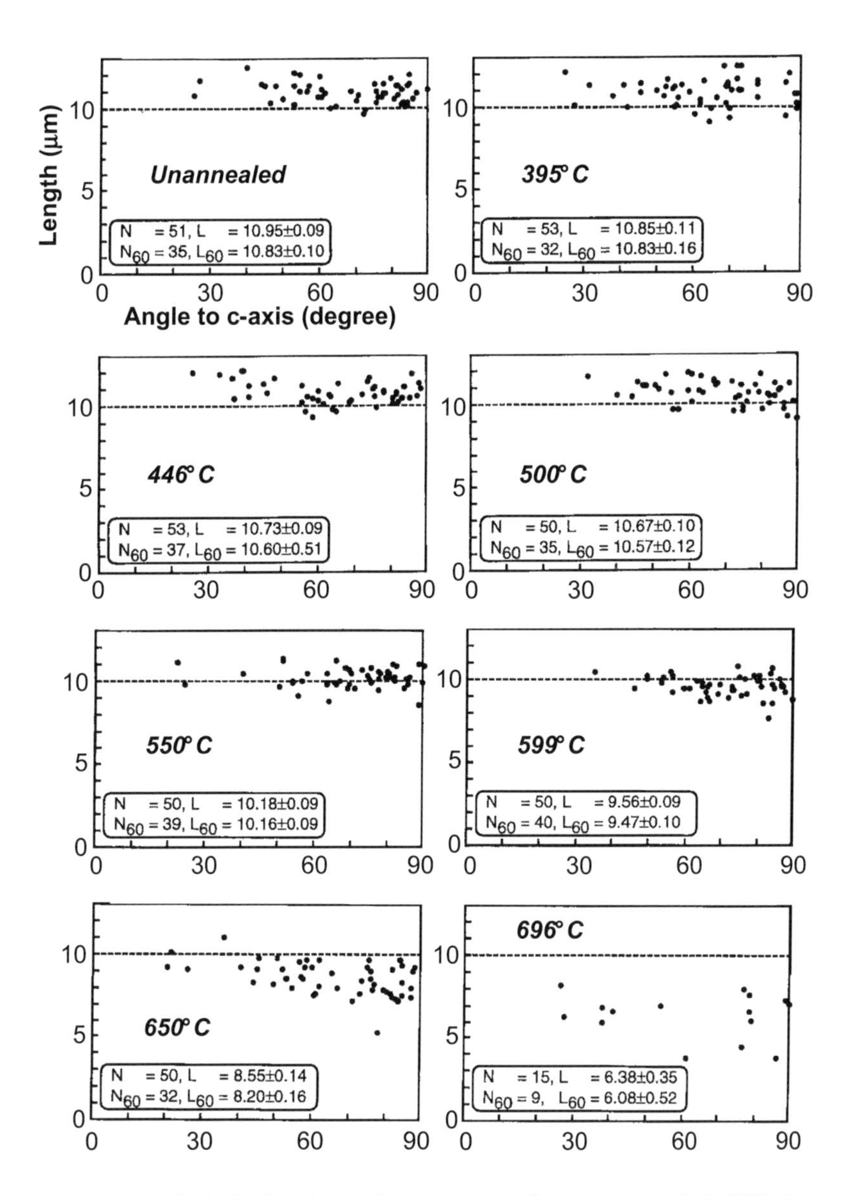

Figure 2. The results of 1-hr isochronal annealing experiments of spontaneous tracks in NST zircons at elevated temperatures (Yamada et al. 1995b). Lengths of horizontal confined tracks are plotted against angles to crystallographic *c*-axis. Note that the angular distribution of measured tracks is highly anisotropic and, at advanced stages of annealing, tracks with greater angles to *c*-axis are slightly shorter than others. *N* and *L* indicate respectively the number and mean length (µm, ± 1 standard error) of measured tracks. The subscript "60" denotes the value for only tracks 60° to the *c*-axis.

- True reduction of spontaneous track density during 1 hour isochronal annealing was measured using both the uranium contents correction (i.e., fission-track dating of annealed zircons) and angular distribution measurement of surface tracks (Tagami et al. 1990). As annealing proceeds, the density of spontaneous tracks decreases gradually and becomes more anisotropic, with the smaller number of tracks for greater azimuth angles to *c*-axis. As a result, reliable density-length relationship was first established for laboratory annealing of spontaneous tracks.
- The optimum track etching time shows a significant increase at annealing temperatures above ~450–600 °C during 1 hour isochronal annealing experiments of spontaneous tracks (Tagami et al. 1990; Yamada et al. 1995b; The threshold temperature may be different from sample to sample). Above those temperatures, the anisotropy of track etching appears further enhanced and the track revelation rates become approximately constant, regardless of the variation of uranium within and between zircon grains (Tagami et al. 1990). These lines of evidence suggest that the accumulated α-recoil damage, which increases the general attack rate of zircon, is removed from grains.
- The effects of various experimental biases are assessed in track length measurement of spontaneous tracks (Yamada et al. 1995a). The results indicate that, when tracks are highly annealed, the observed mean length is significantly affected by the etching temperature, etching criteria, measurement criteria, and crystallographic orientation. Hence, some guidelines are needed to standardize these factors for reliable track length analysis (See the section on "Track length measurement" below).
- Annealing rates are indistinguishable at 2 SE level among samples having spontaneous track densities of $\sim 1 \times 10^6$ to $\sim 3 \times 10^7$ cm^{-2} (Yamada et al. 1998; Matsuura and Tagami 2005.). However, annealing rates appear to increase for young zircons with spontaneous track densities lower than $\sim 4 \times 10^5$ cm^{-2} (Yamada et al. 1998; Matsuura and Tagami 2005), as suggested by their shorter mean lengths for annealing at 700 °C for 1 hour. These results, in conjunction with the data by Kasuya and Naeser (1988) mentioned above, imply that compositional effects do not play a major role in track annealing of zircon.

Quantitative track annealing models. Kinetic parameters that describe fission-track annealing in zircon at laboratory time scales have been established for zircon based on laboratory heating studies of confined track lengths (Yamada et al. 1995b; Tagami et al. 1998). The first reliable annealing model was proposed by Yamada et al (1995b) using spontaneous tracks in the NST zircon crystals annealed under isothermal heating conditions of ~350–750 °C for $\sim 10^{-1}$ to 10^3 hours. The annealing data were then fitted by parallel and fanning models originally developed for track annealing in apatite by Laslett et al. (1987). Galbraith and Laslett (1997) applied the new statistical method of Laslett and Galbraith (1996) to the data in Yamada et al. (1995b) and determined that both the parallel and fanning models are tenable. However, when additional annealing data for 10^4 hours heating are included, a clear statistical preference was found for a fanning model (Tagami et al. 1998). Consequently, the fanning model formula for the mean length (μ; note this parameter is for tracks 60° to *c*-axis) after annealing at temperature *T* Kelvin for time *t* hours was given by:

$$\mu = 11.35\left[1-\exp\left\{-6.502+0.1431\frac{(\ln t+23.515)}{(1000/T-0.4459)}\right\}\right] \quad (1)$$

Figure 3 summarizes the results of the annealing experiments using spontaneous tracks in the NST zircon and iso-length contours given by the fitted fanning model. If we define the lowest and highest temperature limits of the zircon partial annealing zone (ZPAZ) by mean track lengths of 10.5 and 4.0 μm, respectively (these values represent approximately the detectable,

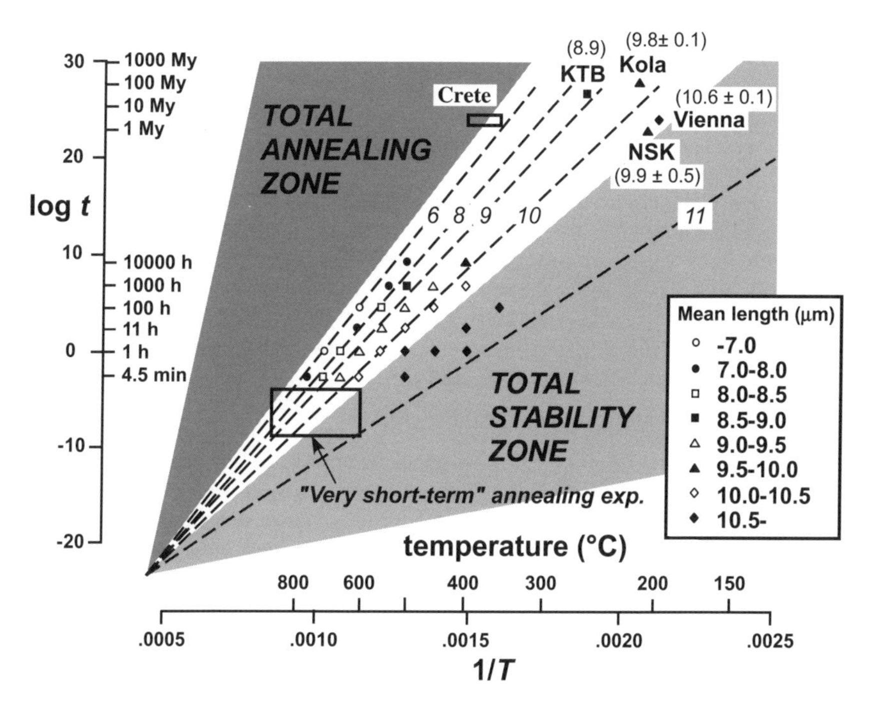

Figure 3. Arrhenius plot showing the design points of the laboratory annealing experiments of spontaneous fission tracks in zircon as well as contour lines for the fitted fanning model extrapolated to geological time scale (Tagami et al. 1998; see also Yamada et al. 1995b; Galbraith and Laslett 1997). Three temperature zones can be defined as first-order approximation of fission-track annealing: the total stability zone (TSZ) where tracks are thermally stable and hence accumulated as time elapsed; the partial annealing zone (PAZ) where tracks are partially stable and slowly annealed and shortened; and the total annealing zone (TAZ) where tracks are unstable and faded soon after their formation. Note that the zircon PAZ is defined here as a zone having mean lengths of ~4 to 10.5 μm and shown as a white region intervened by the TSZ and TAZ. Also shown are the time-temperature region of "very short-term" annealing experiments that simulate the shear heating of fault motion (Murakami et al. 2005); data points of four deep borehole samples subjected to long-term natural annealing, which were used to test the extrapolation of the annealing model (Tagami et al. 1998; Hasebe et al. 2003); and a box that indicates the time-temperature condition estimated for the higher temperature limit of the zircon PAZ on Crete (Brix et al. 2002).

minimum length reduction and total fading of surface tracks; see Yamada et al. 1995b), the extrapolation of Equation (1) predicts the ZPAZ at geological timescales as ~190–380 °C, ~180–350 °C and ~160–330 °C for heating durations of 1, 10 and 100 m.y., respectively.

Very short-term track annealing for fault studies. Fission-track investigations of rocks from fault zones, as will be discussed later, require special consideration of two factors. (1) Fault rocks may have been subjected to hydrothermally-pressurized conditions at some stage during fault development. To evaluate the effect of pressure on track annealing, laboratory heating experiments of zircon were carried out using a hydrothermal synthetic apparatus (Brix et al. 2002; Yamada et al. 2003). Yamada et al. (2003) annealed the same zircon samples, using the same temperature monitor and experimental procedure as those employed in the previous experiment at atmospheric conditions (Yamada et al. 1995b). The observed fission-track annealing characteristics are indistinguishable between the heating carried out at atmospheric and hydrothermally-pressurized conditions. This finding probably validates the application of annealing kinetics based on the experiments at atmospheric conditions to rocks subjected to hydrothermal conditions in nature, such as those in fault zone and plate subduction settings. (2) Frictional heating along a fault is a short-term phenomenon with heating durations of seconds, significantly shorter than conventional laboratory heating of ~10^{-1} to 10^4 h. Thus, high-temperature and short-term annealing experiments were newly designed and conducted using a graphite furnace coupled with infrared radiation thermometry (Murakami et al. 2005). Their results show that the observed track length reduction by 3.6 to 10 s heating at 599 to 912 °C is similar to that predicted by the fission-track annealing kinetics based on the conventional laboratory heating for ~10^{-1} to 10^4 h at ~350 to 750 °C (Yamada et al. 1995b; Tagami et al. 1998). In addition, it was determined that spontaneous tracks in zircon are totally annealed at 850 ± 50°C for ~4 s. These physical conditions are similar to those of some pseudotachylyte formation in nature (e.g. Otsuki et al. 2003, for pseudotachylyte of the Nojima fault). A new zircon fission-track annealing model is in progress by incorporating these high-temperature and short-term annealing data (Murakami et al. 2005).

Long-term track annealing at geological timescales

Fossil partial annealing zone. Track annealing properties of zircon have also been constrained through the fission-track analysis of heating around a granitic pluton (Tagami and Shimada 1996) (Fig. 4). These data come from the contact metamorphic aureole in the Cretaceous Shimanto Belt, southwest Japan, formed by the intrusion of the Takatsukiyama Granite at 15 Ma, for which the heating duration is estimated as ~10^5–10^6 yr using one-

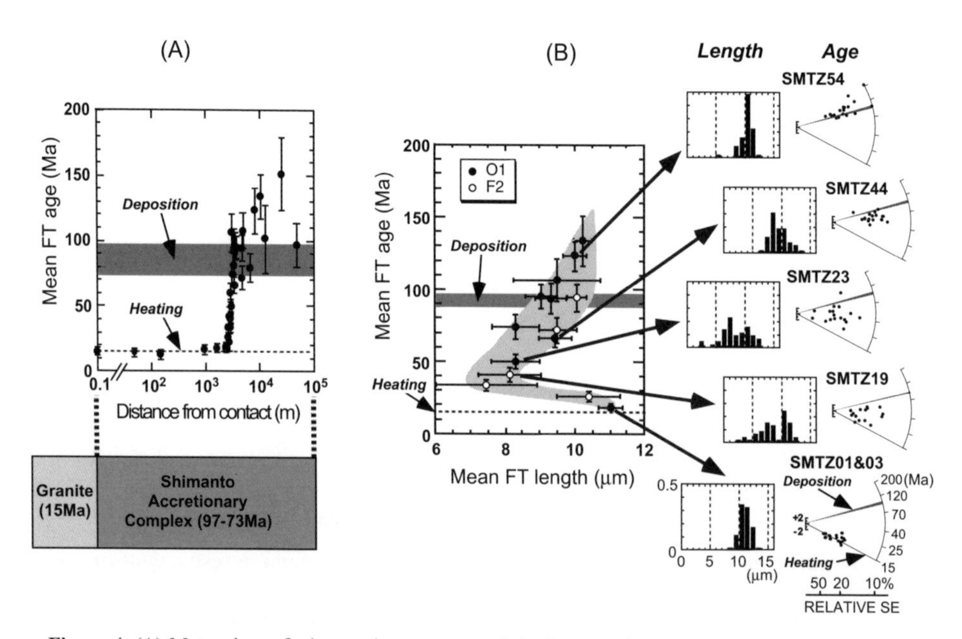

Figure 4. (A) Mean zircon fission-track ages versus their distances from the contact of Takatsukiyama Granite (Tagami and Shimada 1996). The shaded zone and dashed line represent, respectively, ~97–73 Ma deposition of the Shimanto accretionary complex and ~15 Ma rapid cooling of the granite subsequent to its intrusion and thermal overprinting. The transitional nature of track annealing is observed at ~3 km distance from the contact, as a result of 15 Ma heating of Mesozoic zircons in the Shimanto sandstones. (B) Mean zircon fission-track ages versus mean track lengths for the samples from the transitional annealing zone. The plotted data exhibit a characteristic "Boomerang-shape" bend, with the upper end predating deposition and lower end coincident with the granite intrusion event. Also shown are distributions of track lengths and single-grain ages of five representative samples. Error bars are ± 2 standard error. O1, O1 member of the Ogura Formation; F2, F2 member of the Furushiroyama Formation.

dimensional heat-conduction modeling (Tagami and Shimada 1996). Figure 4a presents the relationship between sample-mean zircon fission-track ages and their surface (two-dimensional) distances from the intrusion contact of the Takatsukiyama granite (Tagami and Shimada 1996). The ages are generally similar to or older than Cretaceous depositional ages for samples >4 km away from the contact. Near the contact, however, ages show a continuous reduction from about 100 to 15 Ma at ~ 3 km distance. For distances <2.5 km, the FT ages coincide with cooling ages of the Takatsukiyama Granite (15.0 ± 0.4 Ma). Figure 4b shows representative examples of single-grain age distributions using radial plots (Galbraith 1990) and track-length histograms. Track-length distributions (Fig. 4b) show variations consistent with increasing resetting due to heating at 15 Ma. In unreset samples, track lengths show a unimodal distribution and average ~10 μm. Approaching the contact, samples that experienced the thermal effects of the pluton intrusion were significantly annealed and shortened, thus yielding a bimodal distribution with a greater number of tracks <9 μm in length. Less than 2.5 km from the contact, thermal resetting is complete due to higher paleotemperatures, and the samples have a unimodal distribution of long tracks formed after the 15 Ma event. The mean lengths in un-annealed or totally annealed (reset) samples are close to the reference value of 10.5 ± 0.1 μm determined for zircon age standards from rapidly cooled volcanic rocks (Hasebe et al. 1994).

Figure 5 illustrates a schematic relationship between; (i) three temperature zones that approximately express the thermal stability of the zircon fission-track system, i.e., total

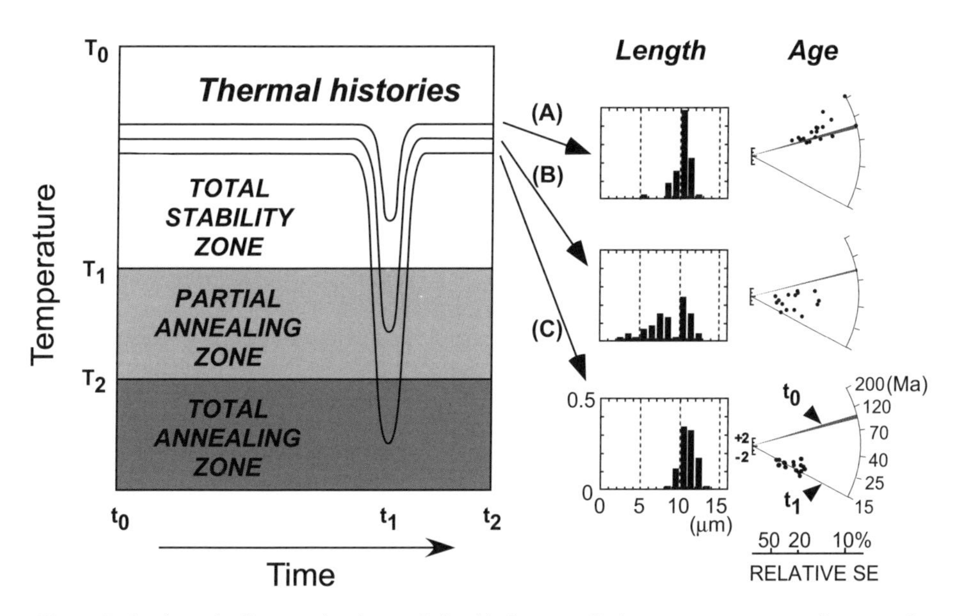

Figure 5. A schematic diagram showing a relationship between (i) three temperature zones in terms of thermal sensitivity of the zircon fission-track system, i.e., total stability zone (TSZ), partial annealing zone (PAZ) and total annealing zone (TAZ), (ii) three hypothetical thermal histories characterized by secondary heating to different paleo-maximum temperatures (T_{max}), and (iii) three sets of zircon fission-track length and age distributions resulted from the three different time-temperature pathways (after Tagami and Murakami 2005b). Using this relationship, we can inversely reconstruct the thermal history of rocks from observed fission-track length and age data. T_0 = temperature at earth's surface, T_1 = temperature at the boundary between TSZ and PAZ, T_2 = temperature at the boundary between PAZ and TAZ, t_0 = time of deposition (or rock formation), t_1 = time of secondary heating, t_2 = present, (A) a thermal history with $T_{max} < T_1$, (B) a thermal history with $T_1 < T_{max} < T_2$, (C) a thermal history with $T_2 < T_{max}$. Note that T_1 and T_2 substantially vary with heating time duration (see Fig. 3).

stability zone, partial annealing zone and total annealing zone; (ii) three hypothetical thermal histories characterized by secondary heating to different paleo-maximum temperatures; and (iii) three sets of length and age distributions resulted from the three different time-temperature pathways. Using this relationship, we can inversely reconstruct the thermal history of rocks from measured fission-track length and age data (for quantitative analysis, see Gallagher et al. 1998). Here we note that the boundary temperatures (i.e., T_1 and T_2 in Fig. 5) for the three temperature zones substantially vary with heating time duration, as shown in Figure 3. This variation in PAZ boundaries should be kept in mind when we analyze a heating phenomenon that has duration significantly shorter (<1 m.y.) than ordinary geological timescales (1–1000 m.y.), the former of which may be the case of thermochronology of fault zones.

Geological test of laboratory-based annealing models. Samples from deep boreholes have provided additional constraints on the fission-track annealing behavior in zircon on geological timescales (Vienna Basin, Austria, Tagami et al. 1996; Kola, Russia, Green et al. 1996; KTB-HB, Germany, Coyle and Wagner 1996; MITI-Nishikubiki, Japan, Hasebe et al. 2003) (Fig. 3). Of these, data from well-studied sedimentary basins (Vienna Basin and MITI-Nishikubiki) are more reliable in constraining the long-term annealing characteristics because thermal histories of rocks from those basins were reconstructed with better confidence using sedimentary records. Overall, the validity of extrapolating the kinetic model to geological timescales is supported by the consistency between the model extrapolations and borehole constraints (Tagami et al. 1998; Hasebe et al. 2003). These data are particularly effective in estimating the lower temperature limit of the zircon partial annealing zone (ZPAZ). In addition, the mineral parageneses and quartz microstructures of high-P – low-T rocks on Crete, Greece, indicate that the higher temperature limit of the zircon partial annealing zone (ZPAZ) is between 350 and 400 °C for a heating duration of 4 ± 2 Ma (Brix et al. 2002). This constraint is consistent with the model extrapolation within errors. The zircon fission-track closure temperature was directly calibrated from cooling histories modeled by $^{40}Ar/^{39}Ar$ multi-domain diffusion theory of K-feldspar, yielding ~260–265 °C and ~235 °C for cooling rates of 30–50 °C/m.y. and ~4 °C/m.y., respectively (Foster et al. 1996). These estimated temperatures are both within the ZPAZ based on the extrapolation (Fig. 3).

A note should be added concerned with other geological approaches to assess the thermal sensitivity of the zircon fission-track system (e.g., Tagami et al. 1996, Brandon et al. 1998). Classically, the sensitivity was expressed by the closure temperature that was estimated by comparing (or interpolating) measured zircon ages with reconstructed cooling curves of the samples, yielding a range of values, i.e., ~175 °C (Harrison et al. 1979) and 240 ± 50 °C (Hurford 1986). The diversity in the two values was derived, at least in part, from the diversity in closure temperatures of the K-Ar biotite system adopted for the reconstruction of the cooling histories: 180–260 °C in the former and 300 ± 50 °C in the latter for similar cooling rates of ~50 °C/m.y. Since these values substantially control the reconstructed cooling curves, they exert a major influence on the closure temperature of zircon fission-track system estimated from their interpolation. More recent studies indicate the closure temperature concept for K-Ar biotite system to be an over-simplification, since it is not possible to describe the Ar transportation process in biotite solely by ideal volume diffusion (e.g., Lee 1994; Lo et al. 2000). In consequence, the former determination of zircon fission-track closure temperature based on such interpolations must be regarded as very approximate estimates. A different approach was adopted by Zaun and Wagner (1985), who analyzed a borehole section in Urach III and recognized a remnant (fossil) ZPAZ. They reported a series of downhole reductions in mean zircon fission-track ages measured for individual samples. From a linear extrapolation of the observed age reduction, and assuming the samples' original age, the magnitude and duration of a Mesozoic subsidence, and paleogeothermal gradient, a zircon fission-track closure temperature was estimated as 195 ± 20 °C with a ZPAZ of 158–224 °C, for a heating duration of 100 m.y. However, unless these assumed parameters were well quantified, this type of approach does not provide direct evidence and, again, only offers an approximation of the annealing temperature.

ANALYTICAL PROCEDURES

The aim of this section is to present briefly the zircon fission-track laboratory procedure using the external detector method (EDM), as practiced in the fission-track laboratory, Kyoto University. See the chapter by Tagami and O'Sullivan (2005) for the principle and outline of the analysis. Also, see Garver (2003) for a recent survey of the experimental procedures used for chemical etching and track counting in zircon.

Zircon fission-track dating

Mounting of zircon crystals. After the concentration and separation of zircon crystals from other minerals, small fraction of zircon grains is spread on a silica glass slide. The glass slide is brought under a stereo binocular microscope for inspection. If the fraction contains small zircons not suitable for analysis (i.e., <40 μm in diameter) or other minerals, it is desirable to remove them from the glass slide by handpicking (this procedure is to achieve reliable polishing appropriate for track observation).

The silica glass is then shifted onto a hot plate having good temperature control. For mounting zircon grains, teflon sheets is generally used because of its stability against the etchant of zircon. FEP (polyfluoroethylene) type teflon (Gleadow et al. 1976) was first used successfully for this purpose, whose mounting temperature is about 280 °C. An alternative teflon, PFA (copolymer of tetrafluoroethylene-perfluoroalkoxyethylene), was later found by T. Danhara and applied to zircons of a wide range of ages (Tagami 1987; Tagami et al. 1988b). The advantage of PFA is in that it softens at higher temperature around 320 °C and, thus, grains tend not to dislodge from the mount even during long etching needed for young zircons (i.e., ~50 h or more; Gleadow et al. 1976; Tagami 1987; Seward and Kohn 1997). In the routine of Kyoto University, we employ a sheet of PFA Teflon of the size 1.5 × 1.5 cm^2 and of thickness 0.50 mm. Although the larger thickness is better for keeping the mount flat, the transparency decreases with thickness which, in turn, causes inconvenience in counting tracks under a transmitted light microscope.

For mounting, a Teflon sheet is held vertically using tweezers on the silica glass slide and heated for about 30 s until its bottom end which touches the glass slide melts slightly so as to stay vertically oriented by itself. The placement of the Teflon sheet is chosen in such a manner that when tilted slightly with tweezers to allow to fall gently on the glass slide, the zircon grains are mounted in the central area. A second silica glass plate smaller in size, preheated on the same hot plate, is put on the Teflon sheet and pressed gently so that the grains are fixed in the Teflon sheet. The entire arrangement is then removed from the hot plate and kept on a metallic plate. It is allowed to cool to the room temperature with a small load on the upper smaller glass slide so as not to produce any curvature of PFA sheet during cooling.

The Teflon mount, after being removed from the glass slide by applying a little alcohol, is then slightly cut asymmetrically at its right top corner. This asymmetric cutting makes it convenient to recognize the grain side of the Teflon sheet by naked eyes and without using a microscope. The sample code is written with a needle pen on the backside of the Teflon mount.

Grinding and polishing. To expose flat internal surfaces of the mounted grains by grinding and polishing, the Teflon mount is fixed on a transparent acrylic block of suitable size using bifacial (double-sided) tape. The grains are first ground using 1500# emery paper wetted with water until the maximum possible area is exposed. Then, all of the exposed areas are recorded by taking their digital images so that later checks are possible to choose appropriate areas having 4π geometry for track counting. The mount is further ground to remove at least a thickness that corresponds to half of etchable track length, ~6 μm for zircon (Krishnaswami et al. 1974; Hasebe et al. 1994). The removed thickness can be measured by choosing some surface feature, such as exposed inclusion, on a grain and noting the removal of surface feature depth by z-motion of microscope. In case of zircon, it is not preferred to grind crystals along their crystallographic *c*-axis because it often produces, in our experience, deep cracks or damages on the surface.

After the grinding, the mount is polished successively with diamond pastes of 7 μm, 2.5 μm and even smaller sizes, if necessary. At every step of polishing, the direction is changed from that of preceding polishing. This makes it easy to distinguish the disappearance of the preceding polishing scratches, ensuring thereby the completion of each polishing step. Since it may be difficult to check the removal of the scratches with transmitted light microscope, reflected light microscope is useful for this purpose. Before switching from one diamond paste to another, the samples are washed with ultrasonic cleaner to make them clean of the preceding diamond paste. The polishing time with each diamond paste should always be about five times longer than that just required to erase the scratches of the preceding paste (otherwise, the scratches will emerge after etching and disturb the observation of tracks).

Etching of spontaneous tracks in zircon. Because the optimum etching time of spontaneous tracks in zircon is negatively correlated with the track density (Gleadow et al. 1976; Hasebe et al. 1994), we need to perform progressive (step) etching and track observation for unknown zircons. It is necessary to continue the step-etching until the thin tracks parallel to the crystallographic *c*-axis are completely revealed (Gleadow 1981; Sumii et al. 1987; Watanabe 1988; see chapter by Tagami and O'Sullivan 2005). For young zircons, it is often difficult to judge the optimum etching condition using a small number of tracks therein (Gleadow 1980; Watanabe 1988). In such a case, etching needs to be continued until the width (diameter) of tracks perpendicular to the *c*-axis becomes 1 μm (Tagami et al. 1990). Using this etching criterion, even induced tracks in a pre-annealed zircon, which represents an extreme state of anisotropic etching, have an approximately isotropic angular distribution. Note that the optimum etching time in zircon can vary, even in one sample, from grain to grain and area to are, due to large variation of spontaneous track density. In such a case, the decision for further etching can be made on the basis of number of grains that are underetched. If the observer feels that further etching makes more grains available for measurement than the number it spoils by etching, one can etch the sample further.

For etching a large number of zircon mounts together, we use an electric hot plate having holes for keeping Teflon beakers (Fig. 6). The depth of these holes are made slightly smaller than the height of the Teflon beaker containing the eutectic NaOH:KOH (1:1) etchant (Gleadow et al. 1976). After the required temperature of the heater (252° ± <1 °C) is attained, the etchant is put into the beakers and kept there for about 12 h with the funnel covers. After this period, no air bubbles are left in the etchant, which is then ready for use. In our routine, 4–5 mounts are put together in each beaker for etching. The mounts float on the surface of etchant during etching, grain side down. In the present system, the temperature of etchant is controlled to 221° ± 2 °C, so that the track etching is reproducible (Note: this temperature is a recalibrated value and was quoted previously as 225 °C for the same setting).

After etching, individual zircon mounts are taken out from the etchant and put in 5% HCL in an ultrasonic cleaner for 10 minutes to remove minute quantities of etchant on the mount. It is not uncommon for the PFA Teflon to curve during etching. Such a curvature makes the

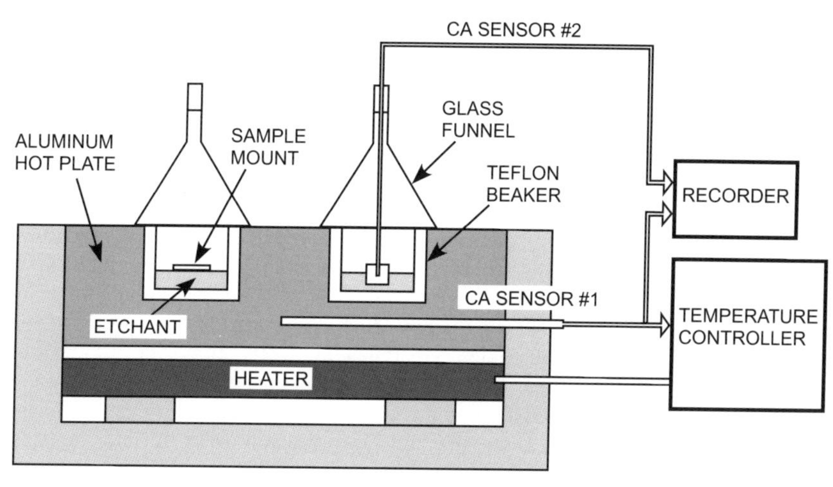

Figure 6. A sketch of track etching system for zircon using the eutectic NaOH:KOH etchant (after Tagami et al. 1988b). Temperature of the etchant is kept at 221° ± 2 °C with the help of a temperature controller and a thermocouple (CA sensor #1) inserted to the center of the Aluminum hot plate, for which temperature is controlled to 252° ± <1 °C. Stability of temperature with time can be monitored by connecting the output of thermocouple to x-t recorder. In order to monitor the temperature of etchant, another similar thermocouple (CA sensor #2, with teflon coating on its end) is dipped in the etchant inside the beaker and its output is also fed to the x-t recorder.

microscopic observation somewhat difficult and will disturb the complete attachment of the muscovite detector during the irradiation. To flatten the mount, the Teflon mount is sandwiched between two silica glass slides and put them, with a small weight, in an electric oven at 250 °C for a few hours, and then allowed to cool. From our experience, the etchant once prepared can be used for a week and then should be replaced by new one.

Neutron irradiation. A uranium-free muscovite sheet, such as Brazilian Ruby Clear (Shin and Park 1989), is cut into square pieces of the size slightly larger than the area of the mounted grains and of suitable thickness (~ 0.1 mm). The sample code is written with a needle pen on the back of the muscovite detector. To distinguish the surface attached to the mineral grains, one corner of it is cut asymmetrically such that when fixed to Teflon sheet, the cuts of both are in the same direction. Before putting the detectors, all mineral mounts are observed under microscope and undesirable particles or dust on the mounts are removed. Each detector is attached firmly to the etched mineral surface for neutron irradiation. In a similar fashion, a muscovite detector is attached to the surface of U-doped standard glass, such as NIST-SRM 612 (or 962a) or Corning CN1 through CN6, which is used as a neutron dosimeter. Then, sample batches are stacked vertically between standard glasses, and the entire set is put into a capsule for neutron irradiation in a nuclear reactor. In recent years, we irradiate samples at the Thermal Column Pneumatic Tube facility of Kyoto University Research Reactor. The facility is well thermalized, with nominal thermal, epithermal and fast neutron fluxes of 4×10^{11}, 6×10^{8} and 8×10^{7} $cm^{-2}s^{-1}$, respectively.

Etching of induced tracks on muscovite detector. After irradiation and radioactive cooling, the detectors are removed from the mounts and fixed individually in the cuts made on a HF-resistant rubber ring. The ring containing the muscovite sheets is put in a plastic beaker having some hole on its walls. For starting the etching, this beaker is put into 47% HF in a constant-temperature water bath kept at 32° ± 1 °C. The optimum etching time for the Brazilian Ruby Clear muscovite is about 4 min using this condition. After etching, the plastic beaker containing the ring is immersed in distilled water in an ultrasonic cleaner for a few minutes. The plastic beaker is then dried in an electric oven for about an hour to evaporate HF vapors completely.

Counting tracks. The sample mount and muscovite detector are fixed on a glass slide using transparent manicure (fingernail polish). Etched tracks are counted in transmitted light using the Nikon Eclipse® microscope with 100× dry objective and 10× eyepieces. A transparent glass sheet having 10 × 10 square-grids, inserted in one of the eyepieces of the microscope, is employed to measure the counted area. For reliable track density measurement in zircon, we should count only (a) grains surfaces of high etching efficiency which are judged by the existence of sharp polishing scratches (Naeser et al. 1980; Gleadow 1981) and (b) optimally etched grains (or areas of the grains) which are judged by the clear revelation of thin tracks parallel to the crystallographic *c*-axis (Gleadow 1981; Sumii et al. 1987; Watanabe 1988). Also, overall magnification of the microscope should be calibrated precisely using a micro-scale for reliable determination of the counted area. To measure the induced track density of U-doped standard glass, the attached area of a muscovite detector is uniformly scanned using an automatic stage. Routinely 1000–2000 tracks are totally counted for each dosimeter glass. It is stressed here that special attention is needed to use an oil-immersion objective for counting tracks because shallow etch pits tend to be obscured with it in some minerals. This will lead to a systematic drift of the ζ factor, which will be given below. See the chapter by Tagami and O'Sullivan (2005) for other analytical criteria.

Age calculation and calibration. In the zeta age calibration, a fission-track age t is given by:

$$t = \frac{1}{\lambda_D} \ln\left[1 + \lambda_D \zeta \rho_D \left(\frac{\rho_S}{\rho_I}\right) QG\right] \quad (2)$$

where λ_D is the α-emission decay constant of ^{238}U (1.55125×10^{-10} yr^{-1}), ζ is the zeta age calibration factor, ρ_D is the induced fission-track density on a U-doped standard glass, ρ_S is the surface density of etched spontaneous fission tracks, ρ_I is the surface density of etched induced fission tracks, Q is the integrated factor of registration and observation efficiency of fission tracks, and G is the integrated geometry factor of etched surface. For derivation of this equation, see Tagami and O'Sullivan (this volume). ζ is determined empirically by analyzing standard minerals of known ages (Hurford 1990). Q and G are the same between the age standards and age-unknown samples as long as they are analyzed using identical experimental procedures and criteria. In the routine analysis using the EDM, therefore, Q is ignorable and G is usually assumed to be 0.5, the ideal ratio between the 2π and 4π geometries for ρ_I and ρ_S measurements, respectively. Thus the equation is simplified to:

$$t = \frac{1}{\lambda_D} \ln\left[1 + 0.5\lambda_D \zeta \rho_D \left(\frac{\rho_S}{\rho_I}\right)\right] \quad (3)$$

ζ is given by inputting the measured data of standard (i.e., ρ_D, ρ_S and ρ_I) and its reference age t_{STD} into:

$$\zeta = \left(\frac{\rho_I}{0.5\lambda_D \rho_D \rho_S}\right)\left\{\exp(\lambda_D t_{STD}) - 1\right\} \quad (4)$$

Table 1 presents age standard minerals widely used for zeta calibration. For reliable calibration, ζ should be determined repeatedly for a set of standards to examine the internal consistency between experimental runs (Hurford 1990). If measured ζ values show a significant scatter over analytical uncertainties, the experimental conditions/procedures adopted should

Table 1. Age standards for fission-track analysis.

Sample	*Mineral*	*Age ±2σ (Ma)*	*References*
Buluk Tuff	zircon	16.3 ± 0.2	[1]
Fish Canyon Tuff	zircon, apatite, sphene	27.9 ± 0.7	[2]
Durango Apatite	apatite	31.4 ± 0.5	[3]
Tardree Rhyolite	zircon	58.4 ± 0.7	[4]
Mt. Dromedary Banatite	zircon, apatite, sphene	98.7 ± 1.1	[3]

References:
[1] McDougall and Watkins 1985; [2] Hurford and Green 1983; [3] Green 1985; [4] Gamble et al. 1999.

be improved/refined before analyzing samples of unknown ages. Routinely 10–30 zircon grains are measured for a single FT analysis of zeta calibration and of age determination. The single-grain data are subjected to the χ^2-test to statistically assess the extra-Poissonian variation(s) (Tagami and O'Sullivan 2005).

Track length measurement

For track-length analysis, zircon grains are mounted, ground and polished in the same way as for track-density measurement, except that the rigorous exposure of 4π geometry is not necessary because confined tracks are used throughout (see Tagami and O'Sullivan 2005). Tracks are etched in KOH-NaOH eutectic etchant at 248 ± 1 °C until the widths of surface-intersecting tracks perpendicular to the *c*-axis reach 2.0 ± 0.5 μm in most grains (Yamada et al. 1995a). (Note that this temperature is a recalibrated value and was quoted previously as 250 °C for the same setting. Also note that, for the higher temperature etching, we use a new track etching system that is a refined version of the first one in Fig. 6. It has a stainless-steel lid—with insulator inside—on top of the hot plate, which substantially decreases the temperature gradient within the system and increases the temperature stability of etchant.) A criterion for etched track widths is needed to standardize the etching state since the general attack rate (V_G) that controls the track revelation is positively correlated with the spontaneous track density. The 2.0 μm criterion generally requires longer etching compared to the optimum etching of surface-intersecting tracks mentioned above, and the difference is larger for zircons of higher track densities. However, such etching is necessary to increase the number of measurable confined tracks, particularly for zircons of low to intermediate spontaneous track densities (i.e., <1 to ~3 × 10^6 cm^{-2}). (This is also the case for zircons annealed by laboratory heating experiments.) Etching is conducted at higher temperature, 248 ± 1 °C, to make the etching time effectively shorter, i.e., ~ 20% of that at 221° ± 2 °C (e.g., 26 hr at 248 ± 1 °C versus 120 hr at 221° ± 2 °C for induced tracks of NST zircon; Yamada et al. 1995a). Prior to the track length measurement, the sample mount is repolished if the track density is too high (say, 10^7 cm^{-2}) for observation of confined tracks.

Track lengths and angles to crystallographic *c*-axis are measured on horizontal confined tracks (HCTs; Laslett et al. 1982), revealed either as tracks-in-tracks (TINTs) or as tracks-in-cleavages (TINCLEs). We adopt HCTs for measurement that have (1) orientations 60° to the *c*-axis because of enhanced anisotropy in annealing and etching for tracks <60° (i.e., etched tracks sub-parallel to *c*-axis appear longer than those sub-perpendicular to *c*-axis) and (2) widths of 1.0 ± 0.5 μm in order to reduce the influences of different etching states on measured lengths (Yamada et al. 1993, 1995a; Hasebe et al. 1994; see chapter by Tagami and O'Sullivan 2005). The lengths of HCTs are measured by taking photographs of tracks using the Nikon® DXM1200 CCD digital camera installed on the Nikon Eclipse® microscope with a 100× dry objective, followed by counting the pixel length of the track using an image analysis software on a computer. In terms of the horizontal criterion, therefore, we measure only confined tracks that are fully in focus. Conversion from pixel length to true length is calibrated by measuring a Nikon microscale with a minimum scale of 0.1 μm. Overall precision and accuracy of measured length is about ± 0.1 μm (1 SE).

Approximately 50–100 confined tracks are routinely measured for each zircon sample. Confined track lengths and track densities (ρ_S) need to be measured on different grains because etching criteria are different between the two analyses. In general, this is not a desirable situation as the two groups of grains may represent different grain populations having different thermal sensitivities. In the particular case of zircon, however, annealing rates are indistinguishable among zircons of different origins and ages, except very young ones on which both length and density can be measured at the same etching condition. Hence, the present issue is probably not a major problem, although further technical developments are strongly needed to measure routinely both the length and density on the same grain.

A note should be added here concerning with the experimental technique to increase the number of measurable confined tracks. Yamada et al. (1998) compared three techniques using annealed and unannealed zircons; namely, 1) irradiation of nuclear fission fragments using the ^{252}Cf source (Donelick and Miller 1991), 2) irradiation of heavy nuclides using the accelerator (Watanabe et al. 1991) and 3) artificial fracturing using a well-controlled grinding machine (Yoshioka et al. 1994). They found that the latter two works well for zircon compared to the first because of the difficulty to implant long host tracks on zircon using the ^{252}Cf source (Note this technique works well for apatite; see chapter by Donelick et al. 2005). Where an accelerator facility is not easily accessible, therefore, the artificial fracturing method is recommended for zircon and outlined below (for details, see Yamada et al. 1998). To make artificial cracks in zircon, the sample mount is ground parallel to the crystallographic *c*-axis using 1500# emery paper wetted with water, followed by polishing. For reproducible, gentle cracking, a steel weight (740 g) is put on the back of the acrylic sample holder to give a constant pressure on the sample during grinding. For the rotation speed of ~2 rps and ~1 m/s, artificial cracks are observed in 40–60% of grains after grinding for ~10 s. The length, width and depth of artificial cracks after etching are 50–100, 1–2, and 2–10 μm, respectively. Note that these values depend on the number, average size and hardness of mounted zircons. Because most cracks are formed parallel to the *c*-axis, the observed TINCLEs are approximately perpendicular to the *c*-axis, satisfying the orientation criterion of measuring confined tracks mentioned above (i.e., 60° to the *c*-axis). In the particular case of this artificial fracturing, tracks are etched until the widths of surface-intersecting tracks perpendicular to the *c*-axis reach 1.0 ± 0.5 μm in most grains, since TINCLEs and surface-intersecting tracks are etched approximately at the same rate (note the 1.0 ± 0.5 μm width criterion of measuring confined tracks mentioned above). This technique is useful for zircons with low track densities (i.e., $<1 \times 10^6$ cm^{-2}), and also for those separated from a limited amount of rocks, such as the case of borehole samples.

APPLICATION TO THE NOJIMA FAULT ZONE

Quantitative assessment of heat generation and transfer along faults is of primary importance in understanding the dynamics and structural history of faulting (Scholz 1996), as well as in constraining the heat budget and thermotectonic evolution of mobile belts (Fukahata and Matsuura 2001). These thermal signatures also provide a tool for constraining the ages of faults. Among a range of methodologies available to investigate thermal regimes around faults, the fission-track analysis of fault rocks has been successfully applied by many researchers to understand the thermal history in or near faults in various regions (see Wagner and Van den haute 1992; Gallagher et al. 1998, for references); e.g., the Median Tectonic Line,

Japan (Tagami et al. 1988a), the Alpine fault, New Zealand (Kamp et al. 1989), the Pejo fault system, Italian Eastern Alps (Viola et al. 2003), Castle Mountain Fault, Alaska (Parry et al. 2001), and San Gabriel Fault, California (d'Alessio et al. 2003).

Here I present zircon fission-track data of the Nojima fault rocks collected from the University Group 500 m (UG-500; or DPRI 500 m, after Murata et al. 2001; Ogura 500 m, Tagami et al. 2001) borehole, Geological Survey of Japan 750 m (GSJ-750) borehole, the fault trench at Hirabayashi, and nearby outcrops on the Awaji Island, Japan (Figs. 7 and 8; Tagami and Murakami 2005b). The Nojima fault, which ruptured during the 1995 Kobe earthquake (Hyogoken-Nanbu earthquake; M7.2), has been investigated multifariously from the viewpoint of geology and geophysics (Ando 2001). Together with such information, I discuss geologic processes producing the observed fission-track profiles around the fault.

Geological setting

Awaji Island is located in the southern area of Inner Zone of Southwest Japan, with the southern tip of the island in contact with the Median Tectonic Line (MTL) (Fig. 7). The basement of the island is comprised by Cretaceous Ryoke Granitic Rocks consisting mainly of granodiorite (Mizuno et al. 1990), with hornblende and biotite K-Ar ages ranging from 88

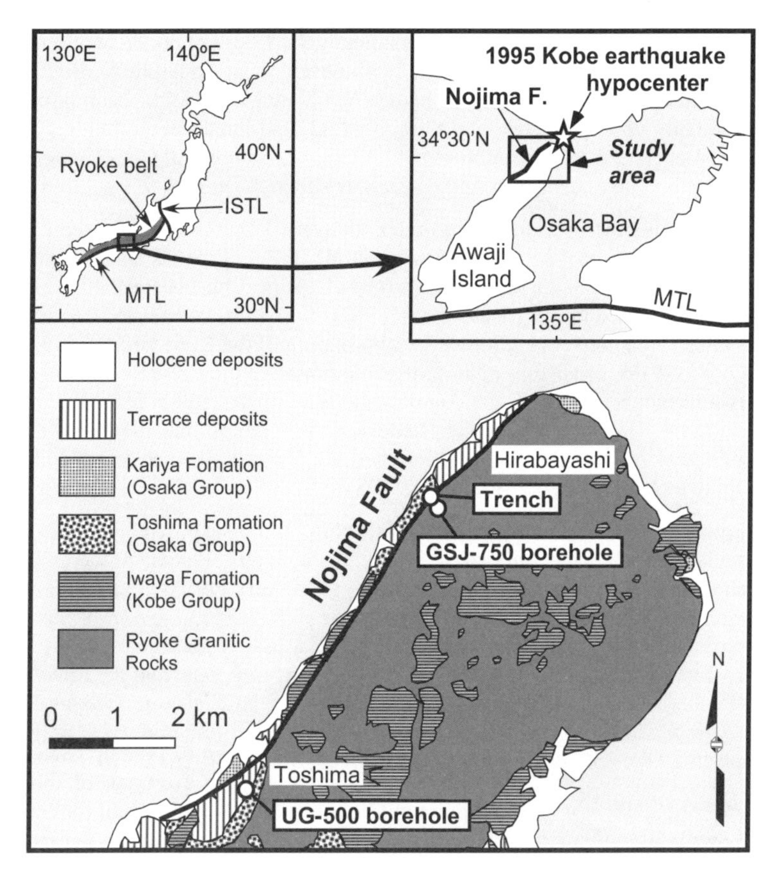

Figure 7. A geologic map showing sampling localities of Nojima fault, which runs along the northwestern coast of Awaji island. MTL, Median Tectonic Line; ISTL, Itoigawa-Shizuoka Tectonic Line. After Murakami et al. (2002).

± 4 (1 SE) to 90 ± 5 Ma and 70 ± 4 to 88 ± 4 Ma, respectively (Takahashi 1992). Zircon FT ages of the Ryoke Granitic Rocks in the Kii Peninsula, located ~80 km to the east of the Awaji Island, are 59 ± 3 to 78 ± 9 Ma (Tagami et al. 1988a and Hasebe and Tagami 2001). In the northern part of the island, the Eocene-Oligocene Iwaya Formation locally overlies basement rocks (Mizuno et al. 1990; Yamamoto et al. 2000). The Plio-Pleistocene Osaka Group is distributed along the northwest coast of the island.

The 1995 Kobe earthquake caused great disaster in the southern parts of Kobe City and the northwest area of Awaji Island. The hypocenter was located in the Akashi strait and the focal depth was reported to be 14 km (Fig. 7). As a result, >10 km long surface rupture occurred along the preexisting NE-SW striking the Nojima fault, which is a high-angle reverse fault dipping 83°SE (Awata et al. 1996; Murata et al. 2001) (Fig. 7). Maximum displacement of the surface rupture was observed as 180 cm right lateral and 130 cm reverse components at Hirabayashi, located at the northern part of the fault. The fault in the area forms a geological boundary between the Osaka Group and Iwaya Formation on the west and the Cretaceous Ryoke Granitic rocks on the east (Fig. 7). To the southwest around Toshima, however, the sedimentary sequences are observed on both sides of the fault, with no granitic basement outcrops nearby.

Sample description

After the 1995 earthquake, three boreholes were drilled, two of which intersected the Nojima fault zone (Fig. 8). One of these boreholes is the GSJ-750 borehole at Hirabayashi, drilled through the Cretaceous granodiorite with seven shear zones bearing fault gouges (Fig. 9). Of the seven shear zones, the one named MSZ is regarded as the central (main) zone of the Nojima fault, with the active fault trace at 625.27 m (Tanaka et al. 2000). The fission-track analysis was carried out on zircon separates from ten samples of GSJ-750 borehole and an outcrop sample ~130 m distant from the borehole site to the east.

The second borehole used for fission-track analysis is the UG-500 borehole (Fig. 10), with the following stratigraphic section (Murata et al. 2001): (1) sands and gravels (0–118.0 m; borehole apparent depth throughout), correlated to the Plio-Pleistocene Osaka Group, (2) arkosic sandstones and silty sandstones (118.0–191.6 m), correlated to the Eocene-Oligocene (after Yamamoto et al. 2000) Iwaya Formation of the Kobe Group and (3) granitic rocks (191.6–389.4 m), consisting mainly of granodiorite of the Cretaceous Ryoke Granitic Rocks. This sequence is observed again on the other side of the fault at 389.4 m depth: sands and gravels (389.4–410.7 m), arkose sandstones and silty sandstones (410.7–494.2 m), and granitic rocks (494.2–550.7 m). All samples used for analysis were collected from granitic rocks, except two from the Osaka Group (Fig. 10).

The Nojima fault was trenched at Hirabayashi (Fig. 7). The exposed fault rocks consist of granitic cataclasite, a 2–10 mm wide pseudotachylyte layer and siltstone of the Osaka Group, from the hanging wall (southeast) to the footwall (northwest). Otsuki et al. (2003) estimated the temperature of the pseudotachylyte formation as ~750–1280°C, based primarily on the observation of melting of K-feldspar and plagioclase. For fission-track analysis, a 50-cm-wide gray fault rock consisting of the following four layers was sampled from the footwall toward the hanging wall (Murakami and Tagami 2004) (Fig. 12); (1) greenish-gray gouge of footwall (NT-LG; ~20 mm wide), (2) pseudotachylyte (NT-PTa; ~2–10 mm wide), (3) gray gouge of hanging wall (NT-UG; ~30 mm wide) and (4) reddish granite (NF-HB1; ~20 mm wide).

Data and interpretation

GSJ-750 borehole. Figure 9 presents zircon fission-track mean age, mean length and length distributions plotted against distance to the fault (Murakami et al. 2002). Most tracks are not shortened in zircons from the outcrop sample (NF-HB2), with a mean length of 10.61 μm and unimodal length distribution with slightly negative skewness. This length is

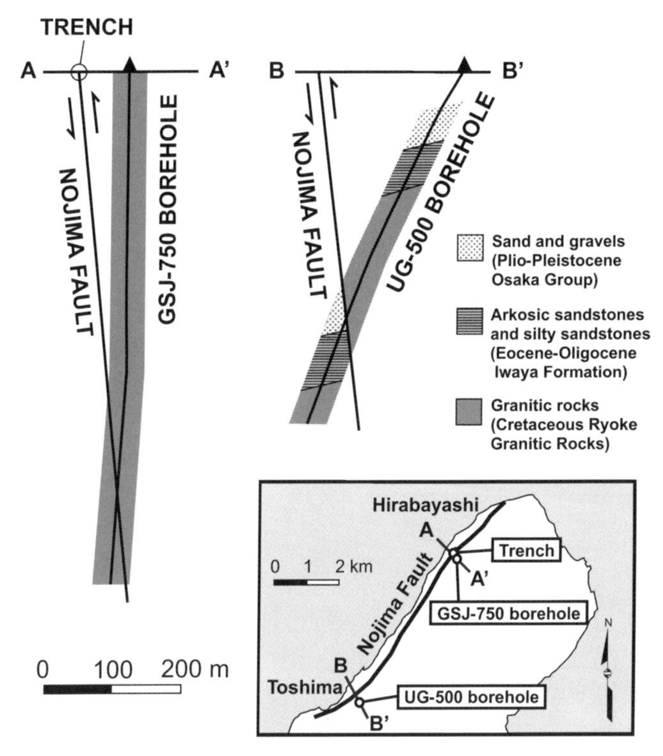

Figure 8. Schematic cross sections of the two boreholes analyzed (after Tagami and Murakami 2005b). The section consists of granitic rocks throughout the GSJ 750 borehole, whereas sedimentary sequences covering the granitic basement are observed on both the hanging wall and footwall of the fault in the UG 500 borehole. The Nojima fault is a high-angle thrust with a dip of 83°SE.

indistinguishable from the reference mean length of 10.5 ± 0.1 (1 SE) μm (Hasebe et al. 1994). The mean age of this sample is 74 ± 3 (1 SE) Ma, consistent with zircon fission-track and biotite and hornblende K-Ar cooling ages of the Ryoke Granitic Rocks in the region (Takahashi 1992; Tagami et al. 1988a; Hasebe and Tagami 2001). These lines of evidence suggest that the zircon fission-track age is interpreted as the time of cooling of granodiorite from high temperatures. Among all borehole samples, GSJ-TH064 at 455 m is the only sample in which the mean fission-track age and length are in agreement with those of outcrop samples. The mean fission-track ages and lengths of the other nine samples are younger and shorter than those of these two samples. The fission-track length distributions of some of nine samples (GSJ-TH001, 023, 093 and 103) are bimodal. This bimodal distribution is interpreted either by (1) residence within the zircon partial annealing zone (ZPAZ) (Fig. 3) temperatures under the regional geothermal regime for an extended period of time, after the initial cooling of granodiorite to within the ZPAZ, or (2) secondary heating up to within the ZPAZ temperatures after the initial cooling below the ZPAZ. The model (2) is further divided into two cases; (2a) secondary heating as a result of tectonic downward motion of the regional geotherm (burial or underthrusting), and (2b) secondary heating by a thermal event that locally perturbed the geothermal structure. The model (2b) is preferred to the others because the mean age and length data that indicate stronger annealing upward in the upper section (GSJTH001 to 064; Fig. 9) are not consistent with the normal geothermal regime, and because the zone of stronger annealing downward (GSJ-TH064 to 099) is too narrow (~25 m in distance orthogonal to the

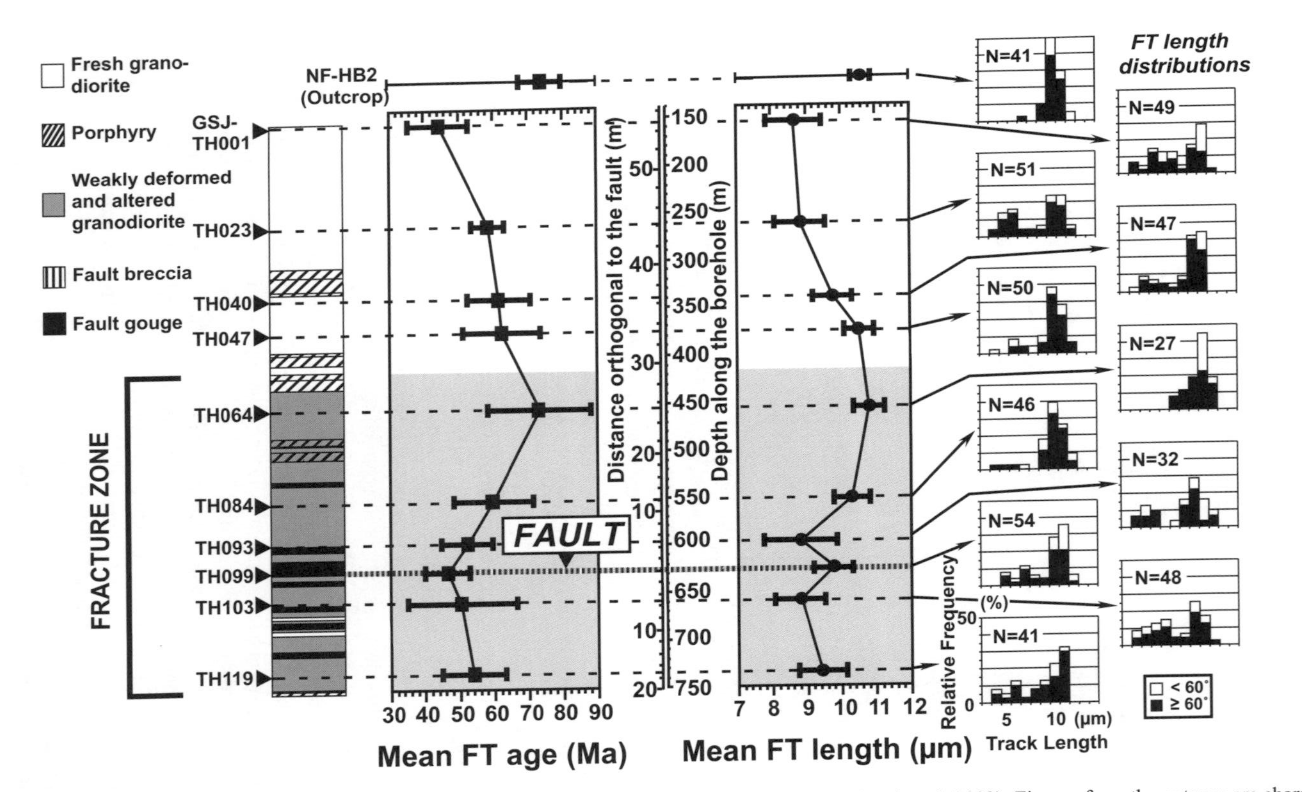

Figure 9. Zircon fission-track length and age versus distance to the fault in the GSJ 750 borehole (after Murakami et al. 2002). Zircons from the outcrop are characterized by mean lengths of ~10–11 μm and unimodal length distributions, showing no signs of appreciable reduction of fission-track length. In contrast, those from adjacent to the fault at depth show significantly reduced means of ~6–8 μm and distributions having two components of long and short tracks. Error bars are ±2 SE. Note that the track length distribution is shown in a histogram with different patterns for tracks marking azimuth angles greater (solid column) and less (open column) than 60° to the crystallographic c-axis, because track lengths in zircon depend on etching and annealing properties which show an angular variation. Also shown are the distribution of fault rocks (Tanaka et al. 1999) and sample localities (Murakami et al. 2002).

fault, which is equal to ~210 m in a vertical direction above the fault; Fig. 8) to be formed by residence within, or reheating into, the ZPAZ under the geothermal gradient in the study area (21–24 °C/km, measured in the NIED 1800 m borehole 200 m southeast of the GSJ 750; Kitajima et al. 1998: Note that there are no geological signs of magmatism-related thermal anomalies around the borehole site in the recent past).

GSJ-TH001 and TH099 have youngest mean ages of ~45–47 Ma. While the former has a mean track length shorter than the neighboring samples, the mean length of the latter is longer than neighboring samples. The lower mean track lengths may have accumulated before the secondary heating and were subjected to stronger annealing and thus almost removed from the sample (i.e., heating into the region lower than the Fig. 3b). Therefore, it is likely that the GSJ 750 borehole rocks record the heat signatures at two different locations, one <25 m perpendicular to the fault and the other at the shallower depth (>25 m).

The age of initiation of last cooling after the secondary heating, which marks the cessation of the secondary heating, was estimated using the Monte Trax program (Gallagher 1995) for zircon fission-track data from the GSJ-750 borehole (Murakami et al. 2002) (Fig. 9). Length distributions and age data of three partially-annealed samples near the fault (GSJ-TH093, 099 and 103) yielded ages of initiation of last cooling as 35 ± 1 (1 SE), 38 ± 2 and 31 ± 1 Ma by averaging eight, eight and nine model runs, respectively (Tagami and Murakami 2005a) (Fig. 11). The accuracy of the age is 16, 14 and 12 Ma, respectively. The ages are indistinguishable from each other averaging ~35 Ma.

UG-500 borehole. Track lengths of nineteen samples were analyzed from the outcrops and borehole section (Tagami et al. 2001; Tagami and Murakami 2005a) (Fig. 10). Fission-track lengths in zircons from localities >25 m away from the fault plane as well as one 0.1 m away from the fault in the footwall (NF51-FTH03M, a sample from the non-deformed Osaka Group sandstone) are characterized by concordant mean values of ~10–11 μm and unimodal distributions with negative skewness, which showed no signs of appreciable reduction in fission-track length. In contrast, those adjacent (<3 m) to the fault at depths on the hanging wall side (e.g., NF51-FTG14) showed significantly reduced mean track lengths of ~6–8 μm and distributions having a peak around 6–7 μm with rather positive skewness. A sample (NF51-FTG11) was likely to yield a distribution of a transient pattern by mixing the two end-member components. The former pattern is similar in mean length to spontaneous tracks in unannealed zircons (Hasebe et al. 1994) and, thus, is interpreted to reflect cooling through the ZPAZ, without later, partial thermal overprints. The latter indicates substantial track shortening as a result of partial annealing and is interpreted either by the three models mentioned in the last section. The width of the observed zone of zircon fission-track partial annealing is as narrow as <3 m in distance orthogonal to the fault (Fig. 10), which is equal to <25 m in a vertical direction above the fault (Fig. 8). On the other hand, the mean fission-track length shifts from >10 μm to <8 μm within the zone (Fig. 10), which requires the change of heating temperature of ~100 °C for a geological timescale (e.g., a heating duration of 10 m.y.). Hence, the heating cannot be explained by models (1) or (2a), both of which are accompanied by the ZPAZ under a subnormal geothermal regime (e.g., ~23 °C/km; Yamano and Goto 2001). In addition, the region is ~200 m beneath the base of the Eocene-Oligocene Iwaya Formation, and this indicates that the post-Eocene uplift is of limited amount in the borehole site, probably less than 5 km. These facts support model (2b) which calls on secondary heating by a thermal event.

The age of initiation of last cooling after the secondary heating was estimated by the inverse modeling of a sample near the fault (NF51-FTG14b) that yielded a partially annealed track length distribution (Fig. 10). The length distribution and apparent age data were modeled, as shown in Figure 11 (Tagami and Murakami 2005a), using the Monte Trax program (Gallagher 1995). The age of initiation of last cooling was estimated as 2.5 ± 0.4 (1 SE, precision) Ma by

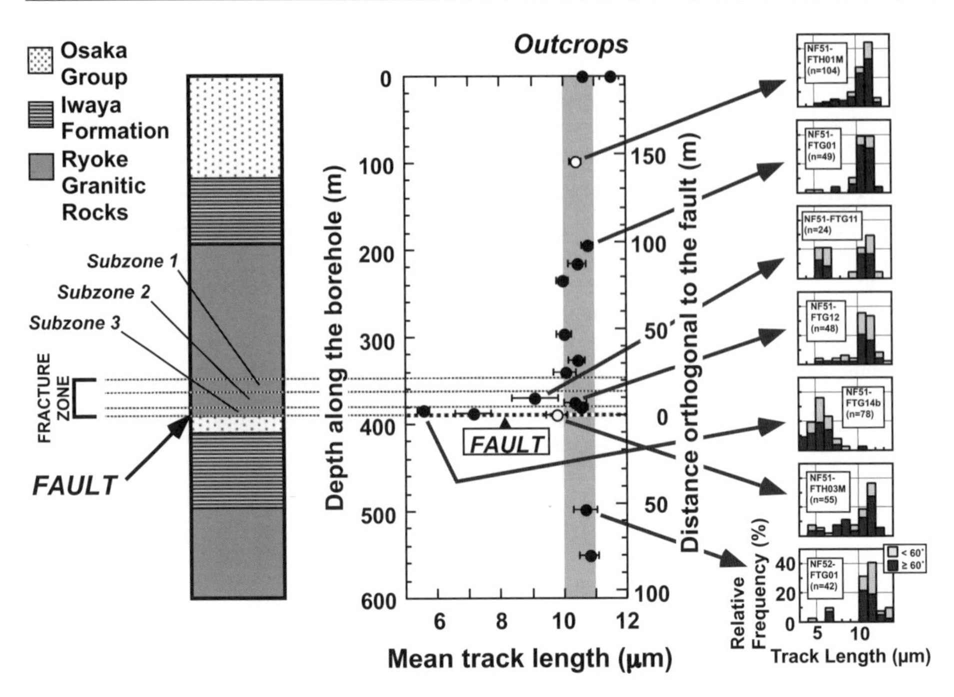

Figure 10. Zircon mean fission-track length versus distance to the fault in the UG 500 borehole, with track length histograms of representative samples (Tagami et al. 2001; Tagami and Murakami 2005a). Zircons from localities >30 m away from the fault plane as well as those from outcrops are characterized by the mean lengths of ~10–11 μm and unimodal distributions with negative skewness, showing no signs of appreciable reduction of fission-track length. In contrast, those from adjacent to the fault at depth show significantly reduced means of ~6–8 μm and distributions having a dominant peak around 6–7 μm with rather positive skewness. Error bars are ±1 SE. Also shown is the lithology of the borehole section (Murata et al. 2001), along with the three subzones within the fracture zone that represent dominant fault rock types (Tanaka et al. 2001); subzone 1 = non- to weakly deformed and altered rocks, subzone 2 = weakly deformed and altered rocks, subzone 3 = fault breccia and fault gouge.

averaging the results of model runs repeated six times. The accuracy of the age was estimated as 1.8 Ma by propagating the errors from length distribution and age analysis. The age is significantly younger than the ~35 Ma age of the GSJ-750 borehole mentioned above.

Fault trench at Hirabayashi. Figure 12 shows zircon fission-track data of the nine samples from the Hirabayashi trench (Murakami and Tagami 2004). The mean age of NF-HB1 is 72 ± 5 Ma, indistinguishable from the initial cooling age of the Ryoke basement of 74 ± 3 Ma (NF-HB2, ~200 m from the fault; Murakami et al. 2002). NT-PTa, however, yielded an age of 56 ± 4 Ma, significantly younger than the initial cooling. In addition, NT-UG4 and NT-LG2 are different in age from NT-PTa at the 95% confidence level. The other pseudotachylyte sample (NT-PTb) also gave a younger age of 59 ± 8 Ma, approximately concordant with the former age. The ages of NT-UG1-4 and NT-LG1-2 are given as from 65 to 76 Ma with continuous reduction toward the pseudotachylyte layer. NF-HB1 and NT-PTa have mean lengths of ~11 μm and unimodal distributions (Fig. 12), approximately indistinguishable from the reference zircon mean length of 10.5 ± 0.1 μm (Hasebe et al. 1994). In contrast, NT-LG1 and NT-UG1-3 yielded mean lengths of 8–9 μm, shorter than the formers. Their fission-track length distributions have components of short tracks of 4–9 μm lengths. This is interpreted as the result of partial annealing of tracks within the ZPAZ at ancient time. Also found were

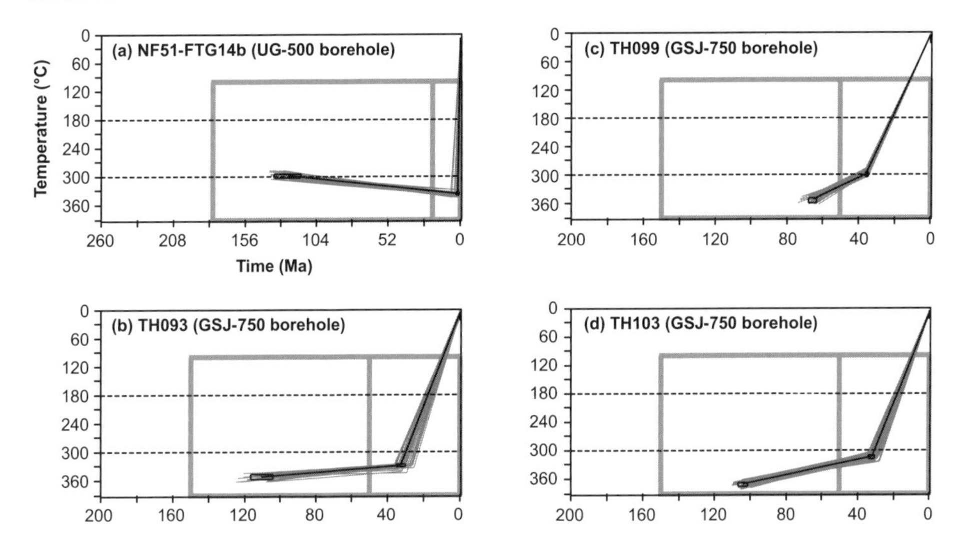

Figure 11. Zircon fission track modeling results for representative runs of the four samples using the Monte Trax program (Gallagher 1995). Two grey boxes for each diagram show initial time-temperature constraints given, from where time-temperature points were selected at random when the modeling started. 50 modeled runs at the final stage of iterations are shown by dark grey lines for each sample. The solid lines represent the average of the modeled thermal histories. The 95% confidence limits for the kink points of the history were shown as open black boxes. After Tagami and Murakami (2005a).

"syringe-shaped" tracks showing the evidence of partial annealing in these samples (e.g., NT-UG1; Fig. 12). NT-LG2 and NT-UG4 have mean lengths of ~10 μm, and their length distributions also have short tracks but their proportions are smaller than those of NT-LG1 and NT-UG1-3. These characteristic age and length profiles suggest that the zircon fission-track system of the pseudotachylyte layer was totally reset (or kept reset) and subsequently cooled at ~56 Ma. It is also inferred that the gray gouge layers on the footwall and hanging wall side were not heated into ZPAZ at that time. Its significantly older ~94 Ma age probably represents a dominant detrital age component of the host rock of the gouge, i.e., Osaka Group siltstone.

Geological implications

Heat source. Two plausible models of the heat source responsible for the observed cooling ages were examined; (a) frictional heating of fault motion, and (b) heat transfer or dispersion via fluids in the fault zone. The range of detectable zircon fission-track annealing formed by the model (a) is calculated as ~1 mm from the center of fault by one-dimensional heat conduction modeling (Murakami and Tagami 2004) (Fig. 13), using the following conditions; (1) thermal diffusivity of 2 mm^2s^{-1}, (2) environmental temperature of 200°C at dry condition, (3) frictional heat of 1000°C (e.g., Otsuki et al. 2003) for 5 seconds (Kikuchi and Kanamori 1996) at the fault plane, followed by heat dispersion due to thermal conduction, (4) zircon fission-track annealing kinetics of Tagami et al. (1998), along with the definition of partial resetting (annealing) zone mentioned above, and (5) frictional heating model of Cardwell et al. (1978). The predicted 1 mm range is approximately concordant with that of the ~56 Ma cooling age observed for the Hirabayashi trench. In the case of GSJ 750 and UG 500 boreholes, however, the spatial range of zircon fission-track annealing reaches at least ~5 m and ~3 m, respectively, from the fault (Figs. 9 and 10). These observed ranges are too wide to be explained by frictional heating even if earthquakes occurred repeatedly over an extended period of time.

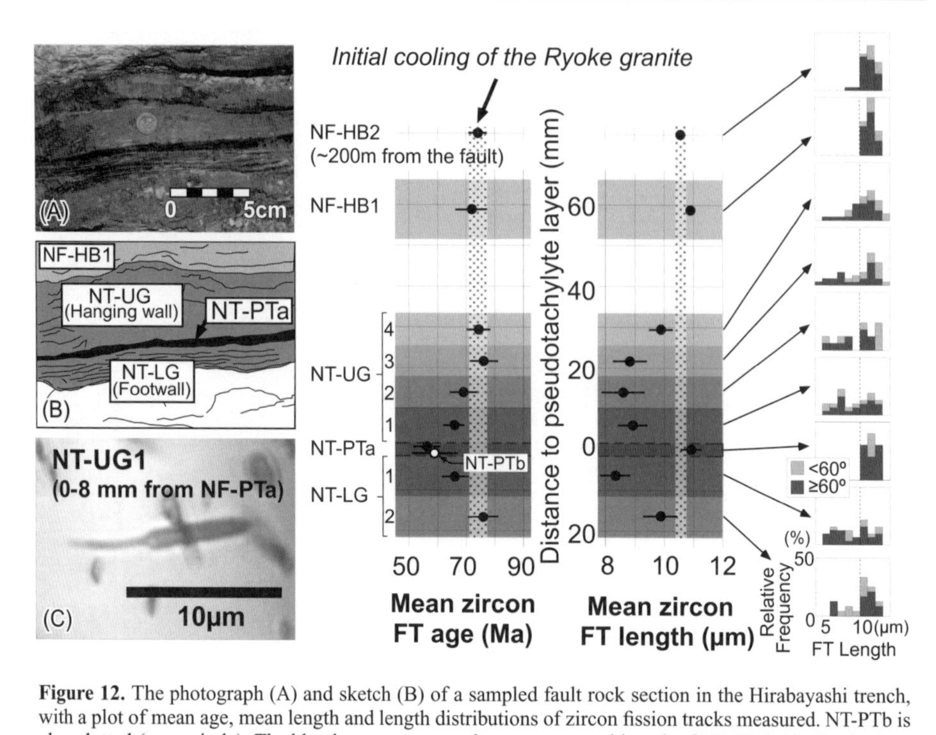

Figure 12. The photograph (A) and sketch (B) of a sampled fault rock section in the Hirabayashi trench, with a plot of mean age, mean length and length distributions of zircon fission tracks measured. NT-PTb is also plotted (open circle). The blue boxes represent the mean age and length of NF-HB2 (the Ryoke host rock sample). The sample from the pseudotachylyte layer (NT-PTa) has an age significantly younger than that of initial cooling of the Ryoke host rock sample (NF-HB1 and 2). The photograph (C) is "syringe-shaped" tracks found in NT-UG1, which shows the evidence of partial annealing. Error bars are ±1 SE. After Murakami and Tagami (2004).

With respect to model (b), four samples <12 m away from the fault in the GSJ 750 borehole (GSJ-TH084, 093, 099 and 103) are from deformed and altered rock. In particular, three of them (GSJ-TH093, 099 and 103) belong to shear zones deformed and altered prominently compared to other parts of the borehole section (Tanaka et al. 1999). Because these three samples show greater degrees of zircon fission-track annealing (Fig. 9), it appears that the degree of annealing and deformation/alteration are related. A similar relationship occurs in the UG 500 borehole: the region of greatest zircon fission-track annealing corresponds to the zone of cataclastic deformation of rocks and dissolution of heavy minerals (i.e., the subzone 3 in Fig. 10; Tagami et al. 2001). These observations favor the model (b) as a heat source of the observed thermal anomalies. In case the frictional heat (1000 °C, 5 sec) is homogeneously and instantaneously dispersed via fluids to 3-m-wide zone, the temperature increase is estimated as ~1 °C using the heat calculation, giving no significant effects on fission tracks in zircon. Hence, it is likely that the heat was transferred by migration of hot fluids along the fault from the deep crustal interior. The paleo-depth of the rocks was probably shallower than ~8 km at the time of last cooling because the zircon fission-track system is closed at <~200 °C for a geological timescale, which corresponds to <~8 km by assuming a geothermal gradient of 23 °C/km.

Age constraints on the Nojima fault. The age data presented in the previous sections place important constraints on the history of Nojima fault. The age of the Nojima fault formation was previously estimated as ~1.2 Ma, on the basis of the spatial distribution of the Osaka Group in and around the Awaji Island (Murata et al. 2001). However, the ~56 Ma age of the

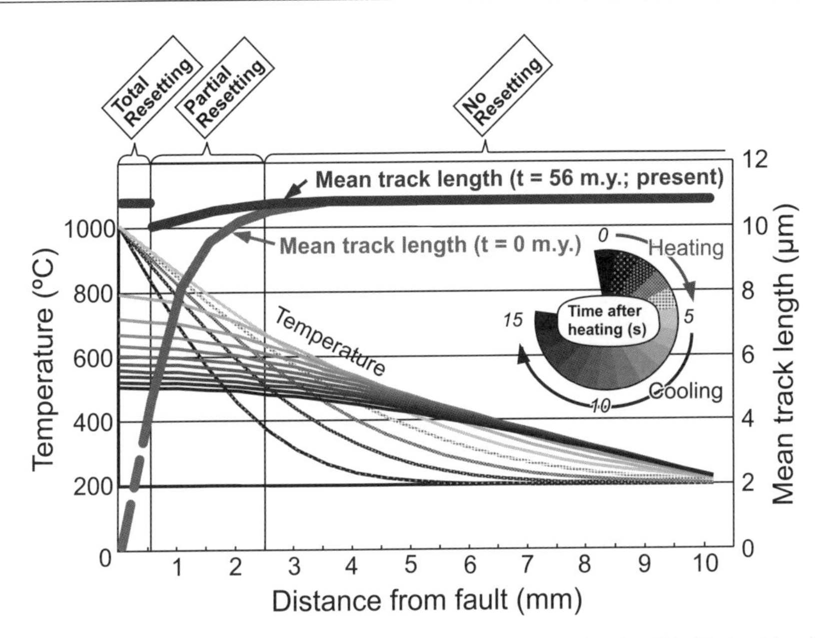

Figure 13. Results of one-dimensional heat conduction modeling for the plausible frictional heating of Nojima fault (Murakami and Tagami 2004). Fifteen temperature profiles are given, as examples, against distance from the fault, with each representing the time elapsed (i.e., 1–15 seconds) since the beginning of heating. Also shown are predicted mean fission-track length profiles at 56 Ma (t = 0 m.y.) and the present (t = 54 m.y.), suggesting that fission tracks are totally annealed at < ~0.5 mm from the fault plane. The following conditions are assumed: (1) thermal diffusivity of 2 mm^2s^{-1}, (2) environmental temperature of 200 °C at dry condition, (3) frictional heat of 1000 °C for 5 seconds at the fault plane, followed by heat dispersion due to thermal conduction, (4) zircon fission-track annealing kinetics after Tagami et al. [1998], along with the definition of partial resetting (annealing) zone mentioned in the previous section.

pseudotachylyte suggests that the Nojima fault had already been initiated at ~56 Ma, much older than the previous estimate. Furthermore, microscopic observation using thin sections suggests that the pseudotachylyte layer was formed during at least two faulting stages (Otsuki et al. 2003). Thus the ~56 Ma age is probably the time of final pseudotachylyte formation. It is likely, therefore, that the present Nojima fault system was formed by the Middle Quaternary reactivation of an ancient fault, which was already initiated at ~56 Ma at the interior of crust. This new reconstruction is consistent with the ~35 Ma secondary heating found near the fault in the GSJ 750 borehole.

Another important aspect is the temporal/spatial variation in reconstructed thermal histories among the three sites:

1. UG 500 borehole site (Toshima); ancient heating into ZPAZ, followed by cooling started at ~2.5 Ma within ~3 m from the fault on the hanging wall side,
2. GSJ 750 borehole (Hirabayashi); ancient heating into ZPAZ, followed by cooling started at ~35 Ma within ~25 m from the fault on both hanging wall and footwall sides,
3. Trench (Hirabayashi); no thermal overprints for a rock from the footwall ~10 mm away from the fault as well as one from the hanging wall ~ 30 mm from the fault, in contrast to the pseudotachylyte layer that was heated above ZPAZ and had last cooling at ~56 Ma.

These lines of evidence demonstrate the heterogeneity of the Nojima fault in terms of ancient heat transfer or dispersion via fluids and their resultant thermal anomalies. In addition, the thermal anomaly observed in the shallower part of the GSJ 750 (i.e. GSJ-TH001, 023, 040, 047) was not found for the UG 500. Because calcite layers and altered K-feldspars are widely observed in the four samples, they may represent a zone of ancient alteration possibly related to the past fault motion at Hirabayashi.

Summary

1. Zircon fission-track age and length data suggest that the GSJ 750 borehole rocks record the heat signatures at two different locations, one <25 m perpendicular to the fault and the other at the shallower depth (>25 m). The age of cooling after the heating event was estimated as ~35 Ma for samples within ~25 m from the fault by modeling fission-track data.
2. Zircon fission-track length data suggest that the UG 500 borehole rocks record a heat signature in the hanging wall <3 m from the fault. The age of cooling after the heating event was estimated as ~2.5 Ma by modeling fission-track data.
3. The plausible heat source of these thermal events is heat transfer via fluids along the fault zone from the deep crustal interior, on the basis of one-dimensional heat conduction modeling, the positive correlation between the degree of fission-track annealing and deformation/alteration of borehole rocks, and the in-situ heat dispersion calculation.
4. Zircon fission-track age and length data suggest that the zircon fission-track system of the pseudotachylyte layer at the Hirabayashi trench site was totally reset (or kept reset) and subsequently cooled at ~56 Ma. However, no signs of such cooling was found for a rock from the footwall ~10 mm away from the fault as well as one from the hanging wall ~30 mm from the fault. The spatial distribution of these data is approximately concordant with that of detectable zircon fission-track annealing predicted by thermal modeling of the frictional heating of fault motion.
5. The present Nojima fault system was probably formed by the Middle Quaternary reactivation of an ancient fault, which was already initiated at ~56 Ma at the interior of crust.
6. The Nojima fault shows a temporal/spatial variation in terms of the thermal anomalies recorded in the fault rocks, implying the heterogeneous heat transfer via fluids migrated along the fault zone from the deep crustal interior.

CONCLUDING REMARKS

As reviewed in the present chapter as well as in Tagami and O'Sullivan (this volume), the zircon fission-track thermochronology has been much advanced in the last decade and are being applied with success to unravel thermal histories of rocks formed under a variety of settings. The ongoing refinements of parameters of the track annealing model will offer further valuable basis to attain more reliable thermal modeling of crustal tectonics. In particular, further studies are strongly needed to better constrain the track annealing rate around the upper temperature limit of partial annealing zone on the geological timescale, as performed by Brix et al. (2002).

With respect to the future perspectives, we could extract more thermal history information from the same zircon sample using a combination of etchants (or etching conditions) because the use of different etchants can give rise to different track annealing rates. The revival of the acid mixture etchant (Krishnaswami et al. 1974), coupled with the alkalic mixture we use, may be a realistic candidate to test this hypothesis using the modern technique of horizontal confined track length analysis. The other potential way to extract more information is to apply a very weak etch to zircons and observe thin tracks under modern high-resolution microscopes. Chemical etching of latent tracks to the optical microscopic scale is indeed a convenient way of observing tracks, but may have obscured the detailed structure of partially annealed tracks. If we could develop this approach, I expect that the effective partial annealing zone will become significantly wider toward both low-temperature and high-temperature sides, offering a more robust tool for thermal history analysis.

ACKNOWLEDGMENTS

I thank members of the Kyoto University Fission Track Research Group who contributed to many issues in the present chapter. I also thank John Garver and Rich Stewart for their constructive reviews of the manuscript. This study has been supported by a Grant-in Aid (no. 12440137) as well as by a Grant-in-Aid for the 21st Century COE Program (Kyoto University, G3) from the Japanese Ministry of Education, Culture, Sports, Science and Technology.

REFERENCES

Ando M (2001) Geological and geophysical studies of the Nojima fault from drilling: an outline of the Nojima fault zone probe. Island Arc 10:206-214

Awata Y, Mizuno K, Sugiyama Y, Imura R, Shimokawa R, Okumura K, Tsukuda E (1996) Surface fault ruptures on the northwest coast of Awaji Island associated with the Hyogoken Nanbu earthquake of 1995, Japan (in Japanese with English abstract). J Seismol Soc Japan 49:113-124

Brandon MT, Roden-Tice MK, Garver JI (1998) Late Cenozoic exhumation of the Cascadia accretionary wedge in the Olympic Mountains, northwest Washington State. Bull Geol Soc Am 110:985-1009

Brix MR, Stockhert B, Seidel E, Theye T, Thomson SN, Kuster M (2002) Thermobarometric data from a fossil zircon partial annealing zone in high pressure-low temperature rocks of eastern and central Crete, Greece. Tectonophys 349:309-326

Cardwell RK, Chinn DS, Moore GF, Turcotte DL (1978) Frictional heating on a fault zone with finite thickness. Geophys J Royal Astronom Soc 52:525-530

Carter A (1990) The thermal history and annealing effects in zircons from the Ordovician of North Wales. Nucl Tracks Radiat Meas 17:309-313

Coyle DA, Wagner GA (1996) Fission-track dating of zircon and titanite from the 9101 m deep KTB: observed fundamentals of track stability and thermal history reconstruction. Abstr International Workshop on Fission Track Dating, Gent 1996, Gent, p 22

d'Alessio MA, Blythe AE, Burgmann R (2003) No frictional heat along the San Gabriel Fault, California; evidence from fission-track thermochronology. Geology 31:541-544

Davis DW, Williams IS, Krogh TE (2003) Historical development of zircon geochronology. Rev Mineral Geochem 53:145-181

Donelick RA, O'Sullivan PB, Ketcham RA (2005) Apatite fission-track analysis. Rev Mineral Geochem 58:49-94

Donelick RA, Miller DS (1991) Enhanced TINT fission track densities in low spontaneous track density apatites using ^{252}Cf-derived fission fragment tracks: a model and experimental observations. Nucl Tracks Radiat Meas 18:301-307

Fleischer RL, Price PB, Walker RM (1964) Fission-track ages of zircons. J Geophys Res 69:4885-4888

Fleischer RL, Price PB, Walker RM (1965) Effects of temperature, pressure, and ionization on the formation and stability of fission tracks in minerals and glasses. J Geophys Res 70:1497-1502

Foster DA, Kohn BP, Gleadow AJW (1996) Sphene and zircon fission track closure temperatures revisited; empirical calibrations from ^{40}Ar/^{39}Ar diffusion studies of K-feldspar and biotite. Abstr International Workshop on Fission Track Dating, Gent 1996, Gent, p 37

Fukahata Y, Matsuura M (2001) Correlation between surface heat flow and elevation and its geophysical implication. Geophys Res Lett 28:2703-2706

Galbraith RF (1990) The radial plot: Graphical assessment of spread in ages. Nucl Tracks Radiat Meas 17: 207-214

Galbraith RF, Laslett GM (1997) Statistical modeling of thermal annealing of fission tracks in zircon. Chem Geol 140:123-135

Gallagher K (1995) Evolving temperature histories from apatite fission-track data. Earth Planet Sci Lett 136: 421-435
Gallagher K, Brown R, Johnson C (1998) Fission track analysis and its applications to geological problems. Ann Rev Earth Planet Sci 26:519-572
Gamble JA, Wysoczanski RJ, Meighan IG (1999) Constraints on the age of the British Tertiary volcanic province from ion microprobe U-Pb (SHRIMP) ages for acid igneous rocks from NE Ireland. J Geol Soc London156:291-299
Garver JI (2003) Etching zircon age standards for fission-track analysis. Radiat Meas 37:47-53
Gleadow AJW (1980) Fission track age of the KBS Tuff and associated hominid remains in northern Kenya. Nature 284:225-230
Gleadow AJW (1981) Fission-track dating methods: What are the real alternatives? Nucl Tracks 5:3-14
Gleadow AJW, Hurford AJ, Quaife RD (1976) Fission track dating of zircon: improved etching techniques. Earth Planet Sci Lett 33:273-276
Green PF (1985) Comparison of zeta calibration baselines for fission-track dating of apatite, zircon and sphene. Chem Geol 58:1-22
Green PF, Hegarty KA, Duddy IR, Foland SS, Gorbachev V (1996) Geological constraints on fission track annealing in zircon. Abstr International Workshop on Fission Track Dating, Gent 1996, Gent, p 44
Harrison TM, Armstrong RL, Naeser CW, Harakal JE (1979) Geochronology and thermal history of the Coast Plutonic Complex, near Prince Rupert, British Columbia. Can J Earth Sci 16:400-410
Hasebe N, Mori S, Tagami T, Matsui R (2003) Geological partial annealing zone of zircon fission-track system: additional constraints from the deep drilling MITI-Nishikubiki and MITI-Mishima. Chem Geol 199: 45-52
Hasebe N, Tagami T (2001) Exhumation of an accretionary prism; results from fission track thermochronology of the Shimanto Belt, Southwest Japan. Tectonophys 331:247-267
Hasebe N, Tagami T, Nishimura S (1994) Towards zircon fission-track thermochronology: reference framework for confined track length measurements. Chem Geol 112:169-178
Hurford AJ (1986) Cooling and uplift patterns in the Lepontine Alps, south-central Switzerland, and an age of vertical movement on the Insubric fault line. Contrib Mineral Petrol 92:413-427
Hurford AJ (1990) Standardization of fission track dating calibration: recommendation by the Fission Track Working Group of the I. U. G. S. Subcommission on Geochronology. Chem Geol 80:171-178
Hurford AJ, Gleadow AJW, Naeser CW (1976) Fission track dating of pumice from the KBS Tuff, East Rudolf, Kenya. Nature 263:738-740
Hurford AJ, Green PF (1983) The zeta age calibration of fission-track dating. Isot Geosci 1:285-317
Kamp PJJ, Green PF, White SH (1989) Fission track analysis reveals character of collisional tectonics in New Zealand. Tectonics 8:169-195
Kasuya M, Naeser CW (1988) The effect of α-damage on fission-track annealing in zircon. Nucl Tracks Radiat Meas 14:477-480
Kikuchi M, Kanamori H (1996) Rupture process of Kobe, Japan, earthquake of determined Jan. 17, 1995, from teleseismic body waves. J Phys Earth 44:429-436
Kitajima T, Kobayashi Y, Ikeda R, Iio Y, Omura K (1998) Heat flow measurement for the borehole at Hirabayashi, Awaji Island (in Japanese). Gekkan Chikyuu 21:108-113
Kohn BP, Pillans B, McGlone, MS (1992) Zircon fission track ages for middle Pleistocene Rangitawa tephra, New Zealand: Stratigraphic and paleoclimatic significance. Palaeogeog Palaleoclim Palaeoecol 95:73-94
Kowallis BJ, Heaton JS, Bringhurst K (1986) Fission-track dating of volcanically derived sedimentary rocks. Geology 14:19-22
Krishnaswami S, Lal D, Prabhu N (1974) Characteristics of fission tracks in zircon: applications to geochronology and cosmology. Earth Planet Sci Lett 22:51-59
Laslett GM, Galbraith RF (1996) Statistical modeling of thermal annealing of fission tracks in apatite. Geochim Cosmochim Acta 60:5117-5131
Laslett GM, Green PF, Duddy IR, Gleadow AJW (1987) Thermal annealing of fission tracks in apatite, 2. A quantitative analysis. Chem Geol 65:1-13
Lee JKW (1994) Multipath diffusion and the influence of microstructure on closure temperatures and cooling ages. Abstr 8th Int Conf Geochronol Cosmochronol Isot Geol, Berkeley, p 189
Lo CH, Lee JKW, Onstott TC (2000) Argon release mechanisms of biotite in vacuo and the role of short-circuit diffusion and recoil. Chem Geol 165:135-166
Matsuura S, Tagami T (2005) Effects of radiation damages on fission-track annealing in zircon: experimental assessment by laboratory isochronal annealing. to be submitted to Chem Geol
McDougall I, Watkins RT (1985) Age of hominoid-bearing sequence at Buluk, northern Kenya. Nature 318: 175-178
Mizuno K, Hattori H, Sangawa A, Takahashi Y (1990) Geology of the Akashi district, quadrangle-series (in Japanese with English abstract). Geol Surv Japan, Tsukuba, scale 1:50,000, 90 p.

Murakami M, Tagami T (2004) Dating pseudotachylyte of the Nojima fault using the zircon fission-track method. Geophys Res Lett 31:10.1029/2004GL020211

Murakami M, Tagami T, Hasebe N (2002) Ancient thermal anomaly of an active fault system: Zircon fission-track evidence from Nojima GSJ 750 m borehole samples. Geophys Res Lett 29:10.1029/2002GL015679

Murakami M, Yamada R, Tagami T (2005) Short-term annealing characteristics of spontaneous fission tracks in zircon: A qualitative description. submitted to Chem Geol

Murata A, Takemura K, Miyata T, Lin A (2001) Quaternary vertical offset and average slip rate of the Nojima Fault on Awaji Island, Japan. Island Arc 10:360-367

Naeser CW (1969) Etching fission tracks in zircons. Science 165:388

Naeser CW, Izett GA, Obradovich JD (1980) Fission-track and K-Ar ages of natural glasses. US Geol Surv Bull 1489:1-31

Naeser CW, Izett GA, Wilcox RE (1973) Zircon fission-track ages of Pearlette family ash beds in Meade County, Kansas. Geology 1:187-189

Nishida T, Takashima Y (1975) Annealing of fission tracks in zircons. Earth Planet Sci Lett 27:257-264

Otsuki K, Monzawa N, Nagase T (2003) Fluidization and melting of fault gouge during seismic slip: Identification in the Nojima fault zone and implications for focal earthquake mechanisms. J Geophys Res 108:10.1029/2001JB001711

Parry WT, Bunds MP, Bruhn RL, Hall CM, Murphy JM (2001) Mineralogy, $^{40}Ar/^{39}Ar$ dating and apatite fission track dating of rocks along the Castle Mountain Fault, Alaska. Tectonophys 337:149-172

Rahn MK, Brandon MT, Batt GE, Garver JI (2004) A zero-damage model for fission-track annealing in zircon. Am Mineral 89:473-484

Reiners PW (2005) Zircon (U-Th)/He thermochronometry. Rev Mineral Geochem 58:151-179

Scholz CH (1996) Faults without friction? Nature 381:556-557

Seward D (1979) Comparison of zircon and glass fission-track ages from tephra horizons. Geology 7:479-482

Seward D, Kohn BP (1997) New zircon fission-track ages from New Zealand Quaternary tephra: an interlaboratory experiment and recommendations for the determination of young ages. Chem Geol 141:127-140

Seward D, Mancktelow NS (1994) Neogene kinematics of the central and western Alps: Evidence from fission-track dating. Geology 22:803-806

Shin SC, Park KS (1989) Determination of low levels of uranium impurity by fission track registration using mica and polycarbonate detectors. Nucl Tracks Radiat Meas 16:271-274

Sorkhabi RB (1993) Time-temperature pathways of Himalayan and Trans-Himalayan crystalline rocks: A comparison of fission-track ages. Nucl Tracks Radiat Meas 21: 535-542

Spikings RA, Winkler W, Seward D, Handler R (2001) Along-strike variations in the thermal and tectonic response of the continental Ecuadorian Andes to the collision with heterogeneous oceanic crust. Earth Planet Sci Lett 186:57-73

Sumii T, Tagami T, Nishimura S (1987) Anisotropic etching character of spontaneous tracks in zircon. Nucl Tracks Radiat Meas 13:275-277

Tagami T (1987) Determination of zeta calibration constant for fission track dating. Nucl Tracks Radiat Meas 13:127-130

Tagami T, Carter A, Hurford AJ (1996) Natural long-term annealing of the zircon fission-track system in Vienna Basin deep borehole samples: constraints upon the partial annealing zone and closure temperature. Chem Geol 130:147-157

Tagami T, Galbraith RF, Yamada R, Laslett GM (1998) Revised annealing kinetics of fission tracks in zircon and geological implications. In: Advances in Fission-track Geochronology. Van den haute P, De Corte F (eds) Kluwer academic publishers, Dordrecht, The Netherlands, p 99-112

Tagami T, Hasebe N, Kamohara H, Takemura K (2001) Thermal anomaly around the Nojima fault as detected by fission-track analysis of Ogura 500 m borehole samples. Island Arc 10:457-464

Tagami T, Ito H, Nishimura S (1990) Thermal annealing characteristics of spontaneous fission tracks in zircon. Chem Geol (Isot Geosci Sect) 80:159-169

Tagami T, Lal N, Sorkhabi RB, Ito H, Nishimura S (1988b) Fission track dating using external detector method: A laboratory procedure. Mem Fac Sci Kyoto Univ 53:14-30

Tagami T, Lal N, Sorkhabi RB, Nishimura S (1988a) Fission track thermochronologic analysis of the Ryoke Belt and the Median Tectonic Line, Southwest Japan. J Geophys Res 93:13705-13715

Tagami T, Murakami M (2005a) Ancient thermal anomaly of an active Nojima fault: additional constraints from zircon fission-track analysis of University Group 500 m borehole. submitted to Tectonophys

Tagami T, Murakami M (2005b) Zircon fission-track thermochronology of the Nojima fault zone, Japan. Am Assoc Petrol Geol Spec Paper in press

Tagami T, O'Sullivan PB (2005) Fundamentals of fission-track thermochronology. Rev Mineral Geochem 58: 19-47

Tagami T, Shimada C (1996) Natural long-term annealing of the zircon fission track system around a granitic pluton. J Geophys Res 101:8245-8255

Takahashi Y (1992) K-Ar ages of the granitic rocks in Awaji Island with an emphasis on timing of mylonitization (in Japanese with English abstract). Gankou 87:291-299

Tanaka H, Higuchi T, Tomida N, Fujimoto K, Ohtani T, Ito H (1999) Distribution, deformation and alteration of fault rocks along the GSJ core penetrating the Nojima Fault, Awaji Island, Southwest Japan (in Japanese with English abstract). J Geol Soc Japan 105:72-85

Tanaka H, Hinoki S, Kosaka K, Lin A, Takemura K, Murata A, Miyata T (2001) Deformation mechanisms and fluid behavior in a shallow, brittle fault zone during coseismic and interseismic periods: Results from drill core penetrating the Nojima Fault, Japan. Island Arc 10:381-391

Tanaka H, Tomida N, Sekiya N, Tsukiyama Y, Fujimoto K, Ohtani T, Ito H (2000) Distribution, deformation and alteration of fault rocks along the GSJ core penetrating the Nojima Fault, Awaji Island, Southwest Japan. In: International Workshop of the Nojima Fault Core and Borehole Data Analysis. Ito H et al. (eds) Geol Surv Japan, Tsukuba, p 81-101

Viola G, Mancktelow NS, Seward D, Meier A, Martin S (2003) The Pejo fault system; an example of multiple tectonic activity in the Italian Eastern Alps. Geol Soc Am Bull 115:515-532

Wagner GA, Van den haute P (1992) Fission-track Dating. Kulwer Academy, Norwell, Massachusetts

Watanabe K (1988) Etching characteristics of fission tracks in Plio-Pleistocene zircon. Nucl Tracks Radiat Meas 15:171-174

Watanabe K, Izawa E, Kuroki K, Honda T, Nakamura H (1991) Detection of confined ^{238}U fission tracks in minerals and its application to geothermal geology. Annual Rep Tandem Accel Lab (Kyushu Univ) 3: 151-155

Yamada K, Tagami T, Shimobayashi N (2003) Experimental study on hydrothermal annealing of fission tracks in zircon. Chem Geol 201:351-357

Yamada R, Tagami T, Nishimura S (1993) Assessment of overetching factor for confined fission track length measurement in zircon. Chem Geol (Isot Geosci Sect) 104:251-259

Yamada R, Tagami T, Nishimura S (1995a) Confined fission-track length measurement of zircon: assessment of factors affecting the paleotemperature estimate. Chem Geol (Isot Geosci Sect) 119:293-306.

Yamada R, Tagami T, Nishimura S, Ito H (1995b) Annealing kinetics of fission tracks in zircon: an experimental study. Chem Geol (Isot Geosci Sect) 122:249-258

Yamada R, Yoshioka T, Watanabe K, Tagami T, Nakamura H, Hashimoto T, Nishimura S (1998) Comparison of experimental techniques to increase the number of measurable confined fission tracks in zircon. Chem Geol (Isot Geosci Sect) 149:99-107

Yamamoto Y, Kurita H, Matsubara T (2000) Eocene calcareous nannofossils and dinoflagellate cysts from the Iwaya Formation in Awajishima Island, Hyogo Prefecture, southwest Japan, and their geologic implications (in Japanese with English abstract). J Geol Soc Japan 106:379-382

Yamano M, Goto S (2001) Long-term temperature monitoring in a borehole drilled into the Nojima Fault, southwest Japan. Island Arc 10:326-335

Yoshioka T, Tagami T, Nishimura S (1994) Experimental technique to increase the TINCLEs in zircon (in Japanese). Fission Track News Lett 7:50

Zaun PE, Wagner GA (1985) Fission-track stability in zircons under geological conditions. Nucl Tracks 10: 303-307

Zeitler PK, Tahirkheli RAK, Naeser CW, Johnson NM (1982) Unroofing history of a suture zone in the Himalaya of Pakistan by means of fission-track annealing ages. Earth Planet Sci Lett 57:227-240

Reviews in Mineralogy & Geochemistry
Vol. 58, pp. 123-149, 2005

Fundamentals of Noble Gas Thermochronometry

T. Mark Harrison

Research School of Earth Sciences
The Australian National University
Canberra, A.C.T. 0200, Australia
director.rses@anu.edu.au

Peter K. Zeitler

Department of Earth and Environmental Sciences
Lehigh University
Bethlehem, Pennsylvania 18015, U.S.A.
peter.zeitler@lehigh.edu

INTRODUCTION

The ideal geochronometer would be a universally stable phase that quantitatively retains both parent and daughter isotopes. Though a few mineral systems such as zircon U-Pb dating come reasonably close to this ideal, most minerals are incompletely retentive of daughter-product nuclides under crustal conditions. The mechanisms by which the daughter product can be lost from minerals include dissolution–reprecipitation reactions (e.g., salt; Obradovich et al. 1982), recrystallization (e.g., micas undergoing deformation; Chopin and Maluski 1980), and diffusive loss (e.g., ^{40}Ar degassing of K-feldspar; Foland 1974). The latter mechanism is perhaps the most common source of discrepancy between a radiometric mineral date and the age of the rock from which it formed.

Geochronologists have learned to turn this non-ideal behavior to their advantage and we now understand that most mineral ages from exhumed crustal rocks act in effect as kinetic thermometers sensitive to geologically-induced thermal effects. Such apparent ages are a measure of the temperature range over which daughter product ceased to be lost from a crystal, with intracrystalline diffusion usually acting as the rate-limiting process.

Consider the case in which a mineral sample containing a radioactive parent element experiences a complex thermal evolution, possibly involving heating as well as cooling. Within the sample, daughter product is continually produced by radioactive decay and lost by diffusion at natural boundaries. Although random at the scale of an individual atom, both diffusion and radioactive decay are highly predictable processes over longer and larger scales involving many particles. Coupled with the strong temperature dependence of diffusion, the elegant mathematics of the production-diffusion relationship make it possible to recover information about the thermal history experienced by such a sample, simply by knowing the amount of daughter product remaining in the mineral following cooling, or even better, by knowing the distribution of daughter product within the mineral. If one can justify the assumption of a monotonic cooling history of simple form, it is possible to assign a specific temperature to a bulk mineral date. By interpolating between several such data, a good approximation of the cooling history can be obtained. In the case of thermal histories involving reheating, if sufficient constraints are available it can be possible to estimate the magnitude or duration of a thermal event. Because natural thermal variations may be too subtle to be revealed by the

1529-6466/05/0058-0005$05.00 DOI: 10.2138/rmg.2005.58.5

interpolation method mentioned, the preferred method is to harness knowledge of the internal concentration distributions where possible (see Harrison et al. 2005). Collectively, such approaches are termed thermochronometry (or thermochronology) (Berger and York 1981). Since geodynamic processes alter the distribution of heat in the crust, the thermal record preserved as isotopic variations in minerals can provide valuable insights regarding timing and rates of tectonic and surficial processes.

The tendency for minerals to lose daughter isotope relative to its parent is due in part to the nature of the transmutations caused by radioactive decay. For example, all three forms of beta decay result in the conversion of the parent element into adjacent isobars, with either one extra (β^+) or one fewer (β^-, electron capture) proton. Thus ^{40}K (an alkaline metal) decays either to ^{40}Ca (an alkaline earth) or ^{40}Ar (an inert gas). Both daughter elements are geochemically incompatible in a host mineral that contains potassium as an essential structural constituent. In the case of the U-Th/He system, ^{4}He is produced by intermediate, radioactive daughter isotopes during the series decay of ^{238}U, ^{235}U, and ^{232}Th into isotopes of Pb. For both these noble-gas systems, incompatibility is enhanced by the inability of inert gases to chemically bond in silicates and so at elevated temperatures in the crust, Ar and He will tend to be lost from potassium- and uranium-bearing minerals, resulting in K-Ar and U-He ages much younger than that of mineral formation. Indeed, for most minerals, quantitative noble-gas retention occurs only in the outer ~0.2% of the Earth, explaining why ^{40}Ar is the third most abundant gas in the atmosphere (He is sufficiently light that it rapidly achieves escape velocity and is lost from the atmosphere on short timescales). In general we expect that under crustal conditions in minerals of appropriate chemistry, the K-Ar and U-Th/He decay systems will behave as diffusively controlled thermochronometers.

In this chapter we focus on the K-Ar and U-Th/He decay systems and show how thermochronological data can be understood using solutions to the radioactive decay and diffusion equations. We review the fundamental principles of noble gas thermochronology with the intention of providing a basic grounding for those new to the field. Recent developments and advances in the field are the main subject of this volume, and references to recent work and discussion about ongoing controversies can be found in the relevant chapters.

BASICS OF NOBLE-GAS GEOCHRONOLOGY

K-Ar and ^{40}Ar/^{39}Ar systematics and analysis

The general age equation is:

$$t = \frac{1}{\lambda}\ln\left(1+\frac{D}{P}\right) \tag{1}$$

where D is the daughter product, P is the parent, λ is the decay constant, and t is age. Because ^{40}K decays to both ^{40}Ar and ^{40}Ca, we must modify the K-Ar age equation to account for the fact that only about one in ten ^{40}K decays yield ^{40}Ar. This is done by dividing the total decay constant for ^{40}K ($\lambda = 5.543 \times 10^{-10}$/yr) by the partial decay constants to the ^{40}Ar branch by electron capture to an excited state ($\lambda_e = 0.572 \times 10^{-10}$/yr) and electron capture to the ground state ($\lambda_e = 0.0088 \times 10^{-10}$/yr) (see McDougall and Harrison 1999). Thus the branching ratio is:

$$\frac{\lambda}{\lambda_e + \lambda_e'} = 9.54 \tag{2}$$

Substituting this expression in Equation (1) yields:

$$t = \frac{1}{\lambda}\ln\left(1+\frac{\lambda}{\lambda_e+\lambda_e'}\frac{^{40}\text{Ar}}{^{40}\text{K}}\right) = \frac{1}{\lambda}\ln\left(1+9.54\frac{^{40}\text{Ar}}{^{40}\text{K}}\right) \tag{3}$$

where $^{40}Ar/^{40}K$ is the present ratio of radiogenic ^{40}Ar to ^{40}K.

In the $^{40}Ar/^{39}Ar$ method, the sample to be dated is irradiated with fast neutrons to transform a proportion of the ^{39}K atoms to ^{39}Ar via the $^{39}K(n,p)^{39}Ar$ reaction. Following irradiation, the sample is placed in an ultrahigh vacuum system and heated to fusion to release argon which is then isotopically analyzed in a mass spectrometer. Following correction of the measured Ar isotope ratios for argon isotopes produced by interfering neutron reactions and atmospheric Ar (assuming that non-nucleogenic ^{36}Ar is associated with ^{40}Ar in a ratio equivalent to the modern atmospheric value of $^{40}Ar/^{36}Ar = 296$), a radiogenic $^{40}Ar/^{39}Ar$ ratio is calculated. Because this ratio reflects the sample $^{40}Ar/^{39}K$ value, it is therefore proportional to age (see Eqn. 9 below). This is because ^{39}Ar is a measure of the amount of ^{39}K in the sample, and the $^{40}K/^{39}K$ ratio is essentially constant in nature (Humayun and Clayton 1997). Rather than determining the absolute dose of fast neutrons the sample has received during irradiation, a standard sample of accurately known K-Ar age is irradiated together with the unknown, and the age of the unknown is then derived by comparison with the $^{40}Ar/^{39}Ar$ of the flux monitor standard.

Thus to derive the $^{40}Ar/^{39}Ar$ age equation, we first rearrange Equation (3) in terms of ^{40}Ar:

$$^{40}\text{Ar} = {}^{40}\text{K}\,\frac{\lambda_e+\lambda_e'}{\lambda}\left[(\exp\lambda t)-1\right] \tag{4}$$

The amount of ^{39}Ar that is produced from ^{39}K during neutron irradiation is given by:

$$^{39}\text{Ar} = {}^{39}\text{K}\,\Delta\;\phi(E)\,\sigma(E)\,dE \tag{5}$$

where ^{39}Ar is the number of atoms of ^{39}Ar produced from ^{39}K in the sample, ^{39}K is the original number of atoms of ^{39}K present, Δ is the duration of the irradiation, $\phi(E)$ is the neutron flux at energy E, and $\sigma(E)$ is the neutron capture cross section at energy E for the $^{39}K(n,p)^{39}Ar$ reaction. Combining Equations (4) and (5) we get:

$$\frac{^{40}\text{Ar}}{^{39}\text{Ar}} = \frac{^{40}\text{K}}{^{39}\text{K}}\,\frac{\lambda_e+\lambda_e'}{\lambda}\,\frac{\left[(\exp\lambda t)-1\right]}{\Delta\int\phi(E)\sigma(E)dE} \tag{6}$$

We then define the irradiation parameter, J, as:

$$J = \frac{^{39}\text{K}}{^{40}\text{K}}\,\frac{\lambda}{\lambda_e+\lambda_e'}\,\Delta\int\phi(E)\sigma(E)dE = \frac{(\exp\lambda t)-1}{\left(^{40}\text{Ar}/^{39}\text{Ar}\right)} \tag{7}$$

If the age of a standard is known from K-Ar age measurements, then J can be determined from Equation (7) by simply measuring the $^{40}Ar/^{39}Ar$ ratio of the standard following irradiation. Substituting Equation (7) in Equation (6) gives:

$$\frac{^{40}\text{Ar}}{^{39}\text{Ar}} = \frac{(\exp\lambda t)-1}{J} \tag{8}$$

Thus the $^{40}Ar/^{39}Ar$ age of an unknown is:

$$t = \frac{1}{\lambda}\ln\left(1+J\frac{^{40}\text{Ar}}{^{39}\text{Ar}}\right) \tag{9}$$

Note from Equation (9) that the ratio of daughter ^{40}Ar to parent ^{40}K (via the ^{39}Ar proxy) is measured in a single isotopic analysis, obviating the need for separate analyses of potassium and argon and thus overcoming problems of sample inhomogeneity. Another advantage is that an $^{40}Ar/^{39}Ar$ ratio can be measured more precisely and on smaller samples than a conventional K-Ar age. However, the major advantage of the $^{40}Ar/^{39}Ar$ method over the K-Ar method is that an irradiated sample can be heated in steps, starting at relatively low temperatures and eventually reaching fusion, permitting a series of apparent ages related to the gas released at that step to be determined on a single sample. This approach, known as the step-heating technique, provides a wealth of additional information that can provide insights into the distribution of ^{40}Ar in the sample relative to the distribution of ^{39}K (and thus ^{40}K).

In the ideal case, release of the argon in the vacuum system occurs by diffusion as the sample is progressively heated. Thus, for a sample that has retained its ^{40}Ar since crystallization, both ^{40}Ar and ^{39}Ar likely occur in similar lattice sites as they have both been derived from potassium (Fig. 1). If the two isotopes have similar transport behavior, they will be degassed in similar proportions thus yielding an essentially constant $^{40}Ar/^{39}Ar$ ratio and age in each gas fraction extracted. A plot of the apparent $^{40}Ar/^{39}Ar$ age for each step against cumulative proportion of ^{39}Ar released (termed an "age spectrum") will yield a flat release pattern called a plateau (Fig. 1a). However, a sample that has lost some of its ^{40}Ar, either during protracted residence at high temperature in the deep crust or during a thermal excursion, will have sites within its lattice that have different ratios of daughter ^{40}Ar to parent ^{40}K. During a step-heating experiment, such differences may be revealed by variations in the $^{40}Ar/^{39}Ar$ ratio measured on the gas fractions successively released from the sample, yielding a staircase-type age spectrum (Fig. 1b,c).

In principle, any potassium-bearing mineral or rock can be used for K-Ar or $^{40}Ar/^{39}Ar$ dating. In practice, dateable samples are limited to those in which potassium is an essential

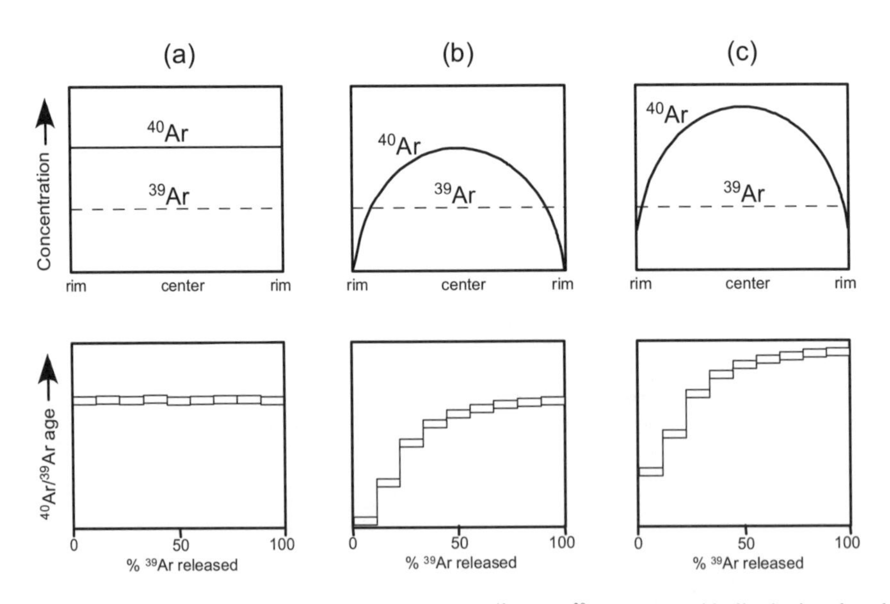

Figure 1. The top diagram portrays the concentration of ^{40}Ar and ^{39}Ar across an idealized mineral, and the lower diagram shows the associated $^{40}Ar/^{39}Ar$ age spectrum. a) Undisturbed subsequent to initial crystallization and rapid cooling yields a flat age spectrum; (b) diffusive loss of ^{40}Ar in recent times; (c) diffusive loss of ^{40}Ar during an ancient event with subsequent uniform accumulation of ^{40}Ar. The age of the diffusion loss event is given by the intercept at 0% ^{39}Ar release and a minimum age for crystal formation by the last measured age (after McDougall and Harrison 1999).

structural constituent, such as micas and feldspars. Because the method is based upon the accumulation of radiogenic ^{40}Ar (^{40}Ar*), the radiogenic signal becomes greater with increasing age, and therefore there is no older limit for the method. The main limitation occurs when the relative proportion of ^{40}Ar* to total ^{40}Ar in a sample tends to zero. As this ratio decreases, the error in its measurement will increase exponentially and come to be quite sensitive to analytical limitations (see McDougall and Harrison 1999).

^{40}Ar/^{39}Ar mineral thermochronometers

White micas. White micas typically contain high (8–10%) potassium concentrations and occur widely in peraluminous granitoids and metamorphic rocks (e.g., they are common in metapelites ranging from chlorite to ultra-high pressure facies). Muscovite is retentive of radiogenic Ar below ~400 °C (Purdy and Jäger 1976) and is generally less susceptible to contamination by uptake of excess radiogenic Ar than biotite. Phengite, however, often contains excess argon (e.g. de Jong 2003). Unlike biotite, muscovite and phengite commonly yield ^{40}Ar/^{39}Ar age spectra that can be interpreted in terms of distribution of radiogenic Ar in the crystals, as the intracrystalline distribution of ^{40}Ar appears to be retained during vacuum step heating.

Biotite. Biotite is virtually ubiquitous in granitoid rocks and common in medium to high-grade metapelites. Its typically high (7–8%) potassium content makes it a common candidate for K-Ar analysis, although it generally contains relatively high atmospheric Ar contents. Its retentivity for Ar is somewhat less than that of muscovite. While a biotite ^{40}Ar/^{39}Ar age will approximate the age of emplacement in the case of rapidly cooled igneous rocks, biotite ages from more slowly cooled igneous and metamorphic rocks will generally reflect the time of cooling. The inability to produce meaningful age spectra from biotites appears to result from its instability during heating *in vacuo,* which tends to destroy radiogenic-Ar diffusion gradients (McDougall and Harrison 1999).

K-feldspar. Because of their abundance in nature, especially in silicic rocks, and because of their relatively high potassium contents (up to 17%), alkali feldspars have been increasingly used for ^{40}Ar/^{39}Ar dating (e.g., Lovera et al. 1989), despite once being viewed as unretentive of argon at room temperature (Faure 1977). K-feldspar is the most prevalent potassium-rich silicate in the crust that is sufficiently stable during *in vacuo* heating to provide a reasonable expectation that natural ^{40}Ar* diffusion properties could be reproduced in laboratory step-heating experiments (Fitz Gerald and Harrison 1993). Although the overall retentivity of K-feldspar for ^{40}Ar* is typically lower than that of the micas, its characteristically complex microstructure results in a broad temperature range for partial argon retention. Because of its stability during *in vacuo* heating, K-feldspar is able to record nearly continuous thermal histories under middle to upper crustal conditions (Harrison et al. 2005).

Amphibole. Amphiboles are common in calc-alkaline igneous and metabasic rocks but are somewhat limited in K-Ar dating due to typically low (0.3–1%) potassium contents which make them susceptible to contamination with excess Ar or intergrowths with less retentive K-rich phases such as biotite (see McDougall and Harrison 1999). Hornblende is highly retentive of radiogenic Ar (Harrison 1981), although there appear to be a compositional dependence on Ar retention. The presence of fine-scale exsolution in metamorphic amphiboles, particularly involving micaceous phases, can have a significant effect on Ar retention in amphiboles.

Principal interpretive methods and analytical issues, ^{40}Ar/^{39}Ar

The age spectrum. As shown in Figure 1, plotting ^{40}Ar/^{39}Ar age as a function of cumulative ^{39}Ar release can in theory create a representation of the intracrystalline distribution of radiogenic Ar. Such a diagram is known as an age spectrum. Assuming that Ar is transported solely by diffusion and that ^{40}Ar and ^{39}Ar originate from similar sites and diffuse at similar rates, we can derive a simple quantitative model that predicts the age spectrum for a given internal distribution of ^{40}Ar (Fig. 5). Thus for a uniform distribution of ^{40}Ar, the model predicts a uniform $^{40}Ar/^{39}Ar$ ratio throughout the laboratory degassing process and a flat age spectrum is interpreted as a closed system since formation (Turner 1968). In many cases, notably the biotite micas, other possible degassing mechanisms (e.g., structural breakdown) dominate, potentially leading to flat age spectra in cases where initially heterogeneous $^{40}Ar/^{39}Ar$ distributions are homogenized (McDougall and Harrison 1999). It is also important to keep in mind that the values plotted on an age-spectrum diagram are not equivalent to a directly sampled concentration profile across a mineral: both the measured ^{40}Ar and ^{39}Ar have been released from the sample by progressive diffusion; indeed, age spectra will always tend to reach apparent plateaus late in the gas release as diffusion brings the ^{40}Ar and ^{39}Ar concentration profiles within the sample into similar forms.

Despite these cautions, for several years the practice of using the flatness of an age spectrum to distinguish between undisturbed and disturbed systems became unfortunately widespread, with any number of *ad hoc* definitions being proposed for what constitutes a plateau and therefore a "good' sample (see discussion in McDougall and Harrison 1999, p. 111). From a thermochronological perspective (as we will show below), a simple rule is that a sample exhibiting a flat release pattern will at best be boring, at worst be misleading, and in any case be of little use.

Total-fusion age. As an alternative to step-heating analysis, which can be quite time-consuming, one has the option of measuring the total gas content of a sample in a single step. The result is termed a total-fusion age, and is equivalent to the sample's conventional K-Ar age. This approach would be inadvisable for samples having recoverable diffusion information, but certainly in the case of biotite, which is notorious for yielding false plateaus even in cases where the mineral must contain diffusion-produced concentration gradients, the argument could be made that for equivalent cost and effort, analysis of a greater number of discrete samples far outweighs any benefits gained by step-heating.

Excess argon. The K-Ar method routinely assumes that all mineral samples are contaminated with non-radiogenic Ar, albeit with Ar of atmospheric composition as noted above. However, this will only be the case if the minerals' immediate surroundings represent a more compatible reservoir for Ar than the host phase (see Baxter et al. 2002). Thus in certain cases, samples can be contaminated with non-radiogenic Ar that greatly exceeds the atmospheric $^{40}Ar/^{36}Ar$ ratio resulting in anomalously old ages that in some cases exceed that of the Earth (Pankhurst et al. 1973; Harrison and McDougall 1981). Early on, it was believed that many age spectra thus contaminated were characterized by "saddle-shaped" behavior, in which high initial ages decrease to a minimum and then rise to high apparent ages in the last portion of the gas release. The high ages early in gas release were thought to be due to diffusive uptake of excess Ar ($^{40}Ar_E$) from a high P_{Ar} reservoir exterior to the grain with the minimum ages yielding an estimate of the true age. Harrison and McDougall (1981) suggested that the anomalously old ages late in release were due to excess Ar trapped in retentive lattice sites. It was subsequently noted that feldspars degassed using duplicate isothermal steps yielded saw-toothed age patterns (Harrison et al. 1993) that reflect decrepitation of Ar_E–bearing fluid inclusions. By linking the saw-tooth age pattern with Cl/K ratios calculated from nucleogenic Ar isotopes, a correction scheme can be devised permitting separation of $^{40}Ar_E$ from *in situ* radiogenic ^{40}Ar, at least for feldspars (Harrison et al. 1994).

Recoil. While we assume that the ^{39}Ar produced from ^{39}K during irradiation is distributed in a manner similar to the ^{40}K in a sample, the $^{39}K(n,p)^{39}Ar$ reaction results in ^{39}Ar nuclei being recoiled with a characteristic lengthscale of 0.08 μm (Turner and Cadogan 1974). Ultra-fine grained samples, such as clays or lunar basalts, exhibit effects of recoil redistribution of ^{39}Ar and can yield highly complex age spectra owing to depletion of ^{39}Ar from some phases and implantation into retentive catcher phases like olivine (Huneke and Smith 1978). However, age spectra of the type routinely used in thermochronometry (i.e., monomineralic separates with particle dimensions much greater than the recoil distance) are unlikely to be affected by recoil redistribution. Note however, that the characteristic recoil lengthscale of 0.08 μm limits the spatial resolution of diffusion studies of irradiated material to ~0.1 μm.

(U-Th)/He systematics and analysis

Shortly after the discovery of radioactivity, it was recognized that the accumulation of ^{4}He from U and Th α-decay could be used as a geochronometer (e.g., Strutt 1908). By the 1950's, however, it was generally recognized that most minerals were extremely "leaky" for He and the approach was essentially abandoned (Hurley 1954) until the method was re-evaluated from a thermochronological perspective (Zeitler et al. 1987).

In the minerals of greatest relevance to thermochronology, ^{4}He and several Pb isotopes are the stable products of the ^{238}U, ^{235}U, and ^{232}Th decay series (^{4}He production from ^{147}Sm decay is usually negligible). The equation for ^{4}He accumulation is:

$$^4He_{all} = 8\left(\frac{137.88}{(1+137.88)}\right)C_U\left(e^{\lambda_{238}t}-1\right)+7\left(\frac{1}{(1+137.88)}\right)C_U\left(e^{\lambda_{235}t}-1\right)+6C_{Th}\left(e^{\lambda_{232}t}-1\right) \quad (10)$$

where C_U and C_{Th} are, respectively, the concentrations of uranium and thorium, λ_{238} (1.55125×10^{-10} yr^{-1}), λ_{235} (9.8485×10^{-10} yr^{-1}), and λ_{232} (4.9475×10^{-11} yr^{-1}) are the relevant U and Th decay constants, and t is age. The expression, which can be solved recursively for age using the Newton-Raphson method, assumes secular equilibrium within the ^{238}U, ^{235}U, and ^{232}Th decay chains. This condition is met provided the host mineral is older than ca. 350 ka; in the case of petrologically old but very recently exhumed samples giving very young He ages, secular equilibrium in the decay series will long ago have been established, the young age being a reflection of ^{4}He loss, not decay-series parents.

Unlike ^{40}Ar/^{39}Ar dating, which is generally applied to rocks and minerals containing the major element potassium, U-Th/He dating involves analysis of accessory minerals in which the radioactive parent elements uranium and thorium are concentrated, but nevertheless present in only trace amounts. Several factors combine to leave helium dating a viable method. First, the lower abundances of U and Th relative to K are to some degree offset by the faster effective production of He compared to Ar: for a given concentration of parent, more than 20 times more ^{4}He is produced per decay of ^{238}U than ^{40}Ar per decay of ^{40}K. Next, while the noble gases have impressively low detection limits in general, the measurement of even small amounts of ^{4}He is straightforward and not complicated by isobaric interferences from hydrocarbons as can be the case for Ar isotopes. Finally, of greatest significance is that background levels of He in the atmosphere are very low, on the order of 1 ppm, compared to the ~1% of air that is ^{40}Ar. Thus, He ages do not require correction for an atmospheric component. In the considerable number of studies carried out in recent years, there is little evidence that geologically trapped, "excess" ^{4}He is an issue. However, ^{4}He is a crustal component that can locally reach high concentrations (e.g., gas wells) and there may be certain environments (e.g., active shear zones, fluid inclusions) in which excess ^{4}He may be present. In any event, given the great range in observed ^{4}He/^{3}He ratios, the magnitude of this ratio (~10^6), and the low abundance of natural ^{3}He, it would not be realistic to attempt corrections for trapped ^{4}He in the way that Ar ages are corrected for atmospheric Ar, and fortunately atmospheric contamination does not appear to be a problem.

Most laboratories today analyze for ^{4}He using vacuum extraction systems similar to that used for ^{40}Ar/^{39}Ar dating. Samples are heated in either a furnace or by laser, the released gases are purified, a ^{3}He spike is added, and the resulting mixture analyzed, in most cases by an inexpensive quadrupole mass spectrometer. Step-heating analysis is possible for the purpose of determining diffusion data from the release of ^{4}He, but there is no way to generate a reference helium isotope from U and Th in the way that ^{39}Ar is used in ^{40}Ar/^{39}Ar dating, and so age spectra cannot be measured. One practical matter unique to helium step-heating analysis is that the low-temperatures at which He is released from some minerals make the traditional double-vacuum furnaces used in many noble-gas laboratories inappropriate, as these furnaces are too slow to equilibrate at these temperatures. Instead, special furnaces need to be constructed that are more responsive at low temperature, like the projector-bulb system described by Farley et al. (1999).

(U-Th)/He mineral thermochronometers

Apatite. Apatite is the principal host of phosphorous in crustal rocks and is therefore virtually ubiquitous. Apatite typically contains 2–20 ppm U and, as other phosphates, tends to anneal radiation damage at low temperatures making it an appealing mineral for (U-Th)/He dating. Ironically, what makes apatite particularly interesting is the characteristic that led to the approach being abandoned in the middle 20th century—the purported "leakiness" of He in minerals. In revisiting applications of (U-Th)/He dating, Zeitler et al. (1987) focused on apatite as a phase likely to be unretentive of He. They proposed a "closure temperature" (see discussion below) of He in apatite of about 100 °C which was subsequently refined by further diffusion measurements to ~70 °C (Farley et al. 1996, 2002; Wolf et al. 1996). This temperature is substantially lower than all other thermochronometers and thus has the greatest sensitivity to changes occurring at the Earth's surface.

Zircon. Zircon is a very common accessory mineral found in most rock types, and the mineral has seen a great deal of use in geochronology, for U-Pb dating of high-temperature and rock-forming events, for provenance and protolith studies in sedimentary and metamorphic rocks, and for thermochronological studies using fission-track dating. Zircons typically contain 100's to 1000's of ppm of U and Th, and thus even very young zircons contain abundant ^{4}He. Zircon is more retentive of He than apatite (Reiners et al. 2002) and appears to overlap the retentivity of K-feldspar for Ar (Reiners et al. 2004).

Titanite (sphene). Titanite is another accessory mineral, fairly common in calc-alkaline igneous and metamorphic rocks, that contains 10's to 100's of ppm of U and Th. Work by Reiners et al. (1999) and Stockli and Farley (2004) have shown that titanite can give reliable and consistent ages, with the mineral having a retentivity for ^{4}He roughly on par with that of zircon. One drawback associated with titanite is that natural crystals often have irregular shapes that make it hard to identify fractured grains and hard to determine recoil-correction factors (see below).

Principal interpretive methods and analytical issues, (U-Th)/He

Bulk age. All helium ages are bulk ages that reflect the total integrated ^{4}He content of the sample. Work is underway to develop means of sampling He concentration gradients, but currently and for the immediate future, helium ages will be equivalent to K-Ar or total-fusion ^{40}Ar/^{39}Ar ages.

^{4}He/^{3}He spectra. Shuster et al. (2003) and Shuster and Farley (2003) describe a means of assessing ^{4}He diffusion kinetics in the context of a uniform distribution of ^{3}He, somewhat analogous to how ^{40}Ar diffusion can be referenced to that of ^{39}Ar. A crucial difference is that the ^{3}He is produced via spallation using proton bombardment. Because most of the major elements in the sample will serve as targets, the ^{3}He that is produced will tend to be uniformly distributed. However, there is no possibility of obtaining age information because the spallation helium is not produced from the U and Th that are the parents for the radiogenic ^{4}He in the sample. Though technically more challenging, this approach does show some promise in providing information about ^{4}He concentration gradients that will be difficult if not impossible to sample directly with laser techniques, given the fine spatial scale of microns over which these gradients are likely to exist.

***Alpha* (α)*-recoil*.** A potentially serious complication inherent to (U-Th)/He dating is that the kinetic energy imparted to α-particles during decay result in their being displaced many microns through the host mineral, leading to a spatial separation between parent and daughter. In the case of minerals whose size is similar to the ejection length-scale, this can result in the ejection of a substantial fraction of the daughter ^{4}He. Typically, the outermost ~20 μm of a crystal is affected, the average alpha-particle range varying slightly depending on parent (Farley et al. 1996). The solution is to estimate the fraction of ejected ^{4}He and make a correction to the observed age (assuming that implantation from adjacent grains is insignificant and that U and Th are uniformly distributed). The most commonly used approach is that of Farley et al. (1996) who provide expressions to calculate the fraction of α-particles retained in crystals of varying geometry (the "F_T" parameter). Of practical significance, Farley et al. (1996) found that the key element in controlling alpha ejection is the surface-to-volume ratio, and they provide an empirical expression between this ratio and F_T, allowing corrections to be readily made for more complex grain geometries for which a simple analytical solution is not available.

The need to correct helium ages for recoil raises some analytical constraints and issues. First, it is important to analyze intact grains, unless pairs of broken grains can be matched, since the correction procedure assumes an intact grain, all of whose surfaces have experienced recoil loss. Next, in the strictest sense, correction for α-recoil is only applicable to samples that have cooled quickly and never lost helium. For samples that have experienced a complex cooling history, the concentration gradient induced near grain ages by the recoil process will modify the concentration profiles produced by diffusion and thus the rate of diffusion itself (Meesters and Dunai 2002a). However, this will be a small effect, and for most samples which spend the greatest part of their lives accumulating a full complement of helium subsequent to cooling, the recoil correction makes sense and yields a far more meaningful result than leaving the age uncorrected.

***U and Th heterogeneity*.** U and Th distributions in accessory minerals tend to be complex. Zoning in apatite is not usually very strong but in zircon in particular, years of experience with U-Pb and fission-track dating have shown that zircons can display extremely complicated high-amplitude zoning variations that are not always of regular pattern and morphology (Meesters and Dunai 2002b). This can greatly complicate the recoil-loss behavior of a sample and render attempts at recoil correction almost worse than useless. Consider the case where a sample is strongly zoned such that a good fraction of its total U content is located within 20 μm of grain edges. Much of the helium generated in such a sample will be ejected from it, and an alpha-recoil correction made on the assumption of uniform U distribution will be inadequate. In practice, this problem can sometimes be overcome using replicate analyses, although if grains in a sample all have similar zoning structure, even replication may not help. Hourigan et al. (2005) describe a method of characterizing zoning patterns in zircons in order to make more appropriate alpha-recoil corrections.

DIFFUSION

Background

In noble-gas geochronology, it is generally assumed that the phase of interest is immersed in an 'infinite reservoir' of zero daughter product concentration and that the rate-determining diffusion process is volume diffusion through the crystal lattice, as opposed, say, to grain-boundary diffusion along crystallographic structures. In this case, no issues related to chemical potential arise and simple concentration is the parameter of interest. Noble-gas diffusion obeys

Fick's First Law: the rate of transfer of mass per unit area is proportional to the concentration gradient (Fick 1855). It is worth noting that Fick's First Law applies to the aggregate behavior of many diffusing particles. In detail, any one particle could move randomly in any direction, but if there is a concentration gradient present, this locally random behavior will result in a net movement of particles down the diffusion gradient. As an example, imagine an immaculate house in autumn surrounded by a garden filled with fallen leaves. As people tramp into and out of the house, they will tend to bring a leaf or two with them, and leaves will flow in both directions. However, because at the start there were more leaves outside than inside, there will be a net flux of leaves into the house. Should the homeowner surrender in despair and not intervene, eventually the concentration of leaves inside and out would become equal, and there would be no net flow, even as leaves continue to enter and exit the house.

The diffusion equation. The general problem of unsteady-state diffusion within a solid involves the prediction of the concentration distribution $C(x,y,z)$ within a solid as a function of the space coordinates and time, t. To derive an equation that can be solved for $C(x,y,z,t)$, conservation of mass and Fick's first law are applied to a differential control volume. The resulting expression is the diffusion equation:

$$\frac{\partial C}{\partial t} = D\left(\frac{\partial^2 C}{\partial x^2} + \frac{\partial^2 C}{\partial y^2} + \frac{\partial^2 C}{\partial z^2}\right) \quad (11)$$

where D is the diffusion coefficient. Solutions of the diffusion equation can be obtained for a number of initial and boundary conditions covering simple concentration distributions and diffusion geometries (Crank 1975). For example, using coordinate transformations, Equation (11) can be modified to describe radial flow in a sphere (see McDougall and Harrison 1999) and solved for the case of a sphere of radius r with an initially uniform concentration, C_0, held in an infinite reservoir of zero concentration. The concentration distribution ($\mathcal{R}$) is given by:

$$C = \frac{C_0 2r}{\pi \mathcal{R}} \sum_{n=1}^{\infty} \frac{(-1)^n}{n} \sin \frac{n\pi \mathcal{R}}{r} \times \exp\left(\frac{-n^2\pi^2 Dt}{r^2}\right) \quad (12)$$

Diffusion mechanisms

Atoms migrate within a solid through a series of random jumps between equilibrium lattice sites. Distortion of the lattice to permit a diffusion jump is affected via local thermal energy and thus the rate of diffusion increases with increasing temperature. Although the rate of this random migration is independent of chemical potential in dilute solutions, the presence of concentration differences results in a net flux down the gradient. When a boundary is encountered beyond which no return of the diffusing species is possible (e.g., the edge of the crystal, sub-grain planar defects), the randomizing process of diffusion results in the movement of mass from regions of high to lower concentration.

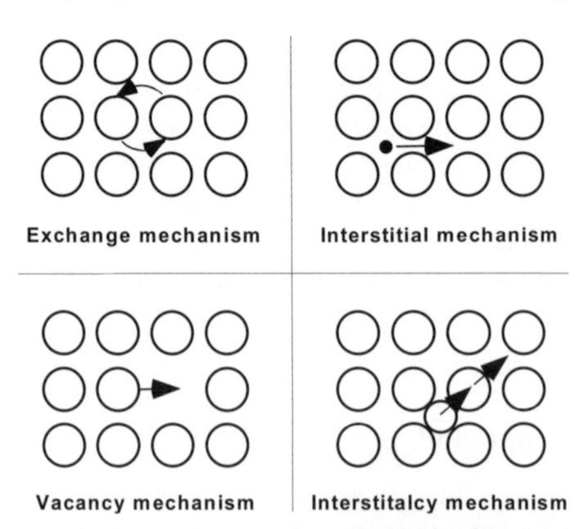

Figure 2. Schematic illustration of the four possible mechanisms of diffusion transport: Exchange, vacancy, interstitial, and interstitalcy.

Four possible transfer mechanisms for atomic diffusion are possible (Fig. 2): an exchange of adjacent atoms, an atom moving into a neighboring vacant lattice site, an interstitial atom sited in between normal lattice sites moving into a new interstitial location by squeezing past atoms in regular sites, and an interstitial atom displacing a normally sited atom into another interstitial location.

The Arrhenius relationship

Point defects form via thermal processes (intrinsic defects) and as a result of chemical impurities (extrinsic defects) which create vacancies in order to conserve charge. Above absolute zero temperature, there is a finite probability of an atom having sufficient local thermal energy to migrate from its current position to an adjacent site by the mechanisms shown in Figure 2. As temperature is raised, the probability of an atom in the Boltzmann distribution acquiring the threshold energy to overcome the potential barrier increases exponentially. Because both the rate of defect formation and migration are exponentially activated, the overall temperature dependence of the diffusion coefficient, D, is given by the Arrhenius relationship:

$$D = D_0 \exp\left(-\frac{E}{RT}\right) \tag{13}$$

where E is the activation energy, R is the gas constant, T is absolute temperature, and D_0 is the frequency factor. Units for D are typically given in cm^2/s.

By taking the base-10 logarithm of both sides of Equation (11) we obtain

$$\log D = \log D_0 - \frac{E}{2.303RT} \tag{14}$$

Note from Figure 3 that both the activation energy (E) and frequency factor (D_0) parameters can be extracted from a linear array of diffusion data on an Arrhenius-type plot. In this example, the logarithm of the diffusion coefficient D is plotted against the reciprocal absolute temperature. Equation (14) is in the form of an equation of a straight line, $y = b + mx$, where y is the log D coordinate, $E/(2.303R)$ is the slope of the line in Figure 3, $1/T$ is the x axis and $\log(D_0)$ is the y axis intercept of the line. Thus the slope of the line in Figure 3 is proportional to the activation energy and the y axis intercept is the logarithm of the frequency factor.

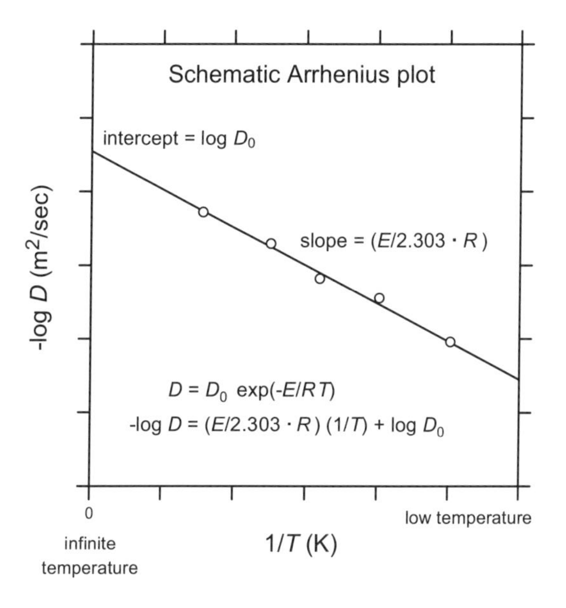

Figure 3. Schematic plot illustrating the relationship of the Arrhenius parameters in terms of the equation of a straight line. The circles represent experimental determinations of diffusion coefficient.

While strongly dependent on temperature, diffusion is predicted to decrease as pressure increases due to both a drop in number of vacancies in response to the crystal relieving internal pressure and the extra work diffusing atoms must perform against the confining pressure to distort the lattice to make a diffusion jump. The modified Arrhenius equation is:

$$D = D_0 \exp\left(-\frac{(E+PV)}{RT}\right) \tag{15}$$

where P is the confining pressure and V is the activation volume (e.g., Harrison et al. 1985; Giletti and Tullis 1977).

Episodic loss

Equation (12), the expression for the concentration distribution within a sphere, is of limited use in $^{40}Ar/^{39}Ar$ or U-Th/He dating as we lack the analytical resolution needed to directly image isotope distributions *in situ*. However, in cases where we can estimate the amount of uptake or loss into or from the solid, solutions relating the degree to which the system has approached equilibrium as a function of the Fourier number ($Dt/r^2 = Fo$) can be useful.

To obtain expressions for the fractional loss (or uptake), the concentration remaining after time t (determined by integrating the concentration distribution between $x = 0$ to $x = r$ at t) is subtracted from the initial uniform concentration, C_0, and this remainder is then normalized to C_0. This fraction represents the approach from zero loss ($f = 0$) at $t = 0$ to total equilibration at $t = \infty$. Thus fractional loss, f, is

$$f = \frac{M_0 - M_t}{M_0} = 1 - \frac{6}{\pi^2}\sum_{n=1}^{\infty}\frac{1}{n^2}\exp\left(\frac{-Dn^2\pi^2 t}{r^2}\right) \tag{16}$$

Graphical results of these equations for f vs. Dt/r^2 for spherical, cylindrical, and plane sheet geometries are shown in Figure 4; note for example, that a value of Dt/r^2 of 0.018 is required to produce a 40% fractional loss from a sphere.

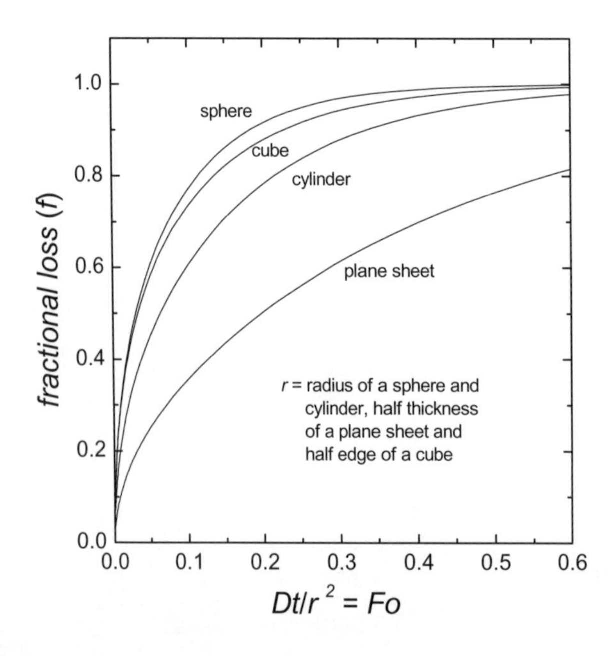

Figure 4. Relationship between fractional loss and the Fourier number (Dt/r^2) for spherical, cylindrical, plane sheet, and cubic geometries.

Coupling fractional loss equations with the Arrhenius relationship

Solutions of the diffusion equation in terms of fractional loss (f) yield expressions in terms of the Fourier number, Dt/r^2 (= Fo) (Fig. 4). Thus a numerical value of the Fo number can be calculated for any estimate of f. Substituting the Arrhenius relationship (Eqn. 13) into the definition of the Fo number yields an expression that relates fractional loss to peak temperature T in a square-pulse type thermal event of duration t:

$$\frac{E}{RT} = \ln\left(\frac{Fo^{-1} t D_0}{r^2}\right) \tag{17}$$

The principal limitation of this solution is that D lacks time dependence, restricting the equation to describing the geological uninteresting case of an isothermal history. However, by substituting:

$$\zeta = \int_0^t \left(\frac{D(t)}{r^2}\right) dt \tag{18}$$

for (Fo^{-1} t) we can then handle any arbitrary thermal history (Brandt 1974; Dodson 1975; Lovera et al. 1989). Understanding that this too-often unheralded substitution can be made is of critical importance in thermochronology. For example, assuming a temperature history of the form $t = 1/T$ permits Equation (17) to be integrated to yield an expression of the form

$$\frac{E}{RT_c} = \ln\left(\frac{A\tau D_0}{r^2}\right) \tag{19}$$

where A is a geometric constant and τ is a constant related to the cooling rate and activation energy (Dodson 1973). The parameter T_c is referred to as the closure temperature. It is in effect the characteristic temperature of retention associated with the age of the bulk geochronological system.

Calculation of age spectra resulting from episodic loss

While we are not yet able to clearly image $^{40}Ar/^{39}Ar$ distributions directly, the $^{40}Ar/^{39}Ar$ age spectrum method offers the potential to make observations at a fine spatial scale. Assuming the presence of a single site for Ar, synthetic $^{40}Ar/^{39}Ar$ age spectra based on diffusion from simple diffusion geometries can be constructed. In the case of radial diffusion from a sphere, the theoretical $^{40}Ar/^{39}Ar$ release pattern (Turner 1968) can be determined by evaluating the relative flux of ^{40}Ar with respect to ^{39}Ar.

For the case where: C_0 is the concentration of ^{40}Ar produced prior to outgassing, Δt_1 is duration of natural outgassing, Δt is the duration of laboratory outgassing, and C_{39} is the unit value of $^{39}Ar_K$ concentration, the $^{40}Ar/^{39}Ar$ ratio for any fractional loss of ^{39}Ar due to a laboratory heating Δt, can be obtained by dividing the function that describes the $^{40}Ar^*$ flux from the sphere by the ^{39}Ar flux equation yielding:

$$\frac{^{40}Ar^*}{^{40}Ar_K} = \frac{C_0}{C_{39}} \left\{ \frac{\left[\left[\sum_{n=1}^{\infty} \exp\left(\frac{-n^2\pi^2 D(\Delta t_1 + \Delta t')}{r^2}\right)\right]\right]}{\left[\sum_{n=1}^{\infty} \exp\left(\frac{-n^2\pi^2 D(\Delta t')}{r^2}\right)\right]} \right\} \tag{20}$$

Synthetic $^{40}Ar/^{39}Ar$ ratios derived from this equation corresponding to 60 and 80% loss are shown in Figure 5 plotted against cumulative % ^{39}Ar. Also plotted in Figure 5 are theoretical age spectra for an infinite cylinder geometry.

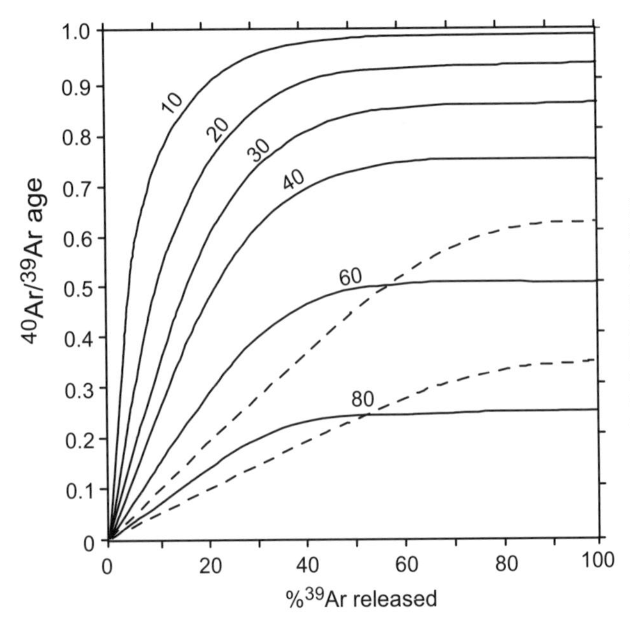

Figure 5. Theoretical $^{40}Ar/^{39}Ar$ age spectra for a sample of plane sheet geometry (solid lines) that has been outgassed today by 10, 20, 30, 40, 60, and 80% of $^{40}Ar^*$ present. For comparison, curves for 60 and 80% loss from spheres of uniform radius are shown as dashed lines.

The basis of this model can be seen in Figure 4. Note that the flux of ^{40}Ar or ^{39}Ar for any value of Dt/r^2 is given by the tangent to the curve. Because of the pronounced curvature of the function representing the sphere (due to the center of mass of the gas being closer to the diffusion boundary compared to the sheet model), a sample that has experienced even minor ^{40}Ar loss will not reach a uniform $^{40}Ar/^{39}Ar$ value until very late in the degassing. This is because even the small offset in Dt/r^2 between the ^{40}Ar and ^{39}Ar distributions places the ^{40}Ar gradient well up the steeply convex curve thus preventing the two tangents from becoming coincident until the sample is essentially outgassed. In contrast, the more linear nature of the plane sheet geometry curve allows the ^{40}Ar and ^{39}Ar gradients to rapidly reach a common value.

Closure temperature

Introduction. In the previous section, an expression was derived relating the Arrhenius relationship with solutions for fractional loss (for a constant D) to yield an expression that relates temperature and heating duration to the degree of isotopic equilibration (Eqn. 17). Of far greater importance is the case in which a sample experiences a complex thermal history, possibly involving heating as well as cooling, during which the daughter product is continually produced by radioactive decay and lost by diffusion. Although a closure temperature is only relevant to samples which that monotonically cooled from high to low temperatures, this condition holds for many samples, and in addition, closure temperature serves as a useful shorthand for describing the retentivity of a system.

Solutions for time-dependent D. Recall that by linking the dimensionless parameter Fo ($\equiv Dt/r^2$) obtained from the fractional loss expressions with the Arrhenius equation yielded a relationship (Eqn. 17) from which the maximum temperature of a square pulse thermal history could be calculated. Although this is of little geological utility, this simple expression does embody the basic form of the thermochronological relationship in that a characteristic temperature of the system is related to the diffusivity and time. The principal limitation of this relationship is that D lacks time dependence, restricting the equation to describing only a square-pulse thermal history. However, by substituting a time-varying diffusion coefficient defined as

$$\zeta = \int_0^t \left(\frac{D(T)}{r^2} \right) dt \qquad (21)$$

into the diffusion equation modified to account for radiogenic ingrowth, then any arbitrarily thermal history can be evaluated yielding an equation of the form:

$$\frac{E}{RT} = \ln\left(\frac{A\tau D_0}{r^2} \right) \qquad (22)$$

where A is a constant related to the geometry of diffusion (e.g., radial diffusion in a sphere) and the nature of the thermal history, τ is a function related to the form of the integrated thermal history, and T is the characteristic temperature of the system. For example, by assuming that the temperature history proceeds linearly in $1/t$, Equation (21) can be integrated to yield a simple relationship for τ that in turn permits a "closure temperature" to be calculated.

Monotonic cooling. Minerals originating at deep crustal levels undergo a transition during slow cooling from temperatures that are sufficiently high that the daughter escapes as fast as it is formed, to temperatures sufficiently low that diffusion is negligible and the retention of radiogenic isotopes by the mineral can be thought of as complete. Between these two states there is a continuous transition over which accumulation eventually balances loss, then exceeds it. Figure 6 (Dodson 1973) shows how a calculated age in this situation relates to the transition interval—the apparent age is the extrapolation of the total accumulation part of the curve to the time axis, which implicitly corresponds to an apparent temperature at which the bulk system became closed. In spite of the fact that closure does not occur at this time, the prior accumulation and subsequent loss balance to give thermal significance to the apparent age. Thus the general problem of isotopic closure during slow cooling is to quantify the shifting balance between accumulation and loss of daughter product. Assuming that a cooling history proceeds linearly in $1/T$, the decrease in the diffusion coefficient has the form of a simple exponential decay (Dodson 1973) with a time constant, τ, which corresponds to the time taken for D to diminish by a factor e^{-1} (i.e., to drop to 37% of its previous value). Thus from the Arrhenius relationship we can write

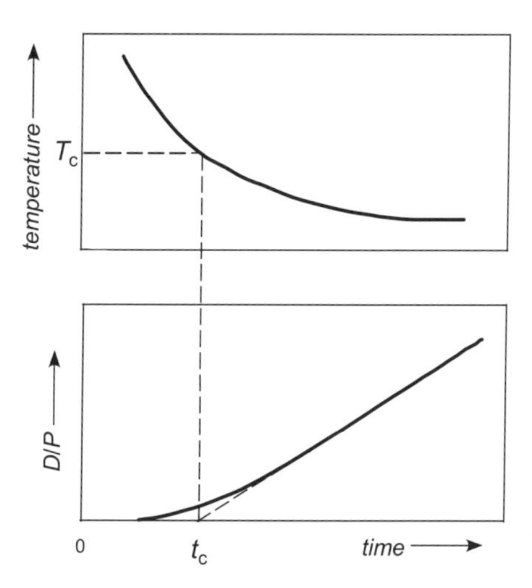

Figure 6. Diagrammatic depiction of closure model, where T_c is closure temperature and t_c is mineral age. During slow cooling from high temperature, the daughter to parent ratio (D/P) passes from a completely open state (i.e., D/P = 0) through a zone of partial accumulation, shown by the curved portion of the curve in the lower diagram, until the temperature is sufficiently low that diffusion loss ceases and the D/P ratio grows at a constant rate with time.

$$D = D_0 \exp\left(-\frac{E}{RT_0} - \frac{t}{\tau} \right) = D(0) \exp\left(-\frac{t}{\tau} \right) \qquad (23)$$

where T_0 and $D(0)$ are the initial temperature and diffusion coefficient, respectively, at time t = 0. For example, when t corresponds to twice the time constant, the exponential coefficient reduces the initial diffusivity to 14% of its original value. Note from Equation (23) that

$$\tau = \frac{R}{E\, dT^{-1}/dt} = \frac{RT^2}{E\, dT/dt} \tag{24}$$

This relationship holds provided that the cooling interval is short with respect to the half-life of the decay system used.

The exponential decay of the loss coefficient with time, a consequence of the form of the Arrhenius law, allows for a closed mathematical solution to the diffusion problem. Dodson (1973) derived a general solution of the accumulation–diffusion–cooling equation for a single diffusion length scale using appropriate substitutions and variable boundary conditions that reduced the problem to an equation identical to Fick's Second Law. This equation can then be solved using general infinite series solutions (Carslaw and Jaeger 1959) for sphere, cylinder, and plane sheet geometries yielding expressions for the concentration distributions. Dodson (1973) also evaluated the coefficients which reduce these expressions to a single characteristic temperature he called the closure temperature (T_c) and provided two heuristic examples to illustrate the nature of the closure concept.

Below we show a non-rigorous calculation to illustrate how linking linear cooling in $1/T$ with diffusive loss via the Arrhenius relationship can derive the general form of the closure equation.

Simplified derivation of the closure temperature equation. Consider the cooling history shown in Figure 6 in which $1/T$ increases linearly. Assuming for the moment that the transition between open and closed system behavior occurs over the relatively narrow temperature interval from T_1 to T_2 (due to the strong temperature dependence of D), we can write two Arrhenius equations:

$$D_1 = D_0 \exp\left(-\frac{E}{RT_1}\right) \tag{25}$$

$$D_2 = D_0 \exp\left(-\frac{E}{RT_2}\right) \tag{26}$$

Dividing (26) by (25) yields

$$\frac{D_1}{D_2} = \exp\left[\left(-\frac{E}{R}\right)\left(\frac{1}{T_2} - \frac{1}{T_1}\right)\right] \tag{27}$$

Thus for a drop in diffusivity from T_1 to T_2 of a factor of e^{-1},

$$\frac{D_1}{D_2} = \frac{E}{R}\left(\frac{1}{T_2} - \frac{1}{T_1}\right) = 1 \tag{28}$$

But because $(T_2^{-1} - T_1^{-1}) \approx (\Delta T/T^2) = \dot{T}\Delta t/T^2$ (where $\dot{T}$ is cooling rate), Equation (28) becomes

$$\frac{E}{R}\frac{\Delta T}{T^2} \approx 1 \approx \frac{E}{R}\frac{\dot{T}\Delta t}{T^2} \tag{29}$$

Recall that fractional loss from a sphere of radius r is associated with a specific value of Dt/r^2. If we arbitrarily state that closure occurs at $Dt/r^2 = 1/A$ and substitute the Arrhenius relationship for D, then the bulk closure temperature of a mineral is given by:

$$\frac{E}{RT_c} = \ln\left(\frac{ART_c^{\,2}D_0/r^2}{E\;dT/dt}\right) \tag{30}$$

Despite the highly oversimplified nature of this derivation, it leads to the form of the closure equation. For the rigorous solution, the values for the geometric constant A can be evaluated and correspond to 55 for the sphere, 27 for the cylinder, and 8.7 for the plane sheet (Dodson 1973). Note that in the spherical case, a value of $A = 55$ corresponds to $Dt/r^2 = 0.018$ which in turn translates to a fractional loss of about 40%.

Note that Equation (30) is iterative in T_c. That is, a trial value inserted into the argument of the logarithm will return a second-order estimate of T_c. Because the logarithm dampens sensitivity to variations in the iterative process, this loop converges rapidly, usually in two iterations.

Assumptions of the closure temperature model. The closure temperature model (Eqn. 30) applies only in those cases where either the simplifying assumptions are met or the results are insensitive to violations of these assumptions. In the case of the slow-cooling model, the time constant (over which the diffusivity drops by a factor e^{-1}) must be much greater than the initial value of the characteristic diffusion time, $r^2/D(0)$. In other words, diffusion must be initially so rapid that the daughter product is not retained on a timescale equivalent to τ. Although the assumption of linear cooling in $1/T$ was adopted in order to make the mathematical solution tractable, a calculated closure temperature is generally highly insensitive to the form of the cooling history (Lovera et al. 1989). Other obvious violations of model interpretations include the presence of excess daughter product due to failure of the infinite reservoir assumption, subsequent open-system behavior, and mineral recrystallization. To expand on this latter point, should the mineral of interest form below the closure temperature, for example, during low grade metamorphism, then the closure model is no longer a valid description of the daughter retention history. Similarly, if cooling is accompanied by differential stress resulting in recrystallization, then the likelihood of diffusion being the rate-limiting daughter product transport mechanism during such structural reconstitution is remote.

Partial retention zone. The transitional interval of temperature (and for cooling geologic systems, time) that is implicit in closure theory finds real expression in crustal rocks. At Earth's surface temperature, all thermochronological systems will be effectively closed, and at lower-crustal depths and temperatures, all of the commonly used minerals used in thermochronometry will be completely open. Clearly with increasing depth and therefore temperature there will be a gradual transition for each system from closed to open behavior. This transition has been observed in deep boreholes, quite commonly for the less retentive systems such as fission-tracks in apatite and zircon (e.g., Naeser and Forbes 1976), but also in other minerals in deeper boreholes like the KTB hole or in exhumed crustal blocks (e.g., Warnock and Zeitler 1998; Stockli and Farley 2004). This transitional interval is referred to the "partial retention zone" or PRZ (Wolf et al. 1998); in fission-track dating the interval is referred to as the "partial annealing zone" or PAZ (Gleadow and Fitz Gerald 1987).

The exact nature of the PRZ depends on local geological conditions and the local tectonic history. In an active, steady-state orogen for example, where erosional exhumation is causing cooling of the crust, the crustal profile of mineral ages will be under the control of the rock uplift which is moving samples through the PRZ. In contrast, in a stable crustal column beneath, say, a craton, samples at various depths are in effect experiencing a prolonged isothermal heating in which daughter-isotope production and diffusional loss trade off to varying degrees, creating potentially large gradients in mineral age with depth. Figure 7 illustrates several forms that the PRZ might take. The accumulation of Ar or He in samples that were stagnant in the PRZ or took a complex path through has been modeled in most laboratories using numerical methods

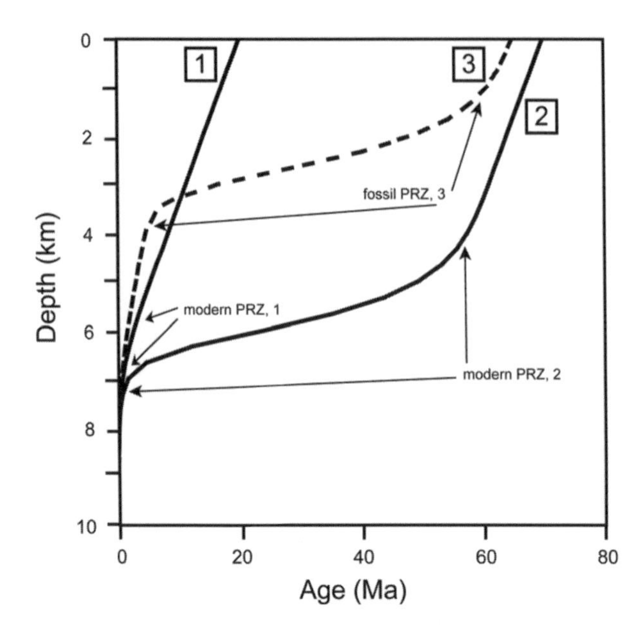

Figure 7. Schematic representation of "partial retention zone." Solid curves show age profile in crust under conditions of continuous erosional cooling at orogenic rates (c1, at left) and after erosional cooling followed by a prolonged interval of stable conditions without erosion or burial (c2, right). Dashed curve (c3) shows crustal age profile after an modest pulse of uplift and erosion is superimposed on case c2. Approximate locations of modern and "fossil" PRZs are also shown.

that account for the general case of radiogenic accumulation and diffusive loss of daughter product (e.g., Wolf et al. 1998; Dunai 2005; Harrison et al. 2005; Ketcham 2005).

In certain circumstances, the PRZ has a sufficiently distinctive form that it can serve as an important marker in tectonic studies, something that has been widely exploited in low-temperature thermochronology in particular, using fission-track dating of apatite, and more recently U-Th/He dating of apatite and zircon (e.g. Stockli et al. 2000). Specifically, if a region has been tectonically stable and then experiences a pulse of rock uplift, it can be possible to identify a "fossil" PRZ in the pattern of ages with structural depth, providing information about both the timing and amount of exhumation (Fig. 7).

It is also very important to keep in mind that thermochronometers that have cooled through the PRZ will preserve an internal record of their passage in their daughter-product concentration profiles. Such "closure profiles" (Dodson 1986) can be inverted to provide a record of the relevant segment of the sample's time-temperature history, provided this profile can be sampled be either indirect means such $^{40}Ar/^{39}Ar$ or $^{4}He/^{3}He$ stepheating (Harrison et al. 2005; Shuster and Farley 2005) or direct sampling by direct means such as depth profiling by ion probe (e.g. Harrison et al. 2005).

EXPERIMENTAL DETERMINATION OF DIFFUSION PARAMETERS

Calculation of diffusion coefficients from bulk loss experiments

Because we are not able to directly observe natural $^{40}Ar/^{39}Ar$ or ^{4}He diffusion profiles, we must assess diffusion behavior by estimating the amount of uptake or loss. One commonly used approach to measure diffusion of noble gases in natural materials is to measure the fractional approach toward equilibration of the radiogenic daughter resulting from a controlled heating experiment.

Approximations for fractional loss (f) from a sphere as a function of Dt/r^2 can be written (Fechtig and Kalbitzer 1966) in terms of the diffusion coefficient, D/r^2. They are:

$$\frac{D}{r^2} = \begin{cases} \frac{1}{\pi^2 t}\left(2\pi - \frac{\pi^2}{3}f - 2\pi\sqrt{1-\frac{\pi}{3}f}\right) & for\ f < 0.85 \\ \frac{1}{\pi^2 t}\ln\left[\frac{\pi^2}{6}(1-f)\right] & for\ f > 0.85 \end{cases} \tag{31}$$

In cases where r is the measured particle size, or can be inferred from experiment, a unique value for D can be determined.

Choosing an appropriate diffusion model for a mineral of complex geometry may not be immediately obvious. As an example, consider the case of biotite. Since diffusion in biotite proceeds much more rapidly parallel to the cleavage compared with perpendicular, an infinite cylinder may be an appropriate model despite the fact that mica is a sheet-like mineral. Planar structures such as perthite lamellae in alkali feldspar can be satisfactorily described as a plane sheet if the lamellar boundaries define the effective diffusion dimension.

Calculation of Ar and He diffusion coefficients from step-heating results

If we make the assumption that the diffusion mechanisms and boundaries that define Ar and He retentivity in nature are the same as those that control degassing during laboratory heating, we can also extract model diffusion coefficients (D/r^2) from stepwise heating of irradiated samples. Because the reactor-produced ^{39}Ar is distributed in a similar fashion to ^{40}K (which is broadly uniform within K-feldspars), it is preferable to use ^{39}Ar for this purpose as the concentration distribution of ^{40}Ar from slowly cooled samples is unknown. In the case of helium, until recently there has been no real choice but to use radiogenic ^4He as measured and, depending on the nature of the sample, either assume a uniform distribution, or attempt a workaround that assumes that there has been He loss or slow cooling sufficient to produce a concentration gradient, with diffusion calculations for later steps being relatively insensitive to the details of this assumption.

For the mth laboratory heating step, $\log(D/r^2)$ can be written as (Lovera et al. 1997)

$$\log\left(\frac{D}{r^2}\right)_m = \log\frac{\Delta\zeta^*_m}{\Delta t_m} = \log\frac{\zeta^*_m - \zeta^*_{m-1}}{\Delta t_m} \tag{32}$$

where ζ_m is obtained from inversion of the approximate expressions for the cumulative fractions of ^{39}Ar or ^4He released. For the case of a plane sheet, $\Delta\zeta^*_m$ is

$$\Delta\zeta^*_m = \begin{cases} \frac{\pi}{4}\left(f_m^2 - f_{m-1}^2\right) & for\ f_m \le 0.6 \\ -\alpha_1^{-2}\ln\left[\frac{(1-f_m)}{(1-f_{m-1})}\right] & for\ f_m \le 0.6 \end{cases} \tag{33}$$

A potential limitation of this approach is that errors introduced at any step m are propagated into subsequent steps. An expression for the propagation of errors in calculation of diffusion coefficients from step heating data is given in Lovera et al. (1997).

Experimental criteria

The requirements of a diffusion study vary depending on the nature of the experiment, but two criteria must be met in all cases: the mineral must remain stable throughout the duration of the experiment and the initial distribution of the diffusing substance must be known. In the specific case of bulk-loss type experiments (e.g., Giletti 1974), several other factors must be known including the shape and effective diffusion length scale of the particles in the aggregate. For purposes of interpretation, diffusion experiments should be designed to address questions regarding chemical vs. tracer diffusion or wet vs. dry conditions. Recrystallization and incongruent dissolution during annealing are concerns that must be evaluated as are possible radiation-induced effects for $^{40}Ar/^{39}Ar$ or $^{4}He/^{3}He$ experiments.

Laboratory diffusion studies - helium

General observations. Studies conducted to date on He diffusion in apatite, zircon, and titanite all suggest that in these accessory minerals, the effective diffusion radius seems to be the physical grain size. All of these three phases show artifacts in their diffusion behavior in the form of non-linearities on their Arrhenius plots, with apatite being the best behaved at the more geologically relevant lower temperatures. In all cases, laboratory diffusion behavior under vacuum seems to be consistent with field calibrations, comparisons to one another, and comparisons to other thermochronometers. Shuster et al. (2003), in conducting their initial work on using proton-induced ^{3}He as a means of studying He diffusion, made the interesting finding that ^{3}He and ^{4}He seem to diffuse at very similar rates, despite the substantial relative difference in their masses. They speculate that the absence of the expected mass dependence of He diffusivity may be indicative of the diffusion mechanism in operation in apatite, with a process like diffusion of lattice defects being the controlling factor.

Apatite. Studies of Durango apatite indicate that He diffusion occurs via a thermally-activated volume diffusion process in the temperature region 100° to 300 °C with an activation parameters of $E = 32.9 \pm 1.6$ kcal/mol and $D_0 = 50$ cm²/s (Farley 2000). D_0/a^2 varies with grain size indicating that the grain size defines the diffusion domain (Farley 2000). These parameters translate into a closure temperature of 70 °C for apatites with a ca. 100 μm diffusion radius at a cooling rate of ca. 10 °C/Ma.

Zircon. Reiners et al. (2002) found that He release from zircon is complex. Temperature cycling experiments suggest that either zircon contains diffusion domains smaller than the physical grain size, or that annealing of radiation damage during the diffusion experiment was altering the diffusion properties of the zircon (Reiners 2005). They suggest a minimum activation energy for geological relevant ^{4}He diffusion of about 44 kcal/mol, and a 10 °C/m.y. closure temperature for typical zircons of about 190 °C. This estimate was supported by comparisons between K-feldspar-derived cooling histories and zircon helium ages, as well as further diffusion experiments (Reiners et al. 2003). These more recent data suggest activation energies of around 40 kcal/mol for portions of the low-temperature gas release, equivalent to closure temperatures of between 170 and 190 °C which show good concordance with the K-feldspar data (10 °C/m.y., 60 μm crystal radius).

Titanite. Reiners et al. (1999) examined He diffusion in titanite. Despite some non-linearity in diffusion behavior early in gas release, overall, He diffusion in titanite seems well-behaved, with much of the gas release following a linear Arrhenius relationship having an activation energy of ~45 kcal/mol. For the samples studied, the diffusion data equate to closure temperatures of about 190 to 220 °C (200 to 800 μm crystal radius, 10 °C/m.y.). In their study of titanites takes from the KTB deep borehole, Stockli and Farley (2004) found that their measured ages agreed very well with predictions for retentivity made from the diffusion data.

Laboratory diffusion studies - argon

Biotite-phlogopite. Using the bulk loss approach, Giletti (1974) measured Ar diffusion from phlogopite (Ann_4) under hydrothermal conditions in the temperature range 900–600 °C and reported Arrhenius parameters of $E = 57.9 \pm 2.6$ kcal/mol and $D_0 = 0.75^{+1.7}_{-0.52}$ cm²/s. Using a similar experimental method, but under controlled oxygen fugacity conditions, Harrison et al. (1985) measured Ar diffusion from the Cooma biotite (Ann_{56}) in the temperature range 750–600 °C. For grain sizes less than 202 μm, they found $E = 47.0 \pm 2.1$ kcal/mol, $D_0 = 0.077^{+0.21}_{-0.06}$ cm²/s. Experiments at 14 kbar indicated an activation volume of 14 cm³/mol.

The fact that the 202 μm radius data plotted above the other data was attributed to the intrinsic effective diffusion radius being smaller than the measured grain radius. If so, this implies an *effective* diffusion radius of ~150 μm for this material, similar to estimates inferred from geological studies (e.g., Wright et al. 1991). Note that inferences from *in situ* laser probe studies that the characteristic Ar diffusion dimension is equivalent to grain size can be incorrect due to both low spatial resolution and the fact that the effective diffusive lengthscale is not equivalent to the largest domain size observed in a crystal (see Grove and Harrison 1996).

Grove and Harrison (1996) revisited the Cooma biotite and obtained $E = 45 \pm 3$ kcal/mol and $D_0 = 0.015^{+0.022}_{-0.005}$ cm²/sec, both values within 1σ of those previously obtained. Combining the two data sets yields an $E = 47.1 \pm 1.5$ kcal/mol and $D_0 = 0.075^{+0.049}_{-0.021}$ cm²/sec (Fig. 8). They also performed bulk loss experiments using an iron-rich biotite (Ann_{71}) which yielded $E = 50.5 \pm 2.2$ kcal/mol and a D_0 of $0.40^{+0.96}_{-0.28}$ cm²/sec—virtually indistinguishable from the Cooma biotite result.

Using an empirical diffusion model (Dowty 1980; Fortier and Gilleti 1989) to predict relationships between Ar diffusivity and chemistry, Grove (1993) identified composition controls on Ar diffusion in the biotite-phlogopite series. Specifically, Ar diffusivities are 1) enhanced by replacement of Mg^{2+} by Fe^{2+}; (2) lowered by incorporation of Al(VI) and/or Fe^{3+} into the octahedral sheet; and (3) significantly lowered by replacement of the hydroxyl group by halogens.

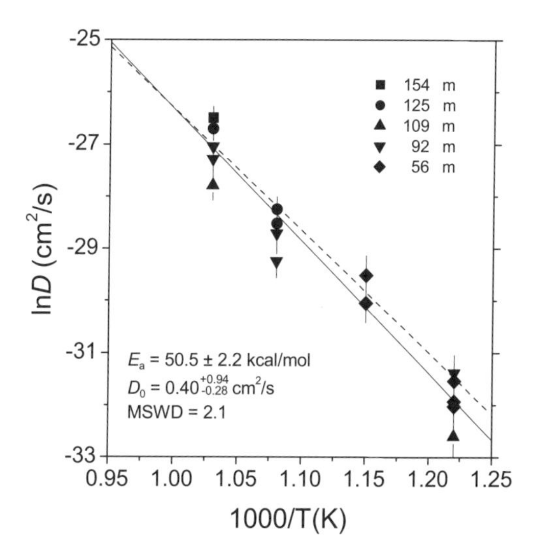

Figure 8. Arrhenius plot showing argon diffusion results for iron-rich [Fe/(Fe + Mg) = 71] Fe-mica from Grove and Harrison (1996). The composite line for the Cooma biotite is shown by the dashed line.

K-feldspar. Foland (1974) undertook an Ar diffusion study of a homogeneous orthoclase (Benson Mines) using hydrothermal and vacuum heating. These data yielded an $E = 43.8 \pm 1.0$ kcal/mol and $D_0 = 0.0098^{+0.0066}_{-0.0037}$ cm^2/s and revealed that ^{40}Ar loss from alkali feldspar proceeds at the same rate whether heated in vacuum or under water pressure. Numerous vacuum step heating studies of irradiated K-feldspars have been undertake as a by-product of $^{40}Ar/^{39}Ar$ dating (e.g., Berger and York 1981). Results of these experiments generally yield an initially linear array that was similar to the E obtained by Foland (1974). However, this simple Arrhenian behavior breaks down at higher temperatures yielding complex patterns. This behavior of the Arrhenius plot is now understood to result from basement K-feldspars typically containing a discrete distribution of diffusion length scales (Lovera et al. 1991).

The approach of using the vacuum step heating approach to establish argon diffusion behavior in K-feldspars was well documented by Lovera et al. (1997) who developed a large data base of analyzed samples. They focused on the low-temperature data which typically yield a linear Arrhenius relationship and statistically analyzed the results using an automated routine with a uniform set of selection criteria (Fig. 9). In general, they found that the propagated uncertainties in the calculation of $\log(D/r^2)$ from step heating data amount to only ± 0.05 log units (in s^{-1}). K-feldspars measured using a consistent extraction schedule (i.e., duplicate isothermal steps from 450–500 °C) yield Arrhenius plots with $\log(D/r^2)$ values that vary by less than two orders of magnitude at any given temperature.

Muscovite. The relatively narrow pressure-temperature field of stability of muscovite and Ar diffusion kinetics severely limits the experimental determination of argon diffusion to between 700 °C and 600 °C. Robbins (1972) undertook a hydrothermal diffusion study of muscovite and calculated Arrhenius parameters for his results using both an infinite cylinder and plane sheet model. A reasonably good fit of all data could be made with the plane sheet model yielding E = 40 kcal/mol and a $D_0 = 6 \times 10^{-7}$ cm^2/s, although an equivalent argument could be

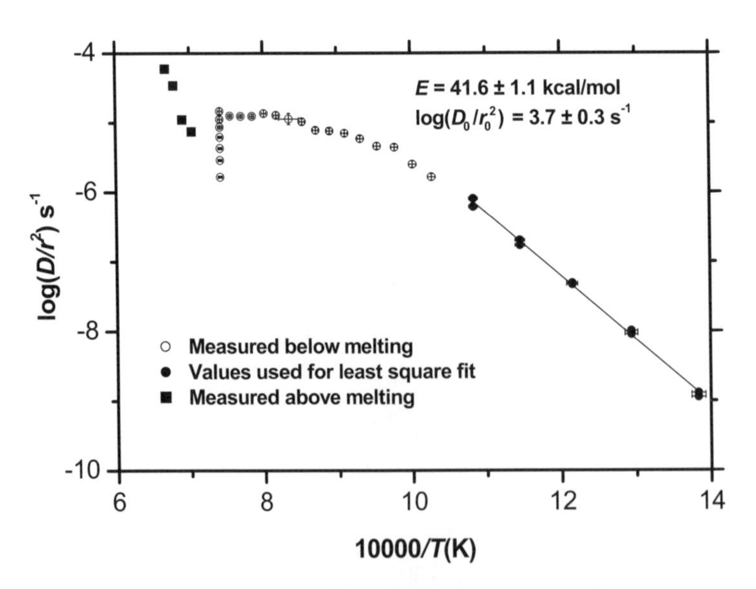

Figure 9. Arrhenius plot for a K-feldspar sample 93-NG-17 illustrating the method to calculate E and $\log(D_0/r_0^2)$. D/r^2 values are indicated by filled symbols and the solid line represents a weighted, least squares fit to these data.

made for the infinite cylinder calculation with a similar activation energy but a frequency factor some 500 times larger. Hames and Bowring (1994) re-evaluated Robbins (1972) results and preferred an infinite cylinder model with E = 43 ± 9 kcal/mol and log D_0 = −3.4 ± 2 (cm^2/s). Lister and Baldwin (1996) instead argued for a sheet geometry with a characteristic thickness of ~12 μm for Robbins (1972) data.

Hornblende. Harrison (1981) used the bulk approach to examine the Ar diffusion behavior of two hornblende samples with Mg#'s of 0.72 and 0.36 in the temperature range 900–750 °C. The combined results defined a line of E = 66 ± 4 kcal/mol and D_0 = $0.06^{+0.4}_{-0.01}$ cm^2/s assuming a spherical model. Harrison et al. (1992) reported an additional diffusion measurement of the Mg-rich hornblende and obtained a somewhat lower value. Baldwin et al. (1990) further investigated the Fe-rich hornblende of Harrison (1981) and obtained results that were consistent with the earlier determined activation energy. They also hydrothermally treated two compositionally contrasting metamorphic amphiboles (Mg# of 0.71 and 0.43) containing complex exsolution structures. Results indicated that the effective diffusion lengthscale was substantially smaller than the measured particle size, probably reflecting the observed phyllosilicate intergrowths and/or exsolution lamellae that partition the hornblende crystals into smaller subdomains. No systematic dependence of argon diffusivity on Mg# has been yet established (Harrison 1981; Baldwin et al. 1990; Cosca and O'Nions 1994). The ionic porosity model (Fortier and Gilleti 1989; Dahl 1996) applied to the compositional dependence of argon retention in amphiboles predicts relatively small effects.

INTERPRETATION OF THERMOCHRONOLOGICAL DATA

It is beyond the scope of this chapter to explore modern approaches in interpreting thermochronological data. Instead we provide a few comments about the nature of such data, how they should be viewed, and what their current limitations are, echoing comments made by Reiners et al. (2005).

Heat transfer

Thermochronological data are inextricably linked to issues of how heat is transferred in the lithosphere and how various geodynamic processes alter the temperature potential field (Ehlers 2005). How exactly a thermochronological study is designed depends very much on the geologic context of the problem (Braun 2005; Gallagher et al. 2005). Interestingly and fortunately for the practitioner, both mass and heat conduction occur by diffusion, and so share a common mathematical description. However, in many cases geological problems are highly underdetermined when it comes to posing exact thermal solutions, and thermochronological data alone cannot make the difference even as they provide critical boundary conditions.

Very generally, higher-temperature systems will be easier to work with and interpret, as higher-temperatures in the crust will usually be less prone to locally complex temperature distributions and rapid changes in temperature. In contrast, at the very lowest temperatures at which thermochronology is currently applicable, below 100 °C, many processes beyond conductive heat transfer can become important (e.g., fluid flow), and topographic, structural, and lithologic factors all begin to play major roles in shaping the temperature field. In the shallow crust such phenomena as isotherm advection in three dimensions become significant in tectonically active regions.

Sampling considerations

In detail, what methods to use and how to distribute samples will depend entirely on the nature of the problem at hand. In general, geochronology has a legacy that emphasizes precise analysis of small numbers of well-characterized samples. While we are certainly not opposed to precision and characterization, the thermal considerations we mentioned above do bring with them some demands. Faced with a complex temperature field that has evolved through time, sampling campaigns for thermochronology need to be wary of aliasing results through undersampling. Modern laboratories have greatly improved their capacities for sample throughput by means of automation and by reducing the amount of material that must be processed to obtain an analysis, but some compromise in method and precision might be required to meet the demands of an extended sampling campaign (e.g., opting for total-fusion rather than step-heating data for biotite samples, or substituting U-Th/He ages on zircon for detailed K-feldspar age spectra). Thinking about how seismic data are acquired and treated, in which multichannel sampling and procedures such as stacking enhance signal-to-noise ratios, we would argue that for many applications thermochronologists would benefit from an analogous approach.

Constraining power

Well-designed thermochronological studies and well-behaved samples can provide remarkably tight bounds on segments of thermal histories. However, it is important to keep in mind that at the other extreme, any single mineral age can be explained by myriad thermal histories, including monotonic cooling at varying rates, and all manner of resetting scenarios including prolonged thermal stagnation. To be most successful, thermochronological studies need to be carried out in conjunction with good structural, stratigraphic, and geochronological constraints.

Intercomparison and accuracy of thermochronological data

To date, there are no thermochronological standards available with which labs can compare results, either at the analytical level (e.g., diffusion data obtained from step-heating) or at a more derived level (e.g., complete thermal histories). In some cases, as in $^{40}Ar/^{39}Ar$ step-heating analysis of K-feldspar, segments of thermal histories can be determined that are probably precise and accurate to within 5–10 °C and well less than 1 m.y., assuming a well-behaved sample and a well-calibrated furnace. On the other hand, for many other mineral systems, much of our knowledge about their behavior comes from comparisons to other thermochronometers, diffusion data are scanty, and what diffusion data are available are frequently applied to unknowns based on mere assumptions about similarities in diffusion kinetics. Thus, determinations of parameters like cooling rates using "mineral pairs" will be fraught with pitfalls, given the assumptions and absolute uncertainties involved, and there is a clear need for more systematic and comprehensive studies of diffusion kinetics in minerals of importance.

CONCLUDING REMARKS

The fundamental nature of noble gas decay systems (i.e., chemically inert daughter products) leads to a high degree of predictability of closure temperature with a minimum of assumptions regarding initial non-radiogenic components. Similarities in the manner in which we can treat diffusive loss of daughter product from K-Ar and U-Th/He decay make it convenient for us to review the underlying theory of thermochronology using these two systems as exemplars. The great versatility of K-Ar and U-Th/He thermochronometers, which provide geological temperature monitoring across the range from 500°C to 60°C, make them the most popular methods for extracting temperature-time histories form crustal rocks.

REFERENCES

Baldwin SL, Harrison TM, Fitz Gerald JD (1990) Diffusion of ^{40}Ar in metamorphic hornblende. Contrib Mineral Petrol 105:691-703

Baxter EF, DePaolo DJ, Renne PR (2002) Spatially correlated anomalous Ar-40/Ar-39 "age" variations in biotites about a lithologic contact near Simplon Pass, Switzerland: A mechanistic explanation for excess Ar. Geochim Cosmochim Acta 66:1067-1083

Berger GW, York D (1981) Geothermometry from ^{40}Ar/^{39}Ar dating experiments. Geochim Cosmochim Acta 45:795-811

Brandt SB (1974) A new approach to the determination of temperatures of intrusions form radiogenic argon loss in contact aureoles. *In:* Geochemical Transport and Kinetics. Hofmann AW, Giletti BJ, Yoder HS, Yund RA (eds), Carnegie Institute, Washington, Publication # 634, p 295-298

Braun J (2005) Quantitative constraints on the rate of landform evolution derived from low-temperature thermochrohonology. Rev Mineral Geochem 58:351-374

Carslaw HS, Jaeger JC (1959). Conduction of heat in solids, 2nd ed. Clarendon, Oxford.

Chopin C, Maluski H (1980) ^{40}Ar/^{39}Ar dating of high-pressure metamorphic micas from the Gran Paradiso area (Western Alps) – Evidence against the blocking temperature concept. Contrib Mineral Petrol 74: 109-122

Cosca MA, O'Nions RK (1994) A re-examination of the influence of composition on argon retentivity in metamorphic calcic amphiboles. Chem Geol 112:39-56

Crank J (1975) The mathematics of diffusion. Oxford University Press, Oxford

Dahl PS (1996) The effects of composition on retentivity of argon and oxygen in hornblende and related amphiboles: a field-tested empirical model. Geochim Cosmochim Acta 60:3687-3700

de Jong K (2003) Very fast exhumation of high-pressure metamorphic rocks with excess ^{40}Ar and inherited ^{87}Sr, Betic Cordilleras, southern Spain. Lithos 70:91-110

Dodson MH (1973) Closure temperature in cooling geochronological and petrological systems. Contrib Mineral Petrol 40:259-274

Dodson MH (1975) Kinetic processes and thermal history of rocks. Ann Rep Dir Dept Terr Magn Carnegie Inst Yearbook 74, p 210-217

Dodson MH (1986) Closure profiles in cooling systems. *In:* Materials Science Forum, Vol 7. Trans Tech Publications, Aedermannsdorf, Switzerland, p 145-153

Dowty E (1980) Crystal-chemical factors affecting the mobility of ions in minerals. Am Mineral 65:174-182

Dunai TJ (2005) Forward modeling and interpretation of (U-Th)/He ages. Rev Mineral Geochem 58:259-274

Ehlers TA (2005) Crustal thermal processes and the interpretation of thermochronometer data. Rev Mineral Geochem 58:315-350

Farley KA (2000) Helium diffusion from apatite: general behavior as illustrated by Duragno fluorapatite. J Geophys Res 105:2903-2914

Farley KA, Reiners PW, Nenow V (1999) An apparatus for measurement of noble gas diffusivities from minerals in vacuum. Anal Chem 71:2059-2061

Farley KA, Wolf RA, Silver LT (1996) The effects of long alpha-stopping distances on (U-Th)/He dates. Geochim Cosmochim Acta 60:4223-4230

Faure G (1977) Principles of isotope geology. Wiley, New York

Fechtig H, Kalbitzer S (1966) The diffusion of argon in potassium-bearing solids. *In:* Potassium-argon Dating. Schaeffer OA, Zähringer J (eds) Springer-Verlag, New York, p 68-107

Fick A (1855) Ueber diffusion. Ann Phys Chem 94:59-86

Fitz Gerald JD, Harrison TM (1993) Argon diffusion domains in K-feldspar I: microstructures in MH-10. Contrib Mineral Petrol 113:367-380

Foland KA (1974) ^{40}Ar diffusion in homogeneous orthoclase and an interpretation of Ar diffusion in K-feldspar. Geochim Cosmochim Acta 38:151-166

Fortier SM, Giletti BJ (1989) An empirical model for predicting diffusion coefficients in silicate minerals. Science 245:1481-1484

Gallagher K, Stephenson J, Brown R, Holmes C, Ballester P (2005) Exploiting 3D spatial sampling in inverse modeling of thermochronological data. Rev Mineral Geochem 58:375-387

Giletti BJ (1974) Diffusion related to geochronology. *In*: Geochemical Transport and Kinetics. Hofmann AW, Giletti BJ, Yoder Jr. HS, Yund RA (eds), Carnegie Inst. of Wash. Publ. 634, p 61-76

Giletti BJ, Tullis J (1977) Studies in diffusion. IV. Pressure dependence of Ar diffusion in phlogopite mica. Earth Planet Sci Lett 35:180-183

Gleadow AJW, Fitzgerald PG (1987) Tectonic history and structure of the Transantarctic Mountains: new evidence from fission track dating in the Dry Valleys area of southern Victoria Land. Earth Planet Sci Lett 82:1-14

Grove M, Harrison TM (1996) ^{40}Ar diffusion in Fe-rich biotite. Am Mineral 81: 940-951

Grove M (1993) Thermal histories of southern California basement terranes. PhD dissertation, University of California, Los Angeles, CA

Hames WE, Bowring SA (1994) An empirical evaluation of the argon diffusion geometry in muscovite. Earth Planet Sci Lett 124:161-167

Harrison TM (1981) Diffusion of ^{40}Ar in hornblende. Contrib Mineral Petrol 78:324-331

Harrison TM, McDougall I (1981) Excess ^{40}Ar in metamorphic rocks from Broken Hill, New South Wales: Implications of ^{40}Ar/^{39}Ar age spectra and the thermal history of the region. Earth Planet Sci Lett 55: 123-149

Harrison TM, Duncan I, McDougall I (1985) Diffusion of ^{40}Ar in biotite: temperature pressure and compositional effects. Geochim Cosmochim Acta 49:2461-2468

Harrison TM, Heizler MT, Grove M, Wartho J (1992) Argon loss from hornblende. EOS Trans Am Geophys Un 73:362

Harrison TM, Grove M, Lovera OM, Zeitler PK (2005) Continuous thermal histories from inversion of closure profiles. Rev Mineral Geochem 58:389-409

Hourigan JK, Reiners PW, Brandon MT (2005) U-Th zonation dependent alpha-ejection in (U-Th)/He chronometry. Geochim Cosmochim Acta 69:3349-3365

Huneke JC, Smith SP (1978) The realities of recoil: ^{39}Ar recoil out of small grains and anomalous age patterns in ^{39}Ar-^{40}Ar dating. Proc 7th Lunar Sci Conf 1987-2008.

Hurley PM (1954) The helium age method and the distribution and migration of helium in rocks. *In:* Nuclear Geology. Faul H (ed) Wiley, New York, p 301-329

Ketcham RA (2005) Forward and inverse modeling of low-temperature thermochronometry data. Rev Mineral Geochem 58:275-314

Lister GS, Baldwin SL (1996) Modelling the effect of arbitrary P-T-t histories on argon diffusion in minerals using the MacArgon program for the Apple Macintosh. Tectonophysics 253:83-109

Lovera OM, Richter FM, Harrison TM (1989) ^{40}Ar/^{39}Ar geothermometry for slowly cooled samples having a distribution of diffusion domain sizes. J Geophys Res 94:17917-17935

Lovera OM, Richter FM, Harrison TM (1991) Diffusion domains determined by ^{39}Ar release during step heating. J Geophys Res 96:2057-2069

Lovera OM, Grove M, Harrison TM, Mahon KI (1997) Systematic analysis of K-feldspar ^{40}Ar/^{39}Ar step-heating results: I Significance of activation energy determinations. Geochim Cosmochim Acta 61:3171-3192

Meesters AGCA, Dunai TJ (2002a) Solving the production-diffusion equation for finite diffusion domains of various shapes - Part I. Implications for low-temperature (U–Th)/He thermochronology. Chem Geol 186: 333-344

Meesters AGCA, Dunai TJ (2002b) Solving the production-diffusion equation for finite diffusion domains of various shapes - Part II. Application to cases with alpha-ejection and nonhomogeneous distribution of the source. Chem Geol 186:57-73

McDougall I, Harrison TM (1999) Geochronology and Thermochronology by the ^{40}Ar/^{39}Ar Method. 2nd ed, Oxford University Press, New York

Obradovich JD, Tatsumoto M, Manuel OK, Mehnert H, Domenick M, Wildman T (1982) K-Ar and K-Ca dating of sylvite from the late Permian Salado Formation New Mexico Implications regarding stability of evaporite minerals. Fifth Int Conf Geochronol Cosmochronol Isotope Geol, Japan, 283-284

Pankhurst RJ, Moorbath S, Rex DC, Turner G (1973) Mineral age patterns in ca. 3700 my old rocks from West Greenland. Earth Planet Sci Lett 20:157-170

Purdy JW, Jäger E (1976) K-Ar ages on rock-forming minerals from the Central Alps. Mem 1st Geol Min Univ Padova 30, 31 pp.

Reiners PW (2005) Zircon (U-Th)/He thermochronometry. Rev Mineral Geochem 58:151-179

Reiners PW, Ehlers TA, Zeitler PW (2005) Past, present, and future of thermochronology. Rev Mineral Geochem 58:1-18

Reiners PW, Farley KA (1999) Helium diffusion and (U-Th)/He thermochronometry of titanite. Geochim Cosmochim Acta 63:3845-3859

Reiners PW, Farley KA, Hickes HJ (2002) He diffusion and (U-Th)/He thermochronometry of zircon: Initial results from Fish Canyon Tuff and Gold Butte, Nevada, Tectonophysics, 349:297-308

Reiners PW, Spell TL, Nicolescu S, Zanetti, KA (2004) Zircon (U-Th)/He thermochronometry: He diffusion and comparisons with ^{40}Ar/^{39}Ar dating, Geochim Cosmochim Acta 68:1857-1887

Robbins GA (1972) Radiogenic argon diffusion in muscovite under hydrothermal conditions. MS thesis, Brown University, Providence RI

Shuster DL, Farley KA (2003) ^{4}He/^{3}He thermochronometry. Earth Planet Sci Lett 217:1-17

Shuster DL, Farley KA (2005) ^{4}He/^{3}He thermochronometry: theory, practice, and potential complications. Rev Mineral Geochem 58:181-203

Shuster DL, Farley KA, Sisterson JM, Burnett DS (2003). Quantifying the diffusion kinetics and spatial distributions of radiogenic ^{4}He in minerals containing proton-induced ^{3}He. Earth Planet Sci Lett 217: 19-32

Stockli DA, Farley KA (2004) Empirical constraints on the titanite (U-Th)/He partial retention zone from the KTB drill hole. Chem Geol 207:223-236

Stockli DA, Farley KA, Dumitru TA (2000) Calibration of the (U-Th)/He thermochronometer on an exhumed normal fault block in the White Mountains, eastern California and western Nevada. Geology 28:983-986

Strutt RJ (1908) The accumulation of helium in geological time. Proc Roy Soc Lond A80:272-277

Turner G (1968) The distribution of potassium and argon in chondrites. *In:* Origin and Distribution of the Elements. Ahrens LH (ed) Pergamon, London, p 387-398

Turner G, Cadogan PH (1974) Possible effects of ^{39}Ar recoil in ^{40}Ar-^{39}Ar dating. Geochim Cosmochim Acta Suppl 5 (Proceedings of the Fifth Lunar Science Conference), 1601-1615

Warnock AC, Zeitler PK (1998) ^{40}Ar/^{39}Ar thermochronometry of K-feldspar from the KTB borehole, Germany. Earth Planet Sci Lett 158:67-79

Wright N, Layer PW, York D (1991) New insights into thermal history from single grain ^{40}Ar/^{39}Ar analysis of biotite. Earth Planet Sci Lett 104:70-79

Wolf RA, Farley KA, Kass DM (1998) Modeling of the temperature sensitivity of the apatite (U-Th)/He thermochronometer. Chem Geol 148:105-114

Wolf RA, Farley KA, Silver LT (1996) Helium diffusion and low-temperature thermochronometry of apatite. Geochim Cosmochim Acta 60:4231-4240

Zeitler PK, Herczig AL, McDougall I, Honda M (1987) U-Th-He dating of apatite: a potential thermochronometer. Geochim Cosmochim Acta 51:2865-2868

Reviews in Mineralogy & Geochemistry
Vol. 58, pp. 151-179, 2005

6

Zircon (U-Th)/He Thermochronometry

Peter W. Reiners

Department of Geology and Geophysics
Yale University
New Haven, Connecticut, 06520, U.S.A.
peter.reiners@yale.edu

INTRODUCTION

A number of features of zircon ($ZrSiO_4$), including high U-Th concentrations, high abundance in a wide range of lithologies, refractory nature under metamorphic and some magmatic conditions, and resistance to physical and chemical weathering, make it highly suitable for geochronology and thermochronology and thus a versatile tool for examining a wide range of earth processes. Like apatite and many other minerals, radioisotopic dating of zircon was first performed using the (U-Th)/He system, but the thermochronologic significance of zircon He ages has emerged only in the last few years. In this chapter, I review the current status of zircon He dating in the earth sciences, primarily as applied to thermochronology, including the controls on He diffusivity, the role of radiation damage, analytical techniques for measuring zircon He ages, special considerations unique to zircon He dating, and a series of case studies. Several examples from the literature are briefly summarized to illustrate the diversity of geologic problems accessible by zircon He dating and highlight the future potential of the system and outstanding unresolved issues. Exemplary applications include determining the timing and rates of orogenic exhumation and constraining provenance, depositional ages, and source terrain histories using He-Pb double dating of detrital zircons.

Historical perspective

Previous geo- and thermochronometric studies of zircon have utilized a wide range of decay schemes, including Pb-α (e.g., Webber et al. 1956), U/Pb, Pb/Pb, Th/Pb (Larsen et al. 1952; Vinogradov et al. 1952; Tilton et al. 1955; Wetherill 1955; Silver and Deutsch 1963; Parrish and Noble 2003; Ireland and Williams 2003; Bowring and Schmitz 2003), U-series (Scharer 1984; Reid et al. 1997), fission-track (Naeser et al. 1981; Brandon and Vance 1992; Bernet and Garver 2005; Tagami 2005), Lu/Hf (in concert with other phases; e.g., Pettingill and Patchett 1981), Sm/Nd (Futa 1986; Wernicke and Getty 1997), and $^{244}Pu/^{136}Xe$ (Turner et al. 2004). As with many other minerals, however, zircon was first dated using the (U-Th)/He system (Strutt 1910a,b). Strutt's pioneering work came not long after Ernest Rutherford reported the first radioisotopic age of any type, using He dating. Along with iron ores, titanite, and other minerals, Strutt measured He ages in zircons from a wide range of localities, reporting dates as young as 100 ka for zircons from Mt. Vesuvius, to as old as 565 Ma for a zircon from Ontario, Canada. Unlike some of his contemporaries, Strutt recognized that both U and Th produced He. Like others at the time, however, he also recognized that He ages were, in general, "minimum values, because He leaks out from the mineral, to what extent it is impossible to say" (Strutt 1910c). Not comfortable with attributing geochronologic significance to these apparent ages, Strutt generally referred to ages determined from relative He and U-Th measurements as "helium ratios."

Other early studies measuring zircon He ages include Holmes and Paneth (1936), who measured Oligo-Miocene ages for xenocrystic zircons in South Africa kimberlites, Larsen

1529-6466/05/0058-0006$05.00 DOI: 10.2138/rmg.2005.58.6

and Keevil (1942), reporting a zircon He age of 23 Ma for the Lakeview tonalite in southern California, and Keevil et al. (1944) with a zircon He age of 260 Ma for the Chelmsford granite in Massachusetts. A number of studies led by Patrick Hurley (Hurley 1952, 1954; Hurley and Fairbairn 1953; Hurley et al. 1956) reported zircon He ages from a wide variety of locations and tectonic settings. Some of the more notable of these include He ages of 435–495 Ma for detrital zircons with relatively low U concentrations from Sri Lanka, and a much younger age of 130 Ma for a single specimen from this suite with much higher U content (Hurley et al. 1956). Hurley (1954) also reported numerous zircon He ages of 63–82 Ma from batholithic rocks in the Sierra Nevada, Idaho batholith, and southern California, and numerous zircon He ages of 650–880 Ma from Ontario. In 1957, Damon and Kulp reported zircon He ages from Ontario and Sri Lankan zircons, though their results on the latter suite were generally older than both previous and more recent studies.

By the 1950s it was generally clear that progress in (U-Th)/He chronometry would require a better understanding of the phenomena leading to natural He loss and ages younger than known or presumed formation ages. For some reason, however, most studies emphasized the role of radiation damage, rather than thermally activated diffusion, in observed variations of natural He loss. Several studies emphasized correlations between indices of radiation damage and apparent He retention among zircons from a cogenetic or geographically localized suite. Holland (1954) showed good correlations between natural radiation dosage and density and a unit cell parameter in Sri Lankan zircons. Holland noted that zircon density changed most rapidly after dosages of about 2×10^{18} α/g. He also noted, on the basis of work by Seitz (1949), that the number of atoms displaced by parent nuclide alpha-decay recoil for a dosage of this magnitude would be about 2×10^{21} displacements/g, about the same as the number of atoms per gram in zircon, thus demonstrating a "satisfactory agreement between the predicted and observed radiation dosage required for disordering the zircon structure completely."

Hurley and coworkers also examined relationships between radiation dosage, He age, refractive index, hardness, specific gravity, birefringence, and other crystallographic indices in zircon, with the hope of using an easily measured indicator of damage as a proxy for age (Hurley 1952; 1954; Hurley and Fairbairn 1953; Hurley et al. 1956). Such uses have not yet been developed, however, partly because it became clear, as was the case later with He retention, that the thermal history of zircons strongly affected the retention of radiation damage. In this work Hurley (1954) foreshadowed the development of (U-Th)/He thermochronometry in general by noting that in at least one region, He ages from a wide variety of minerals seemed to yield very similar ages despite widely varying radiation dosages and known He retentivities. He speculated that in such cases the minerals may "have accumulated helium only since some period of metamorphism."

Although the issue of radiation damage clouded recognition of the thermochronometric potential of (U-Th)/He dating in the early part of the 20th century, there were some early studies that attempted to quantify He diffusion rates in various minerals. Gerling (1939) measured the "heat of diffusion" (activation energy) of He diffusion in several different minerals, including monazite and uraninite, apparently finding increasing values for later stages of the experiments. For some reason, however, despite widespread use of closure concepts and thermochronometry based on other radioisotopic systems, clear recognition of the thermochronometric potential of (U-Th)/He dating did not develop until Zeitler et al.'s (1987) paper on apatite He dating. Subsequent work by Farley and coworkers developed the interpretational and methodological bases of apatite He dating (e.g., Wolf et al. 1996, 1998; Farley et al. 1996; Farley 2000). Subsequent work on other phases has developed experimental bases and methods for (U-Th)/He thermochronometry of titanite (Reiners and Farley 1999) and zircon (Reiners et al. 2002, 2004; Tagami et al. 2003), and shown the utility of these methods in interpreting thermal histories of rocks from a variety of settings (e.g., Reiners et al. 2000, 2003; Stockli et al. 2000; Pik et al. 2003; Stockli and Farley 2004).

Apatite (U-Th)/He (apatite He) dating has received the bulk of attention in the thermochronologic community and in geologic applications, largely because of its uniquely low closure temperature (~60–70 °C for typical crystal sizes and orogenic cooling rates). At the time of writing of this volume, there are three existing reviews focusing on or including significant focus on apatite (U-Th)/He thermochronology (Farley 2002; Farley and Stockli 2002; Ehlers and Farley 2003). Although there have been some important development in apatite He dating since their publication, including insights from $^4He/^3He$ thermochronology (Shuster and Farley 2004, 2005; Shuster et al. 2004), here we focus on zircon He thermochronometry, although we show several examples of thermochronometric results combining apatite He, zircon He, and apatite and zircon fission-track (AFT and ZFT, respectively) data.

HELIUM DIFFUSION IN ZIRCON

Step-heating experiments

As in the case of apatite and titanite, understanding the thermal sensitivity of the zircon (U-Th)/He system relies primarily on results of step-heating experiments to determine He diffusion in zircon. Empirical studies of zircon He age patterns in natural settings with presumably well-understood thermal histories provide important complementary support, however. Reiners et al. (2002, 2004) presented He diffusion results from zircons with a range of ages and U concentrations. In most cases, cycled step-heating experiments on both whole unmodified crystals and interior chips of large gem-quality crystals yielded similar results, in which, following the initial ~5% degassing, activation energies (E_a) were 163–173 kJ/mol (39–41 kcal/mol) and $\log(D_0/a^2)$ were 3.7–4.7 s^{-1} (Fig. 1). These data also provide preliminary indications that, as in the case of apatite and titanite, physical grain size scales with diffusivity (i.e., the crystal size controls the lengthscale of the diffusion domain). Assuming that one half of the minimum dimension of a typical zircon crystal (i.e., the tetragonal prism half-width) corresponds to the diffusion domain lengthscale, then the D_0 derived from these experiments show a much smaller range than the D_0/a^2, with a range between 0.10–1.5 cm^2/s. Reiners et al. (2004) cited means and standard deviations of E_a and D_0 derived from the post-high-temperature heating parts of step-heating experiments as the best estimates for He diffusion in zircon. These were: $E_a = 169 \pm 3.8$ kJ/mol (40.4 ± 0.9 kcal/mol), and $D_0 = 0.46^{+0.87}_{-0.30}$ cm^2/s. For typical igneous zircons with half-widths of ~40–100 μm, these diffusion parameters would yield closure temperatures (Dodson 1973) of 175–193 °C, for a cooling rates of 10 °C/m.y. (Fig. 2).

Reiners et al. (2004) showed that zircons with young He ages (~120 Ma) and low U concentrations (~100 ppm) display essentially identical diffusion characteristics as zircons with old He ages (~440 Ma) and very high U concentrations (~900 ppm). This is important because it suggests that at least in these cases the effects of radiation damage on He diffusivity do not become important until radiation dosages higher than about 2–4 × 10^{18} α/g, a level at which macroscopic and crystallographic characteristics of zircon also show large changes (e.g., Nasdala et al. 2001, 2004).

It is important to note that most available He diffusion experiments for zircon display anomalously high diffusivity in the earliest stages of step-heating. This behavior has also been observed in other minerals and is not well understood, and could have a range of origins, including effects from geometric vagaries of natural grain morphology, localized high-diffusivity from small high radiation-damage zones, crystallographically anisotropic diffusion, or inhomogeneously distributed He. Reiners et al. (2004) compared these non-Arrhenius features of the diffusion experiments with forward models of degassing from

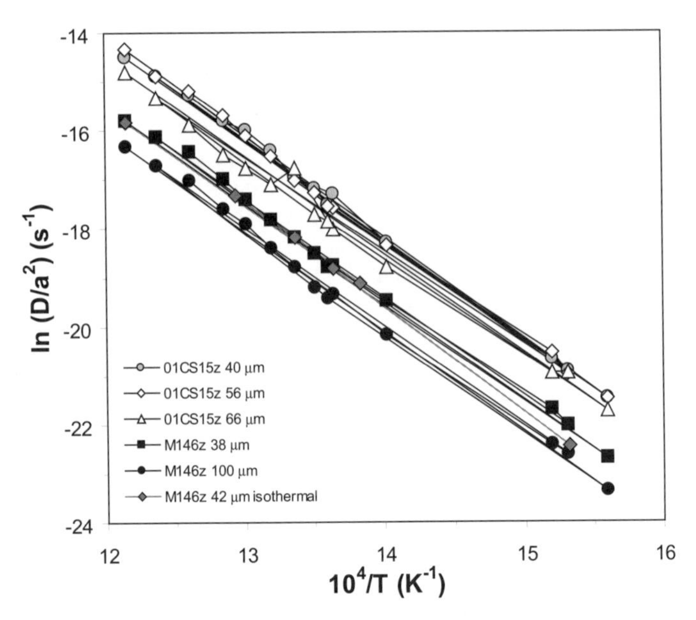

Figure 1. Arrhenius plot of post-high-temperature steps of cycled step-heating He diffusion experiments on different size fractions of two different zircon specimens, after Reiners et al. (2004). Early, low-temperature steps of each experiment displayed anomalously high diffusivity (not shown here), but these effects were absent after the initial steps at higher temperature (425–520 °C). Zircons from sample 01CS15z have a U/Pb age of ~122 Ma and zircon He cooling age of ~100 Ma, and U and Th concentrations of about 175 and 38 ppm, respectively. The sample from which these zircons came is the one at about 100 m from the dike in Figure 6. Specimens from M146z are fragments of a large gem-quality zircon with a U/Pb age of ~570 Ma and zircon He age of ~440 Ma, and U and Th concentrations of 923 and 411 ppm, respectively (Nasdala et al. 2004). Sizes quoted in inset are half-widths of minimum dimensions of individual crystals or crystal fragments. With one exception, data from smaller aliquots are shifted to systematically higher D/a^2, consistent with equivalency of diffusion domain size and grain size in zircon. These experiments indicate E_a = 163–173 kJ/mol (39–41 kcal/mol), and D_0 = 0.09–1.5 cm²/s, with an average E_a of 169 ± 3.8 kJ/mol (40.4 ± 0.9 kcal/mol) and average D_0 of 0.46 $^{+0.87}_{-0.30}$ cm²/s. For an effective grain radius of 60 μm and cooling rate of 10 °C/m.y., these yield closure temperatures, T_c, of 171–196 °C, with an average of 183 °C.

multiple domains of different sizes. These features could be matched well by positing small fractions of gas (2–4%) in domains with length-scales that are a factor of about 25–200 times smaller than the bulk grains themselves (Fig. 3). Zircons with a wide range of ages and radiation dosages exhibited approximately the same degree of non-Arrhenius behavior in initial diffusion steps. Although such modeling does not prove such a mechanism for these non-Arrhenius effects, it suggests that only a small proportion of gas resides in domains that exhibit anomalously high diffusivity, and therefore this phenomenon may not significantly affect the bulk closure temperature or He diffusion properties of most natural zircons.

Although Reiners et al. (2004) suggested that the anomalously high and non-Arrhenius diffusion seen in early stages of step heating experiments can be explained in ways that may not be important for most thermochronometric applications, this has yet to be proven. These and other features of He diffusion in zircon (and in other minerals) such as erratic behavior in some samples, are not easily explained and their origins may yet prove to be important in understanding anomalous ages and model thermal histories. One potential concern for step-heating diffusion studies of the type shown here, and which is common to most other He diffusion studies, is the possibility for crystallographic modifications during experiments

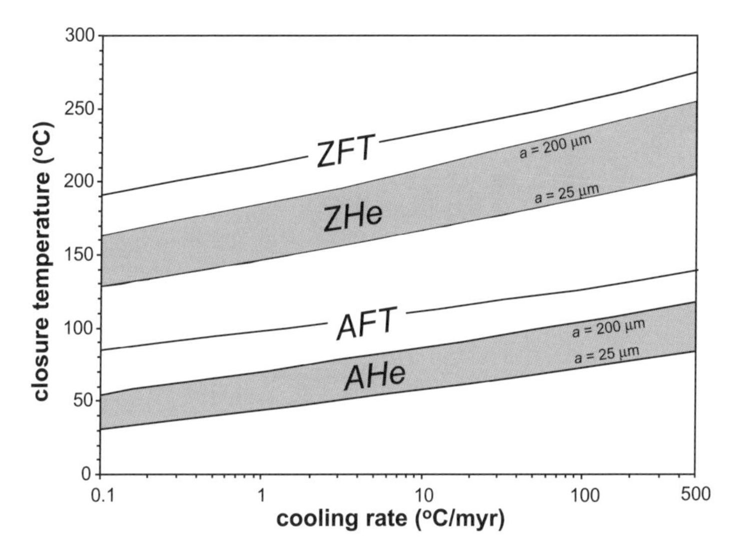

Figure 2. Closure temperatures calculated for apatite He and zircon He (Dodson 1973), using parameters from Farley (2000) and Reiners et al. (2004), respectively, and for AFT and ZFT [following approach of Bernet (2002), after Dodson (1979)] using parameters from Laslett et al. (1987) [$\beta = 9.83 \times 10^{11}$ s^{-1}, E_a = 187 kJ/mol (44.6 kcal/mol)] and compiled data of Brandon and Vance (1992) [$\beta = 1.0 \times 10^8$ s^{-1}, E_a = 208 kJ/mol (49.8 kcal/mol)], respectively. Closure temperatures are calculated for large but not unreasonable potential variations in grain size for apatite He and zircon He, and cooling rates (a = diffusion domain size, corresponding roughly to the radius of a sphere with approximately the same surface-area-to-volume ratio as the crystal). AFT and ZFT closure temperatures are always higher than those for the (U-Th)/He system on the same minerals.

themselves. If heating at experimental temperatures and durations changes crystallographic features that affect diffusivity (e.g., by causing annealing), it is conceivable that such step-heating experiments may underestimate He diffusivity in natural zircons. Other important unresolved questions include the role of localized radiation damage zones in crystals whose bulk compositions would not suggest significant damage, the possibility of anisotropic diffusion, the interaction of microstructures, inclusions, or fission-tracks with migrating He, the possible role of pressure in He diffusion, and He solubility in zircon.

Radiation damage

As indicated by early studies relating radiation dosage and apparent (U-Th)/He ages, there is good evidence that He diffusion in zircon is strongly affected by relatively high degrees of radiation damage. This is most easily seen in suites of old zircons from a single sample (or from samples from a restricted region that experienced similar thermal histories) with a wide range of U concentrations. Detrital zircons from Sri Lanka, which presumably experienced similar thermal histories, show reproducible (U-Th)/He ages of 440 ± 9 Ma (2σ), and He diffusion characteristics similar to much younger and lower U zircons (Reiners et al. 2004), as long as U concentrations are less than ~1000 ppm. At higher U concentrations, zircon He ages decrease rapidly (Fig. 4). Zircons with high U concentrations also liberate He at high rates in vacuum, at low temperature (e.g., we have observed ~0.2 nmol ^{4}He/g/min at room temperature, for a metamict zircon with ~5000 ppm U). The apparent cutoff at which low-temperature He diffusion rates dramatically increase corresponds to a radiation dosage, calculated from either (U-Th)/He age or U/Pb age, of about 2×10^{18} α/g, similar to the dosages at which Holland (1954) and Hurley et al. (1956) observed large changes in He diffusivity.

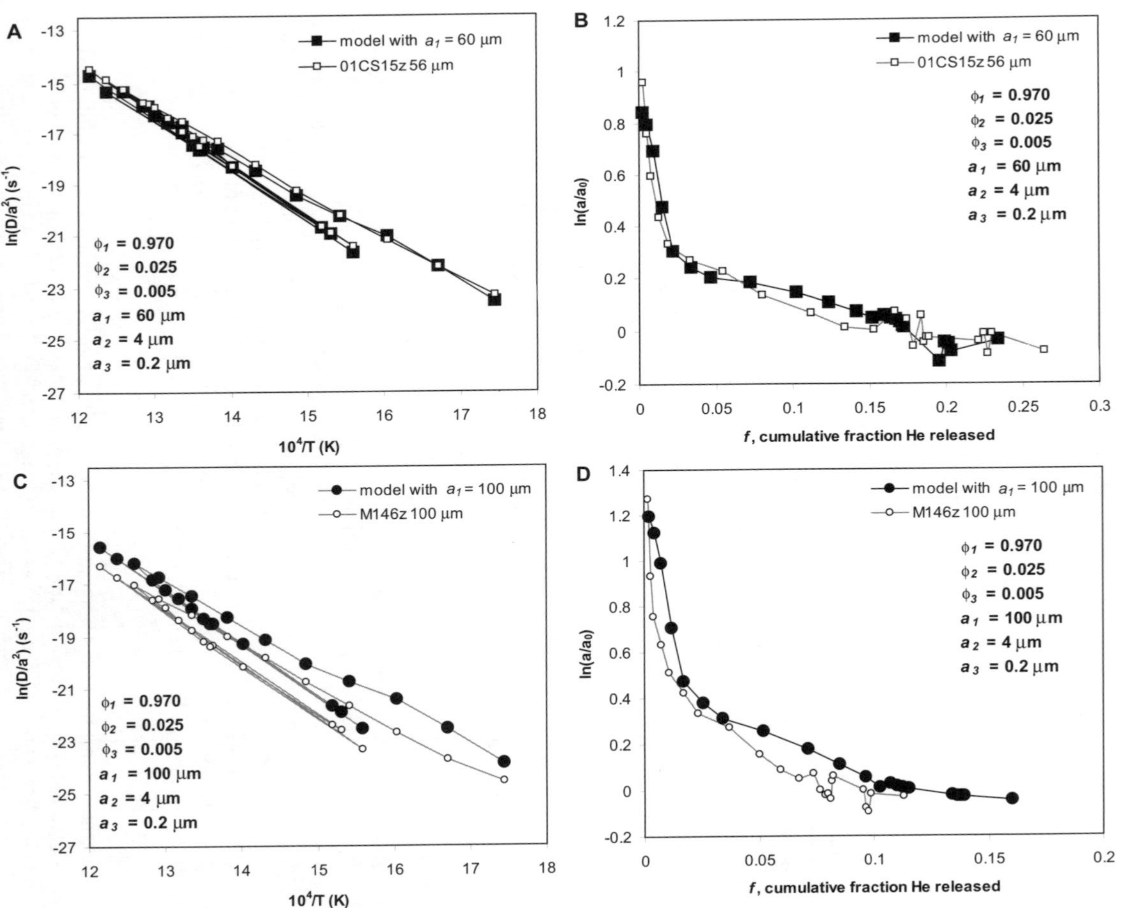

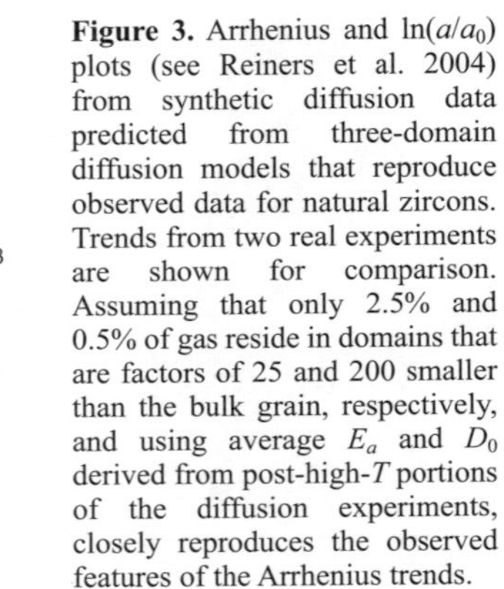

Figure 3. Arrhenius and $\ln(a/a_0)$ plots (see Reiners et al. 2004) from synthetic diffusion data predicted from three-domain diffusion models that reproduce observed data for natural zircons. Trends from two real experiments are shown for comparison. Assuming that only 2.5% and 0.5% of gas reside in domains that are factors of 25 and 200 smaller than the bulk grain, respectively, and using average E_a and D_0 derived from post-high-T portions of the diffusion experiments, closely reproduces the observed features of the Arrhenius trends.

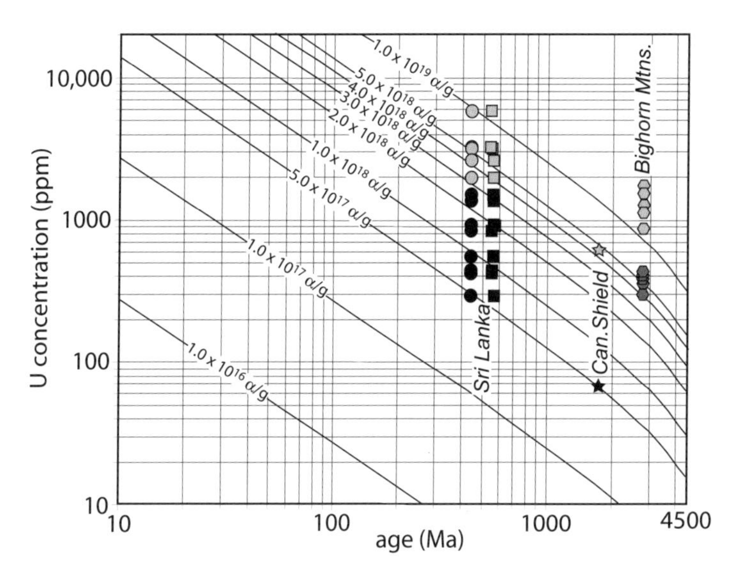

Figure 4. Contours of alpha dosage as a function of age and U concentration, assuming Th/U = 0.5, a typical value for zircon. Also shown are zircons from the detrital Sri Lankan suite studied by Nasdala et al. (2004) (circles), two zircons from the Canadian shield (stars; samples courtesy of Rebecca Flowers, MIT), and a suite of zircons from the Shell Canyon region of the Bighorn Mountains (octagons) (samples BH-12 and BH-17 of Reiners and Farley, 2001). Sri Lankan zircons are shown for both He age (442 Ma) and U/Pb age (555-560 Ma). Canadian shield and Bighorns zircons are shown for $^{40}Ar/^{39}Ar$ biotite cooling ages of 1.76 Ga and 2.8 Ga, respectively. Black symbols denote samples yielding zircon He ages that are reproducible and consistent with other thermochronologic constraints, implying insufficient accumulated radiation damage to affect the He age. Light grey symbols represent samples with anomalously young and unreproducible He ages, attributed to the effects of high radiation damage. Darker grey symbols for the low-U Bighorn samples denote samples with ages between 330–570 Ma; lighter grey symbols denote samples with ages between 7–178 Ma. None of these samples have reproducible zircon He ages at any U concentration, suggesting that they have all accumulated radiation damage sufficient to cause low-temperature He loss and anomalously young zircon He ages. In most suites of zircons, the transition from reproducible old ages to unreproducible young ages appears to occur at approximately $2–4 \times 10^{18}$ α/g.

Nasdala et al. (2004) and others have shown that the alpha-parent-recoil radiation damage responsible for metamictization is annealed at elevated temperatures. Although the kinetics of radiation damage are not well understood, it is clear that many zircons do not retain radiation damage that is simply proportional to their crystallization ages. In the case of the Sri Lankan suite, the retained radiation damage is only about one-half of that which would be accumulated since the U/Pb ages. This suggests that the critical radiation dosage for rapid low-temperature He loss in zircon should be calculated only for the duration of time a zircon has spent below some temperature at which damage is accumulated; in other words, from a cooling age rather than U/Pb age, though it is not yet clear exactly what temperature should be used for the cooling age.

Other suites of zircons with old U/Pb ages and relatively long low-temperature histories show similar results consistent with dramatically reduced He retentivity at higher radiation dosages. In some cases, it is not possible to estimate the extent of retained radiation dosage as a function of U-Th concentration, because no zircons at any U concentration provide reliable and reproducible He ages, and few if any other constraints are available on the low-temperature thermal histories. However in some cases, suites of zircons with wide ranges of U concentrations can be found in rocks with cooling age and thermal history constraints from other systems.

Such cases can be used to place at least some constraints on the effective radiation dosage at which low temperature He loss increases rapidly. Figure 4 shows U concentrations and cooling ages of several zircon suites in which single-crystal age measurements appear to cross a threshold of radiation damage leading to anomalously young and irreproducible He ages. These data also suggest that this low-temperature radiation accumulation limit lies somewhere between about 2–4 × 10^{18} α/g. Assuming a critical retained dosage of 2 × 10^{18} α/g represents an effective upper dosage limit for meaningful (U-Th)/He ages, then typical zircons with U concentrations of about 100-1000 ppm would require full retention at low temperatures for at least 0.6 to 4.0 b.y.. Potentially more insidious effects of smaller degrees of radiation damage at intermediate temperatures are not yet known.

As noted by Nasdala et al. (2004), strong He diffusivity changes in zircon are observed at a critical accumulated radiation dosage corresponding to double-overlapping of alpha recoil damage zones (see also Nasdala et al. 2001). If the relationship between He diffusivity and alpha damage is also systematic at intermediate extents of damage (i.e., not just a critical threshold), it may be possible to use the specific relationship between alpha fluence and apparent (U-Th)/He age among zircons from the same rock to deduce thermal histories. Figure 5 shows the apparent zircon He ages as a function of alpha fluence for zircons from rocks with similar

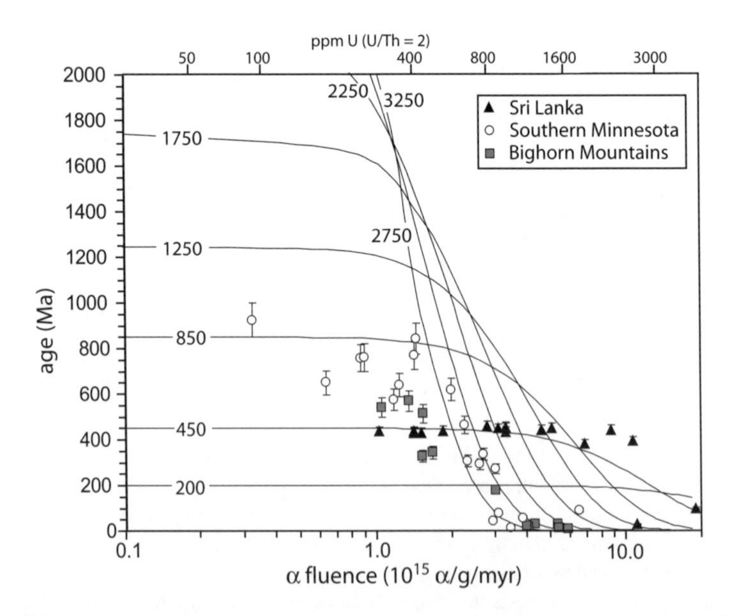

Figure 5. Measured zircon He age versus alpha fluence from measured U and Th concentrations for zircons from a range of settings. The upper *x*-axis is for reference, to denote U concentrations that would produce given alpha fluences for a U/Th of 2, typical of zircon; it does not denote actual U concentrations in these zircons. Zircons shown are: (1) the Sri Lankan suite of Nasdala et al. (2004), with crystallization and cooling ages of ~560 Ma and ~440 Ma, respectively; (2) granitoids of the Shell Canyon section, Bighorn mountains, Wyoming (Reiners and Farley 2001), with a crystallization age of ~2.8 Ga and an unknown age of cooling through the zircon He closure temperature (apatite He ages on the same samples require temperatures less than ~70 °C for at least the last ~600 m.y.); (3) granitoids and metamorphic rocks of the Minnesota river valley, with crystallization ages ranging from ~2.6–3.6 Ga and unknown cooling ages. Model trends are constructed to test the simple hypothesis that the ratio of observed to "actual" zircon He cooling age is equal to the fraction of remaining crystallinity in the zircon, where the latter is determined using the double-overlapping cascade model of Holland and Gottfried (1955). This simple model appears to work for the Sri Lankan zircons, but not for the older groups, suggesting a more complex relationship between He retention and fractional crystallinity from radiation damage.

thermal histories involving long-term residence at low temperatures (note that this analysis makes use of alpha fluence, or rate of production, not cumulative alpha dosage). Also shown in this figure are trends of zircon He ages that would be measured for zircons with the same original He cooling age, but varying radiation dosage, assuming that the ratio of observed zircon He age to actual cooling age is equal to the fraction of remaining crystallinity, where crystallinity is determined by the double-overlapping cascade model (Holland and Gottfried 1955). In the context of such a hypothetical model, zircons with young "actual" cooling ages would preserve geologically significant zircon He ages even at high alpha fluences, but with increasing "actual" cooling age, zircons with progressively lower alpha fluences would show decreasing apparent ages. The Sri Lankan zircon suite shows a trend that may be consistent with this model, with high fluence grains falling to lower ages than the inferred cooling age for these samples in approximate proportion to crystallinity. Other samples, from granitoids of southern Minnesota and the Bighorn Mountains of Wyoming, however, are not consistent with this simple model. At relatively high fluences, these zircons show age-fluence trends similar to the model trends for "actual" cooling ages of ~2.5–3.0 Ga, but at lower fluences, the ages are lower than expected for such trends. This may mean that He diffusivity does not scale linearly with crystallinity and/or alpha damage in zircon. More work is clearly needed to understand the relationships between radiation damage and He diffusivity, and to extract any potentially useful information in age-fluence relationships. It is worth noting again, however, that at least in the cases we have examined such as the Sri Lankan suite, radiation damage effects only become important at fairly high accumulated radiation dosages, requiring old ages, long-term residence at low temperatures, and relatively high U-Th concentrations.

ANALYTICAL AND AGE DETERMINATION TECHNIQUES

Analytical methods

Most (U-Th)/He dating procedures involve measurement of parent and daughter nuclides on the same aliquots, because of the potential for varying U and Th concentrations among crystals. Single aliquot measurements are particularly important for zircon He dating, because U and Th concentrations among zircon crystals often show large variations. Currently, measurement of a reasonably precise He age using routine methods in the Yale (U-Th)/He chronometry lab typically requires ^{4}He contents at least as high as about 0.3 fmol. For a typical zircon of about 200 μm length and 100 μm width, with 350 ppm U and U/Th ~2, this requires only about 20 kyr of ingrowth. Thus, most zircon He applications require only single crystals, though it is conceivable that some applications, such as dating of very young or low-U-Th samples may be better suited to multi-crystal aliquots. Ideally, at least two single crystal replicates of zircons are dated (many more are dated if the sample comprises unreset detrital grains), to check reproducibility of single grain ages. The procedures described here are those employed for routine zircon He dating in the Yale (U-Th)/He chronometry lab, and may differ from protocols in other labs.

Crystal selection, documentation, and alpha-ejection corrections. Dated crystals are selected from heavy mineral separates prepared by standard procedures, on the basis of size, morphology, abundance of inclusions, and clarity. The following selection guidelines apply to non-detrital zircons, where there is no risk of biasing age populations by grain selection criteria. In general, suitable crystals have tetragonal prism widths of at least 75–90 μm. As a general rule of thumb, except in rare circumstances crystals with tetragonal prism widths less than 60 μm are not dated, because uncertainties in the very large alpha-ejection corrections (and inherent assumptions about parent zonation required in conventional analyses) can lead to potentially large errors. Interestingly, however, Hourigan et al. (2005) noted that U-Th zonation effects on age inaccuracy actually decrease at crystal sizes less than ~60 μm, because at such small sizes, all parts of the crystal are affected by alpha ejection roughly equally. The potential disadvantage of extremely large crystals (e.g., ~300 μm widths), however, is that they may require multiple heating and gas extractions to reduce subsequent extractions to less than ~2% of the total, and such large grains occasionally also show dissolution problems. Thus optimally sized crystals for routine zircon He dating procedures have tetragonal prism widths of about 75–150 μm.

Alpha-ejection corrections (described below) require an assumption of a characteristic grain morphology, so in many cases grains with morphologies most similar to an idealized tetragonal prism with bipyramidal terminations are selected. Other suitable morphologies include regular prolate spheroids or tetragonal prisms with broken ends perpendicular to the *c*-axis. In general, however, highly irregular morphologies or grains with obviously fractured surfaces at low angles to the *c*-axis are avoided. In most samples, it is difficult to select zircons without any inclusions, but in order to minimize potential zonation effects, grains with few large or obvious inclusions are selected.

Selected crystals are photographed and their dimensions are measured in at least two mutually perpendicular perspectives parallel to the a_1 and a_2 crystallographic axes. These dimensions and an assigned morphology are used to calculate the alpha-ejection correction, to account for ^{4}He lost from the crystal by long-stopping distances of alpha-particles. In zircon, stopping distances average about 17.0 μm for the ^{238}U series, 19.6 for the ^{235}U series, and 19.3 for the ^{232}Th series (Farley et al. 1996; Hourigan et al. 2005). Farley (2002) provided equations for zircon alpha-ejection corrections based on assumed crystal morphologies of tetragonal prisms with pinacoidal terminations. This approach used Monte-Carlo modeling to parameterize second order polynomial factors as a function of crystal surface-area-to-volume ratio β, to solve for fractions of He retained within the crystal for the U- and Th-series individually. Measured Th/U of individual zircons are then used to weight the series' specific retention factors appropriately. This approach has been used in most zircon He dating studies, and provides accurate ages for standards of known age (Kirby et al. 2002; Reiners et al. 2002, 2003, 2004; Tagami et al. 2003).

As part of a larger study examining intracrystalline U-Th zonation in zircon He dating, Hourigan et al. (2005) examined the effects of more realistic crystal geometries on alpha-ejection correction. They found that bipyramidal terminations generally have the effect of changing alpha-ejection corrections from those calculated assuming pinacoidal terminations by about 1–3%. The magnitude of the discrepancy is proportional to the fraction of the total *c*-axis parallel length of the crystal comprising the pyramidal tips. Typical zircons have pyramidal terminations that are about 10–30% of the total length, although considerable variation exists in natural crystals. Using this more realistic morphology, Hourigan et al. (2005) determined new factors for the polynomial on β for determining fraction of He retained in the crystal. Table 1 shows the U- and Th-series factors determined in that study.

The surface-area-to-volume ratio β, is calculated from equations shown in Table 2. Some natural zircons, especially those in sedimentary environments, are well rounded and bear more resemblance to prolate spheroids than to tetragonal prisms with or without pyramidal terminations. In such cases, measurements of equatorial and polar radii are made (with a mean equatorial radius determined from the average of two mutually perpendicular measurements), and different equations are used for surface-area and volume calculations (Table 2).

Although concentration determinations of neither parent nor daughter are required for He age determinations, most studies report estimated U and Th concentrations of dated crystals. This can be useful in several ways, including assessing radiation damage effects, identifying anomalous crystals, or elucidating zonation from He-Pb double-dating results (Reiners et al. 2005). Because single crystal aliquots are typically too small to precisely weigh by standard

Table 1. Factors A_1 and A_2 for calculating fraction of He retained in crystals from the ^{238}U and ^{232}Th decay series in zircon, for different assumed crystal geometries.

	Tetrahedral prism with pinacoidal terminations				*Tetrahedral prism with pyramidal terminations*	
	(Farley 2002)		(Hourigan et al. 2005)		(Hourigan et al. 2005)	
Parent Nuclide	A_1	A_2	A_1	A_2	A_1	A_2
^{238}U	−4.31	4.92	−4.35	5.47	−4.28	4.37
^{232}Th	−5.00	6.80	−4.94	6.88	−4.87	5.61

Table 2. Surface areas and volumes of zircons for assumed geometries of zircon crystals.

Geometry	*Volume*	*Surface Area*
Tetragonal* prism with pyramidal terminations	$V_z = 4 r_1 r_2\left[(l-h_1-h_2)+\frac{1}{3}(h_1+h_2)\right]$	$SA_z = 4(l-h_1-h_2)(r_1+r_2)+2r_1a+2r_2b$ $a=\sqrt{h_1^2+r_2^2}+\sqrt{h_2^2+r_2^2}$ $b=\sqrt{h_1^2+r_1^2}+\sqrt{h_2^2+r_1^2}$
Prolate Spheroid	$V_{ps} = \frac{2}{3}\pi r^2 l$	$SA_{ps} = 2\pi r^2 + \left[\frac{2\pi r\left(\frac{l}{2}\right)^2}{\sqrt{\left(\frac{l}{2}\right)^2 - r^2}}\right]\sin^{-1}\left[\frac{\sqrt{\left(\frac{l}{2}\right)^2 - r^2}}{\left(\frac{l}{2}\right)}\right]$

Note: l = c-axis-parallel length; h_1, h_2 = pyramidal termination lengths; r_1, r_2 = mutually-perpendicular prism half-widths or average equatorial radius. Prolate spheroid geometry was not modeled for independent polynomial factors, and is assumed to have the same A's as the tetragonal prism with pyramidal terminations.

*Although the Monte Carlo models are derived from tetragonal prism morphologies, differing prism half-widths are typically measured to calculate volumes and surface areas, because typical zircons are commonly significantly orthogonal.

techniques, mass estimates for concentrations are determined from volume estimates, assuming an average density, for which we use 4.6 g/cm^3 for zircon. Given that grain dimension measurements typically include length, widths, and, if pyramidal terminations are considered, their lengths, or else polar and equatorial radii, the morphological assumption used for volume calculation has a large effect on estimated mass. Using realistic morphologies as opposed to simplified tetragonal prisms with pinacoidal terminations results in estimated volumes and masses that are typically 10–50% lower (Reiners et al. 2005).

In summary, alpha-ejection corrections are made in the following way. Following Farley et al. (1996) and Farley (2002), the fraction of He retained in crystals due to alpha-ejection (F_{He}) for ^{238}U and ^{232}Th (^{235}U is similar to ^{232}Th) are calculated from a polynomial fit to surface-area-to-volume ratios and fraction of alphas retained in the crystal, where:

$$F_{He} = 1 + A_1\beta + A_2\beta^2 \tag{1}$$

where A_1 and A_2 are given for different morphologies in Table 1. The F_{He} for each nuclide is then weighted according to the measured Th/U of the sample, to derive the bulk F_{He}, using the equations:

$$\hat{F}_{He} = a_{238}\,{}^{238U}F_{He} + (1 - a_{238})\,{}^{232Th}F_{He} \tag{2}$$

and

$$a_{238} = \left(1.04 + 0.245(Th/U)\right)^{-1} \tag{3}$$

Extraction and measurement of He, U, and Th. Early analytical procedures extracted He from zircons by either *in vacuo* fluxing or heating of crystals in small (~5 mm) stainless steel or Ti capsules in a resistance furnace at ~1200–1300 °C (Reiners et al. 2002). Although the latter method was successful in most cases, disadvantages include potential loss of material for U-Th analyses upon retrieval from the capsules, as well as the relatively high He blanks and time-consuming temperature cycling of furnace heating, compared with laser heating. Routine He extraction techniques in the Yale lab now typically involve placement of a single crystal into a small (~1 mm) Nb foil envelope that is lightly crimped closed, and heating of the foil by direct lasing with a focused 10 μm beam of a 1064-nm Nd:YAG laser. The foil sits in a Cu or stainless steel planchet with a few dozen sample slots, directly underneath a KBr coverslip, which prevents vapor deposition on the underside of a sapphire viewport above the samples, in a high-vacuum sample chamber connected to the He purification/measurement line. Nb foils are heated to approximately 1100–1250 °C for 15 minute extraction intervals; all samples are then subject to at least two and occasionally more extractions and He measurements, to assess the extent of degassing of the crystal. Typically, re-extracts yield less than 0.5% of previous ^4He yields, but some zircons can display stubborn He extraction, with multiple extracts required to reduced successive yields to less than 2–3%. It is not known why some zircons display such reticent He extraction during laser heating (it is not obviously related to radiation dosage, thermal history, U-Th content, for example), or whether this reflects unusual He diffusion properties during natural cooling as well.

Gas extracted from zircons by heating is spiked with approximately 0.1–1.0 pmol ^3He, cryogenically (and/or via gettering) concentrated and purified, and expanded into a small volume with a gas-source quadrupole mass spectrometer. The $^4He/^3He$ is measured for about ten seconds following gas release and nominal equilibration time. The measured ratio is corrected for background and interferences on mass 3 (HD^+ and H_3^+), and compared with the $^4He/^3He$ measured on pipetted aliquots of a manometrically calibrated ^4He standard processed by the same methods. ^4He in the unknown zircon is assumed to be the product of the ^4He content of the standard with the ratio of the $^4He/^3He$ measurements on the unknown and the standard. Linearity of this calibration approach has been confirmed in the Yale He lines over about four orders of magnitude of ^4He signal. Each batch of zircons is processed with several "hot blanks" and "line blanks" to check the measured $^4He/^3He$ of laser extraction procedures on empty Nb foil envelopes, and $^4He/^3He$ of ^3He-only shots. Nominal ^4He blanks from these procedures range between 0.05–0.1 fmol, although reproducibility of blanks, and therefore uncertainty on unknown He contents make precise determination of ^4He contents lower than roughly 0.3 fmol difficult.

Parent nuclide contents of degassed zircons are measured by isotope dilution and solution ICP-MS. This requires spiking with isotopically distinctive U-Th spike, sample-spike equilibration, and dissolution to a final solution suitable for ICP-MS. Zircon requires different dissolution techniques than those developed for apatite or titanite (House et al. 2000; Reiners and Farley 1999), because most zircons do not dissolve in nitric acid alone and require HF-HNO_3 mixtures at higher than ambient temperatures and pressures. Dissolution of zircon directly from Pt foil, as in the case of apatite, is not possible because of Pt dissolution, forming $PtAr^+$ complexes in the ICP-MS with isobaric interferences on U isotopes. In principle,

$^{195}Pt^{40}Ar$ and $^{198}Pt^{40}Ar$ could be resolved from ^{235}U and ^{238}U using high resolution, but the disadvantages of lower sensitivity, mass calibration, and triangular peak shape make avoidance of dissolved Pt in the solution a simpler option. House et al. (2000) used Pd foil for wrapping and laser-heating of titanite, and dissolved titanite directly from the foil in a spiked HCl-HF solution. In our experience however, the lower melting temperature of Pd made it more prone to melting during lasing, compromising quantitative recovery of the U-Th in the zircon. Rather than unwrapping Pt foils and retrieving naked zircons, which also potentially compromises quantitative parent recovery, an alternative procedure is to dissolve the entire Nb foil and contents in Parr pressure digestion vessels (Parr bombs). An alternative procedure, outlined by Tagami et al. (2003), involves flux melting of zircons in Pt foils in the presence of U-Th spike, and subsequent dissolution of the resulting glass for ICP-MS measurement. This procedure avoids time-intensive acid dissolution steps, but, at least as reported in Tagami et al. (2003), yields U-Th blanks about 4–25 times higher than those we observe for acid dissolution.

The first step of our U-Th measurement procedure is spiking of zircon-bearing Nb-foils with 0.4–0.8 ng of ^{233}U and 0.6–1.2 ng of ^{229}Th. In contrast to some apatites, the low Sm content of zircon generally does not lead to significant age contributions from this element. The foil and spike are initially bombed in Teflon vials at 225 °C for 72 hours. Samples are then heated to dryness and then rebombed and dissolved in HCl at 200 °C for 24 hours to redissolve refractory fluoride salts. After a final drydown the sample is redissolved in 6% HNO_3 and 0.8% HF, and an aliquot of this solution is introduced to the ICPMS via an all-PFA samples introduction system with sapphire injector. Ratios of $^{238}U/^{233}U$ and $^{232}Th/^{229}Th$ are quantified by 2000 measurements of the average intensities in the middle 10% of peakwidths in low resolution mode on an Element2 high-resolution ICP-MS. $^{238}U/^{235}U$ is also measured to check for Pt contamination and mass fractionation. U and Th contents of zircons are calculated from multiple determinations of isotope ratios on pure spike and spiked normals containing 1–4 ng of isotopically normal U and Th. Procedural blanks for U and Th are determined by processing empty Nb foil envelopes, and average 2.9 ± 0.9 pg U and 5.6 ± 1.0 pg Th; this is about four to ten times higher than procedural blanks for apatite.

Figure 6 shows 83 single-crystal analyses of zircons from the Fish Canyon Tuff (FCT) analyzed by methods described above; our mean age is 28.3 Ma, and two standard deviations of the population is 2.6 Ma (9.1%). The most precise U/Pb age of FCT zircon is 28.48 ± 0.06 Ma (2σ) (Schmitz and Bowring 2001). There is some debate as to the crystallization and cooling ages of different phases in the FCT (e.g., Lanphere and Baadsgaard 2001; Schmitz and others 2001; Dazé et al. 2003; Schimtz and others 2003). Some chronometers (with nominally lower closure temperatures) yield ages as young as 27.5 Ma (e.g., Lanphere and Baadsgaard 2001), but Schmitz and Bowring (2001) argue against a prolonged magmatic residence origin for this discrepancy, and several studies suggest systematic ^{40}K decay constant errors may be partly responsible (Min et al. 2000; Kwon et al. 2002). In any case, this uncertainty in FCT ages is well within even one standard deviation of the observed zircon He ages (~1.3 Ma).

Age calculation considerations unique to zircon He dating. Several additional considerations may be necessary in zircon He dating. One that has been recognized for some time is the potential for He age inaccuracies via intracrystalline zonation of U and Th (Farley et al. 1996; Reiners et al. 2004; Hourigan et al. 2005). This problem is not unique to zircon, but because intracrystalline zonation in zircon may be particularly strong, it may be particularly important in some cases of systematic zonation. Strongly heterogeneous distribution of U and Th within crystals affects the alpha-ejection correction because, at least as conventionally applied, this correction assumes uniform distribution of parent nuclides. In the extreme, if all parent nuclides are located on the rim of a crystal, the fraction of alphas retained due to ejection would be slightly less than 0.5 (depending on crystal geometry and size). Conversely

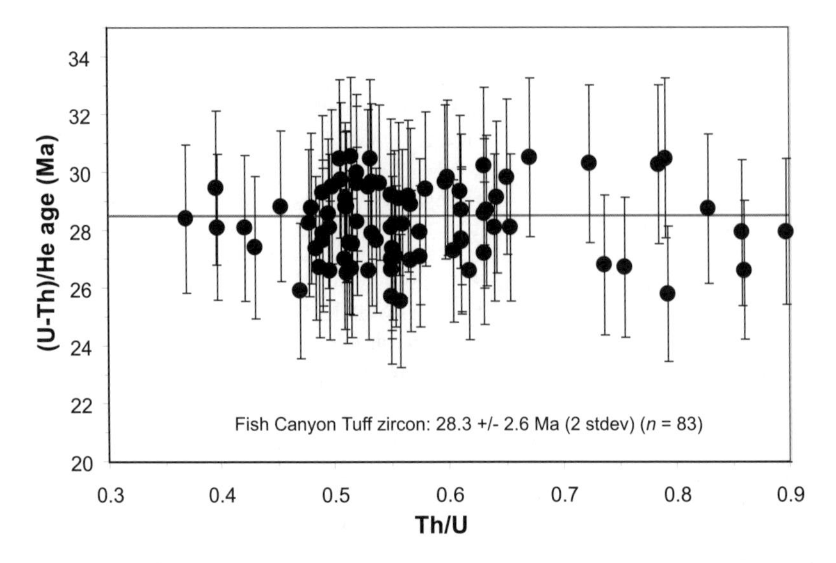

Figure 6. (U-Th)/He ages of 83 single-crystal Fish Canyon Tuff zircons plotted against Th/U, analyzed by procedures described in this chapter, involving Nd:YAG heating of grains in Nb foil, followed by Parr bomb dissolution, and isotope dilution on a high-resolution ICP-MS. The mean age is 28.3 Ma, and two-standard-deviation range of the population is 2.6 Ma (9%). The horizontal grey bar shows the ID-TIMS age and uncertainty of FCT zircon given by Schmitz and Bowring (2001): 28.48 ± 0.06 Ma (2σ).

if all parents are located more than one alpha stopping distance from the rim, no correction would be needed for this effect ($F_{He} = 1$). In practice however, the maximum age inaccuracy that could occur for the most extreme zonation for the vast majority of potential crystal sizes is about 35%, and this decreases with increasing crystal size above about 60 μm prism width.

Hourigan et al. (2005) examined the effects of U-Th zonation on zircon He age inaccuracies for a wide variety of crystal morphologies and styles and extents of zonation. Zircons with U-Th enriched cores produce significantly "too-old" ages (if homogeneous U-Th distributions were assumed in the alpha-ejection correction) with maximum inaccuracies for depleted rims about one stopping distance thick (Fig. 7). The magnitude of this effect however, decreases rapidly as the depleted rim becomes thinner. Enriched rims, however, pose potentially significant age inaccuracies ("too-young" ages) even if they are only 1–2 μm wide. One way to deal with the potential effects of zonation in practice is to examine images of common zonation types and extents in populations of grains exposed in polished mounts and assume similar features in specific grains selected from the population for dating. This approach is often not realistic, however, because of the large variety of zonation exhibited in single samples, especially for detrital samples. Another approach is simply to perform multiple single-crystal age determinations on single samples, though this may not rule out systematic zonation. Hourigan et al. (2005) have developed methods for grain-specific characterization of U-Th zonation by depth profiling in one-dimensional core-to-rim laser ablation pits in crystals selected for dating. These depth-profiles can be converted to zonation models for arbitrary patterns and used to derive customized alpha-ejection corrections specific to each grain. This approach has shown promise for ameliorating systematic age inaccuracies in problematic specimens (Reiners et al. 2004; Hourigan et al. 2005), although complex zonation patterns may still pose problems. Finally, another method under development is characterization of two-dimensional zonation patterns in dated grains by selecting zircons from mounts used for fission-track dating.

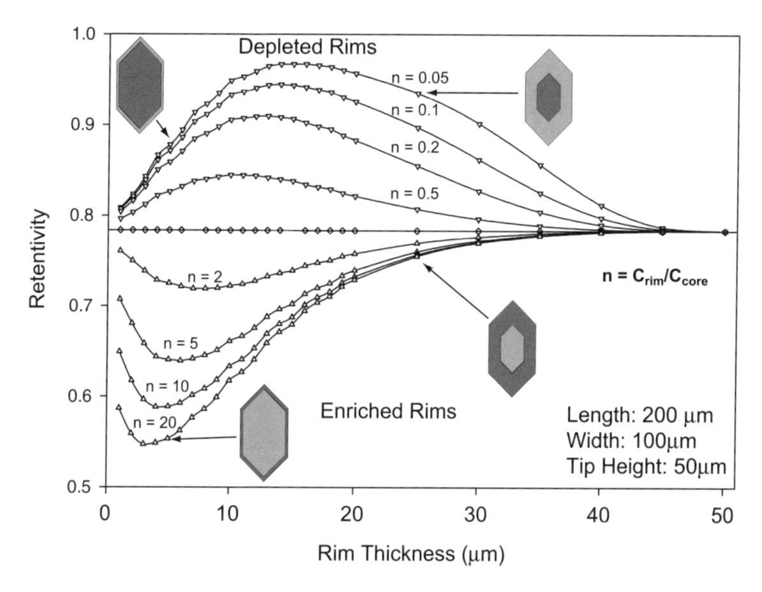

Figure 7. Zoning-dependent bulk retentivity plots for tetragonal model crystals with rims (concentration step functions) of variable width and degree of enrichment or depletion, after Hourigan et al. (2005). The model crystal is a tetragonal prism 200 × 100 μm, with pyramidal terminations (tip height = 50 μm). Although step-function, concentric zonation is obviously a simplification to natural zonation in zircons, this model provides an indication of the magnitudes and styles of zonation required for significant age inaccuracies. Zircons with U-Th enriched rims will show maximum age inaccuracies (as high as about 40% for this morphology) when rims are thin (~2–5 μm), but those with U-Th depleted rims show maximum age inaccuracies (~25%) when rims are about one alpha-stopping-distance thick.

Another consideration that may be important for detrital zircons is the effect of natural abrasion of crystals during transport, and the removal of all or part of the alpha-ejection depleted rim of the crystal. This problem was treated by Rahl et al. (2003) in their study of highly rounded aeolian zircons of the Navajo sandstone. In some cases, abrasion may be so significant that alpha-ejection should only be applied to the post-depositional history of the crystal. Rahl et al.'s equation for a modified post-depositional alpha-ejection corrected age A_c, is:

$$A_c = A_d\left(1 - F_{\mathrm{He}}\right) + A_m \tag{4}$$

where A_d is the depositional age of the host sedimentary rock, F_{He} is the standard alpha-ejection correction factor as described above, and A_m is the measured age as determined from relative abundances of parent and daughter nuclides

The relatively high U-Th concentrations of most zircons make zircon He dating potentially useful for dating young (10^3–10^5 ka) volcanic rocks, in age ranges that may be difficult to access by other techniques, as demonstrated by Farley et al. (2002). In such cases, however, consideration of the effects of secular disequilibrium are necessary. Zircon typically has U/Th greater than unity, and thus may exclude ^{230}Th relative to ^{238}U during crystallization. Establishment of secular equilibrium between ^{238}U and ^{230}Th occurs only after about five half-lives of ^{230}Th, or ~375 kyr. For zircons with initial activity ratios of (^{230}Th/^{238}U) < 1, He will be produced at a slower rate for the first ~375 kyr following crystallization, and zircon He ages in this range may be several percent to several tens of percent too-young, depending on the age. As shown by Farley et al. (2002), correction for this secular disequilibrium requires knowledge of (^{230}Th/^{238}U) of the melt from which the zircon crystallized, which may be constrained by analyzing Th/U of cogenetic phases that fractionate Th/U differently, such as apatite and zircon. An additional important consideration is the magma residence time of crystallized phases, which, for zircon, may be typically about 100–500 kyr (Halliday et al. 1989; Reid et al. 1997; Hawkesworth et al. 2000), comparable with the duration of time required for establishment of secular equilibrium, or longer.

It is worth noting that only when (U-Th)/He ages reflect formation, as opposed to cooling, ages, is consideration of secular equilibrium required. Numerous examples of Holocene through Pleistocene zircon He ages have been documented that reflect resetting by magmatic heating, hydrocarbon burning, or cooling associated with tectonic or erosional exhumation. In such cases, cooling ages are much younger than formation ages, and, assuming no loss of intermediate daughters during cooling or the resetting event, U-series disequilibria considerations are not necessary.

CASE-STUDY EXAMPLES

Comparison with K-feldspar $^{40}Ar/^{39}Ar$ cooling models

K-feldspar $^{40}Ar/^{39}Ar$ age spectra from step-heating degassing experiments are widely used to model thermal histories of rocks between temperatures of about 350–150 °C (Lovera et al. 1989, 1991, 1997, 2002; Richter et al. 1991; Harrison et al. 2005). The ability of K-feldspar to provide continuous time-temperature paths, rather than a single point in the thermal history, arises from the inferred multi-domain or multi-path Ar diffusion behavior exhibited by this mineral. The temperature range of many K-feldspar cooling models overlaps with the closure temperatures for the zircon He system inferred from diffusion experiments. Several studies have presented results of zircon He ages from rocks with K-feldspar $^{40}Ar/^{39}Ar$ cooling models (Kirby et al. 2002; Reiners et al. 2004; Stockli 2005). Although in most cases K-feldspar models suggested relatively rapid cooling rates through the zircon He closure temperatures, the results showed good agreement between the two techniques in nearly all cases (Fig. 8), suggesting that the experimentally determined He diffusion parameters for zircon and its inferred closure temperatures apply in natural settings. In some cases, although mean zircon He ages calculated from single crystal ages still overlapped with K-feldspar cooling models, some single crystals were displaced to older ages. Some proportion of zircons in these samples displayed fairly extreme intracrystalline U-Th zonation (core-to-rim contrasts up to a factor of ~30), which was suggested as a likely cause of some single grain age discrepancies. As discussed subsequently, severe intracrystalline U-Th zonation may lead to scatter in He ages of any mineral, and in the case of zircon, maximum age inaccuracies by the most extreme zonation would not exceed ~35%.

Dike heating

In certain circumstances, wallrocks adjacent to basaltic dikes provide opportunities to compare experimentally determined thermal sensitivities of thermochronometric systems to their natural behavior in a presumably relatively well controlled setting. Characteristic timescales of thermal histories associated with heating from large (~10 m) dikes are typically 10^2–10^4 yr, providing potential calibration tests at an intermediate timescale between the ~10^{-3}–10^{-1} yr of laboratory experiments and ~10^5–10^7 yr natural tectonic processes. Although uncertainties in detailed intrusion history and thermal properties of magma and wallrocks may prevent robust tests or "natural calibrations" based on such results, these settings provide at least first-order consistency tests, and qualitative comparisons of resetting profiles among different thermochronometers with differing thermal sensitivities. Figure 9 shows fission-track and (U-Th)/He ages of zircon and apatite in a horizontal transect of trondhjemitic wallrocks adjacent to a ~9 m wide vertical dike of basaltic andesite in the Wallowa Mountains of

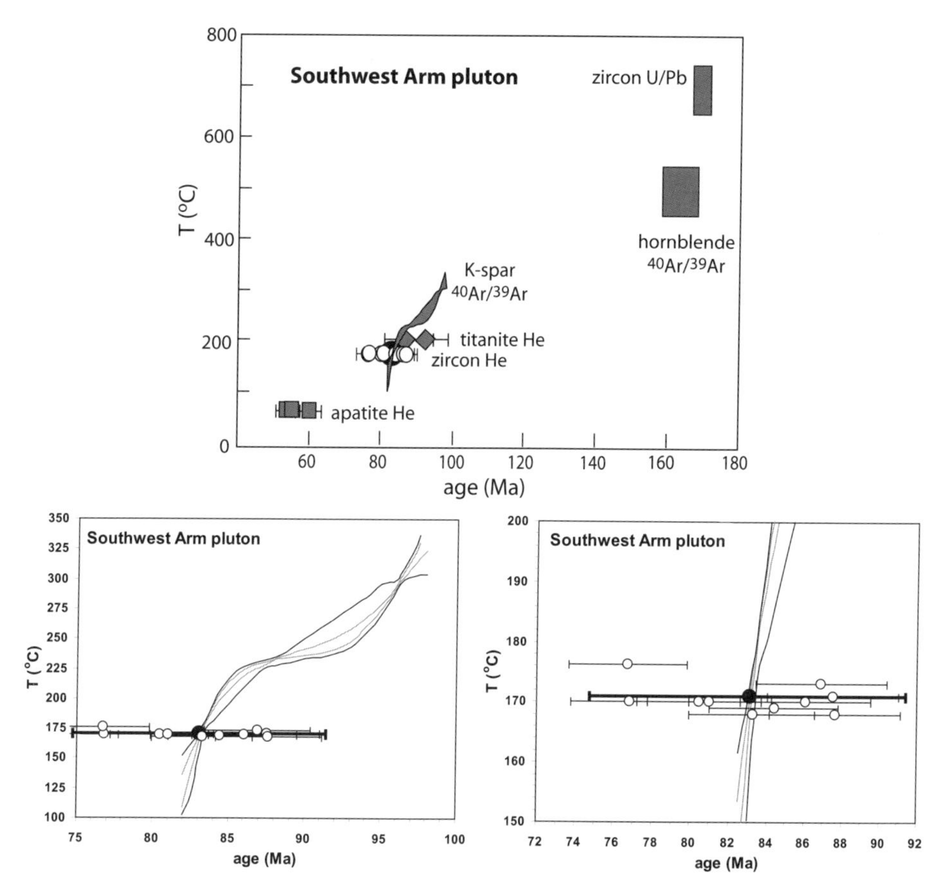

Figure 8. Comparisons between single crystal and averaged zircon He ages and other thermochronometric constraints from the Southwest Arm pluton, Stewart Island, New Zealand (after Reiners et al. 2004). Upper panel: all available data. Lower panels: Detailed comparisons between K-feldspar $^{40}Ar/^{39}Ar$ cooling models and zircon He ages. White circles are single grain ages; black circles are means. Error bars on single grain ages are 8% (2σ) estimates of reproducibility based on multiple analyses of Fish Canyon Tuff zircon. Error bars on mean ages are two standard deviations of the single grain ages.

northeastern Oregon. Comparisons between high temperature (zircon U/Pb and biotite $^{40}Ar/^{39}Ar$) and low-temperature cooling ages in the wallrock at distances greater than ~50 m, as well as other geologic evidence, indicate temperatures less than 40 °C prior to dike intrusion, since the middle Cretaceous. Ages of all four thermochronometers (ZFT, zircon He, AFT, and apatite He) are completely reset to the age of the dike ~17 Ma to a distance of 2 m, and each system shows increasing ages at greater distances, to ~105–120 Ma in unreset portions. To first order, the resetting profiles of each system are similar to those predicted by simple thermal and fractional degassing models based on reasonable thermal properties of the dike and wallrocks. Some discrepancies are observed however, most notably a more complex pattern of resetting in both the ZFT and zircon He systems at distances about 8–15 m from the dike, which requires a more complex thermal evolution at these distance, possibly caused by hydrothermal convection in the wallrock (e.g., Barker et al. 1998), or another heat source out of the plane of exposure. Such a mechanism may also explain why the observed AFT resetting profile is farther from the dike than predicted by simple thermal models. Note, however, that

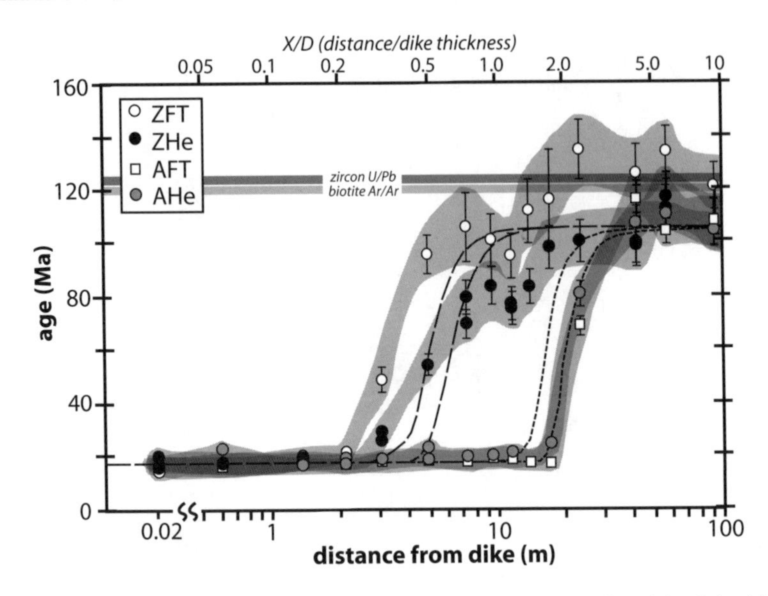

Figure 9. Thermochronometric data for samples adjacent to a ~10 m wide dike of the Columbia River Basalt Group (CRBG), in the Wallowa Mountains of northeastern Oregon. Dashed lines with long and short segments are predicted zircon He and apatite He ages, respectively for dike widths of 8 or 10 m widths (8 m dike predictions are closer to dike margin). Horizontal grey fields show age ranges including 2σ uncertainties on zircon U/Pb (by laser-ablation ICP-MS) ages, and biotite $^{40}Ar/^{39}Ar$ plateau ages. Both ZFT and zircon He ages show decreased ages at about 11–15 m distances from the dike that are not predicted by thermal and diffusion/annealing models. Model assumptions and parameters: Country rock cooling age for zircon He, apatite He, and apatite FT = 105 Ma (based on roughly invariant ages at distances greater than ~40 m from the dike, and other (U-Th)/He ages in the vicinity); dike age = 17 Ma; dike temperature = 1100 °C [typical for CRBG magma; Ho and Cashman (1997)]; country rock temperature between 105 and 17 Ma = 10 °C; thermal diffusivity of country rock and magma = 8.4×10^{-7} m^2/s; heat of fusion of melt in dike = 0.32 MJ/kg; heat capacity of country rock and magma = 1 kJ/kg.

the position of the apatite He profile agrees well with a simple dike heating model, which would not be the case if extra heat were required to explain the ZFT/zircon He patterns or the shifted AFT profile. In summary, dike heating experiments such as this and others (Hart 1964; Wartho et al. 2001) can probably only provide qualitative checks on the thermal sensitivity of thermochronometers, due to uncertainties in a range of parameters, though they are useful as qualitative tests of interchronometer consistency for thermal processes operating at timescales intermediate between laboratory and many tectonic processes.

Exhumed crustal sections

Crustal sections exposed in footwalls of large normal faults provide an opportunity to compare cooling ages of multiple thermochronometers in rocks that have experienced conceivably simple thermal histories that may be simply related by a single geothermal gradient and exhumation history. Typically, these crustal sections are interpreted in the context of long periods of steady-state conditions with little or no cooling/exhumation, producing a partial retention or partial annealing zone (PRZ or PAZ), followed by an episode of rapid tectonic exhumation by hanging wall removal and isostatically-induced tilting/uplift of the footwall.

One example of a tectonically exhumed crustal section is the Gold Butte block in southeastern Nevada, which, on the basis of structural and thermobarometric evidence, has been proposed to be the deepest continuous section of continental crust in the southwestern

U.S. (Fryxell et al. 1992). Apatite fission-track (Fitzgerald et al. 1991), and apatite, titanite, and zircon (U-Th)/He studies (Reiners et al. 2000, 2002) have shown that each thermochronometer records invariant ages at relatively deep paleodepths, and increasingly older ages at shallower paleodepths (Fig. 10). The shallow parts of each trend have been interpreted as partial retention zones developed during a long period of tectonic and erosional quiescence from the Jurassic through Miocene, and the young and invariant ages at depth as recording rapid exhumation during mid-Miocene Basin and Range extension. The break-in-slope in a plot of paleodepth versus zircon He ages is difficult to resolve given zircon abundance in the critical paleodepths, but corresponds to an approximate pre-exhumational temperature of about 180–250 °C, assuming a pre-exhumational geothermal gradient of 20–25 °C/km, consistent with age-paleodepth relations of lower temperature thermochronometers.

At least two potential complications to this simple interpretation of the Gold Butte data exist, however. First, it has been noted (Bernet 2002) that the depth range of the upper section showing a correlation between age and paleodepth is greater than expected for a static PRZ. Bernet (2002) suggested that the upper section instead records slow erosional exhumation of a missing ~4 km section of sedimentary rocks that are no longer preserved in this region. If this is correct, then an apparent break-in-slope would represent the base of a moving, not static PRZ, though it would still correspond approximately to the depth of the closure isotherm at the

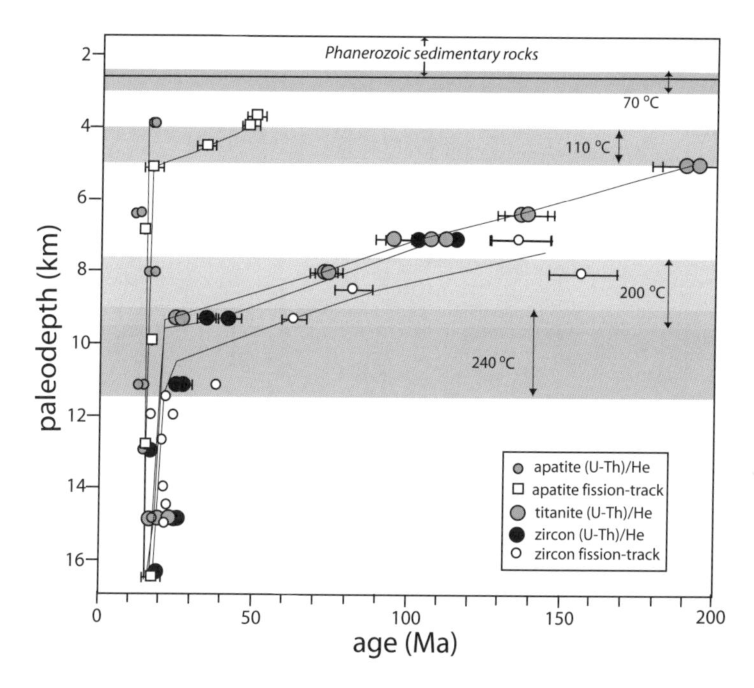

Figure 10. Thermochronologic data from the Gold Butte block, southern Nevada. AFT data are from Fitzgerald et al.(2000), apatite He and titanite He from Reiners et al. (2000), zircon He from Reiners et al. (2002), and ZFT data are from Bernet (2002). Each system except apatite He shows a correlation between age and paleodepth (as determined from distance from and dip of overlying sedimentary rock units) at relatively shallow paleodepths, underlain by young and roughly invariant ages at greater paleodepth. The apatite He system shows roughly constant ages throughout the block. These data have been interpreted to represent slow exhumation-related cooling prior to about 15 Ma, at which time rapid top-down cooling occurred in response to tectonic exhumation of the footwall by normal faulting. Vertical arrows and grey bars show paleodepths of given isotherms, assuming paleogeothermal gradients between 20–25 °C/km.

onset of rapid Miocene extension. The second complication is essentially the possibility that the normal fault that exhumed the Gold Butte footwall changed dip with depth, in which case the apparent paleodepth and temperature at any position in the block may not be easily related by distance from overlying sedimentary units. In this scenario, a listric normal fault would shallow with depth, so paleodepths in the structurally lower part of the block would be less than inferred from their position. Further detailed thermobarometric and higher-temperature thermochronologic work could address this.

Another example of an exhumed crustal section that can be used as either an empirical check on the effective closure depth (and, by inference, closure temperature) of the zircon He system, or else to constrain timing and rate of Miocene exhumation, is the Wassuk range of western Nevada (Tagami et al. 2003; Stockli 2005). This range exposes a paleodepth of approximately 8.5 km, and shows *en echelon* sets of PRZs for the apatite He and zircon He systems, and a PAZ for the AFT system (Fig. 11). The age versus paleodepth data appear to show relatively rapid exhumation-related cooling in the early Paleogene, followed by either slow or no exhumation-related cooling until ~15 Ma, when exhumation rates again increased. Stockli et al. (2002) have estimated pre-15-Ma geothermal gradients of about 26–30 °C/km for this section, which, combined with a paleosurface temperature of ~10 °C and assuming a closure temperature of 180 °C for zircon He, would predict an effective closure depth of the zircon He system of about 5.7–6.5 km. There is a distinct break-in-slope in the Wassuk zircon He data at about 6.5 km (Fig. 11), in good agreement with the prediction based on estimates of the zircon He closure temperature and the pre-exhumation geothermal gradient.

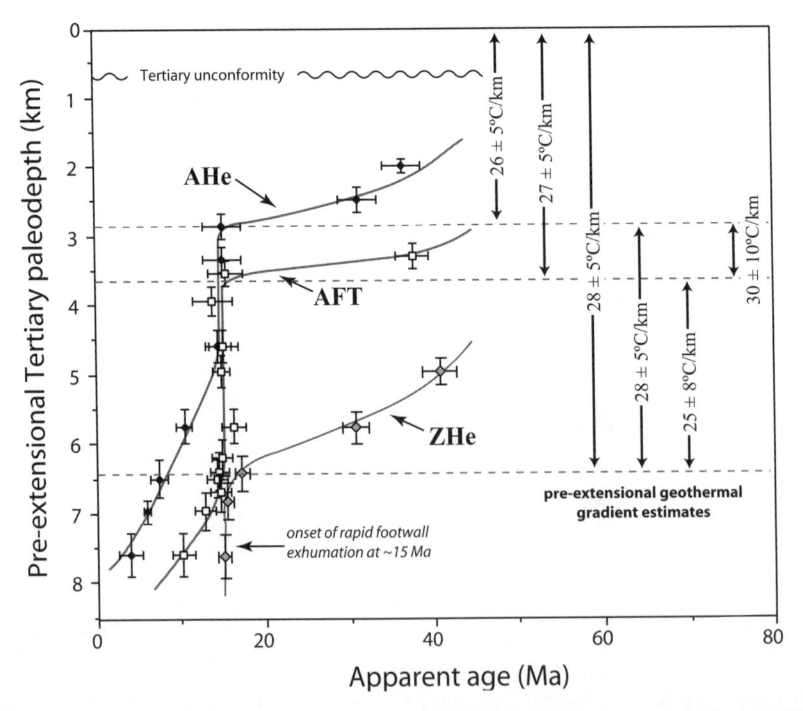

Figure 11. Apatite He, AFT, and zircon He data from the Wassuk crustal section, western Nevada (Stockli 2005). These data show extremely systematic relationships with paleodepth that have been interpreted similarly to those in Figure 10, with slow cooling prior to 15 Ma, at which time the block was rapidly exhumed, quenching thermochronometric ages in all three systems. See Stockli (2005) for further details.

Orogenic exhumation: Dabie Shan

The Dabie Shan was the focus of one of the first regional applications of zircon He thermochronology, combined with other low-temperature thermochronometers (Reiners et al. 2003; see also Kirby et al. 2002). This mountain range is a relatively low-relief (< 1.5 km) orogen in eastern China along part of the early Mesozoic collision zone between the north and south China blocks. The Dabie Shan is famous for its ultra-high pressure (UHP) metamorphic rocks, and its high through intermediate temperature geochronology and thermochronology is well studied by U/Pb and $^{40}Ar/^{39}Ar$ methods. Following collision and exhumation of UHP rocks in the Triassic-Jurassic (Hacker et al. 2000; Grimmer et al. 2003), Cretaceous granitoids were intruded into a large region of the orogen and this was accompanied by locally high-grade metamorphism at the presently exposed crustal depth (Ratschbacher et al. 2000). Post-orogenic evolution of the range since the Cretaceous has been characterized by slow erosion rates and little tectonism, although Grimmer et al. (2002) invoked early-mid-Tertiary exhumation on the east side of the range based on AFT length modeling.

Figure 12 shows two different ways of representing orogen-scale spatial-temporal patterns of exhumation in the Dabie Shan, using zircon He data combined with AFT and apatite He. Figure 12a shows cooling ages of each system projected onto a horizontal transect across the range, along with a topographic profile. Both zircon and apatite He ages show generally younger ages in the center of the orogen, where mean elevations are the highest. This pattern is consistent with higher long-term erosion rates in the core of the range than on the flanks. A simple model to estimate erosion rates from the zircon He, AFT, and apatite He ages in the core of the range assuming a geothermal gradient of 25 °C/km and closure temperatures of 180 °C, 110 °C and 65 °C, respectively, yields long-term average rates of about 0.07 km/m.y., and slower rates, by about a factor of two, on the range flanks. Figure 12b shows ages of all three systems from samples in a vertical transect in the core of the range. For apatite He ages, the ages are plotted against sample elevation. For AFT and zircon He ages, each sample's elevation is adjusted to a "pseudo-elevation," which is sample elevation plus the ratio between the zircon He-apatite He or AFT-apatite He closure temperature difference and geothermal gradient. This effectively converts each system to a common closure temperature and paleodepth, simulating a single age-elevation trend that would be measured for a single system, if such a high-relief vertical transect existed. This plot shows a steep age-elevation trend at the highest pseudo-elevations (oldest zircon He samples) suggesting rapidly exhumation at about 100 Ma, overlying a trend of roughly constant slope to the lowest pseudo-elevations and youngest ages. A best fit line through the apatite He and AFT data has a slope of about 0.07 km/m.y.. A detailed finite-element model by Braun and Robert (in press) also concluded that an average erosion rate of about 0.07 km/m.y. in the core of the range best explains these data. One potential complication to these data, and a factor that was not considered by Reiners et al. (2003) is the possibility for non-monotonic cooling histories for rocks of the Dabie Shan, involving reburial and reheating following early Mesozoic exhumation.

Detrital zircon dating

Zircon's resilience to weathering and alteration during transport and diagenesis makes it particularly useful in detrital settings, for providing constraints on provenance and depositional age of clastic sedimentary rocks, and for deducing long-term orogenic histories of source terranes. This has been well-established by the long history of U/Pb and fission-track dating studies of detrital zircons (e.g., Wilde et al. 2001; Bernet and Garver 2005). In comparison, few (U-Th)/He dating studies of detrital zircon exist. Those that do have focused on combining multiple radioisotopic techniques on single detrital crystals. For example, Rahl et al. (2003) combined U/Pb dating of ~30 μm deep pits on the crystal exteriors by laser ablation ICP-MS, with subsequent bulk-grain zircon He age determination, to perform

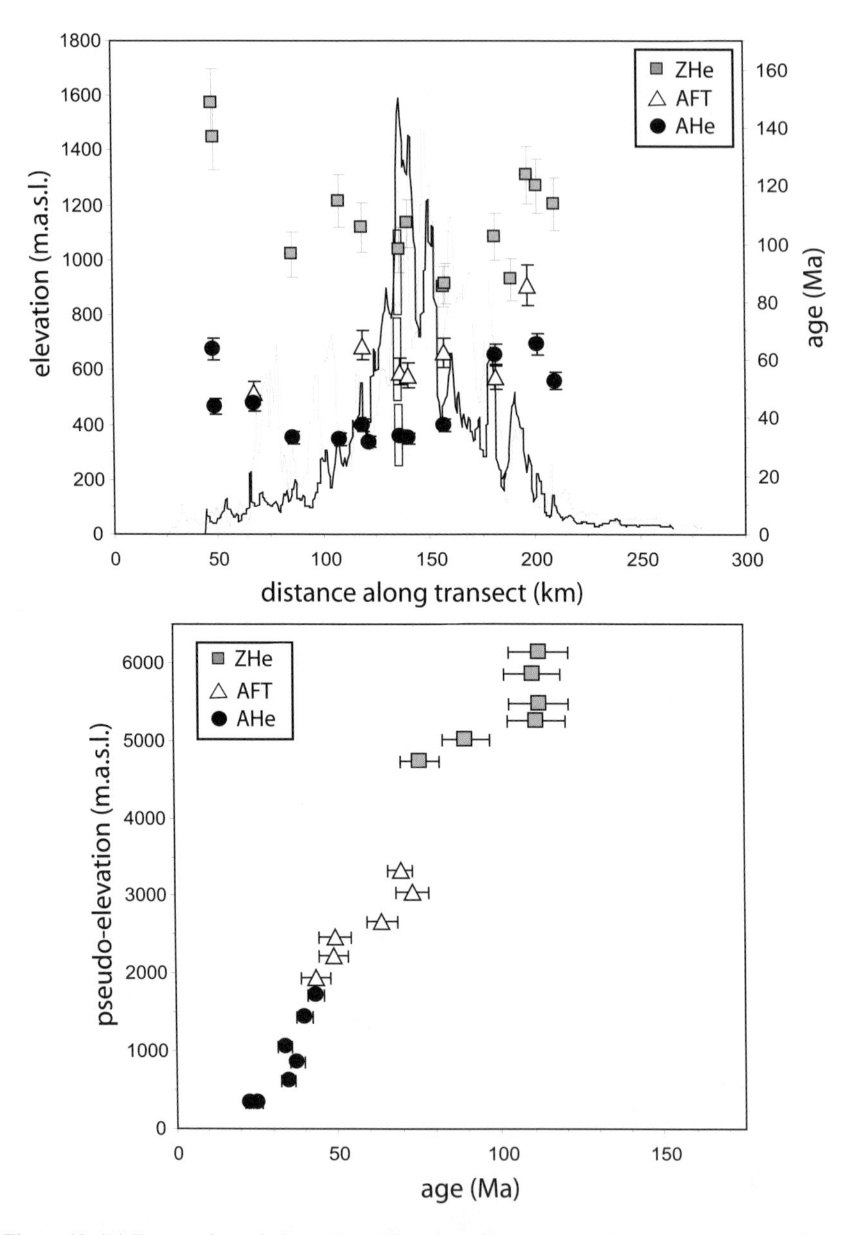

Figure 12. (A) Topography and zircon He, AFT, and apatite He ages projected onto a horizontal transect through the approximate center of the Dabie Shan, from southwest to northeast. Both zircon He and apatite He ages are systematically younger in the core of the range, in the location of the highest topography, consistent with higher long-term erosion rates there, after Reiners et al. (2003). Vertically-stretched boxes in center of range denote age ranges exhibited by samples in a single vertical transect there. (B) Age versus pseudo-elevation plot for zircon He, AFT, and apatite He data from a single vertical transect located in the central part of the Dabie Shan. Pseudo-elevations for the AFT and zircon He systems are shifted in proportion to the difference between their closure temperature and that of the apatite He system. This shows an apparent thermal history involving relatively rapid cooling at ~100 Ma, followed by slow cooling, consistent with an exhumation rate of about 0.07 km/m.y., through the present.

"He-Pb double-dating" of single detrital zircons from the Navajo sandstone in southern Utah. They showed that most of the zircons, and by inference most of the material, in this large erg deposit in the southwestern U.S. was derived from sources with combined U/Pb and He ages most characteristic of the Appalachian-Caledonide orogen of eastern North America (also see Dickinson and Gehrels 2003). In particular, a large population of zircons in these samples had distinctive combinations of ~1.0–1.2 Ga U/Pb ages with ~300–500 Ma He ages. Rahl et al. (2003) suggested that this reflected source rocks formed in the Grenvillian orogeny, but ultimately exhumed and cooled below ~180 °C in the Appalachian-Caledonide orogeny.

Other examples of He-Pb double dating include studies of zircons in active margin flysches in the Olympic Mountains and Kamchatka Peninsula, and Paleogene paleofluvial deposits in northeastern Oregon (Reiners et al. 2005). Campbell et al. (2005) extended He-Pb double-dating to measurement of distinct core and rim U/Pb ages, in addition to bulk grain He ages, on single crystal zircons from the Ganges and Indus rivers. This latter study showed U/Pb ages ranging from ~20–3000 Ma, roughly 70% of which had He ages less than 5 Ma, requiring high exhumation rates of diverse lithologies in the Himalayan source rocks.

Recent developments in detrital zircon dating have expanded to dating of single grains by three separate techniques: U/Pb, ZFT, and zircon He. Figure 13 shows compiled U/Pb, ZFT, and zircon He ages obtained on single grains from the Missouri River, along with He-Pb double-dates on zircons from the Mississippi River and the Navajo Sandstone.

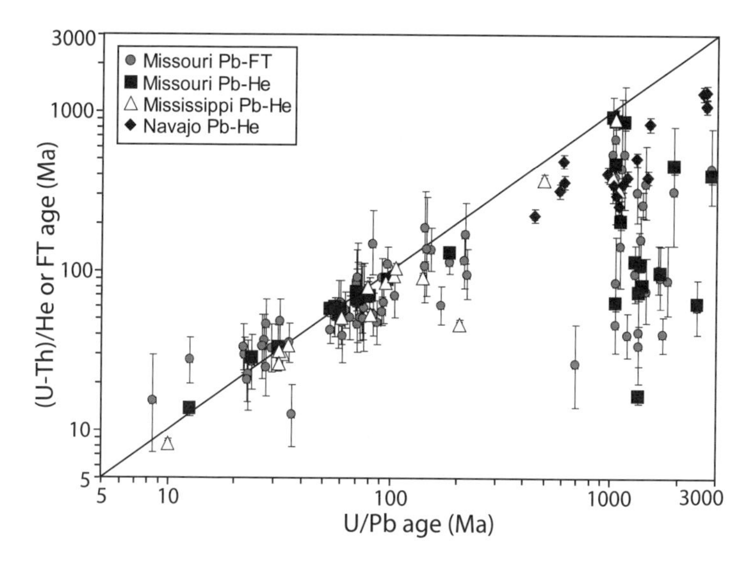

Figure 13. Zircon He and fission-track ages versus U/Pb ages in single zircon crystals from the Missouri and Mississippi rivers, and the Navajo sandstone. Mississippi river and Navajo sandstone zircons are He-Pb double-dates only; Missouri river zircons are He-FT-Pb triple-dates, though not all zircons have been dated by zircon He. Most zircons from both rivers have both crystallization and cooling ages less than 105 Ma, and most of these are close to the first-cycle volcanic trend of identical He and Pb ages. The lack of rocks with such characteristics in the eastern and central parts of North America suggest that most of this detritus in both rivers is derived from the western U.S. Three clusters of young zircons are seen, with potential "magmatic gaps" separating each. A much older group of Precambrian zircons shows a wide range of zircon He ages. Most of the zircons with Grenvillian U/Pb ages show Appalachian zircon He ages, whereas Mesoproterozoic zircons typically have younger cooling ages, corresponding to uplift and exhumation in the Mesozoic-Paleogene Cordilleran orogeny in western North America. Navajo sandstone data are from Rahl et al. (2003).

Most zircons in the modern sediment of these rivers have both U/Pb and He ages less than 105 Ma, and most of these have He ages that are close to or indistinguishable from the U/Pb ages. This is true for the Mississippi, as well as the Missouri, and the near absence of mid-Cretaceous and younger igneous east of the Rockies/Great Plains require that most of the Mississippi's detritus is derived from the western U.S.

Within the dominant population of zircons younger than 105 Ma, there are three clusters of zircons that fall along or near to the first-cycle trend: those with U/Pb ages of about 8–15 Ma, 22–35 Ma, and 55–105 Ma. The lack of zircons with U/Pb ages between 35 and 55 Ma is particularly well pronounced and is likely to represent a "magmatic gap" in the U.S. Cordillera. Of these three younger age groups, the 55–105 Ma group appears to contain the largest number of grains that fall significantly below the first-cycle volcanic trend, with apparent lag times of 20–40 m.y., possibly indicating slow cooling of plutonic rocks or burial and reheating prior to ultimate exhumation for these source rocks.

A fourth group of Phanerozoic zircons is observed in both rivers, with U/Pb ages between about 140 and 220 Ma. Most of these zircons have apparent Pb-He lag times as high as 100–150 m.y. implying slow exhumation or multicycle histories. With the exception of two zircons with U/Pb ages of 0.5 and 0.7 Ga, there is a very pronounced gap in U/Pb ages between ~220 Ma and 1.0 Ga. Precambrian zircons with ages between 1.0–2.8 Ga are abundant in the Missouri and Mississippi rivers. This group shows a hint of an inverse correlation between U/Pb and He ages, caused by the fact that most zircons with 0.9–1.2 Ga U/Pb ages have He ages of ~300–600 Ma, whereas most zircons with U/Pb ages older than ~1.2 Ga have He ages younger than 150 Ma. The zircons with Grenvillian U/Pb ages and Appalachian He ages are almost certainly derived from eastern North America. The presence of these grains in Missouri river sediment is probably due to transport and multi-cycle burial of grains derived from eastern North America, in the western part of the continent, as in the case of the Navajo sandstone. Zircons with Mesoproterozoic U/Pb ages but He ages less than ~150 Ma probably represent sources in the Belt Supergroup and related units, and possibly Ancestral Rockies uplifts, that were ultimately exhumed and cooled in Mesozoic through Paleogene Cordilleran orogenic events.

FUTURE DEVELOPMENTS

Zircon's relatively high abundance in diverse lithologies, high U concentrations, and fairly well-understood thermal sensitivity, make it a promising geo- and thermochronometric target for innovative He dating approaches in a wide range of applications. One example is using zircon He ages to constrain the timing, duration, intensity, and spatial patterns of natural surface or subsurface fires. Zircon He ages have been used to identify the effects of wildfire heating in exposed bedrock (Mitchell and Reiners 2003), and to map the spatial-temporal patterns of natural coalfire that have occurred in the Great Plains since the Pliocene (Heffern and Coates 2004), creating metamorphosed rocks known as clinker. Clinker has been dated by zircon He methods as young as ~10 ka in the Powder River basin of Wyoming and Montana, and the spatial patterns of clinker ages show systematic relationships with respect to topography that hold promise for constraining landscape evolution over timescales of 10^3–10^6 yr. Another promising aspect of using zircon He dating to identify surface reheating signals arises from the relative kinematics of He diffusion and fission-track annealing in zircon. As is also the case in apatite, at relatively short timescales and high temperatures, fission-tracks anneal more rapidly than He diffuses (Fig. 14). This leads to FT ages that are younger than He ages, a diagnostic indicator of short-duration reheating events that holds promise for elucidating volcanic processes (Stockli et al. 2000), shear heating along faults (e.g., Tagami 2005), and the distributions and dynamics of paleowildfire (Reiners and Donelick 2004).

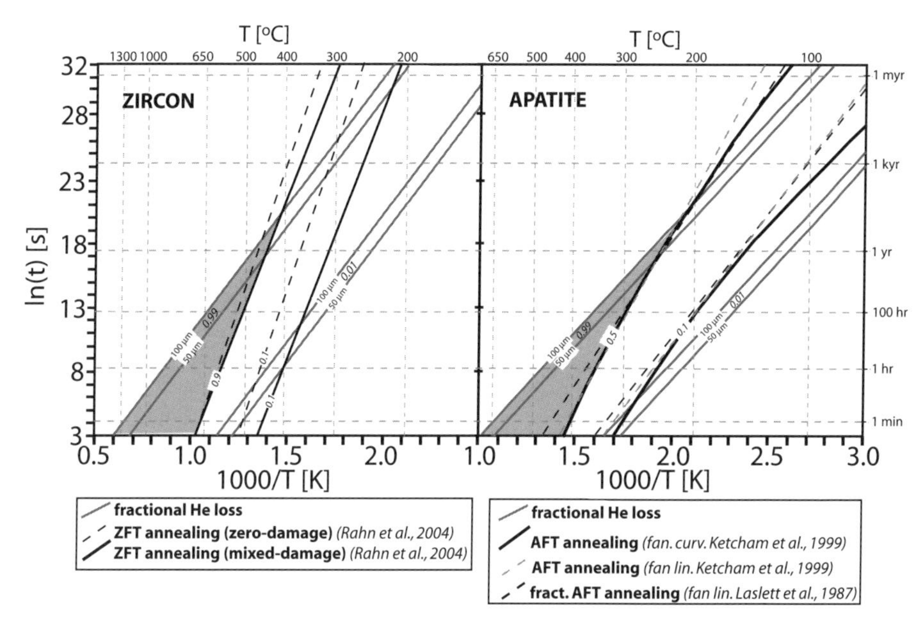

Figure 14. Pseudo-Arrhenius plots for He diffusion and fission-track shortening, relating time, temperature, and fractional degassing or track shortening for stepwise heating events. Solid grey lines denote fractional He degassing of 1.0% and 99% for spherical diffusion domain radii of 50 and 100 μm. AFT and ZFT contours are fractional fission-track shortening of 0.1 and 0.5 (at which point track density rapidly decreases), and 0.1 and 0.9, respectively. He diffusion parameters are from Farley (2000) for apatite, and Reiners et al. (2004) for zircon. Differing activation energies of the He and FT systems lead to "inverted" He-FT age relationships (He age older than FT age, or partially reset He age with fully reset FT age, the latter shown by shaded area) after short duration heating events. This is a diagnostic indicator of wildfire or other short-timescale, near-surface thermal events.

Another particularly promising prospect for zircon He dating is the potential to obtain spatially-resolved He concentrations or ages within single crystals. Just as ion probe and laser-ablation ICP-MS techniques allow discrimination of distinct U/Pb ages in rim and core portions of detrital zircons, advances in He dating techniques may soon allow in situ determinations of intracrystalline He age or concentration variations. Although no technique can currently measure both He and U-Th contents in precisely the same micro-analytical pit, two methods pose considerable promise. One is ^{4}He/^{3}He thermochronometry (e.g., Shuster and Farley 2005), whereby homogeneous ^{3}He distributions are generated throughout crystals by proton bombardment, and the evolution of ^{4}He/^{3}He in gas fractions released during step heating experiments is used to model detailed thermal histories of crystals. This technique has been successfully applied to several minerals, and assuming intracrystalline U-Th zonation issues could be addressed, it should be suitable for zircon as well. Another technique that may be able to directly measure in situ age distributions comes from the use of high-precision and -accuracy excimer laser ablation and He analysis of small (~5–10 μm) pits in core-to-rim traverses across single grains, followed by U-Th analyses in the same, or parallel, traverses, as suggested by Hodges and Boyce (2003). Again assuming that complications arising from the combined effects of U-Th zonation and the ~17 μm alpha-stopping distances can be addressed, this approach should allow estimation of not only bulk crystal closure ages, but also intracrystalline closure profiles (Dodson 1986), providing continuous time-temperature paths and constraints on cooling rates of the source rocks, as is done in K-feldspar ^{40}Ar/^{39}Ar and monazite U/Pb thermochronometry (e.g., Harrison et al. 2005). Zircon may prove particularly useful for this type of approach, because of the potentially high He contents needed to overcome blanks in small laser-ablation pits within crystals.

ACKNOWLEDGMENTS

I gratefully acknowledge the collaboration of those who have worked in the Yale He dating lab, especially Stefan Nicolescu, Jeremy Hourigan, and Kyle Min. I also thank Mark Brandon, John Garver, Rich Ketcham, and Peter Zeitler for helpful discussions. Some of the Wallowa dike heating data composed part of Victoria Lee's 2002 senior thesis at Yale. Thanks to Charlotte Allen for LA-ICP-MS U/Pb dating of the Wallowa zircons, and Peter Zeitler for $^{40}Ar/^{39}Ar$ ages on the biotites. Some of the southern Minnesota zircons were analyzed by Louise Miltich, during *HeDWaY 2004* at Yale. I acknowledge Taka Tagami's illustration of the potential importance of zircon He-FT kinetic "crossovers" in his talk at the 2003 Goldschmidt meeting. Modern He dating in general, and this author's opportunities to contribute to it, owe much to the creativity and insight of Ken Farley. Much of the work discussed here was supported by grants from the U.S. National Science Foundation and the American Chemical Society's Petroleum Research Fund. Constructive reviews by Raphael Pik and Danny Stockli are gratefully acknowledged.

REFERENCES

Barker CE, Bone Y, Lewan MD (1998) Fluid inclusion and vitrinite-reflectance geothermometry compared to heat-flow models of maximum paleotemperature next to dikes, western onshore Gippsland Basin, Australia. Int J Coal Geol 37:73-111

Bernet M (2002) Exhuming the Alps through time: Clues from detrital zircon fission-track ages. PhD Dissertation, Yale University, New Haven, Connecticut

Bernet M, Garver JI (2005) Fission-track analysis of detrital zircon. Rev Mineral Geochem 58:205-238

Brandon MT, Vance JA (1992) Tectonic evolution of the Cenozoic Olympic subduction complex, Washington State, as deduced from fission track ages for detrital zircons. Am J Sci 292:565-636

Bowring SA, Schmitz MD (2003) High-precision U-Pb zircon geochronology and the stratigraphic record. Rev Mineral Geochem 53:305-326

Braun J, Robert X (in press) Constraints on the rate of post-orogenic erosional decay from low temperature thermochronological data: application to the Dabie Shan, China. Earth Surf Proc Land.

Campbell IH, Reiners PW, Allen C, Nicolescu S, Upahdyay R (2005) He-Pb double-dating of detrital zircons from the Ganges and Indus rivers: Implication for quantifying sediment recycling, exhumation rates and provenance studies, Earth Planet Sci Lett, in press

Dazé A, Lee JKW, Villeneuve M (2003) An intercalibration study of the Fish Canyon sanidine and biotite $^{40}Ar/^{39}Ar$ standards and some comments on the age of the Fish Canyon Tuff. Chem Geol 199:111-127

Dickinson WR, Gehrels GE (2003) U–Pb ages of detrital zircons from Permian and Jurassic aeolian sandstones of the Colorado Plateau, USA: paleogeographic implications. Sed Geol 163:29-66

Dodson MH (1973) Closure temperature in cooling geochronological and petrological systems. Contrib Mineral Petrol 40:259-274

Dodson MH (1979) Theory of cooling ages. *In:* Lectures in Isotope Geology. Jäger E, Hunziker JC (eds) Springer-Verlag, Berlin, p 194-202

Dodson MH (1986) Closure profiles in cooling systems. *In:* Materials Science Forum. Vol. 7. Trans Tech Publications, Aedermannsdorf, Switzerland. p 145-153

Ehlers TA, Farley KA (2003) Apatite (U-Th)/He thermochronometry: methods and applications to problems in tectonics and surface processes. Earth Planet Sci Lett 206:1-14

Farley KA (2000) Helium diffusion from apatite: General behavior as illustrated by Durango fluorapatite. J Geophys Res 105:2903-2914

Farley KA (2002) (U-Th)/He dating: techniques, calibrations, and applications. Rev Mineral Geochem 47: 819-844

Farley KA, Stockli DF (2002) (U-Th)/He dating of phosphates: apatite, monazite, and xenotime. Rev Mineral Geochem 48:559-577

Farley KA, Wolf RA, Silver LT (1996) The effects of long alpha-stopping distances on (U-Th)/He ages. Geochim Cosmochim Acta 60:4223-4229

Farley KA, Kohn BP, Pillans B (2002) The effects of secular disequilibrium on (U–Th)/He systematics and dating of Quaternary volcanic zircon and apatite. Earth Planet Sci Lett 201:117-125

Fitzgerald PG, Fryxell JE, Wernicke BP (1991) Miocene crustal extension and uplift in southeastern Nevada: Constraints from fission track analysis. Geology 10:1013-1016

Fryxell JE, Salton GG, Selverstone J, Wernicke B (1992) Gold Butte crustal section, South Virgin Mountains, Nevada. Tectonics 11:1099-1120

Futa K (1986) Sm-Nd systematics of a tonalitic augen gneiss and its constituent minerals from northern Michigan. Geochim Cosmochim Acta 45:1245-1249

Gerling EK (1939) Diffusion temperature of helium as a criterion for the usefulness of minerals for helium age determination. Compt Red Acad Sci URSS 24:570-573

Grimmer JC, Jonckheere R, Enkelmann E, Ratschbacher L, Hacker BR, Blythe AE, Wagner GA, Wu Q, Liu S, Dong S (2002) Cretaceous-Cenozoic history of the southern Tan-Lu fault zone: apatite fission-track and structural constraints from the Dabie Shan (eastern China). Tectonophy 359:225–253

Grimmer JC, Ratschbacher L, McWilliams M, Franz L, Gaitzch I, Tichomirowa M, Hacker BR, Zhang Y (2003) When did the ultra-high pressure rocks reach the surface? A $^{207}Pb/^{206}Pb$ zircon, $^{40}Ar/^{39}Ar$ white mica, Si-in-white mica, single-grain provenance study of Dabie Shan synorogenic foreland sediments. Chem Geol 197:87–110

Hacker BR, Ratschbacher L, Webb L, McWilliams MO, Ireland T, Calvert A, Dong S, Wenk HR, Chateigner D (2000) Exhumation of ultrahigh-pressure continental crust in east central China: Late Triassic-Early Jurassic tectonic unroofing. J Geophys Res 105:13,339–13,364

Halliday AN, Mahood GA, Holden P, Metz JM, Dempster TJ, Davidson JP(1989) Evidence for long residence times of rhyolitic magma in the Long Valley magmatic system—the isotopic record in precaldera lavas of Glass Mountain. Earth Planet Sci Lett 94:274–290

Harrison TM, Grove M, Lovera OM, Zeitler PK (2005) Continuous thermal histories from inversion of closure profiles. Rev Mineral Geochem 58:389-409

Hawkesworth CJ, Blake S, Evans P, Hughes R, MacDonald R, Thomas LE, Turner SP, Zellmer G (2000) Time scales of crystal fractionation in magma chambers—integrating physical, isotopic and geochemical perspectives. J Petrol 41:991–1006

Hart SR (1964) The petrology and isotopic-mineral age relations of a contact zone in the Front Range, Colorado. J Geol 72:493-525

Heffern EL, Coates DA (2004) Geologic history of natural coalbed fires, Powder River Basin, USA. Int J Coal Geol 59:25-47

Ho AM, Cashman KV (1997) Temperature constraints on the Gingko flow of the Columbia River Basalt Group. Geology 25:403-406

Hodges K, Boyce J (2003) Eos, Trans, Am Geophys U 84(46), Fall Meeting Supplement, Abstract V22G-05

Holland HD (1954) Radiation damage and its use in age determination. *In*: Nuclear Geology. Faul H (ed) Wiley, New York, p 175–179

Holland HD, Gottfried D (1955) The effect of nuclear radiation on the structure of zircon. Acta Crystallogr 8: 291–300

Holmes A, Paneth FA (1936) Helium-ratios of rocks and minerals from the diamond pipes of South Africa. Proc Roy Soc Lond A154:385-413

Hourigan JK, Reiners PW, Brandon MT (2005) U-Th zonation dependent alpha-ejection in (U-Th)/He chronometry, Part I: Theory. Geochim Cosmochim Acta 69:3349-3365

House MA, Farley KA, Stockli D (2000) Helium chronometry of apatite and titanite using Nd-YAG laser heating. Earth Planet Sci Lett 183:365-368

Hurley PM (1952) Alpha ionization damage as a cause of low helium ratios. Trans Am Geophys U 33:174-183

Hurley PM (1954) The helium age method and the distribution and migration of helium in rocks. *In*: Nuclear Geology. Faul H (ed) Wiley, New York, p 301-329

Hurley PM, Fairbairn HW (1953) Radiation damage in zircons: a possible age method. Bull Geol Soc Am 64: 659-674

Hurley PM, Larsen ES Jr., Gottfried D (1956) Comparison of radiogenic helium and lead in zircon. Geochim Cosmochim Acta 9:98-102

Ireland TR, Williams IS (2003) Considerations in zircon geochronology by SIMS. Rev Mineral Geochem 53: 215-241

Keevil NB, Larsen ES Jr., Wank FJ (1944) Distribution of helium and radioactivity in rocks. VI: The Ayer granite-migmatite at Chelmsford, Mass. Am J Sci 242:345-353

Ketcham RA, Donelick RA, Carlson WD (1999) Variability of apatite fission-track annealing kinetics: III. Extrapolation to geological time scales. Am Mineral 84:1235-1255

Kirby E, Reiners PW, Krol M, Hodges K, Farley KA, Whipple K, Yiping L, Tang W, Chen Z (2002) Late Cenozoic uplift and landscape evolution along the eastern margin of the Tibetan plateau: Inferences from $^{40}Ar/^{39}Ar$ and U-Th-He thermochronology. Tectonics 10.1029/2000TC001246

Larsen ES Jr., Keevil NB (1942) The distribution of helium and radioactivity in rocks. III: Radioactivity and petrology of some California intrusives. Am J Sci 240:204-215

Larsen ES, Keevil NB, Harrison HC (1952) Method for determining the age of igneous rocks using the accessory minerals. Geol Soc Am Bull 63:1045-1052

Laslett GM, Green PF, Duddy IR, Gleadow AJW (1987) Thermal annealing of fission tracks in apatite. 2. A quantitative analysis. Chem Geol (Isot Geosci Sect) 65:1-13

Lovera OM, Richter FM, Harrison TM (1989) The $^{40}Ar/^{39}Ar$ thermochronometry for slowly cooled samples having a distribution of diffusion domain sizes. J Geophys Res 94:17917-17935

Lovera OM, Richter FM, Harrison TM (1991) Diffusion domains determined by ^{39}Ar released during step heating. J Geophys Res 96:2057-2069

Lovera OM, Grove M, Harrison TM (2002) Systematic analysis of K-feldspar $^{40}Ar/^{39}Ar$ step heating results II: Relevance of laboratory argon diffusion properties to nature. Geochim Cosmochim Acta 66:1237-1255

Lovera OM, Grove M, Harrison TM, Mahon KI (1997) Systematic analysis of K-feldspar $^{40}Ar/^{39}Ar$ step heating results. 1. Significance of activation energy determinations. Geochim Cosmochim Acta 61: 3171-3192

Mitchell SG, Reiners PW (2003) Influence of wildfires on apatite and zircon (U-Th)/He ages. Geology 31: 1025-1028

Naeser CW, Zimmermann RA, Cebula GT (1981) Fission-track dating of apatite and zircon: An interlaboratory comparison. Nucl Tracks 5:65-72

Nasdala L, Reiners PW, Garver JI, Kennedy AK, Stern RA, Balan E, Wirth R (2004) Incomplete retention of radiation damage in zircon from Sri Lanka. Am Mineral 89:219-231

Nasdala L, Wenzel M, Vavra G, Irmer G, Wenzel T, Kober B (2001) Metamictisation of natural zircon: accumulation versus thermal annealing of radioactivity-induced damage. Contrib Mineral Petrol 141: 125–144

Parrish RR, Noble SR (2003) Zircon U-Th-Pb geochronology by isotope dilution—thermal ionization mass spectrometry (ID-TIMS). Rev Mineral Geochem 53:183-213

Pettingill HS, Patchett PJ (1981) Lu-Hf total-rock age for the Amîtsoq gneisses, West Greenland. Earth Planet Sci Lett 55:150-156

Pik R, Marty B, Carignan J, Lave J (2003) Stability of the Upper Nile drainage network (Ethiopia) deduced from (U-Th)/He thermochronometry: implications for uplift and erosion of the Afar plume dome. Earth Planet Sci Lett 215:73-88

Rahl JM, Reiners PW, Campbell IH, Nicolescu S, Allen CM (2003) Combined single-grain (U-Th)/He and U/Pb dating of detrital zircons from the Navajo Sandstone, Utah. Geology 31:761-764

Rahn, MK, Brandon MT, Batt GE, Garver JI (2004) A zero-damage model for fission-track annealing in zircon Am Mineral 89:473-484

Ratschbacher L, Hacker BR, Webb LE, McWilliams M, Ireland T, Dong S, Calvert A, Chateigner D, Wenk HR (2000) Exhumation of the ultrahigh-pressure continental crust in east central China: Cretaceous and Cenozoic unroofing and the Tan-Lu fault. J Geophys Res 105:13,303–13,338

Reid MR, Coath CD, Harrison TM, McKeegan KD (1997) Prolonged residence times for the youngest rhyolites associated with Long Valley Caldera: ^{230}Th-^{238}U ion microprobe dating of young zircons. Earth Planet Sci Lett 150:27–39

Reiners PW, Brady R, Farley KA, Fryxell JE, Wernicke BP, Lux D (2000) Helium and argon thermochronometry of the Gold Butte block, South Virgin Mountains, Nevada. Earth Planet Sci Lett 178:315-326

Reiners PW, Campbell IH, Nicolescu S, Allen CA, Hourigan JK, Garver JI, Mattinson JM, Cowan DS (2005) (U-Th)/(He-Pb) "double-dating" of detrital zircons. Am J Sci 305:259-311

Reiners PW, Donelick RA (2004) Thermochronology of wildfire and fault heating through single grain (U-Th)/He and fission-track double dating. Geol Soc Am Abstr Prog Vol. 36, No. 5, p. 447

Reiners PW, Farley KA (1999) He diffusion and (U-Th)/He thermochronometry of titanite. Geochim Cosmochim Acta 63:3845-3859

Reiners PW, Farley KA (2001) Influence of crystal size on apatite (U-Th)/He thermochronology: an example from the Bighorn Mountains, Wyoming. Earth Planet Sci Lett 188:413-420

Reiners PW, Farley KA, Hickes HJ (2002) He diffusion and (U-Th)/He thermochronometry of zircon: Initial results from Fish Canyon Tuff and Gold Butte. Tectonophys 349:247-308

Reiners PW, Spell TL, Nicolescu S, Zanetti KA (2004) Zircon (U-Th)/He thermochronometry: He diffusion and comparisons with $^{40}Ar/^{39}Ar$ dating. Geochim Cosmochim Acta 68:1857-1887

Reiners PW, Zhou Z, Ehlers TA, Xu C, Brandon MT, Donelick RA, Nicolescu S (2003) Post-orogenic evolution of the Dabie Shan, eastern China, from (U-Th)/He and fission-track dating. Am J Sci 303:489-518

Richter FM, Lovera OM, Harrison TM, Copeland P (1991) Tibetan tectonics from $^{40}Ar/^{39}Ar$ analysis of a single K-feldspar sample. Earth Planet Sci Lett 105:266-278
Schärer U (1984) The effect of initial ^{230}Th disequilibrium on young U-Pb ages: the Makalu case, Himalaya. Earth Planet Sci Lett 67:191–204
Schmitz MD, Bowring SA (2001) U-Pb zircon and titanite systematics of the Fish Canyon Tuff: an assessment of high precision U-Pb geochronology and its application to young volcanic rocks. Geochim Cosmochim Acta 65:2571-2587
Seitz F (1949) On the disordering of solids by action of fast massive particles. Disc Far Soc 5:271-282
Shuster DL, Farley KA (2004) $^{4}He/^{3}He$ thermochronometry. Earth Planet Sci Lett 217:1-17
Shuster DL, Farley KA, Sisterson JM, Burnett DS (2004) Quantifying the diffusion kinetics and spatial distributions of radiogenic ^{4}He in minerals containing proton-induced ^{3}He. Earth Planet Sci Lett 217: 19-32
Silver LT, Deutsch S (1963) Uranium-lead isotopic variations in zircons: a case study. J Geol 71:721-758
Stockli DF (2005) Application of low-temperature thermochronometry to extensional tectonic settings. Rev Mineral Geochem 58:411-448
Stockli DF, Farley KA (2004) Empirical constraints on the titanite (U–Th)/He partial retention zone from the KTB drill hole. Chem Geol 227:223-236
Stockli DF, Farley KA, Dumitru TA (2000) Calibration of the apatite (U-Th)/He thermochronometer on an exhumed fault block, White Mountains, California. Geology 28:983-986
Stockli DF, Surpless BE, Dumitru TA, Farley KA (2002) Thermochronological constraints on the timing and magnitude of Miocene and Pliocene extension in the central Wassuk Range, western Nevada. Tectonics 21:4, doi: 10.1029/2001TC001295.02
Strutt RJ (1910a) The accumulation of helium in geologic time, II. Proc Roy Soc Lond, Ser A 83:96-99
Strutt RJ (1910b) The accumulation of helium in geologic time, III. Proc Roy Soc Lond, Ser A 83:298-301
Strutt RJ (1910c) Measurements of the rate at which helium is produced in thorianite and pitchblende, with a minimum estimate of their antiquity. Proc Roy Soc Lond, Ser A 83:379-388
Tagami T, Farley KA, Stockli DF (2003) (U-Th)/He geochronology of single zircon grains of known Tertiary eruption age. Earth Planet Sci Lett 207:57-67
Tilton GR, Patterson C, Brown H, Inghram M, Hayden R, Hess D, Larsen E (1955) Isotopic composition and distribution of lead, uranium and thorium in a Precambrian granite. Geol Soc Am Bull 66:1131-1148
Turner G, Harrison TM, Holland G, Mojzsis SJ, Gilmour J (2004) Extinct ^{244}Pu in ancient zircons. Science 306:89-91
Vinogradov AP, Zadorozhnyi IK, Zykor SI (1952) Isotopic composition of lead and the age of the earth. Dokl Akad Nauk SSSR 85:1107-1110
Wartho J-A, Kelley SP, Blake S (2001) Magma flow regimes in sills deduced from Ar isotope systematics of host rocks J Geophys Res 106:4017-4035
Webber GR, Hurley PM, Fairbairn HW (1956) Relative ages of eastern Massachusetts granites by total lead ratios in zircon. Am J Sci 254:574-583
Wernicke B, Getty SR (1997) Intracrustal subduction and gravity currents in the deep crust: Sm-Nd, Ar-Ar, and thermobarometric constraints from the Skagit Gneiss Complex, Washington. Geol Soc Am Bull 109: 1149-1166
Wetherill GW (1955) An interpretation of the Rhodesia and Witwatersrand age patterns. Geochim Cosmochim Acta 9:290-292
Wilde SA, Valley JW, Peck WH, Graham CM (2001) Evidence from detrital zircons for the existence of continental crust and oceans on the Earth 4.4 Gyr ago. Nature 409:175-178
Wolf RA, Farley KA, Silver LT (1996) Helium diffusion and low-temperature thermochronometry of apatite. Geochim Cosmochim Acta 60:4231-4240
Wolf RA, Farley KA, Kass DM (1998) Modeling of the temperature sensitivity of the apatite (U-Th)/He thermochronometer. Chem Geol 148:105-114
Zeitler PK, Herczeg AL, McDougall I, Honda M (1987) U-Th-He dating of apatite: a potential thermochronometer. Geochim Cosmochim Acta 51:2865-2868

Reviews in Mineralogy & Geochemistry
Vol. 58, pp. 181-203, 2005

7

^{4}He/^{3}He Thermochronometry: Theory, Practice, and Potential Complications

David L. Shuster and Kenneth A. Farley

Division of Geological and Planetary Sciences
California Institute of Technology
Pasadena, California, 91125, U.S.A.
dshuster@caltech.edu *farley@gps.caltech.edu*

INTRODUCTION

Thermochronometry most often involves the determination of a cooling age from parent and daughter abundances within an entire crystal or population of crystals (Dodson 1973). Complementary information exists in the spatial concentration distribution of the daughter, $C(x,y,z)$, within a single crystal. By combining a bulk cooling age with $C(x,y,z)$ on the same sample, it is possible to place tight limits on the sample's time-temperature (t-T) path. Techniques for this kind of analysis have been developed for several different parent/daughter systems including U-Th-Pb and K-Ar (Harrison et al. 2005). Here we describe how this approach is applied to the (U-Th)/He system. The particular attraction of the (U-Th)/He method is its sensitivity to uniquely low temperatures. For example, the nominal ^{4}He closure temperatures (at 10 °C/Myr) for apatite, zircon and titanite are 70 °C, 180 °C, and 200 °C, respectively (Reiners and Farley 1999, 2001; Farley 2000; Reiners et al. 2002, 2004). In the case of apatite, we will show that significant diffusive mobility of ^{4}He occurs at temperatures just slightly higher than those of the Earth's surface. In this chapter, we present an overview of the *^{4}He/^{3}He thermochronometry* technique in which the natural spatial distribution of ^{4}He is constrained by stepwise degassing ^{4}He/^{3}He analysis of a sample containing synthetic, proton-induced ^{3}He. We present the fundamental theory, assumptions, practical aspects of proton irradiation and stepwise ^{4}He/^{3}He analyses, as well as several example applications of ^{4}He/^{3}He thermochronometry.

In particular, we illustrate how the ^{4}He/^{3}He technique can be used to determine the helium diffusion kinetics *and* constrain the natural ^{4}He distribution within an individual crystal or a small population of crystals, and how this information can be used to constrain the sample's t-T path. We also discuss some of the complications that have arisen and summarize the current state of research on this new thermochronometer.

FUNDAMENTAL CONSIDERATIONS

The basic principles and assumptions of (U-Th)/He dating have been described in detail elsewhere (Farley 2002). Here we concentrate only on aspects particular to the ^{4}He/^{3}He variant of the method. Like other radio-thermochronometers, (U-Th)/He dating involves two physical processes: radiogenic ingrowth of a daughter product (^{4}He) and thermally activated volume diffusion of the daughter. However, a difference between the (U-Th)/He system and other chronometers is that multiple parent nuclides produce a common daughter through α decay. Although α decay of ^{147}Sm also produces ^{4}He, the vast majority of radiogenic ^{4}He in minerals is produced via actinide decay. Once in a state of secular equilibrium, the actinide decay series

1529-6466/05/0058-0007$05.00 DOI: 10.2138/rmg.2005.58.7

emit 8, 7, and 6 α-particles for a single decay of ^{238}U, ^{235}U and ^{232}Th, respectively. The ^4He ingrowth equation can therefore be written:

$$^{4}\mathrm{He} = 8\cdot {}^{238}\mathrm{U}\cdot\left(e^{\lambda_{238}t}-1\right)+7\cdot\left(\frac{^{238}\mathrm{U}}{137.88}\right)\cdot\left(e^{\lambda_{235}t}-1\right)+6\cdot {}^{232}\mathrm{Th}\cdot\left(e^{\lambda_{232}t}-1\right)$$

where ^4He, ^{238}U and ^{232}Th indicate present-day abundances, t is the accumulation time or He age, λ is a radioactive decay constant ($\lambda_{238} = 1.511 \times 10^{-10}$ yr^{-1}, $\lambda_{235} = 9.849 \times 10^{-10}$ yr^{-1}, $\lambda_{232} = 4.948 \times 10^{-11}$ yr^{-1}), and (1/137.88) is the present day $^{235}U/^{238}U$ ratio.

The ^4He spatial distribution

The basis for $^4He/^3He$ thermochronometry is that the spatial distribution of radiogenic ^4He within a U and Th bearing crystal is an evolving function of the sample's t-T path. This can be summarized by the following schematic equation which applies to an individual crystal:

$$\int_{t_o}^{today}\left[Production(x,y,z,t) - Removal(x,y,z,T,t)\right]dt = Distribution(x,y,z,today)$$

where *Production* is the time dependant radiogenic production function of ^4He, *Removal* is the time and temperature dependent diffusive loss function, *Distribution* is the spatial concentration function of ^4He within the sample today, and t_o is the time when ^4He accumulation initiates. With knowledge of these functions, the above expression provides a relationship between measurable quantities and the desired t-T path of the sample. Since the physics which describes and relates these functions is well established, the challenge is to quantify the functional form of each. $^4He/^3He$ thermochronometry provides an analytical technique to constrain (1) the *Distribution* function in a sample today and (2) the *Removal* function (i.e., via the helium diffusion kinetics). For a sample with uniformly distributed parent nuclides, the standard assumptions of (U-Th)/He dating provides (3) the *Production* function through knowledge of the bulk U and Th concentrations in the sample today. Although non-uniform parent distributions could easily be incorporated into the theory, for simplicity we initially consider only the uniform case.

Classical diffusion theory provides the necessary relationships between time, temperature and the spatial distribution of a radiogenic noble gas within a solid matrix (Carslaw and Jaeger 1959; Crank 1975). Using numerical methods, the classical theory can be extended to any arbitrary geometry. However, specific analytic solutions to the production – diffusion equation exist. The spherical solution is the most useful, and one which provides the clearest way to illustrate and conceptualize the relationship between t, T and $C(x,y,z)$ within a crystal. For the purposes of this chapter, we focus on the spherical solution, and discuss below why the spherical model is useful for many geological applications.

Within a spherical diffusion domain, and for known diffusivity (D), the radial concentration distribution of a diffusing substance can be described along a single spatial dimension r, ($0 \le r \le a$). For an initial radial concentration distribution $C_0(r)$ and assuming no ingrowth, the concentration at a later time is given by:

$$C(r,\tau) = \frac{2}{ar}\sum_{k=1}^{\infty} e^{-k^2\pi^2\tau}\sin\left(\frac{k\pi r}{a}\right)\int_0^a r' C_o(r')\sin\left(\frac{k\pi r'}{a}\right)dr' \qquad (1)$$

(Carslaw and Jaeger 1959) if its mobility follows volume diffusion and $C_0(a) = 0$ for all t. Here, we use the non-dimensional diffusion time

$$\tau(T,t) = \int_0^t \frac{D(T,t')}{a^2} \cdot dt' \quad (2)$$

and assume that diffusion is thermally activated. These relationships provide a mechanism by which to compute the ^4He distribution along any arbitrary *t-T* path. By discretization of τ_i to be a piecewise linear quantity

$$\tau_i = \tau(T_i, t_i) \quad (3)$$

the radial distribution after a discrete step of duration t_i at T_i is given by

$$C_i(r) = C(r, \tau_i) \quad (4)$$

These expressions are useful for predicting the evolution of a radial ^4He profile within a spherical diffusion domain over geologic time. Along a discretized *t-T* path, the profile evolution is calculated by first adding a finite "dose" of radiogenic ^4He, then diffusively evolving the profile according to Equation (4). Because individual crystals of the minerals commonly used for (U-Th)/He dating [apatite (Farley 2000; Reiners and Farley 2001), titanite (Reiners and Farley 1999), zircon (Reiners et al. 2002)] act essentially as single diffusion domains, we consider the effect of thermal history on the ^4He distribution within a single diffusion domain rather than a distribution of domains.

To illustrate the above expressions, Figure 1 shows ^4He profiles obtained after eight different *t-T* paths assuming helium diffusion kinetics equivalent to those of Durango apatite (Farley 2000; Shuster and Farley 2004). We have chosen to use the Durango apatite He diffusion kinetics throughout this chapter because they are amongst the best determined of all minerals, and because they demonstrate the low temperature sensitivity available from the apatite $^4He/^3He$ method. Shown in Figure 1a,d are cooling histories; the resultant radial ^4He distributions are shown in Figure 1b,e. The first simulation (Fig. 1a,b) ended when the temperature reached 25 °C whereas the second simulation (Fig. 1d,e) was followed by late stage isothermal accumulation at 25 °C for 5 Myr. Note that these simulations assume temperature only varies with time, and does not vary across the diffusion domain.

Figure 1 shows that slow cooling and prolonged residence at elevated temperatures yields low concentrations of ^4He near the domain edge. In contrast to these "rounded" profiles, those generated by rapid cooling or a long duration at low temperatures where diffusion is very slow have higher concentrations near the edge, they are more "square." The point is that each of the eight distributions in Figure 1 distinctly reflects the *t-T* path on which it was produced. With knowledge of a sample's helium diffusion kinetics, model ^4He distributions can be calculated according to any arbitrary *t-T* path. Alternatively, if the natural ^4He spatial distribution within a sample can be constrained, then a finite set of *t-T* paths consistent with both the sample's ^4He distribution and He age can be identified.

Proton-induced ^3He

As Figure 1 illustrates, most of the difference among the ^4He profiles is found in the outermost few percent of the domain (e.g., $r/a > 0.90$). Because typical accessory minerals have dimensions of only ~100 μm, distinction among profiles like these (and thus *t-T* paths) requires a technique for ^4He detection that has a spatial resolution of better than a few microns. We are unaware of a technique by which to directly measure a ^4He concentration distribution at this resolution. Instead, we use a stepwise degassing approach that simultaneously yields both pieces of information we require: the ^4He concentration profile *and* the helium diffusion kinetics of the same sample. With only a single isotope (^4He) it is impossible to de-convolve the effects of a non-uniform ^4He distribution and unknown helium diffusivity from a set of

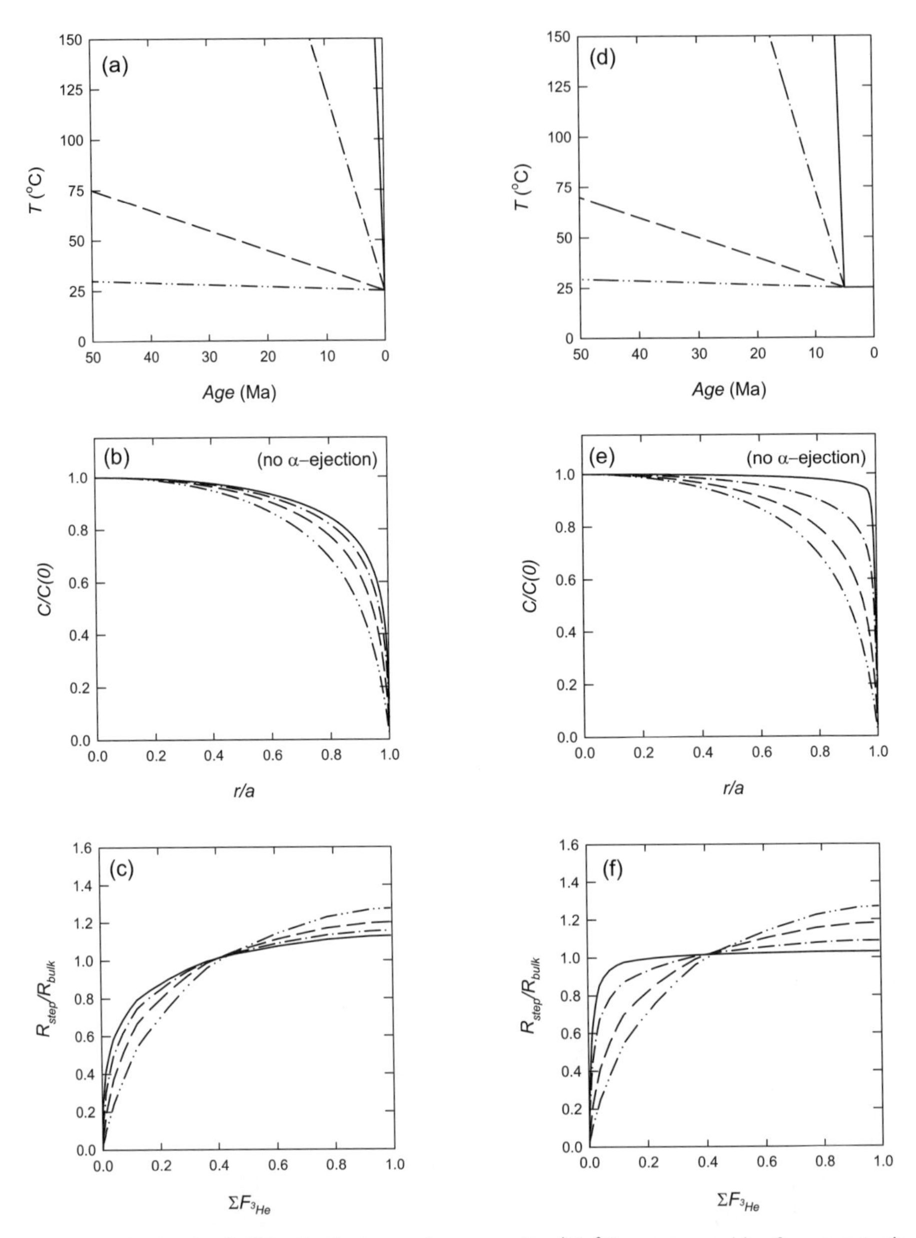

Figure 1. Simulated radial ^{4}He distributions and corresponding ^{4}He/^{3}He spectra resulting from monotonic cooling, modified from Shuster and Farley (2004). The concentration profiles shown in (b) and (e) result from the *t-T* paths shown in (a) and (d), respectively and the helium diffusion kinetics of Durango apatite and no isotopic fractionation. The resultant step-heating simulations are shown as ratio evolution diagrams in (c) and (f), respectively. Shown are the simulated isotope ratios for each release step, R_{step} (R=^{4}He/^{3}He), normalized to the bulk ratio R_{bulk} plotted vs. the cumulative ^{3}He release fraction, $\Sigma F_{^3\text{He}}$. The simulations excluded the effect of α-ejection.

helium release fractions (Shuster and Farley 2004). In 1969, Turner showed that the presence of a second, synthetically generated, and uniformly distribution Ar isotope (^{37}Ar or ^{39}Ar) is useful for constraining the distribution of radiogenic ^{40}Ar in a degassing experiment (Turner 1969). In 1978, Albarède showed how stepwise degassing Ar data can be used to directly invert for the ^{40}Ar distribution (Albarède 1978). The development of ^{4}He/^{3}He thermochronometry has largely been based on these results. Step-heating of a sample that contains a uniform distribution of ^{3}He solves this difficulty: ^{3}He release fractions quantify helium diffusion kinetics, while evolution of the ^{4}He/^{3}He ratio over the course of sequential degassing constrains the natural radiogenic helium profile.

Shuster et al. (2004) demonstrated that sufficient ^{3}He for this application can be generated within minerals via energetic proton irradiation. The ^{3}He nuclei are produced from all atoms in the mineral as spallation products of nuclear reactions initiated by the incident protons. The experiments of Shuster et al. (2004) showed that (i) lattice damage associated with proton irradiation does not affect the helium diffusion properties of at least apatite and titanite, (ii) the proton-induced distribution of ^{3}He is uniform, (iii) diffusion parameters determined from ^{3}He are in excellent agreement with those determined from ^{4}He for both apatite and titanite, (iv) the technique produces ^{4}He at levels orders of magnitude lower than are found in minerals of interest for He dating, and (v) the sample is heated by less than a few °C during irradiation, so helium diffusion during the process is negligible.

A uniform ^{3}He distribution is useful for two reasons: (i) it enables a stepwise degassing experiment in which the ^{4}He release fractions are normalized to the ^{3}He released in the same step; we illustrate below how this *ratio evolution experiment* constrains the natural ^{4}He distribution within the sample, and (ii) it satisfies the initial condition from which diffusion coefficients are easily calculated (Fechtig and Kalbitzer 1966).

Unlike ^{39}Ar, which is induced via neutrons reacting with ^{39}K and thus tracks the parent isotope ^{40}K, ^{3}He induced by energetic protons is not uniquely generated from the parent nuclides of radiogenic ^{4}He. Proton-induced ^{3}He is effectively generated from all atoms that are present in the irradiated mineral. Therefore, unlike the ^{40}Ar/^{39}Ar method, ^{4}He/^{3}He release data do not define a radiometric age for each step. However, if U and Th are uniformly distributed throughout a particular sample, then some of the features in the ratio evolution diagram carry age significance when combined with a bulk He age.

The ^{4}He/^{3}He ratio evolution diagram

Proton-induced ^{3}He provides a means to interrogate the natural ^{4}He distribution within a sample by sequentially measuring ^{4}He/^{3}He ratios during stepwise degassing. During a degassing experiment, the ^{4}He/^{3}He release spectrum or *ratio evolution diagram* is a sensitive function of the natural spatial distribution of ^{4}He (Shuster and Farley 2004). This diagram is a plot of the ^{4}He/^{3}He ratio in each step (R_{step}) normalized to the ^{4}He/^{3}He ratio of the bulk sample (R_{bulk}) as a function of cumulative ^{3}He release fraction, $\Sigma F_{^3\text{He}}$. In effect, each step "mines" deeper into the diffusion domain. We will show how forward model simulations can be compared against an observed ratio evolution diagram to constrain the ^{4}He distribution. Alternatively, Shuster and Farley (2004) described a linear inversion to directly solve for the unknown ^{4}He distribution from the ^{4}He/^{3}He spectrum.

The simulated radial distributions presented in Figure 1 can be used to illustrate the sensitivity that the ^{4}He/^{3}He ratio evolution diagram has for constraining spatial ^{4}He distributions. The same expressions used to model radiogenic ingrowth and diffusion in nature can also be used to model evolution of the ^{4}He and ^{3}He distributions during a degassing experiment. The only difference is that the radiogenic production is negligible over the timescale of an experiment.

For a given diffusion kinetics, $D(T)/a^2$, Equation (4) predicts the piecewise evolution of the radial distribution after discrete steps of duration t_i at T_i of a simulated degassing experiment for any arbitrary initial profile, $C_0(r)$. By integrating the profiles and taking their differences between each step, a set of simulated helium release fractions is calculated for any arbitrary heating schedule. And if the two helium isotopes have known relative diffusivity (see below), Equation (4) can be used to calculate isotope ratios for the concurrent release of a uniformly distributed isotope (^{3}He) and an isotope with an arbitrary natural distribution (^{4}He). This simulates a ^{4}He/^{3}He stepwise degassing experiment.

Figure 1(c,f) shows the simulated ratio evolution diagrams corresponding to each of the distributions shown in Figure 1(b,e), respectively. Here we assume that both helium isotopes have the same diffusivity, an issue which we consider more fully below. The important result is that the various profiles yield easily measurable differences in isotopic ratio (up to factors of a few), especially in the initial few percent of ^{3}He release (i.e., the helium derived from near the domain edge).

The effect of α-ejection

One complication unique to the (U-Th)/He method is that α particles are emitted with sufficient energy that they travel ~20 microns from the site of decay, and some fraction are ejected from grain surfaces. In the case of commonly dated minerals like apatite, titanite, and zircon, the edge of the diffusion domain corresponds to the α-ejection boundary. This effect exerts a strong and predictable spatial effect on the ^{4}He production function that is independent of diffusion (Farley et al. 1996), particularly in smaller grains.

Although α-ejection influences the ^{4}He distribution, Shuster and Farley (2004) demonstrated that if the effect is incorporated into the modeling, it does not significantly diminish the sensitivity of ^{4}He/^{3}He thermochronometry. Figure 2 illustrates the consequences of α-ejection on a ^{4}He profile and its corresponding ^{4}He/^{3}He ratio evolution diagram. This profile was generated by assuming the production function consistent with an α-particle range of 20 microns (see Eqn. 1 of Farley et al. 1996). Shown in Figure 2a is the radial distribution within a 65 μm (radius) spherical domain that experienced 10 °C/Myr cooling and α-ejection. Shown for reference is the profile calculated for the same cooling rate, but without α-ejection. Alpha-ejection clips and flattens the ^{4}He distribution and

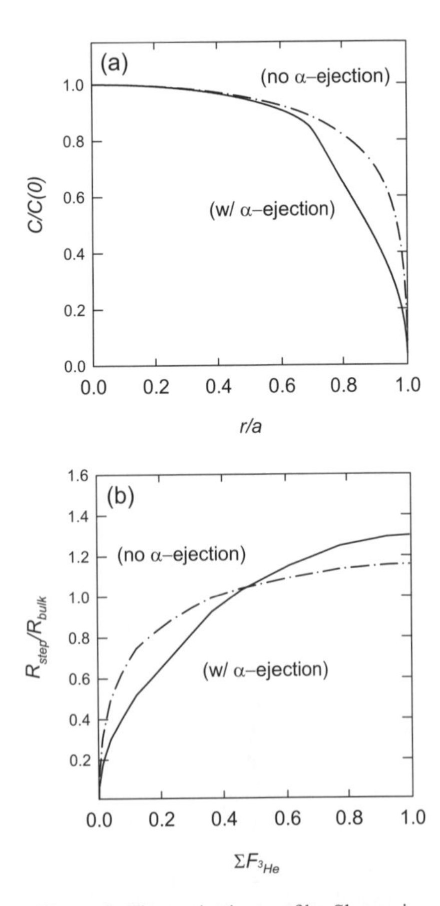

Figure 2. The α-ejection profile. Shown in (a) are two ^{4}He profiles calculated including (solid curve) and excluding (dash-dot curve) the effect of α-ejection, modified from Shuster and Farley (2004). The α-ejection profile was calculated for a spherical diffusion domain with 65 μm radius. These profiles both result from the same *t-T* path shown as a "dash-dot" curve in Figure 1a. Shown in (b) are the corresponding ratio evolution diagrams.

as shown in Figure 2b it also dramatically influences the ratio evolution diagram. The effect introduces a distinct linearity in the first ~30% of the ratio evolution diagram which cannot be ignored in interpretive models.

Note that the overall *shape* of a ratio evolution diagram is independent of the specific diffusion kinetics (i.e., $D(T)/a^2$) and heating schedule of a particular experiment. For a single diffusion domain, the shape of the diagram is controlled only by the initial spatial distribution of ^{4}He (a small potential effect of differences in diffusivity between the two isotopes is discussed below). Therefore, the ^{4}He spatial distribution can be determined *without* knowledge of the helium diffusion kinetics of a particular sample. However, to interpret the profile in terms of the *t-T* path by which it was produced requires knowledge of $D(T)/a^2$ for the helium in that specific sample.

The ^{3}He Arrhenius plot

Calculating diffusion coefficients from a stepwise degassing experiment requires specification of the initial spatial distribution, $C_0(x,y,z)$ of the diffusant (Fechtig and Kalbitzer 1966). Although an initially uniform concentration across a spherical diffusion domain, $C_0(r) = constant$, is typically assumed, this assumption is violated for samples that experienced ^{4}He loss by diffusion and/or α-ejection. Failure to incorporate an appropriately rounded or clipped distribution in the Fetchig and Kelbitzer (1969) computation will yield diffusivities that underestimate the true values, potentially by several orders of magnitude. Therefore, ^{4}He release fractions cannot in general be used to quantify helium diffusion kinetics in most natural samples, despite previous efforts to do so (Lippolt et al. 1994; Wolf et al. 1996; Warnock et al. 1997; Reiners et al. 2002). Laboratory generated ^{3}He, on the other hand, does not pose this problem because the induced distribution is uniform.

An important consideration for this method is whether proton-induced ^{3}He diffusivities are a reliable proxy for radiogenic ^{4}He. It is possible that the two isotopes are sited sufficiently differently that their diffusion behavior differs. In addition, there is a general expectation that the two isotopes will diffuse at slightly different rates given their substantial mass difference.

Shuster et al. (2004) demonstrated that to within analytical uncertainty the diffusion kinetics inferred from ^{3}He are equivalent to those based on radiogenic ^{4}He in Durango apatite and Fish Canyon tuff titanite. The apatite study was performed on an interior aliquot of this large gem quality apatite. Because the analyzed material was obtained at a distance from the α-ejection clipped edge, and because the apatite was quickly cooled it can be assumed to have a uniform distribution of radiogenic ^{4}He. (Note that the variability in He content inferred from the spatial variations in U and Th content reported by (Hodges and Boyce 2003) are far too small to influence this conclusion). In a more detailed study of Durango apatite, Shuster et al. (2004) also demonstrated that diffusive fractionation of helium isotopes can safely be neglected throughout a degassing experiment of that sample. These results indicate that proton-induced ^{3}He is an excellent proxy for ^{4}He in at least this apatite. Additional experiments on more typical apatites are more difficult to interpret because of probable rounding of the ^{4}He profile. In some of these samples diffusive fractionation may be present. Even if the diffusivities vary by the 15% predicted from their masses, a ^{3}He based Arrhenius plot would adequately describe ^{4}He diffusion kinetics for calculating He ages and concentration profiles on specified *t-T* paths. However, small diffusivity differences between the two isotopes *will* influence the shape of the ratio evolution diagram (see below).

An important advantage of ^{4}He/^{3}He thermochronometry over the conventional bulk age approach is that helium diffusion kinetics is determined for each analyzed sample. There is no need to extrapolate diffusion parameters to different grain sizes or to assume that a set of diffusion parameters measured on one sample apply to another sample.

Constraining thermal histories

In this section we consider how results of a $^4He/^3He$ experiment can be used to restrict a sample's allowable *t-T* paths. Three pieces of information are required: (i) the ^4He distribution, (ii) the function $D(T)/a^2$ for helium in the sample, and (iii) the bulk He age. In general a large number of thermal histories will be consistent with these three observations. Like other problems of this nature (e.g., see (Albarède 1978)), to determine a sample's actual thermal history from a measured spatial ^4He distribution is an ill posed problem (Shuster and Farley 2004); a unique solution is not generally possible.

In many ways this situation is analogous to attempts to constrain *t-T* paths from fission track length distributions (Gallagher 1995) and from multi-domain K-feldspar $^{40}Ar/^{39}Ar$ dating (Lovera et al. 1989). As with those techniques, one approach is to generate possible *t-T* paths, forward model the resulting observables, and minimize the mismatch to measurements. Ultimately, we believe this will be the most effective way to interpret $^4He/^3He$ data. Nevertheless, there are several approaches that allow a more intuitive interpretation of the data under certain circumstances.

$^4He/^3He$ age spectra

In many thermochronometry studies the age of a discrete and relatively large magnitude cooling event is sought. For example, rocks may cool rapidly as a consequence of vertical motion on faults or due to river incision associated with surface uplift. The most common approach for dating these events is to obtain an age-elevation transect involving multiple samples (House et al. 1998; Stockli et al. 2000). Provided the event was of sufficient magnitude to cool rocks that were originally above their closure temperature, the age of the event can be determined. However, in cases where the event is of smaller magnitude or where vertical sampling is impractical, it can be difficult or impossible to date the event using a bulk age. The $^4He/^3He$ method provides an alternative and far more sensitive approach that can be performed on a single sample.

Figure 3 illustrates such an application. Consider a forward problem in which geologic evidence indicates a discrete rapid cooling event, arbitrarily assumed to be from 50 °C to 0 °C. If one obtained a (U-Th)/He age of, e.g., 5 Ma on an apatite with $a = 65$ μm, it would not be possible to determine when the cooling event occurred from this sample alone: all six cooling histories in Figure 3 would result in a 5 Ma bulk He age. However, each of the *t-T* paths yields a distinct ratio evolution diagram, especially for the $^4He/^3He$ ratio of the first gas released (e.g., at $\Sigma F_{3He} < 0.01$). The earlier the cooling event the higher the $^4He/^3He$ ratio in the first step, from zero for recent cooling up to the limit of ~ 0.55 dictated by α-ejection alone for the case of cooling at 5 Ma. The intuitive explanation is that prior to cooling, these samples were experiencing ^4He diffusion such that the edge of the grain was effectively at zero age. After rapid cooling and the cessation of diffusion, the edge of the grain began to quantitatively retain ^4He and "age."

This effect is analogous to the age significance in initially derived gas in a $^{40}Ar/^{39}Ar$ age spectrum (McDougall and Harrison 1999). However, an important difference between the Ar and He systems is the influence that α-ejection has upon the distribution. Note that the computations shown in Figure 3 included the effect of α-ejection. This phenomenon is not problematic for the technique, but it does need to be considered. When we account for α-ejection, the initially derived $^4He/^3He$ carries age significance at the limit that essentially no diffusion takes place after a fixed point in time, and as $\Sigma F_{3He} \rightarrow 0$. A second difference between the He and Ar based methods is the need to use the bulk He age to translate the $^4He/^3He$ ratio into a step age. This arises because ^3He is not directly correlated with the parent nuclides of ^4He.

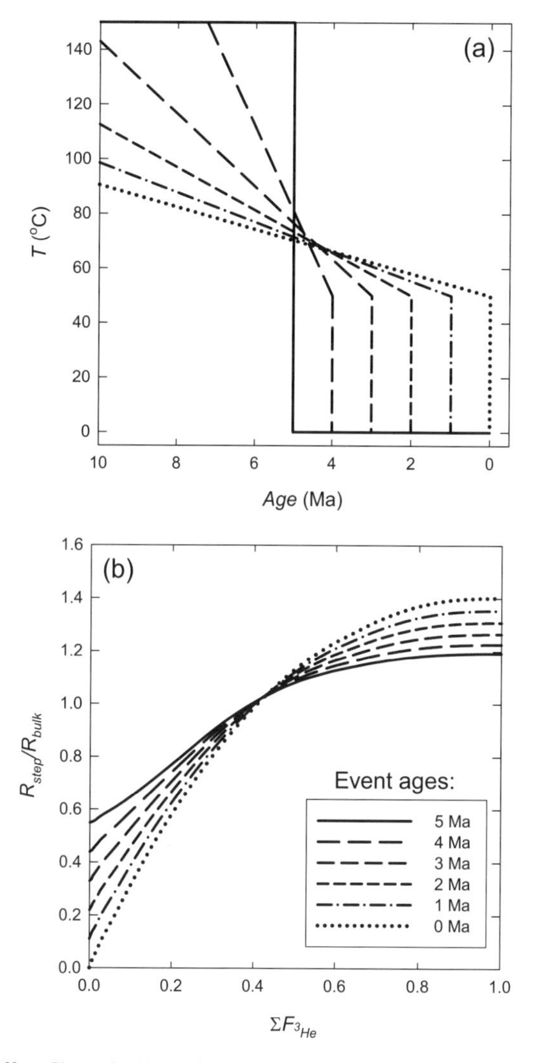

Figure 3. The edge effect. Shown in (a) are six *t*-*T* paths which each result in a bulk He age of 5 Ma. Their corresponding ^4He profiles (not shown) were calculated using Durango apatite helium diffusion kinetics including the effect of α-ejection. The ^4He profiles correspond to the six simulated ratio evolution spectra shown in (b).

On timescales less than ~100 Ma, a model "*Edge Age*" can be calculated simply as a linear scaling of the α-corrected He age ($HeAge_{cor}$) by the ratio of the observed initial $^4He/^3He$ ratio [i.e., $(R_{initial}/R_{bulk})_{measured}$] to that predicted for a profile that expected from α-ejection alone [i.e., $(R_{initial}/R_{bulk})_{\alpha\text{-}ref}$]:

$$\text{"}EdgeAge\text{"} = \frac{\left(\dfrac{R_{initial}}{R_{bulk}}\right)_{measured}}{\left(\dfrac{R_{initial}}{R_{bulk}}\right)_{\alpha-ref}} HeAge_{cor} \tag{5}$$

where

$$\left(\frac{R_{initial}}{R_{bulk}}\right)_{\alpha-ref} = \frac{F_{edge}}{F_T}\sqrt{\frac{D^{^4He}}{D^{^3He}}} \quad (6)$$

F_T is the α-ejection correction factor (Farley et al. 1996):

$$F_T = 1 - \frac{3S}{4a} + \frac{S^3}{16a^3} \quad (7)$$

and F_{edge} is given by:

$$F_{edge} = -\frac{1}{4}\frac{(S-2a)}{a} \quad (8)$$

F_{edge} represents the fraction of α-particles that are retained at the edge of a spherical domain of radius a, and S is the average stopping distance of α-particles along the U and Th decay series (~20 μm). Using these expressions and by extrapolating the curves shown in Figure 3b to $\Sigma F_{3He} \rightarrow 0$, we find that the initially derived $^4He/^3He$ ratios correspond exactly to the ages of rapid cooling in each simulation. The D^{4He}/D^{3He} term accounts for any diffusive fractionation between the helium isotopes upon degassing (see below). In the above simulations, $D^{4He}/D^{3He} = 1$.

The previous example illustrates how a portion of the ratio evolution diagram can have direct age significance. This concept can be generalized to produce a $^4He/^3He$ age spectrum. Because proton-induced ^3He is uniformly distributed, the ^3He release fractions are a measure of the volume of material interrogated by each step. Hence the $^4He/^3He$ ratio of each step is proportional to the ^4He concentration in the interrogated volume. By knowing the ^4He production function within the grain assuming uniform U and Th, we can translate the $^4He/^3He$ ratio for each step into a step age. To make the results more readily interpretable the $^4He/^3He$ release spectrum can be converted into a model age spectrum as follows:

$$\text{"}StepAge\text{"} = \frac{\left(\frac{R_{step}}{R_{bulk}}\right)_{measured}}{\left(\frac{R_{step}}{R_{bulk}}\right)_{\alpha-ref}} \cdot HeAge \quad (9)$$

Here the "α *-ref*" ratio is taken from the ratio evolution diagram computed for a sample of the same grain size as that actually analyzed, but with a ^4He profile dictated solely by α-ejection (the formula for such a profile is given by (Farley et al. 1996)). The "*measured*" and "α *-ref*" ratios are evaluated at the same values of ΣF_{3He}.

An example serves to illustrate how an age spectrum can be used to answer a common geologic question. As discussed by Stockli et al. (2000), a rapid cooling event, for example induced by normal faulting, may be recorded as a fossil helium partial retention zone in an age-elevation plot, where the age of the lower in ection point ("break-in-slope") indicates the age of onset of exhumation. This is illustrated in Figure 4a, which shows a model age-elevation pattern for apatites that experienced thermal quiescence (no cooling) for 55 Myr followed by a rapid exhumation event at 10 Ma. In this example the in ection point would only be revealed after ~3 km of exhumation (assuming a geothermal gradient of 20 °C/km); if only 2 km of exhumation occurred, the He ages would provide no insight to the age of the cooling event (see Fig. 4a, right hand axis). However, the concentration profiles of samples

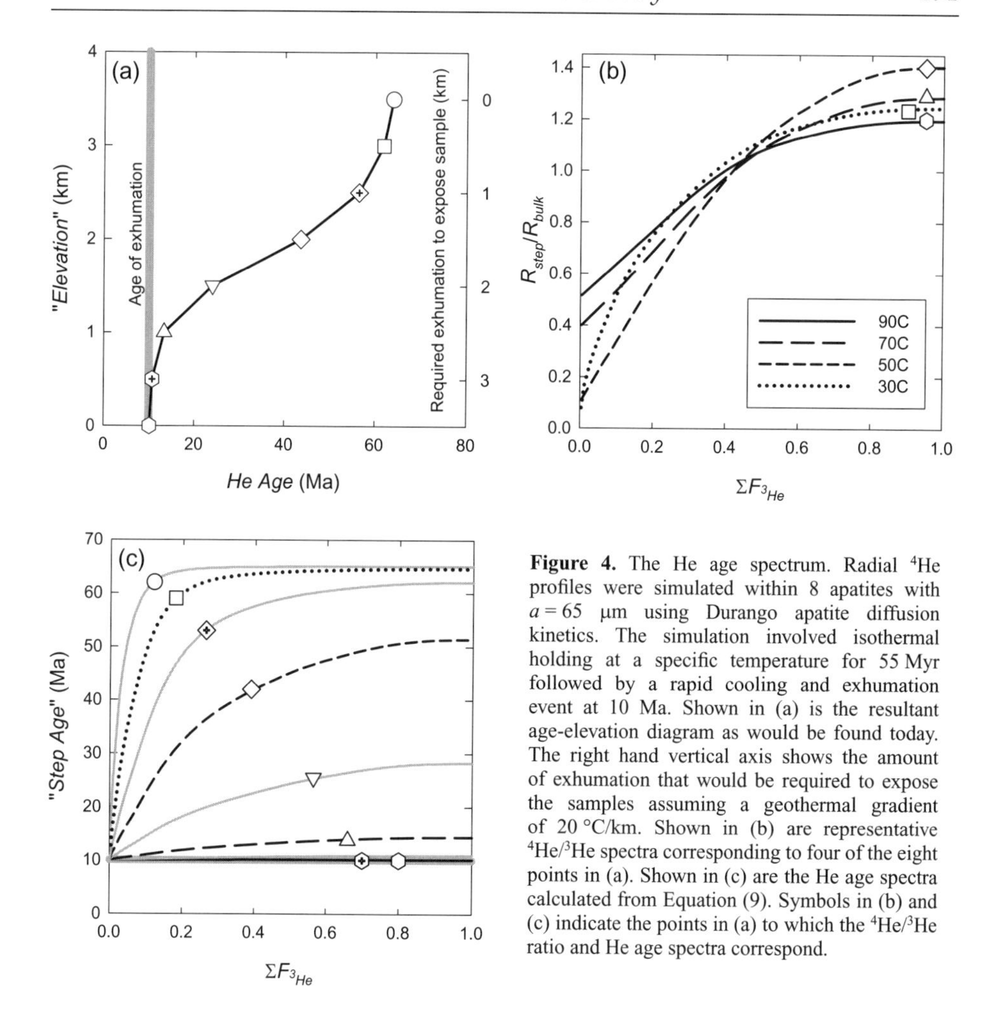

Figure 4. The He age spectrum. Radial ^4He profiles were simulated within 8 apatites with $a = 65$ μm using Durango apatite diffusion kinetics. The simulation involved isothermal holding at a specific temperature for 55 Myr followed by a rapid cooling and exhumation event at 10 Ma. Shown in (a) is the resultant age-elevation diagram as would be found today. The right hand vertical axis shows the amount of exhumation that would be required to expose the samples assuming a geothermal gradient of 20 °C/km. Shown in (b) are representative $^4He/^3He$ spectra corresponding to four of the eight points in (a). Shown in (c) are the He age spectra calculated from Equation (9). Symbols in (b) and (c) indicate the points in (a) to which the $^4He/^3He$ ratio and He age spectra correspond.

high up in the section carry the sought-for timing information, as well as information on their thermal state prior to exhumation. Shown in Figure 4b are the ratio evolution diagrams for 4 representative apatites at 0, 1, 2, and 3 km elevation. In the simulation, these samples resided at 90, 70, 50 and 30 °C, respectively, prior to exhumation.

Figure 4c shows model age spectra for all 8 apatites in this hypothetical sample suite. Each spectrum corresponds to a bulk He age and elevation as indicated by symbols in Figure 4a. As shown in Figure 3, every sample that was sufficiently hot to be diffusing significant He from the grain edge (i.e., above ~30 °C) will have an edge age approximating the age of exhumation. (In principle, the initial He released should be precisely the exhumation age, but in practice one must "mine deeper" into the profile to extract enough He to make the measurement).

The shape of the age spectrum at higher release fractions is also significant. For apatites coming from within the former ^4He partial retention zone (*HePRZ*; ~40–80 °C) the age spectrum is highly curved demonstrating long residence at temperatures where diffusive loss was rapid. Using numerical forward models it would be straightforward to distinguish samples generated in a thermally static *HePRZ* (this model) from, e.g., samples that migrated through the *HePRZ* at some slow rate. More directly, steps from each of the structurally highest two samples achieve a plateau age of 65 Ma. This is because throughout the 65 Ma simulations these two samples were cool enough to quantitatively retain helium in the grain interior. From a practical perspective, such a plateau would demand that the sample had resided at low temperatures (e.g., <30 °C; a more quantitative estimate requires full modeling) for 65 Ma. This would set a limit on the paleodepth of this sample prior to exhumation, and hence an estimate of the total amount of exhumation that had occurred after 65 Ma. Thus by performing a $^{4}He/^{3}He$ analysis on one or several samples it would be possible to determine (i) the time at which a sample first cooled to temperatures where He is quantitatively retained in a grain interior (plateau age), (ii) the total amount of cooling and hence exhumation (from structural relationships involving a sample with a plateau age), (iii) the rate of cooling through the *HePRZ* (from the curvature of the age spectrum), and (iv) the age of onset of a rapid cooling event (from the edge age).

TECHNICAL ASPECTS

Proton irradiation

We have performed four proton irradiations: two at the Harvard Cyclotron Laboratory (now closed) using a 150 MeV proton beam, and two at the Northeast Proton Therapy Center (NPTC) with a 220 MeV beam. The NPTC generates a proton beam with a 4.5 m diameter cyclotron. In our latest irradiation, we generated 10^9 atoms ^{3}He/mg with a fluence of 1×10^{16} protons/cm^2 at 220 MeV, accomplished with a beam current of ~280 nA over a single 8 hour period. This is an enormous concentration of ^{3}He which easily permits single grain analyses.

Energetic proton irradiation produces spallation ^{3}He from nearly all target nuclei in the same way that cosmic rays produce ^{3}He in meteorites in space (Leya et al. 1998; Wieler 2002). Spallation ^{3}He is dominantly produced by a process known as charged particle evaporation. The initial interaction between the incident proton and a target nucleus can leave the residual target nucleus in an excited state, although most of the incident kinetic energy is likely carried off by a scattered nucleon. Upon de-excitation to a new ground state, the residual target nucleus has some probability of emitting a ^{3}He nucleus, e.g., ^{28}Si (p, p) $^{28}Si^{*}$ $^{25}Mg + {}^{3}He$. Because this process emits ^{3}He nuclei along stochastic trajectories at most probable energies between 1–10 MeV, we expect approximately isotropic ^{3}He distributions to be generated within the solid ~1 to ~50 μm surrounding each target atom in a mineral. For this reason the final siting of the ^{3}He nucleus should be crystallographically random and similar to that in which radiogenic ^{4}He resides after nuclear ejection. And since ^{3}He production probabilities are approximately equal from all target elements, the ^{3}He is *a priori* expected to be uniform. We believe that ^{3}He emission from grain edges will be approximately balanced by implantation from surrounding materials, so there is no equivalent to the α-ejection phenomenon with which to contend.

Spallation ^{4}He is produced along with ^{3}He, with a $^{4}He/^{3}He$ ratio of about 10. This amount of ^{4}He is negligible compared to the radiogenic ^{4}He in natural U and Th bearing minerals like apatite at least at the irradiation dosages we have used.

Cross sections for $X(p,x)^{3}He$ reactions are strongly dependent upon proton energy below ~50 MeV, but remain relatively constant above ~100 MeV (Leya et al. 2000). Proton energies exceeding ~100 MeV will not substantially improve ^{3}He yields within a given sample. However because proton energy drops as the beam passes through solid matter, a higher beam energy does allow a thicker stack of samples to be irradiated. The range of protons through a target stack is directly proportional to the energy of the incident beam. To maximize the number of samples simultaneously irradiated, the target stack is constructed such that its length is near the overall range of protons. The range of 150 MeV protons is ~16 cm in Lucite and ~7.5 cm in aluminum. At 220 MeV these ranges are ~34 and ~14.5 cm, respectively. In our latest irradiation, we irradiated ~100 samples at once. This number could easily be doubled while maintaining sufficient ^3He yield in each sample.

To generate a uniform ^3He distribution within a sample, a broad and defocused proton beam is required. This is achieved by first passing the incident beam through a 100 μm Pb foil prior to the samples. The Pb foil, as well as the target stack, causes a scattering of the protons causing the beam intensity profile (normal to the beam axis) to be approximately Gaussian in shape. For individual grains < 150 μm in radius, the resultant ^3He concentration distribution is uniform to within a few percent. This has been observationally verified in a crushed and sieved aliquot of ~180 μm Durango apatite (Shuster et al. 2004). Because the beam intensity varies by as much as 10% normal to its axis (along 15 mm diameter), for samples with diffusion domains >500 μm in radius, uniform ^3He production remains to be established. For such large samples the irradiation can be made more uniform by using a thicker scattering foil (at the expense of overall proton flux) or by continuously moving or spinning the sample relative to the beam.

When working with grains of a size typical of accessory minerals (<150 μm), the container within which the sample is held during the irradiation is critical. For each of the four irradiations, we experimented with a different type of sample container, and found the most success with small packets composed of two pieces of Sn foil cold welded together. Sn foil does not degrade under the high flux of protons, and the cold welded packets ensure quantitative recovery of all irradiated grains.

Sample requirements

The same sample requirements for conventional (U-Th)/He dating also apply for $^4He/^3He$ thermochronometry: euhedral crystals free of fluid and mineral inclusions. Since most information on the *t-T* path of a sample is contained toward the edge of the diffusion domain, it is of particular importance that the analyzed samples contain only original crystal surfaces (see Potential Complications).

As with conventional (U-Th)/He dating, the physical dimensions of each analyzed crystal need to be measured to determine the F_T value (Farley 2002). Since the degassing experiment provides the characteristic diffusive length scale "a" within the function $D(T)/a^2$, the F_T value is only relevant for simulating the effect of α-ejection. For an individual grain, a single F_T value relates the observed ratio evolution to spherical model calculations including α-ejection. However, for samples with low ^4He concentration, it has been necessary to run as many as 20 grains in a single degassing analysis. In such cases we have selected grains within a narrow size range, and calculated F_T values for each; the average F_T value for the population is then used in the model simulations. Because α-ejection adds uncertainty to both the ratio evolution diagram and the conventional age determination, it is desirable to analyze the largest possible grains.

Stepwise degassing analysis

For the reasons discussed above, the objective of a $^4He/^3He$ degassing experiment is two-fold: (i) to measure the ratio evolution diagram and (ii) to determine the function $D(T)/a^2$. These two objectives are not necessarily best achieved with the same heating schedule, so a compromise is required. For instance, a degassing experiment designed solely for the purpose of quantifying diffusion kinetics might incorporate multiple retrograde heating cycles (Farley 2000; Shuster and Farley 2004) and would not require steps at high cumulative helium yields. On the other hand, a well quantified ratio evolution diagram requires an even distribution of points across the budget of ^3He (i.e., on $0 < \Sigma F_{3He} < 1.0$). And, because retrograde degassing steps typically evolve small amounts of helium, they do not necessarily improve the quality of a ratio evolution diagram.

Our $^4He/^3He$ stepwise degassing analyses use the projector-lamp heating device (Farley et al. 1999) and a sector field noble gas mass spectrometer capable of resolving ^3He from HD (MAP 215-50). Because ^3He blanks are typically very low in this system, the ^3He detection limit does not limit the amount of irradiated material from being analyzed. A proton fluence of ~10^{16} p/cm^2 generates sufficient ^3He abundance that individual ~100 μm grains can be studied.

POTENTIAL COMPLICATIONS

Mineral surfaces

Analysis of samples with intact original surfaces is important because most of the thermal information contained in a concentration distribution is located toward the grain edges and because broken surfaces will generate misleading results in the stepwise degassing analysis. A broken surface (or a large crack) will expose a steeper concentration gradient than would otherwise exist in the sample. This would result in initially evolved $^4He/^3He$ ratios that are artificially too high. With careful sample selection, broken and/or cracked grains can be identified and avoided. Fortunately, in samples containing sufficient amounts of natural ^4He, the $^4He/^3He$ degassing analysis can be performed on very few or even single grains. The ^3He blank is sufficiently low that it is not usually the limiting factor in the $^4He/^3He$ analysis. While these are stringent criteria, we have been able to locate at least a few appropriate grains from the modest number of granitic apatite samples we have so far worked on.

Geometry

Throughout this chapter, we used the spherical diffusion domain as an analytically tractable model for actual diffusion domains. This model is clearly an oversimplification; in many cases euhedral (non-spherical) crystals act as the domains for ^4He diffusion (Reiners and Farley 1999, 2001; Farley 2000; Reiners et al. 2002). Although more elaborate calculations could be constructed to incorporate actual crystal geometry, the spherical domain is sufficient for the purposes of bulk (U-Th)/He thermochronometery (Meesters and Dunai 2002a,b). This conclusion applies to the $^4He/^3He$ method as well. For example, Shuster et al. (2004) illustrated that accurate spherical representations of diffusively modified ^4He distributions could be retrieved using $^4He/^3He$ release spectra from degassed Durango apatite despite the fact that the shards were non-spherical. In the case of a non-spherical domain, the Fechtig and Kalbitzer (1966) calculation effectively averages over the geometrically complicated nature of the material and returns a diffusion domain radius, a [i.e., $\ln(D/a^2)$], which on the average describes the characteristic diffusion length scale of the material and describes a sphere with a surface area to volume ratio approximating that of the actual domain.

However, for the above statements to be true, two conditions must be met. The first is that ^3He and ^4He release fractions are determined simultaneously. The ^3He $\ln(D/a^2)$ values and ^4He release fractions are specific to the diffusion domain geometry *and* experimental conditions during an analysis. Second, it is critically important that forward-calculated thermal models use the function $D(T)/a^2$ that is specific to the sample. The experimentally determined diffusivity, D, *and* the characteristic length scale, a, specify the sample. By transforming a problem of profile-model-matching to the spherical domain, the two parameters are inextricably linked. Modeled profiles determined through forward calculation must contain each. The extrapolation of a specific experimentally determined function $D(T)/a^2$ (e.g., for Durango apatite) to an unstudied specimen possibly of a different grain size may not be accurate.

Does proton irradiation affect helium diffusion kinetics?

An important assumption of the $^4He/^3He$ method is that proton irradiation does not modify the helium diffusion kinetics of the material under investigation. Since the helium diffusion kinetics is determined on an irradiated sample, it is important to demonstrate whether or not the irradiation modifies a given mineral in such as way as to alter the otherwise natural diffusion kinetics. Our detailed work on Durango apatite demonstrated that ^{4}He diffusivities of irradiated and un-irradiated aliquots are analytically indistinguishable. Figure 5, taken from Shuster et al. (2004), shows that the ^{4}He Arrhenius relationships determined for an irradiated and a non-irradiated aliquot are equivalent. A similar conclusion can be drawn from work on titanite (Shuster et al. 2004). These experiments clearly indicate that for the proton fluence we used, the irradiation does not significantly modify ^{4}He diffusion kinetics. These observations only apply to the dosage and minerals we investigated; experiments on additional minerals at higher fluences could conceivably yield different result.

Diffusive fractionation of helium isotopes?

Due to the ~25% relative difference in mass between ^{3}He and ^{4}He, the potential exists for a difference between the diffusivities of ^{3}He and ^{4}He. For example, from the kinetic theory of gases the ratio of diffusivities for the two isotopes might be expected to be controlled by the inverse square root of their masses, $D^{4He}/D^{3He} = 0.868$. Shown in Figure 6, detailed experiments on Durango apatite clearly indicate far smaller, possibly zero, isotopic mass fractionation in that sample (Shuster et al. 2004). However results on other apatites (unpublished, see below for examples) hint at some degree of fractionation. Thus it is presently unclear whether isotopic mass fractionation does or does not accompany helium release from apatite in all cases. Hence we evaluate its potential effects in this section.

A difference of 15% in diffusivity would introduce negligible bias to thermochronometric calculations. However, it would fractionate ^{4}He/^{3}He ratios upon stepwise degassing,

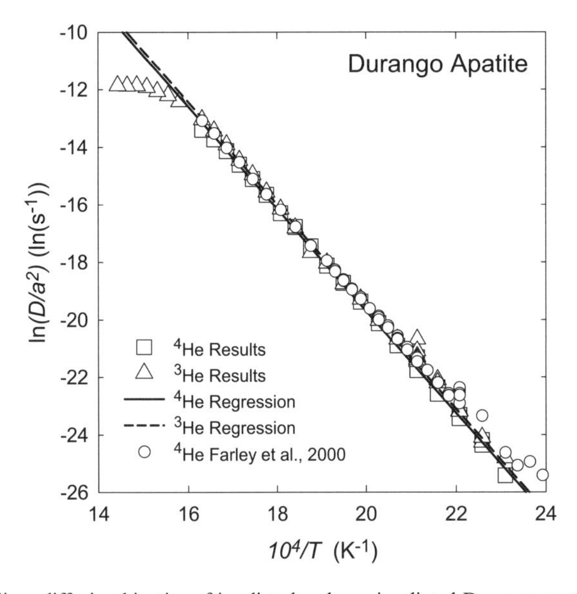

Figure 5. The helium diffusion kinetics of irradiated and non-irradiated Durango apatite, modified from Shuster et al. (2004). Diffusion coefficients (D/a^2) calculated using (Fechtig and Kalbitzer 1966) are plotted against inverse absolute temperature ($10^4/T$). Open triangles are values calculated from proton-induced ^{3}He and open squares calculated from ^{4}He for the irradiated aliquot. The dashed line indicates least squares regression through subsets of the ^{3}He results, and the solid line the ^{4}He results. Shown as open black circles in are ^{4}He results from (Farley 2000) for non-irradiated Durango apatite. Modified from Shuster et al. (2004).

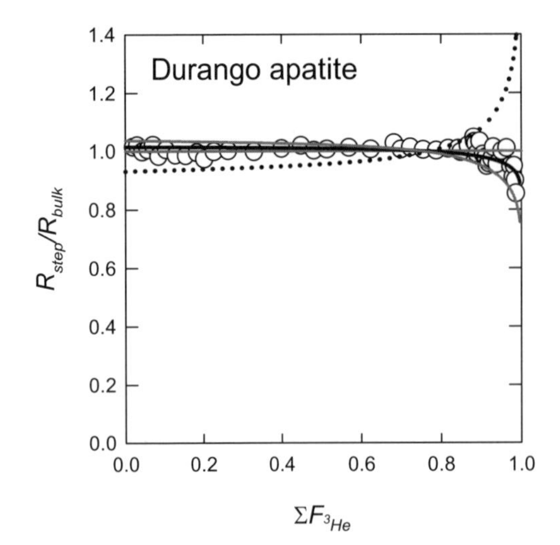

Figure 6. Durango apatite ratio evolution diagram, modified from Shuster et al. (2004). Shown are measured isotope ratios for each release step, R_{step} (R=^{4}He/^{3}He), normalized to the bulk ratio R_{bulk} plotted vs. the cumulative ^{3}He release fraction, ΣF_{3He}. Four diffusion models are also shown. The model that best fits the entire dataset, $D^{4He}/D^{3He} = 1.03$, is shown as a solid black curve with 95% confidence intervals shown as solid gray curves: D^{4He}/D^{3He} = 1.00 and 1.07, respectively. Also shown as a dotted curve is the model corresponding to the inverse root mass relationship: D^{4He}/D^{3He} = SQRT(m_3/m_4) = 0.868. Modified from Shuster et al. (2004).

particularly toward the end of an analysis at high helium yields. Because this would influence the shape of a ^{4}He/^{3}He ratio evolution diagram, the effect needs to be considered when deriving a thermochronometric interpretation.

During sequential degassing, an isotopic difference in diffusivity has a predictable effect on measured ^{4}He/^{3}He ratios. Figure 7 illustrates how this would influence the ^{4}He/^{3}He observations for a typical ^{4}He distribution. For reference, this simulation uses the same ^{4}He profile calculated according to a relatively rapid cooling trajectory as in example 2 (see below). Shown are the resultant ratio evolution spectra for three values of the ratio D^{4He}/D^{3He} between 1.00 and 0.868. Notice that the differences between the curves are small when $\Sigma F_{3He} < 0.25$ and that all three curves are convergent as ΣF_{3He} approaches ~ 0.8. The curves then strongly diverge at values of $\Sigma F_{3He} > 0.8$. The differences between the curves are most pronounced as ΣF_{3He} approaches 1.0. This "distillation" effect is due to preferential diffusion of ^{3}He from the domain over the course of the analysis. As expected, the effect is much less pronounced at the beginning of the analysis. Particularly for samples containing a diffusive ^{4}He profile, the spectral shapes for $\Sigma F_{3He} < 0.25$ are not strongly dependent on the diffusivity ratio.

Because the diffusivity ratio does not appear to be consistent between different samples, its uncertainty will propagate into the t-T paths constrained by a ratio evolution diagram. As shown in Shuster and Farley (2004), most of the information on t-T is expressed in the ratio evolution for $0 < \Sigma F_{3He} < 0.25$. Even if a conservative range in D^{4He}/D^{3He} is assumed (e.g., from 1.00 to 0.868), the resultant t-T uncertainty will be negligible when models are primarily matched to data below $\Sigma F_{3He} < 0.25$. Conversely, the ^{4}He/^{3}He ratios as ΣF_{3He} values approach 1.0 are more strongly dependent upon the diffusivity ratio than on the ^{4}He distribution or its corresponding t-T path. Systematically increasing ratios towards the end of an analysis likely indicate higher ^{3}He diffusivity compared to ^{4}He in a particular sample. Only when models are matched over the entire range of ΣF_{3He} should the influence of diffusive fractionation become strongly relevant.

Non-uniform U and Th distributions

Throughout this chapter, we have assumed a uniform distribution of U and Th throughout the diffusion domain. A strongly heterogeneous distribution of parent nuclides would clearly influence the final spatial distribution of ^{4}He within a particular sample. Recent efforts on zircon involving laser ablation mass spectrometry have attempted to quantify heterogeneous

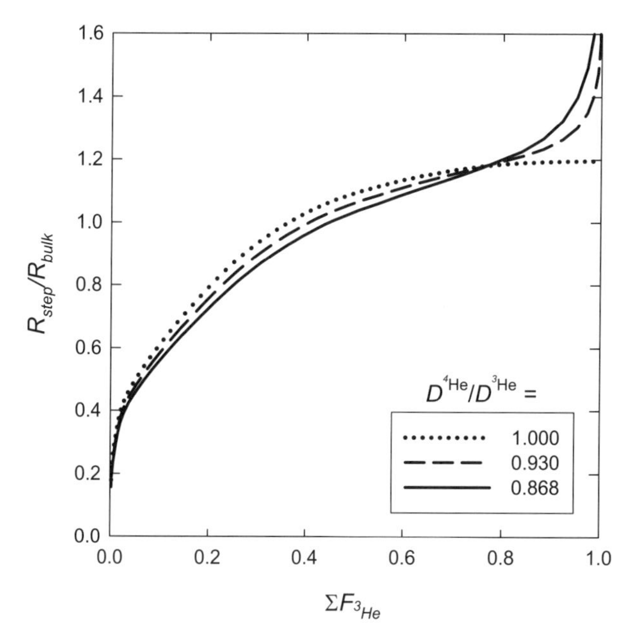

Figure 7. The potential effect of diffusive isotope fractionation upon a ratio evolution diagram. Shown are three $^4He/^3He$ ratio evolution spectra for a common ^4He profile, but for different values of the diffusivity ratio D^{4He}/D^{3He}: 1.00, 0.93 and 0.868. For reference, the ^4He profile was calculated according to the solid black *t-T* path shown in Figure 9(c) and $F_T = 0.81$.

distributions of U and Th on a small spatial scale (Hourigan et al. 2003). If this information is accurately determined for a given sample, it could be easily incorporated into the $^4He/^3He$ method by simply allowing for spatial variability in the ^4He production function. The model ^4He distributions and their corresponding $^4He/^3He$ spectra described in this chapter can easily be calculated for any radially symmetric, but variable ^4He production function.

EXAMPLE APPLICATIONS

In this section we present three example applications of $^4He/^3He$ thermochronometry. These examples are intended to illustrate the quality of $^4He/^3He$ data and *t-T* information that is obtainable using the method rather than to address a particular geological problem. Therefore, the examples are presented without geologic context. The first example is a set of three controlled experiments designed to test the method's accuracy. The other two examples are granitic apatites expected to contain naturally diffusive ^4He distributions which reflect their cooling trajectories. All of these apatites were irradiated simultaneously using a 220 MeV proton beam and total proton fluence of ~1×10^{16} protons/cm^2.

Example 1: controlled ^4He distributions

Three aliquots of Durango apatite were heated under vacuum for different durations in order to generate distinct distributions of ^4He. These experiments were performed on the same interior aliquot of Durango apatite used in previous studies (Farley 2000; Shuster et al. 2004) and which had a uniform distribution of ^4He prior to heating. Since the initial ^4He distribution and helium diffusion kinetics of this sample are sufficiently well known, the three aliquots could be partially degassed for different durations to generate diffusive profiles with known deficit gas fractions (*dgf*). The heated aliquots are analogous to samples which experienced simple, yet known thermal perturbations. Following partial ^{4}He degassing, the aliquots were subjected to proton bombardment and the ^{4}He/^{3}He analysis described above.

The results are shown in Figures 8a-c as ratio evolution spectra. All three diagrams clearly reveal diffusive ^{4}He profiles; each has an initial ^{4}He/^{3}He ratio ~ 0 followed by a systematic increase in ^{4}He/^{3}He to relatively constant values near $R_{step}/R_{bulk} = 1$ when $\Sigma F_{3\text{He}} > 4.0$. It is also clear that each profile is distinct from the others, with the magnitude of diffusive rounding increasing in the order a < b < c. For reference, compare these three results with Figure 6, which shows the observed ^{4}He/^{3}He spectrum for an aliquot containing a uniform ^{4}He distribution within Durango apatite.

Superimposed on each result in Figure 8 is a model ^{4}He/^{3}He spectrum for a diffusive ^{4}He profile within a spherical domain. The models use $D^{4\text{He}}/D^{3\text{He}} = 1$, and each model corresponds to the specific deficit gas fraction (*dgf*) independently measured on each aliquot. We find excellent agreement between the models and the ^{4}He/^{3}He observations. Although not shown, the ^{3}He Arrhenius plots for each experiment yield diffusion parameters that are statistically indistinguishable from the known helium diffusion kinetics of Durango apatite. Coupled with the deficit gas fractions implied by each ratio evolution spectrum, the ^{3}He based diffusion kinetics successfully constrains the actual heating temperatures and durations which were used to partially degas the ^{4}He distributions prior to irradiation.

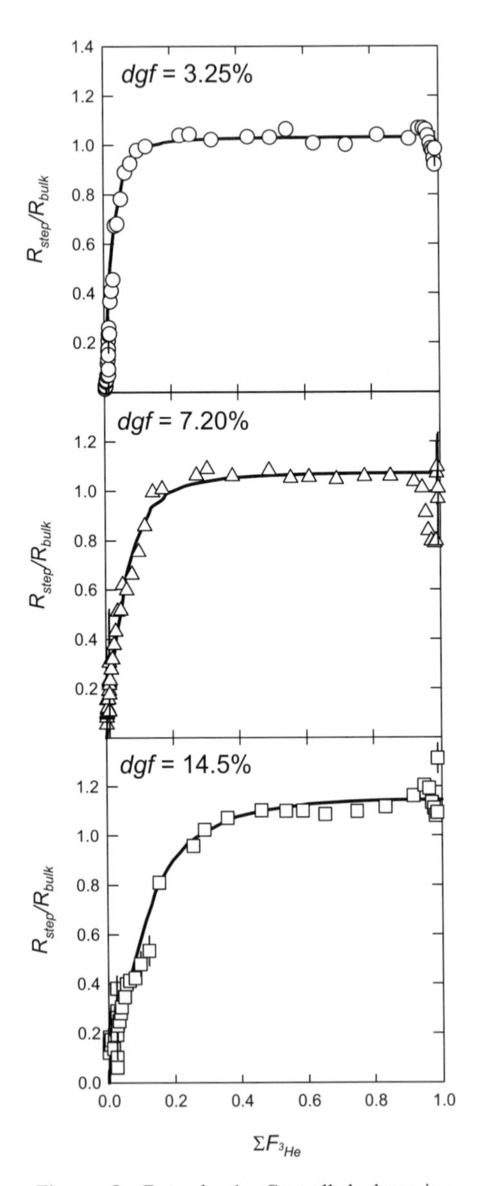

Figure 8. Example 1: Controlled degassing experiments. Shown as points in (a)-(c) are the observed ^{4}He/^{3}He evolution spectra of 3 partially degassed aliquots of Durango apatite. Shown as solid curves are the ^{4}He/^{3}He evolution spectra corresponding to the expected ^{4}He profiles in these aliquots. The deficit gas fractions (*dgf*) of the curves are indicated on the figures.

These experiments demonstrate that (i) the ratio evolution diagrams successfully recover the expected ^{4}He profiles resulting from simple diffusive modification, (ii) $\ln(D/a^2)$ values calculated from proton-induced ^{3}He adequately describe helium diffusion kinetics in the irradiated sample, and (iii) when combined, this information successfully constrains the actual thermal perturbation experienced by each of the three aliquots. Because the grains used were not radially symmetric spheres, these results also illustrate the self-consistency of the spherical model. As long as the domain geometry and diffusion coefficients are self-consistently applied, the profiles obtained by forward model matching can be used to constrain a sample's low-temperature thermal history in nature even if the sample is not spherical.

Example 2: natural apatite

The second example is a population of nine euhedral apatites free of inclusions collected from a single hand specimen of a granitic pluton. The apatites had an average F_T value of 0.81 and an α-ejection corrected bulk He age of ~39 ± 2 Ma. Grain-to-grain variance in F_T was ± 0.02. The apatites were subjected to proton bombardment as a larger population of grains, and then subsequently picked for $^4He/^3He$ analysis. The results are shown in Figure 9.

Figure 9a shows the diffusion coefficients determined from ^3He which yields an Arrhenius relationship with activation energy = 135 kJ/mol and $\ln(D_0/a^2)$ = 12.4. These parameters correspond to a 10 °C/Myr helium closure temperature of ~70 °C. Note that 8 points which clearly deviate from linearity at the highest temperatures express the same phenomenon observed by (Farley 2000), and were excluded from the regression. The observed ratio evolution spectrum is shown in Figure 9b as points. Details of the spectrum on

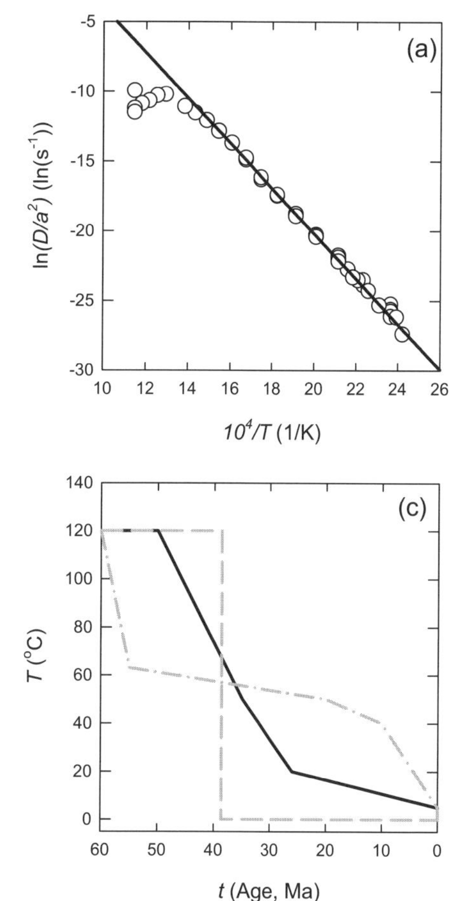

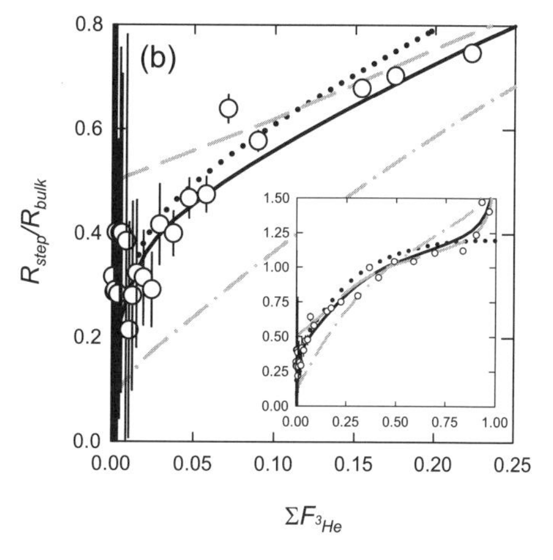

Figure 9. Example 2: Natural granitic apatites. Shown in (a) is the ^3He Arrhenius plot for this analysis calculated from (Fechtig and Kalbitzer 1966). The solid line is a least squares regression through a subset of the points and defines the diffusion kinetics used to calculate models shown in (b) and (c). Shown in (b) as points is the observed $^4He/^3He$ ratio evolution spectrum for $0 < \Sigma F_{3He} < 0.25$. The curves shown in (b) are model spectra which correspond to $F_T = 0.81$ and the t-T paths shown in (c). Each model produces a bulk He age = 39 Ma and was calculated for $D^4He/D^3He = 0.868$. Also shown as a dotted curve in (b) is the same model as the solid curve, but calculated for $D^4He/D^3He = 1.0$. The inset plot in (b) shows the observed spectrum and models over the entire ΣF_{3He} range.

$0 < \Sigma F_{3He} < 0.25$ are presented, and the entire spectrum is shown in the inset. The analytical uncertainty indicated in Figure 9b was typically dominated by ^{4}He blank corrections.

Superimposed on Figure 9b are three model ^{4}He/^{3}He spectra calculated for spherical domains with $F_T = 0.81$ and include the effect of α-ejection. Each model shown as a solid or dashed curve corresponds to a cooling trajectory shown in Figure 9c and spatial ^{4}He distribution which would result in the bulk He age of the sample, 39 Ma. The three models were calculated for $D^{4He}/D^{3He} = 0.868$; notice in the inset that elevated ^{4}He/^{3}He ratios when $\Sigma F_{3He} > 0.80$ hint at diffusive fractionation between proton-induced ^{3}He and radiogenic ^{4}He in this sample. For comparison, the same model shown as a solid black curve is also shown for $D^{4He}/D^{3He} = 1.0$ as a dotted black curve.

The three cooling models clearly do not represent an exhaustive coverage of cooling history space, but are simply intended to illustrate the types of models which can be developed for ^{4}He/^{3}He comparison. The initially observed ^{4}He/^{3}He ratios with value ~ 0.3 require that the sample had accumulated a significant fraction of its ^{4}He below ~30 °C. This and the observed curvature when $\Sigma F_{3He} < 0.25$ clearly eliminate the two models shown as dashed curves as possible *t*-*T* paths for the sample. Instant cooling at 39 Ma would result in initial ^{4}He/^{3}He ratios ~ 0.55, and prolonged residence in the partial retention zone would result in a more diffusive distribution. The data are reasonably consistent with the model shown as a solid black curve corresponding to relatively rapid cooling before 25 Ma, followed by more gradual recent cooling below 30 °C. Note that models calculated for $D^{4He}/D^{3He} = 1.0$ would yield a very similar result, particularly when matching observations between $0 < \Sigma F_{3He} < 0.25$.

Example 3: natural apatite

The third example is a population of 27 apatites collected from a hand specimen of a different granitic pluton. These apatites were also free of inclusions and euhedral with an average F_T value of 0.76 and an α-ejection corrected bulk He age of ~9.7 ± 0.6 Ma. Grain-to-grain variance in F_T was ± 0.03. The apatites were also subjected to proton bombardment as a larger population of grains, and then subsequently selected for ^{4}He/^{3}He analysis. The results are shown in Figure 10.

Diffusion coefficients calculated from ^{3}He are shown in Figure 10a. The best fit Arrhenius relationship (which excludes the 7 highest temperature steps for the same reason as discussed above) is shown as a solid line in Figure 10a and corresponds to an activation energy of 124 kJ/mol and $\ln(D_0/a^2) = 10.2$. These apatites are slightly less helium retentive than the previous example, with a 10 °C/Myr helium closure temperature of ~60 °C. Details of the ^{4}He/^{3}He spectrum are shown in Figure 10b for $\Sigma F_{3He} < 0.25$, and the entire spectrum is shown in the inset. The analytical uncertainties were typically dominated by ^{4}He blank corrections.

As seen in Figure 10b, the ^{4}He/^{3}He ratios measured between $0.05 < \Sigma F_{3He} < 0.25$ were very well constrained and show systematically increasing values. Superimposed on the data are three models corresponding to the cooling trajectories shown in Figure 10c. Each model uses the Arrhenius relationship shown in Figure 10a, and each corresponds to a bulk He age of 9.7 Ma. Sharply increasing ratios for $\Sigma F_{3He} > 0.80$ (Fig. 10b inset) suggest diffusive fractionation of the helium isotopes; the three models were calculated for $D^{4He}/D^{3He} = 0.868$. As with the previous example, the ^{4}He/^{3}He data are not consistent with either immediate cooling at the He age or prolonged residence at elevated temperatures of partial ^{4}He retention. The initially elevated ^{4}He/^{3}He ratios near the grains' edges require that a significant fraction of ^{4}He was accumulated at relatively low temperatures. The data are in excellent agreement with a thermal history involving ~15 °C/Myr cooling to < 30°C by ~ 6 Ma (solid black curve). Interestingly, the mean value $R_{initial}/R_{bulk} = 0.3$ corresponds to an "edge age" ~ 5.5 Ma for this sample, which is in good agreement with the full model calculation.

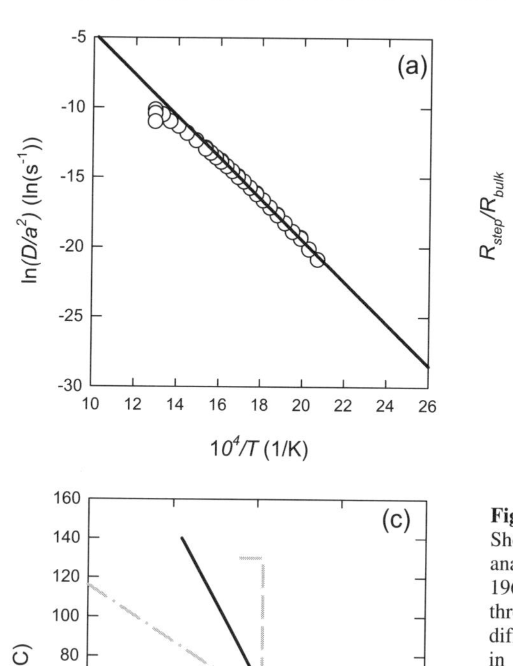

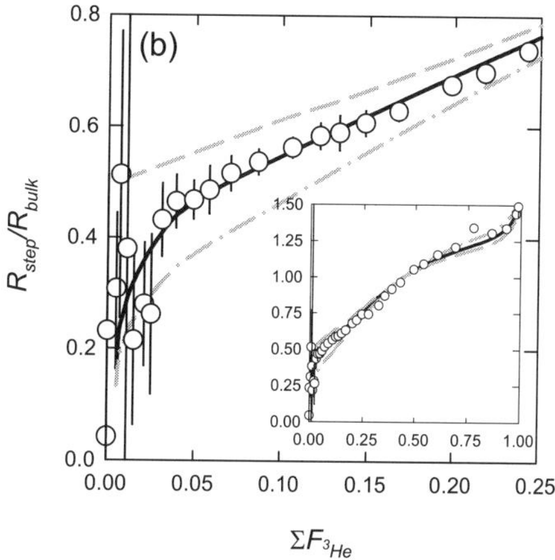

Figure 10. Example 3: Natural granitic apatites. Shown in (a) is the ^3He Arrhenius plot for this analysis calculated from (Fechtig and Kalbitzer 1966). The solid line is a least squares regression through a subset of the points and defines the diffusion kinetics used to calculate models shown in (b) and (c). Shown in (b) as points is the observed $^4He/^3He$ ratio evolution spectrum for $0 < \Sigma F_{3He} < 0.25$. The curves shown in (b) are model spectra which correspond to $F_T = 0.76$ and the t-T paths shown in (c). Each model produces a bulk He age = 9.7 Ma and was calculated for $D^{4He}/D^{3He} = 0.868$. The inset plot in (b) shows the observed spectrum and models over the entire ΣF_{3He} range.

Although the successful models shown in examples 2 and 3 (solid curves Figs. 9b and 10b) do not prove the t-T paths of the samples, the $^4He/^3He$ observations clearly permit elimination of many possible thermal histories in a very low temperature range. The examples illustrate that high-precision $^4He/^3He$ data is attainable on relatively small populations of apatites, and that when combined with the bulk He age, these data can be used to place stringent restrictions on the low temperature t-T paths of the samples.

CONCLUSIONS

A uniform ^3He distribution can be artificially produced within minerals by irradiation with a 220 MeV proton beam. Outgassing of spallation ^3He produced by such irradiation can be used as a proxy for radiogenic ^4He diffusion. The ability to generate a uniform ^3He distribution within minerals permits helium diffusivity measurements on samples in which the natural concentration distribution is non-uniform. Variations in the $^4He/^3He$ ratio over the course of a stepwise heating experiment re ect the initial ^4He distribution within the sample. Both forward and inverse modeling can be used to constrain these profiles, which can in turn be used to constrain the t-T path of the sample. When coupled with the bulk (U-Th)/He age of a sample, this information places stringent limitations on the possible low temperature thermal histories of a particular sample.

ACKNOWLEDGMENTS

We thank D. Burnett and J. Sisterson for helpful discussions and F. Albarède for a helpful review of the manuscript. This work was supported by the National Science Foundation and by a N.S.F. Graduate Research Fellowship to D.L.S.

REFERENCES

Albarède F (1978) The recovery of spatial isotope distributions from stepwise degassing data. Earth Planet Sci Lett 39(3):387-397

Carslaw HS, Jaeger JC (1959) Conduction of Heat in Solids. Oxford University Press, New York

Crank J (1975) The Mathematics of Diffusion. Oxford University Press, New York

Dodson MH (1973) Closure temperatures in cooling geological and petrological systems. *Contrib Mineral Petrol* 40:259-274

Farley K, Reiners P, Nenow V (1999) An apparatus for high-precision helium diffusion measurements from minerals. Anal Chem 71:2059-2061

Farley KA (2000) Helium diffusion from apatite: general behavior as illustrated by Durango fluorapatite. J Geophys Res 105:2903-2914

Farley KA (2002) (U-Th)/He dating: techniques, calibrations, and applications. Rev Mineral Geochem 47: 819-844

Farley KA, Wolf RA, Silver LT (1996) The effects of long alpha-stopping distances on (U-Th)/He ages. Geochim Cosmochim Acta 60:4223-4229

Fechtig H, Kalbitzer S (1966) The diffusion of argon in potassium bearing solids. *In:* Potassium-Argon Dating. Schaeffer OA, Zahringer J (eds) Springer, Heidelberg, p 68-106

Gallagher K (1995) Evolving temperature histories from apatite fission-track data. Earth Planet Sci Lett 136: 421-435

Harrison TM, Grove M, Lovera OM, Zeitler PK (2005) Continuous thermal histories from inversion of closure profiles. Rev Mineral Geochem 58:389-409

Hodges K, Boyce J (2003) Laser-ablation (U-Th)/He geochronology. Eos Trans AGU, Fall Meet Suppl 84(46)

Hourigan JK, Reiners PW, Nicolescu S, Plank T, Kelley K (2003) Zonation-dependent α-ejection correction by laser ablation ICP-MS depth profiling: toward improved precision and accuracy of (U-Th)/He ages. Eos Trans AGU, Fall Meet Suppl 84(46)

House MA, Wernicke BP, Farley KA (1998) Dating topography of the Sierra Nevada, California, using apatite (U-Th)/He ages. Nature 396:66-69

Leya I, Busemann H, Baur H, Wieler R, Gloris M, Neumann S, Michel R, Sudbrock F, Herpers U (1998) Cross sections for the proton-induced production of He and Ne isotopes from magnesium, aluminum, and silicon. Nucl Instr Methods Phys Res B 145:449-458

Leya I, Lange HJ, Neumann S, Wieler R, Michel R (2000) The production of cosmogenic nuclides in stony meteoroids by galactic cosmic-ray particles. Meteor Planet Sci 35(2):259-286

Lippolt HJ, Leitz M, Wernicke RS, Hagedorn B (1994) (U+Th)/He dating of apatite: experience with samples from different geochemical environments. Chem Geol 112:179-191

Lovera O, Richter F, Harrison T (1989) The $^{40}Ar/^{39}Ar$ thermochronometry for slowly cooled samples having a distribution of diffusion domain sizes. J Geophys Res 94:17917-17935

McDougall I, Harrison TM (1999) Geochronology and Thermochronology by the $^{40}Ar/^{39}Ar$ method. Oxford University Press, New York

Meesters AGCA, Dunai TJ (2002a) Solving the production-diffusion equation for finite diffusion domains of various shapes (part I): implications for low-temperature (U-Th)/He thermochronology. Chem Geol 186: 333-344

Meesters AGCA, Dunai TJ (2002b) Solving the production-diffusion equation for finite diffusion domains of various shapes (part II): application to cases with α-ejection and non-homogeneous distribution of the source. Chem Geol 186:347-363

Reiners PW, Farley KA (1999) Helium diffusion and (U-Th)/He thermochronometry of titanite. Geochim Cosmochim Acta 63:3845-3859

Reiners PW, Farley KA (2001) Influence of crystal size on apatite (U-Th)/He thermochronology: an example from the Bighorn mountains, Wyoming. Earth Planet Sci Lett 188:413-420

Reiners PW, Farley KA, Hickes HJ (2002) He diffusion and (U-Th)/He thermochronometry of zircon: Initial results from Fish Canyon Tuff and Gold Butte, Nevada. Tectonophys 349:297-308

Reiners PW, Spell TL, Nicolescu S, Zanetti KA (2004) Zircon (U-Th)/He thermochronometry: He diffusion and comparisons with Ar-40/Ar-39 dating. Geochim Cosmochim Acta 68(8):1857-1887

Shuster DL, Farley KA (2004) $^3He/^4He$ thermochronometry. Earth Planet Sci Lett 217(1-2):1-17

Shuster DL, Farley KA, Sisterson JM, Burnett DS (2004) Quantifying the diffusion kinetics and spatial distributions of radiogenic ^4He in minerals containing proton-induced ^3He. Earth Planet Sci Lett 217(1-2):19-32

Stockli DF, Farley KA, Dumitru TA (2000) Calibration of the (U-Th)/He thermochronometer on an exhumed fault block, White Mountains, California. Geology 28:983-986

Turner G (1969) Thermal histories of meteorites by the $^{40}Ar/^{39}Ar$ method. *In:* Meteorite Research. Millman PM (ed) Reidel, Dordrecht, p 407-417

Warnock AC, Zeitler PK, Wolf RA, Bergman SC (1997) An evaluation of low-temperature apatite U-Th/He thermochronometry. Geochim Cosmochim Acta 61(24):5371-5377

Wieler R (2002) Cosmic-ray-produces noble gases in meteorites. Rev Mineral Geochem 47:125-163

Wolf RA, Farley KA, Silver LT (1996) Helium diffusion and low-temperature thermochronometry of apatite. Geochim Cosmochim Acta 60(21):4231-4240

Reviews in Mineralogy & Geochemistry
Vol. 58, pp. 205-238, 2005

8

Fission-track Analysis of Detrital Zircon

Matthias Bernet

Department of Geosciences
State University of New York
New Paltz, New York, 12561, U.S.A.
matthias.bernet@aya.yale.edu

John I. Garver

Geology Department
Olin Building, Union College
Schenectady, New York, 12308-3107, U.S.A.
garverj@union.edu

INTRODUCTION

Zircon has become one of the most important minerals for studying sediment provenance and the exhumation history of orogenic belts. The reason for this utility is that zircon is common in many igneous, metamorphic, and sedimentary rocks, it is resistant to weathering and abrasion, and it can be dated with various isotopic methods having reasonable high concentrations of uranium and thorium (Fig. 1). Techniques used to date detrital zircon include U/Pb and (U-Th)/He dating, but in this chapter we focus exclusively on fission-track (FT) analysis.

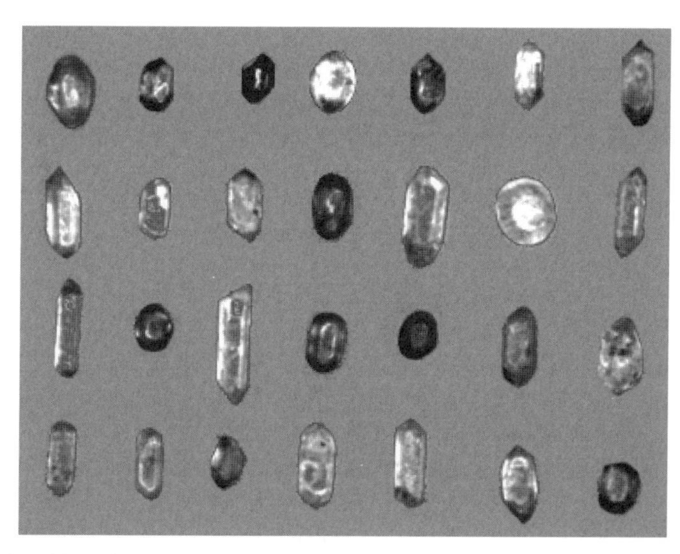

Figure 1. Suite of detrital zircon showing the whole spectrum of zircon shapes and colors that encountered in detrital samples. This particular samples is a suite of zircon from a single sandstone sample of the Eocene Ukelayet Flysch, Northern Kamchatka, Russia. Several end members are worth noting (see text for discussion): Very well-rounded grains are likely to be polycyclic; Colorless with little damage and/or low REE; Colorless and euhedral; Grains of the red series; Grains of the yellow series.

1529-6466/05/0058-0008$05.00 DOI: 10.2138/rmg.2005.58.8

FT analysis allows age determination of single zircon grains that may have cooling ages between several hundred thousand to a billion years or more. The datable range depends on individual uranium content and cooling history of a zircon grain. Fission tracks in zircon have an effective annealing temperature of ~240 °C ± 30 °C in natural systems (Hurford 1986; Brandon et al. 1998; Bernet et al. 2002). Therefore most detrital zircon are fairly resistant to thermal annealing in typical sedimentary basins after deposition, while the other low-temperature thermochronometers anneal at lower temperatures common in sedimentary basins (i.e., Helium dating and apatite FT) and therefore more readily have compromised provenance information (Fig. 2). Consequently, the strength of detrital zircon fission-track (DZFT) analysis lies in the fact that this method provides robust cooling ages of source terrains. The ability of zircon to retain information about the most recent thermal history of a source area is invaluable in elucidating the processes and system response in a range of geodynamic settings, especially the evolution of orogenic belts. This characteristic makes DZFT dating superior to U/Pb dating when the objective is to link sedimentation to the uplift and exhumation history of the source terrain. U/Pb dating of single crystals provides crystallization ages (or zircon growth during metamorphism), which typically pre-date the latest orogenic cycle. This long-term memory is partly due to the fact that zircon is so robust that recycling is common and it is typical for zircons to be polycyclic, even in crustal melts. As such, a U/Pb age on a detrital zircon may have little bearing on the nature of the immediate source rock, but may be ideal for understanding the long-term record of crustal formation. In addition, U/Pb ages of detrital zircon rarely allow the determination of exhumation rates and because of the possibility of multiple recycling of zircon grains their U/Pb ages can only vaguely be assigned to non-distinct source regions. Therefore, FT ages tend to be directly related to actively evolving source terrains.

Consequently, as we explain below, DZFT analysis is a method ideally suited for: a) tracing the provenance of clastic sediments; b) determining stratigraphic ages in volcanically active areas; c) studying the long-term exhumation history of convergent mountain belts with little active volcanism, and d) dating low-temperature thermal events. Some interesting recent

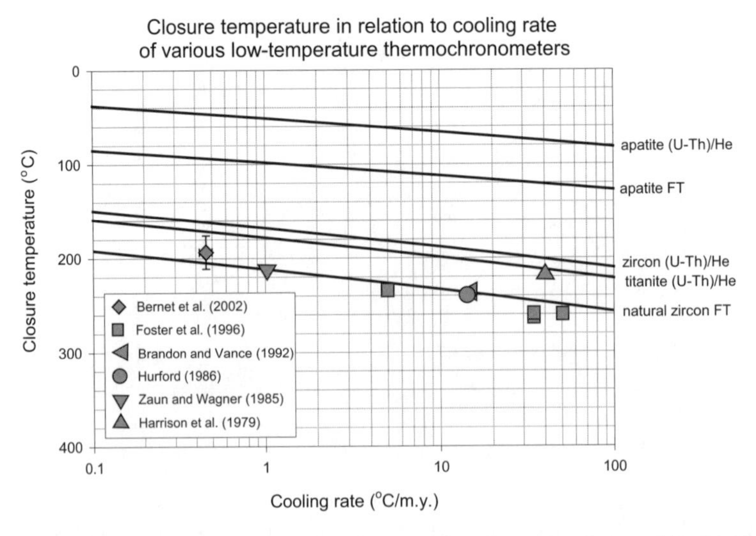

Figure 2. Closure-temperature as a function of cooling rate, given for apatite, zircon and titanite (U-Th)/He and apatite and zircon FT thermochronometers. All curves are calculated after Dodson (1973). Field-based estimates of the zircon fission-track closure temperature are shown from Harrison et al. (1979), Zaun and Wagner (1985), Hurford (1986), Brandon and Vance (1992), Foster et al. (1996), and Bernet (2002).

work has been aimed at combining DZFT with U/Pb or (U-Th)/He dating on the same grains or samples (see below; i.e., Reiners et al. in review).

In this chapter we explain basic aspects of DZFT analysis, and provide some practical considerations on sampling techniques in the field and laboratory analysis. We then show how results can be presented and discuss the interpretation of fission-track grain-age (FTGA) distributions in several different applications. Finally, we give an overview of the current developments in DZFT analysis and end by outlining some outstanding issues that need further attention.

FISSION-TRACK DATING OF DETRITAL ZIRCON

This section is meant to be an introduction to practical and technical aspects of FT analysis on detrital zircon. The basics of ZFT analysis are reviewed elsewhere (i.e., see Tagami et al. 2005), so here we highlight the principal methodological aspects that are unique to analysis of detrital zircon.

Field collection

Fission-track analysis can be preformed on detrital zircon from any clastic sedimentary environment, but most studies have focused on ancient sandstones and a few have investigated zircon from modern environments such as fluvial and beach facies. Sampling techniques are different for both kinds of samples (rock or sediment) and sampling strategies depend on the intended study. For exhumation and provenance studies one should carefully consider where in the field samples are collected. For example, regional studies require samples collected from river deltas of large-scale drainages, from marine turbidite sequences in the outcrop, or from drill-cores, if available. In any case, the most common mistake is that too little sample is collected (see suggestions below) and too few zircons are separated for proper analysis.

Zircon and source rocks. A crucial first step is to understand the nature of the source rock and whether that rock will yield zircon in an appreciable quantity. Geologic maps usually provide reasonable information about the potential zircon yield that can be expected in any given drainage area, but obviously the source for ancient sequences can be more difficult to infer. Zircon is a common accessory mineral in many acidic and sodium rich igneous rocks such as granite, granodiorite, tonalite or rhyolite and their metamorphic equivalents (see Table 1; Poldervaat 1955, 1956; Deer et al. 1992). As such, zircon occurs in siliciclastic deposits derived from such source rocks. In many river drainages the variety of gravels in the riverbed will provide a quick overview of lithology in the source area, but they are likely to be biased towards more resistant lithologies.

Table 1. Relative zircon concentration by source lithology.

Source lithology	*High concentration*	*Intermediate concentration*	*Low to no concentration*
Igneous rocks	granite, granodiorite, tonalite	rhyolite, ignimbrite	gabbro, ultramafic rocks, basalt
Metamorphic rocks	orthogneiss	paragneiss, meta-rhyolite, meta-sandstone, phyllites	marble, eclogite, schist
Sedimentary rocks	arkose	conglomerates, quartz arenite, litharenite, siltstone	claystone, dolomite carbonate rocks

Note: Relative zircon concentrations are based on Poldervaart (1955, 1956) and Deer et al. (1992)

Owing to its stability (hardness of 7.5; lacks distinct cleavage) zircon survives significant weathering and transport while other detrital components are selectively removed. This trend is reflected in the zircon-tourmaline-rutile (ZTR) index for heavy minerals in clastic sedimentary rocks. This index is used to semi-quantitatively evaluate sediment maturity and source rock weathering, and increases when these three very stable minerals are relatively enriched in the heavy mineral fraction of clastic sediment by either transport or dissolution (e.g. Morton 1984; Mange and Maurer 1992). For example quartz arenite and quartzite have a particularly high ZTR index and commonly at least an intermediate zircon yield. Lithologies with unusually low zircon yield include carbonates, mafic rocks, and ultramafic rocks (Table 1).

Recent sediment. Collecting detrital zircon samples from Recent sediment and loosely consolidated sedimentary rock is relatively simple. Zircon, commonly of fine sand size in detrital samples, has a density of ~4.55–4.65 g/cm^3 (Deer et al. 1992), so its settling velocity is similar to quartz grains of medium sand size. For this reason zircon is typically deposited with somewhat coarser grained material, and samples should be preferably collected from sand bars and beaches with coarse- to medium-sand grain sizes. Simple gravity separation in the field (i.e., gold panning), can easily concentrate zircon so that a final density separation in the lab involves only a small quantity of material (200–300 g instead of 2–4 kg). Therefore, loose sediment can be directly processed in gold pans, and panning removes the lighter material (quartz, feldspar, micas etc.) and enriches the heavy minerals such as zircon, garnet, magnetite and even gold. In general, it is sufficient to pan between 12–14 pans of material, but the final outcome depends on zircon yield, panning efficiency, etc.

It is also possible to collect samples from gravel bars. In this case, gravel and all finer grained material can be run through a coarse sieve. The finer fraction (coarse to fine sand and smaller sizes) should be retained and processed further in the gold pan, while all coarse material (> 2 mm) can be discarded. It is worthwhile to look for heavy mineral placer deposits, which can be easily recognized by black and reddish colors from magnetite and garnet. If placer deposits are available it is not even necessary to use gold pans, because the top layers of the placer deposits can be scraped from the surface. If only loose sediment is collected for processing in the lab, without panning in the field, one should collect at least 4–7 kg of sample material. However, even this size of sample may not have sufficient zircon if the source lithology is not zircon bearing.

Ancient sandstone. Collecting samples from sandstone outcrops for DZFT analysis is routine, but there are some important considerations to bear in mind. All sample sizes suggested here are based on our experience in a number of different geodynamic settings of different composition and age. We find that the best samples are medium-grained arkosic sandstones and 2 kg of sample is generally sufficient, but many sandstone compositions are appropriate for collection and zircon extraction. Samples of lithic sandstone samples should be 4–7 kg. The presence of visible quartz is generally a good indicator, because quartz-rich lithologies require smaller samples (Table 2). The target grain size should be medium- to coarse-grained sandstone: fine-grained sandstone should be avoided, but collected only as a last resort (see below: fine-grained sandstones can yield c. 50 μm zircons, which are possible to analyze). For graded beds (i.e., sandy turbidites), this observation requires that in some cases only the base of a bed is sampled.

Table 2. Sandstone sample size for DZFT analysis.

Lithology	*weight*
Arkose	2–4 kg
Quartzo-feldspathic sandstone	~4 kg
Quartz-bearing volcaniclastic sandstone	4–7 kg
Lithic sandstone	4–7 kg
Silicic volcaniclastic sandstone	2–4 kg

The yield of zircon from most sandstone is usually satisfactory because many common lithologies produce appreciable yield and post-depositional modification is not significant. However, detrital apatite is much less predictable because it is more variable in source rock, and it can be severely affected by post-depositional dissolution. If the aim is to analyze detrital apatite as well, then it is important to avoid altered sandstones, especially those with excessive iron oxide and evidence of interstratal dissolution. These strata may have very poor yields of apatite, and there may be significant secondary minerals such as pyrite, siderite, or barite.

Analytical considerations in the lab

Mineral separation. After the samples have been collected in the field it is necessary to extract zircons in the laboratory with standard heavy liquid and magnetic separation techniques (Table 3). When large amounts of kyanite, barite, or pyrite are present in the zircon fraction, it may be necessary to further concentrate the zircon by hand picking, or it is possible to remove pyrite with 5 N HNO_3 over 24 hours, which leaves zircon unaffected.

Table 3. Summary of mineral separation steps.

	Separation step
1)	Crush and pulverize the rock
2)	Separate the sample using a shaking table (e.g., Rogers or Gemeni table).
3)	(optional) Sieving with 0.25 – 0.088 mm sieves. Process only the 0.25 – 0.088 mm fraction. Store the >0.25 and <0.088 mm fractions.
4)	Separate heavy minerals from light-mineral contaminates by passing the sample through heavy liquids (i.e., Sodium polytungstate or Tetrabromoethane).
5)	Pass the heavy fraction through the Frantz magnetic separator stepwise at 0.1 – 1.5 amp. (Possible to loose Fe-rich zircon during this separation step).
6)	Process the nonmagnetic fraction in heavy liquid (i.e., Methylene iodide).

Mounting. Separated zircons are mounted in PFA or FEP Teflon[1], as is routine in ZFT analysis, but there are a few aspects unique to DZFT analysis. Depending on available sample material we like to include 200 to 1000 zircons in a mount ($\sim 2 \times 2$ cm^2) to easily ensure that 50–100 randomly selected grains can be dated per sample. The number of grains on the mount is important because a large fraction will be uncountable due to heterogeneous uranium distribution, high radiation damage, cracks and inclusions, etch quality, and other factors that are typical of zircon on detrital grain mounts. After the grains have been mounted in Teflon, the mounts are polished to expose smooth internal zircon surfaces. One distinctly different approach is that DZFT typically involves making several mounts that then receive different etch times. We recommend at least two mounts per sample that are then etched for different lengths of time (see Naeser et al. 1987; Garver et al. 2000b; Bernet et al. 2004b), but in cases where the grain-age distribution is very large (i.e., grain ages span 1000 m.y.) then up to six mounts may be required to fully capture the whole grain-age distribution (i.e., Meyer

[1] In the past many FT labs have used FEP Teflon, which is available in tape form commercially (i.e., Saunders Inc.). In the mid- to late 1990's many FT labs switched to PFA Teflon, which is composed of tetraflouroethylene-perflouroalkoxyethene. One problem, is that PFA Teflon has limited commercial availability. For details, see Ontrack, v. 2, n.2, p. 17 (November 1992, available on the internet). PFA Teflon has a higher melting temperature, and is more resistant to chemical attack in the etchant. However, it is more difficult to handle while mounting.

and Garver 2000). The reason for the different etch times is that a detrital sample contains a mixture of zircons with different amounts of radiation damage, and therefore different chemical reactivity (or "etchabilities"—discussed below—see Naeser et al. 1987; Garver et al. 2000a; Bernet et al. 2004b). Nevertheless, in most provenance and exhumation studies, grain ages between a few million and several hundred millions of years can be dated (Fig. 3).

Etching. Etching polished mounts is among the most crucial steps in DZFT analysis. Zircon etching is done with a strong acid or base that attacks the polished crystal surface. The increase in crystal disorder in the damaged track is preferentially attacked and the track is fully revealed for optical analysis when etched long enough. Detrital suites typically have a large variation in single-grain radiation damage, which is generally attributed to α-recoil damage from the decay of uranium and thorium (see Garver and Kamp 2002). Accumulated α-damage increases the chemical reactivity of a zircon, so that highly damaged grains (generally older grains all else being equal) are much easier to etch than grains with little damage. This difference is not trivial and typical etch times can vary by about 3 orders of magnitude (1–100 hr for our lab set up). The etchant should be replaced regularly during the etching process (every 24 to 48 hours), especially when working with impure mounts, to maintain etching efficiency.

Note that the etch formula and etching temperature vary from lab to lab and a useful summary of these different conditions is given in Garver (2003). Here we discuss some general etching characteristics using conditions most commonly employed in labs around the world. This typical set up includes an etchant composed of a NaOH:KOH eutectic at 225–230 °C, in a covered Teflon dish heated by a laboratory oven. We are not trying to imply that this is the best approach for etching, but these are the conditions that we are most familiar with. The single

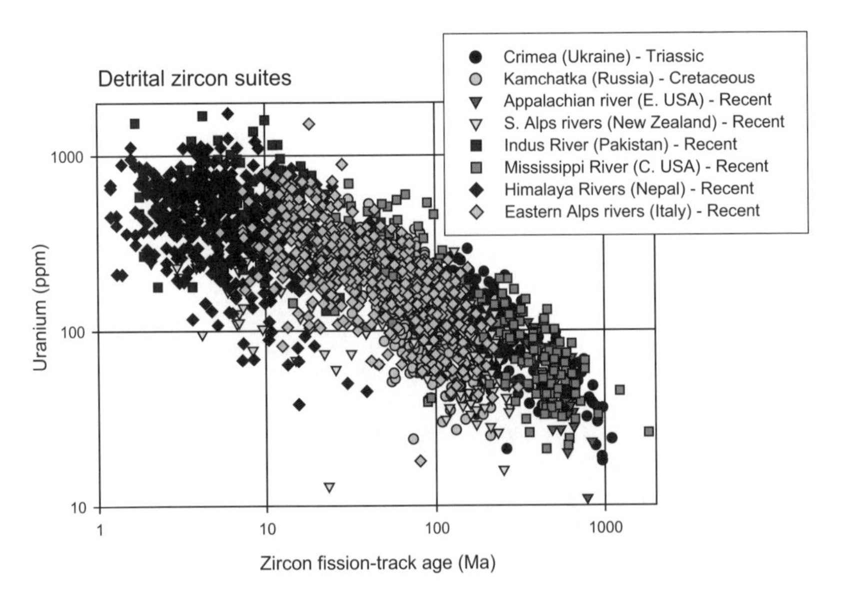

Figure 3. Plot showing the general range of expected grain-age distributions and uranium concentrations in typical detrital zircon suites. Note that the upper right field has no data points because these zircons have track densities that are too dense to count using standard methodologies. The lower left field, largely empty, corresponds to grains that are generally underetched in most analyses, but long etch times (c. 50–150 hr) could have captured them in these cases. Data Sources: Crimea (Soloviev, unpublished); Kamchatka (Garver et al. 2000); Appalachian rivers (Meyer and Garver, 2000); Southern Alps (Garver and Kamp 2002); Indus River (Cerveny et al., 1988); Mississippi River (Meyer and Garver, 2000), Himalaya rivers (Brewer, unpublished); European Alps (Bernet et al. 2004b).

biggest variable that affects etch time is temperature (Garver 2003), so temperature control should be well calibrated, and strictly controlled.

Three approaches have been used to attempt to fully reveal tracks in a Teflon-mounted mixed suite of zircon. The *Muti-Etch technique* assures both random selection and optimal etching by repeatedly etching and counting a single mount at regular intervals (Hasebe et al. 1993). This method provides an unbiased distribution of grain ages, but is very time intensive and operationally difficult (see Hasebe et al. 1993, p. 124). The *Multi-Mount technique* optimizes the total range of countable grains by insuring that all grain populations are well etched by etching several mounts over different lengths of times and counting grains from all mounts (Naeser et al. 1987). The principal advantage of this technique is that it quickly and reliably provides the full FTGA spectrum. However, this approach may result in an inadequate quantitative sampling of the FTGA distribution because of the overlap of the ages dated in the individual mounts certain age groups may be over-represented. The *Optimal Etch technique* attempts to maximize a certain population of grains from a single sample (e.g. Kowallis et al. 1986; Garver and Brandon 1994b). This approach requires that a particular population is optimally etched at the expense of all other populations, and it has largely been used to date the young population of grains.

There is no simple formula for determining etch time required for a suite of samples, but here we offer some general rules that work in our labs using the *Multi-Mount technique.* To illustrate the general variation in etch times we refer to Figure 4, which shows the relation between etch time, U content, and fission-track age of detrital zircon from a number of rivers that drain the Alps (Bernet et al. 2004a, b). This plot demonstrates that zircon with young cooling ages and low accumulation of radiation damage need longer etch times to reveal countable tracks as discussed above. Recall that zircons with high radiation damage etch much

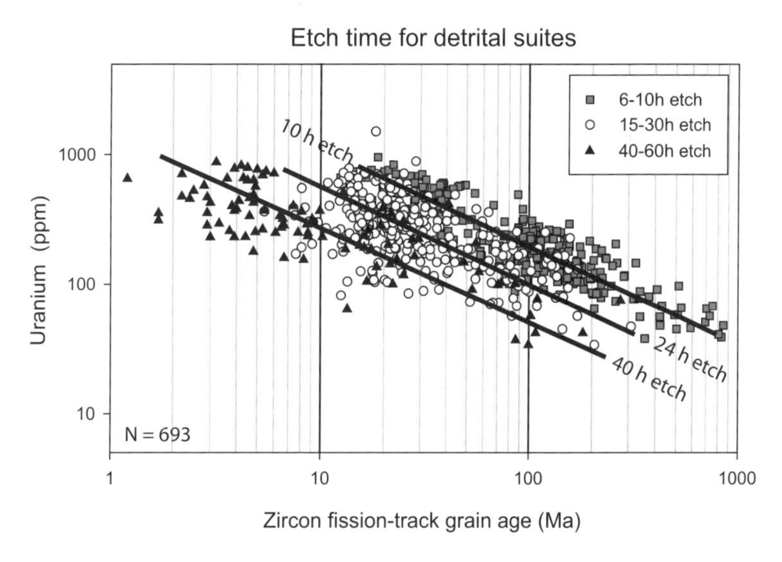

Figure 4. Uranium concentration and FT age correlation of detrital zircon, shown here in relation to etch time duration. Note that long tech times tend to reveal countable tracks in grains with higher U concentrations and younger cooling ages. The reason for this is that grains with younger cooling ages have less radiation damage accumulated and the grains are more pristine, reducing the etching efficiency. Grains with older cooling ages and higher radiation damage etch more easily and therefore have shorter etch times. Etching was done in a NaOH:KOH eutectic in Teflon dishes at 228 °C in a laboratory oven. All data are from modern river sediment (Bernet et al. 2004a,b).

more easily and countable tracks are visible after short etch times. Therefore, a good starting approach is to etch one mount for a few hours, remove from the etchant, clean, and evaluate tracks under the microscope. If the majority of the grains are under-etched, then additional etch-time is needed. We found that in many settings it is good to start with etch times between 8 and 30 hours. The etch time of the second mount can then be selected shorter or longer, depending on the etching response of the first mount. With two mounts, one should attempt to straddle the optimal etch time. When we prepare a series of samples for analysis (20–40 samples), we typically budget about 5–7 days for all etching, well in excess of the 1–2 hr that would be required to etch a comparable suite of apatite mounts.

A unique situation involves samples with both very young grains mixed with older grains. Typical zircons have uranium concentrations between about 200 and 450 ppm (see Garver and Kamp 2002; Reiners et al. in review). Detrital grains with typical uranium concentrations and ZFT ages of less than about 1–3 Ma in age have little radiation damage and require very long etch times (c. 30–100 hr). These low-damage grains have an etching anisotropy that results in a differential rate of track revelation parallel to *c*-axis (slower) compared to perpendicular to the *c*-axis (faster). However, there is an additional problem in that because they are young they may have few if any tracks: it can be difficult to evaluate whether a grain is properly etched if it has no tracks (a zero-track grain) because the quality of the etch is evaluated by most workers by track-pit diameter. If there are no tracks it is difficult to ascertain if the grain has been sufficiently etched. One possible solution to this problem is to etch for a very long time (40–100 hr), and assume that all grains are well etched, and count all grains (even zero-track grains). In this approach, older grains may be sacrificed due to overetching. If zero-track grains are ignored, the data set will be biased, and not representative of the grain-age distribution in the source region. As such, when evaluating detrital zircon with a population of grains $< 1–3$ Ma, one needs to carefully devise the experimental conditions to capture this difficult-to-etch population.

Counting. Counting tracks in zircon for FT analysis is routine, but for DZFT dating there are a few specific procedural aspects that are unique. At issue here is sample bias, and grain countability. Grains with high spontaneous track densities (track densities $> 3 \times 10^7$ tracks/cm^2, usually old grains or those with very high uranium concentrations) or metamict grains cannot be dated with the FT method, because individual tracks cannot be differentiated and counted. On the other end of the spectrum, grains with low track densities may be underetched.

In an attempt to avoid further bias and to obtain representative and reliable results only a random selection of countable zircons should be analyzed, from a randomly mixed suite of zircons in the Teflon mount. That approach differs from routine ZFT analysis where one would select representative grains of the best population of grains for dating. Detrital zircon grains should be selected by their countability and not by shape, size, clarity, or other attributes. Therefore, only grains containing well-etched fission tracks should be counted and under-etched or over-etched grains should be omitted. In addition, grains with strong zoning, uneven surfaces, cracks, inclusions, or very small counting areas should not be counted. The crucial aspect of counting is that specific criteria are determined at the onset of analysis and that these criteria are then strictly followed.

Track-lengths. Measuring track lengths of horizontally confined tracks (HCT) for modeling thermal histories is a standard procedure in apatite FT analysis, but not so for zircon. While several labs do measure track lengths in zircon, routine analysis is hindered by variability in grain-to-grain etch times, which is attributed to variation in alpha damage. Track-lengths measurements have never been reported in DZFT analysis, largely due to the fact that detrital grains have extreme variation in etch times, and therefore single grain measurements are nearly meaningless. Additionally, it is difficult to establish a unique population for a single grain that might have a measurable track length. Note that even with the analysis of detrital apatites, track lengths are rarely done because it is difficult to assign single grains to component populations with distinct thermal histories (see Garver et al. 1999).

Grain-age analysis and data presentation

The results obtained from DZFT analysis can be evaluated in several different ways, but the goal in each approach is to discriminate populations of cooling ages. First, it is important to determine if any grains are younger than the depositional age of the sample. All the youngest grains would naturally fall into the minimum age group of a detrital sample, which has special significance in many studies. The minimum age is determined either by binomial peak fitting or by χ^2 evaluation, and may be of importance in refining the depositional age of the sample or for detecting partial resetting during low-temperature thermal events (see discussion below). Second, it is common to calculate the mean age of the FTGA distribution, which may be of interest when determining average exhumation rates in exhumational studies. A third approach, which is widely used, is to decompose the distribution of grain ages into individual grain-age components through a number of statistical techniques (e.g. see Brandon 1996 for discussion). Currently available software packages for data analysis and graphical or numeric data presentation include BINOMFIT based on Brandon (1996), POPSHARE from Dunkl (unpublished), or MACMIX based on the approach of Sambridge and Compston (1994). We prefer a binomial peak fitting routine from Galbraith and Green (1990), which is used in the BINOMFIT program. This approach involves taking the observed grain-age distribution and then decomposing it into major grain-age components or peaks (labeled P1, P2, P3 etc.).

The full grain-age spectrum and binomial fitted peaks are conveniently presented in histograms, probability density (PD) plots, or radial plots (Fig. 5). If detrital samples of the same stratigraphic age are collected and compared to each other, than it is useful to look for reoccurring peak-age groups (i.e., label P1, P2, P3 and so on, see Table 1 in Bernet et al. 2004b; also see Sircombe and Hazelton 2004). If samples with different depositional ages from a stratigraphic section are presented, than P1, P2, P3 etc. should be assigned just as they occur.

Stewart and Brandon (2004) provide a discussion on the detection limit of the FT method on detrital samples and on how many grains should be counted per sample to identify major grain-age components. It is our experience that counting more than 100 grains per sample does not significantly improve the results and is usually not justified given the amount of time it takes to date 100 grains, but for a different opinion see the discussion by Vermeesch (2004).

INTERPRETATION OF FISSION-TRACK GRAIN-AGE DISTRIBUTIONS

Once individual grains from samples are dated, results can be placed in geologic context. Before we discuss specific results, we review a few important concepts in the literature that frame the context of detrital peak ages. For a start, one could ask the following questions: are all age components or peaks older than the depositional age and reflect exhumation and cooling in the source area? Is there a volcanic component not related to exhumation-driven cooling? If grains have been derived from an orogenic source, which grains are from reset metamorphic sources, and which grains are recycled from sedimentary cover units? If P1 in a sample is younger than the depositional age, is that an indication for partial annealing after deposition?

The partial annealing zone and closure of the ZFT system

Fission tracks in zircon result from the spontaneous fission of ^{238}U and the formation of a track or damage zone in the crystal from these fission events. At elevated temperatures these tracks anneal, which means they shorten and then disappear as fast as they are formed, but at low temperatures all tracks are fully retained. Because detrital zircon in sedimentary strata commonly get buried and heated it is important in any DZFT study to determine if there is

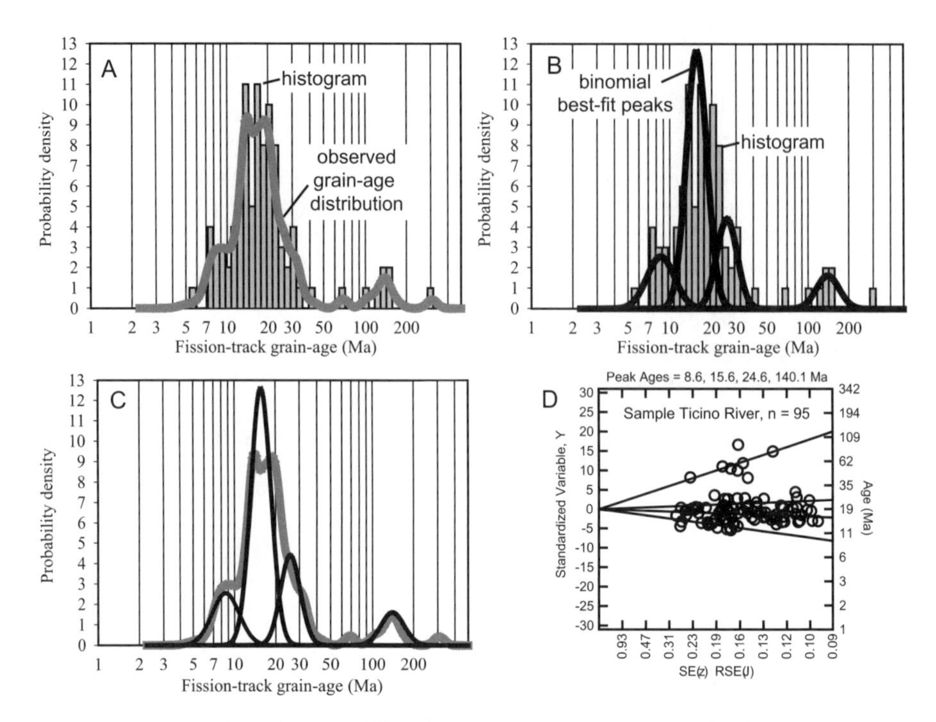

Figure 5. Shown are the various possibilities to present detrital zircon FTGA distributions and best-fit peaks in probability density and radial plots. The data shown here are from the Ticino River in Italy (Bernet et al. 2004b). A) Histogram and curve of the observed grain-age distribution. B) Histogram and curves of binomial best-fit peaks. C) Curve of observed grain-age distribution and curves of binomial best-fit peaks. D) Radial plot with best-fit peaks. Peak fitting after Galbraith and Green (1990) and (Brandon 1996) using BINOMFIT from Brandon.

any evidence of full or partial FT annealing after deposition. To evaluate thermal maturity of sedimentary rocks, we can independently employ techniques such as apatite FT, (U-Th)/He, vitrinite reflectance (R_o), or conodont color-alteration indices (CAI).

The temperature range below which tracks are retained and above which tracks are lost is commonly referred to as the Partial Annealing Zone (PAZ) (see Wagner and van den Haute 1990). For simplicity many workers refer to an effective closure temperature instead of the partial annealing zone, which represents the temperature of nearly full track retention, and therefore closure of the FT system (after Dodson 1973). Even if this is a rather simplified concept when considering the parameters (time, temperature, cooling rate, radiation damage, pressure etc.) that influence partial and full track retention or resetting in zircon, it is a widely used concept that works reasonably well. In most geological settings zircon has an effective closure temperature of about 240°C ± 30 (Brandon et al. 1998; Bernet et al. 2002), but this temperature is sensitive to the rate of cooling and radiation damage in the zircon (Fig. 2; also see Garver et al. 2002, 2005; Rahn et al. 2004). We use the estimates of Brandon and Vance (1992) that suggest the 90% retention temperature in most cases (or $T_{90\%}$) is ~240 °C. Likewise, detrital samples that have preserved unreset zircons are assumed to have resided below temperatures (track retention of greater than 10% or $T_{10\%}$) of ~175–200 °C for heating times between 25 and 1 million years (see discussion in Brandon and Vance 1992). Note that the $T_{90\%}$ (~240 °C) corresponds to a depth of about 7.5–8 km assuming a typical continental geotherm of 30 °C/km and an average surface temperature of ~10 °C.

Resetting for any particular grain is largely a function of internal radiation damage, which affects its annealing properties: low-damage grains are more resistant to annealing than high-damage grains (Garver et al. 2005). These end members can be simplified in general conceptual terms: Low-Retentive zircon (LRZ) has a partly disordered crystalline structure, significant radiation damage and a low temperature of annealing (c. 180–200 °C). High-Retentive zircon (HRZ), which is nearly crystalline, fully anneals at temperatures in excess of ~280–300 °C. At higher temperatures, all grains are reset provided the sample remains at these temperatures for a geologically significant time (>10^6 yr). Most differential annealing occurs in the range of about 180 to 280 °C. Exhumation of rocks that have been buried and heated to this degree commonly have a population of grains that are fully reset and then a wide range of grain ages that are either partially reset or those that represent provenance ages. Consequently, in a number of studies where a young reset population has been identified, it is not clear if the older grain ages are unreset and therefore retain the original provenance information, or if they are partly reset (or both). This is an area of active research, but it is clear that rocks heated to temperatures between 180 °C and about 220 °C (the lower end of resetting and partial resetting) have the potential to record both thermal resetting and original provenance information. A crucial factor in this sort of setting is the amount and range in inter-grain radiation damage.

Lag time

Perhaps the most distinctive aspect of using DZFT analysis on zircon that has not been partially or fully reset after deposition is that cooling ages recorded in the sedimentary detritus can be related to past thermal events in the source terrain. In many cases these cooling events are directly related to uplift and exhumation of source rock, so cooling ages provide a direct link between long-term sediment supply and sediment accumulation. Once a FTGA distribution is determined and peak ages have been fitted, lag times can be determined (Fig. 6).

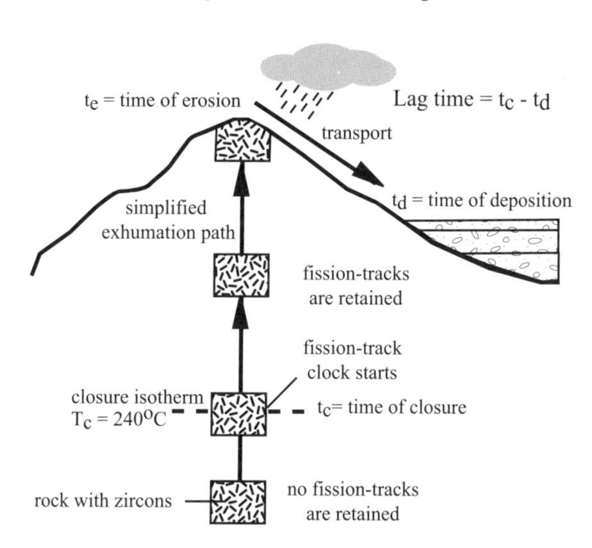

Figure 6. The lag-time of a sample is the time required for the sample to cool, get exhumed to the surface, and then get deposited in a nearby basin. As a rock is exhumed to the surface, the rock cools below the closure temperatures of the different thermochronometers (here only ZFT is shown): when this happens, various isotopic clocks start. Eventually the rock reaches the surface where it is subject to erosion. Apatite, zircon, and mica grains are released into sediment and transported by glaciers and rivers into the adjacent basins, where they are deposited. The time for erosion and sediment transport is generally regarded as geologically instantaneous (Heller et al. 1992; Bernet et al. 2004a), but this is not always the case. Lag time integrates the time between closure and the time of deposition, and mainly represents the time needed to exhume the rock to the surface.

In this case, lag time is defined as the difference between the peak age and the depositional age (Garver and Brandon 1994a; Garver et al. 1999; Bernet et al. 2001), and it represents the lag or difference between closure in the source and deposition in the adjacent basin. In areas of active volcanism, closure occurs during eruption, and erosion may immediately transfer grains to flanking basins, so lag time is nearly zero. In other, non-volcanic cases, rock in the source area is exhumed from depth and the rock passes through a closure isotherm at depth at which time the lag-time clock is set. In this case, the lag time represents the time required for the rock to be exhumed to the surface, eroded, and then the zircon being transported to an adjacent basin. Lag time is then a function of exhumation rate in the source area.

Transformation of lag time to an exhumation rate estimate requires several simplifying assumptions. The basic calculation necessitates that the cooling age can be related to a closure depth, and therefore an estimate of the geothermal gradient and an effective closure temperature is required. In this respect the effect of isotherm advection needs to be considered if exhumation is rapid (> 1 km/m.y. — see Garver et al. 1999). Another simplifying assumption commonly made is that storage of sediment in the orogenic belt is negligible and that sediment is relatively quickly removed from the orogenic belt and deposited in adjacent basins. This latter assumption appears to be valid for sediment shed off active orogenic belts (i.e., see Garver and Kamp 2002, Bernet et al. 2004a). In its simplest form, lag time can be converted to an exhumation rate using the relation presented in Figure 7 (see Garver et al. 1999 for details). It is not uncommon to just focus on the shorter lag times (millions to tens of millions of years) because these zircons have been derived from the fastest and most deeply exhuming areas of the source region. Zircons with longer lag-times (tens to hundreds of millions of years) are typically recycled from sedimentary cover units.

Types of lag-time changes

In principle, three basic lag-time trends can be expected when studying synorogenic samples from a stratigraphically coordinated sequence (Fig. 8). The first trend is a shortening of lag time up-section, which indicates continuous and accelerating exhumation. The FTGA peaks P1, P2 etc. are then regarded as *moving peaks*, because they become continuously younger at a rate faster than change in depositional age (Fig. 8b). The second possibility is that peak ages

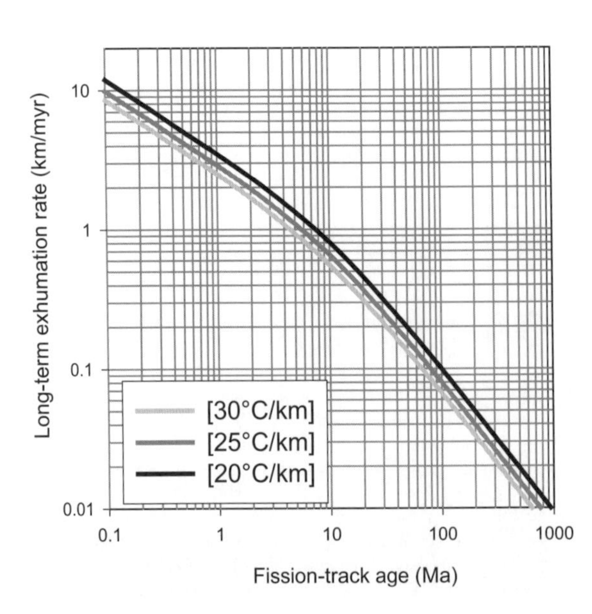

Figure 7. Relationship of FTGA or peak-age and long-term average exhumation rate (for radiation-damaged zircon), shown here for common geothermal gradients of 20 °C, 25 °C and 30 °C (after Garver et al. 1999). Advection of isotherms during fast exhumation has been considered in constructing this graph.

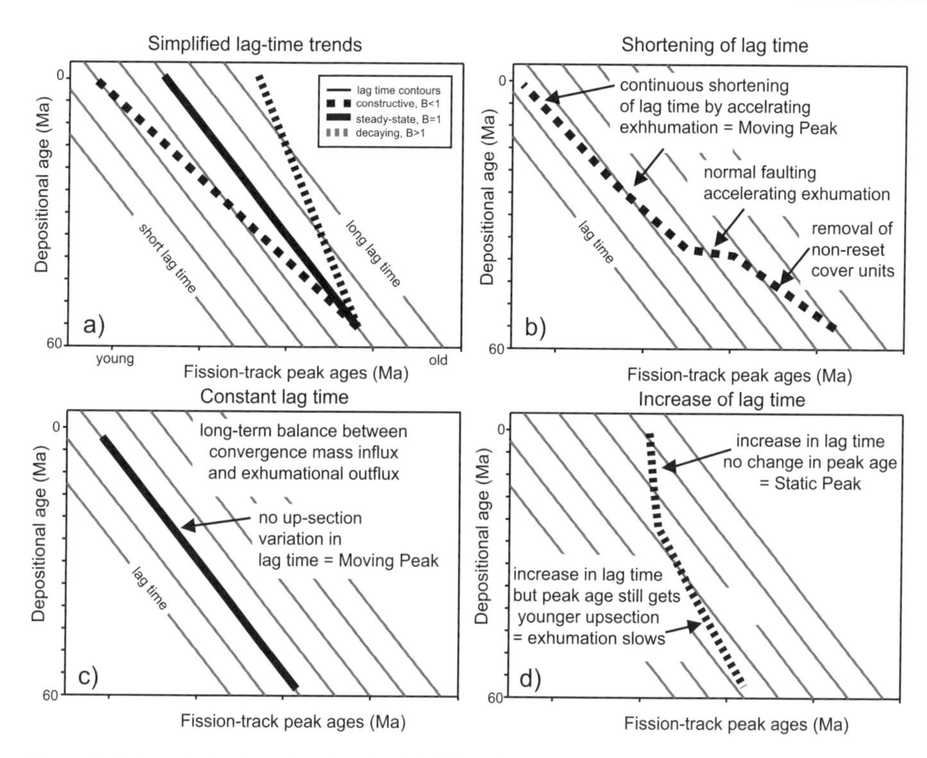

Figure 8. Schematic lag-time plots based on FTGA peak ages and depositional ages. A) The up-section lag-time trend can be approximated with the linear relationship $t_c = A + B\ t_d$ (Bernet et al. 2001). The slope B of the lag-time line is a function of orogenic evolution and can indicate orogenic construction, steady state, or orogenic decay. B) Overall shortening trend of lag time reflects removal of non- or partially reset cover units in the beginning and increase in exhumation rate throughout the record. Peak ages are becoming continuously younger and are therefore *moving peaks*. C) Constant lag times can be observed if zircons always need the same time to pass through the closer temperature, be exhumed, eroded an deposited throughout part of the stratigraphic section. Peak ages also become continuously younger and are also *moving peaks*. D) Increase in lag time indicates a decrease in exhumation rates, which means that the mountain belt or parts of it became inactive and are decaying way. If peak ages do not change up-section, then these peaks are regarded as *static peaks*.

young at the same rate as change in depositional age. In this case, lag time remains constant, but because the peaks young up-section they are also regarded as *moving peaks* (Fig. 8c). This type of lag time would be characteristic of a constantly exhuming source terrain. The third possibility is that lag time increases up-section, which indicates slowing of exhumation rates (Fig. 8d). If peak ages do not change at all up-section, than they are described as *static peaks*. Such peaks reflect a FT source terrain, which has been rapidly cooled in the past, maybe by fast, episodic exhumation (normal faulting or erosion), and was exhumed slowly since.

EXAMPLES AND APPLICATIONS

In this section we highlight a few examples of recent studies and applications of DZFT analysis that we think have made an impact on how we look at and analyze data. We present this review to give the reader some suggestions of what can be done with DZFT analysis, and to point out where we think the future lies. Note that this section is not a historical overview, and as such we leave out and ignore some early pioneering work.

Provenance analysis

Detrital zircon fission-track analysis has a long tradition in provenance analysis (e.g. see review in Hurford and Carter 1991; Carter 1999; Garver et al. 1999). In fact, the earliest use of DZFT was for simple provenance analysis because the technique allows identification of major cooling ages in the source terrain, and this alone is a powerful discriminator of sediment provenance. This approach to provenance analysis—the analysis of a single mineral phase—is commonly referred to as a varietal study because a single mineral phase is used to address sediment provenance (e.g. Haughton et al. 1991). Although powerful, varietal studies have limitations because the unique source terrain indicated by the data only pertains to the specific mineral studied, and there may be a host of other lithologies in the source terrain that are essentially unidentified. Therefore, varietal studies are most effective when combined with other sediment provenance techniques aimed at identifying the full provenance spectrum. In considering sediment provenance and zircon source, it is prudent to consider potential source rock lithologies that could have supplied detrital zircon with the shape, color, and morphology in the sample of interest.

Historically, the young populations of grain ages have received the most attention, because they can be commonly ascribed to active processes in the source terrain. For example, if a young population of euhedral ZFT ages is close or identical to the depositional age, then they are likely derived from a volcanic source (Fig. 9a; see Kowallis et al. 1986; Garver and Brandon 1994b; Garver et al. 2000b; Soloviev et al. 2002; Stewart and Brandon 2004). Otherwise, young age peaks in sediment derived from convergent mountain belts without active volcanism reflect

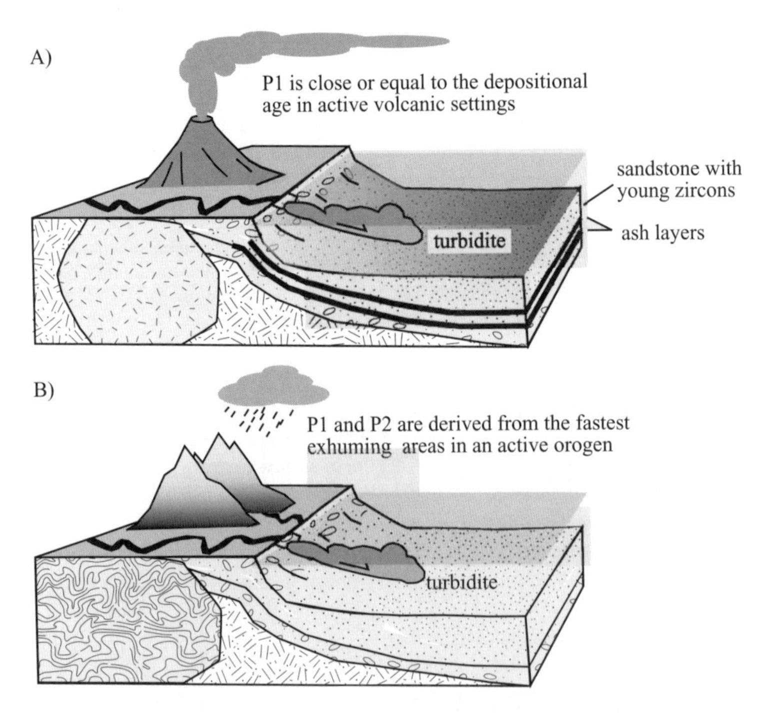

Figure 9. A) In areas with active volcanism ZFT analysis can be used to date ash layers to obtain stratigraphic ages, or by determining P1 in contemporaneous sandstone layers (e.g. Garver et al. 2000b; Soloviev et al. 2001; Stewart and Brandon 2004). B) In orogenic settings without active volcanism, FT ages are related to exhumational cooling and depositional ages must be determined with other means, such as biostratigraphy.

rapid exhumation of deep-seated metamorphic rocks in the core of the orogen (Fig. 9b; also see Brandon and Vance 1992; Garver and Brandon, 1994b; Garver et al. 1999; Bernet et al. 2001). Zircons from such rocks have been fully reset during regional metamorphism and their cooling ages represent the recent thermal history of the source area. Zircons with older cooling ages are usually derived from partially or non-reset cover units (Fig. 10). Non-reset zircons are therefore recycled and re-introduced into the rock cycle. Here we highlight several examples that demonstrate the utility of this technique in provenance analysis.

European Alps. Geologic settings where abundant bedrock ZFT cooling ages are available allow comparison of DZFT peak ages from modern river samples with the bedrock FT age distribution in the drainage area. Such comparative studies were done in the European Alps. DZFT peak ages from rivers that drain the Alps toward their foreland and hinterland, were compared to the dense data set of bedrock ZFT ages available for the Alps (Bernet et al. 2004a,b). These studies helped to improve our understanding of how FTGA distributions can be used to recognize sediment source areas on a local and regional scale, and also demonstrated that detrital samples provide a reliable and representative overview of the bedrock age distribution in their river drainages. Furthermore, it was shown that the provenance signal revealed in the ZFT peak ages is detectable even >500–1000 km away from the source and that sediment transport time from source to sink is essentially geological instantaneous in orogenic systems (Bernet et al. 2004a).

In addition to information contained in the ZFT peak ages, another parameter can be evaluated to better constrain zircon provenance is grain morphology. Detailed zircon morphology classifications have been presented in the past (Pupin 1980), but in simplistic terms, euhedral grains are likely to be derived from igneous sources while rounded grains are likely to be derived from sedimentary or meta-sedimentary sources. The FT peak age – grain

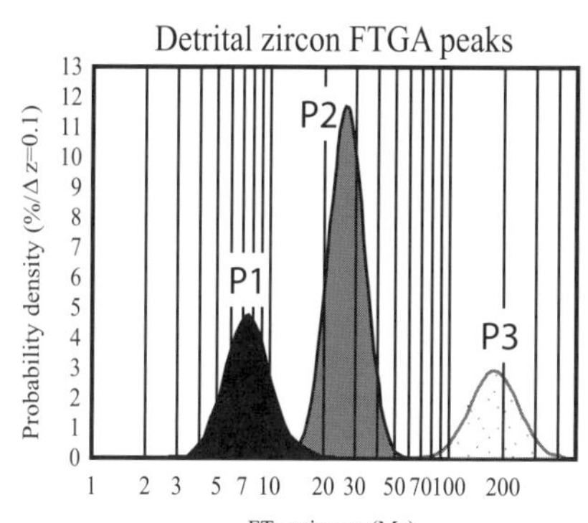

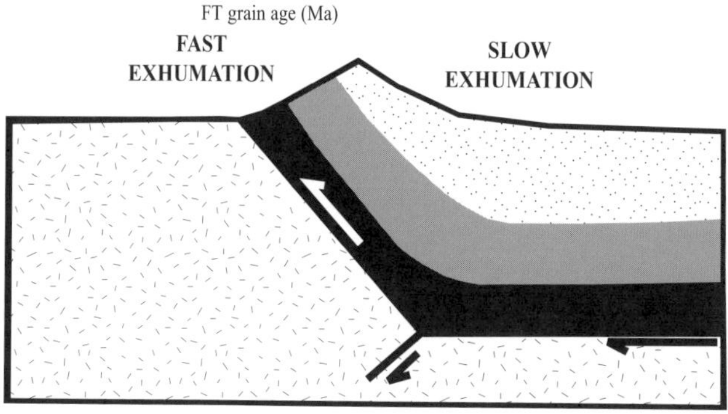

Figure 10. Schematic diagram showing exposure of synorogenic cooling ages in a single vergent mountain belt, similar to the Southern Alps in New Zealand. Older, non-reset cooling ages occur in cover units. The hypothetical probability density plot shows the general distribution of FT age components derived from such a setting.

morphology relation was explored by Dunkl et al. (2001) in a study of the upper Oligocene Macigno Formation in the northern Apennines, Italy. Most sedimentary rocks that occur today in the Apennines were originally derived from the European Alps (e.g. Cibin et al. 2001). Dunkl et al. (2001) showed that some of the zircons belonging to the youngest FTGA component were in fact derived from exhumed metamorphic rocks, while the others were derived from periadriatic igneous rocks in the Alps. The importance of this study lies in the attempt to identify the contribution of igneous zircon in the FTGA distribution of Alpine derived sediment.

In areas where only part of the source lithologies contain zircon, it is sensible to combine DZFT dating with other provenance techniques. For example, Spiegel et al. (2004) used Nd isotope ratios in detrital epidote to trace provenance from non-zircon-bearing basic igneous rocks in the Central Alps in addition to DZFT. Through combination of these two particular provenance techniques these authors were able to propose a detailed picture of sediment source and transport pathways into the foreland basin during the Oligocene and Miocene. This study highlights the important trend towards using multiple provenance techniques to develop a more robust provenance picture.

Southern Alps of New Zealand. Zircon from either side of the Southern Alps have distinctive FT ages and radiation-damaged-induced color that are distinctive and diagnostic on either side of the orogenic belt. In a study where Recent sediment was collected from drainages with known source rocks, Garver and Kamp (2002) mapped the distribution of FT ages and zircon color (Fig. 11). Color in zircon is a function of radiation damage and rare earth element (REE) content. There are two dominant color series in zircon: the pink series ranges between light pink, pink, rose, red, purple (hyacinth) and black; and the yellow series ranges between pale yellow, straw, honey, brown, and black (i.e., Gastil et al. 1967). The color of the pink series gets reset, and the zircon becomes colorless, between ~250–400 °C. In the Southern Alps zircons can be grouped into three categories (from deepest to shallowest crustal levels): 1) reset FT age – reset color; 2) reset FT age – non-reset color; and 3) non-reset FT age – non-reset color, in the order of decreasing temperature ranges (Garver and Kamp 2002). Because uplift and exhumation of the Southern Alps is asymmetric across the range, deeply exhumed rocks occur on the west side, and rocks that have been at shallow crustal levels occur on the east side. This difference is dramatically reflected in the sediment provenance of river sediment. In the west-flowing rivers, 80% of the zircons are colorless and about ~60% of the dated grains have FT ages of less than 22 Ma. This assemblage represents deeply exhumed rocks that have come from depths of at least 10 km. Quite the opposite occurs on the eastern side of the Southern Alps, where ~50–70% of the grains have color and almost all FT ages are older than 100 Ma (Fig. 11). This latter assemblage of zircon represents rocks that had been fed laterally into the orogenic system, and these rocks have resided at shallow crustal levels (<10 km) for about 100 Ma. The important point of this example is that not only FT age, but also other aspects of the zircon can be used to locate crustal material with a specific thermal history.

Ecuadorian Andes. Basins flanking the Andes have an excellent record of the uplift and exhumation of the orogenic belt as well as adjacent continental blocks. The basin strata that flank these crustal blocks provide some of the most important information on the movement history of adjacent crustal blocks. It is difficult, however, to determine the source of the basin fill in some cases, because many crustal blocks have geologic similarities. To solve this problem, Ruiz et al. (2004) studied 24 Cretaceous to Tertiary samples from strata in the Andean Amazon basin in Ecuador. They recognize several important changes in ZFT grain-age distributions in the stratigraphic sequence. In the lower part of the Cretaceous section they discovered that part of the population consists of relatively old DZFT ages (Paleozoic), in zircons that are characteristically dark and rounded. These zircons occur in sedimentary rocks that have a high ZTR index. Because the ZTR index is high for these samples, they attributed the zircons as polycyclic and derived from the Paleozoic platform cover to the craton. Up-section, they

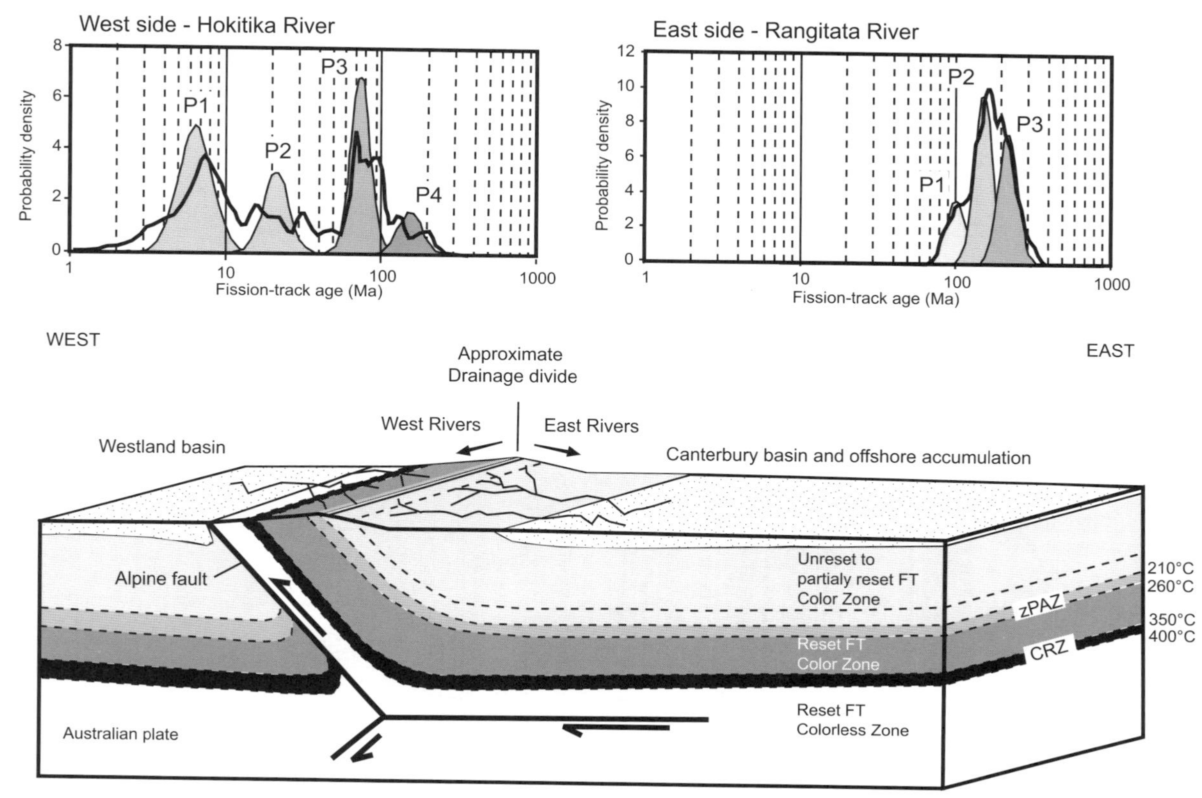

Figure 11. Schematic profile of the Southern Alps of New Zealand, showing zircon color retention and FT annealing zones. Also shown are zircon FTGA distributions and best-fit peaks of the Rangitata and Hokitika rivers that drain the Southern Alps to the east and west respectively.

identified an influx of near-zero-lag-time grains that coincide with an influx of heavy minerals with a decrease in the ZTR index and an increase in the amount of kyanite and sillimanite. The DZFT cooling ages in this part of the section are therefore inferred to record Middle to Late Eocene exhumation of a high-grade metamorphic terrane that may have been affected by collision of the Macuchi Arc terrane to the eastern edge of Ecuador. In the Early Miocene, short lag times are attributed to significant volcanic activity in the source region because the heavy mineral assemblage contains euhedral biotite, hornblende, diopside, apatite and idiomorphic zircon. In addition, the ZFT and apatite FT ages are identical, and therefore it is likely that both reflect cooling of volcanic rocks (Ruiz et al. 2004). This study demonstrates the utility of interpreting the DZFT age patterns using supporting provenance information, especially heavy mineral assemblages that can provide crucial clues as to the nature of the source rock.

Dating strata

In a number of studies FT analysis of detrital zircon has been used to establish maximum depositional ages of poorly dated or undated sedimentary rocks. One approach that is particularly powerful is to date volcanic ash deposits interbedded with a poorly dated sedimentary sequence (i.e., Kowallis et al. 1986), but obviously this approach only works on stratigraphic sequences that have stratified tuffs. There are many sequences with a partial volcanic provenance where a young volcanic component is mixed with detritus from other sources. We focus on this latter setting, in which one needs to rely on the information contained in the detrital constituents of the sandstones that have a heterogeneous provenance. Stratigraphic sequences in these studies where this technique has been applied have several things in common: 1) they have little or no biostratigraphic control; 2) they are thick, monotonous, and commonly internally structurally imbricated; and 3) they have a partial provenance from a volcanic center (mainly continental volcanic arc). The strata have few, if any, internal stratigraphic marker horizons.

Naturally, if sandstone has a population of ZFT cooling ages that represent primary cooling in the source region, then deposition of the sedimentary rock must postdate or equal that cooling age. Where this approach has been most useful is in those instances where the source region contains an active volcanic source that contributes a significant fraction of zircons with nearly syndepositional cooling ages to the basin. In our experience, this generally means that the source included a continental volcanic arc, which produces relatively large volumes of sediment, and many of the volcanic rock types are rich in zircon. Because this young age is strictly a limiting age, it has been referred to as a FT minimum age in the literature because the calculated age is the minimum FT component in the grain-age distribution (i.e., Garver et al. 2000b).

Kamchatka - Forearc strata of the Ukelayet Flysch. The thick, deep-water flysch sequences in the Olutorsky collision zone provide a good case study for this approach because the Kamchatka margin has been volcanically active for the last 100 Ma, and a tremendous thickness of poorly dated strata have accumulated. Work on a number of these sequences has demonstrated how FT dating of detrital zircon can be used to determine depositional ages of terrigenous sequences in a continental arc setting (Garver et al. 2000b; Shapiro et al. 2001; Soloviev et al. 2001, 2002). These researchers carried out detailed analyses of Cretaceous to Eocene turbiditic sandstone along most of the Kamchatka margin and in the southern Koryak upland farther north. They determined that the youngest age component of each of their samples was comprised mainly of euhedral and colorless zircon inferred to be first-cycle volcanic zircons. These first-cycle zircons are inferred to have been derived from active magmatism in the nearby Okhotsk-Chukotka continental arc and the Western Kamchatka-Koryak Volcanic belt between 88 and ~44 Ma. This young population of cooling ages constrains depositional ages in this 10-km-thick package of uniform and monotonous turbidites (Fig. 12). Zircons in the second age component, P2, were associated with continuous exhumation and cooling of basement rocks to Okhotsk-Chukotka continental arc. While many

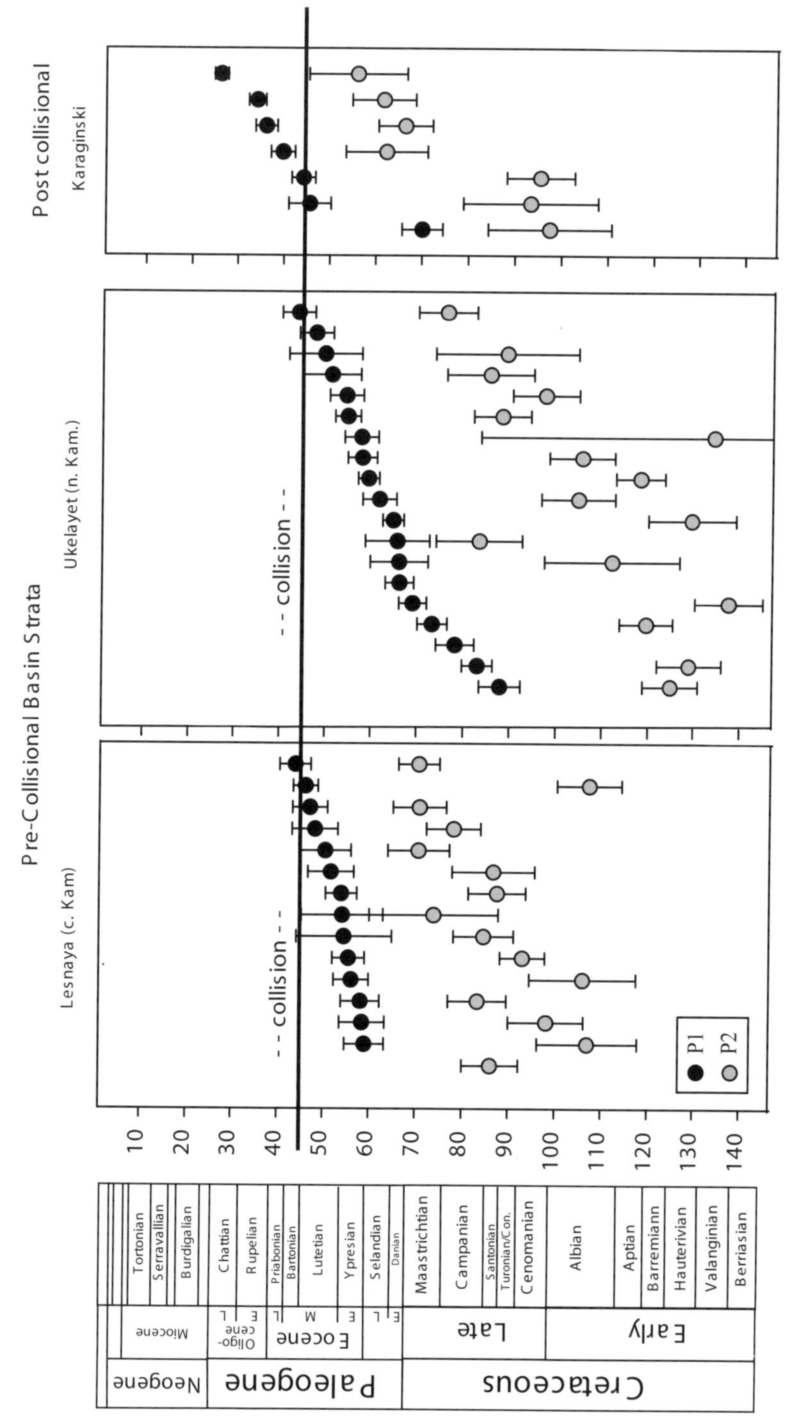

Figure 12. ZFT ages from poorly dated and undated sandstones associated with the Olyutorsky collision zone, Kamchataka. The young populations have been used to constrain the depositional age of sandstones in the pre- and post-collisional basins. These studies fundamentally changed interpretations of Kamchatka geology because many of these sandstone units were originally mapped as Cretaceous, and it is clear that they must be at least Eocene in a number of cases (from Garver et al. 2000a; Garver unpublished).

of the dated sandstones are from sequences that have no fossils, one study area focused on dating detrital zircon from sandstones that had age control from nannofossils in interbedded shales. Without exception, the ZFT minimum ages coincided with the age constraints provided by nannofossils (Shapiro et al. 2001; Soloviev et al. 2001). Note that for the most part, these sandstones are quartzo-feldspathic and arkoses with a relatively minor amount of volcanic detritus (see Shapiro et al. 2001). Despite this lack of obvious volcanic detritus, it is certain that the grains are volcanic (or high level) because those in the young population have U/Pb and ZFT ages that are statistically indistinguishable (Hourigan et al. 2001).

Olympic Subduction Complex, Cascadia forearc. The Olympic subduction complex (OSC) comprises much of the uplifted and exhumed part of the subduction complex to the Cascadia subduction wedge (Brandon et al. 1998). Sedimentary units in the subduction complex are thick, structurally imbricated, and mostly monotonous sequences of Tertiary deep-water turbidites. DZFT dating of these units has fundamentally altered our understanding of the age-distribution of accreted units in the subduction complex and flanking strata (Brandon and Vance 1992; Garver and Brandon 1994a; Brandon et al. 1998; Stewart and Brandon 2004). Dating of sedimentary units of the central part of the OSC has shown that many of the units have significant populations of cooling ages that fall at 43, 57, and 74 Ma and these are related to rapidly cooled crustal blocks in the hinterland, behind the Cascade arc. These populations have the same age regardless of depositional age of the sandstone, and as such they are referred to as *static peaks* (Fig. 8). The sandstones also have a minor population of young ages that is variable and appears to be very close to depositional age, where depositional age is constrained. The authors referred to this young peak as a *moving peak* because it becomes younger with time (Fig. 8). This young moving peak was inferred to represent material from the syn-contemporaneous Cascade arc.

More recent work in the Olympic Subduction Complex confirms earlier conclusions. Stewart and Brandon (2004) conducted a detailed examination of the siliciclastic, lower Miocene "Hoh Formation" of the Coastal OSC. They analyzed 34 sandstone samples and 2 volcanic ash layers of the coastal OSC, and used the young peak age to show that most of the strata are Lower Miocene. They note that most sandstones in the Hoh Formation are variable in composition, but most fall between lithic arkoses or lithic wackes, with volcanic lithic fragments. The young population of cooling ages is inferred to represent material from the syn-contemporaneous Cascade arc, and in a few instances they were able to show that the young population was the same as the paleontologically determined depositional age. They nicely summarize the reasoning behind the assumption that P1 (young peak age) can be used as a proxy for depositional age, which is mainly focused on an analysis of those units that have fossil control.

Exhumation studies

Exhumation studies are aimed at gaining a better understanding of the long-term evolution and thermal structure of an orogen, and determining the rate of exhumation in known source regions. Convergent mountain belts, such as the European Alps, the Southern Alps of New Zealand, or the Himalayas have been successfully studied using this analysis, largely because these mountain belts lack significant volcanic activity, so most, if not all, of the cooling ages are related to tectonic or erosional exhumation and not igneous activity. It is difficult to study exhumation of orogenic systems with significant igneous activity, such as continental arcs, because the thermal structure of the crust is affected by both exhumation and igneous heating.

There are several practical considerations one needs to bear in mind when using the sediment record to understand orogenic exhumation. Recall the objective here is to use cooling ages of zircons in basin strata to make inferences about the long-term evolution of the source area. The typical approach is to isolate and analyze detrital zircon from a number of different stratigraphic levels so that the nature of the source through time can be evaluated. Individual zircons in basin strata may have an uncertain provenance and the inferences need to be made as to original source rock. Additionally, almost all orogenic systems produce a wide variety of ZFT cooling ages, so as discussed above, cooling age populations need to be carefully isolated. Finally, one needs to understand sediment transport in the basin and how that sediment transport might have changed in the basin.

In general, exhumation studies using DZFT ages are based on understanding a prominent peak-age distribution, determining the lag time of that peak age and an inferred exhumation rate, and then evaluating how that exhumation rate changes with time, as described above. Samples should be collected from strata that are stratigraphically well dated: if they are not, the lag time, and hence the calculated exhumation rate, will have a high uncertainty. The possible effects of sediment storage need to be evaluated as well. Significant sediment storage, which may be characteristic of moderate to slow exhumed systems, increases lag time and therefore calculated exhumation rates would be too slow if storage time is significant (on the order of millions of years). However, in most studies, where source-rock exhumation is on the order of 200 m/m.y. or faster, it is assumed that sediment storage is insignificant (i.e., Bernet et al. 2001, 2004a).

Himalayas. The earliest studies aimed at understanding orogenic exhumation were focused on the sedimentary apron at the foot of the Himalayas (Zeitler et al. 1986; Cerveny et al. 1988). In fact, the work by Zeitler et al. (1986) and Cerveny et al. (1988) was ground breaking and of unparalleled importance for DZFT analysis. These authors took the method from being merely useful for provenance analysis to being a powerful tool to study the long-term evolution of convergent mountain belts. In these studies, DZFT analysis was used to improve the understating of exhumation in the Nanga Parbat region in the northwestern Himalayas. The authors analyzed samples from the modern Indus River, as well as from stratigraphic sections of the Miocene to Pliocene Siwalik Formations in Pakistan. They concluded that exhumation rates of 300 m/m.y. and above have existed at least in part of the Himalayan zircon source areas (Zeitler et al. 1986). Cerveney et al. (1988) came to the conclusion that high exhumation rates and high relief were common features in the Nanga Parbat-Haramosh Massif over the past 18 Ma. Their conclusion is based on the occurrence of young zircons, within 1–5 m.y. of the depositional age in each of their stratigraphic samples and in modern Indus River sediment.

It is interesting to reconsider the data from Cerveny et al. (1988) using the lag-time concept outlined above. The results of this re-analysis indicate that lag time becomes shorter up-section for both P1 and P2 age components (*moving peaks*) in the Indus River and Siwalik sediments from the Middle Miocene to the Recent (Fig. 13). This up-section change suggests that this part of the Himalayas has been in a constructional phase with increasing relief and accelerating exhumation rates since the Miocene.

British Columbia Coast Range, Canada. One of the earliest examples of DZFT analysis applied to long-term source rock exhumation was from a well-dated stratigraphic section of the Tofino basin in Washington State and British Columbia, which records the erosional exhumation of the British Columbia Coast plutonic complex that makes up most of the Coast Range (Garver and Brandon 1994b). Eight stratigraphically coordinated samples ranging in age from Middle Eocene to Miocene (40 to 19 Ma) were analyzed using this approach. Known ZFT cooling ages in the modern Coast Plutonic Complex (CPC), the incidental source of the sediment, were also considered in the analysis. This combined record of cooling ages allowed for interpretation of a ~40 m.y. record of lag times that are interpreted to represent the emergence of the CPC and continued exhumation through time. These lag-time data suggest a nearly constant long-term average exhumation rate of 250 m/m.y., a moderate exhumation rate.

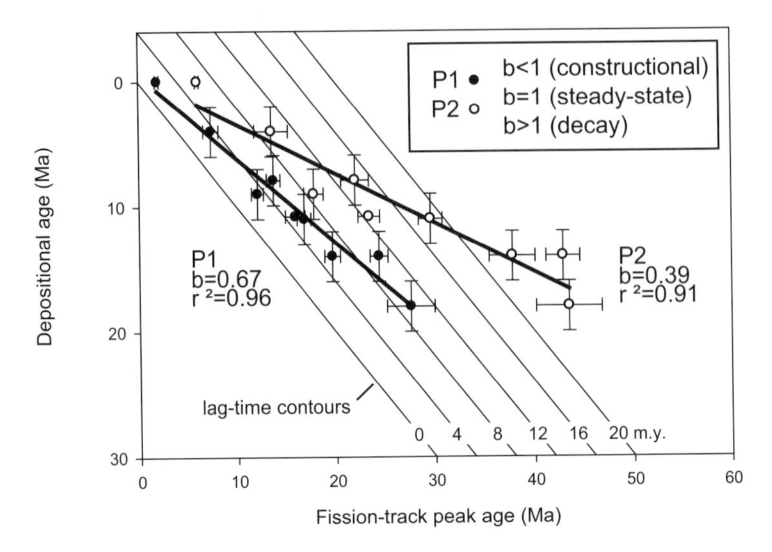

Figure 13. FT peak ages plotted against depositional age of Indus River and Siwalik Group sediments. Contour lines designate lag time. Both P1 and P2 data indicate that the fastest exhuming areas of the Himalayas in NW Pakistan are in a constructional phase since at least the Miocene. Shortening of lag time up-section implies an increase in exhumation rates (data from Cerveny et al. 1988).

In this study, FT ages were interpreted in the context of sediment provenance, paleocurrents, and basin infill history, and there are two distinctive aspects of the sediment provenance in this study. The first is that the oldest sample was derived from metamorphic rocks of the nearby Leech River Schist (not the CPC), but the more quartzo-feldspathic facies were derived from plutonic rocks of the CPC and adjacent cover rocks. The second aspect is that the first detritus shed off the uplifted and exhuming CPC included old basin deposits (Cretaceous), which resulted in a complicated distribution of grain ages, and a lithic feldspathic sandstone composition. The important point of these two examples is that the sediment provenance of this basin sequence plays a crucial role in interpreting the significance of the DZFT ages.

European Alps. The Alps are an excellent mountain belt for exhumation studies because they have evolved without significant volcanism since the Oligocene. Because the orogen lacks significant igneous heating, samples collected from controlled stratigraphic sections of synorogenic sediment of foreland and hinterland basins provide insight into the long-term exhumation history. One of the most important observations in recent studies in the Alps is the up-section evolution of ZFT peak ages (Spiegel et al. 2000; Bernet et al. 2001, in press). The peak ages change at the same rate as the depositional age, and are therefore described as *moving peaks* (Fig. 14), and they indicate relatively fast, continuous exhumation.

Analysis of the youngest peak age in each sample shows that exhumation rates of the fastest exhuming areas in the Alps have remained relatively constant since the Early Miocene at long-term average rates of about 700 m/m.y. Continuous P2 lag times give exhumation rates of 300–400 m/m.y. (Bernet et al. 2001), which are in the range of long-term erosion rate estimates for the Alps in other studies (e.g. Schlunegger et al. 2001; Kuhlemann et al. 2002). Nevertheless, the interpretation of a long-term exhumational steady state of the European Alps by Bernet et al. (2001), on the basis of DZFT lag times, is controversial. This work initiated a debate on the long-term evolution and steady state of mountain belts in general and the Alps in particular, and led to the increased use of the lag-time concept to understand orogenic

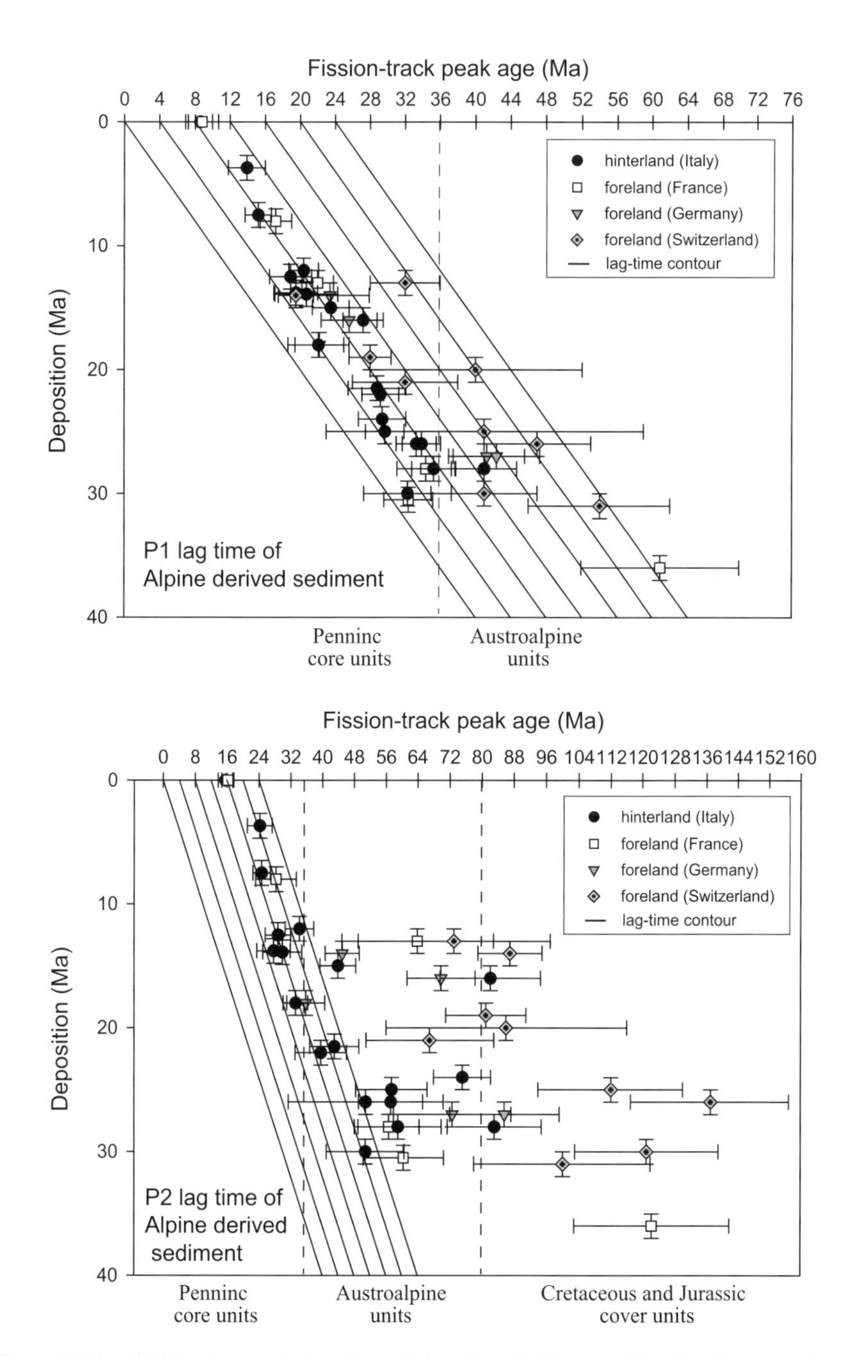

Figure 14. P1 and P2 lag-time trends in sediment derived from the European Alps. Samples were collected in the adjacent foreland and hinterland basins. Samples from Italy, France and Germany are from Bernet et al. (2001) and Bernet (2002). Swiss samples are from Spiegel et al. (2000). Note that most samples show relatively constant P1 and P2 trends (*moving peaks*) since the early Miocene. The main source areas are indicated as Penninic core, Austroalpine, and Cretaceous to Jurassic cover units in the Alps.

exhumation. While exhumation rates determined by Bernet et al. (2001) agree with estimates from other workers (e.g. Clark and Jäger 1969; Hinderer 2001), the exhumational steady-state interpretation is in apparent conflict with a dramatic increase in sediment yield from the Alps since the Pliocene as predicted from sediment budget calculations (Kuhlemann 2000) or apatite FT analysis in drill cores from the North Alpine Foreland basin (Cederbom et al. 2004). Nonetheless, additional work has shown that the same steady lag-time trend and same long-term average exhumation rates can be observed in the Alpine foreland as in the hinterland (Fig. 14), because the Alps are a doubly vergent orogenic wedge that has shed zircon with young cooling ages to both sides of the mountain belt (Bernet et al. in press).

Dating low-temperature thermal events and strata exhumation

We described situations above where detrital zircon occur in sedimentary basins, but in some cases samples may come from deeply buried and heated sequences that may possibly be partially or fully reset. Partial resetting of mixed suites of zircon is most conspicuous when sedimentary zircons with a wide range of radiation damage are brought to elevated temperatures (c. 200 °C), and then allowed to cool. Partially reset samples have LRZ that were partially of fully annealed and HRZ that were not annealed after deposition. In this case, cooling ages are not concordant, and the young population, which is younger than depositional age, corresponds to cooling following the thermal event. Full annealing of both LRZ and HRZ results in cooling ages that are concordant, but requires relatively high temperatures (>300 °C and above). This property of partial annealing can be used to date low-temperature thermal events (<300 °C) and the exhumation of strata (see full discussion in Garver et al. 2005). In this section, we draw attention to several studies that used post-depositional partial resetting of detrital zircon to date low-temperature thermal overprint and exhumation of heated sedimentary rocks.

Olympic Mountains, Western USA. Deeply exhumed strata in the core of the Olympic Mountains were first deposited in the offshore accretionary complex, then accreted into the Olympic subduction complex, and finally exhumed to the surface by erosional processes. FT analysis of detrital zircon from Cenozoic sandstone in the exhumed core of the Olympic Subduction Complex (OSC) and in flanking units, define the timing of deposition, subduction accretion, and exhumation in the core of the Olympic Mountains (Brandon and Vance 1992). Detrital zircons have reset FT ages of ~14 Ma in the core of the OSC, and this cooling age is related to post-metamorphic cooling driven by erosional exhumation. Samples from *unreset* sandstone units that flank the main reset area have preserved their original undisturbed grain-age distributions with several distinct grain-ages populations related to episodes of source terrain cooling (see above, Brandon and Vance 1992).

The *reset zone* in the center of the OSC represents the youngest and most deeply exhumed part of the OSC. This region also coincides with the area of the highest topographic relief in the Olympic Mountains. Subaerial erosion started at ~12 Ma, when the OSC first became emerged. Since then, roughly 12 km of rock has been removed from the core of the OSC, resulting in a long-term exhumation rate of ~1000 m/m.y. (Brandon et al. 1998). All grains in samples from the core were not fully reset during metamorphism, despite the fact that they achieved the highest temperatures of any rocks exposed in the accretionary complex. However, the young fully reset population is clearly geologically meaningful, and therefore the authors report FT minimum ages, which is the youngest population of grains. The FT ages for single grains range in age from 6–36 Ma, but they are resolvable into young peak ages (P1) between 13 and 14.5 Ma, and older peak ages (P2) between about 17 and 25 Ma that are defined by about half the grain ages. Assuming monotonic cooling, they estimate that these samples reached peak temperatures of 239 °C and cooled at rates between 15 and 20 °C/m.y. (see Brandon and Vance 1992 and Brandon et al. 1998 for details).

Taiwanese Alps. Detrital zircon from metamorphosed Eocene to Miocene sedimentary rocks of the Taiwanese Alps record the progressive north-to-south exhumation that has brought meta-sediments in the axial spine of the range to the surface as a consequence of the ongoing oblique collision between the Luzon arc and the Asian mainland (Liu et al. 2001; Willett et al. 2003). These studies show that zircons in the Central Range of the Taiwanese Alps are largely reset with minimum ages of 0.9–2.0 Ma. Such young minimum ages reflect resetting of the least retentive of the zircons in the sample distribution. More retentive zircons remain unreset or partly reset and these occur in most of the samples from the Central Range, and therefore it is unlikely that these samples attained temperatures in excess of 280–300 °C (i.e., see Brandon and Vance 1992).

In contrast, ZFT ages from the Western Foothills and southern Taiwan are consistently older than depositional ages of host strata therefore the grains still retain cooling ages of their source region. Like the Olympics, these unreset samples occur around the deeply exhumed samples and represent rocks with a shallower depth of burial. This restricted spatial extent of reset minimum ZFT ages indicates limited exhumation of the Western Foothills belt and supports an interpretation of the southward propagation of the collision zone (Willett et al. 2003).

Peruvian Andes. The Cordillera Huayhuash and surrounding areas of the Puna surface of this part of the high Andes are underlain by Cretaceous quartzites that have been subjected to moderate temperatures for long intervals of time and therefore they record the effects of reheating and prolonged cooling of high-damaged zircon (Garver et al. 2005). Bedrock is dominated by folded Mesozoic miogeoclinal rocks unconformably overlain by mid-Tertiary volcanics intruded by late Tertiary granitic rocks and silicic dikes. In areas where the rocks are completely unreset, quartzites have late Paleozoic cooling ages and therefore by the time they were heated in the Tertiary, zircons had at least 200 to 300 m.y. of accumulated radiation damage, much more, on average, than the two examples highlighted above.

These Lower Cretaceous quartzites have ZFT ages with a wide range of cooling ages, but almost all are younger than depositional age of the host strata, so resetting has been pervasive (Fig. 15). In this study (Garver et al. 2005), the authors identify LRZ and HRZ depending on single-grain susceptibility to annealing of fission tracks. They discovered that most LRZ have reset ages at c. 27 Ma, and 63 Ma in rocks that probably never attained temperatures higher than c. 180–200 °C (based on vitrinite reflectance values). In this case, the young peak age of 27 Ma can be attributed to cooling following a period of intrusion and widespread volcanism, so there was a readily available heat source at this time. It is not clear if the 63 Ma ages represent a thermal event or if they represent partially reset grain ages that are meaningless with respect to the geologic history of this area.

Hudson Valley, Eastern USA. Lower Paleozoic strata of the lower Hudson Valley in New York State were deposited and shallowly buried (c. 5 km) prior to rifting of the North Atlantic and associated rift-basin formation in eastern North America. Detrital zircons from the Ordovician Austin Glen Formation and the Silurian Shawangunk Conglomerates have a wide spectrum of cooling ages, most of which are younger than depositional ages, so resetting is widespread (Garver et al. 2002). Cooling ages can be divided into three populations: a) reset in the Early Jurassic (~185 Ma); b) reset or partially reset in the late Paleozoic (c. 275–322); and; c) unreset to partially reset in the early Paleozoic (Fig. 16). These FT data clearly show that the Shawangunk Cg. experienced an Early Jurassic thermal event, and it would appear that only the most damaged grains were reset. Rocks in this part of the Hudson Valley experienced temperatures of ~180–220 °C, based on published vitrinite reflectance, CAI, and illite crystallinity values. These data suggest that the zircons were reset during Early Jurassic heating and an elevated geothermal gradient of ~50 °C/km (see Garver et al., 2002).

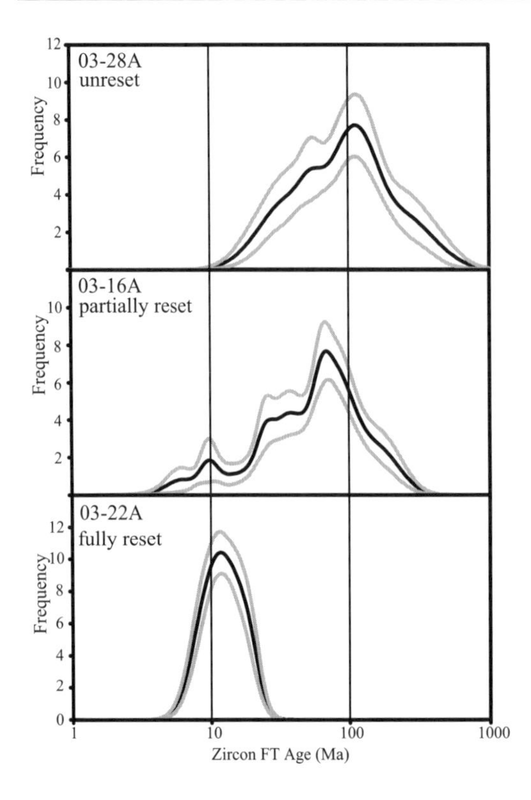

Figure 15. Probability density plots of unreset, partly reset, and fully reset zircons from Cretaceous quartzites in the Cordillera Huayhuash, Perú (from Garver et al. in press). Note that variable resetting results in a complicated grain-age distribution. Once heated, less retentive grains are fully reset, while more retentive grains are unreset or only party reset. A good example of the resulting mixed populations of grains (unreset and fully reset) is shown in 03-16a. This sample retains old grains that are presumably High Retentive Zircon (HRZ)), some grains that are partly reset (mid Tertiary), and a small component of Low Retentive Zircon (LRZ) that are full reset at about 10 Ma. Gray lines represent error envelope. Depositional age is Lower Cretaceous (100–120 Ma).

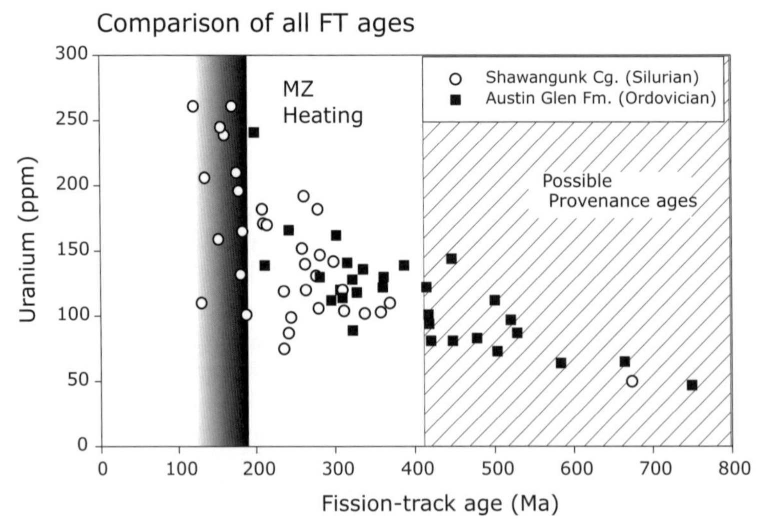

Figure 16. DZFT plot of zircon fission track age and uranium concentration from two units in the mid-Hudson Valley (NY). These Paleozoic sandstones are inferred to have been heated to temperatures in the range of 180–200 °C during the thermal affects associated with opening of the North Atlantic (Garver and Bartholomew 2001).Depositional age of these units is Silurian and Ordovician, so any grain with a possible provenance age (and hence unreset) falls in the diagonally ruled field. Subsequent heating (gray shaded zone) apparently fully reset a number of grains that collectively define cooling at c. 185 Ma. Heating occurred in the early Mesozoic (MZ) and was associated with the opening of the North Atlantic.

Shimanto Belt, SW Japan. Strata of the Shimanto Belt represent an exhumed accretionary complex that accumulated at the leading edge of the Eurasian plate. Similar to the Olympic Subduction Complex discussed above, these rocks are part of a thick imbricated package of sedimentary rocks deformed in a subduction setting. These strata include sandstones with a continental provenance, and zircons from these sandstones have had a wide range of grain ages prior to burial and heating. Resetting of detrital zircon in strata in the Kii and Kyushu regions showed the spatial variability and timing of exhumation in this part of the accretionary complex (Hasebe and Tagami 2001).

The Kii region provides important insight into the effects of widespread thermal resetting of detrital zircon. This region consists of three main belts of interest (inboard to outboard): Ryoke, Sambagawa, and Shimanto, which show widespread resetting, with almost none of the samples passing χ^2, suggesting heterogeneous annealing throughout the belt. Several end members are represented in the data. Some samples are from psammatic schists heated to greenschist conditions, and these have a range of grain ages younger than depositional age but still fail χ^2. Others are from sandstones with grain-age distributions that are nearly representative of provenance ages, and therefore have not been heated to any great degree. The general interpretation is that most of these rocks have been heated to well within the zircon PAZ, but the crucial question is the temperature limit of this heating.

The widespread resetting of most samples resulted in ZFT ages with a young population (c. 55–75 Ma) that can be interpreted as the time of maximum burial and heating. Older component ages, some of which are younger than depositional age and some older than depositional age, cannot be interpreted in any geologically meaningful way. It is important to note that virtually all samples that have been heated and reset, still fail χ^2, which suggests the original population of grain ages had heterogeneous internal radiation damage. Some of these grains must have been quite resistant to annealing: in the Sambagawa belt ZFT ages fail χ^2 but Ar-Ar muscovite ages are reset and the rocks have been metamorphosed to greenschist facies. This setting is instructive because it seems that annealing even at relatively high temperatures produces a wide range of grain ages that reflect heterogeneous annealing due to variation in radiation damage.

Combination with other isotopic dating techniques

If a DZFT age distribution is useful for understanding sediment provenance, dating strata, and exhumation studies, then it is only logical to assume that multiple geochronometers on the same mineral assemblage or multiple dating of the same grains provide an even deeper and more detailed understanding of the source region. Technical and financial issues are the most significant with respect to why this multi-dating approach hasn't been used more often, but it seems likely that these approaches will see greater use in the future due to methodological advances. Here we briefly highlight a couple of different approaches that should see widespread use in the future.

Multi-cooling studies (FT on two phases). Combining apatite and ZFT analysis of detrital grains from the same sandstone can be used to reconstruct the time temperature history of a source region provided the grains are derived from the same source rock (Lonergan and Johnson 1998). This approach was used to reconstruct the exhumation history of the Betic Cordillera, in southeastern Spain (Lonergan and Johnson 1998). An important aspect of this study was that they analyzed apatite and zircon from the same samples collected from synorogenic sediment. They showed that the structurally highest rocks of the Malaguide Complex cooled relatively slowly during the latest Oligocene (Aquitanian), while deep-seated metamorphic units of the Alpujarride complex experienced rapid cooling of up to 300ºC/m.y. between the Burdigalian to Langhian (c. 15–20 Ma). This change in cooling rate coincides with a change from erosional exhumation to predominantly tectonic exhumation (normal faulting) starting at ~21 Ma. Tectonic exhumation is related to a phase of orogenic extension in the internal parts of the mountain belt. Heavy mineral analysis on the same rocks shows an increase in metamorphic minerals (i.e., blue sodic amphibole and Mg-rich chloritiods) since ~18 Ma.

There are several complexities associated with this approach. These authors attempted to measure track lengths on the detrital apatite, but too few tracks were measured for a meaningful analysis. In this case, and in the case of most detrital apatite studies, it is important to ascertain which cooling age population the grain belongs to if a track length is measured. Otherwise track-length measurements will be nearly useless if they represent a mix of populations. Another complication is that the relative precision of single zircon ages is about 10× better than apatite, so fitted peak ages tend to be correspondingly less precise. This lack of precision for the apatite system complicates exhumation estimates (discussed in Garver et al. 1999). In sum, this approach is excellent for those cases where both apatite and zircon are derived from the same source rock, most commonly a granitic source terrane, and in those cases where the change in cooling rate is relatively small.

Fission-track, U/Pb, and Helium dating on detrital zircon. A natural marriage of analytical techniques for zircon dating is U/Pb dating and FT dating of single crystals so that both the crystallization age and the cooling age can be determined (i.e., Carter and Moss 1999; Carter and Bristow 2000, 2003). One of the limiting factors in this sort of analysis is the physical handling of single grains, and the fact that ZFT is partly destructive and U/Pb analysis by TIMMS is fully destructive. Advances in (U-Th)/He dating allows for dating of single zircon grains (see Reiners et al. 2005). Some of the analytical challenges have disappeared with the routine use of the less destructive U/Pb determination by SHRIMP analysis and Eximer LA-ICPMS (i.e., see Reiners et al. 2005). In this regard, the future is bright for double- and triple-dating techniques.

U/Pb and FT dating on a detrital suite from the Khorat Basin in Thailand, helped refine the identification of source rocks by providing cooling age and crystallization ages (Carter and Moss 1999; Carter and Bristow 2000, 2003). In these papers, Carter, Moss, and Bristow argued that determining provenance of zircon solely based on either fission-track or U/Pb dating would lead to ambiguous identification of source terrains. In early analyses with just FT ages, it was not clear if the FT ages represented rock formation ages (volcanic ages) or cooling ages of metamorphic rocks (exhumation ages). Likewise, interpretation of U/Pb data had the problem that crystallization ages can only vaguely be assigned to general source regions but not to distinct source areas, largely because of the propensity of zircon to survive multiple recycling.

The approach to solving this problem in the Khorat Basin was to first analyze two aliquots of zircon from the Mesozoic Phra Wihan Formation, one with the fission-track method and one with the U/Pb method (Carter and Moss 1999). Grains dated with the FT method were removed from their Teflon mounts and dated with the U/Pb method using an ion probe (SHRIMP). FT ages from aliquot one showed two main age components at 114 ± 6 Ma and 175 ± 10 Ma. The U/Pb ages from aliquot two revealed five main age components. Removal of the grains from the Teflon mount is required because the ion probe requires a high-quality Au-coat that is hard to achieve with a grain embedded in Teflon. These new U/Pb ages of the zircons from aliquot one were representative of the ages from each of the five main U/Pb age components. Thus, these results demonstrated that almost all FT ages of zircon from the Phra Wihan Formation are cooling ages related to exhumation of metamorphic rock and not rock formation ages or volcanic eruption ages.

The combination of isotopic dating techniques is an important trend in low-temperature thermochronology that promises to dramatically improve our understanding of source rock evolution. This development comes at the advent of routine (U-Th)/He (herein ZHe) dating of detrital zircon, which compliments DZFT (see Reiners et al., in review). In effect, ZHe dating, like ZFT, provides a low-temperature cooling age (c. 180 °C cf. 240 °C), and therefore this approach can also be effectively combined with U/Pb dating to address similar source regions as discussed above (see Reiners 2005). For example ZHe dating and U/Pb dating done on the same single zircon grains from the Jurassic Navajo sandstone in the southwestern United States allowed a robust interpretation of the source region (Rahl et al. 2003). These authors showed that zircons in the Navajo sandstone were not locally derived from western North America but more likely came from the Appalachians and had crossed the North American continent to be deposited in the southwest.

CONCLUSIONS

In this chapter we provided an introduction to DZFT analysis. We gave some practical and analytical considerations concerning sample and data handling, and showed examples of fission–track dating of detrital zircon. These examples include: a) determining sediment provenance and source rock characterization; b) dating strata; c) establishing exhumation histories of orogenic belts; and d) dating low-temperature thermal events. We also provided a series of examples of these main applications. The interested reader can find the associated publications of these applications in the reference list to obtain further information. We conclude with a summary of main points and the potential for future research directions.

1. The revelation of fission tracks in zircon is routine, but challenges remain with respect to etching detrital suites of zircons. Most natural suites of zircons have a wide range of radiation damage, and therefore a wide range of chemical reactivity that is manifested in different etch times. There is a need for studies aimed at quantifying the etch response associated with varying degrees of chemical reactivity (i.e., Garver 2003). A number of strategies have evolved to fully reveal tracks in a detrital suite of grains with a wide range of etchabilities, we use the multi-mount technique, but there are other approaches that might give a full qualitative representation of grain ages. Particularly difficult are those suites that contain grains <2–5 Ma, because these have a relatively low chemical reactivity.
2. While the general bounds for thermal annealing of zircon are well known. Most laboratory studies have focused on annealed zircon with induced tracks and little to no radiation damage. On the other hand, most studies of the thermal limits of natural fission tracks involve grains that have a moderate level of radiation damage, and they predict annealing temperatures that are somewhat lower. It is clear that the main difference is radiation damage, which lowers the effective closure temperature (Rahn et al. 2004). In light of this situation, it seems clear that more studies are needed to quantify the effective closure temperature of monotonically cooled zircon with low, moderate, and high levels of radiation damage. Essentially this approach involves gaining a better understanding of how grains with different damage become reset and fully annealed in different temperature-time conditions. This avenue of research includes understanding the stability of fission tracks in moderately warm settings (150–200 °C), where it seems clear that full resetting of highly damaged grains can occur. This finding has important implications for what we would except in terms of reset grains in basins and other settings where strata are warmed. We also need a better understanding of how and why some grains appear to retain tracks even at relatively high temperatures as seen in the Taiwan Alps and the Olympics. In these settings, the fully reset population clearly gives geologically significant ages, but it is not clear why particular grains become fully reset and others don't.
3. DZFT analysis is most powerful when combined with other provenance techniques and should not be limited only to isotopic dating methods. The combination with

conventional sediment petrography, heavy mineral and geochemical analysis, or the relatively new SEM-CL of quartz, can provide a detailed picture of an evolving source terrain. We are confident that DZFT analysis will be more and more applied in connection with other techniques in the future, while it also retains its value as a stand-alone tool.

4. DZFT has made important contributions to understanding sediment provenance and the exhumation of source terrains. While the potential and utility of this technique has been explored in a number of publications, it is clear that there is wide scope for future studies aimed at high-resolution evaluation of source rock exhumation. Important is a full characterization of sediment provenance and changes in sediment transport in the basin. However, once stratigraphic sections are well characterized, exhumation can be evaluated. Future studies using double dating of low temperature thermochronometers (ZHe and ZFT) will provide high resolution cooling histories of now-eroded orogenic belts.

5. The combination of ZFT dating with ZHe or U/Pb dating allows for a full characterization of source terrains. So far these double (and triple) dating schemes are not routine and will clearly improve with time as methodological challenges are overcome. The application of Eximer LA-ICP-MS to U/Pb dating is a significant improvement in this respect, because it allows analyzing many grains quickly and inexpensively. The future lies in those studies where a creative approach in combining these techniques allows new insights into poorly understood orogenic belts and poorly resolved tectonic settings.

ACKNOWLEDGMENTS

We want to thank the editors P. Reiners and T. Ehlers for inviting us to write this chapter. We acknowledge useful discussions, paper reviews, and electronic discourse we have had over the years with P.A.M. Andriessen, A. Carter, M. Brix, M.T. Brandon, I. Brewer, D. Burbank, A. Gleadow, R.L. Fleischer, B. Fügenschuh, N. Hasebe, J. Hourigan, N. Hovius, A. Hurford, P.J.J. Kamp, B. Kohn, B. Kowallis, N. Naeser, C.W. Naeser, L. Nasdala, J. Rahl, M.K. Rahn, B.C.D. Riley, P.W. Reiners, D. Seward, E. Sobel, C. Spiegel, R.J. Stewart, A.V. Soloviev, T. Tagami, S.N. Thompson, J.A. Vance, P. van der Beek, B. Ventura, G-C. Wang, G. Xu, and M. Zattin. We would also like to thank students over the years who have made DZFT part of their thesis work which is partly summarized in this paper in some form or another: A. Bartholomew, M.E. Bullen, A.J. Frisbie, S.R. Johnston, J.R. Lederer, N.M. Meyer, B.R. Molitor, M.J. Montario, S. Perry, B.C.D. Riley, C.R. Schiffman, S. J. Shoemaker, and L. J. Walker. Support for part of this research was provided by the US NSF grants EAR 9911910 (Kamchatka) and EAR 9614730 (New Zealand) (both to Garver), as well as by a James Dwight Dana Fellowship (Yale University) and a Marie Curie Fellowship (European Union) (both to Bernet). This manuscript profited from detailed reviews by Andy Carter and Massimilliano Zattin, which we gratefully acknowledge.

REFERENCES

Bernet M, Zattin M, Garver JI, Brandon, MT, Vance, JA (2001) Steady-state exhumation of the European Alps. Geology 29:35-38

Bernet M, Brandon MT, Garver JI, Reiners PW, Fitzgerald PG (2002) Determining the zircon fission-track closure temperature. GSA Cordilleran Section, 98th annual meeting, Abstract with Programs 34:18

Bernet M, Brandon MT, Garver JI, Molitor BR (2004a) Downstream changes of Alpine zircon fission-track ages in the Rhône and Rhine rivers. J Sed Res 74:82-94

Bernet M, Brandon MT, Garver JI, Molitor BR (2004b) Fundamentals of detrital zircon fission-track analysis for provenance and exhumation studies with examples from the European Alps. *In:* Detrital Thermochronology – Provenance Analysis, Exhumation, and Landscape Evolution of Mountain Belts. Bernet M, Spiegel C (eds) GSA Spec Pub 378:25-36

Bernet M, Brandon, MT, Garver JI, Balestieri, ML, Ventura, B, Zattin, M (2005) Exhuming the Alps through time: Clues from detrital zircon fission-track ages. Am J Sci (in press)

Brandon MT (1996) Probability density plot for fission track grain-age samples. Radiation Meas 26:663-676

Brandon MT, Vance JA (1992) New statistical methods for analysis of fission track grain-age distributions with applications to detrital zircon ages from the Olympic subduction complex, western Washington State. Am J Sci 292:565-636

Brandon MT, Roden-Tice MK, Garver JI (1998) Late Cenozoic exhumation of the Cascadia accretionary wedge in the Olympic Mountains, northwest Washington State. GSA Bull 110:985-1009

Carter A (1999) Present status and future avenues of source region discrimination and characterization using fission-track analysis. Sed Geol 124:31-45

Carter A, Moss SJ (1999) Combined detrital-zircon fission-track and U-Pb dating: A new approach to understanding hinterland evolution. Geology 27:235-238

Carter A, Bristow CS (2000) Detrital zircon geochronology: Enhancing the quality of sedimentary source information through improved methodology and combined U-Pb and fission-track techniques. Basin Res 12:47-57

Carter A, Bristow CS (2003) Linking hinterland evolution and continental basin sedimentation by using detrital zircon thermochronology: a study of the Khorat Plateau Basin, eastern Thailand. Basin Res 15:271-285

Cederbom CE, Sinclair H, Schlunegger F, Rahn M (2004) Climate-induced rebound and exhumation of the European Alps. Geology 32:709-712

Cerveny PF, Naeser ND, Zeitler PK, Naeser CW, Johnson NM (1988) History of uplift and relief of the Himalaya during the past 18 million years: Evidence from fission-track ages of detrital zircons from sandstones of the Siwalik Group. *In:* New Perspectives in Basin Analysis. Kleinspehn K, Paola C (eds) Springer-Verlag, New York, p 43-61

Cibin U, Spadafora E, Zuffa, GG, Castellarin A (2001) Continental collision history from arenites to episutural basins in the Northern Apennines, Italy. GSA Bulletin 113:4-19

Clark SP, Jäger E (1969) Denudation rate in the Alps from geochronologic and heat flow data. Am J Sci 267: 1143-1160

Deer WA, Howie RA, Zussman J (1992) An Introduction to the Rock-Forming Minerals. 2nd ed., Longman Scientific and Technical, Essex, England

Dodson MH (1973) Closure Temperature in Cooling Geochronological and Petrological systems. Contrib Mineral Petrol 40:259-274

Dunkl I, Di Gulio A, Kuhlemann J (2001) Combination of single-grain fission-track chronology and morphological analysis of detrital zircon crystals in provenance studies – sources of the Macigno Formation (Apennines, Italy). J Sed Res 71:516-525

Foster DA, Kohn BP, Gleadow, AJW (1996) Sphene and zircon fission track closure temperatures revisited: empirical calibrations from $^{40}Ar/^{39}Ar$ diffusion studies on K-feldspar and biotite. International Workshop on Fission Track Dating, Abstracts, Gent 37

Galbraith RF, Green PF (1990) Estimating the component ages in a finite mixture. Nucl Tracks Radiat Meas 17:197-206

Garver JI (2003) Etching age standards for fission track analysis. Radiat Meas 37:47-54

Garver JI, Brandon MT (1994a) Fission-track ages of detrital zircon from Cretaceous strata, southern British Columbia: Implications for the Baja BC hypothesis. Tectonics 13:401-420

Garver JI, Brandon MT (1994b) Erosional denudation of the British Columbia Coast Ranges as determined from fission-track ages of detrital zircon from the Tofino Basin, Olympic Peninsula, Washington: GSA Bull 106:1398-1412

Garver JI, Bartholomew A (2001) Partial Resetting of fission tracks in detrital zircon: Dating low Temperature events in the Hudson Valley (NY): GSA Abstracts with Programs 33:82

Garver JI, Kamp PJJ (2002) Integration of zircon color and zircon fission track zonation patterns in Orogenic belts: Application of the Southern Alps, New Zealand. Tectonophysics 349:203-219

Garver JI, Brandon MT, Roden-Tice MK, Kamp PJJ (1999) Exhumation history of orogenic highlands determined by detrital fission track thermochronology. *In:* Exhumation processes: Normal faulting, ductile flow, and erosion. Ring U, Brandon MT, Willett SD, Lister GS (eds) Geol Soc London Spec Pub 154:283-304

Garver JI, Brandon MT, Bernet M, Brewer I, Soloviev AV, Kamp PJJ, Meyer N (2000a) Practical considerations for using detrital zircon fission track thermochronology for provenance, exhumation studies, and dating sediments. *In:* The Ninth International Conference of Fission-track Dating and Thermochronology. Noble WP, O'Sullivan PB, Brown RW (eds) Geol Soc Australia - Abstracts 58:109-111

Garver JI, Soloviev AV, Bullen ME, Brandon MT (2000b) Towards a more complete record of magmatism and exhumation in continental arcs using detrital fission-track thermochronometry. Phys Chem Earth 25: 565-570

Garver JI, Riley BCD, Wang G (2002) Partial resetting of fission tracks in detrital zircon. European Fission-track conference, Cadiz, Spain. Geotemas 4:73-75

Garver JI, Reiners PR, Walker LJ, Ramage JR, Perry SE (2005) Implications for timing of Andean uplift based on thermal resetting of radiation-damaged zircon in the Cordillera Huayhuash, northern Perú. J Geol. 113:117-138

Gastil RG, DeLisle M, Morgan J (1967) Some effects of progressive metamorphism on zircons. GSA Bull 78: 879-906

Harrison TM, Armstrong RL, Naeser CW, Harakal JE (1979) Geochronology and thermal history of the Coast Plutonic Complex, near Prince Rupert, British Columbia. Can J Earth Sci 16:400-410

Hasebe N, Tagami T (2001) Exhumation of an accretionary prism results from fission track thermochronology of the Shimanto Belt, southwest Japan. Tectonophysics 331:247-267

Hasebe N, Tagami T, Nishimura S (1993) Evolution of the Shimanto accretionary complex: A fission-track thermochronological study. *In:* Thermal Evolution of the Tertiary Shimanto Belt, Southwest Japan: An example of Ridge-Trench Interaction. Underwood MB (ed) GSA Spec Pub 273:121-136

Haughton PDW, Todd SP, Morton AC (1991) Sedimentary provenance studies. *In:* Developments in Sedimentary Provenance Studies. Morton AC, Todd SP, Haughton PDW (eds) Geol Soc London Spec Pub 57:1-11

Heller PL, Tabor RW, O'Neil, JR, Pevear DR, Shafiquillah M, Winslow NS (1992) Isotopic provenance of Paleogene sandstones from the accretionary core of the Olympic Mountains, Washington: GSA Bull 104: 140-153

Hinderer M (2001) Late Quaternary denudation of the Alps, valley and lake fillings and modern river loads. Geodinamica Acta 14: 231-263

Hourigan JK, Brandon MT, Garver JI, Soloviev AV (2001) A Comparison of the detrital zircon grain-age distributions from the Ukelayat Group and the Kamchatskiy Complex: Implications for the origin of the Sredinniy Range, Kamchatka. Seventh Zoneshain International Conference on plate tectonics, Moscow, Russia 504

Hurford AJ (1986) Cooling and uplift patterns in the Lepontine Alps, South Central Switzerland and an age of vertical movement on the Insubric fault line. Contrib Mineral Petrol 92:413-427

Hurford AJ, Carter A (1991) The role of fission track dating in discrimination of provenance. *In:* Developments in Sedimentary Provenance Studies. Morton AC, Todd SP, Haughton PDW (eds) Geol Soc London Spec Pub 57:67-78

Kowallis BJ, Heaton JS, Bringhurst K (1986) Fission-track dating of volcanically derived sedimentary rocks. Geology 14:19-22

Kuhlemann J (2000) Post-collisional sediment budget of circum-Alpine basins (Central Europe). Memorie degli Istituti di Geologia e Mineralogia dell' Universita di Padova 52:1-91

Kuhlemann J, Frisch W, Székely B, Dunkl I, Kazmér M (2002) Post-collisional sedimenty budget history of the Alps: tectonic versus climatic control. Int J Earth Sci 91:818-837

Liu TK, Hseih S, Chen Y-G, Chen W-S (2001) Thermo-kinematic evolution of the Taiwan oblique-collision mountain belt as revealed by zircon fission track dating. EPSL 186:45–56

Lonergan L, Johnson C (1998) Reconstructing orogenic exhumation histories using synorogenic zircons and apatites: An example from the Betic Cordillera, SE Spain. Basin Research 10:353-364

Mange MA, Maurer HFW (1992) Heavy Minerals in Colour. Chapman and Hall, London

Meyer NR, Garver JI (2000) Zircon fission-track fingerprint of major tributaries to the Mississippi River. GSA Abstracts with Program 32:59

Morton AC (1984) Stability of detrital heavy minerals in Tertiary sandstones of the North Sea Basin. Clay Minerals 19:287-308

Naeser ND, Zeitler PK, Naeser, CW, Cerveny PF (1987) Provenance studies by fission track dating of zircon – Etching and counting procedures. Nucl Tracks Radiat Meas 13:121-126

Poldervaart A (1955) Zircon in rocks 1, Sedimentary rocks. Am J Sci 235:433-461

Poldervaart A (1956) Zircon in rocks 2, Igneous rocks. Am J Sci 234:521-554

Pupin JP (1980) Zircon and Granite petrology. Contrib Mineral Petrol 73:207-220

Rahl JM, Reiners PW, Campbell IH, Nicolescu S, Allen CM (2003) Combined single-grain (U-Th)/He and U/Pb dating of detrital zircons from the Navajo Sandstone, Utah. Geology 31:761-764

Rahn MK, Brandon MT, Batt GE, Garver JI (2004) A zero-damage model for fission-track annealing in zircon. Am Mineral 89:473-484

Reiners PW (2005) Zircon (U-Th)/He thermochronometry. Rev Mineral Geochem 58:151-179

Reiners PW, Campbell IS, Nicolescu S, Allen CA, Hourigan JK, Garver JI, Mattinson, JM, Cowan DS (2005) (U-Th)/(He-Pb) double-dating of detrital zircons. Am J Sci 305:259-311

Ruiz GMH, Seward D, Winkler W (2004) Detrital thermochronology – a new perspective on hinterland tectonics, an example from the Andean Amazon Basin, Ecuador. Basin Res 16:413-430

Sambridge MS, Compston W (1994) Mixture modeling of multi-component data sets with application to ion probe zircon ages. ESPL 128:373-390

Schlunegger F, Melzer J, Tucker GE (2001) Climate, exposed source-rock lithologies, crustal uplift and surface erosion: a theoretical analysis calibrated with data from the Alps/North Alpine Foreland Basin system. Int J Earth Sci 90:484-499

Shapiro MN, Soloviev AV, Garver JI, Brandon MT (2001) Sources of zircons from Cretaceous and Lower Paleogene terrigenous sequences of the southern Koryak upland and western Kamchatka. Lith Mineral Resource 36:322-336

Sircombe KN, Hazelton ML (2004) Comparison of detrital zircon age distributions by kernel functional estimation. Sed Geol 171:91-111

Soloviev AV, Garver JI, Shapiro, MN (2001) Fission-track dating of detrital zircon from sandstone of the Lesnaya Group, northern Kamchatka. Strat Geologic Correlation 9:293-303

Soloviev AV, Shapiro MN, Garver JI, Shcherbinina EA, Kravchenko-Berezhnoy IR (2002) New age data from the Lesnaya Group: A key to understanding the timing of arc-continent collision, Kamchatka, Russia. The Island Arc 11:79-90

Spiegel C, Kuhlemann J, Dunkl I, Frisch W, von Eynatten H, Balogh K (2000) The erosion history of the Central Alps: Evidence from zircon fission-track data of the foreland basin sediments. Terra Nova 12: 163-170

Spiegel C, Siebel W, Kuhlemann J, Frisch W (2004) Towards a comprehensive provenance analysis: a multi-method approach and its implications for the evolution of the Central Alps. *In:* Detrital Thermochronology – Provenance Analysis, Exhumation, and Landscape Evolution of Mountain Belts. Bernet M, Spiegel C (eds) GSA Spec Pub 378:37-50

Stewart RJ, Brandon MT (2004) Detrital zircon fission-track ages for the "Hoh Formation": Implications for late Cenozoic evolution of the Cascadia subduction wedge. GSA Bull 116:60-75

Tagami T (2005) Zircon fission-track thermochronology and applications to fault studies. Rev Mineral Geochem 58:95-122

Vermeesch P (2004) How many grains are needed for a provenance study? EPSL 224:441-451

Wagner G, Van den Haute P (1992) Fission-track dating: Solid Earth Sciences Library. Kluwer Academic Publishers, Amsterdam

Willett SD, Fisher D, Fuller C, En-Chao Y, Chia-Yu L (2003) Erosion rates and orogenic-wedge kinematics in Taiwan inferred from fission-track thermochronometry. Geology 31:945-948

Zaun PE, Wagner GA (1985) Fission-track stability in zircons under geological conditions. Nucl Tracks 10: 303-307

Zeitler PK, Johnson MN, Briggs ND, Naeser CW (1986) Uplift history of the NW Himalaya as recorded by fission-track ages of detrital Siwalik zircons. *In:* Proceedings of the Symposium on Mesozoic and Cenozoic Geology. Jiqing H (ed) Geological Publishing House, Beijing, p 481-494

Reviews in Mineralogy & Geochemistry
Vol. 58, pp. 239-257, 2005

$^{40}Ar/^{39}Ar$ Thermochronology of Detrital Minerals

K.V. Hodges*, K.W. Ruhl, C.W. Wobus, M.S. Pringle

Department of Earth, Atmospheric, and Planetary Sciences
Massachusetts Institute of Technology
Cambridge, Massachusetts, 02139, U.S.A.
*kvhodges@mit.edu

INTRODUCTION

Clastic sediments and sedimentary rocks provide important records of the erosional history of active and ancient orogenic systems. For example, foreland basin deposits are especially rich sources of information about the character of orogenic hinterlands that have long-since eroded away. A popular approach to extracting such information is the U-Pb dating of detrital zircon (e.g., Ross and Bowring 1990; Gehrels 2000); by matching detrital zircon age populations with age patterns in exposed bedrock regions, it is possible to constrain the source regions for basin fill—provided, of course, that what is left of the hinterland is at least representative of what was there at the time of exhumation! One reason that zircon is used for such studies is that it has a very high closure temperature for Pb diffusion, higher that metamorphic temperatures in all but high-temperature granulite settings (Cherniak and Watson 2000), so that detrital U-Pb zircon dates almost always reflect provenance ages. Unfortunately, the refractory nature of the U-Pb zircon system means that it is less useful for understanding the thermal and erosional evolution of the hinterland than other systems with lower closure temperatures. Other chapters in this volume deal with the application of low-temperature (U-Th)/He and fission-track thermochronometers to detrital samples. Here we review the methodology of detrital $^{40}Ar/^{39}Ar$ thermochronology and explore how it has been used in a variety of tectonic studies. Additional perspectives are available in Stuart (2002). We presume a basic familiarity with the fundamentals of $^{40}Ar/^{39}Ar$ geochronology and the laser $^{40}Ar/^{39}Ar$ microprobe; readers who would like a review should consult Harrison and Zeitler (2005) of this volume, McDougall and Harrison (1998), and Hodges (1998).

MOTIVATIONS FOR DETRITAL $^{40}Ar/^{39}Ar$ STUDIES

With the development of laser microanalytical protocols for terrestrial $^{40}Ar/^{39}Ar$ geochronology in the late 1970's and early 1980's, a new dimension was added to studies of detrital minerals. As is the case with the U-Pb zircon method, the $^{40}Ar/^{39}Ar$ method can be used to pinpoint the provenance of detrital minerals. However, it also can be used to establish the cooling histories of source terrains, to constrain the timescales of sedimentary processes, and—in some cases—to define the positions of deformational features that control patterns of uplift in active orogens. While detrital fission-track and (U-Th)/He) thermochronology also offer such insights, the current state of these arts is such that the $^{40}Ar/^{39}Ar$ laser microprobe technique provides considerably higher precision and more rapid sample throughput.

A relatively recent innovation has been the application of detrital $^{40}Ar/^{39}Ar$ thermochronology to modern fluvial sediments. One important motivation for this application is reconnaissance mapping of bedrock cooling ages in modern surface exposures. In active orogenic settings, the method provides a simple and effective way to development regional

1529-6466/05/0058-0009$05.00 DOI: 10.2138/rmg.2005.58.9

maps—catchment by catchment—of exhumation patterns indicative of tectonic evolution. Moreover, when bedrock patterns of cooling ages are known, the technique can be used to explore erosional processes for individual fluvial systems and thus track the geomorphic evolution of mountainous landscapes.

SAMPLING AND SAMPLE PREPARATION

As is the case in other geochemical studies, the sampling of fresh, unaltered materials is important for $^{40}Ar/^{39}Ar$ detrital thermochronology. The best sedimentary rock samples are impure sandstones and arkoses that show no evidence of post-depositional metamorphism. Similarly, the best modern sediments are those with high concentrations of K-rich minerals. In all cases, minerals should be free of obvious signs of alteration. Examining potential samples with a hand lens before collecting them can substantially improve the probability of successful laboratory work. Do K-feldspar grains show signs of authigenic overgrowths that may complicate data interpretation? Are the micas clean, or do they show intergrowths with other minerals that may yield poor $^{40}Ar/^{39}Ar$ results? In some cases, simple ancillary studies—such as the use of illite crystallinity to evaluate the degree of post-depositional metamorphism—can be useful in selecting optimal samples.

Most detrital $^{40}Ar/^{39}Ar$ studies have focused on one mineral, usually either K-feldspar, a dioctahedral mica such as muscovite or phengite, or a trioctahedral mica such as biotite or phlogopite. All are resistant to abrasion and chemical dissolution during transport, and should persist for long distances downstream from the source region (Kowalewski and Rimstidt 2003). K-feldspar yields dates corresponding to the low-temperature evolution of source terrains and thus is especially useful for comparisons with results obtained from detrital zircon fission-track studies (Bernet and Garver 2005). However, feldspars are also very susceptible to alteration, particularly in submarine environments, and most grains derived from metamorphic or intrusive igneous rocks are sufficiently structurally complex that they do not have a single, narrow range of $^{40}Ar/^{39}Ar$ closure temperatures (Parsons et al. 1999; Lovera et al. 2002). Since laser fusion is the most frequently used procedure for $^{40}Ar/^{39}Ar$ detrital geochronology (see below), such complexities can greatly hinder the robust interpretation of K-feldspar data. As a consequence, the majority of modern detrital studies focus on the micas.

Micas are common in clastic sediments, and it is relatively easy to make high-purity separates of them. Because they are elastic at surface conditions, they survive sedimentary transport over long downstream distances with relatively minor grain size reduction. It is possible in some cases to distinguish multiple components within a sample's population of micas on the basis of grain size alone. These may represent different source regions, or different lithologies in the same region. If the provenance cooled very slowly (<1 °C/m.y.), multiple grain size fractions may provide important information about its thermal history due to the grain size dependence of closure temperature (e.g., Markley et al. 2002).

Of the micas, muscovite has proven to be the best mineral for detrital $^{40}Ar/^{39}Ar$ work. The trioctahedral micas can be problematic for several reasons. First, they are highly susceptible to secondary alteration (Mitchell et al. 1988; Dong et al. 1998; Murphy et al. 1998). Small-scale intergrowths of chlorite are common in biotites, and their presence renders a grain unsuitable for $^{40}Ar/^{39}Ar$ dating (Roberts et al. 2001; Di Vincenzo et al. 2003). Second, many biotites contain excess ^{40}Ar not produced by *in situ* decay of ^{40}K and, as a consequence, yield geologically meaningless $^{40}Ar/^{39}Ar$ dates (Kelley 2002). In contrast, dioctahedral micas are less susceptible to alteration and, with the exception of phengite (e.g., Scaillet 1996), only infrequently incorporate excess ^{40}Ar during growth.

Grains can be separated from sands or sedimentary rock samples using a variety of standard mechanical, gravimetric, and magnetic techniques. Because it is desirable to maintain original grain sizes, the best separation protocols involve only as much mechanical crushing as is absolutely necessary. In order to ensure maximum purity, some laboratories leach samples in dilute acids to remove contaminant primary minerals and alteration products. Inevitably, hand picking of individual, optically pure grains is necessary for good analytical results. Most laboratories sieve each sample to two or more grain size ranges. These ranges are analyzed independently in order to test for grain size-dependent, apparent age variations.

Although the sample itself dictates the range of grain sizes available for analysis, the best grain sizes for $^{40}Ar/^{39}Ar$ detrital thermochronology range from approximately 250 to 1000 μm. Substantially smaller grains are difficult to manipulate and, after irradiation, pose significant laboratory safety issues. Moreover, small grains that are very young may not contain sufficient radiogenic ^{40}Ar for a high-precision $^{40}Ar/^{39}Ar$ analysis. Very small grains (<20 μm) may lose ^{39}Ar through recoil during irradiation and yield spurious results (Hunecke and Smith 1976; Onstott et al. 1995; Lin et al. 2000), although methods have been suggested to overcome this limitation (e.g., Hemming et al. 2002).

The number of analyses necessary for a robust result

For all detrital geochronology studies—whether they are based on the $^{40}Ar/^{39}Ar$ method or other techniques—a persistent question is: how many analyses are necessary to characterize adequately the cooling age signal for a particular sample? A typical sedimentary sample is a mechanical mixture of thousands of grains of a datable mineral from an unknown number of source regions. Given that it is practically impossible to date all the grains in such a sample, how many must we date to ensure that we do not miss one or more components altogether? To some extent the answer depends on how many components there are in the sample; if the population is unimodal, a small number of analyses is sufficient. Additionally, the answer depends on the relative proportions of each component; if a sample contains 5000 datable grains but only one is derived from a specific source, all 5000 grains must be dated if we are to be absolutely sure that we do not miss that single component. Unfortunately, we know neither the number of components in a sample nor their relative abundances *a priori*.

In a recent innovative paper, Vermeesch (2004) developed a general statistical relationship among the number of analyzed grains, the fraction of the population represented by the least well-represented component, and the probability that no component was missed in the study. For example, in order to be ~95% confident of not missing a component making up at least 5% of the population, we would have to analyze *at least* 117 grains. Note that this analysis presumes the worst-case scenario: a uniform distribution of components in the sample, all of the same size. Populations with a smaller number of components require fewer analyses to characterize adequately (Ruhl and Hodges 2005), but a good rule of thumb is to analyze 100 or more grains from each sample whenever a sufficient number of optically pure grains is available.

ANALYTICAL TECHNIQUES

Detrital $^{40}Ar/^{39}Ar$ studies are typically done with Ar-ion or CO_2 laser microprobes (Hodges 1998). Ar-ion lasers emit light at a variety of wavelengths, but most is within the blue-green visible spectrum at 514 or 488 nm. CO_2 instruments produce far-infrared light at 10.6 μm, and their lower cost (as compared to Ar-ion lasers) has promoted their widespread use over the past decade. Both blue-green and far-infrared light are readily absorbed by detrital feldspars and micas, and the two microprobes are equally suitable for laser fusion dating of detrital materials. Most Ar-ion or CO_2 laser microprobe systems are fully automated, and enough single-grain analyses to characterize a detrital sample can be done in approximately 48 hours.

Typical gas extractions involve complete fusion of a grain using a defocused laser beam. In some cases, grains are heated incrementally by varying the power output of the laser upward through a series of steps. While time consuming, this approach can help evaluate the possibility of alteration in suspect samples (Najman et al. 2002). In many laboratories, samples suspected of having slight alteration are routinely heated prior to analysis by either: 1) a bakeout of the sample chamber for several hours at low temperatures (e.g., 350 °C); or 2) exposing each grain to the laser set at low power (to achieve a temperature of ~500–600 °C) for a few minutes prior to fusion.

A small number of detrital $^{40}Ar/^{39}Ar$ studies have been done using a focused Nd-YAG laser (1064 nm) to melt 50–100 micron holes in large single crystals (e.g., Kelley and Bluck 1989). More recently, Sherlock et al. (2002) and Haines et al. (2004) have demonstrated the utility of high spatial-resolution mapping of intracrystalline apparent age domains in single detrital micas using an ultraviolet (frequency-quadrupled Nd-YAG) laser microprobe.

Regardless of the laser employed, modern microprobe systems are capable of producing $^{40}Ar/^{39}Ar$ for single, 250–1000 μm grains with a precision of from 0.5 to 1 Ma for smaller crystals and of 200,000 to 500,000 years for larger crystals. (In this paper, all explicit precisions are cited at the 2σ, or ~95% confidence, level.) Some samples that are more fine-grained or contain components of Plio-Pleistocene age yield single-crystal dates of much poorer precision. One solution to this problem is to analyze multigrain aliquots, rather than single grains. However, this approach precludes detailed analysis of the modality of ages because multiple components may be mixed in each aliquot, so it should be used only when single-crystal analyses are impractical.

Data presentation and interpretation

Specific analytical protocols vary from laboratory to laboratory, so it is imperative that papers with detrital $^{40}Ar/^{39}Ar$ data contain a section describing the methods used in some detail. At the very least, the reported information should include: 1) the name of the reactor used for sample irradiations (because different reactors have different neutron flux and thermal characteristics); 2) the duration of irradiation and the specific position in the reactor; 3) the specific values used to correct for interfering nuclear reactions during irradiation; 4) the name and assumed age of the material used as a neutron fluence monitor; 5) isotopic discrimination factors for the mass spectrometer used for the analyses; and 6) a description of the analytical backgrounds or "blanks" during the measurements. Detailed discussions of these parameters may be found in McDougall and Harrison (1998).

Typically, reported uncertainties for $^{40}Ar/^{39}Ar$ dates represent analytical imprecision at the 2σ (or ~ 95% confidence) level. When uncertainties in the age of neutron flux monitors and uncertainties in ^{40}K decay constants are also propagated into the error as well, the quoted uncertainty reflects the accuracy of the measured date (Renne et al. 1998). However, as is the case with all geochronologic data, a $^{40}Ar/^{39}Ar$ date for a detrital mineral grain is not necessarily a geologically meaningful age. For example, alteration may yield dates that are too young, and ^{39}Ar recoil loss may result in dates for very small grains that are too old. Both problems can be avoided by careful sample selection and the use of appropriate experimental protocols. A grain may be contaminated by excess ^{40}Ar and the laser fusion technique may not provide evidence of that contamination. In some of these cases, laser incremental heating of a representative number of grains may help eliminate this possibility or demonstrate the unreliability of some dates (e.g., White et al. 2002). Another complication can be the existence of significant intracrystalline gradients in radiogenic ^{40}Ar related to slow cooling (Hodges and Bowring 1995) or polymetamorphism (Hames and Hodges 1993). For samples containing older grains, this possibility can be assessed by reconnaissance use of ultraviolet laser microprobes to map individual crystals (Sherlock et al. 2002).

Once the possibility of such "geologic error" has been dismissed, most detrital $^{40}Ar/^{39}Ar$ data interpretation is based on analytical imprecision alone. This statistical measure is useful for exploring the range of apparent ages in a sample population irradiated in a single package and analyzed in a single laboratory using the consistent protocols. The frequency distribution of detrital mineral ages in a sample is typically depicted using either histograms or graphs of the synoptic probability density function (SPDF). The familiar histogram is adequate for small datasets with an age dispersion much greater than the uncertainty of individual analyses, but it is far less valuable for data exploration than a graph of the SPDF. Derived from the probability density functions for individual measurements with known uncertainties*, the SPDF is a normalized summation of such functions for each measurement in a population. Assuming that analytical precision follows a Gaussian distribution, the SPDF for a population of n dates can be written as:

$$SPDF = \frac{\sum_{i=1}^{n} \left(\frac{1}{\sigma_i \sqrt{2\pi}} \right) e^{-(t-t_i)^2 / 2\sigma_i^2}}{n} \tag{1}$$

where t is age, t_i is the ith measured age, and σ_i is the standard deviation of the ith age†. Sircombe (2004) recently presented a useful discussion of the relative merits of SPDF graphs and histograms, and he has made available for download a Microsoft Excel spreadsheet for producing both from a user-supplied dataset. For example, Figure 1 is a combined histogram and SPDF plot generated with this software for a set of 111 detrital muscovite $^{40}Ar/^{39}Ar$ dates for sands from the Nyadi River in central Nepal (Ruhl and Hodges 2005).

Inferring population characteristics

The Nyadi River has a relatively small drainage area (~200 km^2), so it comes as no surprise that the SPDF in Figure 1 displays a single mode or "hump"—the detrital micas have a range of $^{40}Ar/^{39}Ar$ dates that reflects the unroofing history of the bedrock in the catchment rather than multiple source regions with significantly different unroofing histories. Figure 2 illustrates a very different distribution of $^{40}Ar/^{39}Ar$ muscovite dates. Analyses from several samples of the Triassic part of the Torlesse Supergroup of New Zealand (Adams and Kelley 1998) reveal a decidedly multimodal SPDF. The two groups of apparent ages comprise ranges from ~150–300 Ma and from ~400–500 Ma, suggesting two different source regions. For the purposes of Adams and Kelley (1998), this distinction was sufficient to draw an important tectonic conclusion regarding the source area for *most* of the detrital muscovites: of all the candidate sources, only the New England and Hodgkinson fold belts of northern New South Wales and eastern Queensland, Australia, have the appropriate range of bedrock cooling ages. But what if there was not such a marked age distinction between the modes, or what if there was a value in developing a quantitative understanding of the distinctive modes that might be present? A more sophisticated statistical approach is needed to address such issues.

One of the most widely used methods of modal analysis is mixture modeling (Galbraith and Green 1990; Sambridge and Compston 1994). After the user specifies a number of assumed components, the modeling routines set about finding the best-fit modes, their uncertainties, and relative proportions. For the Torlesse data shown in Figure 2, a search for two components using the Sambridge and Compston approach results in the definition of components at 448 ± 49 Ma and 235 ± 18 Ma. Clearly, these two components alone cannot explain the total dispersion in the dataset, particularly the large number of apparent ages younger than 225 Ma

* For a review of the underlying statistics, see Bevington and Robinson (1992).

† For ease of reference, variables used throughout this chapter are defined in Table 1.

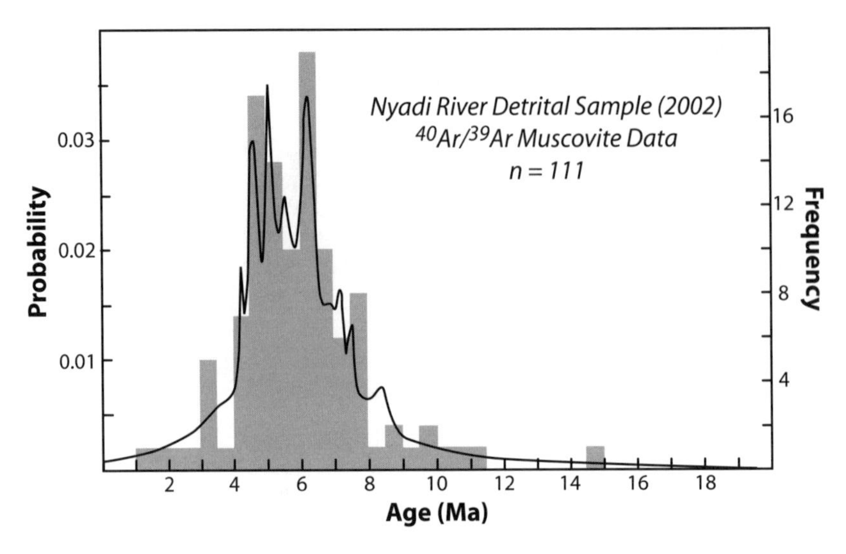

Figure 1. Combined histogram and synoptic probability density function (SPDF) plot for 111 single-crystal, total-fusion, muscovite dates from a modern sand sample collected in 2002 from the Nyadi River, Annapurna Range, central Nepal (Ruhl and Hodges 2005). The histogram (gray boxes) was constructed using a bin size of 500,000 years. A sample collected from the same river bed in 1997 yielded a statistically indistinguishable SPDF plot for 34 multigrain analyses of detrital muscovites (Brewer et al. 2001), implying that there is little year-to-year variability in the age distribution of muscovites in well-mixed sediment samples from this river.

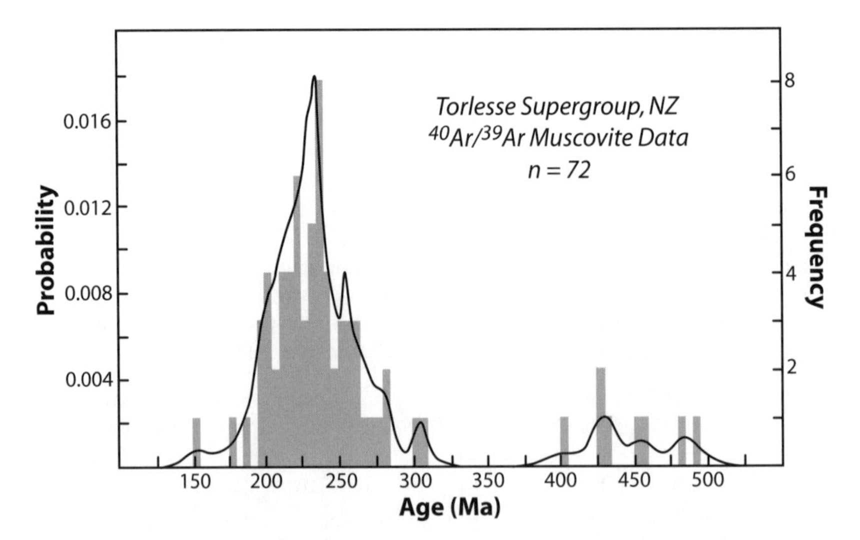

Figure 2. Combined histogram and SPDF plot for 72 $^{40}Ar/^{39}Ar$ muscovite dates from samples of the Triassic Torlesse Supergroup from New Zealand (Adams and Kelley 1998). The bin size for the histogram is 5 million years. Although the age distribution forms two broad groupings (400–500 Ma and 150–300 Ma), at least eight distinct modes are indicated in the dataset by mixture modeling implying numerous source regions with distinctive cooling ages.

Table 1. Symbols used.

Symbol	Definition
A_s	Surface volumetric heat production
C	Specific heat
$CSPDF$	Cumulative probability distribution function
E	Erosion rate
G_o	Nominal geothermal gradient (e.g., 25 °C/km)
h	Depth at which heat production drops off to 1/e of A_s
L	Lag time
L_o	Lag time for a mineral in the oldest studied horizon of a sedimentary section
n	Number of measured cooling ages in a series
R	Relief in a stream catchment
$SPDF$	Synoptic probability density function
t	Age
t_m	Measured cooling age
T_m	Bulk closure temperature for an isotopic chronometer
T_{mi}	Cooling age measurement i in a series of n = 1 to i measurements
t_{range}	Range of cooling ages in a distribution
T_s	Surface temperature
T_λ	Assumed temperature at depth λ (e.g., 750 °C)
z	Depth
z_m	Depth of the bulk closure isotherm for an isotopic chronometer
$\Delta\tau_E$	Time elapsed between cooling of a mineral through Tc_m and its exposure at the surface as a consequence of erosion
$\Delta\tau_{TR}$	Time elapsed between initial surface exposure of a detrital mineral and its sedimentary deposition
κ	Thermal diffusivity
λ	Assumed layer thickness for thermal model (e.g., 30 km)
ρ	density
σ_i	Sample standard deviation of the ith measured cooling age in a series of n measurements

and between ~250 and ~300 Ma. Sambridge and Compston (1994) suggest specifying successively larger numbers of modes for analysis until the modeling results stabilize (e.g., the same mode is picked more than once, or the uncertainties in the modes do not decrease as the number of specified modes increase). The Torlesse dataset is very complicated in this regard. The apparent ages in the older group are sufficiently few that the ~448 Ma mode, while robust, is hard to refine further. Mixture modeling experiments with the younger group define robust modes at ~290 Ma, ~233 Ma, ~210 Ma, and ~170 Ma, but a mixture of at least eight modes would be necessarily to describe the full complexity of the data.

This relatively objective approach to analyzing mixed populations provides a powerful tool for comparative studies of detrital age datasets. For example, Gehrels (2000) and Gehrels et al. (2002) showed how the similarities and distinctions between two distributions of U-Pb detrital zircon data could be quantified. Sircombe and Hazelton (2004) have presented a more sophisticated statistical protocol for testing the similarity of modes in different samples.

APPLICATIONS AND EXAMPLES

In the paragraphs that follow, we examine how $^{40}Ar/^{39}Ar$ geochronology of modern and ancient detrital minerals may be used to address important problems in tectonics and geomorphology.

Determining sediment source regions

The Adams and Kelley (1998) study reviewed above is an example of one of the most common uses of detrital mineral dating: establishing provenance. Among the $^{40}Ar/^{39}Ar$ community, this approach was pioneered by Kelley and Bluck (1989; 1992) in their studies of detrital micas from the Ordovician Southern Uplands sequence of Scotland. Subsequent research demonstrated the applicability of the technique to understanding sedimentary transport in a variety of continental settings (e.g., Renne et al. 1990; Dallmeyer et al. 1997; Stuart et al. 2001; Barbieri et al. 2003; Rahman and Faupl 2003; Haines et al. 2004). In some studies, the fingerprinting of source regions has helped inform large-scale, tectonic reconstructions (e.g., Dallmeyer and Neubauer 1994; von Eynatten et al. 1996; Hutson et al. 1998; Carrapa et al. 2004a) . Changes in provenance with time, as evidenced by variations in cooling age SPDFs from one stratigraphic level to another, have helped to constrain the evolution of drainage patterns (e.g., Aalto et al. 1998; Najman et al. 2003; von Eynatten and Wijbrans 2003; Carrapa et al. 2004b) and the deformational histories of source regions (e.g., White et al. 2002). Although there have been few $^{40}Ar/^{39}Ar$ dating studies of fine-grained terrigenous sediments in oceanic environments, Hemming et al. (2000; 2002) reported impressive results for hornblendes and micas from ice-rafted detritus in Quaternary North Atlantic sediment cores that indicate derivation from the Labrador Sea region.

Constraining minimum depositional ages of ancient sediments

In some cases, sedimentary successions contain abundant detrital minerals suitable for $^{40}Ar/^{39}Ar$ dating but no fossils or volcanogenic strata providing an indication of the age of deposition. Najman et al. (1997) demonstrated how the cooling ages for detrital minerals provide at least a minimum estimate of the depositional age. More recently, Carrapa et al. (2004a) showed how such indirect dating has helped to constrain tectonic models for the Alps-Apennine junction area.

Estimating the timing of source region exhumation

The geochronology of detrital minerals is a powerful way to explore the exhumation histories of orogenic hinterlands. Copeland and Harrison (1990) used $^{40}Ar/^{39}Ar$ data from Ocean Drilling Program sediment cores in the distal Bengal Fan to demonstrate that rocks metamorphosed during the Himalayan orogeny had been uplifted to the surface and were providing detritus to the fan by Middle Miocene time. More detailed investigations of the timing of Himalayan rock uplift have involved dating detrital K-feldspar and micas from more proximal foreland deposits (Harrison et al. 1993; Najman et al. 1997; Najman et al. 2001; Najman et al. 2002). Other case studies of this type include those of Grimmer et al. (2003), who sought to determine the timing of exhumation of ultrahigh-pressure rocks of the Dabie Shan orogen, and Barbieri et al. (2003), who documented the episodicity of uplift in the Ligurian Alps.

Constraining the erosion-transport interval for orogenic detritus

The interval between the cooling age of a detrital mineral and the depositional age of the sedimentary rock in which it is found is frequently referred to as "lag time" (Cerveny et al.

1988; Brandon and Vance 1992; Ruiz et al. 2004). Lag time (L) can be expressed as:

$$L = \Delta\tau_E + \Delta\tau_{TR} \tag{2}$$

where $\Delta\tau_E$ is the time necessary for sufficient erosion to have occurred to bring the mineral to the surface from the position of its closure isotherm at depth, and $\Delta\tau_{TR}$ is the time necessary for transport of the eroded grain through the fluvial system to its final point of deposition.

In a few cases, L is extremely long, suggesting protracted sediment storage and reworking prior to final deposition (Sherlock 2001). However, many studies suggest that transport intervals can be very brief in river networks draining active orogenic systems (e.g., Copeland and Harrison 1990; Heller et al. 1992). This suggests a mechanism by which L might be used to formulate a rough estimate of erosion rate in the source region. If we specify the closure temperature for a mineral (T_m), and calculate the depth of its closure isotherm (z_m) below Earth's surface using a modeled geothermal gradient, then erosion rate (E) is:

$$E = \frac{z_m}{\Delta\tau_E} = \frac{z_m}{(L - \Delta\tau_{TR})} \tag{3}$$

As $\Delta\tau_{TR}$ approaches zero, L approaches $\Delta\tau_E$ and $E \approx z_m/L$. (Note that this rate is averaged over the exhumation interval $\Delta\tau_E$. If that interval is sufficiently long, the averaged rate may not be particularly useful for addressing many tectonic problems.)

An estimate of E made with Equation (3) is only as good as the model used for the geothermal gradient. One approach has been to assume a linear, conductive geotherm (Copeland and Harrison 1990; Renne et al. 1990). For example, an estimated geotherm of 30 °C/km implies that z_m is ~12.2 km for a nominal $^{40}Ar/^{39}Ar$ muscovite bulk closure temperature of 366 °C (Hodges 2003). This approach provides a useful rough estimate of E for regions undergoing slow (<1 km/m.y.) erosional exhumation, but a more widely applicable model must account for advective heat transport. Numerical experiments suggest that well-developed orogenic systems (i.e., those in which the orogenic wedge grows rapidly through accretionary processes) achieve thermal steady states relatively quickly (e.g., Huerta et al. 1998). As a consequence, a first attempt at improving upon Equation (3) might be made simply by adopting a sufficiently sophisticated model for the steady-state geotherm that incorporates both conduction and advection.

By way of example, we might presume that the problem is one-dimensional , involving the vertical conduction of heat as well as the vertical advection of both heat and rock as a consequence of erosion. Moreover, we might postulate an exponential decrease in radiogenic heat production with depth, such that the surface heat production (A_s) falls off to a value of 1/e at depth h. As shown by Manktelow and Grasemann (1997), among others, this situation can be modeled as the thermal structure in a layer of thickness λ with a fixed temperature at depth λ (T_λ). For our purposes, the values we choose for λ and T_λ are not particularly critical as long as λ is substantially greater than z_m. The "steady-state" solution for T_m as a function of z_m is:

$$T_m = T_s + \beta\left[1 - \exp\left(-\frac{z_m}{h}\right)\right] + \gamma\left[1 - \exp\left(-\frac{Ez_m}{\kappa}\right)\right] \tag{4}$$

where

$$\beta = \frac{A_s h^2}{\rho C(\kappa - Eh)} \tag{5}$$

and

$$\gamma = \frac{(T_\lambda - T_s) - \beta\left[1 - \exp\left(-\frac{\lambda}{h}\right)\right]}{1 - \exp\left(-\frac{E\lambda}{\kappa}\right)} \tag{6}$$

If we assume zero sediment transport time ($E \approx z_m/L$), we can modify Equation (6) to relate lag time to source-region erosion rate:

$$T_m \approx T_s + \beta\left[1 - \exp\left(-\frac{EL}{h}\right)\right] + \gamma\left[1 - \exp\left(-\frac{E^2 L}{\kappa}\right)\right] \tag{7}$$

Technically, this is *not* actually a steady-state solution; as erosion progresses, heat producing elements are stripped from the top of the column such that A_s in Equation (5) is time-dependent. Manktelow and Grasemann (1997) suggest how this can be dealt with in an iterative fashion, but ignoring this effect has little impact on our particular application of the model.

Equation (7) could be improved upon still further by taking account of the influence of topography on the form of steady state, near-surface isotherms (Stüwe et al. 1994; Mancktelow and Grasemann 1997). However, this effect is damped with increasing depth and has little effect on the $^{40}Ar/^{39}Ar$ muscovite closure isotherm except when exhumation rates are extremely high (>3 km/my; Brewer et al. 2003). A matter of greater concern is the inability of a one-dimensional model to capture lateral advection of both heat and rock during orogenesis (Ehlers and Farley 2003), and developing more realistic models relating lag time to erosion rate is a topic of on-going research.

Provided that the sediment source region does not change, the comparison of lag times from different stratigraphic levels can be used to monitor changes in source-region cooling rate through time (Ruiz et al. 2004). If there was little variation in the thermal structure of the source region, such changes imply changes in erosion rate. Figure 3 illustrates the implications of variations in L with respect to the lag time of the oldest horizon for which

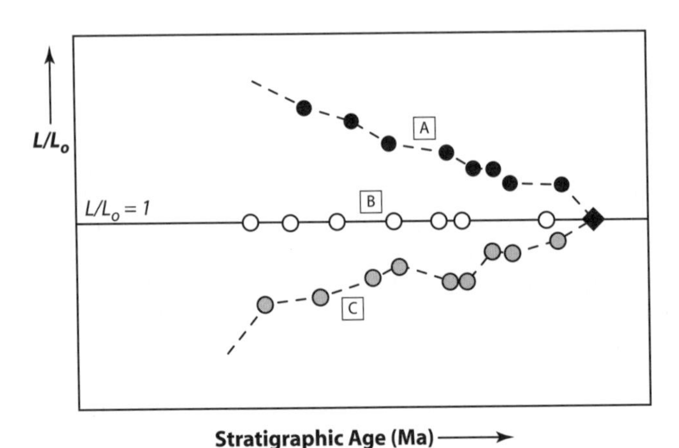

Figure 3. Lag-time evolution diagram. Lag time here is non-dimensionalized by dividing the lag time for a detrital mineral from a horizon of known stratigraphic age (L) by the lag time for that mineral in the oldest studied horizon (L_o). That oldest sample is shown as a filled diamond symbol. Circles indicate lag times for successively younger samples. Curve A implies values of L/L_o >1, implying decreasing source-region erosion rate with time. Curve B (L/L_o = 1) suggests a steady state, whereas Curve C (L/L_o < 1) suggests an increasing erosion rate with time.

there are thermochronologic data (L_o). Constant lag time at all stratigraphic levels ($L/L_o = 1$) indicates a thermal stability in the source region that suggests steady-state erosion. Equation 3 predicts that increasing cooling rates over time would lead to decreasing lag times ($L/L_o < 1$), and that decreasing cooling rates would have the opposite effect ($L/L_o > 1$). With examples from the Himalaya and Western Alps, White et al. (2002), Carrapa et al. (2003) and Najman et al. (2003) showed how comparative studies of lag times from different stratigraphic levels in foreland deposits can help inform tectonic models for orogenic hinterlands.

Most lag time studies have focused on the lag time between the youngest dated grain in a sample and the sample's depositional age. However, a sample containing multimodal mixtures of grains from multiple source regions contains additional information. If the lag times for two different modes are consistently different through time, the implication is that both eroded at similarly different rates. A convergence of lag times suggests instead a change in the erosional character in one or the other source regions, the minimum age of which can be estimated from the cooling age for the youngest mode at the time of the change (Ruiz et al. 2004).

Elucidating modern erosional patterns

Cooling ages for detrital minerals from modern river networks are powerful indicators of the dynamics of erosion in active orogenic systems. The bedloads of river systems that drain vast continental regions are particularly interesting targets of study; their cooling age spectra provide a sense of large-scale variations in unroofing rate as a consequence of regional tectonic processes. One recent investigation of this type was that of Clift et al. (2004), who combined zircon U-Pb, biotite and muscovite $^{40}Ar/^{39}Ar$, and apatite fission track geochronology to explore the source regions for detritus in the lower Indus River of Pakistan. Their data suggest that the high-grade metamorphic rocks exposed in the physiographic Higher Himalaya are the dominant source of modern sediment. A relatively small proportion of detritus comes from the Nanga Parbat syntaxis, despite evidence for remarkably rapid erosion of that region today (Zeitler et al. 2001).

Estimating erosion rates for modern sedimentary catchments

On a more local scale, thermochronologic data from modern river sediments can serve as a valuable proxy for bedrock cooling ages. Consider a simple landscape, with relatively low topographic relief, that erodes very slowly, no more than one or two millimeters each year (Fig. 4). If this landscape is at least several million years old, chances are that conditions of thermal and topographic steady-state may have been achieved or at least closely approached (Whipple 2001; Willett and Brandon 2002). If we furthermore assume that the only material flux from depth to the surface is related to erosion—such that lateral movements related to tectonic activity are ignored—the $^{40}Ar/^{39}Ar$ cooling ages of minerals exposed on the surface of the landscape will show an elevation dependence; samples from the tops of ridges will yield the oldest dates, and those from the bottoms of valleys will yield the youngest dates. This apparent age gradient with respect to elevation (dt_m/dz) is frequently presumed to be a rough proxy for inverse of the time-averaged erosion rate (E) in regional thermochronologic studies (e.g., Wagner and Reimer 1972). As pointed out by Stock and Montgomery (1996), uniform erosion of such a terrain would yield sediments containing minerals with a range of $^{40}Ar/^{39}Ar$ cooling ages (t_{range}) that is related to the total relief (R) through the equation:

$$t_{range} = R\left(\frac{dt_m}{dz}\right) \tag{8}$$

Stock and Montgomery (1996) suggested how this relationship could be used to explore the evolution of relief in the source regions for modern and ancient sediments if the apparent age gradient with respect to elevation in the bedrock is known.

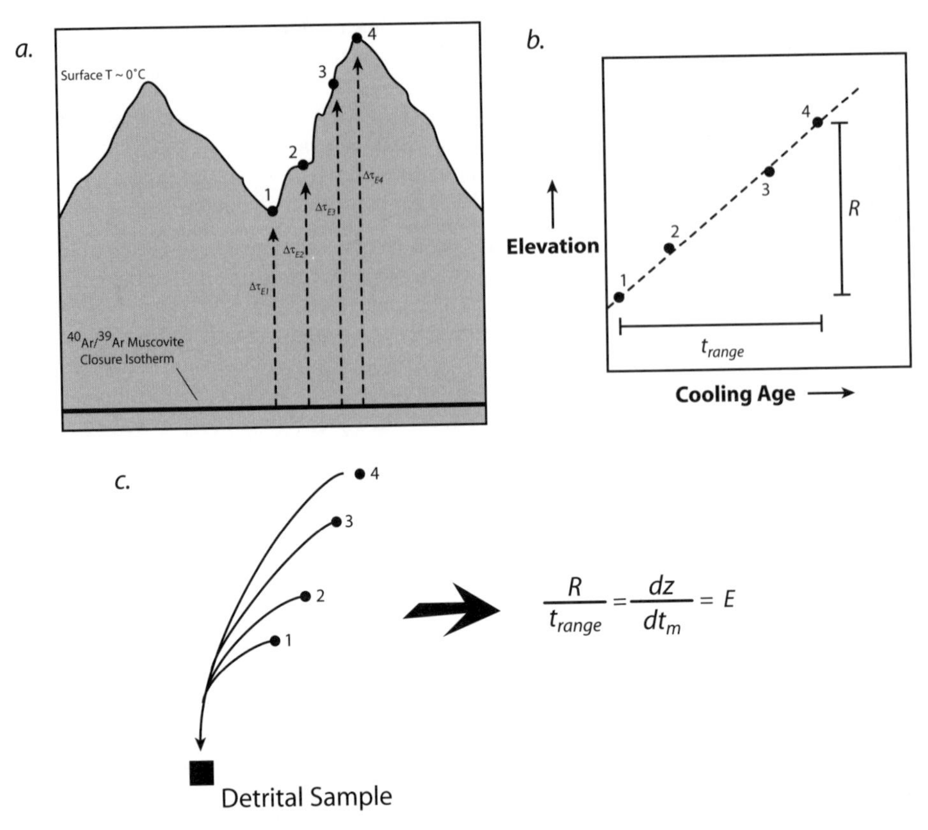

Figure 4. Illustration of how the range of $^{40}Ar/^{39}Ar$ cooling ages for a set of four detrital muscovite samples can be used to estimate source-region erosion rate. (a) Cartoon showing the positions of a range of samples exposed at various elevations in an eroding source terrain. Sample 1 is from the deepest part of the valley and Sample 4 is from the top of the highest ridge. If transport of material from the $^{40}Ar/^{39}Ar$ muscovite isotherm is along vertical paths (shown by dashed lines), Sample 4 should yield the oldest age, having taken τ_{E4} million years to reach the surface, whereas Sample 1 should be the youngest, having taken τ_{E1} million years. The total relief is R and the total range of ages is t_{range}. (b) If Samples 1-4 were collected from bedrock and dated, their $^{40}Ar/^{39}Ar$ cooling ages might be expected to yield a linear relationship with elevation, the slope of which is often interpreted as erosion rate (E). If erosion rate was constant throughout the erosional interval, then E is also equal to R/t_{range}. (c) After muscovite from these samples is eroded and deposited in a river sediment, the range of detrital cooling ages can be used to estimate E *if* we assume that the modern relief of the catchment is the same as the relief during the cooling interval.

For the study of modern catchments with known relief, Equation (2) suggests that the cooling age range also can be used to determine the time-averaged rate of bedrock erosion in the catchment if $E = (dt_m/dz)^{-1}$ (*cf.*, Eqn. 3). An example calculation can be done using the detrital muscovite dataset illustrated in Figure 1 (Ruhl and Hodges 2005), which represents a 13.4 million year range of apparent ages. The total relief in the Nyadi catchment—as measured from a 90-m digital elevation model—is ~6.2 km. As a consequence, we can estimate E as ~0.5 mm/y, although this estimate has large uncertainties related to the high analytical imprecision of very young, single-grain dates in this dataset. If only high-quality dates are used, the apparent age range decreases to ~8.3 million years and the estimate for E increases to 0.75 mm/yr for the period represented by the span of apparent ages (~10.8–2.5 Ma). Note that this rate is only an average. The Nyadi data also imply that the rate of exhumation increased through the ~10.8–2.5 Ma interval because the young end of the range implies a closure-to-surface transport interval ($\Delta\tau_E = 2.5$ Ma) that would require erosion rates of several millimeters per year for any reasonable geothermal gradient.

There are two reasons why this simple way of interpreting the data may lead to erroneous results. The first is that developing orogenic systems have complex and time-variant thermal structures, surface topographies, and patterns of material transport (e.g., Batt and Braun 1997; Braun and Sambridge 1997; Stüwe and Hintermüller 2000; Lague et al. 2003) and are not well-represented by the simple model depicted in Figure 4. The second is that erosion rates may vary substantially over a sediment source region, even in small catchments; Safran (2003) presented a useful analysis of how such issues can influence erosion rate estimates.

Fortunately, there are ways to test some of the critical hypotheses implicit in the method described above. If uniform, steady-state erosion is assumed, the *form* of the cumulative probability distribution function (CSPDF) for a set of detrital cooling ages:

$$CSPDF = \sum_{j=0}^{t} SPDF(j) \tag{9}$$

should be the same as the form of the integral hypsometric curve for the catchment (Ruhl and Hodges 2005). If a statistically based comparison of the integral hypsometric and CSPDF curves suggests a poor match, one or more of the assumptions behind using the detrital age range to estimate erosion rate is unjustified. In the case of the Nyadi Khola data, the two curves match very well, lending additional credence to the 0.75 mm/yr estimate for average erosion rate in the catchment over the closure period.

An alternative approach to estimating erosion rates for active river drainage basins was suggested by Brewer et al. (2003).These authors began with a two-dimensional thermal model to estimate the location of the closure isotherm for Ar retention in muscovite beneath a paleosurface as a function of erosion rate and relief. That result can be used to create a "synthetic" SPDF curve for a specified hypsometry, and different synthetic curves can be generated for different assumed erosion rates. Brewer and co-workers suggested that the positions of major peaks in the modeled SPDF curves could be matched with SPDF curves for actual thermochronologic data to find a best-fit erosion rate. However, when this method is applied to a set of detrital muscovite $^{40}Ar/^{39}Ar$ data from the Nyadi catchment (Brewer et al. 2005), it results in an estimated erosion rate of 2.3 mm/yr, almost three times higher than the Ruhl and Hodges (2005) estimate and inconsistent with the observed range of ages (~8.2 million years) over 6.2 km of relief. One likely explanation for this result is that success of the Brewer et al. (2003) approach depends strongly on the fidelity of the thermal model, which itself depends on parameters (e.g., the distribution of radioactive heat producing elements) that are not well constrained for this region of the Himalaya. Moreover, as mentioned previously, lateral advection of both rocks and heat strongly influence the thermal structure of active orogens. Several research groups are currently developing more realistic models that integrate thermal, mechanical, and erosional processes to achieve a more generally applicable method for deducing exhumation rate from the range of detrital mineral ages in modern sediment samples.

Defining the positions of young deformational features

In some cases, the differences in t_{range} for sediments from modern catchments are large enough to allow the definitions of patterns of differential bedrock uplift that are related to deformational features. For example, Wobus and others (2003) used detrital muscovite $^{40}Ar/^{39}Ar$ dating from the Burhi Gandaki and Trisuli river drainages in central Nepal to define better the nature of the steep, roughly E-W-trending topographic transition from the high Himalayan ranges to their southern foothills to the south. Near the transition, small tributaries enter the major rivers at high angles in a trellised pattern. As a consequence, several of the tributary catchments drain regions north of the transition, whereas others drain regions south of the transition.

Figure 5 shows the SPDF curves for detrital muscovite $^{40}Ar/^{39}Ar$ dates from modern stream sediments collected from these catchments. The curves are markedly different on either side of the transition: almost all grains from tributaries to the north give Miocene or younger apparent ages, while grains from tributaries to the south give Paleozoic or Proterozoic apparent ages. The detrital thermochronologic data suggest that the physiographic transition marks the position of a recently active, south-vergent thrust fault that accommodates uplift of the high Himalayan ranges relative to their foothills. This interpretation gains additional support from cosmogenic ^{10}Be data for sands from the tributary catchments; millennial-timescale erosion rates calculated from those data are four times higher north of the physiographic transition than south of it (Wobus et al. 2005). Although exposures are poor along the transition in the Burhi Gandaki and Trisuli drainages, mapping to the west along strike demonstrates the existence of a system of Pliocene-Quaternary thrust faults with the necessary kinematics to explain the detrital thermochronologic data (Hodges et al. 2004).

FUTURE DIRECTIONS

Detrital mineral $^{40}Ar/^{39}Ar$ thermochronology offers a straightforward, powerful way to explore the tectonic and erosional evolution of orogenic systems and the fluvial networks that transport sediments through them. Because the nominal closure temperatures of $^{40}Ar/^{39}Ar$

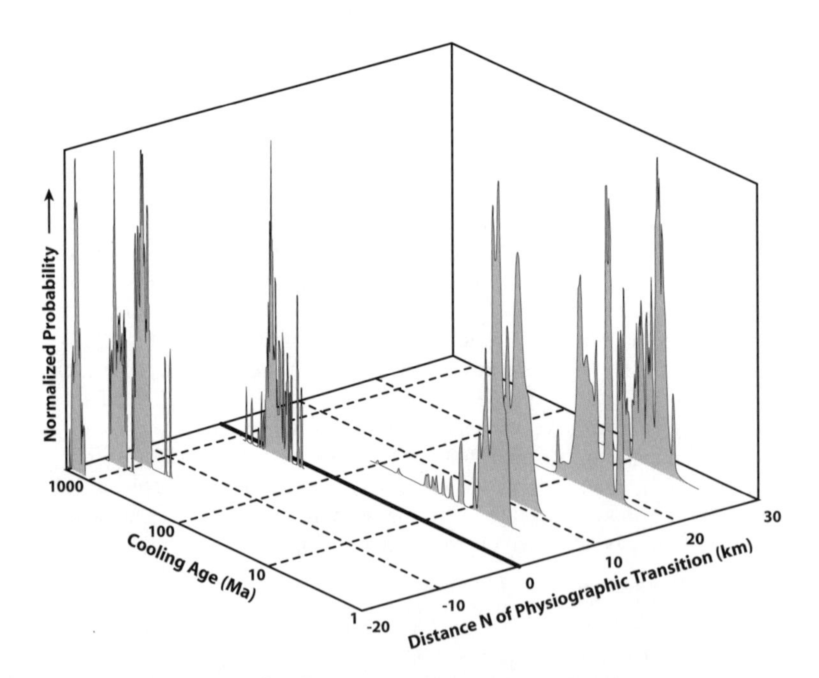

Figure 5. Stacked SPDF plots for $^{40}Ar/^{39}Ar$ muscovite dates from detrital samples from small catchments north and south of the physiographic transition separating the Higher and Lower Himalayan ranges in the Ganesh region of central Nepal, after Wobus et al. (2003). Note the logarithmic cooling age scale of the SPDF plots. The thick solid line at the physiographic transition marks a profound change in detrital cooling ages that is interpreted as a recently active, surface-breaking fault.

thermochronometers occur at mid-crustal levels, their use is particularly informative regarding long-term erosion rates in orogenic settings. Fission-track and (U-Th)/He thermochronometry of detrital accessory minerals, discussed elsewhere in this volume, provides similar information for regarding the cooling of source regions through lower-temperature, near-surface closure isotherms. The U-Pb zircon geochronometer has an extremely high closure temperature and is resistant to thermal resetting; thus, detrital zircon ages are seldom indicative of the thermal evolution of source regions during the most recent phase of mountain building (e.g., DeCelles et al. 2000). There have been some notable attempts to integrate some of these techniques to establish a better understanding of source regions and their thermal and erosional histories (e.g., Adams and Kelley 1998; Clift et al. 2004; Haines et al. 2004), but a systematic integration has yet to be done. A particularly high priority, for example, is the use of low-temperature thermochronometers to evaluate the tectonic significance of significant lag times documented by detrital mica $^{40}Ar/^{39}Ar$ thermochronometry for proximal deposits in the forelands of rapidly unroofed orogens like the Himalaya (White et al. 2002) .

Although the application of $^{40}Ar/^{39}Ar$ dating to ancient sediments has been highly successful, the method has only recently been used in the study of modern sediments. This is a particularly fruitful arena for research. For example, more aggressive dating campaigns in active foreland basins and river systems can help elucidate erosional patterns in developing orogens. Temporal variations in sediment storage in active river systems can be studied by applying the technique to fluvial terrace deposits, the ages of which might be determined by ^{14}C or cosmogenic dating methods. Moreover, the spatial patterns of glacial erosion can be addressed by detailed dating studies of moraine deposits.

As we learn more and more about the relationship between catchment-scale erosional processes and the apparent age distributions found in modern sediments, these lessons can be used to refine how we use data from more sedimentary rocks to study ancient landscapes. Stock and Montgomery (1996) anticipated such studies by suggesting how detrital thermochronologic data could be used to deduce paleorelief. While their method requires the assumption of a known geothermal gradient, a somewhat less dangerous assumption is that changes in geothermal gradient occur at slower rates than changes in relief, such that studies of changes in age ranges from one stratigraphic level to another can provide more reliable information about the evolution of relief in a sediment source region. Such unexploited opportunities ensure that detrital deposits will continue to be an attractive target for $^{40}Ar/^{39}Ar$ thermochronology.

REFERENCES

Aalto KR, Sharp WD, Renne PR (1998) $^{40}Ar/^{39}Ar$ dating of detrital micas from Oligocene-Pleistocene sandstones of the Olympic Peninsula, Klamath Mountains, and northern California Coast Ranges: provenance and paleodrainage patterns. Can J Earth Sci 35(7):735-745

Adams CJ, Kelley S (1998) Provenance of Permian-Triassic and Ordovician metagraywacke terranes in New Zealand; evidence from $^{40}Ar/^{39}Ar$ dating of detrital micas. Geol Soc Am Bull 110:422-432

Barbieri C, Carrapa B, Di Giulio A, Wijbrans J, Murrell GR (2003) Provenance of Oligocene synorogenic sediments of the Ligurian Alps (NW Italy): Inferences on belt age and cooling history. Int J Earth Sci 92: 758-778

Batt GE, Braun J (1997) On the thermomechanical evolution of compressional orogens. Geophys J Int 128: 364-382

Bernet M, Garver JI (2005) Fission-track analysis of detrital zircon. Rev Mineral Geochem 58:205-238

Bevington PR, Robinson DK (1992) Data Reduction and Error Analysis for the Physical Sciences. McGraw-Hill, Inc., New York

Brandon MT, Vance JA (1992) Fission-track ages of detrital zircon grains: Implications for the tectonic evolution of the Cenozoic Olympic subduction complex. Am J Sci 292:565-636

Braun J, Sambridge M (1997) Modelling landscape evolution on geological time scales: A new method based on irregular spatial discretization. Basin Res 9:27-52

Brewer ID (2001) Detrital-Mineral Thermochronology: Investigations of Orogenic Denudation in the Himalaya of Central Nepal, p. 181. The Pennsylvania State University, State College, PA.

Brewer ID, Burbank DW, Hodges KV (2003) Modelling detrital cooling-age populations: insights from two Himalayan catchments. Basin Res 15:305-320

Brewer ID, Burbank DW, Hodges KV (2005) Downstream development of a detrital cooling-age signal: insights from $^{40}Ar/^{39}Ar$ muscovite thermochronology in the Nepalese Himalaya. *In*: Tectonics, Climate, and Landscape Evolution. Geological Society of America Special Paper. Willett SD, Hovius N, Brandon MT, Fisher D (eds) Geological Society of America, Denver, in press

Carrapa B, Di Giulio A, Wijbrans J (2004a) The early stages of the Alpine collision: an image derived from the upper Eocene–lower Oligocene record in the Alps-Apennines junction area. Sed Geol 171:181-203

Carrapa B, Wijbrans J, Bertotti G (2003) Episodic exhumation in the Western Alps. Geology 31(7):601-604

Carrapa B, Wijbrans J, Bertotti G (2004b) Detecting differences in cooling/exhumation patterns within the Western Alpine arc through $^{40}Ar/^{39}Ar$ thermochronology on detrital minerals (Tertiary Piedmont Basin, NW Italy). *In:* Detrital Thermochronology – Provenance Analysis, Exhumation, and Landscape Evolution of Mountain Belts. Geological Society of America Special Paper 378. Bernet M, Spiegel C (eds.) Geological Society of America, Denver, p 67-103

Cerveny PF, Naeser ND, Zeitler PK, Naeser CW, Johnson NM (1988) History of uplift and relief of the Himalaya during the past 18 million years; evidence from sandstones of the Siwalik Group. *In:* New Perspectives in Basin Analysis. Kleinspehn KL, Paola C (eds) Springer-Verlag, New York, p 43-61

Cherniak DJ, Watson EB (2000) Pb diffusion in zircon. Chem Geol 172:5-24

Clift PD, Campbell IH, Pringle MS, Carter A, Zhang X, Hodges KV, Khan AA, Allen CM (2004) Thermochronology of the modern Indus River bedload: new insight into the controls on the marine stratigraphic record. Tectonics 23, doi:10.1029/2003TC001559

Copeland P, Harrison TM (1990) Episodic rapid uplift in the Himalaya revealed by $^{40}Ar/^{39}Ar$ analysis of detrital K-feldspar and muscovite, Bengal Fan. Geology 18:354-357

Dallmeyer RD, Keppie JD, Nance RD (1997) $^{40}Ar/^{39}Ar$ ages of detrital muscovite within Lower Cambrian and Carboniferous clastic sequences in northern Nova Scotia and southern New Brunswick; applications for provenance regions. Can J Earth Sci 34(2):156-168

Dallmeyer RD, Neubauer F (1994) Cadomian $^{40}Ar/^{39}Ar$ apparent age spectra of detrital muscovites from the Eastern Alps. J Geol Soc London 151:591-598

DeCelles PG, Gehrels GE, Quade J, LaReau B, Spurlin M (2000) Tectonic implications of U-Pb zircon ages of the Himalayan orogenic belt in Nepal. Science 288:497-499

Di Vincenzo G, Viti C, Rocchi S (2003) The effect of chlorite interlayering on 40Ar-39Ar biotite dating: an ^{40}Ar-^{39}Ar laser-probe and TEM investigations of variably chloritised biotites. Contrib Mineral Petrol 145: 643-658

Dong H, Peacor DR, Murphy SF (1998) TEM study of progressive alteration of igneous biotite to kaolinite throughout a weathered soild profile. Geochim Cosmochim Acta 62:1881-1887

Ehlers TA, Farley KA (2003) Apatite (U-Th)/He thermochronometry; methods and applications to problems in tectonic and surface processes. Earth Planet Sci Lett 206:1-14

Galbraith RF, Green PF (1990) Estimating the component ages in a finite mixture. Nucl Tracks Rad Meas 17: 197-206

Gehrels GE (2000) Introduction to detrital zircon studies of Paleozoic and Triassic strata in western Nevada and northern California. *In*: Paleozoic and Triassic Paleogeography and Tectonics of Western Nevada and Northern California. Geological Society of America Special Paper 347. Soreghan MJ, Gehrels GE (eds) Geological Society of America, Boulder, p 1-17

Gehrels GE, Stewart JH, Kettner KB (2002) Cordilleran-margin quartzites in Baja California – implications for tectonic transport. Earth Planet Sci Lett 199:201-210

Grimmer JC, Ratschbacher L, McWilliams M, Franz,L, Gaitzsch I, Tichomirowa M, Hacker BR, Yueqiao Z (2003) When did the ultrahigh-pressure rocks reach the surface? A $^{207}Pb/^{206}Pb$ zircon, $^{40}Ar/^{39}Ar$ white mica, Si-in-white mica, single-grain provenance study of Dabie Shan synorogenic foreland sediments. Chem Geol 197(1-4):87-110

Haines PW, Turner SP, Kelley SP, Wartho JA, Sherlock SC (2004) ^{40}Ar-^{39}Ar dating of detrital muscovite in provenance investigations: a case study from the Adelaide Rift Complex, South Australia. Earth Planet Sci Lett 227:297-311

Hames WE, Hodges KV (1993) Laser $^{40}Ar/^{39}Ar$ evaluation of slow cooling and episodic loss of ^{40}Ar from a sample of polymetamorphic muscovite. Science 261:1721-1723

Harrison TM, Copeland P, Hall SA, Quade J, Burner S, Ojha TP, Kidd WSF (1993) Isotopic preservation of Himalayan/Tibetan uplift, denudation, and climatic histories of two molasse deposits. J Geol 101:157-175

Harrison TM, Zeitler PK (2005) Fundamentals of noble gas thermochronometry. Rev Mineral Geochem 58: 123-149

Heller PL, Renne PR, O'Neil JR (1992) River mixing rate, residence time,, and subsidence rates from isotopic indicators: Eocene sandstones of the U.S. Pacific Northwest. Geology 20:1095-1098

Hemming SR, Bond GC, Broecker WS, Sharp WD, Klas-Mendelson M (2000) Evidence from $^{40}Ar/^{39}Ar$ ages of individual hornblende grains for varying Laurentide sources of iceberg discharges 22,000 to 10,500 yr B.P. Quaternary Res 54:372-383

Hemming SR, Hall CM, Biscaye PE, Higgins SM, Bond GC, McManus JF, Barber DC, Andrews JT, Broecker WS (2002) $^{40}Ar/^{39}Ar$ ages and $^{40}Ar^*$ concentrations of fine-grained sediment fractions from North Atlantic Heinrich layers. Chem Geol 182:583-603

Hodges K, Wobus C, Ruhl K, Schildgen T, Whipple K (2004) Quaternary deformation, river steepening, and heavy precipitation at the front of the Higher Himalayan ranges. Earth Planet Sci Lett 220:379-389

Hodges KV (1998) $^{40}Ar/^{39}Ar$ geochronology using the laser microprobe. *In*: Reviews in Economic Geology 7: Applications of Microanalytical Techniques to Understanding Mineralizing Processes. McKibben MA, Shanks WC (eds) Society of Economic Geologists, Tuscaloosa, p 53-72

Hodges KV (2003) Geochronology and Thermochronology in Orogenic Systems. *In*: The Crust, 3. Rudnick RL (ed) Elsevier Science, Amsterdam, p 263-292

Hodges KV, Bowring SA (1995) $^{40}Ar/^{39}Ar$ thermochronology of isotopically zoned micas; insights from the southwestern USA Proterozoic orogen. Geochim Cosmochim Acta 59(15):3205-3220

Huerta AD, Royden LH, Hodges KV (1998) The thermal structure of collisional orogens as a response to accretion, erosion, and radiogenic heating. J Geophys Res-Solid Earth103(B7):15287-15302

Hunecke JC, Smith SP (1976) The realities of recoil: ^{39}Ar recoil out of small grains and anomalous patterns in ^{40}Ar-^{39}Ar dating. Geochim Cosmochim Acta, Suppl 7:1987-2008

Hutson FE, Mann P, Renne PR (1998) $^{40}Ar/^{39}Ar$ dating of single muscovite grains in Jurassic siliciclastic rocks (San Cayetano Formation); constraints on the paleoposition of western Cuba. Geology 26(1):83-86

Kelley S (2002) Excess argon in K–Ar and Ar–Ar geochronology. Chem Geol:188:1-22

Kelley S, Bluck, BJ (1989) Detrital mineral ages from the Southern Uplands using ^{40}Ar-^{39}Ar laser probe. J Geol Soc London146(3):401-403

Kelley SP, Bluck BJ (1992) Laser ^{40}Ar-^{39}Ar ages for individual detrital muscovites in the Southern Uplands of Scotland, U.K. Chem Geol 101(1-2):143-156

Kowalewski M, Rimstidt JD (2003) Average lifetime and age spectra of detrital grains: Toward a unifying theory of sedimentary particles. J Geology 111:427-439

Lague D, Crave A, Davy P (2003) Laboratory experiments simulating the geomorphic response to tectonic uplift. J Geophys Res 108, doi:10.1029/2002JB001785

Lin L-H, Onstott TC, Dong H (2000) Backscattered ^{39}Ar loss in fine-grained minerals: implications for $^{40}Ar/^{39}Ar$ geochronology of clay. Geochim Cosmochim Acta 64:3965-3974

Lovera OM, Grove M, Harrison TM (2002) Systematic analysis of K-feldspar Ar-40/Ar-39 step heating results II: relevance of laboratory argon diffusion properties to nature. Geochim Cosmochim Acta 66(7):1237-1255

Mancktelow NS, Grasemann B (1997) Time-dependent effects of heat advection and topography on cooling histories during erosion. Tectonophysics 270:167-195

Markley MJ, Teyssier C, Cosca M (2002) The relation between grain size and $^{40}Ar/^{39}Ar$ date for Alpine white mica from the Siviez-Mischabel Nappe, Switzerland. J Struct Geol 24(12):1937-1955

McDougall I, Harrison TM (1998) Geochronology and Thermochronology by the $^{40}Ar/^{39}Ar$ Method. Oxford University Press, New York

Mitchell JG, Penven MJ, Ineson PR, Miller JA (1988) Radiogenic argon and major-element loss from biotite during natural weathering – a geochemical approach to the interpretation of potassium-argon ages of detrital biotite. Chem Geol 72:111-126

Murphy SF, Brantley SL, Blum AE, White AF, Dong H (1998) Chemical weathering in a tropical watershed, Luquillo Mountains, Puerto Rico: II. Rate and mechanism of biotite weathering. Geochim Cosmochim Acta 62:227-243

Najman Y, Garzanti E, Pringle M, Bickle M, Stix J, Khan I (2003) Early-middle Miocene paleodrainage and tectonics in the Pakistan Himalaya. Geol Soc Am Bull 115(10):1265-1277

Najman Y, Pringle M, Godin L, Oliver G (2001) Dating of the oldest continental sediments from the Himalayan foreland basin. Nature 410(6825):194-197

Najman Y, Pringle M, Godin L, Oliver G (2002) A reinterpretation of the Balakot Formation: Implications for the tectonics of the NW Himalaya, Pakistan. Tectonics 21(5), doi:10.1029/2001TC001337

Najman YMR, Pringle MS, Johnson MRW, Robertson AHF, Wijbrans JR (1997) Laser Ar-40/Ar-39 dating of single detrital muscovite grains from early foreland-basin sedimentary deposits in India: Implications for early Himalayan evolution. Geology 25(6):535-538

Onstott TC, Miller ML, Ewing RC, Arnold GW, Walsh DS (1995) Recoil refinements: Implications for the $^{40}Ar/^{39}Ar$ dating technique. Geochim Cosmochim Acta 59:1821-1834

Parsons I, Brown WL, Smith JV (1999) $^{40}Ar/^{39}Ar$ thermochronology using alkali feldspars: real thermal history or mathematical mirage of microtexture? Contrib Mineral Petrol 136:92-110

Rahman MJJ, Faupl P (2003) $^{40}Ar/^{39}Ar$ multigrain dating of detrital white mica of sandstones of the Surma Group in the Sylhet Trough, Bengal Basin, Bangladesh. Sed Geol 155(3-4):383-392

Renne PR, Becker TA, Swapp SM (1990) $^{40}Ar/^{39}Ar$ laser-probe dating of detrital micas from the Montgomery Creek Formation, Northern California; clues to provenance, tectonics, and weathering processes. Geology 18(6):563-566

Renne PR, Swisher CC, Deino AL, Karner DB, Owens T, DePaolo DJ (1998) Intercalibration of standards, absolute ages and uncertainties in $^{40}Ar/^{39}Ar$ dating. Chem Geol (Isotope Geosciences) 145:117-152

Roberts HJ, Kelley SP, Dahl PS (2001) Obtaining geologically meaningful ^{40}Ar-^{39}Ar ages from altered biotite. Chem Geol 172(3-4):277-290

Ross GM, Bowring SA (1990) Detrital zircon geochronology of the Windermere Supergroup and the tectonic assembly of the southern Canadian Cordillera. J Geol 98:879-893

Ruhl KW, Hodges KV (2005) The use of detrital mineral cooling ages to evaluate steady-state assumptions in active orogens: an example from the central Nepalese Himalaya. Tectonics, in press

Ruiz GMH, Seward D, Winkler W (2004) Detrital thermochronology – a new perspective on hinterland tectonics, an example from the Andean Amazon Basin, Ecuador. Basin Res 16, doi:10.1111/j.1365-2117.2004.00239.x

Safran EB (2003) Geomorphic interpretation of low-temperature thermochronologic data: Insights from two-dimensional thermal modeling. J Geophys Res 108, doi:10.1029/2002JB001870

Sambridge MS, Compston W (1994) Mixture modeling of multicomponent data sets with application to ion probe zircon ages. Earth Planet Sci Lett 128: 373-390

Scaillet S (1996) Excess ^{40}Ar transport scale and mechanism in high-pressure phengites; a case study from an eclogitized metabasite of the Dora-Maira nappe, Western Alps. Geochim Cosmochim Acta 60:1075-1090

Sherlock SC (2001) Two-stage erosion and deposition in a continental margin setting; an $^{40}Ar/^{39}Ar$ laserprobe study of offshore detrital white micas in the Norwegian Sea. J Geol Soc London 158(5):793-799

Sherlock SC, Jones KA, Kelley SP (2002) Fingerprinting polyorogenic detritus using the $^{40}Ar/^{39}Ar$ ultraviolet laser microprobe. Geology 30(6):515-518

Sircombe KN (2004) AGEDISPLAY: an EXCEL workbook to evaluate and display univariate geochronological data using binned frequency histograms and probability density distributions. Computers & Geosciences 30(1):21-31

Sircombe KN, Hazelton ML (2004) Comparison of detrital zircon age distributions by kernel function estimation. Sed Geol 171:91-111

Stock JD, Montgomery DR (1996) Estimating paleorelief from detrital mineral age ranges. Basin Res 8:317-327

Stuart FM (2002) The exhumation history of orogenic belts from $^{40}Ar/^{39}Ar$ ages of detrital micas. Mineral Mag 66:121-135

Stuart FM, Bluck BJ, Pringle MS (2001) Detrital muscovite $^{40}Ar/^{39}Ar$ ages from Carboniferous sandstones of the British Isles: Provenance and implications for the uplift history of orogenic belts. Tectonics 20(2): 255-267

Stüwe K, Hintermüller M (2000) Topography and isotherms revisited: the influence of laterally migrating drainage divides. Earth Planet Sci Lett 184:287-303

Stüwe K, White L, Brown R (1994) The influence of eroding topography on steady-state isotherms. Application to fission track analysis. Earth Planet Sci Lett 124:63-74

Vermeesch P (2004) How many grains are needed for a provenance study? Earth Planet Sci Lett 224:441-451

von Eynatten H, Gaup R, Wijbrans JR (1996) $^{40}Ar/^{39}Ar$ laser-probe dating of detrital white micas from Cretaceous sedimentary rocks of Eastern Alps: evidence for Variscan high-pressure metamorphism and implications for Alpine Orogeny. Geology 24:691-694

von Eynatten H, Wijbrans J (2003) Precise tracing of exhumation and provenance using $^{40}Ar/^{39}Ar$ geochronology of detrital white mica: the example of the Central Alps. *In*: Tracing Tectonic Deformation Using the Sediment Record. Geological Society Special Publication 208. McCann T, Saintot A (eds) The Geological Society, London, p 289-305

Wagner GA, Reimer GM (1972) Fission track tectonics: The tectonic interpretation of fission track apatite ages. Earth Planet Sci Lett 14:263-268

Whipple KX (2001) Fluvial landscape response time: How plausible is steady-state denudation? Am J Sci 301: 313-325

White NM, Pringle M, Garzanti E, Bickle M, Najman Y, Chapman H, Friend P (2002) Constraints on the exhumation and erosion of the High Himalayan Slab, NW India, from foreland basin deposits. Earth Planet Sci Lett 195(1-2):29-44

Willett SD, Brandon MT (2002) On steady states in mountain belts. Geology 30:175-178

Wobus C, Heimsath A, Whipple K, Hodges K (2005) Active surface thrust faulting in the central Nepalese Himalaya. Nature 434:1008-1010

Wobus CW, Hodges KV, Whipple KX (2003) Has focused denudation sustained active thrusting at the Himalayan topographic front? Geology 31:861-864

Zeitler PK, Meltzer AS, Koons PO, Craw D, Hallet B, Chamberlain CP, Kidd WSF, Park SK, Seeber L, Bishop M, Shroder J (2001) Erosion, Himalayan geodynamics, and the geomorphology of metamorphism. GSA Today 11:4-9

Reviews in Mineralogy & Geochemistry
Vol. 58, pp. 259-274, 2005

10

Forward Modeling and Interpretation of (U-Th)/He Ages

Tibor J. Dunai

Faculty of Earth and Life Sciences
Vrije Universiteit
De Boelelaan 1085, 1081 HV
Amsterdam, The Netherlands
dunt@geo.vu.nl

INTRODUCTION

Soon after the initial studies on He-diffusion in apatite pointed out the use of the (U-Th)/He system for low temperature geochronology (Zeitler et al. 1987; Wolf et al. 1996; Warnock et al. 1997), the potential value of forward modeling for refined interpretation of (U-Th)/He dating was indicated (Wolf et al. 1998). While the first studies provided evidence for the low closure temperature (T_C, Dodson 1973) of the (U-Th)/He system in apatite for monotonously cooling systems, the latter study investigated the sensitivity of the (U-Th)/He system for a wide range of geologically feasible heating/cooling scenarios. Forward modeling of He-accumulation in minerals expanded the use of the (U-Th)/He geochromometer to geological systems that do not fit one of the main conditions for the meaningful application of the closure temperature concept, i.e. monotonous cooling at a constant rate through T_C. Forward modeling can help to give a meaning to apparent (U-Th)/He-ages from rocks with cooling histories with protracted residence time in the helium partial retention zone (HePRZ; e.g., ~40–85 °C in apatite; Wolf et al. 1998), where the closure temperature is undefined (Fig. 1).

There are many examples of successful applications of forward modeling to answering geological questions (House et al. 1997, 1998; Stockli et al. 2000; Batt et al. 2001; Reiners and Farley 2001; House et al. 2002; Persano et al. 2002; Ehlers et al. 2003; Pik et al. 2003). These, and others, are discussed in detail in various chapters of this volume (Armstrong 2005; McInnes et al. 2005; Spotila 2005; Stockli 2005). The scope of this review is to provide a guide to the safe use of forward modeling for the potential user, and to clarify aspects of forward modeling for practitioners in field. This chapter covers the correct translation of geometric sample parameters for the use in forward modeling algorithms. Further it describes the use of a comprehensive, but easy to use forward modeling program (DECOMP) that is included in this volume. Guidelines for the correct choice of geothermal and other boundary conditions for successful forward modeling are provided in other chapters of this volume (Braun 2005; Ehlers 2005).

FORWARD MODELING

General remarks

Forward modeling (FM) in (U-Th)/He thermochronology aims to find numerical solutions for equations describing He-accumulation in a mineral as a result of concurrent He-production by radioactive decay and He-loss by volume diffusion. The desired output in most FM exercises consists of model (U-Th)/He ages that result from a specific temperature history

1529-6466/05/0058-0010$05.00 DOI: 10.2138/rmg.2005.58.10

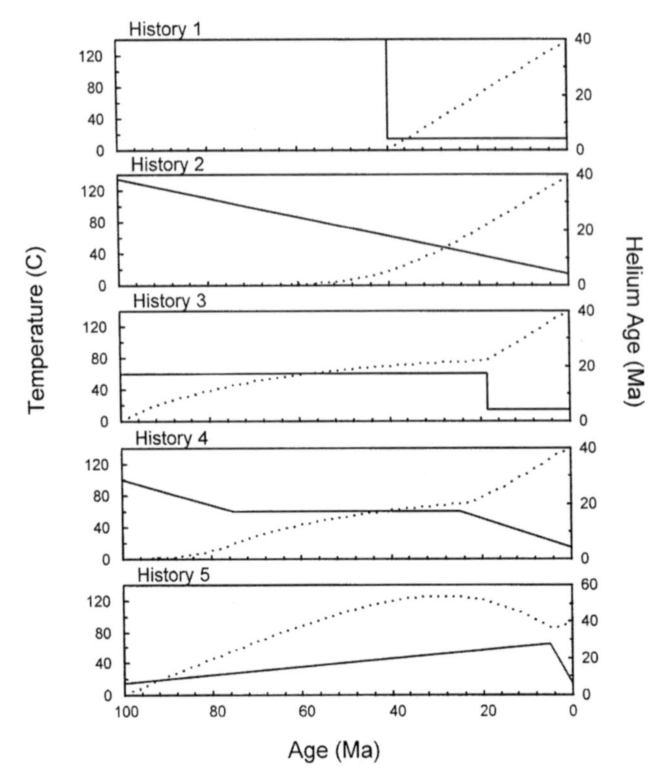

Figure 1. Apatite He-age evolution curves (dotted lines) for several representative time-temperature paths (solid lines). Forward modelling was conducted using the Crank-Nicolson method for spheres of 60 μm radius. Note that only for history 2 the closure temperature concept (Dodson 1973) is defined, and the age obtained relates to passing through a point (T_c) on the time-temperature path. For histories that are similar to 3–5, which know a protracted residence time in the HePRZ, there is no direct geological meaning of He-ages. Forward modelling is required to give a geological meaning to the apparent ages obtained from He and U, Th concentrations in samples. [Used with permission of Elsevier, from Wolf et al. (1998), *Chemical Geology*, 148, Fig.5, p. 109]

(Fig. 1). Usually the aim is to compare these model ages to measured (U-Th)/He ages and to narrow down the number of temperature histories that can explain the data. Helium production by radioactive decay is described by:

$$\left[{}^4He\right] = \sum_{j=1}^{4} \frac{p_j}{\lambda_j}\left(e^{\lambda_j t} - 1\right) \tag{1}$$

with

$$p_1 = 8\,\lambda_1\,[^{238}U]; \quad \lambda_1 = 1.55125 \times 10^{-10}\ a^{-1};$$
$$p_2 = 7\,\lambda_2\,[^{235}U]; \quad \lambda_2 = 9.8485 \times 10^{-10}\ a^{-1};$$
$$p_3 = 6\,\lambda_3\,[^{232}Th]; \quad \lambda_3 = 4.9475 \times 10^{-11}\ a^{-1};$$
$$p_4 = \lambda_4\,[^{147}Sm]; \quad \lambda_4 = 6.539 \times 10^{-12}\ a^{-1},$$

the number before the decay constants ($\lambda_{1\text{-}4}$) denote the number of alpha-particles (^{4}He-nuclei) emitted in the uranium and thorium decay chains. He-loss by volume diffusion can be described by an Arrhenius relationship:

$$D = D_0 e^{-E_a / RT} \tag{2}$$

where D is the diffusion coefficient ($m^2\ s^{-1}$), D_0 is the pre-exponential factor (the theoretical diffusion constant at infinite temperature, or a measure of "conduction" in the mineral), E_a is the activation energy (J), R is the gas constant ($J\ K^{-1}\ mol^{-1}$) and T is temperature (K). The activation energy E_a and the pre-exponential factor D_0 are usually experimentally derived values, and are different for each mineral (e.g., Dunai and Roselieb 1996; Reiners and Farley 1999; Farley 2000; Reiners et al. 2004).

There are several different mathematical recipes that can be used for FM. For the end-user the detailed nature of these differences is usually irrelevant. However, these differences are the reason why certain approaches are more versatile than others. In the following the two main approaches used in FM of production-diffusion of He are briefly discussed.

Finite difference methods such as the Crank-Nicolson method (Crank and Nicolson 1947; Crank 1975; Wolf et al. 1998; Ketcham 2005) are useful to describe coupled production-diffusion of He in systems that can be reduced mathematically to one dimension. This is the case for spheres, *infinite* plane sheets and *infinite* cylinders. For all other forms where such a reduction is impossible finite difference models become comparatively very slow and practically unfeasible (Meesters and Dunai 2002a).

Decomposition into eigenmodes of production-diffusion equations can be applied to the same shapes as finite difference methods (Lovera et al. 1989), however, also to *finite* cylinders and rectangular blocks of any shape (Meesters and Dunai 2002a). Moreover, this method allows the coupled treatment of effects of zonation, alpha-ejection and production-diffusion (Meesters and Dunai 2002b).

While these two principal pathways in FM are different, it is important to note that, when modeling the same system (e.g., spheres) they give the same result (Meesters and Dunai 2002a). When carefully applied, i.e. paying attention to inherent limitations, both methods can be used with advantage. Later in this section it will be explained which aspects require attention.

Effect of shape and surface/volume ratio

The significance of shape and surface/volume (S/V) ratios for the *quantitative* description of diffusion processes was first described in detail by Lagerwall and Zimen (1964). Their systematic assessment shows the overruling importance of the surface to volume ratio for bodies of modest aspect ratios, and a strong influence of shape for bodies of large aspect ratios (e.g., very elongate shape). In a nutshell, their work demonstrates that the characteristics of diffusive loss from two bodies with the same surface/volume ratio are essentially the same, with the exception of elongate shapes. The way this is incorporated in numerical models is crucial, since half of the effort in FM is the quantitative description of He-loss by diffusion.

To gain computational speed and for convenience of the end-user, the shape of crystals is usually approximated by simpler geometric forms. For apatite prisms infinite cylinders (Farley 2000) or spheres of identical radius (Ehlers et al. 2003) or spheres of identical S/V-ratio (Meesters and Dunai 2002a) are used. For the following discussion on the effect of shape and S/V ratio, the sphere is used as reference shape.

Bodies of identical minimum dimension. The diffusion of He from spheres and cylinders of the same diameter, examples of bodies of identical minimum dimension (if length of a prism is equal or greater than its diameter), can differ widely. This is governed by the difference in S/V -ratios (Table 1). Nevertheless, commonly the average radii of apatite prisms are used as radii of spheres applied in FM calculations. Intuitively this simplification is attractive, however, in practice it may have undesirable effects (i.e., wrong model ages). The following examples illustrate the effects of ignoring differences in S/V -ratios.

Table 1. The effect of surface to volume ratio for various shapes with identical minimum dimension

Shape	*Minimum Dimension* [μm]	*Other Dimensions* [μm]	*S/V* [1/μm]	*% Δ age relative to sphere Temperature history # in Fig. 1*				
				1	2	3	4	5
Sphere	120 (Ø)	—	0.05	0	0	0	0	0
Cube	120	120	0.05	0	1	5	5	6
Cylinder	120 (Ø)	240	0.0417	0	4	14	15	21
Inf. Cylinder	120 (Ø)	∞	0.0333	0	8	27	25	37
Block	120	240 × 240	0.0333	0	11	34	30	43

In Table 1 the relative differences between model ages of various shapes relative to those obtained for a sphere are shown. The minimum dimension of these bodies is identical to the spheres used to calculate model ages depicted in Figure 1. Also the same thermal histories (#1–5) as in Figure 1 are used for this comparison. The age differences refer to the final ages (i.e., 40 Ma in Fig. 1). The results in Table 1 have been obtained using the algorithms of Meesters and Dunai (2002a).

Two issues are evident from Table 1. Firstly, all bodies with a lower *S*/*V*-ratio than a sphere have a significantly higher (4–43%) model ages than would be obtained when approximating these bodies to spheres of the same minimum dimension. Such an approximation is, therefore, not recommended. Secondly, the relative difference increases with the duration of the passage through the HePRZ. However, even monotonous cooling (history #2) creates differences between 4–11%. Only for temperature histories with quasi-instantaneous cooling through the HePRZ (temperature histories like #1) produce no difference (because diffusion is irrelevant). Note that even cylinders with a very modest aspect ratios (1:2) behave quite differently (4–21% age difference) from spheres of the same radius. With increasing aspect ratio these differences further increase. The subordinate, but still significant effect of shape can be deduced from the fact that bodies of identical *S*/*V* ratios in the above example produce varying relative age differences.

Bodies of identical S/V-ratio. As seen above, bodies of same *S*/*V*-ratio have a smaller relative age difference than bodies of a lower *S*/*V* ratio when compared to a sphere of the same minimum dimensions. The following experiment illustrates the extent to which this finding can be generalized (Table 2). The shapes and thermal histories used are the same as in the previous experiment; however, *S*/*V*-ratios are equalized to that of the sphere by reducing the minimum dimensions. The size reduction is performed in a way that the aspect ratios remain constant. Note that the reduction in dimensions required to obtain the same *S*/*V*-ratio is generally significant (16–31%).

The results of the FM using identical *S*/*V*-ratios (Table 2) show much-reduced relative age differences as compared to the first experiment (Table 1). The maximum relative age differences range between −7 and 6%. As in the previous experiment, relative differences increase with the duration of the passage through the HePRZ. The *finite* cylinder shows the smallest difference compared to the sphere (max. 2% deviation), whereas the *infinite* cylinder shows the largest deviations of all shapes modeled (−3 to −7%). As a preliminary conclusion of the two numerical experiments (Tables 1 and 2) it can be stated that (i) a sphere can be used to calculate He loss from other shapes of equal *S*/*V*, and (ii) this approximation works particularly well for the finite cylinder.

Table 2. The effect of shape for bodies of identical surface to volume ratio.

Shape	*Minimum Dimension* [μm]	*Other Dimensions* [μm]	*S/V* [1/μm]	*% Δ age relative to sphere Temperature history # in Fig. 1*				
				1	2	3	4	5
Sphere	120 (Ø)	—	0.05	0	0	0	0	0
Cube	120	120	0.05	0	1	5	5	6
Cylinder	100 (Ø)	200	0.05	0	<1	1	1	2
Inf. Cylinder	80 (Ø)	∞	0.05	0	−3	−6	−5	−7
Block	80	160 × 160	0.05	0	<1	3	3	4

Simultaneous treatment of alpha ejection and diffusion

The long stopping distances of α-particles, ~20 μm in the case of apatite (Ziegler 1977; Farley et al. 1996), cause a relative depletion of the outer zones of a crystal since a proportion of the α-particles leave the crystal (Fig. 2). Farley et al. (1996) have discussed in detail how to correct for this effect in the absence of diffusion. In the presence of diffusion, it is currently customary to first perform FM using a (spherical) crystal and, subsequently, multiply the result with a correction factor for α-ejection (F_T; Farley et al. 1996). However, the strong depletion of He in the outer parts of a crystal due to α-ejection decreases the relative degree of diffusive loss of He as compared to a rectangular concentration profile that would be valid in the absence of α-ejection. From this it follows that the two processes (α-ejection and diffusion) are interwoven because the first process affects the spatial distribution of He and the effect of the second process depends on this spatial distribution. Consequently, simultaneous treatment of the concentration profile that is modified by α-ejection and diffusive loss is required (Meesters and Dunai 2002b). Failing to do so, i.e., applying a correction for α-ejection *after* FM (or vice versa), will over-correct ages for samples that spend significant length of time in the HePRZ.

In the following numerical experiment this effect is illustrated. In Table 3 the age differences between the simultaneous and sequential treatment of α-ejection in FM are given. These ages were calculated for spheres using the algorithms of Meesters and Dunai (2002). The results obtained for spheres can be generalized for other shapes (Meesters and Dunai 2002b). The temperature histories and the dimensions of the spheres are the same as in the previous experiments. Only for temperature histories with quasi-instantaneous or monotonous cooling through the HePRZ (histories #1 & 2) there is no resolvable difference between ages obtained. In all other instances the "conventional" sequential treatment results in a 4–10% overcorrection of ages relative to ages obtained by the correct simultaneous treatment of α-ejection and diffusion. Note that this error increases with the *S/V* ratio, i.e. it is larger for bodies that are smaller than the body, a sphere of 120 μm diameter, that serves as a model here (Table 3).

Fortunately a large proportion of thermochronological studies using (U-Th)/He are carried out in areas where rocks, that are now sampled at the surface, moved swiftly through the HePRZ (e.g., in active mountain belts that are at or close to steady state erosion). In these settings there is no resolvable difference between the sequential and the simultaneous treatment of α-ejection and diffusion. However, it also has to be noted that in these situations FM is overkill, as the closure temperature concept of Dodson (1973) is defined in these situations and will provide exactly the same ages as FM for a given cooling-trajectory. In all situations where FM is required to give meaning to apparent (U-Th)/He ages, e.g., temperature histories with periods of heating and isothermal holding in the HePRZ (histories of types 3–5, Fig. 1), simultaneous treatment is required to obtain correct model ages.

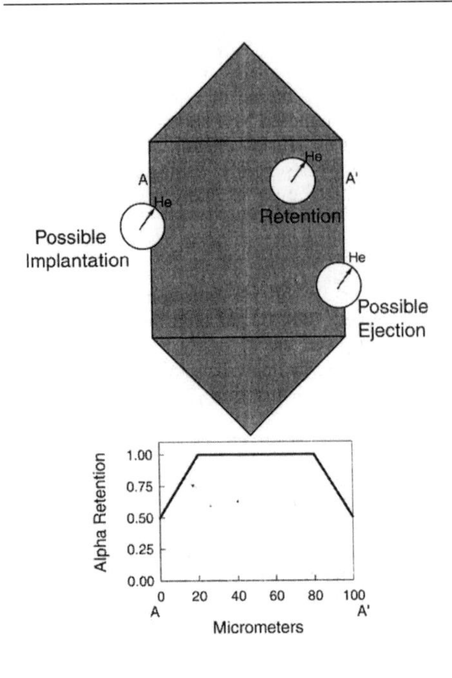

Figure 2. The effects of α-ejection on He-concentration profiles. The upper figure illustrates the three relevant possibilities with a schematic crystal: α-retention, possible α-ejection and possible α implantation. The center of the circle denotes the site of the parent U and Th nuclide, the edge of the white circles denote the locus of the points where α-particles may come to rest. The arrows indicate a possible trajectory. The lower plot shows schematically how α-ejection affects the α-retention along the path A-A'. Without ejection alpha retention would be 1 throughout the crystal, i.e. the He concentration profile would be rectangular, if the parent nuclides are homogenously distributed in the crystal. Most diffusion algorithms intrinsically assume an initially rectangular concentration profile. Exact equations describing the shape of the retention curve can be found in Farley et al. (1996).

Table 3. The effect of accounting for the concentration profile rounding by alpha ejection.

Temperature history # of Fig. 1	*% Δ age relative to "conventional" F_T corrected model age*
1	0
2	0
3	4
4	6
5	10

Considering parent nuclide distribution

Zoned distribution of U (and by inference also Th) is frequently observed in fission track analysis (Meesters and Dunai 2002b). If zonation of parent nuclides has been indicated for an apatite population that is used for (U-Th)/He dating and the measured ages are to be compared to results of FM, the effects of this zonation have to be considered. Zoned parent nuclide distribution has a two-fold influence on (U-Th)/He ages. The first effect is via its influence on the α-ejection correction (in the case of sequential treatment of ejection correction and FM); the second is via longer/shorter *average* diffusion pathways for helium leaving the crystals.

The influence of zonation on the α-correction is purely geometric and has been discussed by Farley et al. (1996). It can be summarized as follows: If the zonation is such that parent nuclides are concentrated in the outer zone of the crystal the standard F_T correction will under-correct ages; if the parent nuclides are concentrated in the core of the crystal the F_T correction will over-correct ages. For instance, if all parent nuclides are in the core and are one emission distance (~20 μm) away from the edge of the crystal no He will be lost by α-ejection and a F_T correction performed ignoring the zonation would introduce and error as large as the correction itself.

Zonation of the parent nuclides, in conjunction with the long stopping ranges of α-particles, leads to a highly irregular concentration profile that has to be considered in FM (Meesters and Dunai 2002b). Depending on where He is relatively more abundant as compared to a homogenous parent nuclide distribution, He loss will be faster or slower. When He is relatively more abundant closer to the edge of the crystal it will be faster and vice versa. Qualitatively this can be explained by the longer/shorter average distances He atoms have to move by diffusion before reaching the edge of the crystal. To assess the influence quantitatively a refined decomposition algorithm is required (Meesters and Dunai 2002b). In the case of a sequential FM/F_T treatment, assuming homogenous distribution of parent nuclides instead of a zoned distribution, the effects on diffusive loss of He can easily double the error of an erroneously conducted F_T correction as the errors work in the same direction. Examples for effects of parent nuclide zonation can be found in Meesters and Dunai (2002b).

It is currently not possible to quantitatively determine the parent nuclide zonation in the apatite grains that are used for (U-Th)/He dating. The information on zonation that can usually be obtained is qualitative only; mostly in the form of: (i) The population from which the analyzed apatite grains analyzed were handpicked contains zoned crystals and (ii) those crystals that are zoned have their parent nuclides mostly at their rims/cores. Therefore the treatment of zonation in FM is currently limited to a qualitative assessment of the consequences of zonation. However, due to the drastic effects parent nuclide zonation can have on measured and modeled ages, even a qualitative assessment can be advantageous to understanding thermal histories. The FM software described in the next section of this chapter can be used for such an assessment.

F_T correction vs. FM an apparent conflict resolved

The realization that the combined effect of α-ejection and diffusion can lead to overcorrection of ages when applying the F_T correction to FM results (Meesters and Dunai 2002b), raises the question whether the F_T correction should be applied to measured ages. The worry would be that the corrected ages of samples could be likewise over-corrected. In the following discussion, however, it will be shown that for commonly encountered cooling histories the F_T correction does not over-correct ages, and generally it should be applied to all reported sample ages.

Above it was established that for cooling histories with monotonous cooling through the HePRZ no discernible difference could be found between results of the two approaches. The same must therefore hold for the F_T correction of sample data of rocks of simple cooling histories. On trying out with DECOMP (program description below) it can be found that at least for cooling rates faster than 1 °C/Ma no significant over-correction does occur. Generally it can be stated that in situations where the closure temperature concept is defined (monotonous cooling through T_C) the F_T correction does not overcorrect ages.

In all other cases, however, where apparent (U-Th)/He ages do not have a simple meaning, because of protracted/non-trivial passage through in the HePRZ, the F_T correction will over-correct ages. Because in the latter cases the apparent ages do not have a direct geological meaning anyway, the over-correction usually will not harm the geological interpretation. In these situations FM can help provide a geological meaning to the measured ages. When the results of these FM exercises are compared to real data, (U-Th)/He ages uncorrected for α-ejection have to be used. For *reporting data*, however, in data tables or in diagrams, F_T corrected ages should be used.

A checklist for FM

It is clear that any FM algorithm/software is at best as good as the parameters it is fed with. In the sections above the main focus has been on *S*/*V*-ratios and shape of geometric forms used to approximate crystals in FMs. This perspective might appear limited, however, the radii of the model sphere or infinite cylinder that is used for actual FM calculations are the only sample dependent input parameters that can be fed into to all currently used FM software. The discussion has aimed to illustrate that (i) these input parameters are important, and (ii) that it is straightforward to choose the correct dimensions and geometry. Thus we have the first points for a short checklist for input parameters to FMs.

- Use spheres of the same *S*/*V* ratio as the real-life crystals of interest (definitely for typical apatites and zircons; also good for titanites and other non-needle like minerals)
- Do not use infinite cylinders to approximate (finite) prisms
- Check whether the expected temperature history permits a sequential treatment of -ejection and diffusion or whether a simultaneous treatment is required. The need of a simultaneous treatment might require the change to a different algorithm.
- If a sequential treatment of -ejection and diffusion is possible you may consider using the closure temperature concept instead of a FM, as it will give the same results.

Other input parameters can be parameters describing the diffusion process.

- Use current or generally accepted determinations of diffusion parameters of the relevant minerals e.g., (Reiners and Farley 1999; Farley 2000; Reiners et al. 2004), if not give a good justification.

The remaining input parameters are temperature histories that are based on geological/geothermal constraints, the output of thermo-kinematic models (Pik et al. 2003; ter Voorde et al. 2004; Braun 2005; Ehlers 2005) or the imagination of the user. The correct choice of these, admittedly most important parameters are beyond the scope of this section and are ultimately the responsibility of the users of FMs. The basis for a responsible treatment of these parameters is provided in various chapters (Armstrong 2005; Ehlers 2005; McInnes et al. 2005; Spotila 2005; Stockli 2005) of this volume. The range of this section is to avoid unnecessarily compromising results of FMs via poor choice of sample parameters.

DECOMP – A USER FRIENDLY FM SOFTWARE

As outlined in the introduction, forward modeling has a wide range of uses in low-temperature geochronology. However, the threshold for use of FM is commonly perceived as high by potential users. To overcome this situation *and* to provide the first program that is capable of simultaneous treatment the combined effects of -ejection and diffusion as well as zoned parent nuclide distribution, DECOMP was created (Bikker et al. 2002). This program uses the algorithms of Meesters and Dunai (2002a, 2002b) and its name is derived from the *decomp*osition into eigenmodes that is used in these algorithms. DECOMP can be downloaded from *http://www.minsocam.org/MSA/RIM*.

The target group for DECOMP are beginners as well as experienced practitioners in thermochronology. The functionality of the software lends itself for following applications:

- Qualitative assessment of how changes of sample parameters (diffusion parameters, *S*/*V*-ratios, emission distance) and model temperature histories affect ages. This aspect is particularly useful for teaching/learning what the most sensitive parameters for a certain system are.

- Quantitative FM of any time temperature history.
- Qualitative assessment of the effect of parent nuclide zonation. The calculation is quantitative, however, input will in most cases be qualitative (see previous section), consequently also the results.
- While DECOMP was originally designed for FM of (U-Th)/He ages, FM of any thermochronological system that is governed by volume diffusion, and of which the diffusion parameters of the radionuclide (D_0; E_a) are known, is possible.

A quick guide to DECOMP

The DECOMP program window is shown on Figure 3. In the following the main functions, input and output options are briefly described.

Button panel. Under the *save* and *open* buttons geometric parameters, temperature histories, constants and annotations can be saved or retrieved. The *quit* button is used to leave the program, temporarily saving geometric parameters, temperature histories, and constants for the next time the program is started.

The *recalculate* button starts the calculation of the age evolution diagram. The *edit constants* button gives access to a menu where the number of eigenvalues used for computation, the activation energy E_a and the pre-exponential factor D_0 can be changed. The units for input of E_a and D_0 are [cal/mol] and [cm^2/s], respectively. These non-SI units are used here as currently most values are still published in [cal/mol] or [cm^2/s] (For conversion 1 joule = 0.239 cal).

Input of parameters describing the geometry of a sample.

- *Sphere radius:* Radius (μm) of a sphere of identical surface to volume ratio as the crystals of interest.

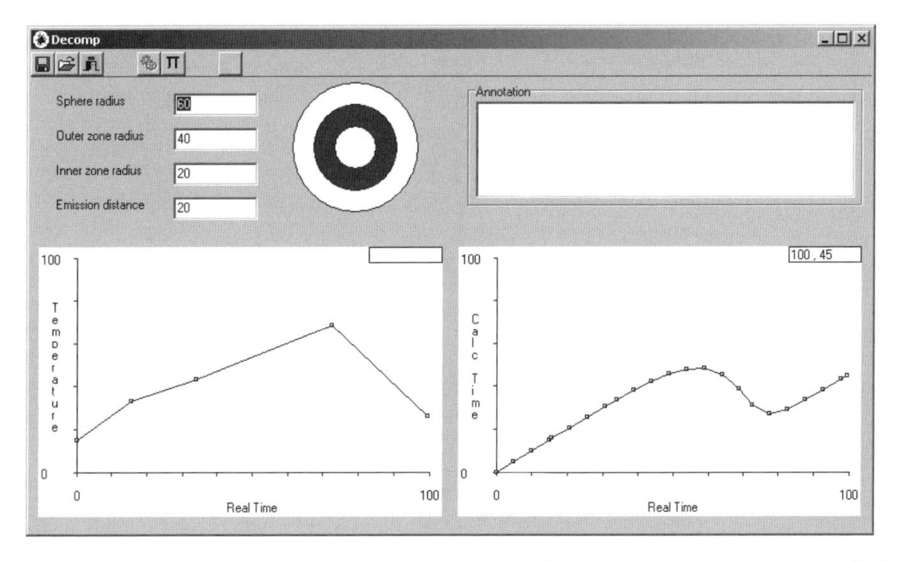

Figure 3. The DECOMP program window. Functionalities and formats for input and output are described in the text.

- *Outer* and *inner zone radius:* Gives the outer and inner radius (μm) of the zone containing the parent nuclides. The shape of the zone containing the parent nuclides is visualized by a red zone in the circle to the right of the input field that changes simultaneously while changing dimensions. Samples with a homogenous parent nuclide distribution have an outer zone radius that is equal to the sphere radius and an inner radius of 0 μm.
- *Emission distance:* Emission distance (μm) of α-particles. In order to conduct FM of thermochronological systems other than the (U-Th)/He system, the emission distance can be set to zero. As the program actually does not accept input of zero in this field a value close to zero (e.g., 0.000001) should be inserted in this case.

Temperature history diagram. In this diagram the user selected time [Ma] vs. temperature [°C] history is shown. The term "real time" signifies the time elapsed since beginning of the model. In this window there is a left and right mouse button functionality.

- The right mouse button gives access to following menu:

  copy the numeric values of the temperature history that can be pasted as a comma delimited table e.g., in a spreadsheet program.

  add and delete points to/from the temperature history

  table edit: edit the numeric values of existing points of the temperature history
- The left mouse button allows dragging of the diagram axis to scale and existing nodes of the temperature history to new values.
- The temperature and "real time" at the position of the tip of the cursor arrow are indicated at the top right of the diagram.

Age evolution diagram. In this diagram real time [Ma] vs. calculated age [Ma] is shown. The calculated age is the age as would be calculated from U, Th and He concentrations in a sample (i.e., raw ages, without F_T correction).

Initially this diagram is blank, on pressing the *recalculate* button the age evolution is calculated for the current temperature history and sample parameters. The numeric values of the curve can be copied (right mouse button) and pasted into a spreadsheet program (table comma delimited). The calculated age and real time at the position of the tip of the cursor arrow are given in the frame at the top right of the diagram. The left mouse button allows dragging the diagram axis to scale.

Annotation field. This field can be used for annotations to be saved with the model parameters (save with the save button).

EVALUATION OF SAMPLE DATA BY FORWARD MODELING

Of the many applications FM already had in (U-Th)/He thermochronology (see Introduction) two exemplary studies are highlighted in the following. These studies serve as examples for the use of FM (i) to *qualitatively* test/falsify predictions of hypothesis (Persano et al. 2002) and (ii) *quantify* rates of processes in a well-defined morphotectonic setting (Ehlers et al. 2003).

Qualitative evaluation of competing hypothesis

Qualitative evaluations are often sufficient to test first order geomorphological concepts. The study of Persano et al. (2002) for instance investigates competing hypothesis on the development of the Great Escarpment on the southeastern Australian passive margin following continental breakup. While it is itself a local study, the hypothesis tested are competing models for passive margin evolution in general.

Two principal hypotheses are put forward to explain the evolution of the Great Escarpment (Bishop and Goldrick 2000). One argues that the outer edge of the margin was tectonically lowered by down-warping and/or faulting soon after breakup and the escarpment retreated into this down-warped/down-faulted plateau (Ollier 1982; Seidl et al. 1996; Ollier and Pain 1997). The alternative hypotheses does not accept any post-break-up tectonic lowering of the margin and, instead, suggests that the escarpment has retreated across a coastal plain, accompanied by flexural isostatic rebound (Bishop and Goldrick 2000). The latter hypothesis knows two sub-scenarios, one is a down-wearing of the rift shoulder the other an escarpment retreat (Fig. 4). Using FM each of the hypotheses allows predictions on the thermal evolution of present day surface rocks and corresponding model (U-Th)/He ages. The *predicted age patterns* (Fig. 4) for transects perpendicular to the escarpment are sufficiently different to allow a distinction between the two principal hypotheses.

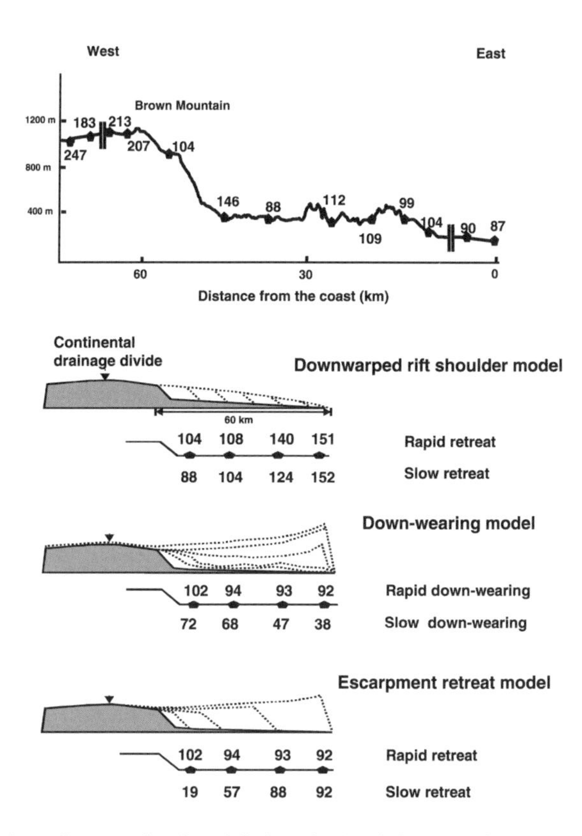

Figure 4. Comparison of measured and modelled apatite (U-Th)/He ages along an E-W transect through the southeast Australian continental margin. The upper panel gives the measured ages, with vertical lines separating samples that were collected along the transect from samples taken along the coast and the plateau hinterland. The lower panel shows schematic cross sections showing different models for the generation of escarpments on passive continental margins. Forward modelled apatite He-ages are given for each model, one for rapid (escarpment formed within the first 5 Ma after rifting 95 Ma ago) and for slow erosion (constant erosion rates since 95 Ma). The dashed lines refer to the amount and style of denudation. Discussion is given in the text. [Used with permission of Elsevier, from Persano et al. (2002) *Earth and Planetary Science Letters*, 200, 79-90, Fig. 2 p. 84 & Fig.3 p. 86]

Comparing the predicted age patterns to measured ages (Fig. 4) it is evident that the He age data are inconsistent with plausible erosion scenarios of a down-warped rift margin, thereby *falsifying* this *hypothesis* for landscape development in the region. Further the data are inconsistent with a constant (slow) post-break-up rate of lateral escarpment retreat across the coastal plain or by constant down-wearing. The data are, however, consistent with either rapid escarpment retreat or rapid in-place excavation of the escarpment commencing soon after continental break-up. Slow escarpment formation can be ruled out in any case, thereby providing valuable kinetic information. The study of Persano et al. (2002) is good example of hypothesis testing, as it is actually possible to rule out one of the competing concepts (Ollier 1982; Seidl et al. 1996; Ollier and Pain 1997). Further the study provides bounds on the process speed of possible alternative concepts.

Quantification of process rates and model parameters

In areas where the principal modes of tectono-morphological evolution have been established, FM can be used to constrain process rates and values of the principal parameters. Ehlers et al. (2003) investigate the Wasatch fault in Utah, USA, marking the boundary between the stable Colorado Plateau to the east and the extending Basin and Range to the west. Aim of the study is to better understand the thermo-kinematic evolution of this fault to eventually provide insights into intracontinental extensional tectonics and deformation processes in other rift zones.

For their study Ehlers et al. (2003) use apatite fission track *and* (U-Th)/He data to quantify process rates and parameters of a 2-D thermokinematic model describing the Wasatch fault. Quantities constrained are (i) the spatial and temporal variability of exhumation and erosion rates, (ii) the geometry of footwall tilt, (iii) the fault dip angle, and (iv) the magnitude and duration of exhumation.

To constrain these quantities, (i) – (iv), both a visual comparison between predicted and measured ages (Fig. 5) as well as a reduced χ^2 measure of fit are used (Fig. 6). The visual, 2-D, comparison is convincing for well fitting or badly fitting model results, however, the results of the reduced χ^2 measure of fit are essential to cover a more comprehensive range of results in the model space. The first is intuitively more accessible and the latter is crucial to obtain well-constrained numeric values. The study of Ehlers et al. (2003) nicely shows the value of the combined use of these two tools.

The reduced χ^2 measure of fit as used by Ehlers et al. (2003) can be generally applied to evaluate the parameters of a FM exercise, fitting predicted and observed ages.

$$\chi^2 = \sqrt{\frac{\sum_{i=1}^{N}\left(\frac{Age_{p_i} - Age_{o_i}}{U_i}\right)^2}{N - M}} \quad (3)$$

Age_{pi} and Age_{oi} are the predicted and observed ages for the ith point, respectively, U_i is the one standard deviation uncertainty in the ith age; N is the number of samples; and M is the number of model parameters. χ^2 provides an unbiased estimator of the root mean square error and a quantified measure of fit.

OUTLOOK FOR FORWARD MODELING

The two examples of FM discussed in the previous section (Persano et al. 2002; Ehlers et al. 2003) used 2D-thermokinetic modeling to predict time-temperature paths and ultimately (U-Th)/He ages. The questions addressed in these studies could be reduced to 2D

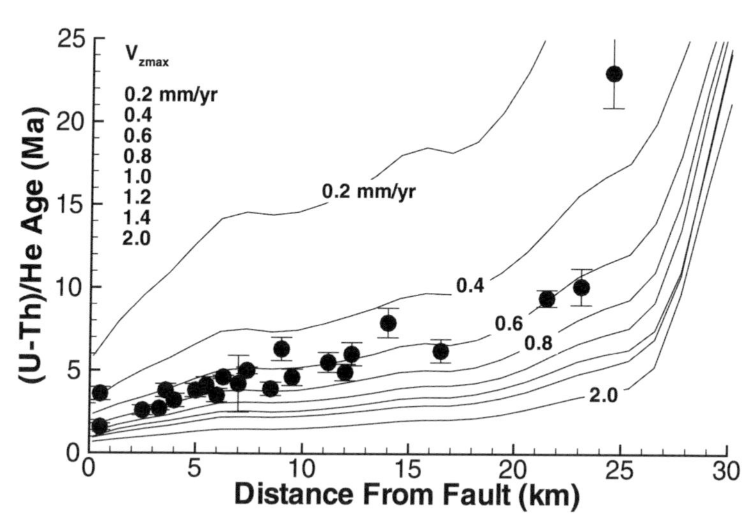

Figure 5. Effect of changing the exhumation rate on predicted model apatite (U-Th)/He ages across an uplifted block in the Wasatch Mountains USA. Solid lines represent predicted He-ages for footwall exhumation rates at the fault. Except for few samples the measured ages agree favourably with predicted ages for 0.6–0.8 mm/yr exhumation. [From Ehlers et al. (2003) *Journal of Geophysical Research*, 108 (B3), 2173-2193, Fig. 6b, p. 2182]

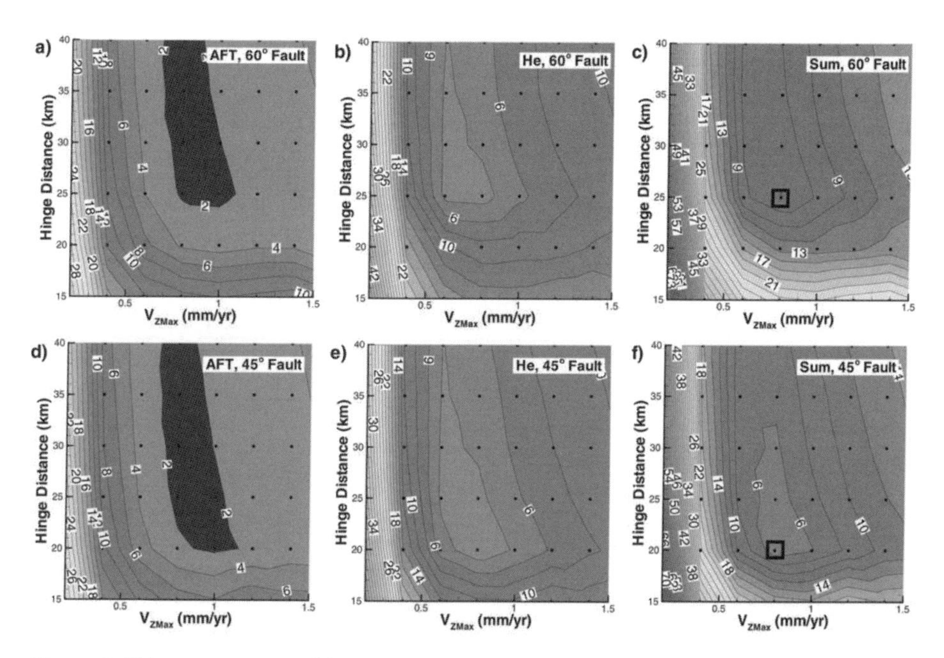

Figure 6. Chi-square measure of fit between predicted and observed thermochronometer ages for the model solution space. (a) AFT, (b) (U-Th)/He, (c) summed AFT and (U-Th)/He misfit for a model with a 60° dipping fault. (d) AFT, (e) (U-Th)/He, (f) summed AFT and (U-Th)/He misfit for a model with a 45° dipping fault. Black boxes in Figures 6c and 6f represent the best fit models [From Ehlers et al. (2003) *Journal of Geophysical Research*, 108 (B3), 2173-2193, Fig. 9 p. 2186].

as both the escarpment and the fault modeled can adequately be described by 2D transects perpendicular to the main geological features. In geological settings, however, where topography is less symmetric *and* is significant, 3D-thermokinematic models are required to predict the subsurface thermal field (Braun 2003; Ehlers and Farley 2003). Studies aiming at understanding the dynamics of relief formation in such areas do need a 3-D approach (Braun 2003; Ehlers and Farley 2003).

The first codes enabling 3D thermo-kinematic modeling are now available for general use (Braun 2003; Braun 2005; Ehlers 2005) and will undoubtedly help to further increase the use of (U-Th)/He dating to study Earth surface processes. These programs and their application/interpretation are discussed in various chapters of this volume (Braun 2005; Ehlers 2005; Gallagher et al. 2005). Important next steps in FM will be the full integration of *numeric* landscape/tectonomorphological models (Braun and Sambridge 1997; Beaumont et al. 2000) and as well as starting to use the unique information that can be derived from scaled *analog* models (Perrson and Sokoutis 2002; Smit et al. 2003).

After a rigorous cross-calibration of the various low-temperature thermochronometers, (U-Th)/He and fission track dating of apatite, titanites and zircons, and K-feldspar Ar-Ar dating, is achieved, the combination of these systems in FM will allow a comprehensive modeling of exhumation of upper crustal rocks (Ehlers and Farley 2003).

These future levels of sophistication in FM, however, will still rely on the same foundations as their beginnings today. Therefore a constant eye has to be cast on whether our computational abilities do not exceed our understanding of the basic physical processes involved, such as He-diffusion (Ehlers and Farley 2003) or annealing of fission tracks (Tagami et al. 2005). Small deviations in diffusion parameters, or a sloppy description of the diffusion domains, which may stand at the very beginning of a FM-exercise, may have as large an influence on the results of FM than changes in geological parameters that we want to understand.

REFERENCES

Armstrong PA (2005) Thermochronometers in sedimentary basins. Rev Mineral Geochem 58:499-526

Batt GE, Brandon MT, Farley KA, Roden-Tice T (2001) Tectonic synthesis of the Olympic Mountains segment of the Cascadia wedge, using two-dimensional thermal and kinematic modeling of thermochronological ages. J. Geophys. Res. - Solid Earth 106(B11):26731-26746

Beaumont C, Kooi H, Willett SD (2000) Coupled tectonic-surface process models with applications to rifted margins and collisional orogens. *In*: Geomorphology and Global Tectonics. Summerfield MA (ed) Wiley, Chichester, p 29-55

Bikker A, Dunai TJ, Meesters AGCA (2002) DECOMP. 1.1 FALW, Vrije Universiteit

Bishop P, Goldrick G (2000) Geomorphological evolution of the East Australian continental margin. *In*: Geomorphology and Global Tectonics. Summerfield M (ed) Wiley, Chichester, p 227-255

Braun J, Sambridge M (1997) Modelling landscape evolution on geological time scales: a new method based on irregular spatial discretization. Basin Res 9:27-52

Braun J (2003) Pecube: a new finite-element code to solve the 3D heat transport equation including the effects of a time-varying, finite amplitude surface topography. Computers & Geosciences 29:787-794

Braun J (2005) Quantitative constraints on the rate of landform evolution derived from low-temperature thermochrohonology. Rev Mineral Geochem 58:351-374

Crank J, Nicolson P (1947) A practical method for numerical evaluation of solutions of partial differential equations of the heat-conduction type. Proc Cambridge Philos Soc 43:50-67

Crank J (1975) The Mathematics of Diffusion. Clarendon, Oxford

Dodson MH (1973) Closure temperature in cooling geochronological and petrological systems. Contr Mineral Petrol 40:259-274

Dunai TJ, Roselieb K (1996) Sorption and diffusion of helium in garnet: implications for volatile tracing and dating. Earth Planet Sci Lett 139:411-421

Ehlers TA, Farley KA (2003) Apatite (U-Th)/He thermochronometry: methods and applications to problems in tectonic and surface processes. Earth Planet Sci Lett 206:1-14

Ehlers TA, Willet SD, Armstrong PA, Chapman DS (2003) Exhumation of the central Wasatch Mountains, Utah: 2. Thermokinematic model of exhumation, erosion and thermochronometer interpretation. J Geophys Res 108(B3):2173, doi:10.1029/2001JB001723

Ehlers TA (2005) Crustal thermal processes and the interpretation of thermochronometer data. Rev Mineral Geochem 58:315-350

Farley KA, Wolf RA, Silver LT (1996) The effects of long alpha-stopping distances on (U-Th)/He ages. Geochim Cosmochim Acta 60:4223-4229

Farley KA (2000) Helium diffusion from apatite: General behaviour as illustrated by Durango fluorapatite. J Geophys Res 105:2903-2914

Gallagher K, Stephenson J, Brown R, Holmes C, Ballester P (2005) Exploiting 3D spatial sampling in inverse modeling of thermochronological data. Rev Mineral Geochem 58:375-387

House MA, Wernicke BP, Farley KA, Dumitru TA (1997) Cenozoic thermal evolution of the central Sierra Nevada, California, from (U-Th)/He thermochronometry. Earth Planet Sci Lett 151:167-179

House MA, Wernicke BP, Farley KA (1998) Dating topography of the Sierra Nevada, California, using apatite (U-Th)/He ages. Nature 396:66-69

House MA, Kohn B, Farley KA, Raza A (2002) Evaluating thermal history models for the Otway Basin, southeastern Australia, using (U-Th)/He and fission-track data from borehole apatites. Tectonophysics 349:277-295

Ketcham RA (2005) Forward and inverse modeling of low-temperature thermochronometry data. Rev Mineral Geochem 58:275-314

Lagerwall T, Zimen KE (1964) The kinetics of rare gas diffusion in solids. EURATOM Report # 772, Brussels

Lovera OM, Richter FM, Harrison TM (1989) The $^{40}Ar/^{39}Ar$ thermochronology for slowly cooled samples having a distribution of diffusion domain sizes. J Geophys Res 94B:17917-17935

McInnes BIA, Evans NJ, Fu FQ, Garwin S (2005) Application of thermochronology to hydrothermal ore deposits. Rev Mineral Geochem 58:467-498

Meesters AGCA, Dunai TJ (2002a) Solving the production-diffusion equation for finite diffusion domains of various shapes (part I): implications for low-temperature (U-Th)/He-thermochronology. Chem Geol 186: 333-344

Meesters AGCA, Dunai TJ (2002b) Solving the production-diffusion equation for finite diffusion domains of various shapes (part II): application to cases with α-ejection and non-homogenous distribution of the source. Chem Geol 186:347-363

Ollier CD (1982) The Great Escarpment of eastern Australia: tectonic and geomorphic significance. J Geol Soc Aust 29:13-23

Ollier CD, Pain CF (1997) Equating the basal unconformity with the plaeoplain: a model for passive margins. Geomorphology 19:1-15

Perrson KS, Sokoutis D (2002) Analogue models of orogenic wedges controlled by erosion. Tectonophysics 356:323-336

Persano C, Stuart FM, Bishop P, Barfod DN (2002) Apatite (U-Th)/He age constraints on the development of the Great Escarpment on the southeastern Australian passive margin. Earth Planet Sci Lett 200:79-90

Pik R, Marty B, Carignan J, Lave J (2003) Stability of the Uppe Nile drainage network (Ethopia) deduced from (U-Th)/He thermochronometry: implications for uplift and erosion of the Afar plume dome. Earth Planet Sci Lett 215:73-88

Reiners PW, Farley KA (1999) Helium diffusion and (U-TH0/He thermochronometry of titanite. Geochim Cosmochim Acta 63:3845-3859

Reiners PW, Farley KA (2001) Influence of crystal size on apatite (U-Th)/He thermochronology: an example from the Bighorn Mountains, Wyoming. Earth Planet Sci Lett 188:413-420

Reiners PW, Spell TL, Nicolescu S, Zanetti KA (2004) Zircon (U-Th)/He thermochronometry: He diffusion and comparisons with Ar-40/Ar-39 dating. Geochim Cosmochim Acta 68:1857-1887

Seidl MA, Weissel JK, Pratson LF (1996) The kinematics and pattern of escarpment retreat across the rifted continental margin of SE Australia. Basin Res 12:301-316

Smit JHM, Brun JP, Sokoutis D (2003) Deformation of brittle-ductile thrust wedges in experiments and nature. J Geophys Res - Solid Earth 108(B10): Art. # 2480

Spotila J (2005) Applications of low-temperature thermochronometry to quantification of recent exhumation in mountain belts. Rev Mineral Geochem 58:449-466

Stockli DF, Farley KA, Dumitru TA (2000) Calibration of the apatite (U-Th)/He thermochronometer on an exhumed fault block, White Mountains, California. Geology 28:983-986

Stockli DF (2005) Application of low-temperature thermochronometry to extensional tectonic settings. Rev Mineral Geochem 58:411-448

Tagami T, O'Sullivan PB (2005) Fundamentals of fission-track thermochronology. Rev Mineral Geochem 58: 19-47

ter Voorde M, de Bruijne CH, Cloetingh SAPL, Andriessen PAM(2004) Thermal consequences of thrust faulting: simultaneous versus successive fault activation and exhumation. Earth Planet Sci Lett 223: 395-413

Warnock AC, Zeitler PK, Wolf RA, Bergman SC (1997) An evaluation of low-temperature apatite U-Th/He thermochronology. Geochim Cosmochim Acta 61:5371-5377

Wolf RA, Farley KA, Silver LT (1996) Helium diffusion and low-temperature thermochronometry of apatite. Geochim Cosmochim Acta 60:4231-4240

Wolf RA, Farley KA, Kass DM (1998) Modeling of the temperature sensitivity of the apatite (U-Th)/He thermochronometer. Chem Geol 148:105-114

Zeitler PK, Herczeg AL, McDougall I, Honda M (1987) U-Th-He dating of apatite: a potential thermochronometer. Geochim Cosmochim Acta 51:2865-2868

Ziegler JF (1977) Helium Stopping Powers and Ranges in all Elemental Matter. Pergamon, New York

Reviews in Mineralogy & Geochemistry
Vol. 58, pp. 275-314, 2005

11

Forward and Inverse Modeling of Low-Temperature Thermochronometry Data

Richard A. Ketcham

Jackson School of Geosciences
The University of Texas at Austin
Austin, Texas, 78712, U.S.A.
ketcham@mail.utexas.edu

INTRODUCTION

The thermochronometric systems discussed in this volume broadly share three features: parent isotopes, daughter products, and one or more time-dependent, temperature-sensitive processes by which daughter products are altered or lost. If these processes can be measured in the laboratory, and their behavior confidently extrapolated to geological time scales, it becomes possible to construct a *forward model* of the system that predicts how a given instance of it will evolve assuming a particular starting arrangement and subsequent time-temperature history.

Once a forward model has been created and verified, it then becomes possible to apply it in the *inverse* sense: given a measured ending condition and an assumed starting one, find the intervening time-temperature history. In general, because of information loss, limited precision of measurements, and lack of system uniqueness, more than one history is consistent with a given ending condition. As a result, an inverse model solution usually consists of a set of thermal histories that are consistent with the measured data, as judged by some statistical criterion.

This chapter will concentrate on two low-temperature thermochronometers: fission-track (primarily for apatite) and (U-Th)/He. All of the calculations described here are implemented in a computer program called "HeFTy," which is available with this volume (see Ehlers et al. 2005).

A theme that will be touched upon throughout this chapter is that forward and inverse models are only as good as the data and assumptions behind them. Although this principle of course holds for all scientific investigations, it is often obscured when such details are packaged in user-friendly software that produces publication-ready graphics.

FORWARD MODELING OF THE FISSION-TRACK SYSTEM

Fission tracks form continuously over time at a rate dependent solely upon the concentration of uranium present. Earlier-formed fission tracks tend to be shorter than later-formed tracks, as they will have had more time to anneal, and may have experienced higher temperatures. The distribution of fission-track lengths observed in an unknown represents a sampling of all tracks formed and annealed during its residence below the total annealing temperature. In essence, they preserve an integrated thermal history, making this thermochronometer a uniquely powerful tool for thermal history reconstruction. The fission-track system has been most thoroughly studied and characterized for apatite; a growing body of work is paving the way for zircon fission track lengths to also be used in this way, although significant complexities remain.

1529-6466/05/0058-0011$05.00 DOI: 10.2138/rmg.2005.58.11

Early attempts to characterize fission-track annealing behavior were based on reduction in track density (number of tracks intersecting a polished surface per unit area) rather than length (e.g., Dakowski et al. 1974; Bertel and Märk 1983). These important first steps were limited in their capabilities, however, in that they could only be used to calculate a model age given a proposed time-temperature history. Insofar as many histories can produce a given age, such models could only constrain the time at which a mineral passed through its approximate closure temperature.

The foundations for forward modeling of fission track lengths in apatite were laid in a series of four papers. In the first, Green et al. (1986) documented length reduction in induced, confined horizontal tracks for Durango apatite at laboratory time scales. Next, Laslett et al. (1987) performed a statistical analysis of these data and proposed an empirical equation that described the laboratory results well and allowed them to be extrapolated to geological time scales. Duddy et al. (1988) then reported on a series of experiments that supported the *equivalent time* hypothesis, which states that the annealing behavior of a track is a function of only its length, and insensitive to the prior time-temperature history by which it obtained that length. These tools allowed Green et al. (1989) to construct a complete forward model describing the evolution of the fission-track length distribution as a function of time and temperature. A fifth paper (Green 1988) provided the basis for relating fission-track length to fission-track density, allowing ages to be modeled as well.

All fission-track forward and inverse models share these fundamental building blocks. However, subsequent work has revealed a number of significant issues in need of further study. The most notable of these is *kinetic variability*, or the fact that different grains of a particular mineral can anneal at different rates (Green et al. 1986; James and Durrani 1986). Another current concern is *calibration and consistency of length measurements*, which not only vary with analytical technique (i.e., etching) but also have been shown to vary between analysts to a degree not accounted for by standard statistical approaches (Barbarand et al. 2003b).

Calibrations

The term *calibration* is used here in two senses. First, it serves as a shorthand for the set of components underlying an equation describing fission-track length reduction: the apatite variety (i.e., composition), the laboratory annealing measurements, the statistical analysis of those data, and the equation form used for fitting. Second, it is used in the usual sense to refer to the means by which an analyst ensures that measurements are acquired correctly and consistently with the model used to interpret those measurements. Other chapters in this volume review in more detail the procedures and principles involved, but they are covered briefly here as an introduction to the issues that underlie the choice of a particular calibration for modeling, and to provide the background for assessing model reliability and precision.

Overview. Laboratory measurements of annealing are characteristically undertaken on populations of induced fission tracks. Naturally occurring tracks are fully annealed in a furnace, and fresh tracks are induced by thermal neutron bombardment, which causes a certain amount of ^{235}U to fission. This procedure ensures that all tracks studied have a common, unannealed starting point, greatly facilitating interpretation of the ensuing measurements. However, it also requires assuming that induced ^{235}U tracks are equivalent to spontaneous ^{238}U tracks, and that the pre-annealing process has not altered the crystal structure in a way that affects annealing rates. The former assumption is considered warranted because the masses and energies of the fission products of both U isotopes are very similar (Crowley 1985). The latter assumption is more difficult to verify, but broad similarities between annealing rates of spontaneous and induced tracks in apatite (e.g., Green 1988) suggest that any effect is small in that mineral. Zircon, which tends to have much higher U and Th concentrations, may be different in this respect, as the accumulation of alpha damage has been shown to decrease resistance to annealing (Kasuya and Naeser 1988; Yamada et al. 1995b; Rahn et al. 2004).

Once fresh fission tracks have been created, aliquots of the mineral being studied are subjected to isothermal annealing conditions. In most cases heating times range from 5 minutes to 10,000 hours (~10^2–10^7 seconds), and temperatures from 150 °C to 425 °C for apatite and 350 °C to 800 °C for zircon. Recently experiments on zircon have been carried out at extremely high temperatures for extremely short durations (3.6 seconds, 1000 °C; Yamada et al. 2003). Such experiments are potentially valuable because they considerably increase the extent of data in log-time units (down to ~10^0 seconds), which may increase confidence when extrapolating trends to geological time scales (10^{13-15} seconds). Particular care must be taken in temperature calibrations for these experiments, especially when different equipment is used for annealing at different time scales, as slight biases can lead to large errors when extrapolated.

Following annealing the mineral grains are mounted and etched, and confined track lengths measured. A variety of etching protocols have been used for both apatite and zircon. Etching technique has frequently been a decisive factor when workers choose an annealing model, as it is desirable that the calibration be based on the same technique used for unknowns. Other procedural factors shown to affect length measurements include whether TINCLEs (track-in-cleavages) are measured or not (Barbarand et al. 2003b; Jonckheere and Wagner 2000) and whether Cf irradiation was employed (Donelick and Miller 1991; Ketcham 2005). Typically, the mean track length (l_m) and associated standard deviation and standard error of the mean are reported. Increasing attention to the role of fission-track angle has led to an increasing degree of reportage of relevant data, as discussed below.

Length calibration. Measurement of track lengths in standards such as Durango and Fish Canyon has long been a part of the routine training regimen for apatite analysis. Such standards tend to have long mean lengths (>14 μm), and the primary method of comparison among workers is via the mean, standard deviation, and standard error. There have been relatively few inter-laboratory comparisons of length measurements, with the most comprehensive being by Miller et al. (1993), involving 33 analysts and 15 laboratories. For two apatites, again with relatively long mean track lengths (14.01 ± 0.36 μm and 13.43 ± 0.35 μm), the results were in fact not very good: for one apatite 9 of 23 analyses were further than 2 standard errors from the mean of all analyses, and for the other apatite 12 of 26 analyses were likewise further from the mean than predicted by their respective 95% confidence intervals. Moreover, if propagated into a simple thermal history forward model, the observed variations translate into a difference in maximum paleotemperature of over 30 °C (Barbarand et al. 2003b).

Partly in response to the paucity of length calibration studies, Barbarand et al. (2003b) performed a systematic investigation of the reproducibility of length measurements, both among different researchers and for one worker over time. Their results mimic the Miller et al. (1993) study, in that they find that the standard error does not adequately predict the degree of variation among fission-track length measurements. They also document that the frequency and degree of excessive dispersion tends to increase with increasing annealing level (Fig. 6 from Barbarand et al. 2003b). In analyzing their data, they found that many cases of poor reproducibility could be traced to the effects of fission-track angle.

Angular effects and biasing. It has long been known that fission tracks at high angles to the apatite crystallographic *c* axis tend to anneal more rapidly than tracks at lower angles (Green and Durrani 1977; Donelick 1991). The divergence between lengths of low-angle and high-angle tracks increases with progressive annealing, and this is responsible for the large rise in the standard deviation of track length populations as annealing progresses. Thus, the mean track length can vary significantly depending on whether high-angle or low-angle tracks are measured preferentially.

Biasing of fission-track length measurements has long been considered to be dominated by track length (Laslett et al. 1982, 1984), as long tracks are more likely to be observed and measured than short ones. Another documented biasing factor is the cross-sectional shape of an inclined track serving as an etchant pathway to a confined horizontal track: the tendency of this cross section to be elongated in the *c*-axis direction means that high-angle tracks have a larger "target" for intersection, and are thus intersected and measured more frequently (Galbraith et al. 1990; Ketcham 2003b). However, a closer examination of length data demonstrates that these biases are in fact secondary to other factors that correlate with angle (Ketcham 2003b). In most apatites, low-angle tracks appear thinner than high-angle tracks, and frequently appear under-etched, with indistinct tips. Such tracks are less likely to be noticed, and when they are observed they are likely to be passed over in favor of better-defined tracks. Whether to pass over a poorly etched low-angle track is a decision made by the measurer, and the data strongly suggest that how this decision is made can vary from analyst to analyst, and for one analyst over time. Since such low-angle tracks are longer than high-angle tracks for a given level of annealing, it is easy to see that these subjective decisions can drive the mean towards larger or smaller values. A similarly subjective decision is how to interpret very short tracks: some workers measure tracks down to 2 μm or below, while others never measure tracks below 5 μm. Again, it is clear that this variable preference can heavily influence the measured mean track length. These shortened tracks tend to be high-angle, so this bias is likewise correlated with track angle.

A potential solution for observer bias in length measurement lies in defining annealing level in terms of both length and angle, rather than only length (Galbraith and Laslett 1988; Galbraith et al. 1990; Donelick 1991). The most actively pursued approach has been based on the observation of Donelick (1991) that fission-track lengths are distributed along an ellipse on a polar plot at low to intermediate levels of annealing (Fig. 1a-c). In such cases the variation of length with angle is well-described simply by the *c*-axis parallel and perpendicular axes of an ellipse (l_c and l_a). Donelick et al. (1999) found that the relationship between l_c and l_a is systematic and linear. At high levels of annealing the elliptical relationship breaks down, with the annealing rate of high-angle tracks accelerating relative to low-angle ones (Fig. 1d-f). Such cases can be adequately described by truncating the ellipse at the angle where the breakdown occurs and using a line segment to describe higher-angle tracks (Donelick et al. 1999; Ketcham 2003b). These elements can be combined to create a complete, empirical *c-axis projection model* that describes the variation of track length with angle at all levels of annealing (Fig. 2). This model allows any track-angle measurement to be projected to a *c*-axis parallel orientation, thus rendering moot the issue of which angular populations are preferentially sampled (e.g., Fig. 3). The *c*-axis projection is intended to remove dispersion effects correlated with track angle while preserving thermal history information. The length-angle relationship likely changes as a function of etching method, and the existing models only apply to the etching protocol used by Carlson et al. (1999), although work is in progress on one applicable to the more widely used Barbarand et al (2003a) protocol (Ketcham et al. 2004).

Annealing data sets. In the years immediately following their publication, the Green et al (1986) data set and Laslett et al. (1987) annealing equation for Durango apatite served as the near-universal basis for modeling of apatite fission-track data. However, even at that time there were indications of significant variations in the fission-track closure temperature among some apatites. Green et al. (1985) noted that chlorine-rich apatites tend to give older ages than fluorine apatites, implying a higher closure temperature; independent experimental data by James and Durrani (1986) also found high Cl content to correlate with increasing resistance to annealing. While Durango is a widely-used standard due to its size and availability, it is not typical of the majority of apatites, as it has an appreciable chlorine content (0.42 wt%) and perhaps other features that make it slightly more resistant to annealing than typical end-member F-apatite. Carlson et al. (1999) found that Durango apatite anneals in the laboratory at

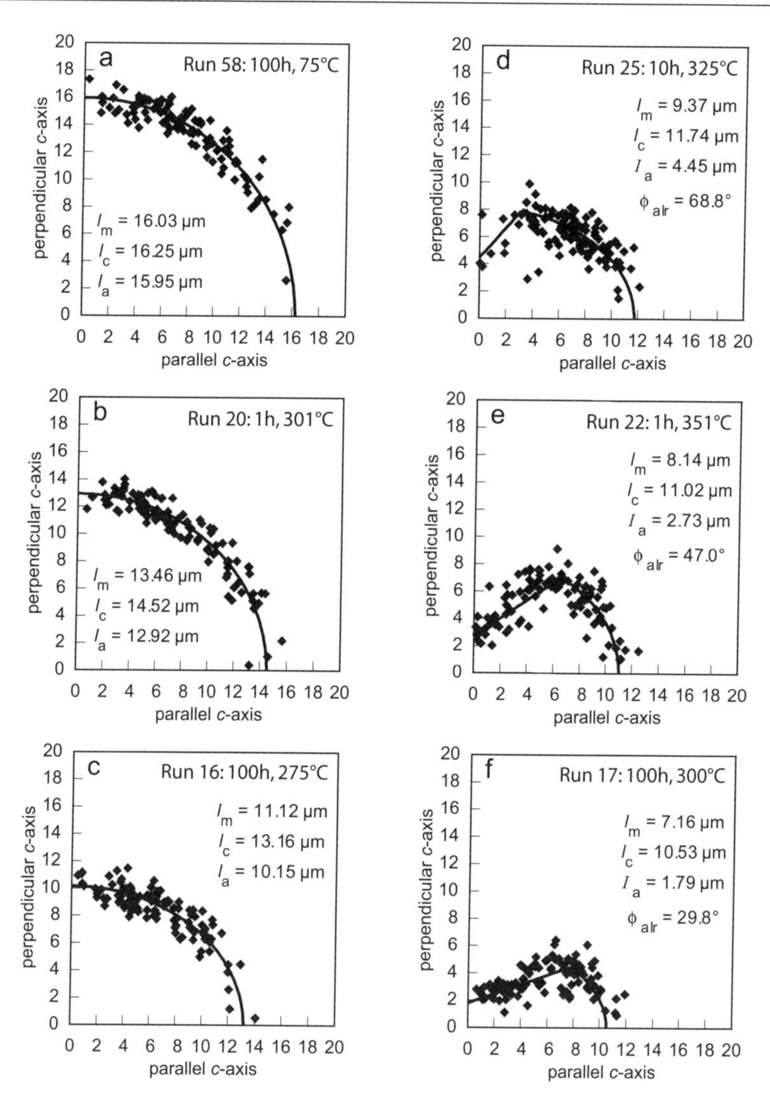

Figure 1. Polar coordinate plots of fission-track length measurements of Durango apatite at progressively higher levels of annealing, with fitted ellipses and line segments. Run refers to experimental annealing run from Carlson et al. (1999); l_m is mean length; l_c and l_a are *c*-axis parallel and perpendicular intercepts of fitted ellipses; I_a is *c*-axis perpendicular intercept of line segment fit to accelerated length reduction tracks; ϕ_{alr} is angle of onset of accelerated length reduction. From Donelick et al. (1999), with variable names changed according to Ketcham (2003b).

temperatures up to ~25 °C higher than end-member F-apatite, and extrapolation to geological time scales predicts that Durango has closure temperatures some 10–12 °C higher than F-apatites (Ketcham et al. 1999). Analysis of the Barbarand et al. (2003a) annealing data set corroborates this divergence (Ketcham et al. 2004).

Table 1 shows the major data sets for apatite fission-track annealing, including which apatites were studied, what etching technique used, and whether and how angular data were reported. Crowley et al. (1991) published the first thorough data set on annealing of an F-

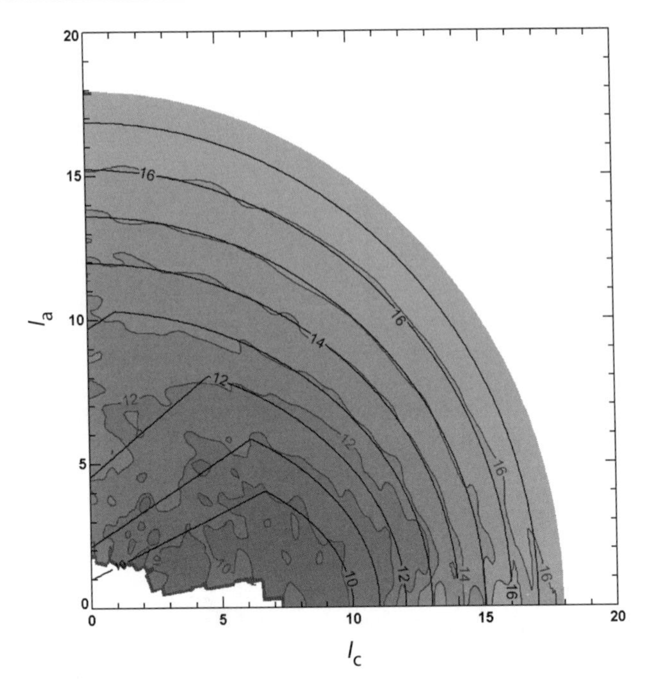

Figure 2. Polar contour plot of distribution of $l_{c,fit}$ values as a function of length and angle for six near-endmember F-apatites in the Carlson et al. (1999) data set (gray contours and shading), and empirical *c*-axis projection model (black contours) allowing any (l,ϕ) measurement to be converted to l_c. From Ketcham (2003b).

apatite, and also studied an F-apatite with significant Sr substitution for Ca. In addition to using a different etching procedure than the Green et al. (1986) data set, this study featured the uncommon utilization of oil-immersion microscopy. Donelick (1991) reports annealing results on the Cl-rich Tioga apatite, which also includes an uncommon degree of Fe substitution. Annealing models based on that data set predict much higher closure temperatures than F-apatite (Ketcham et al. 1999).

The recognition of a diversity of potential compositional factors affecting annealing kinetics led to the creation of a number of data sets featuring multiple varieties of apatite, including those of Carlson et al. (1999), Barbarand et al. (2003a), and Ravenhurst et al. (2003). These data sets corroborate that, along with Cl, various cation substitutions (Mn, Fe, Sr, REE) also affect annealing rates. The frequency with which cation substitutions occur to a kinetically significant degree is unknown, in part because they are rarely analyzed for, and in part because current data are as yet insufficient to confidently map out what amounts are required to have a given effect. It has also been proposed that the OH anion substitution is a significant factor, either in and of itself (Indrelid and Terken 2000) or in the context of mixing and ordering requirements with F and Cl (Hughes et al. 1989; Ketcham et al. 1999).

Reporting of angular data for apatite fission tracks has progressed through time (Table 1). Green et al. (1986) provide none. Crowley et al. (1991) followed the approach of Donelick (1991) and used ellipses to describe the variation of length versus angle with progressive annealing; the fitted ellipse radii parallel and perpendicular to the *c* axis are referred to here as $l_{c,fit}$ and $l_{a,fit}$. Ravenhurst et al. (2003) likewise report $l_{c,fit}$ and $l_{a,fit}$ for their data, with the additional step of eliminating high-angle tracks in highly annealed experiments for which

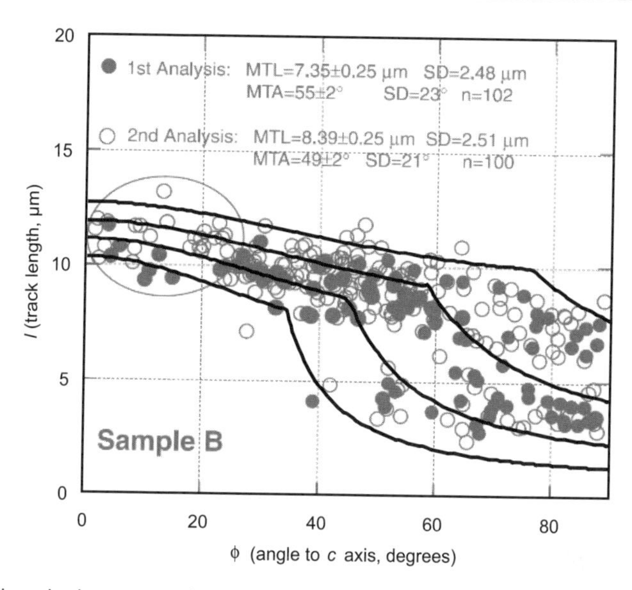

Figure 3. *c*-axis projection contours from Ketcham (2003b) superimposed on Figure 4b from Barbarand et al. (2003b). The mean lengths of replicate analyses of the same mount by the same analyst taken three months apart were more than two standard errors apart, outside the level of hoped-for reproducibility of measurements. The two data sets have similar distributions of length versus angle, but different samplings of angles; removal of this differential sampling density improves data reliability. Contours from top to bottom are for l_c values of 12.7, 11.9, 11.1, and 10.3 µm. Track lengths are converted to their *c*-axis-projected equivalents by tracing along the contours to the $\phi = 0$ axis. [Used with permission of Elsevier from Ketcham (2005) *Rad Meas*, Vol. 39, p. 595-601, Fig. 1.]

Table 1. Apatite fission-track confined length annealing data sets.

Reference	*Apatite*	*Etching method*[1]	*Reporting of angle*
Green et al. (1986)	Durango (low Cl)	1	None
Donelick (1991)	Tioga (Cl, Fe)	2	Ellipse, mean angle
Crowley et al. (1991)	F-apatite, Sr-F-apatite	3	Ellipse, mean angle
Carlson et al. (1999)	15 apatites, including Durango, F-apatites, Fish Canyon, Tioga	4	Partial ellipse, $l_{c,mod}$
Barbarand et al. (2003a)	13 apatites, including Durango, F-apatites, Fish Canyon	1	Mean angle
Ravenhurst et al. (2003)	4 apatites, including Durango, F-apatite	Various	Partial ellipse

[1] *Etching methods:* (1) 5.0 M HNO_3, 20 s, 20 °C; (2) 5.0 M HNO_3, 25 s, 21 °C; (3) 1.6 M HNO_3, 40 s, 20 °C; (4) 5.5 M HNO_3, 20 s, 21 °C.

the elliptical relationship no longer holds. Carlson et al. (1999) also fitted partial ellipses, albeit with slightly different criteria, and also report $l_{c,mod}$, the mean *c*-axis projected length as defined by the model of Donelick et al. (1999). Barbarand et al. (2003a) report only mean angle for their experiments, but further analysis has indicated that *c*-axis projection can be successfully applied to their data (Ketcham et al. 2004).

Comparatively few laboratory studies have been done for fission-track length reduction in zircons. The principal data set is provided by Yamada et al. (1995b), with subsequent additions by Tagami (1998) and Yamada et al (2003). No evidence has been reported for compositional factors having a direct effect on zircon fission-track annealing kinetics, although extent of alpha damage has been shown to affect etching characteristics (Yamada et al. 1995b), and likely annealing as well (Rahn et al. 2004). This feature makes geologically comprehensive calibration of annealing more problematic, as alpha damage is a function of U and Th content, as well as the amount of time spent below the alpha damage annealing temperature.

Yamada et al. (1995a,b) document slightly anisotropic annealing in zircon, and report mean lengths for all tracks and for those lower than 60° to the zircon *c*-axis. The resulting change in the mean is fairly small, however, and arguably dismissible (Rahn et al. 2004). Zircon annealing and etching anisotropy characteristics are similar to apatite, in that both annealing and etching are slower in the *c*-axis direction, and it has a similar relationship between length and density reduction. It is possible that in the future approaches similar to those developed for apatite could be investigated.

Modeling equations. A number of equations have been proposed for fitting fission-track length data. Most of these are empirical, in large part because there is no widely agreed-upon physical model for how fission tracks are configured and how they anneal on the atomic scale. The only fully developed model with a physical basis was created by Carlson (1990), based on the assumption of fission tracks consisting of vacancies and interstitials in the crystal lattice that are annealed by short translations of atoms from abnormal positions back to stable sites. The mechanism is similar to topotactic transformation, which controls the transitions between mineral polymorphs such as calcite and aragonite, except the driving force is release of strain energy rather than a chemical potential difference. The resulting equation, based on the kinetic equations for atomic motions across a coherent interface, is:

$$l_{as} = l_0 - A\left(\frac{\mathrm{k}}{\mathrm{h}}\right)^n \left[\int_0^t T(\tau)\exp\left(\frac{-Q}{\mathrm{R}T}\right)d\tau\right]^n \qquad (1)$$

where l_{as} is the length of an individual track after axial shortening (i.e., shortening from the tips); l_0 is the initial (unannealed) length; k is Boltzmann's constant; h is Planck's constant; T is absolute temperature; t is time since the beginning of the annealing episode; R is the universal gas constant; τ is a dummy variable for time integration; and A, n and Q are empirically determined parameters that respectively reflect a rate constant, the shape of the radial defect distribution, and an activation energy. This equation is not considered appropriate for forward and inverse modeling, as there is structure in the residuals between the fitted equation and the data (Green et al. 1993), which leads to large biases when model predictions are extrapolated to geological time scales. However, the model is interesting because it contains elements of the various empirical models that have been proposed, as discussed below.

In a preliminary analysis of the as yet incomplete Green et al. (1986) Durango apatite annealing data set, Green et al. (1985) found that their length data could be well represented by an equation with the form:

$$\ln(1 - l/l_0) = c_0 + c_1\ln(t) + c_2/T \qquad (2)$$

where l is the mean track length after isothermal annealing for time t (in seconds) at temperature T (in Kelvins), l_0 is the unannealed, or initial mean track length, and the c's are fitted parameters. This equation defines a parallel series of contours of constant annealing on an Arrhenius plot (Fig. 4, short-dashed lines), implying that annealing is controlled by a kinetic process with a single activation energy. A more thorough analysis of the complete Green et al.

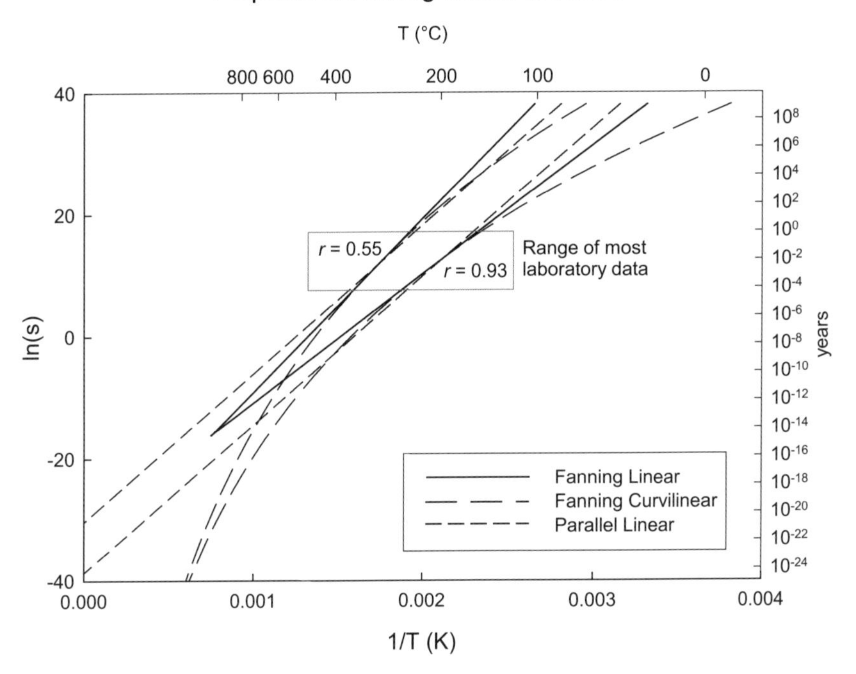

Figure 4. Contour lines of mean length reduction for three different annealing equations fitted by Ketcham et al. (1999) to the Renfrew F-apatite data set of Carlson et al. (1999). For each pair of contours, right shows reduced mean *c*-axis projected length of 0.93, corresponding to the amount inferred to occur in apatite fission tracks at surface temperatures over geological time scales. The left contour is for a reduced mean *c*-axis projected length of 0.55, approximately where density (and thus age) falls to zero. Although these are very similar over the range of time scale of most laboratory experiments, they make very different predictions of geological time-scale behavior.

(1986) data set by Laslett et al. (1987) found that the observations were better described by a model in which annealing contours fan from a single point (Fig. 4, solid lines):

$$\frac{\left[\left(1-r^{\beta}\right)/\beta\right]^{\alpha}-1}{\alpha}=c_0+c_1\left[\frac{\ln(t)-c_2}{(1/T)-c_3}\right] \tag{3}$$

where α and β are additional fitted parameters, r is the normalized mean length (l/l_0), and the fanning point is defined by the coordinates (c_2, c_3). Laslett et al. (1987) found that the Green et al. (1986) data could be adequately explained by omitting c_3 (i.e., setting c_3=0), but others have found that this is not necessarily the case with other data sets. It should be noted that posing annealing contours as fanning from a single point is principally a computational convenience and simplification, and there is no basis for ascribing any physical significance to this point (e.g., Crowley 1993b).

The fanning Arrhenius equation form was also favored by Crowley et al. (1991); however, the Laslett et al. (1987) fitting method was inappropriate for their data set, owing to its disproportionate number of experiments with only modest annealing. Laslett and Galbraith (1996), in their re-analysis of the Crowley et al. (1991) data, suggested an amended form of the fanning Arrhenius equation:

$$\log\left[1-\left(\frac{l}{l_{\max}}\right)^{1/\lambda}\right]=c_0+c_1\left[\frac{\ln(t)-c_2}{(1/T)-c_3}\right] \quad (4)$$

in which λ is a fitted parameter that they propose should always be set to a value of 3, and l_{max} is the "true" (but unknown and therefore to be fitted) mean initial track length. Part of their purpose was to generalize the normalizing quantity from the measured "unannealed" track length, because data presented by Donelick et al. (1990) indicates that some annealing may take place at room temperature in the seconds and minutes following irradiation. This is a reasonable objective, but it is not clear that fitting annealing data is an effective way to find this parameter. Ketcham et al. (1999) found that models fitted to equations (3) and (4) (using a revised merit function) make very similar predictions on geological time scales.

In analyzing the Carlson et al. (1999) data, set, Ketcham et al. (1999) tried several models, and favored one based on mean *c*-axis projected lengths rather than unprojected mean lengths, with a slightly different form on the Arrhenius plot:

$$\frac{\left[\left(1-r^{\beta}\right)/\beta\right]^{\alpha}-1}{\alpha}=c_0+c_1\left[\frac{\ln(t)-c_2}{\ln(1/T)-c_3}\right] \quad (5)$$

This equation form causes lines of constant annealing to have some curvature on the Arrhenius plot (Fig. 4, long-dashed lines). Although it fit the laboratory data slightly worse than Equation (3), it was chosen because it makes predictions of geological time-scale annealing behavior more in accord with expectations from independent evidence; gauging geological time-scale behavior will be discussed further in the next section. Preliminary analysis of the Barbarand et al. (2003a) data set indicates that Equation (5) also produces models that better fit geological time-scale expectations (Ketcham et al. 2004).

In considering the relative merits of the various empirical equations, it is good to bear in mind that the Carlson (1990) model contains a physical basis for both linear and curvilinear Arrhenius terms. For an isothermal annealing interval, Equation (1) can be rearranged to yield:

$$\ln(1-r)=\ln\left(\frac{A}{l_0}\right)+n\ln\left(\frac{\mathrm{k}T}{\mathrm{h}}\right)+n\left(\ln t-\frac{Q}{\mathrm{R}T}\right) \quad (6)$$

If the first temperature term is deleted we see that this corresponds exactly to Equation (2), the parallel linear Arrhenius model. The first temperature term gives rise to slightly curved Arrhenius contours, and corresponds to a frequency factor. Put simply, the first temperature term defines how frequently an activation energy barrier to annealing a given defect is assailed, and the second defines how often an assault is successful. The transition between linear and curvilinear Arrhenius behavior may thus be defined by which of these terms is dominant. Although as noted above the Carlson (1990) model as currently posed is not suitable for extrapolating laboratory data to geological time scales, it is possible that enhancements, such as using a range of activation energies rather than a single one, could bridge this gap.

Given a set of fitted parameters, all of the equations above describe the kinetics for a single class of apatite. Because apatite kinetics vary considerably, some means of accounting for this variation is required. Ketcham et al. (1999) found that laboratory annealing measurements for any two apatites in the Carlson et al. (1999) data set could be described adequately using a fairly simple equation. Give two apatites, one more resistant to annealing and the other less resistant, subjected to the same thermal conditions (i.e., annealed together in the same furnace), their annealed lengths are related by:

$$r_{lr} = \left(\frac{r_{mr} - r_{mr0}}{1 - r_{mr0}} \right)^{\kappa} \tag{7}$$

where r_{mr} and r_{lr} are the reduced lengths of the more-resistant and less-resistant apatites, and r_{mr0} and κ are fitted parameters (Fig. 5a). In particular, the parameter r_{mr0} corresponds to the reduced length of the more-resistant apatite at the time-temperature conditions where the reduced length of the less-resistant apatite falls to zero. Equation (7) can be used for both regular and *c*-axis projected lengths. Using this equation, Ketcham et al. (1999) were able to construct a single model that incorporated all of the different types of apatite in the Carlson et al. (1999) data set.

The set of fitted values for r_{mr0} and κ calculated by Ketcham et al. (1999) suggest a further simplification (Fig. 5b):

$$r_{mr0} + \kappa \cong 1 \tag{8}$$

Using this approximation, it is in principle possible to describe the behavior of any apatite using an annealing equation for the "most" resistant apatite variety combined with Equation (7) and a value for r_{mr0}. A complete annealing model for apatite could thus be constructed by finding a function that relates measurable parameters (hereinafter termed kinetic parameters), such as etch figure length (D_{par}, measured in μm) or chemical composition [Cl or OH atoms per formula unit (apfu), based on apatite formula $Ca_{10}(PO_4)_6(F,Cl,OH)_2$], to r_{mr0}. Etch figure length can be used to infer kinetic behavior (Donelick 1993, 1995; Burtner et al. 1994) but is not entirely robust. Likewise, as discussed above no single chemical compositional variable provides a completely reliable predictor for r_{mr0}. Ketcham et al. (1999) suggest the following two equations to characterize r_{mr0} in terms of D_{par} (in μm) and Cl content (in apfu) for the fifteen apatites they examined:

$$r_{mr0} = 1 - \exp\left[0.647\left(D_{par} - 1.75\right) - 1.834\right] \tag{9a}$$

$$r_{mr0} = 1 - \exp\left\{2.107\left[1 - \mathrm{abs}\left(Cl - 1\right)\right] - 1.834\right\} \tag{9b}$$

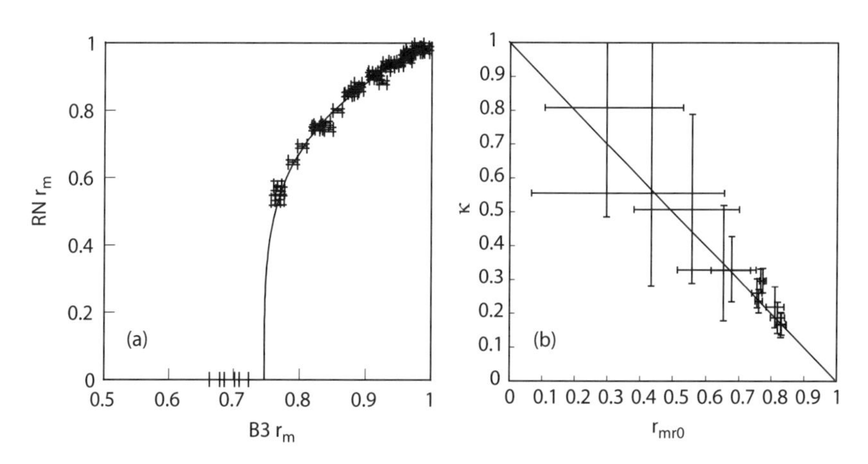

Figure 5. (a) Diagram relating reduced mean lengths of RN F-apatite and B3 Cl-apatite that were annealed at the same time; the systematic relation between the two can be characterized with Equation (7). (b) Values of r_{mr0} and κ, for relating all Carlson et al. (1999) apatites to B2 Cl-OH-Fe apatite, the most resistant in that data set. 95% confidence intervals show them to be within error of the line defined by Equation (8).

It should be kept in mind, however, that these should be considered "80–90% solutions" —sufficient in the majority of cases but inaccurate in a non-negligible minority.

Geological time-scale predictions. Given the uncertainty concerning which equation is best for extrapolating laboratory measurements to geological time scales, it is important to have a means of gauging and comparing their geological time-scale predictions. Ketcham et al. (1999) proposed high-temperature and low-temperature indices. For the former, three index temperatures were proposed: the closure temperature (T_C), the temperature of the system at the time given by the measured age, assuming a linear cooling path (Dodson 1973); the fading temperature (T_F), the temperature at which fission tracks anneal totally after an isothermal annealing episode of a given duration; and the total annealing temperature (T_A), the temperature at which a fission-track population formed at low temperature fully anneals for a given linear heating rate (Issler 1996b), or equivalently the temperature at the time of formation of the oldest remaining track after linear cooling. T_C and T_A vary with heating and cooling rate (in K/Myr), and T_F varies with the duration of the isothermal episode (in My). The proposed low-temperature index was the predicted mean length after the long-duration, low-temperature history provided by Vrolijk et al. (1992) for a deep-sea sample.

The calculations necessary to determine these indices are provided accompanying this volume (see Ehlers et al. 2005) in the computer program FTIndex. Also provided is the C programming language source code for the program which implements the calculations discussed in the next section. The example input file includes all of the constants for the previously discussed annealing models.

Ketcham et al. (1999) proposed two geological benchmarks. For high-temperature behavior, $T_{F,30}$ = 95 °C corresponds to observations of F-apatite in the Flaxmans-1 well in the Otway Basin (Gleadow and Duddy 1981). The low-temperature benchmark was a mean length of 14.6 ± 0.1 μm, as measured by Vrolijk et al. (1992). Table 2 shows indices for the annealing models most in use today. The Ketcham et al. (1999) and Laslett et al. (1987) Durango models give very similar results, although very different methods and equations were employed. The high index temperatures for the Crowley et al. (1991) F-apatite model were lowered somewhat in the improved treatment of the same data by Laslett and Galbraith (1996), but they are still higher than the corresponding results for more-resistant Durango apatite. However, the Crowley et al. (1991) data are congruent with the Carlson et al. (1999) F-apatite data that are fitted the same way (Ketcham et al. 1999). The range of temperatures for the Carlson et al. (1999) F-apatites indicates that there is probably about a 10 °C range in temperature predictions, placing an upper limit on the temperature resolution of the technique. Similarly, the 0.5 μm range in low-temperature index lengths, stemming from the documented variability of the initial track length, provides an indication of the precision of length data and modeling results.

Length distribution calculation

For a given *t-T* history, continuous track formation is approximated by subdividing the *t-T* path into discrete intervals. The length distribution of the track population formed during each interval is calculated based on the estimated amount of annealing over the course of the subsequent thermal history, and the presumed distribution of lengths about the mean. The populations are then summed together, using weighting factors that reflect the relative likelihood of observing each population. Each of these steps is discussed in further detail below.

Subdividing the time-temperature path. A time-temperature path can be defined either as a set of line segments (e.g., Green et al. 1989; Gallagher 1995; Willett 1997; Ketcham et al. 2000) or a continuous function obtained by combining Chebyshev polynomials (Corrigan 1991). The latter form is particularly suitable when the annealing equation is in the form of an integral (e.g., Eqn. 1), which is however not the case with the most-used calibrations.

Table 2. Index temperatures and lengths for various annealing models.

Apatite	*Fit*	$T_{F,100}$	$T_{F,30}$	$T_{F,10}$	$T_{C,1}$	$T_{C,10}$	$T_{C,100}$	$T_{A,10}$	$T_{A,30}$	$T_{A,100}$	$l_{m,low\text{-}T}$[1]
F-apatite B-5	Crowley et al. (1991)	120.7	127.1	133.0	116.9	130.0	144.1	136.3	149.3	163.1	14.40–14.98
F-apatite B-5	Laslett and Galbraith (1996)	102.0	108.7	115.0	103.0	116.8	131.8	118.2	132.0	146.8	14.62–15.21
F-apatites (various)[2]	Ketcham et al. (1999)[3]	81–90	89–99	98–107	81–87	99–105	118–124	100–109	117–127	136–145	14.21–14.70
Durango (Green)	Laslett et al. (1987)	96.8	103.9	110.6	97.6	112.4	128.4	113.6	128.2	144.0	15.24
Durango (Carlson)	Ketcham et al. (1999)[3]	92.2	101.1	109.4	94.2	112.3	131.6	111.5	129.3	147.9	14.72

[1] Range of model values for Vrolijk et al. (1991) thermal history. For F-apatite models, range is based on 15.78-16.42 μm range of unannealed induced track lengths of F-apatites from Carlson et al. (1999). For Durango models, assumed initial length is 16.21 μm.

[2] Results are bounds defined by five near-endmember F-apatites from Carlson et al. (1999): PQ, RN, SC, UN and WK.

[3] Ketcham et al. (1999) results are for model with *c*-axis projected lengths fitted using fanning curvilinear equation; results for $l_{m,low\text{-}T}$ are converted back to non-projected means.

The more segments a time-temperature path is subdivided into, the more accurate the numerical solution will be. Conversely, an excessive number of time steps will slow computation down unnecessarily. The optimal time step size to achieve a desired solution accuracy was examined in detail by Issler (1996b), who demonstrated that time steps should be smaller as the total annealing temperature of apatite is approached. For the Ketcham et al. (1999) annealing model for F-apatite, Ketcham et al. (2000) found that 0.5% precision is assured if there is no step with greater than a 3.5 °C change within 10 °C of the F-apatite total annealing temperature. The total annealing temperature (T_A) for F-apatite for a given heating or cooling rate (R) is given by the equation

$$T_A = 377.67\, R^{0.019837} \tag{10}$$

where T_A is in Kelvins and R is in K/Myr.

Estimation of mean annealed length. The length distribution for a population of fission tracks formed during a particular time interval is typically defined by a mean length (either non-projected or *c*-axis projected) and a dispersion about that mean. The evolution of the mean with time and temperature is described by one of the calibration equations discussed previously. For annealing equations not defined as time integrals, the calculation of the mean requires employing the equivalent time hypothesis, which posits that the annealing behavior of a track is solely a function of its length, and not to the prior time-temperature history (Duddy et al. 1988). It should be noted that, although the equivalent time hypothesis has been verified on laboratory time scales, it cannot be tested on geological ones, and this extrapolation therefore should be added to the long list of assumptions to be kept in mind (Jonckheere 2003).

One way to understand the equivalent time hypothesis is to see how it is used to model track annealing over the course of a time-temperature path. In the time interval in which the population is formed, it undergoes the amount of annealing corresponding to the duration of the interval and the mean temperature experienced (under the simplifying assumption that all tracks formed at the beginning of the interval). For the next *t-T* path segment, the mean temperature is determined, and the time required to achieve the current level of annealing at this current temperature is determined by solving the annealing equation for length. This "equivalent time" is then added to the duration of the interval, which is then used to calculate a new annealed length.

Equivalent time can also be used to speed up the calculation by applying it backwards in time (Crowley 1993a). If the annealed length is calculated for the final time step (i.e., the one that includes the present) first, then this length can be used to calculate an equivalent time that can be applied to the *previous* time step, resulting in the complete estimate of annealing for the population formed in that time step. The equivalent time of that time step can then be applied to the one before, and so on. The entire time-temperature path thus only needs to be traversed once, in the reverse direction, rather than forward from each time step.

Initial track length. Most of the annealing calibrations discussed above characterize reduced length (l/l_0). To arrive at an actual mean length (or mean *c*-axis projected length), the result of these calculations must be multiplied by an "initial" track length. A number of considerations underlie the choice of what value to use. Traditionally, it has been the unannealed induced track length of the apatite upon which the calibration was based, such as Durango. In an ideal sense this value should be measured by the analyst whose data are being modeled, as it will vary somewhat due to differences in etching procedures, microscope equipment, and analyst tendencies. However, given the almost 0.7 μm variation in the mean unannealed induced length of F-apatites reported by Carlson et al. (1999), it is clear that a single value does not adequately describe a random apatite.

Carlson et al. (1999) also demonstrated that initial length is correlated with kinetic variables such as Cl content and, particularly intuitively, D_{par}. They report the relations:

$$l_{0,m}\ (\mu m) = 16.18 + 0.544\ Cl\ (apfu);\ R^2 = 0.63 \quad (11a)$$

$$l_{0,m}\ (\mu m) = 15.63 + 0.283\ D_{par}\ (\mu m);\ R^2 = 0.82 \quad (11b)$$

$$l_{0,c,mod}\ (\mu m) = 16.49 + 0.407\ Cl\ (apfu);\ R^2 = 0.64 \quad (11c)$$

$$l_{0,c,mod}\ (\mu m) = 16.10 + 0.205\ D_{par}\ (\mu m);\ R^2 = 0.81 \quad (11d)$$

Again, these relations are only strictly valid for the analyst who made the Carlson et al. (1999) measurements, and in particular for the etching procedure used for that study. Re-calibration for other laboratories and etching protocols is possible by making induced length measurements of some subset of the Carlson et al. (1999) apatites (e.g., Sobel et al. 2004).

It should also be noted that although the Laslett and Galbraith (1996) annealing equation is based on actual rather than normalized length, its predictions should still be re-normalized to account for differences in etching and measuring procedures among laboratories.

Estimation of length distribution shape. The final component of estimating the characteristics of the track population for a particular time interval is to characterize the dispersion about the mean. This is usually done by assuming that track lengths are normally distributed about the mean, for which the standard deviation is an adequate descriptor. The standard deviation as a function of mean length (Fig. 6) is obtainable from any annealing data set, and similar for all of them.

Figure 6a shows that, for non-projected lengths, standard deviation rises considerably as annealing progresses. This is due to the increasing change in annealing behavior with respect to crystallographic angle. Eventually the change in behavior becomes so large that track length distributions become increasingly kurtotic and skewed, and occasionally even bimodal. This change in distribution shape has two implications. First, the standard error is no longer a good estimate of the uncertainty of the mean. Second, model length distributions calculated using this short-cut are inaccurate. The extent to which this inaccuracy contaminates model predictions is probably minor in most instances, as highly-annealed populations tend to be infrequently observed when mixed with less-annealed ones. However, in certain cases, such as boreholes studies where the deepest samples have exclusively highly-annealed tracks, the mismatch between model and reality may have a more serious impact on model reliability. A reasonable rule of thumb is that whenever tracks less than 10 μm are measured, they probably reflect a non-normal population in the data. In any event, the consequences of this simplification have not been studied in detail.

When lengths are *c*-axis projected, their distributions tend to be much closer to normally distributed and remain so at all stages of annealing, as shown by Figure 6b. For this case, the assumption of a Gaussian distribution is better-supported.

Summation and biasing of populations. Once the track length populations for all time steps have been calculated, they must be combined to create a length distribution for comparison to measured data. However, rather than an estimate of the "true" length distribution, what is required is more indirect: the distribution of likely samples from the true underlying distribution. Because the likelihood of observing a particular track is a function of its length and angle, samples from a heterogeneous distribution will tend to be biased in favor of the more easily observed tracks. Weighting factors thus must be introduced to replicate this effect. These factors can be separated into the relative number of tracks formed during each time step, and the relative probability of observing tracks from each step, or observational bias.

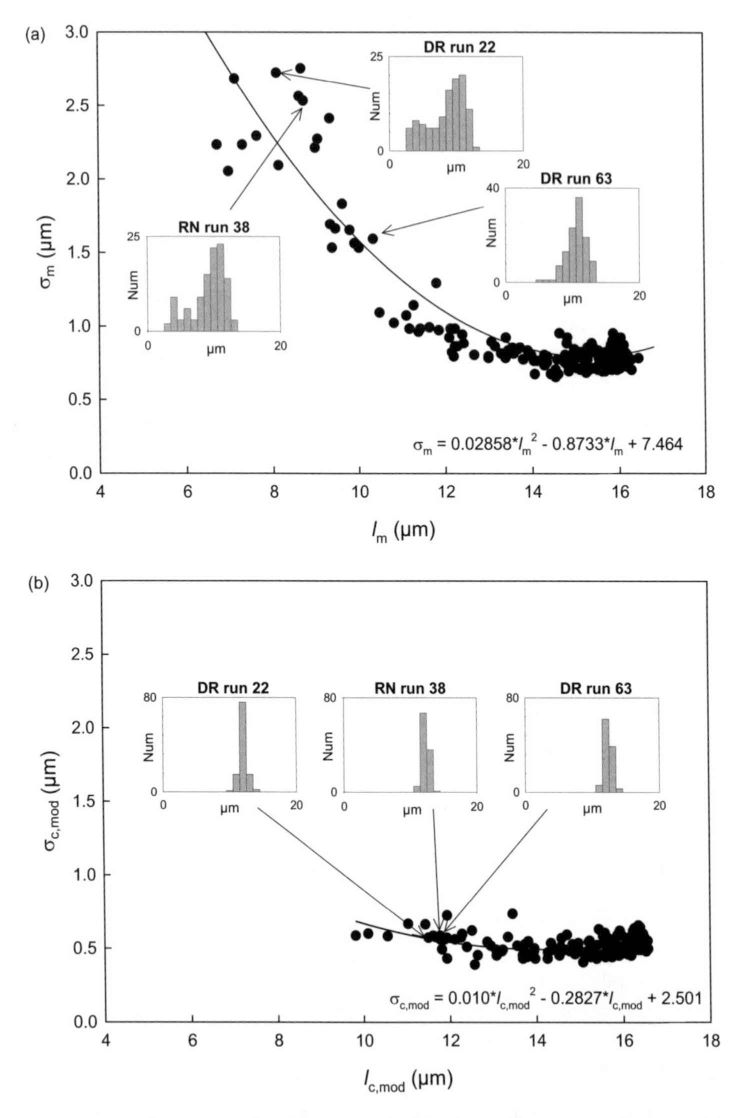

Figure 6. (a) Relationship of non-projected length mean versus standard deviation for F-apatites and Durango apatite from Carlson et al. (1999); histograms show various length distributions, which become skewed and bimodal as annealing progresses. (b) Relationship of *c*-axis projected mean versus standard deviation of same apatites, based on projection model of Ketcham (2003b), with histograms of same experiments showing that the distributions have become more Gaussian, with dispersion growing only very slightly with increasing annealing.

The number of tracks formed during a time step is the product of its duration and its concentration of uranium relative to other time steps. In young samples the concentration of uranium can be approximated as being constant through time but as sample age increases the depletion due to radioactive decay becomes increasingly significant. For example, uranium concentration and thus fission track production in any given apatite was 1% higher at 64.2 Ma. The equation for the weighting factor w is:

$$w(n) = (t_2 - t_1)\frac{\int_{t_1}^{t_2} e^{\lambda t}dt}{t_2 - t_1} = \left(e^{\lambda t_2} - e^{\lambda t_1}\right)/\lambda \tag{12}$$

where time step n is bounded by times t_1 and t_2, in m.y. before present, and λ is the total decay constant for ^{238}U in My^{-1} (Ketcham et al. 2000).

The observational bias quantifies the relative probability of observation among the different fission-track populations calculated by the model. Highly annealed populations are less likely to be detected and measured than less-annealed populations for two primary reasons. First, shorter tracks are less frequently impinged and thus etched (e.g., Laslett et al. 1982). Second, at advanced stages of annealing some proportion of tracks at high angles to the c axis may be lost altogether, even though lower-angle tracks remain long; thus the number of detectable tracks in the more-annealed population diminishes, at a rate disproportionate to measured mean length (Ketcham 2003b). These two factors can be approximated in a general way by using an empirical function that relates measured fission-track length to fission-track density (e.g., Green 1988). For example, the model of Willett (1997) uses a pair of line segments fit to the Green (1988) data. Ketcham et al. (2000) use the similar relation:

$$\rho = 1.600 r_{c,\mathrm{mod}} - 0.600, \quad r_{c,\mathrm{mod}} \geq 0.765 \tag{13a}$$

$$\rho = 9.205 r_{c,\mathrm{mod}}^2 - 9.157 r_{c,\mathrm{mod}} + 2.269, \quad r_{c,\mathrm{mod}} < 0.765 \tag{13b}$$

where ρ is normalized fission-track density, which is used as a proxy for observational frequency, and $r_{c,\mathrm{mod}}$ is reduced c-axis projected length ($l_{c,\mathrm{mod}}/l_{0,c,\mathrm{mod}}$).

When ^{252}Cf irradiation is used to enhance detection of confined tracks (Donelick and Miller 1991), it is more likely that all available tracks will be seen and therefore the observational bias due to length is expected to decrease (Ketcham 2005). Although the effect of bias removal within a single track population can be ameliorated by c-axis projection, the shift in bias between track populations has not yet been quantified.

Age calculation

Age is derived from the length distribution calculation by assuming that each time step of length Δt will contribute Δt to the total fission-track age, modified by the amount of track density reduction of the population in that time step relative to the age standard (Willett 1992). The total age is then the sum of the contributions of each population:

$$age = \frac{1}{\rho_{st}} \sum_i \rho_i \Delta t_i \tag{14}$$

where ρ_{st} is the estimated fission-track density reduction in the age standard, ρ_i is the fission-track density reduction in the population for time step i, and Δt_i is the duration of time step i. The density reduction in the age standard is calculated using its estimated track length reduction, using the assumption that density reduction is proportional to length reduction, and that spontaneous fission tracks are initially as long as induced tracks. For example, the Durango apatite has a measured present-day spontaneous mean track length of ~14.47 μm (Donelick and Miller 1991) and a mean induced track length of ~16.21 μm (Carlson et al. 1999), then ρ_{st} = 14.47/16.21 = 0.893. Density reduction for individual model fission-track populations is estimated using the same conversion between track length reduction and track density reduction used for population observational biasing (Eqn. 13a,b). The correct length value to use is the one at the midpoint of a time step, which can be approximated as the mean of the endpoints (Ketcham et al. 2000).

Oldest track

In addition to reporting an age and track length distribution for a given thermal history, HeFTy also reports the "oldest track," which corresponds to the earliest-formed track population that has not entirely annealed by model's end. This number defines the earliest time constrained by fission-track data in that model. It is often useful for interpretive purposes, especially when interpreting samples from sedimentary settings. For example, in thermal histories featuring reheating, an "oldest track" that post-dates the thermal peak indicates that the peak reheating temperature is not constrained by the FT data. Similarly, if the oldest track predates the provenance age then it indicates that the fission-track data may contain an inherited (pre-depositional) component, which may be a source of extra-Poissonian variation in the data.

Example FT forward models

A few examples serve to illustrate the effects of time-temperature history and kinetic variability on fission-track length distributions. Figure 7a-c illustrates the "classic" fission-track length distributions from fast cooling, slow cooling, and reheating (Green et al. 1989). Fast cooling through the apatite so-called partial-annealing zone (Fig. 7a) results in a relatively long, non-skewed length distribution. Slow cooling (Fig. 7b) brings about a negatively skewed length distribution, as the older tracks have spent more time at high temperatures, resulting in more shortening. If reheating extends into the partial annealing zone, where annealing rates are fast but not so fast as to fully erase tracks (usually cited as 60–110 °C for F-apatite), a bimodal distribution can result (Fig. 7c), with the short-length peak corresponding to tracks that formed before the reheating event and the long-length peak consisting of post-reheating tracks. In all cases, the *c*-axis projected length results (thin curves) are similar to the non-projected ones (thick curves), except that they have higher means and tighter distributions.

Figure 7d demonstrates that kinetic variability can yield bimodal length distributions without a reheating event. The upper distribution is for an apatite with the same kinetics as for the other models (D_{par} = 1.75 μm), and the lower one has a somewhat larger D_{par} of 2.5 μm. According to the Ketcham et al. (1999) annealing model, this results in an upward shift in index temperatures of about 25 °C, and correspondingly raises the partial annealing zone as well. The long residence time in the upper part of the partial annealing zone of the former apatite and the lower part of the partial annealing zone of the latter causes the distribution to have distinctly different modes. If there are roughly equal proportions of each type of apatite, the net distribution would be bimodal.

FORWARD MODELING OF THE (U-Th)/He SYSTEM

The (U-Th)/He system is, in principle, quite simple. In a mineral grain that contain U and Th, these atoms and their decay products undergo alpha decay throughout its history, producing ^{4}He. This He can be retained within the mineral, but can also be lost by diffusion to the grain margin. In addition, because alpha particles do not stop until they have traveled on the order of 20 microns from their parent atom, some proportion of ^{4}He produced near the margin will be ejected from the grain. Modeling this system thus requires quantifying alpha production and stopping combined with diffusion within a solid crystal.

A number of computational approaches to this problem have been put forward. Wolf et al. (1998) provide an analytical solution for isothermal calculations for spherical grains, and use a finite difference scheme for arbitrary histories but do not describe it. Meesters and Dunai (2002a,b) use the eigenmode method with versions for a number of grain geometries, including a sphere, finite cylinder, and rectangular bock of any aspect ratio. Their model also provides for zoning, albeit only for the binary case (i.e., zones with and without U and Th).

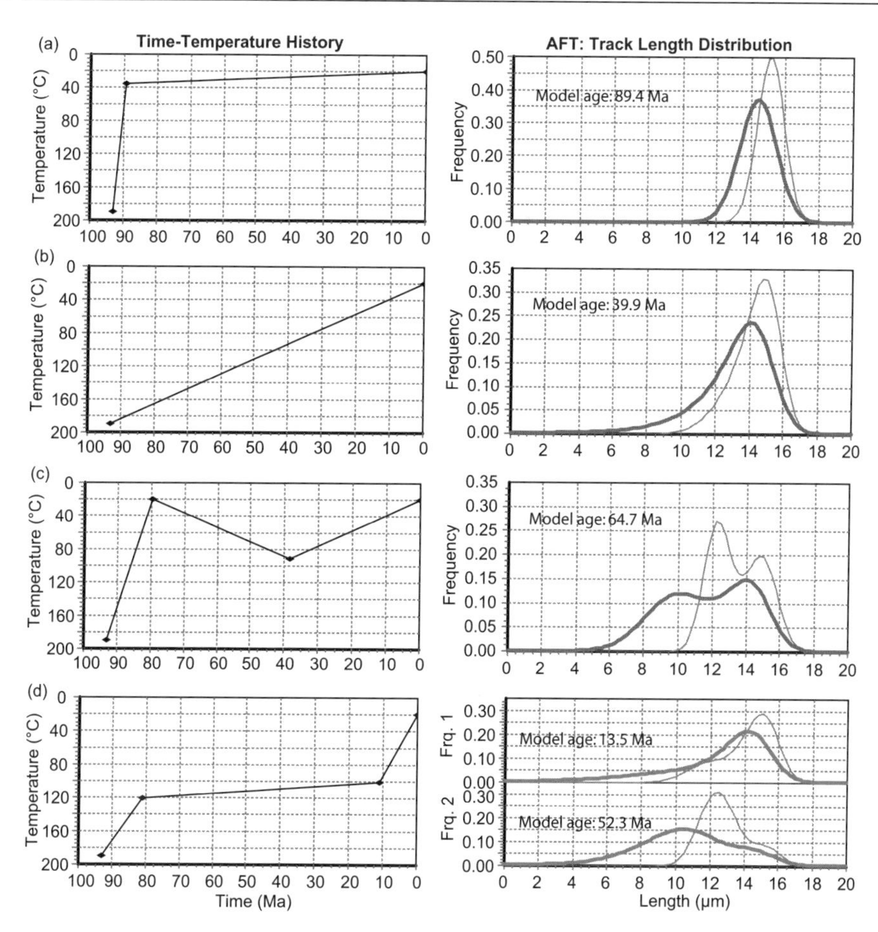

Figure 7. Example fission-track length distributions (right column) for various time-temperatures histories (left column) and apatite kinetics. In length distributions, thick line is for unprojected lengths, and thin curve is for *c*-axis projected lengths. (a) Fast cooling. (b) Constant cooling. (c) Reheating. (d) Cooling only, for two kinetic populations. All models based on Ketcham et al. (1999), D_{par} = 1.75 μm, except (d) lower length distribution, which has D_{par} = 2.50 μm.

Alpha particle ejection is only modeled for spherical grains. Ehlers (2001) provides a finite element solution, but without zoning or alpha ejection.

This section outlines a finite difference (FD) solution for a spherical grain, with provisions for long alpha stopping distances and concentric zoning of U and Th. The FD method, while venerable, has the advantages of speed and stability; the model discussed here is implemented in HeFTy and calculates quickly enough (several solutions per second) that it can be connected to a click-and-drag graphical interface.

Equations defining the (U-Th)/He dating system

The basic equation for age is the total amount of ^4He divided by the production rate. The amount of ^4He generated between a certain time *t* in the past and the present day is:

$$^4He = 8\,^{238}U\left(e^{\lambda_{238}t}-1\right)+7\,^{235}U\left(e^{\lambda_{235}t}-1\right)+6\,^{232}Th\left(e^{\lambda_{232}t}-1\right) \qquad (15)$$

where ^{238}U, ^{235}U and ^{232}Th are the present-day concentrations of these elements, and the λ's are their corresponding decay constants. Wolf et al. (1998) advise simplifying the calculation by omitting ^{235}U, as it produces <0.5% of total system He, and neglecting diminishment of U and Th concentrations over time due to radioactive decay. This latter step is reasonable in many situations because of the low closure temperature of the apatite He system (causing a tendency towards low ages) and long half-lives of the isotopes; however, it is less appropriate for higher-closure-temperature systems. Because a general solution is desired, these simplifications are not employed here.

The computational task at hand can be summarized as follows: generate ^{4}He at the correct rate through time, and convert a total amount of ^{4}He into an age given present-day concentrations of U and Th. In both cases, zoning should be accounted for which alpha ejection is a natural extension. ^{4}He generation at any particular position i in a zoned crystal between two times t_1 and t_2 is represented simply by:

$$^4He_i = 8\,^{238}U_i\left(e^{\lambda_{238}t_2} - e^{\lambda_{238}t_1}\right) + 7\,^{235}U_i\left(e^{\lambda_{235}t_2} - e^{\lambda_{235}t_1}\right) + 6\,^{232}Th_i\left(e^{\lambda_{232}t_2} - e^{\lambda_{232}t_1}\right) \quad (16)$$

To convert a final ^{4}He profile to an age t in a zoned crystal, we substitute 0 for t_1 (making all t_1 exponentials equal to 1), t for t_2, and sum over all positions, giving:

$$S_{\text{He}} + S_{232} + S_{235} + S_{238} = S_{232}e^{\lambda_{232}t} + S_{235}e^{\lambda_{235}t} + S_{238}e^{\lambda_{238}t} \quad (17)$$

where S indicates a summation for that isotope over the whole crystal, weighted by volume (or area, in the case of an infinite cylinder). Note that when calculating volume or area, the constant (4π/3 or π) cancels, so we can omit it. This is problematic to solve for unique t, but the answer can easily be found iteratively.

Calibration

The main parameters required to model the (U-Th)/He system are the variables describing the diffusion of ^{4}He through the crystal structure of the mineral being studied. Widely-used minerals with good calibration data include apatite (Farley 2000), zircon (Reiners et al. 2004), and titanite (Reiners and Farley 1999). This is characteristically done using a series of step-heating experiments, as described elsewhere in this volume. In general it is assumed that only one set of diffusion parameters is necessary to describe diffusion in a crystal, although there is some evidence that diffusion rates in apatite may be faster perpendicular to the c-axis than parallel (Farley 2000). The details of obtaining and interpreting diffusion measurements are discussed elsewhere in this volume.

Finite difference solution

The finite difference solution (and most others) makes several assumptions to simplify and speed up the calculations. One of these is symmetry: three-dimensional shapes such as spheres and infinite cylinders can be reduced to a single dimension (radius), while finite cylinders can be reduced to two dimensions. In this chapter the spherical solution is described. Although most crystal grains that are dated are not terribly spherical, Meesters and Dunai (2002a) find that a spherical solution gives answers roughly equivalent to other shapes if the sphere size is chosen to give it the same surface to volume ratio. However, if diffusion is anisotropic to a significant degree then alternative approximations may be required. For example, Farley (2000) suggests that if diffusion rates are faster perpendicular than parallel to the apatite c axis then an infinite cylinder may be a more appropriate model.

Another assumption is that all ^{4}He diffusing out of a crystal is immediately removed from the vicinity, probably by fast grain boundary diffusion. This allows the edge of the crystal to be treated as an infinite, instantaneous ^{4}He sink. However, there may be cases where this assumption is not true, such as instances where there is excess He in the intergranular fluid (Kohn et al. 2003), or where grain boundary paths are sealed (Belton et al. 2004).

Solving the diffusion equation for a sphere. The basic equation for diffusion in a sphere (Carslaw and Jaeger 1959) is:

$$\frac{\partial v}{\partial t} = \kappa \frac{1}{r^2}\frac{\partial}{\partial r}\left(r^2 \frac{\partial v}{\partial r}\right) + A_0 \tag{18}$$

where v is the diffusing quantity (helium), κ is diffusivity, r is radial position, and A_0 is the rate of ^{4}He generation. The central term is simplified using the u substitution $u = vr$, which results in:

$$\frac{\partial u}{\partial t} = \kappa \frac{\partial^2 u}{\partial r^2} + A_0 r \tag{19}$$

which can be solved on a one-dimensional grid. The Crank-Nicholson finite difference solution for this equation is (e.g., Press et al. 1988):

$$\frac{u_j^{n+1} - u_j^n}{\Delta t} = \frac{\kappa}{2}\frac{\left(u_{j+1}^{n+1} - 2u_j^{n+1} + u_{j-1}^{n+1}\right) + \left(u_{j+1}^n - 2u_j^n + u_{j-1}^n\right)}{\Delta r^2} + A_j r \tag{20}$$

where the j subscript indicates nodes (numbered 1 to *rad*), the n superscript indicates time step number, Δt is time step duration, Δr is grid spacing, and A_j is the He production rate. In this scheme, concentric zoning is implemented simply by radially varying A. Using the common summarization $\beta = 2\Delta r^2/\kappa\Delta t$, the tridiagonal system of equations is obtained:

$$(-2-\beta)u_j^{n+1} + u_{j-1}^{n+1} + u_{j+1}^{n+1} = (2-\beta)u_j^n - u_{j-1}^n - u_{j+1}^n - A_j r \beta \Delta t \tag{21}$$

Once the solution is found, u has to be converted back to v. This provides the He diffusion profile as a function of radius, which may be useful if the shape of the He profile can be measured or inferred.

Setting up the grid and boundary conditions. A one-dimensional grid is set up from the center to the edge of the sphere, as outlined in Figure 8. The boundary condition at the edge simulates constant zero ^{4}He, corresponding to the usually reasonable assumption that any

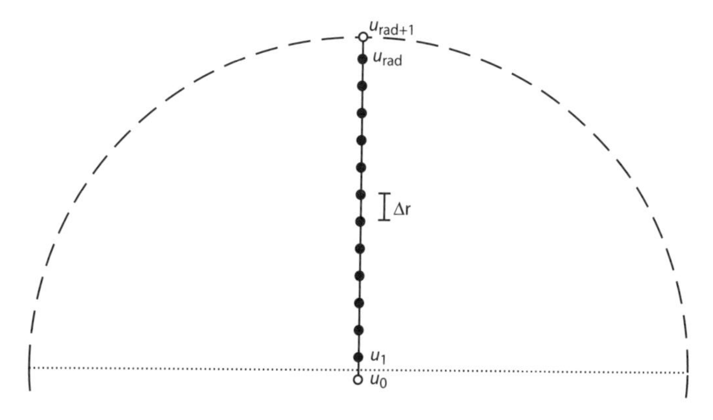

Figure 8. Schematic illustration of finite difference grid for spherical solution. Dashed outline shows sphere radial outline and dotted line shows mid-plane. Closed circles are real grid points, and open circles are imaginary grid points used to implement boundary conditions. In the actual solution, many more grid points (513) are used.

and all ^{4}He that diffuses to the edge is instantly taken away by diffusing quickly along grain boundaries. The *Dirichlet* condition at the edge of the sphere is implemented by:

$$u^n_{rad+1} = u^{n+1}_{rad+1} \tag{22}$$

where *rad* is the outermost calculated node, and *rad*+1 is the (imaginary) node at the position of the sphere radius.

At the center node a zero-flux *Neumann* boundary condition is used, which needs to be changed somewhat because of the variable transformation. The desired condition is:

$$\frac{dv}{dr} = 0 \quad \text{at} \quad r = 0 \tag{23}$$

Because of the transformation to u, this condition is applied as:

$$\frac{du}{dr} = \frac{d(rv)}{dr} = r\frac{dv}{dr} + v\frac{dr}{dr} = \frac{u}{r} \tag{24}$$

To set up the grid, the innermost node (index 1) is placed at position $\Delta r/2$ (where Δr is the node spacing), where it faces imaginary node $-\Delta r/2$ across the center of the sphere (Fig. 8). Using the approximation that $u(\Delta r/2)$ is sufficiently close to $u(0)$ to be considered equal, the finite difference solution is

$$\frac{u_1^n - u_0^n}{\Delta r} = \frac{u_1^n}{r} \tag{25}$$

where subscript 1 indicates the first node and subscript 0 indicates the imaginary node. Because at the first node $r = \Delta r/2$, this solves to the simple (if somewhat unexpected)

$$u_0^n = -u_1^n \tag{26}$$

which is the exact opposite of the normal implementation of the Neumann condition. Testing shows that these approximations are accurate to within a few hundredths of a percent of the exact solution.

Incorporating long alpha stopping distances. During alpha decay the ejected alpha particle travels a certain distance from its parent atom, on the order of 20 microns. This can lead to loss of helium at the edges of crystals, leading to lower net concentrations in the crystal compared to expectation for accumulation with no loss. Farley et al. (1996) analyzed this effect in terms of the fraction of decays lost and retained, and provide a correction that consists simply of dividing the age by the fraction retained. However, this treatment neglects the effect of ejection on diffusion. Farley (Farley 2000, p. 2910) claims that the result of ejection is to lower the edge gradient, reducing diffusion rates and increasing retentivity and closure temperature; this seems to be supported by an unpublished calculation based on Wolf et al. (1998). However, from an alternative viewpoint the effect of ejection can be seen as pushing the effective boundary inwards, increasing the effective diffusion rate.

Alpha ejection can be modeled straightforwardly by reducing the ^{4}He production rate of nodes near the edge of the spherical crystal, based on stopping distance (a function of isotope and mineral). Following Farley et al. (1996), the proportion of ^{4}He lost from decays at a certain position X_i in a crystal of radius R is equal to the fraction of the surface area of a sphere centered at that point with ejection radius S that lies outside of the crystal (Fig. 9). To calculate this, the required parameter is the location of the plane defining the intersection of the two spheres, X^*. Using the fact that the cosine of the angle across from S (θ) is X^*/R, and the law of cosines: $S^2 = Xi^2 + R^2 - 2X_iR \cos\theta$, we get: $X^* = (X_i^2 + R^2 - S^2)/(2X_i)$. The surface area of the partial

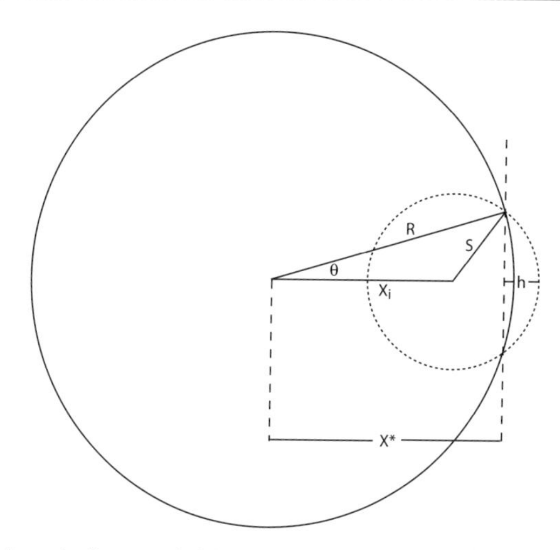

Figure 9. Schematic diagram of alpha ejection calculation. Variables explained in text.

ejection sphere is $2\pi\, S\, h$, where h is given by $(S - (X^* - X_i))$. Dividing through by the surface area of the whole sphere, $4\pi S^2$, the fraction lost is $0.5 - (X^* - X_i)/2S$. The fraction retained is the complement, $0.5 + (X^* - X_i)/2S$. Incorporating this into the numerical simulation is as easy as multiplying the He production rate at edge nodes by this factor. In HeFTy this is currently done individually using averages for ^{238}U, ^{235}U and ^{232}Th, as provided in Farley et al. (1996). Additional accuracy can be obtained by calculating this factor for each decay product; this is likely to be a very minor effect, however.

A wrinkle in this scheme is the fact that long alpha stopping distances also result in a *redistribution* of ^4He within the crystal. If there is no zoning this redistribution cancels out, but if zoning is present then the effect is to smooth things out somewhat, transferring ^4He from high-production areas into nearby low-production areas. To model this, the effective production rate at any particular location needs to reflect the amount of U and Th within a surrounding shell, rather than at that location.

Effective production is calculated by averaging the ^4He contribution in the shells corresponding to each alpha ejection radius. If we take the shell for one decay series and integrate it along X as used above, then the intersection of each infinitesimally thin region subtended by dX and the shell will have the same surface area. Taking ^4He production A as a function of radial position X, the effective production at position i then is

$$A_{eff,i} = \frac{\int_{X_i-S}^{X_i+S} A(X')\,dX}{\int_{X_i-S}^{X_i+S} dX} \tag{27}$$

where X' is the radial position of the shell edge with respect to the spherical crystal. The shell edge position has the Y coordinate $(S^2 - (X - X_i)^2)^{1/2}$, and X' is $(X^2 + Y^2)^{1/2}$. Plugging everything in and simplifying a little bit, we get:

$$A_{eff,i} = \frac{\int_{X_i-S}^{X_i+S} A\left[\left(S^2 + 2XX_i - X_i^2\right)^{1/2}\right]dX}{2S} \tag{28}$$

This can easily be numerically approximated by summing over arbitrarily small dX. Note that each isotope will have its own $A_{eff,i}$, as each will have its own stopping radius S. This value is then converted to a depletion D by dividing by actual production:

$$D_i = \frac{A_{eff,i}}{A_i} \tag{29}$$

These calculations can also be used to estimate an age correction due to alpha ejection. The correction (factor F_T in Farley et al., 1996) corresponds to the alpha ejection corrected production rates divided by the non-corrected production rate. Assuming that the rate can be simplified as an instantaneous quantity, and summing over volumetric shells, this comes to:

$$F_T = \frac{\sum_{i=0}^{rad} V_i \left(8\lambda_{238}\,{}^{238}U_i D_{238,i} + 7\lambda_{235}\,{}^{235}U_i D_{235,i} + 6\lambda_{232}\,{}^{232}Th_i D_{232,i}\right)}{\sum_{i=0}^{rad} V_i \left(8\lambda_{238}\,{}^{238}U_i + 7\lambda_{235}\,{}^{235}U_i + 6\lambda_{232}\,{}^{232}Th_i\right)} \tag{30}$$

where V_i is the volume of shell i, and we've explicitly specified the depletion factors for each isotope. Interestingly, Meesters and Dunai (2002b) also provide a solution for redistribution, using a different approach, and arrive at a different but equivalent result.

Calculating the age. Once the radial He profile has been determined, the age is calculated using Equation (17) above. However, a significant issue is how to calculate the sums, particularly for He. If He production is uniform (the simplest case), then He will vary from core to rim according to a function with first and second derivatives that are always negative. This systematic bias has a bad impact on any linear interpolation scheme; as a corollary, the choice of integration method is quite important. Romberg integration (Press et al. 1988) is an effective antidote, as it uses estimates based on progressively increasing numbers of nodes to extrapolate a "correct" solution. The implementation used requires that the number of nodes be a power of 2 plus 1, but this does not present a problem.

Example He forward models

A few examples serve to illustrate the general effects of time-temperature path shape and zoning on He forward models. Figure 10 illustrates the difference in He radial profiles and ages for a 100-μm radius apatite given three general histories: early-cooling, linear cooling, and late-cooling. The early cooling case (Fig. 10a) shows a distinct corner and slope at the edge, caused by alpha ejection. The slight slope and rounding reflect that even at 20 °C there is some appreciable He loss. As cooling is pushed later and made to proceed more slowly through the apatite He partial retention zone (Fig. 10b,c), the slope becomes more rounded, and the age falls.

Figure 11 illustrates the effect that zoning can have on diffusional profiles and ages. Fairly modest zoning at the rim (Fig. 11a) affects the zoning profile considerably (compare to Fig. 10b), and lowers the age by 3%, owing to the proportionally increased loss in He from alpha ejection. Similar zoning at the core (Fig. 11b) has a more subtle effect on the diffusional profile, and does not appreciably affect the age, as all He is retained. Zoning in zircon can be much more severe than in apatite, and Fig. 11c shows an example zircon profile (J. Hourigan, unpub. data) in which U and Th contents span up to three orders of magnitude. In this case, the zoning raises the model age by 5% (over 13% if ages are not alpha-corrected), presumably because the He has a much greater distance to travel through the crystal lattice to leave the grain. These examples make it clear that any attempt to utilize He zoning to extract thermal history information will require measurement of U and Th zoning as well.

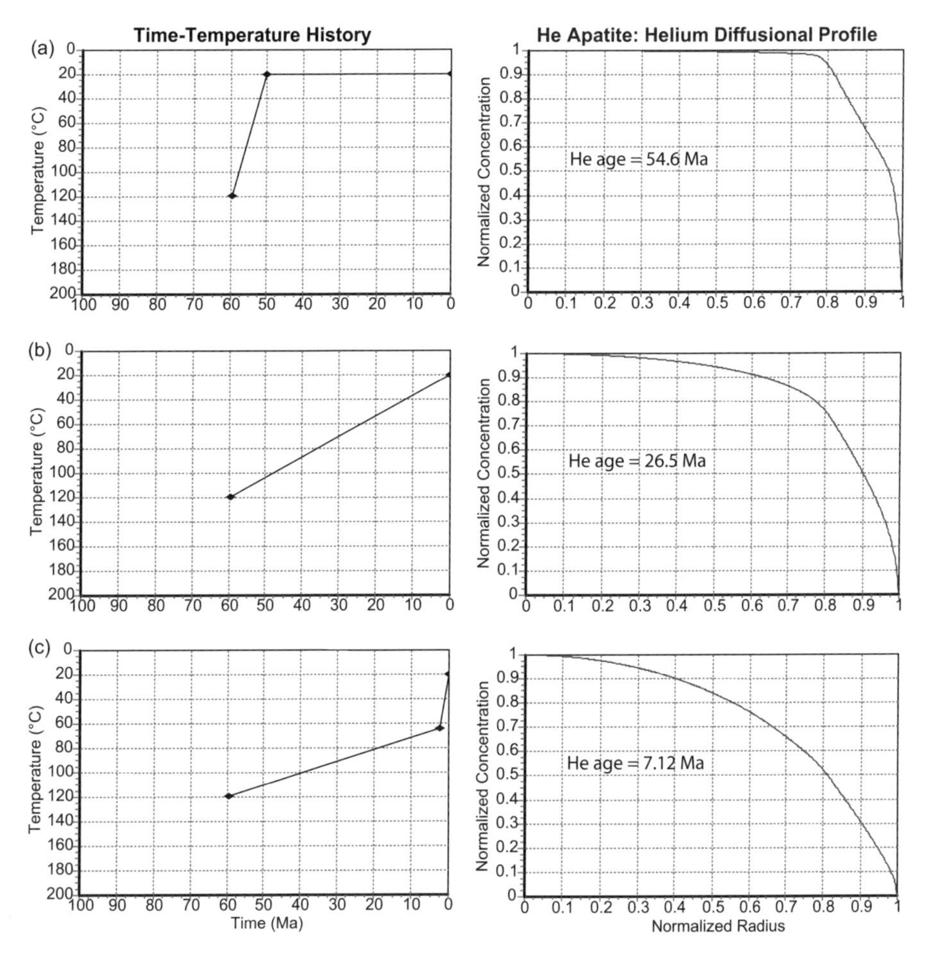

Figure 10. Example He diffusional profiles for various time-temperature histories. All models calculated using the Farley (2000) Durango apatite diffusivity values and a grain diameter of 100 μm. U and Th are uniformly distributed. (a) Early cooling. (b) Linear cooling. (c) Late cooling. Ages are reported using the alpha correction factor of Farley et al. (1996).

INVERSE MODELING

The inverse problem in thermochronology can be stated thus: given one or more measurements describing the present-day condition of a thermochronometric system, and an assumed starting condition, what can be determined about its time-temperature history? A number of computational solutions have been proposed (Corrigan 1991; Crowley 1993a; Gallagher 1995; Hadler et al. 2001; Issler 1996a; Ketcham et al. 2000; Lutz and Omar 1991; Willett 1997), all of which share these five basic features:

- A theoretical annealing model that predicts how the system evolves as a function of time and temperature
- An algorithmic means of calculating model evolution continuously over a time-temperature path
- A statistical means for comparing the model calculations to measurements

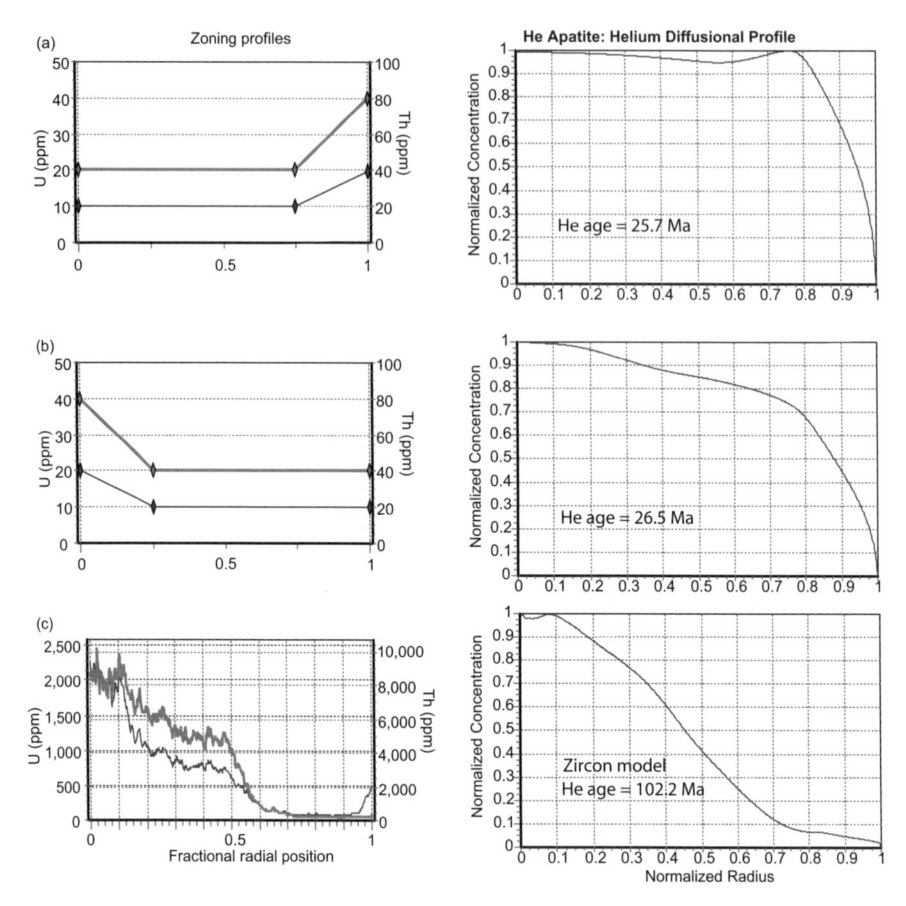

Figure 11. Example He diffusional profiles for different zoning patterns of U (thin lines) and Th (thick lines), given a linear cooling history (Fig. 10b). (a) Zoning at the rim, starting at 0.75 fraction radius and rising linearly to twice core values. (b) Zoning at the core, starting at 0.25 fractional radius and rising to twice rim values. (c) Zoning as measured in a zircon grain by LA-ICPMS (J. Hourigan, unpub. data). For this run zircon diffusivity values were used (Reiners et al. 2004) and linear cooling began at 140 Ma and 250 °C. Age without zoning = 97.2 Ma, a difference of 5% (corrected) and over 13% (uncorrected).

- A strategy for posing candidate thermal histories to be tested against the data, and possibly searching for optimal fits between model and data
- A means of showing the range of possible thermal histories that are consistent with the measured data

The first two points have been dealt with in previous sections for the fission-track and (U-Th)/He systems, and the final three will be discussed here.

Statistical tests

Two approaches to statistically comparing fission-track model predictions to measurements have been proposed. One is to use separate statistics to evaluate the goodness of fit for the age and the length distribution (e.g., Corrigan 1991; Ketcham et al. 2000; Lutz and Omar 1991; Willett 1997), whereas the other evaluates the joint likelihood of obtaining the individual lengths and counts measured (Gallagher 1995). The former approach is adopted and described in detail here, but is compared to the latter at the end of this section. The former approach has three components: a goodness-of-fit measure for the age (which can also be used for (U-Th)/He ages), a goodness-of-fit measure for the length distribution, and a means of combining multiple measures into a single merit function value.

Length distribution goodness of fit. The simplest measure for evaluating the degree of fit between fission-track length distributions is the Kolmogorov-Smirnov (K-S) test (Press et al. 1988; Willett 1997). The K-S test relies on two parameters: the maximum separation between two cumulative distribution functions (cdf) representing the measured and model track lengths, and the number of observations comprising the measured cdf. In the version of the K-S test used here, it is assumed that the model cdf describes a continuous, completely known distribution, and the measured data are a finite sample from some unknown, underlying cdf. The result of the test is the probability that a set of samples taken randomly from the model distribution would have a greater maximum separation from it on a cdf plot than is the case for the data distribution. For example, a K-S probability of 0.05 means that, if N random samples were taken from the distribution described by the calculation result, where N is the number of fission track lengths actually measured, there would be a 5% chance that the resulting distribution would have a greater maximum separation from the model on a cdf plot than is observed between the data and the model. The 0.05 value thus describes the traditional 95% confidence boundary. The expected value for the case in which the measured track lengths are in fact samples from the model distribution is 0.5: 50% of sample populations from the model distribution would have a greater separation, and 50% would have a lesser one. The 50% limit thus marks a logical boundary of statistical precision for the track-length distribution according to the K-S test.

Age goodness of fit. Most of the merit functions that have been proposed for evaluating ages are based on the assumption that the uncertainty of the age measurement is normally distributed (and thus symmetric) about the measured value (τ_{meas}), and defined by estimated standard deviation (σ_{meas}). Ketcham et al. (2000) use for their goodness-of-fit function the proportion of samples from this distribution that would be further away from the measured than the model age (τ_{mod}):

$$GOF_{age} = 1 - \int_{\tau_{meas} - |\tau_{meas} - \tau_{mod}|}^{\tau_{meas} + |\tau_{meas} - \tau_{mod}|} \frac{1}{\sqrt{2\pi}} e^{-(x-\tau_{meas})^2 / 2\sigma_{meas}^2} dx \tag{31}$$

where x is an integration variable. An advantage of this scheme is that the GOF value has an easily interpretable meaning that is analogous to the K-S test result, only we have reversed the "known" and "sample" entities. A value of 0.05 means that 5% of possible random samples from the distribution described by the data are further away from the measured age than the model age, and the expected value for a random sample taken from the data distribution is 0.5.

This approach can be improved upon by recognizing that the uncertainty about the measured age in fact cannot be completely symmetric, insofar as it can never go below zero. If the uncertainty in the age determination is large, this asymmetry can be appreciable. This is particularly the case in AFT age determinations of young samples, as the uncertainty is proportional to the square root of the relatively few counts.

A more rigorous method for estimating and reporting uncertainties is given by Galbraith (1984) and Galbraith and Laslett (1985). Following their example, recast the fission-track age equation as:

$$\tau = a \ln\left(1 + b e^{\gamma}\right); \quad \gamma = \ln\left(\zeta \rho_d \frac{\rho_s}{\rho_i}\right) \tag{32}$$

where τ is the age, ζ is the analyst-dependent zeta calibration factor determined using age standards, ρ_s, ρ_i and ρ_d are the spontaneous, induced and dosimeter densities respectively, and a and b are constants. All of the uncertainties in the age determination are within the γ term. The motivation behind this recasting is that the sampling distribution of the estimate of γ should be closer to normal (Gaussian) than the sampling distribution of the estimate of τ. Using standard error propagation techniques, it can be shown that the total standard error in the γ term is:

$$\sigma_\gamma = \sqrt{\frac{1}{N_s} + \frac{1}{N_i} + \frac{1}{N_d} + SE_\zeta^2} \tag{33}$$

where SE_ζ is equal to σ_ζ/ζ. The 95% confidence limits for τ are then calculated by adding and subtracting $2\sigma_\gamma$ from γ in Equation (32). To estimate the goodness of fit of a model result (τ_{mod}) to a measured age (τ_{meas}) in the asymmetric case, we calculate the difference in γ,

$$\gamma_{mod} - \gamma_{meas} = \ln\left(\frac{e^{\frac{\tau_{mod}}{a}} - 1}{e^{\frac{\tau_{meas}}{a}} - 1}\right) \tag{34}$$

and compare it to σ_γ using an analogue to Equation (31) above.

Combined merit function. The pair of tests described above has the advantage that they both describe essentially the same quantity: the "probability of a worse fit." In other words, each describes the probability that, were the model *t-T* path truly correct and the annealing calculation accurate, a set of fission-track lengths or an age measured from a population described by the model would be less similar to the model than the data. The general equivalence in meaning suggests that there should be no need to weight these statistics to ascribe more importance to one than the other (e.g., Corrigan 1991; Lutz and Omar 1991).

In order to combine these two statistics into a single merit function for evaluating thermal histories against each other, one can simply take their minimum (Willett 1997). When multiple kinetic populations or thermochronometers are being evaluated, the minimum statistic value across all populations is used (Ketcham et al. 2000). When the minimum is above 0.05, all statistics pass the 95% confidence test, and by convention the model is termed *acceptable*. When the minimum is above 0.5, the statistical precision limit, the model is termed *good*.

Comparison with likelihood method. Gallagher (1995) proposed using the following log-likelihood function to evaluate goodness-of-fit:

$$L = \sum_{j=i}^{N_c} \left\{ N_s^j \ln[\theta] + N_i^j \ln[1-\theta] \right\} + \sum_{k=1}^{N_l} \ln\left[P(l_k)\right] \tag{35}$$

where $\theta = \rho_s/(\rho_s + \rho_i)$, with ρ_s and ρ_i estimated from the model calculations; N_c and N_l are the number of grains counted and tracks measured, N_s, N_i, and l are the individual measurements, and:

$$P(l_k) = \int_{l_k - 0.1}^{l_k + 0.1} f(L)\, dL \tag{36}$$

where $f(L)$ is the model-generated length distribution. This scheme is clearly "closer to the data," in that the statistics are based on equations specifically describing the physics and geometry of track generation, revelation, and measurement, albeit in an idealized and simplified way. However, it has the possible weakness that the log-likelihood function contains no *a priori* expectation of results; values do not have the same straightforward interpretation as the statistics discussed previously. The log-likelihood is simply a number to be maximized, and model resolution is defined in terms of closeness to the maximum likelihood value. It is thus conceivable that a problem in the inversion process, whether it be data that do not fit the assumptions behind the model and statistics (bad measurements, kinetic variability, multiple provenance), or improper user input (annealing calibration choice, initial length specification, thermal history constraints) could escape notice, whereas the same problem may prevent the statistics with expected values from ever signifying a "good" result, raising a warning flag. Overall, both statistical schemes are based on similar underlying principles, and should give broadly similar results, although no direct comparisons have been made.

Defining and searching candidate thermal histories

A number of schemes have been proposed for defining the range of time-temperature paths that can be used to attempt to match the measured data. They can be broadly classified based on form, degrees of freedom, and how they map out the confidence interval of the solution.

The two main forms are sets of line segments (Lutz and Omar 1991; Gallagher 1995; Issler 1996a; Willett 1997; Ketcham et al. 2000; Hadler et al. 2001) and sets of smooth polynomial functions (Corrigan 1991). In practical terms the difference between these approaches is minor, with the main potential difference being that line segment models may predict slightly higher peak temperatures in thermal histories featuring reheating. Ideally, there should be a means to constrain them to pass through certain regions of time-temperature space, and/or be monotonic (heating or cooling only) when passing through certain regions of time-temperature space; an example from HeFTy is shown in Figure 12.

The differences brought about by varying degrees of freedom, however, are more significant. One set of inversion schemes employs relatively few (<10) time-temperature nodes (Lutz and Omar 1991; Gallagher 1995; Hadler et al. 2001) or polynomials (Corrigan 1991; Lovera et al. 1997; Quidelleur et al. 1997), whereas the other set (Issler 1996a; Willett 1997; Ketcham et al. 2000) generally uses more (10–50+). This divergence reflects a shift in philosophy that is quite influential. The former schemes embody the principle that the

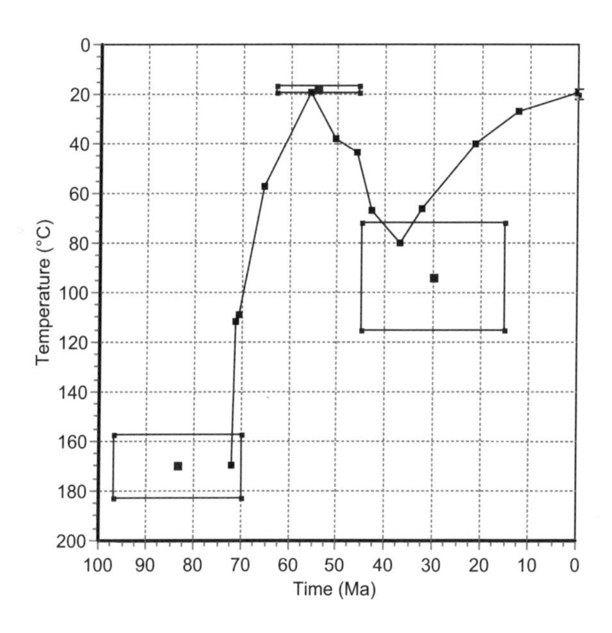

Figure 12. Example of time-temperature constraints from HeFTy. Boxes define regions in *t-T* space through which thermal history must pass, and all paths between constraints are constrained to be monotonic (heating- or cooling-only), as shown in example path.

simplest answer is the best one, and thus the thermal history that fits the data while using the fewest nodal points is to be preferred over more complex ones. The latter approaches take the viewpoint that geological histories are likely to have a certain amount of complexity, and it is the job of the inversion scheme to map out the uncertainty that this complexity implies. This issue will be discussed further in the final section.

There are also two overall approaches to searching among the candidate thermal histories for adequately-fitting solutions. The first is to use a simple Monte Carlo approach, generating and evaluating a large number (usually some tens of thousands) of independent time-temperature paths. The second is to use an optimization algorithm that iteratively uses already-tried thermal histories to eventually arrive at one with the best goodness-of-fit statistic value; useful algorithms have included simulated annealing (Corrigan 1991), the genetic algorithm and Powell's method (Gallagher 1995), and the simplex method (Lutz and Omar 1991) and related algorithms such as constrained random search (Willett 1997; Ketcham et al. 2000). Another upshot of the number of time-temperature nodes permitted is that these algorithms tend to work best when there are fewer degrees of freedom, as many-node models are more likely to trap optimization algorithms into local minima or maxima. Because of this, in such cases the Monte Carlo method is usually the most appropriate.

The last consideration in thermal history searching is how the confidence interval surrounding the solution is determined. When the Monte Carlo method is used, the software keeps track of each path tested, and reports the set of paths that pass whatever statistical criteria are being utilized. In such cases, it is almost certain that this mapping of permissible solutions will not be complete, and will change subtly every time an inversion is run. When optimization algorithms are used, the similar approach of recording close-fitting paths calculated on the way to the best solution (e.g., Corrigan 1991; Willett 1997; Ketcham et al. 2000) can be used. However, this procedure results in even more incomplete mapping of the solution space than the Monte Carlo method, as algorithm convergence will tend to restrict the extent of time-temperature space tested. A better alternative is to calculate divergences from the best solution in some systematic way. Gallagher (1995), for example, proposes various methods for calculating the amount that each node in the final solution can vary, although the most complete solutions are computationally expensive.

Presentation of inversion results

Generally, a completed inversion model will consist of either a set of paths passing one or more statistical criteria, or a single path with confidence intervals around each node. In the latter case, presentation is self-evident: the best path is plotted and the confidence intervals delineated. In the former case, presentation options include plotting all paths, plotting all nodal points corresponding to temperature minima and maxima, and envelopes surrounding or contours depicting some aspect of the range of solutions. Examples are shown in Figure 13.

Envelopes and contours can be the cleanest way of showing a solution set, but can be susceptible to misinterpretation. Envelopes bound all models passing a given statistical test (e.g., Ketcham et al. 2000); at any one time, the temperature bounds are defined by the highest and lowest temperatures experienced by any accepted t-T path at that time. Not all paths that stay within an envelope are valid, however. In particular, paths that follow the top or bottom margin of an envelope almost certainly would not pass the criterion in question. As a result, envelopes can give a somewhat overstated picture of the spread of solutions. Ketcham et al. (2000) use nested envelopes surrounding all "acceptable" and "good" solutions.

Contours have been the graphic of choice when polynomials are used to define t-T paths (Corrigan 1991). Contours reflect the number of valid paths passing through a given region of t-T space, with the idea that areas where more paths path through are more likely to be

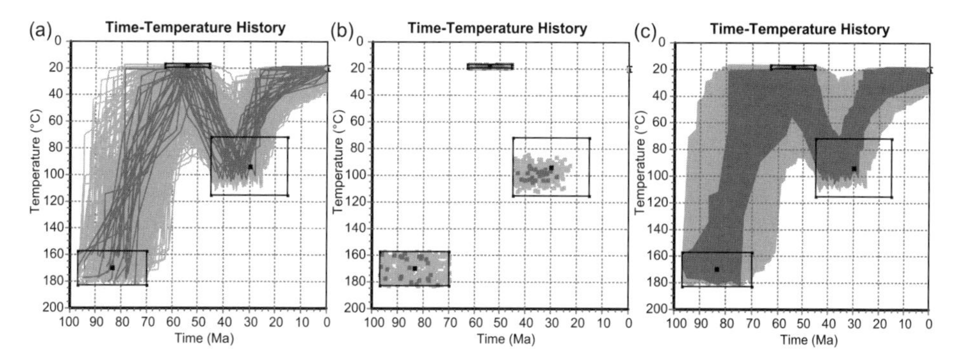

Figure 13. Various methods for displaying inversion results. In all cases, light = acceptable fit, dark = good fit. (a) All paths. (b) Minimum and maximum *t-T* points. (c) Envelopes around all paths fitting each criterion.

correct: more roads lead there than elsewhere. This interpretation requires tolerating a certain amount of circularity, however, as the distribution of paths tested depends in part on how they are generated.

EXECUTION AND INTERPRETATION OF INVERSE MODELING

Using inverse modeling software is part art and part science, at least judging by the range of ways these techniques have been employed. This is probably inevitable, as the best use of inverse modeling will depend on the geological situation, the geological problem being solved, and the thermochronologic data themselves. Given the range of objectives and circumstances, there are probably a number of ways of doing inverse modeling "right." However, there are also many more ways of doing it wrong, which should of course be avoided.

This final section lists some guidelines for approaching inverse modeling and reporting the results, generally following recommendations outlined by Ketcham et al. (2000), adapted for use of HeFTy. First and foremost, the data must be evaluated to see if they are worth modeling, or what additional work might be necessary to make them so. Next, fitting conditions must be selected carefully, in particular with regard to user-imposed constraints on the form and course of the thermal history. Finally, the results should be interpreted judiciously (humbly) and parameters reported completely.

Data requirements. The rule of thumb here is that the data should be as simple as the assumptions used to model them. If the data show some sign of complexity, such as multiple kinetic populations in apatite fission-track or significant zoning in (U-Th)/He, don't use a modeling approach that assumes that such complexity is absent. Either get whatever additional data are necessary to document these complexities, such as measuring a kinetic parameter for every fission-track grain or measuring zoning with LA-ICPMS, or leave that particular sample out of the modeling effort.

For the fission-track method, simple first-cut checks are the chi-squared test (Green 1981) and the radial plot (Galbraith and Laslett 1993). If these indicate that single-grain ages are not drawn from a single population, then it is likely that this sample has multiple inheritance and/or multiple kinetic populations. For apatite fission-track data it is highly advisable that some kinetic information be acquired even if there is no evidence of variability, so that one can choose an appropriate annealing calibration. An interesting variation on this theme is to use binomial peak fitting (e.g., Galbraith and Green 1990; Brandon 1996) to segregate single-grain ages into populations, and then use some measured characteristics of those populations to estimate their kinetic properties. In this case, the use of length data would be limited to those measured in grains on which ages were determined as well.

Specification of constraints, and "allowed complexity." In general, all time-temperature histories should begin at a sufficiently high temperature to ensure that there is total annealing (i.e., no fission tracks present) and/or no retained He as an initial condition. Thus, the earliest *t-T* constraint should have a minimum temperature above the total annealing temperature of the most resistant apatite being modeled, or significantly above the closure temperature of the most retentive (U-Th)/He thermochronometer. An exception to this principle might be the modeling of volcanic rocks where it is known *a priori* that the initial condition is represented by no daughter products present. The time of the initial constraint should be somewhat earlier than the fission-track age of the most resistant kinetic population being modeled, to account for the phenomenon of age reduction by partial annealing.

The final constraint in time should correspond to the temperature at which the sample was collected. Another constraint might correspond to the depositional time and assumed temperature of the sedimentary sequence from which the sample was obtained.

In HeFTy, constraints are entered as a box surrounding a section of time-temperature space. Candidate paths are constructed by randomizing a point within each constraint, and connecting these points with line segments. These line segments can either be straight, or they can be broken down into sub-segments to allow greater variability or "noise" in the path. The purpose in doing so is not so much to try to extract additional detail from the data, but rather to admit that one doesn't know how complex the true history was. This concept is referred to as *allowed complexity* (Ketcham 2003a).

An example demonstrates this idea. Figure 14a shows a bimodal fission-track length distribution, and four constraints for fitting it: a high-temperature initial condition, low-temperature conditions corresponding to deposition and present-day, between which a high-temperature condition is added to allow reheating. This reheating event can be surmised from the fact that the fission-track age is younger than the depositional age, indicating partial annealing. Figure 14b shows the range of solutions if uninterrupted line segments are used to connect randomly generated points within the constraints. From this model, it appears that the time and temperature of peak reheating is fairly well constrained by the data. (The alert reader may notice that this distribution is truncated on the right-hand side of this constraint. This indicates that additional acceptable solutions would likely be found if the range of permissible paths is expanded. In general this should be done, unless the position of the constraint is controlled by independent information.)

The situation changes markedly if we allow variation in the shape of the heating and/or cooling paths (Fig. 14c,d). For these models, one additional degree of freedom is allowed on the respective path segments, with the only condition being that the paths must remain monotonic (cooling-only or reheating-only) in that area. Each additional degree of freedom expands considerably the range of statistically good locations for the reheating peak. If we add more complexity all along the path (Fig. 14e), the model appears to have lost its ability to constrain the heating peak time and location at all—the fact that the solution space extends to all boundaries of the reheating constraint indicates that the complete solution space is considerably larger.

How should this seeming loss of focus be interpreted? One line of argument holds that the extra degrees of freedom are undesirable because the data can be adequately explained with "simpler" models having fewer nodes, invoking Occam's Razor ("*Pluralitas non est ponenda sine neccesitate*"). A contrary viewpoint is that leaving out these additional nodes amounts

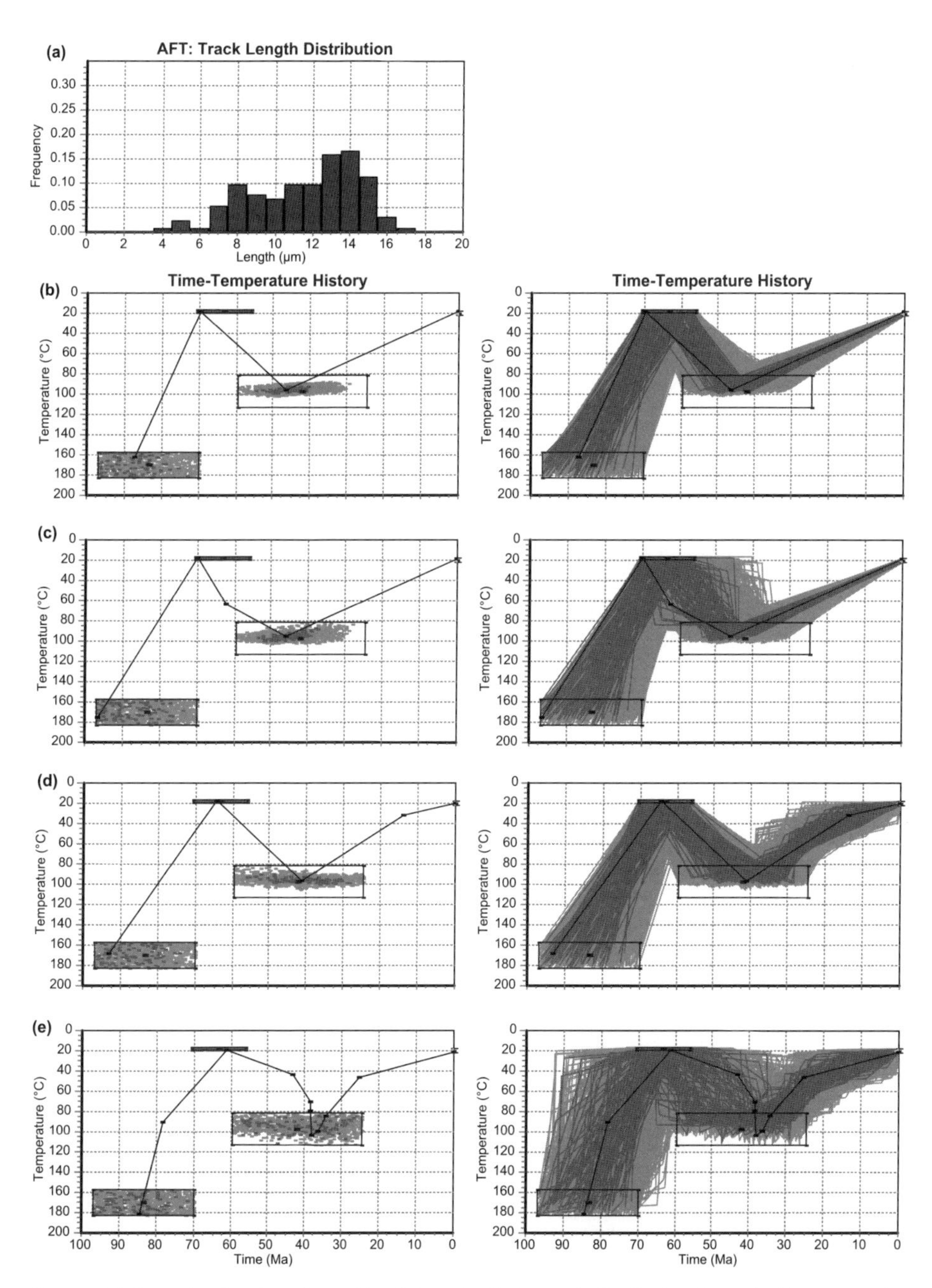

Figure 14. Example of various fits to bimodal apatite fission-track distribution shown in (a); measured age = 58.5 ± 4.8 Ma. (b) Only nodal points within constraints allowed. (c) One extra nodal point along reheating path. (d) One extra nodal point along final cooling path. (e) One extra node along initial cooling paths, and three along reheating and final cooling paths. For each inversion, random paths were generated until 100 "good" paths were found.

to making a very strong and not easily justifiable assumption: that heating and cooling paths must be strictly linear. The additional nodes relax this condition, replacing it with the weaker assumption that we simply don't know the shape of the heating or cooling paths.

Which viewpoint is preferable? This question might be answered by looking at the objective of inverse modeling. If one is seeking the simplest, *single* thermal history that explains the data best, then fewer degrees of freedom are preferable. However, insofar as the usual goal is to document the *range* of statistically consistent thermal histories, assuming strict linearity of long segments of the *t-T* path really amounts to a very influential but essentially hidden prior condition being imposed on the solution set.

How much allowed complexity is sufficient to reasonably document the range of variability in thermal histories? This question is being addressed by work in progress, but ultimately the answer is not so much to have the "correct" amount of complexity as to explicitly state and defend the amount of complexity the one allows into the inversion process. In other words, allowed complexity (or lack thereof) should be elevated from an unstated implicit assumption to an explicitly stated decision that influences how one's results should be interpreted.

One way that complexity can be constrained is to limit how additional nodes are randomized. HeFTy defines three scales of complexity that can be imposed on a *t-T* path segment, which are denoted by corresponding suggested geological interpretations: *episodic*, *intermediate*, and *gradual*. Episodic segments allow sudden changes in heating and cooling rates, and correspond to geological mechanisms that might cause relatively rapid thermal effects, such as faulting, magmatism, and changes in fluid flow conditions. Gradual segments would correspond to geological mechanisms such as slow burial or erosional unroofing. Which type to choose depends of course on the geological environment that one is studying, or at least one's assumptions about it that are presumably based on independent information. The number of additional nodes is also settable, and should probably correspond in some sense to the presumed time scale of geological events, such as the duration of sedimentation vs. quiescent intervals, or the interval between faulting episodes. Figure 15a shows an example of how these conditions can impact the range of solutions. By limiting the degree of randomness on the reheating and late cooling paths, the solution space has been limited somewhat in comparison to Figure 14e. Further constraints might serve to impose maximum heating and cooling rates over selected parts of the history.

Multiple thermal indicators. A potentially powerful capability being developed in current modeling software is the ability to work with multiple thermal indicators simultaneously. This can range from modeling multiple kinetic populations in the apatite fission-track system to combining different systems such as (U-Th)/He or vitrinite reflectance (%Ro). Each new indicator can constrain a different aspect of the thermal history. For example, (U-Th)/He can restrict the range of acceptable thermal histories in the region of the closure temperature of the mineral used, while vitrinite can help pinpoint the peak reheating temperature.

As an example, Figure 15b replicates the conditions of Figure 15a, with the additional datum of a 29.6 ± 1.4 Ma (U-Th)/He determination from a 100 μm radius apatite grain. The timing and temperature of the reheating peak is somewhat better resolved, and the cooling path is much better constrained in the ~60–70 °C region. In Figure 15c a vitrinite determination of 0.59 ± 0.04 %Ro has been added to the mix. Here the cooling path is no better defined, but the peak temperature is limited to about the upper 60% of the range defined in the previous inversion. The apparent expansion in the range of "acceptable" solution stems from the inversion setup: in each case the model stopped after 100 "good" paths were found. Because each additional datum makes such solutions harder to find, more Monte Carlo iterations are required to generate them, and more "acceptable" paths are found along the way, as listed in the figure caption.

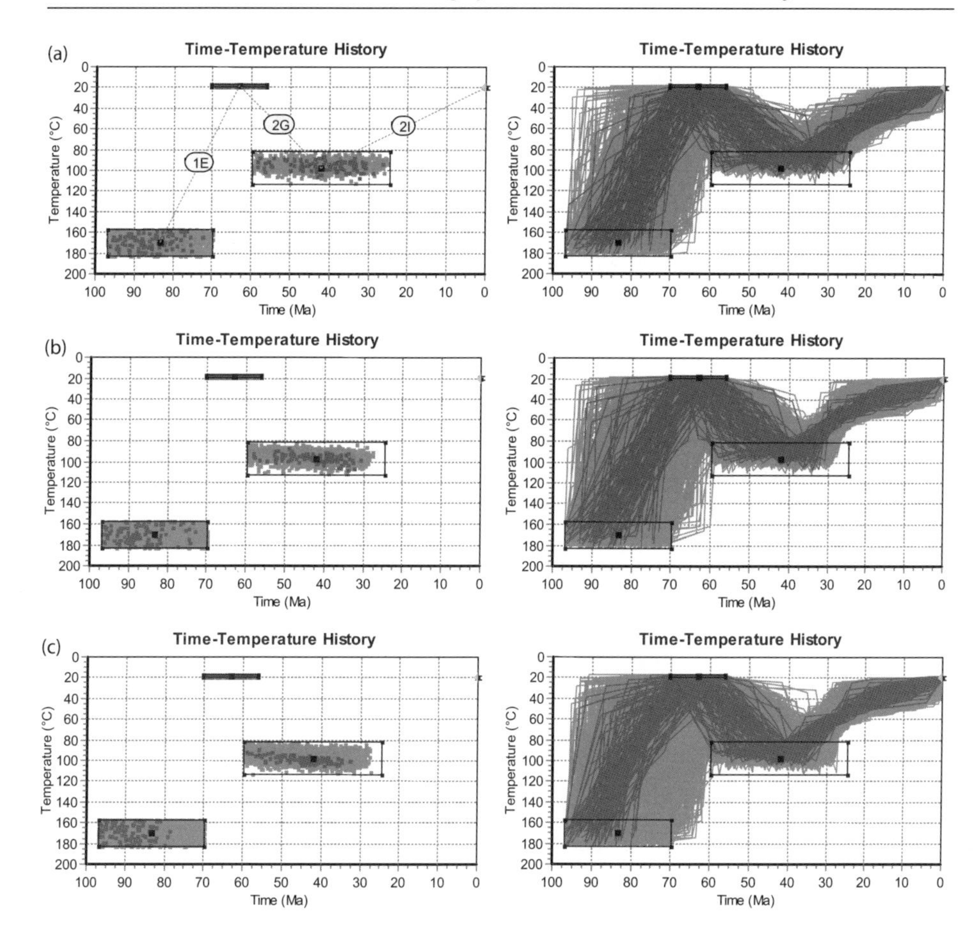

Figure 15. Part (a) shows a fit to apatite fission track data with similar conditions to Figure 14e except degree of complexity along different parts of history has been limited. In left diagram, numbers indicated number of times path between constraints is halved, and letter indicates randomizer settings of episodic ("E"), intermediate ("I") and gradual ("G"). Part (b) is the same as (a), with addition of a single (U-Th)/He apatite datum to further constrain range of solution. Part (c) is the same as (b), with addition of vitrinite reflectance datum. In order to obtain 100 paths with statistically "good" fits, for (a) 11,223 paths were generated and 1,165 "acceptable" paths found; for (b) 42,563 paths were generated and 1,771 acceptable paths found; for (c) 86,178 paths were generated and 3,700 acceptable paths found.

Reporting and interpreting results. Given the impact that inversion model setup has on results, it is abundantly clear that investigators who utilize these models should report all of the inputs used. If nothing else, this is needed to satisfy the requirement of reporting sufficient details to make one's work reproducible. In HeFTy, time-temperature constraints and allowed complexity might be reported using a diagram showing these conditions (e.g., Fig. 15a, left), either independently or with the model result superimposed. Additional parameters that should be provided for fission-track analysis include the annealing calibration, kinetic parameters measured and/or assumed, and initial track length; for (U-Th)/He, parameters include the diffusion calibration, solution geometry and dimension (e.g., sphere of a given radius), type of alpha correction, and whether and how zoning was included.

Finally we come to interpretation. In the opinion of this author, the healthiest way of looking at modeling is as the product of a very extensive series of assumptions. These include the nature of the data, the depiction of the natural system and the calibration of that depiction, and what range of geological histories should be tested. A successful fit constitutes a confirmation that these assumptions are mutually consistent, and then shows what subset of thermal histories is consistent with them. It does not constitute a proof these thermal histories are the only ones possible, however. Inverse modeling does not prove that a particular version of events is required by the data, but rather corroborates and refines a hypothesis about them.

In addition, it pays to keep in mind that there are sources of uncertainty that are not included in the statistics that we use to compare model and data. For example, the uncertainties in the parameters in the Ketcham et al. (1999) calibrations imply a 1-sigma range of up to 2 °C in the temperature predictions of the various annealing models, to say nothing of the potential biases embedded in the empirical simplification of the annealing equations. The uncertainty in the initial track length probably leads to similar or even greater uncertainties in annealing temperatures. The poor reproducibility of length measurements is another significant concern: its effect, and the extent to which *c*-axis projection or some other scheme can correct for it, have also not been studied in detail. At this time there is no way to incorporate these uncertainties into modeling results or interpretations other than to mentally expand the inversion envelopes or contours by a few degrees.

AVAILABLE SOFTWARE

In addition to HeFTy, there are a number of computer programs that implement one or more of the systems discussed here. Table 3 lists some of the modeling software that is currently available to academic users, along with some of their attributes. Note that this list is not exhaustive, as most program authors are willing to share their software in some way, shape or form.

As discussed in the previous sections, these programs vary not only in how they implement their forward and inverse modeling calculations, but in their overall approaches to problem-solving as well. As a result, different geological problems may be best studied with different programs. It is left to the researcher to decide which tool best serves her or his needs and temperament.

Table 3. Partial list of publicly available FT and (U-Th)/He modeling software.

Software	*References*	*Platform*	*Inversion methods*[1]	*Systems modeled*
AFTINV	Issler (1996a), Willett (1997)	WIN	MC	AFT, Ro%
AFTSolve	Ketcham et al. (2000)	WIN	MC, CRS	AFT, Ro%
DECOMP	Meesters and Dunai (2002a,b)	WIN	None	(U-Th)/He
HeFTy	This chapter	WIN	MC	AFT, (U-Th)/He, Ro%
Monte Trax	Gallagher (1995)	MAC	GA	AFT
THA (Thermal History Analyzer)	Hadler et al. (2001)	WIN	MC	AFT

[1] Inversion methods: CRS – Constrained Random Search; GA – Genetic Algorithm; MC – Monte Carlo.

CLOSING THOUGHTS

An old rule of thumb in computer modeling is that any model is only as good as the data and assumptions that go into it; this principle is also succinctly stated "garbage in, garbage out." A corollary cliché is that one gets out of modeling what one puts into it, which in this case can refer to the care and effort an analyst takes in calibrating measurements and checking for sources of natural divergence from simplifying assumptions, such as kinetic variability for fission-track analysis or the presence of inclusions or zoning for (U-Th)/He. An additional facet of this train of thought, and probably less appreciated, is that modeling also embodies assumptions about more philosophical issues, such as what "simplicity" means and what a time-temperature history "should" look like. To the extent that a practitioner is conversant with this range of assumptions, inverse modeling is a powerful tool for extracting the maximum amount of information from thermochronologic data. To the extent that these assumptions are ignored, misunderstood, or taken for granted, it becomes increasingly possible to use inverse modeling to make unsupported or outright erroneous claims. Unfortunately, the statistics seem just as authoritative and the graphics look just as good either way.

ACKNOWLEDGMENTS

Thoughtful reviews by D. Issler and E. Sobel helped improve this chapter. Creation of HeFTy was supported in part by Apatite to Zircon, Inc., and P. Reiners and Yale University.

REFERENCES

Barbarand J, Carter A, Wood I, Hurford AJ (2003a) Compositional and structural control of fission-track annealing in apatite. Chem Geol 198:107-137

Barbarand J, Hurford AJ, Carter A (2003b) Variation in apatite fission-track length measurement: implications for thermal history modelling. Chem Geol 198:77-106

Belton DX, Kohn BP, Gleadow AJW (2004) Quantifying "excess Helium": some of the issues and assumptions in combined (U-Th)/He and fission track analysis. 10th International Conference on Fission-Track Dating and Thermochronology, Abstract code DVL-12-O

Bertel E, Märk TD (1983) Fission tracks in minerals: Annealing kinetics, track structure and age correction. Phys Chem Min 9:197-204

Brandon MT (1996) Probability density plot for fission-track grain-age samples. Rad Meas 26:663-676

Burtner RL, Nigrini A, Donelick RA (1994) Thermochronology of lower Cretaceous source rocks in the Idaho-Wyoming thrust belt. Am Assoc Petrol Geol Bull 78:1613-1636

Carlson WD (1990) Mechanisms and kinetics of apatite fission-track annealing. Am Mineral 75:1120-1139

Carlson WD, Donelick RA, Ketcham RA (1999) Variability of apatite fission-track annealing kinetics I: Experimental results. Am Mineral 84:1213-1223

Carslaw HS, Jaeger JC (1959) Conduction of Heat in Solids. Oxford University Press, Oxford

Corrigan JD (1991) Inversion of apatite fission track data for thermal history information. J Geophys Res 96: 10347-10360

Crowley KD (1985) Thermal significance of fission-track length distributions. Nucl Tracks 10:311-322

Crowley KD (1993a) Lenmodel: a forward model for calculating length distributions and fission-track ages in apatite. Computers Geosci 19:619-626

Crowley KD (1993b) Mechanisms and kinetics of apatite fission-track annealing - Discussion. Am Mineral 78:210-212

Crowley KD, Cameron M, Schaefer RL (1991) Experimental studies of annealing etched fission tracks in fluorapatite. Geochim Cosmochim Acta 55:1449-1465

Dakowski M, Burchart J, Galazka J (1974) Experimental formula for thermal fading of fission tracks in minerals and natural glasses. Bull Acad Pol Sci Ser Sci Terr 22:11-19

Dodson MH (1973) Closure temperature in cool geochronological and petrological systems. Contrib Mineral Petrol 40:259-274

Donelick RA (1991) Crystallographic orientation dependence of mean etchable fission track length in apatite: An empirical model and experimental observations. Am Mineral 76:83-91

Donelick RA (1993) A method of fission track analysis utilizing bulk chemical etching of apatite. U.S. Patent # 5,267,274
Donelick RA (1995) A method of fission track analysis utilizing bulk chemical etching of apatite. Australia Patent # 658,800
Donelick RA, Ketcham RA, Carlson WD (1999) Variability of apatite fission-track annealing kinetics II: Crystallographic orientation effects. Am Mineral 84:1224-1234
Donelick RA, Miller DS (1991) Enhanced TINT fission track densities in low spontaneous track density apatites using 252Cf-derived fission fragment tracks: A model and experimental observations. Nucl Tracks 18:301-307
Donelick RA, Roden MK, Mooers JD, Carpenter BS, Miller DS (1990) Etchable length reduction of induced fission tracks in apatite at room temperature (~23 °C): Crystallographic orientation effects and "initial" mean lengths. Nucl Tracks 17:261-265
Duddy IR, Green PF, Laslett GM (1988) Thermal annealing of fission tracks in apatite 3. Variable temperture behaviour. Chem Geol 73:25-38
Ehlers TA (2001) Geothermics of Exhumation and Erosion in the Wasatch Mountains, Utah. PhD Dissertation, University of Utah, Salt Lake City, Utah
Ehlers TA, Chaudhri T, Kumar S, Fuller CW, Willett SD, Ketcham RA, Brandon MT, Belton DX, Kohn BP, Gleadow AJW, Dunai TJ, Fu FQ (2005) Computational tools for low-temperature thermochronometer interpretation. Rev Mineral Geochem 58:589-
Farley KA (2000) Helium diffusion from apatite: General behavior as illustrated by Durango fluorapatite. J Geophys Res 105:2903-2914
Farley KA, Wolf RA, Silver LT (1996) The effects of long alpha-stopping distances on (U-Th)/He ages. Geochim Cosmochim Acta 60:4223-4229
Galbraith RF (1984) On statistical estimation in fission track dating. Math Geol 16:653-669
Galbraith RF, Green PF (1990) Estimating the component ages in a finite mixture. Nucl Tracks 17:197-206
Galbraith RF, Laslett GM (1985) Some remarks on statistical estimation in fission-track dating. Nucl Tracks 10:361-363
Galbraith RF, Laslett GM (1988) Some calculations relevant to thermal annealing of fission tracks in apatite. Proc R Soc Lond A 419:305-321
Galbraith RF, Laslett GM (1993) Statistical models for mixed fission track ages. Nucl Tracks 21:459-470
Galbraith RF, Laslett GM, Green PF, Duddy IR (1990) Apatite fission track analysis: geological thermal history analysis based on a three-dimensional random process of linear radiation damage. Phil Trans R Soc Lond A 332:419-438
Gallagher K (1995) Evolving temperature histories from apatite fission-track data. Earth Planet Sci Lett 136: 421-435
Gleadow AJW, Duddy IR (1981) A natural long-term track annealing experiment for apatite. Nucl Tracks 5: 169-174
Green PF (1981) A new look at statistics in fission-track dating. Nucl Tracks 5:77-86
Green PF (1988) The relationship between track shortening and fission track age reduction in apatite: Combined influences of inherent instability, annealing anisotropy, length bias and system calibration. Earth Planet Sci Lett 89:335-352
Green PF, Duddy IR, Gleadow AJW, Tingate PR, Laslett GM (1985) Fission-track annealing in apatite: track length measurements and the form of the Arrhenius plot. Nucl Tracks 10:323-328
Green PF, Duddy IR, Gleadow AJW, Tingate PR, Laslett GM (1986) Thermal annealing of fission tracks in apatite 1. A qualitative description. Chem Geol 59:237-253
Green PF, Duddy IR, Laslett GM, Hegarty KA, Gleadow AJW, Lovering JF (1989) Thermal annealing of fission tracks in apatite 4. Quantitative modeling techniques and extension to geological time scales. Chem Geol 79:155-182
Green PF, Durrani SA (1977) Annealing studies of tracks in crystals. Nucl Tracks 1:33-39
Green PF, Laslett GM, Duddy IR (1993) Mechanisms and kinetics of apatite fission-track annealing - Discussion. Am Mineral 78:441-445
Hadler JC, Paulo SR, Iunes PJ, Tello SCA, Balestrieri ML, Bigazzi G, Curvo EAC, Hackspacher P (2001) A PC compatible Brazilian software for obtaining thermal histories using apatite fission track analysis. Rad Meas 34:149-154
Hughes JM, Cameron M, Crowley KD (1989) Structural variations in natural F, OH, and Cl apatites. Am Mineral 74:870-876
Indrelid SL, Terken JMJ (2000) Constraints on the thermal history of the interior basins of the Sultanate of Oman using apatite fission track analysis (AFTA). 9th International Conference on Fission Track Dating and Thermochronology, Geological Society of Australia Abstracts Series 58:177-180
Issler DR (1996a) An inverse model for extracting thermal histories from apatite fission track data: instructions and software for the Windows 95 environment. Geological Survey of Canada, Open File 2325

Issler DR (1996b) Optimizing time step size for apatite fission track annealing models. Computers Geosci 22: 67-74

James K, Durrani SA (1986) The effect of crystal composition on fission-track annealing and closure temperatures in geological materials: Implications for the cooling rates of terrestrial and extraterrestrial rocks. Nucl Tracks 11:277-282

Jonckheere R (2003) On methodical problems in estimating geological temperature and time from measurements of fission tracks in apatite. Rad Meas 36:43-55

Jonckheere R, Wagner GA (2000) On the occurrence of anomalous fission tracks in apatite and titanite. Am Mineral 85:1744-1753

Kasuya M, Naeser CW (1988) The effect of α-damage on fission-track annealing in zircon. Nucl Tracks 14: 477-480

Ketcham RA (2003a) Effects of allowable complexity and multiple chronometers on thermal history inversion. Geochim Cosmochim Acta 67:A213

Ketcham RA (2003b) Observations on the relationship between crystallographic orientation and biasing in apatite fission-track measurements. Am Mineral 88:817-829

Ketcham RA (2005) The role of crystallographic angle in characterizing and modeling apatite fission-track length data. Rad Meas 39:595-601

Ketcham RA, Carter A, Barbarand J, Hurford AJ (2004) Analysis of the UCL apatite annealing data set. 10th International Conference on Fission-Track Dating and Thermochronology, Abstract code DVL-17-O

Ketcham RA, Donelick RA, Carlson WD (1999) Variability of apatite fission-track annealing kinetics III: Extrapolation to geological time scales. Am Mineral 84:1235-1255

Ketcham RA, Donelick RA, Donelick MB (2000) AFTSolve: A program for multi-kinetic modeling of apatite fission-track data. Geol Mat Res 2:(electronic)

Kohn BP, Gleadow AJW, Lorencak M, Belton DX (2003) Combined low temperature thermochronology in slowly-cooled terranes: Challenges and strategies. Geochim Cosmochim Acta 67:A226

Laslett GM, Galbraith RF (1996) Statistical modelling of thermal annealing of fission tracks in apatite. Geochim Cosmochim Acta 60:5117-5131

Laslett GM, Gleadow AJW, Duddy IR (1984) The relationship between fission track length and track density in apatite. Nucl Tracks 9:29-38

Laslett GM, Green PF, Duddy IR, Gleadow AJW (1987) Thermal annealing of fission tracks in apatite 2. A quantitative analysis. Chem Geol 65:1-13

Laslett GM, Kendall WS, Gleadow AJW, Duddy IR (1982) Bias in measurement of fission-track length distributions. Nucl Tracks 6:79-85

Lovera OM, Grove M, Harrison TM, Mahon KI (1997) Systematic analysis of K-feldspar $^{40}Ar/^{39}Ar$ step heating results I: Relevance of laboratory argon diffusion properties to nature. Geochim Cosmochim Acta 61:3171-3192

Lutz TM, Omar G (1991) An inverse method of modeling thermal histories from apatite fission-track data. Earth Planet Sci Lett 104:181-195

Meesters AGCA, Dunai TJ (2002a) Solving the production-diffusion equation for finite diffusion domains of various shapes Part I. Implications for low-temperature (U-Th)/He thermochronology. Chem Geol 186: 333-344

Meesters AGCA, Dunai TJ (2002b) Solving the production-diffusion equation for finite diffusion domains of various shapes Part II. Application to cases with α-ejection and nonhomogeneous distribution of source. Chem Geol 186:347-363

Miller DA, Crowley KD, Dokka RK, Galbraith RF, Kowallis BJ, Naeser CW (1993) Results of interlaboratory comparison of fission track ages for 1992 Fission Track Workshop. Nucl Tracks 21:565-573

Press WH, Flannery BP, Teukolsky SA, Vettering WT (1988) Numerical Recipes in C. Cambridge University Press, Cambridge

Quidelleur X, Grove M, Lovera OM, Harrison TM, Yin A (1997) The thermal evolution and slip history of the Renbu Zedong thrust, southeastern Tibet. J Geophys Res 102:2659-2679

Rahn MK, Brandon MT, Batt GE, Garver JI (2004) A zero-damage model for fission-track annealing in zircon. Am Mineral 89:473-484

Ravenhurst CE, Roden-Tice MK, Miller DS (2003) Thermal annealing of fission tracks in fluorapatite, chlorapatite, manganoapatite, and Durango apatite: experimental results. Can J Earth Sci 40:995-1007

Reiners PW, Farley KA (1999) Helium diffusion and (U-Th)/He thermochronometry of titanite. Geochim Cosmochim Acta 63:3845-3859

Reiners PW, Spell TL, Nicolescu S, Zanetti KA (2004) Zircon (U-Th)/He thermochronometry: He diffusion and comparisons with $^{40}Ar/^{39}Ar$ dating. Geochim Cosmochim Acta 68:1857-1887

Sobel EM, Seward D, Ruiz G, Kuonov A, Ege H, Wipf M, Krugh C (2004) Influence of etching conditions on D_{par} measurements: Implications for thermal modeling. 10th International Conference on Fission-Track Dating and Thermochronology, Abstract code DVL-15-O

Tagami T, Galbraith RF, Yamada R, Laslett GM (1998) Revised annealing kinetics of fission tracks in zircon and geological implications. *In*: Advances in Fission-Track Geochronology. Van Den Haute P, De Corte F (eds) Kluwer Academic Publishers, Netherlands, p 99-112

Vrolijk P, Donelick RA, Queng J, Cloos M (1992) Testing models of fission track annealing in apatite in a simple thermal setting: site 800, leg 129. *In*: Proceedings of the Ocean Drilling Program, Scientific Results. Vol 129. Larson RL, Lancelot Y (eds) Ocean Drilling Program, College Station, TX, p 169-176

Willett SD (1992) Modelling thermal annealing of fission tracks in apatite. *In*: Short Course Handbook on Low Temperature Thermochronology. Zentilli M, Reynolds PH (eds) Mineralogical Association of Canada, p 43-72

Willett SD (1997) Inverse modeling of annealing of fission tracks in apatite 1: A controlled random search method. Am J Sci 297:939-969

Wolf RA, Farley KA, Kass DM (1998) Modeling the temperature sensitivity of the apatite (U-Th)/He thermochronometer. Chem Geol 148:105-114

Yamada R, Murakami M, Tagami T (2003) Zircon fission track annealing: Short-term heating experiment toward the detection of frictional heat along active faults. Geochim Cosmochim Acta 67:A548

Yamada R, Tagami T, Nishimura S (1995a) Confined fission-track length measurement of zircon: assessment of factors affecting the paleotemperature estimate. Chem Geol 119:293-306

Yamada R, Tagami T, Nishimura S, Ito H (1995b) Annealing kinetics of fission tracks in zircon. Chem Geol 122:249-258

Reviews in Mineralogy & Geochemistry
Vol. 58, pp. 315-350, 2005

Crustal Thermal Processes and the Interpretation of Thermochronometer Data

Todd A. Ehlers

Department of Geological Sciences
University of Michigan
Ann Arbor, Michigan, 48109-1063, U.S.A.
tehlers@umich.edu

INTRODUCTION

Many thermochronology studies focus on extracting the thermal history of a sample using fission-track length distributions and track annealing models (e.g., Ketcham 2005, and references therein). Recent developments in measuring the $^3He/^4He$ concentration profile across apatite grains (e.g., Shuster and Farley 2005) offer a new approach for quantifying sample cooling histories over a broader range of temperatures with the (U-Th)/He system. The motivation behind calculating thermal histories from thermochronometer data has traditionally been to date events such as the onset of exhumation or fault motion, erosion, or for quantifying hydrocarbon maturation in sedimentary basins. The calculation of thermal histories from thermochronometer data is a routine aspect of most thermochronology studies and produces valuable geologic information.

A relevant question for thermochronology studies is what can be learned from forward modeling crustal thermal fields from principles of heat conduction and advection beyond what is already learned from the thermal history extracted from the data? Forward modeling of crustal thermal fields requires a physically based model for heat transfer in the geologic setting of interest. Specific geologic processes such as magmatism, fault motion, fluid flow, as well as the kinematic, topographic, and erosional evolution of an orogen can significantly influence the thermal history of thermochronometer samples. These processes can be simulated with thermal models and then compared to thermochronometer derived cooling histories and ages. Thus, predicted thermal histories and thermochronometer ages generated under a known set of conditions can be compared to observed thermal histories and ages to quantify the geologic processes associated with sample cooling. To answer the question posed earlier, forward modeling crustal thermal fields allows robust, or at least constrained, interpretations of what geologic events could have occurred to produce an observed suite of thermochronometer ages. Hence, comparison of thermochronometer data with model predicted thermal histories and thermochronometer ages allows quantification of different geologic processes.

What is the state of our knowledge concerning crustal thermal fields and processes? A rich body of geothermics literature exists documenting the present day thermal state of the crust from borehole measurements of thermal gradients and thermophysical properties. Excellent compilations of global heat flow determinations are available (Chapman and Pollack 1975; Pollack and Chapman 1977; Sass et al. 1981; Pollack et al. 1993). The influence of different geologic processes on subsurface temperatures has been quantified in previous studies using analytic and/or numerical solutions to differential equations for heat transfer (e.g., Benfield 1949a,b; Carslaw and Jaeger 1959; Lachenbruch 1968; Lachenbruch and Sass 1978; Chapman 1986; Philpotts 1990; Chapman and Furlong 1992). Analytic, or exact, solutions to differential

1529-6466/05/0058-0012$05.00 DOI: 10.2138/rmg.2005.58.12

equations are abundant in the literature and provide a starting point for quantifying the thermal field of many geologic settings under a simplified set of assumptions. More complicated, and more realistic, treatments of crustal thermal processes have been conducted with numerical solutions using techniques such as the finite difference and finite element methods. Numerical methods have the advantage of quantifying transient subsurface temperatures in two or three dimensions, tectonically and topographically complicated areas, and in regions where material properties are variable. However, the implementation of numerical techniques is more complicated than application of analytic solutions and the ability to simulate more complex situations introduces additional free parameters that may be poorly constrained. Which approach—analytic or numerical—to use and when is not easy to determine and often times more simple analytic solutions are sufficient for quantifying a particular process and interpreting a thermochronometer data set. This chapter emphasizes the use of analytic solutions to evaluate the influence of geologic processes on subsurface thermal fields so that readers can acquire an intuitive sense for how to quantify thermochronometer cooling histories. The application of numerical models to selected problems is also highlighted along with addition references describing these methods.

Much of the geothermics literature pertaining to measurements of the crustal thermal field and/or crustal thermal processes is under-utilized by the thermochronology community when attempting to estimate crustal thermal gradients for calculation of sample exhumation rates. The objectives of this chapter are to provide the reader with an understanding of: (1) natural variability in terrestrial heat flow and the conditions under which different geologic processes such as erosion, sedimentation, faulting, magmatism, and fluid flow influence rock thermal histories; (2) how different analytic solutions can be applied to thermochronometer data to quantify geologic processes; and (3) future directions for quantifying geologic processes with coupled low-temperature thermochronometers and thermal models. The closure temperature concept will be used throughout this chapter to simplify the discussion and highlight the effect of geologic processes on subsurface temperatures. Note, however, that a rigorous interpretation of thermochronometer ages necessitates consideration of sample cooling histories and composition. Detailed discussions of how to calculate cooling rate dependent thermochronometer ages are covered in complementary chapters in this volume (e.g., Ketcham 2005; Dunai 2005; Shuster and Farley 2005; and Harrison et al. 2005).

Finally, the supplemental online software archive for this volume includes software which simulates many of the processes discussed here and calculates cooling rate dependent thermochronometer ages (*http://www.minsocam.org/MSA/RIM*). Matlab programs that calculate some of the equations and figures presented here are also available via this online resource. These programs are provided as learning tools to help readers explore how different aspects of the Earth influence subsurface thermal fields and thermochronometer cooling histories. The appendix to this chapter provides a compilation of thermophysical property measurements (e.g., thermal conductivity, specific heat, radiogenic heat production) of common Earth materials.

NATURAL VARIABILITY IN TERRESTRIAL HEAT FLOW

Surface heat flow is the expression of geothermal processes at depth and the analysis of these data provides insight into the nature of these processes. Over 25,000 global heat flow determinations are currently freely available (Commision 2004). Figure 1a shows a global representation of these data (Pollack et al. 1993). This figure was generated using a 12th order spherical harmonic representation of the data, meaning the data set is smoothed over spatial scales of about 3300 km. On spatial scales of this size (10^3 km) global surface heat flow ranges between ~0 and 350 mWm^{-2}. Low values are associated with cratons, very old oceanic crust, and regions with high sediment accumulation rates. High values are associated with

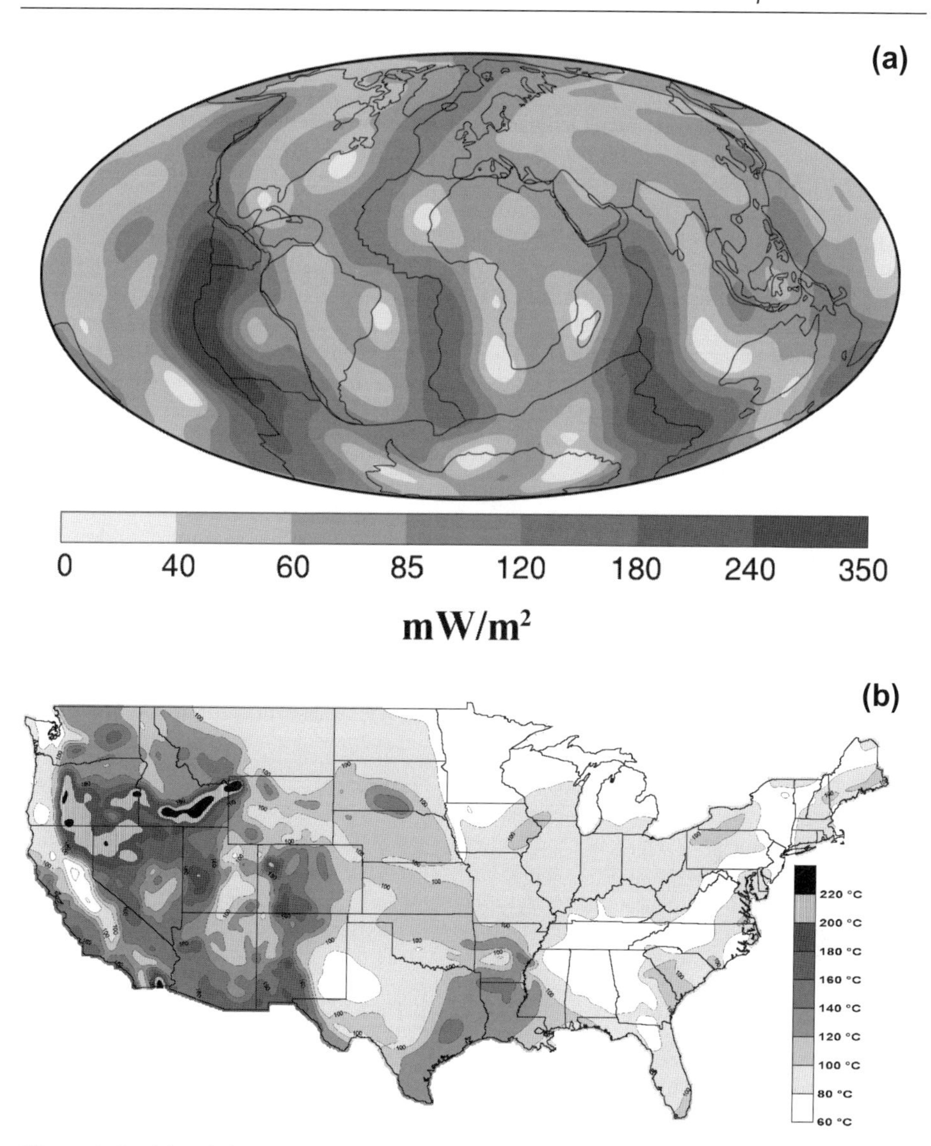

Figure 1. Spatial variations in present day surface heat flow and subsurface temperatures. (a) Long wavelength variations in global heat flow and thermal field contoured with a 12th order spherical harmonic (see text for details). Modified from Pollack et al (1993). (b) Shorter wavelength variations in crustal thermal field for North America. Approximated temperatures at 4 km depth are contoured to highlight spatial variations in subsurface temperatures within individual mountain belts. Modified from Blackwell et al. (1994, 1996).

mid-ocean ridges where new ocean crust is forming. On average terrestrial surface heat flow typically ranges between about 25 and 120 mWm^{-2}. On spatial scales of 10^3 km, variations in terrestrial surface heat flow mimic tectonic provinces (Chapman and Rybach 1985; Furlong and Chapman 1987). Heat flow variations on spatial scales of 10^1–10^2 km (not clearly visible in Fig. 1a) are typically associated with crustal thermal processes such as tectonics, magmatism, or hydrology that affect crustal temperatures. It is on these smaller spatial scales that thermochronometer data are often collected and awareness to natural variations in surface heat flow and thermal gradients are needed.

Surface heat flow (q) is calculated using Fourier's law such that $q = -kdT/dz$. Parameters in this equation include the measured thermal conductivity of rock (k) and the thermal gradient (dT/dz) measured in a borehole. Thus, thermal gradients and depth to a particular closure temperature can be readily approximated from surface heat flow determinations by dividing the surface heat flow by the thermal conductivity. The thermal conductivity of crustal rocks typically averages around 2.5 $Wm^{-1}k^{-1}$ (see appendix A for details). Dividing the previous range in terrestrial surface heat flow values by this average thermal conductivity provides an estimate of continental variations in thermal gradients. This estimated range in continental thermal gradients is about 10 to 50 °C/km. Thus, significant variation is crustal thermal gradients exist and care must be taken when assuming a thermal gradient to interpret thermochronometer cooling histories. The relevant question at this point is whether or not thermal gradients are significantly variable on the scale of 10^1–10^2 km, the scale at which many thermochronometer sampling campaigns are conducted?

Figure 1b illustrates the variability of thermal gradients in the United States on spatial scales greater than ~ 10^1 km. The figure provides an estimate of temperatures at 4 km depth (Blackwell et al. 1994, 1996). Temperatures at 4 km depth range between 60 and 220 °C, which span the closure temperatures of several thermochronometer systems (e.g., apatite and zircon (U-Th)/He and apatite fission-track data) and highlights a significant natural variability in crustal thermal gradients. A second feature of interest in Figure 1b are the spatial scales over which temperatures vary. For example, individual tectonic provinces (e.g., Rocky Mountain of Colorado, Basin Range of Utah and Nevada, and Sierra Nevada Mountains of Colorado) have significant variations in temperature at 4 km depth on spatial scales of tens of km.

The natural variability in surface heat flow underscores the need to quantify crustal thermal processes when interpreting data sensitive to these variations. A prudent interpretation of thermochronometer data not only considers the present day thermal state of a study area from available heat flow determinations, but also an evaluation of geologic processes influencing the thermal history of thermochronometer data. The remainder of this chapter addresses common geologic processes influencing the thermal history of thermochronometer data.

AGE-ELEVATION PLOTS AND SUBSURFACE TEMPERATURES

A common approach to interpreting thermochronometer data is to plot the sample age versus elevation. This approach is often used because information about the sample exhumation/erosion rate can be inferred from the slope of a best-fitting line through the data. However, several important assumptions about the thermal state of crust underlie the interpretation of exhumation rates from age-elevation plots and care must be taken when applying this technique. The most important assumption commonly made is that all samples pass through the closure temperature at the same elevation. The following section discusses settings where this assumption is valid as well as invalid. In settings where samples do not pass through the closure temperature at the same elevation more sophisticated modeling techniques are required to interpret exhumation rates.

Figure 2 illustrates end-member thermal models associated with the interpretation of thermochronometer data from age-elevation plots (Stuwe et al. 1994; Mancktelow and Grasemann 1997). One end-member model, the horizontal isotherm model (Fig. 2a), assumes the closure isotherm is located at a constant elevation (with respect to sea level) and that samples collected at the surface passed through the closure temperature at the same elevation. In this model, *samples could be collected anywhere across the topography* and, as discussed

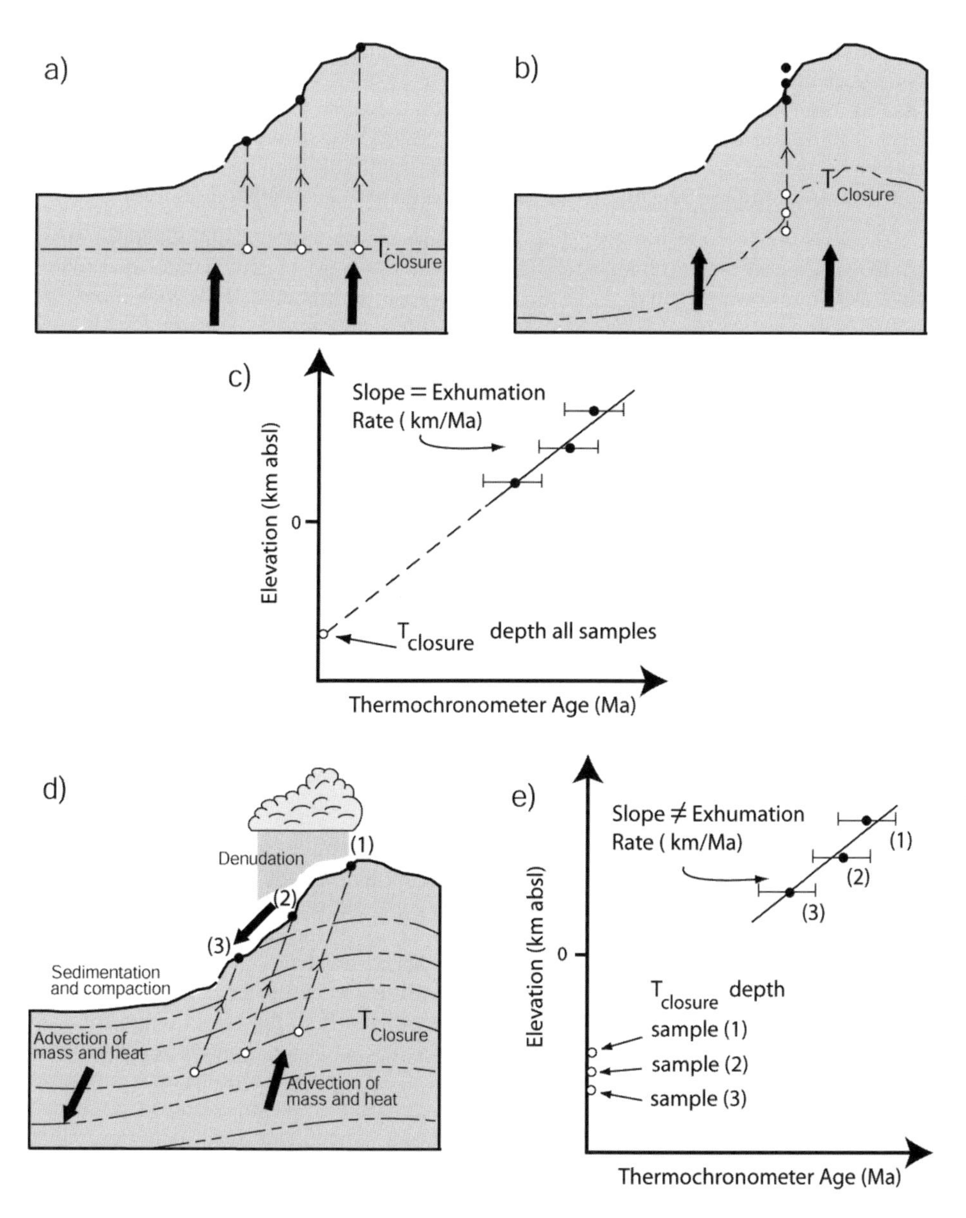

Figure 2. Implication of assumed thermal model on the interpretation of thermochronometer data in age-elevation plots. (a) 1D, horizontal isotherm, thermal model with samples (black circles) collected up a range front. (b) 1D, draped isotherm, thermal model with samples collect up a vertical cliff face. (c) Schematic plot of thermochronometer sample age vs. elevation for the thermal models depicted in (a) and (b). Slope of best fit line equals the sample exhumation rate and depth to closure temperature equals the y-intercept value. (d) Schematic 2D thermal model with variable thermal gradients across range front due to processes illustrated. Each exhumed sample (black circles) pass through the closure temperature at different depths and travel different distances to the surface before exposure. (e) Schematic plot of thermochronometer sample age vs elevation for thermal model shown in (d). Closure depth for each sample is variable and slope of best-fit line through sample ages does not represent the exhumation rate.

below, used to interpret an exhumation rate from a age-elevation plot. This model is applicable for interpreting low-temperature thermochronometer data (e.g., apatite (U-Th)/He or apatite fission track data) in regions with short wavelength (< ~10 km) topography and low erosion rates (e.g., see Stuwe et al. 1994 and Braun 2005 for complete discussion). Alternatively, if higher temperature thermochronometer systems (e.g., biotite, muscovite $^{40}Ar/^{39}Ar$ cooling ages) are being used then this model is also potentially applicable because closure temperature depths are larger and less sensitive to topography (e.g., Stock and Montgomery 1996).

The second end-member model (Fig. 2b) assumes closure isotherms mimic topography. This model assumes the closure temperature is located at a constant depth below the surface, rather than a constant elevation as in the previous case. In this model, *samples should be collected in a purely vertical profile* (e.g., from a cliff face or borehole as shown, Fig. 2b) and would have passed through the closure temperature at the same depth. Geologic settings where this model is relevant include locations where low-temperature thermochronometers (e.g., apatite (U-Th)/He and AFT) with low closure temperatures are investigated, erosion rates are high, and/or long-wavelength (> 40 km) topography is present.

In the previous end-member models the samples in each model pass through the closure temperature at a constant elevation (Fig. 2a) or constant depth (Fig. 2b). Both of these models and sampling strategies represent situations where meaningful information can be interpreted from a plot of the sample age versus elevation. Figure 2c illustrates how an exhumation rate can be calculated from samples if they are plotted in age-elevation space. For example, in both end-member models all the samples pass through the closure temperature at a constant elevation (open circle) and are collected at different elevations on the surface (filled circles). If the assumed thermal model is correct, then a best-fit line through the data should have a y-intercept corresponding to the elevation of the closure temperature and the slope of such a line yields the exhumation/erosion rate. This approach is called the altitude-dependence method (Mancktelow and Grasemann 1997). In practice, these end-member thermal models are not always common and care must be taken before assuming a particular closure isotherm geometry.

Deficiencies in the previous models are several fold and a more realistic model is depicted in Figure 2d where the depth to the closure isotherm and thermal gradients are spatially variable. Thermo-tectonic processes which affect the depth of the closure isotherm include: (1) lateral heat flow across large, range-bounding faults, (2) uplift and erosion, (3) sedimentation and burial, (4) lateral heat refraction around low thermal conductivity sediments deposited in a range front basin, (5) 3D temperature variations due to high-relief topography, (6) magmatism coeval with sample exhumation, (7) topographically driven fluid flow from range crests to valley bottoms, and (8) changes in surface temperature due to the atmospheric adiabatic lapse rate. Low-temperature thermochronometers are sensitive to these thermo-tectonic processes and, as shown in the remainder of this chapter, significant errors can result in calculated exhumation rates due to oversimplified thermal model assumptions such as the examples shown in Figure 2a,b (Grasemann and Mancktelow 1993; Mancktelow and Grasemann 1997; Gleadow and Brown 2000).

Figure 2e illustrates the implication of a non-constant closure temperature depth for samples collected across an orogen as illustrated in Figure 2d. For each of the samples shown in Figure 2e the depth at which the samples passed through the closure temperature is variable because thermal gradients across the range are spatially variable. The sample collected from the highest elevation (sample 1) passed through the closure temperature at a higher elevation than the sample collected at the lowest elevation (sample 3). The implication of variable closure temperature depths is that, unlike previous examples (Fig. 2a,b), there is no physical basis for interpreting an exhumation rate from these samples by taking the slope of a best-fit line through the data because each sample has had a different cooling history.

A more rigorous interpretation of thermochronometer data requires an assessment of the degree to which sample cooling histories depart from the scenarios depicted in Figure 2a,b. Transient and 2D and 3D perturbations to the thermal field can often influence sample cooling ages and result in the situation depicted in Figure 2d,e. Determining if sample cooling histories can be accurately quantified using a 1D constant thermal gradient requires a stepwise evaluation of 2D, 3D, and transient thermal processes using analytic solutions for heat transfer and, if warranted, numerical models. The remainder of this chapter discusses different geologic processes that influence thermochronometer cooling ages and methods to quantify the potential effect of these processes on subsurface temperatures.

GEOLOGIC PROCESSES INFLUENCING THERMOCHRONOMETER AGES

The thermal field of continental crust is influenced by geologic processes that are active on different spatial and temporal scales. For example, sedimentary basin formation occurs on spatial scales of 10^1–10^3 km and although basin infilling in some cases may only occur over 10–20 Ma thermal equilibrium of the crust can take significantly longer and up to 50 Ma or longer. This example highlights the need to quantify the spatial and temporal scale over which the thermal effects of different geologic processes occur and the impact these processes can have on rock thermal histories. In the following section examples of the influence various geologic processes have on the thermal history of thermochronometer samples is discussed.

Background thermal state of the crust

Both conductive and advective heat transfer can be prominent in determining the distribution of temperatures within the crust. Advective heat transfer and spatial variations in heat flow are prevalent in tectonically active areas where mass redistribution by erosion, sedimentation, and fluid flow is common (Blackwell et al. 1989). Conductive heat transfer is more typical of tectonically stable areas and is the starting point for this discussion on the variability of continental geotherms.

The background conductive thermal state of the crust depends on the thermal conductivity, distribution of heat producing radioactive elements, and heat flow into the base of the crust. Prediction of crustal geotherms for different observed surface heat flows is possible using the partial differential equation for time-dependent conductive heat transfer (Carslaw and Jaeger 1959):

$$-div(-k\nabla T) + A = \rho c \frac{\partial T}{\partial t} \tag{1}$$

Where T is temperature, t is time, k is thermal conductivity, A is volumetric heat production and ρ and c are the density and specific heat, respectively. The simplest approach for calculating a crustal geotherm is to consider the case of one-dimensional heat transfer in a homogeneous and isotropic medium with radiogenic heat production. In this case, Equation (1) can be expressed as:

$$\frac{\partial^2 T}{\partial z^2} + \frac{A}{k} = \frac{1}{\alpha}\frac{\partial T}{\partial t} \tag{2}$$

where α is the thermal diffusivity and defined by $\alpha = k/\rho c$. Further simplification of this equation to steady-state heat transfer produces the ordinary differential equation known as Poisson's equation:

$$\frac{d^2 T}{\partial z^2} = -\frac{A}{k} \tag{3}$$

The solution to Poisson's equation in 1D is:

$$T(z) = T_0 + \frac{q_0}{k} z - \frac{Az^2}{2k} \tag{4}$$

Where T_0 and q_0 are the temperature and heat flow at Earth's surface, respectively.

The final step required for calculating steady-state geotherms is assignment of material properties such as thermal conductivity and heat production. Thermal conductivity is primarily determined by composition, but also influenced by temperature and pressure (Cermak and Rybach 1982). Thermal conductivity of common crustal rocks ranges between 1.5 (shale) and 6.0 (quartz arenite) $Wm^{-1}K^{-1}$. A value of 3.0 $Wm^{-1}K^{-1}$ is used here and is an appropriate average value for upper crustal rocks (Roy et al. 1981, see also appendix A). Radiogenic heat production is highly variable in the upper crust and ranges between 0.3 (basalt) to 2.5 (granite/rhyolite) μWm^{-3}, with an average continental crust value of 0.8 μWm^{-3} (Taylor and McLennan 1981; Ryback and Cermak 1982, see also appendix A). The distribution of heat production in the crust varies both laterally and with depth, thereby significantly influencing subsurface temperatures (Pollack 1982). For the upper crust, an exponential decreasing heat production is used such that:

$$A(z) = A_0 \exp\left(\frac{-z}{D}\right) \tag{5}$$

where A_0 is the surface heat generation, z is depth, and D is the characteristic depth over which heat producing elements are distributed. Average crustal values of A_0 and D are 2.0 μWm^{-3} and 10 km, respectively.

Figure 3 shows a suite of steady-state continental geotherms calculated for surface heat flow variations between 40 and 90 mWm^{-2}. The geotherms were calculated using Equations (4) and (5), and the interval method of (Chapman 1986). The geotherms illustrate how subsurface temperatures and thermal gradients vary for a range in terrestrial heat flow. The most important aspect of this figure is the divergence of temperatures with depth from the common surface temperature of 0 °C. For example, at 20 km depth (Fig. 3a) temperatures vary from 220 °C for the 40 mWm^{-2} geotherm to 475 °C for the 90 mWm^{-2} geotherm. A more subtle aspect of each geotherm is the slight downward curvature. The curvature of geotherms is due to heat production and the decrease of heat producing elements with depth (Eqn. 5).

Variation in the depth at which a particular temperature occurs can influence the interpretation of thermochronometer cooling ages. For example, apatite (U-Th)/He and apatite fission-track (AFT) samples are sensitive to the thermal history between 50–90 °C, and 90–150 °C, respectively. Using the 100 °C isotherm for reference, Figure 3b suggests the depth of this isotherm ranges between ~3.4 and 8.2 km for surface heat flow between 90 and 40 mWm^{-2}, respectively. The corresponding average thermal gradients over these depths are 30 and 12 °C/km for the same range in surface heat flow. This variation in thermal gradients is significant because many thermochronometer studies use an average crustal thermal gradient of ~25 °C/km to calculate a depth to closure and an exhumation or burial rate. The point emphasized here is that terrestrial surface heat flow is variable around the world (Fig. 1a), and even within individual orogens (Fig. 1b). Therefore, the notion that exhumation and burial rates can be calculated from an average continental thermal gradient must be discarded. The natural variability in thermal gradients is large enough that significant errors can be introduced into exhumation rate calculations.

Equation (4) provides the basis for calculating the background thermal state of the crust. The thermal consequence of the other geologic processes discussed below are all superimposed upon the heat transfer processes and concepts illustrated in this section. Using

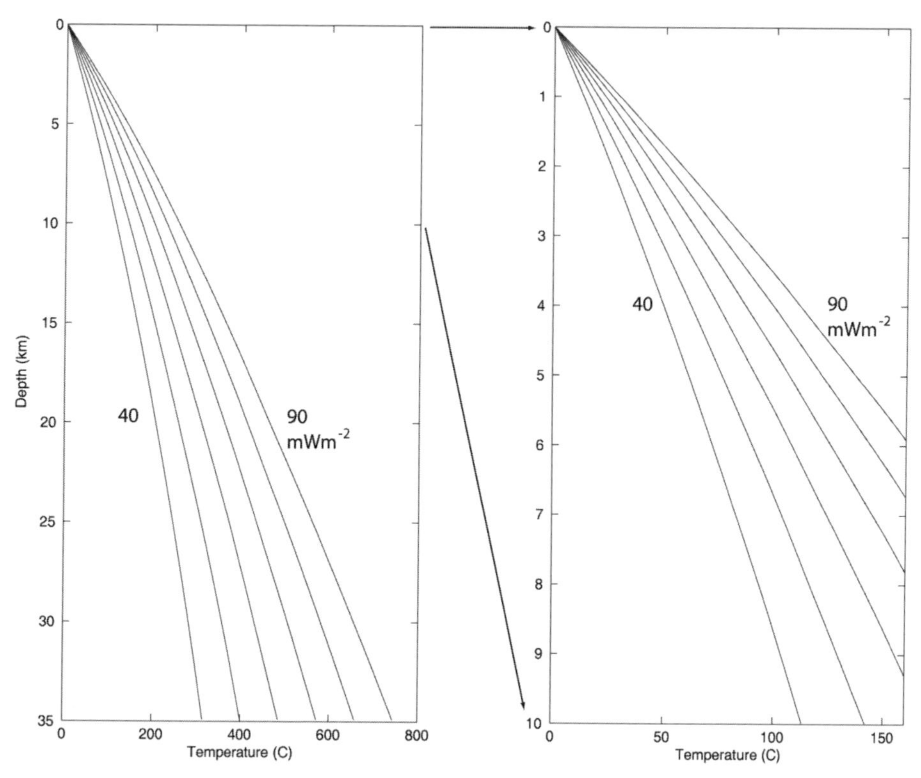

Figure 3. Steady-state crustal geotherms calculated as a function of surface heat flow. Geotherms were calculated for surface heat flow values of 40, 50, 60, 70, 80, and 90 mWm^{-2} using Equation (4). Material properties used include a constant thermal conductivity of 3.0 $Wm^{-1}k^{-1}$ and variable heat production with depth (Eqn. 5) assuming A_o = 2.0 uWm^{-3} and D = 10 km. A Matlab program that reproduces Figure 3 is provided in the supplemental software archive.

thermochronometers to interpret thermal histories from geotherms predicted by Equation (4) should be done only in tectonically stable areas with very slow erosion or sedimentation and minimal topographic influence at closure temperature depths.

Erosion and sedimentation

Significant interest has emerged in the last decade to quantify rates of landscape evolution using low-temperature thermochronometry (House et al. 1998; House et al. 2001; Braun 2002; Persono et al. 2002; Ehlers and Farley 2003). Thermochronometer data are particularly well suited to quantify the rates of landscape and sedimentary basin evolution over million year timescales, which is a timescale where other geochronology techniques (e.g., cosmogenic ^{10}Be, ^{26}Al nuclides) start to lose sensitivity to these processes. Landscape evolution results in the redistribution of mass through the processes of erosion and sedimentation. The redistribution of mass by these processes also results in the redistribution of heat. For example, erosion removes material from Earth's surface resulting in the upward movement of warmer rocks. This upward movement of rocks causes a net increase in temperature at any given depth relative to the initial temperature prior to erosion. Sedimentation can have the opposite effect on subsurface temperatures. As sediment is deposited on the surface of Earth the cool surface temperature is advected downward, thereby causing a net decrease in temperatures at any depth below the surface relative to temperatures prior to sedimentation. If erosion and sedimentation occur at sufficiently high rates (i.e., > ~0.1–0.2 mm/yr) then heat transfer is no longer purely by conduction and advective transfer can be important.

One caveat related to material properties needs mention here, namely that thermal gradients in sedimentary basins can also increase if the density and thermal conductivity of the sediment are sufficiently low. Although the following discussion does not quantify this effect, realize that advection of mass by sedimentation decreases thermal gradients whereas the material properties of the sediment deposited can sometimes increase the gradients. Thus, the effect of these two processes on thermal gradients are opposite in direction.

Calculation of subsurface temperatures. The previously discussed thermal effects of erosion and sedimentation on subsurface temperatures can be quantified with the transient advection-diffusion partial differential equation. Application of the advection-diffusion equation to problems in landform evolution typically requires a 2D or 3D form of the equation to quantify rates and magnitudes of topographic change (e.g., Stuwe et al. 1994; Mancktelow and Grasemann 1997). The 1D form of the advection-diffusion equation will be discussed here to illustrate the physical consequences of sedimentation and erosion on subsurface temperatures, and more complicated 2D solutions will be discussed later. In 1D with no heat production, the advection-diffusion equation is given by (Carslaw and Jaeger 1959):

$$\alpha \frac{\partial^2 T}{\partial z^2} = \frac{\partial T}{\partial t} + v \frac{\partial T}{\partial z} \tag{6}$$

where α is the thermal diffusivity, T is temperature, t is time, z is depth, and v is the velocity of the medium relative to Earth's surface. If z is defined in a coordinate system such that it is positive downward into Earth, then a negative value of v represents erosion and a positive value of v represents sedimentation. In Equation (6), the left-hand term describes conductive heat transfer, the middle term transient heat transfer, and the right hand term advective heat transfer by erosion or sedimentation.

Equation (6) can be solved by assuming a constant surface temperature (T_0) at $t = 0$, $z = 0$ and equilibrium initial thermal gradient (Γ_b) prior to erosion or sedimentation at time $t = 0$. The solution for transient subsurface temperature is then (Powell et al. 1988):

$$T(z,t) = T_0 + \Gamma_b(z - vt) + \frac{1}{2}\Gamma_b \times \left[(z + vt)\exp\left(\frac{vz}{\alpha}\right) erfc\left(\frac{z + vt}{2(\alpha t)^{\frac{1}{2}}}\right) - (z - vt)\, erfc\left(\frac{z - vt}{2(\alpha t)^{\frac{1}{2}}}\right) \right] \tag{7}$$

where all variables are the same as before and *erfc* is the complementary error function (Abramowitz and Stegun 1970). Equation (7) is an appropriate starting point for understanding the effect of sedimentation and erosion of subsurface temperatures.

Figure 4 shows a set of transient crustal geotherms calculated using Equation (7). These geotherms were calculated assuming a constant sedimentation or erosion rate of 1 mm/yr. For the case of sedimentation (Fig. 4a), subsurface temperatures at any depth get cooler from their initial temperature as the duration of sedimentation increases. For example, at 4 km depth (Fig. 4a) the initial temperature ($t = 0$) is 100 °C. After 5, 10, and 15 Ma of sedimentation at 1 mm/yr the temperature at 4 km depth decreases to 70, 55, and 46 °C, respectively. Note from this example that the rate of temperature change decreases with time such that between 0 and 5 Ma the temperature at 4 km depth changes 30 °C per 5 Ma, whereas between 5-10 Ma and 10-15 Ma the temperature change is 15 and 9 °C per 5 Ma, respectively. This decrease in temperature change with time is a result of the thermal field approaching steady-state.

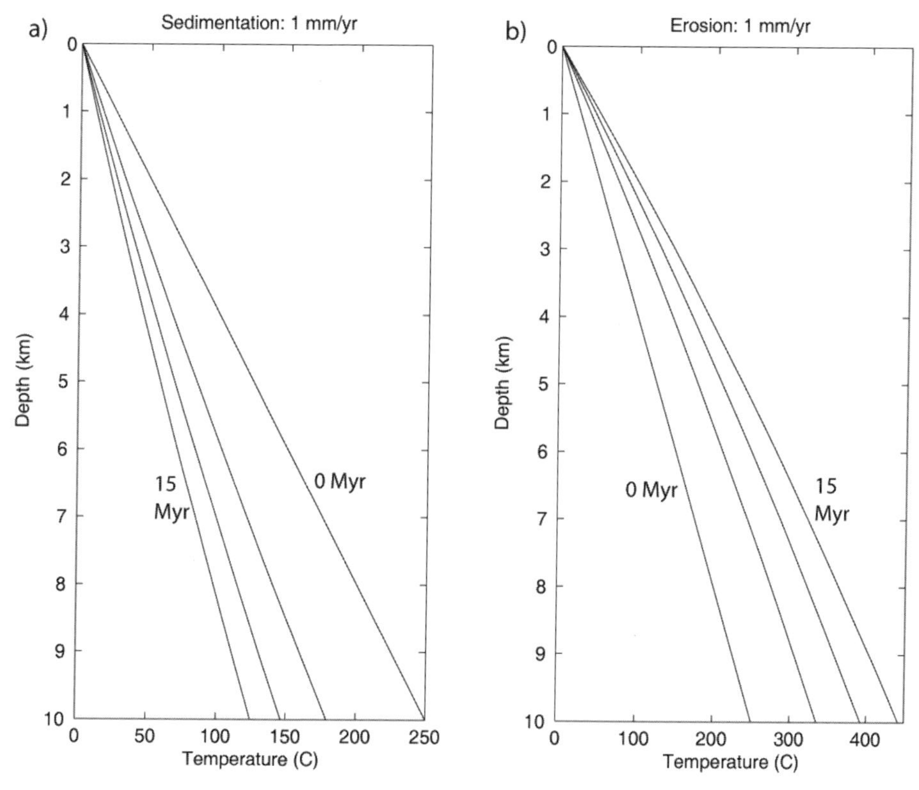

Figure 4. Transient crustal geotherms calculated as a function of erosion and sedimentation. Prescribed erosion and sedimentation rates are 1 mm/yr, with geotherms shown after 0 (initial condition), 5, 10, and 15 Ma of erosion or sedimentation. (a) Influence of sedimentation on geotherms. (b) Influence of erosion on geotherms. Results were calculated using Equation (7) and the following parameters, surface temperature of 0 °C, initial thermal gradient of 25 °C/km, and thermal diffusivity of 32 km^2Ma^{-1}. A Matlab program that reproduces Figure 4 is provided in the supplemental software archive.

The thermal effect of sedimentation decreasing subsurface temperatures must be taken into account when interpreting thermochronometer data collected from sedimentary basins (e.g., Armstrong and Chapman 1999).

Erosion has the opposite effect on subsurface temperatures than sedimentation and rock temperatures at a particular depth increase relative to their initial temperature. For example, in Figure 4b after 5, 10, and 15 Ma of erosion temperatures at 4 km depth increase from 100 °C (at $t = 0$) to 140, 170, and 195 °C, respectively. Thus, upper crustal temperatures progressively increase, and thermal gradients are enhanced in response to erosion. The magnitudes of temperature change between the time intervals of 0–5, 5–10, and 10–15 Ma are 40, 30, and 15 °C per 5 Ma. Note that, as with the case of sedimentation, the magnitude of temperature change decreases with time but that with erosion the change is larger. For example, between 0–5 Ma temperatures at 4 km depth change by 30 °C during sedimentation and 40 °C for erosion. The larger magnitude of temperature change due to erosion compared to sedimentation is a consequence of the direction material is moving with respect to the boundary conditions used to solve Equation (6). For sedimentation, material is moving downward and away from a constant surface temperature, thereby advecting heat downward. For erosion, material is moving upward and away from a constant gradient boundary condition where temperature is free to increase as long as the gradient stays constant, thereby resulting in a larger magnitude of temperature change. The warming of temperatures at depth and enhanced thermal gradients due to erosion must be taken into account in exhumation and erosion studies in active orogenic settings as rocks cool along their trajectory to the surface.

Thermal gradient perturbations. A common problem encountered by Earth scientists working with thermochronometer data is knowing when a process such as erosion or sedimentation has occurred at a high enough rate and for long enough time to influence their interpretation of data. For example, the interpretation of exhumation rates from age-elevation plots (e.g., Fig. 2a,b) assumes thermochronometer samples cool through a temporally and spatially constant thermal gradient. However, in settings where erosion rates are low and/or erosion durations short the perturbation to thermal gradients at closure temperature depths can be negligible. Furthermore, quantifying changes in thermal gradients as a function of an erosion or sedimentation rate is important to understanding how the cooling rate of samples changes. Changes in a cooling rate (dT/dt) are proportional to the product of the thermal gradient (dT/dz) and the erosion or sedimentation rate (dz/dt), where T is temperature, z is depth, and t is time. Thus, quantifying changes in thermal gradients is a necessary step to quantifying changes in cooling rates and calculation of cooling rate dependent closure temperatures (e.g., Dodson 1973,1986; Farley 2000). The following discussion highlights a 1D approach to quantifying first order effects of erosion and sedimentation on thermal gradients and cooling rates.

The previous 1D advection-diffusion Equation (6) can be used to estimate the influence of erosion and sedimentation on crustal thermal gradients. The solution of Equation (6) for corresponding gradients is given by (Benfield 1949b; Kappelmeyer and Haenel 1974; Powell et al. 1988):

$$\frac{\partial T(z,t)}{\partial z} = \Gamma_b + \frac{1}{2}\Gamma_b \bullet \left[-erfc\left(\frac{z-vt}{2(\alpha t)^{\frac{1}{2}}}\right) - \frac{z+vt}{(\pi\alpha t)^{\frac{1}{2}}}\exp\left(\frac{vz}{\alpha}\right)\exp\left(-\left(\frac{z+vt}{2(\alpha t)^{\frac{1}{2}}}\right)^2\right) \right.$$
$$\left. + \frac{z-vt}{(\pi\alpha t)^{\frac{1}{2}}}\exp\left(-\left(\frac{z-vt}{2(\alpha t)^{\frac{1}{2}}}\right)^2\right) + \left(1+\frac{vz}{\alpha}+\frac{v^2 t}{\alpha}\right)\exp\left(\frac{vz}{\alpha}\right) erfc\left(\frac{z+vt}{2(\alpha t)^{\frac{1}{2}}}\right)\right] \tag{8}$$

Equation (8) can be used to quantify the effect of sedimentation and erosion on thermal gradients at Earth's surface and at the depths and temperatures sensitive to low-temperature thermochronometers. Figure 5 uses Equation (8) to calculate thermal gradient changes as a function of different erosion/sedimentation rates and durations. The parameters assumed for making Figure 5 are identical to those used in Figure 4, including the initial thermal gradient (Γ_b = 25 °C/km). On typical time-scales of orogenic development (10^6–10^7 years) thermal gradients at Earth's surface can significantly change from their initial value at erosion and sedimentation rates greater than ~0.1 mm/yr (Fig. 5a,b). For example, at a sedimentation rate of 1 mm/yr (Fig. 5a) the initial thermal gradient of 25 °C/km remains unaffected until about 10^3 years, at which time the gradient starts to decrease. After 10^6 and 10^7 years of sedimentation at the same rate the thermal gradient will decrease to 20 and ~12 °C/km, respectively. Erosion has the opposite effect (Fig. 5b) and causes an increase in thermal gradients with increased erosion duration. Figure 5a,b also clearly shows that the disturbance to thermal gradients from an initial value is more pronounced at higher rates. For example, after 10^5 years of erosion at 10 mm/yr the thermal gradient doubles from 25 to 50 °C/km (Fig. 5b). Thus, the magnitude of change in thermal gradients within active orogenic settings is highly sensitive to both the rate and duration of sedimentation or erosion. Furthermore, the previous discussion highlights

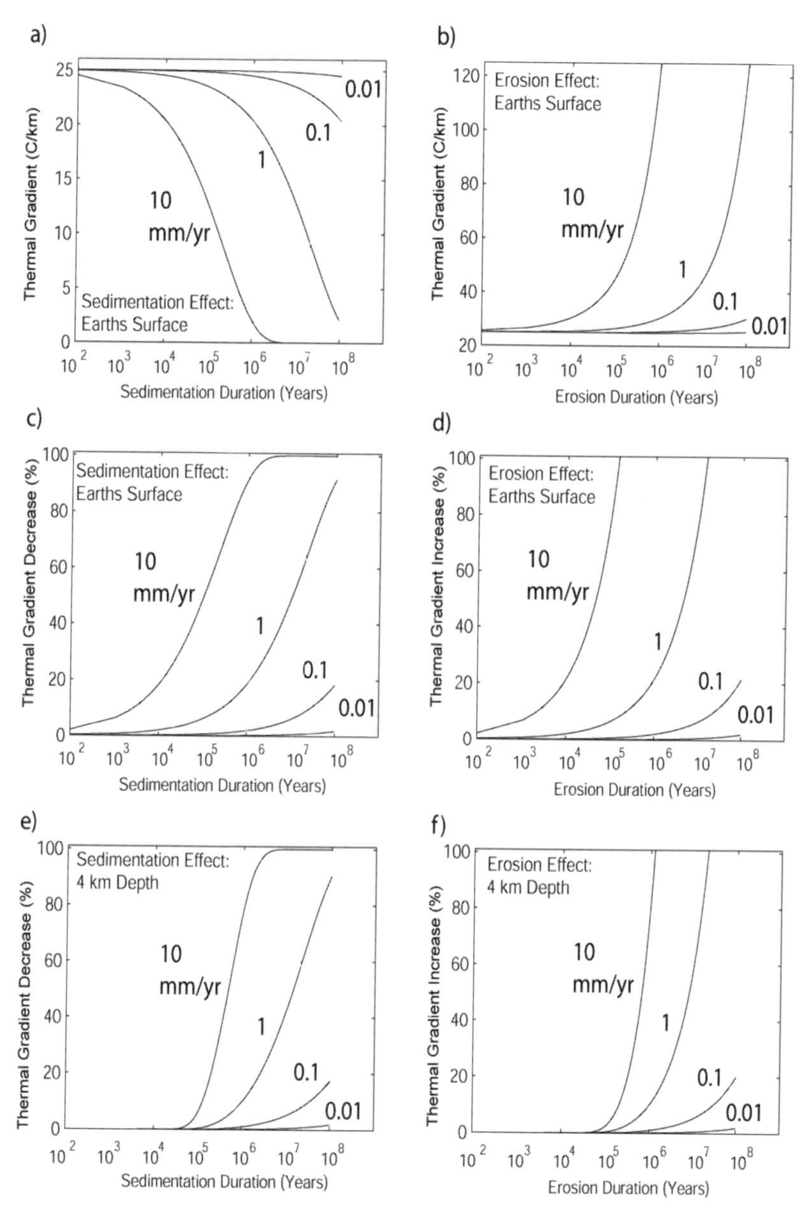

Figure 5. The effect of erosion and sedimentation rate and duration on thermal gradients. Affect of sedimentation (a) and erosion (b) on thermal gradients at Earth's surface for an initial thermal gradient of 25 °C/km. Generalized % effect of sedimentation (c) and erosion (d) on thermal gradients at Earth's surface. Effect of sedimentation (e) and erosion (f) on thermal gradients at 4 km depth. All results were calculated using Equation (8) and the same parameters as in Figure 4. Percent gradient increase or decrease (y-axis) in (c) thru (f) was calculated as percent effect with respect to the initial thermal gradient prior to onset of erosion or sedimentation. The calculated % increase or decrease is applicable to any initial thermal gradient. A Matlab program that reproduces Figure 5 is provided in the supplemental software archive.

the transient evolution of thermal gradients and draws into question assumptions of constant cooling rates often used to interpret thermochronometer data (e.g., Fig. 2).

Present day near surface thermal gradients are spatially variable (e.g., Fig. 1). The remainder of this section generalizes the results of Figure 5a,b, which assumed an initial thermal gradient of 25 °C/km, to show how erosion and sedimentation influence thermal gradients for any initial thermal gradient. Figure 5c-f use Equation (8) to show the *percent thermal gradient change relative to any initial thermal gradient* (Γ_b) as a function of erosion and sedimentation duration and rate. At Earth's surface (Fig. 5c,d) the effect of sedimentation and erosion on thermal gradients increases with the duration of activity and with the rate at which the process occurs. For example, after 1 Ma ($t = 10^6$ years) of sedimentation the thermal gradient is decreased by ~3%, 20%, and 90% for rates of 0.1, 1.0, and 10 mm/yr, respectively (Fig. 5a). After 1 Ma of erosion (Fig. 5d) the thermal gradient is perturbed 5%, 30%, and 200% (not shown) at rates of 0.1, 1.0, and 10 mm/yr. The magnitude of gradient perturbation is greater for erosion than sedimentation because of the direction material is moving with respect to boundary conditions as previously discussed.

At depths around 4 km where low-temperature thermochronometer data such as apatite (U-Th)/He or fission-track are at or near closure, thermal gradients are also influenced by the processes of sedimentation and erosion (Fig. 5e,f). The magnitude of perturbation these processes have on thermal gradients is comparable to those previously discussed. However, at 4 km depth the gradient perturbation is delayed from the time at which it occurs at the surface. For example, for a sedimentation rate of 1 mm/yr, perturbation to the thermal gradient at the Earth's surface starts at ~1,000 years (Fig. 5a) after the onset of sedimentation whereas at 4 km depth an effect is not noticed until ~100,000 years (Fig. 5e). A similar delay in gradient perturbation occurs for the case of erosion (compare Fig. 5d and 5f). The time delay in thermal gradient perturbations at depth is due to the extra time it takes processes occurring at Earth's surface to penetrate downward. The primary controlling factors on the magnitude of the time delay are the rate at which the process is occurring and the thermal diffusivity of the rock.

When taking into account the influence of sedimentation and erosion on thermal gradients and the interpretation of thermochronometer data it is important to not only consider the rate but also the time span over which the sedimentary basin or orogen has been evolving. Figure 5 provides a means to quickly assess the impact of these processes on, for example, the calculation of sedimentation or erosion rates. If erosion or sedimentation rates in a study area are estimated from other geologic constraints to be <0.1 mm/yr then disturbances to the background thermal field will be less than ~10% for durations on order 10 Ma or less. However, if rates are estimated to be higher or durations longer than perturbations to the background gradient could be significantly larger and a more sophisticated treatment of the data to quantify sedimentation or erosion rates may be warranted.

Tectonics and faulting

Thermochronometer data are often used to quantify the timing and rates of fault motion in orogenic belts. In this section the influence of fault motion on thermochronometer data is discussed and references to more detailed observational and modeling studies are presented. In tectonically active areas several processes influence the cooling history of thermochronometer data and cause departures from the simple 1D background thermal field represented in Equation (4) (Fig. 6). The largest disturbance to the background thermal field typically arises from erosion and sedimentation transporting mass from one side of the fault to the other. In extensional, normal fault bounded, mountain ranges erosion of the footwall causes an enhanced thermal gradient in the footwall (e.g., Fig. 4b, Fig. 6a) whereas sedimentation on the hanging wall results in a depressed thermal gradient (e.g., Fig. 4a). In extensional settings where erosion and sedimentation are minimal and rocks are exposed at the surface by tectonic exhumation in

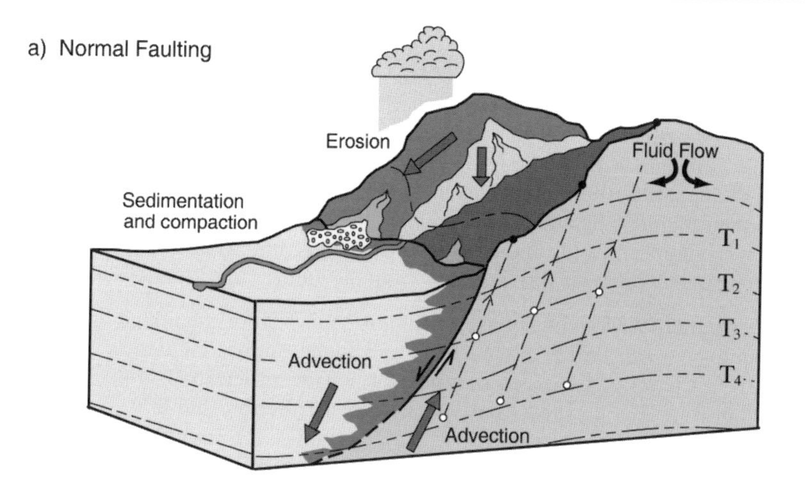

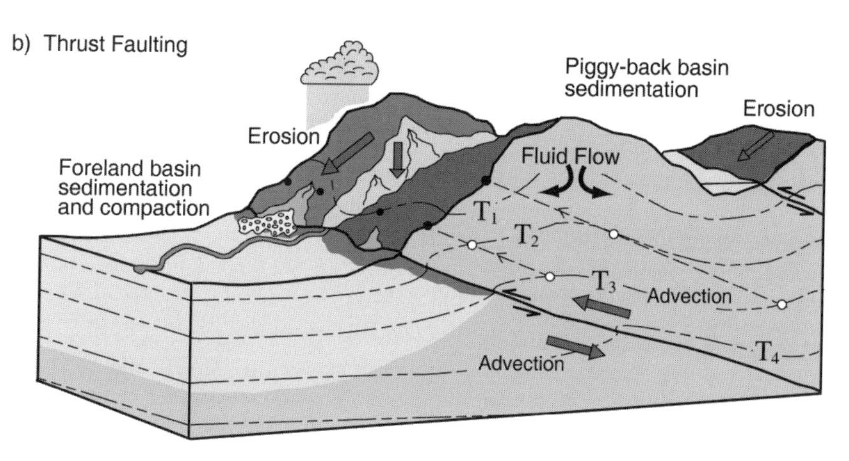

Figure 6. Schematic of thermo-tectonic processes influencing thermochronometer interpretation. (a) Processes influencing thermochronometer interpretation in normal fault settings modified from Ehlers et al (2003). (b) Processes influencing thermochronometer interpretation in thrust tectonic settings modified from Ehlers and Farley (2003). Subsurface isotherms (dashed lines) are indicated with temperatures increasing from T_1 to T_4. Open circles represent position and trajectory of rocks eventually exposed at the surface (filled circles).

the footwall a similarly enhanced footwall thermal gradient can also occur. The reverse effect occurs in compressional, or thrust tectonic, mountain ranges (Fig. 6b) where erosion of the hanging wall enhances thermal gradients and foreland basin sedimentation on the footwall depresses thermal gradients. There are several important thermal consequences to the previous description of mass redistribution across faults (Fig. 6): (1) rates of erosion and sedimentation on each side of the fault can cause significant departures from the background thermal field as discussed in the previous section, (2) the juxtaposition of enhanced and depressed thermal fields across the fault causes lateral heat flow across the fault and curved isotherms at thermochronometer closure temperature depths, and (3) sedimentation and basin formation can result in thermal conductivity contrasts across the fault and lateral heat flow. If the rates of fault motion, erosion, and sedimentation are sufficiently fast (e.g., > ~0.1 mm/yr, Fig. 5d,f) then a significant contrast in thermal gradients across the fault can violate assumptions of vertical, 1D, heat flow (e.g., Eqns. 4 and 7) and 2D or even 3D heat transfer should be considered.

Several other thermal processes associated with mountain building can result in lateral variations in thermal gradients and multidimensional heat flow. These processes include: (1) frictional heating on faults with slip rates of > ~1 cm/yr (e.g., Lachenbruch and Sass 1980), (2) thickening of radiogenic heat producing layers in thrust tectonic settings, (3) stripping off or removal of high heat producing layers from erosion of the uplifted block, (4) displacement on adjacent or underlying thrusts, (5) topographic development and 3D heat flow in the uplifted block, and (6) topographically driven fluid flow from the uplifted block to the valley floor.

Normal faulting. Several numerical modeling studies have quantified the influence of normal faulting on subsurface temperatures and found normal faulting to significantly influence subsurface temperatures when rates of extension are sufficiently fast enough (Furlong and Londe 1986; Ruppel et al. 1988; van Wees et al. 1992; Bertotti and ter Voorde 1994; ter Voorde and Bertotti 1994; Govers and Wortel 1995; Bertotti et al. 1999; Ehlers and Chapman 1999; Ehlers et al. 2001; Armstrong et al. 2003; Ehlers and Farley 2003). Figure 7 highlights the effect of normal faulting and footwall tilt on the interpretation of thermochronometers in age-elevation plots (Ehlers et al. 2001). Samples collected on a traverse between a fault and range crest (open circles in Fig. 7a) will, among other things, be sensitive to the exhumation rate at the fault and the magnitude of footwall tilt. This type of sampling strategy is useful when trying to determine the timing, rate, magnitude, and structural tilt of normal fault bounded ranges. Synthetic (U-Th)/He and AFT data were simulated for this setting using two fault-adjacent exhumation rates of 1.1 (solid lines) and 0.5 (dashed lines) mm/yr (Fig. 7b), Footwall tilt causes range crest samples (2 km elevation) to exhume at slower rates of ~0.9 and 0.3 mm/yr. The predicted sample age after 10 km of exhumation is shown with the open circles and a 1D thermal model (e.g., Fig. 2a,b) based best-fit line through each data set is shown (V_{1D} values). Note that the 1D thermal model does not account for footwall tilt so only one exhumation rate can be calculated from the data.

Two general trends are visible in the predicted (U-Th)/He and AFT ages (Fig. 7b). First, with an increase in exhumation rate, the difference between (U-Th)/He and AFT ages will decrease. This is evident by comparing the age difference between the two fast and slow exhumation rate pairs of curves. This decrease in age differences with increased exhumation rate is a direct result of material moving faster towards the surface and higher exhumation rates shifting isotherms closer to the surface and decreasing the distance between apatite (U-Th)/He and fission track closure temperatures. A second visible trend in the predicted ages is that for either pair of (U-Th)/He and AFT samples, both exhumed at an identical exhumation rate, the slope of the best fit line through the data is different, with the (U-Th)/He samples having a steeper slope than the AFT samples. A 1D thermal model would yield the incorrect interpretation that the steeper slope of the (U-Th)/He data suggests an increase in the exhumation rate relative to the AFT data. However, the different slopes between each pair of (U-Th)/He and AFT data is a result of an increase in the thermal gradient at shallow (U-Th)/He closure temperature depths and the 2D nature of the thermal field (e.g., Fig. 2d,e).

Differences between the 2D and 1D fault perpendicular exhumation rates are as follows. For samples exhumed at 1.1 mm/yr adjacent to the fault (solid lines in Fig. 7b), the 1D thermal model predicts average (U-Th)/He and AFT exhumation rates of 1.2 and 0.8 mm/yr, respectively. Neither 1D exhumation rate is correct with respect to the simulated exhumation rates (v_{exh}) and errors are between 0.1 and 0.3 mm/yr (~10–30%) compared to the fault-adjacent exhumation rate. The (U-Th)/He and AFT samples exhumed at the slower, fault-adjacent rate of 0.5 mm/yr (dashed lines in Fig. 7b) have predicted 1D exhumation rates of 0.4 and 0.3 mm/yr, respectively. The (U-Th)/He and AFT 2D and 1D exhumation rates differ by 0.1 mm/yr (20%)

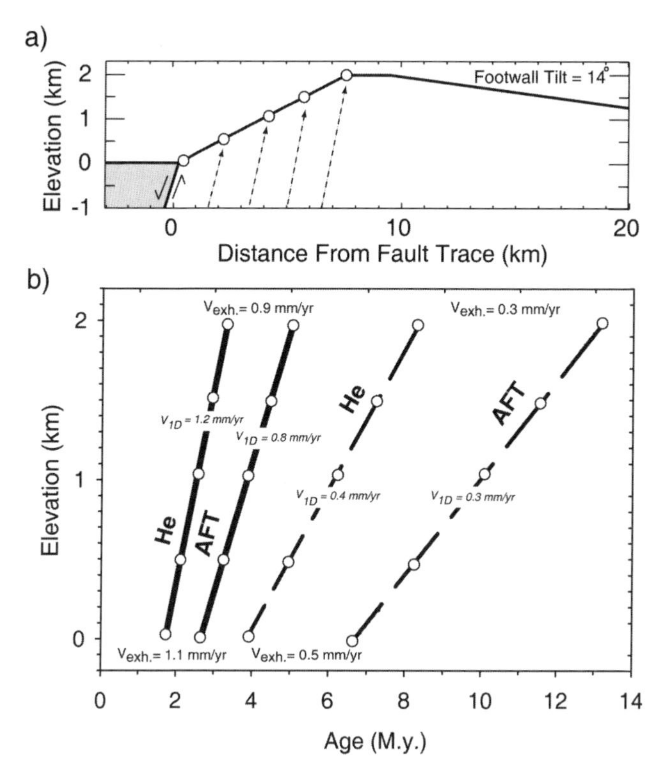

Figure 7. Numerical model predicted sensitivity of interpreted exhumation rates to the assumed thermal model in normal fault bounded ranges. (a) Location of samples collected up the front of a normal fault bounded range. Circles represent sample collection locations, dashed lines represent sample exhumation trajectories. (b) Predicted sample ages (open circles) versus elevation for sample collected along profile shown in (a). The 1D exhumation rates (V_{1D}) refer to rates calculated using the altitude-dependence method (e.g., Figure 2c). The 2D exhumation rates (V_{exh}) refer to the exhumation rate along the samples exhumation trajectory (e.g., Figure 2e). Sample ages in (b) were calculated using a spherical ingrowth diffusion model, and fission-track annealing model for apatite He and apatite fission track ages, respectively (methods described in Ehlers et al. 2004). Modified from Ehlers et al. (2001).

and 0.2 mm/yr (40%), respectively, at the range front. The magnitude of error in 1D exhumation rates generally increases with increased exhumation rate. The difference, or error, between 1D and simulated (v_{exh}) exhumation rates exceed typical uncertainties in measured (U-Th)/He and AFT data and should be of concern when interpreting thermochronometer data.

Thrust faulting. Extensive work has been directed towards delineating the structural evolution of fold and thrust belts (e.g., Suppe 1976, 1980, 1983; Boyer and Elliot 1982; Mitra 1986; Homza and Wallace 1995; Thorbjornsen and Dunne 1997; Martel 1999). Many of these studies have focused on structural styles of deformation and described specific structures such as fault-bend folds, imbricate fans, and duplexes (e.g., Boyer and Elliot 1982; Suppe 1983; Mitra 1986; Allmendinger and Shaw 2000; Salvini et al. 2001) or have described kinematic patterns of structural thickening that occur with underplating (Platt et al. 1985). Large-scale kinematic descriptions of fold and thrust belts reached their apogee with the construction of balanced cross-sections which facilitate the restoration of pre-deformational strata (e.g., Dahlstrom 1969; Jones 1971; Elliott 1976, 1977, 1983; Suppe 1980, 1983; Boyer and Elliot 1982).

Several studies have made significant advances in understanding thrust fault kinematics using apatite fission-track thermochronometry (e.g., Cerveny and Steidtmann 1993; Burtner and Nigrini 1994; Burtner et al. 1994; Omar et al. 1994; O'Sullivan et al. 1998; Quidelleur et al. 1997; Rahn and Grasemann 1999). The focus of most of these studies has been to determine the chronology of thrusting on different faults, exhumation magnitude, and in some cases rates of faulting on a single thrust fault. Most of these studies implement one or more of the following approaches: (1) use one-dimensional (1D) thermal models to deduce exhumation parameters, (2) assume 1D (vertical) kinematics for sample exhumation pathways, (3) ignore sample cooling due to erosion (versus structural cooling and subsequent warming) or prescribed erosion histories, and (4) neglect the 2D thermal, kinematic, and erosional implications that laterally adjacent imbricate or underlying (duplex) thrust faults may have on exhumed thermochronometer sample ages. To the best of my knowledge, no published studies have investigated the thrust belt kinematic processes influencing apatite (U-Th)/He cooling ages.

However, a rich body of literature is available discussing the thermal processes of thrust faulting (e.g., Edman and Surdam 1984; Furlong and Edman 1984, 1989; Edman and Furlong 1987). Studies of individual thrust sheets represent the vast majority of these studies. The simplest of these thermal models consider 1D heat flow through a thrust sheet emplaced instantaneously on top of a relatively cool footwall (e.g., Oxburgh and Turcotte 1974; Graham and England 1976; Brewer 1981; Angevine and Turcotte 1983; Edman and Surdam 1984; England and Thompson 1984; Davy and Gillett 1986; Mailhe et al. 1986; England and Molnar 1993; Husson and Moretti 2002). At the next level of sophistication, 1D models were stacked laterally to approximate 2D heat flow or finite thrust emplacement rates (Karabinos and Ketcham 1988). Fully 2D models were also applied to the thrust-sheet emplacement problem (Shi and Wang 1987; Molnar and England 1990, 1993; Ruppel and Hodges 1994; Quidelleur et al. 1997; Rahn and Grasemann 1999; Husson and Moretti 2002). Recent work by (ter Voorde et al. 2004) has used a 2D thermal model with prescribed erosion histories to investigate the thermal consequence of adjacent thrust faults and thickening of radiogenic heat producing layers. Future work in modeling thrust belt thermal processes is needed to quantify the thermal effects of multiple thrusts, timing and rates of thrust sheet erosion, and foreland and piggy-back basin formation.

Figure 8 illustrates the effect of crustal-scale thrust faulting and fault-bend fold formation on subsurface temperatures (Husson and Moretti 2002). The steady-state thermal field is shown for variable thrust displacement rates, a basal heat flow of 50 mWm^{-2}, and constant surface temperature of 0 °C. Erosion is not accounted for and the depth to detachment is 30 km below the surface. Although the geometry of thrusts and depth to detachment are often less (~10 km deep) in fold and thrust belts the example shown illustrates how variations in displacement rates and advective heat transfer by thrust faulting influence subsurface temperatures. For example, at very fast displacement rates of 2 m/yr (Fig. 8a) isotherms become inverted across the fault and could result in an inverted metamorphic grade of rocks exposed at the surface later. As the displacement rate decreases so does the curvature of subsurface isotherms across the fault. At displacement rates of 20 and 2.5 mm/yr (Fig. 8b,c) isotherms are not inverted as in Figure 8a although they are still notably curved across the fault indicating that advective heat transfer due to thrust faulting influences subsurface temperatures and could influence the cooling history of thermochronometer samples (see also ter Voorde et al. 2004). Future work in quantifying thrust tectonic thermal processes is needed to investigate the combined influence of faulting, erosion, shear heating, and crustal thickening on thermochronometer cooling histories. Furthermore, the influence of adjacent imbricate and underlying duplex faulting on rock cooling histories is not well understood.

A key concept illustrated by the previous discussion of normal and thrust fault thermal processes is that lateral heat transfer between fault blocks is commonplace and that a 2D or 3D

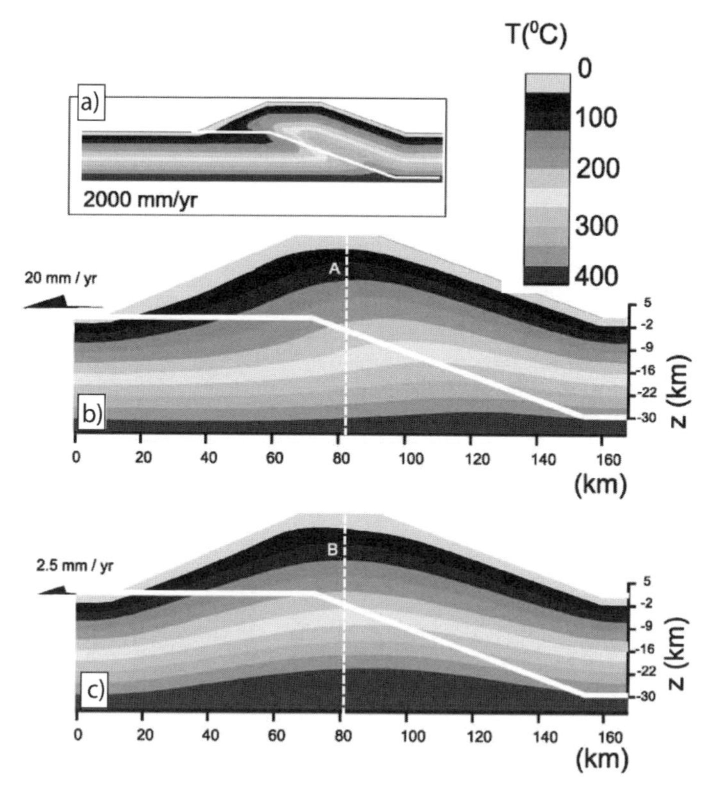

Figure 8. Numerical model predicted sensitivity of steady-state subsurface temperatures to the rate of fault-bend fold formation. (a) Thermal field with a high fault slip rate of 2 m/yr. (b) Thermal field with a fault slip rate of 20 mm/yr. (c) Thermal field with a low fault slip rate of 2.5 mm/yr. Location of fault is shown with solid white line cutting through model. Modified from Husson and Moretti (2002).

representation of the transient thermal field is often needed to make kinematic interpretations out of thermochronometer data. Unfortunately, unlike the previous discussion on the influence of erosion and sedimentation on thermal fields, simple analytic solutions to the advection-diffusion equation are not available in 2D or 3D for realistic fault geometries and inhomogeneous thermal properties of rocks. Thus, numerical solutions (e.g., finite element and finite difference methods) must be applied to quantify thermochronometer cooling history.

Magmatism

The rate of heat loss from magmatic bodies can, in some settings, play an important role in the interpretation of thermochronometer data. High-temperature chronometers (e.g., U/Pb, $^{40}Ar/^{39}Ar$ from hornblende or biotite) collected from within an intrusion often reflect the cooling age of the intrusion and represent the inherited or parent age of a pluton. The parent age of a pluton is useful to know when interpreting low-temperature thermochronometers (e.g., apatite and zircon fission-track or (U-Th)/He methods) that can contain post-emplacement information on the exhumation history of the pluton. Furthermore, the emplacement of magmatic bodies can reheat surrounding country rock to sufficient temperatures to reset low-temperature thermochronometer ages (e.g., Tagami and Shimada 1996, Reiners 2005). Numerous petrologic and heat flow studies have investigated the rate of heat loss from magmatic bodies (e.g., see Furlong et al. 1991; Peacock 1989b for a complementary discussion and overview). The aim of this section is to summarize previous work and the physics of heat loss around magmatic bodies so that readers have a practical guide to quantifying this process and when it may or may not influence thermochronometer ages.

Heat loss from crustal magmatic bodies primarily occurs through the process of conduction (Philpotts 1990; Spear 1993). Although magmatic bodies have a finite shape the thermal evolution of a magmatic body and its surrounding area can be quantified in 1D with several simplifying assumptions (Peacock 1989a; Furlong et al. 1991; Stuwe 2002). First, the emplacement of magmatic bodies occurs rapidly compared to the post-emplacement thermal equilibration of the surrounding country rock. With this in mind, emplacement can be thought of as occurring instantaneously and solutions to the 1D diffusion equation for a step change in heating can be utilized. Second, the geometry of magmatic bodies are small (typically < 10 km or significantly smaller) relative to the distance to Earth's surface or base of the lithosphere where boundary conditions must be imposed to solve the diffusion equation. Hence, the boundary conditions can be assumed to lie at infinite distance from the intrusion. Finally, the latent heat of fusion from conversion of melt to rock will be neglected to quantify the first-order consequence of magmatism.

Using the previous assumptions, the thermal evolution of an intrusion and the surrounding country rock can be quantified with the 1D transient diffusion equation (Eqn. 2) after removing the heat production term which has negligible influence over the timescales of interest for this problem. Thus, the partial differential equation (Eqn. 2) simplifies to the following form:

$$\alpha\frac{\partial^2 T}{\partial z^2}=\frac{\partial T}{\partial t} \tag{9}$$

where the variables were previously defined. Equation (9) can be solved for cooling history of a finite width intrusion using the boundary and initial conditions shown in Figure 9. The intrusion is assumed to be planar with width L, and a coordinate system centered in the intrusion and perpendicular to the intrusion walls. Initial conditions used are that at time $t = 0$ the temperature T equals the intrusion temperature T_i within and at the edges of the intrusion [$-(L/2) < z < (L/2)$] and the temperature T_b equals the background country rock temperature at distances $(L/2) < z < -(L/2)$. Boundary conditions for this problem include the temperature $T = T_b$ at $z =$ and $T = T_i$ at $z = -$ for $t > 0$. With these conditions the solution to Equation (9) is (Carslaw and Jaeger 1959):

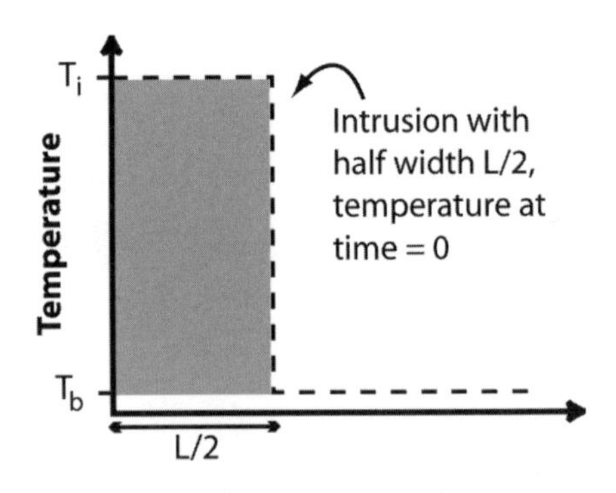

Figure 9. Model setup and initial conditions for 1D thermal model of magmatism (Eqn. 10). Instantaneous intrusion of a body with width L, and intrusion temperature T_i, occurs within country rock with a background temperature T_b.

$$T(z,t)=T_b+\frac{T_i-T_b}{2}\left[erf\left(\frac{L/2-z}{2\sqrt{\alpha t}}\right)+erf\left(\frac{L/2+z}{2\sqrt{\alpha t}}\right)\right] \tag{10}$$

where erf is the error function (Abramowitz and Stegun 1970) and z is measured as the distance from the center of the intrusion. Equation (10) can be used to calculate temperatures on each side of the intrusion; however because the boundary and initial conditions imposed are symmetric around $z = 0$ the solution is also symmetric and only half of the domain need be considered (i.e., from $z = 0$ to $z > 0$).

Figure 10 shows the thermal evolution of intrusions with thicknesses of 100 m, 1 km, and 5 km (Fig. 10a, 10b, 10c, respectively) as well as the thermal evolution of neighboring country rock. An initial intrusion temperature of 700 °C and country rock temperature of 50 °C were chosen to simulate the influence of shallow intrusions on low-temperature thermochronometers in the country rock. A dominant trend in each of the plots is the relaxation, or decrease, in the intrusion temperature and a temporary increase in country rock temperatures. For example,

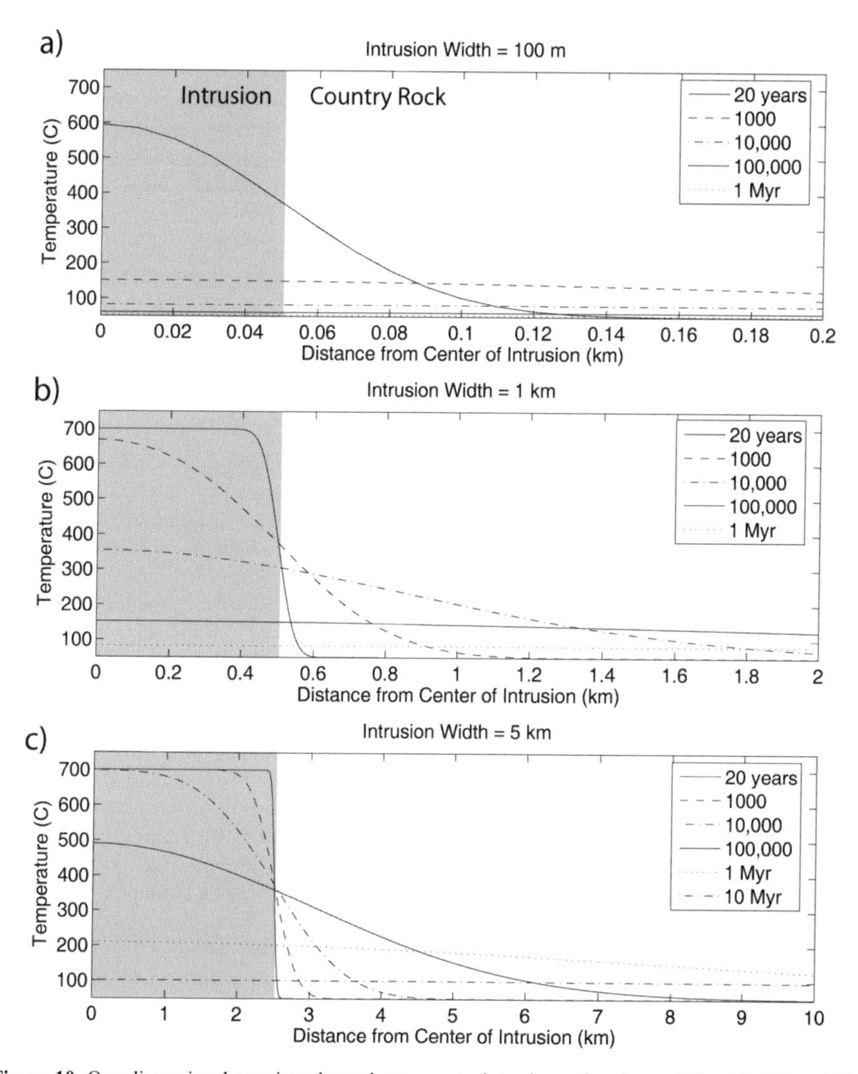

Figure 10. One dimensional transient thermal response to intrusions of various widths. (a), (b), and (c) represent the transient cooling of intrusions and surrounding country rock with intrusion thicknesses of 100 m, 1 km, and 5 km. Temperature profiles were calculated using Equation (10). Intrusion depth was assumed to be shallow and into country rock at background temperature of 50 °C. Other parameters include: initial intrusion temperature of 700 °C, and thermal diffusivity of 32 km^2Ma^{-1}. A Matlab program that reproduces Figure 10 is provided in the supplemental software archive.

with a 1 km wide intrusion (Fig. 10b) between 20 and 1000 years after emplacement of the intrusion the intrusion temperatures decrease from 700 to ~400 °C between the center and wall of the intrusion (0–0.5 km distance). Meanwhile, country rock temperatures (distance > 0.5 km) increase from the background temperature of 50 °C to ~400 °C at the contact and decrease back towards the background temperature at 1 km distance. With increased time since emplacement, temperatures continue to decrease within the intrusion and increase with greater distance from the intrusion. One Ma after emplacement, temperatures are still slightly elevated (~70 °C) above the background temperature between the center of the intrusion and 2 km distance. Thus, the emplacement of intrusions results in a temporary heat pulse that propagates into the country rock and decreases with time.

A second important aspect of Figure 10 is the time required for a heat pulse from an intrusion to decay into the surrounding country rock. The time required for the heat content of an intrusion to dissipate into the surrounding country rock is dependent upon the thermal diffusivity of the rock and the size of the intrusion. A constant thermal conductivity was used in Figure 10 and the discussion here will focus on the effect of intrusion width on the rate of heat dissipation. The smaller the intrusion thickness the faster the temperatures equilibrate back to the background temperature. For example, after 1 Ma with an intrusion thickness of 100 m (Fig. 10a) country rock temperatures (distance > 0.05 km) have equilibrated back to the initial background temperature of 50 °C. However, for an intrusion thicknesses of 5 km (Fig. 10c), 1 Ma after intrusion emplacement country rock temperatures have not equilibrated and are 200–125 °C and decrease with increased distance from the contact. Furthermore, 10 Ma after emplacement (Fig. 10c) country rock temperatures are still elevated to 100 °C. Thus, as intrusion size increases so does the magnitude and duration of heating in adjacent country rock and at greater distances from the intrusion (e.g., Armstrong et al. 1997).

There are several implications of Figure 10 for thermochronometer data collected in country rock adjacent to shallow intrusive bodies. Thermochronometer ages from the country rock will have older, or parent, ages than the intrusion. If samples are collected in the country rock close to the intrusion then the heat pulse associated with intrusion emplacement might reset thermochronometer ages to the intrusion age. Because the magnitude and duration of country rock heating decreases with increased distance from the contact it is possible for thermochronometer ages close to the contact to be reset to the intrusion age whereas samples collected at greater distances will have the parent age of the country rock. Whether or not thermochronometer samples will be reset depends on the magnitude and duration of the heat pulse and the kinetics of He diffusion, Ar loss, and/or fission-track annealing. A detailed analysis of what magnitude and duration of heating is required to reset any given thermochronometer age is beyond the scope of this discussion, and readers should see recent work by Armstrong et al. (1997), Tagami and Shimada (1996) and Reiners (2005). Thus, if the intent of the thermochronometer sampling is to study the magnitude of heating and kinetics of thermochronometer ages then samples should be collected with increasing distance from the intrusion with a finer sampling interval close to the contact. Alternatively, if the objective of a study is to quantify the exhumation history of the country rock then sampling far away from the intrusion is desirable and Equation (10) and Figure 10 can be used to approximate, for different size intrusions, what distance is sufficiently large enough to record cooling associated with country rock exhumation rather than reheating and cooling associated with magmatism.

A thorough study modeling intrusive thermal processes and thermochronometer age responses with distance from the intrusion has not been conducted. Additional work is needed along the lines of (Reiners 2005) to not only measure multiple thermochronometer system ages with increased distance from the intrusion of different sizes and compositions, but also to quantify intrusion thermal processes and evaluate if other associated heat transfer processes such as hydrothermal fluid flow along intrusion contacts can significantly influence intrusion thermal processes (e.g., Cook and Bowman 2000; Cui et al. 2001).

Topography

The temperature field of the upper crust is sensitive to overlying topography (e.g., Fig. 6). Hence, the cooling history of exhumed thermochronometer samples will vary from point to point beneath topography. Unlike the previous examples (Figs. 7 and 8) in which variations in cooling histories and thermochronometer ages resulted from spatial variability in exhumation rate, here cooling histories and thermochronometer sample age variations can be produced in a region experiencing uniform erosion but beneath topography. The distribution of ages at the surface may thus be inverted for paleotopography to provide important limits on paleoelevation. Several studies have highlighted the utility of apatite (U-Th)/He ages for this approach (e.g., House et al. 1998; House et al. 2001; Persono et al. 2002).

The effect of topography on subsurface temperatures has long been known (Lees 1910), and more recent studies have investigated how topography affects cooling histories using 2D and 3D thermal models (Stuwe et al. 1994; Mancktelow and Grasemann 1997; Braun 2002, 2005; Stuwe 2002; Ehlers and Farley 2003). These studies conclude that spatial and temporal variations in cooling rate depend on the wavelength and amplitude of the topography, as well as the exhumation rate and duration. The implications of topography for low temperature cooling ages in exhuming terrains are not generally appreciated and is an area of active research. For additional information on this topic the reader is referred to a thorough synthesis of recent techniques for quantifying topographic change in a companion chapter in this volume by (Braun 2005).

Fluid flow

The previous discussion of crustal thermal processes emphasized conductive heat transfer in stable cratonal settings and conductive and advective heat transfer due to erosion, sedimentation, and faulting. Fluid flow within the upper crust can also significantly perturb the background conductive thermal state of the crust (e.g., Fig. 3) by advective transport of heat by water. A rich body of literature is available quantifying regional scale fluid flow and thermal processes in sedimentary aquifers (e.g., Smith and Chapman 1983 and references therein). Fluid flow in alpine settings, where thermochronometer data are often collected, is unfortunately less well studied. Furthermore, to date no study has addressed the influence of alpine fluid flow on the interpretation of thermochronometer data in age-elevation plots. The purpose of this section is to highlight the relevant hydrologic and thermal processes controlling alpine fluid flow and to draw the readers attention to how these processes can influence thermochronometer cooling ages in age-elevation plots.

Several quantitative studies have investigated the influence of mountain topography on fluid flow (e.g., Freeze and Witherspoon 1967; Jamieson and Freeze 1983; Toth and Millar 1983; Ingebritsen and Sorey 1985; Forster and Smith 1989). Tell-tale signs of active fluid flow in mountainous settings are warm and hot springs discharging at valley bottoms. Paleosprings can are often indicated by hydrothermally altered rock near fractures. It is important to note that springs are ephemeral features and last on the order of 10^4–10^5 years and change location (Elder 1965). Thus, the present-day distribution and thermal power output from springs is not necessarily a good indicator of ancient fluid flow patterns and careful attention to the field geology is also important.

Mountain fluid and heat flow are related through several processes (Forster and Smith 1988a). In mountainous settings, topography causes lateral variations in pressure that drive fluid flow from range crests to valley bottoms (Fig. 6). This type of flow can be thought of as topographically-driven fluid flow and is often responsible for groundwater discharge into mountain streams and hot springs (e.g., Ehlers and Chapman 1999; Manning and Solomon 2003). Thus, mountain topography can enhance groundwater circulation to greater depths where temperatures are warmer and the water will capture heat prior to discharge at the surface. Topographically induced lateral temperature variations can influence fluid density and viscosity, thereby altering the rates and patterns of groundwater flow. Thermally induced variations in fluid density can produce a buoyancy-driven vertical fluid flow. Rates of fluid flow can increase from the reduced viscosity of water at higher temperatures. Although elevated subsurface temperatures under mountains can increase rates of fluid flow it is important to note that rocks in mountainous terrains often have significantly lower permeabilities than sedimentary aquifers. Lower permeabilities in mountain settings can result in longer timescales for fluid flow between ridges and valleys (Forster and Smith 1988a). Fluid flow can discharge as hot springs along range fronts if fracture zones associated with range-bounding faults or sedimentary basins have higher permeability than the range.

Figure 11 illustrates how topographically driven fluid flow can influence crustal thermal fields. Meteoric water in the form of rain and snow melt recharges groundwater near the range crest. As the groundwater flows toward the topographic low (basin) it warms and captures heat from the surrounding bedrock. The warm groundwater eventually rises toward the valley floor due to thermal buoyancy and/or higher permeability in the range-bounding fault and basin sediments. Some groundwater can bypass the fault and recharge a basin aquifer or exit as a hot spring further out in the valley. The influence of this groundwater path on the subsurface thermal field is as follows. Downward flowing groundwater in the "zone of heat capture" (Fig. 11) depresses thermal gradients and heat flow over the range crest. Warm water exiting at the hot spring or bypassing the fault to the basin can elevate ground temperatures in those areas, as well as increasing thermal gradients and heat flow. If all the water that enters at the range crest exits at the hot spring then a plot of heat flow normalized by the background heat flow (Fig. 11, top) will have equal area above and below the background heat flow value due to conservation of energy (e.g., Ehlers and Chapman 1999).

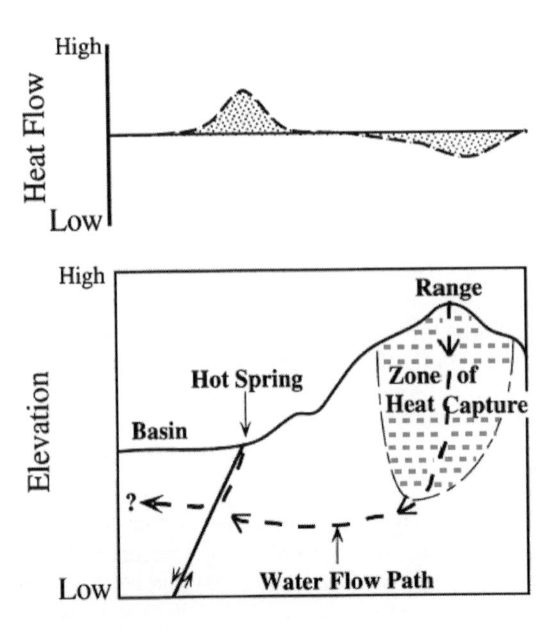

Figure 11. Schematic representation of topographically driven fluid flow and surface heat flow response. Lower panel illustrates groundwater flow paths driven by topography. Heat is captured along the descending path toward the valley and liberated in hot springs or transferred to the adjoining basin. Top panel illustrates how heat flow (and thermal gradients) could be depressed under the range crest and enhanced near the hot spring. Modified from Ehlers and Chapman 1999.

The fluid flow driven depression of thermal gradients under the range crest and elevation of thermal gradients at the valley floor will influence thermochronometer cooling histories of samples collected up the range front. The effect of fluid flow on subsurface temperatures is superimposed upon other relevant thermal processes in this setting, such as topographic, erosional, and faulting effects on subsurface temperatures. Note that although erosion of the range in Figure 11 could enhance thermal gradients, fluid flow could have the opposite effect and depress thermal gradients.

Figure 12 demonstrates the influence of mountain fluid flow on subsurface temperatures. The model accounts for coupled fluid flow and heat transfer for a mountain range with 2 km relief and convex and concave topographic profiles. The models share a common basal heat flow (60 mWm^{-2}), thermal conductivity (2.5 $Wm^{-1}k^{-1}$), impermeable lower and side boundary

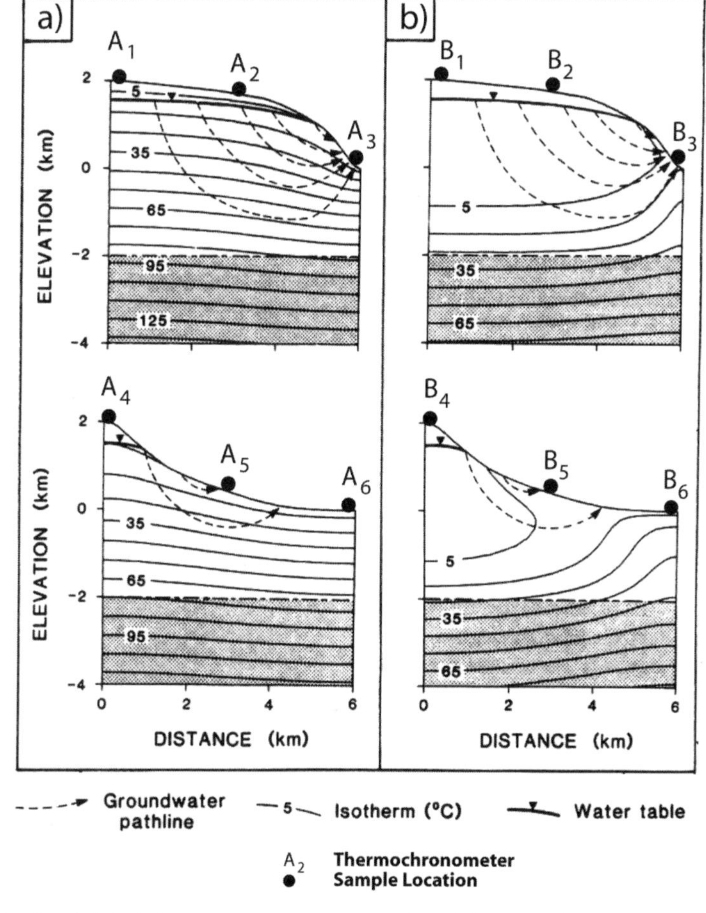

Figure 12. Coupled hydrologic and thermal numerical model results demonstrating the influence of topographically driven fluid flow on subsurface temperatures. See text for model description and parameters. (a) Concave and convex topographic geometries. Fluid flow (dashed lines) is minimal in these simulations, the thermal field is conductive, and not hydrologically disturbed. (b) Same topographies as (a) but thermal field hydrologically disturbed and dominated by advective heat transfer (note change in position of isotherms. Black circles at surface represent hypothetical thermochronometer sample locations discussed in Figure 13. Modified from Forster and Smith (1995).

conditions, and a low-permeability layer below −2 km depth (1.0×10^{-22} m^2). Remaining material properties are described in (Forster and Smith 1988a). The models differ in the permeability of the upper layer above −2 km depth and the infiltration rate at the surface.

The models in Figure 12 demonstrate the transition between conduction and fluid advection dominated thermal fields as a function of increasing permeability in the upper unit. Permeabilities less than 10^{-18} m^2 result in a purely conductive thermal field where shallow isotherms mimic the overlying topography (Fig. 12a). At permeabilities of 10^{-15} m^2 fluid flow significantly perturbs subsurface temperatures and the thermal field is dominated by advective fluid heat transport (Fig. 12b). In a fluid advection dominated thermal field isotherms no longer mimic the overlying topography but rather have a shape that is the inverse of the topography (e.g., compare 35 °C isotherm in Fig. 12a,b). The transition, or threshold, between a conductive and fluid advective thermal field occurs with a permeability around 10^{-16} m^2 (not shown) (Forster and Smith 1988a). The previous threshold permeability corresponds to the lower end of permeabilities associated with rocks present in mountainous settings such as sandstone, limestone and dolomite, and fractured igneous and metamorphic rocks (Freeze and Cherry 1979). The key point illustrated here is that the hydraulic permeability of the upper most crust can have significant influence on subsurface temperatures under certain conditions. Hyrdraulic permeability in mountain settings is controlled not only by lithology but also by secondary fractures generated by removal of overburden and tectonic stresses. A more detailed suite of similar model simulations are presented in (Smith and Chapman 1983; Forster and Smith 1988b,1989).

The transition from a conductive thermal field to one dominated by fluid advection can significantly influence the interpretation of thermochronometer data in age-elevation plots. Figure 12b illustrates how advective heat transport from fluid influences lateral variations in the depth to an assumed closure temperature of 65 °C [e.g., apatite (U-Th)/He closure]. Figure 13 schematically shows how these variations in closure depth could influence ages as a function of elevation. For this thought experiment, uniform erosion of the surface is assumed to occur at a sufficiently low erosion rate (~0.1–0.2 mm/yr) that advective heat transport from erosion is negligible. For a conduction dominated thermal field with concave and convex shaped topography (Fig.12a, filled circles) apatite (U-Th)/He ages will increase with elevation due to the greater distance samples from higher elevations travel from the closure temperature depth (Fig. 13). In this conduction dominated thermal field the samples plot on a straight line. For a fluid advection dominated thermal field with erosion at the same rate as the conductive field the geometry and depth of the 65 °C closure isotherm is perturbed from the conductive field. In the fluid advective model, all samples spend a greater amount of time traveling between the closure temperature and surface and therefore have older ages (Fig. 13, open circles). The increased sample ages in the fluid advective model could be falsely interpreted as exhumation at slower rate when in fact the erosion rate could be the same in both the conductive and fluid advective models. The previous discussion assumes that hydrologic conditions remain constant for the entire time rocks are exhumed.

A second important characteristic of fluid flow on sample ages is that ages do not plot on a straight line as a function of elevation (Fig. 13, open circles). For both topographic geometries considered a concave up curve connects sample ages plotted as a function of elevation. A concave up trend in thermochronometer ages plotted as a function of elevation can also result from a recent change in topographic relief due to differential erosion rates between ridge crests and valley bottoms (Braun 2002). However, in the examples used here a similar concave up trend can occur across topographies with uniform erosion rates and fluid flow between the ridge crest and valley bottom. Thus, caution must be taken when interpreting paleotopography from thermochronometer data that advective heat transport from fluid flow is not biasing the interpretation.

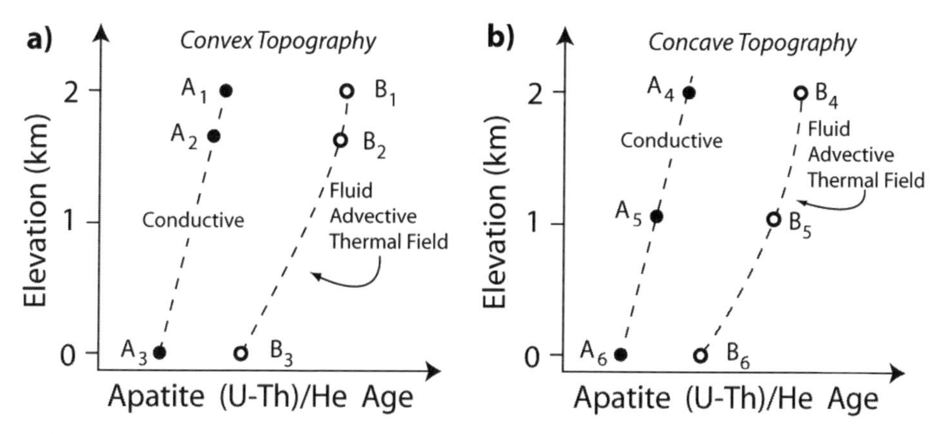

Figure 13. Influence of topographically driven fluid flow on thermochronometer age-elevation plots. (a) Thermochronometer age-elevation relationships for the concave topographies shown in Figure 12. For a conductive (Figure 12a) thermal field (solid circles) thermochronometer samples would plot along a straight line. In an advective fluid flow thermal field (Figure 12b) sample ages would be older (for the same erosion rate as in a conductive simulation) and would have a curved, concave-up geometry that could be falsely interpreted as changing topographic relief. (b) Thermochronometer age-elevation relationships for the convex topographies shown in Figure 12. As with (a) a conductive thermal field produces exhumed sample ages that lie on a straight line. An advective thermal field results in ages that are older and lie along a concave up curve. The relative sample ages shown here were estimated by assuming a constant erosion rate and measuring distances between exhumed sample elevations and the 65 °C isotherms shown in Figure 12.

In summary, the effect of fluid flow on thermochronometer data are as follows: (1) at permeabilities of greater than $\sim 10^{-16}$ m^2 the thermal field can be dominated by fluid advection. (2) all else being equal, fluid dominated thermal fields result in older thermochronometer ages than samples exhumed through a conductive thermal field. Increased thermochronometer ages from fluid flow can result in an overestimate of the timing of the exhumation, and (3) fluid flow can result in a concave up geometry to data plotted as a function of elevation. This concave up pattern could be falsely interpreted as a change in topographic relief. Additional work is needed to quantify the effect of fluid flow on thermochronometer ages. In particular, modeling studies are needed to quantify if fluid flow induced variations in thermochronometer ages significantly influence interpretations of the data when uncertainties in sample age are taken into account. Furthermore, the permeability of mountainous regions with variable fracture density is poorly understood and although the presence or lack of significant fluid flow can be documented with other geochemical or geophysical techniques (e.g., Blackwell et al. 1989; Ehlers and Chapman 1999; Manning and Solomon 2003) studies linking these observations to models that quantify erosion rates and paleotopographic change needs to be conducted.

CONCLUDING REMARKS

This chapter discussed natural variability in crustal thermal gradients and different underlying geologic processes responsible for the observed variability. Mathematic and physical constructs for evaluating thermochronometer thermal histories were presented for the cases of steady-state and transient geotherms, erosion and sedimentation, normal and thrust faulting, magmatism, topography, and hydrothermal fluid flow. In most applications, geologic interpretations of thermochronometer data can be improved by quantifying the thermal consequence of geologic processes responsible for a suite of measured ages. Although thermal modeling of geologic processes is time consuming, the benefits can be rewarding in as much that additional information (e.g., exhumation rates, fault kinematics, etc.) can be obtained from a data set. Thermal modeling geologic processes prior to collecting samples can provide valuable insight into the magnitude of signal potentially preserved in thermochronometer data as well as the optimal sampling strategy required to capture the predicted signal.

Modeling crustal thermal processes requires careful attention to how different model parameters influence the cooling history. For example, if forward modeling the cooling history in an eroding medium (e.g., Eqn. 7), numerous simulations should be conducted exploring how variations in thermal diffusivity, erosion rate, and erosion duration influence the thermal history of exhumed samples. Predicted thermal histories can be compared to data by calculating cooling rate dependent ages using one of several freely available software packages (e.g., see online software archive). Thus, for each of the tens, hundreds, or thousands of predicted thermal histories, a direct comparison to predicted and observed ages can be made. Statistical comparisons of predicted and observed ages such as a Chi-squared misfit (e.g., Batt et al. 2001; Ehlers et al. 2003, Braun 2005), are efficient ways to identify what combination of model parameters produces the best fit predicted ages to the data.

In an ideal world a perfect and unique model fit to the data will be identified. In reality, individual data points will be outliers to model predicted ages or systematic misfits to the data will be predicted regardless of what combination of model parameters are used. If the later occurs, then the model must be re-evaluated to determine if it is appropriately capturing the relevant thermal processes and additional processes will need to be considered with a different model. Alternatively, several model predicted ages may produce an equally good fit to the data and some uncertainty in the interpretation of the data will be unavoidable.

In many cases, the use of the analytic solutions described here will contain restrictive assumptions (e.g., constant thermal properties, or 1D heat transfer) that require the use of more complicated numerical models to circumvent. The interpretation of geologic events using thermal models and thermochronometer data should be thought of as an iterative process that may take significant effort. The Earth is a complicated environment and multiple combinations of model parameters may produce an equally good fit to the data. Thus, the approach taken should not be to find *the* solution that satisfies the data but rather *the range* of solutions that provide a good fit to the data. The pay-off from the time invested in forward modeling thermal histories is a rigorous and quantitative understanding of the underlying geologic processes responsible for an observed data set.

As a summary to the topics covered in this chapter, the following generalized steps are suggested as guidelines for quantifying the influence of crustal thermal processes on thermochronometer sample collection and interpretation:

Step 1: Compile available surface heat-flow determinations to determine spatial variation in the present day thermal field (e.g., Fig. 1) across the study area. This step is primarily relevant for interpreting samples collected from an active orogenic or magmatic setting and is less relevant for the study of ancient (e.g., Paleozoic) mountain belts. Surface heat-flow determinations should be used to help constrain predicted subsurface temperatures and surface heat flow calculated in steps 3 and 4.

Step 2: Compile relevant thermo-physical properties (e.g., thermal conductivity, heat production, etc) for lithologies present in the study area. Ideally material properties are measured on rocks from within a study region because the range values for any given rock type can sometimes be large (e.g., Appendix A). If measured values from the study area are not possible then table 'lookup' values such as those in appendix A provide a good starting point.

Step 3: Identify the dominant thermal processes influencing thermochronometer thermal histories and quantify the spatial dimensions over which these processes are relevant. This step requires application of the relevant differential equation(s) for heat transfer discussed in this chapter, or in referenced material. A step-wise increase is model complexity is recommended by starting with simpler 1D equations and increasing to 2D and 3D formulations if needed. The test of whether 1D, 2D, or 3D models are sufficient hinges on the dominant direction(s) of heat flow in the study over the time span that samples cooled. For example, if heat flow is believed to be predominantly vertical then a 1D simulation of heat transfer is sufficient. Alternatively, if significant lateral heat flow occurs in association with samples collected close to active structures (e.g., Ehlers et al. 2003) or long-wavelength topography (e.g., Stuwe et al. 1994; Braun 2002; Ehlers and Farley 2003) then 2D or 3D models should be applied.

Step 4: Quantify if temporal changes in the crustal thermal field influenced the sample thermal history. An important concept illustrated earlier in this chapter (e.g., Figs. 3, 4, 5, 10) is characterization of if the thermal field was in steady state or evolving at the time samples cooled. Various 1D equations and figures in this chapter were presented for determination of the time scales over which subsurface temperatures evolve in magmatic, or erosional and depositional settings (see also Mancktelow and Grasemann 1997). If the thermal field is determined to be transient then some form of the time-dependent diffusion equation should be applied for sample interpretation.

Step 5: After the spatial and temporal variability in the thermal field is characterized then the relevant governing equation should be used to forward model sample cooling histories. These predicted thermal histories can then be used to compute predicted thermochronometer ages, track length distributions, etc, using the thermal histories generated from steps 2 to 4 (see Dunai 2005; Harrison et al. 2005; Ketcham 2005 for a detailed discussions) that can then be compared to the data.

Step 6: Quantify the range of plausible material properties, boundary conditions, and model free parameters that minimize differences between predicted and observed thermal histories and/or thermochronometer ages. This final step requires identification and exploration of all free parameters within the model being used and iteratively applying the model (e.g., step 5) to compare predicted and observed ages and thermal histories. A statistical comparison between model predicted and observed ages and thermal histories is recommended to identify the range of plausible geologic processes that could have produced the observed sample ages and thermal histories.

Unfortunately, every data set and geologic setting requires a different treatment and thermal modeling analysis for a robust interpretation of thermochronometer data. The previous steps are provided mainly as guidelines for assessing the influence of crustal thermal processes on thermochronometer data and additional, or fewer, steps may be required depending on the problem addressed.

ACKNOWLEDGMENTS

This chapter benefited from discussions with the Surface Processes Research Group, Jason Barnes, and other curious graduate students at the University of Michigan. Phillip Armstrong and Kevin Furlong are thanked for thorough and constructed reviews. This chapter resulted from US National Science Foundation funding to the author (EAR 0409289, 0309779, and 0196414).

REFERENCES

Abramowitz M, Stegun I (eds) (1970) Handbook of Mathematical Functions. Dover Publications, New York

Allmendinger R, Shaw J (2000) Estimation of fault propagation distance from fold shape; implications for earthquake hazard assessment. Geology 28(12):1099-1102

Angevine CL, Turcotte DL (1983) Oil generation in overthrust belts. AAPG Bulletin 67:235-241

Armstrong P, Ehlers T, Chapman D, Farley K, Kamp P (2003) Exhumation of the central Wasatch Mountains, Utah: 1. Patterns and timing of exhumation deduced from low-temperature thermochronology data. J Geophys Res 108(3):2172, doi:10.1029/2001JB001708

Armstrong PA, Chapman DS (1999) Beyond surface heat flow: An example from a tectonically active sedimentary basin. Geology 26:183-186

Armstrong PA, Kamp PJ, Allis RG, Chapman DS (1997) Timing of the heat flow high on Taranaki Peninsula (New Zealand): evidence from combined apatite fission track age and vitrinite reflectance data. Basin Res 9:151-169

Batt G, Brandon M, Farley K, Roden-Tice M (2001) Tectonic synthesis of the Olympic Mountains segment of the Cascadia wedge, using two-dimensional thermal and kinematic modeling of thermochronological ages. J Geophys Res 106(B11):26731-26746

Benfield A (1949a) The effect of uplift and denudation on underground temperatures. J Appl Phys 20:66-70

Benfield A (1949b) A problem of the temperature distribution in a moving medium. Quart Appl Math 6:439-443

Bertotti G, Seward D, Wijbrans J, ter Voorde M, Hurford A (1999) Crustal thermal regime prior to, during, and after rifting; a geochronological modeling study of the Mesozoic South Alpine rifted margin. Tectonics 18(2):185-200

Bertotti G, ter Voorde M (1994) Thermal effects of normal faulting during rifted basin formation; 2, The Lugano-Val Grande normal fault and the role of pre-existing thermal anomalies. Tectonophysics 240(1-4):145-157

Blackwell DD, Steele JL, Wisian KW (1994), Results of geothermal resource evaluation for the eastern United States. Geothermal Resources Council Trans 18:161-164

Blackwell DD, Wisian KW, Richards M (1996) Geothermal resources of the United States based on heat flow and gradient information, DOE Report Contract #C91-103450

Blackwell D, Steele J, Brott C (1989) Heat flow in the Pacific Northwest. *In:* Physical Properties of Rocks and Minerals. Touloukian Y, Judd W, Roy R (eds) McGraw-Hill, New York, p 495-502

Boyer S, Elliot D (1982) Thrust systems. AAPG Bulletin 66(9):1196-1230

Braun J (2002) Quantifying the effect of Recent relief changes on age-elevation relationships. Earth Planet Sci Lett 200(3-4):331-343

Braun J (2005) Quantitative constraints on the rate of landform evolution derived from low-temperature thermochrohonology. Rev Mineral Geochem 58:351-374

Brewer J (1981) Thermal effects of thrust faulting. Earth Planet Sci Lett 56:233-244

Burtner R, Nigrini A (1994) Thermochronology of the Idaho-Wyoming thrust belt during the Sevier Orogeny; a new, calibrated, multiprocess thermal model. AAPG Bulletin 78(10):1586-1612

Burtner R, Nigrini A, Donelick R (1994) Thermochronology of lower Cretaceous source rocks in the Idaho-Wyoming thrust belt. AAPG Bulletin 78(10):1613-1636

Carslaw H, Jaeger J (1959) Conduction of Heat in Solids. Clarendon Press, Oxford

Cermak V, Rybach L (1982) Thermal conductivity and specific heat of minerals and rocks. *In:* Landolt-Bornstein Numerical Data and Functional Relationships in Science and Technology, New Series, Group V. Angenheister G (ed), Springer-Verlag, Berlin, p. 89-134

Cerveny PF, Steidtmann JR (1993) Fission track thermochronology of the Wind River Range, Wyoming; evidence for timing and magnitude of Laramide exhumation. Tectonics 12(1):77-92

Chapman D (1986) Thermal gradients in the continental crust. Geol Soc Spec Publ 24:63-70

Chapman D, Pollack H (1975) Global heat flow: a new look. Earth Planet Sci Lett 28:23-32

Chapman D, Rybach L (1985) Heat flow anomalies and their interpretation. J Geodynamics 4(1-4):3-37

Chapman DS, Furlong KP (1992) The thermal state of the lower crust, in continental lower crust. *In:* Development in Geotectonics. Volume 23. Fountain DM, Arculus RJ, Kay RM (eds) Elsevier, Amsterdam, p 179-199

Clark S (1966) Handbook of Physcial Constants. Memoir Geological Society of America, 97. Geological Society of America, New York

Commision IHF (2004) Global heat flow data base. *http://www.heatflow.und.edu/index2.html*

Condon EU, Odishaw H (1967) Handbook of Physics. McGraw-Hill, New York

Cook SJ, Bowman JR (2000) Contact metamorphism surrounding the Alta stock fluid rock interaction accompanying metamorphism of siliceous dolomites. J Petrology 41:739-757

Cui X, Nabelek PI, Liu M (2001) Controls of layered and transient permeability on fluid flow and thermal structure in contact metamorphic aureals, with application to the Notch Peak aureole, Utah. J Geophys Res 106:6477-6491

Dahlstrom CDA (1969) Balanced cross sections. Can J Earth Sci 6:743-757

Davy P, Gillett P (1986) The stacking of thrust slices in collision zones and its thermal consequences. Tectonics 5:913-929

Dodson MH (1973) Closure temperature in cooing geochronological and petrological systems. Contrib Mineral Petrol 40:259-274

Dodson MH (ed) (1986) Closure Profiles in Cooling Systems. Materials Science Forum, 7. Trans Tech Publications, Aedermannsdorf, Switzerland, 145-153 pp

Dunai TJ (2005) Forward modeling and interpretation of (U-Th)/He ages. Rev Mineral Geochem 58:259-274

Edman JD, Furlong KP (1987) Thrust faulting and hydrocarbon generation: A reply. AAPG Bulletin 71:890-896

Edman JD, Surdam RC (1984) Influence of overthrusting on maturation of hydrocarbons in Phosphoria Formation, Wayoming-Idaho-Utah overthrust belt. AAPG Bulletin 68:1803-1817

Ehlers T, Chapman D (1999) Normal fault thermal regimes; conductive and hydrothermal heat transfer surrounding the Wasatch Fault, Utah. Tectonophysics 312(2-4):217-234

Ehlers T, Farley K (2003) Apatite (U-Th)/He thermochronometry; methods and applications to problems in tectonic and surface processes. Earth Planet Sci Lett 206(1-2):1-14

Ehlers TA, Armstrong PA, Chapman D (2001) Normal fault thermal regimes and the interpretation of low-temperature thermochronometers. Phys Earth Planet Interiors 126:179-194

Ehlers TA, Willett SD, Armstrong PA, Chapman DS (2003) Exhumation of the central Wasatch Mountains, Utah: 2. Thermokinematic model of exhumation, erosion, and thermochronometer interpretation. J Geophys Res 108(B3):2173, doi:10.1029/2001JB001723

Elder J (ed) (1965) Physcial Processes in Geothermal Areas. Terrestrial Heat Flow, Geophysical Monograph 8. American Geophysical Union, Washington D.C.

Elliott D (1976) The energy balance and deformation mechanisms of thrust sheets. Phil Trans Royal Soc London 283:289-312

Elliott D (1977) Some aspects of the geometry and mechanics of thrust belts, Parts 1 and 2. 8th Annual Seminar Canadian Society of Petroleum Geology. University of Calgary, Calgary

Elliott D (1983) The construction of balanced cross-sections. J Struct Geol 5:101-126

England PC, Molnar P (1993) The interpretation of inverted metamorphic isograds using simple physical calculations. Tectonics 12:145-158

England PC, Thompson AB (1984) Pressure-Temperature-time paths of regional metamorphism I. Heat transfer during the evolution of regions of thickened continental crust. J Petrology 25:894-928

Farley K (2000) Helium diffusion from apatite; general behavior as illustrated by Durango fluorapatite. J Geophys Res 105(B2):2903-2914

Forster C, Smith L (1988a) Groundwater flow systems in mountainous terrain; 1, Numerical modeling technique. Water Resour Res 24(7):999-1010

Forster C, Smith L (1988b) Groundwater flow systems in mountainous terrain; 2, Controlling factors. Water Resour Res 24(7):1011-1023

Forster C, Smith L (1989) The influence of groundwater flow on thermal regimes in mountainous terrain: a model study. J Geophys Res 94(B7):9439-9451

Freeze AR, Cherry JA (1979) Groundwater. Prentice Hall, Englewood Cliffs, NJ

Freeze RA, Witherspoon PA (1967) Theoretical analysis of regional groundwater flow, 2. Effect of water table configuration and subsurface permeability variations. Water Resour Res 3:623-634

Furlong K, Chapman D (1987) Crustal heterogeneities and the thermal structure of the continental crust. Geophys Res Lett 14(3):314-317

Furlong KP, Edman JD (1984) Graphical approach to the determination of hydrocarbon maturation in overthrust terrains. AAPG Bulletin 68:1818-1824

Furlong KP, Edman JD (1989) Hydrocarbon maturation in thrust belts; thermal considerations. *In:* Origin and Evolution of Sedimentary Basins and Their Energy and Mineral Resources. Geophysical Monograph, vol. 48, IUGG vol. 3. Price RA (ed) American Geophysical Union, Washington, DC, p 137-144

Furlong KP, Hanson RB, Bowers JB (1991) Modeling thermal regimes. Rev Mineral 26:437-506

Furlong KP, Londe MD (eds) (1986) Thermal-mechanical consequences of basin and range extension. *In:* Extensional Tectonics of the Basin and Range Province: A Perspective. GSA Special Paper 208. Mayer L (ed) Geological Society of America, Boulder, p 23-30

Govers R, Wortel M (1995) Extension of stable continental lithosphere and the initiation of lithosphere scale faults. Tectonics 14(4):1041-1055

Graham CM, England PC (1976) Thermal regimes and regional metamorphism in the vicinity of overthrust faults: an example of shear heating and inverted metamorphic zonation from southern California. Earth Planet Sci Lett 31:142-152

Haenel R, Rybach L, Stegena L (1988) Handbook of Terrestrial Heat-flow Density Determination; with guidelines and recommendations of the international Heat Flow Commission. Kluwer-Acad., Dordrecht

Harrison TM, Grove M, Lovera OM, Zeitler PK (2005) Continuous thermal histories from inversion of closure profiles. Rev Mineral Geochem 58:389-409

Homza TX, Wallace WK (1995) Geometric and kinematic models for detachment folds with fixed and variable detachment depths. J Struct Geol 17:575-588

House M, Wernicke B, Farley K (1998) Dating topography of the Sierra Nevada, California, using apatite (U-Th)/He ages. Nature 396(6706):66-69

House MA, Wernickey KA, Farley KA (2001) Paleo-geomorphology of the Sierra Nevada, California, from (U-Th)/He ages in apatite. Am J Sci 301:77-102

Husson L, Moretti I (2002) Thermal regime of fold and thrust belts - an application to the Bolivian sub Andean zone. Tectonophysics 345:253-280

Ingebritsen SE, Sorey ML (1985) A quantitative analysis of the Lassen hydrothermal system, North Central California. Water Resour Res 21:853-868

Jamieson GR, Freeze RA (1983) Determining hydraulic conductivity distribution in a mountainous area using mathematical modeling. Groundwater 21:168-177

Jones PB (1971) Folded faults and sequences of thrusting in Alberta foothills. AAPG Bulletin 55:292-306

Kappelmeyer O, Haenel R (1974) Geothermics, with Special Reference to Application. Gebruder Borntraeger, Berlin

Karabinos P, Ketcham R (1988) Thermal Structure of Active Thrust Belts. J Metamorph Geol 6(5):559-570

Ketcham RA (2005) Forward and inverse modeling of low-temperature thermochronometry data. Rev Mineral Geochem 58:275-314

Lachenbruch A (1968) Rapid estimation of the topographic disturbance to superficial thermal gradients. Rev Geophys 6(3):365-400

Lachenbruch A, Sass J (1978) Models of an extending lithosphere and heat flow in the Basin and Range Province. GSA Memoir 152:209-250

Lachenbruch A, Sass J (1980) Heat flow and energetics of the San Andreas fault zone. J Geophys Res 85(B11): 6185-6222

Lee Y, Deming D (1999) Heat flow and thermal history of the Anadarko Basin and the western Oklahoma platform. Tectonophysics 313(4):399-410

Lees CH (1910) On the isogeotherms under mountain ranges in radioactive districts. Proc Royal Soc 83:339-346

Mailhe D, Lucazeau F, Vasseur G (1986) Uplift history of thrust belts; an approach based on fission track data and thermal modelization. Tectonophysics 124(1-2):177-191

Mancktelow N, Grasemann B (1997) Time-dependent effects of heat advection and topography on cooling histories during erosion. Tectonophysics 270(3-4):167-195

Manning A, Solomon D (2003) Using noble gases to investigate mountain-front recharge. J Hydrology 275(3-4):194-207

Martel SJ (1999) Mechanical controls on fault geometry. J Struct Geol 21:585-596

Mitra S (1986) Duplex structures and imbricate thrust systems; geometry, structural position, and hydrocarbon potential. AAPG Bulletin 70(9):1087-1112

Molnar P, England PC (1990) Temperatures, heat flux, and frictional stress near major thrust faults. J Geophys Res 95:4833-4856

Omar GI, Lutz TM, Giegengack R (1994) Apatite fission-track evidence for Laramide and post-Laramide uplift and anomalous thermal regime at the Beartooth overthrust, Montana-Wyoming. Geol Soc Am Bull 106(1):74-85

O'Sullivan PB, Wallace WK, Murphy JM (1998) Fission-track evidence for apparent out-of-sequence Cenozoic deformation along the Philip Smith Mountain Front, northeastern Brooks Range, Alaska. Earth Planet Sci Lett 164(3-4):435-449

Oxburgh ER, Turcotte DL (1974) Thermal gradients and regional metamorphism in overthrust terrains with special reference to the Eastern Alps. Schweiz Mineral. Petrogra Mitt 54:641-662

Peacock SM (1989a) Numerical constraints on rates of metamorphism, fluid production, and fluid flux during regional metamorphism. Geol Soc Am Bull 101:476-485

Peacock SM (1989b) Thermal modeling of metamorphic P-T-t paths. *In:* Metamorphic P-T-t Paths. Spear FS, Peacock SM (eds), American Geophysical Union, p 57-102

Persono C, Stuart F, Bishop P, Barfod D (2002) Apatite (U-Th)/He age constraints on the development of the Great Escarpment on the southeastern Australian passive margin. Earth Planet Sci Lett 200:79-90

Philpotts AR (1990) Principles of Igneous and Metamorphic Petrology. Prentice Hall, Englewood Cliffs

Platt JP, Leggett JK, Young J, Raza H, Alam S (1985) Large-scale sediment underplating in the Makran accretionary prism, southwest Pakistan. Geology 13:507-511

Pollack H (1982) The heat flow from the continents. Annual Rev Earth Planet Sci 10:459-481

Pollack H, Chapman D (1977) On the regional variation of heat flow, geotherms, and lithospheric thickness. Tectonophysics 38(3-4):279-296

Pollack H, Hurter S, Johnson J (1993) Heat flow from the Earth's interior; analysis of the global data set. Rev Geophys 31(3):267-280

Popov Y, Pribnow D, Sass J, Williams C, Burkhardt H (1999) Characterization of rock thermal conductivity by high-resolution optical scanning. Geothermics 28:253-276

Powell W, Chapman D, Balling N, Beck AE (1988) Continental heat-flow density. *In:* Handbook of Terrestrial Heat-Flow Density Determination. Haenel R, Rybach L, Stegena L (eds), Kluwer Academic Publishers, Dordrecht p 167-222

Pribnow D, Sass J (1995) Determination of thermal conductivity for deep boreholes. J Geophys Res 100(B6): 9981-9994

Quidelleur X, Grove M, Lovera O, Harrison T, Yin A (1997) Thermal evolution and slip history of the Renbu Zedong Thrust, southeastern Tibet. J Geophys Res 102(B2):2659-2679

Rahn M, Grasemann B (1999) Fission track and numerical thermal modeling of differential exhumation of the Glarus thrust plane (Switzerland). Earth Planet Sci Lett 169(3-4):245-259

Reiners PW (2005) Zircon (U-Th)/He thermochronometry. Rev Mineral Geochem 58:151-179

Roy R, Beck A, Touloukian Y (1981) Thermo-physical properties of rocks. *In:* Physical Properties of Rocks and Minerals. Touloukian Y, Judd W,Roy R (eds), McGraw-Hill, New York, p 409-502

Ruppel C, Hodges K (1994) Role of horizontal thermal conduction and finite time thrust emplacement in simulation of pressure-temperature-time paths. Earth Planet Sci Lett 123(1-4):49-60

Ruppel C, Royden L, Hodges K (1988) Thermal modeling of extensional tectonics; application to pressure-temperature-time histories of metamorphic rocks. Tectonics 7(5):947-957

Ryback L, Cermak V (1982) Radioactive heat generation in rocks. *In:* Landolt-Bornstein Numerical Data and Functional Relationships in Science and Technology, Group V. Angenheister G (ed), Springer-Verlang, Berlin, pp. 353-371

Salvini F, Storti F, McClay K (2001) Self-determining numerical modeling of compressional fault-bend folding. Geology 29(9):839-842

Sass JH, Blackwell DD, Chapman DS, Costain JK, Decker ER, Lawyer LA, Swanberg CA, Blackstone DL, Brott CA, Heasier HP, Lachenbruch AH, Marshall BV, Morgan PM, Robert J, Steele JL (1981) Heat flow from the crust of the United States. *In:* Physical Properties of Rocks and Minerals. Touloukian Y, Judd W, Roy R (eds), McGraw-Hill, New York, p 503-548

Shi Y, Wang C (1987) Two-dimensional modeling of the P-T-t paths of regional metamorphism in simple overthrust terrains. Geology 15(11):1048-1051

Shuster DL, Farley KA (2005) $^4He/^3He$ thermochronometry: theory, practice, and potential complications. Rev Mineral Geochem 58:181-203

Smith L, Chapman D (1983) On the thermal effects of groundwater flow; 1. regional scale systems. J Geophys Res 88(B1):593-608

Somerton WH (1992) Thermal Properties and Temperature-related Behavior of Rock/Fluid Systems. Developments in petroleum science, 37. Elsevier, Amsterdam

Spear FS (1993) Metamorphic Phase Equilibria and Pressure-Temperature-Time Paths. Mineralogical Society of America, Washington D.C.

Stock J, Montgomery D (1996) Estimating palaeorelief from detrital mineral age ranges. Basin Res 8(3): 317-327

Stuwe K (2002) Introduction to the Geodynamics of the Lithosphere: Quantative Description of Geological Problems. Springer-Verlag, Berlin

Stuwe K, White L, Brown R (1994) The influence of eroding topography on steady-state isotherms; application to fission track analysis. Earth Planet Sci Lett 124(1-4):63-74

Suppe J (1976) Decollement folding in southwestern Taiwan. Pet Geol Taiwan 13:25-35

Suppe J (1980) Imbricated structure of Western Foothills Belt, southcentral Taiwan. Pet Geol Taiwan 17:1-16

Suppe J (1983) Geometry and kinematics of fault-bend folding. Am J Sci 283(7):684-721

Tagami T, Shimada C (1996) Natural long-term annealing of the zircon fission track system around a granitic pluton. J Geophys Res 101(4):8245-8255

Taylor S, McLennan S (1981) The composition and evolution of the continental crust: rare earth element evidence from sedimentary rocks. Phil Trans Royal Soc London 301:381-399

ter Voorde M, Bertotti G (1994) Thermal effects of normal faulting during rifted basin formation. 1. A finite difference model. Tectonophysics 240:133-144 ter Voorde M, de Bruijne CH, Cloetingh SAPL, Andriessen PAM (2004) Thermal consequences of thrust faulting: simultaneous versus successive fault activation and exhumation. Earth Planet Sci Lett 223: 395-413

Thorbjornsen KL, Dunne WM (1997) Origin of a thrust-related fold: geometric vs kinematic tests. J Struct Geol 19:309-319

Toth J, Millar R (1983) Possible effects of erosional changes of the topographic relief on pore pressures at depth. Water Resour Res 19(6):1585-1597

Touloukian YS, Judd WR, Roy RF (eds) (1989) Physical Properties of Rocks and Minerals. CINDAS data series on material properties, Group II, Properties of special materials, II-2. McGraw-Hill, New York

van Wees J, de Jong K, Cloetingh S (1992) Two-dimensional P-T-t modelling and the dynamics of extension and inversion in the Betic Zone (SE Spain). Tectonophysics 203(1-4):305-324

APPENDIX A:
THERMOPHYSICAL PROPERTIES OF EARTH MATERIALS

The appendix tables contain compilations of thermophysical properties for common geologic materials. These tables are provided to give readers an appreciation for the natural variability, and average values, of thermal conductivity (Table A1), specific heat (Table A2), and volumetric radiogenic heat production (Table A3). The values presented here can be used for forward modeling crustal thermal properties using equations presented earlier.

Table A1. Thermal conductivity of common geologic materials.

Material Type	*Mean Value* W/m°K	*Range of values* W/m°K	*# of Samples*	*Ref.*	*Comments*
<u>*Unconsolidated Material and Water*</u>					
Clay/Mud	2.1	1.7-3.2	na	[3]	At least 40% saturated
Soil	0.69	0.4-0.9	na	[3]	Saturated
	3.9	2.1-5.3	na	[3]	Unsaturated
Water	1.4	na	na	[4]	At 0 °C
<u>*Sedimentary Rocks*</u>					
Sandstone	4.7	3.5-7.7	7	[3]	
Limestone	3.3	1.3-6.26	445	[1]	
Shale	2.48	1.00-3.96	na	[2]	
	3.88	2.8-5.6	31	[3]	
Sandstone+Shale	1.67	na	223	[6]	*In situ* Measurements
Limestone+Shale	2.08	1.90-2.61	385	[6]	*In situ* Measurements
Limestone+Sandstone+Shale	2.55	na	44	[6]	*In situ* Measurements
<u>*Igneous Rocks*</u>					
Granite	3.3	1.78-5.02	130	[1]	
	7.8	6.2-9.0	12	[3],[4]	High quartz content
Rhyolite	1.3	0.92-1.68	na	[3]	
Basalt	1.7	1.12-2.38	72	[1]	
	1.8	1.6-2.0	na	[2]	
	4	na	na	[4]	
Gabbro	5.6	na	na	[3]	At 0 °C
<u>*Metamorphic Rocks*</u>					
Gneiss	2.95	1.7-5.8	379	[1]	
	6.58	4.6-8.0	37	[3]	High quartz content
	3.49	3.7-3.5	54	[5]	
	3.38	1.9-4.6	30	[7]	
Schist	3.6	2.1-4.6	96	[1]	
	6.6	4.1-8.9	15	[3]	High quartz content
Marble	2.8	2.02-5.59	218	[1], [4]	
Amphibolite	3.4	1.8-4.7	66	[1]	
	2.5	na	27	[5]	
	2.6	2.4-2.8	6	[7]	

Notes: All measurements made at ~20°C unless stated, na = information not available

References: [1] Touloukian et al. 1989; [2] Somerton 1992; [3] Clark 1966; [4] Condon and Odishaw 1967; [5] Pribnow and Sass 1995; [6] Lee and Deming 1999; [7] Popov et al. 1999

Table A2. Specific heat of common geologic materials.

Material Type	*Mean Value* J/kg°K	*Range of values* J/kg°K	*Ref.*
Soil, Air, Water			
Air	1006	na	[2]
Soil	922	796-1005	[2]
Water	4190	na	[2]
Sedimentary Rocks			
Sandstone	964	na	[2]
Limestone	860	910-810	[1]
Igneous Rocks			
Granite	880	na	[1]
Basalt	788	700-875	[1]
Gabbro	775	835-715	[1]
Metamorphic Rocks			
Marble	883	903-863	[1]

Notes: All measurements made at ~20 C; na = information not available
References: [1] Touloukian et al. 1989; [2] Carslaw and Jaeger 1959

Table A3. Volumetric radiogenic heat production of common geologic materials.

Material Type	*Mean Value* $\mu W/m^3$	*Range of values* $\mu W/m^3$	*Ref.*
Sedimentary Rocks			
Sandstone	0.66	0.32-0.99	[1]
Limestone	0.62	na	[1]
Shale	1.80	1.8-5.5	[1]
Igneous Rocks			
Granite	2.45	na	[1]
Dacite	1.48	na	[1]
Diorite	1.08	na	[1]
Basalt	0.31	na	[1]
Gabbro	0.31	na	[1]

Notes: na = information not available
Reference = [1] Haenel et al. 1988

Reviews in Mineralogy & Geochemistry
Vol. 58, pp. 351-374, 2005

13

Quantitative Constraints on the Rate of Landform Evolution Derived from Low-Temperature Thermochrohonology

Jean Braun

Research School of Earth Sciences
The Australian National University
Canberra ACT 0200, Australia
(*Now at Géosciences-Rennes, Université de Rennes 1, Rennes 35042, France*)
Jean.Braun@univ-rennes1.fr

INTRODUCTION

In recent years, the nature of the complex interactions between the solid earth and the overlying atmosphere has been the subject of much debate (England and Molnar 1990; Whipple et al. 1999; Braun et al. 1999). Does erosion play an important role in the dynamics of mountain building? Is there a strong feedback into the balance of forces at the crustal scale from mass transport by erosional and depositional processes at the Earth's surface? Or does erosion passively react to changes in the Earth's surface topography to shape mountains with tectonic forces, originating in the underlying convecting mantle, dominating the dynamics of plate interactions and crustal deformation at convergent plate margins?

The answers to these fundamental questions lie in our ability to determine the rate at which surface processes can react to tectonic forcing (Whipple et al. 1999). Where erosion is efficient, the coupling will be strong as the rate of change of surface loading by mass transport will be similar to the rate of creation of topography by crustal deformation. This may lead to a concentration of deformation and thus exhumation in regions of intense erosion (Koons 1990; Willett et al. 1993; Beaumont et al. 2001; Koons et al. 2002, 2003; Braun and Pauselli 2004). Where erosion is relatively inefficient, mountain building is the result of a balance between internal driving forces and those originating from the deflection of an upper free surface; there is no feedback from erosional processes.

Low-temperature thermochronology, such as (U-Th)/He (He) or fission track (FT) dating in apatite and zircon, provides information on the rate at which rocks cool as they are exhumed towards the "cold" Earth surface (Duddy et al. 1988; Warnock et al. 1997; Farley 2000; Reiners and Farley 1999). Because the temperature structure of the uppermost crust, to which the low-temperature systems will be most sensitive[1], is perturbed by the presence of finite amplitude surface topography (Turcotte and Schubert 1982; Stüwe et al. 1994; Mancktelow and Grasemann 1997; Braun 2002b), one can postulate that He and FT dating can be used to provide constraints on the shape of the surface topography in the past and, consequently, on the rate at which it evolved to reach its present-day form (Braun 2002b).

There have been several attempts to extract such information from low-temperature age datasets (Brown et al. 1994; House et al. 1997, 1998; Braun 2002a,b; Ehlers et al. 2003;

[1] The closure temperature of He in apatite is in the range, 55–75 °C, depending on cooling rate and grain size. The annealing temperature range for FT in apatite is in the range 100–120 °C

1529-6466/05/0058-0013$05.00 DOI: 10.2138/rmg.2005.58.13

Reiners et al. 2003b), most of which used the relationship between age and present-day topographic elevation to make inferences on the shape of the landform in the past, typically at the time at which the low-temperature system closed[2]. In this chapter, I will attempt to define how much information is contained in age-elevation datasets that can be used to constrain the rate at which landform evolves on tectonic timescale, the assumptions that are at the core of this interpretation and the methods to extract this information. More importantly, I will try to define a sampling strategy for collecting age-elevation datasets that optimizes the information about the rate of change of topography. I will then succinctly demonstrate how erosion-driven flexural isostasy can affect age-elevation relationships and how thermochronological data can be used to extract information on the strength of the continental lithosphere (i.e., its effective elastic thickness). I will illustrate each of these points with examples based on several thermochronological datasets collected in a range of tectonic and erosional environments.

TOPOGRAPHY AND TEMPERATURE

To illustrate the effect of surface topography on the thermal structure of the underlying crust, let's consider that the shape of the Earth's surface can be approximated by the sum of a finite number of periodic (cosine and sine) functions. In other words, we assume that surface topography can be expanded in an infinite Fourier series:

$$h(x) = \sum_{i=0}^{\infty} h_{0,i}^{c} \cos(2\pi x/\omega) + \sum_{i=0}^{\infty} h_{0,i}^{s} \sin(2\pi x/\omega) \tag{1}$$

Because heat conduction in solids can be approximated as a linear physical process, the total perturbation caused by any topographic surface can be adequately approximated by the sum of the perturbations caused by each of the components of the Fourier series.

The perturbation caused by a periodic topographic function:

$$h_i^c(x) = h_{0,i}^c \cos(2\pi x/\omega) \tag{2}$$

has been shown (Turcotte and Schubert 1982) to be in phase with the topographic signal, directly proportional to the amplitude of the topography, and to decay exponentially with depth at a rate that is proportional to the wavelength of the topographic signal (ω):

$$T(x,z) = h_{0,i}^c G_0 e^{-z/\omega} \cos(2\pi x/\omega) \tag{3}$$

where G_0 is the regional vertical temperature gradient. The total perturbation caused by a generic surface topography can therefore be expressed as:

$$T(x,z) = \sum_{i=0}^{\infty} h_{0,i}^c G_0 e^{-z/\omega} \cos(2\pi x/\omega) + \sum_{i=0}^{\infty} h_{0,i}^s G_0 e^{-z/\omega} \sin(2\pi x/\omega) \tag{4}$$

Note that this relationship is only valid for small amplitude topography (in comparison with

[2] In most of the following discussion, we will assume that each thermochronological system is characterized by a closure temperature, T_c (Dodson 1973), that corresponds to the temperature at which the system "closed," i.e., the temperature at which solid-state diffusion of the daughter product (He) becomes negligible or when fission tracks are retained in the system as annealing becomes negligible. We recognized, however, that the closure of thermochronological systems is a process that takes place over a range of temperature; where accurate estimates of a rock age are determined from a synthetic temperature history, a more sophisticated approach will be used that takes into account the dependence of solid-state diffusion processes on the rate at which the system cools and the size of the diffusing domain (Dodson 1973; Wolf et al. 1996, 1998).

the wavelength) and assumes that the temperature structure is in steady-state. Furthermore, it assumes that conduction is the dominant mechanism at transporting the heat from the Earth's interior towards the upper free surface, advection by tectonic movement and exhumation has been neglected. More complex semi-analytical expressions have been derived that take into account the advection of heat (Stüwe et al. 1994) and provide first-order estimates of the transient solution (Mancktelow and Grasemann 1997), i.e., following the onset or cessation of a tectonic event. An accurate solution of the finite amplitude topography problem, including the effects of a time-varying surface topography can only be obtained by solving numerically the heat transport equation (Braun 2002b).

The approximate solution given by Equation (4) can be used to illustrate the first-order effect of topography on the underlying temperature structure. The conduction of heat acts as a low-pass filter on the topographic signal (Fig. 1): perturbation caused by the short wavelength features of the topography (km scale) is strongly damped and, assuming a "normal" geothermal gradient of 20 °C/km, does not affect the shape of isotherms above a few tens of degree Celsius. At the other end of the spectrum, the long-wavelength features of the topography (10 km scale and larger) affect the temperature structure of the entire crust. Note that in regions of active tectonic exhumation (i.e., where tectonic uplift is compensated by erosion), the geothermal gradient may reach values of up to 100 °C/km or more[3]. In these situations, the perturbation of short wavelength topography can affect higher temperature isotherms.

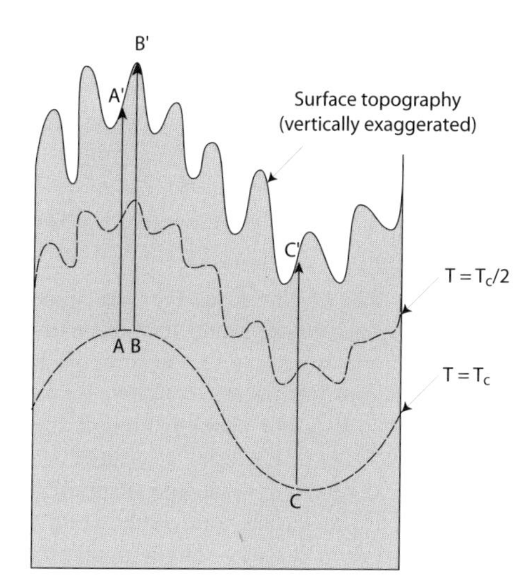

Figure 1. Effect of finite amplitude surface topography on the temperature structure of the underlying crust, using the closure temperature (T_c) of an isotopic dating system for reference. The temperature perturbation decreases exponentially with depth in proportion to the wavelength of the surface topography. Small-scale topographic features do not penetrate sufficiently deep to affect the shape of the closure temperature isotherm, whereas large-scale topographic features causes the closure temperature isotherm to follow the shape of the topography. Small scale relief is ideally suited to extract information about exhumation rate from age-elevation datasets (points A-A' and B-B'), whereas the same data collected across the long-wavelength relief provides constraints on the rate of landform evolution with time (points A-A' and C-C').

This simple, first-order response of a conductive system to perturbation caused by the topography of its upper, cold surface demonstrates that any information that may be contained in age datasets on the past shape of the Earth's surface is limited to a spectral range (a range of wavelengths) that is determined by the local geothermal gradient and the closure temperature of the system. The information is only available for the long wavelength component(s) of the topography, i.e., above a critical wavelength the value of which is determined (to first-order)

[3] The perturbation to the surface geothermal gradient can be shown to be directly proportional to the value of the Peclet number, $\mathbf{Pe} = uL/\kappa$ where u is the advection velocity of rocks towards the surface, L the thickness of the layer being exhumed and κ heat diffusivity, characterizing the balance between conductive and advective heat transport in a given tectonic province.

by the ratio of the closure temperature of the dating system and the local geothermal gradient (Braun 2002a):

$$\omega_c = \frac{T_c}{G_0} \tag{5}$$

AGE-ELEVATION DATASETS

Consider an age dataset that has been collected at a range of elevations but in a region characterized by very high topographic relief, or, more exactly, where the wavelength of the topography is short in comparison with the critical wavelength, ω_c. One can then safely postulate that the isotherm corresponding to the closure temperature of the system is not perturbed by the topography (at least locally) and is therefore horizontal. In this situation, depicted by points A-A and B-B in Figure 1, observed age should be directly proportional to elevation. In fact, one can easily show that the slope of the age-elevation relationship observed along very steep topography is a direct measure of the local mean exhumation rate. Note, that this is true regardless of whether the topography changes with time, as long as the closure temperature isotherm remains "flat" at the scale at which the age sample is collected. Note also that this measure of the local exhumation rate does not rely on an a-priori knowledge of the local geothermal gradient. It is for this reason that age-elevation relationships are commonly collected (e.g., Brown et al. 1994). Unfortunately, in many cases, it is not possible, or simply practical, to sample rocks along a steep topographic feature, either because it does not exist in the area of interest to the investigator, or because it is of very difficult access. One must therefore correct the slope of observed age-elevation relationships for the effect that the topography may have on the shape of the closure temperature isotherm, as suggested by Stüwe et al. (1994). This correction however requires an approximate knowledge of the local geothermal gradient, which is commonly difficult to measure and may lead to inaccurate estimates of the local exhumation rate.

Let's then consider an age dataset collected at a range of elevations but over a very large horizontal distance (points A-A and C-C in Fig. 1), i.e., much greater than the critical wavelength, ω_c[4]. Assuming that the sampling has been done at a range of wavelengths and that the resulting signal has been properly decomposed into its spectral components, one can easily show that, in the situation where topography does not change through time, observed age values should be independent of elevation. This is a simple consequence of the fact that, at wavelengths much greater than the critical wavelength ω_c, the shape of the closure temperature isotherm follows exactly the shape of the surface topography (Fig. 1). One can then state that any finite variation in age with elevation at a wavelength greater than the critical wavelength ω_c must be related to changes in the shape of the topography at that wavelength. In fact, Braun (2002a) has shown that the ratio of the slope of the age-elevation relationship at long wavelength, normalized by the value of the local exhumation rate (which, in turn, can be obtained from the slope of the age-elevation relationship at short wavelength), is a direct measure of the amount of topographic relief change at the long wavelength over a period of time of the order of the mean value of the age datasets. More importantly, one can easily show that this is the ONLY information contained in an age-elevation dataset on the evolution of the surface topography with time (i.e., with respect to its present-day form).

[4] Care must be taken in collecting such a dataset to avoid aliasing effects that arise when, due to poor sampling, some of the information contained in the relationship between age and elevation is transferred from the short wavelength components of the system to its longer wavelength components.

SPECTRAL ANALYSIS

This dependence of the thermal perturbation of surface topography on the wavelength of the topography and its effect on age-elevation distributions can be formalized by making use of traditional spectral analysis methods. Considering the age, ($a = a_i$, where $i = 1,\ldots,n$ and corresponding elevation, ($h = h_i$, where $i = 1,\ldots,n$) datasets as input and output signals of a dynamic, linear system (equivalent in our case to the process of conductive heat transport in the Earth's crust), one can compute the gain function, $G(\omega_i)$, of the system from the Fourier components of the input and output signals—$F(a)$ and $F(h)$, respectively. Note that the Fourier components and the gain function are complex functions, characterized by a real $R(F)$ and imaginary $I(F)$ parts. Because we know that, to first-order, the thermal perturbation caused by the surface topography is in phase with the surface topographic signal, one can easily show (Braun 2002a) that the imaginary part of the gain function should be nil (or at least much smaller than the real part and can therefore be neglected). The real part of the gain function is given by:

$$R(G) = \frac{C_{12}}{C_{11}}$$

$$C_{12} = R[F(a)]R[F(h)] + I[F(a)]I[F(h)]$$

$$C_{11} = R[F(h)]^2 + I[F(h)]^2$$

Note that the gain function has the dimensions of m.y./km (or s/m in SI units). This relates to our choice of what is the input and the output signal. It may seem more intuitive to use the opposite definition such that the gain function had the dimensions of an exhumation rate or a rate of change of topography with time (km/m.y.). Unfortunately, and as I will show below, in cases where the topography does not change with time, this would lead to infinite values for the gain function at long wavelength, a rather difficult measurement and/or inaccurate calculation to perform.

At short wavelengths, the value of the gain function tends towards the inverse of the mean exhumation rate; at long wavelengths, the value of the gain function tends towards zero if the topography has not changed in the interval of time corresponding to the mean age of the age dataset; it is positive if the topographic relief has increased and it is negative if the topographic relief has decreased. The exact value of the amount of relief change, β, i.e., the ratio of the relief then to the relief now, is given by the following relationship (Braun 2002a):

$$\beta = \frac{1}{1 - G_L / G_S} \tag{6}$$

where G_L and G_S are the asymptotic values of the real part of the gain function at long and short wavelengths, respectively.

These concepts and the spectral method they use can be illustrated if one apply them to a set of "synthetic" age-elevation datasets produced by solving the heat transport equation by conduction (and to a lesser degree by tectonic advection) in a region of the crust that is characterized by finite amplitude topography. The method used is briefly described in the following section and in greater details in Braun (2003). Particles are followed that are exhumed at the surface at a rate of 0.3 km/m.y. Their temperature history is extracted from the solution of the heat transport equation and used to compute the synthetic age-elevation datasets. Three cases are envisaged: one in which topography does not change through time, one in which the topographic relief increases by a factor two over the time span of the "experiment" and one in which the topographic relief decreases by a factor two.

The results, shown in Figure 2, demonstrate that in all three cases, the value of the gain function at short wavelengths always tends asymptotically towards the inverse of the exhumation rate (≈3 m.y./km) while the long wavelength value of the gain function tends towards zero in the case where there is no change in topography, is positive when the topographic relief increases and negative when the topographic relief decreases. Furthermore, estimates of the change in surface relief obtained by applying Equation (6) to each of the three cases (0, 2.5 and 0.55, respectively) are first-order accurate, in comparison to the imposed values (0, 2 and 0.5, respectively).

Note that the definition of what are "long" and "short" wavelengths components of the topographic signal depends on the value of the critical wavelength, ω_c, which is a function of the assumed geothermal gradient and the closure temperature of the dating system used. In the situation considered in the synthetic example (Fig. 2), the critical wavelength is approximately 5 km for apatite-He and 20 km for biotite-Ar. Thus, in an active orogen, characterized by a mean exhumation rate of a few hundreds of meters per million years, apatite-He ages contain information on the rate of relief evolution at wavelengths of 10 km or more; Biotite-Ar ages

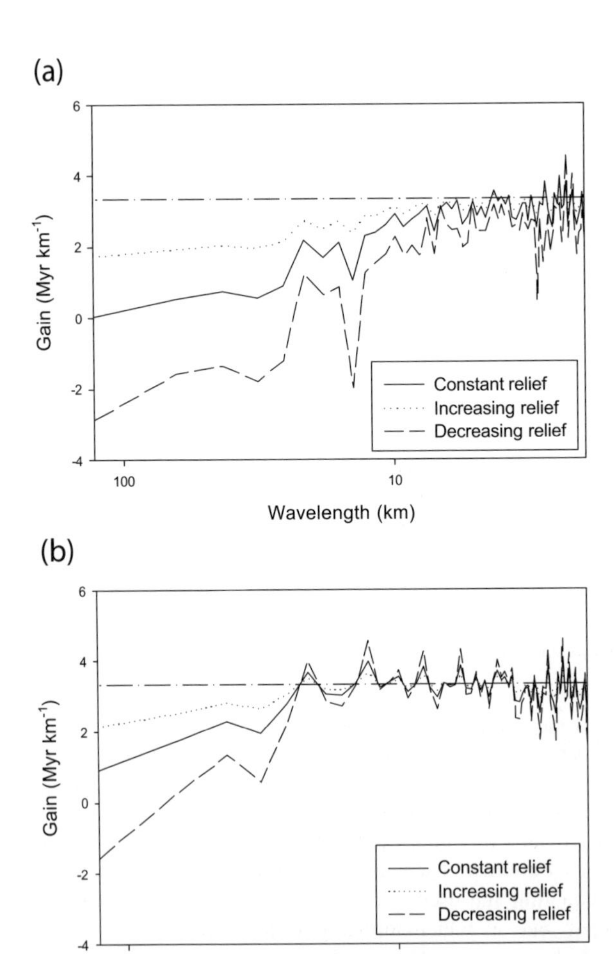

Figure 2. Real part of the gain function between elevation and age from synthetic (a) apatite-He ages and (b) biotite-K/Ar ages computed under the assumption that the surface relief remained constant through time (solid line), increased by a factor of 2 over the last 3 m.y. (dotted line) or decreased by a factor of two over the same time interval (dashed line). The topography has been extracted from GTOPO30 in an area of the Southern Alps, New Zealand, adjacent to Mount Cook. The mixed dashed line represents the gain value corresponding to the inverse of the imposed mean exhumation rate (0.3 km/m.y.). Modified after Braun (2002a).

can only be used to constrain the evolution of the relief at a scale of 20–50 km, which, in most cases, becomes comparable to the scale of the orogen itself. This implies that high-temperature systems (i.e., characterized by a closure temperature greater than 200 °C) can only be used to estimate the rate of change of the orogen-scale topography, in short, the height of the entire mountain belt.

In theory, the spectral method can be used to estimate the value of the gain function at all wavelengths. However, this estimate is accurate only for the spectral components (i.e., the wavelengths) present in the topographic signal. Typically, landforms are dissected by rivers that form valleys. The main trunks of the river system will form valleys that are relatively similar in width and spacing; the tributaries of these main trunks will develop secondary valleys that are more closely spaced and narrower. At the very long wavelength, topography is affected by faulting and differential block uplift which is likely to take place at a wavelength that is determined by the mechanical properties of rocks and the thickness of the layer involved in the deformation (past or present). This implies that accurate measures of the gain function can only be obtained at a small number of critical wavelengths, the value of which is determined by the spectral content of the topographic signal. It is therefore recommended to inspect the power spectrum of the topographic signal and determine which are the "dominant" wavelengths of the topography and focus the spectral analysis, and, optimally, the data collection strategy, at those wavelengths to ensure that the values of the gain function are accurate and meaningful. One can also compute error bars (or confidence intervals) on the gain function estimates by performing the spectral analysis on subsamples of the age-elevation dataset, each half the size of the complete dataset. A statistically significant value for the gain function can be obtained at each wavelength from the mean of the values computed for each of the subsamples and a confidence interval can be extracted from the variance around the mean value (Jenkins and Watts 1968).

Finally, one must note that for the spectral method to be accurate in estimating the mean exhumation rate and the rate of change of surface topographic relief, a specific sampling strategy must be followed, including:

- rocks must be collected along a linear transect at approximately uniform spacing (the exact value of the spacing will depend on the value of the critical wavelength, ω_c; see Braun (2002a) for further details on this);
- rocks must be collected in an area characterized by a uniform exhumation rate, i.e., the transect must be parallel to the strike of the main structures controlling the uplift and related exhumation; this will ensure that the age and elevation signals are stationary, a necessary condition for any Fourier-based spectral method to be accurate;
- horizontal heat advection by movement along low angle structures must be small; one way to avoid this problem is to sample relief produced by large rivers that runs perpendicular to the strike of the main thrust structures and thus parallel to the tectonic transport direction;
- relief must be sampled at all significant wavelengths and at all possible elevations; this may prove rather difficult in many tectonically active areas where local topographic relief may reach several thousands meters.

One can also easily argue that the sampling conditions that are ideal for the use of the spectral method are also those that will yield the most accurate estimate of relief change, i.e., one that does not rely on independent estimates of the local geothermal gradient and/or the mean exhumation rate. In cases where the above sampling guidelines cannot be followed, the spectral method must be abandoned in favor of a less direct, but more flexible approach which we describe in the following section.

Finally, it is worth noting that because it relies on the correlation between elevation and age along a linear transect, the spectral method does not take into account three dimensional effects that might arise from the finite length of valleys and ridges in a direction perpendicular to the transect. In most situations, the thermal perturbation caused by a finite length valley is less significant. This also highlights the need to use a more flexible method to interpret age-elevation datasets in terms of the information they contain on the rate of landform evolution, as proposed in the next section.

3D THERMAL MODELING: PECUBE

Interpreting geological observations to extract meaningful information about the Earth and the dynamic processes that are responsible for its evolution through time is now commonly achieved by using sophisticated numerical models to solve a set of basic differential equations describing the physical processes at play. To interpret thermochronological ages, one needs to predict the temperature history of rocks as they travel through the Earth and are exhumed at the surface. The problem to solve is that of heat transport in a conducting solid where advection by mass transport and production by the decay of radioactive elements may be locally important. The basic partial differential equation to be solved is based on Fourier's law for conductive heat transfer and can be written in the following way (Carslaw and Jaeger 1959):

$$\rho c(\frac{\partial T}{\partial t} + \dot{E}\frac{\partial T}{\partial z}) = \frac{\partial}{\partial x}k\frac{\partial T}{\partial x} + \frac{\partial}{\partial y}k\frac{\partial T}{\partial y} + \frac{\partial}{\partial z}k\frac{\partial T}{\partial z} + \rho H \tag{7}$$

where T is temperature, t is time, x, y and z are the spatial coordinates, ρ is density, c is heat capacity, is mean exhumation rate, k is heat conductivity and H is heat production per unit mass. Boundary and initial conditions must also be imposed for this equation to a have a unique, well-defined solution in both space and time. Here we will assume that the base of the crust is held at a constant temperature, T_1, that no heat is lost through the side boundaries and that the temperature at the surface is imposed to follow a typical atmospheric lapse rate γ (i.e., the rate at which the temperature decreases with elevation in the atmosphere):

$$\begin{aligned} T(z=L) &= T_1 \\ \frac{\partial T}{\partial n} &= 0 \quad \textit{on all four vertical sides} \\ T\left[x,y,z=S(x,y,t)\right] &= T_{msl} + \gamma z \end{aligned}$$

The initial temperature distribution is always assumed to correspond to the steady-state solution of the equation ($\partial T/\partial t = 0$).

I have recently developed a numerical method, based on the finite element discretization of space, to solve this equation, including the effect on the heat transfer balance of a time-varying, finite amplitude surface topography. All parameters are imposed, including the rate of exhumation and the geometry of the surface topography at a finite number of times in the past. In between these times, the topography is assumed to change linearly from one configuration to the next. The method is accurate, stable and efficient; it is fully described in Braun (2003) where estimates of its accuracy are also provided. The code is freely available from my web site (*http://rses.anu.edu.au/~jean/PROJECTS/PECUBE/*). The code, named **Pecube**, was used to estimate the synthetic age-elevation datasets used in the previous section to illustrate the accuracy of the spectral method. During the computations, the position and temperature of a set of rock particles that will end up at the surface of the model at the end of the computations are stored. These temperature and depth estimates are used to compute a *P-T-t* (pressure-temperature-time) path for each of the particles which are then used to compute an age, following the procedure described by Wolf et al. (1998) for (U-Th)/He dating in apatite and by van der Beek et al. (1995) for fission track dating in apatite.

EXAMPLE FROM THE SIERRA NEVADA

One of the most successful attempts to directly measure the "antiquity" of a landform by low-temperature thermochronology was undertaken by House et al. (1998) in the Sierra Nevada of northern California. Two opposing scenarios had been proposed for the evolution of the topographic relief in the area: the first assumes that the present-day relief in the Sierra Nevada was the result of long-term erosion of a landform that was created during the Cretaceous Laramide orogeny (Small and Anderson 1995); the competing scenario assumes that the present-day relief is much younger and resulted from Tertiary surface uplift and dissection, driven by renewed tectonic activity in the area, potentially related to subduction along the nearby plate margin, or to an abrupt climate change some 5 m.y. ago that would have resulted in an amplification of any paleo-relief and an isostatically-driven uplift of the area (England and Molnar 1990).

Interpreting the Sierra Nevada data using the spectral method

Rocks were collected along a linear transect crossing several km deep canyons at an approximately constant elevation of 2000 m. Rocks were dated by apatite-He. The data is shown in Figure 3. This dataset is still the only one to-date that comes close to being suited for the spectral analysis. Unfortunately, it did not sample the relationship between age and elevation at all possible elevations; furthermore it was not collected with uniform spacing between the sites. This will cause the short-wavelength estimates of the gain function to be highly inaccurate. Fortunately a few short transects can be constructed from this dataset and the one published earlier by House et al. (1997) in the same area from which a consistent, and therefore reliable, estimate of the mean exhumation rate can be derived, i.e., 0.04 km/m.y.

To perform the spectral analysis of the age-elevation dataset, both elevation and age values were linearly interpolated to form a 128 long series of equally spaced "data points."

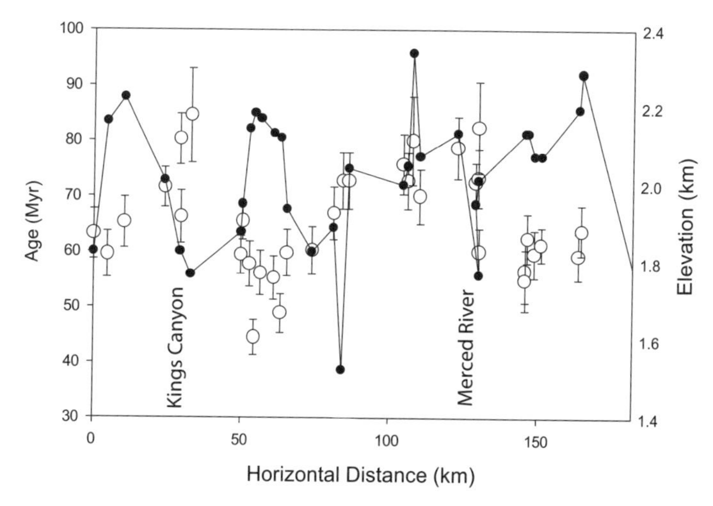

Figure 3. Elevation (open circles) and apatite-He ages (black circles) as a function of distance along a linear transect in the Sierra Nevada, California. See House et al. (1998) for exact location of transect and detailed description of data.

These series were then used to estimate the components of the gain function at wavelengths ranging from 2 to 128 km. The result is shown in Figure 4 and clearly demonstrates that, because the long wavelength estimates of the gain function are negative, the topographic relief at a scale greater than 10–20 km has been decreasing in the last 60–70 m.y. (the mean value of the age data set). Combining the long-wavelength gain estimates with the value of the mean exhumation rate derived from the short transects, leads to the conclusion that the relief amplitude has decreased by a factor of approximately 2. This supports the conclusion at which House et al. (1998) arrived that the large scale topographic features of the area, i.e., the large east-west trending canyons, are indeed the remnants of a much larger amplitude topographic relief that was formed during the Laramide Orogeny. Unlike House et al. (1998), the relief change estimate derived from the spectral analysis is not dependent on any assumption regarding the local past or present geothermal gradient. In fact, that relief has been decreasing for a very long period of time could be concluded from the simple and obvious anti-correlation that exists between age and elevation at the scale of the large valleys (Fig. 3) whereas the strong positive correlation between age and elevation widely documented along several short transects is indicative of the mean exhumation rate of the area driven by the post-orogenic, broad isostatic response to erosional unloading.

Interpreting the Sierra Nevada using Pecube

To illustrate the sensitivity of the relationship between age and elevation for low-temperature systems to landform evolution, we can use **Pecube** to predict the distribution of ages under the two different landform evolution scenarios. I first selected an area around Kings Canyon for which I extracted a 1 km DEM from GTOPO30. Both scenarios have the same initial conditions: it is postulated that the surface relief was twice as large as today's. Numerically I imposed that the surface topography has the same shape as todays' but I doubled its amplitude. In the first scenario, it is assumed that within 20 m.y. of the end of the orogenic event relief has been completely eroded away; in other words the area has been "peneplained." It remains at sea level between 50 and 5 m.y.; finally, over the last 5 m.y. of the experiment the present-day topography is formed by a modern, unspecified process. In the second, simpler

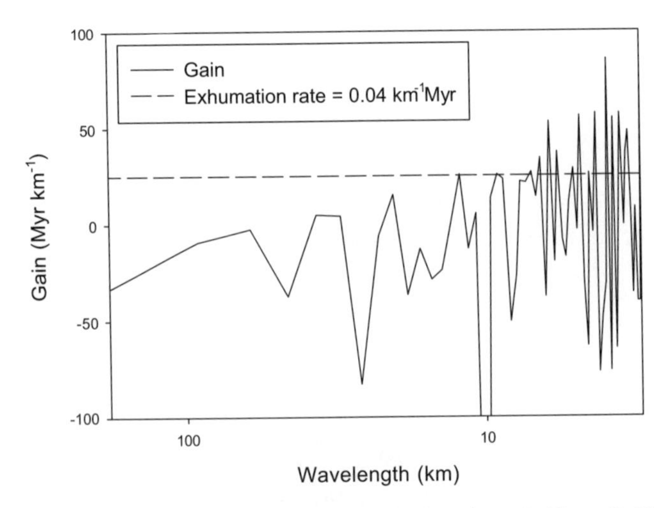

Figure 4. Gain function calculated from the Sierra Nevada data shown in Figure 3. The dashed line corresponds to the gain value equal to the inverse of the exhumation rate calculated from the slope of a well defined age-elevation relationship observed along a very short profile within the area (House et al. 1997). Modified after Braun (2002a).

scenario, relief amplitude decreases linearly from twice its present-day value to todays' value. In both scenarios, I have imposed a mean, slow and steady exhumation rate of 40 m/m.y.

The results of the computations are shown in Figure 5 in terms of predicted age-elevation datasets for each of the two scenarios. In the first scenario, there is a clear linear and positive relation between age and elevation. In the second scenario, there is no apparent correlation between age and elevation. In the first scenario, because there was no surface relief at the time the ages were set, the closure temperature isotherm was perfectly horizontal. That the geometry of the surface topography changed during the late stages of the experiment had no effect on the setting of the ages. Creating the topography simply allowed us to sample the age dataset at various heights. The slope of the age-elevation relationship is exactly equal to the value of the imposed exhumation rate (40 m/m.y.).

In the scenario where the relief decreases linearly over a long period of time, the ages appear to have no correlation to elevation. In fact, two opposite correlations exist at both end of the topographic spectrum: at short wavelengths, there is a positive age-elevation relationship that is proportional to the mean exhumation rate; at long wavelengths, the correlation is negative, reflecting the decrease in topographic relief with time since the rocks passed through their closure temperature. When the entire dataset is plotted in an age-elevation diagram, no clear trend can be extracted. Comparing the results of these two experiments to the distribution of apatite-He ages collected by House et al. (1998) in the Sierra Nevada area (Fig. 5), one is led to conclude that the slow decrease scenario is much more likely than the rapid rejuvenation one (as it predicts a well defined age-elevation relationship at short wavelength only). More importantly, these computations clearly demonstrate how sensitive age datasets are to relief evolution scenarios.

In this and previous sections, we have clearly demonstrated that there is great potential for low-temperature thermochronological datasets to contain important information about landform evolution. Great care must be taken, however, to extract this information from the data and convert it to a meaningful measure of the rate of landform evolution.

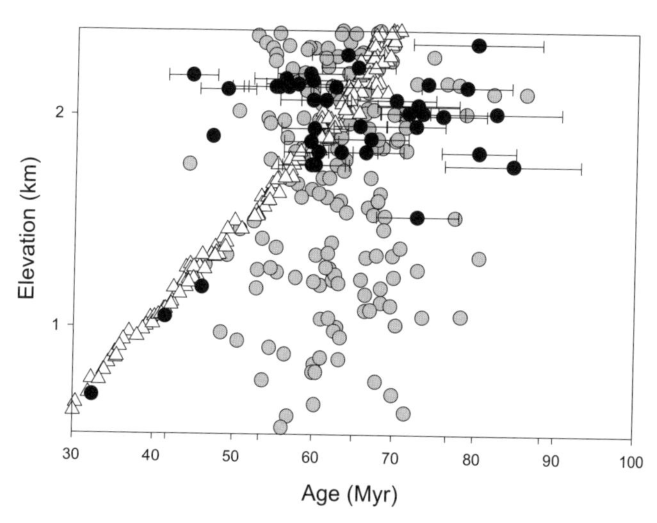

Figure 5. Observed ages (black circles), and predicted ages obtained from the temperature histories computed by **Pecube** assuming that the relief has been decreasing steadily for the last 70 m.y. (grey circles) or that the present-day relief is the product of rejuvenation of the landscape in the last 5 m.y. (white triangle). Modified after Braun (2003).

SLOW EROSIONAL SETTINGS

During the life cycle of an orogenic belt, the post-orogenic phase is commonly accompanied by a slow but steady decrease in surface topographic relief. The processes that act as to decrease the relief (erosion, deposition and transport along the Earth's surface) are still active while the processes that act as to increase the relief (tectonic uplift and/or differential movement on crustal faults) have been turned off. As shown by Kooi and Beaumont (1996), the response of surface processes to an abrupt change in tectonic uplift offers us with an opportunity to estimate the intrinsic rate at which surface processes operate. Most parameterizations of surface processes imply a quasi-linear dependence of erosion and transport rates on local slope (Whipple and Tucker 1999). This, in turn, causes the response of the system to an abrupt cessation in tectonic uplift to be exponential in its time evolution. The e-folding time of this response is regarded as a good measure of the intrinsic rate at which land-forming processes operate in given climatic conditions and for a given set of rock types or lithologies (Kooi and Beaumont 1996).

There are other environments where the geomorphic system is excited by a rapid, quasi-instantaneous event. For example, the rifting between two continents occurring in a region of anomalously high elevation is experienced by both continental fragments as a rapid drop in base level along one of its margins (Gilchrist and Summerfield 1990). The resulting escarpment separating the low-elevation coastal plain from the adjacent continental plateau area has the potential to evolve according to a range of potential scenarios (Gilchrist et al. 1994; Kooi and Beaumont 1994). Some include the retreat of the escarpment as a steep geomorphic feature, others include the slow down-wearing of the escarpment into a gently sloping coastal plain. Because of the absence of tectonic forcing, such environments are thus very useful to isolate the intrinsic response of land-forming processes and have been the target of many attempts to estimate the rate of down-wearing/propagation, both functions of the local rate of stream incision and transport (Gilchrist and Summerfield 1990; Gilchrist et al. 1994; Seidl et al. 1996; Weissel and Seidl 1998; van der Beek et al. 2001, 2002; Brown et al. 2002).

Several studies (Gallagher et al. 1995; Brown et al. 2002; van der Beek et al. 2002; Persano et al. 2002; Braun and van der Beek 2004) have been conducted in recent years to document the rate and nature of escarpment evolution at passive plate margins using thermochronological data collected within the coastal plain at the base of the escarpment. Similarly, several studies have recently focused on constraining the rate of erosional decay of ancient orogenic belts (House et al. 1997, 1998, 2001; Reiners et al. 2003a,b; Braun and Robert 2005).

Isostasy

To state that, in the absence of tectonic forcing, erosional processes can be studied in isolation is, however, an overstatement. Mass transport by erosion and deposition results in unloading/loading of the lithosphere which, in turn, responds by isostatically-driven uplift/subsidence (Turcotte and Schubert 1982). The principle of isostasy states that there is a region within the Earth's interior where, on geological time scales, the strength of rocks is so reduced that no horizontal pressure gradient can be sustained. This region is thought to be located just beneath the lithosphere and is often referred to as the "isostatic compensation depth." The weight of any lithospheric column measured from the compensation depth to the surface must be constant. This is often called the principle of local isostasy and is commonly used to put large-scale constraints on the thickness of the crust and/or the thermal structure of the lithosphere. Furthermore, the lithosphere is characterized by a finite lateral, or flexural, strength (Turcotte 1979) resulting in a damping of the isostatic response to surface loading/unloading for wavelengths that are shorter than the flexural wavelength of the lithosphere. This flexural wavelength can be regarded as the wavelength of deformation of a thin elastic plate subjected to a point load. The effective elastic thickness of continents (T_e) is a function of their composition (especially crustal thickness) and thermal state and varies typically between 10 and 100 km.

The post-orogenic erosional decay of an orogen and/or the evolution of passive margin escarpment will thus be accompanied by an isostatically-driven rock uplift which may cause an amplification of the erosional processes. For example, assuming local isostasy, the lowering of surface topography by 1 km causes approximately 5 km of isostatic uplift and therefore requires erosion of 6 km of rocks.

Because isostasy controls uplift and, potentially, erosion and exhumation, one must therefore consider the effect that it has on the thermal structure of the lithosphere (this is likely to be small as isostatically driven uplift and exhumation rates are slow) and on the paths of rock particles through that thermal structure. For infinitely small values of the effective elastic plate thickness, the isostatic rebound takes place exactly at the same wavelength as the erosional unloading. In this unlikely situation (the Earth's continental lithosphere always has some finite strength), erosional decay is everywhere amplified by isostatically-driven exhumation. As relief decreases, exhumation is greatest along mountain tops than in valley bottoms and ages are predicted to be anti-correlated to elevation at all wavelengths (T_e = 0 km case in Fig. 6).

For large values of the elastic plate thickness, isostatic rebound is small and uniform across the entire mountain belt. Age is negatively correlated to elevation at long wavelengths, reflecting the decrease in surface topographic relief accompanying the erosional decay of the orogen (Fig. 6); at short wavelengths, a positive relationship exists between age and elevation, the slope of which is proportional to the rate of isostatic rebound experienced by the entire system (T_e = 40 km case in Fig. 6). In this situation, the spectral method should be used as it will provide accurate estimates of both the mean, isostatically-driven erosion rate and the amount of relief change.

At intermediate values of the elastic plate thickness, i.e., similar to the wavelength of the eroding topography, the situation is rather complex and the perturbation from the isostatic rebound on the relationship between age and elevation can be substantial, especially at the

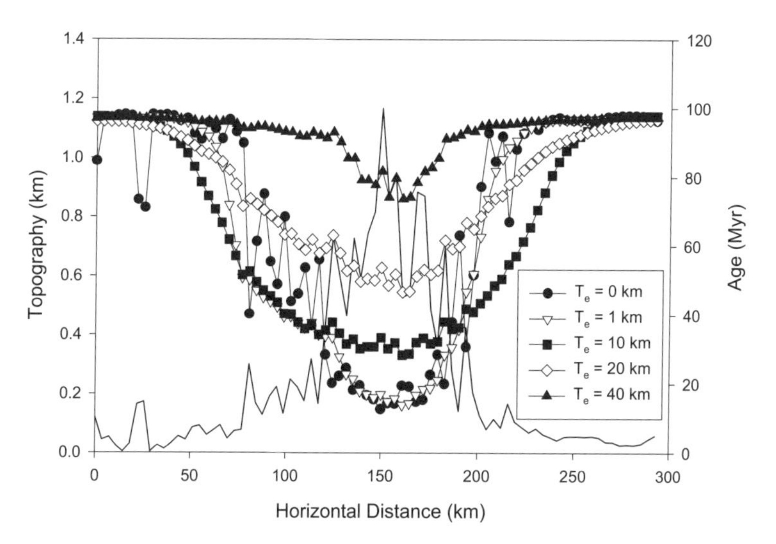

Figure 6. Topography (thin black line) used in **Pecube** to calculate age distributions assuming various values for the effective elastic thickness, Te, for the underlying lithosphere. In all cases it is assumed that relief has decreased steadily by a factor of 4 over the last 100 m.y. Modified after Braun and Robert (2005).

wavelengths that are sensitive to the rate of change of surface relief (T_e = 10 km case in Fig. 6). In this situation, the use of the spectral method is not recommended and a more rigorous, yet computationally costly inversion of the thermochronological data is needed. This is what I propose to describe in the next section and illustrate through a couple of examples.

INVERSION OF AGE-ELEVATION DATASETS

Many existing thermochronological datasets have not been collected with the purpose of using the spectral method to extract from them independent information about the rate of landform evolution. For this reason, they do not fulfill the basic requirements for the method to be accurate: they are not collected along a linear profile perpendicular to the direction of tectonic convergence, they do not sample the relationship between age and elevation at all wavelengths (or at least at the small and long wavelengths contained in the landform) and, most frequently, they do not sample the variation of age across the main geomorphic features of the landscape (the valleys). In fact, for practical reasons, most age datasets are collected along the floor of the main valleys which usually trend in a direction parallel to the tectonic convergence direction and do not sample the main features of the relief, i.e., the relief across the valleys themselves.

For this reason and because the system under investigation may be rendered more complex by the effect of isostasy or horizontal tectonic transport, it is often necessary to perform a more rigorous inversion of the dataset, using for example **Pecube** as a predictor tool. **Pecube** may be regarded as a physically sound method to predict the distribution of ages measured at the Earth's surface for a given erosional scenario and isostatic scenario. The predicted distribution of ages will depend on the imposed erosional scenario, on the assumed degree of flexural compensation of the isostatic response to erosion (through the assumed effective elastic plate thickness), but also on a range of model parameters, including the geothermal gradient, the amount of heat generated in the crust by radioactive decay of K, U and Th, the thermal diffusivity, etc.

One can envisage to run a large number of predictor model runs each corresponding to a different set of model parameters. By comparing the difference between the predicted ages and the observations, one can easily define a degree of "goodness of fit" for each model parameter set. Automated methods have been developed, including the Neighborhood Algorithm (NA) of Sambridge (1999a,b), that efficiently search through parameter space to either map the shape of the misfit function or find the global minimum in misfit which, in turn, should provide the most likely set of model parameters.

However, three situations can arise:

- the misfit cannot be satisfactorily minimized by any combination of parameters; in this case, one or several physical processes that are at play in the natural system have not been incorporated in the predictor model;
- many combinations of the model parameters lead to an acceptable misfit between observations and predictions; no well-defined global minimum can be found; in this case, one can state that the observations do not contain sufficient information to constrain the value of all model parameters; in some cases, only combinations of the model parameters can be constrained;
- the misfit between observations and predictions is clearly minimized for a unique set of model parameters; in this situation, the inversion has led to a clear definition of the "best" or "more likely" model parameters which, in turn, can be translated to determining the contribution/importance of each physical process built into the model.

To illustrate this point and, more generally, the use of inversion methods in interpreting thermochronological datasets to constrain the rate of landform evolution, I will use two examples: one from the Dabie Shan area in eastern China and one from the great coastal escarpment of southeastern Australia. Both examples come from tectonically quiet areas where exhumation (and the subsequent resetting of thermochronological ages) is controlled by the erosional decay of an ancient topography and the associated isostatic uplift.

Post-orogenic erosional decay, example from the Dabie Shan

The Dabie Shan is the result of the collision between the northern edge of the Yangtze craton and the southeastern corner of the Sino-Korean craton that took place during a series of subduction-related episodes of crustal shortening, from the late Paleozoic to the mid-Cretaceous (Schmid et al. 2001). There is debate on whether the orogen was reactivated during the on-going Indo-Asian collision and whether some of the present-day topographic relief is the result of this Cenozoic reactivation (Reiners et al. 2003b). Alternatively, the topography is the erosional remnant of a much larger amplitude relief that was entirely formed in the Cretaceous.

This second hypothesis appears to be supported by apatite (U-Th)/He and FT data recently collected by Reiners et al. (2003b) across the ranges and along a narrow profile near the centre of the orogen (Fig. 7). Today's elevation peaks at approximately 2000 m but most mountain tops do not reach 1500 m (Fig. 7). The thermochronological dataset has been used to constrain the rate evolution of surface relief since the end of the Cretaceous compressional event as well as the effective elastic thickness of the lithosphere, the geothermal gradient and whether the whole area experienced broad-scale (uniform) exhumation during its post-orogenic erosional decay (Braun and Robert 2005). It is the results of this later study that are reported in this section.

Unfortunately, the distribution of the data is such that the spectral method cannot be used to determine the change in surface relief with accuracy. Consequently, we used **Pecube** to solve the 3D heat transport equation and predict the temperature history of rock particles that

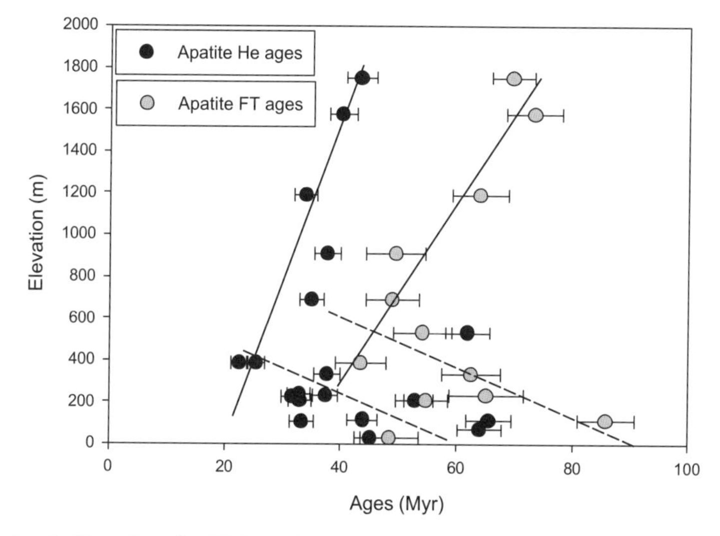

Figure 7. Apatite-He and apatite-FT data collected in the Dabie Shan of eastern China by Reiners et al. (2003). The overall trend is for ages to decrease towards the centre of the orogen (dashed lines); the clear linear trend in age with elevation has been measured along a narrow profile near the centre of the orogen (solid lines). Modified after Braun and Robert (2005).

are exhumed at the surface at the locations where the age data was collected (Braun and Robert 2005). To include the effects of flexural isostasy, we computed the surface load corresponding to each imposed increment in surface topography and computed the incremental thin elastic plate deflection resulting from the application of the load using the two-dimensional spectral method described in Nunn and Aires (1988). We also used a sophisticated inverse method, the Neighborhood Algorithm (NA) (Sambridge 1999a,b) to find the best set of parameters to use in **Pecube** that would predict an age distribution compatible with the observed distribution. NA is used to search through parameter space to map the behavior of a so-called misfit function, that measures the "difference" between the predictions and observations (usually, the L_2 norm of the difference between the observed and predicted age arrays). NA is guided in its mapping by a tessellation of parameter space into Voronoi cells (or neighborhoods) around each point in parameter space representing a forward, **Pecube** model run. NA is designed to find minima in the misfit function and is therefore efficient even for problems involving a large number of parameters.

Result of an NA search are shown in Figure 8. In this case, the Dabie Shan dataset is used as constraints to search in a 6D parameter space, composed of (1) the geothermal gradient (or the temperature at the base of the crust), (2) the timing of the end of the orogenic phase (time at which the erosional decay phase started), (3) the effective elastic thickness of the lithosphere, (4) the amount of relief loss since the end of the orogenic phase, (5) a mean exhumation rate (spatially uniform and constant through time) and (6) the erosional e-folding time scale (the time scale over which half of the original surface relief was eroded away assuming that relief has decreased exponentially with time during the erosional decay phase of the orogen). The results demonstrate that the data does not contain sufficient information to constrain all model parameters, in particular, the basal temperature and erosional e-folding time scale are not constrained at all. The duration of the post-orogenic erosional phase is well constrained at 70 m.y. ago; there appears to be a positive linear relationship between acceptable values for the elastic plate thickness and the amount of relief loss, and a negative linear relationship between acceptable values for the mean exhumation rate and both the effective elastic plate thickness and the amount of relief loss. The predictions of the best fitting model are shown in Figure 9 and corresponds to an elastic plate thickness of ~9 km, a duration of ~70 m.y. for the erosional decay phase, an e-folding time scale for erosion of ~250 m.y., a basal temperature of ~650 °C, a relief loss of ~2.5 and a mean exhumation rate of ~0.02 km/m.y. The fit is very good as the model predictions reproduce not only the orogen-scale trend of ages increasing with decreasing elevation towards the margins of the orogen but also the smaller-scale trend of a strong positive age-elevation relationship along a steep and narrow vertical profile near the centre of the orogen. The results are also consistent with the conclusions reached by Reiners et al. (2003b) using a less sophisticated approach to data inversion. The relatively good fit between observations and model predictions (Fig. 9) demonstrates that the pattern of cooling ages is compatible with a quasi-radially symmetric exhumation caused by erosion-driven isostatic rebound superimposed on a uniform, possibly tectonically-driven, exhumation pattern. This conclusion further illustrate the difficulty to differentiate between the two types of exhumation patterns based on thermochronological data only as already shown by Brandon and Vance (1992) and Brandon et al. (1998).

The information concerning the rate of landform evolution that can be extracted from this dataset is interesting: there has been a decrease in surface relief by a factor of 2 or 3 since the end of the orogenic event some 70 m.y. ago; but whether this relief loss took place very rapidly or has been evolving continuously over the past 70 m.y. cannot be said. The best fitting model suggests that the system has evolved rather slowly, i.e., at a uniform rate. It is also interesting to note that the thermochronological dataset can not only be used to derive constraints on the rate of rock exhumation and amount of relief loss, but also on the effective elastic thickness of

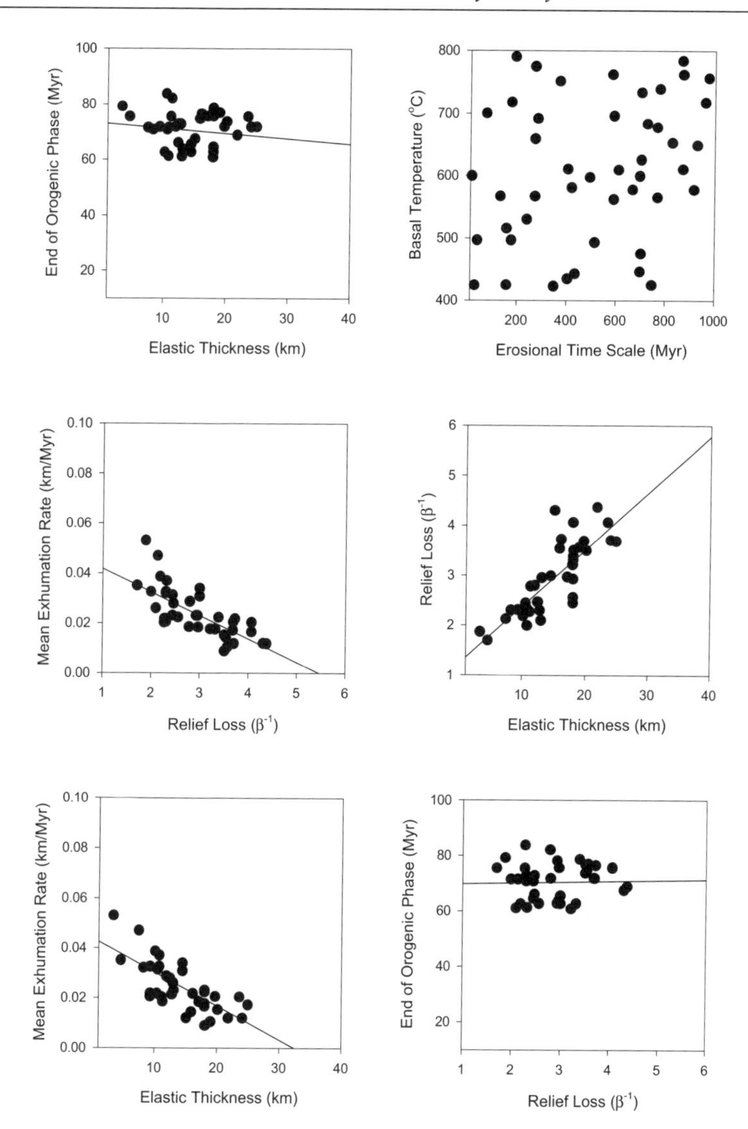

Figure 8. Results of the NA inversion of the Dabie Shan data. Space parameter is of dimension six but results can only be shown by projection onto selected two dimensional sections. Each black dot represent a **Pecube** model run. 17,744 model runs were performed during the search; the 50 best models (i.e., characterized by the lowest misfits) only are shown. Some parameters are well constrained by the data (such as the timing for the end of the orogenic phase); others (such as the basal temperature or the erosional time scale) are not constrained by the data; also shown (black lines) are linear constraints determined by linear regressions between pairs of model parameters (such as elastic thickness and relief loss, or mean exhumation and relief loss). Modified after Braun and Robert (2005).

the lithosphere beneath the area. It is one of the strengths of modern-day inverse methods that they permit to extract from the available data a broader range of constraints on the behavior of the Earth system. They are also very efficient at demonstrating that, sometimes contrary to our intuition, the data does not contain information/constraints on other aspects of the system, in this case, the background geothermal gradient or the erosional e-folding time scale.

(a)

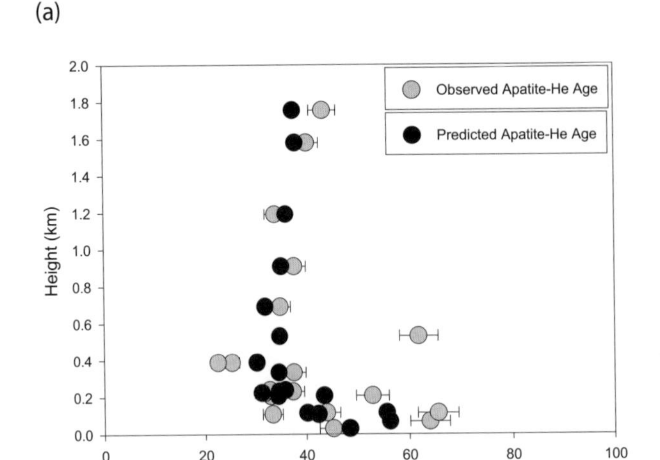

(b)

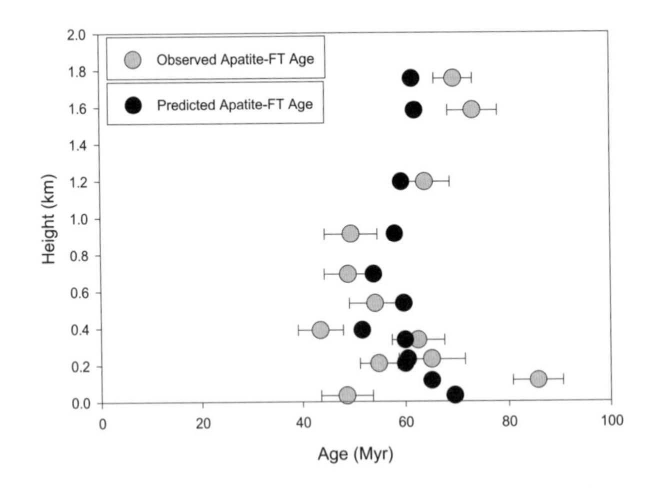

Figure 9. Comparison between the observed ages from the Dabie Shan area Reiner et al. (2003) and the predictions from the **Pecube** run characterized by the smallest misfit. The model predictions reproduce the broad younging of the ages towards the centre of the orogen as well as the well-defined age-elevation relationship along the narrow profile near the centre of the orogen. Modified after Braun and Robert (2005).

The main limitation of the inversion approach/method is its computational requirements. Each of the forward model run performed using **Pecube** took approximately 20 minutes of CPU time on a 2.4 Ghz Pentium IV with very fast access memory. The computations shown in Figure 8 would have taken approximately 246 days to perform if we did not have access to a 128 node cluster of PC's on which the computations took just under 2 days. Clearly, the applicability of this method to invert thermochronological datasets by solving the full 3D heat transfer equation and taking into account the effect of a finite amplitude, time evolving upper surface is limited by the availability of high performance computing facilities. Fortunately, clusters of PC's are now becoming more widely available.

Rate and nature of passive margin escarpment evolution, example from SE Australia

Passive margin escarpments form as a result of continental rifting in an area of anomalously high elevation, such as observed in several continental interiors (Gilchrist and Summerfield 1990). Rifting causes an abrupt drop in base level to which surface processes react. Several hypotheses exist on how escarpments evolve from their original position at the plate margin to their present-day location some tens to hundreds of kilometers inland. Examples include the escarpment surrounding the southern parts of the African continent (King 1951) and the escarpment flanking the eastern margin of the Australian continent (Ollier 1982).

Two conflicting scenarios for escarpment evolution have received much attention in recent years (Gilchrist et al. 1994; Kooi and Beaumont 1994; Seidl et al. 1996; Weissel and Seidl 1998; van der Beek et al. 2001, 2002; Persano et al. 2002; Braun and van der Beek 2004). One postulates that escarpments retreat landward while keeping their steep morphology, the other postulates that escarpments are rapidly worn down and re-establish themselves at the location of a pre-existing inland drainage divide (Fig. 10). The first "model" for escarpment evolution is called the Escarpment Retreat (ER) model, the alternative model is called the Plateau Down-wearing (PD) model (van der Beek et al. 2001).

Several low-temperature datasets have been collected recently along the coastal plains adjacent to well-defined passive margin escarpments (Morley et al. 1980; Gallagher et al. 1995; Brown et al. 2002; Persano et al. 2002), in an attempt to constrain the nature of the escarpment evolution process and the rate at which it took place. Recently, apatite-He ages have been measured from samples collected along a transect running perpendicular to the escarpment of southeastern Australia, in the coastal plains of the Bega Valley (Persano et al. 2002). The data suggests a rapid evolution of the escarpment close to its present-day position and morphology, soon after rifting and opening of the Tasman Sea (Persano et al. 2002). By inverse modeling, Braun and van der Beek (2004) demonstrated that the data is indeed indicative of a rapid evolution in the mid-Cretaceous but that it cannot be used to differentiate between the two competing scenarios.

In this inversion, NA was used to search through parameter space and **Pecube** was used to relate geomorphic scenarios to predicted thermal histories and age distributions. The data was used to constrain 4 parameters (1) the nature of the process by which the escarpment evolved to its present-day position (i.e., following the ER or PD scenario), (2) the rate of evolution of

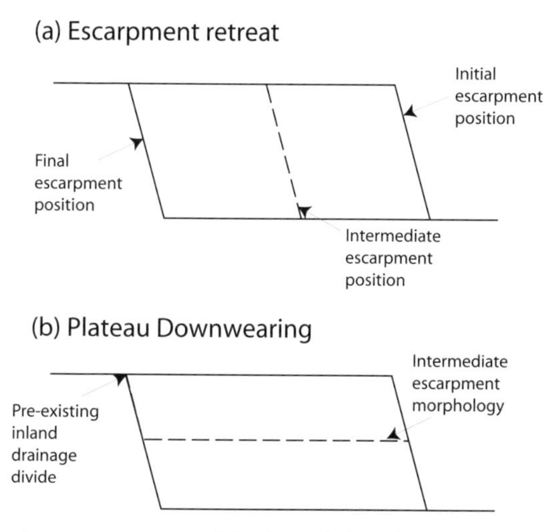

Figure 10. Two scenarios currently proposed for the evolution of a passive margin escarpment. In the escarpment retreat scenario (panel a), the escarpment migrates as a clear topographic step from its initial position near the coast to its present-day position inland; in the plateau down-wearing scenario (panel b), the region comprised between the original escarpment and an pre-existing inland drainage divide is progressively and uniformly eroded away.

the escarpment, (3) the local geothermal gradient, and (4) the effective elastic thickness of the underlying lithosphere. The results show that the best fit to the data was indeed obtained for models characterized by a rapid evolution of the escarpment (i.e., within 15 m.y. of the end of rifting). More interestingly, the inversion also demonstrated that equally acceptable predictions could be made following either of the two scenarios (Fig. 11). The inversion also suggested that the area must be characterized by a very weak lithosphere and/or high geothermal gradient.

Taking the two best fitting model experiments, one for each of the two scenarios and comparing their predictions in terms of the distribution of predicted ages on the natural topography (Fig. 12), one can deduce from the modeling what are the areas where rocks should be collected and dated to provide the best constraints to discriminate between the two scenarios. Consistently, models based on the escarpment retreat scenario predict younger, reset ages near the base of the escarpment. Braun and van der Beek (2004) argue that vertical profiles taken along the face of the escarpment should provide the best constraints on the mode of evolution of the escarpment. As demonstrated earlier, the slope of age-elevation datasets along vertical or near-vertical profiles provides an accurate estimate of the local rate of exhumation. The PD scenario predicts significantly slower local erosion rate than the ER scenario; this difference should be measurable, especially along the face of the escarpment where significant relief exists. Alternatively, a linear transect running parallel to the escarpment and sampling the long-wavelength relief along the face of the escarpment would provide direct estimates of the rate of relief evolution which is markedly different under the two proposed scenarios (Braun and van der Beek 2004).

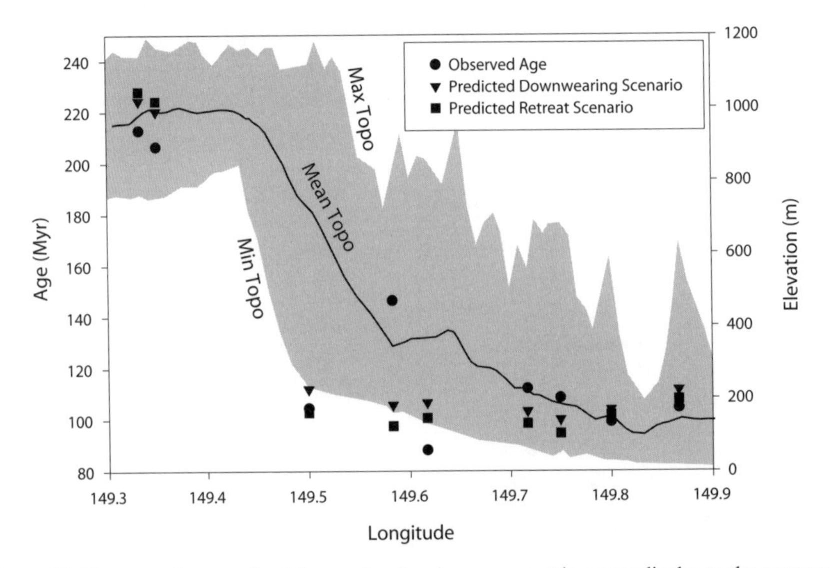

Figure 11. Mean, maximum and minimum elevation in a narrow strip perpendicular to the escarpment in the Bega Valley, where (Persano et al. 2002) collected rocks for apatite-He dating. The observed ages (black circles) are compared to the predictions of two **Pecube** model runs; one model run is based on the escarpment retreat scenario (black squares), the other on the plateau down-wearing scenario (black triangle). That the predictions of the two model runs are equally compatible with the observed ages demonstrates that this thermochronological dataset does not contain information on the mode of evolution of the escarpment in southeastern Australia. The parameters of these two model runs were found by NA search; the search demonstrated that the evolution of the escarpment towards its present-day position took place soon after the opening of the Tasman Sea and that the area must be characterized by a low effective elastic thickness and/or geothermal gradient. Modified after Braun and van der Beek (2004).

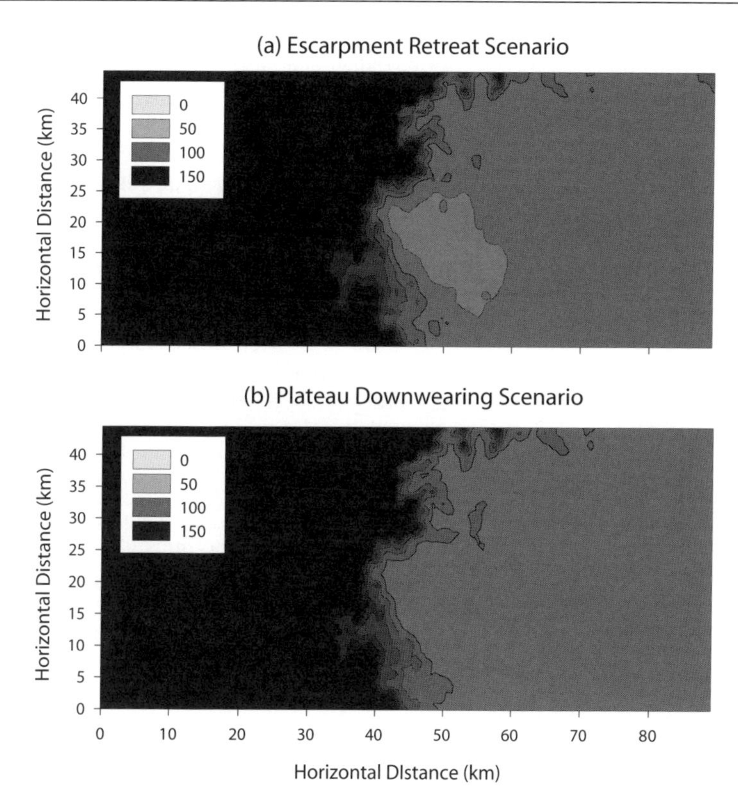

Figure 12. Contour plots of age distributions (in m.y.) predicted from the two "best-fitting" **Pecube** model runs shown in Figure 11. The main difference lies near the bottom of the escarpment. This suggests that this area should be the target of future data collection that could provide the necessary constraints to discriminate between the two hypothesized scenarios. Modified after Braun and van der Beek (2004).

This last point demonstrates how mathematical modeling and the existence of pre-existing data can, and potentially should, be used to define targets for (future) data collection. Mathematical modeling is often regarded as a way to interpret data where, in fact, it should be involved at all stages of investigation, including prior to data collection. In this way, one can make sure that the data collection strategy has been designed to provide optimal constraints on the system being studied. This statement simply reinforces the one made earlier about how to design a sampling strategy that will provide the most useful and independent constraints on the rate of landform evolution through the application of the spectral/gain method. Of course, one cannot avoid the limitations brought to this theoretical consideration by the nature of the geological record: highly spatially variable and incomplete.

CONCLUSIONS AND FUTURE WORK

Recently, much has been learnt about the use and limitations of thermochronological methods in providing constraints on the rate of landform evolution. Sophisticated numerical models have been developed, such as **Pecube**, that have made possible the accurate prediction of the effect of landscape change on the thermal structure of the crust and the distribution of low-temperature age measurements from surface rocks (Braun 2003; Ehlers et al. 2003). It has been clearly demonstrated that the effect is of finite amplitude (i.e., it is much larger than the uncertainty on age determinations) and strongly variable—it depends on the rate at which landform evolved (Stüwe et al. 1994; Mancktelow and Grasemann 1997; Braun 2002b). In particular, thermochronology has been used to demonstrate the antiquity of some landforms—such as the large-scale relief in the Sierra Nevada (House et al. 1998; Braun 2002a) and/or the Dabie Shan (Reiners et al. 2003b; Braun and Robert 2005). It has also been used to show that some geomorphic features, such as passive margin escarpments, are very dynamic (Persano et al. 2002; Braun and van der Beek 2004): their past history is punctuated by short periods of rapid evolution followed by much longer periods of quiescence.

Much work remains to be done, however, to constrain the potential complex dynamical feedbacks between erosion and tectonics. Much has been said about the potential for erosional processes to focus crustal deformation in regions or high rainfall, thereby linking the evolution of tectonic processes to variations in climate on a local to regional scale (Willett et al. 1993; Beaumont et al. 1999). These ideas remain hypotheses that have been suggested by mathematical models and/or first-order field observations such as the relationship between rainfall and rock exhumation along the western slopes of the Southern Alps of New Zealand (Batt and Braun 1999; Koons et al. 2003), Taiwan (Dadson et al. 2003) or the Himalayan front (Beaumont et al. 2001; Kirby et al. 2003). Much work is still required to understand the complexity of this interaction, including the rate at which landforms can adapt to changing tectonic/uplift conditions (Kooi and Beaumont 1996), how or whether landforms are advected by horizontal tectonic transport (Willett et al. 2001), how the rate of land-forming processes has been affected by the advent of widespread Quaternary glaciations (England and Molnar 1990) or how do these coupled systems behave when subjected to cyclic climatic and therefore erosional conditions as glaciers wax and wane over most of the planet active mountain belts (Braun et al. 1999).

ACKNOWLEDGMENTS

Some of the computations shown in this chapter were performed on the TerraWulf cluster at the Centre for Advanced Data Inference of the Australian National University. The author wishes to thank P. Upton and M. Brandon for useful comments they made on this manuscript.

REFERENCES

Batt GE, Braun J (1999) The tectonic evolution of the Southern Alps, New Zealand: insights from fully thermally coupled dynamical modelling. Geophys J Int 136:403–420

Beaumont C, Jamieson R, Nguyen M, Lee B (2001) Himalayan tectonics explained by extrusion of a low-viscosity crustal channel coupled to focused surface denudation. Nature 414:738–742

Beaumont C, Kooi H, Willett S (1999) Coupled tectonic-surface process models with applications to rifted margins and collisional orogens. In: Geomorphology and Global Tectonics. Summerfield M (ed) John Wiley and Sons Ltd, New York, p 29–55

Brandon M, Roden-Tice M, Garver J (1998) Late Cenozoic exhumation of the Cascadia accretionary wedge in the Olympic Mountains, northwest Washington State. Geol Soc Am Bull 110:985–1009

Brandon M, Vance J (1992) Tectonic evolution of the Cenozoic Olympic subduction complex, Washington State, as deduced from fission track ages for detrital zircon. Am J Sci 292:565—636

Braun J (2002a) Estimating exhumation rate and relief evolution by spectral analysis of age-elevation datasets. Terra Nova 14:210–214

Braun J (2002b) Quantifying the effect of recent relief changes on age-elevation relationships. Earth Planet Sci Lett 200:331–343

Braun J (2003) Pecube: A new finite element code to solve the heat transport equation in three dimensions in the Earth's crust including the effects of a time-varying, finite amplitude surface topography. Comput Geosci 29:787–794

Braun J, Pauselli C (2004) Tectonic evolution of the Lachlan Fold Belt (Southeastern Australia): constraints from coupled numerical models of crustal deformation and surface erosion driven by subduction of the underlying mantle. Phys Earth Planet Int 141:281–301

Braun J, Robert X (2005) Constraints on the rate of post-orogenic erosional decay from thermochronological data: example from the Dabie Shan, China. Earth Surf Proc Land, in press

Braun J, van der Beek P (2004) Evolution of passive margin escarpments: what can we learn from low-temperature thermochronology? J Geophys Res109:F04009, doi:10.1029/2004JF000147

Braun J, Zwartz D, Tomkin J (1999) A new surface processes model combining glacial and fluvial erosion. An Glaciol 28:282–290

Brown R, Summerfield M, Gleadow A (1994) Apatite fission track analysis: its potential for the estimation of denudation rates and implications for models of long-term landscape development. *In:* Process Models and Theoretical Geomorphology. Kirby M (ed) John Wiley and Sons Ltd, New York, p 23–53

Brown R, Summerfield M, Gleadow A (2002) Denudational history along a transect across the Drakensberg Escarpment of southern Africa derived from apatite fission track thermochronology. J Geophys Res 107: 2350, doi:10.1029/2001JB000745

Carslaw H, Jaeger C (1959) Conduction of Heat in Solids. Clarendon, Oxford, third edition

Dadson S, Hovius N, Chen H, Dade W, Hsieh M-L, Willett S, Hu J-C, Horng M-J, Chen M-C, Stark C, Lague D, Lin J-C (2003) Links between erosion, runoff variability and seismicity in the Taiwan orogen. Nature 426:648–651

Dodson MH (1973) Closure temperature in cooling geochronological and petrological systems. Contrib Mineral Petrol 40:259–274

Duddy IR, Green PF, Laslett GM (1988) Thermal annealing of fission tracks in apatite 3. Variable temperature behaviour. Chem Geol 75:25–38

Ehlers T, Willett S, Armstrong P, Chapman D (2003) Exhumation of the central Wasatch Mountains: 2 thermo-kinematic models of exhumation, erosion and low-temperature thermochronometer interpretation. J Geophys Res 108:2173, doi:10.1029/2001JB001723

England P, Molnar P (1990) Surface uplift, uplift of rocks and exhumation of rocks. Geology 18:1173–1177

Farley KA (2000) Helium diffusion from apatite: general behavior as illustrated by Durango fluorapatite. J Geophys Res 105:2903–2914

Gallagher K, Hawkesworth C, Mantovani M (1995) The denudation history of the onshore continental margin of southeastern brazil inferred from fission track data. J Geophys Res 99:18,117–18,145

Gilchrist AR, Kooi H, Beaumont C (1994) Post-Gondwana geomorphic evolution of southeastern Africa: implications for the controls on landscape development from observations and numerical experiments. J Geophys Res 99:12,221–12,228

Gilchrist AR, Summerfield MA (1990) Differential denudation and flexural isostasy in the formation of rifted-margin upwarps. Nature 346:739–742

House M, Wernicke B, Farley K (2001) Paleo-geomorphology of the Sierra Nevada, California, from (U-Th)/He ages in apatite. Am J Sci 301:77–102

House MA, Wernicke BP, Farley KA (1998) Dating topography of the Sierra Nevada, California, using apatite (U-Th)/He ages. Nature 396:66–69

House MA, Wernicke BP, Farley KA, Dumitru TA (1997) Cenozoic thermal evolution of the central Sierra Nevada, California, from (U-Th)/He thermochronometry. Earth Planet Sci Lett 151:167–169

Jenkins GM, Watts DG (1968) Spectral Analysis and its Applications. Holden-Day, Oakland, California, first edition

King LC (1951) South African Scenery. Oliver and Boyd, White Plains, NY

Kirby E, Whipple K, Tang W, Chen Z (2003) Distribution of active rock uplift along the eastern margin of the Tibetan Plateau: inferences from bedrock channel longitudinal profiles. J Geophys Res 108(B4):2217, doi:10.1029/2001JB000861

Kooi H, Beaumont C (1994) Escarpment evolution on high-elevation rifted margins: insights derived from a surface processes model that combines diffusion, advection and reaction. J Geophys Res 99:12,191–12,209

Kooi H, Beaumont C (1996) Large-scale geomorphology: classical concepts reconciled and integrated with contemporary ideas via a surface processes model. J Geophys Res 101:3361–3386

Koons P, Norris R, Craw D, Cooper A (2003) Influence of exhumation on the structural evolution of transpressional plate boundaries: an example from the Southern Alps, New Zealand. Geology 31:3–6

Koons P, Zeitler P, Chamberlain C, Craw D, Melzer A (2002) Mechanical links between erosion and metamorphism in Nanga Parbat, Pakistan Himalaya. Am J Sci 302:749–773

Koons PO (1990) Two-sided orogen: collision and erosion from the sandbox to the Southern Alps, New Zealand. Geology 18:679–682

Mancktelow NS, Grasemann B (1997) Time-dependent effects of heat advection and topography on cooling histories during erosion. Earth Planet Sci Lett 270:167–195

Morley ME, Gleadow AJW, Lovering JF (1980) Evolution of the Tasman Rift: Apatite fission track dating evidence from the southeastern Australian continental margin. *In:* Fifth International Gondwana Symposium, pages 289–293, Wellington, New Zealand

Nunn J, Aires J (1988) Gravity anomalies and flexure of the lithosphere at the Middle Amazon Basin, Brazil. J Geophys Res 93:415–428

Ollier CD (1982) The Great Escarpment of eastern Australia: tectonic and geomorphic significance. J Geol Soc Aust 29:13–23

Persano C, Stuart FM, Bishop P, Barford DN (2002) Apatite (U-Th)/He age constraints on the development of the Great Escarpment on the southeastern Australian passive margin. Earth Planet Sci Lett 200:79–90

Reiners P, Ehlers T, Mitchell S, Montgomery D (2003a) Coupled spatial variations in precipitation and long-term erosion rates across the Washington Cascades. Nature 426:645–647

Reiners P, Farley K (1999) Helium diffusion and (U-Th)/He thermochronometry of titanite. Geochim Cosmochim Acta 63:3845–3859

Reiners P, Zhou Z, Ehlers T, Xu C, Brandon M, Donelick R, Nicolescu S (2003b) Post-orogenic evolution of the Dabie Shan, eastern China, from (U-Th)/He and fission-track thermochronology. Am J Sci 303: 489–518

Sambridge M (1999a) Geophysical Inversion with a Neighbourhood Algorithm -I. Searching a parameter space. Geophys J Int 138:479–494

Sambridge M (1999b) Geophysical Inversion with a Neighbourhood Algorithm -II. Appraising the ensemble. Geophys J Int 138:727–746

Schmid R, Ryberg T, Ratschbacher L, Schulze A, Franz L, Oberhansli R, Dong S (2001) Crustal structure of the eastern Dabie Shan interpreted from deep seismic reflection and shallow tomographic data. Tectonophysics 333:347–359

Seidl MA, Weissel JK, Pratson LF (1996) The kinematics and pattern of escarpment retreat across the rifted continental margin of SE Australia. Basin Res 12:301–316

Small E, Anderson R (1995) Geomorphically driven late Cenozoic rock uplift in the Sierra Nevada, California. Science 270:277–280

Stüwe K, White L, Brown R (1994) The influence of eroding topography on steady–state isotherms. Application to fission track analysis. Earth Planet Sci Lett 124:63–74

Turcotte D (1979) Flexure. Adv Geophys 21:51–86

Turcotte DL, Schubert G (1982) Geodynamics: Applications of Continuum Physics to Geological Problems. John Wiley and Sons, New York, first edition

van der Beek P, Andriessen P, Cloetingh S (1995) Morpho-tectonic evolution of rifted continental margins: Inferences from a coupled tectonic-surface processes model and fission-track thermochronology. Tectonics 14:406–421

van der Beek P, Pulford A, Braun J (2001) Cenozoic landscape evolution in the Blue Mountains (SE Australia): Lithological and tectonic controls on rifted margin morphology. J Geol 109:35–56

van der Beek P, Summerfield M, Braun J, Brown R, Fleming A (2002) Modelling post-breakup landscape development and denudational history across the southeast African (Drakensberg Escarpment) margin. J Geophys Res 107(B4):2351, doi:10.1029/2001JB000744

Warnock AC, Zeitler PK, Wolf RA, Bergman SC (1997) An evaluation of low-temperature apatite U-Th/He thermochronometry. Geochim Cosmochim Acta 61:5371–5377

Weissel JK, Seidl MA (1998) Inland propagation of erosional escarpments and river profile evolution across the southeast Australian passive continental margin. *In*: Rivers over Rock: Fluvial Processes in Bedrock Channels. Geophysical Monograph 107. Tinkler KJ, Wohl EE (ed) American Geophysical Union, Washington, p 189–206

Whipple KX, Kirby E, Brocklehurst SH (1999) Geomorphic limits to climate-induced increases in topographic relief. Nature 401:39–43

Whipple KX, Tucker G (1999) Dynamics of the stream-power incision model: implications for height limits of mountain ranges, landscape response timescales and research needs. J Geophys Res 104:17,661–17,674

Willett S, Beaumont C, Fullsack P (1993) Mechanical model for the tectonics of doubly-vergent compressional orogens. Geology 21:371–374

Willett SD, Slingerland R, Hovius N (2001) Uplift, shortening, steady state topography in active mountain belts. Am J Sci 301:455–485

Wolf RA, Farley KA, Kass DM (1998) Modeling of the temperature sensitivity of the apatite (U-Th)/He thermochronometer. Comput Geosci 148:105–114

Wolf RA, Farley KA, Silver LT (1996) Helium diffusion and low-temperature thermochronometry of apatite. Geochim Cosmochim Acta 60:4231–4240

Reviews in Mineralogy & Geochemistry
Vol. 58, pp. 375-387, 2005

Exploiting 3D Spatial Sampling in Inverse Modeling of Thermochronological Data

Kerry Gallagher[1], John Stephenson[1], Roderick Brown[2], Chris Holmes[3], Pedro Ballester[1]

*[1]Dept. of Earth Sciences and Engineering
Imperial College London
South Kensington, London, SW7 2AS, England*

*[2]Division of Earth Sciences
Gregory Building
University of Glasgow
Glasgow, G12 8QQ, Scotland*

*[3]Dept. of Statistics
University of Oxford
1 South Parks Road
Oxford, OX1 3TG, England*

INTRODUCTION

The development of quantitative models for fission track annealing (Laslett et al. 1987; Carlson 1990; Laslett and Galbraith 1996; Ketcham et al. 1999) and more recently, helium diffusion in apatite (Wolf et al. 1996; Farley 2000), has allowed direct inference of the temperature history of the host rocks, and a more indirect inference of denudation chronologies (see Kohn et al. this volume, and references therein). An example of a model prediction of AFT parameter and (U-Th)/He age for a specified thermal history is given in Figure 1. Various approaches exist to extract a thermal history model directly from the data, and these focus around inverse modeling (Corrigan 1991; Gallagher 1995; Issler 1996; Willett 1997; Ketcham et. al. 2000). The user specifies some constraints on the thermal history (e.g., upper and lower bounds on the temperature time, and heating/cooling rate), and then typically some form of stochastic sampling is adopted to infer either the most likely thermal history (ideally with some measure of the uncertainty of the solution), and/or a family of acceptable thermal histories. In both the forward and inverse approaches, the thermal history is typically parameterized as nodes in time-temperature space, with some form of interpolation between the nodes.

Over recent years, one of the major applications of low temperature thermochronology has been the study of long term denudation as recorded in the cooling history of surface samples. More recently, some studies have specifically tried to link relatively short term, local estimates of denudation (e.g., from cosmogenic surface exposure dating) to these longer term estimates (Cockburn et al. 2000; Brown et al. 2001; Reiners et al. 2003). The step from the thermal history to denudation chronology is less direct that inferring the thermal history from the data, in that we need to make some assumptions in order to convert temperature to depth. This may involve an assumption that a 1D steady state with a constant temperature gradient over time is appropriate, or alternatively that a full 3D diffusion-advection model is required. The latter situation is not particularly amenable to an inversion approach, although recent applications have been made with a restricted number of parameters, to identify plausible solutions to relatively specific questions, such as the timing of relief development (Braun

1529-6466/05/0058-0014$05.00 DOI: 10.2138/rmg.2005.58.14

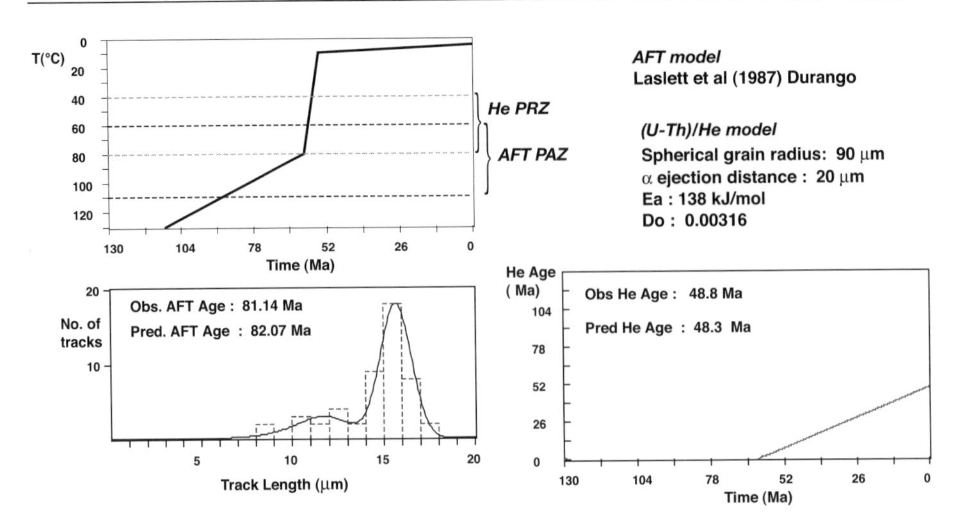

Figure 1. A typical forward model—the thermal history is specified, and having chosen and annealing/diffusion model, we can predict the apatite fission track parameters (age, length distribution), and (U-Th)/He data. PRZ and PAZ are the partial retention zone, and partial annealing zones, over which the He and AFT systems are most sensitive on geological timescales.

2005). In practice, heat transfer can vary spatially (in both horizontal and vertical dimemsions) as a consequence of variations in thermal properties, in the mode of heat transfer (conduction and advection) and spatial variations in erosion rate and surface topography.

In the simplest case, assuming a constant gradient, it is common to adopt a "representative" geotherm of around 25–30 °C and to additionally specify that this is constant over time. Often, we do not know the present day gradient in crystalline basement areas, although it may be possible to adopt a local value from a global heat flow database (Pollack et al. 1993), which may also require assumptions regarding the thermal conductivity of the material that that has been removed by erosion. The role of thermal conductivity is often neglected, but the importance lies in the fact that thermal conductivities of common rocks can vary by a factor of 2–3 (Somerton 1992), and so for a constant heat flow, the geothermal gradient will vary by a similar factor. However, the thermal conductivity or rocks is reasonably predictable, in terms of lithology, mineralogy and porosity. Consequently, if it is possible to infer the nature of the eroded material, then it is possible to make an informed judgement of the thermal conductivity.

Here we consider some aspects related to inverse modeling of the thermal history and describe a strategy which aims to identify good, but simple, thermal history models. Such models are found by jointly fitting data from multiple samples, rather than taking each data set independently. The method has been developed to identify spatial variations in the thermal history, particularly in the context of identifying boundaries (such as faults), across which the thermal history may vary significantly.

What is a good but simple thermal history model?

Although it is relatively straightforward to find a thermal history that fits the observed thermochronological data, a more difficult, but significant stage in modeling is to understand how good this thermal history is. Intuitively, we can argue a good model is one that fits the observations satisfactorily without being overly complex, i.e., having structure that is not supported or required by the available data. Thus, the important criteria are the measure of the data fit and also a measure of the model complexity. For fission track data, a natural choice of data fit is the log likelihood function given by Gallagher (1995). This is defined in terms of the observed spontaneous and induced track counts, N_s^j and N_i^j for each crystal j of a total of N_c, and the N_t individual track length measurements, l_k, $k = 1,N_t$ and is given as

$$L = \sum_{j=1}^{N_c} \left\{ N_s^j \ln(\theta) + N_i^j \ln(1-\theta) \right\} + \sum_{k=1}^{N_t} \ln[P(l_k)] \quad (1)$$

where θ is a function of the predicted spontaneous and induced track densities (ρ_s, ρ_i), given as

$$\theta = \frac{\rho_s}{\rho_s + \rho_i} \quad (2)$$

$P(l_k)$ is the probability of having a track of length l_k in the observed distribution, given that we have predicted the track lengths distribution for a particular thermal history (for details see Gallagher 1995). A common form of likelihood function, probably appropriate for (U-Th)/He dating, is based on a sum of squares statistic between observed and predicted ages, weighted by the error, which for N (U-Th)/He ages, is given as

$$L = \ln\left(\sum_{j=1}^{N} \left(\frac{t_i^{obs} - t_i^{pred}}{\sigma_i} \right)^2 \right) \quad (3)$$

where t^{obs} and t^{pred} are the observed and predicted He ages. This form, interpreted as a log-likelihood, implicitly assumes normally distributed errors.

In practice, the log-likelihood is a negative number, and we look for the thermal history that produces the maximum value of the log-likelihood (i.e., closest to zero), This is equivalent to the thermal history that has the maximum probability of producing the observed data. It is clear from Equation (1), that the value of the log-likelihood will depend on the number of data. As already alluded to above, the likelihood value will also depend on the complexity of the model in that we expect a model with more parameters to provide a better fit to the observed data. However, the issue then is whether the improvement in the data fit is sufficient to justify the additional model parameters. One straightforward way of assessing this is through the Bayesian Information Criterion (Schwartz 1978), which is defined for a model, m_i, as

$$BIC(m_i) = -2L(m_i) + \nu_{m_i} \log(N) \quad (4)$$

where L is the log-likelihood, ν_{mi} is the number of model parameters in the current model, and N is the number of data (observations). The second term in Equation (4) penalizes the improvement in the data fit as a consequence of increasing the complexity of the model. If we consider two models, m_1 and m_2, where m_2 has more model parameters than m_1, then if BIC(m_1) < BIC(m_2), then we infer that model m_1 is preferable to m_2. The BIC is useful for model choice when we use the same number of data for all models and an implicit assumption is that the true model is contained in all the models we consider.

Figure 2 shows three thermal history models inferred from the same set of synthetic data in which the parameterization captures the true model (which has 5 parameters, 3 temperatures and 2 times—we know the present day time). The first model is under-parameterized (3 parameters), and the third model is over-parameterized (7 parameters). As we expect, the 7 parameter model provides the best fit to the data, but the BIC implies that the improvement over the 5 parameter model is not significant, while the 5 parameter model is significantly better than the 3 parameter model. Heuristically, we could infer this by looking at the form of the thermal histories. The 7 parameter model does not really introduce any significantly

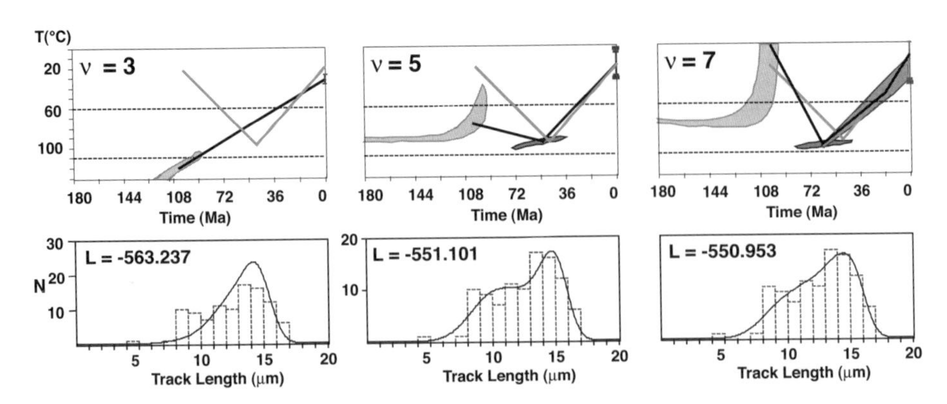

Figure 2. Thermal histories (black lines) derived from fitting AFT synthetic data and ν is the number of model parameters (time and temperature nodes). The original thermal history has ν = 5 and is shown as the grey line (BIC = 1129.4). The shaded regions are the approximate 95% confidence regions (see Gallagher 1995 for details). L is the log-likelihood, and although the model with ν = 7 yields the maximum likelihood, the BIC of 1140.0 implies the improvement over the model with ν = 5 does not warrant the extra model parameters. The model with ν = 3 has BIC = 1142.7.

new features compared with the 5 parameter model, in that the extra time-temperature node effectively falls on the cooling trajectory from the maximum temperature to the present day for the 5 parameter model.

Another aspect of modeling that is relevant to the approach advocated in this contribution is the role of the number of data used. Although there is likely to be redundancy in thermochronological data, incorporating more data to constrain a thermal history model generally leads to less variability in the acceptable solutions, or smaller confidence regions about the inferred thermal history. This is illustrated in Figure 3 where 2 synthetic data sets were generated by sampling the predicted parameters for the same thermal history shown in Figure 2. We calculated the 95% confidence regions about the best thermal history using the methods outlined in Gallagher (1995), and the results show that the inferred thermal history is better constrained with a larger amount of data than with relatively few data. This then implies that if we can group the data from multiple samples and model them jointly with a common thermal history, then the resolution of the inferred thermal history should be better than if we model each sample independently. The likelihood function for the collective samples is just the sum of the log-likelihoods for the individual samples, each calculated using the same, common thermal history. This approach will also tend to produce simpler models, as there will be a degree of compromise in jointly fitting multiple data sets.

The philosophy behind our preferred strategy to modeling thermochronological data can be summarized as follows: we aim to incorporate multiple data sets into a common model, and try to find the simplest thermal history models that can satisfy the observed data. The BIC can be used to address the second aspect, but a remaining issue in addressing the first aspect is how best to group data together from a suite of irregularly distributed spatial samples. In some cases, there are natural groupings. For example, a suite of samples from a borehole, or vertical profiles, in which a suite of samples is collected over a range of elevation at effectively on location. In these cases, the spatial relationship between the samples is the vertical offset and this can be regarded as a 1D geometry (i.e., the vertical dimension). Provided there have not been thermal perturbations within the section (e.g., due to fluid flow or faulting), this can be directly translated into a temperature offset, such that samples at great depth in a borehole,

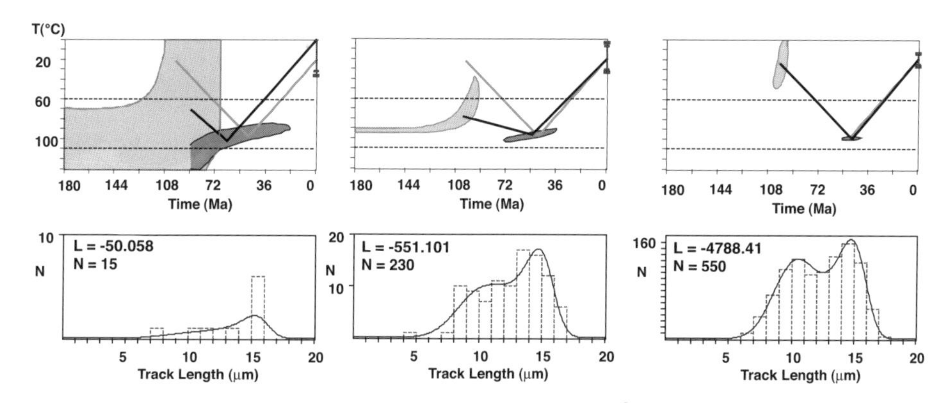

Figure 3. Inferred thermal histories based on different amounts of AFT data. The left panel has 10 track lengths and 5 single grain, ages, the central panel 200 lengths and 30 single grain ages, and the right panel has 500 lengths and 50 single grain ages. The absolute value of the likelihood depends on the number of data. The approximate 95% confidence regions are based on differences in the likelihood and, when more data are used, these are smaller (i.e., the thermal history is more well resolved).

or shallower elevation for a vertical profile, will have been at higher temperatures than the shallower depth (or higher elevation) samples. In practice, the thermochronological definition of a vertical profile does not require the samples to be aligned vertically, but does imply that the dominant direction of heat transfer is vertical. This means that factors such as spatial variations in thermal properties, erosion rate or surface topography, leading to lateral heat transfer, do not significantly influence the thermal history of the samples (vertical profile) being considered. When dealing with sedimentary basins, there may also be complications in terms of preserving provenance signatures (reflecting the pre-depositional thermal history of detrital grains), which can complicate the inference of the post-depositional thermal history (Carter and Gallagher 2004). However, it is straightforward to allow for this, by incorporating extra model parameters to account for the pre-depositional thermal history (which may or may not be specified to be independent between samples).

When dealing with samples irregularly distributed in two spatial dimensions (e.g., latitude and longitude), there is not such an obvious way to group samples in order to share a common thermal histories. One approach, that underlies the geostatistical method of kriging (e.g., Isaaks and Srivastava 1989), is to assume that samples which are close in space will have experienced similar thermal histories. As the distance between samples becomes greater, this requirement is relaxed. One problem with this assumption is that two nearby samples may be separated by a fault (which is a spatial discontinuity). Furthermore, such a fault may or may not have been active over the time span of the thermal history retrievable from the data, or the presence of a discontinuity may not have even been recognized. The most general case, in 3D, incorporates the features of the 1D vertical offset case, and the irregularly distributed 2D samples, which may be separated by unknown discontinuities.

In the next sections, we review a general strategy to deal with these situations, and demonstrate the application to synthetic apatite fission track data, although the basic approach is completely general in terms of application to other thermochronological systems, and to combinations of different types of data, provided suitable likelihood functions can be defined. In all cases, we parameterize the thermal history as a series of time-temperature nodes, and specify bounds on the possible values of the temperature and time as described earlier. To find the thermal history models, we use stochastic sampling methods, primarily genetic algorithms

(GA) and Markov chain Monte Carlo (MCMC). The former method is an efficient optimizer, i.e., for rapidly identifying the better data fitting models. The latter method provides reliable estimates of the joint and marginal probability density functions of the model parameters, from which it is to examine correlation between parameters, and to quantify the uncertainty in terms of, for example, the 95% credible range on individual model parameters. More detail on the methodology and applications to real data sets can be found in Gallagher et al. (2005) and Stephenson et al. (2005). We first consider the 1D vertical profile case, then the 2D case with an unknown number of spatial discontinuities, and finally demonstrate the generalization to 3D.

1D modeling

Here we want to exploit the spatial relationship of samples in the vertical dimension, in which we implicitly assume the lowermost sample was always the hottest and the uppermost sample was always the coolest. The situation we consider is shown in Figure 4 where samples are collected from a vertical profile (e.g., a borehole or up the side of a valley). We specified a thermal history and generated synthetic data for a suite of such samples. These "synthetic samples" were first modeled independently and then modeled jointly. In the second case, the parameters for the thermal history model were specified in the same way as adopted for modeling the samples independently, with additional model parameters which deal with the temperature offset between the upper and lower samples. We consider two cases in which we use a constant temperature offset and a time-varying offset. In both cases, we choose the pale-offset to be independent of the present day offset as many vertical profiles are collected on surface samples which are often at similar present day temperatures (or the temperature offset is effectively the atmospheric temperature lapse rate, typically 5–6 °C/km).

The results are shown in Figure 5. Modeling the samples independently leads to a better log-likelihood (L = 7514.60), as we expect, but there are 72 model parameters required (9 parameters for 8 samples). There are some common features, such as the rapid cooling recorded in the deepest samples, but generally the individual thermal histories show little coherence. When treating the samples jointly, and assuming a constant temperature offset, the inferred thermal history model is much simpler, with only 11 model parameters, and the

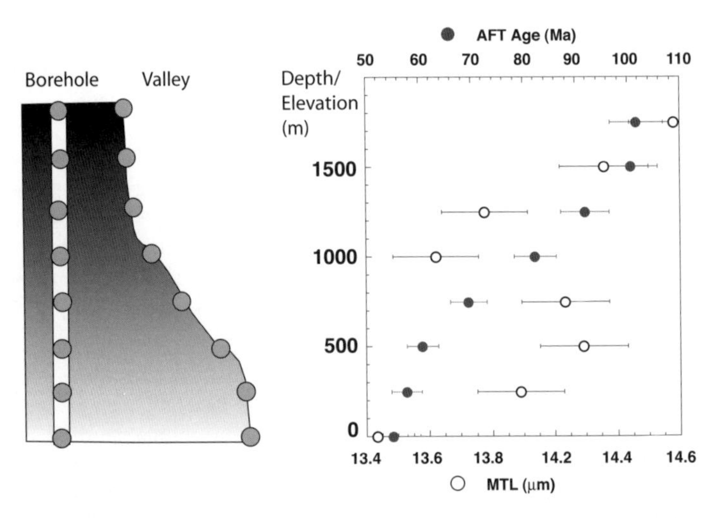

Figure 4. A vertical profile is obtained by sampling from different depths in a borehole, or different elevations in a valley. The distribution of fission track age and mean length with elevation is characteristic of the thermal history.

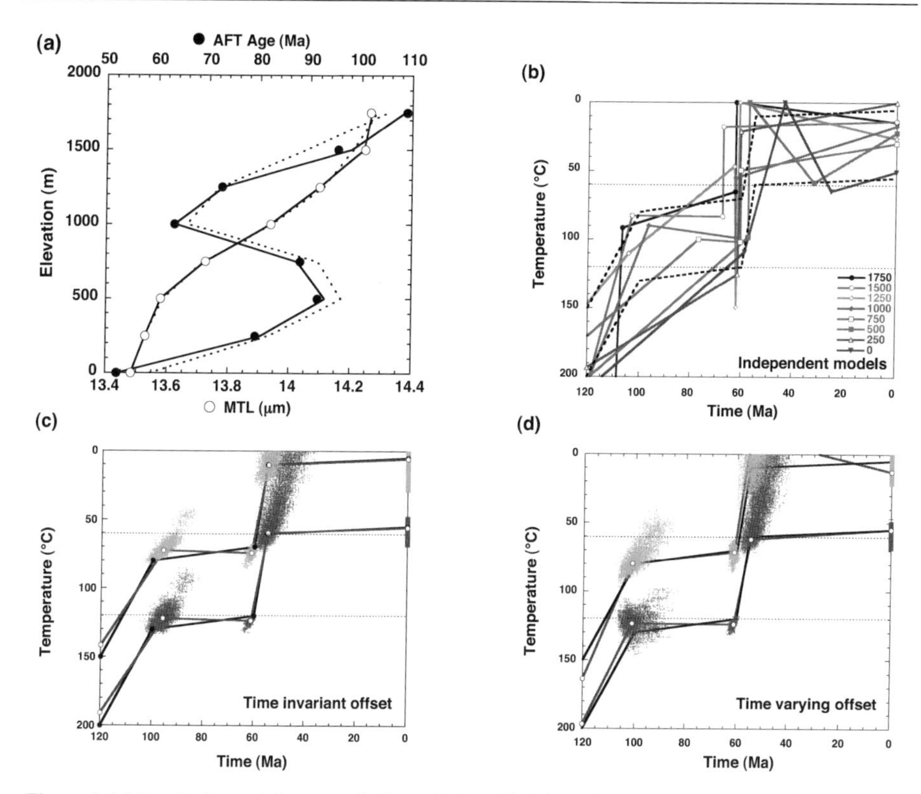

Figure 5. (a) Results for modeling a synthetic vertical profile. The "observed" data and the predictions are shown as symbols, and lines, respectively as a function of elevation. The solid lines are the predictions when modeling each sample independently, and the dashed lines are the predictions when the samples are modeled jointly (on this scale there is no difference between the predictions from the models shown in panels c and d). (b) Inferred thermal histories when modeling the samples independently, labeled according to the present day elevation (BIC = 15569.5). The true thermal history for the uppermost and lowermost samples are shown as the dashed lines, and the intermediate samples are all parallel to these. (c) Thermal histories inferred by modeling the samples jointly, with a constant temperature offset over time (BIC = 15161.9). The grey shaded areas around each time-temperature point are the distributions obtained from MCMC sampling and approximate the 95% confidence regions. The lighter grey regions (around the lower temperature thermal history) incorporate the uncertainty on the temperature offset between the 2 thermal histories. (d) Thermal histories inferred by modeling the samples jointly, but allowing the temperature offset to vary over time (BIC=15175.9). This model is only marginally better than the constant offset model in terms of the likelihood, but the extra model parameters are not justified when assessed with the BIC.

95% credible regions about each time-temperature point imply the thermal history is well resolved. While we do not fit the data quite as well ($L = -7539.60$), the BIC tells us that this simpler model is readily acceptable (the difference in log-likelihood is 25). In fact the difference in the log-likelihood would need to be about an order of magnitude greater before we reject the simpler model. Allowing for a variable temperature gradient over time produces a slightly better model ($L = -7539.04$), the incorporation of the 3 extra model parameters is not warranted, based on the BIC.

As part of the model formulation, we infer the temperature offset over time, which then gives a estimate of the temperature gradient directly from the thermochronological data. Moreover, as we use MCMC to characterize the model parameter space, we also obtain the probability distribution on the temperature gradient (Fig. 6). As mentioned earlier, the

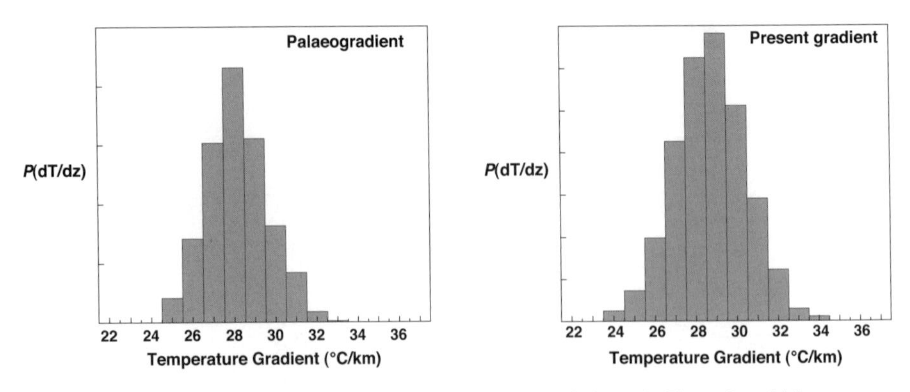

Figure 6. Distributions on the temperature gradient for the model shown in Figure 5c, which assumes a constant palaeogradient. The true solution for both the present and palaeogradient is ~28.6 °C km^{-1}. These distributions can be sampled to produce uncertainty estimates for denudation.

temperature gradient is a key requirement to convert the thermal history to a equivalent depth or denudation chronology. The probability distributions on the thermal history and the temperature offset can be readily sampled to construct the probability distribution of the denudation estimates. If the temperature gradient changed over time, as a consequence of rapid denudation, and there is information in the thermochronological data, then this approach should extract that information (and the uncertainty). However, from our experience, introducing a time-varying offset tends to introduce too much variation, and we would certainly recommend exploring whether the difference in data fit in comparison with a constant offset model is justified.

2D modeling

In this situation, the spatial relationship is nearness, i.e., samples close together are likely to have similar thermal histories. As mentioned earlier, in the real world, there are discontinuities (e.g., faults). The problem then is how to group samples spatially, allowing for the presence of unknown discontinuities. Here we classify the samples into different sub-groups defined by discrete spatial regions or partitions,, such that the thermal history is the same for a given partition,, but varies between partitions. Also, we do not know how many partitions we should look for, i.e., one of the unknown parameters is the number of parameters. The problem as formulated here does not allow for lateral variations in the thermal history within a partition, although this is not a major problem to implement (it just requires some form of interpolation across a partition). Another requirement is that samples do not move laterally relative to each other, which may limit the application to active mountain belts involving large scale lateral transfer (e.g., the southern Alps in New Zealand).

This is solved with a form of Bayesian Partition Modeling (BPM), which is more formally described by Denison et al. (2002). In essence, BPM provides a method for spatial clustering of different samples, according to the spatial structure of the data. In our case, we have an additional complication in that we are interesting in spatial clustering based on the thermal history inferred from the data for particular samples. The 2D space is parameterized with a Voronoi tessellation (Okabe et al. 2000), which are polygonal regions defined by an internal point, such that any sample location that falls within a given Voronoi cell is closer to that internal point. The boundaries of the Voronoi cells are drawn as the perpendicular bisectors of the internal points in each cell (Fig. 7). It is the boundaries of the partitions that are our proxy for geological discontinuities, such as faults, where the thermal history may change rapidly over a small distance.

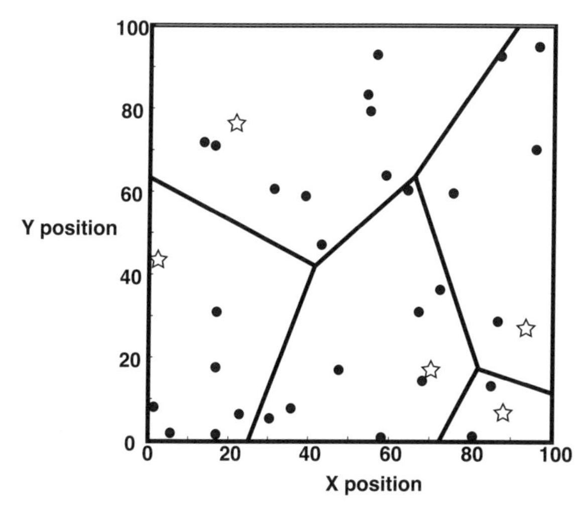

Figure 7. The geometry of 2D Voronoi cells, and their centers (indicated by the stars). The boundaries of each cell is defined as the perpendicular bisectors of the lines joining its centre to all other centers. Any sample location (filled circles) that falls within a given cell is closer to the centre of that cell than any other centre. The linear boundaries are used in our modeling approach to characterize spatial discontinuities, although their number and positions are unknown.

The implementation of BPM we adopt uses a dimension changing version of MCMC, known as Reversible Jump (RJ) MCMC (Green 1995), as we need to deal with an unknown number of partitions. In this approach, we can specify the minimum and maximum number of partitions we allow *a priori*. The maximum range is from 1 to the number of samples, but we typically choose to set the maximum number to value less than the number of samples. Otherwise, we can just model all the samples independently. In order to deal with the unknown thermal histories in each partition, we use the GA described by Ballester and Carter (2004) to find the optimal thermal history with each partition for a given partition configuration generated during the MCMC run. When a given partition configuration is repeated during the MCMC (in that the sample groupings have previously been considered), we take the earlier best GA thermal history model for the partitions in that configuration. This particular approach can lead to the algorithm becoming somewhat static and sub-optimal. However, we can modify the algorithm to run another MCMC run on the thermal history in each partition, for a given partition configuration, which improves the combined sampling of the model space for the thermal histories and partitions (Stephenson et al. 2005). The examples we consider in this paper are based on synthetic data, and here the resolution on the thermal history is not our primary objective. Rather we want to demonstrate the concept of implementing the partition model approach to irregularly distributed spatial samples.

Figure 8 shows the method applied to a 3 partition problem, with 15 sample locations, where the RJ-MCMC was run allowing for up to 7 partitions. The results show that we can recover the correct number of partitions with high probability, with the correct allocation of samples in each partition, and also a good representation of the thermal history within each partition. Note that here we chose thermal histories for each partition that are distinct and relatively easy for the method to identify, as our aim is to demonstrate the ability of the methodology to identify the form and spread in the inferred partition structure. The spread in the solutions for 3 partitions is also an indication of the uncertainty about the location of the partition boundaries. With the implementation we have used here, all partitions geometries that correctly allocate the samples will use the same thermal histories (and have the same likelihood), then it is clear that the range of the location of the boundaries is determined by the location of the sample locations, subject to the requirement that the boundaries are straight. So in the top right of Figure 8. there is a relatively large spread in the location of the boundaries, as there are no sample locations there, but the spread is constrained by the samples around $x = 50$, and $y = 70–80$. Similarly, the

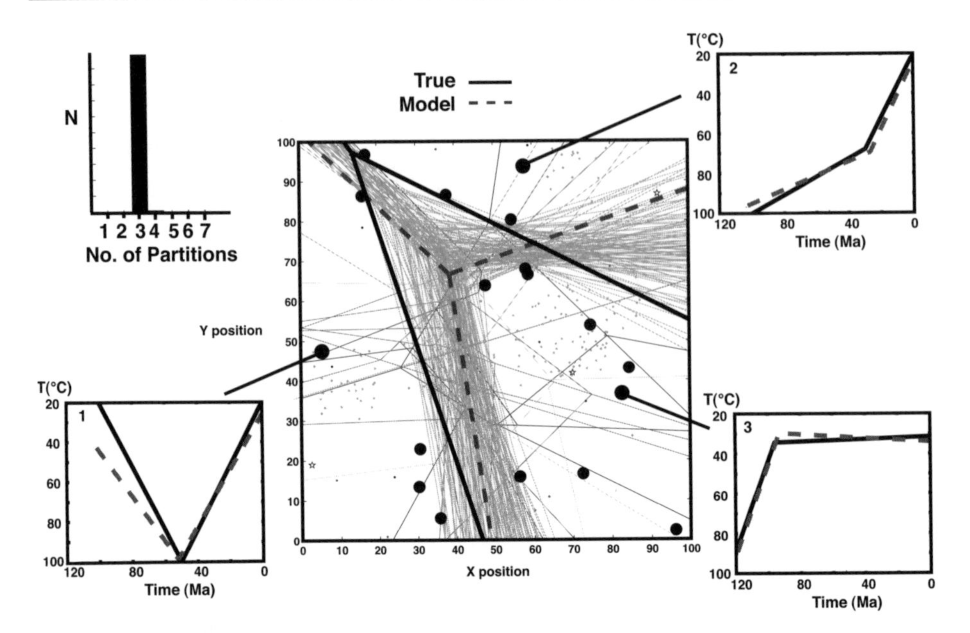

Figure 8. Example of modeling 2D spatial samples, in 3 Voronoi cells, or partitions, with different thermal histories (the true and model results are shown as solid and dashed lines, respectively, for each partition). We used 15 samples (filled circles), randomly distributed around the 3 partitions, and all samples within a given partition have synthetic data generated from the same thermal history. Also, shown is the inferred distribution on the number of partitions, and the location of all combinations of partitions sampled during the model run. There are various partitions locations with the correct allocation of samples, and the variation in these different locations is a function of the sample distribution.

samples between $y = 10$ and 30, and $x = 30$–60, and those $x = 20$, $y = 85$–95, determine the range on the possible partition boundaries. Therefore, if the objective of a sampling campaign is to identify differential cooling (due to fault movement, for example), then it is clearly strategic to sample close to where the boundaries may be.

3D modeling

To extend the approach to 3D, we combine the 1D vertical profile method and 2D partition modeling method, i.e., use the fact that within a partition, we may have samples at different elevations. We do not need to assume that the samples within a partition are at the same location, but merely have not been offset relative to each other (Fig. 9). This lets us combine samples exploiting the 3D sampling geometry, and also lets us estimate the temperature gradient within each partition. In Figure 10, we show the result of a 5 partition model, with 30 samples, with a temperature offset in 2 of the partitions. We allowed up to 10 partitions in the model space. In general, we infer thermal histories which represent well those used to generate the synthetic data. However, we do not recover the same partition structure, converging to 6 with high probability, rather than 5. However, the discrepancies are not too serious. For example, although partition A is subdivided, these 2 partitions do not involve any samples from outside partition A. Moreover, the thermal histories inferred for the 2 partitions are very similar, in terms of the predicted fission track parameters, and so qualitatively, we could ignore the sub-division. Partition C has also been subdivided, and partition D has been merged with one of the subdivided partitions. Again the inferred thermal histories are similar (as are those used to generate the synthetic data for partitions C and D). The fact that the GA converged on different

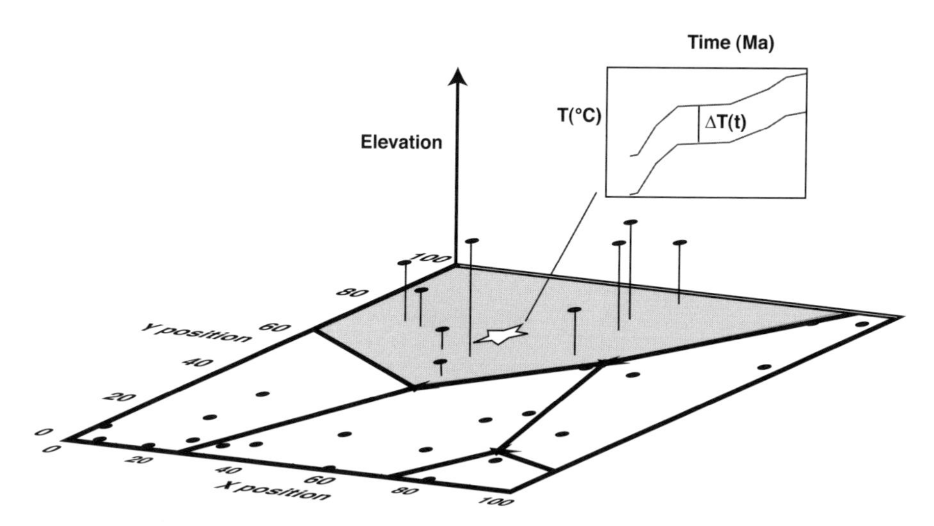

Figure 9. The geometry of the 3D modeling approach which uses 2D Voronoi cells, and within a cell, the 1D vertical profile approach (a temperature offset as a function of sample elevation).

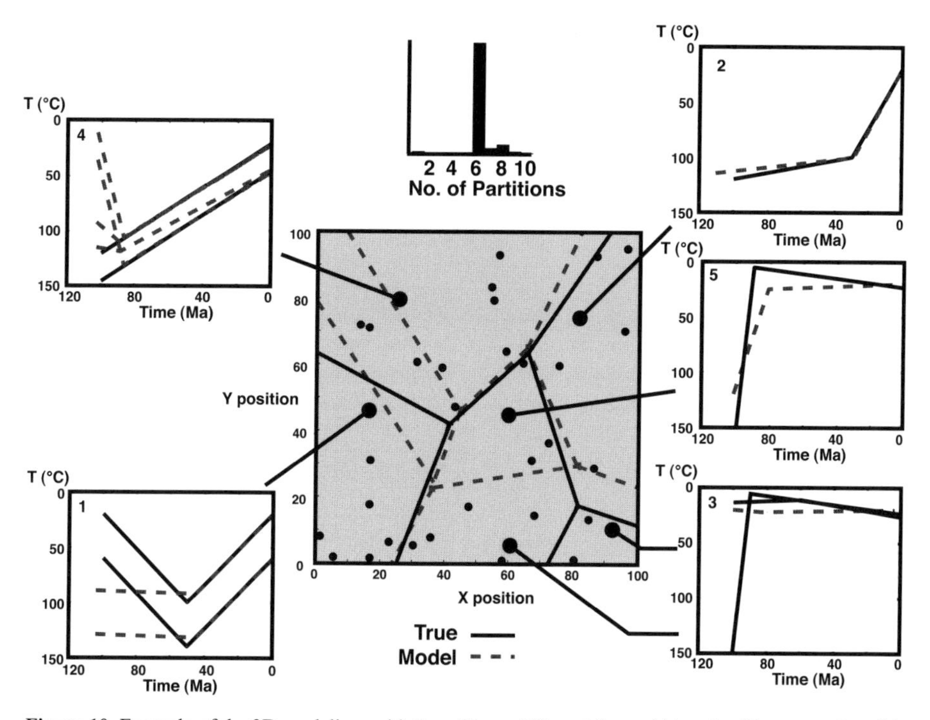

Figure 10. Example of the 3D modeling, with 5 partitions, different thermal histories (the true and model results are shown for each partition and 30 samples (filled circles), randomly distributed around the 5 partitions. Partitions A and E have temperature offsets, while B, C and D do not. The model infers 6 partitions and does not always correctly allocate the samples within partitions. However, this reflects the fact that the thermal histories are similar, and we select optimal thermal histories for a given partition configuration (see the text for details).

solutions in these situations and did not subsequently move appears to be a consequence of this particular implementation not allowing new thermal history models for a given partition configuration (although this is a relatively straightforward modification; see Stephenson et al. 2005). However, the other partition boundaries are well identified (given the distribution of sample locations). For example, the boundary that runs SW-NW is well resolved as there are several samples located close to the boundary. The same comments about the resolution of the boundaries made for the 2D case also applies in 3D. Thus, the boundaries between A and E, and B and D, are as well resolved as they can be, given the sample distributions.

SUMMARY

We have given an overview of a modeling strategy aimed at exploiting the spatial geometry of the sample distributions in order to maximize the retrieval of thermal history information from thermochronological data. Philosophically, we aim to find thermal history solutions that fit the observations well, but do not have unwarranted complexity. These two requirements are quantified through the Bayesian Information Criterion, which combines the data likelihood and the number of model parameters. The overall approach relies on exploiting the spatial geometry of the sample locations to combine data from individual samples and identify a common thermal history. The combination of different data sets has the advantage of improving the resolution on the inferred thermal history, and also reducing the complexity. Markov chain Monte Carlo sampling provides a means of constructing reliable representations on the probability distributions for the model parameters. 1D modeling is relevant to vertical profiles, and provides an estimate of the paleotemperature gradient directly from the data. The 2D approach relies on a partition model, in which each partition contains a subgroup of the samples with a common thermal history. The partition model approach allows for an unknown number of discontinuities, whose locations are also unknown. The extension to 3D combines the 1D and 2D approaches to find partitions in which samples at different elevations have experienced a common form of thermal history, but the actual temperatures depend on the elevation. As presented here, it is implicit that the spatial relationship between samples has not changed over time, at least not in a way that will lead to different thermal histories.

The approach presented here is different to 3D thermal models (Braun 2003, 2005; Ehlers 2005) but complementary. Thus, we infer the thermal history directly from the data, while the other 3D models are specified and certain parameters are adjusted to match the observed data. Both approaches assume that the predictive models for fission track annealing or helium diffusion in apatite are correct. In principle, this assumption can be relaxed and appropriate predictive model parameters can be estimated as part of the modeling process. However, this will lead to significant trade-off between annealing of diffusion parameters and the thermal history (Gallagher and Evans 1991). Future modifications to this approach will include more generalized sampling of the thermal histories during the MCMC sampling of the partition structure, incorporation of multiple data types (e.g., apatite fission track and (U-Th)/He data) and potentially allowing for irregularly shaped partition boundaries.

REFERENCES

Ballester PJ, Carter JN (2004) "An Effective Real-Parameter Genetic Algorithms with Parent Centric Normal Crossover for Multimodal Optimization." Genetic and Evolutionary Computation Conference (GECCO-04, Seattle, USA). Lecture Notes in Computer Science 3102, Springer, p. 901-91

Braun J (2003) Pecube: a new finite element code to solve the 3D heat transport equation including the effects of a time-varying, finite amplitude surface topography. Comp Geosci 29:787-794

Braun J (2005) Quantitative constraints on the rate of landform evolution derived from low-temperature thermochrohonology. Rev Mineral Geochem 58:351-374

Brown RW, Summerfield MA, Gleadow AJW (2002) Denudational history along a transect across the Drakensberg Escarpment of southern Africa derived from apatite fission track thermochronology. J Geophys Res 107(B12):2350, doi:10.1029/2001JB000745

Carlson WD (1990) Mechanisms and kinetics of apatite fission-track annealing. Am Mineral 75:1120-1139

Carter A, Gallagher K (2004) Provenance signatures and the inference of thermal history models from apatite fission track data – A synthetic data study. Geol Soc Am Spec Publ 378:7-23

Cockburn HAP, Brown RW, Summerfield MA, Seidl MA (2000) Quantifying passive margin denudation and landscape development using a combined fission-track thermochronology and cosmogenic isotope analysis approach. Earth Planet Sci Lett 179:429-435

Corrigan J (1991) Inversion of apatite fission track data for thermal history information. J Geophys Res 96(B6):10347–10360, doi:10.1029/91JB00514

Denison DGT, Holmes CC, Mallick BK, Smith AFM (2000) Bayesian Methods for Nonlinear Classification and Regression. John Wiley & Sons, Chichester

Ehlers TA (2005) Crustal thermal processes and the interpretation of thermochronometer data. Rev Mineral Geochem 58:315-350

Farley KA (2000) Helium diffusion from apatite I: General behavior as illustrated by Durango fluorapatite. J Geophys Res 105:2903-2914

Gallagher K (1995) Evolving thermal histories from fission track data. Earth Planet Sci Lett 136:421-435

Gallagher K, Evans E (1991) Estimating kinetic parameters for organic reactions from geological data: an example from the Gippsland Basin, Australia. Appl Geochem 6:653-664

Gallagher K, Stephenson J, Brown R, Holmes C, Fitzgerald P (2005) Low temperature thermochronology and strategies for multiple samples. 1 : vertical profiles, Earth Planet Sci Lett, in press

Green PJ (1995) Reversible jump Markov chain Monte Carlo computation and Baeysin model determination. Biometrika 82:711-732

Isaaks EH, Srivastava RM (1989) Introduction to Applied Geostatistics. Oxford University Press, Oxford

Issler DR (1996) An inverse model for extracting thermal histories from apatite fission track data: Instructions and software for the Windows 95 environment. Geological Survey of Canada, Open File Report 2325

Ketcham RA (2005) Forward and inverse modeling of low-temperature thermochronometry data. Rev Mineral Geochem 58:275-314

Ketcham R, Donelick R, Carlson W (1999) Variability of apatite fission-track annealing kinetics: III. Extrapolation to geological timescales. Am Mineral 84:1235-1255

Ketcham R, Donelick R, Donelick M (2000) AFTSolve: a program for multi-kinetic modeling of apatite fission track data. Geol Mater Res 2:1-32

Laslett GM, Galbraith R (1996) Statistical modeling of thermal annealing of fission tracks in apatite. Geochim Cosmochim Acta 60:5117-5131

Laslett GM, Green PF, Duddy IR, Gleadow AJW (1987) Thermal annealing of fission tracks in apatite. 2. A quantitative analysis. Chem Geol (Isot. Geosci. section) 65:1-13

Okabe A, Boots B, Sugihara K, Chin S-N (2000) Spatial Tessellations: Concepts and Applications of Voronoi Diagrams, 2nd edition. John Wiley & Sons, Chichester

Pollack HN, Hurter SJ, Johnson JR (1993) Heat flow from the earth's interior: analysis of the global data set. Rev Geophys 31(3):267-280

Reiners PW, Zuyi Z, Elhers TA, Xu C, Brandon MT, Donelick RA, Nicolescu S (2003) Post-orogenic evolution of the Dabie Shan, eastern China, from (U-Th)/He and fission-track thermochronology. Am J Sci 303: 489-518

Schwartz G (1978) Estimating the dimension of a model. Ann Statistics 6:461-646

Somerton WH (1992) Thermal Properties and Temperature-related Behaviour of Fluid Rock Systems. Elsevier, Amsterdam

Stephenson J, Gallagher K, Holmes C (2005) Low temperature thermochronology and strategies for multiple samples 2: partition modeling for 2/3D distributions with disontinuities. To be submitted to Earth Planet Sci Lett

Willett SD (1997) Inverse modeling of annealing of fission tracks in apatite 1: A Controlled Random Search method. Am J Sci 297:939-969

Wolf RA, Farley KA, Silver LT (1996) Helium diffusion and low temperature thermchrononmetry of apatite. Geochim Cosmochim Acta 60:4231-4240

Reviews in Mineralogy & Geochemistry
Vol. 58, pp. 389-409, 2005

Continuous Thermal Histories from Inversion of Closure Profiles

T. Mark Harrison

Research School of Earth Sciences
The Australian National University
Canberra, A.C.T. 0200, Australia
director.rses@anu.edu.au

Marty Grove, Oscar M. Lovera

Department of Earth and Space Sciences
University of California, Los Angeles
Los Angeles, California, 90095, U.S.A.

Peter K. Zeitler

Department of Earth and Environmental Sciences
Lehigh University
Bethlehem, Pennsylvania, 18015, U.S.A.

INTRODUCTION

Background

Most geophysical processes impart a characteristic thermal signature to the crust that can be preserved in the form of isotopic variations in radiogenic minerals. Reading the record of these events using thermochronology permits unprecedented insights into the timing and rates of key dynamic processes, such as rifting, thrust faulting, tectonic denudation, erosion/incision, and magmatism, that may otherwise go unnoticed (McDougall and Harrison 1999). However, thermal disturbances are often too subtle to be revealed by conventional thermochronometric methods; i.e., interpolation of discrete temperature-time (T-t) points from bulk analyses using "nominal" closure temperatures. Rather, the highest resolution thermal histories require harnessing knowledge of the concentration distribution of the daughter product in the mineral of interest.

In previous chapters, the case has been explored in which a mineral cooling within the crust transitions from being open to loss of daughter product to closed system behavior. Assuming a monotonic thermal history of simple form (Dodson 1973), it is then possible to use the balance between radiogenic accumulation and loss to assign a bulk closure temperature, T_c, which is given by:

$$\frac{E}{RT_c} = \ln\left(\frac{ART_c^2 D_0 / r^2}{E\, dT/dt}\right) \tag{1}$$

where E is the activation energy, D_0 is the frequency factor, R is the gas constant, T is absolute temperature, A is a geometry factor (sphere = 55, cylinder = 27, and plane sheet = 8.7), r is the effective diffusion length scale (radius or half-width), and dT/dt is cooling rate. When the T_c and age of a number of coexisting mineral thermochronometers are correlated, an estimate of

1529-6466/05/0058-0015$05.00 DOI: 10.2138/rmg.2005.58.15

the temperature history can be interpolated. This method, termed the bulk closure approach, has been used for nearly 30 years (Purdy and Jäger 1976; Mattinson 1978; Berger et al. 1979; Harrison et al. 1979).

However, use of Equation (1) carries several stringent requirements. These include: knowledge of the activation energy, frequency factor, and compositional dependence of daughter product diffusion in the mineral of interest, *a priori* knowledge of the effective diffusion dimension, and assurance that volume diffusion was the rate limiting natural transport mechanism. Although adequate diffusion data are available for most commonly used minerals, it is only in rare cases that all these requirements are met. This represents a significant limitation of thermochronometry.

An example: the bulk closure temperature of biotite

Consider the example of Ar closure in "biotite," which is typically taken to be 300 ± 50 °C (Purdy and Jäger 1976; Mattinson 1978; Harrison et al. 1979; Hodges 1991). This represents a nearly ideal case as the biotite-phlogopite solid solution is experimentally well-characterized in terms of Ar diffusion (Giletti 1974; Giletti and Tullis 1977; Harrison et al. 1985; Grove and Harrison 1996). Nevertheless, rigorous determination of a closure temperature requires that we know: 1) the effective diffusion length scale, 2) important compositional parameters such as Fe/Mg, Al(VI) occupancy, and halogen content, and 3) the approximate cooling rate and pressure during Ar closure (see Harrison and Zeitler 2005). For example, a biotite with an Fe/Mg of 0.6, an Al(VI) occupancy of 1.0, a total halogen content of 0.2%, and r = 150 μm, is characterized by a T_c of 350 °C (assuming P = 200 MPa and dT/dt = 100 °C/Ma). However, increasing the total halogen content to 0.8% and decreasing Fe/Mg to 0.4 increases T_c to about 450 °C (Grove 1993). Together with published estimates of r that vary by a factor of 6, extreme variations in compositional parameters yield a range in "biotite" T_c of over 300 °C.

The magnitude of this range would be of less concern if it could be reported that the compositional and diffusion size parameters were routinely measured. However, to our knowledge, only a handful of the many hundreds of studies involving Ar closure in biotite have actually attempted to do so (e.g., Harrison et al. 1985; Copeland et al. 1987).

Bulk mineral thermochronometry

Implicit in the example just discussed is that the underlying assumptions of the closure temperature model have been met [i.e., $\tau = (RT_c^2/E)\cdot dT/dt >> r^2/D(0)$, linear cooling in $1/T$, no inherited daughter, no subsequent open system behavior, no recrystallization; see Harrison and Zeitler (2005) for definitions]. If so, then the bulk T_c calculated for the biotite can be combined with the measured K-Ar (or bulk $^{40}Ar/^{39}Ar$) age to yield a temperature-time (T-t) datum on a thermal history plot. Figure 1(a) shows a schematic T-t plot with three mineral thermochronometers of contrasting closure temperature. As an example, the input thermal history (solid curve) shown is meant to represent initially slow cooling in the mid-crust, followed by rapid cooling in the footwall of a rapidly slipping normal fault, and then slow cooling due to erosional denudation.

In such a case, it is possible to use the transition from slow to rapid cooling to determine both the initiation age of faulting and the slip rate by fitting the form of the thermal history to an appropriate thermo-kinematic model (Harrison et al. 1995, 1996). Note, however, that the resolution of the bulk closure data set is insufficient to reveal the complex nature of the temperature evolution (Fig. 1a). Indeed, there is no justification to infer anything other than the linear history, shown by the dashed line, which misses the essential character of the T-t path.

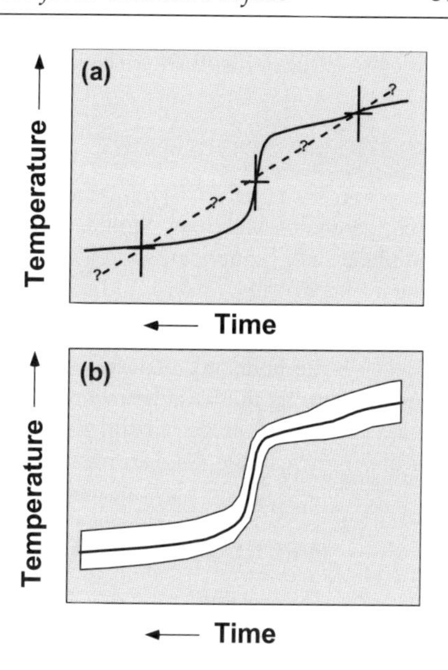

Figure 1. Schematic thermal history (solid curve) showing initial slow cooling, followed by a rapid cooling segment, and then a second slow cooling phase. Crosses in (a) represent three thermochronometers with varying T_c. Note that uncertainties in T_c permit a wide range of possible thermal history interpretations, including the linear path shown by the dashed line. (b) shows a continuous thermal history obtained by knowing the concentration distribution of the daughter product.

How do we obtain the highest accuracy and resolution thermal histories?

We have, thus far, emphasized that bulk closure temperatures are only rarely rigorously constrained and that *T-t* histories generated from such data are characterized by relatively low resolution. These limitations can be appreciated by inspection of Equation (1). Because T_c varies as the natural logarithm of a term that typically has a value in the range 10^{16}–10^{18}, it is highly insensitive to variations in the constituent parameters, including dT/dt. It stands to reason that inverting the problem to solve for time-dependent variations in dT/dt would yield far greater sensitivity, provided T_c could be precisely determined. In fact, Dodson (1973) did just that. The full accumulation-diffusion-cooling equation for a time-dependent diffusion coefficient was solved to yield expressions for the distribution of the daughter product within solids of various geometries. When volume averaged, they reduce to the form given in Equation (1).

In cases where thermal variations are too subtle to be revealed by the interpolation method, they may still be documented by harnessing the full concentration distribution equation. This general approach is shown schematically in Figure 1(b) where knowledge of the daughter product distribution has permitted determination of a continuous thermal history of complex form (the bounding envelope represents an objective estimate of the range of uncertainty associated with the *T-t* path). In this case, the true nature of the thermal history is revealed permitting detailed insights into cooling mechanisms to be drawn.

IN SITU CLOSURE PROFILES

The closure profile equation

To restate the preceding paragraph, although a mineral can be characterized by a bulk T_c, each radial position within the crystal has a unique closure temperature, the volume average of which is the parameter described by Equation (1). Dodson (1973) presented, but did not evaluate, expressions for the limiting concentration distributions within minerals following slow cooling. Dodson (1986) later obtained solutions for the distribution of daughter/parent as a function of position within a cooling solid for sphere, cylinder and plane sheet geometries.

The closure profile equation is given by:

$$\frac{E}{RT_c} = \ln\left(\frac{\gamma\tau D_0}{r^2}\right) + 4S_2(x) \tag{2}$$

where γ = 1.78, $\tau = (RT_c^2/E)\cdot dT/dt$, and $4S_2(x)$ describes the concentration distribution for different geometric solutions. Summations of $4S_2(x)$ (Dodson 1986) as a function of position within plane sheet, spherical, and cylindrical solids are given in Table 1. Note that when the volume averaged value of $4S_2(x)$ (Table 1) is inserted into Equation (2), it reverts to the form of Equation (1).

Because the boundary concentration is maintained at zero (i.e., it is always open), the T_c tends to zero as the boundaries are approached (i.e., $x \rightarrow \pm 1$). A broader implication of Equation (2) is that a single mineral sample can potentially provide a continuous cooling curve rather than only a single *T–t* datum.

INFERING CLOSURE PROFILES FROM $^{40}Ar/^{39}Ar$ DATA

The $^{40}Ar/^{39}Ar$ method is among the most versatile thermochronologic tools (McDougall and Harrison 1999; see also Harrison and Zeitler 2005). It offers both the potential of constraining continuous thermal histories by direct determination of intracrystalline age profiles and indirect analysis based upon step-heating of bulk materials. The indirect nature of the step-heating approach stems from the fact that the age distribution is deduced from measurement of the relative flux of ^{40}Ar and ^{39}Ar from all portions of the bulk material as a function of temperature (see Harrison and Zeitler 2005).

Direct measurement of $^{40}Ar/^{39}Ar$ age profiles in single crystals is a promising approach that is primarily limited by analytical considerations. Consider the case of a biotite (Fe/Mg = 0.6, E = 47 kcal/mol, D_0/r^2 = 350/s; Harrison et al. 1985) that cooled at 5 °C/Ma through the Ar closure interval, which equates to a bulk closure temperature of 300 °C (infinite cylinder geometry). The radial distribution of closure temperature using Equation (2) varies from 325 °C at $x = 0$ to 240 °C at $x = 0.99$, in accord with our expectation of a high T_c in the center and a lower T_c near the diffusion boundary. Thus if the spatially varying age distribution is known, we can in theory determine the variation of τ, which is directly related to dT/dt, as a function of time.

A significant amount of effort has gone into laser probe analysis of millimeter-scale single crystals of mica and other phases (Lee et al. 1990; Scaillet et al. 1990; Onstott et al. 1991; Hodges et al. 1994; Arnaud and Kelley 1995; Reddy et al. 1996; Pickles et al. 1997). Results from these studies have demonstrated that laser ablation is capable of revealing significant age variation in mm-scale crystals that could, under favorable circumstances, lead to thermal history constraints (Hodges et al. 1994; Hames and Andresen 1996; Hames and Cheney 1997; Kelley and Wartho 2000; Wartho and Kelley 2003). Alternatively, *in situ* laser analysis has proven exceedingly useful for revealing problematic excess radiogenic ^{40}Ar ($^{40}Ar_E$) distributions. Some of the best examples occur in phengitic micas from relatively dry, ultra high-pressure rocks (Kelley et al. 1994; Arnaud and Kelley 1995; Sherlock et al. 1999).

In spite of its potential, the major limitation for *in situ* laser age profiling of grains remains the fact that some (not all) intracrystalline features that control Ar transport in $^{40}Ar/^{39}Ar$ thermochronometers appear to be very small (i.e., < 1 μm; e.g., Foland 1974) relative to the dimensions of abalation pits that liberate sufficient argon to accurately resolve age variations (~10 μm for >1 Ga materials and much larger for more youthful minerals). Thus even though depth profiling approaches have been developed to enhance spatial resolution in close proximity to grain boundaries (e.g., Arnaud and Kelley 1997; Wartho et al. 1999), key information remains out of reach for spot profiling approaches that lack the spatial resolution to detect <10 μm scale features. Because of this limitation, the most significant developments have come in the realm of interpreting age spectra obtained from bulk step-heating of anhydrous materials that remain structurally stable over a broad range of conditions during *in vacuo* heating (e.g., Lovera et al. 1993). Such experiments are sensitive to all length scales larger than those characteristic of ^{39}Ar recoil (~0.1 μm) and are capable of simultaneously revealing both the age and kinetic properties of Ar in the thermochronometer.

Table 1. Closure function $4S_2(x)$

x	*Plane sheet*	*Cylinder*	*Sphere*
0.00	0.41194	1.02439	1.38629
0.05	0.41653	1.03831	1.38980
0.10	0.43036	1.03990	1.40039
0.15	0.45367	1.05944	1.41826
0.20	0.48685	1.08728	1.44371
0.25	0.53047	1.12393	1.47723
0.30	0.58535	1.17011	1.51946
0.35	0.65254	1.22676	1.57131
0.40	0.73347	1.29515	1.63392
0.45	0.82999	1.37695	1.70887
0.50	0.94458	1.47439	1.79824
0.55	1.08057	1.59051	1.90484
0.60	1.24254	1.72949	2.03262
0.65	1.43699	1.89732	2.18721
0.70	1.67352	2.10291	2.37704
0.75	1.96710	2.36026	2.61543
0.80	2.34291	2.69317	2.92511
0.85	2.84826	3.14674	3.34950
0.90	3.58951	3.82350	3.98796
0.95	4.90636	5.05471	5.16455
0.96	5.33878	5.46577	5.56120
0.97	5.90027	6.00366	6.08275
0.98	6.69734	6.77401	6.83402
0.99	8.06977	8.11483	8.16128
0.995	9.44913	9.47507	9.49661
Volume average	1.58611	2.71862	3.43012

^{40}Ar/^{39}Ar step-heating of K-feldspar

By far the best opportunity to obtain detailed thermal histories from the K-Ar system is by application of the ^{40}Ar/^{39}Ar step-heating method to K-feldspars. K-feldspar is ideal in this role as it is widespread, ^{40}K rich, and generally stable during laboratory heating up to temperatures near its melting point (~1100 °C). Two distinct sources of information are available from a K-feldspar ^{40}Ar/^{39}Ar step-heating experiment: the age spectrum (Fig. 2a) and the Arrhenius plot (Fig. 2b). The age spectrum is calculated from the flux of radiogenic argon (^{40}Ar*) relative to the reactor produced argon (^{39}Ar$_K$) that is released during discrete laboratory heating steps (see Harrison and Zeitler 2005). The Arrhenius plot is derived by plotting diffusion coefficients (calculated from inversion of the ^{39}Ar release function assuming a single diffusion length scale) against the inverse absolute temperature of laboratory heating. Because the shape of the Arrhenius plot varies with laboratory heating schedule for samples containing a distribution of diffusion domain sizes, an alternate form of data display termed the log (r/r_0) plot is often used (Richter et al. 1991; Fig. 2a). Log (r/r_0) spectra are constructed by plotting the deviation of the measured diffusivities (D/r^2) from a reference diffusion law $(D/r_0^2 = D_0/r_0^2{\cdot}\exp[E/RT])$ at a given temperature T as a function of cumulative % ^{39}Ar released (Fig. 2b). Because the intrinsic diffusivity D from the reference diffusion law is arbitrarily assigned to the sample, the log (r/r_0) value is given simply by the expression $0.5{\cdot}(\log D/r_0^2 - \log D/r^2)$.

Without exception, ^{40}Ar/^{39}Ar age spectra of basement K-feldspars yield age spectra and Arrhenius plots that are inconsistent with the presence of a single diffusion dimension. For example, rather than yielding a single linear array, K-feldspar Arrhenius plots show complex departures from an initial straight line segment (e.g., Fig. 2b). These behaviors undoubtedly

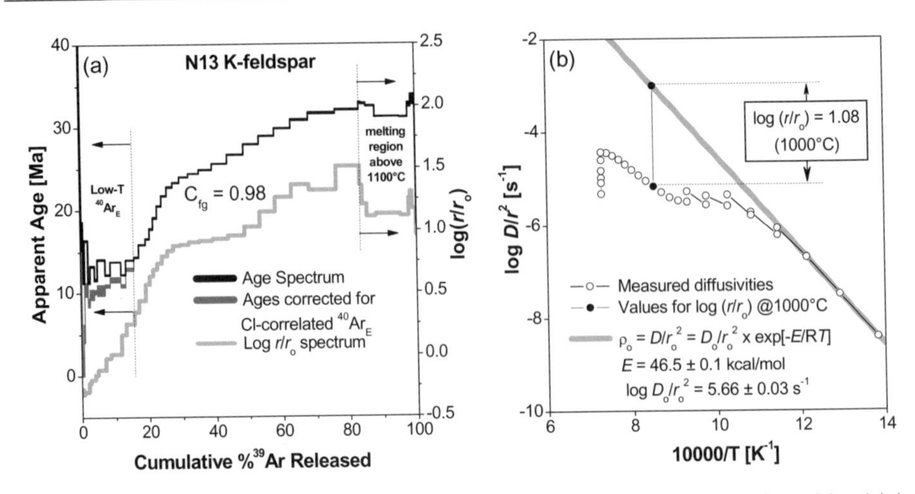

Figure 2. Typical K-feldspar age and Arrhenius properties. (a) Age spectrum (left axis) and log (r/r_0) spectrum (right axis) for N13 K-feldspar (Harrison et al. 2000). Note correlated behavior, particular over the interval of gas release between the disappearance of low-temperature Cl-correlated $^{40}Ar_E$ and the onset of melting above 1100 °C. C_{fg} refers to correlation coefficient calculated by Lovera et al. (2002). (b) Arrhenius plot showing measured diffusivities, reference Arrhenius law (ρ_0) defined by initial gas release (see text), and sample calculation of log (r/r_0) value at 1000 °C (see text).

reflect discrete Ar retentivities within K-feldspar, whether due to varying size, energetics, nested diffusion domains, or other phenomena.

Fundamental assumptions for recovering thermal history information

Two fundamental assumptions *must* be satisfied to permit estimation of crustal thermal histories from K-feldspar $^{40}Ar/^{39}Ar$ step-heating data. These are: (1) that both $^{40}Ar^*$ and ^{39}Ar loss from K-feldspar are governed by volume diffusion; and (2) laboratory Ar release adequately mimics the natural diffusion boundaries and mechanisms. Failure of either of these assumptions precludes recovery of useful thermal history data. By comparison, other commonly made assumptions are second order issues (i.e., uniform $^{39}Ar_K$ distribution, prescribed diffusion geometry (slab, cylinder, sphere), zero $^{40}Ar^*$ boundary conditions, etc.) that can be dealt with by appropriately modifying the model. The reader may be aware that there is continued discussion of the relative importance of these and other factors in controlling Ar diffusion in K-feldspars (e.g., Parsons et al. 1999; Lovera et al. 2002). However, it is a truism that the *only* fundamental requirement for recovering thermal histories from $^{40}Ar/^{39}Ar$ step-heating results is knowledge that Ar loss proceeds by volume diffusion and that laboratory Ar release is controlled by the natural diffusion mechanisms and boundaries.

Evaluation of fundamental assumptions. Examination of the similarity of the age and log (r/r_0) spectra yielded by a K-feldspar provides a first order assessment of the degree to which fundamental assumptions have been upheld. The age spectrum reflects the natural Ar retentivity of the sample over 10's of millions of years of cooling, whereas the Arrhenius data from which the log (r/r_0) plot is developed is generated on timescales of minutes to days. When $^{40}Ar^*$ concentration along diffusion boundaries is negligible, simple volume diffusion theory predicts that age spectra should increase monotonically with progressive ^{39}Ar release because natural $^{40}Ar^*$ concentrations should be highest in the most retentive sites. Exhibition of this predicted behavior represents the initial criterion for selecting samples for thermal history analysis.

The useful (i.e., most readily interpreted) fraction of gas release in a K-feldspar $^{40}Ar/^{39}Ar$ step-heating experiment is almost always limited at low-temperature by weakly bound (fluid inclusion-hosted?) $^{40}Ar_E$ and at high-temperature by the onset of melting (Fig. 2a). Within this interval, we often see correlated behavior between K-feldspar age and log (r/r_0) spectra. Correlated behavior is expected only if Ar diffusion occurs by the same mechanisms and has access to the same diffusion boundaries in nature as it does in the laboratory heating. In the case of N13 K-feldspar in Figure 2, the age and log (r/r_0) spectra show correlated inflections at ~15%, 25%, and 50% ^{39}Ar released (Fig. 2a). We regard this as confirming evidence that, for this sample, diffusion properties obtained via step-heating experiments can be extrapolated to conditions attending natural Ar loss within the crust.

Lovera et al. (2002) developed routines to quantitatively evaluate the extent of correlation between age and log (r/r_0) spectra. About two-thirds of K-feldspars they analyzed gave correlation coefficients >0.9 and thus are well-suited for thermal history analysis. N13 K-feldspar, the example in Figure 2, yielded a correlation coefficient of 0.98. While some K-feldspars yield low (<0.9) correlation coefficients, this is generally due to contamination by $^{40}Ar_E$ or presence of intermediate age maxima (see below). In contrast, hydrous phases (hornblende, biotite, muscovite) characteristically yield poorly—or even negatively—correlated age and log (r/r_0) spectra (Lovera et al. 2002).

Recognition of problematic behavior in K-feldspar $^{40}Ar/^{39}Ar$ age spectra

High temperature $^{40}Ar_E$ contamination. While most samples are adversely affected by at least some $^{40}Ar_E$ at low temperatures (Fig. 2a), its existence does not preclude thermal history modeling since the effect can often be corrected for if step-heating experiments are designed appropriately (e.g., Harrison et al. 1994). When this is done, systematically varying age spectra similar to those shown in Figure 3a are often obtained when samples represent varying positions relative to a major structure such as a fault. Geologically significant age variations can be obscured, however, when the high-temperature $^{40}Ar_E$ contamination is pervasive (Fig. 3b). In general, such samples will be highly unfavorable candidates for thermochronology since reliable correction schemes are unknown.

The hallmark of $^{40}Ar_E$ contamination is highly erratic ^{40}Ar release (Fig. 3b). This generally has a large negative impact upon the degree of correlation between age and log (r/r_0) spectra and characteristic "U-shaped" spectra are typically obtained (e.g., Zeitler and Fitzgerald 1986). Independent geologic constraints often indicate that the least affected, and hence geologically most meaningful portion of the age spectrum occurs after low-temperature $^{40}Ar_E$ has been exhausted (by about ~600–800 °C in typical step-heating sequences) and release of high-temperature $^{40}Ar_E$ is still minimal. This comparatively unaffected portion of gas release typically occurs at about 15–35 cumulative % $^{39}Ar_K$ release in relatively high resolution step-heating experiments. Release of high-temperature $^{40}Ar_E$ generally begins in earnest above ~900–1000 °C or at temperatures that approach those required for melting K-feldspar *in vacuo*. The onset of $^{40}Ar_E$ is often easily recognized by rapidly increasing and highly erratic apparent ages that can exceed maximum ages permitted by independent geologic constraints, such as crystallization ages. For such samples, it is only possible to interpret the lower temperature gas release provided that adequate corrections for low-temperature $^{40}Ar_E$ can be performed.

Intermediate age maxima (IAM). Intermediate age maxima are readily distinguished from the effects of $^{40}Ar_E$ in that they are typically expressed by smoothly varying "humps" over the portion of age spectrum where $^{40}Ar_E$ contamination is minimal (Fig. 3c). Such features are characteristic of K-feldspars that have experienced low-temperature alteration by authigenic adularia (Girard 1991; Foland 1994; Warnock 1999). While the underlying causes of IAM are not well understood, they can potentially be explained by recrystallization,

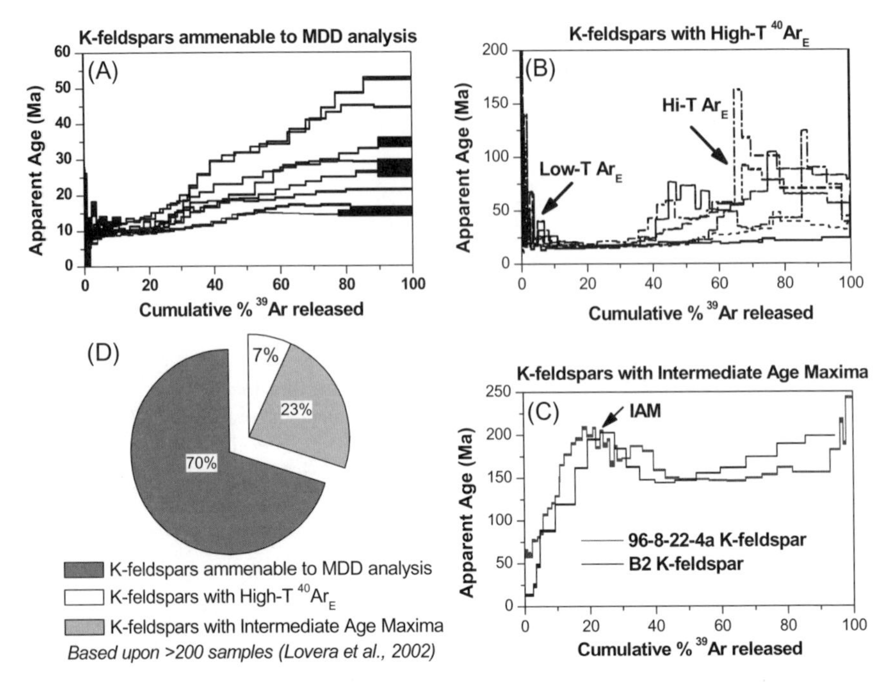

Figure 3. Variability of K-feldspar age spectra in nature. (a) Well-behaved monotonically increasing K-feldspar age spectra after correction for Cl-correlated excess ^{40}Ar ($^{40}Ar_E$) from the Renbu Zedong thrust (Quidelleur et al. 1997). (b) K-feldspar age spectra heavily contaminated with both low- and high-temperature $^{40}Ar_E$. Note U-shaped age spectra with least affected gas release at 15-35 cumulative % $^{39}Ar_K$ release. (c) K-feldspar age spectra exhibiting intermediate age maxima (IAM). These are smoothly continuous and are characteristically developed over the 15-35 cumulative % $^{39}Ar_K$ interval of gas release where $^{40}Ar_E$ contamination tends to be minimal. (d) Survey of more than 200 K-feldspars from Lovera et al. (2002) indicating proportions of "well-behaved" samples, samples exhibiting high-temperature $^{40}Ar_E$ contamination, and samples with IAM.

exsolution or other related phenomenon that has occurred at sufficiently low-temperatures to remobilize previously "locked in" $^{40}Ar^*$ that had resided within the most retentive crystalline domains that are expected to yield the oldest ages. We will revisit the implications of such misbehavior once we have introduced the multi-diffusion domain model and the techniques employed to recover thermal history information from K-feldspar.

To summarize, $^{40}Ar_E$ contamination generally appears to reflect incorporation of $^{40}Ar^*$ from external sources that is either loosely bound (i.e., within fluid inclusions) or tightly held (potentially in higher order defects or similar features in crystalline K-feldspar). Alternatively, existence of IAM most likely reflects problematic redistribution of intrinsic $^{40}Ar^*$ (i.e., derived from *in situ* radiogenic decay of ^{40}K) from high retentivity to low retentivity domains in K-feldspar. Finally, although Figure 3d implies that these types of misbehavior are manifested in the minority of K-feldspars that we have examined, samples that have experienced similar histories will tend to exhibit similar age spectra. *In other words, in settings where problematic behavior is manifested, it tends to be prevalent.*

The multi-diffusion domain model

The multi-diffusion domain (MDD) model (Lovera et al. 1989, 1991) assumes that the characteristic form of the Arrhenius plot and age spectrum in K-feldspars is due to the presence of a discrete distribution of diffusion domain sizes. Thus the form of both plots is a function of the diffusion (E and D_0) and domain distribution (domain size, ρ, and volume fraction, ϕ) parameters and the form of the thermal history. The Arrhenius plot is then a convolution of the parameters that characterize the individual diffusion domains. Since the diffusion parameters are obtained directly from the Arrhenius plot and we have two independent measures of ρ and ϕ (i.e., the log (r/r_0) plot and age spectrum), in theory we have sufficient information to invert the results to obtain a unique, monotonic cooling history.

As noted earlier, there is a diversity of opinion regarding the cause of discrete Ar retentivities in K-feldspar, with some workers disputing the role of diffusion domain size as the primary cause. In fact, a faithful depiction of the intrinsic structure of K-feldspar is not required to calculate a meaningful thermal history. Any diffusion-based model, for example one in which nested diffusion domains interact (e.g., the heterogeneous diffusion model; Lovera et al. 2002) will accurately reproduce the same thermal history as predicted by the MDD model provided the conservation of diffusion mechanism and boundary assumptions are met.

Thermal history calculations based upon the MDD model begin by estimating the diffusion parameters E and $\log(D_0/r_0^2)$ for the K-feldspar in question. This may be accomplished by least squares fitting to the linear initial low-temperature Arrhenius data (Lovera et al. 1997) (Fig. 4a). The slope defined by this low-temperature data is expected to be proportional to $-E/R$ provided that none of the domains contributing ^{39}Ar have been outgassed by more than ~60% (Lovera et al. 1991). It is also possible to employ average values (~46 kcal/mol; see Lovera et al. 1997) when the initial gas release fails to adequately define E and $\log(D_0/r_0^2)$. The next step is to find a set of distribution parameters, volume fraction (ϕ) and the relative size (ρ) for each domain, that best fit the measured Arrhenius data. The domain distribution for N13 K-feldspar is indicated schematically on Figure 4a. Once the MDD diffusion model has been calibrated for the K-feldspar in question, it is possible to perform forward modeling by iteratively inputting trial thermal histories (Fig. 4b) to calculate model age spectra (Fig. 4c). Note that initial temperatures must be high enough to exceed the closure temperatures of the largest domains (i.e., >> 400 °C).

To be sure, one can do very well using the equations of Lovera et al. (1989) to conduct forward modeling and it is relatively straightforward to find a solution in this way. However, this can become tedious with larger numbers of samples, and leaves open the nagging concern that, even when applying available geologic constraints, the solution arrived at is only one of a number of possibilities. Intuitively, given the good precision of most $^{40}Ar/^{39}Ar$ measurements and some experience with forward modeling, one expects that for most cases of simple monotonic cooling, K-feldspar age spectra provide tight constraints on thermal history. Experiments with the inverse models described below bear this out. However, given that it appears that most multidomain samples can be described using a single activation energy, the time-temperature response of different domains will be similar. Thus in the case of more complicated thermal histories involving thermal stagnation or reheating, the range of possible solutions will be substantially larger.

Inversion of ^{40}Ar/^{39}Ar results to thermal history data

In a manner analogous to the way in which information about Earth structure is obtained—using observed seismic data to invert for velocity structure and thus obtain tomographic images of the Earth's interior—we wish to find the set of all temperature histories that can explain our observed data, which take the form of the flux ratio $^{40}Ar/^{39}Ar$ and the absolute quantity of ^{39}Ar degassed from K-feldspar diffusion boundaries.

Calculating a continuous thermal history directly from a measured closure profile entails relatively straightforward and direct inversion using the closure-profile equation (Eqn. 2).

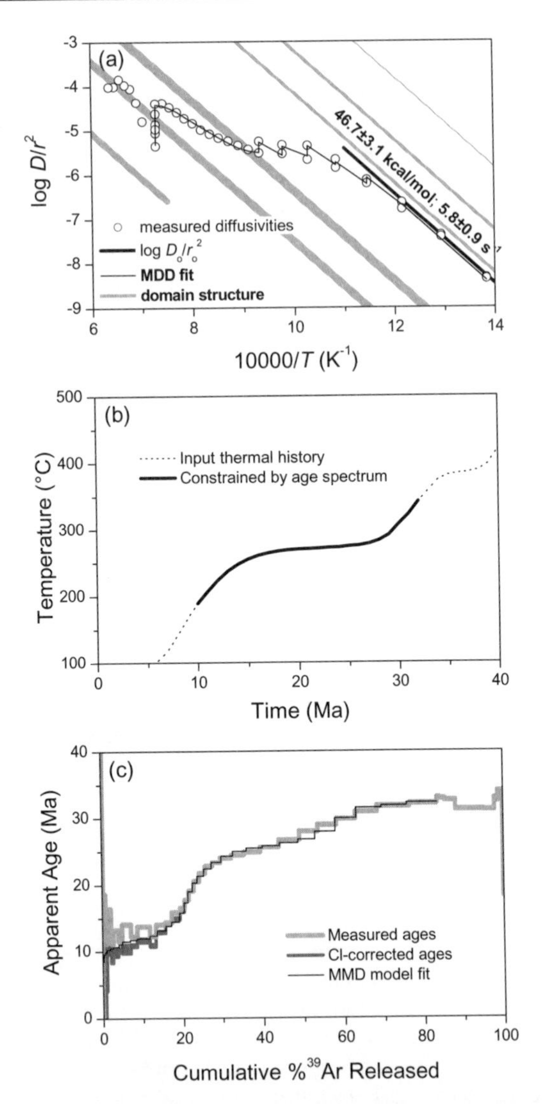

Figure 4. Application of the multi diffusion domain (MDD) model to recover crustal thermal histories from K-feldspar. After calibrating the MDD model with the diffusion properties of the samples, the thermal history is forward modeled by fitting the age spectrum. (a) MDD fit to measured Arrhenius data. Domain structure is indicated schematically by the lines of variable thickness, where thickness is proportional to concentration. (b) Input thermal history corresponding to calculated age spectrum (below). Solid line indicates portion of temperature-time path relevant to sample. (c) Calculated MDD age spectrum.

Observation of the distribution of ^{40}Ar in K-feldspars is indirect, owing to the diffusion-domain structure prevalent in feldspars and the method used in assessing this distribution via measurements of ^{40}Ar and ^{39}Ar diffusive fluxes during step-heating. In general, inversion of age spectra is no more or less complicated than inversion of other data sets. One's goal is to minimize an objective function which amounts to the difference between observed and predicted values. This in practice involves code which can be used to calculate predicted values from parameters (i.e., a forward model) and then some recipe for manipulating parameters so that they are progressively shaped to minimize the value of the objective function. Different schemes have their advantages and disadvantages in terms of overall efficiency, speed towards a solution, and predilection for becoming trapped in a local, rather than global, minima. For instance, a purely Monte Carlo approach is quite simple to implement and generally good at exploring parameter space without becoming trapped, but is impossibly inefficient for use with thermochronological data given the range of thermal histories to explore and the temperature sensitivity of diffusion. Basic simplex methods (see Press et al. 1998) are fairly efficient but can quite readily crawl into and be trapped in a "false minimum." Below, we describe two approaches that seem to do quite well in balancing efficiency with adequate investigation of parameter space.

Controlled random search method. For the purpose of inverting apatite fission-track data for thermal history, Willett (1997) described an implementation of the controlled random search (CRS) algorithm of Price (1977). This algorithm retains the advantages of a Monte Carlo approach in searching parameter space for true minima, while converging far more rapidly due to the "learning" component inherent in the CRS method. Zeitler (1993, 2004) adapted this approach for use with K-feldspar age spectra under the assumptions of the MDD model described above, most recently adding the ability to include other thermochronometers in the inversion process.

As applied to thermochronological data, the CRS algorithm manipulates a starting pool of ~150 randomly generated thermal histories, generating a new history from a subset of some ~10 histories randomly chosen from the master pool, testing it and accepting it into the pool if it provides a better fit between the observed and calculated age spectra (and other ages). In detail, the new history is made as follows: the histories in the subset are averaged, and then a new history is made by reflecting the additional selected history through the averaged values, subject to an amplification factor that might range between 1.1 and 1.5. Thus, as the master pool of histories learns about better solutions and is improved, new histories inherit something of this learning, although incompletely, since they are made from only a small subset. In addition, the reflection and amplification process serves to introduce diversity into the pool and together these two factors serve to explore parameter space while helping the model avoid traps in the form of false minima.

The code which implements the CRS algorithm, Arvert, uses at its core the equations of Lovera et al. (1989) to calculate K-feldspar age spectra from thermal histories, and uses a finite-difference scheme to model results from other systems like U-Th/He. The fit between observed and calculated data can be assessed using either a simple mean percent deviation determined from all relevant steps and other mineral ages, or using a chi-square parameter that takes into account uncertainties on the observed data. The user has the option of providing both explicit constraints in the form of temperature boundaries at various times in the model run, as well as implicit rate constraints for maximum heating and cooling rates.

Figure 5 shows a typical Arvert convergence sequence for synthetic data, determined for linear cooling of a multidomain sample at 10 °C/Ma. For these synthetic data, the convergence criterion was arbitrarily chosen at a mean deviation of one percent. It is clear from the convergence sequence that the model very quickly learns about impossible portions of parameter space (old and cold, young and hot), and then begins to converge on the correct cooling rate over that portion of temperature-time space where the age spectrum has constraining power. Figure 6 shows that the model has little trouble recovering more complex, non-linear cooling histories, and that addition of other mineral data into the inversion, such as U-Th/He apatite data, can significantly extend and improve the quality of the result.

Beyond the intrinsic value and convenience of having an automated, objective means of assessing K-feldspar age spectra and the thermal histories that might be compatible with them, use of such models brings into focus several technical details that can be overlooked.

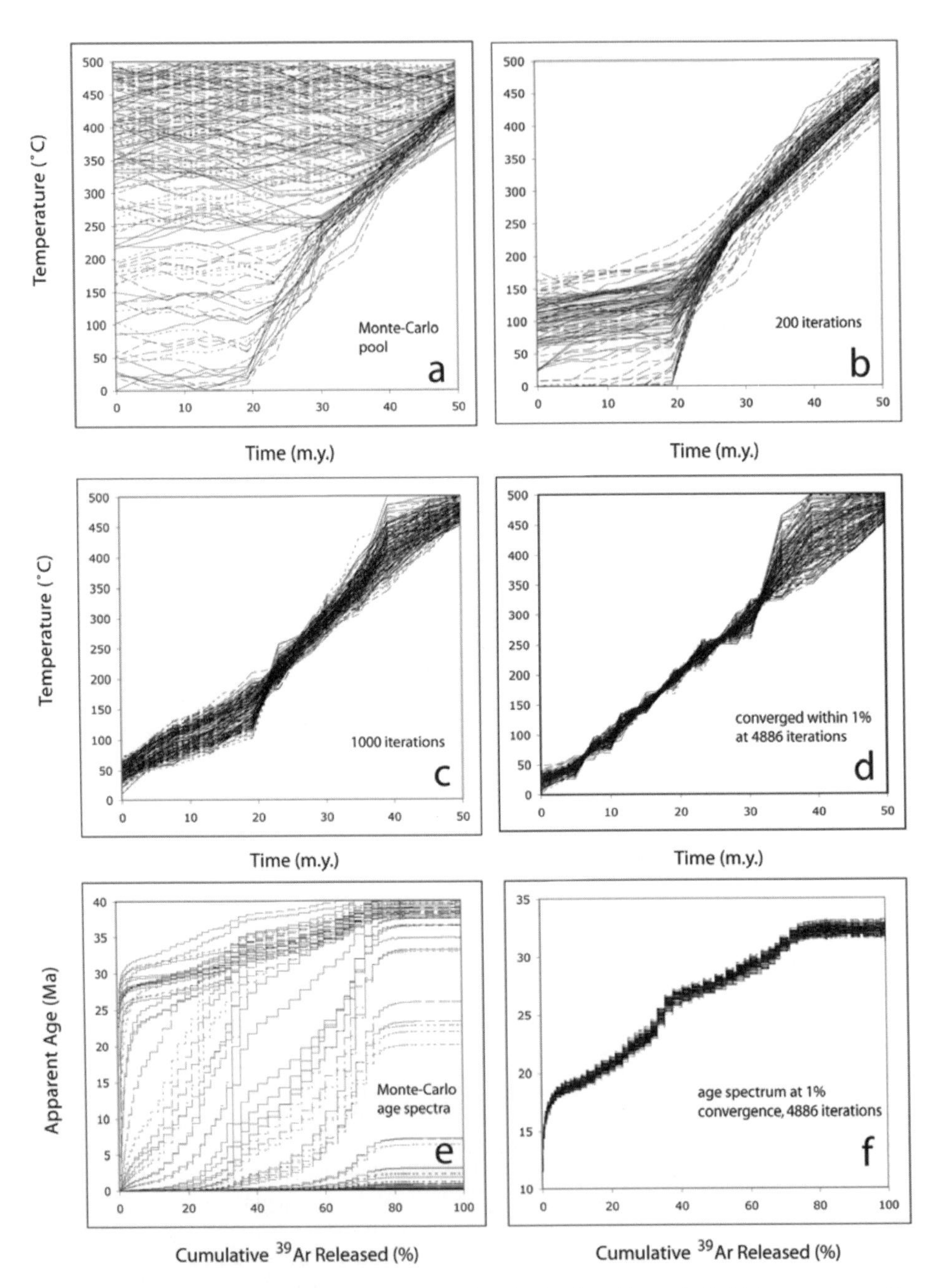

Figure 5. Top: (a)-(d), CRS convergence sequence for simulated multi-domain K-feldspar cooling at 10 °C/Ma, Monte Carlo, 200, 1000, and 4886 iterations. Middle, (e) and (f), predicted age spectra at start and end of model. Bottom: summary of final results after convergence. Box shows time span where age spectrum has constraining power.

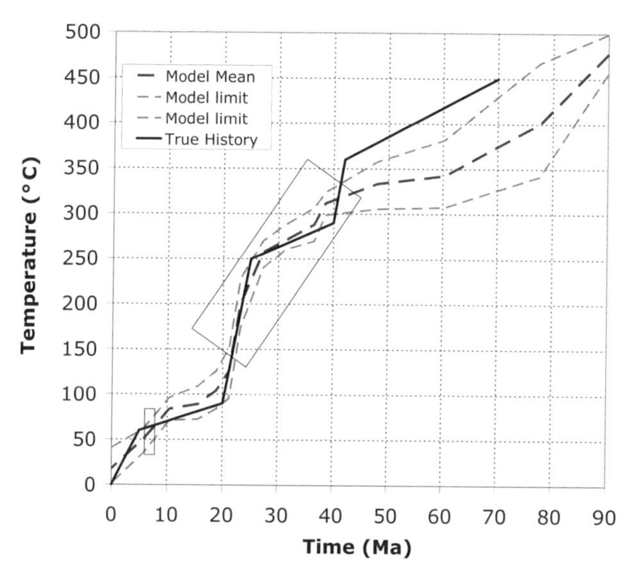

Figure 6. Summary of CRS model for multi-domain K-feldspar having a more complex non-linear cooling history. Boxes shows time span where age spectrum has constraining power, and time where U-Th/He apatite age provides a constraint.

One is the great importance of adequately and accurately determining a sample's domain structure. "Extra" domains do no harm but working with an insufficient number that does not truly represent the domain distribution leads to poor fits and poor convergence because one is violating a fundamental underlying assumption. A second item to keep in mind is that the 150°C or more of thermal history that can be recorded by a K-feldspar is not in most cases uniformly distributed across an age spectrum: the smaller domains usually have a lesser volume fraction but record a significant part of the lower-temperature thermal history. It is this part of the history that is most often obscured by (often Cl-correlated) $^{40}Ar_E$, and most often ignored by beginners as they focus on an overall age-spectrum fit, and in fact can be the last portion of an age spectrum that the inversion code improves. Care is needed with dealing with initial parts of an age spectrum, and this is an added motivation for including other low-temperature thermochronometers in the inversion to provide additional constraints.

The variational approach. An alternative method utilized by Lovera et al. (1997) to fit both the Arrhenius and age data is the Levenberg-Marquardt variational approach (Press et al. 1988). The determination of the domain distribution parameters, ρ and ϕ, involves the fitting of the cumulative ^{39}Ar released from the step-heating experiment. The Levenberg-Marquardt scheme has proven proficient in searching ρ and ϕ parameter space for the set of values that produced the best fit to the measured ^{39}Ar data (Lovera et al. 1997). While the domain distribution (ρ,ϕ) that produces the "best-fit" solution is non-unique, all potential solutions tend to produce broadly equivalent results in terms of diffusion properties. Certainly what differences arise are minor when compared to the impact of varying activation energy (Lovera et al. 1997). Because of the importance of E on thermal history calculations, Lovera et al. (1997) found it desirable to perform calculations in a manner in which this parameter is explicitly varied. To accomplish this, a number of equivalent domain distributions are determined by randomly selecting values of E and log D_0/r_0^2 from a normal distribution centered upon the measured values [see Lovera et al. (1997) for details; MDD programs and instructions are available at *http://argon.ess.ucla.edu*].

A similar Levenberg-Marquardt method can also be employed to model the $^{40}Ar/^{39}Ar$ age spectrum. Since the diffusion and domain distributions are constrained by the ^{39}Ar release data, the only degree of freedom in modeling the age spectrum is given by the thermal history. Thus, a parametric expression for the thermal history must be specified. One approach is to employ a Chebyshev polynomial function to represent the thermal history. The phase of this expansion can be conveniently varied to iterate from a random initial history towards solutions that fit the observed age spectrum to a specified degree. For each set of MDD parameters, a number of best-fit thermal history solutions to the measured age spectrum are then sought beginning with initial, random thermal histories. In practice, calculating 10 equivalent domain distributions and seeking five best-fit solutions to the age spectrum for each (50 total best-fit solutions) appears to adequately explore the impact of varying E.

Results obtained by applying this approach to 96MX-001 K-feldspar are shown in Figure 7. Correction of the low-temperature steps for Cl-correlated $^{40}Ar_E$ allowed us to fit the age spectrum between 10 and 58 Ma (Fig. 7a). The corresponding cooling histories reflect propagation of a ±2 kcal uncertainty in E. This translated into a ±17 °C temperature uncertainty (Fig. 7b). The 90% confidence limits of the mean and median of the resulting distribution of cooling histories are shown in Figure 7c. Note that the median confidence bands are limited to the portion of the thermal history that was constrained by age data from the K-feldspar. The Late Eocene-Early Oligocene rapid-cooling indicated by this sample corresponds with localized basement denudation that diverted extraregional drainages that had developed across the erosionally beveled terrane in the Eocene (Axen et al. 2000). Low-temperature constraints from K-feldspar constrain the timing and magnitude of Late Miocene and Pliocene extensional faulting. Note that Paleocene rapid-cooling is not constrained by age data from the sample and is simply an artifact of the starting conditions.

Figure 8 compares the variational results for sample MX-100 with those obtained using the CRS approach. As indicated, both models yield very similar thermal histories for this sample in spite of using significantly different computational approaches and trial input thermal histories (compare Fig. 8c and 8d).

Numerical simulation of domain instability during slow-cooling

All in all, inversion is not a panacea and is no substitute for careful assessment of data and understanding of multi-domain systematics. Users must be ever vigilant in assessing potentially problematic behavior. While samples affected by significant $^{40}Ar_E$ contamination are generally easily identified and avoided, samples exhibiting IAM often present a more subtle and hence insidious problem. As previously stated, IAM appear to reflect problematic redistribution of $^{40}Ar^*$ from larger domains to smaller domains. To quantify the effects of instantaneous domain reorganization during slow cooling, Lovera et al. (2002) performed numerical simulations based upon a finite difference algorithm based on the Crank-Nicholson implicit method (Crank 1975) to simulate diffusion through plane slabs with zero boundary conditions. A subset of their calculations is presented here. At a specified temperature, Lovera et al. (2002) instantaneously formed small domains that inherited Ar concentrations from their larger predecessors. In order to simplify interpretation, the sample was assumed to undergo linear slow-cooling and the initial and final domain distributions were characterized by the same set of diffusion length scales. Repartitioning of the domains in the eight domain sample was accomplished by readjusting the volume concentrations of the individual length scales. At a given temperature, a portion of the population of the smallest four domains was created at the expense of the four largest domains. The smallest domains inherited a uniform distributed concentration equivalent to the mean $^{40}Ar^*$ concentration that had been accumulated within the largest domains until that time. Because parent domains were 20–50 times larger than daughter domains, use of uniform $^{40}Ar^*$ concentrations within the newly created domains was justified.

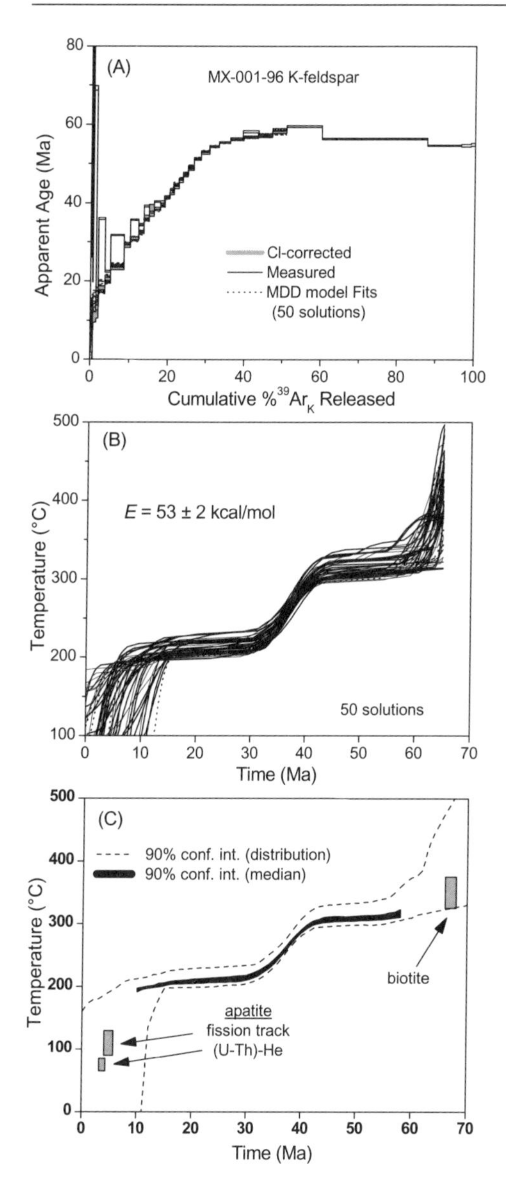

Figure 7. Application of the variational approach to MDD analysis of 96MX-001 K-feldspar (Axen et al. 2000). (a) Measured age spectrum and 50 best-fit MDD age spectra calculated from ten equivalent MDD domain distributions with activation energy (E) values defining a normal distribution about the measured activation energy for the sample (53 ± 2 kcal/mol). See text and Lovera et al. (2002) for additional details. (b) Set of 50 thermal histories corresponding to best-fit age spectra. Note that ±17 °C variation in temperature-time histories results from propagation of ±2 kcal/mol error in E. (c) Confidence intervals (90%) for the overall distribution and median of the 50 thermal histories. Note that median shows only portion of thermal history constrained by age data from the sample. Additional thermochronologic constraints for 96MX-001 from Axen et al. (2000) are shown for reference.

In the most extreme domain transformation examined by Lovera et al. (2002), the proportion of the very smallest domains was increased at the expense of the very largest domains. Thirty-two percent of the bulk K-feldspar was affected. Calculated age spectrum resulting from transformation of the domain structure at four different temperatures (150, 200, 250, and 300 °C) of the imposed linear cooling history are shown in Figure 9. As shown, the age spectrum remains unaffected when breakage occurs at or above 250 °C. This is because even when the four smallest domains comprise more than 60% of the sample, they are completely open to Ar loss at that temperature. In such a case, MDD analysis produces an accurate estimate of the imposed thermal history. This is not the case for the 200 °C and 150 °C transformations however. An IAM that is just barely developed at 200 °C becomes prominently

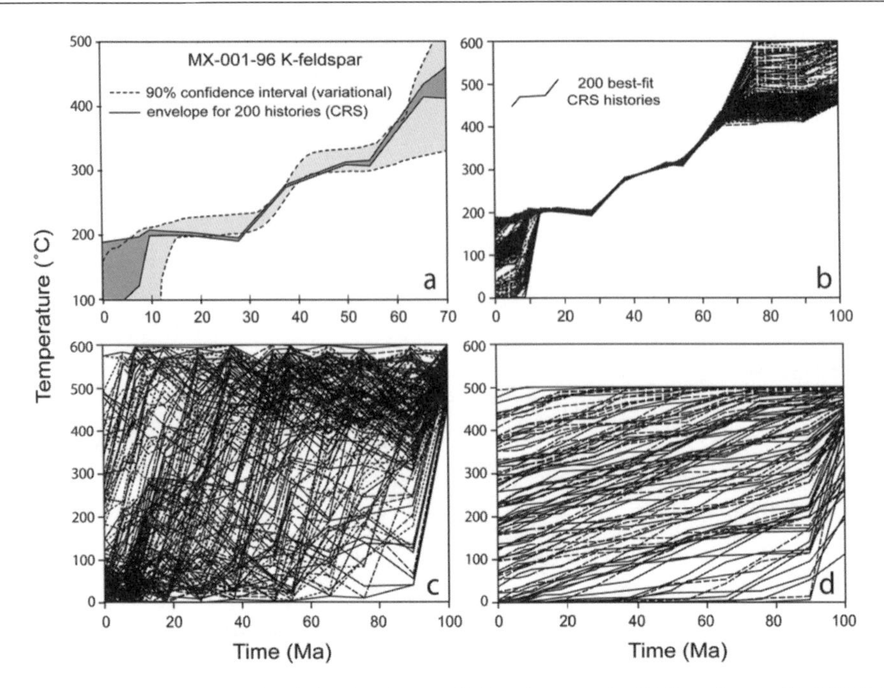

Figure 8. (a) Comparison of variational and controlled-random search (CRS) analyses of 96MX-001 K-feldspar (Axen et al. 2000). (b) CRS model results are composite of 200 best-fit thermal histories calculated from two different starting pools of random thermal histories, shown in (c) and (d), respectively. CRS models based on single domain distribution having same activation energy as the mean of used in variational models (53 kcal/mol).

developed for the 150 °C solution (Fig. 9a). Concomitantly, the calculated thermal histories become increasingly nonsensical (Fig. 9b). It should be noted that it became impossible to obtain good MDD fits to the 200 and 150 °C model age spectra.

Less extreme domain transformations considered by Lovera et al. (2002) produced more subtle effects whose impact was further decreased as the fraction of the bulk feldspar involved was reduced. In spite of this, it was clear that significant restructuring of the domain structure under <250 °C is capable of producing problematic artifacts in calculated thermal histories. Hence, extreme caution should be exerted in analyzing results from K-feldspars that have been either cataclastically deformed or partially altered to adularia under low-temperature (<250 °C) conditions.

Other applications: Th-Pb dating of monazite

Background. Since the development of U-Pb ion microprobe dating 20 years ago, (Compston et al. 1984), the general approach has been to focus a primary beam of oxygen ions on a polished surface and analyze the secondary ions of Pb, U, and UO emitted. By comparison of Pb^+/U^+ vs. UO^+/U^+ of the unknown with a standard of known age, it is possible to date accessory minerals in this way with ±1% accuracy. The spatial resolution of such a measurement is defined by the diameter of the primary ion beam spot. Given the trace concentration levels of Pb usually encountered, typical ion probe spot diameters of 10–30 μm are generally required.

However, because virtually all secondary ions originate from the first or second atomic layer of the instantaneous sample surface, atomic mixing due to the impacting primary

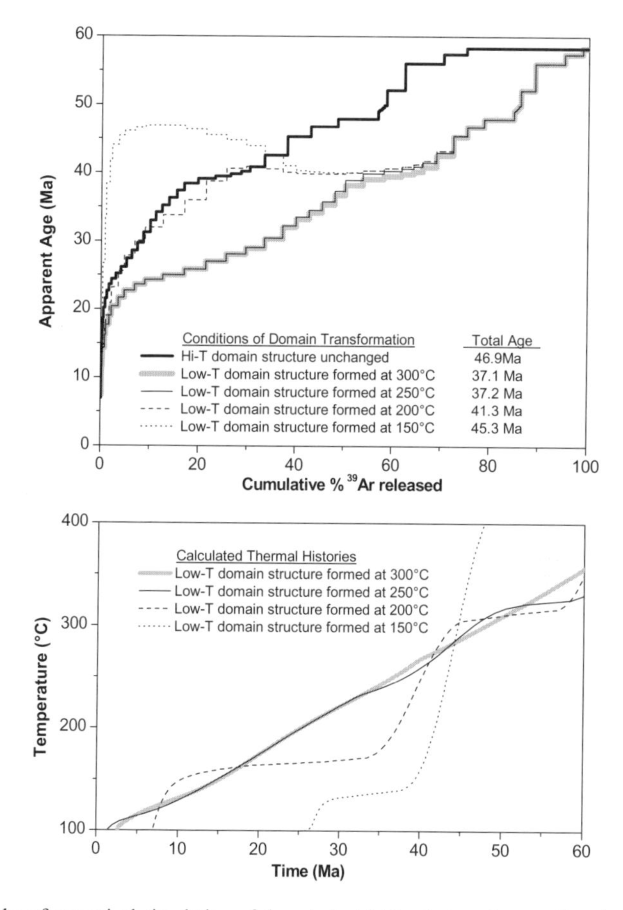

Figure 9. Results of numerical simulation of domain instability during slow-cooling from Lovera et al. (2002). An domain structure is instantly converted to less retentive domain structure by forming small domains from pre-existing larger domains at temperatures between 150–300 °C. All runs have the same imposed linearly decreasing thermal history. See text and Lovera et al. (2002) for additional details.(a) Age spectra for original high-temperature domain distribution and transformed samples. Note formation of intermediate age maxima for transformations taking place at 200 °C or lower. (b) calculated thermal histories for 150–300 °C model runs. Note that there is essentially no impact at or above 250 °C for the conditions explored in these calculations. At 200 °C and below, calculated thermal histories and progressively nonsensical.

ions and geometric effects on crater production are the essential limiting factors on depth resolution. Provided the crystal contains symmetrical overgrowths, it is possible to use the ion microprobe in depth profiling mode on an unpolished surface to increase spatial resolution by up to two orders of magnitude relative to spot analysis. Care must be taken in this analysis mode to ensure material sputtered from the crater walls is not analyzed, typically by placing an aperture in front of the emergent ion beam that restricts entry into the mass spectrometer to only those ions originating on the crater floor. Once the sputtered crater shape and depth have been determined, usually using a surface profilimeter, a continuous age profile as a function of depth is obtained. This works routinely in cases where the age profile is revealed over distances on the order of a few microns – the depth over which inter-element calibrations are stable. For deeper profiles, the sample can be re-polished after initial sputtering and the depth of the residual pit measured by surface profilometry to accurately calculate the thickness of sample removed. This approach has been applied to zircon and monazite (Grove and Harrison 1999; Mojzsis and Harrison 2002).

In cases where observed Pb loss can be attributed to volume diffusion, such gradients contain potentially valuable thermochronological information that can be extracted if the diffusion behavior of Pb were known with confidence.

Pb diffusion in monazite. Pb diffusion in monazite has been measured by Smith and Giletti (1997) and Cherniak et al. (2004), who obtained strikingly different results. Smith and Giletti (1997) measured the tracer diffusion of Pb in natural monazites using ion microprobe depth profiling and obtained an activation energy of 43 kcal/mol. Cherniak et al. (2004) measured Pb diffusion in synthetic and natural monazites under dry, 1-atm conditions using both Rutherford Backscattering Spectroscopy and ion microprobe depth-profiling methods. Their activation energy of 140 kcal/mol is more than three times higher than that reported by Smith and Giletti (1997). Although the source of this discrepancy remains unclear, it is generally accepted that the lowest value of diffusion coefficient measured is the best estimate as diffusion is the rate limiting transport mechanism in solids (i.e., any non-diffusive effect would increase D). While this would seem to favor the Cherniak et al. (2004) study, they noted that linear extrapolation of their diffusion law for Pb in monazite to geologic conditions would preclude the existence of μm-scale profiles as seen in Figure 10a. They allowed that there might be a change in diffusion mechanism between the conditions attainable in the laboratory and that in nature and used the Grove and Harrison (1999) datum to estimate an activation energy of 82 kcal/mol. Although containing an element of circularity, we use this value of E in the subsequent discussion for the purposes of illustrating the approach.

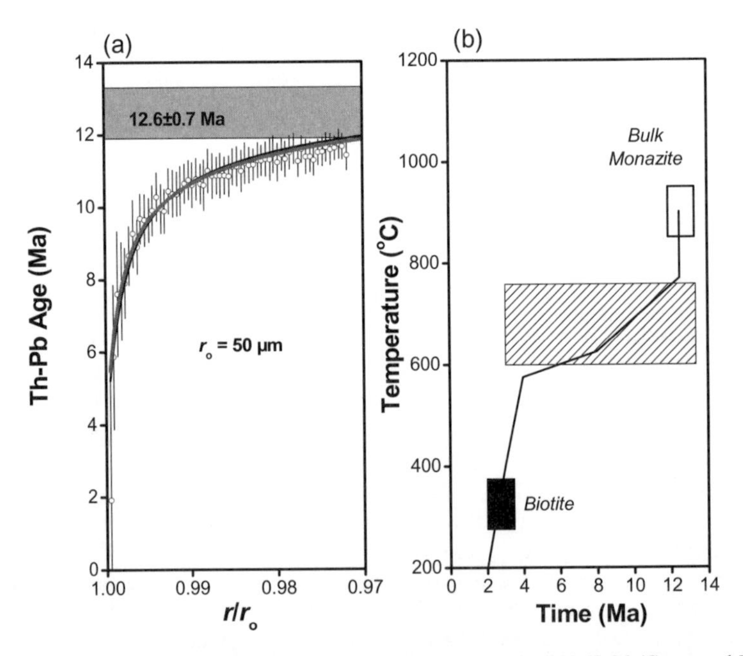

Figure 10. (a) Th-Pb age profile in the near surface region of monazite DH-68-96 (Grove and Harrison 1999). The curve fit to the data reflects the diffusion calibration described in the text and thermal history shown in (b), the thermal history required to fit the Th-Pb age gradient. Hatched region shows the T-t region that are consistent with the thermochronometric and thermobarometric data and other geologic constraints.

Th-Pb age profile of monazite DH-68-96. Gneisses of the crystalline nappe that form the hanging wall of the Himalayan Main Central Thrust experienced Miocene peak metamorphic temperatures of 600–750 °C, but cooled below ca. 400 °C by 3 Ma (Harrison et al. 1999). Th-Pb depth profiling of monazites from sample DH-68-96, obtained from a 13 Ma pegmatite, revealed an age gradient within 1 μm of the crystal surface (Grove and Harrison 1999) (Fig. 10a). Secondary ion intensities were sufficient to permit ~500 Å depth resolution. The continuous nature of these profiles, the relative conformity of their shape to that expected for radiogenic ingrowth/diffusion loss (Eqn. 2), and their inter-grain consistency effectively rules out protracted growth as an explanation for their formation. Thus it was concluded that the age gradient shown in Figure 10a is a closure profile. Given knowledge of Pb diffusion in monazite, the closure profile could then be used to extract a continuous thermal history using Equation (2).

Continuous thermal history. Given a diffusion law for Pb in monazite, albeit a problematic one, it remains only to fit the closure profile (Fig. 10a) using Equation (2). Repeated forward modeling to fit the age profile (Fig. 10a) leads to the thermal history shown in Figure 10b, which is characterized by initial rapid cooling to about 750 °C followed by cooling at ~20 °C/Ma until 4 Ma when a second phase of rapid cooling begins. This history is consistent with thermometric data from the encompassing metamorphic rocks and the age of the coexisting biotite. Furthermore, the form of the thermal history (Pliocene rapid cooling in the MCT hanging wall) was predicted based on the pattern of recrystallization ages in the MCT footwall (Harrison et al. 1998).

CONCLUSIONS

The conventional approach to thermochronology, the bulk closure method, is limited by a set of underlying requirements that are rarely met. Obtaining the highest possible resolution thermal histories requires resolving spatially varying, intra-grain concentration distributions of the daughter product.

K-feldspar provides an extraordinary opportunity to obtain continuous thermal histories by application of $^{40}Ar/^{39}Ar$ step-heating method. This approach yields two distinct sources of kinetic information (the age and Arrhenius spectra) from which quantitative measures of discrete Ar retentivity can be obtained. Provided a high correlation is observed between these two spectra, unique thermal history data can be obtained, regardless of choice of diffusion model.

With the advent of ion microprobe depth profiling, measurable gradients of Pb/Th and Pb/U have been observed over sub-μm length scales in accessory minerals. Ion microprobe depth profiling analysis is a highly underutilized method of thermochronologic analysis. Where Th-Pb gradients in monazite can be shown to be due to diffusive Pb loss, they contain potentially valuable thermochronological. Routine application of this approach using monazite is at present limited by uncertainty regarding extrapolation of kinetic models, but the method has considerable promise for the future.

REFERENCES

Arnaud NO, Kelley SP (1995) Evidence for excess argon during high-pressure metamorphism in the Dora Maira Massif (western Alps, Italy) using a ultra-violet laser ablation ion microprobe ^{40}Ar-^{39}Ar technique. Contrib Mineral Petrol 121:1-11

Arnaud NO, Kelley SP (1997) Argon behavior in gem quality orthoclase from Madagascar: experiments and consequences for geochronology. Geochim Cosmochim Acta 61:3227-3255

Axen GJ, Grove M, Stockli D, Lovera OM, Rothstein DA, Fletcher JM, Farley K, Abbott PL (2000) Thermal evolution of Monte Blanco Dome; low-angle normal faulting during Gulf of California rifting and late Eocene denudation of the eastern Peninsular Ranges. Tectonics 19:197-212

Berger GW, York D, Dunlop DJ (1979) Calibration of Grenvillian palaeopoles by $^{40}Ar/^{39}Ar$ dating. Nature 277:46-48

Cherniak DJ, Watson EB, Grove M, Harrison TM (2004) Pb diffusion in monazite: A combined RBS/SIMS study. Geochim Cosmochim Acta 68:829-840

Compston W, Williams IS, Meyer C (1984) U-Pb geochronology of zircons from Lunar Breccia 73217 using a sensitive, high mass resolution ion microprobe. Proceed 14th Lunar Planet Sci Conf Part 2. J Geophys Res 89:8525-8534

Copeland P, Harrison TM, Kidd WSF, Ronghua X, Yuquan Z (1987) Rapid early Miocene acceleration of uplift in the Gandese Belt, Xizang-southern Tibet, and its bearing on accommodation mechanisms of the India-Asia collision. Earth Planet Sci Lett 86:240-252

Crank J (1975) The mathematics of diffusion, 2nd edition, Oxford University Press, London

Dodson MH (1973) Closure temperature in cooling geochronological and petrological systems. Contrib Mineral Petrol 40:259-274

Dodson MH (1986) Closure profiles in cooling systems. *In*: Materials Science Forum, Vol 7, Trans Tech Publications, Aedermannsdorf, Switzerland, p 145-153

Foland KA (1974) ^{40}Ar diffusion in homogeneous orthoclase and an interpretation of Ar diffusion in K-feldspar. Geochim Cosmochim Acta 38:151-166

Foland KA (1994) Argon diffusion in feldspars. *In*: Feldspars and Their Reactions. Parsons I (ed) Kluwer, Dordrecht, p. 415-447

Giletti BJ (1974) Studies in diffusion I: Argon in phlogopite mica. *In*: Geochemical Transport and Kinetics. AW Hofmann, BJ Giletti, HS Yoder Jr, RA Yund (eds) Carnegie Inst. Wash. Publ 634, p 107-115

Giletti BJ, Tullis, J (1977). Studies in diffusion. IV. Pressure dependence of Ar diffusion in phlogopite mica. Earth Planet Sci Lett 35:180-183

Girard JP, Onstott TC (1991) Application of $^{40}Ar/^{39}Ar$ laser probe and step-heating techniques to the dating of authigenic K-feldspar overgrowths. Geochim Cosmochim Acta 55:3777-3793

Grove M (1993) Thermal histories of southern California basement terranes. PhD dissertation, University of California, Los Angeles, CA

Grove M, Harrison TM (1996) ^{40}Ar diffusion in Fe-rich biotite. Am Mineral 81: 940-951

Grove M, Harrison TM (1999) Monazite Th-Pb age depth profiling. Geology 27: 487–490

Hames WE, Andresen A (1996) Timing of Paleozoic orogeny and extension in the continental shelf of north-central Norway as indicated by laser $^{40}Ar/^{39}Ar$ muscovite dating. Geology 24:1005-1008

Harrison TM, Zeitler PK (2005) Fundamentals of noble gas thermochronometry. Rev Mineral Geochem 58: 123-149

Harrison TM, Yin A, Grove M, Lovera OM, Ryerson FJ (2000) The Zedong Window: A record of superposed Tertiary convergence in southeastern Tibet. J Geophys Res 105:19,211-19,230

Harrison TM, Grove M, Lovera OM, Catlos EJ, D'Andrea J (1999) The origin of Himalayan anatexis and inverted metamorphism: models and constraints. J Asian Earth Sci 17:755-772

Harrison TM, Grove M, Lovera OM, Catlos EJ (1998) A model for the origin of Himalayan anatexis and inverted metamorphism. J Geophys Res 103:27,017-27,032

Harrison TM, Leloup PH, Ryerson FJ, Tapponnier P, Lacassin R, Chen W (1996) Diachronous initiation of transtension along the Ailao Shan-Red River Shear Zone, Yunnan and Vietnam. *In*: The Tectonic Evolution of Asia. Yin A, Harrison TM (eds), Cambridge Press, London, p 205-226

Harrison TM, Copeland P, Kidd WSF, Lovera OM (1995) Activation of the Nyainqentanghla shear zone: implications for uplift of the southern Tibetan Plateau. Tectonics 14: 658-676

Harrison TM, Heizler MT, Lovera OM, Chen W, Grove M (1994) A chlorine disinfectant for excess argon released from K-feldspar during step-heating. Earth Planet Sci Lett 123:95-104

Harrison TM, Duncan I, McDougall I (1985) Diffusion of ^{40}Ar in biotite: temperature, pressure and compositional effects. Geochim Cosmochim Acta 49: 2461-2468

Harrison TM, Armstrong RL, Naeser CW, Harakal JE (1979) Geochronology and thermal history of the Coast Plutonic Complex, near Prince Rupert, British Columbia. Canadian J Earth Sci 16: 400-410

Hodges KV (1991) Pressure-temperature-time paths. Ann Rev Earth Planet Sci 19:207-236

Hodges KV, Hames WE, Bowring SA (1994) $^{40}Ar/^{39}Ar$ age gradients in micas from a high-temperature/low-pressure metamorphic terrain: Evidence for very slow-cooling and implications for the interpretation of age spectra. Geology 22:55-58

Kelley SP, Arnaud NO, Okay AI (1994) Anomalously old Ar-Ar ages in high pressure metamorphic terrains. Mineral Mag 58A:468-469

Kelley SP, Wartho JA (2000) Rapid kimberlite ascent and the significance of Ar-Ar ages in xenolith phlogopites. Science 289:609-611

Lee JKW, Onstoot TC, and Hames JA (1990) An $^{40}Ar/^{39}Ar$ investigation of the contact effects of a dyke intrusion, Kapuskasing Structural Zone, Ontario. A comparison of laser microprobe and furnace extraction techniques. Contrib Mineral Petrol 105:87-105

Lovera OM, Grove M, Harrison TM (2002) Systematic analysis of K-feldspar $^{40}Ar/^{39}Ar$ step-heating experiments II: Relevance of laboratory K-feldspar argon diffusion properties to Nature. Geochim Cosmochim Acta 66:1237-1255

Lovera OM, Grove M, Harrison TM, Mahon KI (1997) Systematic analysis of K-feldspar $^{40}Ar/^{39}Ar$ step-heating experiments I: Significance of activation energy determinations. Geochim Cosmochim Acta 61: 3171-3192

Lovera OM, Heizler MT, Harrison TM (1993) Argon diffusion domains in K-feldspar II: kinetic properties of MH-10. Contrib Mineral Petrol 113:381-393

Lovera OM, Richter FM, Harrison TM (1991) Diffusion domains determined by ^{39}Ar release during step heating. J Geophys Res 96:2057-2069

Lovera OM, Richter FM, Harrison TM (1989) $^{40}Ar/^{39}Ar$ geothermometry for slowly cooled samples having a distribution of diffusion domain sizes. J Geophys Res 94: 17917-17935

Mattinson JM (1978) Age, origin, and thermal histories of some plutonic rocks from Salinian Block of California. Contrib Mineral Petrol 67:233-245

McDougall I, Harrison TM (1999) Geochronology and Thermochronology by the $^{40}Ar/^{39}Ar$ Method. 2nd ed., Oxford University Press, New York

Mojzsis SJ, Harrison TM (2002) Establishment of a 3.83-Ga magmatic age for the Akilia tonalite (southern West Greenland). Earth Planet Sci Lett 202:563-576

Hames, WE, Cheney, JT (1997) On the loss of ^{40}Ar from muscovite during polymetamorphism. Geochim Cosmochim Acta 61:3863-3872

Onstott TC, Phillips D, Pringle-Goodell L (1991) Laser microprobe measurement of chlorine and argon zonation in biotite. Chem Geol 90:145-168

Parsons I, Brown W L, Smith JV (1999) $^{40}Ar/^{39}Ar$ thermochronology using alkali feldspars; real thermal history or mathematical mirage of microtexture? Contrib Mineral Petrol 136:92-110

Pickles CS, Kelley SP, Reddy SM, Wheeler J (1997) Determination of high spatial resolution argon isotope variations in metamorphic biotites. Geochim Cosmochim Acta 61:3809-3833

Press WH, Flannery BP, Peukolsky SA, Vetterling WT (1988) Numerical Recipes: The Art of Scientific Computing. Cambridge University Press, New York

Price WL (1977) A controlled random search procedure for global optimization. Computer J 20:367-370

Purdy JW, Jäger E (1976) K-Ar ages on rock-forming minerals from the Central Alps. Mem 1st Geol Min Univ Padova 30, 31 pp

Quidelleur X, Grove M, Lovera OM, Harrison TM, Yin A, Ryerson FJ (1997) The thermal evolution and slip history of the Renbu Zedong Thrust, southeastern Tibet. J Geophys Res 102:2659-2679

Reddy SM, Kelley SP, and Wheeler J (1996) A $^{40}Ar/^{39}Ar$ laser probe study of micas from the Sesia Zone, Italian Alps: implications for metamorphic and deformation histories. J Metamorph Geol 14:493-508

Richter FM, Lovera OM, Richter FM, Copeland P (1991) Tibetan tectonics from a single feldspar sample: An application of the $^{40}Ar/^{39}Ar$ method. Earth Planet Sci Lett 105:266-276

Scaillet S, Féraud G, Lagabrielle Y, Ballèvre M, Ruffet G (1990) $^{40}Ar/^{39}Ar$ laser probe dating by step heating and spot fusion of phengites from the Dora Maira nappe of the western Alps, Italy. Geology 18:741-744

Sherlock S, Kelley S, Inger S, Harris N, Okay A (1999) $^{40}Ar–^{39}Ar$ and Rb–Sr geochronology of high-pressure metamorphism and exhumation history of the Tavsanli Zone, NW Turkey. Contrib Mineral Petrol 137: 46–58

Smith HA, Giletti BJ (1997) Lead diffusion in monazite. Geochim Cosmochim Acta 61:1047-1055

Warnock AC, van de Kamp PC (1999) Hump-shaped $^{40}Ar/^{39}Ar$ age spectra in K-feldspar and evidence for Cretaceous authigenesis in the Fountain Formation near Eldorado Springs, Colorado. Earth Planet Sci Lett 174:99-111

Wartho J, Kelley SP (2003) $^{40}Ar/^{39}Ar$ ages in mantle xenolith phlogopites: determining the ages of multiple lithospheric mantle events and diatreme ascent rates in Southern Africa and Malaita, Solomon Islands. Geol Soc Spec Pub 220:231-248

Wartho J, Kelley SP, Brooker RA, Carroll MR, Villa IM, Lee MR (1999) Direct measurement of Ar diffusion profiles in a gem-quality Madagascar K-feldspar using the ultra-violet laser ablation microprobe (UVLAMP). Earth Planet Sci Lett 170:141-153

Willett SD (1997) Inverse modeling of annealing of fission tracks in apatite; 1, A controlled random search method. Am J Sci 297:939-969

Zeitler PK, Fitz Gerald JD (1986). Saddle-shaped $^{40}Ar/^{39}Ar$ age spectra from young, microstructurally complex potassium feldspars. Geochim Cosmochim Acta 50:1185-1199

Zeitler PK (1993) Inversion of $^{40}Ar/^{39}Ar$ age spectra using the controlled-random-search method. EOS 74: 650

Zeitler PK (2004) Arvert 4.0.1. Inversion of $^{40}Ar/^{39}Ar$ age spectra. User's Manual (*http://www.ees.lehigh.edu/EESdocs/geochron/downloads/arvert401guide-US.pdf*).

Reviews in Mineralogy & Geochemistry
Vol. 58, pp. 411-448, 2005

Application of Low-Temperature Thermochronometry to Extensional Tectonic Settings

Daniel F. Stockli

Department of Geology
University of Kansas
Lawrence, Kansas, 66045, U.S.A.
stockli@ku.edu

INTRODUCTION

Extension and associated tectonic exhumation have been described across a diverse spectrum of geodynamic environments, ranging from lithospheric-scale extension along divergent plate margins to extensional faulting within compressional orogenic belts. Normal faulting structurally dominates extensional tectonic environments such as passive continental margins, intra-continental extensional provinces and rifts, extensional back-arc basins, and mid-oceanic ridge inside corner extensional systems. In addition, structural domains characterized by normal faulting or extensional crustal attenuation commonly occur within collisional orogens and within transcurrent deformation belts. Numerous studies over the past decades have investigated extensional faulting in these different extensional tectonic environments to elucidate the architecture of extensional fault systems, the driving forces associated with crustal or lithospheric extension, and the timing and rates of extensional tectonic processes.

Many important aspects of extensional tectonics remain controversial, such as (1) the forces driving initiation of extension and rifting and their variation through time (e.g., Jones et al. 1996; Atwater and Stock 1998), (2) the parameters controlling the temporal and spatial distribution of extensional strain in the lithosphere (e.g., Buck 1991; Kusznir and Park 2002), (3) the factors influencing the geometry, kinematics, and mechanics of extensional fault systems (e.g., Wernicke and Burchfiel 1982; Jackson and White 1989; Buck 1991), (4) the role of magmatism during extensional faulting (e.g., Sengor and Burke 1978; Gans et al. 1989; Armstrong and Ward 1991; Axen et al. 1993), and (5) the interplay between extensional tectonism, erosion, and sedimentation in evolving extensional provinces (e.g., Ehlers et al. 2001, 2003).

One of the primary hurdles in resolving many of these questions has been the lack of constraints on the timing and rates of geological processes accommodating extension. Traditionally, the most direct way to assess the timing of normal faulting has been to date stratigraphic or intrusive/extrusive geological marker units that bracket the timing of onset, duration, and rates of extensional faulting and exhumation. However, in many cases there are not enough datable units available to reconstruct the faulting history, since syn-extensional basin fill deposits are often not present or appropriate datable marker beds and magmatic cross-cutting relationships may not be present. Low-temperature thermochronology is a powerful alternative approach to directly date the exhumational cooling of footwall rocks that accompany major normal fault slip in the crust during extension. Footwall rocks move relatively upwards during major slip on a normal fault, leading to exhumation and cooling of the footwall such that the timing of fault slip can be estimated from the age of this cooling (e.g., Gans et al. 1991, Ehlers and Chapman 1999). However, quantitative interpretation of exhumation rates in extensional settings may be more complicated since footwall cooling can

1529-6466/05/0058-0016$05.00 DOI: 10.2138/rmg.2005.58.16

result from a complex interplay between structural and erosional processes (e.g., Stuwe et al. 1994; Mancktelow and Grasemann 1997; Ring et al. 1999; Ehlers et al. 2003; Ehlers 2005).

The application of thermochronological methods such as $^{40}Ar/^{39}Ar$, fission-track, and (U-Th)/He dating has significantly advanced our understanding of tectonic unroofing and associated erosional exhumation of upper and middle crustal rocks during extension. Thermochronological techniques have been extensively used to investigate the timing of inception and duration of extensional tectonic exhumation, rates and rate variations of fault slip, magnitude of exhumation and crustal tilting, initial geometry of extensional fault systems during normal faulting, as well as thermal conditions of the crust prior to, during, and following major crustal extension (Foster et al. 1990; Fitzgerald et al. 1991; Gans et al. 1991; John and Foster 1993; John and Howard 1995; Harrison et al. 1995; Howard and Foster 1996; Miller et al. 1999; Foster and John 1999; Wells et al. 2000; Axen et al. 2001; Stockli et al. 2001, 2002, 2003; Wang and Gans 2004; Colgan et al. 2004).

$^{40}Ar/^{39}Ar$ and fission-track dating have been widely applied to reconstruct cooling histories of footwall rocks from ~350 to 110 °C or ~15 to 4 km depth (assuming a geothermal gradient of ~25 °C/km). More recently, with the development of low-temperature (U-Th)/He dating methods, the application of thermochronology has been expanded to include upper-crustal and near-surface extensional faulting. Furthermore, the combination of multiple (U-Th)/He thermochronometers (apatite, titanite, zircon, etc.) with $^{40}Ar/^{39}Ar$ and fission-track dating methods has made it possible to resolve multi-stage cooling histories in detail. This review paper provides an overview of modern thermochronometry as applied to extensional tectonic settings, discussing styles and mechanisms of extensional exhumation, thermochronological techniques and sampling strategies, interpretation of data sets with case studies, and important assumptions and limitations. The paper concludes with thoughts about future challenges and potential areas of research.

PROCESSES OF EXTENSIONAL UNROOFING AND EXHUMATION

There is abundant structural, stratigraphic, and geo-thermochronological evidence that normal faulting is an effective mechanism to exhume rocks from considerable crustal or even mantle depth along oceanic and continental margin core complexes in a variety of tectonic settings. Large-magnitude extensional fault systems commonly juxtapose mid-crustal mylonitic rocks of the footwalls against brittle upper-crustal hanging-wall rocks. Structural displacement accommodated by normal faults can reach tens of kilometers, although the magnitude, rate, and efficiency of structural exhumation are a function of fault slip, fault geometry, and in particular fault initiation angle.

Extensional tectonic environments are characterized by significant structural and geometric complexities that are largely controlled by the style of faulting and the symmetry or asymmetry of large-scale deformation (e.g., McKenzie 1978; Wernicke and Burchfiel 1982; Wernicke et al. 1985; Buck 1991). The lithospheric-scale two-dimensional deformation geometries of passive continental margins and large-magnitude intra-continental extensional provinces can vary from symmetrical to distinctly asymmetrical. Whereas symmetrical extension often results in conjugate upper-crustal high-angle normal faulting and lithospheric thinning, asymmetric stretching and extension commonly result in the formation of localized low-angle normal fault zones with distinctly different footwall and hanging wall characteristics (e.g., Wernicke 1981, Wernicke et al. 1985; Lister et al. 1986, 1991). However, strongly localized asymmetrical low-angle normal faulting can form within an overall large-scale symmetrical extension regime. For example, Vanderhaege and others (1999) showed that gneiss domes in the Canadian Rocky Mountains and the Massif Central in France have an overall symmetrical deformation pattern expressed by two strongly asymmetrical extensional faults with opposite shear sense.

For both symmetric and asymmetric deformation, horizontal extension and vertical tectonic exhumation of footwall rocks beneath major normal fault systems can be investigated using thermochronological methods, since the thermal history and spatial distribution and magnitude of exhumation in extensional provinces vary dramatically as a function of structural style (e.g., Buck et al. 1988). However, given the complexities of extensional tectonic regimes, it is crucial to fully integrate structural geometry and fault slip kinematics for the proper interpretation of thermochronological data. In particular, the presence of pre-extensional intrusive, volcanic, or sedimentary markers can serve as a paleohorizontal datum. Such geological constraints are critical to the selection of an appropriate thermochronometer, aid in the design of sampling strategy, and provide a sound geological context for proper interpretation of the thermochronological data.

Extensional faults have traditionally been classified into two end-members groups based on their geometry, namely high-angle normal faults and low-angle detachment faults. High-angle normal faults are the most commonly observed structure in extensional provinces, accommodating crustal extension by slip along faults with an initial dip of ~60°. High-angle normal faults can be further subdivided into rotational or non-rotational faults, and planar or listric faults. At depth, high-angle normal faults turn either into high-angle ductile shear zones or interact/merge with other high-angle or low-angle normal faults or middle-crustal shear zones (e.g., Wernicke and Burchfiel 1982; Gans 1987). In highly-extensional regimes, multiple generations of rotational high-angle normal faults are commonly observed; inactive earlier faults undergo passive rotation after initiation of younger, more favorably oriented normal faults (e.g., Profett 1977; Gans and Miller 1983; Jackson et al. 1988; White 1990). Early generation normal faults have been described that underwent passive rotation beyond horizontal, finally resembling thrust faults. In many extensional provinces, such as the Basin and Range province of western North America, the North Sea, or the South China Sea, high-angle normal faults bound major crustal blocks that have experienced asymmetric tilting during faulting and extension (e.g., Proffett 1977; Miller and Gans 1983; Jackson et al. 1988). In the footwalls of such extensional faults, rocks have been exhumed from substantial depths (>3 km) by a combination of structural unroofing and footwall erosion. Progressive footwall tilting of up to ~60–90° has been observed in many highly-extended provinces, exposing entire upper crustal sections (up to 10–15 km) within exhumed fault blocks (e.g., Gans and Miller 1982; Fryxell et al. 1992; Howard and Foster 1996; Surpless et al. 2001; Stockli et al. 2002).

Low-angle normal faults or detachment faults are characterized by shallowly inclined extensional fault planes with a dip of ~10–35°. Until the 1980s, low-angle normal faults were interpreted as thrust faults, despite structurally juxtaposing younger, low-grade rocks over older, higher-grade rocks and structurally omitting substantial crustal sections (e.g., Crittenden et al. 1980; Coney 1980; Davis et al. 1980; Armstrong 1982). However, Ring and others (1999) discussed in detail how the structural juxtaposition of younger, low-grade rocks against older, high-grade rocks is not necessarily indicative of normal faulting, since out-of-sequence thrust faulting can result in the same arrangement. Footwall rocks from below major detachment faults typically have experienced major tectonic unroofing and exhibit penetrative deformation and mylonitization. The occurrence of syn-extensional magmatism is fairly widespread in many extensional settings (e.g., Lister and Baldwin 1993), but is not a unique characteristic. Many detachment faults, such as all detachment faults in the Alps, lack any syn-extensional magmatism (e.g., Selverstone 1988; Mancktelow 1992). The same is true for large-scale detachment-associated subduction complexes, as for example the Cretan detachment (e.g., Ring et al. 2001).

Although Andersonian faulting theory predicts that low-angle normal faults cannot slip in their low-angle orientation, low-angle normal faults have been observed in a wide range of extensional tectonic regimes, including divergent plate boundaries, highly extended intracontinental extensional provinces, back arc extensional domains, contractional orogens characterized by syn- and post-collisional tectonism, and regional releasing bends in large-scale transtensional strike-slip systems (e.g., Selverstone 1988; Froitzheim and Eberli 1990; Malavieille et al. 1990; Burchfiel et al. 1992; Wernicke 1992; Mancktelow 1992; Oldow et al. 1994; Ranero and Reston et al. 1999; Axen et al. 2000; Manatschal 2004). Low-angle normal faults commonly exhume mid-crustal rocks (>10 km), forming metamorphic core complexes that provide tectonic windows into deeper crustal levels, geometrically requiring very large-magnitude fault slip (>20 km). For example, Eocene to Late Miocene extension in western North America formed a semi-continuous belt of detachment faults and metamorphic core complexes stretching from Sonora, Mexico to British Columbia, Canada (e.g., Crittenden et al. 1980, Coney 1980; Armstrong 1982; Axen et al. 1993).

The debate over the mechanical viability of low-angle initiation and fault motion along low-angle normal faulting continues, despite evidence for seismogenic and active low-angle normal faulting, for example in the D'Entrecasteaux islands (e.g. Hill et al. 1992; Abers et al. 1997) and the Gulf of Corinth (e.g. Rietbrock et al. 1996; Rigo et al. 1996; Sorel 2000). Mechanical models reconciling the existence of low-angle normal faults with Andersonian mechanics include the rolling hinge model, which has the detachment rotate passively into a low-angle orientation (Buck 1988; Wernicke and Axen 1988). In this model, footwall rocks are exhumed along a high-angle master fault that soles into the mid-crustal brittle ductile transition. During progressive exhumation these rocks and inactive fault strands are isostatically rotated into a low-angle geometry, while the active master fault remains at high angles and allows for continued exhumation of footwall rocks. In the case of fossil low-angle normal faults, the most compelling data supporting low-angle initiation and slip along extensional detachment faults as well as the existence of rolling-hinge type extensional faults have been provided by combined structural and thermochronometric or paleomagnetic studies (e.g., John and Howard 1995; Lee 1995; Livaccari et al. 1995; John and Foster 1999; Stockli et al. 2001).

Regardless of their initial geometry, extensional faults and fault systems are characterized by abrupt differences in the amount of exhumation between the hanging wall and footwall. Extensional faults in conjunction with syn-tectonic erosion are very effective at exhuming rocks, leading to cooling of footwall rocks during ascent to the surface. This is a process that can be directly dated utilizing a variety of thermochronological techniques with different or overlapping thermal sensitivity windows. High-angle normal faults exhume deeper crustal rocks after only modest amounts of crustal extension, since these faults cut and intersect isotherms at high angles. In comparison, low-angle detachment faults are a very effective mechanism for accommodating large amounts of horizontal extension, but geometrically require large amounts of fault slip to exhume deep rocks. For example, 5 km of fault slip will exhume rocks from ~4.3 km depth along a 60°-dipping normal fault, but only from ~2.5 km depth along a 30°-dipping fault, ignoring footwall rotation and isostatic rebound.

Cooling and exhumation of extensional footwall rocks occurs by a complex combination of tectonic unroofing and erosion. The conceptual role of syn-extensional footwall erosion and hanging wall sedimentation on interpretation of thermochronometric data with respect to exhumation and exhumation rates in extensional settings is discussed in detail by Ehlers (2005). The quantitative interpretation of low-temperature thermochronometric data and in particular the estimation of exhumation rates is significantly affected by perturbations of the three-dimensional temperature field across extensional structures (Ehlers et al. 2003). Tectonic, surficial, and hydrological processes complicating this quantitative interpretation of thermochronometric data include: (1) lateral thermal gradients across range-bounding normal faults due to the juxtaposition of a cool hanging wall and a warm footwall, (2) rock uplift and erosion of the footwall, (3) geothermal fluid circulation, (4) sedimentation and burial of the hanging wall, (5) lateral heat refraction around low thermal conductivity sediments deposited in the hanging wall basin, and (6) topographically-induced temperature variations in the footwall (for summary see Fig. 6a in Ehlers 2005).

LOW-TEMPERATURE THERMOCHRONOMETRIC TECHNIQUES

A well-established approach to assessing the timing and rates of faulting is to determine the low-temperature cooling histories of rocks in the footwalls of normal faults (e.g., Fitzgerald et al. 1986; Foster and John 1999; Miller et al. 1999; Ehlers and Chapman 1999; Ehlers et al. 2001; Stockli et al. 2001, 2002). For example, major slip on a normal fault leads to exhumation and cooling of the footwall such that the timing of fault slip can be estimated from the age of this cooling. This approach is limited only by whether exhumation was of sufficient magnitude to expose rocks that resided at temperatures above the closure temperatures of the minerals in question.

Although there are a multitude of thermochronometric techniques with thermal sensitivity windows ranging from >700 °C to <100 °C, this paper will focus on thermochronometers with closure temperatures and thermal sensitivity windows below 300 °C as applicable to exhumation of middle and upper crustal rocks. Several intermediate and high-temperature Th-U/Pb-based thermochronometers have been developed, although closure temperature estimates have been contested in light of contrasting results from experimental Pb diffusion and natural calibration studies. In particular, Th-U/Pb techniques using minerals such as monazite (closure temperature, T_c = 700–750 °C, e.g., Cherniak et al. 2004 and references therein; Harrison et al. 2002 and references therein), rutile (T_c = >600 °C, e.g., Cherniak 2000; Vry et al. 2003), titanite (T_c = 660–700 °C, e.g., Scott and St. Onge 1995), and apatite (T_c = 450–550 °C, Harrison et al. 2002), and Th/Pb monazite ion-probe depth-profiling (modeled T_c = 700–400 °C, Grove and Harrison 1999; and Harrison et al. 2005) have successfully started to bridge the gap between high- and low-temperature thermochronometry. These intermediate-temperature thermochronometers have helped to elucidate structural processes and extensional exhumation from deeper crustal levels and to directly date the cooling of ductile rocks from contractional and extensional tectonic settings. The following sections, roughly organized by decreasing temperature sensitivity, summarize the fundamentals of low-temperature thermochronological techniques commonly utilized in extensional tectonic settings. For a more detailed treatment of these and other techniques the reader is referred to specific chapters in this volume and references therein.

^{40}Ar/^{39}Ar thermochronometry

^{40}Ar/^{39}Ar geo- and thermochronometry of multiple rock-forming K-bearing mineral phases has been widely utilized to constrain cooling histories in both orogenic belts and highly-extended terrains (e.g., Harrison et al. 2005). Some of the most common ^{40}Ar/^{39}Ar thermochronometers include hornblende (T_c = 500–550 °C, Harrison 1981), mica (T_c = 300–370 °C; Harrison et al. 1985; Grove and Harrison 1996; Lister and Baldwin 1996), and K-feldspar (modeled T_c = 150–350 °C; Lovera et al. 1989). A detailed discussion of methodological and analytical aspects of ^{40}Ar/^{39}Ar dating can be found in McDougall and Harrison (1999) and references therein.

K-feldspar ^{40}Ar/^{39}Ar thermochronometry was developed as a powerful tool to decipher low-temperature cooling histories. ^{40}Ar/^{39}Ar dating in combination with step-heating techniques have shown that K-feldspars lose argon at low temperatures primarily by complicated multi-domain volume diffusion apparently controlled by microstructural domains of variable size rather than on the whole-grain scale. Lovera and others (1989) showed that ^{39}Ar release from K-feldspar during detailed step-heating experiments is controlled by distinct diffusion domains that are characterized by different retentivities and closure temperatures. K-feldspars typically have diffusion domains with a range of sizes that are characterized by identical activation energies. Using the Multi-Domain Diffusion (MDD) approach of Lovera and others (e.g. 1989, 1997) enables iterative modeling of the K-feldspar release data and the deduction of thermal histories between ~350 °C and ~150 °C (Harrison et al. 2005).

However, several studies have questioned the validity of thermal histories recovered from MDD K-feldspar thermochronology, since the smallest estimated diffusion domain size appears to be similar to the ^{39}Ar recoil distance, multi-diffusion domains cannot be identified structurally, and microstructures of K-feldspar are thought to be metastable during step-heating experiments (e.g., Parsons et al. 1999). Although the mineralogical and microstructural nature of K-feldspar diffusion domains remains contested, the fact that ages from the lowest temperature steps are typically only slightly older than apatite fission track ages and that the oldest ages are usually concordant with $^{40}Ar/^{39}Ar$ ages of coexisting biotite, lend strong support to the validity of the MDD approach. Moreover, several case studies have used empirical evidence to demonstrate that cooling histories based on MDD models from K-Feldspar $^{39}Ar/^{40}Ar$ data are in good agreement with other thermochronometers and independent geological evidence (e.g., Lee 1995; Axen et al. 2001; Kirby et al. 2002).

Fission-track thermochronometry

Apatite and zircon fission-track dating methods have been used extensively to constrain the low-temperature cooling histories of rocks in exhumed footwall blocks of major normal faults (e.g., Dokka et al. 1986; Foster et al. 1990, 1994; Fitzgerald et al. 1991; Gans et al. 1991; Howard and Foster 1996; Miller et al. 1999; Stockli et al. 2000, 2002). Fission track thermochronometry is based on the fact that tracks are partially or entirely erased at elevated temperatures, resulting in easily measured reductions in both track lengths and fission track densities. Kinetic models for thermally-activated annealing of fission tracks in apatite are well developed (e.g., Naeser 1979; Gleadow et al. 1986; Green et al. 1989a,b; Carlson 1990; Corrigan 1991; Crowley et al. 1991; Ketcham et al. 1999). The annealing behavior of fission tracks in apatite and the temperature at which all tracks are totally annealed is influenced by apatite compositional variations, principally Cl-content (Green et al. 1985, 1986, 1989b; Gleadow et al. 2002), and crystallographic anisotropy (Donelick et al. 1999; Carlson et al. 1999; Ketcham et al. 1999). Theoretical and natural calibration studies of pre-exhumation apatite fission track data patterns demonstrate how apparent ages and track lengths vary systematically with depth and burial temperature (e.g., Green et al. 1989a, 1989b; Dumitru 2000, Gleadow et al. 2002). In apatite, all fission tracks are totally annealed above ~110 °C and partially annealed between ~60 °C and ~110 °C, a temperature range termed the partial annealing zone (e.g., Gleadow et al. 1986; Green et al. 1989a). Below ~60 °C, fission tracks in apatite are effectively stable, annealing only at very slow rates (e.g., Fitzgerald and Gleadow 1990). If a sample rapidly cooled from temperatures >~110 °C due to tectonic unroofing, then the measured age will record the age of extensional faulting and exhumation. The quantitative understanding of fission track annealing kinetics allow partially annealed apatite fission track apparent ages and confined length data to be used to constrain the thermal evolution (~60–110 °C) of samples using stochastic inverse modeling approaches (e.g., Lutz and Omar 1991; Corrigan 1991; Gallagher 1995; Willett 1997; Ketcham et al. 2000, 2003; Ketcham 2005).

Fission tracks in zircon anneal at higher temperatures than those in apatite and thus constrain higher temperature parts of the cooling history. Based on comparison with K-feldspar $^{40}Ar/^{39}Ar$ results, the zircon fission track closure temperature has been estimated to be ~250 °C (Foster et al. 1996; Wells et al. 2000). This estimate appears to be in good agreement with experimental data suggesting a closure temperature for fission tracks in zircon of ~240 °C (Tagami 2005). Conservative estimates of the temperature bounds of the zircon partial annealing zone are between ~230 °C and ~310 °C (Yamada et al. 1995; Tagami and Dumitru 1996; Tagami 2005). However, little is known about the impact of compositional variations on annealing behavior in zircon.

Detailed descriptions of the fundamentals of apatite and zircon fission track dating have been published (e.g., Gallagher et al. 1998; Dumitru 2000; Gleadow et al. 2002; Tagami et al. 2005; Donelick et al. 2005; Tagami 2005).

(U-Th)/He thermochronometry

(U-Th)/He dating of apatite is now a well-established thermochronological technique and is widely applied in geological, tectonic, and geomorphologic studies (e.g., Zeitler et al. 1987; Lippolt et al. 1994; Wolf et al. 1996, 1998; House et al. 1997, 1999; Farley 2000; Reiners et al. 2000; Stockli et al. 2000; Reiners 2002; Farley and Stockli 2002; Ehlers and Farley 2003; Carter et al. 2004). (U-Th)/He dating of apatite is based on the decay of ^{235}U, ^{238}U, and ^{232}Th by alpha (^{4}He nucleus) emission. ^{4}He is completely expelled from apatite at temperatures above ~80 °C and almost totally retained below ~40 °C (termed the He partial retention zone) (Wolf et al. 1996, 1998; House et al. 1999; Stockli et al. 2000). Helium diffusivity appears to correlate with the physical dimensions of the apatite crystal, indicating that the diffusion domain is the grain itself, so grain size has a small effect on the closure temperature (Farley 2000). The thermal sensitivity of this system is lower than that of any other isotopic thermochronometer. Assuming a mean annual surface temperature of 10 ± 5° C and a geothermal gradient of 25 °C/km, the relevant temperature range is equivalent to depths of ~1 to 3 km. Thus the apatite (U-Th)/He system can be applied to investigate a variety of geologic processes in the uppermost part of the crust, such as rifting, mountain building, erosional exhumation, and landscape evolution. For a more in-depth summary of apatite (U-Th)/He dating, the reader is referred to systematic overview articles (e.g., Farley 2002; Reiners 2002; Farley and Stockli 2002; Ehlers and Farley 2003).

More recently, other U- and Th-bearing minerals, such as titanite, zircon, monazite and xenotime have attracted interest as potential geo- and thermochronometers (e.g., Reiners and Farley 1999; Pik and Marty 1999; Reiners et al. 2002; Farley and Stockli 2002; Tagami et al. 2003; Reiners 2005; Stockli and Farley in review). Each of these mineral phases is characterized by distinct He diffusion kinetics and closure temperatures, ranging from ~120 °C to ~240 °C, offering the possibility of constraining different portions of low-temperature thermal histories (e.g., Reiners et al. 2000; Kirby et al. 2002; Farley and Stockli 2002).

THERMOCHRONOMETRY AND EXTENSIONAL TECTONICS

The combination of biotite and K-feldspar $^{40}Ar/^{39}Ar$ (including MDD modeling), zircon and apatite fission track, and apatite, zircon, titanite, and monazite (U-Th)/He thermochronometers provide a powerful means to decipher complete low-temperature cooling histories of exhumed footwalls in extensional tectonic environments. These methods allow us to either monitor the thermal evolution of entire exhumed sections and/or to track individual rocks samples along their entire temperature-time (*T-t*) path during exhumation from mid-crustal to near surface structural levels. The following sections provide a detailed summary of tectonic questions that can be addressed, sampling strategies, interpretation of thermochronological data sets with case studies, and a discussion of important assumptions and limitations.

Since the 1980's, innumerable studies have utilized thermochronometry in conjunction with structural, petrographic, stratigraphic, and geomorphologic techniques to investigate the timing and dynamics of extensional tectonic settings in different regions of the world. Initially many of these efforts were motivated by scientific and economic interest in understanding the tectonic and thermal evolution of continental margins (e.g., Gallagher and Brown 1997; Gallagher et al. 1998; Gleadow et al. 2002 and references therein). Thermochronometric studies since those early efforts have been expanded to include a multitude of different extensional settings, ranging from faulting along intra-continental rift systems, back-arc basins, transtensional strike-slip regimes, intra-oceanic extensional provinces, contractional orogenic systems characterized by syn- and post-orogenic extension throughout the Phanerozoic in the Aegean (e.g., John and Howard 1995; Baldwin and Lister 1998; Thomson et al. 1999; Gessner et al. 2001; Hejl et al. 2002; Ring et al. 2003), Antarctica (e.g., Fitzgerald and Gleadow 1990, Fitzgerald 1992; Lisker and Olesch 1998; Lisker et al. 2002, 2003), the Alps (e.g., Staufenberg 1987; Seward and Mancktelow 1994, Dunkl and Demeny 1997; Fugenschuh et al. 1997, 2000; Dunkl et al. 2003; Fugenschuh and Schmid 2003), the western Mediterranean, including the Alboran, Ligurian and Tyrrhenian seas (e.g., Thomson 1994; Platt et al. 1998, 2003a,b; Zarki-Jakni 2004), eastern Australia (e.g., Moore et al. 1986; Dumitru et al. 1991; Gallagher et al. 1994; O'Sullivan et al. 1995; Kohn et al. 1999; Kohn and Bishop 1999; Persano et al. 2002), the Basin and Range Province (e.g., Dokka et al. 1986; Davy et al. 1989; Foster et al. 1990; Fitzgerald et al. 1991; Gans et al. 1991; John and Foster 1993; Howard and Foster 1996; Miller et al. 1999; Stockli et al. 2000, 2001, 2002, 2003; Axen et al. 2000; Fletcher et al. 2000; Wells et al. 2000; Coleman et al. 2001; Surpless et al. 2002; Armstrong et al. 2003; Egger et al. 2003; Colgan et al. 2004; Carter et al. 2004), the Canadian Cordillera (e.g., Lorencak et al. 2001; Vanderhaeghe et al. 2003), the Greenland and Norway Atlantic margins (e.g., Rohrman et al. 1995; Clift et al. 1996; Thomson et al. 1999; Hansen and Brooks 2002), the East Africa rift system (e.g., Foster and Gleadow 1992; Noble et al. 1997; Van der Beek et al. 1998), the Red Sea (e.g., Davison et al. 1994; Omar and Steckler 1995; Kohn et al. 1997; Menzies et al. 1997; Abbate et al. 2002; Ghebreab et al. 2002), Lake Baikal (van der Beek 1997), New Zealand (e.g., Seward et al. 1989; Spell et al. 2000), the South West Pacific (e.g., Baldwin et al. 1993; Monteleone et al. 2001), Tibet (e.g., Harrison et al. 1995; Stockli et al. 2002), the South China Sea (e.g., Carter et al. 2000; Hutchison et al. 2000), etc. This rapidly growing list is by no means comprehensive and is simply intended to illustrate the thematic and geographic diversity of thermochronological investigations in extensional settings.

Timing of extensional faulting and exhumation

The timing of onset of normal faulting in extensional settings has traditionally been constrained by detailed documentation of the chronostratigraphy of pre-, syn, and post-kinematic sedimentary and volcanic strata. However, rifting and extension-related strata are often either deeply buried within hanging-wall basins or are poorly dated due to the scarcity of datable strata, especially in non-marine or non-volcanic extensional systems, thus hampering attempts to adequately reconstruct extensional faulting histories. In their summary article, Foster and John (1999) pointed out that there is considerable uncertainty in the timing of onset of extensional exhumation of metamorphic core complexes in the southwest U.S.A. based on stratigraphic constraints. They attribute this uncertainty to the limited availability of reliable dates for the oldest syn-kinematic strata and the non-uniform spatial distribution of these syn-extensional deposits. Th-U/Pb geochronological dating of pre-, syn, and post-kinematic plutons and dikes often provides useful chronometric brackets on the timing of extensional tectonic faulting, but many extensional provinces lack extension-related magmatism or sufficiently high-frequency intrusive events needed to provide precise age constraints on the onset of faulting.

Low-temperature thermochronometric methods allow the determination of timing, duration, rate, and magnitude of extensional faulting from low-temperature cooling histories of exhumed footwall rocks (e.g., Fitzgerald and Gleadow 1988; Fitzgerald et al. 1991; Omar and Steckler 1995; Harrison et al. 1995; Ehlers and Chapman 1999; Miller et al. 1999; Foster and John 1999; Axen et al. 2000; Wells et al. 2000; Stockli et al. 2000, 2002; Wong and Gans 2004). This approach requires rocks to be exhumed from temperatures above the closure temperatures of the mineral phases dated. Strategies for thermochronological investigations to accurately determine the onset of cooling and thus the onset of extensional faulting and exhumation can be broadly subdivided into two fundamentally different approaches.

Age – temperature. This approach utilizes multiple thermochronometric dating techniques with different thermal sensitivity windows to determine complete cooling histories from individual samples or from an array of samples from different structural levels. The construction of complete *T-t* paths for samples uses two types of thermochronometric data sets: (1) measured cooling ages and corresponding closure temperatures of different thermochronometers, and (2) segments of *T-t* paths derived, for example, from MDD modeling of K-feldspar $^{40}Ar/^{39}Ar$ data or modeling of apatite fission track age and confined track length data (e.g., Gallagher 1995; Ketcham 2005) (Fig. 1). The resulting *T-t* reconstructions define discrete cooling episodes, cooling rates, and rate changes and can be interpreted in terms of the magmatic or tectonic evolution of a sample. Whereas slow cooling or isothermal holding of a sample is commonly interpreted as a period of tectonic quiescence, abrupt increases in cooling rate are thought to represent tectonic exhumation in response to normal faulting. The timing of accelerated cooling dates the onset of major normal faulting and extensional exhumation. However, several workers have pointed out that the onset of accelerated cooling in samples from structurally deeper levels can be delayed with respect to the actual onset of extensional

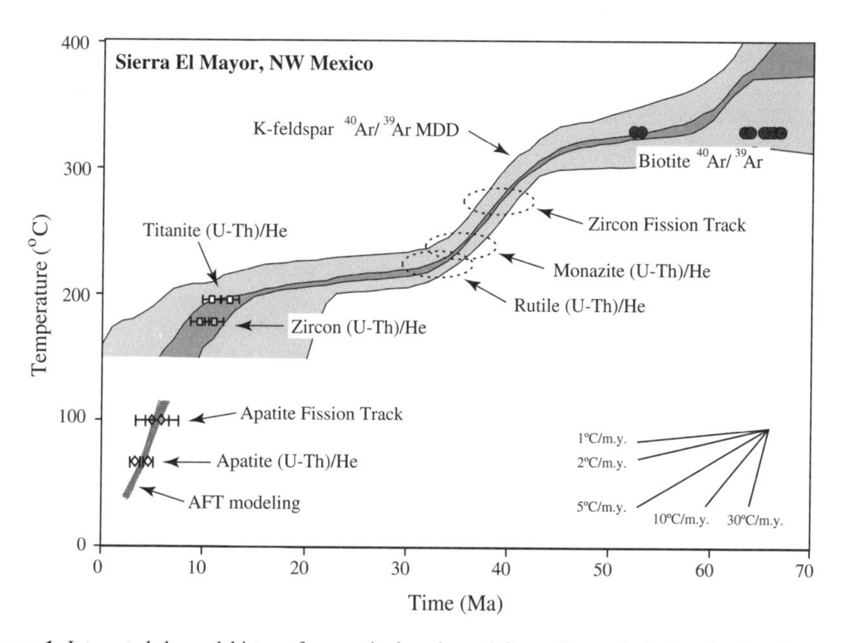

Figure 1. Integrated thermal history from a single exhumed footwall sample below the Canada David Detachment in the Sierra El Mayor, Mexico, derived from complementary and overlapping temperature sensitivity windows of different $^{40}Ar/^{39}Ar$, (U-Th)/He, and fission track thermochronometers. K-feldspar $^{40}Ar/^{39}Ar$ MDD modeling results are shown as median (dark gray) and 90% confidence interval for distribution (light gray). Continuous low-temperature thermal history is derived from MonteTrax fission track age and track length modeling runs (Gallagher 1995). Zircon fission track and monazite and rutile (U-Th)/He closure temperature windows are shown as dashed areas, but no data are available for this sample (Axen et al. 2000; Stockli et al. in prep.).

faulting because of footwall advection and compression of isotherms during rapid tectonic exhumation (e.g., Ruppel et al. 1988; Ketchum 1996; Foster and John 1999). This approach is most commonly employed on deeply exhumed footwall rocks in highly extended extensional terrains and in particular on samples from metamorphic core complexes (e.g., Foster et al. 1993; Harrison et al. 1995; Axen et al. 2001). However, it is clearly not limited to extensional tectonics and has been successfully employed in contractional and transcurrent orogenic systems (e.g., Spotilla 2005).

Low-temperature thermochronometric data such as (U-Th)/He apparent ages cannot necessarily be interpreted as directly dating when a sample cooled through a specific closure temperature. Instead, such data might represent partially reset ages that could have experienced partial diffusive loss and retention of He. Therefore, caution is required when reconstructing *T-t* paths using closure temperatures, which are themselves a function of cooling rate.

Figure 1 shows a *T-t* history derived from integrated $^{40}Ar/^{39}Ar$, fission-track and (U-Th)/He thermochronometry of a single granitic sample from the footwall of the Sierra El Mayor metamorphic core complex in the northern Gulf of California (Axen et al. 2000; Stockli et al. in prep.). Biotite $^{40}Ar/^{39}Ar$ ages of ~65 Ma reflect slow cooling through ~350 °C. MDD modeling of K-feldspar $^{40}Ar/^{39}Ar$ release spectra indicate very slow cooling (~1 °C/m.y.) from ~65 to ~45 Ma. Accelerated cooling from ~315 °C to ~215 °C between ~45 and ~33 Ma records at least 3–4 km of Eocene denudation that is likely related to early Tertiary tectonism. Footwall rocks remained nearly isothermal from ~30 Ma until ~12–10 Ma, when renewed rapid cooling (33 ± 17 °C/m.y.) began. The onset of this rapid cooling event is also recorded by titanite and zircon (U-Th)/He ages of ~ 10–12 Ma. Apatite fission-track and (U-Th)/He ages of ~5 Ma and ~4 Ma, respectively, record continued progressive cooling through ~110 °C to ~70 °C. The integrated cooling history shows a clear break in slope and acceleration of cooling rate at ~10–12 Ma. Rapid middle Miocene cooling is interpreted to date the onset of low-angle normal faulting associated with the formation of the Sierra El Mayor metamorphic core complex that accommodated rift-related extension in what eventually became the Gulf of California (Axen and Fletcher 1998; Axen et al. 2000).

The integration of different isotopic thermochronometers and the interpretive modeling of thermochronometric data to reconstruct thermal histories has been a commonly utilized approach to dating tectonic exhumation of footwall rocks from beneath low-angle normal faults (e.g., Foster et al. 1993; Baldwin et al. 1993; Fitzgerald et al. 1994; Harrison et al. 1995; Forster and John 1999; Miller et al. 1999; Thomson et al. 1999; Fuegenschuh et al. 2000; Wells et al. 2000; Stockli et al. 2001; Lorencak et al. 2001; Foster and Raza 2002; Vanderhaeghe et al. 2003). Forster and John (1999) described a slight variation of this well-established approach by combining different $^{40}Ar/^{39}Ar$ thermochronometers and apatite fission track dating on multiple samples from different structural levels from the footwall of metamorphic core complexes in the Basin and Range province. These vertically stacked *T-t* histories show temporally coincident inflection points dating the onset of extensional exhumation with a slightly delayed advent of cooling recorded by the structurally deepest footwall rocks. Lee (1995) employed a similar approach using continuous thermal histories from K-feldspar $^{40}Ar/^{39}Ar$ MDD modeling to demonstrate diachronous footwall cooling in response to the migration of a rolling hinge in the Snake Range in the Basin and Range province.

Age – depth/distance. A second approach for constraining the timing of onset of extensional exhumation uses single or multiple-method thermochronometric data collected from arrays of samples from the maximum exposed range of paleodepths in the footwall of extensional fault systems. Rather than reconstructing complete *T-t* paths based on closure temperatures and *T-t* modeling results, this technique evaluates the apparent age pattern of thermochronometric data as a function of structural paleodepth or elevation using the concept of partial retention (Ar and He) and annealing zones (fission track). This approach uses sample transects to identify inflection points in age versus paleodepth/elevation arrays which define the bounds of partial retention zones. The shape of apparent age-paleodepth/elevation trends is analogous to systematic down-hole thermochronometric age variations observed in deep borehole studies (e.g., Warnock et al. 1997; Coyle and Wagner 1998; House et al. 1999; Stockli and Farley 2004; Gleadow et al. 2002). In exhumed extensional fault blocks using empirically or numerically derived estimates for the temperature bounds of the different partial retention zones, observed inflection points can be interpreted as paleoisotherms that were exhumed when extensional faulting started. This approach is commonly employed in low-temperature thermochronological studies determining the timing of fault initiation using fission track and (U-Th)/He dating, but can also be applied utilizing higher temperature thermochronometers. However, it requires that thermochronometric data from samples span a significant range of paleodepths/elevation and demands a sound structural understanding of extensional fault geometries and fault kinematics. In particular, the derivation of paleodepth estimates in highly extended terrains commonly requires detailed structural and geological reconstructions of the pre-extensional crust for the proper interpretation of age data (e.g., Fitzgerald and Gleadow 1988; Fitzgerald et al. 1991; Howard and Foster 1996; Stockli et al. 2000, 2002).

In all cases, samples should be systematically collected from an array of structural levels in transects parallel to the extensional fault slip direction. There are, however, subtle differences in sampling strategy and data interpretation depending on structural style and fault geometry. In the case of non-rotational high-angle fault blocks or faults accommodating small to moderate amounts of extension, apparent age data should be collected from vertical transects and evaluated against elevation or paleodepth. For rotational high-angle faults or highly extended terrains, carefully reconstructed paleodepth estimates are required to evaluate the apparent age data. In the case of low-angle detachment faults, sample arrays are commonly collected along near-horizontal transects parallel to the extensional fault slip direction.

Figure 2 illustrates the sampling strategy used for rotational high-angle faults and shows a highly simplified structural model of the exhumation and cooling of extensional crustal fault blocks. Rotational high-angle normal faults accommodate major crustal extension with

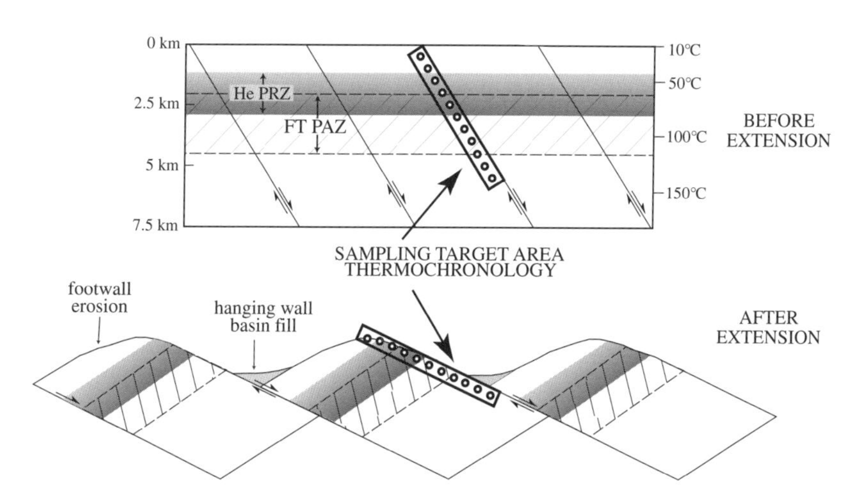

Figure 2. Conceptual model illustrating the exhumation of footwalls during major high-angle normal faulting, exposing a range of pre-extensional paleodepths in intact, tilted fault blocks (~50% extension). Apatite fission track and (U-Th)/He thermochronologic sampling transects are devised to obtain samples from the largest possible range of footwall paleodepths (adapted from Miller et al. 1999 and Stockli et al. 2000).

mountain ranges corresponding to major crustal blocks that underwent asymmetric tilting during normal faulting. In the footwalls of such extensional faults, rocks have been exhumed from substantial depths. If fault slip has been rapid and of sufficient magnitude to exhume samples from just below the zero-retention isotherm, these thermochronometric ages will date the onset of extensional faulting and range formation. At increasingly shallow paleodepths, apparent ages will trend older, because the radiometric clocks were not completely reset before exhumation commenced.

The Wassuk Range and Grey Hills in western Nevada represent Miocene tilted fault blocks that display ~60° of footwall rotation and expose pre-extensional paleodepths of up to ~8.5 km, based on the structural reconstruction of tilted pre-extensional volcanic rocks (e.g., Surpless 1999; Stockli et al. 2003) (Figs. 3 and 4). Figure 5 shows low-temperature apatite fission track and apatite and zircon (U-Th)/He data plotted against pre-extensional paleodepth from the Wassuk Range and Grey Hills (Stockli et al. 2002; Tagami et al. 2003). The data record rapid cooling and exhumation of the upper crust as a consequence of large-magnitude extension and tilting of the Wassuk fault block starting at ~15 Ma. The timing of onset of extension and fault block tilting agrees well with the tilt history bracketed by Tertiary volcanic and sedimentary deposits. The fission track and (U-Th)/He age versus paleodepth curves show clear and distinct inflection points at ~6.4 km (zircon (U-Th)/He), ~3.6 km (apatite fission track) and ~ 2.9 km (apatite (U-Th)/He) (Fig. 5). These curves resemble profiles from other tilted, intact fault blocks (e.g., Reiners et al. 2000; Stockli et al. 2000) and borehole studies (e.g., Green et al. 1989a; House et al. 1999; Stockli and Farley 2004) in which the structurally shallow part of the curve records progressive age reduction in a zone of partial annealing/retention. Structurally deeper samples record rapid cooling and exhumation of rocks that were previously residing below the partial retention/annealing zones (e.g., Bryant and Naeser 1980; Fitzgerald et al. 1991; Gleadow and Fitzgerald 1987; Howard and Foster 1996; Stockli et al. 2000; Reiners et al. 2000, 2002; Stockli et al. 2002, 2003). The inflections in apparent age against pre-extensional paleodepth curves are interpreted to represent the base of the fission track partial annealing zone for apatite (~110 °C) and (U-Th)/He partial retention zones for apatite (~80 °C) and zircon (~180–200 °C) prior to major fault slip and resultant footwall exhumation and cooling (Fig. 5).

In metamorphic core complexes or in footwalls to low-angle detachment faults, the approach of identifying thermochronometric inflection points along horizontal sample traverses has also been used (e.g., Foster et al. 1993; John and Howard 1995; Ketchum 1996; Miller et al. 1999; Foster and John 1999; Wells et al. 2000). For metamorphic core complexes and detachment faults, apparent age data are plotted against distance in slip direction (Fig. 6). The resulting age-distance curves are commonly characterized by four distinct elements (from shallow to deeper structural levels): (1) a zone of ages that are unaffected by extensional exhumation, reflecting an older episode of cooling that might be tectonic or post-magmatic in nature, (2) a zone of partially reset ages from samples that resided within the partial retention zone prior to onset of detachment faulting, (3) a sharp inflection point that marks the base of the partial retention/annealing zone and directly constrains the timing of onset of rapid extensional exhumation, and (4) ages below the inflection point that trend systematically younger in down-dip direction, recording progressive cooling during fault slip and structural unroofing (Fig. 6). The temporal initiation of tectonic exhumation can be deduced from the break in slope directly, since samples just below the partial retention zone cooled very rapidly as a consequence of extensional slip along the low-angle detachment fault. Therefore an individual thermochronometric sample age from down-dip of the inflection point without the context of a systematic sample array would always underestimate the onset of extensional unroofing (Foster and John 1999) (Fig. 6). Syn-extensional plutonism or structural complexities such as the existence of secondary break-away faults, multiple low-angle detachment faults,

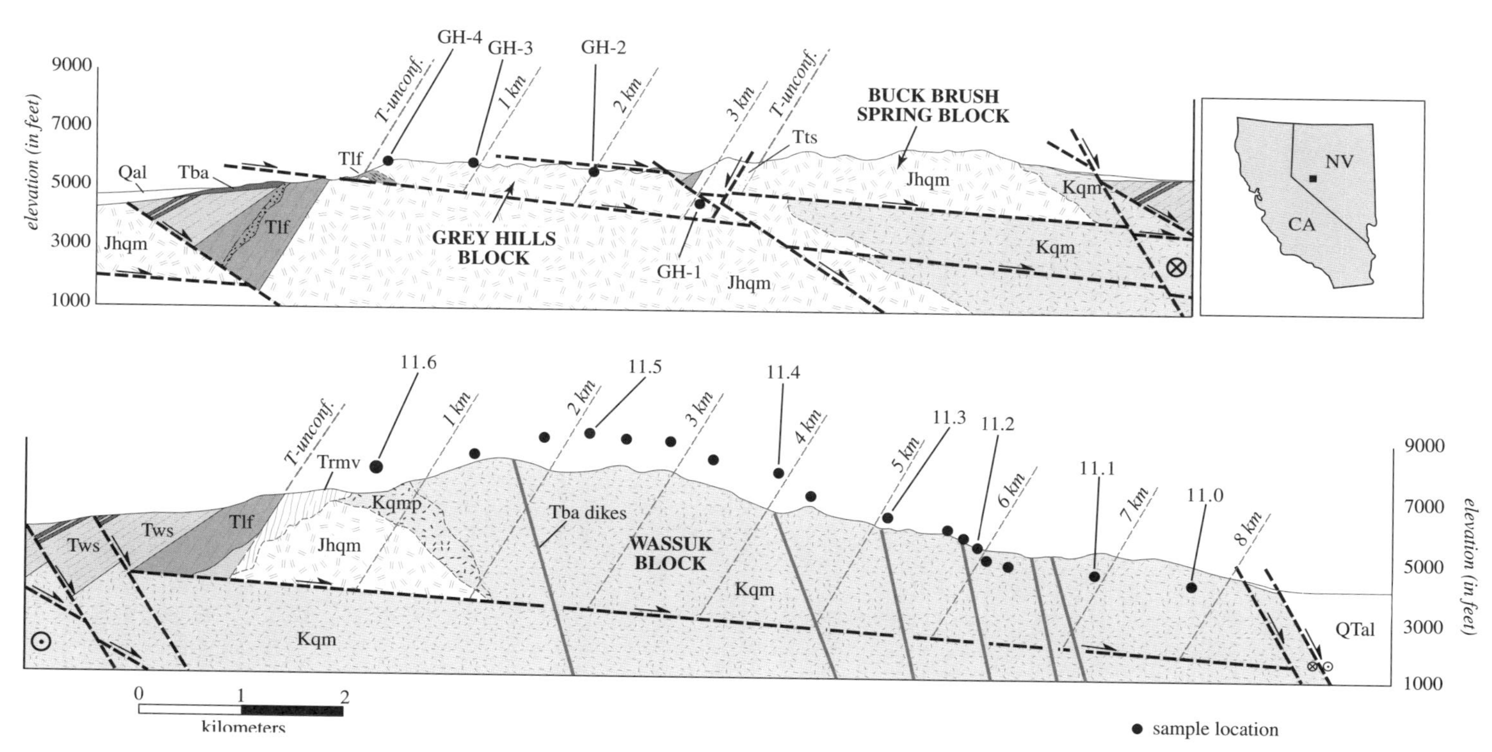

Figure 3. Simplified cross-sections of the central Wassuk Range and Grey Hills, Nevada (from Stockli et al. 2002). Fault blocks are tilted to the west ~60° and expose a cross-sectional view of the upper crust. Thermochronological samples were collected from a range of paleodepths in the Grey Hills and Wassuk blocks. No samples were collected from the Buck Brush Spring block due to extensive hydrothermal alteration of the Jurassic quartz monzonite (Surpless 1999). Rock unit legend given in Figure 4.

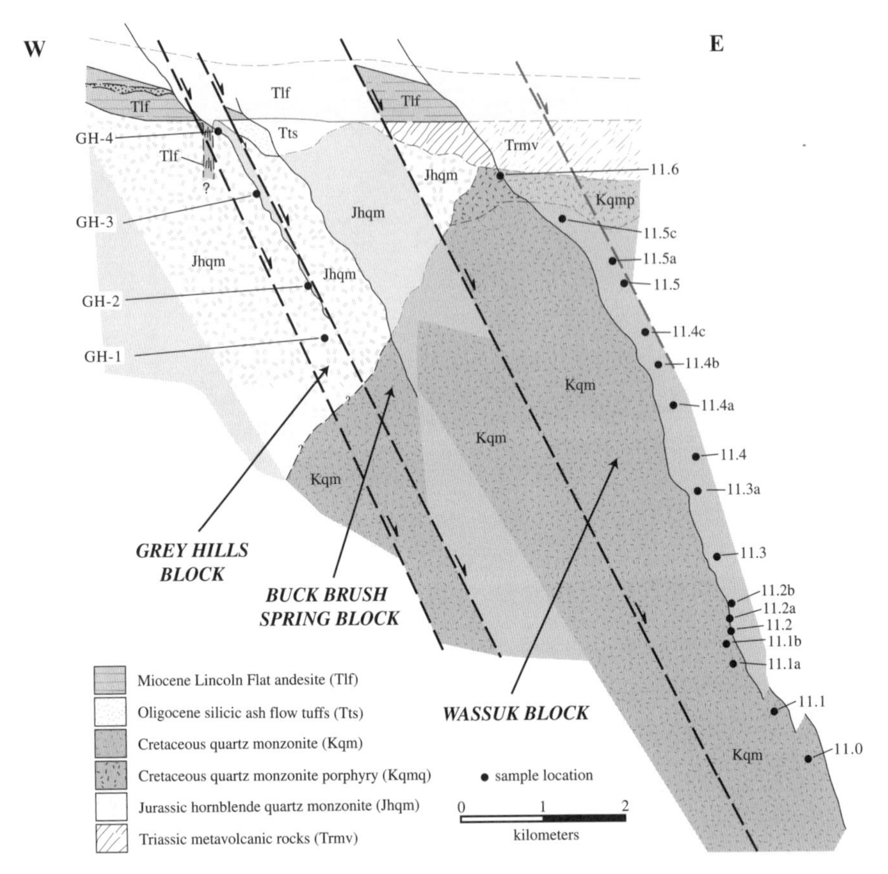

Figure 4. Detailed structural reconstruction of the central Wassuk Range area prior to the onset of large-magnitude extensional faulting at ~15 Ma, based on the youngest preextensional stratigraphic marker unit, the Miocene Lincoln Flat andesite (Tlf) (adapted from Stockli et al. 2002). The total extension derived from this palinspastic restoration is >200% (Surpless 1999). Structural reconstruction of pre-extensional crustal geometries of the fault blocks permits estimation of preextensional paleodepths necessary for the proper interpretation of the thermochronological data. Sample numbers (11.0-11.6 and GH-1-GH-4) correspond to sample numbers and locations in Figure 3, illustrating the current spatial sample distribution.

depth-dependent changes in fault geometry, or complicated multi-stage extensional histories (e.g., diachronous detachment faulting or detachments with opposite kinematic vergence) can complicate the thermal history and interpretation of thermochronometric data from the footwall of metamorphic core complexes. Therefore, as for high-angle normal faulting, a detailed structural and geological context is necessary for the proper interpretation of these age-distance relationships.

Estimation of fault slip rates

Thermochronometric data from both high-angle and low-angle normal faults have been used to constrain rates of exhumation and extensional fault slip (e.g., Naeser et al. 1983; Foster et al. 1993; Foster and John 1999; Wells et al. 2000; Ehlers et al. 2001; Ehlers et al. 2003; Brichau 2004; Carter et al. 2004). Such an approach assumes that cooling ages record the time at which a footwall rock is exhumed along a normal fault through a specific near-horizontal isotherm. Rate estimates are derived from the inverse slope of age-paleodepth and age-

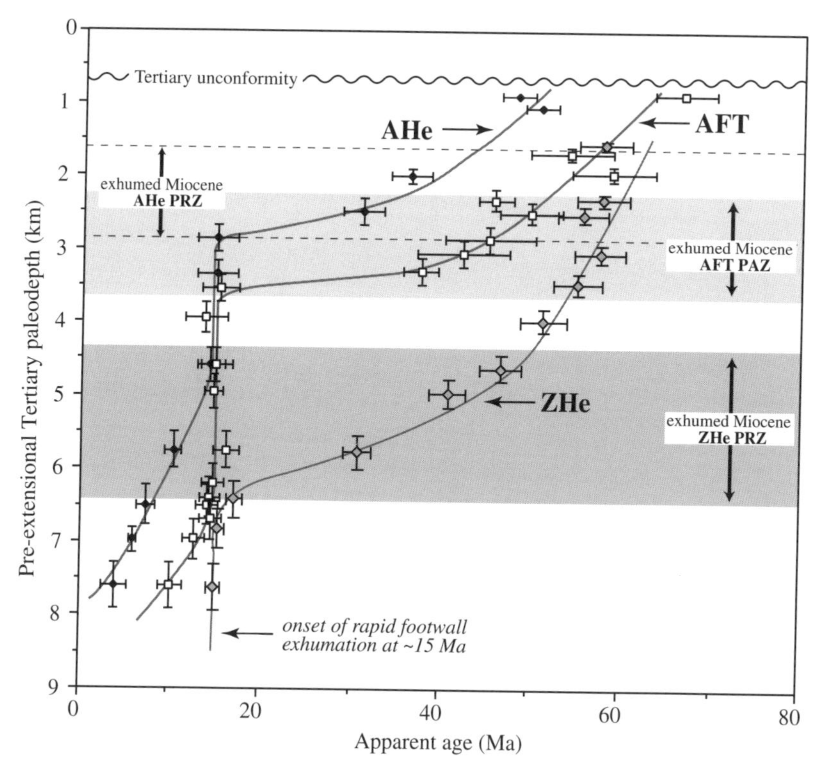

Figure 5. Apatite fission track (AFT) and apatite and zircon (U-Th)/He (AHe and ZHe) thermochronological data from the central Wassuk Range and Grey Hills displayed against preextensional Miocene paleodepths. The data show clearly defined partial annealing/retention zones and inflection points at ~15 Ma, marking the onset of very rapid cooling and exhumation. Cooling ages below the inflection points are invariant over a wide range of structural relief indicating very rapid cooling rates (apatite fission track and (U-Th)/He data from Stockli et al. 2002; zircon (U-Th)/He data from Tagami et al. 2003).

distance relationships from samples that resided at structural levels below inflection points of the different thermochronometers. Using age data from samples that resided within partial retention zones will result in erroneous rate estimates and thus it is imperative to ensure that thermochronometric ages are true cooling ages.

High-angle normal faults. In the case of high-angle normal faulting, the inverse of the slope of a best fit line through the cooling age trend in an age against paleodepth/elevation diagram has traditionally been used to estimate apparent exhumation rates, assuming 60° dipping normal fault planes (e.g., Naeser et al. 1983; Kowallis et al. 1990). However, this approach is only useful to obtain approximate apparent exhumation and fault slip rates and generally cannot be used to quantify true exhumation rates, variations in exhumation rates, fault dip angle, footwall tilting, or exhumation duration (Ehlers et al. 2003). In fact, the inverse slope method commonly underestimates true exhumation and fault slip rates, since isotherms are dynamically advected in the footwall as a result of tectonic and erosional exhumation and the creation of topography, and are depressed in the hanging wall due to sedimentation, resulting in curved isotherms and zero-age thermochronometer surfaces (e.g., Ehlers and Chapman 1999; Ehlers et al. 2001, 2003; Ehlers 2005). Ehlers and others (2003) presented an elegant two-dimensional thermo-kinematic numerical model to translate apparent exhumation

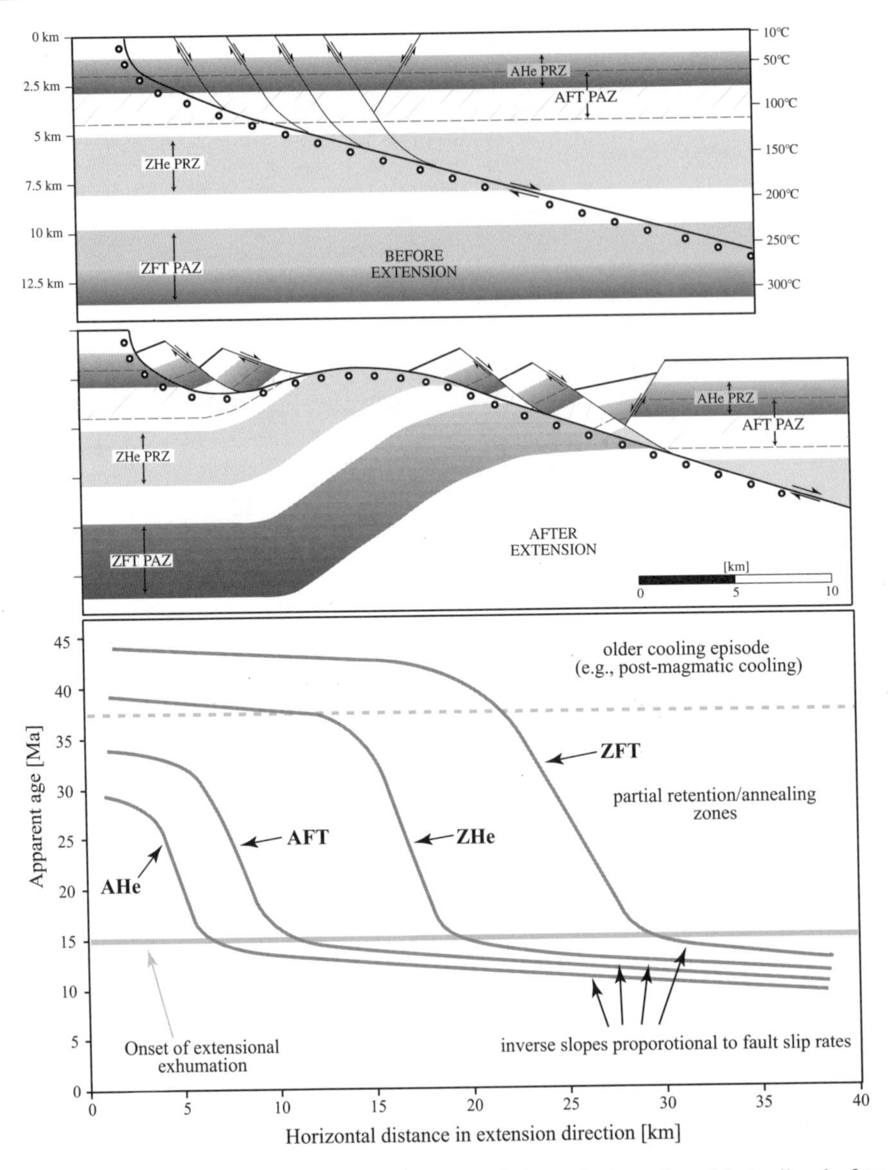

Figure 6. Conceptual model illustrating the structural evolution and exhumation of footwall rocks from beneath low-angle normal faults before and after extension. Sampling transects are devised to obtain samples from the maximum exposed distance parallel to slip direction of low-angle normal faults. Sample locations shown at depth are laterally projected into cross-section view and reflect differential exposure perpendicular to the extension direction (e.g., 3D fault plane corrugations). For discussion of structural complexities and their impact on timing and rates of tectonic exhumation (e.g., isostatic flexure, secondary breakaways, or hanging wall extension) see text. Lower diagram depicts schematic age-distance relationships for different thermochronometers (AHe, apatite (U-Th)/He; AFT, apatite fission track; ZHe, zircon (U-Th)/He; ZFT, zircon fission track) from a sample transect across a metamorphic core complex. Systematic relationships display exhumed fossil partial retention/annealing zones and marked inflection points, constraining the onset of tectonic exhumation and slip along a low-angle normal fault. Ages from below inflection points trend systematically younger in down-dip direction as a result of progressive exhumation (e.g., Foster and John 1999). Inverse slope of regression lines can be used to estimate fault slip rates (see text for detailed discussion).

rates into true exhumation and fault slip rates, and to estimate fault dip angles and the duration of exhumation inferred from thermochronometric data sets. This numerical method is most effective for determining exhumation and fault slip rates from brittle high-angle fault-bounded crustal blocks such as in the Basin and Range province.

Low-angle normal faults. During extension, footwall rocks progressively cool during exhumation along a detachment, resulting in a systematic relationship between cooling ages and distance in slip direction along the fault plane (e.g., Foster et al. 1993) (Fig. 6). Thermochronological slip rate studies of low angle normal faults have been largely restricted to the Basin and Range province (Foster et al. 1993; Fitzgerald et al. 1994; Ketcham 1996; Scott et al. 1998; Foster and John 1999; Fayon et al. 2000; Wells et al. 2000; Stockli et al. 2001; Brady 2002; Carter et al. 2004) and the Greek Cycladic Islands in the Aegean Sea (John and Howard 1995; Brichau 2004).

Fault slip rates can be estimated from thermochronological age trends from low-angle normal fault footwall traverses as long as (1) isotherms are near-horizontal and stationary (Ketcham 1996) and (2) cooling of footwall rocks is only attributable to vertical exhumation caused by fault slip (e.g., Wells et al. 2000). Foster and John (1999) pointed out that for accurate slip rate estimates, closure temperature isotherms have to remain near-horizontal or in a dynamic steady-state during the time interval recorded by specific thermochronometers. Using two-dimensional numerical models investigating the conductive thermal evolution of footwalls beneath low-angle normal faults, Ketcham (1996) demonstrated that despite significant initial advection, isotherms reach a steady-state configuration within <1 m.y. after the onset of extensional fault slip. However, the models also showed that deeper, hotter isotherms are advected more significantly than shallower, cooler isotherms, leading to isotherm compression and possibly causing an underestimation of derived slip rates. Derived slip rates should be regarded as minimum estimates due to isotherm advection and are more significantly underestimated for higher temperature thermochronometers. Therefore, thermochronometers with the lowest possible closure temperatures tend to minimize the uncertainty attributable to isotherm advection. However, the lower the thermal sensitivity of a thermochronometer, the more likely it is that observed exhumation rates are not purely tectonic exhumation rates, but are affected by isostatic footwall doming, erosional exhumation, and topographic cooling.

On the other hand, isostatic footwall rebound and significant extension and structural thinning of the hanging wall can contribute an additional component of vertical exhumation that is not related to simple fault slip, leading to an overestimation of the fault slip rate (e.g., Wells et al. 2000). Other processes including fault dip changes with depth (rolling hinge), the existence of secondary breakaway faults, slip along multiple detachments, or active amplification of fault plane corrugations need to be critically evaluated to ensure that estimated slip rates can be assumed to truly reflect actual fault slip rates. As is the case for determining the timing of onset of extensional faulting, the derivation of meaningful and accurate fault slip rates from age-distance relationships directly depends on a sound understanding of structural and tectonic parameters as well as an awareness of the assumptions and limitations of the technique.

As discussed in detail above, isotherms must be near-horizontal or in a dynamic advective equilibrium controlled by tectonic exhumation for estimated slip rates to be meaningful. The thermal field can be locally perturbed by syn-extensional plutonism or hydrothermal activity along detachment faults. Syn-extensional intrusions have been documented in many metamorphic core complexes and might play an important role in the formation of low angle normal faults (e.g., Lister and Baldwin 1993). It is, therefore, important to establish whether the applied thermochronometer records post-magmatic cooling or fault-induced tectonic exhumation of the intrusion and the adjacent country rock. Morrison and Anderson (1998) suggested that advective heating or cooling by fluid flow along detachment faults might result in cooling ages that do not reflect simple conductive cooling. However, thermochronometric ages do not appear to support this hypothesis, since cooling histories do not vary drastically over short distances from the fault plane (e.g., Brady 2002).

It is worth noting that footwall rocks in metamorphic core complexes are commonly characterized by strong penetrative deformation that affects mineral phases used for geo- and thermochronometry. In particular, dynamic recrystallization, mineralogical phase changes or alteration, and brittle fracturing of datable minerals can significantly affect analytical results and data interpretation. For example, dynamic recrystallization of muscovite can lead to a spectrum of effective $^{40}Ar/^{39}Ar$ closure temperatures as a function of a newly formed grain size distribution (e.g., Hames and Bowring 1994; Wells et al. 2000; Lovera et al. 2002). Similarly, anomalous $^{40}Ar/^{39}Ar$ age spectra observed in K-feldspar have been attributed to brittle low-temperature fracturing and alteration (Lovera et al. 2002). Although many of these complexities might render interpretation of thermochronometric data difficult, at the same time these data can be used to address and quantify structural, microstructural, magmatic, or hydrothermal phenomena. In any case, thermochronometric determinations of slip rates should be cautiously interpreted as apparent slip rate estimates (Wells et al. 2000).

Thermochronometrically constrained slip rates are time-averaged estimates and do not preclude faster or slower rates over shorter time intervals (Foster and John 1999). Single thermochronometer sample densities and data precision (2σ) generally do not allow for derivation of time-variable estimates of fault slip rates. For this reason a linear regression approach is used to calculate time-averaged slip rates. Regression uncertainties of time-averaged fault slip rates can be quite large (e.g., Foster and John 1999), but can be reduced by maximizing the total horizontal distance of sample locations and by selection of higher precision thermochronometers (e.g., Brady 2002).

The application of thermochronology to metamorphic core complexes and low-angle normal faults has greatly increased our understanding of tectonic exhumation of middle crustal rocks, rates of slip, and geometry of detachments (see Foster and John 1999 and references therein). Foster and John (1999) discussed very compelling case studies using fission track and K-feldspar $^{40}Ar/^{39}Ar$ data to estimate fault slip rates for low angle detachment faults from several metamorphic core complexes in the Colorado Extensional Corridor of the southwest U.S.A. Apparent slip rates determined from these regionally-extensive low angle normal faults range from ~3 mm/yr to about 7-9 mm/yr and appear to be supported by geological slip rate estimates derived from geochronological constraints on restored structural markers. In the same area, Brady (2002) successfully demonstrated the utility of apatite (U-Th)/He thermochronometry to calculate higher precision fault slip rates for the Buckskin-Rawhide metamorphic core complex.

Besides the Basin and Range province, the island of Naxos and other Cycladic islands in the Aegean Sea have attracted considerable interest as natural laboratories for the study of extensional processes, including ductile extensional shearing, synextensional magmatism, and brittle low-angle detachment faulting (e.g., Lister et al. 1984, Buick 1991; Gautier et al. 1993; John and Howard 1995; Lister and Forster 1996; Keay et al. 2001). Post-orogenic extension in the Aegean is thought to have started in the Early to Middle Miocene as a result of back-arc extension driven by roll-back of the north-vergent subducting slab and southward retreat of the Hellenic arc (e.g., Jackson 1994; Jolivet et al. 1994). Metamorphic and syn-extensional plutonic rocks exhumed along the Mountsouna shear zone/detachment on the island of Naxos have been the subject of several thermochronological studies investigating footwall cooing/exhumation and fault slip rates (e.g., Andriessen et al. 1979; Wijbrans and McDougall 1988; John and Howard 1995; Brichau 2004). John and Howard (1995) used published $^{40}Ar/^{39}Ar$

biotite, white mica, and hornblende ages to demonstrate a systematic decrease in footwall cooling ages from south to north in extension direction and to estimate fault slip rates of ~7.6–4.7 mm/yr. Apatite fission track analyses from Naxos range in age from ~13.4 Ma to ~8.2 Ma and indicate rapid cooling in the Middle/Late Miocene at maximum rates of ~130 °C/m.y. (e.g., Altherr et al. 1982; Hejl et al. 2002).

More recently, Brichau (2004) systematically investigated the progressive cooling of footwall rocks during exhumation accommodated by the brittle Mountsouna detachment fault using apatite and zircon fission track and (U-Th)/He dating. The ages obtained from each of the four methods yielded internally consistent results that systematically decrease northwards in the slip direction. Their results indicate very rapid tectonically-controlled footwall exhumation and suggest a minimum detachment fault slip rate of ~16–8 mm/yr between 12–9 Ma, implying a minimum displacement of ~25 km along the Mountsouna detachment on the island of Naxos (Fig. 7) (Brichau 2004).

Thermochronometric constraints on fault dip angles

Some of the most polarizing and contested questions in the field of extensional tectonics concern the mechanical viability of fault motion along low-angle detachment faults, initiation angles of low-angle normal faults, and alternative models attempting to reconcile the phenomenon of low-angle normal faults with Andersonian mechanics (e.g., Buck 1988; Wernicke 1995). Much of this debate has centered on whether low-angle normal faults can move in low-angle geometry or if these faults initiate as high-angle normal faults and subsequently rotate as inactive or active faults into a low angle configuration. Not surprisingly, the application of thermochronometry to extensional tectonics has focused on these critical questions, using variations in cooling histories and paleo-temperature along detachment faults to shed light on the initial geometry and the active geometry during tectonic footwall exhumation. Several studies have used thermochronometric results in support of the different

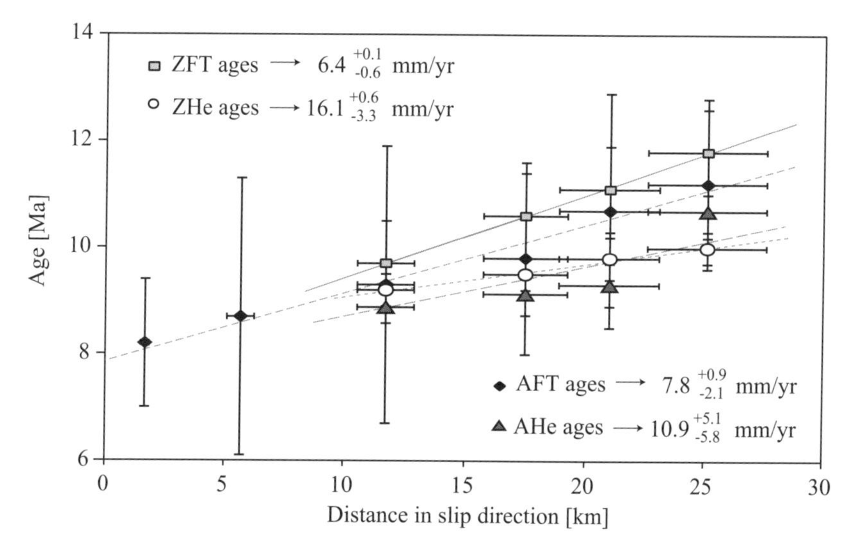

Figure 7. Case study illustrating the systematic decrease of fission track and (U-Th)/He ages in slip direction (~25 km) in the footwall of the Mountsouna detachment on the island of Naxos in the Aegean Sea (Brichau 2004). Average minimum fault slip rates for the Mountsouna detachment are calculated from linear regression of age-distance arrays of fission track and apatite (U-Th)/He ages (Brichau 2004) and zircon (U-Th)/He ages (Brichau and Stockli, unpublished). All error bars are 2 sigma.

structural models for the geometric evolution of low-angle normal faults (e.g., Foster et al. 1990; Holm et al. 1992; Lee 1995, Foster and John 1999; Stockli et al. 2001). Foster and John (1999) pointed out that in principle, thermochronometric methods are capable of constraining the angle of low-angle detachment faults during tectonic exhumation as long as the following criteria are satisfied: (1) the thermal gradient during fault exhumation can be estimated or constrained (see geothermal gradient section below), (2) there is no significant horizontal thermal gradient within the footwall such as can be caused by syn-extensional plutonism, and (3) the cooling of samples from the footwall reflects exhumation in response to a single detachment system.

There are three principle thermochronometric approaches that can be used to elucidate the slip geometry of low-angle normal faults. A first technique uses the systematic geographical distribution of cooling ages from the footwall immediately below the detachment in direction of fault slip (e.g., John and Foster 1993). Contoured cooling ages from different thermochronometers provide systematic estimates of paleotemperatures and the position of paleoisotherms within the footwall immediately prior to the onset of extensional faulting. This estimated paleotemperature field gradient in the footwall of the low-angle normal fault combined with a pre-extensional geothermal gradient estimate can be used to calculate an initiation angle for the fault (John and Foster 1993; Foster and John 1999). The geothermal gradient can be used to deduce estimates of paleodepth for individual samples based on the paleotemperature field gradient in the footwall. These paleodepth or paleodepth ranges for estimated geothermal gradients can be plotted against horizontal distance of samples in fault slip direction to estimate the fault initiation angle of the detachment fault. This approach yielded detachment initiation angle estimates of 15–30° and <30° for the Chemehuevi detachment and Harquahala metamorphic core complex in the Colorado River Extensional Corridor of the southwest U.S.A., respectively, using a wide range of geothermal gradients (30 ± 10 °C/km) (John and Foster 1993; Foster and John 1999).

A second approach uses thermochronometric data to constrain the magnitude of footwall tilting and the known modern fault geometry to reconstruct the initial fault geometry. This technique is not limited to low-angle normal faults, but can also be employed in reconstructing fault geometries of rotational high-angle normal faults (e.g., Stockli et al. 2001; Wong and Gans 2004). Again, the amount of footwall rotation can be estimated from the systematic increase in maximum temperatures in fault slip direction and the horizontal distance over which this paleotemperature field gradient is measured. Stockli and others (2001) demonstrated this approach on footwall rocks below the Sevier Desert detachment fault, documenting ~15–20° of footwall tilting and rotation of the detachment fault since onset of faulting, assuming a geothermal gradient of 25 °C/km (Fig. 8).

A third approach to determine the slip geometry of low-angle normal faults uses thermal histories of multiple samples to constrain paleotemperatures at different structural levels of the extensional footwall at specific times. Using thermal histories derived either from an age-time multi-technique approach or from $^{40}Ar/^{39}Ar$ K-feldspar MDD or fission track modeling, time-specific paleotemperature field gradients along a detachment fault can be reconstructed. The advantage of this technique is that it allows for paleotemperature contouring of rocks below extensional faults before, during, or after extensional faulting as long as thermal histories are well constrained. Assuming a range of geothermal gradients, one can constrain the dip of detachment faults and determine how the dip has changed through time. Lee (1995) used thermal histories based on $^{40}Ar/^{39}Ar$ K-feldspar MDD modeling from footwall rocks in the Snake Range, Nevada to reconstruct fault dip changes in both time and space. Based on these data, he argued that the fault plane dip changed in slip direction through time consistent with a rolling hinge type geometric evolution of the Snake Range decollement.

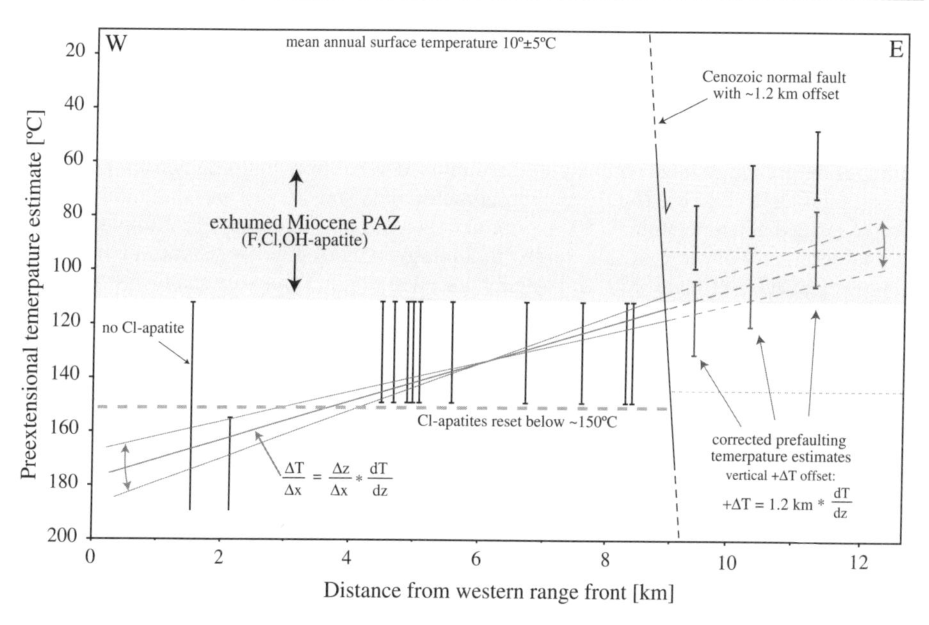

Figure 8. Pre-extensional temperature estimates derived from multi-compositional apatite fission track data displayed against horizontal distance from across the Canyon Range parallel to the slip direction of the Sevier Desert detachment (from Stockli et al. 2000). Temperature estimates for the three easternmost samples are corrected for ~1.2 km of offset along the eastern range-bounding normal fault. Assuming a geothermal gradient, pre-extensional temperature estimates serve as a proxy for structural depth and can be used to calculate the amount of footwall tilting. Given a geothermal gradient (dT/dz) of ~25 °C/km, the systematic spatial trend in temperatures ($\Delta T/\Delta x$) translates into ~15–20° of footwall tilt.

Regardless of which approach is used, estimates of fault geometry are subject to similar problems and uncertainties as fault slip rate calculations (see section above). In particular, differential isotherm compression and syn-extensional plutonism can lead to non-linear thermal field gradients that will affect the data. On the other hand, thermochronometric data can also be used to evaluate and image such phenomena, since down-dip trends would be expected to deviate considerably from linear temperature-depth relationships.

Estimation of crustal tilting and footwall rotation

Thermochronometric data from exhumed footwall blocks can be used to constrain the amount of horizontal-axis rotation due to isostatic rebound or subsequent normal faulting and crustal tilting. This is accomplished by investigating the systematic increase in maximum pre-extensional temperatures along a roughly horizontal transect perpendicular to the tilt axis, which is commonly orthogonal to the fault slip direction, assuming a steady-state geothermal gradient and near-horizontal pre-extensional isotherms (Fig. 8). Significant topography and elevation differences between individual samples can lead to erroneous results, although small differences in elevation between samples can usually be ignored. This is often a reasonable assumption in metamorphic core complexes with limited topographic relief along sampling transects, but has to be taken into account in areas with significant tectonic relief. In such cases, two-dimensional thermo-kinetic modeling (e.g., Ehlers et al. 2003) can be used to constrain initial faults angles and amounts of footwall tilting.

Assuming a reasonable geothermal gradient or a range of geothermal gradients, these pre-extensional temperature estimates can be used as a proxy for structural depth exposed in the footwall block. Thus, the maximum range of structural depths as a function of horizontal footwall distance in fault slip direction, assuming a geothermal gradient, can easily be used to calculate the amount of footwall tilting (e.g., Stockli et al. 2001; Wong and Gans 2004).

Stockli and others (2001) employed this approach to estimate the amount of footwall tilting using pre-extensional temperature constraints derived from multi-compositional apatite fission track data (Fig. 8). The thermochronometric data showed that the structurally lowest sample resided at temperatures 150 °C prior to exhumation, whereas samples from the central and western parts of the footwall resided in a relatively narrow pre-extensional temperature window between ~110 and 150 °C. Partially annealed samples from the shallowest structural portion of the footwall suggest pre-extensional temperatures decreasing from ~80 ± 10 °C to about 60 ± 10 °C. These pre-extensional paleotemperature estimates, assuming a geothermal gradient of 25 °C/km, are indicative of ~15–20° of footwall tilting as a result of rapid footwall exhumation; an estimate supported by independent geological constraints. However, in many extensional provinces, fault block tilt estimates can be more easily derived from tilted pre-extensional sedimentary or volcanic strata (e.g., Stockli et al. 2002).

Estimation of normal fault offset magnitude

Besides providing constraints on the timing and rate of extensional faulting, thermochronometric cooling ages from rocks exposed in footwalls of normal faults have traditionally been used to estimate the magnitude of vertical tectonic exhumation. This approach has predominantly been employed in the absence of preserved hanging wall rocks with offset geological markers across the normal fault and the lack of other geological constraints on fault slip magnitude. Thermochronometric cooling age information from the structurally lowest sample can yield a pre-extensional maximum paleotemperature that in turn can be used to estimate vertical exhumation, assuming a geothermal gradient. These maximum vertical exhumation estimates can be translated into fault slip magnitudes assuming specific fault geometries. In continental extensional and divergent margin tectonic settings, the structural grain is often dominated by asymmetric, high-angle normal fault-bounded tilted crustal blocks and intervening half-grabens with syn- and post-tectonic sedimentary fill. In these cases, estimates of total apparent extensional fault displacement have to be regarded as minimum estimates since syn- and post-tectonic hanging wall sedimentation can obscure portions of exhumed footwall rocks. Geological and geophysical constraints providing estimates for the thickness of sedimentary fill can be combined with thermochronometric exhumation estimates to arrive at more accurate total exhumation and fault slip estimate (e.g., Stockli et al. 2002, 2003).

Relatively small-scale (<2–5 km) displacement across normal faults in the upper crust generally is insufficient to exhume rocks from below partial retention zones or closure isotherms, providing limited direct information on timing and magnitude of fault exhumation. Apparent normal fault throw across such normal faults can often be difficult to assess in geological settings that lack identifiable geologic offset markers, for example in plutonic provinces. Thermochronometric data from both upper crustal footwall and hanging wall blocks can help constrain the magnitude of offset across normal faults. Repetition of thermochronometric ages and age patterns that vary with depth or the repetition and offset of a portion of fossil apparent age against paleodepth/elevation profiles can be used as passive structural markers to estimate apparent fault displacements across individual normal faults (Fig. 9). Invariant cooling ages from below partial retention zones cannot be used as structural markers. Combined with fault kinematic data, thermochronometric offset estimates can constrain structural throw across a normal fault.

Several studies have utilized low-temperature thermochronometric data to quantify the magnitude of apparent fault displacement (e.g., Fitzgerald et al. 1994; Kamp 1997; McInnes et al. 1999; Stockli et al. 2002, 2003). Stockli and others (2003) illustrated this technique using

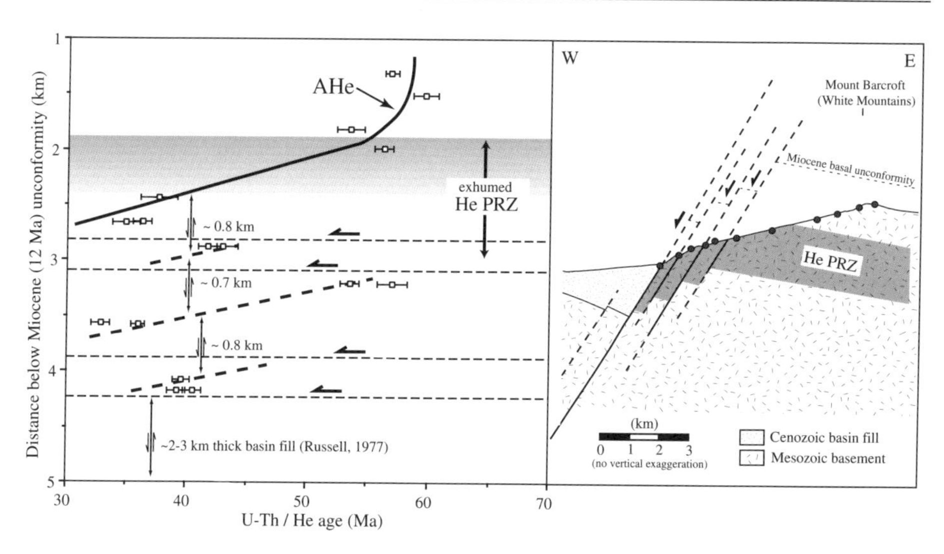

Figure 9. Apatite (U-Th)/He data from the central White Mountains, California, illustrating the use of structurally offset and repeated age-depth patters to estimate normal fault displacement. Diagram shows a systematic offset of apatite (U-Th)/He partial retention across four normal fault zones that progressively down-drop hanging wall fault slivers. Structural duplication and offsets in the (U-Th)/He apparent age versus paleodepth pattern illustrate the down-to-the-west normal faulting and allow the estimation of fault throw across individual normal faults using the pre-extensional thermal state within the crustal fault block as a structural marker. Simplified cross-section shows location of samples and normal fault strand along western flank of the White Mountains (after Stockli et al. 2003).

apatite fission track and (U-Th)/He data from the central White Mountains in California to estimate fault offsets across several normal fault strands sequentially down-dropping granitic hanging wall rocks. The (U-Th)/He age-paleodepth profile exhibits multiple systematic offsets across these normal faults, juxtaposing older apparent ages in the hanging wall with younger ages in the footwall (Fig. 9). These data show that the three major normal faults within the range accommodate ~2.5 km of cumulative normal displacement, without assuming any geothermal gradient information. Gravity and seismic studies estimate that hanging wall basin fill is ~2-3 km thick, resulting in a minimum of ~4.5–5.5 km of Cenozoic normal displacement across the White Mountains normal fault. This estimate is roughly supported by estimates derived from structural reconstruction of tilted volcanic stratigraphy indicating a maximum of ~6 km of Cenozoic normal displacement.

Geothermal gradient estimates

Constraints on the thermal structure of the crust prior to the onset of extensional faulting are important for understanding the external and internal driving forces leading to extensional faulting, especially when trying to evaluate the role of thermal weakening and magmatism in triggering extensional faulting (e.g., Kusznir and Park 2002). Highly extended extensional tectonic provinces are characterized by deeply exhumed crustal rocks that are exhumed by either low-angle normal faults or rotational high-angle normal faults or a combination of both. In particular, rotational high-angle normal faults often result in the tectonic exhumation of intact crustal-scale fault blocks (e.g., Proffett 1977) and the exposure of tilted crustal sections. These rigid crustal-scale fault blocks undergo large-magnitude horizontal axis rotation and expose up to ~15 km of pre-extensional upper crust, providing important insights into the structural and thermal evolution of the crust prior to and during large-scale crustal extension (e.g., Davis et al. 1980; Holm and Wernicke 1990; Fryxell et al. 1992; Mueller and Snoke 1993; Howard and Foster 1996; Surpless et al. 2001; Stockli et al. 2002; Brady et al. 2002). Detailed structural reconstruction of tilted crustal fault blocks and the determination of a pre-extensional paleodatum from pre-extensional strata, unconformities, and magmatic constraints are needed to determine paleodepths within these exhumed crustal sections (Figs. 3 and 4). Several factors including uncertainties in estimating the amount of crustal tilting, paleotopography of unconformities, and preextensional volcanic and sedimentary overburden above the structural paleohorizontal datum need to be considered in order to accurately estimate the preextensional depths of thermochronometric samples (e.g., Stockli et al. 2002). Paleodepth measurements are commonly based on structural depth below a paleohorizontal reference plane, such as Mesozoic or Tertiary unconformities in western North America (Fig. 4). Several studies have used these thermochronometric techniques to estimate paleogeothermal gradients from exhumed crustal sections exposed in the footwalls of major normal faults in the Basin and Range province (e.g., Foster et al. 1991, 1994; Fitzgerald et al. 1991; Gans et al. 1991; Howard and Foster 1996; Foster and John 1999; Stockli 1999; Stockli et al. 2003).

Stockli and others (2002) utilized several thermochronometric methods to estimate the geothermal gradient immediately prior to the onset of middle Miocene extension in the central Wassuk Range of the western Basin and Range province. They determined the temperature difference between the Tertiary paleosurface and the depth to the observed inflection points of different thermochronometers, assuming a mean annual surface temperature of 10 ± 5 °C (Fig. 10). The paleodepth of the apatite fission track inflection point (~3.6 km) yields a geothermal gradient estimate of 27 ± 5 °C/km at the onset of Tertiary extension. The distance from the Tertiary unconformity to the base of the He partial retention zones can similarly be used to estimate the preextensional Miocene geothermal gradient. Paleodepth estimates for the apatite (~2.9 km) and zircon (~6.4 km) (U-Th)/He inflection points suggests a preextensional geothermal gradient of 26 ± 5 °C/km and 28 ± 5 °C/km, respectively (Fig. 10), in excellent agreement with the estimate derived from the fission track data. In addition, the difference in paleodepths between the base of the partial retention/annealing zones can be used to estimate a geothermal gradient, although associated uncertainties are large due to the relatively small difference in paleodepths and temperatures. For example, the paleodepth difference between apatite fission track and (U-Th)/He inflection points yields a rough estimate of 31 ± 10 °C/km for the preextensional Miocene geothermal gradient.

A slightly different approach for evaluating paleogeothermal gradients uses thermochronological modeling results to estimate the temperature of a sample immediately prior to the onset of extension as constrained by the age of the inflection point. Linear regression of the systematic paleotemperature trend as a function of paleodepth yields a pre-extensional geothermal gradient. Stockli and others (2002) used apatite fission track modeling results from a wide range of paleodepths to estimate the paleogeothermal gradient prior to the onset of extensional faulting at 15 Ma in the Wassuk Range (Fig. 11). The linear relationship between modeled paleotemperatures and paleodepths yields a preextensional paleogeothermal gradient of 27 ± 3 °C/km, which is in good agreement with estimates derived from other thermochronological lines of evidence (Fig. 11). Furthermore, the upper intercept of the linear regression of the modeled preextensional temperatures suggests ~0.8 km of pre-extensional overburden above the basal Tertiary unconformity, assuming a 10 ± 5 °C mean annual surface temperature. This estimate is consistent with the average regional thickness estimates of the pre-extensional Tertiary andesite.

Spatial and temporal distribution of extension

Despite decades of study, many of the fundamental processes that govern continental rifting or extensional rupturing of continental lithosphere remain poorly understood. Specifically, our

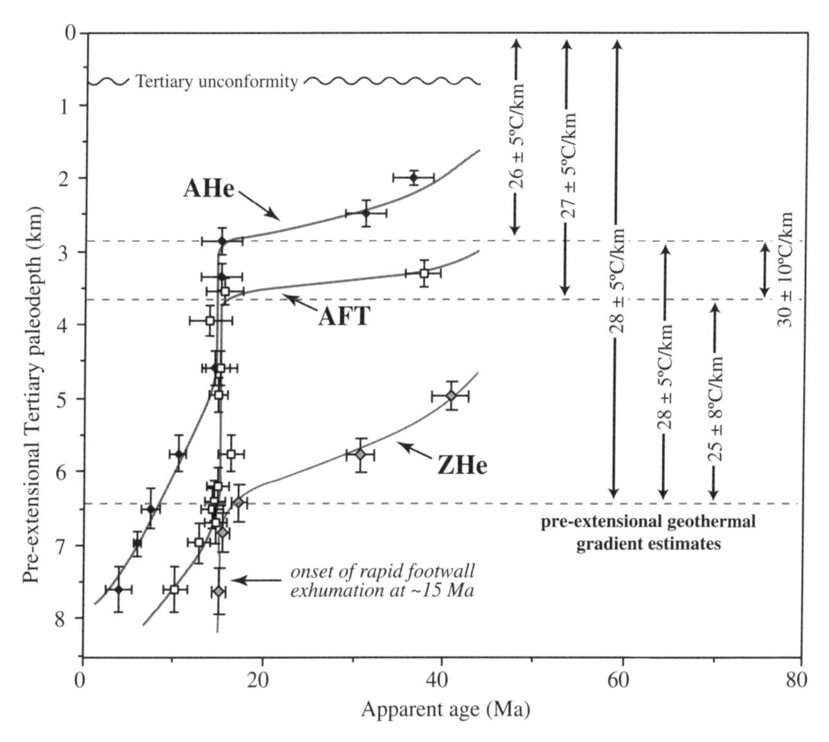

Figure 10. Apatite fission track (AFT) and apatite and zircon (U-Th)/He (AHe and ZHe) thermochronological data from the Wassuk Range area (Fig. 5) constrain the paleodepths of pre-extensional inflection points (paleoisotherms), allowing estimation of geothermal gradients prior to the onset of middle Miocene extension. See text for detailed description of geothermal gradient calculations and estimates (apatite fission track and (U-Th)/He data from Stockli et al. 2002; zircon (U-Th)/He data from Tagami et al. 2003).

limited knowledge of how extensional strain is spatially and temporally distributed has made it difficult to adequately evaluate and test models for the dynamic evolution of extensional fault systems. The spatial and temporal distribution of rifting, the partitioning of extensional faulting, the strain distribution and architectural evolution of extensional provinces and divergent continental margins, and the pre- and syn-extensional thermal structure of the crust in extensional provinces are often poorly constrained. Nevertheless, they are important to our geodynamic understanding of extensional systems, and are key to successful modeling of extensional tectonic processes such as continental rifting and transition to sea-floor spreading or the role of extension in collisional orogenic belts.

Along passive continental margins, the origin, longevity and erosional or tectonic destruction of rift flanks represent first-order constraints on the tectonic, geophysical, and geomorphic framework of rift systems and the evolution of divergent continental margins. Knowledge of the temporal and spatial distribution of crustal extension and the timing of flexural amplification and exhumation of a rift flank is important in evaluating the role of processes such as active vs. passive asthenospheric upwelling (Sengör and Burke 1978), secondary convection (Buck 1986), and flexural unloading of the crust (Weissel and Karner 1989), as well as the distribution of subcrustal lithospheric extension relative to crustal thinning (e.g., Lister et al. 1991).

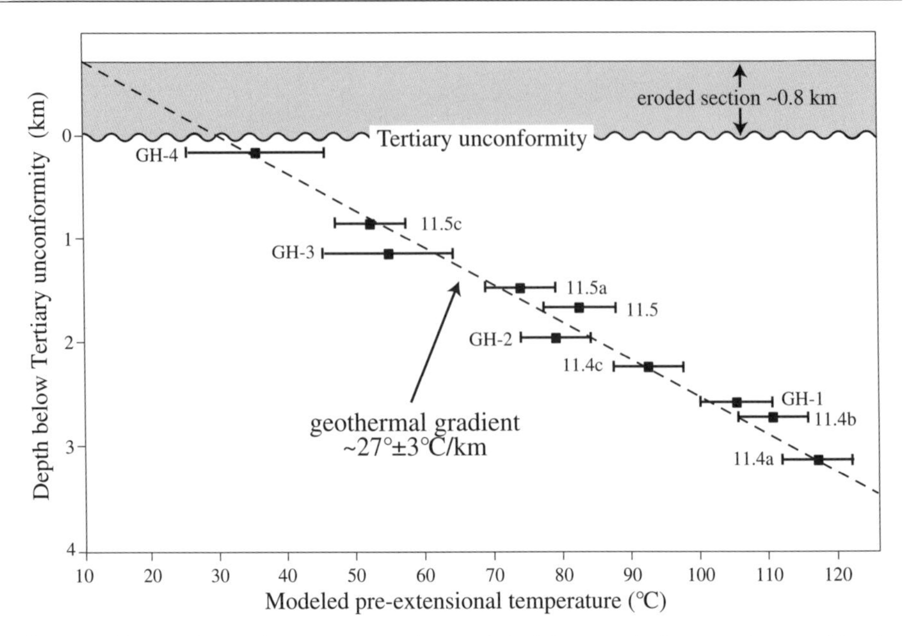

Figure 11. Preextensional paleotemperature estimates derived from apatite fission track age and length modeling (MoneTrax) plotted against depth below Tertiary unconformity in the Wassuk Range, Nevada. Linear regression of temperature estimates intercepts the assumed preextensional mean annual surface temperature (10° ± 5 °C) at ~0.8 km, suggesting ~0.8 km of Tertiary overburden prior to extensional faulting. Slope of regression line yields a preextensional geothermal gradient estimate of 27 ± 3 °C/km (after Stockli et al. 2002).

The most exhaustive thermochronometric regional database for any extensional province has been assembled for the northern and central Basin and Range province. Foster and John (1999) provide an excellent overview of thermochronological applications to crustal extension in the Colorado River extensional corridor of the central Basin and Range province. Several recent studies have shed further light on the timing of extensional faulting in different portions of the central and southern Basin and Range province of the southwest U.S.A. and Mexico (e.g., Foster et al. 1993; Fitzgerald et al. 1994; Fayon et al. 2000; Reiners et al. 2000; Axen et al. 2000; Wong and Gans 2004).

The northern Basin and Range province is characterized by a complex, multi-stage history of Tertiary extensional faulting, but the location, timing and magnitude of the various periods are disputed. As an additional complication, the western portion of the province has been overprinted by late Tertiary transcurrent and transtensional faulting within the Walker Lane belt. Over the past several decades, many studies have elucidated the timing of extensional faulting associated with different structural settings in different areas of the northern Basin and Range (e.g., Naeser et al. 1983; Kowallis et al. 1990; Lee 1995; Miller et al. 1999; Stockli 1999; Wells et al. 2000; Stockli et al. 2000, 2001, 2002, 2003; Coleman et al. 2001; Armstrong et al. 2003; Egger et al. 2003; Colgan et al. 2004). Compilation of thermochronological data shows that most of the northern Basin and Range province was affected by a very-large-magnitude extensional episode between 19–13 Ma. Figure 12 shows the spatial distribution of extensional faulting across the central portion of the northern Basin and Range province and the transition zone to the stable Sierra Nevada as constrained by systematic regional apatite fission-track and (U-Th)/He data (Stockli 1999; Stockli et al. in review). These data point to the early to middle Miocene as a period of major

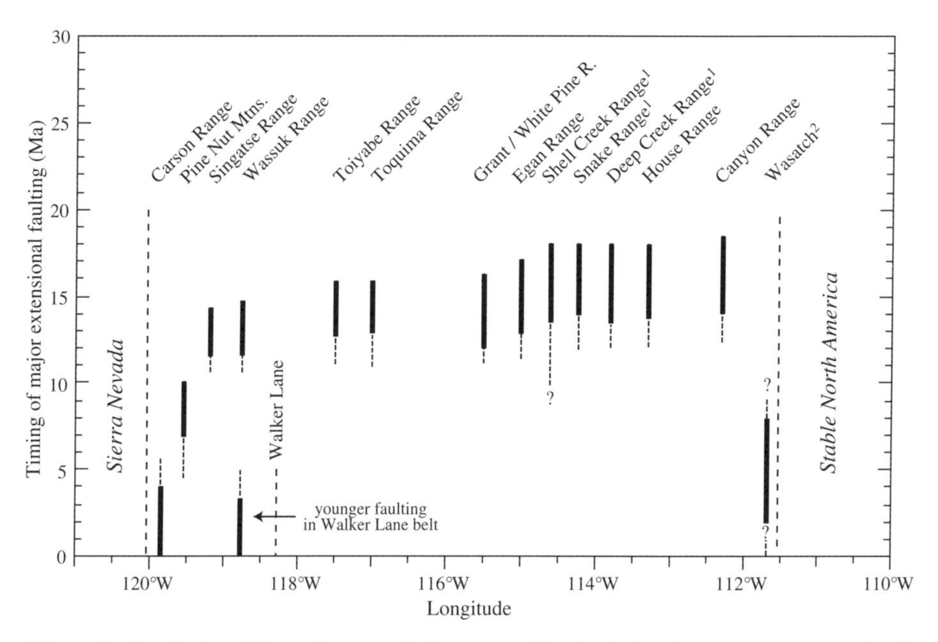

Figure 12. Compilation of thermochronometric data constraining the timing of major Miocene extensional fault slip in the northern Basin and Range province at the latitude of ~39°N. The regional data point to relatively synchronous extensional faulting throughout the central portion of the northern Basin and Range province between 19–13 Ma. In post-middle Miocene times, faulting appears to be localized along the margins of the northern Basin and Range province. In the Late Miocene and Pliocene, renewed rapid exhumation and faulting in the western part of the province is associated with right-lateral transtension within the Walker Lane Belt (e.g., Surpless et al. 2001; Stockli et al. 2003). Data from Stockli 1999 and Stockli and others (in review). Superscripts indicate published data from (1) Miller et al. (1999) and (2) Naeser et al. (1983).

extensional faulting starting between about 19–13 Ma. Although relatively short-lived, this event produced large-magnitude extension at the scale of the entire province. More detailed analysis of regional data patterns points to the existence of regional structural domains that extended simultaneously, exhibiting slightly different extensional histories compared to adjacent extensional domains (e.g., Dumitru et al. 2000).

Systematically younger apatite fission track and (U-Th)/He ages in the transition zone between the northern Basin and Range province and Sierra Nevada document a widening of the extensional province and progressive encroachment of extensional faulting into the unextended Sierra Nevada since the middle Miocene (e.g., Surpless et al. 2001) (Fig. 12). Furthermore, thermochronological data from the footwall of rhomboidal pull-apart basins and other extensional structures associated with right-lateral transcurrent deformation within the Walker Lane belt started <10–6 Ma and migrated systematically westward (e.g., Oldow et al. 1994; Stockli et al. 2003). This very large thermochronometric data set from the northern Basin and Range province constrains the temporal and spatial distribution of extensional faulting, long-term fault slip rates, extensional fault geometries, and pre- and post-extensional geothermal gradients. The constrains on the temporal, spatial, and thermal evolution of the Basin and Range extensional province are crucial for an improved understanding the geodynamic evolution and internal and external driving forces of Tertiary intra-continental extension and transtension in western North America.

CONCLUSIONS AND FUTURE DIRECTIONS

Our understanding of many fundamental structural and thermal processes governing the initiation of normal faulting and rifting, temporal variations in rift architecture, rates of extensional processes, and the geometric evolution of extensional fault systems remains incomplete. For example, along divergent continental margins the processes that control initial rift geometry and ultimately lead to the rupture of continental lithosphere and to the creation of oceanic crust are poorly understood. Specifically, our limited knowledge of how extensional strain is spatially and temporally distributed has made it difficult to adequately evaluate and test models for the dynamic evolution of normal faulting in divergent margin, intra-continental, syn-collisional, and transcurrent tectonic environments. Integrated low temperature $^{40}Ar/^{39}Ar$, fission track, and U-Th/He dating of different mineral phases allows us to reconstruct cooling histories of rocks from exhumed footwall fault blocks between ~350 °C and ~40 °C. These techniques provide a powerful means to reconstruct cooling histories of exhumed rocks and to quantify the timing and rates of extensional processes responsible for the tectonic unroofing of rocks from the footwalls beneath major normal faults in extensional tectonic environments. Many studies have successfully utilized thermochronometric data and have significantly advanced our understanding of fundamental processes in extensional tectonic settings by elucidating important aspects such as (1) the timing of onset of extensional faulting and footwall exhumation, (2) the duration of extensional faulting, (3) the rates and variations in rates of vertical tectonic exhumation and fault slip, (4) the magnitude of total exhumation and fault throw, (5) the initial fault geometry and evolution of fault geometries during progressive extension, (6) the thermal structure and geothermal gradients prior to, during, and after extensional faulting, and (7) the role and impact of magmatism during the development of extensional fault systems. However, to ensure the proper interpretation of low-temperature thermochronometric data and the accurate determination of temporal, spatial, and thermal aspects of extensional faulting, it is critical to interpret the data in a sound geological, structural, and magmatic context. Furthermore, it is worth noting that relatively large thermochronometric data sets are required to adequately constrain temporal and thermal aspects of regional extensional faulting (e.g., Foster and John 1999; Stockli et al. in review). For example, in order to determine the geographic distribution of the timing of inception of faulting within an extensional tectonic province, detailed thermochronological sample transects from multiple areas are required.

The application and quantitative interpretation of low-temperature thermochronometry to elucidate the temporal and thermal aspects of tectonic processes represents a rapidly growing field of research. New and future applications of these methods are not restricted to extensional tectonics, but encompass such field as convergent and transcurrent mountain building and landscape evolution.

As the development of apatite (U-Th)/He dating over the past decade has demonstrated, new and developing thermochronometric techniques will continue to have significant impact on research addressing upper crustal and near surface tectonic problems. The development and application of new (U-Th)/He thermochronometers are likely to further expand the thermal sensitivity window, opening new avenues of investigation. Innovation and advances in analytical techniques and numerical modeling of closure profiles and diffusive concentration gradients will allow recovery of continuous thermal histories from individual samples (e.g., Shuster and Farley 2005; Harrison et al. 2005). Furthermore, integrated modeling of multiple thermochronometer data sets, such as combined fission track and (U-Th)/He modeling (e.g., Ketcham 2005) or integrated $^{40}Ar/^{39}Ar$ MDD and (U-Th)/He modeling, will significantly enhance the ability to derive comprehensive and self-consistent thermal histories from single samples. For example, since both He and Ar are lost from crystals by thermally-activated volume diffusion, it should be possible to model K-feldspar $^{40}Ar/^{39}Ar$ and zircon (U-Th)/He data in a single fully integrated MDD model. Besides this combination of multiple thermochronometers, the systematic combination of low-temperature thermochronometric methods with other analytical techniques, such as cosmogenic nuclide dating or Th-U/Pb depth profiling techniques (e.g., Harrison et al. 2005) will allow an integrated investigation of middle crustal to surficial processes during extensional faulting.

Although in many extensional provinces thermochronometric studies have shed light on the timing and rates of large-scale tectonic processes, strategic collection of high-resolution thermochronometric data sets are necessary to address detailed questions. Specifically, an improved understanding of tectonic processes during extensional faulting requires quantitative interpretation of age data in concert with detailed structural, geological and geophysical models and reconstructions. In extensional provinces, such cooling/exhumation processes include extensional exhumation of footwall rocks, vertical exhumation due to hanging wall attenuation, differential isotherm advection, magmatic heat input, generation of topography, sedimentary burial, hydrothermal heating or refrigeration, topographic cooling, climatic processes, et cetera. Ehlers and others (2001 2003) pointed out that in light of these complexities, 1D thermal models are not adequate to precisely quantify cooling and exhumation rates, necessitating more sophisticated 2D models. Several tectonic studies have developed and utilized 2D thermo-kinetic models to interpret thermochronometric data or to investigate the effect of specific tectonic parameters (e.g., Gans et al. 1991; Harrison et al. 1995; Ketcham 1996, Ehlers et al. 2001; Ehlers et al. 2003). In reality, however, many of these processes, such as creation of topography and erosion, are 3D in nature, creating 3D isotherm deflections and lateral thermal gradients. Therefore, quantification of 3D exhumation processes would require high-resolution sampling along strike, quantification of processes and rates governing erosion, sedimentation, topographic evolution, et cetera, and more computationally intensive 3D thermal models (e.g., Braun 2005).

ACKNOWLEDGMENTS

The author would like to thank T. Dumitru, K. Farley, N. Mancktelow, D. Seward, E. Miller, B. Fügenschuh, B. Surpless, G. Axen, M. Grove, T.M. Harrison, and S. Brichau for many informative discussions about thermochronology applied to extensional tectonics. I would like to acknowledge financial support for aspects of this work by NSF grants EAR-9417939 and EAR-9725371 (Miller), EAR- 0125879 (Stockli), and a Caltech postdoctoral fellowship to Stockli. I also would like to thank L. Stockli and J.D. Walker for improving various versions of the manuscript, U. Ring and A. Fayon for very careful and constructive reviews, and T. Ehlers and P. Reiners for review and editorial handling.

REFERENCES

Abbate E, Balestrieri ML, Bigazzi G (2002) Morphostructural development of the Eritrean Rift flank (southern Red Sea) inferred from apatite fission track analysis. J Geophys Res 107(B11):2319, doi:10.1029/2001JB001009

Abers GA, Mutter CZ, Fang J (1997) Shallow dips of normal faults during rapid extension; earthquakes in the Woodlark-D'Entrecasteaux rift system, Papua New Guinea. J Geophys Res 102:15,301-15,317

Altherr R, Kreuzer H, Wendt I, Lenz H, Wagner GH, Keller J, Harre W, Höhndorf A (1982) A late Oligocene/early Miocene high temperature belt in the Attic-Cycladic crystalline complex (SE Pelagonian, Greece). Geologisches Jahrbuch E23:97-164

Andriessen PAM, Boelrijk NAIM, Hebeda EH, Priem EH, Verdurmen T, Verschure RH (1979) Dating the events of metamorphism and granitic magmatism in the Alpine Orogen of Naxos (Cyclades, Greece). Contrib Mineral Petrol 69:215-225

Armstrong RL (1982) Cordilleran metamorphic core complexes - from Arizona to southern Canada. Ann Rev Earth Planet Sci 10:129-154

Armstrong RL, Ward P (1991) Evolving geographic patterns of Cenozoic magmatism in the North American Cordillera: the temporal and spatial association of magmatism and metamorphic core complexes. J Geophys Res 96:13,201-13,224

Armstrong PA, Ehlers TA, Chapman DS, Farley KA, Kamp PJJ (2003) Exhumation of the Central Wasatch Mountains, 1: Patterns and timing deduced from Low-temperature Thermochronometry data. J Geophys Res 108:3

Atwater T, Stock J (1998) Pacific-North America plate tectonics of the south-western United States - An update. Int Geol Rev 40:375-402

Axen GJ (1993) Ramp-flat detachment faulting and low-angle normal reactivation of the Tule Springs Thrust, southern Nevada. Geol Soc Am Bull 105:1076-1090

Axen GJ, Fletcher JM (1998) Late Miocene-Pleistocene extensional faulting, northern Gulf of California, Mexico and Salton Trough, California. Int Geol Rev 40:217-244

Axen GJ, Stockli DF, Grove M, Lovera OM, Rothstein DA, Fletcher JM, Farley KA, Abbott PL (2000) Thermal evolution of Monte Blanco dome: Late Neogene low-angle normal faulting during Gulf of California rifting and late Eocene denudation of the eastern Peninsular Ranges. Tectonics 19:197-212

Axen GJ, Lam PS, Grove M, Stockli DF, Hassanzadeh J (2001) Exhumation of the west-central Alborz Mountains, Iran, Caspian subsidence, collision-related tectonics. Geology 29:559–562

Baldwin SL, Lister GS, Hill EJ, Foster DA, McDougall I (1993) Thermochronologic constraints on the tectonic evolution of active metamorphic core complexes, D'Entrecasteaux Islands, Papua New Guinea. Tectonics 12:611-628

Baldwin SL, Lister GS (1998) Thermochronology of the South Cyclades shear zone, Ios, Greece; effects of ductile shear in the argon partial retention zone. J Geophys Res 103:7315-7336

Brady RJ (2002) Very high slip rates on continental extensional faults: new evidence from (U–Th)/He thermochronometry of the Buckskin Mountains, Arizona. Earth Planet Sci Lett 197:95-104

Braun J (2005) Quantitative constraints on the rate of landform evolution derived from low-temperature thermochrohonology. Rev Mineral Geochem 58:351-374

Brichau S (2004) Constraining the tectonic evolution of extensional fault systems in the Cyclades (Greece) using low-temperature thermochronology. PhD Dissertation, University of Mainz and University of Montpellier

Bryant B, Naeser CW (1980) The significance of fission-track ages of apatite in relation to the tectonic history of the Front and Sawatch ranges, Colorado. Geol Soc Am Bull 91:I 156-I 164

Buck WR (1986) Small-scale convection induced by passive rifting; the cause for uplift of rift shoulders. Earth Planet Sci Lett 77:362-372

Buck WR, Martinez F, Steckler MS, Cochran JR (1988) Thermal consequences of lithospheric extension: Pure and simple. Tectonics 7:213-234

Buck WR (1988) Flexural rotation of normal faults. Tectonics 7:959-973

Buck WR (1991) Mode of continental lithospheric extension. J Geophys Res 96:20161-20178

Buick IS, Holland TJB (1991) The nature and distribution of fluids during amphibolite facies metamorphism, Naxos (Greece). J Metamor Geol 9:301-314

Burchfiel BC, Chen Z, Hodges KV, Liu Y, Royden LH, Deng C, Xu J (1992) The South Tibetan detachment system, Himalayan Orogen; extension contemporaneous with and parallel to shortening in a collisional mountain belt. Geol Soc Am Special Paper 269

Carlson WD (1990) Mechanisms and kinetics of apatite fission-track annealing. Am Mineral 75:1120-1139

Carlson WD, Donelick RA, Ketcham RA (1999) Variability of apatite fission track kinetics: I Experimental results. Am Mineral 84:1213-1223

Carter A, Roques D, Bristow CS (2000) Denudation history of onshore central Vietnam; constraints on the Cenozoic evolution of the western margin of the South China Sea. Tectonophysics 322:265-277

Carter TJ, Kohn BP, Foster DA, Gleadow AJW (2004) How the Harcuvar Mountains metamorphic core complex became cool: Evidence from apatite (U-Th)/He thermochronometry. Geology 32:985–988

Cherniak DJ (2000) Pb diffusion in rutile. Contrib Mineral Petrol 139:198-207

Cherniak DJ, Watson EB, Grove M, Harrison TM (2004) Pb diffusion in monazite: a combined RBS/SIMS study. Geochim Cosmochim Acta 68:829-840

Clift PD, Carter A, Hurford AJ (1996) Constraints on the evolution of the East Greenland margin; evidence from detrital apatite in offshore sediments. Geology 24:1013-1016

Colgan JP, Dumitru TA, Miller EL (2004) Diachroneity of Basin and Range extension and Yellowstone Hotspot volcanism in northwestern Nevada. Geology 32:121-124

Coleman DS, Walker JD, Bartley JM, Hodges KV (2001) Thermochronologic evidence for footwall deformation during extensional core complex development, Mineral Mountains, Utah. Utah Geological Association Publication 30 - Pacific Section American Association of Petroleum Geologists Publication GB78:155-168

Coney PJ (1980) Cordilleran metamorphic core complexes: An overview. *In:* Cordilleran Metamorphic Core Complexes, Memoir 153. Crittenden MD, Coney PJ, Davis GH (eds), Geological Society of America, Boulder Colorado, p 7-31

Corrigan JD (1991) Inversion of apatite fission track data for thermal history information. J Geophys Res 96: 10,347-10,360

Coyle DA, Wagner GA (1998) Positioning the titanite fission-track partial annealing zone. Chem Geol 149: 117-125

Crittenden MD, Coney PJ, Davis GH (eds) 1980 Cordilleran Metamorphic Core Complexes. Geological Society of America Memoir 153

Crowley KD, Cameron M, Schaefer RL (1991) Experimental studies of annealing of etched fission tracks in fluorapatite. Geochim Cosmochim Acta 55:1449-1465

Davis GA, Anderson JL, Frost EG, Shackelford TJ (1980) Mylonitization and detachment faulting in the Whipple-Buckskin-Rawhide Mountains terrane, southeastern California and western Arizona. *In* Cordilleran metamorphic core complexes, Memoir 153. Crittenden MD, Coney PJ, Davis GH (eds), Geological Society of America, Boulder Colorado, p 79-130

Davison I, Al-Kadasi M, Al-Khirbash S, Al-Subbary AK, Baker J, Blakey S, Bosence D, Dart C, Heaton R, McClay K, Menzies M, Nichols G, Owen L, Yelland A (1994) Geological evolution of the southeastern Red Sea Rift margin, Republic of Yemen. Geol Soc Am Bull 106:1474-1493

Davy P, Guerin G, Brun JP (1989) Thermal constraints on the tectonics evolution of a metamorphic core complex (Santa Catalina Mountains, Arizona). Earth Planet Sci Lett 94:425-440

Dokka RK, ,Mahaffie MJ, Snoke AW (1986) Thermochronologic evidence of major tectonic denudation associated with detachment faulting, northern Ruby Mountains-East Humboldt Range, Nevada. Tectonics 7:995-1006

Donelick RA, Ketcham RA, Carlson WD (1999) Variability of apatite fission track annealing kinetics II: Crystallographic orientation effects. Am Mineral 84:1224-1234

Donelick RA, O'Sullivan PB, Ketcham RA (2005) Apatite fission-track analysis. Rev Mineral Geochem 58:49-94

Dumitru TA, Hill KC, Coyle DA, Duddy IR, Foster DA, Gleadow AJW, Green PF, Kohn BH, Laslett GM, O'Sullivan AB (1991) Fission track thermochronology: Application to continental rifting in southeastern Australia. Aust Petrol Explor Assoc J 31:131-142

Dumitru TA (2000) Fission-track geochronology in Quaternary geology. *In*: Quaternary geochronology: Methods and applications. Noller JS, Sowers JM, Lettis WR (eds) American Geophysical Union Reference Shelf 4:131-156

Dumitru TA, Miller EL, Surpless BE, Martinez CM, Egger A, Stockli DF (2000) Large structural domains of synchronous Miocene extension in the northern Basin and Range Province. Abstract with Program, Geol Soc of Am 32:43

Dunkl I, Demeny A (1997) Exhumation of the Rechnitz Window at the border of the Eastern Alps and Pannonian Basin during Neogene extension. Tectonophysics 272:197-211

Dunkl I, Frisch W, Grundmann G (2003) Zircon fission track thermochronology of the southeastern part of the Tauern Window and the adjacent Austroalpine margin, Eastern Alps. Eclogae Geologicae Helvetiae 96:209-217

Egger AE, Dumitru TA, Miller EL, Savage CFI, Wooden JL (2003) Timing and Nature of Tertiary Plutonism and Extension in the Grouse Creek Mountains, Utah. Int Geol Rev 45:497-532

Ehlers TA, Chapman DS (1999) Normal fault thermal regimes; conductive and hydrothermal heat transfer surrounding the Wasatch Fault, Utah. Tectonophysics 312:217-234

Ehlers TA, Armstrong PA, Chapman DS (2001) Normal fault thermal regimes and the interpretation of low-temperature thermochronometers. Phys Earth Planet Int 126:179–194

Ehlers TA, Willett SD, Armstrong PA, Chapman DS (2003) Exhumation of the Central Wasatch Mountains, 2: Thermo-kinematics of exhumation, erosion, and thermochronometer interpretation. J Geophys Res 108: 3, doi:101029/2001JB001723

Ehlers TA, Farley FA (2003) Apatite (U-Th)/He thermochronometry: methods and applications to problems in tectonic and surface processes. Earth Planet Sci Lett 206:1–14

Ehlers TA (2005) Crustal thermal processes and the interpretation of thermochronometer data. Rev Mineral Geochem 58:315-350

Farley KA (2000) Helium diffusion from apatite: General behavior as illustrated by Durango fluorapatite. J Geophys Res 105:2903-2914

Farley KA (2002) (U-Th)/He Dating: Techniques, calibrations, and applications. Rev Mineral Geochem 47: 819-844

Farley KA, Stockli DF (2002) (U-Th)/He Dating of phosphates: apatite, monazite, and xenotime. Rev Mineral Geochem 48:559-577
Fayon AK, Peacock SM, Stump E, Reynolds SJ (2000) Fission track analysis of the footwall of the Catalina detachment fault, Arizona; tectonic denudation, magmatism, and erosion. J Geophys Res 105:11047-11062
Fitzgerald PG, Gleadow AJW (1988) Fission track geochronology, tectonics and structure of the Transantarctic Mountains in Northern Victoria Land, Antarctica. Isotope Geosci 73:169-198
Fitzgerald PG, Gleadow AJW (1990) New approaches in fission track geochronology as a tectonic tool: Examples from the Transantarctic Mountains. Nucl Tracks Rad Meas 17:351-357
Fitzgerald PG, Fryxell JE, Wernicke BP (1991) Miocene crustal extension and uplift in southeastern Nevada: Constraints from fission track analysis. Geology 19:1013-1016
Fitzgerald PG, Reynolds SJ, Stump E, Foster DA, Gleadow AJW (1994) Thermochronologic evidence for timing of denudation and rate of crustal extension of the South Mountains metamorphic core complex and Sierra Estrella, Arizona. Nucl Tracks Rad Meas 21:555-563
Fletcher JM, Kohn BP, Foster DA, Gleadow AJW, (2000) Heterogeneous Neogene cooling and uplift of the Los Cabos block, southern Baja California: Evidence from fission track thermochronology. Geology 28: 107-110
Foster DA, Harrison TM, Miller CF, Howard KA (1990) The $^{40}Ar/^{39}Ar$ thermochronology of the eastern Mojave Desert, California, and adjacent western Arizona: with implications for the evolution of metamorphic core complexes. J Geophys Res 95:20,005-20,024
Foster DA, Miller DS, Miller CF (1991) Tertiary extension in the Old Woman Mountains area, California: Evidence from apatite fission-track analysis. Tectonics 10:875-886
Foster DA, Gleadow AJW (1992) The morphotectonic evolution of rift-margin mountains in central Kenya; constraints from apatite fission-track thermochronology. Earth Planet Sci Lett 113:157-171
Foster DA, Gleadow AJW, Reynolds SJ, Fitzgerald PG (1993) The denudation of metamorphic core complexes and the reconstruction of the Transition Zone, west-central Arizona: constraints from apatite fission-track thermochronology. J Geophys Res 98:2167-2185
Foster DA, Howard KA, John BE (1994) Thermochronological constraints on the development of metamorphic core complexes in the lower Colorado River area. *In*: Eighth International Conference on Geochronology, Cosmochronology, and Isotope Geology. Volume 1107. Lanphere MA, Dalrymple GB, Turrin BD (eds) United States Geological Survey Circular:103
Foster DA, Kohn BP and Gleadow AJW (1996) Sphene and zircon fission track closure temperatures revisited: Empirical calibrations from $^{40}Ar/^{39}Ar$ diffusion studies on K-feldspar and biotite. International Workshop on Fission Track Dating Abstracts, University of Gent, Gent, Belgium
Foster DA, John BE (1999) Quantifying tectonic exhumation in an extensional orogen with thermochronology: examples from the southern Basin and Range Province. *In*: Exhumation processes: normal faulting, ductile flow and erosion. Ring U, Brandon MT, Lister GS, Willett SD (eds) Geological Society, London, Special Publication, 154:343-364
Foster DA, Raza A (2002) Low-temperature thermochronological record of exhumation of the Bitterroot metamorphic core complex, northern Cordilleran Orogen. Tectonophysics 349:23-36
Froitzheim N, Eberli GP (1990) Extensional detachment faulting in the evolution of a Tethys passive continental margin, Eastern Alps, Switzerland. Geol Soc Am Bulletin 102:1297-1308
Fryxell JE, Salton GG, Selverstone J, Wernicke B (1992) Gold Butte crustal section, South Virgin Mountains, Nevada. Tectonics 11:1099-1120
Fugenschuh B, Seward D, Mancktelow N (1997) Exhumation in a convergent orogen; the western Tauern Window. Terra Nova 9:213-217
Fugenschuh B, Mancktelow N, Seward D (2000) Cooling and exhumation history of the Oetztal-Stubai basement complex, eastern Alps: a structural and fission-track study. Tectonics 19:905-918
Fuegenschuh B, Schmid SM (2003) Late stages of deformation and exhumation of an orogen constrained by fission-track data; a case study in the Western Alps. Geol Soc Am Bull 115:1425-1440
Gallagher K, Dumitru T, Gleadow A (1994) Constraints on the vertical motions of eastern Australia during the Mesozoic. Basin Res 6:77-94
Gallagher K (1995) Evolving temperature histories from apatite fission-track data. Earth Planet Sci Lett 136: 421-443
Gallagher K, Brown R (1997) The onshore record of passive margin evolution. J Geol Soc 154:451-457
Gallagher K, Brown RW, Johnson CJ (1998) Geological Applications of Fission Track Analysis. Ann Rev Earth Planet Sci 26:519-572
Gans PB (1987) An open-system, two-layer crustal stretching model for the eastern Great Basin. Tectonics 6: 1-12

Gans PB, Miller EL (1983) Style of mid-Tertiary extension in east-central Nevada. *In*: Geologic Excursions in the Overthrust Belt and Metamorphic Core Complexes of the Intermontain Region, Nevada Geological Society of America Field Trip Guidebook, Utah Geological and Mineral Survey Special Studies 59: 107-160

Gans PB, Mahood GA, Schermer E (1989) Synextensional magmatism in the Basin and Range province: A case study from the eastern Great Basin. Geol Soc Am Special Paper 233

Gans PB, Miller EL, Brown R, Housman G, Lister GS (1991) Assessing the amount, rate, and timing of tilting in normal fault blocks: A case study of tilted granites in the Kern-Deep Creek Mountains, Utah. Geological Society of America Memoir 23

Gautier P, Brun J-P, Jolivet L (1993) Structure and Kinematics of upper Cenozoic extensional detachment on Naxos and Paros (Cyclades Islands, Greece). Tectonics 12:1180-1194

Gessner K, Ring U, Johnson C, Hetzel R, Passchier CW, Gungor T (2001) An active bivergent rolling-hinge detachment system; central Menderes metamorphic core complex in western Turkey. Geology 29:611-614

Ghebreab W, Carter A, Hurford AJ, Talbot CJ (2002) Constraints for timing of extensional tectonics in the western margin of the Red Sea in Eritrea. Earth Planet Sci Lett 200:107-119

Gleadow AJW, Duddy IR, Green F, Lovering JF (1986) Confined fission track lengths in apatite: A diagnostic tool for thermal history analysis. Contrib Mineral Petrol 94:405-415

Gleadow AJW, Fitzgerald PG (1987) Uplift history and structure of the Transantarctic Mountains; new evidence from fission track dating of basement apatites in the Dry Valleys area, southern Victoria Land. Earth Planet Sci Lett 82:1-14

Gleadow AJW, Belton DX, Kohn BP, Brown RW (2002) Fission track dating of phosphate minerals and the thermochronology of apatite. Rev Mineral Geochem 48:579-611

Green PF, Duddy IR, Gleadow AJW, Tingate PR (1985) Fission track annealing in apatite: Track length measurements and the form of the Arrhenius plot. Nucl Tracks Rad Meas 10:323-328

Green PF, Duddy IR, Gleadow AJW, Tingate PR, Laslett GM (1986) Thermal annealing of fission tracks in apatite, 1, A quantitative description. Chem Geol 59:237-253

Green PF, Duddy IR, Gleadow AJW, Lovering JF (1989a) Apatite fission track analysis as a paleotemperature indicator for hydrocarbon exploration. *In*: Thermal History of Sedimentary Basins: Methods and Case Histories. Naeser ND, McCullogh TH (eds) Springer-Verlag, New York, p 81-195

Green PF, Duddy IR, Laslett GM, Hegarty KA, Gleadow AJW, Lovering JF (1989b) Thermal annealing of fission tracks in apatite, 4, Quantitative modeling techniques and extension to geological timescales. Chem Geol 79:155-182

Grove M, Harrison TM (1996) $^{40}Ar^*$ diffusion in Fe-rich biotite. Am Mineral 81:940-951

Grove M, Harrison TM (1999) Monazite Th-Pb age depth profiling. Geology 27:487-490

Hames WE, Bowring SA (1994) An empirical evaluation of the argon diffusion geometry in muscovite. Earth Planet Sci Lett 124:161-169

Hansen K, Brooks CK (2002) The evolution of the East Greenland margin as revealed from fission-track studies. Tectonophysics 349:93-111

Harrison TM, Duncan I, McDougall I (1985) Diffusion of ^{40}Ar in biotite: Temperature, pressure and compositional effects. Geochim Cosmochim Acta 49, 2461-2468

Harrison TM, Copeland P, Kidd WSF, Lovera OM (1995) Activation of the Nyainquentanghla shear zone; implications for uplift of the southern Tibetan Plateau. Tectonics 14:658-676

Harrison TM, Catlos EJ, Montel JM (2002) U-Th-Pb dating of phosphate minerals. Rev Mineral Geochem 48: 523-558

Harrison TM, Grove M, Lovera OM, Zeitler PK (2005) Continuous thermal histories from inversion of closure profiles. Rev Mineral Geochem 58:389-409

Hejl E, Riedl H, Weingartner H (2002) Post-plutonic unroofing and morphogenesis of the Attic-Cycladic complex (Aegea, Greece). Tectonophysics 349:37-56

Hill EJ, Baldwin SL, Lister GS (1992) Unroofing of active metamorphic core complexes in the D'Entrecasteaux Islands, Papua New Guinea. Geology 20:907-910

Holm DK, Wernicke BP (1990) Black Mountains crustal section, Death Valley extended terrain. California Geology 18:520-523

Holm DK, Snow JK, Lux DR (1992) Thermal and barometric constraints on the intrusive and unroofing history of the Black Mountains; implications for timing, initial dip, and kinematics of detachment faulting in the Death Valley region, California. Tectonics 11:507-522

House MA, Wernicke BP, Farley KA, Dumitru TA (1997) Cenozoic thermal evolution of the central Sierra Nevada, California, from (U-Th)/He thermochronometry. Earth Planet Sci Lett 151:167-179

House MA, Farley KA, Kohn BP (1999) An empirical test of helium diffusion in apatite: borehole data from the Otway Basin Australia. Earth Planet Sci Lett 170:463-474

Howard KA, Foster DA (1996) Thermal and unroofing history of a thick, tilted Basin-and-Range crustal section in the Tortilla Mountains, Arizona. J Geophys Res 101:511-522

Hutchison CS, Bergman SC, Swauger DA, Graves JE (2000) A Miocene collisional belt in North Borneo; uplift mechanism and isostatic adjustment quantified by thermochronology. J Geol Soc London 157:783-793

John BE, Foster DA (1993) Structural and thermal constraints on the initiation angle of detachment faulting in the southern Basin and Range; the Chemehuevi Mountains case study. Geol Soc Am Bull 105:1091-1108

John BE, Howard KA (1995) Rapid extension recorded by cooling-age patterns and brittle deformation, Naxos, Greece. J Geophys Res 100:9,969-9,979

Jones CH, Unruh JR, Sonder LJ (1996) The role of gravitational potential energy in active deformation in the southwestern United States. Nature 381:37-41

Jackson JA, White NJ, Garfunkel Z, Anderson HJ (1988) Relations between normal fault geometry, tilting and vertical movements in extensional terranes: an example from the Southern Gulf of Suez. J Struct Geol 10:155-170

Jackson JA, White NJ (1989) Normal faulting in the upper continental crust: observations from regions of active extension. J Struct Geol 11:15-36

Jackson JA (1994) Active tectonics of the Aegean region. Ann Rev Earth Planet Sci 22:239-271

Jolivet L, Brun J-P, Lallemant S, Patriat M (1994) 3D-kinematics of extension in the Aegean region from the early Miocene to the Present, insights from the ductile crust. Bull Soc Géol France 165:185-209

Kamp PJ (1997) Paleogeothermal gradient and deformation style, Pacific front of the Southern Alps Orogen; constraints from fission track thermochronology. Tectonophysics 271:37-58

Keay S, Lister G, Buick I (2001) The timing of partial melting, Barrovian metamorphism and granite intrusion in the Naxos metamorphic core complex, Cyclades, Aegean Sea, Greece. Tectonophysics 342:275-312

Ketcham RA (1996) Distribution of heat-producing elements in the upper and middle crust of southern and west central Arizona: Evidence from the core complexes. J Geophys Res 101:13,611-13,632

Ketcham RA, Donelick RA, Carlson WD (1999) Variability of apatite fission track annealing kinetics III: Extrapolation to geological time scales. Am Mineral 84:1235-1255

Ketcham RA, Donelick RA, Donelick MB (2000) AFTSolve: A program for multi-kinetic modeling of apatite fission-track data Geological Materials Research, v2, n1

Ketcham RA, Donelick RA, Donelick MB (2003) AFTSolve; a program for multi-kinetic modeling of apatite fission-track data. Am Mineral 88:929

Ketcham RA (2005) Forward and inverse modeling of low-temperature thermochronometry data. Rev Mineral Geochem 58:275-314

Kirby E, Reiners PW, Krol MA, Whipple KX, Hodges KV, Farley KA, Tang W, Chen Z (2002) Late Cenozoic evolution of the eastern margin of the Tibetan Plateau: Inferences from $^{40}Ar/^{39}Ar$ and (U-Th)/He thermochronology. Tectonics 21:1, doi:101029/2000TC001246

Kohn BP, Feinstein S, Foster DA, Steckler MS, Eyal M (1997) History of the eastern Gulf of Suez; II, Reconstruction from apatite fission track and $^{40}Ar/^{39}Ar$ K-feldspar measurements. Tectonophysics 283: 219-239

Kohn BP, Bishop P (1999) Long-term landscape evolution of the southeastern Australian margin: apatite fission track thermochronology and geomorphology. Aust J Earth Sci 46: 155-156

Kohn BP, Gleadow AJW, Cox SJD (1999) Denudation history of the Snowy Mountains: constraints from apatite fission track thermochronology. Aust J Earth Sci 46:181-198

Kowallis BJ, Ferguson J, Jorgensen GJ (1990) Uplift along the Salt Lake segment of the Wasatch fault from apatite and zircon fission track dating in the Little Cottonwood stock. Nucl Tracks Rad Meas 17:325-329

Kusznir NJ, Park RG (2002) The extension strength of the continental lithosphere; its dependence on geothermal gradient, and crust composition and thickness. *In*: Extensional Tectonics; Regional-scale Processes. Holdsworth RE, Turner JP (eds) Geological Society Pub House, London, 2: 97-114

Lee J (1995) Rapid uplift and rotation of mylonitic rocks from beneath a detachment fault; insights from potassium feldspar $^{40}Ar/^{39}Ar$ thermochronology, northern Snake Range, Nevada. Tectonics 14:54-77

Lippolt HJ, Leitz M, Wernicke RS, Hagedorn B (1994) (U+Th)/He dating of apatite: experience with samples from different geochemical environments. Chem Geol 112:179-191

Lisker F, Olesch M (1998) Cooling and denudation history of western Marie Byrd Land, Antarctica, based on apatite fission-tracks. *In*: Advances in Fission-track Geochronology. Van den haute P, De Corte F (eds) Springer, Dordrecht, p 225-240

Lisker F (2002) Review of fission track studies in northern Victoria Land – passive margin evolution versus uplift of the Transantarctic Mountains. Tectonophysics 349:57-73

Lisker F, Brown R, Fabel D (2003) Denudational and thermal history along a transect across the Lambert Graben, northern Prince Charles Mountains, Antarctica, derived from apatite fission track thermochronology. Tectonics 22:5

Lister GS, Banga G, Feenstra A (1984) Metamorphic core complexes of Cordilleran type in the Cyclades, Aegean Sea, Greece. Geology 12:221-225
Lister GS, Etheridge MA, Symonds PA (1986) Detachment faulting and the evolution of passive continental margins. Geology14:246-250
Lister GS, Etheridge MA, Symonds PA (1991) Detachment models for the formation of passive continental margins. Tectonics 10:1038-1064
Lister GS, Baldwin SL (1993) Plutonism and the origin of metamorphic core. Geology 21:607-610
Lister GS, Baldwin SL (1996) Modelling the effect of arbitrary P-*T*-*t* histories on argon diffusion in minerals using the MacArgon program for the Apple Macintosh. Tectonophysics 253:83-109
Lister GS, Forster M (1996) Inside the Aegean metamorphic core complexes. Technical Publications of the Australian Crustal Research Centre 45, 110p
Livaccari RF, Geissman JW, Reynolds SJ (1995) Large-magnitude extensional deformation in the South Mountains metamorphic core complex, Arizona; evaluation with paleomagnetism. Geol Soc Am Bulletin 107:877-894
Lorencak M, Seward D, Vanderhaeghe O, Teyssier C, Burg J-P (2001) Low-temperature cooling history of the Shuswap metamorphic core complex, British Columbia; constraints from apatite and zircon fission-track ages. Can J Earth Sci 38:1615-1625
Lovera OM, Richter FM, Harrison TM (1989) $^{40}Ar/^{39}Ar$ thermochronometry for slowly cooled samples having a distribution of diffusion domain sizes. J Geophys Res 94:17,917-17,935
Lovera OM, Grove M, Harrison TM, Mahon KI (1997) Systematic analysis of K-feldspar $^{40}Ar/^{39}Ar$ step-heating experiments I: Significance of activation energy determinations. Geochim Cosmochim Acta 61: 3171-3192
Lovera OM, Grove M, Harrison TM (2002) Systematic analysis of K-feldspar $^{40}Ar/^{39}Ar$ step heating results II: relevance of laboratory argon diffusion properties to nature. Geochim Cosmochim Acta 66:1237–1255
Lutz TM, Omar GI (1991) An inverse method of modeling thermal histories from apatite fission-track data. Earth Planet Sci Lett 104:181-195
Malavieille J, Guihot P, Costa S, Lardeaux JM, Gardien V (1990) Collapse of the thickened Variscan crust in the French Massif Central; Mont Pilat extensional shear zone and St Etienne Late Carboniferous basin. Tectonophysics 177:139-149
Manatschal G (2004) New models for evolution of magma-poor rifted margins based on a review of data and concepts from West Iberia and the Alps. Int J Earth Sci 93:432-466
Mancktelow NS (1992) Neogene lateral extension during convergence in the Central Alps; evidence from interrelated faulting and backfolding around the Simplonpass; Switzerland. Tectonophysics 215:295-317
Mancktelow NS, Grasemann B (1997) Time-dependent effects of heat advection and topography on cooling histories during erosion. Tectonophysics 270:167-195
McKenzie D (1978) Some remarks on the development of sedimentary basins. Earth Planet Sci Lett 40:25-32
McDougall I, Harrison TM (1999) Geochronology and Thermochronology by the $^{40}Ar/^{39}Ar$ Method, 2nd edition, Oxford University Press, Oxford
McInnes BIA, Farley KA, Sillitoe RH, Kohn BP (1999) Application of apatite (U-Th)/He thermochronometry to the determination of the sense and amount of vertical fault displacement at the Chuquicamata porphyry copper deposit, Chile. Econ Geol Bulle Soc Econ Geologists 94:937-947
Menzies M, Gallagher K, Yelland A, Hurford AJ (1997) Volcanic and nonvolcanic rifted margins of the Red Sea and Gulf of Aden; crustal cooling and margin evolution in Yemen. Geochim Cosmochim Acta 61: 2511-2527
Miller EL, Dumitru TA, Brown R, Gans PB (1999) Rapid Miocene slip on the Snake Range-Deep Creek Range fault system, east-central Nevada. Geol Soc Am Bull 111:886-905
Monteleone BD, Baldwin SL, Ireland TR, Fitzgerald PG (2001) Thermochronologic constraints for the tectonic evolution of the Woodlark Basin, Papua New Guinea. *In:* Proceedings of the Ocean Drilling Program. Huchon P, Taylor B, Klaus A (eds) Sci Results 180:1-34
Moore ME, Gleadow AJW, Lovering JF (1986) Thermal evolution of rifted continental nargins: new evidence from fission tracks in basement apatites from southeastern Australia. Earth Planet Sci Lett 78:255-270
Morrison J, Anderson JL (1998) Footwall refrigeration along a detachment fault: implications for thermal evolution of core complexes. Science 279:63-66
Mueller KJ, Snoke AW (1993) Progressive overprinting of normal fault systems and their role in Tertiary exhumation of the East Humboldt-Wood Hills metamorphic complex, Northeast Nevada. Tectonics 12: 361-371
Naeser CW (1979) Fission track dating and geological annealing of fission tracks. *In:* Lectures in isotope geology. Jäger E, Hunziker JC (eds) Springer-Verlag, New York, p154-169
Naeser CW, Bryant B, Crittenden MD (1983) Fission-track ages of apatite in the Wasatch Mountains, Utah: An uplift study. Geol Soc Am Memoir 157:29-36

Noble WP, Foster DA, Gleadow AJW (1997) The post-Pan-African thermal and extensional history of crystalline basement rocks in eastern Tanzania. Tectonophysics 275:331-350
Oldow JS, Kohler G, Donelick RA (1994) Late Cenozoic extensional transfer in the Walker Lane strike-slip belt, Nevada. Geology 22:637-640
Omar GI, Steckler MS (1995) Fission track evidence on the initial rifting of the Red Sea; two pulses, no propagation. Science 270:1341-1344
O'Sullivan PB, Kohn BP, Foster DA, Gleadow AJW (1995) Fission track data from the Bathurst Batholith: evidence for rapid middle Cretaceous uplift and erosion within the eastern highlands of Australia. Aust J Earth Sci 42:597-607
Parsons I, Brown WL, Smith JV (1999) $^{40}Ar/^{39}Ar$ thermochronology using alkali feldspars: real thermal history or mathematical mirage of microtexture? Contrib Mineral Petrol 136:92–110
Persano C, Stuart FM, Bishop P, Barfod DN (2002) Apatite (U-Th)/ He age constraints on the development of the Great Escarpment on the southeastern Australian passive margin. Earth Planet Sci Lett 200:79-90
Pik R, Marty B (1999) (U-Th)/He Thermochronometry: Extension of the Method to More U-Bearing Minerals. EOS Trans, Am Geophys Union 80:1169
Platt JP, Soto JI, Whitehouse MJ, Hurford AJ, Kelley SP (1998) Thermal evolution, rate of exhumation, and tectonic significance of metamorphic rocks from the floor of the Alboran extensional basin, western Mediterranean. Tectonics 17:671-689
Platt JP, Whitehouse MJ, Kelley SP, Carter A, Hollick L (2003) Simultaneous extensional exhumation across the Alboran Basin: Implications for the causes of late orogenic extension. Geology 31:251-254
Platt JP, Argles TW, Carter A, Kelley SP, Whitehouse MJ, Lonergan L (2003) Exhumation of the Ronda Peridotite and its crustal envelope; constraints from thermal modelling of a P-T-time array. J Geol Soc London 160:655-676
Proffett JM (1977) Cenozoic geology of the Yerington district, Nevada, and implications for the nature of Basin and Range faulting. Geol Soc Am Bull 88:247-266
Ranero CR, Reston TJ (1999) Detachment faulting at ocean core complexes. Geology 27:983-986
Reiners PW (2002) (U-Th)/He chronometry experiences a renaissance. EOS Trans, Am Geophys Union 83: 21-27
Reiners P, Farley K (1999) Helium diffusion and (U-Th)/He thermochronometry of titanite. Geochim Cosmochim Acta 63:3845-3859
Reiners PW, Brady R, Farley KA, Fryxell JE, Wernicke BP, Lux D (2000) Helium and argon thermochronometry of the Gold Butte block, South Virgin Mountains, Nevada. Earth Planet Sci Lett 178:315-326
Reiners PW, Farley KA, Hickes HJ (2002) He diffusion and (U–Th)/He thermochronometry of zircon: initial results from Fish Canyon Tuff and Gold Butte. Tectonophysics 349:297–308
Reiners PW (2005) Zircon (U-Th)/He thermochronometry. Rev Mineral Geochem 58:151-179
Rietbrock A, Tiberi C, Scherbaum F, Lyon-Caen H (1996) Seismic slip on a low angle normal fault in the Gulf of Corinth; evidence from high-resolution cluster analysis of microearthquakes. Geophys Res Lett 23: 1817-1820
Rigo A, Lyon-Caen H, Armijo R, Deschamps A, Hatzfeld D, Makropoulos K, Papadimitriou P, Kassaras I (1996) A microseismic study in the western part of the Gulf of Corinth (Greece); implications for large-scale normal faulting mechanisms. Geophys J Int 126:663-688
Ring U, Brandon MT, Lister GS, Willett SD (1999) Exhumation processes. *In*: Exhumation processes: normal faulting, ductile flow and erosion. Ring U, Brandon MT, Lister GS, Willett SD (eds) Geological Society, London, Special Publication 154:1-27
Ring U, Layer PW, Reischmann T (2001) Miocene high-pressure metamorphism in the Cyclades and Crete, Aegean Sea, Greece: Evidence for large-magnitude displacement on the Cretan detachment. Geology 29: 395–398
Ring U, Thomson SN, Broecker M (2003) Fast extension but little exhumation; the Vari detachment in the Cyclades, Greece. Geol Mag 140:245-252
Rohrman M, van der Beek P, Andriessen P, Cloetingh S (1995) Meso-Cenozoic morphotectonic evolution of southern Norway; Neogene domal uplift inferred from apatite fission track thermochronology. Tectonics 14:704-718
Ruppel C, Royden L, Hodges KV (1988) Thermal modeling of extensional tectonics: application to pressure-temperature-time histories of metamorphic rocks. Tectonics 7:947–957
Scott DJ, St Onge MR (1995) Constraints on Pb closure temperature in titanite based on rocks from the Ungava Orogen, Canada; implications for U-Pb geochronology and P-T-t path determinations. Geology 23:1123-1126
Scott RJ, Foster DA, Lister GS (1998) Tectonic implications of rapid cooling of lower plate rocks from the Buckskin-Rawhide metamorphic core complex, west-central Arizona. Geol Soc Am Bull 110:588-614
Selverstone J (1988) Evidence for east-west crustal extension in the Eastern Alps: implications for the unroofing history of the Tauern Window. Tectonics 7:87-105

Sengör AHC, Burke K (1978) Relative timing of rifting and volcanism on earth and its tectonic implications. Geophys Res Lett 5:419-421
Seward D, Tulloch AJ, White P (1989) Cenozoic tectonics of the Victoria and Paparoa ranges, South Island; a fission-track study. Geol Soc New Zealand Misc Pub 43:86
Seward D, Mancktelow N (1994) Neogene kinematics of the central and western Alps: evidence from fission-track dating. Geology 22:803-806
Sorel D (2000) A Pleistocene and still-active detachment fault and the origin of the Corinth-Patras Rift, Greece. Geology l28:83-86
Spell TL, McDougall I, Tulloch AJ (2000) Thermochronologic constraints on the breakup of the Pacific Gondwana margin; the Paparoa metamorphic core complex, South Island, New Zealand. Tectonics 19: 433-451
Spotila J (2005) Applications of low-temperature thermochronometry to quantification of recent exhumation in mountain belts. Rev Mineral Geochem 58:449-466
Staufenberg H (1987) Apatite fission-track evidence for post-metamorphic uplift and cooling history of the eastern Tauern Window and the surrounding Austroalpine (central Eastern Alps, Austria). Jahrbuch der Geologischen Bundesanstalt Wien 130:571-586
Stockli DF (1999) Regional timing and spatial distribution of Miocene extension in the northern Basin and Range Province. PhD Dissertation, Stanford University
Stockli DF, Farley KA, Dumitru TA (2000) Calibration of the apatite (U-Th)/He thermochronometer on an exhumed fault block, White Mountains, California. Geology 28:983–986
Stockli DF, Linn JK, Walker JD, Dumitru TA (2001) Miocene unroofing of the Canyon Range during extension along the Sevier Desert Detachment, west-central Utah. Tectonics 20:289-307
Stockli DF, Surpless BE, Dumitru TA, Farley KA (2002) Thermochronological constraints on the timing and magnitude of Miocene and Pliocene extension in the central Wassuk Range, western Nevada. Tectonics 21:4, 101029/2001TC00129502
Stockli DF, Taylor M, Yin A, Harrison TM, D'Andera J, Lin D, Kapp P (2002) Miocene-Pliocene inception of E-W extension in Tibet as evidenced by apatite (U-Th)/He data. Abstracts with Program, Geol Soc Am 34:411
Stockli DF, Dumitru TA, McWilliams MO, Farley KA (2003) Cenozoic tectonic evolution of the White Mountains, California and Nevada. Geol Soc Am Bulletin 115:788–816
Stockli DF, Farley KA (2004) Empirical constraints on the titanite (U-Th)/He partial retention zone from the KTB drill hole. Chem Geol 207:223-236
Stockli DF, Farley KA (in review) Helium diffusivity and (U-Th)/He dating of monazite. Submitted to Geochimica Cosmochimica Acta
Stockli DF, Dumitru TA, Miller EL, Surpless BE (in review) Regional timing and distribution of Miocene extension across the northern Basin and Range Province. Submitted to Geol Soc Am Bulletin
Stuwe K, White L, Brown R (1994) The influence of eroding topography on steady-state isotherms Applications to fission track analysis. Earth Planet Sci Lett 124:63-74
Surpless BE (1999) A structural, magmatic, and thermochronological study of the central Wassuk Range, western Nevada. PhD Dissertation, Stanford University
Surpless BE, Stockli DF, Dumitru TA, Miller EL (2002) Two phase westward encroachment of Basin and Range extension into the northern Sierra Nevada. Tectonics 21:1, 101029/2000TC001257
Tagami T, Dumitru TA (1996) Provenance and thermal history of the Franciscan accretionary complex: Constraints from zircon fission track thermochronology. J Geophys Res 101:11,353-11,364
Tagami T, Farley KA, Stockli DF (2003a) (U-Th)/He geochronology of zircon using Nd-YAG laser heating. Earth Planet Sci Lett 207:57-67
Tagami T, Farley KA, Stockli DF (2003b) Thermal sensitivities of zircon (U-Th)/He and fission-track systems. *In:* Abstracts of the 13th annual V M Goldschmidt conference, Kyoto, Japan, Geochim Cosmochim Acta 68; 15A:466
Tagami T (2005) Zircon fission-track thermochronology and applications to fault studies. Rev Mineral Geochem 58:95-122
Tagami T, O'Sullivan PB (2005) Fundamentals of fission-track thermochronology. Rev Mineral Geochem 58: 19-47
Thomson SN (1994) Fission track analysis of the crystalline basement rocks of the Calabrian Arc, southern Italy: evidence of Oligo-Miocene late-orogenic extension and erosion. Tectonophysics 238:331-352
Thomson SN, Stoeckhert B, Brix MR (1999) Miocene high-pressure metamorphic rocks of Crete, Greece; rapid exhumation by buoyant escape. *In*: Exhumation processes: normal faulting, ductile flow and erosion. Ring U, Brandon MT, Lister GS, Willett SD (eds) Geological Society, London, Special Publication 154:87-107
Thomson K, Green PF, Whitham AG, Price SP, Underhill JR (1999) New constraints on the thermal history of North-East Greenland from apatite fission-track analysis. Geol Soc Am Bulletin 111:1054-1068

Van der Beek PA (1997) Flank uplift and topography at the central Baikal Rift (SE Siberia): A test of kinematic models for continental extension. Tectonics 16:122-136

Van der Beek PA, Mbede E, Andriessen PAM, Delvaux D (1998) Denudation history of the Malawi and Rukwa Rift flanks (East African Rift System) from apatite fission track thermochronology. J African Earth Sci 26:363-385

Vanderhaeghe O, Teyssier C, McDougall I, Dunlap WJ (2003) Cooling and exhumation of the Shuswap metamorphic core complex constrained by $^{40}Ar/^{39}Ar$ thermochronology. Geol Soc Am Bull 115:200-216

Vry J, Baker J, Waight TE (2003) *In situ* Pb-Pb dating of rutile from slowly cooled granulites by LA-MC-ICP-MS: Confirmation of the high closure temperature (>600 °C) for Pb diffusion in rutile. Geophys Res Abst 5:30

Warnock AC, Zeitler PK, Wolf RA, Bergman SC (1997) An evaluation of low-temperature apatite U-Th/He thermochronometry. Geochim Cosmochim Acta 61:5371-5377

Weissel JK, Karner GD (1989) Flexural uplift of rift flanks due to mechanical unloading of the lithosphere during extension. J Geophys Res 94:13,919-13,950

Wells ML, Snee LW, Blythe AE (2000) Dating of major normal fault systems using thermochronology: An example from the Raft River detachment, Basin and Range, western United States. J Geophys Res 105: 16,303-16,327

Wernicke BP (1981) Low-angle normal faults in the Basin and Range province: nappe tectonics in an extending orogen. Nature 291:645-648

Wernicke BP (1992) Cenozoic extensional tectonics of the US Cordillera. *In:* The Cordilleran Orogen Conterminous US, Volume G3, Decade of North American Geology. Burchfiel BC, Lipman PW, Zoback M L (eds) Geological Society of America, Boulder, Colorado, 553-581

Wernicke B (1985) Uniform-sense normal simple shear of the continental lithosphere. Can J Earth Sci 22: 108-125

Wernicke B, Burchfiel BC (1982) Modes of extensional tectonics. J Struct Geol 4:105-115

Wernicke BP, Walker JD, Beaufait MS (1985) Structural discordance between Neogene detachments and frontal Sevier thrusts, central Mormon Mountains, southern Nevada. Tectonics 4:213-246

Wernicke B, Axen GJ (1988) On the role of isostasy in the evolution of normal fault systems. Geology 16: 848-851

White N (1990) Does the uniform stretching model work in the North Sea? *In*: Tectonic evolution of the North Sea Rift. Blundell DJ, Gibbs AD (eds) Oxford University Press, Oxford, p 213-235

Wijbrans JR, McDougall I (1988) Metamorphic evolution of the Attic Cycladic metamorphic belt on Naxos (Cyclades, Greece) utilizing $^{40}Ar/^{39}Ar$ age spectrum measurements. J Metamor Geol 6:571-594

Willett SD (1997) Inverse modeling of annealing of fission tracks in apatite; 1, A controlled random search method. Am J Sci 297:939-969

Wolf RA, Farley KA, Silver LT (1996) Helium diffusion and low temperature thermochronometry of apatite. Geochim Cosmochim Acta 60:4231-4240

Wolf RA, Farley KA, Kass DM (1998) Modeling of the temperature sensitivity of the apatite (U-Th)/He thermochronometer. Chem Geol 148:105-114

Wong M, Gans PB (2004) Tectonic implications of early Miocene extensional unroofing of the Sierra Mazatán metamorphic core complex, Sonora, Mexico. Geology 31:953-956

Yamada R, Tagami T, Nishimura S, Ito H (1995) Annealing kinetics of fission tracks in zircon: an experimental study. Chem Geol 122:249–258

Zarki-Jakni B, van der Beek PA, Poupeau G, Sosson M, Labrin E, Rossi P, Ferrandini J (2004) Cenozoic denudation of Corsica in response to Ligurian and Tyrrhenian extension: results from apatite fission-track thermochronology. Tectonics 23, doi: 101029/2003TC001535

Zeitler PK, Herczeg AL, McDougall I, Honda M (1987) U-Th-He dating of apatite: A potential thermochronometer. Geochim Cosmochim Acta 51:2865-2868

Reviews in Mineralogy & Geochemistry
Vol. 58, pp. 449-466, 2005

Applications of Low-Temperature Thermochronometry to Quantification of Recent Exhumation in Mountain Belts

James A. Spotila

Department of Geosciences
Virginia Polytechnic Institute & State University
Blacksburg, Virginia, 24061, U.S.A.
spotila@vt.edu

INTRODUCTION

Exhumation is a primary phenomenon used to characterize geomorphic and tectonic histories. The motion of rock with respect to the earth's surface is produced by erosion and tectonic denudation. Unroofing by erosion redistributes crustal mass and is a major accommodator of shortening in orogenic belts (Brandon et al. 1998; Zeitler et al. 2001; Willett et al. 2003). Tectonic denudation, typically associated with crustal extension, brings rocks to the surface at rapid rates from significant crustal depths (Hodges et al. 1998; Hacker et al. 2003; Vanderhaeghe et al. 2003). Quantifying exhumation using thermochronology is thus an important element in investigations of long-term, crustal-scale geomorphic and tectonic processes.

The thermochronometers most appropriate for estimating exhumation related to landscape or neotectonic history are apatite (U-Th)/He and fission-track dating. Given their low closure temperatures and the typical geothermal gradient of the continents, these cooling ages generally record exhumation from ~1–4 km depths (Farley 2002; Ehlers and Farley 2003; Naeser 1979; Gleadow et al. 1986). The accuracy of the exhumation history inferred from these data is limited by how well the spatially and temporally variant geothermal gradient is known or modeled (Mancktelow and Grasemann 1997; Ehlers and Farley 2003). Beyond these limitations, thermochronology is often hindered by what the rocks in a given study area record. Cooling ages may be "too old" and predate the tectonic or erosional history of interest. Or, topography, accessibility, bedrock lithology, and exposure may hinder collection of a robust sample set. In many studies, experimental design is determined by logistics, available funding and time, and prior analytical training of the investigator. Thus, some studies may not obtain the optimal results to address a given question. Although many papers deal with technical aspects of thermochronometry, the purpose of this paper is to explore guidelines for experimental design for a range of geologic conditions.

Exhumational environment varies substantially across the continents, as a consequence of the varying style and scale of deformation associated with tectonic environment, regional climate, and the local history or stage of geomorphic development. Study locations can range from an isolated, subdued crystalline block from which the cover has been removed (e.g., Bighorn Mountains; Crowley et al. 2002), to an extensive area of extreme relief and rapid denudation (e.g., Himalaya; Burbank et al. 2003). Thermochronometric studies of exhumation thus also vary, in terms of data collection and analysis strategies. Several techniques of bedrock thermochronology have been utilized successfully to estimate exhumation, such as the relief transect, which seeks age-elevation relationships (e.g., the "break in slope"). Because approaches may involve different sample collection strategies or laboratory techniques, it is

1529-6466/05/0058-0017$05.00 DOI: 10.2138/rmg.2005.58.17

advantageous to know what methods are appropriate prior to data collection. In this paper, I explore these methodologies, using cases studies and following a systematic framework of "denudational maturity."

DENUDATIONAL MATURITY

Cooling history and exhumational regimes of continental surfaces vary in their relative degrees of exhumation relative to rock uplift. To place case studies into a logical context, I have devised a denudational maturity classification (Fig. 1). This system is meant as a useful pedagogy to review experimental approaches of thermochronology, much as Davis' (1899) classification system served to communicate differences in landscape morphology. However, just as a topographic surface is dynamic and represents the sum of instantaneous and gradual processes that have acted over a prolonged period, thermal and unroofing histories of bedrock represent the net result of erosion, tectonic rock uplift, isostatic rebound, and other phenomena integrated over the long term (~10^4–10^7 yr). This classification system is thus imprecise and nonunique.

Groups of denudational maturity are defined on the basis of relative rates of rock uplift and exhumation (Fig. 1), where rock uplift is the sum of exhumation and surface uplift following England and Molnar (1990). I use rates, rather than cumulative magnitudes, so that this classification can be applied instantaneously to examples of modern orogens. The difference in

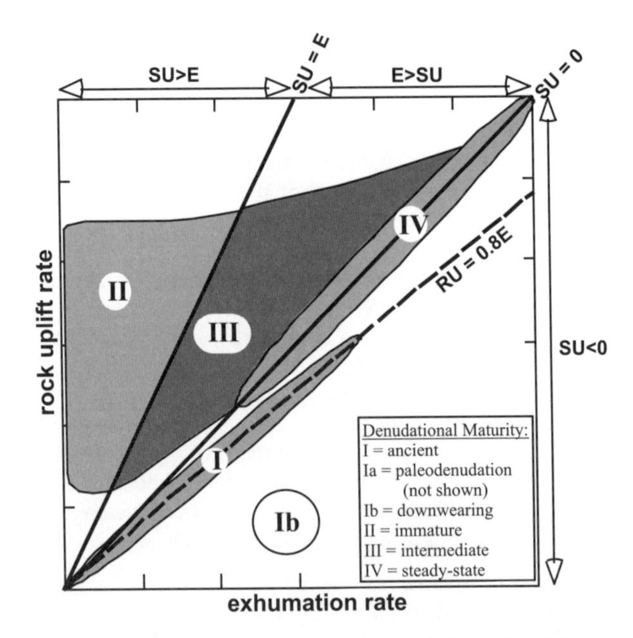

Figure 1. Classification of denudational maturity, based on relative rock uplift and exhumation rate. Case I, for ancient mountain belts, is a narrow zone along a slope of 0.8, representing slow downwearing (negative surface uplift) and isostatic adjustment to exhumation of an orogenic root. Case II represents early (immature) orogenesis, in which surface uplift exceeds exhumation. The line of surface uplift equal to exhumation (slope = 2) is defined as an arbitrary boundary between Cases II and III. Case III represents more mature mountain belts, in which erosional systems have adjusted to rock uplift. The upper and lower limits of Case III are shown with positive slopes, to represent a positive feedback between exhumation and rock uplift (zero surface uplift). Case IV is for orogens with steady-state topography, in which exhumation keeps pace with rock uplift. This occurs for high rates of rock uplift. An additional case is defined for extreme downwearing of local landforms (Ib). Case Ia, for paleodenudation, is not plotted.

approaches is illustrated in Figure 2, where rock uplift and exhumation are plotted separately as histories of cumulative magnitude and rate for evolution of the same fictional mountain belt. For the cumulative case, a convex curve is defined, in which rock uplift initially accumulates faster than exhumation, but which eventually reaches a point of equal cumulative magnitude. The entire active period of the orogen's evolution is confined to the lower quarter of the diagram, and subtleties in denudational development are represented only as slight changes in the curvature of the cumulative line. In the rate case, however, the same stages of development define a clockwise trajectory that returns to the origin when the cycle is complete. This enables

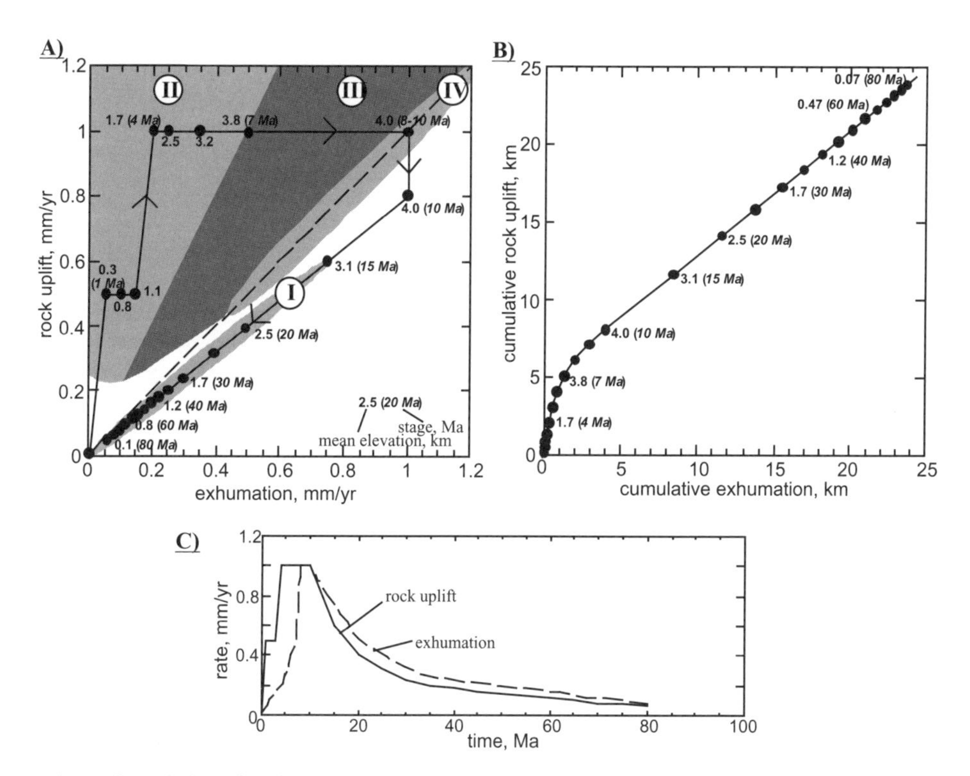

Figure 2. Variation of rock uplift and exhumation for a hypothetical orogenic history. This model illustrates the difference in approaches using relative rates versus cumulative magnitudes. Rates are used in the denudational maturity classification (Fig. 1), because they more clearly define the relative stages of denudational development during active orogenesis. A) Variations in rock uplift and exhumation rate are tracked for an 80-m.y. cycle of the fictional orogen. The first 10 Ma are shown in time-steps of 1 m.y., the latter 70 Ma are shown in decreasing intervals. The mean elevation is labeled at each step (in km), assuming a starting elevation of 100 m. Rock uplift history and surface uplift to exhumation ratio are imposed for the first 10 Ma (see part C), based on the qualitative model of denudational maturity. Isostatic uplift is not specifically calculated for the first 10 Ma, but thereafter the isostatic adjustment to exhumation implicity determines the surface uplift to exhumation ratio. Tectonic rock uplift quickly climbs to 0.5 mm/yr and is stable for 3 m.y. before rising to 1 mm/yr, representing a hypothetical structural adjustment. Rock uplift is then constant until 10 Ma, but exhumation increases at the expense of surface uplift. The orogen is at steady state (zero surface uplift) for 2 m.y., until tectonic rock uplift quickly shuts off at 10 Ma. It then falls to the slope of 0.8, as rock uplift adjusts isostatically to slowly-decreasing exhumation. The four stages of denudational maturity (as in Fig. 1) are shown in open circles. B) Cumulative magnitudes of exhumation and rock uplift for the same orogenic history. Rather than a loop with a clockwise trajectory, the cumulative plot increases steadily, with the bulk of the rock uplift history occurring after the end of tectonism (10 Ma). C) Variation in rock uplift and exhumation rates during the orogen's 80-m.y. history.

a much clearer discrimination of denudational development for the orogen's history. Relative rates of rock uplift and exhumation are therefore used, although for one of the cases described below (Case III), the cumulative magnitude approach is more intuitive.

For all cases of denudational maturity, only positive values are used for rock uplift and exhumation (i.e., no subsidence or burial are permitted), while surface uplift may be negative (i.e., erosional downwearing is permitted). The classification is presented as unitless, to be intentionally ambiguous and avoid exclusion based on absolute numbers, as well as because of the impracticality of estimating surface uplift rate in real mountain ranges. Note that to classify any orogenic system, some measure of surface uplift must be known. In some cases, geomorphic markers such as marine terraces (e.g., Santa Cruz Mountains; Anderson 1990) or assumptions about initial topography (e.g., Alaska Range; Fitzgerald et al. 1993) may provide quantitative estimates of surface uplift, whereas in others, qualitative arguments based on landforms may be all that is available. Although surface uplift is a key parameter in this classification scheme, it is impractical to discuss methodologies or details of data for each case study in this chapter.

The first stage of denudational maturity, labeled Case I, can be considered an "ancient" stage of denudational development. This case is typified by old orogens such as the Appalachian Mountains, in which steady, post-tectonic erosional denudation gradually diminishes a thickened orogenic root (Fischer 2002; Baldwin et al. 2003). It is assumed that regional denudation is isostatically compensated at long wavelengths (~500 km), such that rock uplift recovers ~0.8 of exhumation (Montgomery 1994). The field defining Case I is thus a narrow zone with a slope of 0.8, such that surface uplift is negative (Fig. 1). Given that exhumation rates tend to be slow in ancient orogens, the field is confined near the origin. Two related cases are also defined. Case Ia is defined when thermochronometry records previous tectonic or erosional histories, rather than modern or continuing denudation. Case Ib is defined for the rare occurrence of erosional downwearing, in which local denudation is not locally isostatically balanced. This field, which applies to isolated landforms such as escarpments and volcanoes, is labeled as a poorly-defined circle on the "surface uplift << exhumation" section of the plot (Fig. 1).

The early onset of rock uplift during orogenesis is defined as Case II and is analogous to the "immature" stage of topographic development envisioned by Davis (1899). Surface uplift rate exceeds exhumation, such that the nascent orogen has mountainous topography, despite minimal denudation. Relict topographic features and pre-orogenic cooling ages are present, as the landscape has yet to respond to rock uplift. Examples include fault blocks in the San Bernardino Mountains of California (Spotila et al. 1998) and the southeastern margin of the Tibetan Plateau (Schoenbohm et al. 2004). The Tibet example is the extreme case, in which denudation is so slow, given arid conditions and large areas of internal drainage, that the surface uplift has reached a geodynamic limit (Clark and Royden 2000). Elsewhere, this case should have a finite duration, as landscapes respond via drainage reorganization and knickpoint propagation (Whipple 2001). Note that this is not a universal stage of orogenesis, as it may require certain structural geometry (i.e., large width), low erodibility, or rapid initial rock uplift.

After continued rock uplift and denudation, the cumulative magnitude of exhumation should exceed surface uplift. Case III is defined as a broad field that initiates when surface uplift rate is equaled by exhumation rate. Although this case is defined arbitarily, it is designed to illustrate an intermediate stage between the end members and is suited specifically to host common natural examples. Case III represents actively deforming mountain belts that experience moderate to high rates of denudation and have been active for a significant period (i.e., a few million years). Examples are the San Gabriel Mountains of California (Blythe et al. 2000) and Tien Shan (Bullen et al. 2003). In these cases, topography is locally still rising, such that denudation rate has yet to equal rock uplift rate. At Case IV, however, an orogenic landscape has reached equilibrium and can no longer gain mean elevation, due to rapid denudation. For example, hillslopes may steepen and reach a threshold for bedrock landsliding, such that erosion can keep pace with rock uplift (Burbank et al. 1996; Schmidt and Montgomery 1995). When steady-state topography is reached, surface uplift stalls (i.e., rock uplift rate equals exhumation rate), relief is near constant, the landscape is extremely rugged, rock uplift rates are rapid, and tectonic influx may be balanced by exhumational efflux (Brandon et al. 1998; Willett and Brandon 2002). This case is thus defined in a narrow field along a slope of one in the upper right of Figure 1. Onset of this case is variable, depending on duration and rates of denudation.

Cases I and IV are the most clearly defined fields, whereas Cases II and III are more contrived and gradational stages within a continuum. Nonetheless, these groupings are a systematic way to organize and present cases of denudational development. Each case has certain characteristics and is best studied using different thermochronometric approaches. In the sections that follow, I explore time-temperature manifestations of denudational evolution and appropriate experimental approaches and methods of interpretation using case studies. However, the goal is not rigid classification of orogenic systems. The highlighted examples include local studies within mountain belts, where denudational maturity may vary spatially across tectonic and topographic domains (e.g., Tibetan plateau vs. western Himalayan syntaxis (Kirby et al. 2002; Burbank et al. 1996)). Potential limitations to interpretations in these case studies due to variations in geothermal gradient are also largely ignored. The effects of topography and advection on shallow geothermal gradient are explored elsewhere in this volume (Ehlers 2005; Braun 2005). Long-term variations in geothermal gradient are also not considered, although an assumption that such changes are unimportant is illusory given differences in heat flow among modern and ancient orogens (Ehlers 2005). I adopt these simplifications for the sake of brevity, to allow primary focus on orogenic development and associated thermochronometric signals.

CASE I: ANCIENT OROGENS AND PALEODENUDATION

In the absence of active tectonic uplift, denudation in ancient mountain belts should slowly decelerate, as topographic relief decreases over tens to hundreds of million years (Pinet and Souriau 1988; Fischer 2002; Baldwin et al. 2003). Topography and local geomorphic history in old mountain belts should primarily reflect variations in erodibility (Hack 1980), although denudation may be periodically rejuvenated by tectonic reactivation, drainage reorganization, or climate change (Fitzgerald et al. 1999; Zhang et al. 2001). Exhumation histories of old orogens are relevant for understanding long term landscape dynamics of the continents, but are only loosely constrained by thermochronometry.

Unroofing from the depths corresponding to apatite (U-Th)/He and fission-track closure can take on the order of 100 million years in old, slowly-eroding mountain belts (Roden 1991; Glasmacher et al. 2002). A simple closure-temperature approach to inferring exhumation from individual cooling ages, in which a known or assumed geothermal gradient is inverted for depth of unroofing, thus smoothes out protracted, complex histories of denudation. For example, exhumation rates of 0.02–0.04 mm/yr have been estimated from cooling ages for several physiogeologic provinces of the Appalachian Mountains (Boettcher and Milliken 1994; Spotila et al. 2004a). These are comparable to erosion rates measured over shorter time scales (Hack 1980; Matmon et al. 2003), implying slow denudation for a prolonged period following cessation of orogenesis at ~300 Ma. Individual samples may also be useful for documenting lateral variations in exhumation. Exhumation rates in some ancient orogens, such as the Dabie Shan (China) and Uralides (Russia), increase from range flanks to range cores, presumably because of variations in long-term mean elevation associated with crustal roots (Glasmacher et al. 2002; Reiners et al. 2003).

More detailed exhumation histories can be inferred from single samples using thermal models of helium production/diffusion or fission-track length annealing, which may take into account the effect of evolving mean topography on isotherm shape (Reiners et al. 2003). For example, apatite fission-track length models suggest acceleration of rock cooling in the late Cenozoic in numerous old orogens (Roden 1991; Blackmer et al. 1994; Boettcher and Milliken 1994; Sanders et al. 1999; Glasmacher et al. 2002). Another improvement over using individual cooling ages is the relief transect. Unlike single ages, from which rates of unroofing from closure depths can be inferred, age-elevation relationships can be used to estimate exhumation through closure depths. Note that this requires correction for admittance ratio (the difference in relief of closure isotherm and topography; Braun 2002) when the wavelength of topography is sufficient (>5 km; Reiners et al. 2003). For example, Reiners et al. (2003) used age-elevation relationships for zircon and apatite from a 1.4-km relief transect in the Dabie Shan (eastern China), to document denudation rates as rocks passed through helium partial retention and fission-track partial annealing depths. This provided an estimate of exhumation rate for the mid-Tertiary that was comparable to the long-term average rate estimated using a closure temperature approach on individual ages, supporting a postulation of gradual, steady-state denudation for the ancient mountain range.

Although relief is generally low in old mountain ranges, vertical sample transects from minimal elevation spans are useful and should be sought after. In the absence of adequate relief, pseudo-age-elevation transects can be constructed from paired mineral dating in individual samples. Ages from higher closure temperatures can essentially be translated into imaginary ages for lower temperature systems from higher elevation, extending the record of denudation further back in time (Reiners et al. 2003) (Fig. 3). Inferring exhumation history from thermochronometry that spans long periods (e.g., ~100 Ma) is risky, however, given the likelihood that geothermal gradients are not constant. Systematic changes in geothermal gradient, such as expected for a decrease in heat flow as an orogen becomes tectonically inactive, will depress relative closure depths for individual cooling systems. Although age-elevation gradients should be unaffected by a shift in the closure isotherm depth, the closure-temperature approach in interpreting exhumation rates for individual ages will be inaccurate if changes in the geothermal gradient are ignored. For example, a decrease in geothermal gradient would have the same effect as a deceleration in exhumation. Independent controls on geothermal gradient (e.g., fluid inclusions) can thus be advantageous in studying exhumation in ancient mountain belts.

In some orogens, cooling ages, particularly those from higher closure temperatures, may reflect previous denudational or tectonic regimes (Case Ia). Thermochronometry may thus document geologic events that occurred prior to onset of the active erosional regime. For example, Rahn et al. (1997) used apatite fission-track ages to constrain the rate of isotherm motion relative to the now-inactive Glarus thrust plane in the Swiss Alps, thereby inferring the fault paleo-orientation relative to assumed surface-parallel isotherms. Long-term cooling histories inferred from paired mineral systems or fission-track length modeling can also record complexities through multiple geologic events, such as periods of tectonic activity, quiescence, and rejuvenation (Fletcher et al. 2000; Lisker et al. 2003). It is important to note that such complexities and transitions in exhumational regime are also common in orogens that are still active or that have been active only in the late Cenozoic, such as in the High Himalayas or the Central and Western Alps (Searle et al. 1997; Fugenschuh and Schmid 2003). This implies that long-term averaged cooling histories based on cooling ages in ancient orogens may miss substantial changes in erosional dynamics.

Under unique circumstances, thermochronometry may also record a history of exhumation that greatly exceeds rock uplift (Case Ib). Example landforms that are etched out by erosional downwearing are great escarpments along passive margins. Rift-flank uplift may lead to

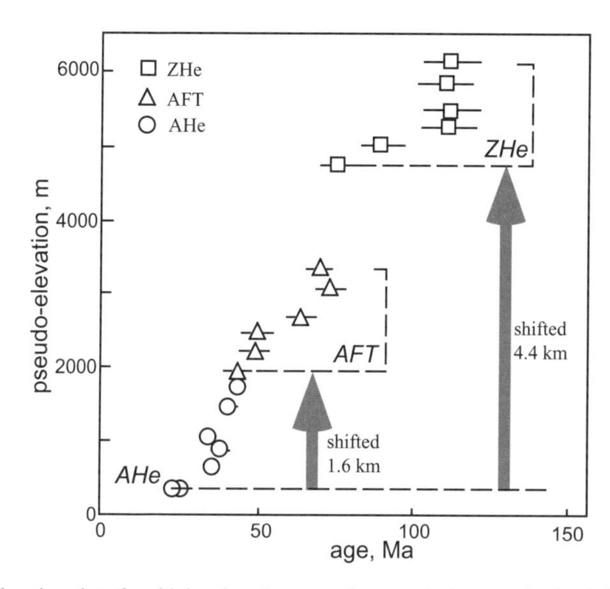

Figure 3. Age-elevation plot of multiple mineral systems for a vertical transect in the Dabie Shan, China, based on data from Reiners et al. (2003). Apatite helium ages are plotted versus sample elevation. Ages representing higher closure temperature for the same six samples are plotted with elevations corrected in proportion to their difference in closure temperature with apatite helium and an assumed 25 °C/km geothermal gradient. For example, the 40 °C difference in closure temperature for apatite fission-track and helium ages requires all apatite fission track ages (AFT) to be shifted upwards by 1.6 km. The zircon helium age (ZHe) closure temperature is 110 °C higher than for apatite helium, requiring a 4.4 km increase in elevation. The result is a pseudo-age elevation transect that indicates steady exhumation rate over a protracted (>100 Ma) period in this ancient orogen, assuming a steady geothermal gradient. Slight changes in slope between the older ages may reflect changes in geothermal gradient or regional tectonic events (Reiners et al. 2003). The basic interpretation of this data set illustrates how paired mineral dating from individual samples can provide a similar type of information as extensive relief transects in old, Case I orogens.

erosional downwearing and retreat of a rugged escarpment into an elevated continental margin (Moore et al. 1986; Steckler and Omar 1994; Gallagher et al. 1994). A nearly universal feature of such escarpments is younging of low-temperature cooling ages away from the escarpment face (Gallagher and Brown 1997). If the magnitude of exhumation is sufficient, the youngest ages may record cooling associated with lateral escarpment retreat. In other cases, the youngest cooling ages may pre-date erosional retreat (i.e., not reset), and are thus analogous to frozen paleo-isochrons exhumed by lateral and vertical erosion (Persano et al. 2002; Spotila et al. 2004a). In either case, exhumation may far exceed rock uplift along escarpments, because the isostatic response to local erosion is distributed over a much longer wavelength.

CASE II: EARLY DENUDATION

Disequilibrium orogenic landscapes, or those that display relict characteristics of pre-tectonic topography and cooling history and in which surface uplift rate exceeds exhumation, are not uncommon in areas of active tectonics (Spotila et al. 1998; Schoenbohm et al. 2004). In a fluvial landscape, the preservation of relict characteristics depends on how quickly the landscape responds to surface uplift via bedrock incision, knickpoint propagation, and drainage network reorganization. Where present, however, this immature denudation case poses a challenge for thermochronometry.

Because of the minimal exhumation (both rate and cumulative magnitude) in immature orogenic belts, cooling ages should predate the onset of tectonism. Therefore, it should be expected that the timing of orogenesis and rate of exhumation will not be constrained using thermochronometry. However, pre-orogenic cooling ages can be used to document an upper limit of exhumation magnitude from a young mountain range. For example, apatite helium ages from atop the eastern Tibetan Plateau are relatively old (~20 Ma) and document minimal (1–2 km) denudation since the onset of plateau growth (Kirby et al. 2002). The Tibetan Plateau is arid and largely internally drained, such that erosion is hindered despite high elevation. Similarly, old apatite helium and fission-track ages (50–70 Ma) identify a horizon of minimal denudation in the recently-uplifted (<2–3 Ma) San Bernardino Mountains of southern California (Spotila et al. 1998; Blythe et al. 2000). These ages confirmed that portions of the range were not in erosional equilibrium with their current elevation, consistent with geologic indicators of slow erosion, including deep weathering and preserved, pre-uplift deposits.

Although cooling ages in immature orogens predate tectonism, isochronous horizons of old ages may be used as markers to document the kinematics of deformation. This can be useful in crystalline bedrock, where other marker horizons, such as stratigraphy, are absent. For example, apatite (U-Th)/He isochrons record gentle block tilting in the San Jacinto Mountains at the northern end of the Peninsular Ranges in southern California (Wolf et al. 1997). In the San Bernardino Mountains, ~50–60 Ma isochrons within the Big Bear and San Gorgonio blocks, in concert with a widespread, geomorphically-defined erosion surface, similarly proved useful for constraining the distribution of relative surface uplift associated with thrust faulting (Spotila et al. 1998) (Fig. 4). Intact fossil helium partial retention or fission-track partial annealing zones may also record rock deformation, such as in the Nevada Basin and Range and Wyoming's Bighorn Mountains (Crowley et al. 2002; Stockli et al. 2002). To establish a 3D pattern of isochronous surfaces, sampling strategy must aim for both wide areal coverage and significant span of relief. Note that using isochrons in such a way is predicated on the validity of an assumption that long-wavelength variations in topography

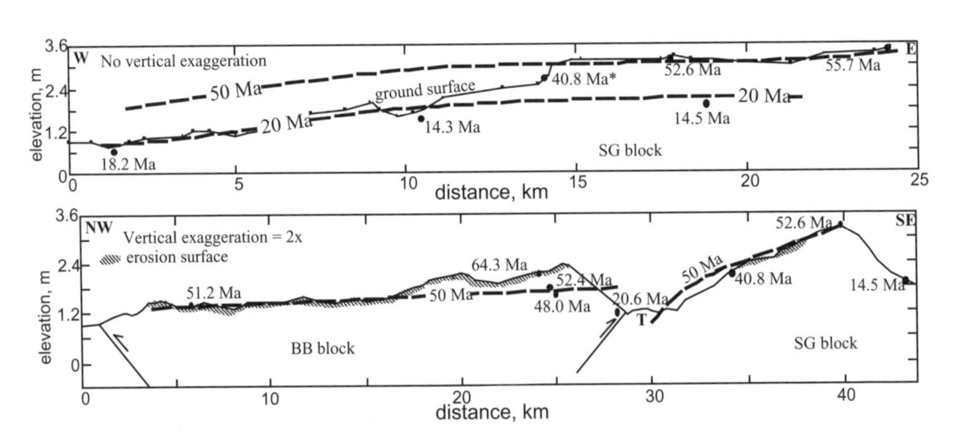

Figure 4. Cross sections from the San Bernardino Mountains, southern California, illustrating how isochrons can be used as structural markers in crystalline rock (adapted from Spotila et al. 1998). Six apatite helium ages from the San Gorgonio (SG) block (above) define 20 and 50 Ma isochrons, which dip gently westward. These define a broad, antiformal arch of this fault-bounded block, as well as document a westward increase in post-cooling incision (i.e., the 50 Ma isochron must be projected over the ground surface). In the lower plot, the 50 Ma isochron is defined by helium ages across the Big Bear (BB) and San Gorgonio (SG) blocks. In this case, the isochron is roughly co-parallel with a deeply-weathered granitic erosion surface. The drop in isochrons between the blocks must occur, because Tertiary sediments (labeled "T") indicate the valley is a structural, fault-bounded low.

and geothermal gradient were minimal at the time rocks closed through closure temperature. In actuality, isochrons will not represent an exact horizon of uniform cooling history. In the absence of other markers, however, isochrons may at least provide a basic constrain on relative uplift that would not otherwise be documented.

Thermochronometry may also be applied to infer paleogeomorphology in some Case II ranges. In a unique application, House et al. (1998) used apatite helium ages along orogen-parallel, single elevation transects to infer long-lived stability of major canyons in the Sierra Nevada, California. This range has experienced Neogene tilting and uplift associated with normal faulting on the east, but is also thought to have high elevation since it was an active magmatic arc in the Cretaceous. Helium ages along horizontal transects are older (~70–80 Ma) at the positions of modern river canyons, below which isotherms should be compressed due to the effect of long wavelength topography, than on less-eroded interfluves (~50–60 Ma). From this, House et al. (1998) inferred significant Sierran relief and elevation since the Mesozoic. Estimating paleogeomorphology in this way is possible, because the western flank of the Sierra Nevada has not experienced enough Cenozoic denudation to erase the record of Cretaceous and early Tertiary cooling.

CASE III: INTERMEDIATE DENUDATION

The rate and cumulative magnitude of exhumation in an orogen will inevitably equal or exceed surface uplift after sufficient time has passed (Case III). During this intermediate stage, between the "immature" and "steady-state" denudational cases, some cooling ages record exhumation associated with active mountain building (i.e., are "reset"), whereas others may still be relicts of cooling prior to onset of orogenesis. The onset of Case III depends on the specific erosional dynamics in a mountain belt, such as climate, geography, and kinematics of deformation. Once this stage has been reached, however, exhumation history is effectively constrained by classic techniques of thermochronometry.

A characteristic feature of intermediate denudation is a break in slope in age-elevation relationship for relief transects. At elevations above the break, pre-orogenic cooling ages with a large range and variance represent slow denudation or crustal stasis at helium partial retention or fission-track partial annealing depths (Gleadow and Fitzgerald 1987; House et al. 2003) (Fig. 5). Although these exhumed fossil retention/annealing zones limit cumulative exhumation magnitude and constrain pre-orogenic geothermal gradient, they do not reflect individual geologic events or rates of orogenic cooling. Younger ages occur below these, which change less with elevation and thereby appear steeper on the age-elevation plot (Fig. 5). These younger ages represent rock uplift through closure depths in the modern orogenic system. For example, apatite fission-track ages from the Alaska Range define a clear break in slope at ~4.5 km elevation, between old ages (6–16 Ma) of a partial annealing zone and young ages (4–6 Ma) that define onset of orogenic denudation (Fitzgerald et al. 1993) (Fig. 6). Fission-track lengths for older ages are variable, consistent with annealing during crustal stasis at depth, whereas tracks are uniformly long below the break in slope, indicating more rapid cooling. The age-elevation gradient defined by young ages (0.7 Ma/km) represents the rate of rock uplift (~1.5 mm/yr) through closure depths as the samples cooled, assuming the same closure temperature for all samples and a fixed closure isotherm depth. Although preservation of the partial annealing zone indicates that the Denali massif (~6.2 km elevation) has experienced minimal denudation during uplift, the Alaska Range as a whole has experienced ~6 km of exhumation and ~3 km of surface uplift (Fitzgerald et al. 1993).

An additional estimate of exhumation rate can be inferred from age-elevation correlations, by extrapolating for the "zero age" (Fig. 5). Although the slope of a steep age-elevation

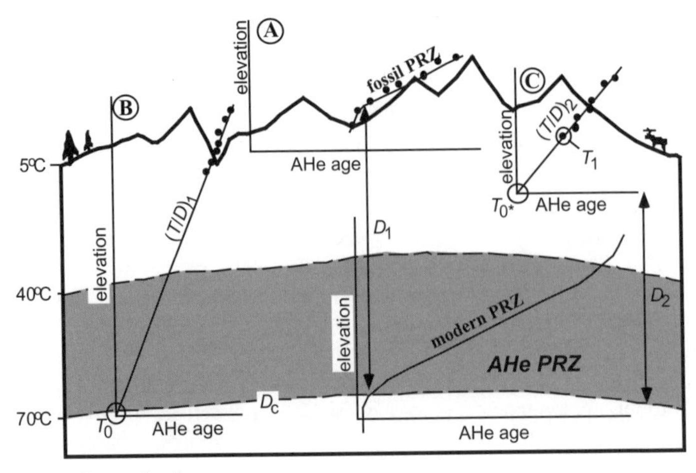

D_c = closure depth
D_1 = bedrock uplift of fossil PRZ
$(T/D)_1$ = age-elevation gradient for active exhumation, gives exhumation rate
T_0 = zero age, expected at closure depth D_c
$(T/D)_2$ = age-elevation gradient for previous period of exhumation
T_{0*} = zero age extrapolated based on age-elevation gradient $(T/D)_2$
D_2 = bedrock uplift after exhumation period $(T/D)_2$, occurs sometime after T_1

Figure 5. Meaning of age-elevation relationships for several exhumation histories shown as a cross section through a fictional orogen, based in part on ideas in House et al. (2003). This case is drawn for apatite helium ages (AHe) with a partial helium retention zone (PRZ) shown from 40–70 °C (Ehlers and Farley 2003). In graph A), a fossil PRZ is preserved as a gentle elevation-age relationship. This records the magnitude of rock uplift (D_1), assuming the geothermal gradient has been constant and that the fossil PRZ formed in the same depth range as the modern PRZ. For graph B), meant to represent a lower elevation section of the same mountain range as A), a steep elevation-age gradient ($[T/D]_1$) indicates rapid exhumation. If the exhumation rate and geothermal gradient have been constant since rapid exhumation initiated, the slope of this elevation-age gradient should predict a zero age at the modern closure depth (T_0). Modern closure depth (D_c) is shown as the base of the PRZ in this case, although this need not be the case, depending on apatite grain size and cooling history (Ehlers and Farley 2003). In graph C), which is meant to represent a different mountain range from A) and B), a steep elevation-age gradient ($[T/D]_2$) similarly implies rapid exhumation. However, this gradient predicts a zero age will occur at a much shallower depth (T_{0*}). This implies additional bedrock uplift (D_2) from closure depth following the period of exhumation at the rate implied by the elevation-age gradient (i.e., sometime after the youngest age, T_1).

relationship, such as for the young ages in the Alaska Range (Fig. 6), represents the rate of exhumation through closure depth, the exhumation of these rocks to the surface is also constrained. If the rate of exhumation through closure depth and the geothermal gradient have been approximately constant to the present, the age-elevation gradient should roughly extrapolate to an age of 0 Ma at the depth of the closure isotherm (Bohannon et al. 1989). If the zero age is predicted at different depth, the rate of exhumation has changed since the time of youngest age. For example, a steep age-elevation gradient in the Santa Lucia Range (California), corresponding to exhumation at 0.35 mm/yr from 6.1–2.3 Ma, predicts a zero apatite helium age at only 0.64 km depth (Ducea et al. 2003). This implies exhumation rate increased to ~0.9 mm/yr since the Pliocene. Similarly, apatite helium age versus elevation in the Coast Mountains (British Columbia) implies an acceleration of exhumation since samples cooled (Farley et al. 2001). Exhumation at 0.21 mm/yr, implied by an age-elevation trend from 10–4 Ma, must have increased after ~4 Ma, because the extrapolated zero age is ~0.7 km higher than the estimated closure depth (Fig. 7). In contrast, extrapolation of the zero age for apatite

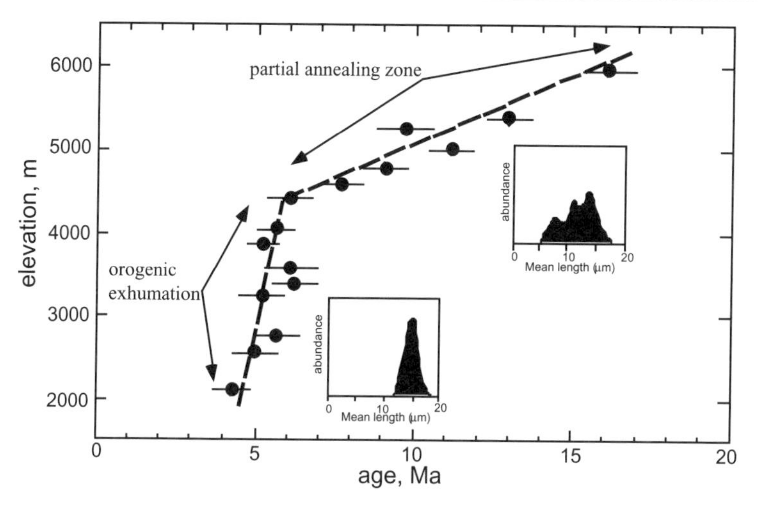

Figure 6. Age-elevation plot for apatite fission-track ages from the Alaska Range, based on data from Fitzgerald et al. (1993). Ages are from the western flank of Denali (Mt. McKinley) and are shown with 2σ error bars. The gently-dipping, older age-elevation correlation is interpreted as a partial annealing zone. The steeper section, in which ages decrease less-rapidly with elevation, represents active, rapid exhumation associated with tectonic uplift of the Alaska Range. The slope of the lower age-elevation correlation implies exhumation through fission-track closure depth of ~0.85 mm/yr (R^2 = 0.50). This rate is close to that predicted for total exhumation using the partial annealing zone as a marker and taking into account mean elevation of the range (Fitzgerald et al. 1993). The small graphs illustrate the general distribution of fission-track lengths in each age group. Older ages from the partial annealing zone have more variable and smaller, annealed tracks, whereas those from the younger group have longer, more uniform-length tracks.

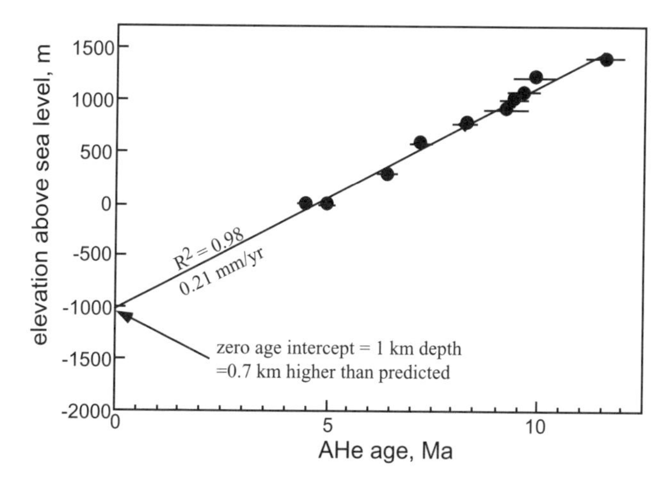

Figure 7. Age-elevation plot of apatite helium ages (AHe) from the Coast Ranges, British Columbia, based on data from Farley et al. (2001), illustrating the "zero age" concept. Ages are shown with 2σ error bars. Ten ages ranging from ~4–11 Ma from Foch Mountain are shown, with a good age-elevation correlation. Ages from higher elevations (>1.5 km) nearby are much older and define a helium partial retention zone. The age-elevation correlation of the young ages implies exhumation through closure depth at ~0.21 mm/yr. However, at this gradient, the zero age should occur at 1 km depth. In contrast, the helium closure depth should be ~1.7 km depth, based on a 68 °C closure temperature, 28 °C/km geothermal gradient, 6 °C mean surface temperature, and mean topography of the Foch Mountain area (Farley et al. 2001). This implies an additional 0.7 km of exhumation since 4 Ma to bring rocks from their closure depth, resulting in a higher post-cooling exhumation rate of 0.39 mm/yr.

fission-track ages in the Tien Shan implies denudation has decelerated (Bullen et al. 2003). An age-elevation relationship from ~10–11 Ma defines an exhumation rate of ~1 mm/yr, whereas the ages themselves predict a long-term average rate since 10 Ma of <0.5 mm/yr. This static interpretation of age groups that correlate with elevation thus provides an additional estimate of exhumation rate, although potential changes in geothermal gradient should also be considered in any individual case.

To maximize the likelihood of measuring reset ages in Case III ranges, it is often advantageous to use apatite helium dating, with its lower closure temperature, rather than apatite fission track. However, fission-track dating has the added benefit of constraining low-temperature history with track-length models (e.g., Blythe and Kleinspehn 1998). A multiple closure temperature approach may also work well in locations of significant denudation, such as along the eastern margin of the Tibetan Plateau (Longmen Shan), which has experienced 8–10 km of denudation since plateau uplift initiated (Kirby et al. 2002). The Tien Shan (China), which has experienced ~4.5 km exhumation and <3 km surface uplift since the Middle Miocene, provides another excellent example of how multiple thermochronometers may be applied to track changes in exhumation rate through the history of a Case-III orogen (Bullen et al. 2003).

Sampling strategy in intermediate mountain ranges should maximize the elevation coverage in relief transects. Lateral sample distribution may depend additionally on deformation style and climatic geography. For example, in transpressional settings, such as the San Gabriel Mountains of southern California, high-angle splay faults may dissect a range into small (~10–50 km) fault blocks with independent exhumation histories, requiring an areal, high-density, "shot-gun" sampling approach (Blythe et al. 2000). Oblique strike-slip regimes in general can result in a complex array of individual crustal blocks that spans the entire spectrum of denudational maturity, resulting in narrow zones of reset and unreset cooling ages (Burgmann et al. 1994; Thomson 2002). In convergent orogens, cross-strike transects of mini-relief sample profiles may effectively document variation in exhumation with climatic (e.g., windward-leeward precipitation variation) or structural (e.g., fault location and orientation) gradients (Burbank et al. 2003). Sampling strategy should also take into consideration potential variations in closure isotherm depth, by covering a range of topographic positions and recording the topographic characteristics of sample sites (e.g., ridge vs. valley, topographic wavelength, relief, etc.). This aids in modeling the effects of topography and advection on evolving geothermal gradient, which is critical when inferring exhumation from reset cooling ages (Mancktelow and Grasemann 1997; Braun 2002; Reiners et al. 2003).

CASE IV: STEADY-STATE

Erosional systems attain an equilibrium with rock uplift, with the onset of steady-state topography (Case IV). Remnants of pre-orogenic topography and bedrock cooling are removed and the rate of rock uplift determines erosional exhumation rate, once an erosional threshold of topographic ruggedness has been attained. The western Himalaya offer two examples of erosional thresholds that may result in steady-state topography and coupling between denudation and rock uplift; the mechanical limit of steep (~30–35°), landslide-prone hillslopes and glacial erosion at equilibrium line altitudes (Burbank et al. 1996; Brozovic et al. 1997; Zeitler et al. 2001). Where steady-state topography occurs, a steady-state flux may also develop, in which tectonic influx is balanced by exhumational efflux and orogenic convergence is accommodated by surficial mass transfer, rather than crustal thickening (Brandon et al. 1998; Willett and Brandon 2002; Spotila et al. 2004b). Given that erosional exhumation is a primary means of partitioning deformation under such conditions, thermochronometry is particularly useful in steady-state orogens.

Age-elevation relationships from relief transects may be useful for documenting rapid exhumation where steady-state topography exists, such as for crustal slivers along the transpressive San Andreas fault (Spotila et al. 2001). In the Yucaipa Ridge block of the San Bernardino Mountains, nearly invariant apatite helium ages over ~1 km relief imply exhumation of ~5 mm/yr for a short period. Hillslopes at the angle of repose and narrow block width enabled rapid erosion and steady-state topography, such that exhumation rate was probably set by vertical displacement rate on bounding faults. Other examples in which minimal variation in age with elevation results from rapid denudation associated with tectonic forcing include Fiordland, New Zealand (House et al. 2002) and the High Himalayas at Gangotri and Shisha Pangma (Sorkhabi et al. 1996; Searle et al. 1997). However, age-elevation invariance does not necessarily mean rapid exhumation. If exhumation is laterally constant and isotherms are roughly parallel to topography (i.e., admittance ratio = 1, for very long topographic wavelength), cooling ages should be invariant with elevation. For example, apatite fission-track ages across the Annapurna High Himalayas do not change with elevation, because exhumation rate is constant across strike due to uniform rock uplift along a crustal ramp (Burbank et al. 2003). Age-elevation relationship in this case does not represent exhumation rate, although exhumation is rapid, as inferred individually from young ages and assumed closure depth.

Because exhumation rates are rapid in steady-state orogens, paired mineral thermochronometry from individual samples may be more useful than age-elevation relationships. Mature histories of exhumation from significant depth translate to cooling through multiple closure temperatures (e.g., Searle 1996; Grafe et al. 2002). Exhumation histories derived from individual samples should be constrained using 1D thermal models that account for advection of isotherms, which is linked with the rapid exhumation typical of steady-state orogens. The use of individual samples further aids in documenting spatial patterns of exhumation, which may vary systematically along or across strike. This is true in orogenic accretionary wedges, in which lateral variations in steady-state exhumation may be caused by particle crustal trajectories and reflected by mapable zones of time-invariant cooling ages for minerals of different closure temperature (Willett and Brandon 2002). For example, zones of reset apatite and zircon fission-track ages are defined in Taiwan, where denudation is focused on the windward flank of the arc-continent collision (Willett et al. 2003). In the Southern Alps of New Zealand, exhumation is concentrated along the oblique-sinistral Alpine fault and decreases towards the range crest (Main Divide), as apparent in belts of reset ages (Batt et al. 2000). Apatite fission-track ages are reset as far away as ~40 km from the fault, while zircon fission-track and higher-closure temperature ages are resent only within ~5 km of the fault (Tippett and Kamp 1993) (Fig. 8). Attainment of steady-state exhumation in these cases is facilitated by erosive conditions (e.g., subtropical monsoons, heavy glaciation, oceanic geography) and narrow orogen width.

DISCUSSION AND CONCLUSIONS

This review of case studies of different exhumational environments reveals guidelines for investigative approaches. For Case I settings of ancient orogens, thermochronometry can generally place first-order constraints on exhumation. Age-elevation transects are useful, but may not be possible given low relief. These must also be corrected for admittance ratio. A closure-temperature approach to individual cooling ages is of limited value, given that ages from slow exhumation rates will average out variations in unroofing history. Paired mineral dating extends the record of cooling and essentially creates pseudo-age-elevation transects from individual samples (Fig. 3), but is limited because higher closure temperatures average exhumation history over even longer periods. Due to slow cooling, secondary techniques, such as fission-track length and helium ingrowth-diffusion modeling, may be more useful than

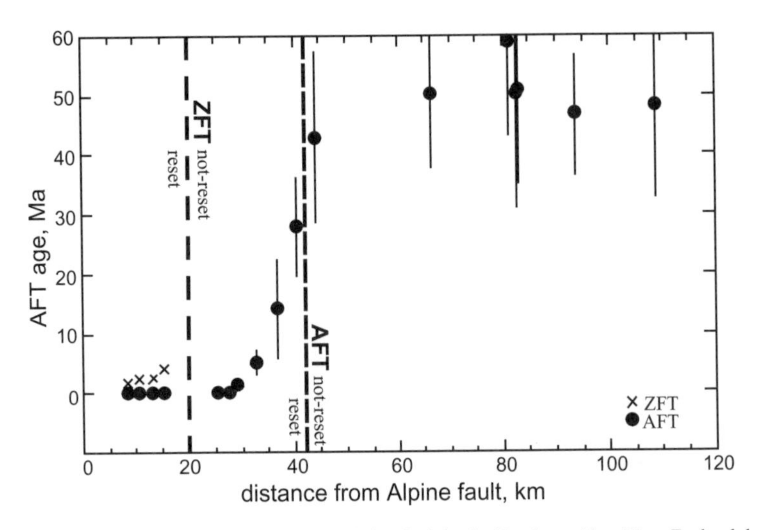

Figure 8. Cooling age versus distance from the Alpine fault in the Southern Alps, New Zealand, based on apatite (AFT) and zircon (ZFT) fission-track data from Tippett and Kamp (1993) for the Whataroa-Godley region (with 2σ error bars). Young, reset apatite ages occur within ~40 km of the fault. If exhumation near the fault is steady-state, these ages should be time invariant. Zircon ages are reset within ~20 km of the fault. At greater distances, zircon fission track ages are >100 Ma (not plotted). The belt of reset zircon ages is smaller than for apatite, because of the higher zircon blocking temperature. Concentration of these belts along the Alpine fault indicates structural control on exhumation. The high rates of exhumation implied by these young cooling ages (Tippett and Kamp 1993) are likely maintained because erosional exhumation keeps pace with rock uplift (i.e., steady-state topography).

the cooling ages themselves. Sampling strategy in old orogens should focus on cross-strike transects or the "shot-gun" approach.

In immature orogens (Case II), denudation has been insufficient to exhume reset cooling ages. Thermochronometry is thus useful as a null test to identify mountain ranges of such minimal denudation. When not reset, old ages may still be useful tectonically. For example, 3D isochrons can be documented using old ages of exhumed partial annealing or retention zones, that can be employed as marker horizons in crystalline rock (Fig. 4). This approach requires both horizontally and vertically-distributed samples. An additional use of thermochronometry where cooling rates are slow is paleogeomorphic reconstruction via the effect of topography on closure isotherm depth.

For Case III mountain belts, it is generally possible to identify the transition between old, unreset ages and young ages that record exhumation associated with the modern denudational regime. This is expressed as a break in slope on age-elevation plots (Figs. 5 and 6). The steeper section of these curves (younger ages) can provide a direct estimate of exhumation rate through the closure isotherms. Additionally, extrapolation of the "zero age" can provide an estimate of exhumation rate from the closure depths (Figs. 5 and 7). The vertical sampling transect is critical in this case, although paired mineral cooling ages can also provide a useful history of cooling. When cooling ages directly represent rapid exhumation associated with orogenesis, it is also important to use caution with regards to the effect of topography and advection on closure isotherm depth. Thermal models can greatly enhance the accuracy of inferred exhumation history. Other considerations for sampling strategy include local deformation kinematics. This is particularly true in transform environments, in which localized uplift and subsidence are so closely controlled by structural geometry.

Finally, a paired mineral approach is most useful for constraining exhumation in steady-state (Case IV) mountain belts. Cooling ages may show invariance to elevation, due to a combination of both rapid exhumation and the high admittance ratio that is common where exhumation rates are high. Like for Case III, thermal models are critical for assessing the shape of isotherms and the effect of isotherm advection associated with rapid exhumation. Given that both along-strike and across-strike belts of reset cooling ages from different closure temperatures may occur (Fig. 8), a "shot-gun" sampling strategy is generally useful.

Effective sampling strategy and investigative approach thus vary depending on the denudation and uplift history of a mountain belt. With existing thermochronometric techniques and recent advances in understanding the thermal behavior of the upper crust, it is possible to place at least first-order constraints on the exhumation of any mountain belt. The accuracy of inferred exhumation histories will improve, as further enhancements are made to analytical techniques. Laser gas extraction now enables rapid analysis of individual mineral grains for (U-Th)/He, thus making paired mineral dating more tractable (e.g., Reiners et al. 2003). Application of both fission track and (U-Th)/He dating is also a significant improvement over use of a single technique, given that these follow independent analytical procedures that are not affected by the same sample limitations (e.g., the effect of U-zonation or U/Th-rich inclusions on helium ages; Ehlers and Farley 2003). When made possible by good mineral separation yields, the grain-size dependence of closure temperatures for helium ages may also be exploited to gain additional information on cooling histories. Greater use of complex (but more user-friendly) thermal models that take into account 3D topography will further improve the accuracy of inferred exhumation histories.

ACKNOWLEDGMENTS

Important ideas were contributed by a number of colleagues. Andrew Meigs and Laura Webb are thanked for careful and thoughtful reviews. The writing of this paper was supported by NSF grant EAR02229628.

REFERENCES

Anderson RS (1990) Evolution of the northern Santa Cruz Mountains by advection of crust past a San Andreas fault bend. Science 249:397-401

Baldwin JA, Whipple KX, Tucker GE (2003) Implications of the shear stress river incision model for the timescale of postorogenic decay of topography. J Geophys Res 108(B3):7:1-17

Batt GE, Braun J, Kohn BP, McDougall I (2000) Thermochronological analysis of the dynamics of the Southern Alps, New Zealand. Geol Soc Am Bull 112:250-266

Blackmer GC, Omar GI, Gold DP (1994) Post-Alleghanian unroofing history of the Appalachian Basin, Pennsylvania, from apatite fission track analysis and thermal models. Tectonics 13:1259-1276

Blythe AE, Burbank DW, Farley KA, Fielding EJ (2000) Structural and topographic evolution of the central Transverse Ranges, California, from apatite fission-track, (U-Th)/He and digital elevation model analyses. Basin Res 12:97-114

Blythe AE, Kleinspehn KL (1998) Tectonically versus climatically driven Cenozoic exhumation of the Eurasian plate margin, Svalbard: Fission track analyses. Tectonics 17:621-639

Boettcher SS, Milliken KL (1994) Mesozoic-Cenozoic unroofing of the southern Appalachian Basin: Apatite fission track evidence from Middle Pennsylvanian sandstone. J Geol 102:655-663

Bohannon RG, Naeser CW, Schmidt DL, Zimmerman RA (1989) The timing of uplift, volcanism, and rifting peripheral to the Red Sea: A case for passive rifting? J Geophys Res 94:1683-1701

Brandon MT, Roden-Tice MK, Garver JI (1998) Late Cenozoic exhumation of the Cascadia accretionary wedge in the Olympic Mountains, northwest Washington State. Geol Soc Am Bull 110:985-1009

Braun J (2005) Quantitative constraints on the rate of landform evolution derived from low-temperature thermochrohonology. Rev Mineral Geochem 58:351-374

Braun J (2002) Quantifying the effect of recent relief changes on age-elevation relationships. Earth Planet Sci Lett 200:331-343

Brozovic N, Burbank DW, Meigs AJ (1997) Climatic limits on landscape development in the northwestern Himalaya. Science 276:571-574

Bullen ME, Burbank DW, Garver JI (2003) Building the Northern Tien Shan: Integrated thermal, structural, and topographic constraints. J Geology 111:149-165

Burbank DW, Blythe AE, Putkonen J, Pratt-Sitaula B, Gabet E, Oskin M, Barros A, Ojha TP (2003) Decoupling of erosion and precipitation in the Himalayas. Nature 426:652-655

Burbank DW, Leland J, Fielding E, Anderson RS, Brozovic N, Reid MR, Duncan C (1996) Bedrock incision, rock uplift, and threshold hillslopes in the northwestern Himalayas. Nature 379:505-510

Burgmann R, Arrowsmith R, Dumitru T (1994) Rise and fall of the southern Santa Cruz Mountains, California, from fission tracks, geomorphology, and geodesy. J Geophys Res 99:20181-20202

Clark MK, Royden LH (2000) Topographic ooze: Building the eastern margin of Tibet by lower crustal flow. Geology 28:703-706

Crowley PD, Reiners PW, Reuter JM, Kaye GD (2002) Laramide exhumation of the Bighorn Mountains, Wyoming: An apatite (U-Th)/He thermochronology study. Geology 30:27-30

Davis WM (1899) The geographical cycle. Geog. Jour 14:481-504

Ducea MD, House MA, Kidder S (2003) Late Cenozoic denudation and uplift rates in the Santa Lucia Mountains, California. Geology 31:139-142

Ehlers TA (2005) Crustal thermal processes and the interpretation of thermochronometer data. Rev Mineral Geochem 58:315-350

Ehlers TA, Farley KA (2003) Apatite (U-Th)/He thermochronometry: methods and applications to problems in tectonic and surface processes. Earth Planet Sci Lett 206:1-14

England P, Molnar P (1990) Surface uplift, uplift of rocks, and exhumation of rocks. Geology 18:1173-1177

Farley KA (2002) (U-Th)/He dating: Techniques, calibrations, and applications. Rev Mineral Geochem 47: 819-843

Farley KA, Rusmore ME, Bogue SW (2001) Post-10 Ma uplift and exhumation of the northern Coast Mountains, British Columbia. Geology 29:99-102

Fischer KM (2002) Waning buoyancy in the crustal roots of old mountains. Nature 417:933-936

Fitzgerald PG, Muñoz JA, Coney PJ, Baldwin SL (1999) Asymmetric exhumation across the Pyrenean orogen: Implications for the tectonic evolution of a collisional orogen. Earth Planet Sci Lett 173:157-170

Fitzgerald PG, Stump E, Redfield TF (1993) Late Cenozoic uplift of Denali and its relation to relative plate motion and fault morphology. Science 259:497-499

Fletcher JM, Kohn BP, Foster DA, Gleadow AJW (2000) Heterogeneous Neogene cooling and exhumation of the Los Cabos block, southern Baja California: Evidence from fission-track thermochronology. Geology 28:107-110

Fugenschuh B, Schmid SM (2003) Late stages of deformation and exhumation of an orogen constrained by fission-track data: A case study in the Western Alps. Geol Soc Am Bull 115:1425-1440

Gallagher K, Brown R (1997) The onshore record of passive margin evolution. J Geol Soc London 154:451-457

Gallagher K, Hawkesworth CJ, Mantovani MSM (1994) The denudation history of the onshore continental margin of SE Brazil inferred from apatite fission track data. J Geophys Res 99:18,117-18,145

Glasmacher UA, Wagner GA, Puchkov VN (2002) Thermotectonic evolution of the western fold-and-thrust belt, southern Uralide, Russ, as revealed by apatite fission track data. Tectonophysics 354:25-48

Gleadow AJW, Duddy IR, Green PF, Lovering JF (1986) Confined fission track lengths in apatite: A diagnostic tool for thermal history analysis. Contrib Mineral Petrol 94:405-415

Gleadow AJW, Fitzgerald PG (1987) Uplift history and structure of the Transantarctic Mountains: A new evidence from fission track dating of basement apatites in the Dry Valleys area, southern Victoria Land. Earth Planet Sci Lett 82:1-14

Grafe K, Frisch W, Villa IM, Meschede M (2002) Geodynamic evolution of southern Costa Rica related to low-angle subduction of the Cocos Ridge: constraints from thermochronology. Tectonophysics 348:187-204

Hack JT (1980) Rock control and tectonism: their importance in shaping the Appalachian highlands. US Geol Survey Professional Paper 1126-B:1-17

Hacker BR, Andersen TB, Root DB, Mehl L, Mattinson JM, Wooden JL (2003) Exhumation of high-pressure rocks beneath the Solund Basin, Western Gneiss region of Norway. J Metamorphic Geol 21:613-629

Hodges KV, Bowring S, Davidek K, Hawkins D, Krol M (1998) Evidence for rapid displacement on Himalayan normal faults and the importance of tectonic denudation in the evolution of mountain ranges. Geology 6:483-486

House MA, Gurnis M, Kamp PJJ, Sutherland R (2002) Uplift in Fiordland region, New Zealand: Implications for incipient subduction. Science 297:2038-2041

House MA, Kelley SA, Roy M (2003) Refining the footwall cooling history of a rift flank uplift, Rio Grande rift, New Mexico. Tectonics 22:5:1060

House MA, Wernicke BP, Farley KA (1998) Dating topography of the Sierra Nevada, California, using apatite (U-Th)/He ages. Nature 396:66-69

Kirby E, Reiners PW, Krol MA, Whipple KX, Hodges KV, Farley KA, Tang W, Chen Z (2002) Late Cenozoic evolution of the eastern margin of the Tibetan Plateau: Inferences from $^{40}Ar/^{39}Ar$ and (U-Th)/He thermochronology. Tectonics 21:1:1001

Lisker F, Brown R, Fabel D (2003) Denudational and thermal history along a transect across the Lambert Graben, northern Prince Charles Mountains, Antartica, derived from apatite fission track thermochronology. Tectonics 22:5:1055

Mancktelow NS, Grasemann B (1997) Time-dependent effects of heat advection on topography on cooling histories during erosion. Tectonophysics 270:167-195

Matmon A, Bierman PR, Larsen J, Southworth S, Pavich MJ, Caffee MW (2003) Temporally and spatially uniform rates of erosion in the southern Appalachian Great Smoky Mountains. Geology 31:155-158

Montgomery DR (1994) Valley incision and the uplift of mountain peaks. J Geophys Res 99:13913-13921

Moore ME, Gleadow A, Lovering JF (1986) Thermal evolution of rifted continental margins: new evidence from fission tracks in basement apatites from southeastern Australia. Earth Planet Sci Lett 78:255-270

Naeser CW (1979) Fission-track dating and geologic annealing of fission tracks. *In:* Lectures in Isotope Geology. Jager E, Hunziker JC (ed) Springer-Verlag, New York, p 154-169

Persano C, Stuart FM, Bishop P, Barfod DN (2002) Apatite (U-Th)/He age constraints on the development of the Great Escarpment on the southeastern Australian passive margin. Earth Planet Sci Lett 200:79-90

Pinet P, Souriau M (1988) Continental erosion and large-scale relief. Tectonics 7:563-582

Rahn MK, Hurfurd AJ, Frey M (1997) Rotation and exhumation of a thrust plane: Apatite fission-track data from the Glarus thrust, Switzerland. Geology 25:599-602

Reiners PW, Zhou Z, Ehlers TA, Xu C, Brandon MT, Donelick RA, Nicolescu S (2003) Post-orogenic evolution of the Dabie Shan, Eastern China, from (U-Th)/He and fission track thermochronology. Am J Sci 303:489-518

Roden MK (1991) Apatite fission-track thermochronology of the southern Appalachian Basin: Maryland, West Virginia, and Virginia. J Geology 99:41-53

Sanders C, Andriessen P, Cloetingh S (1999) Life cycle of the East Carpathian orogen: Erosion history of a doubly vergent critical wedge assessed by fission track thermochronology. J Geophys Res 104:29095-29112

Schmidt KM, Montgomery DR (1995) Limits to relief. Science 270:617-620

Schoenbohm LM, Whipple KX, Burchfiel BC, Chen L (2004) Geomorphic constrains on surface uplift, exhumation, and plateau growth in the Red River region, Yunnan Province, China. Geol Soc Am Bull 116:895-909

Searle MP (1996) Cooling history, erosion, exhumation, and kinematics of the Himalaya-Karakorum-Tibet orogenic belt. *In* The Tectonic Evolution of Asia. Yin A, Harrison M (ed) Cambridge Univ. Press, p 110-137

Searle MP, Parrish RE, Hodges KV, Hurford A, Ayres MW, Whitehouse MJ (1997) Shisha Pangma leucogranite, South Tibetan Himalaya: Field relations, geochemistry, age, origin and emplacement. J Geology 105:295-317

Sorkhabi RB, Stump E, Foland KA, Jain AK (1996) Fission-track and 40Ar/39Ar evidence for episodic denudation of the Gangotri granites in the Garhwal Higher Himalaya, India. Tectonophysics 260:87-199

Spotila JA, Bank GC, Naeser CW, Reiners PW, Naeser ND, Henika BS (2004a) Origin of the Blue Ridge escarpment along the passive margin of eastern North America. Basin Res 16:41-63

Spotila JA, Buscher JT, Meigs A, Reiners PW (2004b) Long-term glacial erosion of active mountain belts: The Chugach/St. Elias Range, Alaska. Geology 32:501-504

Spotila JA, Farley KA, Sieh K (1998) Uplift and erosion of the San Bernardino Mountains, associated with transpression along the San Andreas fault, CA, as constrained by radiogenic helium thermochronometry. Tectonics 17:360-378

Spotila JA, Farley KA, Yule JD, Reiners PW (2001) Near-field transpressive deformation along the San Andreas fault zone in southern California, based on exhumation constrained by (U-Th)/He dating. J Geophys Res 106:30909-30922

Steckler MS, Omar GI (1994) Controls on erosional retreat of the uplifted rift flanks at the Gulf of Suez and northern Red Sea. J Geophys Res 99:12,159-12,173

Stockli DF, Surpless BE, Dumitru TA, Farley KA (2002) Thermochronological constraints on the timing and magnitude of Miocene and Pliocene extension in the central Wassuk Range, western Nevada. Tectonics 21:4:19

Tippett JM, Kamp PJJ (1993) Fission track analysis of the late Cenozoic vertical kinematics of continental crust, South Island, New Zealand. J Geophys Res 98:16,119-16,148

Thomson SN (2002) Late Cenozoic geomorphic and tectonic evolution of the Patagonian Andes between latitudes 42°S and 46°S: An appraisal based on fission-track results from the transpressional intra-arc Liquiñe-Ofqui fault zone. Geol Soc Am Bull 114:1159-1173

Vanderhaeghe O, Teyssier C, McDougall I, Dunlap JW (2003) Cooling and exhumation of the Shuswap metamorphic core complex constrained by $^{40}Ar/^{30}Ar$ thermochronology. Geol Soc Am Bull 115:200-216

Whipple KX (2001) Fluvial landscape response time; how plausible is steady-state denudation? Am J Sci 301: 313-325

Willett SD, Fisher D, Fuller C, En-Chao Y, Chia-Yu L (2003) Erosion rates and orogenic wedge kinematics in Taiwan inferred from fission-track thermochronometry. Geology 31:945-948

Willett SD, Brandon MT (2002) On steady state in mountain belts. Geology, 30:175-178

Wolf RA, Farley KA, Silver LT (1997) Assessment of (U-Th)/He thermochronometry: The low-temperature history of the San Jacinto Mountains, California. Geology 25:65-68

Zeitler PK, Meltzer AS, Koons PO, Craw D, Hallet B, Chamberlain CP, Park SK, Seeber L, Bishop M, Shroder J (2001) Erosion, Himalayan geodynamics, and the geomorphology of metamorphism. Geol Soc Am Today 11:1:4-9

Zhang P, Molnar P, Downs WR (2001) Increased sedimentation rates and grain sizes 2-4 Myr ago due to the influence of climate change on erosion rates. Nature 410:891-897

Reviews in Mineralogy & Geochemistry
Vol. 58, pp. 467-498, 2005

Application of Thermochronology to Hydrothermal Ore Deposits

Brent I. A. McInnes, Noreen J. Evans

CSIRO Exploration and Mining
P.O. Box 1130
Bentley, WA, Australia
brent.mcinnes@csiro.au noreen.evans@csiro.au

Frank Q. Fu

School of Geosciences
University of Sydney, NSW 2006, Australia

Steve Garwin

Geoinformatics Exploration, 57 Havelock St.
West Perth, WA, 6005, Australia

Centre for Exploration Targeting
School of Earth and Geographical Sciences
University of Western Australia, Crawley, WA, 6009, Australia

INTRODUCTION

Thermochronology finds many applications in economic geology. Utilizing temperature-sensitive radiometric dating techniques to reveal low-temperature, upper crustal processes can elucidate many aspects of deposit genesis, including timing and duration of mineralization processes, rate of exhumation and erosion of intrusive ore deposits and comparative preservation potential. The tools are utilized to best advantage when combined with other thermochronometry techniques that provide complementary information. In addition, when thermochronometers are combined with geochronometers (e.g., zircon U/Pb), over 800 °C of thermal history is revealed from emplacement to erosion (Fig. 1). With the advent of computational algorithms that provide more accurate and detailed models of thermochronology results, the economic geologist has a powerful tool to use when assessing economic favorability of a region or prospect.

This chapter summarizes the various ways that low-temperature thermochronometers have been utilized in studies of economic mineralization and primarily focuses on studies of porphyry ore deposits. These deposits are well characterized and provide a good platform from which to demonstrate the extended understanding of ore formation processes that is provided by thermochronology studies.

THERMOCHRONOLOGY AND MINERALIZED SYSTEMS – AN INTRODUCTION

In this section we review the fundamental application of (U-Th)/He, fission track and $^{40}Ar/^{39}Ar$ thermochronometry methods to mineralized systems. We then present a synopsis of how these techniques can be used in thermal history analysis, either alone or in combination with

1529-6466/05/0058-0018$05.00 DOI: 10.2138/rmg.2005.58.18

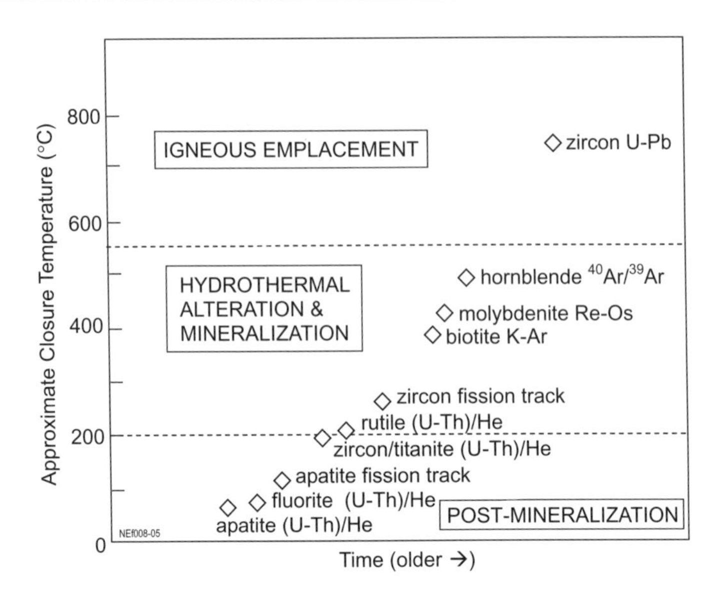

Figure 1. Schematic diagram showing the time-temperature relationship between various chronometers and geologic processes. Nominal closure temperatures are plotted but, in the case of the lowest temperature thermochronometers, closure temperature is dependant on various assumptions regarding grain size and shape, chemical zonation and cooling rate. In mineralized systems, high closure temperature systems like zircon U/Pb delineate igneous emplacement while hornblende $^{40}Ar/^{39}Ar$ and zircon fission track is applied to dating the timing of hydrothermal alteration and mineralization. The timing of post-mineralization processes like exhumation and erosion is revealed using low closure temperature systems like (U-Th)/He and apatite fission track.

other chronometers, in order to provide more information on deposit genesis. The discussion is presented from low through to higher temperature thermochronometers, corresponding to their presentation on Figure 1.

(U-Th)/He thermochronology

As documented earlier in the volume, (U-Th)/He thermochronology is based on measuring the accumulation of radiogenic ^4He produced from U and Th decay. The daughter helium is retained until the mineral is heated to a temperature at which its structure and helium retentivity change. This "closure temperature" varies from mineral to mineral, providing a powerful way to track the low temperature thermal history of an ore deposit if more than one mineral phase from a single sample can be dated (Fig. 1). However, the closure temperature for a given mineral is dependant on a series of assumptions regarding grain size and shape, chemical zonation and cooling rate (e.g., apatite has a closure temperature of 75 °C assuming a cooling rate of 10 °C/m.y.) and interpretation of the age depends on the complexity of the cooling history (how long the mineral was held at a given temperature). This caveat on thermochronometry interpretation is particularly relevant to apatite (U-Th)/He ages and apatite fission track ages where low temperature processes (tectonic or erosional) can result in samples spending long periods in the helium partial retention (Wolf et al. 1998) or fission track partial annealing zone (Fleischer et al. 1975). Careful interpretation of thermochronology ages is critical (Wolf et al. 1998) as an age may reflect: (i) rapid monotonic cooling of the sample over a very short interval on a geological timescale. In this case, the (U-Th)/He age reflects cooling of the sample through its closure temperature and can often be related to a particular geological event (Dodson 1973); or (ii) slow monotonic cooling (or more complex nonmonotonic net cooling) over a long time interval with the final age not necessarily related to any specific geological event. A good example is the use of apatite fission-track and (U-Th)/He techniques in areas with low denudation rates (as is the case for some porphyry deposits in South America and Carlin deposits in Nevada) where rocks may spend prolonged periods in a partial annealing/retention zone. Interpretation of thermochronometry ages is reviewed in more detail in other chapters of this volume (Donelick et al. 2005; Dunai 2005; Gallagher et al. 2005; Harrison and Zeitler 2005; Ketchum 2005).

Apatite (U-Th)/He. The most frequently utilized mineral in (U-Th)/He thermochronometry is the common accessory phase, apatite (see comprehensive reviews by Farley 2002 and Ehlers and Farley 2003). Apatite makes an ideal (U-Th)/He chronometer because it is usually well crystallized, enriched in U and Th, commonly >60 μm in diameter and transparent (making it relatively easy to detect the presence of fluid and mineral inclusions). Apatite is also ideal for other thermochronometry applications (e.g., fission track dating) and is therefore useful for cross-calibration purposes and complementary low-T history studies. Due to the low closure temperature for He in apatite (75–100 °C; Zeitler et al. 1987; Wolf et al. 1996; Farley et al. 1998; House et al. 1998; Farley 2002 and references therein), it is generally assumed (within limits which will be discussed in a later section) that the age documents the time when rocks pass through the upper 1–3 km of the crust. It follows that the apatite (U-Th)/He age can help constrain the post-mineralization uplift (Fig. 1) and exhumation history of a deposit (McInnes et al. 1999) with implications for ore preservation and supergene enrichment processes (McInnes et al. 2003). Another potential use for (U-Th)/He in apatite is to define the timing and scale of low temperature hydrothermal systems like the Carlin-type gold deposits.

Fluorite (U-Th)/He. While the application of (U-Th)/He thermochronometry to fluorite is in its preliminary stages, the closure temperature of vein fluorite from Yucca Mountain, Nevada has been determined to be 80–100 °C based on a 10 °C/m.y. cooling rate and a 200–300 μm diameter (see discussion in Evans et al. 2005). This suggests potential applications for constraining the low-temperature history of hydrothermal ore deposits. Fission-track ages of fluorite mineralization from late hydrothermal veins in Norway have been used to constrain the minimum age of Late Cretaceous/Early Tertiary hydrothermal activity (Grønlie et al. 1990) although the uncertainty on the ages ranges from 35–55%.

Zircon (U-Th)/He. While the first attempts at zircon (U-Th)/He thermochronometry were performed early last century (Strutt 1910a,b), more recent studies (Reiners et al. 2002; Tagami et al. 2003; Reiners et al. 2004) have recognized the value of thermally-dependant helium diffusion from zircon to thermal history studies. Reiners et al. (2004) have identified the closure temperature as being between 171–196 °C. Because zircon is resistant to weathering and is found in magmatic, metamorphic and pegmatitic settings, zircon (U-Th)/He data has wide-ranging application in mineralized systems. For example, phenocrystic zircon (U-Th)/He dating of a porphyry system records the cooling of the rock below 200 °C which corresponds to lower temperature alteration (e.g., argillic alteration stage). As the solubility of Cu in hydrothermal fluids is limited below 200 °C, high temperature Cu transport and mineralization can be constrained as occurring between the magmatic (U/Pb and $^{40}Ar/^{39}Ar$) and zircon (U-Th)/He ages.

Oxide (magnetite, hematite, rutile) (U-Th)/He. Since Fanale and Kulp's (1962) early work on magnetite, the development of (U-Th)/He methods for Fe and Mn-oxides has been primarily aimed at dating magnetite-hematite and base metal vein mineralization (Lippolt and Weigel 1988; Wernicke and Lippolt 1992, 1994a, 1997; Lippolt et al. 1993). While the effect of grain size and compositional variations on diffusion have not been investigated fully, closure temperatures in the range of 180–250 °C for large (5mm) specularite grains and >90–160 °C for botryoidal hematite (>10 μm diameter) have been suggested (Bähr et al. 1994). If the diffusion characteristics of iron oxides can be better understood, numerous applications can be envisioned in iron ore exploration and in understanding the thermal and depositional histories of weathered and altered deposits.

Preliminary development of rutile (U-Th)/He thermochronometry (Crowhurst et al. 2002) has suggested a closure temperature of >180–200 °C. Further development of this chronometer is desirable as it presents yet another resistate mineral with a high closure temperature that can be utilized in metamorphic regimes. In mineralized systems, rutile is a common hydrothermal alteration product of titanite. Preliminary work dating rutile and zircon from the Darrezhar porphyry Cu prospect in Iran yielded identical (U-Th)/He ages within error, (17.0 ± 0.63 Ma and 16.0 ± 0.68 Ma) supporting the predicted similarity in rutile and zircon He closure temperature (B. McInnes, unpublished data).

Fission track

As described in earlier chapters (Donelick et al. 2005; Tagami and O'Sullivan 2005), fission track dating is based on the accumulation of radiation damage tracks from spontaneous nuclear fission of ^{238}U within mineral lattices (Gleadow et al. 2002). The thermally induced annealing of the tracks in minerals such as apatite and zircon provides the basis of the use of fission track dating as a mineral thermochronometer and quantitative predictive models of the temperature dependence of annealing significantly advanced the field (e.g., Crowley 1985; Gleadow et al. 1986; Laslett et al 1987; Duddy et al. 1988; Corrigan 1991; Lutz and Omar 1991; Gallagher 1995; Yamada et al. 1995; Ketcham et al. 1999, 2000; Green et al. 1999). In apatite, the process of fission track annealing is thermally resilient relative to He diffusivity (helium partial retention zone lies at temperatures 35 °C cooler than the analogous fission track partial annealing zone; Wolf et al. 1998) and therefore a slowly cooled sample (10 °C/m.y.) should have a fission track age older than its (U-Th)/He age. One method is often used to corroborate the other. For example, excess helium might be an issue if the apatite (U-Th)/He age is greater than the apatite fission track age. Because fission track lengths in apatite are shortened within the helium partial retention zone, apatite (U-Th)/He ages can be used to test track-length thermal models.

Fission tracks in apatite will anneal above about 125 °C (about 4–5 km depth under a normal geothermal gradient), producing an altered distribution of track lengths. A field based estimate of the zircon fission track closure temperature for a cooling rate of 15 °C/m.y. is 240 °C (zircon from Gold Butte, Nevada; Bernet et al. 2004). Whereas absolute ages can be obtained from fission track analysis, this is only possible for samples that have cooled rapidly and have remained undisturbed at, or close to the surface since formation (Gleadow et al. 2002). More commonly, apparent ages are obtained which reflect aspects of the thermal history (uplift, denudation). In terms of mineralized systems, fission track dating has long been used in conjunction with other techniques to resolve the timing of mineralization (Fig. 1) and geodynamic setting of ore deposits (Banks and Stuckless 1973; Lipman et al. 1976; Shawe et al. 1986; Koski et al. 1990; Naeser et al. 1990; Arne 1992; Hill et al. 2002; Arehart et al. 2003; Suzuki et al. 2004), identify potential buried stocks and potential associated ore deposits (Naeser et al. 1980; Cunningham et al. 1984; Steven et al. 1984; Beaty et al. 1987), constrain the timing and amount of exhumation and erosion (Maksaev and Zentilli 1999), aid in resolution of the sense and amount of fault offset (McInnes et al. 1999) and reveal the thermotectonic history of petroleum basins (Sutriyono 1998; Osadetz et al. 2002).

$^{40}Ar/^{39}Ar$

Based on the natural decay of ^{40}K to ^{40}Ar and the induced decay of ^{39}K to ^{39}Ar, $^{40}Ar/^{39}Ar$ thermochronometry assumes that in the case of partial radiogenic Ar loss, some domains in mineral grains remain unaffected and that excess Ar incorporated during mineral formation, has a different distribution in the crystal than the Ar produced by in situ ^{40}K decay (McDougall and Harrison 1999; Williams 2004). Similar to the (U-Th)/He system, the highly predictable time dependant nature of radioactive decay, in conjunction with the temperature dependence of Ar diffusion can be used to reveal the thermal history of the sample (McDougall and Harrison 1999). Unique to $^{40}Ar/^{39}Ar$ is the step heating technique, where the irradiated mineral is progressively and incrementally heated in order to release the contained Ar. This process accounts for both excess Ar and Ar loss and can yield both the minimum age of crystallization (high temperature steps) and the timing of thermal events (lower temperature steps) on a single sample (e.g., Tomkins et al. 2004).

The recent improvement in precision of isotope ratio measurement and low sample size requirements permit $^{40}Ar/^{39}Ar$ techniques to decipher overprinted hydrothermal and igneous events that are not typically resolved by K-Ar methods (Reynolds et al. 1998). By analyzing a suite of commonly occurring K-bearing minerals (hornblende, micas, potassic alkali feldspars) with a range of diffusion properties (closure temperatures), it is possible to elucidate the timing of thermal events and cooling rates (e.g., Snee 1999) over a higher temperature range than the (U-Th)/He system alone (>200 °C). In combination with other thermochronometry methods (commonly fission track and apatite (U-Th)/He), even more information is obtained. Some applications of $^{40}Ar/^{39}Ar$ thermochronometry in mineralized systems include resolving the timing of mineralization (Fig. 1) and alteration/metamorphic events (e.g., Fullagar et al. 1980; Goldfarb et al. 1991; Arehart et al. 1993, 2003; Dammer et al. 1996; Bierlein et al. 1999; Snee et al. 1999; Chan et al. 2001; Mote et al. 2001; Smith et al. 2001; Hill et al. 2002; Lund et al. 2002; Wilson et al. 2003; Mauk and Hall 2004) and documenting the duration of hydrothermal systems (e.g., Silberman et al. 1979; Arribas et al. 1995; Henry et al. 1995; Marsh et al. 1997; Maksaev and Zentilli 1999; Garwin 2002).

Using thermochronometry in thermal history studies

There are several ways that thermochronometry has been applied to exploring the thermal history of mineralized systems. The first involves use of a single chronometer and can provide meaningful data under some conditions. Of more value is the use of multiple chronometers that can potentially reveal the full thermal history of the region. Finally, modeling of thermochronometry data adds value to thermal history analysis by refining the interpretation of the age data.

Single chronometer. A single thermochronometer will yield an apparent age but interpretation of the significance of that age will require independent geological evidence of the cooling history. For example, an apatite (U-Th)/He age of 60 Ma might record a short-term, low temperature hydrothermal pulse which reset the "helium clock" 60 m.y. ago, or it may suggest that the apatite has experienced a more complex history and has been held within the helium partial retention zone for some extended period. More evidence would be required to ascertain which history is more likely (see Wolf et al. 1998 and Farley 2002 for more discussion). Similar scenarios have been discussed for fission track ages (e.g., Green et al. 1989; Gallagher 1995).

Multiple chronometers. Diffusion experiments on fluorite, apatite, titanite, zircon and rutile have determined a progressive increase in minimum He closure temperature (T_c) from ~75 °C to ~200 °C (Crowhurst et al. 2002; Farley 2002; Reiners et al. 2002; Evans et al. 2005). By combining U/Pb (T_c >900 °C; Lee et al. 1997; Cherniak and Watson 2000), K-Ar, Re-Os (T_c range from 300–500 °C; McDougall and Harrison 1999; Suzuki et al. 1996), fission track and (U-Th)/He techniques, an interval of over 800 °C of thermal history of an ore deposit or mineral district can be elucidated. It should be noted that one may still have a problem deciding on likely cooling paths at low temperature and hence what event is represented by the age determined using low temperature thermochronometers.

Rahl et al (2003) demonstrated for the first time that a single grain of zircon could be analyzed by excimer laser ablation inductively coupled mass spectrometry (ELA ICP-MS) to obtain a U/Pb age and then analyzed by conventional (U-Th)/He methods to yield two ages on a single crystal. McInnes et al. (2003ab, 2004, 2005) have performed apatite (U-Th)/He, zircon (U-Th)/He and U/Pb dating on samples from porphyry deposits in Iran, Chile and Indonesia in order to investigate their genesis, exhumation and preservation potential. Advantages of this approach are that:

1.) a single radioactive decay scheme is utilized U + Th → Pb + He,
2.) apatite and zircon are both usually obtainable in significant quantities for analysis from a 1 kg sample of igneous rock,
3.) apatite and zircon are stable in all hypogene alteration assemblages in porphyry deposits,
4.) the coupled use of zircon U/Pb and zircon (U-Th)/He dating determines both the emplacement age of the porphyry deposit and the interval where the bulk of base metal transport and deposition occurs in magmatic-hydrothermal systems (750–200 °C),
5.) the coupled use of zircon and apatite (U-Th)/He dating determines the post-mineralization uplift and exhumation history of the deposit with implications for ore preservation and supergene remobilization (200–90 °C).

Time-temperature curves are one way of graphically displaying the data outputs from multiple thermochronometers because all the radiometric ages are tied to nominal closure temperatures. Multiple age determinations produce time-temperature curves delineating the thermal history of a porphyry deposit from the time of its emplacement to the time of its thermal decline to ambient conditions. Graphical analysis of combined age-temperature data show a range of cooling profiles varying from a subvertical line for a rapidly cooled intrusion emplaced in the upper 1–2 km of the crust (see right inset in Fig. 2) to a "hockey stick" pattern for a slowly cooled pluton emplaced at depths greater than 2–3 km in the crust (see left inset in Fig. 2). Using the "hockey stick" as a process model for the thermal histories of porphyry copper systems, the handle of the stick (represented by higher temperature chronometers like zircon U/Pb and K-Ar) constrains the high-temperature history of the pluton and its

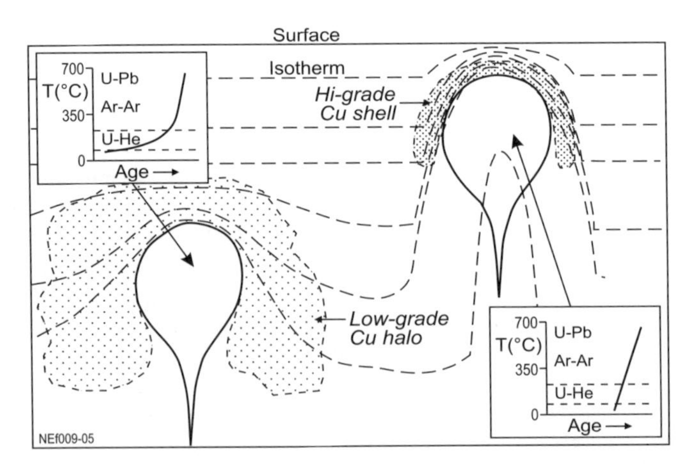

Figure 2. A schematic representation of pluton emplacement at shallow (right) and deep (left) crustal levels and the respective time-temperature histories (insets). If Cu transport and heat transfer are treated as diffusive processes, then intrusions emplaced within the uppermost crust should experience greater thermal gradients and more effective Cu transport than those emplaced in mid-crustal regions where temperature regimes are moderated by the Earth's geotherm.

associated Cu ore shell, while the blade of the stick (zircon (U-Th)/He and apatite (U-Th)/He chronometers) constrain the post-mineralization uplift and exhumation history of the deposit with implications for ore preservation and supergene remobilization.

Modeling of thermochronometry data. Using numerical models to derive potential cooling histories directly from thermochronometry data provides researchers with a powerful tool, particularly when modeling is based on multiple chronometers. As discussed earlier, modeling of apatite and zircon fission track data is particularly advanced (e.g., Crowley 1985; Gleadow et al. 1986; Laslett et al 1987; Duddy et al. 1988; Corrigan 1991; Lutz and Omar 1991; Gallagher 1995; Yamada et al. 1995; Green et al. 1999; Ketcham et al. 1999, 2000) and provides a means to assess denudation histories, timing of hydrothermal activity and the low temperature cooling history of igneous intrusions. Modeling of (U-Th)/He ages has progressed in the past few years (e.g., Wolf et al. 1998; Ehlers et al. 2003 and references therein) and now allows quantification of a number of parameters related to the dynamic processes of magmatic-hydrothermal cooling, exhumation and erosion of igneous intrusions (e.g., Fu et al. 2005). Further detail on forward and inverse modeling is provided in other chapters (Dunai et al. 2005; Gallagher et al. 2005; Ketchum 2005).

One such inverse modeling package (4DTherm v.1.1; Fu et al. 2005) is used later in this work to analyze multiple age data (primarily (U-Th)/He and U/Pb) obtained on selected Indonesian and Iranian porphyry deposits. Physical parameters and assumed initial conditions for porphyry deposits are listed in Table 1 and as the detailed description of the model will be published after this volume goes to press, background information on the modeling technique is provided in Appendix II. Although future versions of 4DTherm will include advective and convective cooling scenarios, the current version inverts thermochronometry data assuming conductive heat transfer, which is the least efficient heat transfer mechanism. The outputs of the modeling runs provided in this paper should, therefore, be considered as end members. For example, if the depth of porphyry emplacement is constrained by the model to be 5 km, then that result is a minimum depth.

Table 1. Main parameters and their initial values used in 4DTherm v.1.1 (Fu et al. 2005).

Parameter	*Initial Value*
Surface temperature	10 °C
Geothermal gradient	50 °C/km
Initial temperature of magma	1000 °C
Thermal diffusivity	1.0×10^{-6} m^2/s
Latent heat of crystallization	100 cal/g
Constant heat flow from bottom	65 mW/m°C

4DTherm v.1.1 treats the cooling history of igneous bodies from their assumed emplacement temperature at 1000 °C to an ambient surface temperature of 10 °C. Throughout the cooling history, two distinct phases were defined (Table 2): (i) *Magmatic-hydrothermal cooling* begins at intrusion emplacement and continues until both igneous and country rocks reach a final thermal equilibrium under a steady-state geothermal gradient. This phase is shown as R_1 on the schematic time-temperature plot (Fig. 3). Initial cooling is rapid but towards the end of the magmatic-hydrothermal cooling stage, the igneous and country rocks cool more slowly until both reach a final thermal equilibration and the geothermal gradient returns to pre-intrusion thermal conditions (defined as the "cooled" state), (ii) *Exhumation cooling* (R_2 in Fig. 3) begins when the intrusion reaches the "cooled" state and continues until the body reaches the surface (at 10 °C). The rate of cooling through this stage is primarily controlled by exhumation and erosion processes. The hypogene deposition of most economic minerals (e.g., Cu, Au) mainly occurs during the early magmatic-hydrothermal cooling stage while supergene remobilization and precipitation processes occur mainly during the late exhumation cooling stage.

Table 2. Division and main characteristics of cooling stages of mineralized porphyry deposits.

Cooling stages	*Magmatic-Hydrothermal Cooling*		*Exhumation Cooling*
	Stage I	*Stage II*	
Duration	1/4 of magmatic-hydrothermal cooling	3/4 of magmatic-hydrothermal cooling	Any length of time (depends on size and emplacement depth)
Relative Cooling Rate	Very high	Relatively high - medium	Very low
Thermal relationship to country rock	High temperature difference (Thermal anomaly)	Low temperature difference (Local thermal equilibrium)	No difference (Regional thermal equilibrium)
Cooling mechanism	Conductive and Hydrothermal circulation cooling	Conductive cooling (Geothermal)	Conductive cooling (Exhumation/erosion)
Geo-/thermo-chronometric constraints	zircon U/Pb, Rb-Sr, Ar-Ar, Re-Os	Rb-Sr, Ar-Ar, Re-Os, zircon fission track, zircon U-He	zircon U-He, apatite fission track, apatite U-He

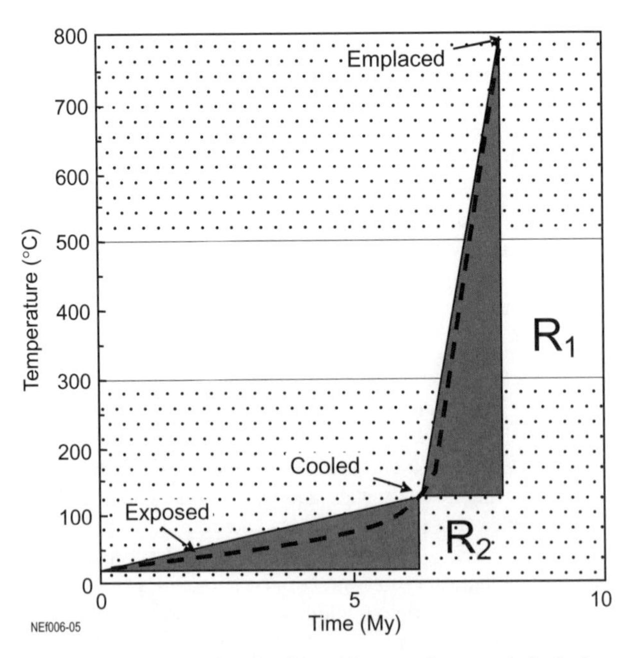

Figure 3. Schematic time-temperature plot showing with rates of magmatic-hydrothermal cooling (R_1) and exhumation cooling (R_2) as defined in 4DTherm v.1.1 (see text). The timing of emplacement, cooling and exposure of a theoretical intrusion are also shown. The white region brackets the thermal regime of hypogene ore formation (500–300 °C).

APPLICATIONS OF THERMOCHRONOMETRY TO GOLD MINERALIZATION

Carlin-type gold deposits

In order to better understand the timing of mineralization and hydrothermal events in Carlin deposits, fission track, (U-Th)/He and $^{40}Ar/^{39}Ar$ methods have all been applied to studies of the sediment-hosted gold deposits (Carlin-type) in the Great Basin of western North America (Arehart et al. 1993, 1995, 2003; Ilchik 1995; Chakurian et al. 2003; Hickey et al. 2003). These giant deposits typically consist of finely (submicron) disseminated gold within hydrothermal arsenian pyrite, hosted in altered, silty carbonate rocks. Ascertaining the timing of gold mineralization is critical to the development of genetic models that aid exploration targeting.

Rb/Sr dates on galkhaite and $^{40}Ar/^{39}Ar$ on hydrothermal illite in dykes and on magmatic biotite in syn-mineral dikes establish that the Carlin-type deposits of the Carlin Trend, Jerritt Canyon district, Getchell district are Eocene in age (Hofstra et al. 1999; Hofstra and Cline 2000; Ressel et al. 2000; Hickey et al. 2003). The reliability of $^{40}Ar/^{39}Ar$ methods for determining the mineralization age of altered rocks associated with sediment-hosted gold deposits has been questioned due to problems of partial resetting of sericite ages after heating by hydrothermal fluids below the sericite closure temperature (Arehart et al. 2003; Chakurian et al. 2003). The most reliable apatite fission track data on several deposits in the Carlin Trend indicate mineralization ages of between 33–42 Ma (Hofstra et al. 1999; Ressel et al. 2000; Arehart et al. 2003; Hickey et al. 2003). Two apatite (U-Th)/He ages on the Carlin East deposit were significantly younger than the fission track ages (31.0 ± 1.9 Ma and 21.4 ± 1.3 Ma) and may be the result of thermal resetting associated with Miocene volcanism. Landscape restoration based on thermal modeling of apatite fission track data suggests that Carlin deposits in the northern Carlin trend were emplaced at paleodepths of between <2–3km for geothermal gradients of 20–30 °C/km, (Cline et al. 2005).

Thermochronology has helped resolve the genetic relationship between the Barney's Canyon and Melco disseminated gold deposits and the giant Bingham Canyon porphyry Cu-Mo-Au deposit in Utah. K-Ar and $^{40}Ar/^{39}Ar$ ages for Bingham Canyon mineralization range from 37 to 40 Ma (Warnaars et al. 1978, Kendrick et al. 2001; and Parry et al. 2001), whereas Arehart et al. (2003) reported an apatite (U-Th)/He age of 34.9 ± 1.8 Ma, indicating rapid cooling and relatively shallow emplacement (<4 km). Apatite from sandstone formations hosting gold at the Barney's Canyon deposit, 8 km distal from Bingham, was found to have a (U-Th)/He age of 33.1 ± 1.6 Ma. This suggests a genetic relationship whereby auriferous low temperature fluids driven by the Bingham Canyon thermal system precipitated gold in sedimentary host rocks at Barney's Canyon. This was confirmed in a major paleothermal study (Cunningham et al. 2004) that also demonstrated that apatite fission track ages in Permian sediments proximal to Bingham Canyon were completely reset whereas zircon tracks were not significantly annealed.

Epithermal gold deposits

Thermochronology studies can be used in conjunction with geochronology methods (refer to discussion in Reiners et al. 2005 to understand the distinction between the two) to determine the temporal relationship between mineralization and magmatism and to explore the thermal history of epithermal deposits. For example, the Porgera epithermal gold deposit in Papua New Guinea was studied by hornblende $^{40}Ar/^{39}Ar$, roscoelite $^{40}Ar/^{39}Ar$ and biotite K-Ar methods to constrain the timing of mineralization to within 0.1 m.y. of emplacement of associated igneous intrusions at around 6 Ma (Richards and McDougall 1991; Ronacher et al. 2002).

Belhadi et al. (1999) combined zircon fission track ages with previously published whole rock, biotite and hornblende K-Ar thermochronometry (Sawai et al. 1998) at quartz vein-hosted gold deposits in the Hoshino gold region, Japan. The results indicate that volcanic activity began at 4.3 Ma with a second phase of activity extending from 3.5 Ma to 2.6 Ma. The zircon fission track ages (2.8 Ma) from the youngest volcanic unit (Takeyama andesite) were within analytical uncertainty of whole rock K-Ar ages from hydrothermally altered samples associated with gold deposits (Sawai et al 1998). The authors postulate that the eruption of the Takeyama andesite was responsible for the hydrothermal alteration and that field relationships also suggest a contemporaneous relationship between the Takeyama andesite and gold metallogenesis.

Archean lode gold deposits

Archean lode gold deposits occur in complex metamorphosed terranes, and the majority of geochronology research has been focused on unraveling the timing of mineralization and deposit genesis. The application of low temperature thermochronometers to the study of the exhumation history of these systems is still in the early stages. Kent and McDougall (1995) applied $^{40}Ar/^{39}Ar$ to hydrothermal muscovite samples from the Kalgoorlie gold field (located in the Archean Yilgarn block, Western Australia) in order to determine the timing of various styles of gold mineralization (see also Witt et al. 1996 and Kent and McDougall 1996). The gold stockwork system at the Mount Charlotte deposit was formed at least 10 m.y. before that at Golden Mile suggesting separate hydrothermal episodes formed the stockwork and shear-hosted mineralization. However, Kent and McCuaig (1997) highlight the need for caution when interpreting $^{40}Ar/^{39}Ar$ ages in hydrothermal ore deposits where regional post-mineralization fluid movement may result in Ar loss. In another study (Napier et al. 1998), amphibole and biotite $^{40}Ar/^{39}Ar$ step-heating was utilized to reveal the complex, post-metamorphic thermal history of the Southern Cross area of the Yilgarn. The results suggest that after main gold mineralization (~2620 Ma), temperatures remained at 500 °C for 20–70 m.y. suggesting longer-lived regional tectonic activity than previously predicted. Tomkins et al. (2004) provide a time-temperature history for the Challenger Deposit, Gawler Craton, South Australia based on an integrated study of U/Pb, Sm-Nd, Rb-Sr and $^{40}Ar/^{39}Ar$ ages. Three thermal events spanning over 1800 Ma were identified including a very low temperature event (~150–200 °C) at about 1531 Ma.

Shale-hosted lode gold deposits

Muscovite, illite and amphibole from argillite-hosted vein gold deposits of the Meguma Terrane, Nova Scotia were studied by $^{40}Ar/^{39}Ar$ to reveal the timing of vein formation (ca. 380–405 Ma), which is about 10–15 m.y. after the peak of regional metamorphism (Kontak et al. 1998). Combined fluid inclusion and thermochronology data led the authors to conclude that the gold was precipitated from metamorphic hydrothermal fluids at 400–450 °C in response to cooling. The auriferous fluids are interpreted to have originated from a reservoir at 18–20 km depth, and gold precipitated at depths of around 12 km as the fluids hydrofractured their way through the crust. Studies of similar deposits in the Lachlan Fold Belt of Australia used $^{40}Ar/^{39}Ar$ to resolve two distinct phases of gold mineralization at about 460–440 Ma and 380–360 Ma (Arne et al. 1998; Foster et al. 1998; Bierlein et al. 1999).

APPLICATION OF THERMOCHRONOMETRY TO PORPHYRY COPPER-MOLYBDENUM-GOLD MINERALIZATION

Reynolds et al (1998) define porphyry deposits as "large (1–5km diameter), low grade (typically <1% Cu) copper (± Mo, Au, Ag) concentrations in which sulfide minerals occur disseminated in a network of veinlets and breccias distributed within, and more or less concentrically around supracrustal porphyritic stocks of intermediate to felsic composition". The cooling rate in the porphyry system changes with time in response to the various processes operating over the life of the system. For example, initial rapid cooling (800 °C → 350 °C) occurs during post-intrusion rupture of the rock column and is typically followed by more moderate cooling and thermal collapse of the system (see "hockey-stick" pattern in left inset, Fig. 2). The timing of this cooling is constrained by the U/Pb and K-Ar or $^{40}Ar/^{39}Ar$ ages. Early cooling of the magmatic fluid plume is largely a result of decompression (from lithostatic to hydrostatic conditions) and phase separation (producing coexisting brine and vapor-rich fluid inclusions) (e.g., Ulrich et al. 2001; Harris et al. 2003). These cooling fluids are responsible for the formation of potassic alteration assemblages and the deposition of the bulk of the mineralization in these deposits. Phyllic and argillic alteration results from magmatic and/or mixed magmatic-meteoric fluid circulation at temperatures ranging from 350 °C to ~200 °C (Hedenquist and Richards 1998) and may develop episodically over a period of a million years or more (e.g., Chuquicamata; Reynolds et al. 1998). The zircon (U-Th)/He age probably records the timing of the lowest temperature hydrothermal event in the deposit whereas the apatite (U-Th)/He age records the thermal collapse of the hydrothermal system and/or unroofing of the system.

Porphyry deposits are found worldwide and have been the focus of numerous thermal history studies employing multiple chronometers, originally based on K-Ar and fission track dating and more recently evolving to (U-Th)/He, U/Pb and $^{40}Ar/^{39}Ar$ methods. This multiple chronometer approach allows discrimination between the intrusion age and the age of later hydrothermal alteration. The (U-Th)/He method is particularly amenable to application in porphyry deposit research, because many of the minerals suited to radiometric dating occur as accessory mineral phases in intrusions known to host disseminated mineralization. In addition, precise low temperature thermochronometers can better resolve the timing of different alteration stages (e.g., potassic versus phyllic) and the duration of the magmatic-hydrothermal system.

Pioneering porphyry thermochronology work was conducted in the southwest U.S.A. by Lipman et al. (1976) and Naeser et al. (1980). Their studies employed K-Ar and fission track dating to demonstrate that mineralization in the San Juan Mountains in the southern Rocky Mountains was episodic, spanning a period of at least 25 m.y.. Cunningham et al. (1987, 2004) suggested that low temperature thermochronometry methods such as fission track dating could be used as a potential exploration tool for porphyry systems under cover. Other work by Naeser et al. (1980) and Larson et al. (1994) predicted the presence of a buried mineralized porphyry system before it was drilled and discovered based on apatite and zircon paleothermal anomalies.

Despite the fact that porphyry deposits have been exploited for 100 years (open pit mining at Bingham began in 1905), a number of time- and temperature-related variables involved in their genesis remain poorly understood:

1.) longevity of the ore precipitation event during the thermal decline of the magmatic-hydrothermal system,

2.) depth of emplacement,

3.) preservation potential of hypogene ores during orogenic uplift and exhumation, and

4.) formation potential of supergene ores from eroded hypogene precursors.

In the following discussion, we review the available data for the world's largest Cu-Mo porphyry deposit, Chuquicamata, Chile. We present new thermochronology data (Table 3, Fig. 4) for samples from the potassic alteration zone of selected porphyries in Indonesia and Iran where multiple chronometers have been applied to yield complete thermal histories. We utilize inverse modeling of the thermal history data for these deposits to demonstrate how such analysis can enhance our understanding of deposit genesis and conclude this section with a summary of how thermochronology studies can address the issues listed above.

Table 3. Thermochronology and geochronology data for porphyry deposits in Indonesia and Iran.

Deposit/Location	*Age (Ma) ± 2σ*	*Method*
SCP Sar Cheshmeh, Iran	13.6 ± 0.1	Zircon U/Pb[1]
	12.2 ± 1.2	Rb-Sr (3-point isochron)[2]
	12.5 ± 0.5	Biotite K-Ar[2]
	10.9 ± 0.4	Zircon U-He[1]
	7.2 ± 0.4	Apatite U-He[1]
Miduk, Iran	12.5 ± 0.1	Zircon U/Pb[1]
	12.4 ± 0.5	Rb-Sr (3 point isochron)[3]
	11.2 ± 0.5	Biotite Ar-Ar[3]
	10.8 ± 0.4	Sericite Ar-Ar[3]
	12.5 ± 0.5	Zircon U-He[1]
	9.5	Apatite U-He*
Adbar, Kuh-eMasahim, Iran[4]	7.5 ± 0.1	Zircon U/Pb[1]
	6.8 ± 0.4	Biotite Ar-Ar[5]
	6.4 ± 0.8; 6.3 ± 0.9	Hornblende Ar-Ar[5]
	7.3 ± 0.3	Zircon U-He[1]
	4.9 ± 0.4	Apatite U-He[1]
Grasberg, Indonesia	2.97± 0.57	U/Pb**
	2.9 ± 0.3	Re-Os[6]
	2.67 ± 0.03–3.33 ± 0.12	Biotite Ar-Ar[7]
	3.1–2.9 ± 0.1	Apatite U-He[1]
Batu Hijau, Indonesia	3.74 ± 0.14	Zircon U/Pb[8]
Young Tonalite	3.73 ± 0.08	Biotite Ar-Ar[8]
	2.23 ± 0.09	Apatite U-He[1]
Ciemas, Indonesia	17.8 ± 0.4	Zircon U/Pb [1]
	15.2 ± 2.7	Sulfide Re-Os[9]
	7.2 ± 0.24	Apatite U-He[1]

U-He and Ar-Ar represent (U-Th)/He and $^{40}Ar/^{39}Ar$, respectively.

Notes: [1]this work (see Appendix I for methodology); [2]Shahabpour 1982; [3]Hassanzadeh 1993; [4]Abdar diorite subvolcanic intrusion in caldera of Kuh-e-Masahim volcano; [5]Hassanzadeh 1993, lava flows on flanks of Kuh-e-Masahim volcano; [6]Mathur et al. 2000b; [7]Pollard et al. 2004; [8]Garwin 2000; [9]McInnes et al. 2000; *No apatite was recovered from the Miduk samples. As Abdar is located only 18 km away and the erosion rate in the area is thought to be similar on this geographic scale, the difference between the zircon and apatite (U-Th)/He ages at Abdar was applied to Miduk resulting in a proxy apatite (U-Th)/He age of 9.5 Ma. ** Gibbins et al. 2003 age for the Ertsberg intrusive.

Selected porphyry deposits

Chuquicamata, Chile. The thermal history of the world's largest porphyry Cu-Mo deposit (15,000 Mt @ 0.71% Cu and 0.01% Mo) at Chuquicamata, northern Chile has been studied extensively and from various perspectives. Ballard et al. (2001) obtained zircon U/Pb ages (excimer laser ablation-inductively coupled plasma-mass spectrometry (ELA-ICP-MS) and sensitive high-resolution ion microprobe (SHRIMP)) and identified two temporally distinct porphyry intrusions, one at 34.6 ± 0.2 Ma and a second emplaced 0.9–1.5 m.y. later (33.3 ± 0.3 Ma and 33.5 ± 0.2 Ma;). The age of the first intrusion (East Porphyry) correlates well with the $^{40}Ar/^{39}Ar$ and Re-Os ages determined for the earliest alteration (Reynolds et al. 1998; Mathur et al. 2000a). Subsequent intrusions (referred to as the Bench and West porphyries)

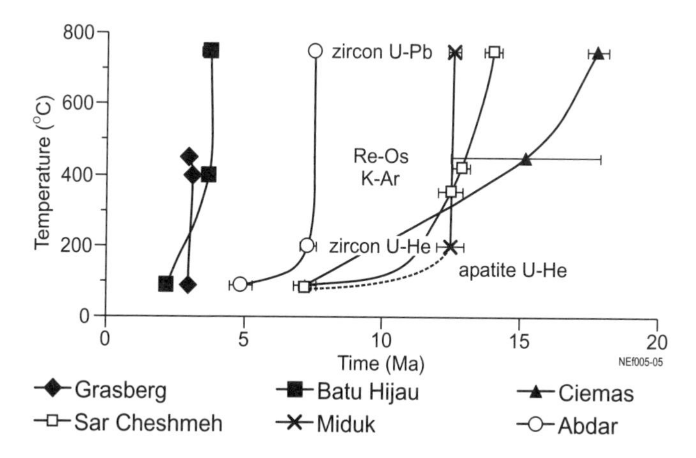

Figure 4. Graphical depiction of time-temperature histories for Indonesian and Iranian porphyries from selected data presented in Table 3. The cooling rate of the various deposits is reflected in the steepness of the curve. Early fast cooling (typical of porphyry deposits) is normally followed by slower cooling until thermal collapse of the system, resulting in the typical "hockey-stick" shaped curve. The contrast between fast cooled systems (e.g., Abdar, Batu Hijau) and more slowly cooled systems (e.g., Ciemas) is clearly evident in this plot.

can be correlated with Cu deposition associated with potassic alteration that has a K-feldspar $^{40}Ar/^{39}Ar$ age of 33.4 ± 0.3 Ma (Reynolds et al. 1998). The Cu-enriched phyllic alteration zone associated with the West Fault structure was developed at least 2 m.y. after the emplacement of the last igneous intrusion at Chuquicamata (sericite $^{40}Ar/^{39}Ar$ age of 31.1 ± 0.3 Ma, Reynolds et al. 1998; pyrite Re-Os age of 31.0 ± 0.3 Ma, Mathur et al. 2000) and cooled rapidly below ~200°C as recorded by the apatite fission track and (U-Th)/He ages of near surface samples (about 31 Ma; McInnes et al. 1999).

A unique feature of the Chuquicamata Cu-Mo deposit is that a significant portion of the western side of the ore body was dislocated by post-mineralization motion along the West Fault, a major N-S strike-slip feature that cuts through Northern Chile. The vertical displacement along the fault was determined by a low temperature thermochronology (apatite (U-Th)/He and fission track) study based on comparative age dating of samples taken from both sides of the fault (McInnes et al. 1999). The study revealed a 600 m vertical offset of the western Fortuna block and determined that the "missing" portion of Chuquicamata was probably eroded during tectonic uplift and denudation of the western crustal block.

Grasberg porphyry Cu-Au deposit, Indonesia. The Grasberg Cu-Au deposit is a giant ore system in Irian Jaya with proven and probable reserve estimates of around 2700 Mt @ 1.08% Cu and 0.98 g/t Au. The economic porphyry is the Main Grasberg Intrusion (MGI), a quartz monzodiorite that intruded the core of the Dalam Diatreme. The final stage of magmatic activity at Grasberg was the emplacement of the dyke-like Kali intrusion. Although U/Pb dating has not been carried out, combined $^{40}Ar/^{39}Ar$ and (U-Th/He) dating on the same samples from the MGI and Kali intrusions shows extremely rapid cooling with apatite (U-Th)/He ages (2.9 to 3.1 (± 0.1 Ma); McInnes et al. 2004) nearly identical to biotite $^{40}Ar/^{39}Ar$ ages (from 2.7–3.3 Ma; Pollard et al. 2005). These ages overlap with a sulfide Re-Os age of 2.9 ± 0.3 Ma (Mathur et al. 2000b) and a more recent molybdenite Re-Os age of 2.88 ± 0.01 Ma (Mathur et al. 2005) indicating that ore-related intrusions in the district underwent extremely rapid monotonic cooling as a consequence of shallow emplacement. These studies corroborate earlier interpretations based on apatite fission track dating that the MGI was emplaced within ~1 km of the paleosurface (Weiland and Cloos 1996). Zircon U/Pb analysis at Grasberg is underway but for the purposes of inverse modeling, the zircon U/Pb age (2.97 ± 0.57 Ma; Gibbins et al. 2003) from the nearby Ertsberg intrusive (1 km to the SE of Grasberg) is adopted as a minimum U/Pb age for the MGI.

Batu Hijau porphyry Cu-Au deposit, Indonesia. The Batu Hijau porphyry Cu-Au deposit is located in SW Sumbawa Island, Nusa Tenggara with proven and probable reserve estimates of around 920 Mt @ 0.55% Cu and 0.41 g/t Au. Mineralization is associated with a multi-phase tonalite porphyry complex hosted in quartz-dioritic and andesitic wallrocks (Garwin 2002). Zircon U/Pb and apatite (U-Th)/He ages for the late-mineralization Young Tonalite (collected at 150 m ASL) are 3.74 ± 0.14 (2σ) Ma and 2.23 ± 0.09 (2σ) Ma, respectively (Garwin 2000). Hydrothermal biotite from the Young Tonalite (3 samples collected from 150–350 m ASL) yields a mean plateau $^{40}Ar/^{39}Ar$ age of 3.73 ± 0.08 (2σ) Ma, which is indistinguishable from the zircon U/Pb age (Garwin 2002). Previous studies using amphibole-plagioclase thermobarometry indicate that the tonalitic magmas began to crystallize at 9km depth with final crystallization and stock emplacement occurring at $\leq$2 km ($\pm$ 0.5 km) depth and between 710–780 °C (Garwin 2000; 2002).

Ciemas Cu-Au porphyry prospect, Indonesia. The Ciemas porphyry Cu-Au prospect is located near the southern coast of western Java, about 150 km south of Jakarta. The Ciemas prospect has seen limited drilling (8 diamond holes) and development, but the results to date suggest extensive zones of sub-economic metal grades (~0.2% Cu, ~0.2 g/t Au) associated with a quartz diorite porphyry intrusion hosted principally by andesitic volcanic rocks. Zircon U/Pb, sulfide Re-Os (McInnes et al. 2000) and apatite (U-Th)/He dating (McInnes et al. 2004) has returned ages of 17.8, 15.2 and 7.2 Ma, respectively. It should be noted that there is a large error bar associated with the Re-Os age (Fig. 4) which places some uncertainty on the thermal history. The geochronology and thermchronology of the Ciemas prospect is included to facilitate thermal history comparison between an apparent "failed" porphyry and the other giant porphyry deposits of Indonesia.

Sar Cheshmeh, Iran. The Kerman Belt, located in southeastern Iran, is an elongated NNW-SSE mountain belt, 500 km long and 100 km wide. It is principally composed of a folded and faulted early Tertiary volcano-sedimentary complex and is bordered to the southwest by a major thrust zone and the Tertiary and Paleozoic sedimentary rocks of the Zagros Mountains (Waterman and Hamilton 1975). The Sar Cheshmeh and Miduk copper deposits (mined by the National Iranian Copper Industries Company, NICICO) are two of the largest known porphyry Cu deposits in the Kerman district. Sar Cheshmeh and Miduk are associated with a high-K calc-alkaline Eocene volcanic arc formed after cessation of subduction of Tethyan oceanic lithosphere at the Zagros suture zone. (Sengor and Kidd 1979). Post-collisional compression and mantle buoyancy forces led to the uplift of the Iranian plateau, with middle Miocene marine sediments occurring at elevations greater than 3000 m in the Kerman belt (Hassanzadeh 1993). The preservation of porphyry Cu deposits in the Kerman belt is therefore dependent on the original depth of emplacement and the rate of exhumation in response to this tectonically driven uplift and erosion.

No previous thermochronology studies have been carried out in the Kerman District. We have determined the zircon U/Pb, zircon (U-Th)/He and apatite (U-Th)/He ages of igneous units from the potassic alteration zone of three Iranian porphyry Cu systems (Table 3, Fig. 4) and have integrated this data with pre-existing whole rock Rb-Sr, biotite K-Ar and $^{40}Ar/^{39}Ar$ geochronology ages (Shahabpour 1982; Hassanzadeh 1993). These ages constrain the maximum period of longevity of hydrothermal mineralization in porphyry-epithermal environments. The thermal histories for Sar Cheshmeh and Miduk are compared to that of the

Abdar Cu-Au prospect hosted within the collapsed and partially eroded caldera of the Kuh-e-Masahim stratovolcano located between the two porphyry copper deposits.

The Sar Cheshmeh Cu deposit (1100 Mt @ 0.64% Cu, 0.03% Mo) is contained within an ovoid (2.5 km ×1 km) Cu shell surrounding a Cu-poor granodiorite to quartz monzonite intrusion known as the Sar Cheshmeh porphyry (SCP) that was emplaced within Eocene to Oligocene volcanic rocks of andesitic composition. The alteration halo and satellite intrusions extend for 7 km. During Cu ore formation the deposit was intruded by 3 intrusions interpreted by Ghorashi-Zadeh (1979) to be fractional crystallization products of the magma that produced the SCP. Cross-cutting relationships define the order of emplacement of the intramineral intrusions as: (1) Late fine porphyry (quartz monzonite), (2) Early hornblende porphyry (dacite) and (3) Late hornblende porphyry dyke swarms (latite). Although these intrusions played a role in redistributing Cu throughout the deposit, their net contribution to ore genesis has been to dilute the initial Cu content of the SCP.

Shahabpour (1982) determined whole rock Rb-Sr and biotite K-Ar ages for the Sar Cheshmeh porphyry of 12.2 ± 1.2 Ma and 12.5 ± 0.5 Ma, respectively. Zircon U/Pb/(U-Th)/He and apatite (U-Th)/He dating was conducted on samples of the SCP and from the open pit in August 2002 (McInnes et al. 2003a,b).

Miduk, Iran. The Miduk Cu deposit (>170 Mt @ 0.82% Cu) is a circular body about 400m in diameter centered over an intrusive quartz diorite stock known as the Miduk porphyry. The main intrusion hosts 90% of the Cu mineralization and is cross-cut by multiple NNE trending dykes called the Miduk fine porphyry. The dykes are of similar composition to the main intrusion and are interpreted as comagmatic (Hassanzadeh 1993).

Zircon U/Pb and zircon (U-Th)/He ages of 12.5 Ma for the Miduk porphyry (McInnes et al. 2003a,b) are essentially identical to mineral-whole rock Rb-Sr ages (12.4 ± 0.5 Ma) reported in Hassanzadeh (1993). However, Hassanzadeh (1993) notes that the Rb-Sr age is considered unreliable due to Rb loss and Sr addition during alteration. Hassanzadeh (1993) also determined $^{40}Ar/^{39}Ar$ isochron ages of 11.2 ± 0.5 Ma for biotite in potassic alteration assemblages and 10.8 ± 0.4 Ma for sericite in phyllic alteration zones. Similar to Sar Cheshmeh, the progressively decreasing ages for the Miduk porphyry in the U/Pb and $^{40}Ar/^{39}Ar$ systems reflects the cooling history of the deposit through the temperature interval 750 °C to 300–350 °C, yielding a cooling rate of 250–350 °C/m.y.. Surprisingly, the zircon (U-Th)/He age is identical to the zircon U/Pb age for the same sample of Miduk porphyry and the $^{40}Ar/^{39}Ar$ isochron ages are younger than the zircon (U-Th)/He age, implying that either: (1) the Miduk sample dated for zircon (U-Th)/He has been emplaced near a contact with cool country rock and therefore has had a more rapid cooling history than the samples in the $^{40}Ar/^{39}Ar$ study, or (2) that zircon (U-Th)/He in rapidly cooled, high level intrusions acts as a *geo*chronometer, rather than a *thermo*chronometer (see Reiners et al. 2005 for definition). No apatite was recovered from the Miduk samples and in order to facilitate modeling, a value of 9.5 Ma was assigned as the apatite (U-Th)/He age. As Abdar is located only 18km away and the erosion rate in the area is thought to be similar to that at Miduk, the difference between the zircon and apatite (U-Th)/He ages at Abdar was subtracted from the zircon (U-Th)/He age for Miduk resulting in a proxy apatite (U-Th)/He age of 9.5 Ma.

Abdar Cu-Au prospect, Kuh-e-Masahim volcano, Iran. The Abdar Cu-Au prospect, located approximately 15 km SE of Miduk, is associated with a subvolcanic dioritic intrusion hosted within the partially eroded caldera of the Kuh-e-Masahim stratovolcano (35 km basal diameter, 3500 m asl total elevation, 1500 m above surrounding plateau). Epithermal high-sulfidation Au-Ag-base metal veins are exposed in the caldera peripheral to the Abdar Cu prospect. Reconnaissance scale drilling of the prospect has detected anomalous yet uneconomic concentrations of Cu (values ranging from 0.1–0.25% Cu). Samples for geochronology investigation were taken from potassic alteration zones from mineralized drill core. Similar to Miduk, the zircon U/Pb and zircon (U-Th)/He age data for the Abdar diorite (Table 3) are identical within error. This further supports the suggestion that the zircon (U-Th)/He system acts as a *geo*chronometer for shallow, rapidly cooled subvolcanic intrusions. The relatively young $^{40}Ar/^{39}Ar$ ages for lava flows on the flanks of the volcano indicate that the feeder conduits did not thermally reset the zircon U/Pb and zircon (U-Th)/He ages of the diorite intrusion. The apatite (U-Th)/He age of 4.9 Ma for the Abdar diorite indicates the time when the subvolcanic intrusion cooled below 90 °C due to caldera collapse and rapid erosion of the overlying volcanic pile.

Duration of hypogene ore formation: measured vs. modeled

It has been shown that the overall duration of magmatic-hydrothermal activity in some porphyry ore deposits (e.g., Divide, Silberman et al. 1979; Far Southeast-Lepanto, Arribas et al. 1995; Potrerillos, Marsh et al. 1997) is within the resolution of K/Ar and $^{40}Ar/^{39}Ar$ techniques (0.1–0.3 Ma; Sillitoe 2000). Through the use of multiple geochronology methods and the age dating of intrusions that cross-cut mineralization, some studies have shown that intrusion-related hydrothermal mineralization takes place within hundred thousand year time frames: Sar Cheshmeh, ~160 Ka (McInnes et al. 2003); Batu Hijau, ~80 Ka (Garwin 2002); Grasberg, ~100 Ka (Pollard et al. 2005); Lepanto-Far South East, 100–300 Ka. (Arribas et al. 1995); Round Mountain, ~100 Ka (Henry et al. 1997). These findings are consistent with heat flow modeling of cooling in and around the small, high-level intrusions typically found associated with porphyry copper deposits (e.g., Norton and Knight 1977; Norton and Cathles 1979; Smith and Shaw 1979; Cathles 1981; Cathles et al. 1997). It is probable that these magmatic-hydrothermal activity duration estimates are maximum values, taking into account the resolution achievable using radiometric dating methods. Zircon U/Pb and zircon (U-Th)/He ages of intrusion-related ore deposits can potentially be used to constrain the maximum duration of hypogene ore formation because their closure temperatures bracket the magmatic-hydrothermal temperature interval of 750° to 200 °C. Taking into account the uncertainties on the determined ages, the measured age differentials suggest a maximum period of ore deposition of 3 Ma for Sar Cheshmeh, and around 0.5 Ma for Abdar and Miduk (Table 3).

The measured geochronology data (Table 3, Fig. 4), intrusion size estimates and a range of emplacement depths (Table 4) were input into 4DTherm v.1.1 to iteratively reproduce an idealized time-temperature history for each deposit (Fig. 5). Because the solubility of chalcopyrite, the main hypogene ore mineral in porphyry deposits, decreases by over two orders of magnitude in hydrothermal fluids during a temperature reduction from 500–300 °C (McPhail and Liu 2002; Liu and McPhail 2005), it is possible to more precisely estimate the duration of hypogene ore formation for each deposit studied by extracting inverse thermal modeling outputs through this narrow temperature interval (Table 4). The modeled ore formation duration results are consistent with intervals previously determined for deposits for which we have comparative data: Sar Cheshmeh 270 k.y. (this study) vs. ~160 k.y. (McInnes et al. 2003a,b); Batu Hijau 10.5 k.y. (this study) vs. ~80 k.y. (Garwin 2002); Grasberg 15 k.y. (this study, assuming the zircon U/Pb age is similar to the biotite $^{40}Ar/^{39}Ar$ age) vs. ~100 k.y. (Pollard et al. 2005). Without wishing to put too fine a point on the modeling outputs, those deposits with ore formation durations less than 100 k.y. scale are Grasberg, Batu Hijau, Abdar and Miduk, whereas Sar Cheshmeh falls into an intermediate category (100–1,000 k.y.), and Ciemas took more than 1,000 k.y. to cool through the 500–300 °C interval (Fig. 5).

Emplacement depth

The main parameters controlling the cooling rate of an intrusion are size; emplacement depth and heat transfer efficiency (conductive vs. advective cooling regime). For similar sized intrusions, deep emplacement and conductive thermal regimes will produce the lowest

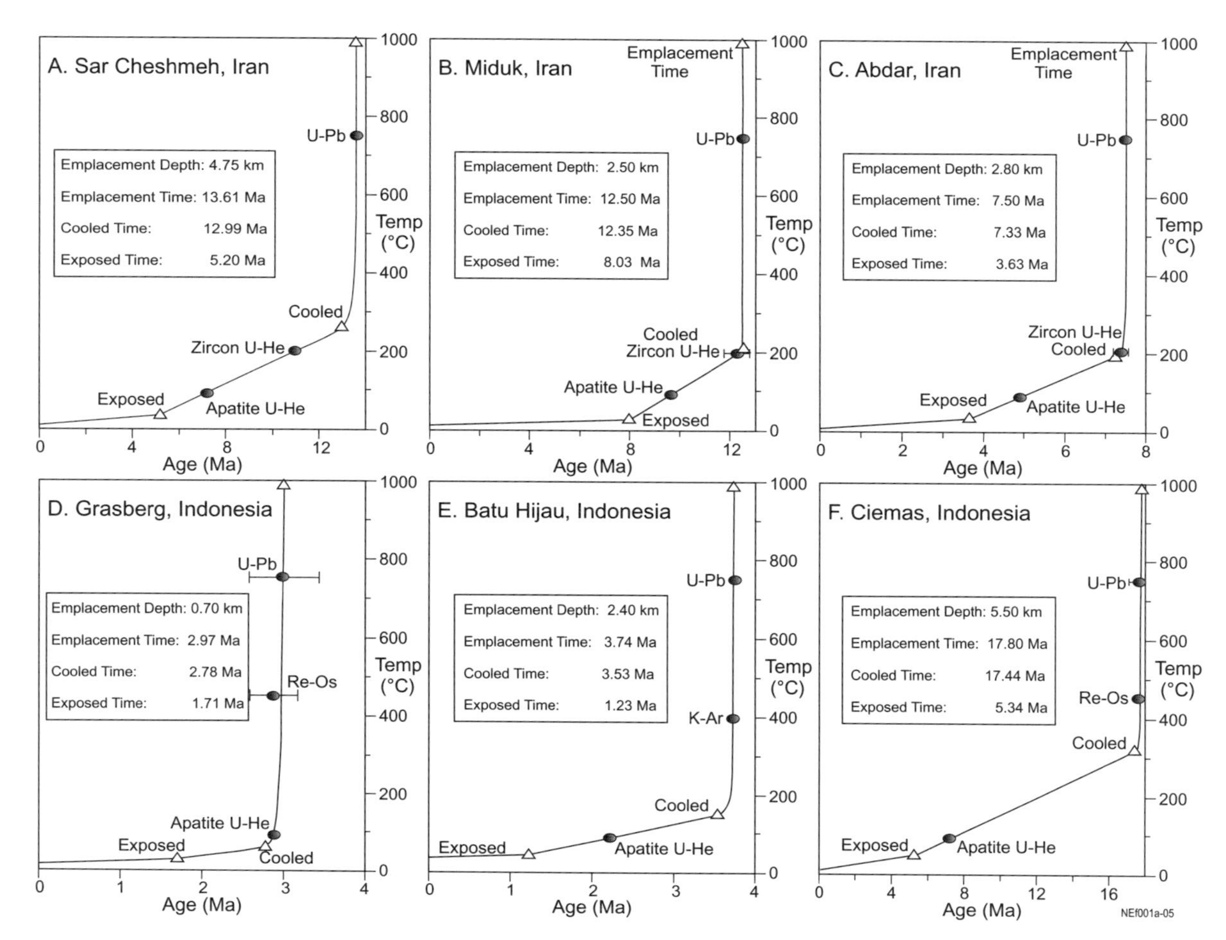

Figure 5. Graphical depiction of modeling results the Indonesian and Iranian porphyry deposits. While the thermo- and geochronology data provides an average cooling rate over temperatures constrained by the closure temperatures, the modeling component adds considerable detail to the interpretation of thermal history including emplacement depth and time, cooling rate during the two main stages of cooling (see text) and time of exposure at the surface (Fu et al. 2005). Errors on age are within the boundaries of the symbol unless otherwise indicated by a horizontal bar.

overall cooling rates whereas shallow emplacement and advective regimes will produce the fastest cooling rates. Raw average cooling rates for the porphyry deposits were calculated (Table 4) based on the age data in Table 3. Of the deposits studied, Grasberg has the fastest (>1,000 °C/m.y.) and Ciemas the slowest (<100 °C/m.y.) raw average cooling rate, with the remaining deposits falling within 100–300 °C/m.y. range (Table 4). Overall average cooling rates calculated by 4DTherm v.1.1 (Table 4) are consistent with the raw results.

As discussed in an earlier section, 4DTherm v.1.1 solves for emplacement depth during cooling in a conductive thermal regime, and therefore the minimum depths of emplacement can be determined for intrusions of a given size. A range of possible emplacement depths along with a best-fit model depth are provided in Table 4, and the results are schematically presented in Figure 6. The model depth of emplacement for the Iranian porphyry deposits indicate that Sar Cheshmeh was emplaced at deeper levels than the Miduk and Abdar intrusions (Table 4, Fig. 6). Geological reconstruction of the Kuh-e-Masahim volcano indicates that approximately 2 km of volcanic cover overlying the Abdar intrusion has been eroded (McInnes, unpublished data), consistent with the minimum possible emplacement depth generated by the model (Table 4). Similarly in Indonesia, model depths of emplacement for Grasberg of 700 m are supported by independently determined geological data that the Main Grasberg Intrusion was emplaced into a volcanic edifice within 1 km of the paleosurface (MacDonald and Arnold 1994; Weiland and Cloos 1996). At Batu Hijau, the modeled depth of emplacement of 2400 m (−400/+600 m) is consistent with paleodepth reconstruction by Garwin (2002) of 2000 ± 500 m. The depth estimate for the Ciemas intrusion is 5500 m below the paleosurface, however, due to the paucity of information for this prospect, this depth cannot be corroborated by other geological data.

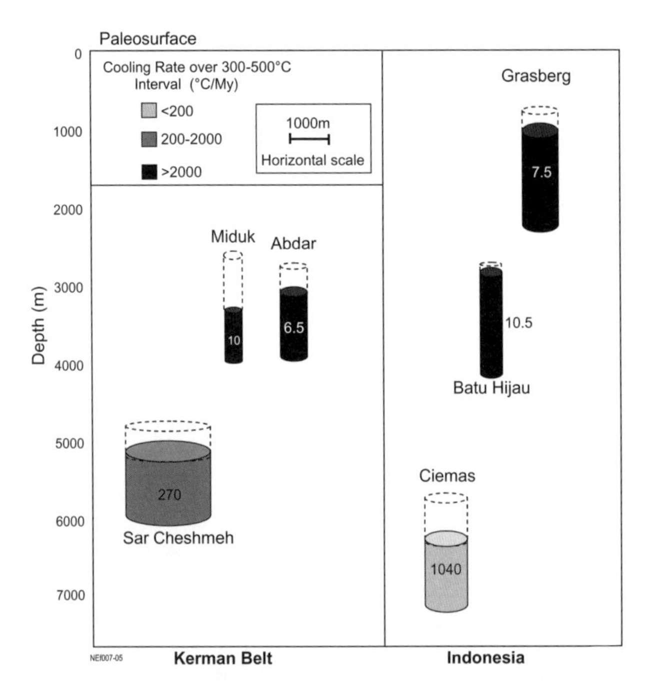

Figure 6. Schematic representation of emplacement depth as determined by 4DTherm. Intrusion diameters and heights are those used in the model (Table 4). Dotted cylinder represents the amount of material eroded from each intrusion since exposure at surface (Table 4). Numbers inside cylinders represent the modeled duration of the 300–500 °C cooling interval for each intrusion (Ka). The higher the cooling rate over this same temperature interval, the darker the shading of the cylinder.

Table 4. Summary of inverse thermal modeling results using 4DTherm v.1.1

Porphyry Deposit		Kerman Belt, Iran			Indonesia		
		SCP Sar Cheshmeh	Miduk	Abdar	Ciemas	Batu Hijau	Grasberg
Range of Possible Emplacement Depths (m)*		3800–5500	2000–3200	2000–3200	4000–6500	2000–3000	500–1000
Best Fit Emplacement Depth (m)**		4750	2500	2800	5500	2400	700
Age of Emplacement (Ma)		13.61	12.50	7.50	17.80	3.74	2.97
Age deposit cooled *** (Ma)		12.99	12.35	7.33	17.40	3.53	2.78
Age of Exposure at Surface**** (Ma)		5.20	8.03	3.63	5.34	1.23	1.71
Cooling Rate# (°C/m.y.)	Raw Average Cooling Rate from Age Data	103	-	253	62	172	1800
	Modeled Overall Average	74	80	133	56	267	337
	Magmatic –Hydrothermal Cooling	1121	742	4721	1890	4295	4817
	Exhumation Cooling	24	13	27	18	43	20
300~500 °C Interval	Duration (k.y.)	270	10	6.5	1040	10.5	7.5
	Cooling Rate (°C/m.y.)	741	20000	30769	192	18868	26666
Average Exhumation Rate (km/m.y.)		0.39	0.26	0.43	0.35	0.72	0.37
Porphyry eroded since exposure (m)		312	800	435	587	300	397
Modeling Parameters	Shape	Cylinder (stock)	Ellipsoid (stock)	Cylinder (stock)	Cylinder (stock)	Cylinder (stock)	Cylinder (stock)
	Dimension (m) of Deposit##	D1=2200 D2=1000 H=1000	D1=400 D2=700 H=700	D=750 H=900	D=1000 H=1000	D=500 H=1700	D=950 H=1700

Assumptions and Notes: Constant surface temperature = 10.0 °C; Constant heat flow from bottom = 65 mW/m°C; Thermal gradient = 50.0 °C/km (Abdar: 60 °C/km); Initial temperature = 1000.0 °C; Diffusivity = 10^{-6} m^2/s. *Range of possible valid values for emplacement depth (m) for each intrusion. A process of iteration and elimination is used to find the best fit value (see Appendix II). ** Emplacement depth is defined in the model as the distance from the paleosurface to the top of the igneous body. ***The igneous body has "cooled" when it reaches the same temperature as the surrounding country rock under normal geothermal gradient conditions. ****Age of Exposure is defined as the time when the top of the igneous body is exposed at surface and begins to be eroded. # The raw average cooling rate derived from the age data is calculated as [(closure temperature of U/Pb)-(closure temperature of apatite (U-Th)/He)] / [(age determined by U/Pb) – (age determined by apatite (U-Th)/He)]. "Overall Average" cooling rate refers to the cooling rate calculated by 4DTherm v.1.1 over the complete cooling history from 1000 °C to surface temperature of 10 °C. Magmatic-hydrothermal cooling refers to the cooling from emplacement to the "cooled" state as defined above. Exhumation cooling refers to the interval from the "cooled" state through to cooling to surface temperature of 10 °C. ## D=stock diameter, H=stock height. Although the horizontal dimension of an igneous body is either known or can be closely estimated from geological evidence, its height (H) is often very difficult to determine because most drilling has only partially penetrated the ore body. Extension of the intrusion at depth would decrease cooling rate. Because the height is often poorly constrained, the emplacement depth, eroded thickness and height of the igneous body are three variables that have to be assumed for each run of the model. These variables are independent and their values will affect the cooling history of the sample. The current solution to this three-variable problem is to make estimates of the variables and then iterate the calculation until the model produces a curve that passes through all the age data points (see Appendix II). Using this method, the erosion rate at Sar Cheshmeh, for example, was determined to be 0.06 mm/yr.

Hypogene copper grade as a function of cooling rate

Temperature is one of the fundamental variables controlling the solubility of copper in magmatic-hydrothermal systems (McPhail and Liu 2002; Liu and McPhail 2005) and therefore thermal history analysis may prove useful in understanding processes that produce high-grade hypogene ores. One way to rapidly deposit Cu-sulfide minerals within small rock volumes is to pass a hydrothermal fluid through a steeply declining thermal gradient. In contrast, weak thermal gradients should generate more diffuse haloes of Cu mineralization. A schematic representation of this concept is provided in Figure 2 where Cu transport and heat transfer are treated as diffusive processes. Under these conditions, intrusions emplaced within the uppermost crust should experience greater thermal gradients than those emplaced in mid-crustal regions where temperature regimes are moderated by the Earth's geotherm. Support for the cooling rate hypothesis can be found in the Indonesian study set. Grasberg, the most shallowly emplaced intrusion-related hydrothermal system with a high average Cu grade of 1.08% experienced the highest rate of cooling of any porphyry system studied. In contrast, the Ciemas porphyry with an average grade of 0.2% Cu experienced the slowest cooling rate over the hypogene temperature interval (almost 200 °C/m.y.) and is interpreted to be the most deeply emplaced intrusion (5.5 km). It took over 1 m.y. for Ciemas to cool through the hypogene Cu window, and it is possible that the low Cu grades for the deposit might be explained by the fact that the original magmatic Cu was distributed over a larger volume of country rock. It is suggested therefore that in the Indonesian examples, Grasberg experienced the thermal gradient conditions of the pluton depicted in the right of Figure 2 and Ciemas represents the pluton on the left. This is also supported by their contrasting patterns in Figures 4 and 5.

Although these preliminary data suggest a correlation might exist between cooling rate and hypogene copper grade in porphyry Cu deposits, it is not known whether thermal history analysis can be applied successfully to mineral exploration. Other factors such as total metal availability, reactivity of wallrocks (e.g., carbonates at Grasberg) and periodic pulses of metal emplacement cannot be assessed through thermal history analysis. More studies of intrusions on a regional scale are needed to increase the data density and definitively assess the relationships between cooling rate and hypogene Cu grade.

Preservation potential of hypogene ores and potential formation of supergene ores

As discussed earlier, 4DTherm v.1.1 treats the cooling of intrusive bodies from magmatic temperatures to surface temperatures (in this study nominated as 10 °C) as a two-stage process: (i) magmatic-hydrothermal cooling (R_1 in Fig. 3) and (ii) exhumation cooling (R_2 in Fig. 3). Exhumation cooling is controlled by the rate of removal of cover material overlying the sample by tectonic (e.g., extension) and/or erosional processes (e.g., glaciation). If a post-emplacement exhumation rate can be determined for an intrusion, then an assessment of the preservation potential of associated hypogene mineralization can be made. Exhumation rates for the porphyry deposits studied range from 0.26 to 0.72 km/m.y. (Table 4). The highest overall exhumation rate was found at Batu Hijau (0.72 km/m.y.), where a combination of collision-driven Pliocene uplift (Garwin 2002 and references therein) and pluvial processes occurred. Collision-driven uplift is also a feature of the Kerman Belt where the majority of uplift is occurring along the actively deforming Zagros Thrust Zone. The exhumation rates determined for the Sar Cheshmeh, Miduk and Abdar deposits are within a narrow range of 0.3–0.4 km/m.y., presumably because they are located co-parallel and 80 km distant from the Zagros Thrust Zone. Grasberg has a lower calculated exhumation rate (0.37 km/m.y.). Other workers (Weiland and Cloos 1996; Hill et al. 2002) have argued that although rapid denudation (0.7–1.0 km/m.y.) is occurring along major thrust fronts actively forming the New Guinea Fold Belt, these fronts are 50 km distant from the Grasberg deposit and the peak uplift forces have not yet transitioned to the Grasberg area.

However, it should be noted that adding the effects of advective and convective cooling to 4DTherm will likely result in increased emplacement depths and hence increased erosion rates, so the 0.37 km/m.y. exhumation rate for Grasberg may be a minimum.

Understanding the rate of exhumation for a mineral district or a metallogenic belt has implications for area selection during mineral exploration. The determination of an exhumation rate permits an assessment of the erosion potential of hypogene ore deposits and an evaluation of the potential for supergene ore formation from eroded hypogene ores. Figure 6 portrays the amount of erosion experienced by each intrusion as a dashed cylinder. In Indonesia, Batu Hijau has experienced the least amount of erosion since exposure whereas Ciemas has experienced the most. Although the deposits of the Kerman belt have similar exhumation rates, differences in their exposure age indicate substantially different amounts of potential hypogene mineralization have been eroded. Taking Sar Cheshmeh as an example, the porphyry copper deposit was exposed 5.2 million years ago and was exhumed at a rate of 0.39 km/m.y. (Table 4), while the erosion rate for the porphyry since exposure is estimated to be about 0.06 km/m.y. (see notes, Table 4). These calculations infer that 312 m of porphyry Cu mineralization have eroded since exposure of the porphyry at surface. Assuming an average copper shell thickness of 100 m and a rock density of 2.7 g/cm^3, the total amount of rock eroded from the Sar Cheshmeh porphyry system was about 810 million tons (Mt). The Sar Cheshmeh porphyry was intruded by three copper-poor igneous dike units that account for about 1/3 of the deposit volume, so the total amount of eroded ore is reduced to 540 Mt. At a minimum copper grade of 0.64%, the total amount of copper eroded equates to around 3.5 Mt, which is nearly half of the remaining reserve as estimated in 1998. If the amount of supergene Cu contained at Sar Cheshmeh is less than the amount of Cu eroded, then the unaccounted Cu may be contained in Exotica-type deposits below sedimentary and volcanic cover in the region. Similar calculations can be performed for the other deposits to assess both the preservation potential of hypogene shells and the potential for formation of supergene ore deposits.

CURRENT TRENDS, FUTURE DIRECTIONS

Thermochronology has made a fundamental contribution to economic geology through the provision of data needed to construct genetic models for hydrothermal ore deposits. Mineral exploration has traditionally involved the search for the measurable anomalies in the Earth's crust produced by hydrothermal systems. Although detection of visual (e.g., Fe-oxide staining or white mica formation), chemical (e.g., trace element enrichments) and physical (e.g., magnetite formation or destruction) anomalies will always be important, economic geologists are increasingly turning to thermochronology data to reveal the paleothermal anomalies produced by hydrothermal systems (e.g., Naeser et al. 1980; Cunningham et al. 1987). Because the search for new ore deposits involves looking deeper below the surface, thermochronology will potentially serve a greater role in testing exploration concepts prior to the expensive stage of drilling.

The power of thermochronology to the explorationist, is the provision of the 4th dimension, vital to the assessment of the metallogenic evolution of a mineral district. The challenge for thermochronologists will be to develop new, low-cost techniques that enable the construction of high-density, temperature-integrated, regional paleothermometry data sets. As outlined in this and other chapters within this review volume, the inversion of thermochronology data to produce 4-D thermal evolution and landscape evolution models is well underway. The coupling of thermal inversion with geochemical and geophysical inversion techniques is still in its nascent stages, but by 2010 the explorationist will be using deterministic, fully coupled thermal-chemical transport models as part of their discovery toolkit.

With respect to porphyry deposit in particular, the following conclusions are offered:

1.) Combining multiple chronometers and in particular apatite (U-Th)/He, zircon (U-Th)/He and zircon U/Pb, provides a thermal history for porphyry deposits over a temperature range of >700 °C. Information that can be obtained from thermal history analysis includes the timing and depth of emplacement of igneous units, the cooling rate during hypogene copper deposition and the exhumation rate of the porphyry deposit.
2.) The disruption of the steady state geothermal gradient during the emplacement of igneous intrusions places limitations on the direct usage of (U-Th)/He age dating in the determination of emplacement depth and exhumation rates. Numerical modeling techniques provide an effective and complementary tool for quantifying cooling and emplacement parameters.
3.) We postulate that strong thermal gradients present the ideal conditions for the generation of high-grade hypogene Cu ores, whereas more diffuse mineralization haloes would be expected for more slowly cooled igneous intrusions. Thermal history analysis using "triple dating" U/Pb-He techniques provides some support for a positive correlation between short duration, rapid cooling from 500 to 300 °C and emplacement depth, however the database is limited and additional district-scale studies are required.
4.) Understanding the rate of exhumation for a mineral district or a metallogenic belt permits an assessment of the erosion potential of hypogene ores, as well as the formation potential of supergene ores from eroded hypogene deposits.

ACKNOWLEDGMENTS

The management of NICICO is gratefully acknowledged for providing access to Sar Cheshmeh, Miduk and Abdar. Thanks to Jeff Davis and Ahmed Ali for mineral separations, Ratih Woodhouse and Marcus Gregson for hand-picking/quality control, Lesley Dotter for zircon dissolutions, Peter Pollard for provision of samples and images from the Grasberg deposit, and to Travis Naughton and Angelo Vartesi for drafting. We are grateful to Rio Tinto Mining and Exploration Ltd, in particular Ross Andrew, Neil McLaurin and John Bartram, for supporting this work. Field activities in Indonesia were supported by AusAID and the Indonesian Department of Energy and Mineral Resources. Jo-Ann Wortho and Barry Kohn provided helpful comments and suggested additional references. The thoughtful reviews of Anthony Harris and Ken Hickey greatly improved the chapter.

REFERENCES

Arehart GB, Foland KA, Naeser CW, Kesler SE (1993) $^{40}Ar/^{39}Ar$, K/Ar, and fission track geochronology of sediment-hosted disseminated gold deposits at Post-Betze, Carlin Trend, Northeastern Nevada. Econ Geol 88:622-646

Arehart GB, Foland KA, Naeser CW, Kesler SE (1995) $^{40}Ar/^{39}Ar$, K/Ar, and fission track geochronology of sediment-hosted disseminated gold deposits at Post-Betze, Carlin Trend, Northeastern Nevada - A Reply. Econ Geol 90:210-212

Arehart GB, Chakurian AM, Tretbar DR, Christensen JN, McInnes BIA, Donelick RA (2003) Evaluation of radioisotope dating of Carlin-type deposits in the Great Basin, Western North America, and implications for deposit genesis. Econ Geol 98:235-248

Arne DC (1992) The application of fission track thermochronology to the study of ore deposits. *In:* Short Course Handbook on Low Temperature Thermochronology. Zentilli M, Reynolds PM (eds) Mineral Assoc Can, Ottawa, p 75-96

Arne DC, Rierlein FP, McNaughton N, Wilson CJL, Morand VJ (1998) Timing of gold mineralization in western and central Victoria, Australia: new constraints from SHRIMP II analysis of zircon from felsic intrusive rocks. Ore Geol Rev 13:251-273

Art JG, Hanson GL (1972) Quartz diorites derived by partial melting of eclogite or amphibolite at mantle depths. Contrib Mineral Petrol 37:161-174

Arribas A Jr., Hedenquist JW, Itaya T, Okada T, Concepcion RA, Garcia JS Jr. (1995) Contemporaneous formation of adjacent porphyry and epithermal Cu-Au deposits over 300ka in northern Luzon, Philippines. Geology 23:337-340

Bähr R, Lippolt HJ, Wernicke RS (1994) Temperature-induced ^{4}He degassing of specularity and botryoidal hematite: a ^{4}He retentivity study. J Geophys Res 99:17695-17707

Ballard JR, Palin JM, Williams IS, Campbell IH, Faunes A (2001) Two ages of phorphyry intrusion resolved for the super-giant Chuquicamata copper deposit of northern Chile by ELA-ICP-MS and SHRIMP. Geology 29:383-86

Banks NG, Stuckless JS (1973) Chronology of intrusion and ore deposition at Ray, Arizona: Part II, Fission track ages. Econ Geol 68:657-664

Beaty DW, Naeser CW, Lynch WC (1987) Recent rapid uplift in the Bolivian Andes: Evidence from fission track dating. Geology 15:680-683

Belhadi A, Himeno O, Watanabe K, Izawa E (1999) Geology and zircon fission track ages of volcanic rocks in the western part of Hoshino gold area, Fukuoka Prefecture, Japan J Min Petr Econ Geol 94:482-494

Bernet MB, Brandon M, Garver J, Reiners P, Fitzgerald P (2004) The zircon fission-track closure temperature. Abstracts, 10th International Fission Track Dating and Thermochronology Meeting: 37-38

Bierlein FP, Foster DA, Mcknight S, Arne DC (1999) Timing of gold mineralization in the Ballarat goldfields, central Victoria: constraints from ^{40}Ar/^{39}Ar results. Aus J Earth Sci 46:301-309

Bray CJ, Spooner ETC, Hall CM, York D, Bills TM, Krueger HW (1987) Laser probe ^{40}Ar/^{39}Ar and conventional K/Ar dating of illites associated with the McClean unconformity-related uranium deposits, north Saskatchewan, Canada. Can J Earth Sci 24:10-23

Cathles LM (1981) Fluid flow and genesis of hydrothermal ore deposits. *In:* Economic Geology 75th anniversary volume; 1905-1980. Skinner BJ (ed) Society of Economic Geologists Publication, p 424-457

Cathles LM, Erendi AHJ, Barrie T (1997) How long can a hydrothermal system be sustained by a single intrusive event? Econ Geol 92:766-771

Cartwright T (1998) Petrology report on 22 samples from Ciemas, West Java for Meekatharra Minerals, Kinston Morrison Mineral Services. 1-38

Chakurian AM, Arehart GB, Donelick RA, Zhang X, Reiners PW (2003) Timing constraints of gold mineralization along the Carlin Trend utilizing apatite fission-track, ^{40}Ar/^{39}Ar, and apatite (U-Th)/He methods. Econ Geol 98:1159-1171

Chan MA, Parry WT, Petersen EU, Hall CM (2001) ^{40}Ar/^{39}Ar age and chemistry of manganese mineralization in the Moab and Lisbon fault systems, southeastern Utah. Geology 29:331-334

Cherniak DJ, Watson EB (2000) Pb diffusion in zircon. Chem Geol 172:5-24

Cline JS, Hofstra AH, Muntean JL, Tosdal RM, Hickey KA (2005) Carlin-type gold deposits in Nevada: critical geologic characteristics and viable models. Economic Geology 100th Anniversary Volume (in press)

Corrigan J (1991) Inversion of apatite fission track data for thermal history information. J Geophys Res 96: 10347-10360

Crowhurst PVC, Farley KA, Ryan C, Duddy I, Blacklock K (2002) Potential of rutile as a (U-Th)-He thermochronometer. Geochim Cosmochim Acta 66:A-158

Crowley KD (1985) Thermal significance of fission track lengths. Nucl Tracks 10: 311-322

Cunningham CG, Steven TA, Campbell DL, Naeser CW, Pitkin JA, Duval JS (1984) Multiple episodes of igneous activity, mineralization and alteration in the western Tushar mountains, Utah. *In:* Igneous Activity and Related Ore Deposits in the Western and Southern Tushar Mountains, Maryvale Volcanic Field, West-Central Utah. Steven TA (ed) US Geol Surv Prof Paper 1299-A,B, p 1-22

Cunningham CG, Naeser CW, Cameron DE, Barrett LF, Wilson JC, Larson PB (1987) The Pliocene paleothermal anomaly at Rico, Colorado is related to a major molybdenum deposit. Geol Soc Am Abstracts with Program 19:268-269

Cunningham CG, Austin GW, Naeser CW, Rye RO, Ballantyne GH, Stamm RG, Barker CE (2004) Formation of a paleothermal anomaly and disseminated gold deposits associated with the Bingham Canyon Porphyry Cu Au Mo Sytem, Utah Econ Geol 99:780-806

Dammer D, Chivas AR, McDougall I (1996) Isotopic dating of supergene manganese oxides from the Groote Eylandt Deposit, Northern Territory, Australia. Econ Geol 91:386-401

Dodson MH (1973) Closure temperatures in cooling geological and petrological systems. Contrib Mineral Petrol 40:259-274

Donelick RA, O'Sullivan PB, Ketcham RA (2005) Apatite fission-track analysis. Rev Mineral Geochem 58: 49-94

Duddy IR, Green PF, Laslett GM (1988) Thermal annealing of fission tracks in apatite 3. Variable temperature behaviour. Chem Geol Isot Geosci Sect 73:25-38
Dunai TJ (2005) Forward modeling and interpretation of (U-Th)/He ages. Rev Mineral Geochem 58:259-274
Ehlers TA, Farley KA (2003) Apatite (U-Th)/He thermochronometry: methods and applications to problems in tectonic and surface processes. Earth Planet Sci Lett 206:1-14
Evans NJ, Wilson NSF, Cline JS, McInnes BIA, Farley KA, Byrne J (2005) (U-Th)/He thermochronology of fluorite and the low temperature history of Yucca Mountain, Nevada. Appl Geochem 20(6):1099-1105
Fanale FP, Kulp JL (1962) The helium method and the age of the Cornwall, Pennsylvania magnetite ore. Econ Geol 57:735-746
Farley KA, House MA, Kohn BP (1998) Laboratory and natural diffusivity calibrations for apatite (U-Th)/He thermochronology. Mineral Mag 62A:436–437
Farley KA (2002) U-He dating: Techniques, calibrations and applications. Rev Mineral Geochem 47:819-844
Fleischer RL, Price PB, Walker RM (1975) Nuclear Tracks in Solids: Principles and Techniques. University of California Press, Berkeley, 605p
Foster DA, Gray DR, Kwak TAP, Bucher M (1998) Chronology and tectonic framework of turbidite-hosted gold deposits in the Western Lachlan Fold Belt, Victoria: $^{40}Ar/^{39}Ar$ results. Ore Geol Rev 13:229-250
Fu FQ, McInnes BIA, Davies PJ, Evans NJ (2005) Numerical modeling of the thermal and exhumation histories of ore systems as constrained by (U-Th)/He thermochronology and U-Pb dating. To be submitted to Econ Geol.
Fullagar PD, Kish SA, Odom AL, Dallmeyer RD, Bottino ML (1980) Possible excess 40Ar in hornblende and biotite from the Appalachian massive sulfide deposits at Ore Knob, North Carolina, and Ducktown, Tennessee. Econ Geol 75:329-334
Gallagher K (1995) Evolving temperature histories from apatite fission-track data. Earth Planet Sci Lett 136: 421-435
Gallagher K, Stephenson J, Brown R, Holmes C, Ballester P (2005) Exploiting 3D spatial sampling in inverse modeling of thermochronological data. Rev Mineral Geochem 58:375-387
Garwin SL (2000) The setting, geometry and timing of intrusion-related hydrothermal systems in the vicinity of the Batu Hijau porphyry copper-gold deposit, Sumbawa, Indonesia. PhD Dissertation, University of Western Australia, Nedlands, Western Australia
Garwin SL (2002). The geologic setting of intrusion-related hydrothermal systems near the Batu Hijau porphyry copper-gold deposit, Sumbawa, Indonesia. *In:* Integrated Methods for Discovery: Global Exploration in the 21st Century. Goldfarb RJ, Nielsen RL (eds) Society of Economic Geologists, Special Publication 9:333-366
Ghorashi-Zadeh M (1979) Development of hypogene and supergene alteration and copper mineralisation patterns, Sar Cheshmeh porphyry copper deposit, Iran. MSc Dissertation, Brock University, St. Catharines, Ontario, Canada
Gibbins S, Titley S, Friehauf K (2003) Age, origin, petrology and petrography of the Ertsberg Diorite, west Papua, Indonesia. Geol Soc Am 36/6:400
Gleadow AJW, Duddy IR, Green PF, Lovering JF (1986) Confined fission track lengths in apataite: A diagnostic tool for thermal history analyses. Contrib Mineral Petrol 94:405-415
Gleadow AJW, Belton DX, Kohn BP, Brown RW (2002) Fission track dating of phosphate minerals and the thermochronology of apatite. Rev Mineral Geochem 48:579-630
Goldfarb RJ, Snee LW, Miller LD, Newberry RJ (1991) Rapid dewatering of the crust deduced from ages of mesothermal gold deposits. Nature 354:296-298
Green PF, Duddy GM, Laslett GM, Hegarty KA, Gleadow AJW, Lovering JF (1989) Thermal annealing of fission tracks in apatite 4. Quantitative modeling and extension to geological timescales. Chem Geol Isot Geosci Sect 79:155-182
Grønlie A, Harder V, Roberts D (1990) Preliminary fission-track ages of fluorite mineralization along fracture zones, inner Trondheimsfjord, Central Norway. Norsk Geologisk Tidsskrift 70:173-178
Harris AC, Kemenetsky VS, White NC, Van Achterbergh E, Ryan CG (2003) Melt inclusions in veins: linking magmas and porphyry Cu deposits. Science 302:2109-2111
Harris AC, Allen CM, Bryan SE, Campbell IH, Holcombe RJ, Palin JM. (2004) ELA-ICP-MS U-Pb zircon geochronology of regional volcanism hosting the Bajo de la Alumbrera Cu-Au deposit: implications for porphyry-related mineralization. Mineral Dep 39:46-67
Harrison TM, Zeitler PK (2005) Fundamentals of noble gas thermochronometry. Rev Mineral Geochem 58: 123-149
Hassanzadeh J (1993) Metallogenic and tectonomagmatic events in the SE sector of the Cenozoic active continental margin of central Iran (Shahr e Babak area, Kerman Province). PhD Dissertation, University of California Los Angeles, California

Hedenquist JW, Richards JP (1998) The influence of geochemical techniques on the development of genetic models for porphyry copper deposits. *In*: Techniques in Hydrothermal Ore Deposits Geology. Richards JP, Larson PB (eds) Rev Econ Geol 10:235-256

Henry CD, Elson HB, Heizler MT, Castor SB (1995) Brief duration of hydrothermal activity at Round Mountain, Nevada determined from $^{40}Ar/^{39}Ar$ geochronology. Geol Soc Am 27/6:329

Hickey KA, Donelick RA, Tosdal RM, McInnes BIA (2003) Restoration of the Eocene landscape in the Carlin-Jerritt Canyon mining district: constraining depth of mineralization for Carlin-type Au deposits using low temperature apatite thermochronology. Geol Soc Am 35:358

Hill KC Kendrick RD, Crowhurst PV, Gow PA (2002) Copper-gold mineralization in New Guinea: tectonics, lineaments, thermochronology and structure. Aus J Earth Sci 49:737-752.

Hofstra, AH, Snee LW, Rye RO, Folger HW, Phinisey JD, Loranger RJ, Dahl AR, Naeser CW, Stein HJ, Lewchuk M (1999) Age constraints on Jerritt Canyon and other Carlin-type gold deposits in the western United States -relationship to mid-Tertiary extension and magmatism. Econ Geol 94:769-802

Hofstra, AH, Cline JS (2000) Characteristics and models for Carlin-type gold deposits: Rev Econ Geol 13: 163-220

House MA, Wernicke BP, Farley KA (1998) Dating topography of the Sierra Nevada, California, using apatite (U-Th)/He ages. Nature 396:66-69

Ilchik RP (1995) $^{40}Ar/^{39}Ar$, K/Ar, and fission track geochronology of sediment-hosted disseminated gold deposits at Post-Betze, Carlin Trend, Northeastern Nevada - A Discussion. Econ Geol 90:210-212

Kendrick M A, Burgess R, Pattrick RAD, Turner G (2001) Halogen and Ar-Ar age determinations of inclusions within quartz veins from porphyry copper deposits using complementary noble gas extraction techniques. Chem Geol 177:351-370

Kent AJR, McDougall I (1995) $^{40}Ar/^{39}Ar$ and U/Pb age constraints on the timing and gold mineralization in the Kalgoorlie gold field, Western Australia. Econ Geol 90:845-859

Kent AJR, McDougall I (1996) $^{40}Ar/^{39}Ar$ and U/Pb age constraints on the timing and gold mineralization in the Kalgoorlie gold field, Western Australia -A Reply. Econ Geol 91:795-798

Kent AJR, McCuaig TC (1997) Disturbed ^{40}Ar-^{39}Ar systematics in hydrothermal biotite and hornblende at the Scotia gold mine, Western Australia: evidence for argon loss associated with post-mineralisation fluid movement. Geochim Cosmochim Acta 61:4655-4669

Ketcham RA, Donelick RA, Carlson WD (1999) Variability of apatite fission-track annealing kinetics: III. Extrapolation to geological time scales. Am Mineral 84:1235-1255

Ketcham RA, Donelick RA, Donelick MB (2000) AFTSolve: A program for multi-kinetic modeling of apatite fission-track data. Geol Mater Res 2

Ketcham RA (2005) Forward and inverse modeling of low-temperature thermochronometry data. Rev Mineral Geochem 58:275-314

Kontak DJ, Horne RJ, Sandeman H, Archibald D, Lee JKW (1998) $^{40}Ar/^{39}Ar$ dating of ribbon-textured veins and wall-rock material from Meguma lode gold deposits, Nova Scotia: implications for timing and duration of vein formation in slate-belt hosted vein gold deposits. Can J Earth Sci 35:746-761

Koski RA, Hein JR, Bouse RM, Evarts RC, Pickthorn LB (1990) K-Ar and fission-track ages of tuff beds at the Three Kids Mine, Clark County, Nevada: Implications for manganese mineralization. Isochron/West 56:30-32

Larson PB, Cunningham CG, Naeser CW (1994) Large-scale alteration effects in the Rico paleothermal anomaly, southwest Colorado. Econ Geol 89:1769-1779

Lee JKW, Williams IS, Ellis DJ (1997) Pb, U and Th diffusion in natural zircon. Nature 390:159-162

Lipman PW, Fisher FS, Mehnert HH, Naeser CW, Luedke RG, Steven TA (1976) Multiple ages of mid-Tertiary mineralization and alteration in the western San Juan Mountains, Colorado. Econ Geol 71:571-588

Lippolt HJ, Weigel E (1988) ^{4}He diffusion in ^{40}Ar retentive minerals. Geochim Cosmochim Acta 52:1449-1458

Lippolt HJ, Leitz M, Wernicke RS, Hagedorn B (1994) (U+Th)/He dating of apatite: Experience with samples from different geochemical environments. Chem Geol 112:179-191

Lippolt HJ, Wernicke RS (1997) (U+Th)-He evidence of Jurassic continuous hydrothermal activity in the Schwarzald basement, Germany. Chem Geol 138:273-285

Liu W, McPhail DC (2005) Thermodynamic properties of copper chloride complexes and copper transport in magmatic hydrothermal solutions. Chem Geol, in press

Lutz TM, Omar G (1991) An inverse method of modeling thermal histories from apatite fission-track data. Earth Planet Sci Lett 104:181-195

Maksaev V, Zentilli M (1999) Fission track thermochronology of the Domeyko Cordillera, Northern Chile: implications for Andean tectonics and porphyry copper metallogenesis. Explor Mining Geol 8:65-89

Maksaev V, Munizaga F, McWilliams M, Mathur R, Ruiz J, Fanning MC, Zentilli M (2003) New Timeframe for El Teniente Cu-Mo Giant Porphyry Deposit: U-Pb, 40Ar-39Ar, Re-Os and Fission Track Dating. South American Symposium on Isotope Geology, IV SSAGI:736-739

Marsh TM, Einaudi MT, McWilliams M (1997) $^{40}Ar/^{39}Ar$ geochronology of Cu-Au and Au-Ag mineralization in the Potrerillos district, Chile. Econ Geol 92:784-806

Mathur R, Ruiz J, Munizaga F (2000a) Relationship between copper tonnage of Chilean base metal phrohyry deposits and Os isotope ratios. Geology 28:555-558

Mathur R, Ruiz J, Titley S, Gibbons S, Margotomo W (2000b) Different crustal sources for Au-rich and Au-poor ores of the Grasberg Cu-Au porphyry deposit. Earth Planet Sci Lett 183:7-14

Mathur R, Titley R, Ruiz J, Gibbins S, Friehauf K (2005) A Re-Os isotope study of sedimentary rocks and copper-gold ores from the Ertsberg District, West Papua, Indonesia. Ore Geol Rev 26(3-4):207-226

Mauk JL, Hall CM (2004) $^{40}Ar/^{39}Ar$ ages of adularia from the Golden Cross, Neavesville and Komata epithermal deposits, Hauraki Goldfield, New Zealand. J Geol Geophys New Zealand 47:227-231

McDougall I, Harrison MT (1999) Geochronology and Thermochronology by the $^{40}Ar/^{39}Ar$ Method. 2nd edition, Oxford University Press, New York

McDowell FW, McMahon TP, Warren PQ, Cloos M (1996) Pliocene Cu-Au bearing intrusions of the Gunung Bihj (Ertsberg) district, Irian Jaya, Indonesia: K-Ar geochronology. J Geol 104:327-340

McInnes BIA, Farley KA, Sillitoe RH, Kohn BP (1999) Application of apatite U-He thermochronology to the determination of the sense and amount of vertical fault displacement at the Chuquicamata porphyry copper deposit, Chile. Econ Geol 94:937-945

McInnes BIA, Evans NJ, McBride J, Keays R, Lambert D (2000) Metallogenic fertility and Re-Os geochronology of ore systems. CSIRO Exploration and Mining Confidential Report 921C

McInnes BIA, Evans NJ, Belousova E and Griffin WG (2003a) Porphyry copper deposits of the Kerman Belt, Iran: timing of mineralization and exhumation processes. CSIRO scientific research report:41

McInnes BIA, Evans NJ, Belasouva E, Griffin WG, Andrew RL (2003b) Timing of mineralization and preservation processes at the SarCheshmeh and Miduk porphyry Cu deposits, Kerman belt, Iran. 7th Biennial SGA Meeting, Athens, Millpress Rotterdam: 1197-1200

McInnes BIA, Evans NJ, Sukarna D, Permanadewi S, Garwin S, Belousova E, Griffin WL, Fu FQ (2004) Thermal histories of Indonesian porphyry Cu-Au deposits determined by U-Pb-He and K-Ar methods. *In:* Predictive Mineral Discovery Under Cover: Extended Abstracts. Muhling J, Goldfarb R, Vielreicher N, Bierlein F, Stumpff E, Groves DI, Kenworthy S (eds) Centre for Global Metallogeny, The University of Western Australia 33:343-346

McInnes BIA, Evans NJ, Fu FQ, Belousova E, Griffin WL, Bertens A, Sukarna D, Permanadewi S, Andrew RL, Deckart K (2005) Thermal history analysis of selected Chilean, Indonesian and Iranian porphyry Cu-Mo-Au deposits. *In:* Super Porphyry Copper and Gold Deposits: A Global Perspective. Porter TM (ed) PGC Publishing, Adelaide, p 27-42

McPhail DC, Liu W (2002) Metal transport in hypersaline brines. Geol Soc Australia Abstracts 67:295

Mote TI, Becker TA, Renne P, Brimhall GH (2001) Chronology of exotic mineralization at El Salvador, Chile, by $^{40}Ar/^{39}Ar$- dating of copper wad and supergene alunite. Econ Geol 96:351-366.

Naeser CW, Cunningham CG, Marvin RF, Obradovich JD (1980) Pliocene intrusive rocks and mineralization near Rico, Colorado. Econ Geol 75:122-133.

Naeser CW, Cunningham CG, Beaty DW (1990) Origin of the ore deposits at Gilman, Colorado; Part III, Fission track and fluid inclusion studies. *In:* Carbonate-hosted Sulfide Deposits of the Central Colorado Mineral Belt. Beaty DW, Landis GP, Thomspon TB (ed) Ichnos 7:219-228

Napier RW, Guise PG, Rex DC (1998) $^{40}Ar/^{39}Ar$ constraints on the timing and history of amphibolite facies gold mineralisation in the Southern Cross area, Western Australia. Aust J Earth Sci 45:285-296

Norton D, Cathles LM (1979) Thermal aspects of ore deposits. *In:* Geochemistry of Hydrothermal Ore Deposits. Barnes HL (ed) Wiley, New York, p 611-631

Norton D, Knight J (1977) Transport phenomena in hydrothermal systems: cooling plutons. Am J Sci 277: 937-981

Osadetz KG, Kohn BP, Feinstein S, O'Sullivan PB (2001) Thermal history of Canadian Williston basin from apatite fission-track thermochronology-implications for petroleum systems and geodynamic history. Tectonophysics 349:221-249

Parry WT, Wilson, PN, Moser, D, Heizier MT (2001) U-Pb dating of zircon and $^{40}Ar/^{39}Ar$ dating of biotite at Bingham, Utah. Econ Geol 96:1671-1684

Pollard PJ, Taylor RG, Peters L (2005) Ages of intrusion, alteration and mineralization at the Grasberg Cu-Au deposit, Irain Jaya, Indonesia. Special issue on "Giant Porphyry Cu-Mo deposits of the Andean and PNG/Irian Jaya foldbelts." Econ Geol, in press

Rahl JM, Reiners PW, Campbell IH, Nicolescu S, Allen CM (2003) Combined single-grain (U-Th)/He and U/Pb dating of detrital zircons from the Navajo Sandstone, Utah. Geology 31:761-764

Ressel MW, Noble DC, Heizler MT, Volk JA, Lamb JB, Park DE, Conrad JE, Mortensen JK (2000). Gold-mineralized Eocene dikes at Griffin and Meikle: Bearing on the age and origin of deposits of the Carlin Trend. *In:* Geology and Ore Deposits 2000: The Great Basin and beyond. Geol Soc Nevada, Symposium Proceedings:79-101

Reiners PW, Ehlers TA, Zeitler PW (2005) Past, present, and future of thermochronology. Rev Mineral Geochem 58:1-18
Reiners PW, Farley KA, Hickes HJ (2002) He diffusion and U-He thermochronology of zircon: initial results from Fish Canyon Tuff and Gold Butte. Tectonophysics 349:297-308
Reiners PW, Spell TL, Nicolescu S, Zanetti KA (2004) Zircon (U-Th)/He thermochonometry: He diffusion and comparisons with $^{40}Ar^{39}Ar$ dating. Geochim Cosmochim Acta 68:1857-1887
Reynolds P, Ravenhurst C, Zentilli M, Lindsay D (1998) High-precision $^{40}Ar/^{39}Ar$ dating of two consecutive hydrothermal events in the Chuquicamata phrphyry copper system, Chile. Chem Geol 148:45-60
Richards JP, McDougall I (1991) Geochronology of the Porgera gold deposit, Papua New Guinea: Resolving the effects of excess Ar on K-$^{40}Ar/^{39}Ar$ and $^{40}Ar/^{39}Ar$ age estimates for magmatism and mineralization. Geochim Cosmochim Acta 54:1397-1415
Ronacher E. Richards JP, Villeneuve ME, Johnston MD (2002) Short life-span of the ore-forming system at the Porgera gold deposit, Papua New Guinea: laser $^{40}Ar/^{39}Ar$dates for roscoeilite, biotite and hornblende. Mineral Dep 37:75-86
Sawai O, Matsumura N, Itaya T (1998) K-Ar ages of volcanic rocks and gold deposits in the Hoshino gold area, Northern-Central Kyushu, Japan. J Geol Soc Japan 104:377-386
Sengor AMC, Kidd WSF (1979) Post-collisional tectonics of the Turkish-Iranian plateau and a comparison with Tibet. Tectonophysics 55:361–376
Shahabpour J (1982) Aspects of alteration and mineralization at the Sar-Cheshmeh copper-molybdenum deposit, Kerman, Iran. PhD Dissertation, Univ. Leeds, London
Shawe DR, Marvin RF, Andriessen PAM, Mehnert HH, Merritt VM (1986) Ages of igneous and hydrothermal events in the Round Mountain and Manhattan Gold Districts, Nye County, Nevada. Econ Geol 81:388-407
Silberman ML, White DE, Keith TEC, Dockter RD (1979) Duration of hydrothermal activity at Steamboat Springs, Nevada, from ages of spatially associated volcanic rocks. USGS Prof Pap P458-D:1-14
Sillitoe RH (2000) Gold-rich porphyry deposits: descriptive and genetic models an dtheir role in exploration and discovery. *In*: Gold in 2000. Hagemann SG, Brown PE (eds) Rev Econ Geol 13:314-345
Smith RL, Shaw HR (1979) Igneous-related geothermal systems. US Geol Surv Circ 790:12-17
Smith PE, Evensen NM, York D, Szatmari P, de Oliviera DC (2001) Single-crystal ^{40}Ar-^{39}Ar dating of pyrite: no fool's clock. Geology 29:403-406.
Snee L (1999) Thermal history of the Butte Porphyry system, Montana. Geol Soc Am 31:A380.
Steven TA, Cunningham CG, Anderson JJ (1984) Geologic history of uranium potential of the Big John Calders, Southern Tushar Mountains, Utah. *In:* Igneous activity and related ore deposits in the western and southern Tushar Mountains, Maryvale Volcanic Field, West-Central Utah. US Geol Surv Prof Paper 1299-A,B, p 25-33
Strutt RJ (1910a) The accumulation of helium in geologic time, II. Proc Roy Soc Lond Ser A83:96-99
Strutt RJ (1910b) The accumulation of helium in geologic time, III. Proc Roy Soc Lond Ser A83:298-301
Sutriyono E (1998) Cenozoic thermotectonic history of the Sunda-Asri basin, southeast Sumatra: new insights from apatite fission track thermochronology. J Asian Earth Sci 16:485-500
Suzuki K, Shimizu H, Masuada A (1996) Re-Os dating of molybdenites from ore deposits in Japan: Implication for the closure temperature of the Re-Os system for molybdenite and the cooling history of molybdenum in ore deposits. Geochim Cosmochim Acta 60:3151-3159
Tagami T, Farley KA, Stokli DF (2003) (U-Th)/He geochronology of single zircon grains of known Tertiary eruption age. Earth Planet Sci Lett 207:57-67
Tomkins AG, Dunlap WJ, Mavrogenes JA (2004) Geochronological constraints on the polymetamorphic evolution of the granulite-hosted Challenger gold deposit: implications for assembly of the northwest Gawler Craton. Aus J Earth Sci 51:1-14
Ulrich T, Gunthur D, Heinrich CA (2001) The evolution of a porphyry Cu-Au deposit, based on LA-ICP-MS analysis of fluid inclusions: Bajo de la Alumbrera, Argentina. Econ Geol 96:1743-1774
Warnaars FW, Smith WH, Bray RE, Lanier G, Shafiqullah M (1978) Geochronology of igneous intrusions and porphyry copper mineralization at Bingham, Utah. Econ Geol 73:1242-1249
Waterman GC, Hamilton RL (1975) The Sar Cheshmeh porphyry copper deposit. Econ Geol 70:568-576
Weiland RJ, Cloos M (1996) Pliocene-Pleistocene asymmetric unroofing of the Irian fold belt, Irian Jaya, Indonesia: apatite fission-track thermochronology. Geol Soc America Bull 108:1438-1449
Wernicke RS, Lippolt HJ (1992) Botryoidal hematite from the Schwarzwald (Germany): heterogeneous uranium distributions and their bearing on the helium dating method. Earth Planet Sci Lett 114:287-300
Wernicke RS, Lippolt HJ (1994a) ^{4}He age discordance and release behaviour of a double shell botryoidal hematite from the Schwarzwald, Germany. Geochim Cosmochim Acta 58:321-429
Wernicke RS, Lippolt HJ (1994b) Dating of vein specularite using internal (U+Th)/^{4}He isochrones. Geophys Res Lett 21:345-347

Williams IS (2004) Measuring the ages of granites: the challenge to get it right. *In:* Magmas to Mineralisation: The Ishihara Symposium. Geoscience Australia 2003/14:141-143

Wilson NSF, Zentilli M, Reynolds PH, Boric R (2003) Age of mineralization by basinal fluids at the El Soldado manto-type copper deposit, Chile: $^{40}Ar/^{39}Ar$ geochronology of K-feldspar. Chem Geol 197:161-176

Witt KW, Swager CP, Nelson DR (1996) $^{40}Ar/^{39}Ar$ and U/Pb age constraints on the timing and gold mineralization in the Kalgoorlie gold field, Western Australia. Econ Geol 91:792-795

Wolf RA, Farley KA, Silver LT (1996) Helium diffusion and low temperature thermochronology of apatite. Geochim Cosmochim Acta 60:4231–4240

Wolf RA, Farley KA, Kass DM (1998) Modeling of the temperature sensitivity of the apatite (U-Th)/He thermochronometer. Chem Geol 148:105-114

Yamada R, Tagami T, Nishimura S, Ito H (1995) Annealing of fission tracks in zircon: an experimental study. Chem Geol Isot Geosci Sect 122:249-258

Zeitler PK, Herczig AL, McDougall I, Honda M (1987) U-Th-He dating of apatite: A potential thermochronometer. Geochim Cosmochim Acta 51:2865-2868

APPENDIX I:
U/PB AND (U-TH)/HE ANALYTICAL PROCEDURES

Apatite and zircon grains for (U-Th)/He thermochronology were selected by hand picking in order to avoid U- and Th-rich mineral inclusions that may contribute excess helium. Images of selected grains were recorded digitally and grain measurements were taken for the calculation of an alpha correction factor (F_t; Farley et al. 1996). Helium is thermally extracted from single crystals, loaded into platinum micro-crucibles and heated using a 1064 nm Nd-YAG laser. ^4He abundances were determined by isotope dilution using a pure ^3He spike, calibrated daily against an independent ^4He standard tank. The uncertainty in the sample ^4He measurement is <1%. The U and Th content of degassed apatite is determined by isotope dilution using ^{235}U and ^{230}Th spikes. Apatite is digested in 7 M HNO_3. Zircon is digested in Parr bombs using HF. Standard solutions containing the same spike amounts as samples were treated identically as were a series of unspiked reagent blanks. For single crystals digested in small volumes (0.3–0.5 ml), U and Th isotope ratios were measured to a precision of <3%. Overall the (U-Th)/He thermochronology method at CSIRO has a precision of 2.5% for apatite, based on multiple age determinations (n = 70) of Durango standard which produce an average age of 31.5 ± 1.6 (2σ) Ma.

Zircon U/Pb dating at Macquarie University utilizes a LA-ICP-MS facility combining a New Wave/Merchantek 213 nm UV laser ablation (LA) system and a HP 4500 ICP-MS, with analytical methods detailed by Jackson et al. (2004). A split sample of zircon grains from the (U-Th)/He study was mounted in epoxy discs and polished to expose the grains. The mounts were examined using back-scattered electron/cathodoluminescence microprobe imaging to record internal zonation features and external morphology prior to selecting grains for analysis. U/Pb geochronology results were based on the analysis of $^{207}Pb/^{235}U$, $^{208}Pb/^{232}Th$ and $^{206}Pb/^{238}U$ on between 14 and 20 grains per sample. The analysis of the sample zircons was bracketed by multiple analyses of the gem quality GJ-1 zircon, and other in-house standards 91500 and Mud Tank zircon (Black and Gulson 1978; Wiedenbeck et al. 1995) were analyzed in every run as an independent control on reproducibility and instrument stability.

APPENDIX II: EXPLANATIONS AND CALCULATIONS OF MODELED PARAMETERS

1. Sample position, eroded thickness of the porphyry, and initial sample depth

In order to run the algorithm, the position of the dated sample within the porphyry unit must be known or assumed. In most cases, it was assumed that the sample was taken from an exposure or outcrop, which was not the original "top" of the porphyry. If there had been significant erosion since exposure or a portion of the porphyry had been removed by mining, the sample position was somewhat deeper in the body. As shown in Figure 7, the distance from the "top" of the porphyry to the pre-mine topography surface is defined as the "Eroded Thickness" of the porphyry since exposure, a variable we attempt to estimate.

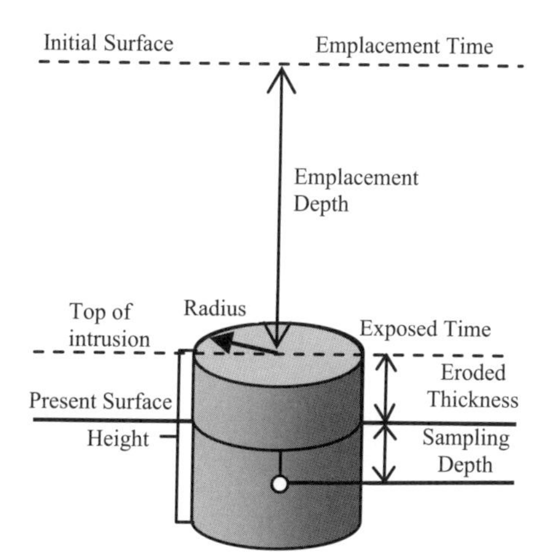

Figure 7. Terminology and algorithms for the calculation of erosion rates in the model defined by Fu et al. (2005)

The small circle within the porphyry as shown in Figure 7 represents the position of the samples, and the "*Sampling Depth*" is defined as the vertical distance between the pre-mine surface and the position of samples. If the samples were taken from the outcrop, the *Sampling Depth* equals zero; if they were either from the mine pits or from drill holes, the *Sampling Depth* is some value larger than zero. The position of the samples at the time of emplacement (called *Initial Sample Depth* for simplicity) is defined as the *Emplacement depth* + *Eroded Thickness* + *Sample Depth.*

In the case of the Sar Cheshmeh Porphyry (SCP), the *Sample Depth* is set to be 100 m because that about 100 m of porphyry has been previously removed by mining. In this case, the position of the sample from SCP at the emplacement time (*Initial Sample Depth*) = *Emplacement Depth* + *Eroded Thickness* of the porphyry + 100 m.

Iterations of the model indicated that pluton emplacement depth and dimension are the key factors controlling the cooling histories of igneous bodies. It can also be proved that different positions within an igneous body have different cooling histories, although symmetry points may have same cooling histories if the igneous body is of regular shape. So the *Initial Sample Depth* (sample position) is very important.

2. Determination of emplacement depth

Determination of emplacement depth and calculation of erosion rates are based on the following two assumptions:

i. Weathering and erosion processes have been occurring at the surface since the intrusion of the porphyry. This infers erosion is occurring during the intrusion and cooling of the body as well as during and after exposure;

ii. The average erosion rate remains constant for a given rock type. In our model, eroded country rocks above the intrusion are assumed to be a single (mixed) rock. We have assumed that the erosion rate for the intrusion is slower than that for the country rock which is assumed to contain sediments and/or sedimentary rock.

For each run of the model, we assign initial values for the *Emplacement Depth* and *Eroded Thickness* of stock and then calculate a cooling curve for the sample. If this cooling curve passes through all age data points determined during dating of the sample, this run of the model is considered to be successful, and the *Emplacement Depth* and *Eroded Thickness* are potentially valid values. The word "potentially" is used here because there are a number of other pairs of *Emplacement Depth* and *Eroded Thickness* values that can also satisfy the model and result in successful cases. This introduces some uncertainty in the final result. Fortunately, the *Emplacement Depth* can be limited to a certain range. Beyond this range, no matter what value of *Eroded Thickness* is taken, the modeling will fail. Similarly, the *Eroded Thickness* can also be limited to a narrow range.

We can further reduce the uncertainties by considering the erosion rates for both country rock and igneous stock separately and by comparing the modeling results with those from other deposits in the region.

3. Calculation of exhumation rates

The apatite (U-Th)/He age was used to calculate a depth (called *He Depth* for simplicity) which corresponds to the position of the sample at the time of closure for apatite (U-Th)/He. For example, by assuming that the apatite (U-Th)/He age is 1 Ma with a closure temperature = 90 °C, surface temperature = 10 °C, thermal gradient = 50 °C/km, then the depth of the sample at 1 Ma:

$$He\ Depth = (90-10)/50 = 1.6\ \text{km}$$

So, we can say, under the above conditions, that the sample was at the depth of 1.6 km below the surface at 1 Ma.

The erosion rate before exposure (*rate* 1) is defined as:

$$rate\ 1 = \frac{(Initial\ Sample\ Depth\ -\ He\ Depth)}{(Emplacement\ Time\ -\ He\ Age)} \quad (1)$$

The erosion rate (*rate* 2) after exposure of the intrusion is defined as:

$$rate\ 2 = \frac{(Eroded\ Thickness\ of\ Intrusion)}{(Exposure\ Age)} \quad (2)$$

where *Eroded Thickness of Intrusion* = *Initial Sample Depth* − (*rate* 1) × (*Emplacement Time* − *Exposure Age*).

The *Exposure Age* will be generated automatically by the model when the top of the intrusion is exposed and begins to erode.

The calculation of *He Depth* is based on the assumption that the igneous stock is already "cooled" (as defined in Table 4 and the text) before it passes through the closure temperature of apatite (U-Th)/He. Most deposits in this work fall into this category. If it is not "cooled" (like Grasberg), the calculation is more complicated but follows the same general assumptions as outlined above.

4. Example: determination of emplacement depth and exhumation rate for the Batu Hijau Porphyry

Thermochronology	*Closure temp*	*Age (± 2 Ma)*	*Note*
Apatite (U-Th)/He	90 °C	2.23 ± 0.09	*He Depth* = 1600 m
K-Ar	400 °C	3.73 ± 0.08	
U/Pb	750 °C	3.74 ± 0.14	*Emplacement Time* = 3.74 Ma

Batu Hijau was modeled as a single composite stock with a cylindrical shape where the width = 500 m and height = 1700 m (residual size). The U/Pb and (U-Th)/He samples of late-mineralization Young Tonalite were taken from drill core at about 350m below the pre-mine topography surface. Thus, the *Sampling Depth* = 350 m. Theoretical modeling studies show that igneous bodies of 500 m width and about 2000 m height would be "cooled" within <1.0 m.y. if the *Emplacement Depth* was <10 km. So, the Batu Hijau tonalite porphyry body would have "cooled" before its temperature passed through the apatite (U-Th)/He closure temperature (90 °C).

We can limit the valid range of the emplacement depths for Batu Hijau before running the model. For example, if *Emplacement Depth* = 2.0 km and *Eroded Thickness* of porphyry = 0 m, the erosion rate before exposure is about 0.89 mm/yr using Equation (1).

$$\begin{aligned} \textit{Initial Sample Depth} &= \textit{Emplacement Depth} + \textit{Eroded Thickness} + \textit{Sampling Depth} \\ &= 2.0 + 0 + 0.35 = 2.35\ km \end{aligned}$$

$$Rate\ 1 = \frac{(2.35\ km - 1.6\ km)}{3.74\ m.y. - 2.23\ m.y.} = \frac{0.75\ km}{1.51\ m.y.} = 0.497\ km/m.y.\ (or\ mm/yr)$$

At this rate, Batu Hijau would still not be exposed because the eroded thickness of country rock is 1.86 km (0.497 km/m.y. × 3.74 m.y.) which is less that the emplacement depth (2.0 km). Therefore, the emplacement depth must be deeper than 2.0 km.

i. If we assume the *Eroded Thickness* of the porphyry is 0~500 m, the minimum *Emplacement Depth* for Batu Hijau is 1.5~2.0 km;

ii. Due to the absence of zircon (U-Th)/He age data, the maximum emplacement depth for Batu Hijau could be up to 5.0 km. However, if we consider reasonable values for the erosion rates of the porphyry and country rock, the maximum emplacement depth can be limited <5 km. For example, assuming *Emplacement Depth* = 3.5 km, if *Eroded Thickness* of porphyry = 100 m, the model will generate an erosion rate before exposure of about 1.56 mm/yr and an erosion rate after exposure of about 0.05 mm/yr. If we increase the *Eroded Thickness* of the porphyry to 1000 m, the erosion rates before and after exposure would be 2.15 mm/yr and 0.46 mm/yr. However, because the erosion rate before exposure is too fast and the difference between the two rates is too large, this is an unlikely scenario and the maximum emplacement depth is 3.5 km.

iii. If the emplacement depth can be determined, we can also limit the *Eroded Thickness* of the porphyry to a reasonable range by considering the erosion rate of the overlying country rock and porphyry rock units.

Then, based on the ranges of *Emplacement Depth* and *Eroded Thickness* obtained, we can generate a cooling curve that matches all the real age data and produces reasonable erosion rates. An *Emplacement Depth* of 2.4 km and *Eroded Thickness* of about 300 m are geologically reasonable and produce a cooling curve that successfully passes through all age data points (Fig. 5e). The calculated average erosion rate for the porphyry is 0.24 mm/yr, consistent with other porphyry units in the region—the calculated erosion rates for the Grasberg porphyry and for Ciemas porphyry are 0.23 mm/yr and 0.11 mm/yr, respectively.

5. Limitations and future improvements

Due to the limited information on sample position and the plethora of uncertainties, the above algorithms are not ideal. However, a feasible solution is yielded under the current conditions and improvements are constantly being made. For example, if two or more samples (with as large an age difference as possible) are obtained from same porphyry deposit for dating, we can generate a unique solution to the porphyry emplacement depth. A range of emplacement depths that satisfy the cooling curve constraints of the real age data for each sample can be identified. At a certain point, the "possible" emplacement depth ranges for both samples will intersect and we can solve for the true emplacement depth for the porphyry.

Reviews in Mineralogy & Geochemistry
Vol. 58, pp. 499-525, 2005

Thermochronometers in Sedimentary Basins

Phillip A. Armstrong

Dept. of Geological Sciences
California State University, Fullerton
800 N. State College Blvd.
Fullerton, CA 92831
parmstrong@fullerton.edu

INTRODUCTION

Sedimentary basins, both modern and ancient, cover most of Earth's land and subsea surface, and provide some of the best natural laboratories for studying and constraining geologic processes. The record of sedimentation, burial, erosion, and uplift provides a rich history that can be combined with various analytical and modeling techniques to evaluate: (1) processes that lead to basin formation; (2) deformation of regions related to plate tectonic effects; (3) timing and duration of hydrocarbon generation, migration, and trapping; (4) past and present effects of fluid flow in basin deposits; and (5) past climate change. Sedimentary basins have been classified by many workers (e.g., Bally and Snelson 1980; Dickinson 1993; Ingersoll and Busby 1995). Ingersoll and Busby (1995) generally classify the basins as forming in divergent settings, intraplate regions, convergent settings, transform settings, and hybrid settings. Many mechanisms have been proposed for the formation of the different styles of basins including, but not limited to, crustal thinning, sedimentary and tectonic loading, subcrustal loading, and mantle lithospheric thickening (e.g., Ingersoll and Busby 1995).

Thermochronometers have played an increasingly important role in the evaluation of both intra- and interbasin processes. Thermochronometers, especially the lower-temperature apatite fission-track and apatite (U-Th)/He dating, are now commonly combined with burial and thermal history analysis and modeling to provide important constraints on the timing and duration of heating/cooling events that can be used to evaluate hydrocarbon systems as well as structural and basin-forming mechanisms.

Past reviews of thermochronometer use in sedimentary basins are given by Naeser et al. (1989), Green et al. (1989a), Naeser (1993), and Giles and Indrelid (1998). Rather than review each study of thermochronometer use in sedimentary basins during the last decade, in this chapter some of the fundamental concepts used to evaluate the present-day thermal field for sedimentary basins, how a basin analyst would evaluate and combine burial and thermal histories, and the main thermochronometers used in sedimentary basin analysis are summarized. Finally, several recent case studies that illustrate the modern uses of thermochronometers in sedimentary basins are reviewed.

PROCESSES THAT AFFECT BASIN TEMPERATURES – THE HEAT BUDGET

Present-day temperatures and paleotemperatures help constrain models of processes that occur within and below sedimentary basins. These temperatures, and their thermochronologic signatures, are the net result of several different thermal and tectonic processes taking place at different spatial and temporal scales. Armstrong and Chapman (1998) broadly group these

1529-6466/05/0058-0019$05.00 DOI: 10.2138/rmg.2005.58.19

processes into six categories: (1) variations in basal heat flow that might be due to lateral temperature changes in the mantle; (2) vertical and/or lateral changes in rock properties such as heat production or thermal conductivity within or below the basin; (3) heat-flow perturbations resulting from exhumation or burial; (4) thermal effects of fluid flow; (5) anomalous heat flow related to magmatic processes such as intrusion emplacement; and (6) tectonic heating or cooling caused by lithospheric thinning and thickening.

Any combination of these processes can perturb the thermal field of a sedimentary basin at any point in its history. Standard basin analysis techniques (e.g., Allen and Allen 1990) make quantifying the thermal effects of some of these processes possible (i.e., exhumation and burial), thus making interpretations of other processes possible if indicators that constrain past temperatures are available.

PRESENT-DAY THERMAL FIELD

Any thermal history modeling exercise in a sedimentary basin must honor the present-day temperature constraints for that basin. Among the principal locations for making the temperature measurements are oil and gas exploration wells. Uncorrected bottom-hole temperatures (BHT) are measured as part of the routine geophysical well logging and typically range from 25–175 °C at depths of 1–5 km. These BHTs must then be corrected for the thermal disturbance caused by drilling and well circulation. There are several reviews of methods for correcting BHTs (e.g., Deming 1989; Funnell et al. 1996), but the method generally includes extrapolating temperatures to infinite time based on multiple temperature measurements at different times following the stoppage of drilling at a particular depth.

The temperature profile between discrete BHTs is computed and depends on rock thermophysical properties such as thermal conductivity and heat production. Thermal conductivity data can be measured in the lab on available drill core or drill cuttings, however, these values relate to discrete sections of the well and specific rock types. One strategy is to establish end-member thermal conductivity values for the different rock types present in a particular basin, then apply these values to similar rock types within the basin based on well log data. For example, Funnell et al. (1996) measured sandstone and shale thermal conductivities of 4.0 and 2.7 $Wm^{-1}K^{-1}$, respectively, for the Taranaki Basin in New Zealand. Heat production within the sedimentary column itself is typically small and often a neglected contributor to the basin's overall heat budget. Typical specific heat production (mass specific) values are $\sim 10–100 \times 10^{-5}$ μW/kg for different sedimentary rock types; this heat production amounts to a heat flow increase of ~1 mW/m^2 for every 1 km of sedimentary section in the basin. All thermal conductivity and specific heat production estimates must be corrected for porosity prior to using them in subsurface temperature and heat flow computations (see below). Porosity is often considered to decrease exponentially with depth (e.g., Magara 1976; Funnell et al. 1996) in sedimentary basins, though many other models have been proposed (e.g., Issler 1992).

Once thermophysical properties of the section are determined and corrected, variations in temperature between discrete BHTs can be computed. The temperature as a function of depth for steady-state heat conduction in a horizontally layered sedimentary section is given by:

$$T(z) = T_0 + \sum_{i=1}^{n}\left[\frac{q_{i-1}\cdot\Delta z_i}{\lambda_i} - \frac{A_i\Delta z_i^{\,2}}{2\lambda_i}\right] \tag{1}$$

where

$$q_i = q_{i-1} - A_i\Delta z_i \tag{2}$$

and T_0 is the surface temperature, λ_i, Δz_i, and A_i are thermal conductivity, thickness, and volumetric heat production, respectively, for the ith depth interval. q_{i-1} is the heat flow at the top of the ith interval and q_0 is the surface heat flow. The temperature-depth profile can then be determined by iterating on the surface heat flow until the difference between BHTs and model temperatures at those depths is minimized (Funnell et al. 1996).

BUILDING A BURIAL AND THERMAL HISTORY

Because sedimentary basins retain a record of burial and exhumation, it is possible to build a burial history and couple that to a thermal history based on the physics of heat transport. Hermanrud (1993), Lerche (1993), and Poelchau et al. (1997) provide reviews on the concepts of basin simulation. The basin analyst generally is concerned with burial, erosion, thermal history, hydrocarbon generation, expulsion, and migration, trap formation, fluid accumulations, and overpressuring. We restrict our discussion to the burial, erosion, and thermal history aspects because they are the parts primarily constrained by thermochronometry. A general overview is given below, and the basic scheme of implementation is shown in Figure 1.

Input parameters for the burial/erosion history include known stratigraphic thicknesses and ages, age bounds on erosional unconformities, and porosity-depth relations. The thickness of each stratigraphic unit is decompacted based on the present-day thickness and an assumed porosity-depth relation (e.g., Schlater and Christie 1980). Input sedimentation rates then are estimated from the thicknesses of the decompacted sediments and their ages. Once timing and rates of burial, and erosion, are determined for each stratigraphic unit, the units can be redeposited or eroded sequentially to construct the burial history. At each time step, material properties are adjusted for depth changes and the temperatures are computed, usually based

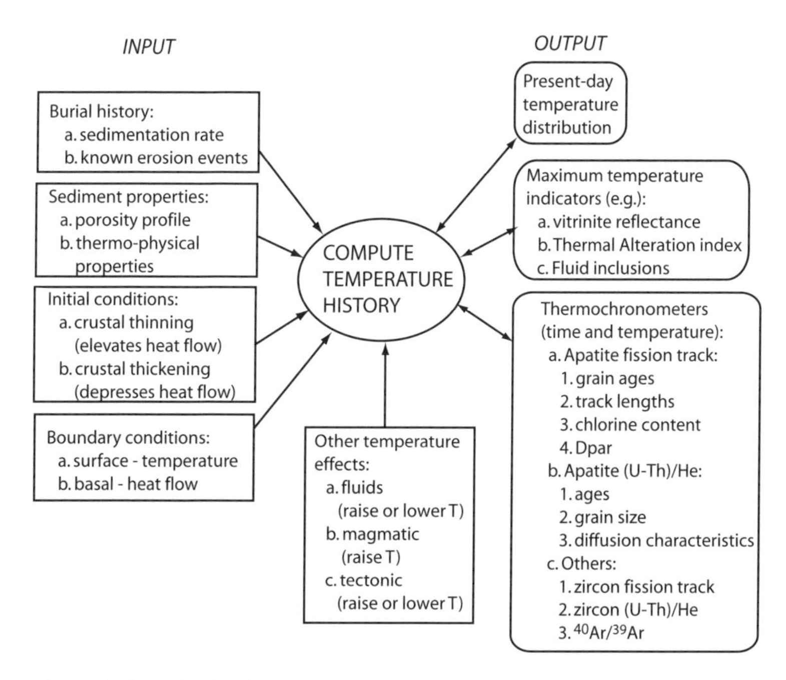

Figure 1. General schematic showing the basic input and output information used to combine burial and thermal histories with thermochronometer data.

on transient conductive heat flow laws. At the end of the simulation, model temperatures and thermochronologic information are compared with measured values and the model is iterated until good fits are obtained. Poor fits to these data sets can then be evaluated and remodeled by incorporating more complex histories that include alternative temperature effects such as those caused by intrusion emplacement, fluid flow, and/or tectonic perturbations (i.e., crustal thinning or thickening; Armstrong and Chapman 1998).

Most of the coupled burial-thermal history simulators assume one-dimensional conductive heat flow. However, large potential temperature variations in sedimentary basins can be caused by lateral thermal conductivity variations and fault-related heat advection. In such cases, heat will be partially refracted across thermal conductivity boundaries such as faults and into high-conductivity regions, which causes temperature to increase in the high-conductivity regions and decrease in the low-conductivity regions. Weir and Furlong (1987) demonstrated that 2D simulations are necessary to account for the heat refraction problem when the width to depth ratio of a basin is less than approximately 5 to 1. Therefore, except for very narrow and deep basins, a 1D solution for these types of burial histories is generally good enough, especially given the uncertainties in sedimentation rates, erosion rates, and thermochronologic modeling.

The boundary conditions used in most thermal modeling routines include an upper surface temperature boundary and a lower heat flow boundary. Mean annual surface temperatures generally are estimated from known environmental conditions at the time of burial (paleoclimate, paleogeography, paleobathymetry). These temperatures can vary substantially and propagate rapidly into the basin sediments. In the Taranaki Basin, surface temperatures based on marine fauna, land flora, and oxygen isotopic data indicate surface temperature changes of ~15 °C in the late Cenozoic (Armstrong et al. 1996), which can affect apatite fission-track ages in the subsurface (e.g., O'Sullivan and Brown 1998). The lower heat flow boundary condition should be placed at sufficient depth to allow the lithosphere to be affected by near surface transient affects. It is unrealistic to place the lower boundary condition at the base of the sediments in the basin because that would imply that the sediments are decoupled thermally from the basement rocks below. Nielsen and Balling (1990) showed that excluding transient thermal coupling between basin strata and the underlying basement can lead to inaccurate estimates of heat flow and temperature structure of a basin. The lower heat flow boundary should be placed at depths of 10's of kilometers. In many models, it is the heat flow at the bottom boundary that is adjusted during the burial–thermal history simulation to account for the temperatures at the end of the simulation (Armstrong and Chapman 1998).

All thermal history simulations must begin with an assumed set of initial conditions. Normally the initial condition is based on presumed heat flow conditions at the time of initial basin formation. For example, rift-related basins will include initially high heat flow related to thinning of the lithosphere (e.g., McKenzie 1978) where the stretching factor is computed from analysis of tectonic subsidence. Alternatively, initial heat flow might be lower than present-day in cases where basins form in response to crustal thickening at convergent plate margins. The transient effects of these lithospheric-scale processes can last for 10^7 years or more and should be included in thermal history analyses if possible. Thermal modeling programs generally allow the user to apply stretching factors with implicit heat flow variations or to apply explicit high or low heat flow during the simulations. Unfortunately, low-temperature thermochronometers are less sensitive to the early heat flow conditions than they are to maximum temperature and burial conditions that usually occur after the initial thermal conditions wane. Nielsen (1996) and Ferrero and Gallagher (2002) used stochastic thermal modeling to constrain heat flow changes through the basin history. These stochastic models utilize proposed burial histories, vitrinite reflectance, and present-day temperatures to place bounds on paleo heat flow values.

If input parameters such as burial history, sediment thermophysical properties, and initial conditions can be determined, or at least reasonably bracketed, then potential temperature histories can be computed for a basin. The suite of computed thermal histories and their output can then be compared to present-day temperature conditions, maximum temperature indicators such as vitrinite reflectance, and to computed thermal histories for the different thermochronometers (mainly apatite fission-track and apatite (U-Th)/He—see later sections) to determine if the temperature history from burial information is acceptable (Fig. 1). Unacceptable temperature histories might require additional thermal effects to explain the thermal data. Such effects might include fluid flow, which can raise or lower subsurface temperatures, or localized magmatic intrusion, which raises subsurface temperatures.

THERMOCHRONOMETERS USED IN SEDIMENTARY BASINS

Several thermochronometers can be used in the study of sedimentary basins, including apatite fission-track (AFT), zircon fission-track (ZFT), apatite and zircon (U-Th)/He dating, $^{40}Ar/^{39}Ar$ in white mica and potassium feldspar. These thermochronometers have different and sometimes overlapping closure temperatures ranging from ~70 to >200 °C. In this review, the lower-temperature thermochronometers of apatite fission-track and apatite (U-Th)/He are mainly illustrated. In the following sections, the dating methods as they pertain to sedimentary basins are outlined; details of the measurement techniques and general interpretations are given in earlier chapters of this volume.

Apatite fission-track dating

The fundamentals and theory of AFT thermochronology are discussed in this volume by Donelick et al. (2005) and Tagami et al. (2005) and by other recent reviews (e.g., Gleadow et al. (2002). Furthermore, the application of AFT thermochronology to sedimentary basins in particular has been outlined by several papers in the last couple decades (Gleadow et al. 1983; Green et al. 1989a; Naeser et al. 1989; Naeser 1993; Giles and Indrelid 1998).

AFT analysis is particularly important to basin analysis and hydrocarbon exploration because its range of annealing temperatures, which generally has been considered to be between 60 and 120 °C in most studies, approximately corresponds with liquid hydrocarbon generation temperatures at geologic time scales. However, this generally used annealing temperature range (60–120 °C) does not reflect the annealing range of apatite grains encountered in many samples, which has bounds of approximately 50–150 °C depending on apatite composition—see below and Donelick et al. (2005) in this volume; apatite kinetic data should be determined for all dated grains.

Naeser (1979) provides a good example of the expected effects of AFT age with temperature in a sedimentary basin succession that is currently at maximum temperature at any depth. Figure 2 shows that the shallowest/coolest samples should have relatively old AFT ages, primarily due to containing detrital AFT grain ages that are at least as old as the depositional age of the strata. The distribution of the detrital AFT grain ages within a single sample in the shallow section may show considerable scatter reflecting variations in apatite chemistry (see below), variations in source regions with different exhumation histories, or both. In the shallow-section, AFT sample ages can either increase or decrease down section. An increase in the shallow AFT sample ages down section usually reflects detrital input from progressively younger denuded source terrains within which the apatites had been partially or totally reset. Alternatively, a decrease in the shallow AFT sample ages down-section usually reflects sourcing from regions where the apatites were not reset.

Below the depth where temperature is ~50–70 °C (depending on composition), fission-track annealing begins and AFT ages start to decrease. For apatites with the same annealing

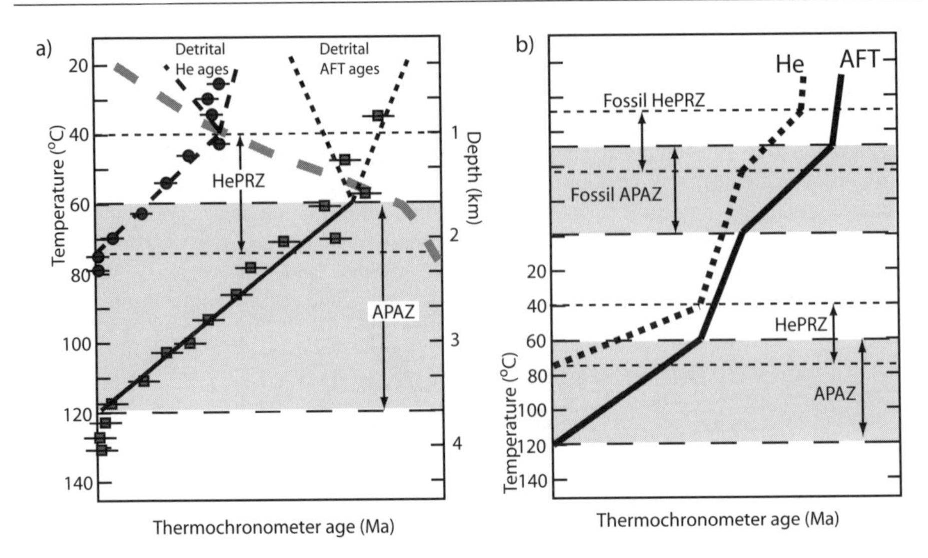

Figure 2. Idealized low-temperature thermochronometer versus temperature profiles in a well. Plot a) shows scenario where no uplift and exhumation has occurred. Apatite fission-track ages (filled squares) in this case are all older than apatite (U-Th)/He ages (filled circles). The thick dashed gray line is stratigraphic age. APAZ represents apatite partial annealing zone. HePRZ represents the apatite helium partial retention zone. In b), the sequence has been uplifted toward surface to give fossil APAZ and HePRZ (see text). Note that in b), temperature scale is condensed. Modified after Naeser (1979).

properties, the sample ages should decrease systematically to a depth that corresponds to the total annealing temperature for that apatite composition (~100–150 °C). This region between no annealing and total annealing is termed the partial annealing zone (PAZ). At greater depth and at higher temperatures, fission-tracks should be totally annealed and AFT ages should be zero (Fig. 2a). Examples of borehole AFT age profiles that exhibit equilibrated maximum temperature profiles include the Otway Basin, Australia (Gleadow and Duddy 1981; Gleadow et al. 1983; Green et al. 1989a) and the San Joaquin Valley, California (Naeser et al. 1990).

If a sedimentary section is uplifted toward the erosional surface (exhumed) after maximum burial, the former zone of total annealing and the PAZ are uplifted as well. Uplifted apatites from former totally annealed (zero age) and partially annealed depths will begin to accumulate tracks thus increasing the ages in these sections to form a fossil total annealing zone and a fossil partial annealing zone (Fig. 2b) (Naeser 1979; Gleadow et al. 1983; Gleadow and Fitzgerald 1987; Fitzgerald and Gleadow 1990). It should be noted that it is rare to have both a fossil and modern PAZ preserved in a borehole profile because most wells are too shallow. Fossil PAZs and the break in slope generally are observed in surface sample age versus elevation transects (e.g., Fitzgerald and Gleadow 1990) whereas the base of the fossil total annealing zone is preserved in well sections.

Fission-track length data provide important constraints in the thermal history assessment of sedimentary basins. Track length distributions, in combination with apparent fission-track ages, allow a basin analyst to discriminate between different thermal history scenarios. Although extremely simplistic, Figure 3 illustrates seven possible example thermal histories for sedimentary strata derived from a single monocompositional source and their characteristic track length distributions. Paths 1–3 show progressive burial to maximum depths below the top of the PAZ followed by near isothermal conditions. The shallowest sample (path 1) shows long track lengths with narrow distributions indicative of that sample's grains not being annealed

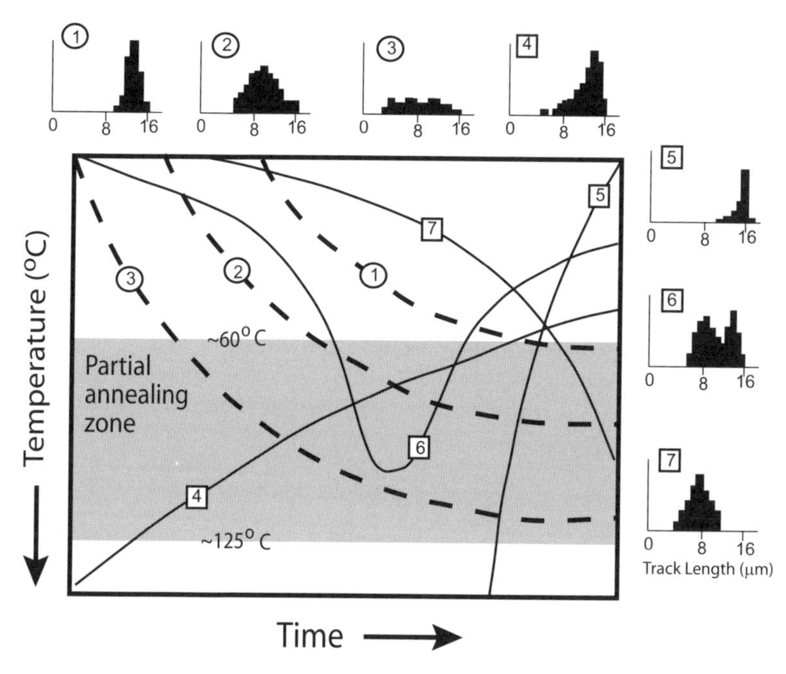

Figure 3. Schematic burial/temperature histories and representative track length distributions. Paths 1-3 show heating followed by nearly isothermal conditions in the apatite partial annealing zone. Paths 4 and 5 show heating from temperatures greater than the base of the partial annealing zone. Path 6 shows heating into the partial annealing zone followed by cooling out of it. Path 7 shows rapid heating to temperatures in the partial annealing zone. Modified after Gleadow et al. (1983).

after deposition. With greater depth (and temperature) the average track lengths progressively get shorter and the distribution of tracks spreads out reflecting the partial annealing. The amount of spread in these paths could be more than illustrated if detrital grains were from multiple sources or had different compositions. Paths 4 and 5 show cooling from depths within or below the PAZ. With cooling from temperatures above the base of the PAZ for this apatite species, average track lengths tend to be long but with negative skewness (path 4 and 5), but the distribution gets narrower with more rapid cooling through the PAZ—compare path 4 with 5. For a sample that is heated to PAZ temperatures and then cooled (path 6), a bimodal distribution can be expected because tracks that were shortened while the sample was in the PAZ are mixed with longer tracks that formed after cooling out of the PAZ. For samples that have experienced relatively recent and rapid heating to maximum temperatures (path 7), average track lengths will be relatively short but not as spread out as in the case of longer periods of time in the PAZ (paths 2 and 3).

In the last decade and a half, several computer programs have been developed (e.g., Green et al. 1989b; Corrigan 1991; Crowley 1993; Gallagher 1995; Willett 1997; Ketcham et al. 2000; Issler 2004) to compute thermal histories from AFT age and track length data. Most of these programs are based on annealing models that assume a single composition for apatite grains (e.g., Laslett et al. 1987; Crowley et al. 1991), but the program (AFTSolve) of Ketcham et al. (2000) is based on multi-kinetic and/or multi-compositional models—see Donelick et al. (2005) and Ketcham (2005) in this volume for detailed discussions.

Apatite composition strongly affects the annealing characteristics of individual apatite grains (e.g., Green et al. 1985). Chlorine-rich apatites tend to anneal more slowly than fluorine-rich apatites, thus the temperature for total annealing is greater for chlorine-rich apatites than for fluorine-rich. This is particularly important in sedimentary basins because grains can come from a variety of sources with different compositions. A substantial advancement in our ability to utilize apatite fission-track data in the last decade has been the realization that apatite fission track analyses should be done in combination with some measurement of composition. In some studies, each dated grain in AFT samples have been microprobed for composition to make direct comparisons between grain ages and chemistry (e.g., O'Sullivan and Brown 1998; Crowhurst et al. 2002; Lorencak et al. 2004). Recently, labs have started to routinely assess relative composition between apatites by measuring the etch pit width parallel to the crystallographic *c*-axis (D_{par}) (e.g., Burtner et al. 1994; Ketcham et al. 1999; Donelick et al. 2005). The utility of measuring D_{par} is that it is straightforward to measure on dated grains, is less expensive than microprobe analysis, and there is publicly available software that allows easy incorporation of D_{par} data—see Donelick et al. (2005) and Ketcham (2005) in this volume for review. Calibrated kinetics (composition or D_{par}) can now routinely be incorporated into quantitative thermal history models (e.g., Ketcham et al. 1999, 2000; Ketcham 2005). The ease of generating compositional information from D_{par} and its incorporation in modeling programs allows the basin analyst to more rigorously evaluate potential thermal histories from AFT data.

Because apatite grains in sedimentary basins can be derived from multiple sources, from a single source with a complicated cooling history, and/or have a range of compositions, it is useful to evaluate the sample for heterogeneous age distributions, especially in samples from the upper part or above the PAZ. Radial plots (Galbraith 1990) are useful for visualizing different age populations in a set of single grain ages. Figure 4 shows an example radial plot from a surface sample in northern Scotland (Carter 1999). O'Sullivan and Parrish (1995) illustrate simple cases (their Fig. 7) of single grain age distributions on radial plots for different positions relative to the PAZ. They show that samples from the PAZ tend to have a fanning out pattern of ages, or "open jaw", reflecting a wide spread in single grain ages. Samples from below or above the PAZ would tend to have narrower distributions of single grain ages. In the example shown in Figure 4, a central age of 320 Ma is shown by the horizontal line extending to the radial part of the plot at the right. Extrapolation of a line from the origin to the radial arc gives the age of each grain. Single grain ages that lie far from the horizontal line, especially outside of the ± 2σ intervals on the *y*-axis, may belong to other age modes than the one suggested by the central age. Several methods for decomposing fission-track ages using peak-fitting methods are available (e.g., Galbraith and Green 1990; Brandon 1992; Sambridge and Compston 1994). In the case of Figure 4, two additional age components of 206 and 557 Ma were recognized by Carter (1999), who used annealing models to interpret the age components as representing slow cooling of the source region through the PAZ starting prior to 500 Ma, and rapid cooling through the apatite PAZ at about 200 Ma. This example is one

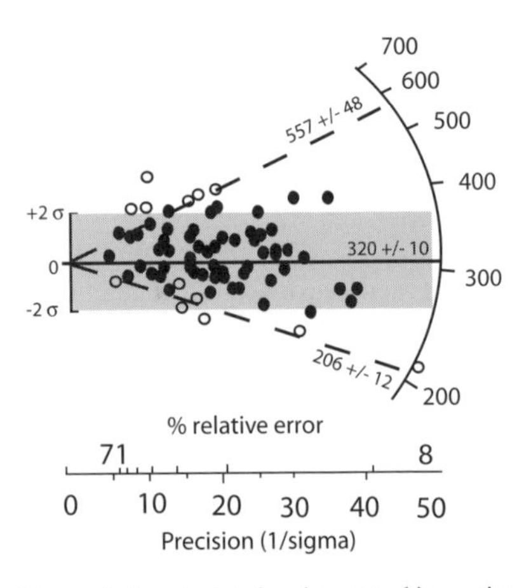

Figure 4. Radial plot showing spread in apatite single grain age data. A central age (Galbraith and Laslett 1993) of 320 Ma is shown by the horizontal line extending to the radial part of the plot at the right. The *x*-axis shows the normalized uncertainty (% error) of each grain age and the y-axis shows the 2σ uncertainty normalized to the length of the scale on the *y*-axis. Filled dots represent ages that fit one age population. Trends of open circles that bound the dots are interpreted to represent additional age populations (see text). Modified after Carter (1999).

that shows that sedimentary basin AFT data can provide powerful constraints not only on the thermal evolution (burial and exhumation) of the depositional basin, but also on source region exhumation rates and timing (see later section) and on tectonic processes that lead to the basin formation.

Apatite (U-Th)/He dating

One of the major advancements in low-temperature thermochronology over the last decade has been use of apatite (U-Th)/He dating. Reviews of the methods and uses of apatite helium are given by Farley (2002) and Ehlers and Farley (2003). The method is based on the production of helium from isotopes of uranium and thorium, but the partial loss of helium due to diffusion out of the crystals has lead to the realization that the method can be used as a thermochronometer (e.g., Zeitler et al. 1987; Lippolt et al. 1994; Wolf et al. 1996). Systematic diffusion studies (Wolf et al. 1996; Farley 2000) indicate that apatite He ages provide thermochronologic information for temperatures between approximately 40–75 °C; at geologic time scales helium is completely diffused from apatite at >75 °C and nearly all the helium is retained at <40 °C. Thus apatite He ages potentially provide lower temperature thermochronologic constraints than that of apatite fission-track ages.

The expected apatite He age profile in a borehole sequence where maximum temperatures have been attained is similar to that of the AFT profile, except that temperatures for any age greater than zero is lower for He ages than for AFT ages (Fig. 2a). The region between about 40–75 °C is referred to as the helium partial retention zone (HePRZ). He ages at depths below the base of the HePRZ are zero and above the HePRZ they are governed by the detrital He ages, which can show considerable spread if there are variable sources. Like the AFT age profile in Figure 2b, during exhumation of the section the He profile is shifted upward so that a new HePRZ and a fossil HePRZ are established as helium accumulates. In one example, paired AFT and apatite He ages from a tilted fault block in the White Mountains (California) confirm a paired fossil HePRZ and PAZ (Stockli et al. 2000).

Wolf et al. (1998) developed a mathematical model to evaluate the time-temperature sensitivity of helium diffusion. The size of grains affects the He age such that larger grains are expected to give older ages, especially in cases of slow cooling (Farley 2000; Reiners and Farley 2001). Helium age modeling with grain size effects is now routinely used to compute model He ages from temperature histories; forward and inverse modeling schemes are discussed by Ketcham (2005) in this volume.

A study in the Otway Basin of Australia (House et al. 1999) showed that He ages from boreholes decrease from about 75 Ma at the surface to nearly zero at depths where temperature is approximately 80 °C. This He age versus temperature profile is consistent with a HePRZ discussed above and with predictions from laboratory data (Wolf et al. 1996). However, the measured He age profile is not consistent with one predicted from a thermal history based mainly on published AFT and vitrinite reflectance data. House et al. (2002) integrated detailed apatite He age data with new and old AFT data from Otway Basin to address these discrepancies and to reevaluate its thermal history. Figure 5 shows that Otway Basin He and AFT ages decrease with increasing temperature and that at any temperature the AFT ages are greater than the He ages. There is considerable scatter in the data, but a positive correlation between grain size and He age suggests that the scatter in the He ages may be caused by grain size effects and the scatter in the AFT ages is shown to be caused by variations in apatite grain composition (House et al. 2002). Model He age versus depth profiles computed from published thermal histories, which are based on burial models and AFT and vitrinite reflectance constraints (e.g., Duddy 1994), are generally similar to measured He age versus depth profiles for eastern part of the Otway Basin. In the western part of the basin, the model He ages are as much as 40 Ma older than measured ages at the same temperature (Fig. 5). House et al. (2002)

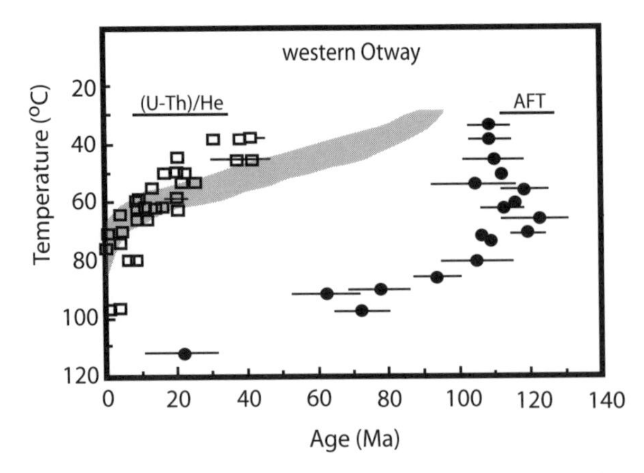

Figure 5. Apatite fission-track (dots) and (U-Th)/He (squares) ages versus present-day temperature for wells from the western part of the Otway Basin in southeast Australia. Shaded region represents model (U-Th)/He ages based on published thermal histories from fission-track data. The misfit of the model and measured He ages suggests that temperatures were greater in the past than indicated by fission-track data alone. Modified after House et al. (2002).

interpret the mismatch of the model and predicted He data to be the result of slightly higher temperatures due to higher geothermal gradients during the Cenozoic, perhaps related to fluid flow. The He age data in this case allowed the thermal history the Otway Basin to be refined and better constrained when compared to models based purely on the higher temperature apatite fission-track data. This study, as well as others (e.g., Crowhurst et al. 2002; Lorancak et al. 2004), points to the effectiveness of extracting more information from sedimentary basins by combining multiple low-temperature thermochronometer data sets.

Combining apatite fission-track and other thermal indicators

In the last decade AFT analysis has been combined with maximum paleotemperature indicators such as vitrinite reflectance (e.g., Bray et al. 1992; Arne and Zentilli 1994; Green et al. 1995; Kamp et al. 1996; O'Sullivan 1999; Marshallsea et al. 2000; Ventura et al. 2001; O'Sullivan and Wallace 2002), fluid inclusions (e.g., Pagell et al. 1997; Parnell et al. 1999), or $^{40}Ar/^{39}Ar$ (e.g., Kohn et al. 1997) to provide stronger and more detailed constraints on thermal histories of sedimentary basins.

Thermochronometers, especially when combined with maximum paleotemperature techniques as listed above, provide some of the best indicators of both magnitude and timing of basin inversion in sedimentary basins (e.g., Kamp and Green 1990; Bray et al. 1992; Green et al. 1995; Hill et al. 1995). Basin inversion is the process of changing from general basin subsidence to local or regional uplift of the sedimentary section. If the overlying strata are removed by erosion during uplift, then the once deeper strata provide a record of the inversion. Bray et al. (1992) outlined a method for using combined AFT and vitrinite reflectance data to estimate amounts of past burial and subsequent inversion. Vitrinite reflectance is a measure of coalification rank of sedimentary organic matter (e.g., Lopatin 1971; Tissot and Welte 1978) and provides a measure of relative maturity that is generally related to depth of burial. More recently, vitrinite reflectance has been quantified so that maximum temperature information can be estimated (Burnham and Sweeney 1989). However, vitrinite reflectance data alone can not provide estimates of the timing of maximum temperatures whereas AFT data can in some cases provide both maximum temperature and its timing. To estimate past burial amounts, computed maximum temperatures from vitrinite reflectance and/or AFT data are plotted as a function of depth and compared to the present-day geothermal gradient (Fig. 6). Projection of the line through the maximum temperature data to the present-day surface temperature provides an estimate of denudation amount. Bray et al. (1992) argue that if the paleotemperature gradient is parallel to the present-day gradient, then paleoheat flow at the time of maximum burial was the

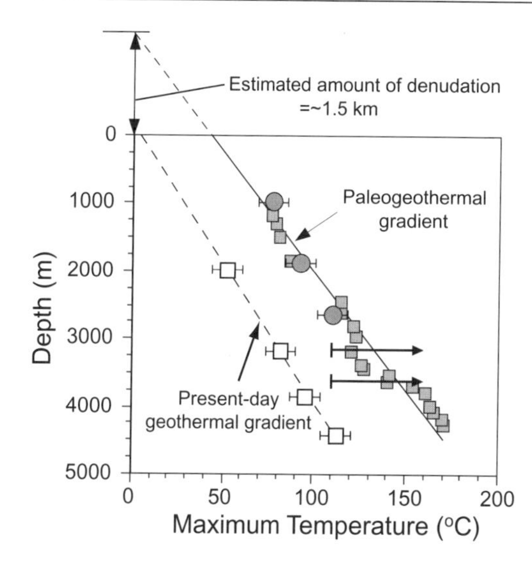

Figure 6. Temperature versus depth plot showing present-day temperatures in a well (open squares) and computed maximum paleotemperatures from vitrinite reflectance (gray squares) and apatite fission-track (gray circles) data. Horizontal arrows indicate that temperatures based on apatite fission-track data were greater than about 110 °C. Lines through the data approximate present-day and paleo temperature gradients. Linear paleotemperature gradients that are parallel to the present gradient suggest paleoheat flow that was similar to heat flow today. In this example, paleoheat flow was similar to present-day. Projection of the paleotemperature gradient to the zero temperature axis gives estimate of vertical uplift and denudation (Bray et al. 1992). Modified after Bray et al. (1992) and O'Sullivan and Wallace (2002).

same as today. If the paleotemperature gradient is greater or less than the present-day profile, then paleoheat flows of greater or less than today, respectively, can be interpreted. Non-linear paleotemperature gradients may indicate higher or lower heat flow through the sediments due to magmatic intrusion or fluid flow (Bray et al. 1992; Duddy et al. 1994).

Higher temperature thermochronometers

Higher temperature thermochronometers such as zircon fission-track dating, $^{40}Ar/^{39}Ar$ dating, and zircon (U-Th)/He dating have been used less extensively than the lower closure temperature systems (apatite fission-track and (U-Th)/He) in evaluating the thermal history of sedimentary basins. The closure temperatures for these systems are ~185–240 °C for zircon fission-track (e.g., Brandon et al. 1998; Tagami et al. 2005), ~350–420 °C for white mica $^{40}Ar/^{39}Ar$ (McDougall and Harrison 1999; Hodges et al. 2005), ~150–200 °C for k-feldspar $^{40}Ar/^{39}Ar$ (Lovera et al. 1997; Harrison et al. 2005), and 180–200 °C for zircon (U-Th)/He (Reiners et al. 2002; Reiners 2005). These higher closure temperatures are less sensitive to typical temperature ranges in most sedimentary basins thus making them ideal for using detrital grains in evaluating source region parameters. A later section of this paper outlines the use of some of these methods.

EXAMPLES OF THERMOCHRONOMETER USE IN SEDIMENTARY BASINS

Thermochronometer data have been used in many studies to evaluate the thermal history of sedimentary basins in the last decade (e.g., Ravenhurst et al. 1994; Green et al. 1995; Sobel and Dumitru 1997; Issler et al. 1999; O'Sullivan 1999; Ventura et al. 2001; Arne et al. 2002; Lim et al. 2003). The intent is not to review all of these, but to offer a few example studies.

Example of a sedimentary basin thermal history – the Williston Basin

Williston Basin is a continental interior epicratonic basin (~800 km diameter) that straddles the Canada-United states boundary and that formed dominantly in the Paleozoic (Osadetz et al. 2002 and references therein). Subsidence in the basin has been described as monotonic and persistent (e.g., Ahern and Mrkvicka 1984) as well as containing rapid subsidence phases (e.g., Burgess et al. 1997). Rocks of the Williston basin and surrounding region consist of Precambrian basement overlain by Paleozoic, Mesozoic, and Cenozoic sediments. The Paleozoic strata consist of carbonates, clastics, and organic-rich mudstone and shale. The preserved cumulative stratigraphic thickness ranges up to about 5 km. The Canadian Williston basin is an important hydrocarbon basin and is the region where the petroleum system concept was first described (Dow 1974).

Osadetz et al. (2002) evaluated apatite fission-track data from Precambrian basement rocks collected from drill holes distributed throughout the Canadian part of the basin (Fig. 7). The basement rocks underlie the Phanerazoic strata at depths as great as 3 km. In most of the study region, the temperature at the top of the Precambrian basement is <70 °C, well below the total annealing temperature of fission tracks in apatite. Surface heat flow typically ranges from about 40–60 mW/m^2. Two regions on the east and west sides of the basin give AFT ages ~350–450 Ma with an area of 250–300 Ma ages in between. The southern part of the basin has ages of <100 Ma. The chlorine content of nearly all probed grains was <0.2 wt% Cl indicating that the apatites are mostly fluorine-rich. Osadetz et al. (2002) argue that the annealing model of Laslett et al. (1987) could be used to evaluate thermal histories, though that particular annealing model is based on a higher chlorine content of ~0.4 wt% for Durango apatite. Thus, the estimated paleotemperatures described below should be considered maximum estimates; the overall interpretations for elevated paleoheat flow in the late Paleozoic discussed below should not change.

Figure 8 shows two representative thermal histories for samples collected in different parts of the basin (Fig. 7). These thermal histories were generated using the program "Monte Trax" described by Gallagher (1995), which uses a Monte Carlo approach to invert for thermal histories that satisfy the fission-track ages and length distributions. The main features of these two simulations, which are similar to those from other wells, are the cooling events from elevated temperatures in the late Paleozoic (~245–365 Ma) and in the Late Cretaceous-Paleogene (~75–50 Ma). In the Lashburn well (Fig. 8), temperatures reached ~85 °C and well within the apatite partial annealing zone in the late Paleozoic. This heating was followed by

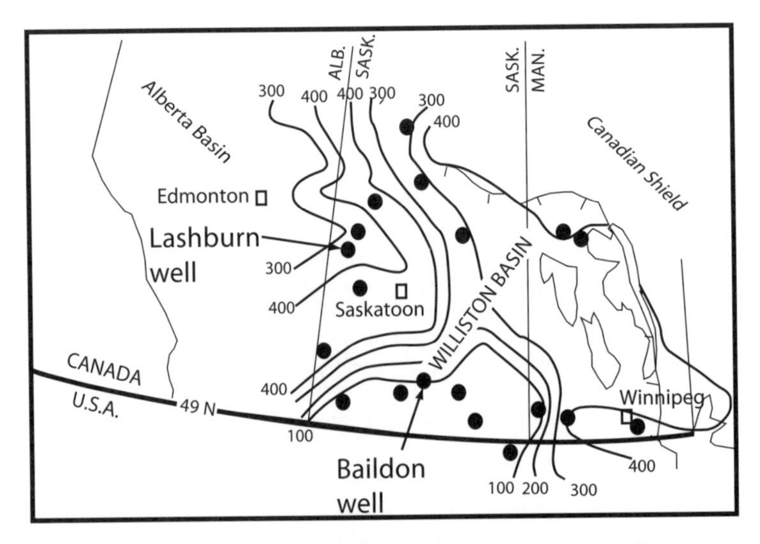

Figure 7. Map showing location of Williston Basin in Canada and United States. Dots show locations of wells that penetrate to basement. Wells "Lashburn" and "Baildon" are discussed in text and in Figure 8. Contours are apatite fission-track ages for basement samples. Modified after Osadetz et al. (2002).

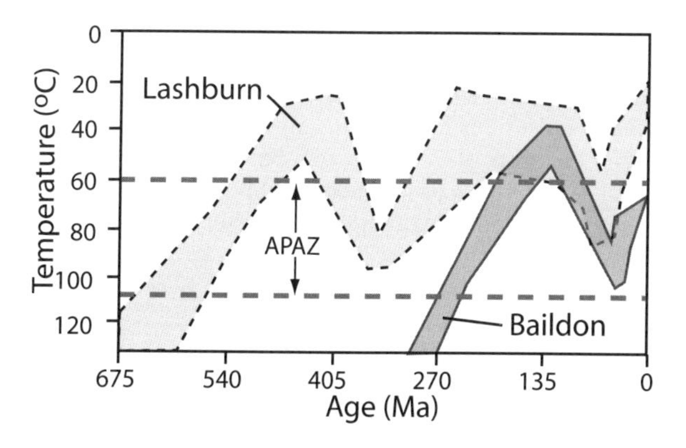

Figure 8. Thermal histories based on apatite fission-track data from basement samples in "Lashburn" and "Baildon" wells in Williston Basin—see Figure 7 for locations. APAZ is apatite partial annealing zone. Thermal histories computed using Monte Trax program described by Gallagher (1995). Modified after Osadetz et al. (2002).

an episode of cooling, then additional heating in the Late Cretaceous. In the Baildon well, temperatures were greater than 120 °C in the late Paleozoic, decreased to about 50 °C in the Cretaceous, and increased again at the end of the Cretaceous. Thus, in both wells the basement samples would have been partially or completely annealed during the late Paleozoic.

A burial history was constructed from known regional stratigraphic thicknesses for the area around the Baildon well (Fig. 9). Only the top of the basement horizon is shown in Figure 9, but it is clear that the maximum burial occurred in the Late Cretaceous – Early Tertiary when basin depths reached about 3 km. During the time period when fission-track models indicate temperatures >120 °C in the late Paleozoic, maximum burial depths were <2 km. For heat

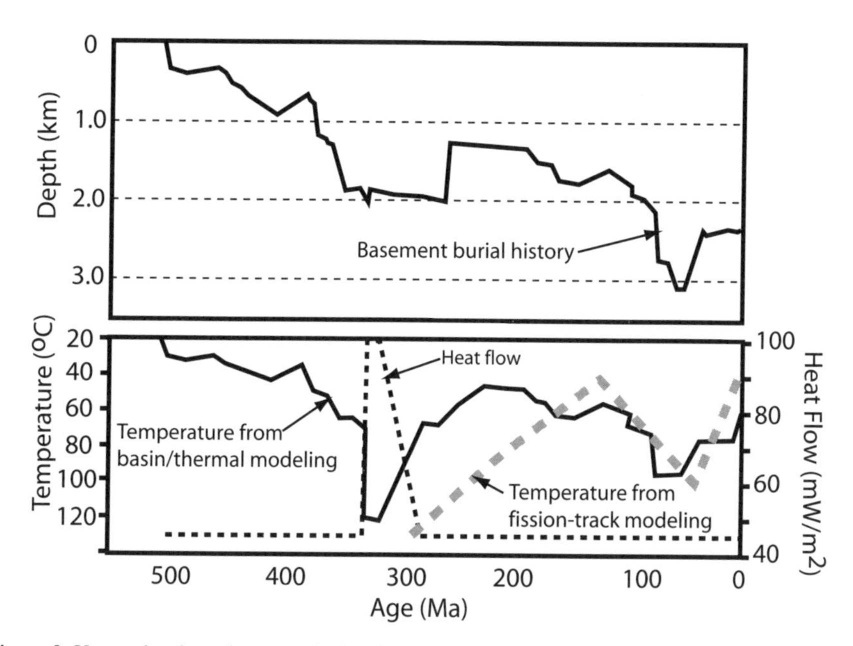

Figure 9. Upper plot shows basement burial history for the Baildon well in Williston Basin. Lower plot shows temperature history based on the burial history and the heat flow condition that includes high heat flow in late Paleozoic (dashed thin line). Thick gray dashed curve in lower plot is acceptable temperature history from fission-track modeling. Modified after Osadetz et al. (2002).

flow values similar to those at present, temperatures must have been on the order of 50–60 °C at these depths. Clearly, the fission-track data require higher heat flow and higher geothermal gradients during the late Paleozoic; nearly 100 mW/m^2, or nearly twice the present-day heat flow, is necessary in the late Paleozoic in order to increase basin temperatures at this time and to satisfy the fission-track constraints. Osadetz et al. (2002) used a basin modeling program (BasinMod™) to construct a temperature history based on the burial history of the well and the proposed heat flow history (Fig. 9). The computed temperature history (solid curve in lower plot of Fig. 9) shows temperatures of ~120 °C in the late Paleozoic and ~90 °C in the Late Cretaceous-Early Tertiary; this temperature history is very similar to that required by the fission-track data in the well. Therefore a varying heat flow history that was significantly elevated in the late Paleozoic was interpreted for the Williston Basin by Osadetz et al. (2002).

The elevated late Paleozoic heat flow has significant implications for the petroleum system in the basin as well as its geodynamic history. The fission-track data and modeling imply that the lower Paleozoic source rocks in the basin would have been in the oil window in the late Paleozoic, which is quite different from previous models that place the earliest hydrocarbon generation in the Late Cretaceous. Osadetz et al. (2002) proposed that the high heat flow was caused by dynamic mantle upwelling starting in the middle Devonian, which caused lithospheric weakening that lead to the rapid subsidence seen at ~400 Ma in Figure 9.

Example integrating burial history with AFT data in an active-margin basin

Taranaki Basin is located both onshore and offshore along the western side of the North Island of New Zealand (Fig. 10), straddles the plate-boundary zone between the Indo-Australian plate and the Pacific plate, and has been variably affected by plate-boundary deformation during Neogene to Recent times. The basin initially developed during the Cretaceous by a rifting event that separated New Zealand from Australia resulting in extensive deposition in fault-controlled sub-basins (King and Thrasher 1996). The deep (>5.5 km) petroleum well Kapuni Deep-1 is located onshore in Taranaki Basin (Fig. 10) and is chosen as a test case for discussion of thermal history modeling combined with thermochronometer data and detailed basin modeling.

Figure 10 shows a burial-thermal history for Kapuni Deep-1 using input constraints that include: (1) burial history from detailed well and seismic data; (2) thermal properties for all sedimentary units; (3) porosity-depth relations; and (4) estimates of paleo-surface temperature based on oxygen isotope and paleontologic evidence. A one-dimensional finite element model ("Bassim" modified from Willett 1988) was used to compute the transient conductive thermal field during sediment deposition and erosion, sediment compaction, sill intrusion, thrusting, and syndepositional and synerosional crustal thickening and thinning (Armstrong et al. 1996). The lower model boundary condition (basal heat flow) was placed at a depth of 40 km to couple basin thermal effects to deeper thermal conditions.

Initial conditions include Late Cretaceous rifting (Weissel and Hayes 1977; Armstrong et al. 1996), which caused initial rapid subsidence and as much as 5 km of Upper Cretaceous-Paleocene sediments to be deposited (Fig. 10). Slower subsidence followed until the late Miocene, when ~300 m of denudation occurred due to uplift along local faults. Rapid subsidence ensued during the Pliocene, followed by ~500 m of rapid Pleistocene erosion during North Island uplift.

Bottom-hole temperatures in the well increase systematically to a depth of about 4.5 km where the down-hole gradient increases. Apatite fission-track ages show a very rapid change from >80 Ma at depths of about 3.5 km to <1 Ma at 5.2 km depth (Fig. 10). Chlorine contents from the three deepest samples average 0.3 wt.%, with 75% having less than 0.4 wt.%, but not all dated grains were probed; if this study were done today, each grain would be analyzed

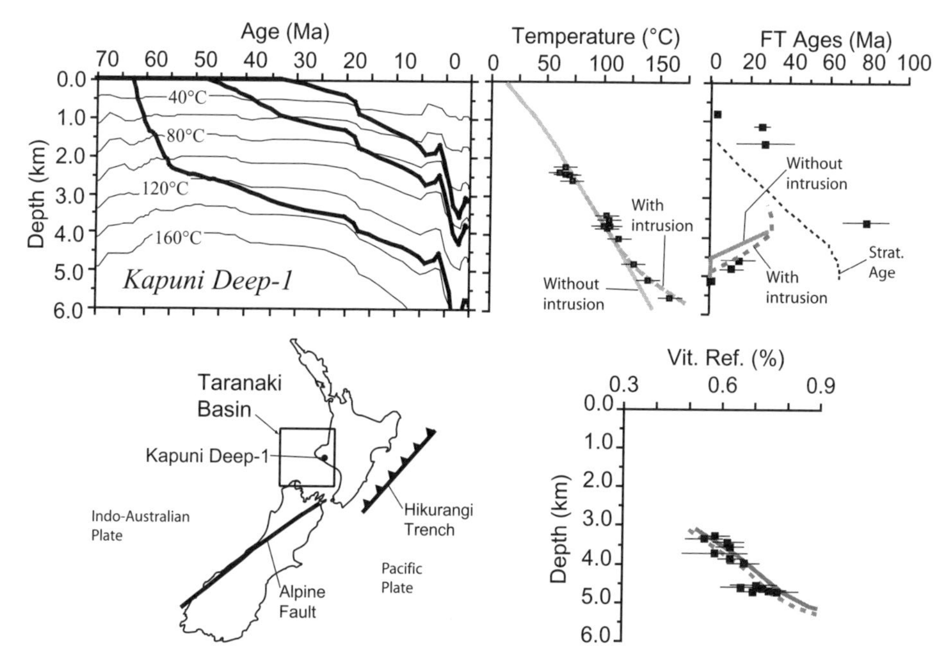

Figure 10. Burial/thermal history and output parameters temperature, apatite fission-track, and vitrinite reflectance data for well Kapuni Deep-1 in Taranaki Basin, New Zealand. Burial history shows three stratigraphic horizons (bold curves) and isotherms (thin curves). Solid curves in the temperature, fission-track, and vitrinite reflectance plots are model values for case of tracking thermal history with burial and erosion only. Thick dashed curves are for case of including sill intrusion below the sediments in the last <1 M.y. Map shows Taranaki Basin in New Zealand in relation to the plate tectonic configuration and well location discussed in this paper. Modified after Armstrong et al. (1996).

using D_{par} as a proxy for composition as discussed earlier. Thus, the interpretations listed below are acceptable, but may not be the only acceptable ones. The >80 Ma sample is older than the stratigraphic age indicating that its age reflects a significant provenance signal and that it has not been substantially reset since the Late Cretaceous. At depths of 4.7 km the AFT ages are ~15 Ma, yet the temperature is 120–125 °C; apatites typical of most grains in Taranaki Basin should be completely annealed if temperature is held at these temperatures for significant time. The three deepest samples may be part of a relatively narrow partial annealing zone (<1000 m wide); this interpretation is consistent with a high geothermal gradient in the deeper parts of the section, as is presently the case in the well. Vitrinite reflectance increases systematically with depth from 0.5% to 0.8%.

The annealing model of Willett (1992), which uses the Arrhenius model of Crowley et al. (1991), was used to predict the AFT ages and the model of Burnham and Sweeney (1989) was used to predict vitrinite reflectance for the burial history. The solid curves in the temperature, AFT age, and vitrinite reflectance plots (Fig. 10) show the model results for the case of tracking temperature change with burial and erosion only. The modeled present-day temperatures do not show the increase in gradient in the deeper parts of the well. Modeled fission-track ages are completely annealed (zero age) at 4.6 km where measured age is 16 Ma. Apatite compositions suggest that the apparently old AFT ages are not related to high chlorine composition.

To account for increased temperature gradient and the poor fit to the AFT data, a simulation was ran that includes shallow intrusion in the last <1 Ma (Armstrong et al. 1997). An intrusion model is reasonable for this area because the Kapuni Deep-1 well is in line with and in the direction of a younging set of Quaternary volcanic edifices on the Taranaki Peninsula. Intrusion associated with the volcanic activity may be responsible for locally high heat flow farther north on the peninsula (Funnell et al. 1996). The intrusion model includes the transient thermal effects of instantaneous horizontal sill intrusion as outlined by Ehlers (2005) and shows good fits to both the temperature and AFT data (Fig. 10). Vitrinite reflectance data show essentially the same fit with both the intrusion and non-intrusion model, indicating that the vitrinite reflectance data are less sensitive to the short-term transient thermal effects than are the AFT data. The interpretation is that both data sets (temperature and AFT) are being affected by a thermal wave propagating up from below; given enough time, the increased temperature gradient will smooth out and the AFT samples at >4.5 km will be completely annealed.

This example illustrates the power and importance of utilizing multiple data sets and basin modeling techniques to extract a detailed thermal history from sedimentary basins. However, additional low-temperature thermochronometer data from apatite (U-Th)/He dating and additional kinetic data on all AFT-dated grains would help constrain the timing of potential intrusion heating further, or provide insight into other potential interpretations that may not include intrusion at all.

A complex history example – constraining structures with outcrop and well data

Low-temperature thermochronometry data from sedimentary basins are commonly used to back-out complex structural histories from both well and outcrop samples. One such case history, based on the work of O'Sullivan and Wallace (2002), in the E-W trending Sadlerochit Mountains of the northeastern Brooks Range (Fig. 11), northern Alaska is illustrated below. The northeastern Brooks Range has been intensely folded by a series of deformational events probably associated with convergence between the North America and Pacific Plates about 1200 km to the south.

Deformation in the main part of the Brooks Range included shortening associated with multiple allochthon emplacements in the late Mesozoic. This shortening is thought to have propogated northward into the Sadlerochit Mountains area in the Paleocene and Eocene. Structures of the northeastern Brooks Range generally comprise a set of north-verging folds and thrusts that form a complex array of fault-bend folds, fault-propagation folds, duplexes, and horses. However, the sequence and timing of thrusting in the Sadlerochit Mountains are poorly constrained because syn- and post-tectonic deposits that could be used to date the deformation are lacking. The goal of evaluating fission-track data from deformed sedimentary basin strata, in combination with complex fault/fold relations and vitrinite reflectance data, is to place constraints on the timing and sequence of fault/fold development and their relations to basement involvement in the deformation style.

O'Sullivan and Wallace (2002) present an extensive fission-track data set for the region around the Sadlerochit Mountains and from one well located ~20 km west of and along trend with the Sadlerochit Mountains. Only a subset of the outcrop data are discussed here. Figure 12 shows apatite and zircon fission-track ages for from Triassic–Tertiary age clastic sediments from the north flank of the Sadlerochit Mountains. The data are plotted to represent a stack of upward-younging sediments as if they were collected from a well. The apatite fission-track ages are all younger than their respective stratigraphic ages, but converge toward the stratigraphic age up-section. The upper part of the section yields AFT ages of 44–51 Ma and the lower section yields ages of 26–31 Ma (Fig. 12). Mean track lengths are 14.8 ± 0.1 to 13.9 ± 0.1 μm. Track length distributions are very narrow, with slight tails to shorter track lengths (Fig. 12), indicating that all samples cooled quickly through the apatite partial annealing zone (Fig. 3). Apatite composition was determined via microprobe on a subset of the grains; all the grains probed had <0.4 wt.% Cl indicating a dominantly fluorine-rich apatite

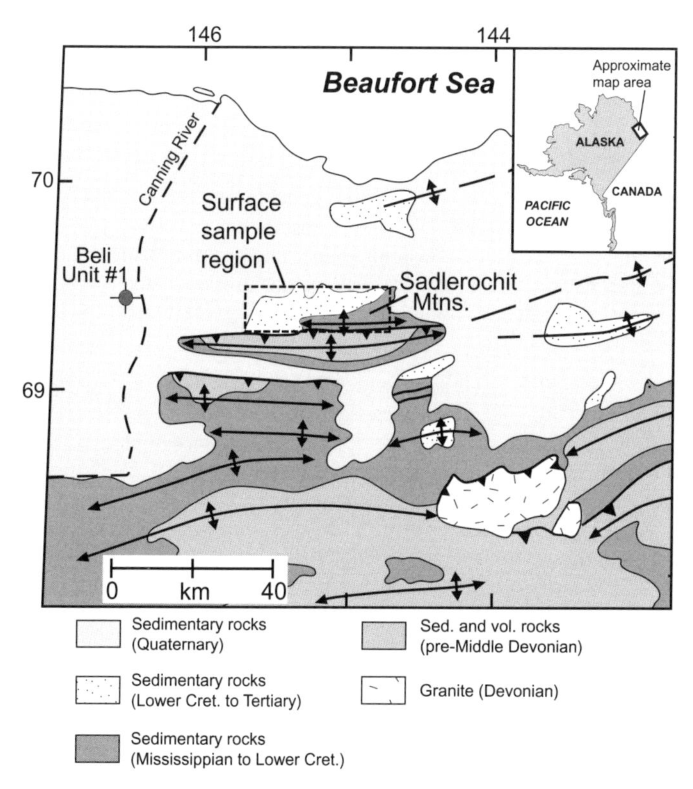

Figure 11. Generalized geologic map showing the locations of the Sadlerochit Mountains and well Beli Unit#1 in northern Alaska. Arrowed curves show locations of major anticlines. Modified after O'Sullivan and Wallace (2002).

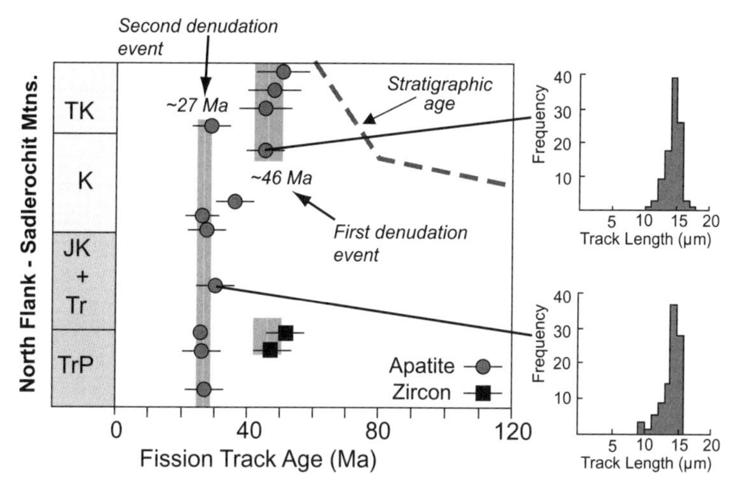

Figure 12. Fission-track data for the north flank of the Sadlerochit Mountains. See Figure 11 for sampling area. T=Tertiary, K=Cretaceous, J=Jurassic, Tr=Triassic, P=Permian. Modified after O'Sullivan and Wallace (2002).

population. Modeling of the age and track length distributions using the annealing model of Laslett et al. (1987) for fluorine-rich apatites indicates that the upper section cooled rapidly to temperatures of less than 50 °C at ~45 Ma. The deeper samples at that time were still hotter than ~110 °C until 27 Ma when they rapidly cooled to less than 50 °C. As the authors point out, the estimated paleotemperatures should be interpreted as maximum values because the actual chlorine contents are less than the 0.4 wt.% used in the Laslett et al. (1987) model.

Two samples in the deeper part of the section yield zircon fission-track ages of 51 Ma and 47 Ma. These ages are much younger than their stratigraphic ages (Fig. 12), thus indicating at least partial resetting of some grains. Assuming a partial annealing temperature range of ~180 to 225 °C for zircon fission tracks, the samples deeper in the section would have cooled to between 180–225 °C and ~110 °C about the 47–50 Ma, approximately the same time as the upper samples cooled to less than 50 °C.

Samples from the well Beli Unit #1, which is located west of and along strike with the Sadlerochit Mountains (Fig. 11) yield ages that decrease with increasing depth. The three deepest samples are younger than the stratigraphic ages, decrease with increasing depth from 52–8 Ma, and are all within the partial annealing zone. Mean track lengths for two of the deepest samples are ~10.6–11.8 μm and distributions are fairly wide, which is characteristic of samples that are within, and perhaps near the base of, the partial annealing zone. Annealing modeling reveals that these samples rapidly cooled 35–45 °C from maximum temperatures of 110 °C about 40–50 Ma. This cooling amount corresponds to ~1.3–1.8 km of depth difference assuming typical geothermal gradient of 27 °C/km. The timing of cooling is concordant with the early phase of cooling in the Sadlerochit Mountains discussed earlier based on surface samples.

Maximum paleotemperatures from the vitrinite reflectance data combined with estimates of paleotemperature from apatite fission-track data in well Beli Unit #1 clearly show a linear trend offset from the present-day temperature gradient (Fig. 6). The offset paleogeothermal gradient line suggests ~1.5 km of exhumation at the well location west of the Sadlerochit Mountains.

The combined apatite fission-track, zircon fission-track, and vitrinite reflectance results from outcrop and well samples indicates two major rapid cooling periods that are interpreted to be the result of major denudation events. These include: (1) approximately 2 km of unroofing at ~45 Ma in the Sadlerochit Mountains and (2) a second rapid cooling event that occurred ~27 Ma as indicated by the samples from deeper in the section. Assuming a constant geothermal gradient, the deeper samples would have passed rapidly through depths of 2–4 km implying more that 2 km of unroofing during this event.

The preferred model of O'Sullivan and Wallace (2002) is that there were two levels of structural detachment that emplaced basement duplexes at ~45 Ma. This duplexing would have elevated the Sadlerochit region enough to cause >2 km of unroofing above the duplex zone. The later unroofing event at ~27 Ma is interpreted to be caused by additional out-of-sequence back-thrusting above the already established duplex system (see O'Sullivan and Wallace 2002). The results of this study indicate between 4 and 8 km of structural thickening probably due to basement-involved thrust stacking occurred in the middle Tertiary. Thus, the thermochronologic data help to identify structural style and timing in a very complexly deformed sedimentary basin.

Additional illustrative examples of AFT analysis in sedimentary basins

Burtner and Negrini (1994) used apatite fission-track data in combination with organic maturation data (vitrinite reflectance and Rock-Eval pyrolysis) to evaluate the thermal history of the Idaho-Wyoming (western U.S.A.) thrust belt, which formed during the Sevier Orogeny in the late Mesozoic and early Cenozoic. Burtner and Negrini (1994) used the distribution of AFT ages from outcrop and well samples to show that gravity-driven fluid flow was responsible for advecting large amounts of heat from the foreland basin towards the east. The relatively hot fluids disturbed the thermal field and caused early hydrocarbon generation. Later, thrust stacking caused meteoric water recharge in high regions which drove cooling of samples along the western part of the foreland basin. Burtner et al. (1994) used refinements to apatite fission-track methods, mainly by correcting ages based on composition and etch pit width (D_{par}) data, to show that Lower Cretaceous source rocks were heated sufficiently to generate hydrocarbons prior to thrust faulting to the west; thrust faulting generally had been assumed to be the cause of most heating and trap formation in the foreland basin.

Cederbom et al. (2004) used apatite fission-track age data to infer climate-induced exhumation of the European Alps. They evaluated detrital apatite from several foreland basin boreholes located in the North Alpine Foreland Basin of Switzerland. Individual grain ages from samples at any depth in the wells vary considerably, but the pooled ages show a consistent decrease with increasing depth. The pooled ages recorded in the deepest samples are interpreted to reflect the bases of exhumed partial annealing zones. The bases of the exhumed PAZs have mean ages of about 5 Ma suggesting that uplift and exhumation of the section began then. This age corresponds to the timing of increased sediment accumulation in nearby depositional centers and with an intensification of the Atlantic Gulf Stream. The erosional power of the climate intensification increased and led to isostatic rebound and uplift of the Swiss Alps and the proximal foreland basin (Cederbom et al. 2004).

O'Sullivan and Brown (1998) were able to relate known past surface temperature changes to AFT ages near the base of a well on the North Slope of Alaska. Their analysis showed that late Cenozoic mean annual surface temperature changes of 10–20 °C affected AFT grain ages at depths of ~3–4 km. One of the important implications of this study is that the subsurface effects of long-term surface temperature changes must be accounted for, if possible, when estimating exhumation magnitudes from AFT and apatite (U-Th)/He data.

Higher-temperature thermochronometers in sedimentary basins

Higher-temperature thermochronometers, those with closure temperatures of greater than about 150 °C, have been and will continue to be used in sedimentary basin analysis. These methods generally include zircon fission-track (Zeitler et al. 1986; Cerveny et al. 1988; Brandon and Vance 1992; Carter 1999; Garver et al. 1999; Spiegel et al. 2000; Bernet et al. 2001; 2004; Bernet and Garver 2005), $^{40}Ar/^{39}Ar$ on K-feldspar (e,g., Harrison and Be' 1983; Harrison and Burke 1989; Lovera et al. 1997; Mahon et al. 1998; Harrison et al. 2005), $^{40}Ar/^{39}Ar$ on white mica (e.g., Copeland and Harrison 1990; Najman et al. 1997; White et al. 2002; Carrapa et al. 2003; 2004; Hodges et al. 2005) , or zircon (U-Th)/He dating (e.g., Rahl et al. 2003).

Harrison and Burke (1989) provide early examples of using $^{40}Ar/^{39}Ar$ thermochronology on microcline, based on age spectrum work of Harrison and Be' (1983), to demonstrate the utility of extracting thermal history constrains for geodynamic models. More recently, Mahon et al. (1998) applied $^{40}Ar/^{39}Ar$ thermochronometry, using a multi-diffusion domain model and corrected for excess ^{40}Ar, to evaluate detrital K-feldspars from a deep well in the San Joaquin Valley, California. Their results are consistent with roughly linear heating of the sedimentary section during the early and middle Miocene, followed by more rapid heating between 9–6 Ma. The thermal model generated from their $^{40}Ar/^{39}Ar$ study allowed assessment of porosity and insights into the pore fluid evolution in the overlying sediments, which are major hydrocarbon producers in the San Joaquin Valley.

A major advance in the last decade in sedimentary basin thermochronometer utility has been in higher-temperature detrital thermochronometry, which mainly provide information on the provenance region(s) for the sediments. Detrital thermochronometry methods are reviewed in earlier chapters by Bernet and Garver (2005) and Hodges et al. (2005). Apatite fission-track data generally are less useful for detrital studies because of their lower annealing temperatures and larger relative errors on ages, but Carter and Gallagher (2004) argue that useful source region thermal history information can be recovered from samples heated to as high as 100 °C after burial.

Information gleaned from higher-temperature thermochronometry studies generally includes the numbers of source region exhumation episodes as well as their timing and rates of exhumation (e.g., Bernet et al. 2001, Ruiz et al. 2004). The data also can provide information on long-term stability (i.e., steady-state) of landscape in the source region (e.g., Willett and Brandon 2002; Ruiz et al. 2004). Regardless of the method (fission-track, $^{40}Ar/^{39}Ar$, helium), the technique generally involves measuring ages on many detrital grains (50–100) and decomposing the ages into peaks (e.g., Brandon 1992;1996). Figure 13 shows an example from the Olympic sudduction complex where the distribution of ages shows a main peak at 20 Ma and a broad composite peak centered around ~50 Ma. Two smaller peaks are superimposed on the broad peak at ~45 and 60 Ma. The main peaks then are fitted using Gaussian peak fitting to show the best-fit peaks (Fig. 13).

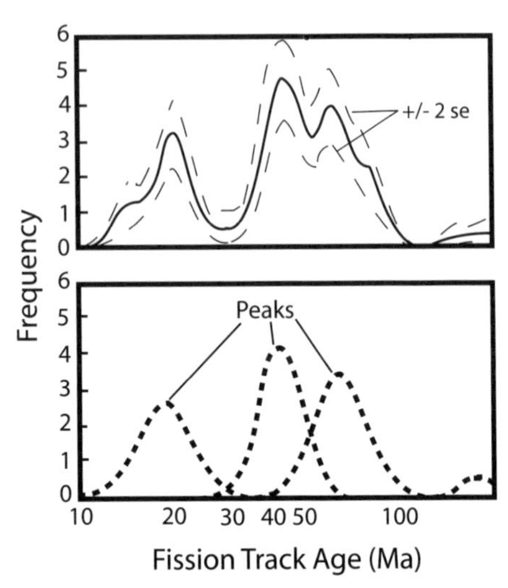

Figure 13. Zircon fission-track age frequency distributions from the Olympic subduction complex with two standard error bounds (upper plot). Lower plot show the age distribution decomposed using Gaussian peak fitting (e.g., Brandon 1992; 1996). Three peaks are present in this example.

Once the peaks are determined from each sample, they can be plotted against known depositional age to determine the "lag time" between when the mineral groups passed through their respective closure temperatures and were deposited in the sedimentary basin. In general, detrital ages increase with increasing depositional age reflecting the sequential exhumation of a source region and burial in the sedimentary basin. Ruiz et al. (2004) define five lag time paths for the relations between detrital age and depositional age. These include the following. (1) Decreasing lag time where the detrital and depositional ages converge up-section. This path indicates increasing source region exhumation with time. (2) Constant lag time where difference in age is constant throughout the section. This path indicates constant source region exhumation rate. (3) Increasing lag time up section, but detrital ages get younger up section. This scenario indicates slowing of exhumation with time. (4) Invariant detrital age up section – indicates major component of volcanic detritus. (5) Detrital ages increase up section. This path indicates possible volcanic contributions, cannibalism of non-reset grains that were deposited previously and re-eroded, and/or changes in source region tectonics.

Figure 14 is an example of detrital zircon fission-track peak ages and their respective depositional ages from the Northern Apennines, Italy (Bernet et al. 2001). The detritus in the sedimentary basin deposits is known to have been derived from the central and western Alps. The angled lines in the plot represent iso-lag time lines; ZFT peak age versus depositional age pairs that plot on one these lines will give the lag time for that sample's time of deposition. In this example, two main source regions are interpreted with lag times of 8 and 16 m.y. The

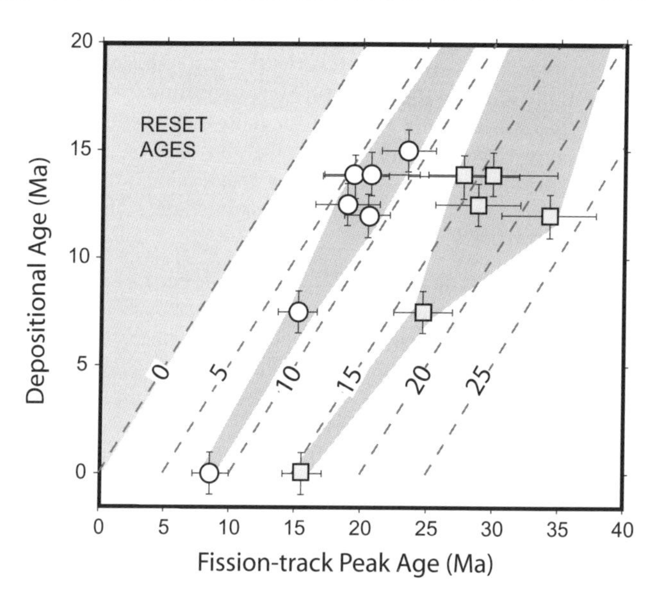

Figure 14. Detrital zircon fission-track peak ages versus depositional age of the stratigraphic unit from which they were extracted in the Northern Appenines, Italy. The area labeled "RESET AGES" represent area where fission-track ages are younger than stratigraphic ages. Diagonal dashed lines represent iso-lag times in million years. In this example, two peak sets are evident at lag times of 8 and 16 m.y. Modified after Bernet et al. (2001).

constant lag time with increasing depositional age for both of the peak sets suggests that the Alps are maintaining exhumational steady-state (Bernet et al. 2001).

Multi-method approaches using different mineral or dating methods are commonly being used to glean more information from detritus in sedimentary basins. Spiegel et al. (2004) use zircon fission-track data in combination with Nd isotope ratios in epidote to evaluate the exhumation history of the central Alps. Carter and Moss (1999) combined U/Pb and zircon fission-track dating to link exhumation and sedimentation in Thailand. Rahl et al. (2003) use zircon (U-Th)/He and U/Pb dating of the same detrital crystals to evaluate source regions for the Navajo Sandstone in Utah.

CONCLUSIONS AND FUTURE DIRECTIONS

Sedimentary basins and their deposits cover most of the solid surface of Earth. The sedimentary layers contain a history of depositional and erosional magnitudes, rates, and timing, which can be combined with modern modeling and thermochronologic techniques to evaluate other factors that influence the basin's history. Low-temperature thermochronometers, mainly apatite fission-track dating, have been used for a couple decades to help constrain thermal histories of sedimentary basins. The last ten years has seen the increased use of apatite fission-track dating in combination with vitrinite reflectance data to provide powerful constraints on past temperature and heat flow changes and magnitude and timing of burial and unroofing events. The temperature changes recorded by the thermochronometers allow assessment of the timing and duration of hydrocarbon production, expulsion, and migration to greatly enhance prospectivity of hydrocarbon systems. Paleoheat flow changes, which can be identified by apatite fission-track parameters, can be used to evaluate tectonic processes responsible for the basin's formation (e.g., Williston Basin). Thermochronometers provide important checks on modeled basin/thermal histories and allow assessment of past temperatures not evident in the present-day temperature field alone. For example, low-temperature thermochronometry data can provide evidence of past localized intrusion or fluid flow events that may have waned after causing basin temperatures to rise or heating that may still be propagating through the basin system. Low-temperature thermochronometers provide important constraints on basin inversion magnitude and timing as well as constraints on timing and style of deformation associated with major tectonic events.

A major advancement in the field of AFT dating has been the appreciation that apatites span a wide range of compositions and that this range in composition leads to a larger range in annealing temperatures than assumed in many studies, from perhaps ~50 °C to ~150 °C. The lower and upper bounds on the PAZ depend on each apatite's composition and kinetic properties. Thus, to extract the most information out of AFT data sets, especially those from sedimentary basins, future studies should include composition/kinetic data on each grain for which an age is determined. Labs have recently begun to assess relative composition/kinetics of apatites by measuring D_{par} (etch pit width parallel the crystallographic *c*-axis). These data should continue to be collected and used in modeling steps to assess acceptable thermal histories.

A significant advance in the last decade has been the combined use of thermochronometers, especially the lowest temperature methods of apatite fission-track and (U-Th)/He dating. The combination of these techniques is already showing great promise in constraining the lowest temperature part of the temperature history. Much effort is being put into combining these techniques, and future directions certainly should include continued efforts to find new ways to integrate these two powerful methods in terms of modeling temperature histories from the collective data sets and finding novel ways of interpreting them. Future directions should also include combined seamless thermochronometer modeling with burial/thermal history modeling in software packages.

The last decade also has seen a growth in using the rich stratigraphic record in sedimentary basins with thermochronometers to assess processes other than the thermal history of the basin itself. The use of higher-temperature thermochronometers such as zircon fission-track, zircon (U-Th)/He, $^{40}Ar/^{39}Ar$, and U-Pb has become increasingly more powerful for assessing source region exhumation events and their causes. These methods are, and will continue to be, used in combination with one another to take advantage of each method's strengths and differences in closure temperatures.

ACKNOWLEDGMENTS

I thank the donors of The Petroleum Research Fund, administered by the American Chemical Society, for support of this effort. S. Perry helped with figure and reference preparation. T. Ehlers, P. Kamp, and, especially, P. O'Sullivan provided reviews of this chapter.

REFERENCES

Ahern JL, Mrkvicka SR (1984) A mechanical and thermal model for the evolution of the Williston Basin. Tectonics 3:79-102

Allen PA, Allen JR (1990) Basin Analysis - Principles and Applications. Blackwell, London

Armstrong PA, Chapman DS (1998) Beyond surface heat flow: An example from a tectonically active sedimentary basin. Geology 26:183-186

Armstrong PA, Chapman DS, Funnell RH, Allis RG, Kamp PJJ (1996) Thermal modelling and hydrocarbon generation in an active-margin basin: The Taranaki basin, New Zealand. Am Assoc Petrol Geol Bull 80: 1216-1241

Armstrong PA, Kamp PJJ, Allis RG, Chapman DS (1997) Thermal effects of intrusion below the Taranaki basin (New Zealand): Evidence from combined apatite fission track age and vitrinite reflectance data. Basin Research 9:151-169

Arne D, Zentilli M (1994) Apatite fission track thermochronology integrated with vitrinite reflectance. *In*: Vitrinite reflectance as a maturity parameter. Mukhopadhyay PK,Dow WG (eds) American Chemical Society, Washington, DC, p 250-268

Arne DC, Grist AM, Zentilli M, Collins M, Embry A, Gentzis T (2002) Cooling of the Sverdrup Basin during Tertiary basin inversion: implications for hydrocarbon exploration. Basin Res 14:183-206

Bally AW, Snelson S (1980) Realms of subsidence. *In*: Facts and Prinicples of World Petroleum Occurrence. Miall AD (ed) Canadian Society of Petroleum Geologists, p 9-79

Bernet M, Brandon MT, Garver JI, Molitor BR (2004) Fundamentals of detrital zircon fission-track analysis for provenance and exhumation studies with examples from the European Alps. *In*: Detrital Thermochronology: Provenance Analysis, Exhumation, and Landscape Evolution of Mountain Belts. Bernet M, Spiegel C (eds) The Geological Society of America, p 25-36

Bernet M, Garver JI (2005) Fission-track analysis of detrital zircon. Rev Mineral Geochem 58:205-238

Bernet M, Zattin M, Garver JI, Brandon MT, Vance JA (2001) Steady-state exhumation of the European Alps. Geology 29:35-38

Brandon MT (1992) Decomposition of fission-track grain age distributions. Am J Sci 292:535-564

Brandon MT (1996) Probability density plot for fission-track grain-age samples. Rad Meas 26:663-676

Brandon MT, Roden-Tice MK, Garver JI (1998) Late Cenozoic exhumation of the Cascadia accretionary wedge in the Olympic Mountains, northwest Washington State. Geol Soc Am Bull 110:985-1009

Brandon MT, Vance JA (1992) New statistical methods for analysis of FT grain age distributions with applications to detrital zircon ages from the Olympic subduction complex, western Washington State. Am J Sci 292:565-636

Bray RJ, Green PF, Duddy IR (1992) Thermal history reconstruction using apatite fission track analysis and vitrinite reflectance: a case study from the UK East Midlands and Southern North Sea. *In*: Exploration Britain: Geological insights for the next decade. Hardman RFP (ed) Geological Society, p 3-25

Burgess P, Gurnis M, Moresi L (1997) Formation of sequences in the cratonic interior of North America by interaction between mantle, eustatic, and stratigraphic process. Geol Soc Am Bull 108:1515-1535

Burnham AK, Sweeney JJ (1989) A chemical kinetic model of vitrinite maturation and reflectance. Geochim Cosmochim Acta 53:2649-2657

Burtner RL, Negrini A (1994) Thermochronology of the Idaho-Wyoming thrust belt during the Sevier Orogeny: A new, calibrated, multiprocess model. Am Assoc Petrol Geol Bull 78:1586-1612

Burtner RL, Nigrini A, Donelick RA (1994) Thermochronology of Lower Cretaceous source rocks in the Idaho-Wyoming Thrust belt. Am Assoc Petrol Geol Bull 78:1613-1636

Carrapa B, Wijbrans J, Bertotti G (2003) Episodic exhumation in the Western Alps. Geology 31:601-604

Carrapa B, Wijbrans J, Bertotti G (2004) Detecting provenance variations and cooling patterns within the western Alpine orogen through 40Ar/39Ar geochronology on detrital sediments: The Tertiary Piedmont Basin, northwest Italy. *In*: Detrital Thermochronology: Provenance Analysis, Exhumation, and Landscape Evolution of Mountain Belts. Bernet M, Spiegel C (eds) Geological Society of America, p 67-84

Carter A (1999) Present status and future avenues of source region discrimination and characterization using fission track analysis. J Sed Geol 124:31-45

Carter A, Gallagher K (2004) Characterizing the significance of provenance on the inference of thermal history models from apatite fission-track data-A synthetic data study. *In*: Detrital Thermochronology: Provenance Analysis, Exhumation, and Landscape Evolution of Mountain Belts. Bernet M, Spiegel C (eds) Geological Society of America, p 7-24

Carter A, Moss SJ (1999) Combined detrital-zircon fission-track and U-Pb dating: a new approach to understanding hinterland evolution. Geology 27:235-238

Cederbom CE, Sinclair HD, Schlunegger F, Rahn MK (2004) Climate-induced rebound and exhumation of the European Alps. Geology 32:709-712

Cerveny PF, Naeser ND, Zeitler PK, Naeser CW, Johnson NM (1988) History of uplift and relief of the Himalaya during the past 18 million years; evidence from sandstones of the Siwalik Group. *In*: New Perspectives in Basin Analysis. Kleinspehn KL, Paola C (eds) Springer-Verlag, New York, p 43-61

Copeland P, Harrison TM (1990) Episodic rapid uplift in the Himalaya revealed by $^{40}Ar/^{39}Ar$ analysis of detrital K-feldspar and muscovite, Bengal fan. Geology 18:354-357

Corrigan J (1991) Inversion of apatite fission track data for thermal history information. J Geophys Res 96: 10,347-10,360

Crowhurst PV, Green PF, Kamp PJJ (2002) Appraisal of (U+Th)/He apatite thermochronology as a thermal history tool for hydrocarbon exploration: An example from the Taranaki Basin, New Zealand. Am Assoc Petrol Geol Bull 86:1801-1819

Crowley KD (1993) Lenmodel: a forward model for calculating length distributions and fission-track ages in apatite. Computers and Geosciences 19:619-626

Crowley KD, Cameron M, Schaefer RL (1991) Experimental studies of annealing of etched fission tracks in fluorapatite. Geochim Cosmochim Acta 55:1449-1465

Deming D (1989) Application of bottom-hole temperature corrections in geothermal studies. Geothermics 18: 775-786

Dickinson WR (1993) Basin geodynamics. Basin Research 5:196-197

Donelick RA, O'Sullivan PB, Ketcham RA (2005) Apatite fission-track analysis. Rev Mineral Geochem 58: 49-94

Dow WG (1974) Application of oil-correlation and source-rock data to exploration in the Williston Basin. Am Assoc Petrol Geol Bull 80:1253-1262

Duddy IR (1994) The Otway Basin: thermal, structural, tectonic and hydrocarbon generation histories. *In*: NGMA/PESA Otway Basin Symposium, Extended Abstracts Record. Finalyson DM (ed) Australian Geological Survey Organisation, Canberra, p 35-42

Duddy IR, Green PF, Bray RJ, Hegarty KA (1994) Recognition of the thermal effects of fluid flow in sedimentary basins. *In*: Geofluids: Origin, Migration and Evolution of Fluids in Sedimentary Basins. Parnell J (ed) Geological Society Special Publication # 78, p 325-345

Ehlers TA (2005) Crustal thermal processes and the interpretation of thermochronometer data. Rev Mineral Geochem 58:315-350

Ehlers TA, Farley KA (2003) Apatite (U-Th)/He thermochronometry: methods and applications to problems in tectonic and surface processes. Earth Planet Sci Lett 206:1-14

Farley KA (2000) Helium diffusion from apatite: General behavior as illustrated by Durango fluorapatite. J Geophys Res 105:2909-2914

Farley KA (2002) (U-Th)/He dating: Techniques, calibrations, and applications. Rev Mineral Geochem 47: 819-843

Farley KA, Wolf RA, Silver LT (1996) The effects of long alpha-stopping distances on (U-Th)/He ages. Geochim Cosmochim Acta 60:4223-4229

Ferrero C, Gallagher K (2002) Stochastic thermal history modelling. 1. Constraining heat flow histories and their uncertainty. Mar Petrol Geol 19:633-648

Fitzgerald P, Gleadow AJW (1990) New approaches in fission track geochronology as a tectonic tool: Examples from the Transantarctic Mountains. Nucl Tracks Radiat Meas. 17:351-357

Funnell RH, Allis RG, Chapman DS, Armstrong PA (1996) Thermal regime of the Taranaki basin, New Zealand. J Geophys Res 101:25197-25215

Galbraith RF (1990) The radial plot: Graphical assessment of spread in ages. Nucl Tracks Radiat. Meas 17: 207-214

Galbraith RF, Green PF (1990) Estimating the component ages in a finite mixture. Nucl Tracks Radiat Meas 17:197-206

Galbraith RF, Laslett GM (1993) Statistical models for mixed fission track ages. Nucl Tracks Radiat Meas 21: 459-470

Gallagher K (1995) Evolving temperature histories from apatite fission-track data. Earth Planet Sci Lett 136: 421-435

Gallagher K, Sambridge M (1992) The resolution of past heat flow in sedimentary basins from non-linear inversion of geochemical data: the smoothest model approach with synthetic examples. Geophys J Int 109:78-95

Garver JI, Brandon MT, Roden-Tice M, Kamp PJJ (1999) Exhumation history of orogenic highlands determined by detrital fision-track thermochonology. *In*: Exhumation processes: Normal faulting, ductile flow and erosion. Ring U, Brandon MT, Lister GS, Willett SD (eds) Geological Society, London, p 283-304

Giles MR, Indrelid SL (1998) Divining burial and thermal histories from indicator data: application and limitations: An example from the Irish Sea and Cheshire Basins. *In*: Advances in Fission-Track Geochronology. Van den haute P, De Corte F (eds) Kluwer Academic Publishers, Dordrecht, p 115-150

Gleadow AJW, Belton DX, Kohn BP, Brown RW (2002) Fission track dating of phosphate minerals and the thermochronology of apatite. Rev Mineral Geochem 48:579-630

Gleadow AJW, Duddy IR (1981) A natural long-term annealing experiment for apatite. Nucl Tracks Radiat Meas 5:169-174

Gleadow AJW, Duddy IR, Lovering JF (1983) Fission track analysis; a new tool for the evaluation of thermal histories and hydrocarbon potential. APEA Journal 23:93-102

Gleadow AJW, Fitzgerald PG (1987) Uplift history and structure of the Transantarctic Mountains: New evidence from fission-track dating of basement apatites in the Dry Valleys area, southern Victoria Land. Earth Planet Sci Lett 82:1-14

Green PF, Duddy IR, Bray RJ (1995) Applications of thermal history reconstruction in inverted basins. *In*: Basin Inversion. Buchanan JG,Buchanan PG (eds) Geological Society of London, p 149-165

Green PF, Duddy IR, Gleadow AJW, Lovering JF (1989a) Apatite fission track analysis as a paleotemperature indicator for hydrocarbon exploration. *In*: Thermal History of Sedimentary Basins - Methods and Case Histories. Naeser ND,McCulloh T (eds) Springer-Verlag, New York, p 181-195

Green PF, Duddy IR, Gleadow AJW, Tingate PR (1985) Fission track annealing in apatite: track length measurements and the form of the Arrhenius plot. Nucl Tracks Radiat Meas 79:155-182

Green PF, Duddy IR, Laslett GM, Hegarty KA, Gleadow AJW, Lovering JF (1989b) Thermal annealing of fission tracks in apatite 4: quantitative modelling techniques and extension to geological timescales. Chem Geol (Isotope Geoscience Section) 79:155-182

Harrison TM, Be' K (1983) $^{40}Ar/^{39}Ar$ thermochronology of detrital microcline from the southern San Joaquin basin, California: an approach to determining the thermal evolution of sedimentary basins. Earth Planet Sci Lett 64:244-256

Harrison TM, Burke K (1989) $^{40}Ar/^{39}Ar$ thermochronology of sedimentary basins using detrital K-feldspars: Examplesfrom the San Joaquin Valley, California, Rio Grande Rift, New Mexico, and North Sea. *In*: Thermal History of Sedimentary Basins - Methods and Case Histories. Naeser ND, McCulloh TH (eds) Springer-Verlag, New York, p 141-155

Harrison TM, Zeitler PK (2005) Fundamentals of noble gas thermochronometry. Rev Mineral Geochem 58: 123-149

Hermanrud C (1993) Basin modelling techniques-an overview. *In*: Basin Modelling: Advances and Applications. Dore AG,Auguston JH,Hermanrud C,Stewart DJ,Sylta O (eds) Norwegian Petroleum Society (NPF), p 1-34

Hill KC, Hill KA, Cooper GT, O'Sullivan AJ, O'Sullivan PB, Richardson MJ (1995) Inversion around the Bass Basin, SE Australia. *In*: Basin Inversion. Buchanan JG, Buchanan PG (eds) Geological Society of London, p 525-547

Hodges KV, Ruhl KW, Wobus CW, Pringle MS (2005) $^{40}Ar/^{39}Ar$ thermochronology of detrital minerals. Rev Mineral Geochem 58:239-257

House MA, Farley KA, Kohn BP (1999) An empirical test of helium diffusion in apatite: borehole data from the Otway basin, Australia. Earth Planet Sci Lett 170:463-474

House MA, Kohn BP, Farley KA, Raza A (2002) Evaluating thermal history models for the Otway Basin, southeastern Australia, using (U+Th)/He and fission-track data from borehole apatites. Tectonophysics 349:277-295

Hu S, O'Sullivan PB, Raza A, Kohn BP (2001) Thermal history and tectonic subsidence of the Bohai Basin, northern China: a Cenozoic rifted and local pull-apart basin. Phys Earth Planet Intl 126:231-245

Ingersoll RV, Busby CJ (1995) Tectonics of sedimentary basins. *In*: Tectonics of sedimentary basins. Busby CJ, Ingersoll RV (eds) Blackwell Science, Oxford, p 1-51

Issler DR (1992) A new approach to shale compaction and stratigraphic restoration, Beaufort-Mackenzie basin and Mackenzie Corridor, northern Canada. Am Assoc Petrol Geol Bull 76:1170-1189

Issler DR (2004) AFTINV: An inverse multi-kinetic annealing model for apatite fission track thermal history reconstruction. 10th International Fission Track Dating and Thermochronology: 48

Issler DR, Willett SD, Beaumont C, Donelick RA, Grist AM (1999) Paleotemperature history of two transects across the Western Canada Sedimentary Basin: Constraints from apatite fission track analysis. Bull Can Petrol Geol 47:475-486

Kamp PJ, Green PF (1990) Thermal and tectonic history of selected Taranaki basin (New Zealand) wells assessed by apatite fission track analysis. Am Assoc Petrol Geol Bull 74:1401-1419

Kamp PJJ, Webster KS, Nathan S (1996) Thermal history analysis by integrated modelling of apatite fission track and vitrinite reflectance data: application to an inverted basin (Buller Coalfield, New Zealand). Basin Res 8:383-402

Ketcham RA (2005) Forward and inverse modeling of low-temperature thermochronometry data. Rev Mineral Geochem 58:275-314

Ketcham RA, Donelick RA, Carlson WD (1999) Variability of apatite fission-track annealing kinetics: III. Extrapolation to geological time scales. Am Mineral 84:1235-1255

Ketcham RA, Donelick RA, Donelick MB (2000) AFTSolve: A program for multi-kinematic modeling of apatite fission-track data. Geol Mat Res 2:1-32

King PR, Thrasher GP (1996) Cretaceous-Cenozoic geology and petroleum systems of the Taranaki Basin, New Zealand. Institute of Geological and Nuclear Sciences Limited, Lower Hutt, 6 enclosures

Kohn BP, Feinstein S, Foster DA, Steckler MS, Eyal M (1997) Thermal history of the eastern Gulf of Suez, II. Reconstruction from apatite fission track and $^{40}Ar/^{39}Ar$ K-feldspar measurements. Tectonophysics 283: 219-239

Laslett GM, Gleadow AJW, Duddy IR (1987) Thermal annealing of fission tracks in apatite 2: a quantitative analysis. Chem Geol (Isotope Geoscience Section) 65:1-13

Lerche I (1993) Theoretical aspects of problems in basin modelling. *In*: Basin Modelling: Advances and Applications. Dore AG, Auguston JH, Hermanrud C, Stewart DJ, Sylta O (eds) Norwegian Petroleum Society (NPF), p 35-65

Lim HS, Lee Y, Min KD (2003) Thermal history of the Cretaceous Sindong Group, Gyeongsang Basin, Korea based on fission track analysis. Basin Res 15:139-152

Lippolt HJ, Leitz M, Wernicke RS, Hagedorn B (1994) (U+Th)/He dating of apatite: experience with samples from different geochemical environments. Chem Geol 112:179-191

Lopatin NV (1971) Temperature and geologic time as factors in coalification. Izv Akad Nauk SSSR 3:95-106

Lorencak M, Kohn BP, Osadetz KG, Gleadow AJW (2004) Combined apatite fission track and (U-Th)/He thermochronometry in a slowly cooled terrane: results from a 3440-m-deep drill hole in the southern Canadian Shield. Earth Planet Sci Lett 227:87-104

Lovera OM, Grove M, Harrison TM, Mahon KI (1997) Systematic analysis of K-Feldspar 40Ar/39Ar step-heating experiments: I. Significance of activation energy determinations. Geochim Cosmochim Acta 61: 3171-3192

Magara K (1976) Thickness of removed sedimentary rocks, Paleopore pressure and paleotemperatures, southwestern part of western Canada Basin. Am Assoc Petrol Geol Bull 60:554-565

Mahon KI, Harrison TM, Grove M (1998) The thermal and cementation histories of a sandstone petroleum reservoir, Elk Hills, California. Chem Geol 152:227-256

Marshallsea SJ, Green PF, Webb J (2000) Thermal history of the Hodgkinson Province and Laura Basin, Queensland: Mutiple cooling episodes identified from apatite fission track analysis and vitrinite reflectance data. Austral J Earth Sci 47:779-797

McDougall I, Harrison TM (1999) Geochronology and Thermochronology by the ^{40}Ar/^{39}Ar Method. Oxford University Press, Oxford

McKenzie D (1978) Some remarks on the development of sedimentary basins. Earth Planet Sci Lett 40:25-32

Naeser CW (1979) Thermal history of sedimentary basins in fission track dating of subsurface rocks. SEPM Special Publication 26:109-112

Naeser CW (1993) Apatite fission-track analysis in sedimentary basins-a critical appraisal. *In*: Basin Modeling: Advances and applications. Dore AG, Auguston JH, Hermanrud C, Stewart DJ, Sylta O (eds) Norwegian Petroleum Society, p 147-160

Naeser ND, Naeser CW, McCulloh TH (1989) The application of fission-track dating to depositional and thermal history of rocks in sedimentary basins. *In*: Thermal History of Sedimentary Basins - Methods and Case Histories. Naeser ND, McCulloh TH (eds) Springer-Verlag, New York, p 157-180

Naeser ND, Naeser CW, McCulloh TH (1990) Thermal history of rocks in southern San Joaquin Valley, California: Evidence from fission-track analysis. Am Assoc Petrol Geol Bull 74:13-29

Najman YMR, Pringle MS, Johnson MRW, Robertson AHF, Wijbrans JR (1997) Laser ^{40}Ar/^{39}Ar dating of single detrital muscovite grains from early foreland-basin sedimentary deposits in India: Implications for early Himalayan evolution. Geology 25:535-538

Nielsen SB (1996) Sensitivity analysis in thermal and maturity modelling. Mar Petrol Geol 13:415-425

Nielsen SB, Balling N (1990) Modelling subsidence, heat flow, and hydrocarbon generation in extensional basins. First Break 8:23-31

Osadetz KG, Kohn BP, Feinstein PB, O'Sullivan PB (2002) Thermal history of Canadian Williston basin from apatite fission-track thermochronology-implications for petroleum systems and geodynamic history. Tectonophysics 349:221-249

O'Sullivan PB (1999) Thermochronology, denudation and variations in paleosurface temperature: a case study from the North Slope foreland basin, Alaska. Basin Res 11:191-204

O'Sullivan PB, Brown RW (1998) Effects of surface cooling on apatite fission track data: Evidence or Miocene climatic change, North Slope, Alaska. *In*: Advances in fission track geochronology. Van den haute P, De Corte F (eds) Kluwer Academic Publishers, Dordrecht, p 255-267

O'Sullivan PB, Parrish RR (1995) The importance of apatite composition and single-grain ages when interpreting fission track data from plutonic rocks: A case study from the Coast Ranges, British Columbia. Earth Planet Sci Lett 132:213-224

O'Sullivan PB, Wallace WK (2002) Out-of-sequence, basement-involved structures in the Sadlerochit Mountains region of the Arctic National Wildlife Refuge, Alaska: Evidence and implications from fission-track thermochronology. Geol Soc Am Bull 114:1356-1378

Pagell M, Braun JJ, Disnar JR, Martinez L, Renac C, Vasseur G (1997) Thermal history constraints from studies of organic matter, clay minerals, fluid inclusions and apatite fission tracks at the Ardeche paleo-margin (BAI Drill Hole, CPF Program). J Sed Res 67:235-245

Parnell J, Carey PF, Green PF, Duncan W (1999) Hydrocarbon migration history, West of Shetland: Integrated fluid inclusion and fission track studies. *In*: Petroleum Geology of Northwest Europe. Fleet JJ, Boldy SAR (eds) Proc 5th Conf. Geological Society of London, p 613-625

Poelchau HS, Baker DR, Hantschel T, Horsfield B, Wygrala B (1997) Basin simulation and the design of the conceptual basin model. *In*: Petroleum and Basin Evolution. Welte DH, Horsfield B, Baker DR (eds) Springer, p 5-70

Rahl JM, Reiners PW, Campbell IH, Nicolescu S, Allen CM (2003) Combined single-grain (U-Th)/He and U/Pb dating of detrital zircons from the Navajo Sandstone, Utah. Geology 31:761-764

Ravenhurst CE, Willett SD, Donelick RA, Beaumont C (1994) Apatite fission track thermochronometry from central Alberta: Implications for the thermal history of the Western Canada Sedimentary Basin. J Geophys Res 99:20023-20041

Reiners PW (2005) Zircon (U-Th)/He thermochronometry. Rev Mineral Geochem 58:151-179

Reiners PW, Farley KA (2001) Influence of crystal size on apatite (U-Th)/He thermochronology: an example from the Bighorn Mountains, Wyoming. Earth Planet Sci Lett 188:413-420

Reiners PW, Farley KA, Hickes HJ (2002) He diffusion and (U-Th)/He thermochronology of zircon: Initial results from Fish Canyon Tuff (Colorado) and Gold Butte (Nevada). Tectonophysics 349:247-308

Rorhman M, Andreissen P, van der Beck P (1996) The relationship between basin and margin thermal evolution assessed by FT thermochronology: an application to offshore southern Norway. Basin Res 8:45-63

Ruiz GMH, Seward D, Winkler W (2004) Detrital thermochronology-a new perspective on hinterland tectonics, an example from the Andean Amazon Basin, Ecuador. Basin Res 16:413-430

Sambridge MS, Compston W (1994) Mixture modelling of multi-component data sets with application to ion probe zircon ages. Earth Planet Sci Let 128:373-390

Sclater JG, Christie PAF (1980) Continental stretching: An explanation of the post-mid-cretaceous subsidence of the central North Sea Basin. J Geophys Res 85:3711-3739

Sobel ER, Dumitru TA (1997) Thrusting and exhumation around the margins of the western Tarim basin during the India-Asia collision. J Geophys Res 102:5043-5063

Spiegel C, Kuhlemann J, Dunkl I, Frisch W, von Eynatten H, Kadosa B (2000) Erosion history of the Central Alps: evidence from zircon fission track data of the foreland basin sediments. Terra Nova 12:163-170

Spiegel C, Siebel W, Kuhlemann J, Frisch W (2004) Toward a more comprehensive provenance analysis: A multi-method approach and its implications for the evolution of the Central Alps. *In*: Detrital Thermochronology: Provenance Analysis, Exhumation, and Landscape Evolution of Mountain Belts. Bernet M, Spiegel C (eds) Geological Society of America, p 37-50

Stockli DF, Farley KA, Dumitru TA (2000) Calibration of the apatite (U-Th)/He thermochronometer on an exhumed fault block, White Mountains, California. Geology 28:983-986

Tagami T (2005) Zircon fission-track thermochronology and applications to fault studies. Rev Mineral Geochem 58:95-122

Tagami T, O'Sullivan PB (2005) Fundamentals of fission-track thermochronology. Rev Mineral Geochem 58: 19-47

Tissot BP, Welte DH (1978) Petroleum Formation and Occurrence: A New Approach to Oil and Gas Exploration. Springer-Verlag, Berlin

Ventura B, Pini GA, Zuffa GG (2001) Thermal history and exhumation of the Northern Apennines (Italy): evidence from combined apatite fission track and vitrinite reflectance data from foreland basin sediments. Basin Res 13:435-448

Weir LA, Furlong KP (1987) Thermal regimes of small basins: Effects of intrabasinal conductive and advective heat transport. *In*: Sedimentary Basins and Basin Forming Mechanisms. Beaumont C, Tankard AJ (eds) Can Soc Petrol Geol p 351-362

Weissel JK, Hayes DE (1977) Evolution of the Tasman Sea reappraised. Earth Planet. Sci Lett 36:77-84

White NM, Pringle MS, Garzanti E, Bickle M, Najman Y, Chapman H, Friend P (2002) Constraints on the exhumation and erosion of the High Himalayan Slab, NW India, from foreland basin deposits. Earth Planet Sci Lett 195:29-44

Willett S, Brandon MT (2002) On steady states in mountain belts. Geology 30:175-178

Willett SD (1988) Spatial variation of temperature and thermal history of the Uinta Basin Type thesis, Univ. of Utah

Willett SD (1992) Modelling thermal annealing of fission tracks in apatite. *In*: Short Course on Low Temperature Thermochronology. Zentilli M,Reynolds PH (eds) p 42-74

Willett SD (1997) Inverse modeling of annealing of fission tracks in apatite 1:A controlled random search method. Am J Sci 297:939-969

Wolf RA, Farley KA, Kass DM (1998) Modeling of the temperature sensitivity of the apatite (U-Th_/He thermochronometer. Chem Geol 148:105-114

Wolf RA, Farley KA, Silver LT (1996) Helium diffusion and low-temperature thermochronometry of apatite. Geochim Cosmochim Acta 60:4231-4240

Zeitler PK, Herczeg AL, McDougall I, Honda M (1987) U-Th-He dating of apatite: A potential thermochronometer. Geochim Cosmochim Acta 51:2865-2868

Zeitler PK, Johnson NM, Briggs ND, Naeser CW (1986) Uplift history of the NW Himalaya as recorded by fission-track ages of detrital Siwalik zircons. *In*: Proceedings of the Symposium on Mesozoic and Cenozoic Geology. Jiqing H (ed) Geological Publishing House, Beijing, p 481-494

Reviews in Mineralogy & Geochemistry
Vol. 58, pp. 527-565, 2005

Visualizing Thermotectonic and Denudation Histories Using Apatite Fission Track Thermochronology

Barry P. Kohn[1], Andrew J.W. Gleadow[1], Roderick W. Brown[2], Kerry Gallagher[3], Matevz Lorencak[1] and Wayne P. Noble[1]

*[1]School of Earth Sciences
The University of Melbourne
Melbourne, Australia 3010*

*[2]Division of Earth Sciences, Centre for Geosciences
University of Glasgow
Glasgow G12 8QQ, Scotland*

*[3]Department of Earth Science and Engineering
Imperial College London, South Kensington
London SW7 2AS, England*

INTRODUCTION

Thermochronology, the use of temperature-sensitive radiometric dating methods to reconstruct the time-temperature histories of rocks, has proved to be an important means of constraining a variety of geological processes. In general, different depths within the Earth's crust are characterized by different temperature regimes and processes. Within the upper crustal environment, temperature can often be used as a proxy for depth, so that reconstructed cooling histories may reveal a record of rock movement towards the surface. That portion of this process which involves temperature variations within the uppermost ~150–200 °C of crustal depth has been the basis for the application of low temperature thermochronology to a range of interdisciplinary problems in the Earth Sciences. The last fifteen or so years have sparked widespread interest in this field and this proliferation has been driven in part by advances in analytical techniques, numerical modeling, and fundamental changes of perspectives on the significance of radioisotopic ages (e.g., McDougall and Harrison 1999; Gleadow et al. 2002a; Farley 2002). One area of rapidly growing interest, which has provided unprecedented insights in this regard, has been the quantification in time and space of surface processes and shallow crustal tectonism using low temperature thermochronology, often combined with complementary techniques structural analysis, geomorphic, numerical modeling, and cosmogenic isotope studies (e.g., House et al. 1998; Ehlers and Farley 2003; Belton et al. 2004; Ehlers 2005).

One of the best established and most sensitive low temperature thermochronology methods available for reconstructing such histories in the upper ~3–5 km of the continental crust, over time scales of millions to hundreds of millions of years, is apatite fission track (AFT) thermochronology which responds to temperatures of typically <110 ± 10 °C.

As for other thermochronological methods, fission track analysis involves a geological dating technique in which the retention of radioactive decay products is sensitive to elevated temperatures. Monitoring the degree to which a particular dating system has remained closed with respect to retention of the daughter products enables the history of exposure to elevated temperatures in the geological environment to be quantified. In many cases, such thermochronometers give rise to apparent ages, which only rarely relate to the time the system

1529-6466/05/0058-0020$05.00 DOI: 10.2138/rmg.2005.58.20

was initiated. These apparent ages reflect the record of the thermal and tectonic processes which have controlled the evolution of such environments and the resulting long-term denudation patterns at the Earth's surface, rather than the original formation or depositional ages of the rocks involved. In most cases, the apparent AFT ages obtained are "mixed" ages, which reflect some integrated product of the low temperature thermal history of the crust. Only in relatively few situations do they directly date a particular discrete geological event involving rapid cooling. Therefore, the significance of regional AFT patterns is not always obvious and non-specialists have often found such seemingly intractable and unwieldy data difficult to interpret, resulting in an inability to fully visualize the implications of the results.

In this paper we show how large regional AFT data sets assembled from surface samples collected from southeastern Canada, southern and eastern Africa, and southeastern Australia, can be presented in ways that their patterns of variation can be readily understood. This allows useful geological information to be extracted in a format that can be readily combined with other large-scale data sets, e.g., digital elevation and heat flow. The terranes investigated mainly comprise crystalline rocks where conventional stratigraphic markers and cross-cutting relationships which might be useful for reconstructing their regional Phanerozoic tectonic and exhumation history are largely absent. Imaging and visualizing that part of the thermal history information contained in the AFT data therefore provides a regional framework, for quantifying the spatial coherence and variability in the timing and magnitude of cooling and crustal denudation, through a part of geological time, hitherto largely unconstrained in these terranes.

APATITE FISSION TRACK THERMOCHRONOLOGY

The general principles of AFT thermochronology, the interpretation of data and their application to geological problems have been outlined in several works (e.g., Wagner and Van den haute 1992; Brown et al. 1994b; Gallagher et al. 1998; Gleadow and Brown 2000; Gleadow et al. 2002a; Tagami and O'Sullivan 2005; Donelick et al. 2005). The process of annealing and the ability to adequately constrain the thermal response of that behavior through experiments on laboratory timescales is the key to the investigation of thermal histories by fission track studies. Briefly, AFT annealing is a thermally activated process occurring over a range of temperatures typically up to ~100–120 °C over geological time scales. With increased levels of annealing, fission tracks become progressively shorter and once a rock cools to the temperature range of relative track stability the tracks retain most of their full initial length. During annealing, tracks will shorten to lengths largely controlled by the maximum paleotemperature to which they have been exposed, so that fission track lengths can be used to provide a measure of the amount of annealing that has occurred. Each individual track is added by a radioactive decay event at a different time, and thus experiences a different fraction of the thermal history. Hence, the lengths of individual tracks are related to the paleotemperatures experienced by samples over different time intervals. Because of the numerous possible time-temperature paths experienced by a particular sample, it is clear that an AFT age alone can be interpreted in a number of ways (e.g., Gleadow and Brown 2000). Considering the AFT age and length data together however, reflect a combination of the time over which tracks have been retained as well as the thermal history of the host rock over that period. Integration of the age and track length parameters can therefore place rigorous constraints on the history of cooling through the fission track annealing zone, e.g., fast or slow cooling or more complex types (Gleadow et al. 1986). Since apatite is the mineral for which annealing systematics are best understood and because it typically contains uranium in the 1–100 ppm range and is a common accessory mineral in many rock types, AFT thermochronology is almost universally applicable to large areas of the Earth's continental crust.

THERMAL HISTORY MODELING

The kinetics of fission track annealing can be studied at higher temperatures for times ranging from hours to years in the laboratory using controlled heating experiments on fresh, neutron-induced ^{235}U fission tracks. These are essentially identical to the natural ^{238}U tracks used in geological dating. Such laboratory annealing studies have given rise to quantitative models of fission track annealing in apatite (e.g., Laslett et al. 1987; Corrigan 1991; Crowley et al. 1991; Carlson et al. 1999; Donelick et al. 1999). These annealing models can in turn be used to calculate the AFT age and track length distribution with the least amount of variance that would result from any given thermal history on a geological time scale (e.g., Green et al. 1989; Ketcham et al. 1999). In order to extract the most plausible thermal histories from the observed AFT data inversion modeling procedures are used. For modeling purposes various mathematical approaches for sampling time-temperature space have been described (e.g., Corrigan, 1991; Lutz and Omar 1991; Gallagher 1995; Willett 1997) and several software applications have been developed to automate the procedure (e.g., Crowley 1993; Gallagher 1995; Issler 1996; Ketcham et al. 2000; Ketcham 2005). Where possible any additional geological information and temperature information can also be incorporated into models to provide more relevant time-temperature constraints.

For the studies described here we have adopted the approach of Gallagher (1995) which uses the algorithm reported by Laslett et al. (1987) to simulate the time-temperature dependence of fission track annealing in apatite as determined from a detailed set of laboratory experiments. The modeling procedure uses a stochastic search method for exploring a wide range of possible thermal histories with statistical testing of the predicted fission track age and length parameters against the observed values. Since the possible solutions that satisfactorily match the observed data are not necessarily unique, a guided search by means of a genetic algorithm (Gallagher and Sambridge 1994) is used to sort through a large search (typically thousands) of potential thermal history histories. The maximum likelihood or probability of each time-temperature path is assessed, providing rapid convergence towards an optimal fit of the observed data. The model thermal history procedure can be refined to be locally optimal and it is then possible to also define the confidence limits around a path (Fig. 1). The application of such inverse approaches to AFT modeling has generally focused on the thermal history inference for individual samples. Increasingly however, there is a necessity to consider the results of thermal history modeling in a more regional context, using larger sample arrays.

REGIONAL APATITE FISSION TRACK DATA ARRAYS

An important consequence of fission track annealing is that fission track ages in general, gradually decrease from some observed value at the Earth's surface to an apparent value of zero at the depth where no fission tracks are retained. The depth to the base of this fission track annealing zone will depend on the geothermal gradient and the annealing properties of the particular apatites being studied (see below). The shape of an AFT age profile, such as may be obtained in an area of high relief or from a deep borehole, will reflect the thermal history of the rocks as they cooled through the annealing zone. Such profiles will vary for different thermal history styles (Gleadow and Brown 2000).

The importance of such vertical arrays of samples is that they contain more information than that which can be obtained from any individual sample alone (Gallagher et al. 2005). Because of the fixed geometric relationship that the samples have to each other, in most cases they are constrained to have followed essentially parallel temperature-time histories. Such a sampling approach does not lend itself so readily to large continental regions where there

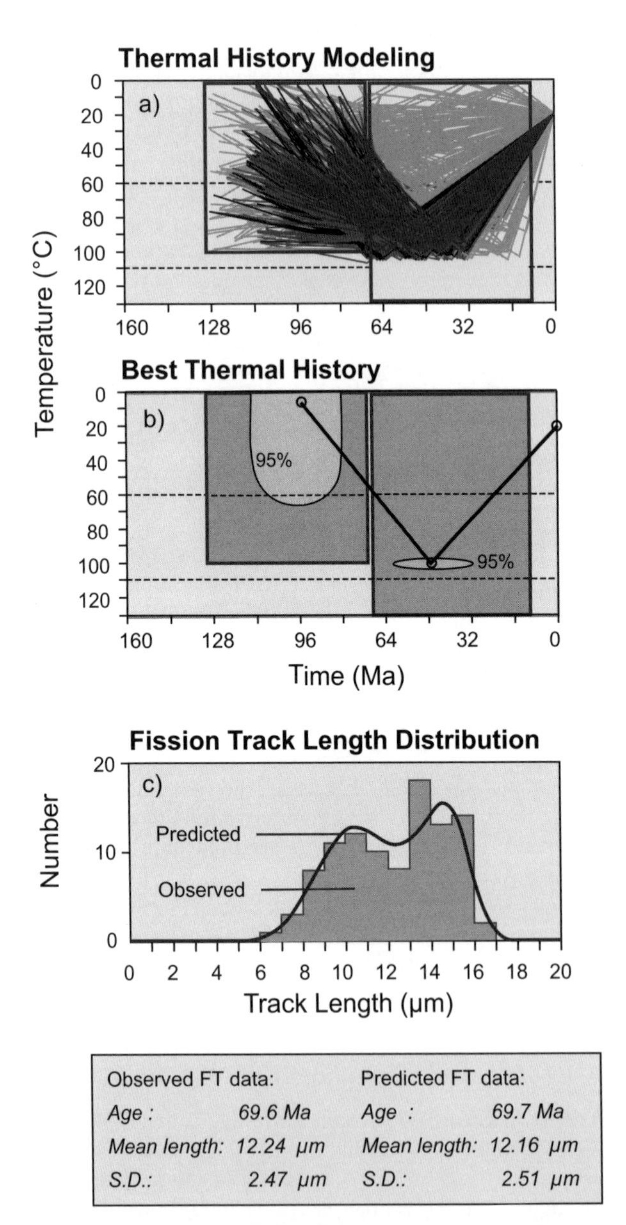

Figure 1. Time-temperature inversion modeling of apatite fission track data. Panel (a) searching time-temperature space, here defined by two boxes, using a Monte Carlo approach for a best-fit two stage thermal history from some 10,000 randomly generated paths. The dark grey paths are those that satisfactorily match the measured apatite fission track data (age, mean confined track length and confined track length distribution). Panel (b) shows the optimal-fit thermal history path determined by employing a guided search by means of a genetic algorithm (Gallagher and Sambridge 1994) and assessment of the maximum likelihood of each path, with the 95% confidence limits around key points on the path. Panel (c) histogram showing the observed track-length distribution compared with that predicted from the optimal-fit thermal history path. The lowermost panel shows a numerical comparison of the input observed fission track data and the predicted estimates of fission-track age, mean track length and standard deviation of the track length distribution arising from the best-fit thermal history path shown in (b) above.

is relatively little relief (or minimal deep borehole data) and for which the assumption of a common cooling history is clearly not appropriate. An alternative approach, suitable for the rapid interpretation of large regional data sets of outcrop samples, is to sequentially model the thermal histories for all samples in an array using a common search strategy. In this case the thermal histories are not constrained by neighboring samples, as they would be in a vertical profile but are free to vary independently from each other. However, using common search parameters (e.g., fission track age, mean track length and track length distribution) and model time-temperature space encourages consistency between the thermal histories for different samples, and will reveal similarities if such information is implicit in the data. The results of this modeling approach can then be interpolated spatially to link the paleotemperatures for individual samples over a consistent set of time steps. Note that this approach does not quantitatively link the thermal histories of nearby samples as in the partition modeling approach described by Gallagher et al. (2005). However, this does represent an end member case in that each sample is effectively allocated a separate partition.

QUANTIFYING LONG-TERM DENUDATION

AFT thermal history information may be related to thermal relaxation following increased heat-flow (for example related to rifting), to localized magmatism, hot fluid flow or to denudation at the land surface. Thermal modeling studies have thus far indicated however, that the direct thermal effects of rifting are unlikely to be significant within the shallow crust environment (~10 km) of the onshore regions of margins (e.g., Buck et al. 1988, Gallagher et al. 1994b, Brown et al. 1994a). The movement of hydrothermal fluids in former cover successions or structural pathways in crystalline basement may influence the AFT pattern in some cases but generally this is viewed as a more localized effect (e.g., Steckler et al. 1993; Duddy et al. 1994; Gleadow and Brown 2000; Gleadow et al. 2002b). Magmatism, also, is mostly restricted locally rather than regionally in the areas reported here. Hence, most cooling in the near-surface environment is dominated by tectonic and erosional denudation. Therefore, a principal assumption in our studies is that it is usually the amount of denudation and the pattern of tectonic offsets that causes the variation in apparent AFT ages at the land surface. Consequently, AFT data can be used to reconstruct regional denudation patterns (e.g., Gallagher and Brown 1999a,b).

Assumptions and uncertainties

Paleotemperatures, heat flow and thermal conductivities. Estimates of long-term denudation are made by converting temperature histories (typically estimated to have an uncertainty of ~10 °C for the paleotemperature at any given point) to an equivalent depth history by making assumptions regarding past heat flow and surface temperatures, as well as the thermal conductivity of the material eroded (Gallagher and Brown 1999a; Gleadow and Brown 2000; Brown et al. 2002). Where vertical profiles are available paleogeothermal gradients may be estimated for time modeled maximum paleotemperatures prior to the onset of cooling (e.g., Brown et al. 2002; Gallagher et al. 2005). For most situations however, where only surface samples are available, it is difficult to constrain past paleogeothermal gradients and the thermal conductivity of the missing section explicitly, unless some form of joint modeling can be used (Gallagher et al. 2005). Region specific geothermal gradients may be extracted from the global heat flow data set of Pollack et al. (1993) and these data suggest that the range of continental gradients in Precambrian and Paleozoic crystalline terranes is relatively restricted (with a mean between ~20–30 °C·km^{-1}) and that large anomalies are usually localized. Further, errors arising from anomalous transient thermal gradients could be significant, but even in extreme cases are unlikely to be greater than a factor of ~2 (Gleadow and Brown 2000). In sedimentary basins where thermal conductivities are generally lower however, geothermal gradients may be more variable and it may also be possible to estimate past gradients in conjunction with vitrinite reflectance studies (e.g., Bray et al. 1992).

Results using a constant as opposed to a spatially variable present day heat flow show that in general, the timing of enhanced episodes of denudation do not differ markedly, although the magnitude may vary (Kohn et al. 2002a; Gunnell et al. 2003; see also Fig. 2). This assumption along with others (e.g., assumed constant thermal conductivity of the eroded section, surface temperature and paleogeothermal gradient) clearly limit the accuracy of long-term denudation rate estimates, which are considered here only as a first-order approximation (see also Brown and Summerfield 1997).

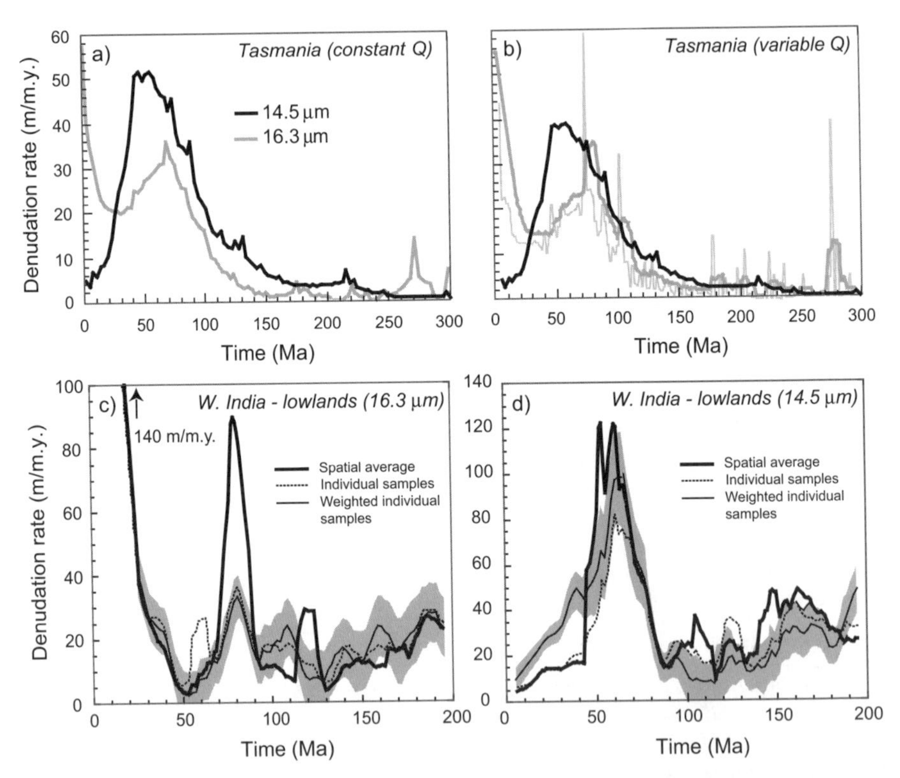

Figure 2. Long-term smoothed (spatially averaged) denudation chronology plots for Tasmania (a and b, modified after Kohn et al. 2002) and the lowlands of western peninsular India (c and d, modified after Gunnell et al. 2003). The spatial average (bold curve in all plots and light curve in plots a and b) is based on interpolating the results over each study area; the individual sample curve (very light curve in plot b and dashed curve, plots c and d) is the unweighted mean of the denudation chronology at individual locations; the weighted sample curve (plots c and d, thin solid line) is the mean denudation at individual locations, weighted by the uncertainty in the inferred thermal history for each location. The shaded area (plots c and d) is the standard error on the weighted mean estimate and is considered to indicate the magnitude of uncertainty inherent in the modeling procedure. Plots (a) and (b) also compare the effects of spatially and temporally constant heat flow (60 mW m^{-2}) and spatially variable but temporally constant (i.e., present day pattern) heat flow (Q). Use of different heat flow parameters leads to differences in magnitude of denudation rates but with no marked change in the timing of periods of accelerated denudation. The effect on denudation chronology of using initial track lengths of 16.3 μm (light line) and 14.5 μm (bold line) are also shown (plots a and b, and plot c versus d). The utilization of initial track lengths of 16.3 μm for thermal history modeling leads to the inference of major cooling and hence dramatically increased denudation rates during the mid to late Tertiary (see text for further discussion).

If paleogeothermal gradients were elevated at the time of a particular period of denudation compared to those of the present day then the actual magnitude of denudation would be lower (as would the calculated long-term denudation rate). For the terranes reported here we have not considered variations of heat flow with time, as we have no constraints on how this may have occurred, except perhaps at some of the rifted continental margins. As mentioned previously however, the influence of rift-related heat flow variations tend to be relatively minor in the rift flanks where some of the onshore samples studied were collected (e.g., southeastern Australia, southern and eastern Africa) and is expected to be even less significant elsewhere.

Compositional variations. Annealing properties of fission tracks in apatite vary with duration of heating, chemical composition (Gleadow and Duddy 1981; Green et al. 1985; Barbarand et al. 2003) and mineralogical properties (Carlson et al. 1999). The total annealing temperature for chlorine-rich apatites for example, occur at higher temperatures ~110–150 °C compared to that in the more common fluorine-rich apatites ~90–100 °C (e.g., Green et al. 1985; Burtner et al. 1994). Although chlorine substitution probably exerts the most important effect, the possible influence of other trace elements (including rare earths) has also been reported (Barbarand et al. 2003).

In the regions studied the rocks sampled are of limited compositional range (mostly granites and granodiorites) and the apatites they contain are mainly fluorine-rich apatite. This has been confirmed by electron microprobe analyses of a representative sampling of grains (mostly <0.2 wt% chlorine) and the qualitative consideration of apatite solubility in that the track etching rate and etch pit size are known to correlate with chlorine content (Donelick 1993; Barbarand et al. 2003). Fast etching grains, likely to be of more extreme composition and hence displaying different annealing properties (Burtner et al. 1994; Carlson et al. 1999), were generally avoided for the AFT analyses presented here. However, in the case study from southern Canada some higher chlorine content apatites were observed and their effect on the measured AFT parameters can be clearly seen (Plate I). For other areas studied we consider that the annealing properties of the apatites analyzed represent a coherent set in terms of their annealing properties and do not depart to a significant degree from the Durango apatite composition of ~0.4 wt% chlorine upon which the Laslett et al. (1987) annealing model is based. It is probable that the average chlorine content will be lower than this Durango apatite value suggesting that paleotemperatures reconstructed on this basis may be too high in some cases, possibly by as much as ~10–15 °C, but are unlikely to be too low.

Modeling strategies. The annealing model of Laslett et al. (1987), the model adopted here to obtain long-term denudation rates, is formulated in terms of the current measured track length normalized to an initial track length. This is referred to as the reduced track length and is defined as $r = l/l_0$, where l and l_0 are the annealed and "unannealed" track lengths, respectively. In applications of the original model formulation, using an initial track length of 16.3 μm, the lack of low temperature annealing typically leads to the inference of major cooling in the geologically recent past (Kohn et al. 2002; Gunnell et al. 2003). This can be alleviated partly by considering what the initial track length parameter represents in these annealing models. Mean spontaneous track lengths from rapidly cooled geological samples rarely exceed ~14.5 to 15 μm while the mean lengths from "unannealed" induced tracks are typically 16.3 ± 0.9 μm (Gleadow et al. 1986; Green 1988). Donelick et al. (1990) showed that the initial length of induced tracks is not constant over very short times and that room temperature annealing occurs in a matter of days. It is clear then, that the generally assumed initial track length is itself a variable, implying that 16.3 μm may not be the relevant unannealed track length over geological timescales. Hence, for geological timescales, spontaneous tracks may effectively be ~10% shorter than that observed on laboratory timescales for induced tracks (Gleadow et al. 1986; Laslett and Galbraith 1996, Ketcham et al. 1999). As a consequence, the Laslett et al. (1987) model, based on the laboratory determined initial length does not appear to predict sufficient annealing at temperatures lower than ~50–60 °C.

As a model calibrated for reduced track lengths depends strongly on the assumed initial track length we have compared the effect of two initial track length values, i.e., 16.3 and 14.5 μm on the modeled thermal histories of two different terranes (Fig. 2). The former value is that inferred for induced tracks, while the latter is consistent with the maximum value typically observed from spontaneous tracks in surface geological samples, assumed to have undergone little post-formation thermal disturbance. In considering regional denudation in this work we have used the latter value. It is acknowledged that this is a departure from an initial length of 16.3 μm upon which the Laslett et al. (1987) model is based and that further refinement and treatment of initial length estimates to account for the observed amount of annealing is required. Strategic approaches to tackle this problem have been outlined by Gunnell et al. (2003).

Data for each sample from southern Africa, eastern Africa and southeastern Australia were modeled to produce optimal data fitting thermal histories. As we are dealing entirely with surface or very shallow samples, we used a scheme to encourage cooling in the thermal history by starting models at high temperature and fixing the present day surface temperature appropriate for the relevant study area. The points in-between can show cooling followed by reheating (e.g., a saw tooth pattern). Rapid temperature variations are damped out as these are poorly constrained by the original data and cannot resolve well the amount of cooling below the subsequent reheating event. Further, any temperature points less than the maximum of a more recent reheating event are moved so that the cooling only proceeds to 5 °C below the subsequent reheating maximum, but such that the rate of the previous cooling event is maintained. If there is more than one point in the cooling episode it is removed, as it makes no difference to the data fit relative to the damped thermal history. This approach implicitly minimizes variations in the thermal history that are unconstrained by the observed data, e.g., multiple episodes of heating and cooling.

Regional-scale imaging

A flow chart showing the various possible inputs and outputs which can be used in applying fission track modeling to the imaging of thermal history, denudation and paleotopography estimates is shown in Figure 3. By combining denudation information with digital elevation data, it is possible to model the evolution of paleotopography. The paleotopography is estimated by "backstacking" the amount of section removed by denudation in a given time period onto the current surface elevation at that location and allowing the back-stacked column to regain isostatic equilibrium (Brown 1991). This is achieved by using a regional flexural isostasy using a thin plate model with an effective elastic thickness of 25 km (Gallagher and Brown 1999b). Estimating the paleotopography between data points requires adding the inferred overburden to a smoothed topographic surface and applying an isostatic correction. Such reconstructed "paleoelevation" estimates need to be interpreted with some caution as they only reflect the passive response to denudation unloading and do not take into account any possible transient episodes of tectonic uplift, subsidence relative to the present land surface or correction for local deformation and/or faulting.

As a consequence of propagating multiple sources of uncertainty from earlier stages, the farther removed the information sought is from the primary fission track data (Fig. 3), the greater will be the cumulative uncertainties associated with them. These include the analytical uncertainties inherent in the data and the annealing model adopted, as well as uncertainties in assumptions regarding paleoheat flow, thermal conductivity, surface temperatures, thermal equilibrium in the crust when converting temperature to depth/denudation and isostatic mechanisms when calculating paleoelevation. Subject to an awareness of these various uncertainties a set of images can be constructed for any particular time-slice for which the

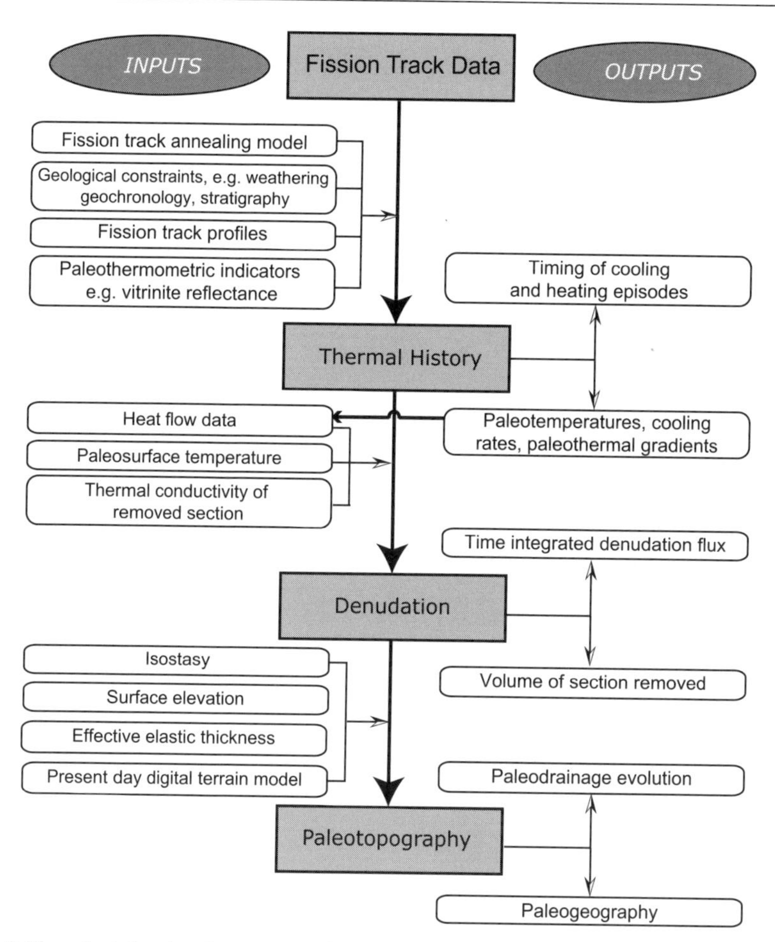

Figure 3. Flow chart showing the sequence of steps and possible inputs which can be used to determine geologically useful outputs from regional apatite fission track data. It is also possible to display as images the regional thermal history, magnitude of denudation and paleotopography for different time slices (see text and Fig. 4 and Plates II-IV). Sources of error are cumulative so that uncertainties increase with each step away from the original apatite fission track data (see text for further discussion on sources of uncertainty) (modified after Kohn et al. 2002; Gleadow et al. 2002a).

modeled temperature remains within the AFT annealing window (Fig. 4). The resulting individual time-temperature solutions can then be "stacked" and visualized as a sequence of regional time-slice images which, when combined into a computer animation depict how temperature, denudation and elevation of present day surface rocks have varied during their passage through the upper crust (e.g., Gleadow et al. 1996; Gallagher and Brown 1999a,b; Kohn et al. 2002). Although such images are extremely useful for visualizing the evolving thermal history for large regional data sets, they can only be considered as broad estimations because the process of bulk modeling and interpolation removes some of the detail that can be obtained during an assessment of the thermal histories of individual samples.

To produce the images, interpolation of the modeled data was carried out with Generic Mapping Tools software (Wessel and Smith 1991) using an adjustable tension, continuous curvature surface gridding algorithm (Smith and Wessel 1990). Interpolation was performed

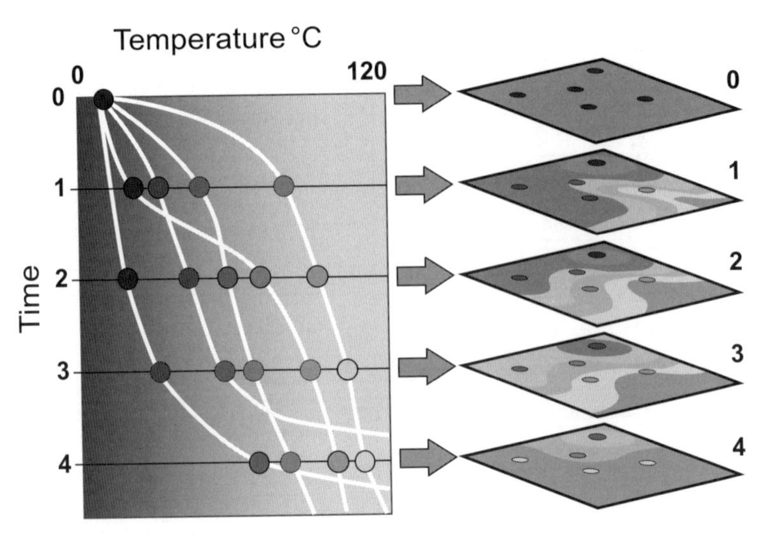

Figure 4. Derivation of paleotemperature images. Following the sequential thermal history modeling of samples within a regional subset, a series of time-temperature curves (white lines, left panel) are derived. Then for different nominal time slices (1-4) paleotemperatures for different samples (filled circles) can be spatially interpolated on a regional map and compiled as a series of time-slice image stacks. Such stacks can then be visualized as a time sequence (and also as a digital movie) showing how temperature of present day surface rocks has varied through time. Such stacks can also be extended to image denudation and paleotopography (see Fig. 3 and text).

across regional data sets for any particular parameter in a time-slice, e.g., paleotopography, but masked to exclude regions where no data were available within a specifically defined distance.

Denudation chronologies

Thermal and denudation histories inferred from AFT data can also be summarized and made more accessible by presenting the results in terms of their spatially integrated denudation rate history. Whereas results from previously published work tend to emphasize the detailed aspects of cooling histories of individual samples, the more regional representation quantifies the average denudation rate as a function of time from multiple samples. This approach takes the denudation chronology inferred for each sample and uses a nearest neighbor interpolation method (Sambridge et al. 1995) to produce a spatial grid of denudation at each time scale. This grid is then integrated spatially at successive time-slices to derive a regional denudation chronology. This scheme satisfies the observed data exactly, and uses weighting based on the spatial distribution of the data points to perform interpolation. This method does not rely on splines (e.g., Mitas and Mitasova 1995), and so does not suffer from the common problem of unconstrained features in the interpolated field. For the case histories presented from Tasmania, southeastern Australia and the lowlands seaward of the Western Ghats escarpment, India (Fig. 2) we interpolated the model results (e.g., paleotemperature, denudation) onto a grid, with a spacing of approximately 10 km. We generated these spatial grids for the results at 2 Ma intervals back to 300 Ma.

One easy way of representing the results is to integrate the spatial denudation information for each time interval, to produce the locally averaged denudation rate as a function of time, or denudation chronology (e.g., Gallagher and Brown 1999a,b). In Figure 2 we represent the denudation chronologies both as the raw integrated estimates and five point smoothing (i.e., over 10 Ma) on these estimates. The smoothing (the spatial average) is used because, although discrete short cooling episodes (and by inference denudation) can occur, there is some uncertainty on the timing, which is not incorporated into the interpolation/integration process. Thus, the amplitude of such rapid cooling/denudation may not be as high as implied by the raw estimates, or as rapid (Fig. 2b-d). The smoothed curves serve to highlight this implicit uncertainty.

Clearly, as we are using interpolation, there will be some uncertainty and possible artifacts in the denudation chronology as a consequence of the interpolation procedure (e.g., Brown et al. 2001). There are various ways to assess this procedure, for example cross-validation and bootstrapping (Efron and Tibshirani 1993) or using kriging as the interpolation scheme (Isaaks and Srivastava 1989). We illustrate one very simple approach for western peninsular India (Fig. 2c,d) where we compare the denudation chronology determined from spatial interpolation and integration, with the denudation chronology averaged over the individual samples, with no interpolation. We also show a weighted average denudation curve, where the weighting is equivalent to the uncertainty on the individual thermal histories for each sample. Our motivation for this is that features, which consistently appear in the different estimates of the denudation chronology, are not likely to be artifacts of the interpolation process.

In making such first-order long-term denudation reconstructions some of the assumptions as detailed above should be borne in mind, as all of them will introduce uncertainties. Such assumptions clearly limit the accuracy of the denudation magnitude and rates, although the timing of enhanced denudation phases appears to be robust to physically reasonable variations in these parameters.

REGIONAL APATITE FISSION TRACK DATA ARRAYS

Southern Canadian Shield – record of a foreland basin across a craton

Geological overview. The Canadian Shield in Ontario consists mainly of the Archaean Superior Province and Proterozoic Southern Province (Fig. 5). To the southeast, the craton is bordered by the 1–1.3 Ga Grenville Province. The geological development of the shield has been discussed by Hoffman (1988), Thurston et al. (1991) and Lucas and St-Onge (1998).

Southeastern Ontario was subjected to rifting during the Late Proterozoic, resulting in the formation of graben structures such as the Ottawa-Bonnechère and Lake Timiskaming grabens (e.g., Kumarapelli 1985 and Fig. 5). Transition to a compressional regime followed during the Paleozoic, characterized by accretion of the Appalachian Orogen to the southeast. Three principal orogenic phases are usually distinguished: the Late Ordovician Taconic orogeny, the Late Devonian Acadian orogeny and the Carboniferous-Permian Alleghenian orogeny. On the shield itself, a number of structural arches developed during the Paleozoic. These represent areas of repeated cratonic uplift and criss-cross the craton in dominant northeast and northwest trends; basement arch movements may have been triggered and controlled by plate motions and related orogenic activity at or beyond the margins of the craton (e.g., Sanford et al. 1985 and Fig. 5). Paleozoic intracratonic basins; Moose River Basin to the north, Williston Basin to the west and Michigan Basin to the south are located in depressions between the arches and surround the craton (Fig. 5), their formation is however poorly understood. While today the basins are deeply eroded, outliers of Mid-Ordovician sediments in Eastern Ontario suggest a greater paleo-extent of sediments and burial of parts of the craton during the Paleozoic.

Present day heat flow of both the Superior and Grenville Provinces is similar, with average values of 42 ± 10 $mW{\cdot}m^{-2}$ and 41±11 $mW{\cdot}m^{-2}$, respectively (Guillou-Frottier et al. 1995; Mareschal et al. 2000). Major spatial and temporal variations are not observed, and the study area can be considered as a single heat flow province (Guillou-Frottier et al. 1995).

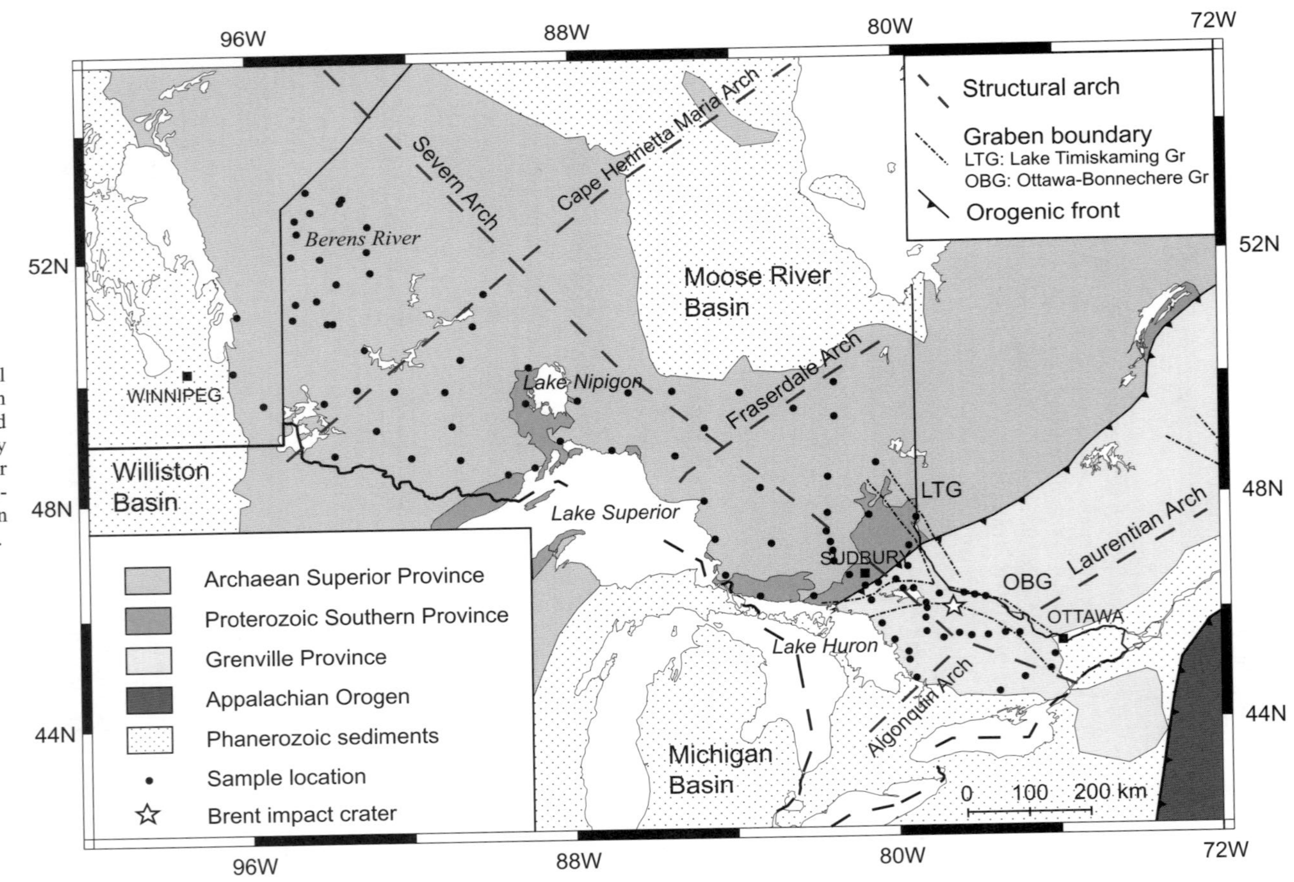

Figure 5. Geological map of the southeastern Canadian shield and adjacent sedimentary basins, showing major tectonic units, structures and apatite fission track sample locations.

Sampling for this regional reconnaissance study focused on the exposed part of the Canadian Shield and the Grenville Province across Ontario, following major roads and extending from easternmost Manitoba to Ottawa (Fig. 5). Fernando Corfu also provided a series of apatite samples from the Berens River province in western Ontario (Fig. 5), from which U-Pb apatite studies had been previously carried out (Corfu and Stone 1998). Regional surface sampling is complemented by 19 samples from a 3440 m deep drillhole in the Sudbury Igneous Complex (Lorencak et al. 2004).

Fission track results. AFT analysis was carried out on 93 samples and the regional distribution of central fission track ages and mean track lengths for apatites are shown in Plate Ia and b respectively. Apparent AFT ages range from ~600–140 Ma and all are considerably younger than the age of crystallization or metamorphism of their host rocks. The oldest ages are found north of Lake Superior, mainly ~500 Ma and these decrease to ~350–400 Ma towards western Ontario and the Berens River area. A similar decrease in age is observed towards the east; northeast of Lake Superior and north of Lake Huron, where most apparent AFT ages fall around ~350–400 Ma. This pattern changes progressively towards southeastern Ontario. There, a relatively rapid decrease in apparent AFT ages across the southern Superior and Grenville Provinces, with the youngest ages of ~140−160 Ma are observed in the vicinity of the present-day sedimentary cover of the shield.

Apatite chemistry was determined by electron microprobe analysis on representative samples and also estimated qualitatively from etch pit diameters in other samples. Most samples are fluorapatite with only a trace of chlorine (up to 0.03 wt%). The exception is from a few samples surrounding Lake Nipigon in central Ontario (Plate I) from the Nipigon diabase, where apatite chlorine content of up to 1.0 wt% was measured.

Mean horizontal confined track lengths (HCTL) range from 13.8–10.5 μm. Most fall into the range of ~11.5–12.2 μm across much of central and western Ontario. One noticeable exception is the few samples from the relatively chlorine-rich apatites of the Nipigon diabase. A second group of relatively long mean HCTL is noticeable in the northeast of the study area along the Fraserdale Arch. Similarly to the apparent AFT ages however, the most prominent change in the mean HCTL pattern is observed towards the southeast, where the increase in mean track lengths mirrors the decrease of AFT ages.

Thermotectonic history. Thermal histories of areas in the study region featuring the oldest apparent AFT ages, immediately north of Lake Superior and between the Cape Henrietta Maria and Fraserdale Arches (Plate Ia) are difficult to constrain due to the lack of independent geological observations. Time-temperature models suggest a mid-Paleozoic heating-cooling event, during which some of the tracks were partially annealed, followed by cooling during the Late Paleozoic. Further north and west, where the apparent AFT ages decrease to ~350−400 Ma, cooling was probably significantly slower and extended into the Mesozoic.

By contrast, in the eastern half of Ontario geological information provides more independent controls for the modeled thermal histories. In the Lake Timiskaming Graben in eastern Ontario, Mid-Ordovician clastic sediments directly overlie the crystalline basement, implying that the present day outcrops of the shield must have been close to the surface in early Paleozoic time. Similarly, undeformed sediments of the same age range fill the ~450 Ma Brent impact crater in southeastern Ontario (Fig. 5). Such constraints can be included into time-temperature models, such as applied to the Sudbury profile (Figs. 5 and 6) (Lorencak et al. 2004).

In the northeastern corner of the study area, along the Fraserdale Arch, a few samples display greater mean track lengths, but with apparent AFT ages in a range similar to the surrounding region (Plate Ia). Models suggest peak paleotemperatures of near-total annealing during Silurian-Early Devonian time, followed by Carboniferous cooling.

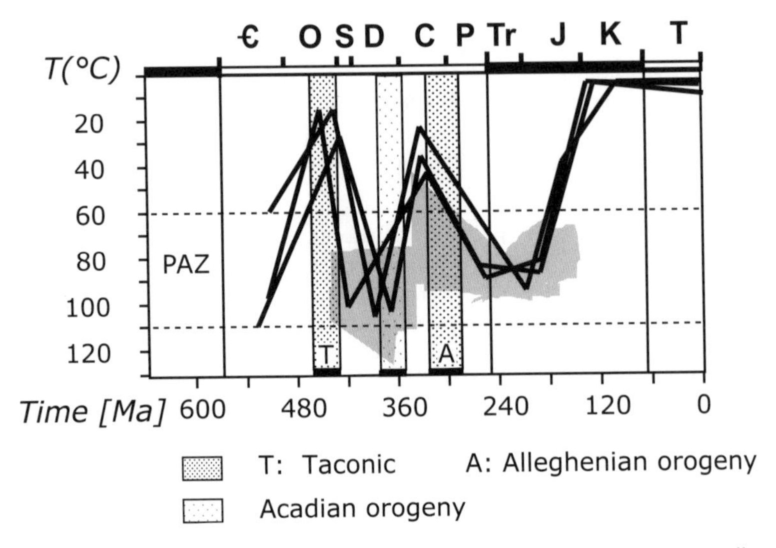

Figure 6. Modeled *t-T* paths from a near-surface sample in the Sudbury drill hole. Three equally possible best-fit paths are shown, all overlapping within 95% confidence intervals (represented by shaded areas) and each determined from 1500 model iterations, following the procedure of Gallagher (1995). Timing of the three main stages of the Appalachian orogeny are shown for comparison. PAZ = apatite partial annealing zone (modified after Lorencak et al. 2004).

Progressing towards the south however, observed track length distributions change. While most of the region north of Lake Huron retains the record of a Paleozoic thermal event, progressive younger thermal overprinting is observed. This is documented by the decrease of apparent AFT ages towards the southeast (Plate Ia), until complete resetting is attained in the Grenville Province, as indicated by Mesozoic AFT ages and a significant increase in mean HCTL (Plate Ib).

Geological implications of the thermal event, reflected in the AFT data from southeast Ontario, can be found in the Michigan Basin immediately south of the study area. The basin contains ~4.5 km of Paleozoic sediments ranging in age from Late Cambrian to Pennsylvanian (Fischer et al. 1988). Only ~200 m of Pennsylvanian age section is preserved and this is directly overlain by local, thin Jurassic red beds. Nevertheless, evidence from organic indicators suggests a greater overburden existed in late Paleozoic time. Vitrinite reflectance values (R_0) of the surface rocks are >0.55 and the Thermal Alteration Index values (TAI) >2.5 (Cercone 1984; Cercone and Pollack 1991). In the central and northern part of the basin, the oil window (R_0 >0.65 or TAI >0.65) extends up to 500 m below the present surface (Cercone 1984). Hydrocarbons from Silurian and Devonian strata similarly require either higher paleotemperatures for their in-situ generation or alternatively their upward migration by up to 2 km, in some cases through impermeable salt beds (Nunn et al. 1984). On the shield itself, AFT studies carried out on a vertical profile from a 3440 m deep drill hole in Sudbury suggest Permo-Triassic heating followed by Late Triassic-Early Jurassic cooling (Lorencak et al. 2004). This cycle has also been previously observed using AFT analysis in other parts of the shield and the Michigan Basin (Crowley 1991).

The present data set also records Permo-Triassic heating followed by Late Triassic-Early Jurassic cooling; this is in accord with the geological observations described above and with the timing suggested by previous work. The most likely cause for a regionally coherent AFT

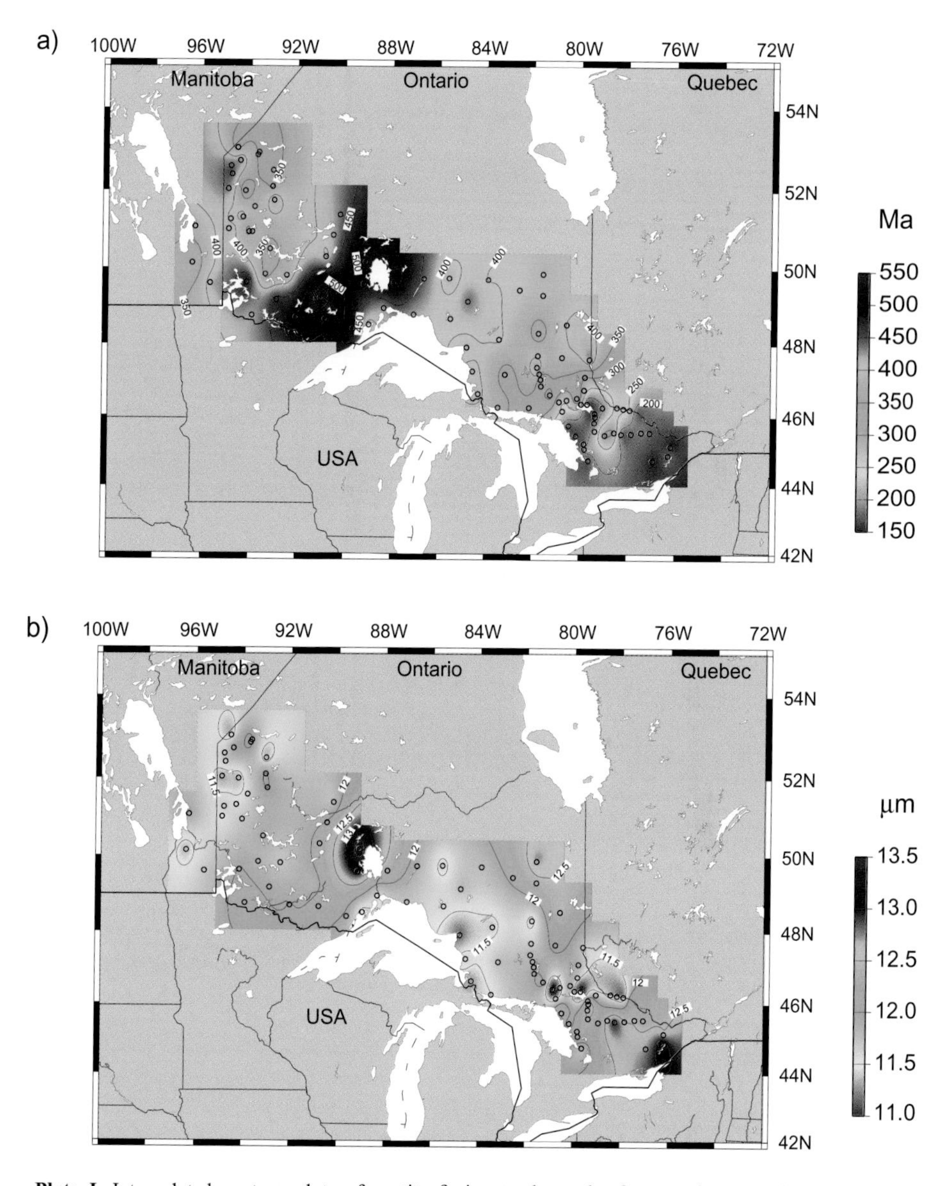

Plate I. Interpolated contour plots of apatite fission track results from surface samples across the southeastern Canadian shield showing: (a) the distribution of apparent apatite fission track "central" ages, and (b) the mean horizontal confined fission track lengths (HCTL) in microns (modified after Lorencak 2003). See also caption for Plate III for information on contouring procedure.

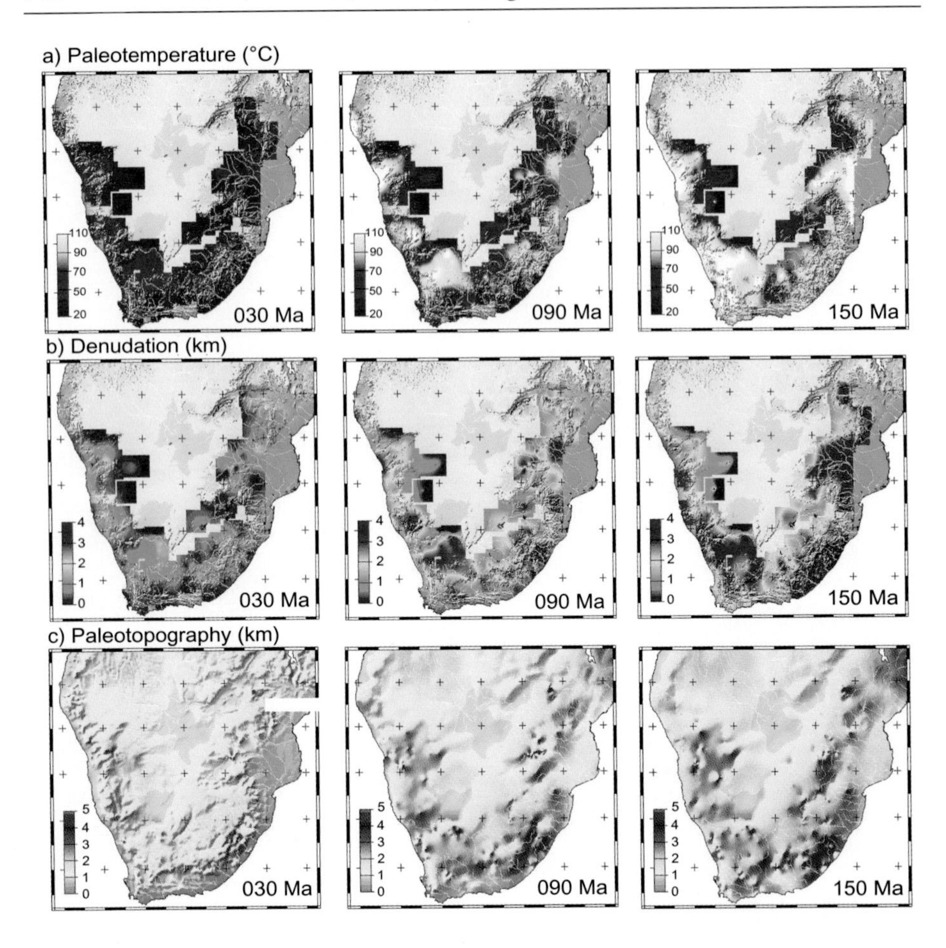

Plate II. Images showing the reconstructed (a) paleotemperatures experienced by rocks now at the surface, (b) the amount of denudation, and (c) an estimated reconstruction of paleotopography for southern Africa for four three separate time-slices; Late Jurassic (150 Ma), mid Cretaceous (90 Ma) and Oligocene (30 Ma). Paleotopography is based on "back-stacking" the present topography using the digital elevation data with the estimated amount of denudation and adjusting for an effective elastic thickness (Te) of 25 km (see Fig. 3 and Gallagher and Brown 1999a; Gleadow and Brown 2000). See also caption for Plate III for information on contouring procedure.

Plate III. *(caption continued from facing page)*
Data Center, Boulder. Colorado, USA) Global Heat Flow data set and is compiled from the work by Pollack et al. (1993). White circles indicate measurement localities, (e) contour images of post Early Jurassic temperature evolution in five time slices (180 Ma, 120 Ma, 90 Ma, 60 Ma and 30 Ma) of apatite fission track samples from eastern Africa and (f) contour images of the amount of cumulative post Early Jurassic denudation occurring across eastern Africa. The slices were generated by contouring the product of paleotemperature estimates (e) geothermal gradient (based on estimates displayed in d). Contouring of data shown in (a) to (f) was accomplished using GMT-3 (Generic Mapping Tools version 3.0 (Wessel and Smith 1991) using the commands of Surface and Blockmean with a contour interval of 30 (background topography as for Fig. 7). Areas of no data control have been masked out using the command psmask (GMT-3.0) using a confidence interval of 1°. Note that only samples with age, track length and standard deviation information were used to generate the images shown in (e) and (f), therefore not all samples show in (a) were used for image compilation.

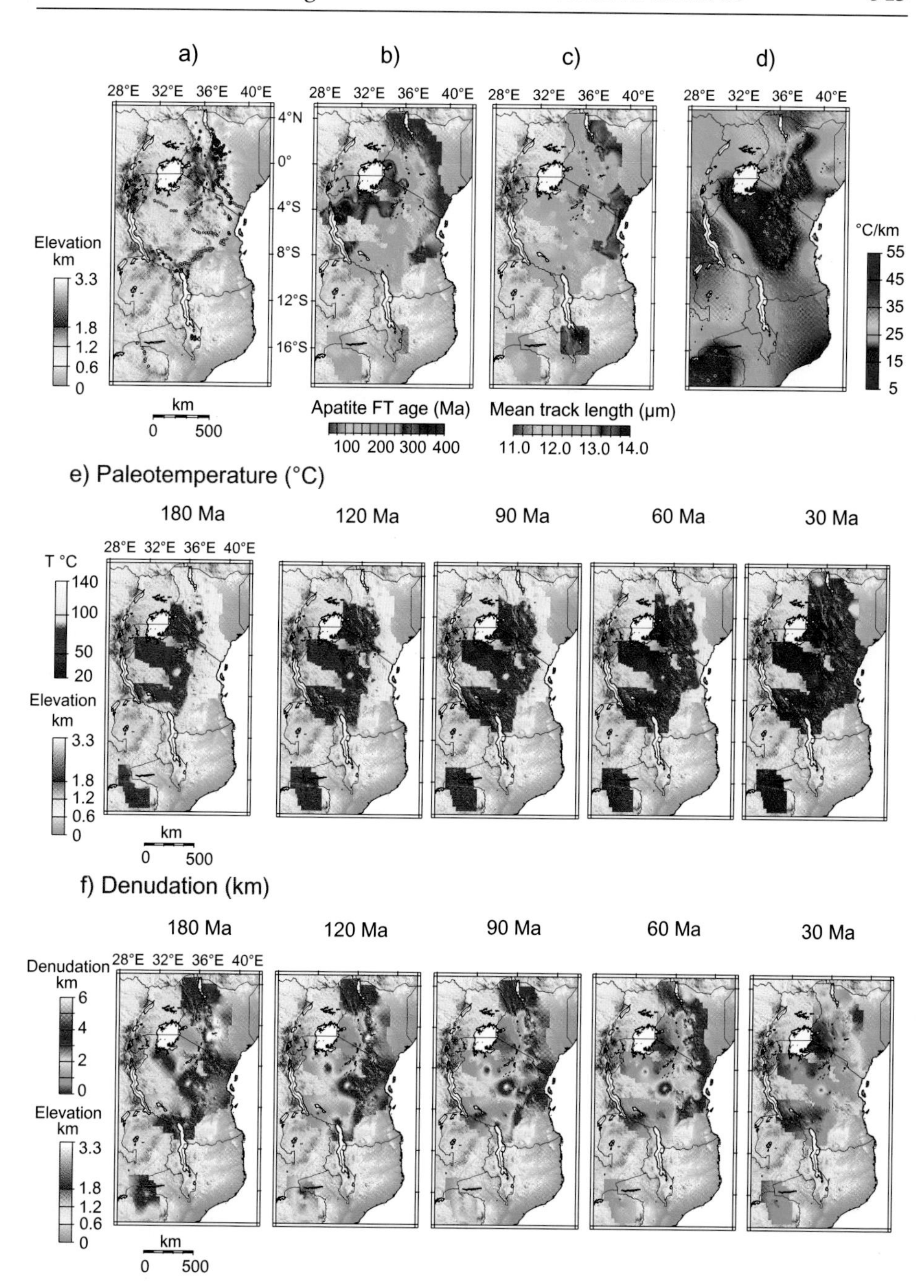

Plate III. Contour images of the AFT data collected in eastern Africa (generated using GMT-3, Wessel and Smith 1991); (a) sample locality map – green dots for samples collected in studies reported by Noble et al. (1997) and Noble (1997), red dots for samples collected in previous studies to that of Noble and co-workers (see text for further details), (b) contour image of apatite fission track ages, (c) contour image of apatite fission track mean lengths, (d) contoured present geothermal gradient values across eastern Africa used to derive the denudation estimates (see f). Data is from NGDRC's (National Geophysical

(caption continued on previous page)

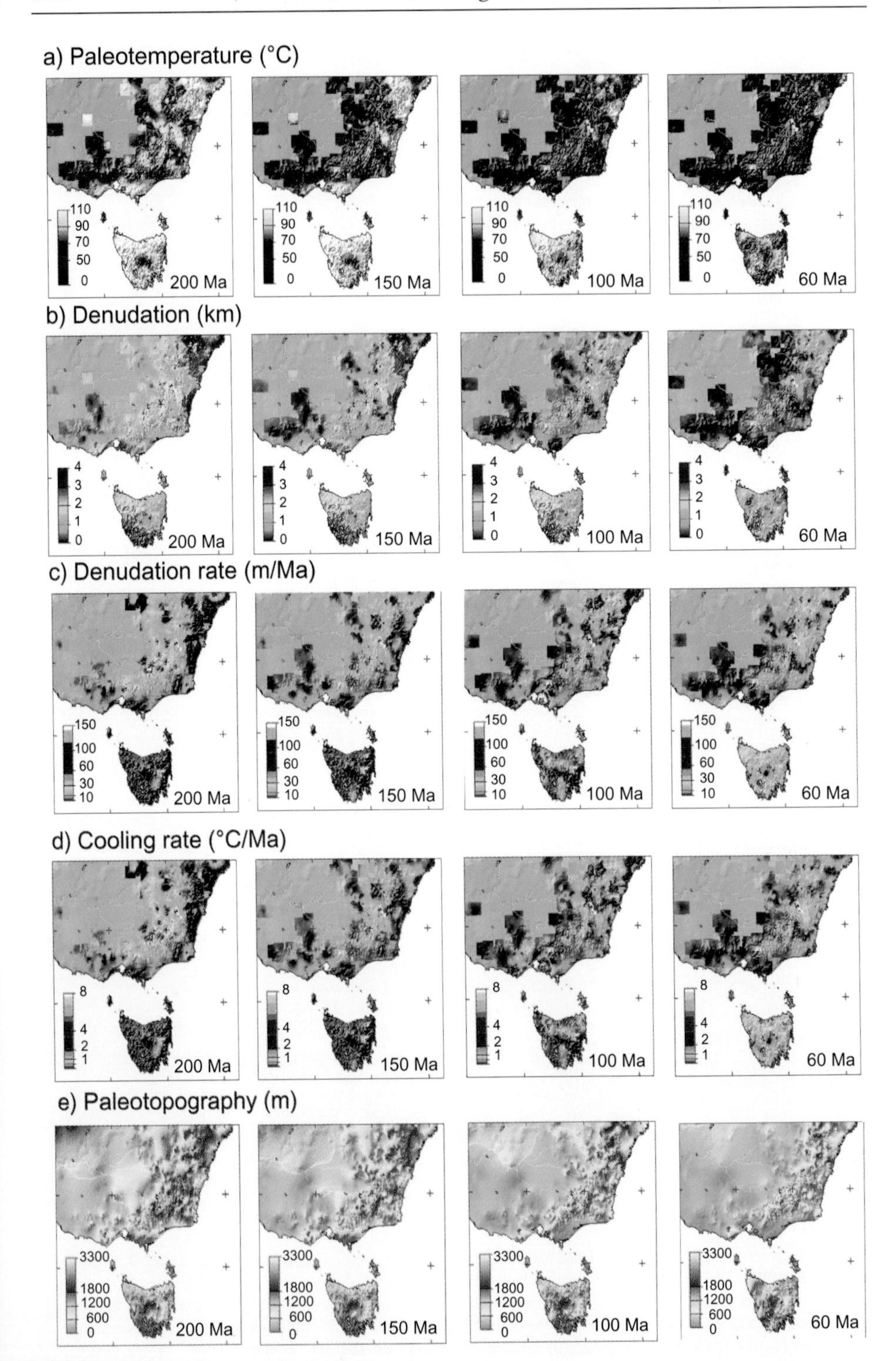

Plate IV. (caption on facing page)

pattern is burial under a sedimentary cover. We suggest that burial by foreland basin sediments shed from the Alleghenian Orogen has left a thermal imprint on the underlying southern Ontario shield rocks. The thickness of the sediments was sufficient to totally reset the AFT clocks (>~110 °C) in most of the Grenville Province. The thickness however rapidly decreased away from the orogen towards the north, resulting in only partial annealing of shield rocks on the southwestern flank of the Severn Arch, north of Lake Huron and east of Lake Superior. Along the Fraserdale Arch, and to the north of Lake Superior, no discernable thermal effect by the proposed burial is recorded by the AFT data.

In summary, the entire transect across the Canadian Shield in Ontario records a Paleozoic heating-cooling event, at least for the eastern half of the province, where the modeled time-temperature paths are well constrained. This heating is most likely associated with burial under sedimentary cover related to the Taconic orogeny, which remains preserved in outliers on the shield. The thermal signature of this event is progressively overprinted in the southeast Ontario shield by later burial under Alleghenian foreland sediments. No evidence for thermal effects related to the Acadian orogeny has been observed. This may reflect the possibility that none ever existed, that they are too subtle to be differentiated within the limits of modeling precision or that the effects of later thermal overprinting has removed such a record. It is very likely however, that large parts of the shield remained buried beneath a sheet of sediments throughout the Paleozoic and parts of the Mesozoic, as also implied independently by the work Patchett et al. (2004).

Southern Africa – formation and evolution of a continental interior

Geological overview. The pre-break-up geological evolution of southern Africa was dominated by the development of the Paleozoic-Mesozoic Karoo Basin. The Karoo Basin in South Africa formed as an extensive foreland basin ahead of the Cape Fold Belt (CFB) during the Early Permian (Tankard et al. 1982; Söhnge and Hälbich 1983). Thick sedimentary sequences accumulated along the southern margin of the CFB, but the sequence thins rapidly northwards forming a relatively thin cover (c. 1–2 km) over the Archaean and Proterozoic basement rocks in the cratonic interior. Remnants of the Permian Dwyka Formation tillites (basal Karoo) occur in exhumed paleovalleys where they underlie elevated river terraces (Martin 1953, 1973). These glacial deposits and striated bedrock surfaces, such as those exposed at Nooitgedagt (near Kimberly, South Africa), indicate that the present land surface represents an exhumed Permo-Carboniferous landscape in places (Visser 1987, 1995). The Clarens Formation aeoliantites (top Karoo) in South Africa (Dingle et al. 1983) have a maximum thickness of 300 m in the upper Orange River valley, but are generally about 100 m thick. This unit indicates terrestrial, semi-arid paleoclimatic conditions across southwestern Gondwana from the latest Triassic to earliest Cretaceous. Sedimentation within the Karoo Basin was terminated abruptly by the eruption of voluminous and extensive lavas of the Karoo continental flood basalts (~183 Ma) and the Paraná-Etendeka CFB (~132 Ma), which are up to 1.4 km thick in places (Erlank et al. 1984; Hawkesworth et al. 1992; Renne et al. 1996). After eruption of the Paraná-Etendeka flood basalts the surface geology is dominated by the thin (generally <200 m) Kalahari basin, which covers much of central southern Africa (Thomas

Plate IV (*on facing page*). Images showing the reconstructed paleotemperatures experienced by rocks now at the surface, their cooling rate, amount of denudation, rate of denudation and an estimated reconstruction of paleotopography for southeastern Australia for four separate time-slices; Early Jurassic (200 Ma), Late Jurassic (150 Ma), mid Cretaceous (100 Ma) and Paleocene (60 Ma). Paleotopography is based on "back-stacking" the present topography using the digital elevation data with the estimated amount of denudation and adjusting for an effective elastic thickness (Te) of 25 km (see Fig. 3; Gallagher and Brown 1999a; Gleadow and Brown 2000). See also caption for Late III for information on contouring procedure. Note samples from the Murray Basin area (Fig. 9) have not been included for generation of the image.

and Shaw 1990). The age of the base of the Kalahari basin sequence is thought to be Late Cretaceous to earliest Tertiary (Ward 1988; Thomas and Shaw 1990; Partridge 1993).

Continental break-up. Continental rifting between South America and Africa began during the Middle Jurassic (~150 Ma) (Nürnberg and Müller 1991). The rifting seems to have propagated south from the Falklands Agulhas fracture zone northwards towards the Walvis Ridge-Rio Grande Rise. The oldest magnetic anomaly clearly identifiable on oceanic crust on both the African and South American plates is M4 (130 ± 1 Ma), while further north near the Walvis Ridge the oldest identified anomaly is M0 (~125 Ma). Continental break-up was accompanied by syn- and post-breakup reactivation of pre-existing basement features such as the Central African and Mwembeshi shear zones (Coward and Potgieter 1983; Coward and Daly 1984; Daly et al. 1989) and involved significant intraplate deformation within west, central and southern Africa (Fairhead 1988; Unternehr et al. 1988; Fairhead and Binks 1991; Binks and Fairhead 1992; Brown et al. 2000). This later period of intracontinental deformation has been ascribed to shear stresses related to major changes in the geometry and relative motions of the plates involved in the opening of the Central and South Atlantic ocean basins.

Although rifting began in the Middle Jurassic and break-up finally occurred during the Early Cretaceous the major volume of sediment within the Orange and Walvis basins was deposited during the Late Cretaceous-early Tertiary (Brown et al. 1995; Rust and Summerfield 1990). Rust and Summerfield (1990) determined that this volume of ~2.8×10^6 km^3 (adjusted to equivalent rock volume) is equivalent to an average depth of denudation of 1.8 km over the whole of the Orange River catchment (an area of 1.55×10^6 km^2). The sedimentary record in the offshore basins clearly indicates that the continent has experienced very significant amounts of denudation (an average of 1.8 km) since the Middle Jurassic-Early Cretaceous. However, the chronology and spatial distribution of onshore denudation are likely to have been highly variable, depending on the post break-up tectonics, the pattern of drainage development, style of landscape evolution, lithological heterogeneity and long-term climatic variations.

Fission track results. A substantial set of AFT data (several hundred samples in total) has been collected from southern Africa. While the coverage is still sparse over large areas, the available data do provide some important new insights into the timing and distribution of long-term denudation at a sub-continental scale. The stratigraphic ages of samples analyzed range from Precambrian (Namaqua metamorphic belt) to Late Triassic (Stormberg Group, Upper Karoo Sequence). Despite this wide range of stratigraphic ages virtually all samples analyzed yielded AFT ages ranging between 166 ± 6 Ma and 70 ± 5 Ma (i.e., Cretaceous), with a conspicuous lack of younger AFT ages. AFT ages predating break-up at ~134 Ma were only obtained from samples in the interior regions of the continent and at elevations in excess of ~1500 m. Significantly, however, this is not a general characteristic of the continental interior, as some of the youngest ages (~70 Ma) were obtained from samples 600 km inland. AFT ages generally increase systematically with increasing elevation for specific localities.

The fact that all of the AFT ages are significantly younger than the stratigraphic age of the host rocks indicates that all the sampled rocks have been subjected to substantially higher temperatures in the past (mostly >~110 °C). Almost all the samples with Cretaceous AFT ages have mean confined track lengths >~12.5 to ~13 μm. The distributions of track lengths within these samples are generally unimodal with the mode between ~13 and 14 μm, and are generally negatively skewed with "tails" of shorter tracks (<~10 μm). This shows that most of the tracks have experienced only a moderate degree of thermal annealing (shortening) at temperatures <~70 °C. The majority of the samples must therefore have cooled from maximum paleotemperatures close to or greater than ~110 °C during the Cretaceous.

Quantitative images. Here we assumed that the eroded rock had an average thermal conductivity of 2.2 W·m·K^{-1} and used surface heat flow data from Brazil and southern

Africa (Pollack et al. 1993) to derive estimates of the near surface temperature gradient. This approach accounts for spatially varying thermal gradients. Temporally varying heat flow and conductivity values, as derived from an independent thermal model for continental rifting for example, could easily be incorporated into the methodology but have not been in this paper.

In Plate IIa, we present maps showing the estimated paleotemperature at three times, 150 Ma, 90 Ma and 30 Ma of rocks presently outcropping on the surface. The earliest time broadly represents the time of continental break-up around southern Africa. For regions covered by the Late Cretaceous-Tertiary Kalahari basin sediments within southern Africa (Thomas and Shaw, 1990) paleotemperatures were set to the surface temperature during periods of deposition. The paleotemperature estimates were converted to equivalent depth as described above and maps representing the amount of denudation for each time are shown in Plate IIb and the estimated paleotopography in Plate IIc.

Thermotectonic history. At a regional scale the long-wavelength geomorphic response to continental rifting and break-up, indicated by the chronology and magnitude of denudation, varied significantly along the margins. For example, the substantial amounts of post-rift denudation indicated for the southwestern African margin probably reflects the geometry and timing of post-rift tectonic reactivation of major intracontinental structures (e.g., Raab et al. 2002). Overall the AFT data from southern Africa are consistent with models of landscape development which predict a major phase of denudation following continental rifting (e.g., Kooi and Beaumont 1994; Gilchrist et al. 1994; Brown et al. 2002). The chronology and rates of denudation inferred from the AFT results are also broadly similar to estimates derived from the offshore sedimentary record (Brown et al. 1990; Rust and Summerfield 1990). However, the timing and distribution of denudation is not compatible with simple escarpment retreat models following break-up, which predict only moderate amounts ($\leq$~1 km) of post-rift denudation inland of the margin escarpments. This is particularly true for southwestern Africa, and is partly a consequence of the post break-up tectonic history of the continental interior. A possible explanation of this discrepancy is that discrete tectonic episodes, inferred to have occurred during the Late Cretaceous and which included reactivation of major intracontinental structures, caused locally accelerated phases of denudation to be superimposed on the secular regional pattern. Alternative explanations for the observed pattern and history of denudation across the sub-continent which incorporate the recently documented dynamic uplift history of the African Superswell (e.g., Lithgow-Bertelloni and Silver 1998; Gurnis et al. 2000; Conrad and Gurnis 2003; Behn et al. 2004) during Cretaceous-Tertiary time will likely provide a more complete explanation for the spatially and temporally variable geomorphic history as documented using the regional imaging approach to analyzing thermochronologic data sets.

Eastern Africa – development of an intracontinental rift system

Geological overview. The crustal-scale mobile belts in East Africa formed during different Precambrian and early Paleozoic orogenic episodes (Shackelton 1986; Muhongo 1989; Stern 1994; Noble et al. 1997). Following the latest of these events, the Pan-African (~900–550 Ma), eastern Africa underwent a period of relative quiescence. This resulted in a phase of extensive peneplanation (Stagman 1978; Wopfner 1986), which was finally disrupted in Late Carboniferous-Early Permian time when the embryonic motions of the break-up of Gondwanaland commenced and sediments started to accumulate in basins of eastern Africa (e.g., Stagman 1978; Lambiase 1989; Kreuser et al. 1990; Wopfner and Kaaya, 1991).

The subsequent Phanerozoic geological history of eastern and central Africa has been dominated by continental extension (Daly et al. 1989; Lambiase 1989). This process has strongly influenced the regional geomorphology and led to the formation of widely recognized rift basins throughout East Africa (Reeves et al. 1987; Fairhead 1988; Daly et al. 1989; Lambiase 1989). It is also well documented that many post-Proterozoic faults that define these basins and adjacent horst blocks were formed by the reactivation of pre-existing structures in the mobile belts (e.g., Gregory 1896; Smith and Mosley 1993; Smith 1994).

In Jurassic time the development of a triple junction centred on what is now the Lamu Embayment (Fig. 7) led to seafloor spreading between Africa and Madagascar and the development of the East African passive margin (Reeves et al. 1987). The Anza Rift, which extends northwest from the Lamu Embayment as far as Lake Turkana (Fig. 7), is the failed arm of this Jurassic triple junction (Reeves et al. 1987). The Anza Rift was periodically reactivated with major periods of extension and sedimentation in the Early and Late Cretaceous and continuing through to Oligocene time (Reeves et al. 1987; Greene et al. 1991; Bosworth and Mosely 1994). The Cretaceous extensional episodes marked the eastern limit of intracontinental deformation in the west and central African rift system, which forms a series of rift basins and transform faults that developed in response to differential movement between continental fragments during opening of the South Atlantic Ocean (Fairhead 1988; Daly et al. 1989). Each of the rifting episodes in the Anza Rift led to regionally extensive erosional denudation of basement rocks in East Africa (Foster and Gleadow 1992a, 1993a, 1996; Noble et al. 1997).

In Late Oligocene to Early Miocene time, extension related to the Kenya Rift began in north Kenya (e.g., Morley et al. 1992). The present morphology of the Kenya Rift itself is a result of continental extension that has been taking place from Miocene to Recent times (Baker et al. 1972). A nascent continuation of the eastern branch of the East African Rift System, which represents an initial stage of development for a propagating rift, is observed in eastern Tanzania (Fig. 7). Some of the present topography and structural architecture of the Kenya Rift area is probably related to older rifts, such as the Anza Rift and Lamu Embayment (Reeves et al. 1987).

Fission track results. Over 430 AFT ages have been determined across eastern Africa and these are reported in several studies (Gleadow 1980; van den Haute 1984; Wagner et al. 1992; Foster and Gleadow 1992a, 1993a, 1996; Mbede et al. 1993; van der Beek 1995; Eby et al. 1995; Noble 1997; Noble et al. 1997; van der Beek et al. 1998). About 350 of these results also have accompanying track length determinations and the localities for these samples are shown in Plate IIIa, along with the AFT data summarized in interpolated images presented in Plate IIIb-c. Sampling strategies varied according to the particular study, but where quantitative images were constructed (Plate IIIe-f) most samples used were based on data reported by Foster and Gleadow (1992a, 1993a, 1996), Noble (1997) and Noble et al. (1997). These samples were mainly derived from suites collected systematically with elevation or across important structural blocks and regional trends, and also sampled to provide insights into the low temperature thermal effects of rift propagation and the response of cratonic crust during Phanerozoic rifting episodes (Plate IIIa).

AFT ages are generally <250 Ma, with two prominent groupings (Plate IIIb): <100 Ma along the coastal area and in northern Kenya and ~100 to 250 Ma inland in Kenya (E39°:S5° and E37°:N4°), Tanzania (E35°:S7°), Rwanda (E30°:S9°) and northern Zimbabwe (E30°: S16°). The youngest ages (≤50 Ma) are confined to the margins of Phanerozoic basins, i.e., the Anza Rift of Kenya (E34°:N4° to E40°:S4°), and the western branch of the East African rift system in southern Tanzania (E34°:S9°), and Malawi (E36°:S15°). The oldest ages (>300 Ma) are generally restricted to the Tanzanian craton of central Tanzania and southwestern Kenya (E33°:S5°).

There are two significant departures from this general pattern. Firstly, AFT ages in Burundi, adjacent to the western branch of the East African rift system and adjacent to Lake Rukwa in southwestern Tanzania are significantly older than any other samples lying in proximity to extensional basins. Secondly, one AFT age on the eastern margin of the

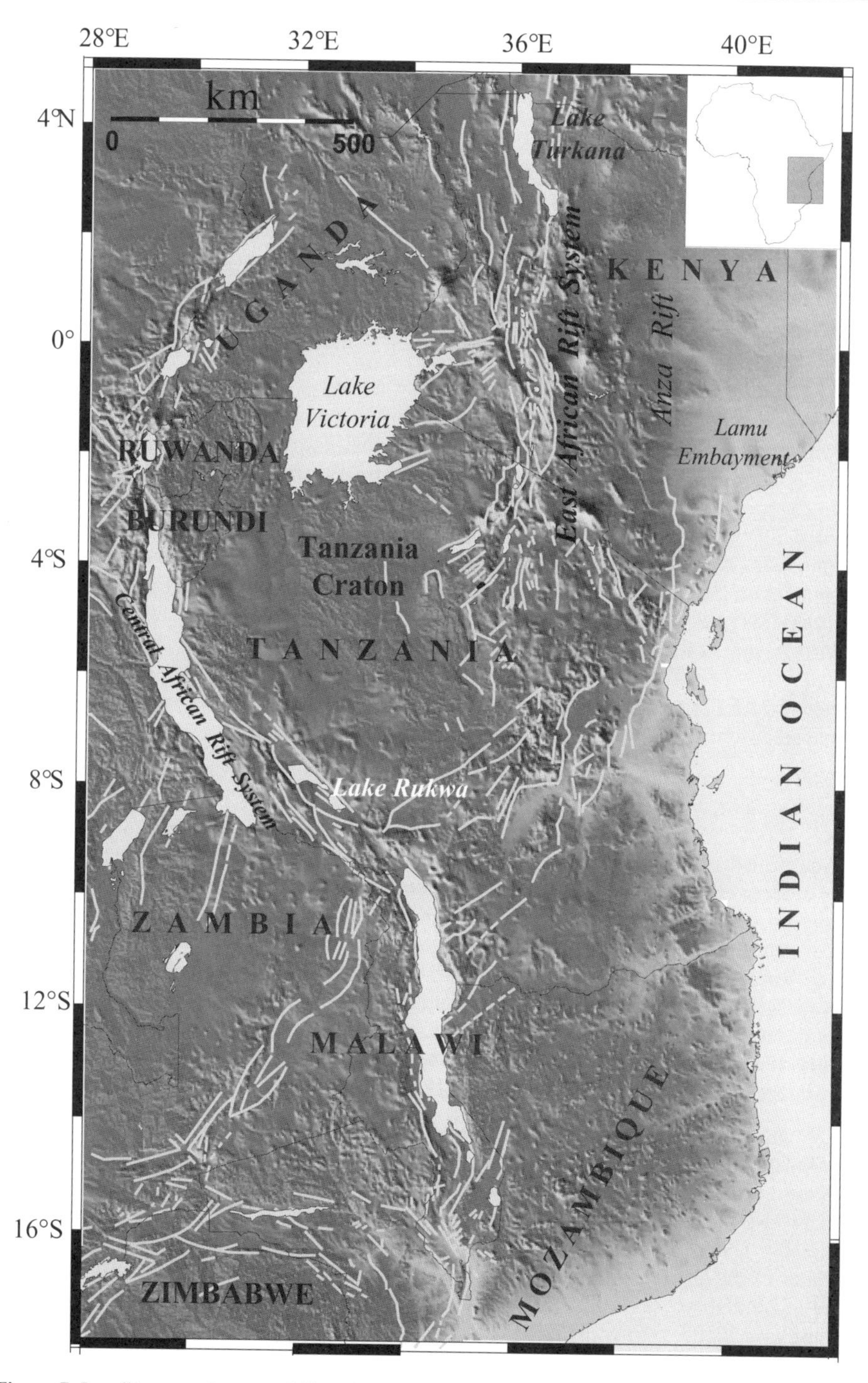

Figure 7. Locality map of eastern Africa showing topography, some geological elements and faults (in white), which are related to Phanerozoic rifting (faults after Rosendahl 1987), generated using GMT-3, version 3.0 (Wessel and Smith 1991). Shaded topography uses the 30 DEM from the US Geological Survey EROS Data Centre.

Tanzanian craton (E35°:S6°) yields a significantly younger age (~60 Ma) compared to all other cratonic samples.

In addition, the range of AFT ages can also be linked with changes in elevation. In general, youngest ages are found at lowest elevations, i.e., the coastal region (~35 to70 Ma) and increase towards the higher elevations of the continental interior (375 to 400 Ma).

Mean HCTL range predominantly between 12 to 13 μm (Plate IIIc). A smaller number of samples located along the flanks of the Anza Graben, in isolated areas on the Tanzanian craton and in southern Malawi yield longer mean HCTL between 13 to 14 μm. The lowest mean HCTL (~11 μm) are found in areas adjacent to the east-bounding fault of the Tanzania Craton, and in southeastern Kenya. The regional pattern of HCTL also appears to vary with elevation showing a decrease away from low-lying coastal areas (~13 to 14 μm) towards the higher interior of the continent (12–13 μm).

Quantitative images. The evolving temperature history of samples from eastern Africa is shown in Plate IIIe. Regional modeling suggests that during Early Jurassic time (180 Ma) rocks currently exposed along the eastern margin of the sampled area and northeastern margin of the Zimbabwe Craton (E30°:S16°) were at temperatures of >110 °C. It is important to note that modeling parameters have a predefined upper default value of ~110 °C, thus until samples record cooling below this temperature the modeling protocol assigns a value of 110 °C. By 180 Ma samples on the eastern margin of the craton (E35°:S7°), in SW Kenya (E36°:S1°), southern Tanzania (E34°:S9°), and northern Zimbabwe (E30°:S16°) had cooled to below ~80 °C. At the same time, some areas of southwest Kenya (E35°:S1°) and central Tanzania (E34°:S4°) are predicted to have cooled to surface temperatures. Prior to the Early Jurassic the information provided by AFT data is limited and only the oldest rocks exposed in the western portion of the sample area were at or below 110 °C.

During the period between 180 Ma and 90 Ma cooling continued, most notably around central Tanzania and southwestern Kenya, expanding the area over which rocks had cooled below ~100 °C. Also during this time interval the eastern margin of the study area and around the Malawi rift (E34°:S10°) does not appear to have undergone any cooling. This area most likely underwent some cooling at this time but remained at a temperature >110 °C. The cooling history between middle Cretaceous and the middle Tertiary (90 to 30 Ma) is much more pronounced along the coastal region, with a significant number of samples cooling more rapidly from >~110 °C to <80 °C. At the same time the samples from southwestern Kenya (E35°:S1°) and central Tanzania (E34°:S4°) underwent more subdued cooling. By 30 Ma most rocks in the study area were close to surface temperatures however a number of important exceptions in Kenya (E35°:N4°, E38°:N1°and E39°:S4°) and central Tanzania (E35°:S6°) suggest that some areas have undergone significant cooling in the last 30 Ma.

As shown previously, thermal histories can be used to estimate the amount and timing of denudation if a paleothermal gradient, paleosurface temperature and rock conductivity of the removed section is assumed (see also Fig. 3). It is noteworthy here that as the geothermal gradient varies across the area (Plate IIId) the resulting pattern of denudation will differ from the cooling histories shown in Plate IIIe (i.e., lower geothermal gradients result in higher estimates of the amount of denudation and vice-versa). The lowest geothermal gradients are found in Zimbabwe, central Tanzania and eastern Kenya, whereas the areas adjacent to rift margins have higher estimates. Time-slices for the denudation history of samples collected in eastern Africa are shown in Plate IIIf.

Areas in southwestern Kenya (E35°:S1°), central Tanzania (E34°:S4°) and Zimbabwe (E30°:S17°) are predicted to have had ~2 to 4 km of section removed since Late Jurassic time while the same areas experienced only 1 to 2 km of denudation since the mid Cretaceous (Plate

IIIf). By contrast, the coastal region of eastern Africa appears to have undergone ~6 km of denudation over the last 90 Ma.

Thermotectonic history. Previous geochronology studies for exposed rocks in eastern Africa generally yield Precambrian ages and therefore do not provide direct information on events occurring over the past ~500 Ma of geological history (e.g., Frisch and Pohl 1986; Munyanyiwa 1993; Möller 1995). The AFT data however, provide tectonothermal information for this significant time period. The range and distribution of AFT parameters across eastern Africa indicate that the Phanerozoic regional history is complex and closely related to the development and reactivation of sedimentary basins (e.g., Foster and Gleadow 1992a, 1993a, 1996). Another important feature is that the fission track parameters determined for the Precambrian mobile belts and Archaean cratons (Tanzanian and northern Zimbabwe) are, at least in some areas, remarkably similar. This suggests that these areas also share segments of a common Phanerozoic low temperature thermal history.

AFT data show that the post Pan African development of eastern Africa was characterized by long periods of slow cooling punctuated by at least four accelerated cooling events, commencing in Triassic (>220 Ma), Early Cretaceous (~140–120 Ma), Late Cretaceous-Early Tertiary (~80–60 Ma) and Middle to Late Tertiary time (Wagner et al., 1992; Foster and Gleadow, 1992a, 1993a, 1996; Mbede et al., 1993; van der Beek, 1995; Noble 1997; Noble et al., 1997). For the most part the relatively rapid cooling is interpreted as resulting from episodes of increased denudation related to the formation and reactivation of high angle fault blocks that moved in response to intraplate stresses. The episodes of denudation are also broadly contemporaneous with the deposition of packages of clastic sedimentary rocks in the basins of eastern Africa (Fig. 8). The periods of relatively rapid cooling are likely to be due to denudation at rates >30 m/Ma because of the preservation of relatively long mean track lengths. The actual rates of denudation probably ranged between 30 to 100 m/Ma during episodes of

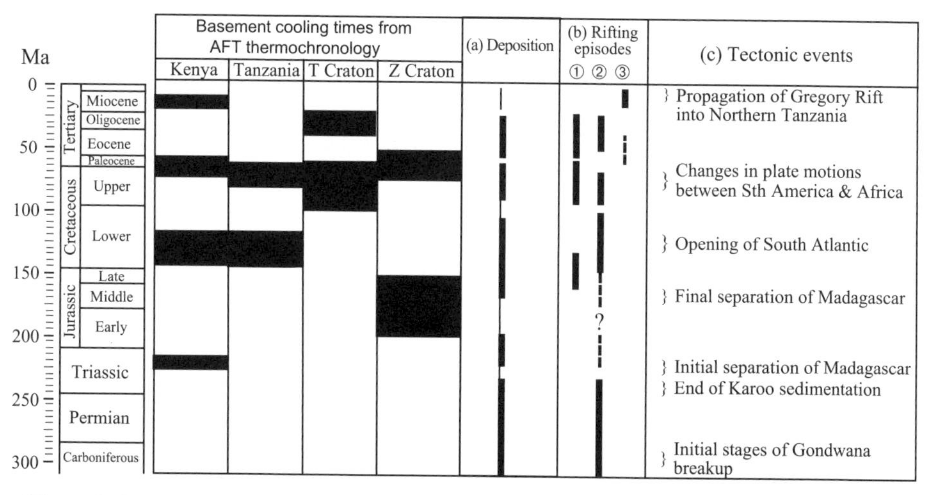

Figure 8. Comparison of periods of rapid denudation/cooling and fault reactivation revealed through AFT studies in Eastern Africa, with periods of: (a) sediment deposition into East African basins; (b) rifting episodes in the Anza Rift, Central African Rift System and East Africa Rift System; (c) regional tectonic events (modified after Winn et al. 1993). Periods of accelerated cooling defined from AFT thermochronology studies (Foster and Gleadow 1992a, 1993a, 1996; Noble 1997; Noble et al. 1997). Thick lines for deposition and rifting episodes denote time periods when both rate and style of sedimentation changed as a result of periods of rifting). T Craton = Tanzania Craton, Z Craton = Zimbabwe Craton (modified after Noble 1997 and Noble et al. 1997).

accelerated erosion (over periods of 10 to 20 Ma) and <5 m/Ma for the intervening times. This calculation is based on age versus elevation gradients and the timing of model histories and the geothermal gradients (Foster and Gleadow 1992a, 1993a, 1996; Noble 1997).

The last two episodes appear to be regionally more extensive. In most cases this is due to the removal of evidence for the older events by later denudation during the younger events. This is especially true for areas adjacent to the younger rifts, e.g., southeast coastal Kenya. The prominence of the last episode during Middle to Late Tertiary may also have been influenced by the regional interaction of plumes in eastern Africa during the later part of the Phanerozoic.

During Phanerozoic rifting in eastern Africa, it has previously been assumed that the Tanzanian and Zimbabwe Cratons have remained relatively inert (e.g., Muhongo 1989; Rach and Rosendahl 1989). The quantitative images however, clearly show that this is not the case and that periods of accelerated denudation and fault reactivation punctuated the tectonothermal history of the cratons at least since Mesozoic time (Plate IIIf) and even earlier (Noble et al. 1997). The timing of periods of denudation and fault reactivation are related to the tectonic evolution of East Africa and are contemporaneous with tectonic reactivation in the adjacent structural belts.

Southeastern Australia – evolution of a complex rifted passive margin

Geological overview. During Paleozoic to mid Cretaceous time, Australia, Antarctica and New Zealand were joined together in eastern Gondwanaland. Throughout much of this time eastern Gondwanaland was a convergent plate margin whose architecture was shaped mainly by its convergence with oceanic plates driven from the Pacific region (Veevers 1984). As a result the Paleozoic fold belts and basins of eastern Australia record a series of subduction-related deformational events. These include Early to Middle Paleozoic episodes responsible for formation and deformation of the Lachlan Fold Belt (Veevers 1984; Coney et al. 1990; Fergusson and Coney 1992; Gray 1997; Foster and Gray 2000) and the Late Permian to Early Triassic episodes which created the New England Fold Belt (Fig. 9) to the north and east (e.g., Harrington and Korsch 1985). The Lachlan Fold Belt is characterized by early to middle Paleozoic rocks including metamorphosed Cambrian through Devonian (primarily Ordovician) volcanic and cratonic-derived deep marine sedimentary rocks and extensive Early Silurian, Early Devonian and Late Carboniferous granitic rocks. By the Early to Middle Carboniferous all regional deformation within the Lachlan Fold Belt had ceased (Foster and Gray 2000).

The New England Fold Belt is composed predominantly of the deformed and metamorphosed remnants of an accretionary complex initially formed during Late Devonian-Early Carboniferous time. By the Late Carboniferous an extensional tectonic regime became predominant and numerous synkinematic S-type granites were intruded, mostly into the lower crust. Extension had ceased by the Late Permian and a major thrust-dominant contractional deformational event (Hunter-Bowen Orogeny) occurred accompanied by the intrusion of significant volumes of Late Permian to Triassic magma (Collins 1991). The Sydney Basin is a foreland basin overlying basement rocks, in part consisting of the Lachlan Fold Belt to the west and the New England Fold Belt to the north (Fig. 9). Subsidence and deposition and subsidence in the basin of a thick sequence of marine and non-marine sediments commenced in the Early Permian and continued through to the Jurassic, with a possible hiatus in the Late Triassic (Mayne et al. 1974). Permian sedimentary rocks within the Sydney Basin were deformed during the Late Permian-Early Triassic Hunter-Bowen Orogeny.

In the Late Jurassic, extension between Australia and Antarctica initiated from west to east (Johnson and Veevers 1984; Norvick and Smith 2001), followed in the Early Cretaceous by rifting along the southern margin (Veevers et al. 1991). Break-up between the Australian and Antarctic plates subsequently occurred in the middle Cretaceous (~95 Ma) and is marked

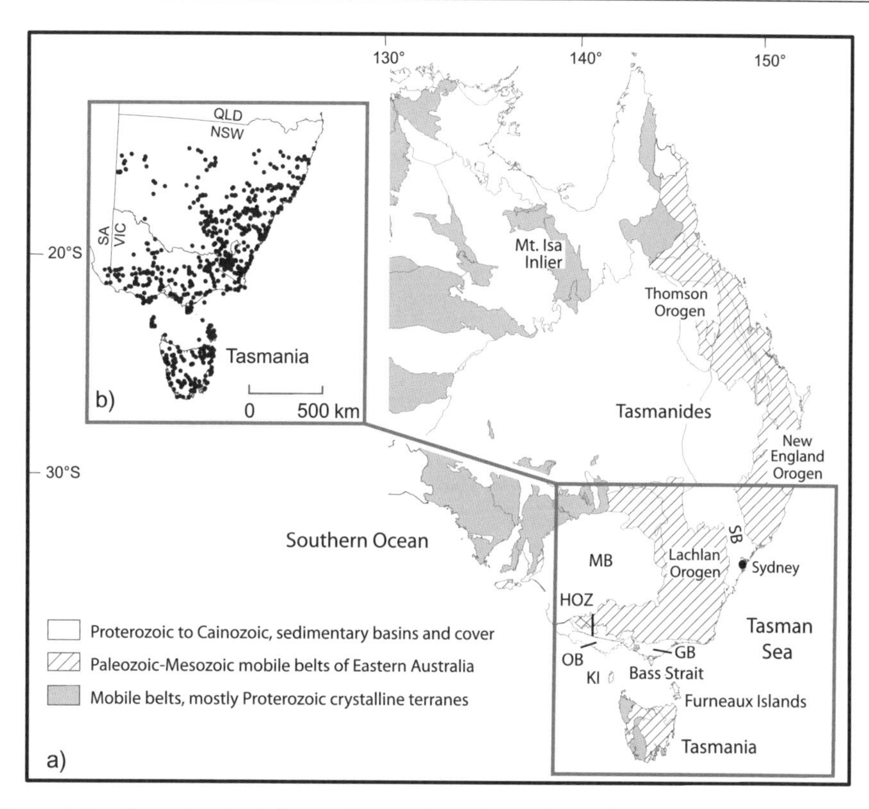

Figure 9. Locality and geological map of eastern Australia (modified after Palfreyman 1984) showing: (a) main crystalline terranes and grey inset delineating the southeastern Australia study area. Sedimentary cover, shown in white, was largely excluded from this study. Map (b) shows apatite fission track sample localities (~850) most of which were used for this study to construct the images shown in Plate IV. *GB* = Gippsland Basin, *KI* = King Island, *MB* = Murray Basin, *SB* = Sydney Basin, *HOZ* = hybrid orogenic zone (separating the Lachlan and Delamerian orogenic belts) and *OB* = Otway Basin.

by a major unconformity throughout the rift-related basins (Veevers 2000), e.g., Otway and Gippsland Basins located along the southern margin of Australia (Fig. 9a). However, the final separation to open the Southern Ocean did not propagate through Bass Strait, but was offset to the south of Tasmania. The magnetic anomalies in the Southern Ocean suggest that spreading was very slow until the Eocene (Cande and Mutter 1982; Veevers 2000). During the mid-Cretaceous (at ~105 Ma), subduction to the east of Australia ceased (Cande and Kent 1995). Subsequently, at ~100 Ma south- to north-directed continental rifting between Australia and the Lord Howe Rise/New Zealand began along what is now the eastern Australian margin. The timing of passive margin rifting along eastern Australia is well constrained by ocean floor magnetic anomalies and seismic data from the Tasman Sea rift (e.g., Weissel and Hays 1977; Veevers et al. 1991) which suggest spreading commenced at ~86 Ma and continued until ~62 Ma (Cande and Kent 1995). The onset of fast spreading in the Southern Ocean leading to the final separation of Australia and Antarctica in middle Eocene time is related to a significant global plate rearrangement (e.g., Cande and Mutter 1982). Structure and topography of the present-day southeastern margin are believed to be predominantly controlled by Late Mesozoic-Early Tertiary rifting, further details of which are summarized by Johnson and Veevers (1984) and Lister and Etheridge (1989).

Fission track results. Most samples for AFT studies were collected from exposed granitic rocks of the Paleozoic to Mesozoic mobile belts on the mainland, as well as in Tasmania and some offshore islands, which are part of the continuous continental crust (Fig. 9a). A small number were collected from basement rocks intersected in shallow borehole occurrences (<100 m depth) beneath shallow cover. In some areas samples included various metamorphic and sedimentary lithologies, and in Tasmania, Jurassic dolerites, but overall, some 90% of the samples studied are rocks of granitic composition with a relatively limited compositional range. AFT data from ~830 samples (see Fig. 9b) were judged to be of sufficient quality in both age and length data to be included in this study. Samples were excluded on the basis of having too few length measurements (<40) or too few grains counted (<6), or where obvious local geological disturbance, reflected in anomalous AFT data had occurred, such as by young volcanic activity.

Over the past ~25 years, several AFT studies have been carried out in areas bordering the southeastern continental margin (e.g., Gleadow and Lovering, 1978a,b; Moore et al. 1986; Dumitru et al. 1991; Foster and Gleadow 1992b, 1993b; O'Sullivan et al. 1995a,b, 1996a,b, 1999a,b, 2000a,b,c; O'Sullivan and Kohn 1997; Gleadow et al. 1996, 2002b; Kohn et al. 1999, 2002; Weber et al. 2004). Although the apparent AFT ages reported from these areas may range from mid-Paleozoic to Tertiary, they only rarely reflect the formation or depositional ages of the host rocks sampled, which are mostly of Paleozoic age. Further, the AFT data often show broad spatial variations that reflect their thermal and denudation histories. These have been interpreted in terms of a regional late Paleozoic cooling over the area modified by later cooling events associated with continental rifting and break-up on the eastern and southern margins (Dumitru et al. 1991; Gleadow et al. 1996, 2002b; Kohn et al. 2002). The most important regional AFT variations include:

1.) A tendency for the youngest ages ranging between ~50 to 100 Ma to be concentrated on and around the southeastern and eastern rifted margin of the continent. Many mean track lengths from these same areas are very long and often exceed 14 μm indicating that the apparent ages are actually dating the time of episodes of rapid cooling. Inland of the "young" age trend there is generally a belt of older ages and much reduced mean track lengths (<12.5 μm). Some of the individual length distributions for this region appear bimodal or unusually broad in character, indicative of mixed ages, intermediate between an older and younger value.

2.) A distinctly different pattern for Tasmania, south of the mainland, with ages across the island mostly <100 Ma and ranging between 30 and 250 Ma. Many of these young ages are also associated with long mean track lengths and narrow unimodal track length distributions, indicating that the ages reflect episodes of rapid cooling (O'Sullivan and Kohn 1997). Some older ages and shorter mean track lengths similar to those observed in the inland trend on the mainland are found in the centre of Tasmania and on King and the Furneaux Islands within Bass Strait (see Fig. 9a and O'Sullivan et al. 2000c).

3.) A relatively abrupt transition to much older ages (~300–400 Ma) in western Victoria going westwards along the southern margin of the mainland. This was first reported by Foster and Gleadow (1992, 1993) who suggested that the transition coincided approximately with the terrane boundary between the early Paleozoic Delamerian (to the west) and Lachlan orogenic belts (to the east) and reflected a change in the style of rifting along the southern margin to the west and east of a hybrid orogenic zone (Miller et al. 2004; also see Fig. 9a). The mean track lengths in the region of old ages are of generally intermediate values (12.5–13.5 μm) indicative of more prolonged cooling histories.

Southeastern Australia AFT age versus mean track length plots all show a characteristic "boomerang" trend with an upwards trend of longer track lengths to young ages (Gleadow et al. 2002b). This implies that the region has been widely affected by rapid cooling episodes with distinct differences in the timing of these rapid cooling episodes (defined by the track lengths clusters >14 μm) between different regions, with Victoria older than the remaining southeastern margin which in turn is older than Tasmania.

Quantitative images. Using the approach described above (see also Fig. 3) the southeastern Australia AFT data set has been modeled to construct quantitative paleotemperature, cooling rate, amount of denudation, denudation rate and paleotopography images as a function of time and space. A series of four time-slice images at 200 Ma, 150 Ma, 100 Ma and 60 Ma is presented in Plate IV. For the construction of the images we have used a spatially constant present day heat flow and surface temperature, and a constant thermal conductivity of 2.5 $Wm^{-1}K^{-1}$ (as used by Sass and Lachenbruch 1979 for the Eastern Heat Flow Province of Australia) to convert temperature to depth. It should be noted that samples from the Murray Basin area (Fig. 9b) were excluded in the construction of the quantitative images.

A remarkable feature of the time slice images presented is the sizeable area of present day surface rocks which were at paleotemperatures close to or greater than 110 °C at 200 Ma. Many of these areas had cooled to lower temperatures by 100 Ma, although parts of coastal eastern Australia and Tasmania still remained at relatively high temperatures into Early Tertiary time. In this analysis western Victoria stands out as one area where Paleozoic rocks have experienced relatively little thermal disturbance and have remained at relatively low temperatures since ~200 Ma (Plate IV).

Converting the paleotemperature data into denudation estimates and taking into account the assumptions described previously, suggests that cumulative denudation in most parts of the study region was between ~2–4 km since 200 Ma. In some mainland coastal areas and over much of Tasmania however, this amount of denudation was mainly achieved over the past ~60 Ma.

The visual pattern observed from the images displays clearly in time and space some of the main points arising from earlier studies, which indicate that since the end of orogeny, different regions of the mobile belt rocks of southeastern Australia record distinct episodes of accelerated denudation (e.g., O'Sullivan et al. 1996a, 2000b; Gleadow et al. 2002b; Kohn et al. 1999, 2002). Such reported episodes occurred during Late Permian to Early Triassic (~265–230 Ma), middle Cretaceous (~100–85 Ma) and Paleocene to Middle Eocene (~60–45 Ma) time.

Thermotectonic history. Late Permian to Early Triassic cooling in the Lachlan Orogen has been related to a far-field denudational response associated with the Hunter-Bowen Orogeny (e.g., O'Sullivan et al. 1996a, 1999a; Kohn et al. 1999, 2002). By contrast, later mid Cretaceous cooling may have been caused by the cessation of dynamic platform tilting due to subduction in early Late Cretaceous time (e.g., Gallagher et al. 1994a; Waschbusch et al. 1999). This would have resulted in rebound, leading to km-scale denudation. Added to this is the effect of contractional deformation (~90–95 Ma) associated with inversion and reactivation along the eastern margin of the continent and the onset of rifting in the Southern Ocean and Tasman Sea (Hill et al. 1995).

Van der Beek et al. (1999) analyzed the present day drainage pattern and denudation history of southeastern Australia and suggested that regional km-scale mid-Cretaceous uplift may not have taken place. Rather, they proposed that the observed cooling may have resulted from denudation related to base-level drops associated with rifting in the Bass-Gippsland basins to the south and the Tasman Sea to the east (Fig. 9a). The isolated occurrences of mid Cretaceous igneous rocks in southeastern Australia rule out the possibility that mid-

Cretaceous cooling may be the result of cooling following a period of elevated heat flow linked to a magmatic event (Kohn et al. 2003).

In general, coastal plain areas exhibit a greater amount of denudation than those inland. But the complex nature of events associated with continental break-up and the formation of new-rifted margins is highlighted by fact that times of rapid cooling may pre-date and postdate the actual time of seafloor spreading onset in the adjacent ocean basins (Plate IV).

The difference in timing of cooling episodes between different areas is also demonstrated by surface denudation rates through time presented for six specific sub-regions of southeastern Australia (Fig. 10). These may be equated with the volumes of sediment derived from the respective landscapes over the time intervals indicated and allow for the predictions to be tested by matching against the stratigraphic record of various sedimentary basins, provided these are preserved (e.g., Weber et al. 2004). It is emphasized that the regional denudation chronology estimates, derived by the integration of spatial denudation information for a particular time interval, produces the "locally-averaged" denudation rate for a specified region as a function of time. Hence, distinctive cooling patterns seen in some individual samples or in samples from a restricted area may not necessarily be highlighted prominently within the overall "average" regional denudation pattern.

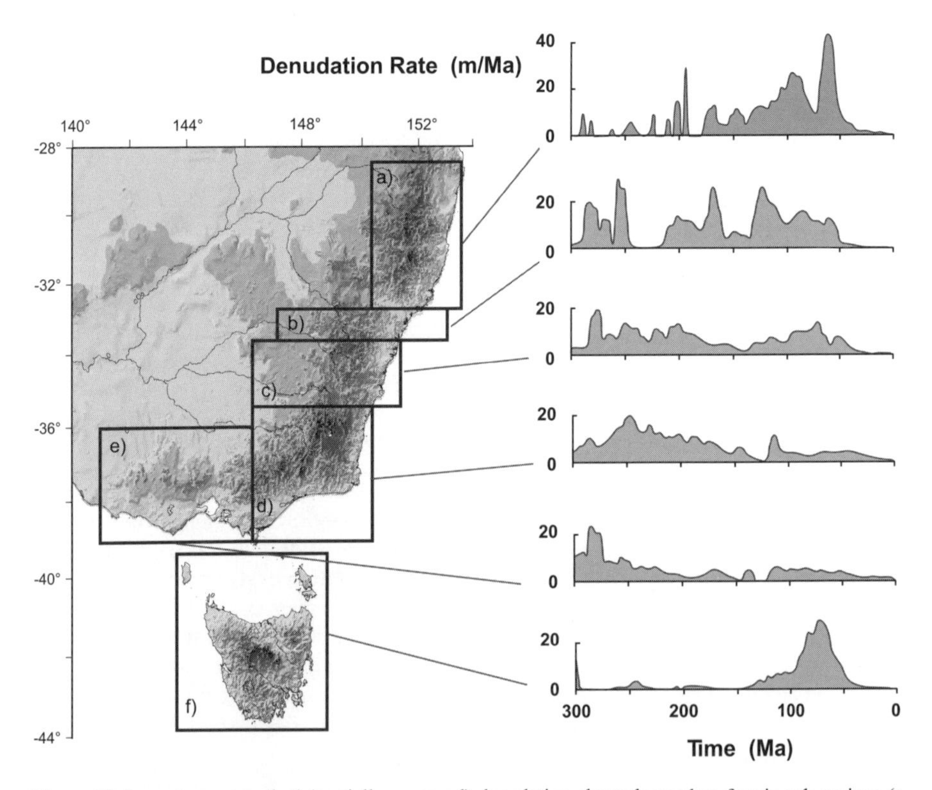

Figure 10. Long-term smoothed (spatially averaged) denudation chronology plots for six sub-regions (a to f) in southeastern Australia based on an initial track length of 14.5 μm. Note the marked differences in denudation history apparent across the different areas. Areas (a) and (f) show pronounced Cretaceous to Early Tertiary denudation, areas (b)-(d) show accelerated denudation in Late Paleozoic-Early Triassic time, whereas area (e) shows particularly low rates of denudation from Mesozoic to Present (see text for further discussion).

Tasmania shows a markedly different and significantly younger cooling history to the mainland. The denudation chronology for Tasmania reveals a steady increase in Early Cretaceous time to a maximum peak in mid Paleocene to mid Eocene time and then a decrease to the Present (Figs. 2b and 10). This is indicative of its continued tectonic emergence and denudation throughout much of the Cenozoic and probably reflects the proximate position of Tasmania, with its very narrow continental shelves and the evolving rift systems between Australia and Antarctica and the Lord Howe Rise/New Zealand.

CONCLUDING REMARKS

AFT thermochronology is a well-established tool for reconstructing the low temperature thermal and tectonic evolution of the continental crust. From the earliest studies of continental terranes it has been apparent that AFT data show broad regionally consistent patterns of variation. The variations are primarily controlled by cooling, which may be initiated by earth movements and denudation at the Earth's surface and/or by changes in the thermal regime. As such the data frequently bear little or no relationship to the original formation ages of the rocks involved. However, the significance of the regional patterns is not always obvious and there have often been difficulties in interpreting and integrating the results of such studies with other sets of geological observations. Using numerical forward modeling procedures the measured AFT parameters can be matched with time-temperature paths using an optimal data fitting procedure, which enables thermal and tectonic processes to be mapped out in considerable detail. Large regional arrays of data can be modeled sequentially and inverted into time-temperature solutions for visualizing the evolution of paleotemperatures, denudation rates and paleotopography of present day surface rocks. These can then be viewed as a series of time-slice images and the regional spatially integrated denudation rate chronology. The images provide a striking new quantitative perspective on crustal processes and landscape evolution and allow important tectonic and denudation events over time scales up to hundreds of million years to be readily visualized in a variety of ways and integrated with other regional data sets such as digital terrain models, heat flow, etc. This approach provides a readily accessible framework for quantifying the often undetectable, timing and magnitude of long-term crustal denudation in many terranes, for a part of the geological record often largely unconstrained.

The images are not only valuable for visualizing the thermochronological information but also allow a new range of quantitative measurements to be made on the virtual landscapes constructed. For example, a direct consequence of the denudation models is to predict sediment volumes and to trace the evolution of drainage basins, at least on a broad scale. This opens up the possibility of making a new range of mass-balance calculations on the amounts and nature of eroded material and sediment accumulation in appropriate depocenters (e.g., Weber et al. 2004). Similarly, predictions of long-term surface denudation rates can be tested against more recent estimates based on cosmogenic isotope analyses (e.g., Belton et al. 2004), at least for the most recent part of the record. The acquisition of apatite (U-Th)/He data (e.g., Farley 2002) on a similar regional scale should provide more robust information on the lower temperature portion of the thermal history (<~60–70 °C), which is poorly constrained by AFT data. In addition, information derived from higher temperature thermochronometric systems e.g., $^{40}Ar/^{39}Ar$ K-feldspar (McDougall and Harrison 1999; Harrison et al. 2005), (U-Th)/He zircon and titanite (Reiners and Farley 1999; Reiners 2005) and zircon fission track (Tagami 2005) could potentially also be integrated into imaging schemes, providing for a more comprehensive visualization of intermediate to low temperature thermal histories.

With increasingly large AFT datasets from regional studies becoming available it will be necessary to make data collection more efficient. One promising direction in this regard is through emergence of Laser-Ablation-Microprobe Inductively Coupled Plasma Mass

Spectrometry (LAM-ICP-MS). This approach allows for the analysis of trace elements in small areas (10–30 μm) within mineral grains and has opened the way to a radically different approach to fission track analysis that does not require neutron irradiation for the analysis of uranium content (Hasebe et al. 2004). As such LAM-ICP-MS promises a drastic reduction in sample turn-around time and improved laboratory safety due to the elimination of the need for neutron irradiations, a requirement, which now imposes relatively long delays on sample processing.

ACKNOWLEDGMENTS

The Australian Research Council, the Australian Institute of Nuclear Science and Engineering, and the former Australian Geodynamics Cooperative Research Centre funded this research. We acknowledge the various contributions of many colleagues related to the acquisition of regional fission track data sets, particularly former members of the fission track groups at the School of Earth Sciences at La Trobe University and more recently at the University of Melbourne, particularly to Paul O'Sullivan, David Foster and Asaf Raza. We are also grateful to Kirk Osadetz and Richard Everitt for their assistance in acquiring samples for the Canadian data set. We thank Paul Andriessen, Geoff Batt, Ann Blythe and Tim Carter for their helpful reviews of this chapter.

REFERENCES

Baker BH, Mohr PA, Williams LAJ (1972) Geology of the Eastern rift system of Africa. Geol Soc Am Spec Paper 136:1-67

Barbarand J, Carter A, Wood I, Hurford A (2003) Compositional and structural control of fission-track annealing in apatite. Chem Geol 198:107-137

Behn MD, Conrad CP, Silver PG (2004) Detection of upper mantle flow associated with the African Superplume, Earth Planet Sci Lett 224:259-274

Belton DX, Brown RW, Kohn BP, Fink D, Farley KA (2004) Quantitative resolution of the debate over antiquity of central Australian landscapes and implications for the tectonic and geomorphic stability of cratonic interiors. Earth Planet Sci Lett 219:21-34

Binks RM, Fairhead JD (1992) A plate tectonic setting for the Mesozoic rifts of West and Central Africa. Tectonophys 213:141-151

Bosworth W, Morley CK (1994) Structural and stratigraphical evolution of the Anza rift, Kenya. Tectonophys 236:93-115

Bray RJ, Green PF, Duddy IR (1992) Thermal history reconstruction using apatite fission track analysis and vitrinite reflectance: a case study from the UK East Midlands and the Southern North Sea. *In:* Exploration Britain: Into the Next Decade. Hardman RFP (ed) Geol Soc Lond Spec Pubs 67, p 3-55

Brown LF Jr, Benson JM, Brink GJ, Doherty S, Jollands A, Jungslager EHA, Keenen JHG, Muntingh A, van Wyk NJS (1995) Sequence Stratigraphy in Offshore South African Divergent Basins. An Atlas on Exploration for Cretaceous Lowstand Traps, Soekor (Pty) Ltd, Am Assoc Petrol Geol Studies in Geol 41, 184 p

Brown RW (1991) Backstacking apatite fission track "stratigraphy": a method for resolving the erosional and isostatic rebound components of tectonic uplift histories. Geology 19:74-77

Brown RW, Summerfield MA (1997) Some uncertainties in the derivation of rates of denudation from thermochronologic data. Ear Surf Proc Landforms 22:239-248

Brown RW, Gallagher K, Duane M (1994a) A quantitative assessment of the effects of magmatism on the thermal history of the Karoo sedimentary sequence. J Afr Ear Sci 18:227-243

Brown RW, Summerfield MA, Gleadow AJW (1994b) Apatite fission track analysis: Its potential for the estimation of denudation rates and implications for models of long term landscape development. *In:* Process Models and Theoretical Geomorphology. Kirkby MJ (ed), John Wiley and Sons Ltd, Chichester, p 23-53

Brown RW, Summerfield MA, Gleadow AJW (2002) Denudational history along a transect across the Drakensberg Escarpment of southern Africa derived from apatite fission track thermochronology. J Geophys Res 107: doi:10.1029/2001JB000745

Brown RW, Gallagher K, Gleadow AJW, Summerfield MA (2000) Morphotectonic evolution of the South Atlantic margins of Africa and South America. *In:* Geomorphology and Global Tectonics. Summerfield MA (ed) John Wiley and Sons Ltd, Chichester, p 257-283

Brown RW, Gallagher K, Johnson K, Cockburn HAP, Summerfield M.A, Gleadow AJW (2001) All landscapes, great and small: problems and strategies for deriving regional denudation histories from sparse data. Earth System Processes, Abst Geol Soc Amer and Geol Soc Lond, Edinburgh, Scotland, p 81

Brown RW, Rust DJ, Summerfield MA, Gleadow AJW, De Wit MCJ (1990) An accelerated phase of denudation on the south-western margin of Africa: evidence from apatite fission track analysis and the offshore sedimentary record. Nucl Tracks Rad Meas 17:339-350

Buck WR, Martinez F, Steckler MS, Cochran JR (1988) Thermal consequences of lithospheric extension: pure and simple. Tectonics 7:213-234

Burtner RL, Nigrini A, Donelick RA (1994) Thermochronology of Lower Cretaceous source rocks in the Idaho-Wyoming Thrust Belt. Bull Am Assoc Petrol Geol 78:1613-1636

Cande SC, Kent DV (1995) A new geomagnetic polarity time scale for the Late Cretaceous and Cenozoic. J Geophy Res 97:13917-13951

Cande SC, Mutter JC (1982) A revised identification of the oldest sea-floor spreading anomalies between Australia and Antarctica. Earth Planet Sci Lett 58:151-160

Carlson WD, Donelick RA, Ketcham RA (1999) Variability of apatite fission-track annealing kinetics: I. Experimental results. Am Mineral 84:1213-1223

Cercone KR (1984) Thermal history of Michigan Basin. Am Assoc Petrol Geol Bull 68:130–136

Cercone KR, Pollack HN (1991) Thermal maturity of the Michigan Basin. Geol Soc Am Spec Paper 256: 1-11

Collins WJ (1991) A reassessment of the 'Hunter-Bowen Orogeny': Tectonic implications for the southern New England Fold Belt. Aust J Earth Sci 38:409-423

Coney PJ, Edwards A, Hine R, Morrison F, Windrim D (1990) The regional tectonics of the Tasman orogenic system, eastern Australia. J Struct Geol 12: 519-543

Conrad CP, Gurnis M (2003) Seismic tomography, surface uplift, and the breakup of Gondwanaland: Integrating mantle convection backwards in time. Geochem Geophys Geosyst 4:1031, doi:10.1029/ 2001GC000299

Corfu F, Stone D (1998) The significance of titanite and apatite U-Pb ages: Constraints for the post-magmatic thermal-hydrothermal evolution of a batholitic complex, Berens River area, northwestern Superior Province, Canada. Geochim Cosmochim Acta 62:2979-2995

Corrigan J (1991) Inversion of apatite fission track data for thermal history information. J Geophys Res 96: 10374-10360

Coward MP, Daly MC (1984) Crustal lineaments and shear zones in Africa: their relationship to plate movements. Precamb Res 24:27-45

Coward MP, Potgieter R. (1983) Thrust zones and shear zones of the margin of the Namaqua and Kheis mobile belts, Southern Africa. Precamb Res 24:27-45

Crowley KD (1991) Thermal history of Michigan Basin and Southern Canadian Shield from apatite fission track analysis. J Geophys Res 96:697-711

Crowley KD (1993) Lenmodel – a forward model for calculating length distributions and fission-track ages in apatite. Computer Geosci 19:619-626

Crowley KD, Cameron M, Schaffer LR (1991) Experimental studies of annealing of etched fission tracks in fluorapatite. Geochim Cosmochim Acta 55:1449-1465

Daly MC, Chorowicz J, Fairhead JD (1989) Rift basin evolution in Africa: The influence of reactivated steep basement shear zones. *In:* Inversion Tectonics. Cooper MA, Williams GC (eds) Geol Soc Lond Spec Pub 44, p 309-334

Dingle RV, Siesser WG, Newton AR (1983) Mesozoic and Tertiary geology of southern Africa. AA Balkema, Rotterdam

Donelick RA (1993) Apatite etching characteristics versus chemical composition. Nucl Tracks Radiat Meas 21:604

Donelick RA, Ketcham RA, Carlson WD (1999) Variability of apatite fission-track annealing kinetics: II. Crystallographic orientation effects. Am Mineral 84:1224-1234

Donelick RA, O'Sullivan PB, Ketcham RA (2005) Apatite fission-track analysis. Rev Mineral Geochem 58: 49-94

Donelick RA, Roden MK, Mooers JD, Carpenter BS, Miller DS (1990) Etchable length reduction of induced fission tracks in apatite at room temperature (~23°C): crystallographic orientation effects and "initial" mean lengths. Nucl Tracks Radiat Meas 17:261-265

Duddy IR, Green PF, Bray RJ, Hegarty KA (1994) Recognition of the thermal effects of fluid flow in sedimentary basins. *In*: Geofluids; Origin. Migration and Evolution of Fluids in Sedimentary Basins. Parnell J (ed) Geol Soc Lond Spec Pub 78, p 325-345

Dumitru TA, Hill KC, Coyle DA, Duddy IR, Foster DA, Gleadow AJW, Green PF, Laslett GM, Kohn BP, O'Sullivan AB (1991) Fission track thermochronology: Application to continental rifting of southeastern Australia. Aust Petrol Exploration Assoc J 31:131-142

Eby GN, Roden-Tice M, Krueger HL, Ewing W, Faxon EH, Woolley AR (1995) Geochronology and cooling history of the northern part of the Chilwa Alkaline Province, Malawi. J Afr Earth Sci 20:275-288

Efron B, Tibshirani RJ (1993) An Introduction to the Bootstrap. Chapman and Hall, New York

Ehlers TA (2005) Crustal thermal processes and the interpretation of thermochronometer data. Rev Mineral Geochem 58:315-350

Ehlers TA, Farley KA (2003) Apatite (U-Th)/He thermochronometry: methods and applications to problems in tectonics and surface processes. Earth Planet Sci Lett 206:1-14

Erlank AJ, Marsh JS, Duncan AR, Miller RmcG, Hawkesworth CJ, Betton PJ, Rex DC (1984) Geochemistry and petrogenesis of the Etendaka volcanic rocks from SWA/Namibia. Spec Pub Geol Soc S Afr 13:195-245

Fairhead JD (1988) Mesozoic plate tectonic reconstructions of the Central-South Atlantic Ocean: the role of the west and central African Rift System. Tectonophys 155:181-191

Fairhead JD, Binks RM (1991) Differential opening of the Central and South Atlantic Ocean and the opening of the west African Rift System. Tectonophys 187:191-203

Farley KA (2002) (U-Th)/He dating: techniques, calibrations and applications. Rev Mineral Geochem 47: 819-843

Fergusson CL, Coney PJ (1992) Convergence and intraplate deformation in the Lachlan Fold Belt of southeastern Australia. Tectonophysics 214:417-439

Fischer JH, Barratt MW, Droste JB, Shaver R (1988). Michigan Basin. *In:* Sedimentary Cover - North American Craton: US. The Geology of North America vol D-2. Sloss LL (ed) Geol Soc Amer, Boulder, p 361-382

Foster DA, Gleadow AJW (1992a) The morphotectonic evolution of rift-margin mountains in central Kenya: Constraints from apatite fission track analysis. Earth Planet Sci Lett 113:157-171

Foster DA, Gleadow AJW (1992b) Reactivated tectonic boundaries and implications for the reconstruction of southeastern Australia and northern Victoria Land, Antarctica. Geology 20:267-270

Foster DA, Gleadow AJW (1993a) Episodic denudation in East Africa: A legacy of intracontinental tectonism. Geophys Res Lett 20:2395-2398

Foster DA, Gleadow AJW (1993b) The architecture of Gondwana rifting in southeastern Australia: evidence from apatite fission track thermochronology. *In:* Gondwana 8 - Assembly, Evolution and Dispersal. Findlay R, Unrug R, Banks RH, Veevers, JJ (eds) Balkema, Rotterdam, p 597-603

Foster DA, Gleadow AJW (1996) Structural framework and denudation history of the flanks of the Kenya and Anza Rifts. East Africa. Tectonics 15:258-271

Foster DA, Gray DR (2000) Evolution and structure of the Lachlan Fold Belt (Orogen) of Eastern Australia. Ann Rev Earth Planet Sci 28:47-80

Frisch W, Pohl W (1986) Petrochemistry of some mafic and ultramafic rocks from the Mozambique Belt, SE Kenya. Mitt Österr Geol Ges 78:97-114

Gallagher K (1995) Evolving temperature histories from apatite fission-track data. Earth Planet Sci Lett 136: 421-435

Gallagher K, Brown RW (1999a) Denudation and uplift at passive margins: the record on the Atlantic Margin of southern Africa. Phil Trans Roy Soc London A 357:835-859

Gallagher K, Brown RW (1999b) The Mesozoic denudation history of the Atlantic margins of southern Africa and southeast Brazil and the relationship to offshore sedimentation. *In:* The Oil and Gas Habitats of the South Atlantic. Cameron N, Bate R, Clure V (eds) Geol Soc London Special Pub 153, p 41-53

Gallagher K, Sambridge M (1994) Genetic algorithms: a powerful method for large scale non-linear optimisation problems. Computer Geosci 20:1229-1236

Gallagher K, Brown RW, Johnson C (1998) Fission track analysis and its applications to geological problems. Ann Rev Earth Planet Sci 26:519-572

Gallagher K, Dumitru TA, Gleadow AJW (1994a) Constraints on the vertical motion of eastern Australia during the Mesozoic. Basin Res 6:77-94

Gallagher K, Hawkesworth CJ, Mantovani MJ (1994b) The denudation history of the onshore continental margin of SE Brazil inferred from apatite fission track data. J Geophys Res 99:18,117-18,145

Gallagher K, Stephenson J, Brown R, Holmes C, Ballester P (2005) Exploiting 3D spatial sampling in inverse modeling of thermochronological data. Rev Mineral Geochem 58:375-387

Gilchrist AR, Summerfield MA (1994) Tectonic models of passive margin evolution and their implications for theories of long-term landscape development. *In:* Process Models and Theoretical Geomorphology. Kirkby MJ (ed) John Wiley and Sons, Chichester, p 55-84

Gleadow AJW (1980) Fission track age of the KBSTuff and associated hominid remains in northern Kenya. Nature 284:225-230

Gleadow AJW, Brown RW (2000) Fission track thermochronology and the long-term denudational response to tectonics. *In:* Geomorphology and Global Tectonics. Summerfield MA (ed) John Wiley and Sons Ltd, Chichester, p 57-75

Gleadow AJW, Duddy IR (1981) Early Cretaceous volcanism and early breakup history of southeastern Australia: Evidence from fission track dating of volcaniclastic sediments. *In:* Gondwana Five - Proceedings of the Fifth International Gondwana Symposium. Cresswell MM, Vella P (eds) Balkema, Rotterdam, p 295-300

Gleadow AJW, Lovering JF (1978a) Fission track geochronology of King Island, Bass Strait, Australia; relationship to continental rifting. Earth Planet Sci Lett 37: 429-437

Gleadow AJW, Lovering JF (1978b) Thermal history of granitic rocks from western Victoria : a fission track dating study. J Geol Soc Australia 25:323-340

Gleadow AJW, Belton DX, Kohn BP, Brown RW (2002a) Fission track dating of phosphate minerals and the thermochronology of apatite. Rev Mineral Geochem 48:579-630

Gleadow AJW, Duddy IR, Green PF, Lovering JF (1986) Confined fission track lengths in apatite: a diagnostic tool for thermal history analysis. Contrib Mineral Petrol 94:405-415

Gleadow A, Kohn B, Gallagher K, Cox S (1996) Imaging the thermotectonic evolution of eastern Australia during the Mesozoic from fission track dating of apatites. *In:* Mesozoic Geology of the Eastern Australia Plate Conference, Geol Soc Aust Abstr 43, p 195-204

Gleadow AJW, Kohn BP, Brown RW, O'Sullivan PB, Raza, A (2002b) Fission track thermotectonic imaging of the Australian continent. Tectonophysics 349:5-21

Gray DR (1997) Tectonics of the southeastern Australian Lachlan Fold Belt: structural and thermal aspects. *In:* Orogeny Through Time. Burg JP, Ford M (eds) Geol Soc Lond Spec Pub 121, p 149-177

Green PF (1988) The relationship between track shortening and fission track reduction in apatite: combined influences of inherent instability, annealing anisotropy, length bias and system calibration. Earth Sci Planet Lett 89:335-352

Green PF, Duddy IR, Gleadow AJW, Tingate PR, Laslett GM (1985) Fission track annealing in apatite: Track length measurements and the forms of the Arrhenius Plot. Nucl Tracks 10:323-328

Green PF, Duddy IR, Laslett GM, Hegarty KA, Gleadow AJW, Lovering JF (1989) Thermal annealing of fission tracks in apatite, 4. Quantitative modeling techniques and extension to geological timescales. Chem Geol 79:155-182

Greene LC, Richards DR, Johnson RA (1991) Crustal structure and tectonic evolution of the Anza Rift, Northern Kenya. Tectonophys 197:203-211

Gregory JW (1896) The Great Rift Valley. London, John Murray

Guillou-Frottier L, Mareschal JC, Jaupart C, Gariepy C, Lapointe R, Bienfait G (1995) Heat flow variations in the Grenville Province, Canada. Earth Planet Sci Lett 136:447-460

Gunnell Y, Gallagher K, Carter A, Widdowson M, Hurford AJ (2003) Denudation history of the continental margin of western penisular India since the early Mesozoic – reconciling apatite fission-track data with geomorphology. Earth Sci Planet Lett 6761:1-15

Gurnis M, Mitrovica JX, Ritsema J, van Heijst H-J (2000) Constraining mantle density structure using geological evidence of past uplift rates: the case of the African Superplume. Geochem Geophys Geosyst 1, doi:10.1029/1999GC000035

Harrington HJ, Korsch RJ (1985) Tectonics of the New England Orogen. Aust J Earth Sci 32:163-179

Harrison TM, Grove M, Lovera O (2005) Time-temperature paths from K-feldspar $^{40}Ar/^{39}Ar$ and monazite U/Pb dating. *In:* Thermochronology. Reiners PW, Ehlers TA (eds), Rev Min Geochem XX:xxx-xxx

Hasebe N, Barbarand J, Jarvis K, Carter A, Hurford AJ (2004) Apatite fission track chronometry using laser ablation-ICP-MS. Chem Geol 207:135-145

Hawkesworth CJ, Gallagher K, Kelley S, Mantovani M, Peate D.W, Regelous M, Rogers N (1992) Paraná magmatism and the opening of the South Atlantic. *In* Magmatism and the Causes of Continental Break-Up. Storey BC, Alabaster T, Pankhurst RJ (eds) Geol Soc Spec Pub 68, p 221-240

Hill KC, Hill KA, Cooper GT, O'Sullivan AJ, O'Sullivan PB, Richardson J (1995) Inversion around the Bass Basin, SE Australia. *In:* Basin Inversion. Buchanan JG, Buchanan PG (eds) Geol Soc Lond Spec Pub 88, p 525-547

Hoffman PF (1988) United Plates of America, the birth of a craton: Early Proterozoic assembly and growth of Laurentia. Ann Rev Earth Planet Sci 16:543-603

House MA, Wernicke BP, Farley KA (1998) Dating topography of the Sierra Nevada, California, using apatite (U-Th)/He ages. Nature 396:66-69

Isaaks EH, Srivastava RM (1989) Introduction to Applied Geostatistics, Oxford University Press

Issler DR (1996) Optimising time-step size for apatite fission-track annealing models. Computer Geosci 22: 67-74

Johnson BD, Veevers JJ (1984) Oceanic paleomagnetism. *In:* Phanerozoic History of Australia. Veevers JJ (ed) Clarendon Press, Oxford, p 17-38

Ketcham RA (2005) Forward and inverse modeling of low-temperature thermochronometry data. Rev Mineral Geochem 58:275-314

Ketcham RA, Donelick RA, Carlson WD (1999) Variability of apatite fission-track annealing kinetics: III. Extrapolation to geological time scales. Am Mineral 84:1235-1255

Ketcham RA, Donelick RA, Donelick MB (2000) AFTSolve: A program for multi-kinetic modeling of apatite fission-track data. Geol Mater Res 2

Kohn BP, Gleadow AJW, Cox, SJ (1999) Denudation history of the Snowy Mountains: constraints from apatite fission track thermochronology Aust J Earth Sci 46:181- 198

Kohn BP, Gleadow, AJW, Brown RW, Gallagher K, O'Sullivan PB, Foster DA (2002) Shaping the Australian crust over the last 300 million years: Insights from fission track thermotectonic and denudation studies of key terranes. Aust J Earth Sci 49:697-717

Kohn BP, O'Sullivan PB, Gleadow AJW, Brown RW, Gallagher K, Foster DA (2003) Fission track thermotectonic imaging and denudation history of Tasmania: Reply to Discussion by D.E. Leaman. Aust J Earth Sci 50:646-650

Kooi H, Beaumont C (1994) Escarpment evolution on high-elevation rifted margins: Insights derived from a surface processes model that combines diffusion, advection and reaction. J Geophys Res 99:12191-12209

Kreuser T, Wopfner H, Kaaya CZ, Markwort S, Semkiwa PM, Aslanidis P (1990) Depositional evolution of Permo-Triassic Karoo basins in Tanzania with reference to their economic potential. J Afr Earth Sci 10: 151-167

Kumarapelli PS (1985) Vestiges of Iapetan rifting in the craton west of the Northern Appalachians. Geosci Canada 12:54-59

Lambiase JJ (1989) The framework of African rifting during the Phanerozoic. J Afr Earth Sci 8:183-190

Laslett GM, Galbraith RF (1996) Statistical modeling of thermal annealing of fission tracks in apatite. Geochim Cosmochim Acta 60:5117-5131

Laslett GM, Green PF, Duddy IR, Gleadow AJW (1987) Thermal annealing of fission tracks in apatite, 2. A quantitative analysis. Chem Geol 65:1-13

Lister GS, Etheridge MA (1989) Detachment model for the uplift and volcanism of the Eastern Highlands. *In:* Intraplate Volcanism in Eastern Australia and New Zealand. Johnson W (ed) Cambridge Univ Press, New York, p 297-312

Lithgow-Bertelloni C, Silver PG (1998) Dynamic topography, plate driving forces and the African superswell. Nature 395:296-272

Lorencak M (2003) Low temperature thermochronology of the Canadian and Fennoscandian Shields: Integration of apatite fission track and (U-Th)/He methods. PhD thesis University of Melbourne

Lorencak M, Kohn BP, Osadetz KG, Gleadow AJW (2004) Combined apatite fission track and (U-Th)/He thermochronometry in a slowly cooled terrane: Results from a 3440 m deep drill hole in the southern Canadian Shield. Earth Planet Sci Lett 227:87-104

Lucas SB, St-Onge MR (eds) (1998) Geology of the Precambrian Superior and Grenville Provinces and Precambrian fossils in North America. Geol Canada No 7, Geol Surv Canada (also Geol Soc Amer, The Geology of North America, Decade Nth Amer Proj Vol C1)

Lutz TM, Omar GI (1991) An inverse method of modeling thermal histories from apatite fission-track data. Earth Planet Sci Lett 104:181-195

Mareschal JC, Jaupart C, Gariepy C, Cheng LZ, Guillou-Frottier L, Bienfait G, Lapointe R (2000) Heat flow and deep thermal structure near the southeastern edge of the Canadian Shield. Can J Earth Sci 37:399-414

Martin H (1953) Notes on the Dwyka succession and on some pre-Dwyka valleys in South West Africa. Trans Geol Soc S Afr 56:37-43

Martin H (1973) The Atlantic margin of southern Africa between latitude 17° south and The Cape of Good Hope. *In:* The Ocean Basins and Margins Vol 1, The South Atlantic. Nairn AEM, Stehli F (eds) Plenum Press, NY, p 277-300

Mayne SJ, Nicholas E, Bigg-Wither AL, Rasidi JS, Raine MJ (1974) Geology of the Sydney Basin – A Review. Bur Min Res Aust Bull 149

Mbede EI, Hurford A, Ebinger CJ, Schandelmeier H (1993) A constraint on the uplift history of the Rukwa Rift, SW. Tanzania: A reconnaissance apatite fission track analysis. Mus Roy Afr Cent Tervuren (Belg), Dept Geol Min Rapp Ann 1991-1992:99-108

McDougall I, Harrison TM (1999) Geochronology and Thermochronology by the $^{40}Ar/^{39}Ar$ Method. Oxford Monographs on Geology and Geophysics 9 (2nd Ed) Oxford University Press, New York

Miller JMcL, Phillips D, Wilson CJL, Dugdale LD (2004) A new tectonic model for the Delamerian and western Lachlan Orogens. Geol Soc. Austral Abst 73: p 174

Mitas L, Mitasova H (1995) Interpolation by regularized spline with tension: 1. theory and implementation. Math Geol 25:641-655

Möller A (1995) Pan- African granulites and Early Proterozoic eclogites in the Precambrian basement of eastern Tanzania: P-T-t history and crustal evolution of the complex Mozambique Belt. PhD thesis, Kiel Univ

Moore ME, Gleadow AJW, Lovering JF (1986) Thermal evolution of rifted continental margins: new evidence from fission track dating of apatites from southeastern Australia. Earth Planet Sci Lett 78:255-270

Morley CK, Wescott WA, Stone DM, Harper RM, Wigger ST, Karanja FM (1992) Tectonic evolution of the northern Kenyan Rift. J Geol Soc Lond 149:333-34

Muhongo S (1989) Tectonic setting of the Proterozoic metamorphic terrains in eastern Tanzania and their bearing on the evolution of the Mozambique Belt. IGCP No. 255 Newslett Bull 2:43-50

Munyanyiwa H (1993) Thermobarometry of mafic rocks within the Zambezi Mobile Belt, northern Zimbabwe. *In:* Proc Eighth Gondwana Symp. Finlay RH, Unrug R, Banks MR, Veevers JJ (eds) Balkema, Rotterdam, p 83-95

Noble PW (1997) Post Pan African tectonic evolution of Eastern Africa: an apatite fission track study. PhD thesis, La Trobe University

Noble WP, Foster DA, Gleadow AJW (1997) The post Pan African tectonothermal development of the Mozambique belt in Eastern Tanzania. Tectonophys 275:331-350

Norvick MS, Smith MA (2001) Mapping the plate tectonic reconstruction of southern and southeastern Australia and implications for petroleum systems. APPEA J 41:15-35

Nunn JA, Sleep NH, Moore WE (1984) Thermal subsidence and generation of hydrocarbons in Michigan Basin. Am Assoc Petrol Geol Bull 68:296-315

Nürnburg D, Müller RD (1991) The tectonic evolution of the South Atlantic from late Jurassic to present. Tectonophys 191:27-53

O'Sullivan PB, Kohn BP (1997) Apatite fission track thermochronology of Tasmania. Aust Geol Surv Org Record 1997/35

O'Sullivan PB, Belton DX, Orr M (2000a) Post-orogenic thermotectonic history of the Mount Buffalo region, Lachlan fold belt, Australia; evidence for Mesozoic to Cenozoic. Tectonophysics. 317:1-26

O'Sullivan PB, Kohn BP, Cranfield L (1999b) Fission track constraints on the Mesozoic to Recent thermotectonic history of the northern New England Orogen, southeastern Queensland. *In*: New England Orogen, Proceedings of the NEO Conference. Flood PG (ed) University of New England Press, Armidale, p 285-293

O'Sullivan PB, Coyle DA, Gleadow AJW, Kohn BP (1996b) Late Mesozoic to Early Cenozoic thermotectonic history of the Sydney Basin and the eastern Lachlan Fold Belt, Australia. *In* Mesozoic Geology of the Eastern Australia Plate Conference, Geol Soc Aust Abst 43, p. 424-432

O'Sullivan PB, Foster DA, Kohn BP, Gleadow AJW (1996a) Multiple post orogenic denudation events: An example from the eastern Lachlan fold belt, Australia. Geology 24:563-566

O'Sullivan PB, Kohn BP, Foster DA, Gleadow AJW (1995b) Fission track data from the Bathurst batholith: Evidence for rapid mid-Cretaceous uplift and erosion within the eastern highlands of Australia. Aust J Earth Sci 42:597-607

O'Sullivan PB, Orr M, O'Sullivan AJ, Gleadow AJW (1999a) Episodic Late Paleozoic to Recent denudation of the Eastern Highlands of Australia: evidence from the Bogong High Plains, Victoria. Austral J Earth Sci 46:199-216

O'Sullivan PB, Foster DA, Kohn BP, Gleadow AJW, Raza A (1995a) Constraints on the dynamics of rifting and denudation on the eastern margin of Australia: Fission track evidence for two discrete causes of rock cooling. Aust Inst Min Metal Publ 9/95:441-446

O'Sullivan PB, Gibson DL, Kohn BP, Pillans B, Pain CF (2000b) Long-term landscape evolution of the Northparkes region of the Lachlan Fold Belt, Australia: Constraints from fission track and omagnetic data. J Geol 108:1-16

O'Sullivan PB, Mitchell MM, O'Sullivan AJ, Kohn BP, Gleadow AJW (2000c) Thermotectonic history of the Bassian Rise, Australia: Implications for the breakup of eastern Gondwana along Australia's southeastern margins. Earth Planet Sci Lett 182:31-47

Patchett PJ, Embry AF, Ross GM, Beauchamp B, Harrison JC, Mayr U, Isachsen CE, Rosenberg EJ, Spence GO (2004) Sedimentary cover of the Canadian Shield through Mesozoic time reflected by Nd isotopic and geochemical results for the Sverdrup Basin, Arctic Canada. J Geol 112:39-57

Palfreyman WD (1984) Guide to the Geology of Australia. Bur Min Res Bull 181:111p.

Partridge TC (1993) The evidence for Cainozoic aridification in southern Africa. Quat Int 17:105-110

Pollack HN, Hurter S J, Johnson JR (1993) Heat flow from the Earth's interior: Analysis of the global data set. Rev Geophys 31:267-280

Raab MJ, Brown RW, Gallagher K, Carter A, Weber K (2002) Late Cretaceous reactivation of major crustal shear zones in northern Namibia: constraints from apatite fission track analysis. Tectonophys 349:75-92

Rach NM, Rosendahl BR (1989) Tectonic controls of the Speke Gulf. J Afr Ear Sci 8:471-488

Reeves CV, Karanja FM, MacLeod IN (1987) Geophysical evidence for a failed Jurassic rift and triple junction in Kenya. Earth Planet Sci Lett 81:299-311

Reiners PW (2005) Zircon (U-Th)/He thermochronometry. Rev Mineral Geochem 58:151-179

Reiners PW, Farley KA (1999) Helium diffusion and (U-Th)/He thermochronometry of titanite. Geochim Cosmochim Acta 63:3845-3859

Renne PR, Glen JM, Milner SC, Duncan AR (1996) Age of Etendeka flood volcanism and associated intrusions in southwestern Africa. Geology 24:659-662

Rosendahl BR (1987) Architecture of continental rifts with special reference to East Africa. Ann Rev Earth Planet Sci 15:445-503

Rust DJ Summerfield MA (1990) Isopach and borehole data as indicators of rifted margin evolution in southwestern Africa. Mar Pet Geol 7:277-287

Sambridge M, Braun J, McQueen H (1995) Geophysical parameterization and interpolation of irregular data using natural neighbor. Geophys J Int 122:837-857

Sanford BV, Thompson FJ, McFall GH (1985) Plate tectonics - a possible controlling mechanism in the development of hydrocarbon traps in southwest Ontario. Bull Canad Petrol Geologists 33:52-71

Sass JH, Lachenbruch AH (1979) The thermal regime of the Australian continental crust. *In*: The Earth – Its Origin, Structure and Evolution. McElhinny MW (ed) Academic Press, New York, p 301-352

Shackleton RM (1986) Precambrian collision tectonics in Africa. *In*: Collision Tectonics. Coward MP, Ries AC (eds) Geol Soc Lond Spec Pub 19, p 329-349

Smith M (1994) Stratigraphic and structural constraints on mechanisms of active rifting in the Gregory rift, Kenya. Tectonophys 236:1-2

Smith M, Mosley P (1993) Crustal heterogeneity and basement influence on the development of the Kenya rift, East Africa. Tectonics 12:591-606

Smith WHF, Wessel P (1990) Gridding with continuous splines in tension, Geophysics 55:293-305

Söhnge APG, Hälbich IW (1983) Geodynamics of the Cape Fold Belt. Geol Soc S Afr Spec Pub 12

Stagman JG (1978) An Outline of the Geology of Rhodesia. Rhodesia Geol Surv Bull 80

Steckler MS, Omar GI, Karner GD, Kohn BP (1993) Pattern of hydrothermal circulation within the Newark Basin from fission-track analysis. Geology 21:735-738

Stern RJ (1994) Arc assembly and continental collision in the Neoproterozoic East African Orogeny. Ann Rev Earth Plan Sci 22:319-351

Tagami T (2005) Zircon fission-track thermochronology and applications to fault studies. Rev Mineral Geochem 58:95-122

Tagami T, O'Sullivan PB (2005) Fundamentals of fission-track thermochronology. Rev Mineral Geochem 58: 19-47

Tankard AJ, Jackson MP, Eriksson KA, Hobday DK, Hunter DR, Minter WEL (1982) Crustal Evolution of Southern Africa: 3.8 Billion Years of Earth History, Springer-Verlag, New York

Thomas DSG Shaw PA (1990) The deposition and development of the Kalahari Group sediments, Central Southern Africa, J Afr Earth Sci 10:187-197

Thurston PC, Williams HR, Sutcliffe RH, Stott GM (eds) (1991) Geology of Ontario. Ontario Geol Surv Spec Vol 4, Parts I and II

Unternehr P, Curie D, Olivet JL, Goslin J, Beuzart P (1988) South Atlantic fits and intraplate boundaries in Africa and South America. Tectonophysics 155:169-179

Van den haute P (1984) Fission track ages of apatites from the Precambrian of Rwanda and Burundi: relationship to East African rift tectonics. Earth Planet Sci Lett 71:129-140

van der Beek P (1995) Tectonic evolution of rifts: inferences from numerical modeling and fission track thermochronology. PhD thesis Vrije Universiteit

van der Beek P, Mbede E, Andriessen PAM, Delvaux, D (1998) Denudation of the Malawi and Rukwa rift flanks (East African System) from apatite fission track thermochronology. J Afr Earth Sci 26:363-385

van der Beek PA, Braun J, Lambeck K (1999) Post-Paleozoic uplift history of southeastern Australia revisited: results from a process-based model of landscape evolution. Aust J Earth Sci 46:157-172

Veevers JJ (ed) (1984) Phanerozoic Earth History of Australia. Clarendon Press Oxford

Veevers JJ, Powell C McA, Roots SR (1991) Review of seafloor spreading around Australia. I. Synthesis of the patterns of spreading. Aust J Earth Sci 38:373-389

Veevers JJ (2000) Changes of tectono-stratigraphic regime in the Australian plate during the 99 Ma (mid-Cretaceous) and 43 Ma (mid-Eocene) swerves of the Pacific. Geology 28:47-50

Visser JNJ (1987) The paleogeography of part of southwestern Gondwana during the Permo-Carboniferous glaciation. Paleogeog Paleoclim Paleoecol 61:205-219

Visser JN (1995) Post-glacial Permian stratigraphy and geography of southern and central Africa: boundary conditions for climatic modeling. Paleogeog Paleoclim Paleoecol 118:213-243

Viviers MC, de Azevedo RLM (1988) The southeastern area of the Brazilian continental margin: its evolution during the middle and late Cretaceous as indicated by paleoecological data. Revista Bras Geosci 18: 291-298

Wagner GA, Van den haute P (1992) Fission–Track Dating. Enke Verlag – Kluwer Academic Publishers, Dordrecht

Wagner M, Altherr R, Van den haute P (1992) Apatite fission-track analysis of Kenyan basement rocks: Constraints on the thermotectonic evolution of the Kenya dome. A reconnaissance study. Tectonophys 204:93-110

Ward JD (1988) Geology of the Tsondab Sandstone Formation. J Sed Geol 55:143-162

Waschbusch P, Beaumont C, Korsch RJ (1999) Geodynamic modeling of aspects of the New England Orogen and adjacent Bowen, Gunnedah and Surat basins. *In:* New England Orogen. Proceedings of the NEO Conference. Flood PG (ed) University of New England Press, Armidale, p. 204-210

Weber UD, Hill KC, Brown RW, Gallagher K, Kohn BP, Gleadow AJW, Foster DA (2004) Sediment supply to the Gippsland Basin from thermal history analysis: constraints on Emperor-Golden Beach reservoir composition. APPEA J 44:397-415

Weissel JK, Hayes DE (1977) Evolution of the Tasman Sea re-appraised. Earth Planet Sci Lett 36:77-84

Wessel P, Smith WHF (1991) Free software helps map and display data, EOS Trans Am Geophys Union 72: 445-446

Willett SD (1997) Inverse modeling of annealing of fission tracks in apatite 1: A Controlled random search method. Am J Sci 297:939-969

Winn RD, Steinmetz JC, Kerekgyarto WL (1993) Stratigraphy and rifting history of the Mesozoic-Cenozoic Anza Rift, Kenya. Am Assoc Petrol Geol Bull 77:1989-2005

Wopfner H (1986) Evidence for Late Paleozoic glaciation in southern Tanzania. Palaegeog Paleoclim Paleoecol 56:259-275

Wopfner H, Kaaya CZ (1991) Stratigraphy and morphotectonics of Karoo deposits of the Northern Selous Basin, Tanzania. Geol Mag 128:319-334

Reviews in Mineralogy & Geochemistry
Vol. 58, pp. 567-588, 2005

Low-Temperature Thermochronometry of Meteorites

Kyoungwon Min

Department of Geology and Geophysics
Yale University, 210 Whitney Ave.
New Haven, Connecticut, 06511, U.S.A.
kyle.min@yale.edu

INTRODUCTION

Timing of accretion of the primary planetesimals in the early solar system has been investigated using short-lived (e.g., ^{26}Al, ^{53}Mn, ^{182}Hf and ^{129}I) and long-lived (e.g., ^{238}U, ^{235}U, ^{232}Th and ^{147}Sm) radionuclides. The first condensed objects in the solar system are known to be the calcium-aluminum-rich inclusions (CAIs) in chondrites (undifferentiated meteorites mainly composed of Fe- and Mg- bearing silicates), the most abundant meteorites found on Earth. The formation age of the CAIs was precisely determined by ^{207}Pb/^{206}Pb dating at ~4.56–4.57 Ga (Tilton 1988; Allègre et al. 1995; Amelin et al. 2002). Following the formation of the CAIs, more solids were rapidly accreted. Bodies grown to relatively large size (e.g., large terrestrial planets) experienced differentiation forming metallic cores, whereas smaller bodies of asteroids (e.g., parent bodies of chondritic meteorites) underwent relatively mild thermal metamorphism over the next ~15–20 Myr (summarized in Gilmour and Saxton 2001), followed by cooling to the their present thermal status.

Thermal histories of asteroids and planets after primary crystallization or metamorphism have been the subject of considerable attention. One reason for this attention is because the deduced thermal histories provide key information on accretion processes, physical dimensions and heat budget of the early solar bodies. Using Ni distribution between two metallic phases, kamacite and taenite, Wood (1967) deduced cooling rates of ~2–10 °C/Ma at ~500 °C for ordinary chondrites, and showed how this information can be used to estimate the size of the parent body and original depth of individual meteorites in the parent body. Turner et al. (1978) deduced cooling rates at ~240 ± 120 °C by studying ^{40}Ar/^{39}Ar system in 16 unshocked chondrites, and suggested that these are from parent bodies with radii less than 30 km, or from at shallow depths in larger bodies. Pellas and Storzer (1981) re-estimated the cooling rates and extended the thermal histories into the low-temperature regime for individual chondritic and iron meteorites using ^{244}Pu fission track method, finding a strong correlation between the metamorphic grade and cooling rate of H-chondrites (a subgroup of the ordinary chondrites with relatively high Fe/Si). Based on these data, they concluded that the parent body of the H-chondrites had a layered structure with less-metamorphosed, rapidly cooled meteorites (H4) at the margin and more-metamorphosed, slowly cooled meteorites (H6) at the deeper parts of the parent body. Although this "onion shell" model is questioned by several authors (Scott and Rajan 1981; Grimm 1985; Taylor et al. 1987), it is supported by later studies on thermal histories estimated from the same type of meteorites (Göpel et al. 1994; Trieloff et al. 2003). More sophisticated modeling of thermal histories has revealed internal structures and size of the chondrite parent bodies (Miyamoto et al. 1981; Bennett and McSween 1996).

Thermal histories can be also used to infer dynamic effects of impact between different solid bodies. Impacts are (1) the major surface-(re)shaping processes for most solid bodies in the solar system and (2) the dominant processes by which solid materials can be transported among

1529-6466/05/0058-0021$05.00 DOI: 10.2138/rmg.2005.58.21

different asteroids or planets. The timing, intensity and post-shock cooling histories of the impact metamorphism have been evaluated by $^{40}Ar/^{39}Ar$ and (U-Th)/He thermochronometers for various types of chondrites (Heymann 1967; Bogard and Hirsch 1980; Wasson and Wang 1991; Bogard 1995), Lunar samples (Turner et al. 1971; Schaeffer and Schaeffer 1977; Bogard et al. 1987; Bogard 1995; Cohen et al. 2000; Culler et al. 2000) and Martian meteorites (Ash et al. 1996; Turner et al. 1997; Bogard and Garrison 1999; Weiss et al. 2002; Min et al. 2004; Min and Reiners 2005).

Of the fundamental importance, estimated meteorite thermal histories can be used in calibrating decay constants of radionuclides. The accuracy of isotopic ages is seriously limited by the accuracy of decay constants of the radionuclides (e.g., Min et al. 2000), and recent publication called for improvement in the accuracy of decay constants (Begemann et al. 2001). One method to better constrain the decay constants is through comparison of different isotopic systems from the same samples. Because meteorites are normally old, they have enough daughter nuclides to be measured with high precision and accuracy, thus providing an excellent opportunity to cross-calibrate decay constants of different parent nuclides (provided the radioisotopic systems are not decoupled geochemically). The decay constant of ^{87}Rb was calibrated by comparing the isochrons obtained from chondrites with U-Th-Pb ages (Minster et al. 1982). The same kind of approach was applied to eucrites providing a ^{187}Lu decay constant calibration (Patchett and Tatsumoto 1980). The decay product of ^{40}K, ^{40}Ar, is noble gas and diffuses more readily than the daughter elements produced from ^{238}U, ^{235}U or ^{187}Lu. Therefore, the thermal history of target samples must be well known in order to calibrate the ^{40}K decay constant. The decay constant of the ^{40}K was estimated using Acapulco meteorite with an assumed rapid cooling history (Kwon et al. 2002), supported by (U-Th)/He method (Min et al. 2003), although the low-temperature thermal history of the Acapulco remains debated (Pellas et al. 1997; Trieloff et al. 2001).

Medium- to high- temperature (> ~300 °C) thermal histories of meteorites were relatively well constrained from various methods. Absolute timing when meteorites experienced certain temperatures was estimated using $^{40}Ar/^{39}Ar$ dating (references above). Cooling rates were deduced by examining microstructures and Ni contents in metallic phases of taenite and kamacite (Wood 1967; Moren and Goldstein 1978, 1979; Rasmussen 1981, 1982, 1989; Rasmussen et al. 1995, 2001; Haack et al. 1996; Yang et al. 1997a,b; Reisener and Goldstein 2003a,b) and by studying Fe^{2+}-Mg ordering states in orthopyroxene crystals (Ganguly et al. 1994; Zema et al. 1996; Ganguly and Stimpfl 2000). However, low-temperature (< ~300 °C) or short duration thermal events are poorly understood probably because of a paucity of appropriate thermochronometric constraints as well as high sensitivity of those thermochronometers to later stage thermal disturbances. The whole rock (U-Th)/He and ^{244}Pu fission track methods have been applied to constrain low-temperature histories, but many of the results have large uncertainties. The recently improved single-grain (U-Th)/He and ^{244}Pu fission methods have high potential for wide application to various meteorites. These thermochronometers, combined with other higher-temperature tools, may allow us to understand more complete thermal histories of extraterrestrial samples as well as to address some of the issues discussed above. In this chapter, I will review the (U-Th)/He and ^{244}Pu fission track methods, which are the two main low-temperature thermochronometers applicable to extraterrestrial materials.

(U-Th)/He METHOD

Fundamentals

The fundamentals of the (U-Th)/He method as applied to terrestrial materials are summarized by Harrison and Zeitler (2005). For application to meteoritic samples, four more factors need to be considered: (1) common target materials are meteoritic apatite and merrillite (also called as whitlockite) crystals, (2) the density of cosmic rays in space is much higher than on earth surface, thus cosmic ray-induced ^{4}He should be considered, (3) radiogenic ^{4}He produced from extinct radionuclides, such as ^{244}Pu, may be present in the old (>4.0 Ga) samples, and (4) trapped ^{4}He may exist.

History

In the early history of (U-Th)/He dating, iron meteorites (meteorites mainly composed of Fe and Ni in the form of Fe-Ni alloys) were the most common target materials. Among various meteorites, the iron meteorites were known to have very slow He diffusion (Paneth et al. 1930), thus they were believed to be more robust to diffusive He loss presumably yielding reliable formation ages of the parent asteroids. Paneth et al. (1930) and Arrol et al. (1942) analyzed large amounts (>10 mg) of iron meteorites and reported a wide range of (U-Th)/He ages (60–7000 Ma) with large uncertainties. Cosmic ray-induced ^{4}He, however, was not considered in the age calculations until Baur (1947) discovered a negative correlation between age and the size of iron meteorites and suggested the presence of cosmogenic ^{4}He to explain such relationship. Subsequent work revealed more direct evidence of cosmic ray-induced He production (Huntley 1948), as well as data providing quantitative evaluation of the cosmogenic He production rates and their depth dependency in iron meteorites (Baur 1948a,b). Although there was debate on the precise amount of cosmogenic ^{4}He in meteorites (Chackett et al. 1950), it was generally accepted that the cosmogenic ^{4}He should be corrected for (U-Th)/He age determinations. When the cosmogenic component is considered, the corrected (U-Th)/He ages of the iron meteorites would become much younger as the iron meteorites' U- and Th- contents are very low, and produce only small amounts of radiogenic ^{4}He compared to the cosmogenic ^{4}He. Without knowing accurate and precise production rates of ^{4}He, it is impossible to obtain reliable (U-Th)/He ages from such meteorites. To separate the cosmogenic component from the radiogenic one, Paneth et al. (1952) used ^{3}He/^{4}He of the samples with a reasonable assumption that all the ^{3}He is produced by cosmic rays. If cosmic ray-induced (^{3}He/^{4}He) is known and trapped ^{4}He is negligible, the radiogenic ^{4}He can be calculated with following equation:

$$^4\mathrm{He}_{rad} = {}^4\mathrm{He}_{meas} - {}^4\mathrm{He}_{cos} = {}^4\mathrm{He}_{meas} - {}^3\mathrm{He}_{cos} \times ({}^4\mathrm{He}/{}^3\mathrm{He})_{cos} \quad (1)$$

This method essentially provided a basis for further sophistication, and many later studies have been devoted to estimations of $(^3\mathrm{He}/^4\mathrm{He})_{cos}$. Since the early 60's, the cosmogenic noble gas isotopes of He, Ne, Ar, Kr and Xe were intensively studied to understand exposure histories of extraterrestrial materials. The production rates of cosmogenic noble gas isotopes, such as ^{3}He, ^{21}Ne, ^{38}Ar and ^{81}Kr, are now better constrained by inter-calibration among different isotopic systems in meteorites, proton-bombardment experiments, and theoretical calculations (reviewed by Wieler 2002).

U, Th and He compositions of various meteorites have been measured and (U-Th)/He ages frequently calculated. Detailed interpretations of these calculated ages, however, were rarely addressed in many studies probably because the calculated ages, commonly younger than K/Ar or ^{40}Ar/^{39}Ar ages, were believed to result from partial loss of ^{4}He during thermal disturbance(s) after formation of the parent body. Limited reproducibility of the (U-Th)/He ages undermined further vigorous applications of this method to diverse meteorites. Nevertheless, a few studies such as Heymann (1967) and Alexeev (1998) were devoted to (U-Th)/He dating on chondrites. Wasson and Wang (1991) calculated (U-Th)/He ages for ~300 ordinary chondrites based on previously published He data combined with the average U and Th compositions of the meteorite group.

In the late 80's, Zeitler et al. (1987) observed volume diffusion behavior of He in Durango apatite suggesting its possible application for (U-Th)/He thermochronology. The He diffusion properties were further examined for the (U,Th)-rich minerals titanite (Reiners et al. 1999), apatite (Farley 2000) and zircon (Reiners et al. 2002). The revised techniques are widely applied to understand low-temperature thermal histories of terrestrial samples, but rarely applied to meteorites. From single phosphate grains in Acapulco meteorite, Min et al. (2003) obtained (U-Th)/He ages of ~4.5 Ga, essentially indistinguishable from the formation age of the solar system. This study showed that the (U-Th)/He method can yield reliable results, even for such old samples. Another example of single grain study was on Los Angeles (LA) Martian meteorite (Min et al. 2004). The (U-Th)/He ages from the LA merrillites are consistent with cosmic ray exposure ages suggesting that the impact metamorphism and ejection of the LA from Mars happened in a very short time interval. The main differences between these two studies and the previous ones are (1) the amount of samples analyzed was significantly reduced to the scale of single phosphate grains having enriched U and Th, and (2) the U, Th and ^{4}He compositions were measured in the same material excluding possible age bias due to sample heterogeneity.

Sample preparation

The most common U, Th reservoirs in meteorites are phosphates such as apatite [$Ca_5(PO_4)_3(F,OH,Cl)$] and merrillite [$Ca_{18}Na_2(Mg,Fe)_2(PO_4)_{14}$]. Although phosphates are minor phases (< 1 modal %), these minerals are present in chondrites (Crozaz 1979; Crozaz et al. 1989), Acapulcoites (Zipfel et al. 1995; Mittlefehldt et al. 1996), mesosiderites (Crozaz et al. 1982), Martian meteorites, lunar samples and iron meteorites (Crozaz et al. 1982; McCoy et al. 1993). Other phosphates (brianite and panethite) were also reported in the Dayton and Carlton iron meteorites (McCoy et al. 1993).

In (U-Th)/He dating, knowing the original morphologies of individual grains is important in order to correct for alpha recoil effects (discussed in the following section). Most phosphates in meteorites are small and anhedral, thus it is almost impossible to separate grains preserving their original morphologies. Optical microscopy, commonly employed to examine the quality and size of the terrestrial samples, does not provide enough information for meteoritic samples. To infer the original morphologies, or to estimate the alpha ejection factors of individual grains, it is important to characterize each grain with non-destructive methods before (U-Th)/He analysis. Compositional mapping and Energy Dispersive Spectra (EDS) are recommended for meteorite studies. More details of sample preparation and analytical procedures are available in the Appendix.

Age corrections

Trapped and cosmogenic components of ^{4}He. Three major sources of ^{4}He in meteorites are (1) trapped, (2) cosmogenic and (3) radiogenic components. It is necessary to distinguish the radiogenic component from the other two in order to calculate (U-Th)/He ages.

The trapped component can be further divided into planetary gas and solar gas components. The planetary gas component is found in Martian meteorites as well as in primary meteorites such as carbonaceous chondrites (a group of chondrites having high carbon contents; believed to be more primary than other meteorites as characterized by their presolar signatures), but rarely reported in other meteorite groups (Ott 2002). Light noble gases in the planetary component are extremely depleted relative to the heavy noble gases (reviewed by Ott 2002; Swindle 2002), therefore planetary ^{4}He components in most meteorites (except primary meteorites) are negligible. Even in primary meteorites, most of the planetary component is contained in carbonaceous phases such as nanodiamond, SiC or graphite (e.g., Ott et al. 1981), and it is unlikely that phosphates harbor any significant amount of planetary ^{4}He.

The solar gas component, however, cannot be ignored because it is enriched in light noble gases relative to heavy ones. One way to discriminate between the trapped solar gas and cosmogenic components is through $^{20}Ne/^{22}Ne$ and $^{21}Ne/^{22}Ne$ values. Because there is no radiogenic Ne, all the Ne isotopes can be explained by solar and cosmogenic components, as described by Wieler (1998) or Lorenzetti et al. (2003) (Fig. 1). The trapped solar wind $^{20}Ne/^{22}Ne$ is ~12 whereas the cosmic ray-produced $^{20}Ne/^{22}Ne$ is close to 1. The two components also have different $^{21}Ne/^{22}Ne$ values, although the cosmogenic $^{21}Ne/^{22}Ne$ is variable depending on the duration of exposure to the cosmic rays. The solar $^{4}He/^{3}He$ is ~3500 or higher for most meteorites (Eugster 1988; Lorenzetti et al. 2003). Although the majority of meteorites are known to have negligible amounts of trapped ^{4}He, there are exceptions in aubrites and some chondrites (Wasson and Wang 1991). It is still difficult to quantitatively estimate trapped ^{4}He in U- and Th-bearing meteorites, and thus it is recommended to check if the target meteorites have $^{20}Ne/^{22}Ne$ of < ~1.2 to avoid any samples with trapped solar gas component.

Cosmic rays, which are composed mainly of protons with high energy (~10 GeV), induce interactions with nuclei in meteorites producing ^{3}He and ^{4}He. Most cosmic rays originate outside of the solar system, and are called galactic cosmic rays. To properly estimate cosmogenic component, it is necessary to obtain $(^{4}He/^{3}He)_{cos}$ (=cosmogenic isotopic ratio) or the ratio of production rates for ^{3}He (P_3) and ^{4}He (P_4). The cosmogenic ^{4}He can be estimated using the equation (1) with an assumption that $^{3}He_{meas} = {}^{3}He_{cos}$. The $(^{4}He/^{3}He)_{cos}$ has been determined for various meteorite samples. Heymann (1967) obtained He isotopic data from L-chondrites (a subgroup of the ordinary chondrites with relatively low Fe/Si; LL-chondrites have even lower Fe/Si) bearing similar petrologic features and presumably originating from the same parent asteroid. If these meteorites experienced complete degassing of He by a single event, the data should show a linear trend (isochron) in the ^{3}He vs. ^{4}He diagram (Fig. 2).

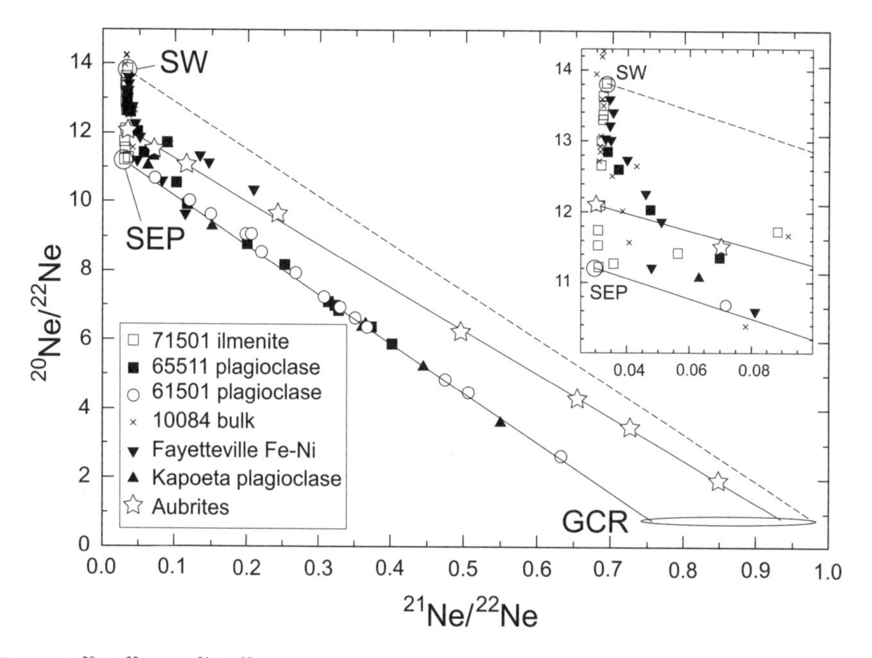

Figure 1. $^{20}Ne/^{22}Ne$ vs. $^{21}Ne/^{22}Ne$ plot showing three components of SW (solar wind), SEP (second solar; explained by Wieler 2002) and GCR (produced by galactic cosmic rays), demonstrating how the trapped solar components can be detected using Ne isotope systems. All the samples are within the field defined by these three components. The trapped solar components (SW and SEP) have higher $^{20}Ne/^{22}Ne$ values (~ 11–14) than the cosmic ray-produced component (~0.9). Data are from lunar samples (71501, 65511, 61501, 10084), H-chondrite (Fayetteville), howardite (Kapoeta) and aubrites (Wieler 1998; Lorenzetti et al. 2003). The linear trend is well established for aubrites.

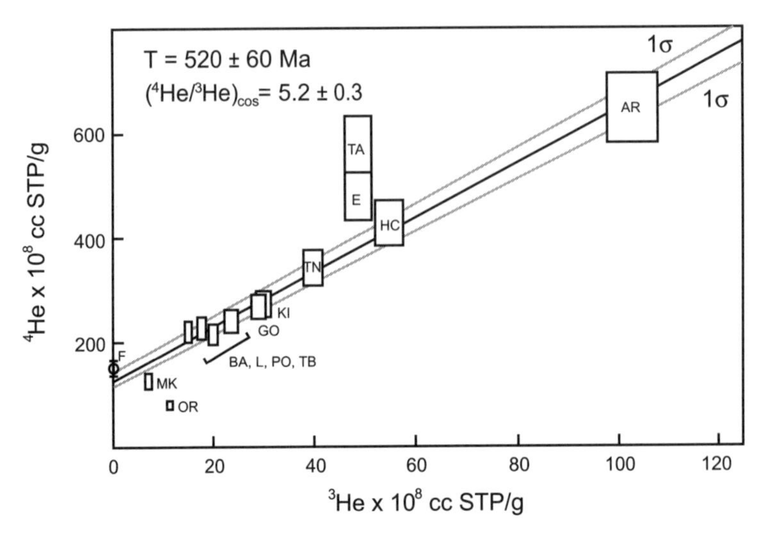

Figure 2. ^{4}He vs. ^{3}He diagram showing the linear trend of isotopic data from 14 L-chondrites. The slope and *y*-intercept of the isochron represent $(^4He/^3He)_{cos}$ and $^4He_{rad}$, respectively. Abbreviations are: AR, Arapahoe; BA, Barratta; E, Ergheo; F, Farmington; GO, Goodland; HC, Hayes Center; KI, Kingfisher; L, Lubbock; MK, McKinney; OR, Orvinio; PO, Potter; TA, Tadjera; TN, Taiban I; TB, Taiban II. [Used with permission of Elsevier, from Heymann (1967) *Icarus*, Vol. 6, p. 189-221, Fig. 12]

The slope and y-intercept of the isochron represents $(^4He/^3He)_{cos}$ and $^4He_{rad}$ (radiogenic ^{4}He), respectively. From this plot, Heymann (1967) deduced a $(^4He/^3He)_{cos}$ of 5.2 ± 0.3 (1σ). Alexeev (1998) employed essentially the same approach but used updated isotopic data for L-chondrites suggesting a higher $(^4He/^3He)_{cos}$ value of 6.1 ± 0.3 (1σ).

Iron meteorites are known to produce lower $(^4He/^3He)_{cos}$ than the chondrites values. The He production rate varies as a function of chemical composition and location (depth) of the sample in the meteorite body. Because iron meteorites do not have considerable radiogenic ^{4}He, it is possible to obtain reliable $(^4He/^3He)_{cos}$ by measuring $(^4He/^3He)$ and abundances of U and Th. For He-rich iron meteorites, Schultz and Franke (2000) deduced $(^4He/^3He)_{cos}$ of 3.2–4.4, generally consistent with the previous results (~4) obtained from proton-bombardment experiments with iron targets (Schaeffer and Zähringer 1959).

To use the Equation (1), the cosmogenic ^{3}He content should be measured for the target sample. Alternatively, the cosmogenic ^{3}He can be estimated if the production rate of ^{3}He (P_3), and the mass and exposure age of the sample are known:

$$^3He_{cos}\,(\text{atoms}) = P_3(\text{atoms/g} - \text{Ma}) \times \textit{mass of sample}\ (\text{g}) \times \textit{exposure age}\ (\text{Ma}) \tag{2}$$

The P_3 is better constrained than the P_4 simply because there is no radiogenic ^{3}He.

Eugster (1988) estimated production rates for noble gas isotopes of ^{3}He, ^{21}Ne, ^{38}Ar and ^{83}Kr by comparing ^{81}Kr-Kr exposure ages from various chondrites. The following general equation was suggested for P_3 calculation:

$$P_3 = F\left[2.09 - 0.43\left(^{22}Ne/^{21}Ne\right)\right] \tag{3}$$

where F is a composition factor ranging from 0.97 to 1.01 for different kinds of chondrites. $^{22}Ne/^{21}Ne$ is used to correct for shielding (the depth effect). A P_3 of 1.61 × 10^{-8} cm^3 STP/g-Ma was suggested for average chondrite composition at average depth (average $^{22}Ne/^{21}Ne$ =

1.11). The production rate of ^{3}He has also been studied by performing proton-bombardment experiments for various target materials (e.g., Leya et al. 2000) and by applying a pure physical model (Masarik et al. 2001; reviewed by Wieler 2002). One of the advantages of the proton-bombardment experiments is that an elemental production rate can be deduced, thus the results eventually can be generalized to any compositions of target materials. The proton-bombardment experiments combined with physical model yield production rates consistent with but slightly lower than the Eugster's (1988) estimates. In combining the estimated $(^4\mathrm{He}/^3\mathrm{He})_{cos}$ and P_3 (1.61×10^{-8} cm^3 STP/g-Ma) values, P_4 of $8 \sim 10 \times 10^{-8}$ cm^3 STP/g-Ma is deduced for bulk chondrites at average depth.

One may question if the production rate estimated for bulk chondrite is comparable with the production rate in specific target minerals, i.e., phosphates. The phosphates are mainly composed of O, Ca and P, significantly different from the bulk composition of chondrites. Elemental P_3 values were estimated for six targets of O, Mg, Al, Si, Fe and Ni (Leya et al. 2000). Unfortunately, P_3 for the main phosphate constituents of Ca and P are not available yet. It is known, however, that the production rate generally decreases as the mass difference between the target and product elements increases (Wieler 2002). If this rule is applicable to the ^{3}He production in P and Ca, then the P_3 for phosphates should be comparable with production rates for bulk chondrites.

The proportion of $^4\mathrm{He}_{cos}$ to the total ^{4}He is a function of several factors including U and Th concentrations, ^{4}He retention age, sample exposure age and the size of the samples. Heymann (1967) noted that the calculated cosmogenic ^{4}He contents in bulk samples are commonly more than 50% of the total ^{4}He, causing large uncertainties in resultant (U-Th)/He ages. As noted by Alexeev (1998), an increase of ~20% in the $(^4\mathrm{He}/^3\mathrm{He})_{cos}$ parameter causes ~25% decrease in the (U-Th)/He ages for the L-chondrites. This is true for many other chondrites, and even more serious for U- and Th- poor iron meteorites. The contribution of $^4\mathrm{He}_{cos}$ can be significantly reduced by choosing U- and Th- concentrated phases such as phosphates. Single grains of apatite with mass of 5 μg and U and Th concentrations of 5 ppm and 1 ppm, respectively, would yield less than ~0.1% $^4\mathrm{He}_{cos}$ for given conditions (He retention age = 4.5 Ga, exposure age < 50 Ma; as shown in Fig. 3a). The $^4\mathrm{He}_{cos}$ contribution increases as U and Th contents decrease. For meteorites with young He retention ages, the cosmogenic component becomes significant because of low radiogenic ^{4}He abundance (lines 4, 5 and 6 in Fig. 3a).

Single grain (U-Th)/He ages are much more robust to the cosmogenic ^{4}He correction than those from bulk samples, one of the many advantages using U-, Th- rich single grains for (U-Th)/He dating.

Effects of ^{147}Sm and ^{244}Pu. ^{147}Sm decays to ^{143}Nd producing radiogenic ^{4}He. Because of the relatively long half life (106 b.y.), low natural abundance (^{147}Sm/Sm = 15%) and low ^{4}He productivity (one ^{4}He atom per one ^{147}Sm atom), ^{147}Sm produces much less ^{4}He than U and Th do in meteorites. However, this effect may be significant if the target material has very low U- and Th- abundances. Figure 3b shows the ^{147}Sm effect for different U- and Th- contents, and cooling ages. The ^{147}Sm contents in phosphates from St Sevérin and Richardton chondrites were measured recently in the Yale (U-Th)/He lab, and the effect on the final ages is generally less than 0.5% for apatites, but up to 4–5 % for merrillites. This effect should be considered for accurate age determinations particularly for merrillites having low U and Th.

Extinct-^{244}Pu (half life = 82 m.y.) is another source of ^{4}He. Because ^{244}Pu was present during the early solar history, this correction is needed for meteorites older than ~4.4 Ga. The original amount of ^{244}Pu can be roughly inferred using fossil fission tracks and fissiogenic Xe isotopes. For various chondrites, ^{244}Pu/^{238}U values of ~0.004–0.017 were inferred with relatively large uncertainties (Hagee et al. 1990). Similar values were estimated for apatites (0.004) and merrillites (0.011) from the Acapulco meteorite (Pellas et al. 1997). The

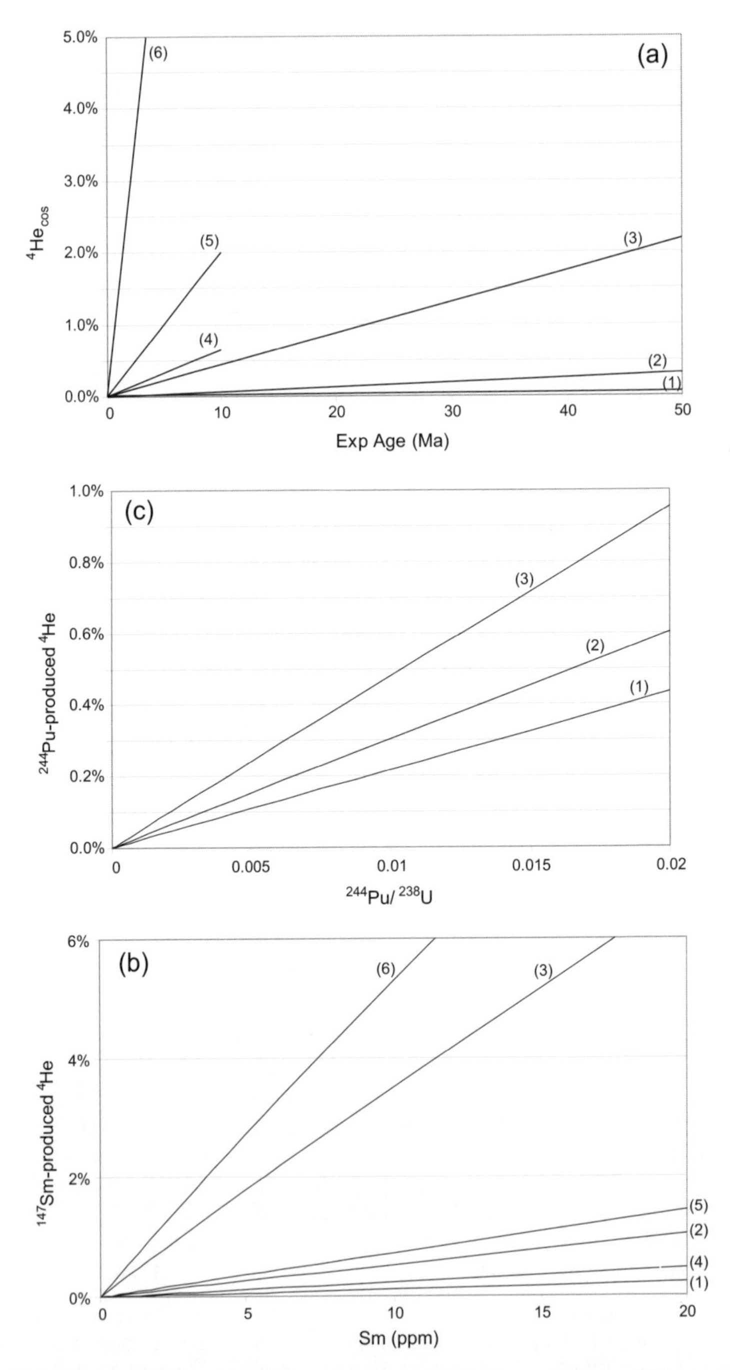

Figure 3. Contributions of ^4He produced by cosmic rays (a), ^{147}Sm (b) and extinct-^{144}Pu (c). The ^4He components were calculated for U and Th concentrations: (1, 4) 5 ppm U, 1 ppm Th; (2, 5) 0.5 ppm U, 5 ppm Th; (3, 6) – 0.1 ppm U, 0.5 ppm Th. He retention ages used were (1–3) 4.5 Ga; and (4–6) 200 Ma.

contribution of ^{244}Pu is generally less than 1% for ~4.5 Ga meteorites (Fig. 3c). For Acapulco apatites with ages older than ~4.5 Ga, the number of ^{4}He atoms produced by ^{147}Sm and extinct-^{244}Pu is calculated less than 0.15% of the total ^{4}He (Min et al. 2003).

Alpha recoil correction. When alpha particles (^{4}He atoms) are produced from parent nuclides, the alpha particles become energized and move (recoil) from their original positions. The recoil distance is generally in the range of 10–30 μm depending on the parent nuclide and the medium the alpha particles pass through. If some of alpha particles are ejected from the system, (U-Th)/He ages can appear younger than their true ages. The details of alpha recoil corrections are available in Farley et al. (1996) and Hourigan et al. (2005).

To correct for alpha ejection loss, it is important to know natural morphologies of the separated samples. Practically, it is quite difficult to estimate the original morphologies of the meteoritic phosphate samples because many of the phosphates are small, irregularly shaped and sometime intensively cracked. It is almost impossible to extract phosphate grains retaining their original crystal forms. Instead, it may be possible to indirectly infer the alpha ejection correction factors by carefully examining natural shapes (*in situ*) and separated grains of phosphates. Min et al. (2003) used phosphate crystals separated from the Acapulco meteorite and obtained alpha ejection-uncorrected (U-Th)/He ages of ~4.5 Ga, essentially same as the metamorphic age deduced from ^{207}Pb/^{206}Pb method. Based on the petrographic observations, they concluded that the phosphate grains used for the analyses were probably derived from larger grains (Fig. 4a), and the marginal parts of the original grains were removed during sample preparation procedure. For these samples, alpha ejection correction factor was estimated to be ~1.

Another example is from Los Angeles, shocked Martian meteorite. By examining thin sections and mineral separates of the phosphates from Los Angeles, Min et al. (2004) suggested that phosphate separates with other phases attached are probably from marginal parts of originally larger grains (Fig. 4b). The original phosphates have many cracks, and the morphology of separated grains was probably determined by these cracks. For such samples, alpha particles ejected from the phosphate fragment may be retained in the neighboring phases, if the thickness of the attached material is comparable with the He recoil distance,

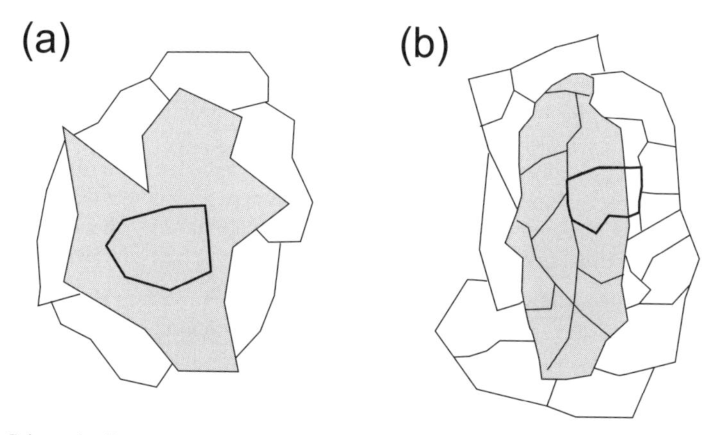

Figure 4. Schematic diagram showing the possible original locations of separated phosphates: (a) from inner part of a large grain, and (b) from the margin of a large grain containing many fractures. The shaded areas represent the original phosphate crystals. The separated grains (areas with bold curves) may retain most of the alpha particles produced inside if the abrasion thickness (for (a)) or the thickness of attached phases (for (b)) is larger than the alpha recoil distance, ~20 μm.

~20 μm. Based on the morphologies of the samples, it was concluded that the alpha ejection factors of phosphates used for analyses was ~1. They obtained alpha ejection-uncorrected (U-Th)/He ages identical with cosmic ray exposure ages within their uncertainties.

Generally it is difficult, if not impossible, to quantitatively estimate alpha ejection factors for individual grains from meteorites and the best method may be identifying samples with least potential of alpha ejection. Selecting large grains is certainly helpful in reducing the alpha ejection problems. Because of these difficulties, the single grain (U-Th)/He ages obtained from the Acapulco are regarded as minimum ages (Min et al. 2003).

Diffusion properties

Diffusion parameters. The diffusion properties of the extraterrestrial phosphates are poorly understood because phosphate phases are generally rare and small, thus contain very little He gas, causing technical difficulties in performing diffusion experiments. In addition, many meteoritic phosphates have cracks and euhedral crystal forms undermining precise estimate of the diffusion parameters. Min et al. (2003) performed stepped heating experiments for two large apatite crystals from the unshocked Acapulco meteorite and obtained $E_a = 44.2$ kcal/mol and $D_0/a^2 = 24.1\ s^{-1}$. The closure temperatures calculated from these data are slightly higher than those estimated from the data for terrestrial apatites [$E_a = 33 \pm 1.0$ kcal/mol (2σ) and $\log D_0 = 1.5 \pm 1.2\ cm^2/s$ (2σ); Farley 2000].

For merrillites, which typically have less radiogenic ^4He than apatites, there are no published estimates of He diffusion characteristics. Nevertheless, Min et al. (2004) suggested that a similarity between the He diffusion properties for apatite and merrillite can be indirectly supported by the systematic relationships between He diffusion and ^{244}Pu fission track annealing kinetics in the two minerals. The He closure temperatures for terrestrial minerals are in the order of (apatite < titanite   zircon), and a similar trend is observed for fission track annealing temperatures in these phases. The retention temperatures (temperatures at which 50% of fission tracks are retained) in extraterrestrial merrillites are comparable or slightly higher than that of apatites (see ^{244}Pu FISSION TRACK DATING). If the He diffusion and fission track annealing in merrillites also follow the general trend found in the terrestrial apatite, titanite and zircon, the He closure temperature of merrillite is expected to be similar to or slightly higher than that of apatite.

The best material for direct diffusion experiments is large phosphate crystals from old and unshocked (thus having enough ^4He) meteorites. However, such samples are rare, and the Acapulco apatite is probably one of the best samples for this purpose. Alternatively, it may be possible to measure diffusion properties using the recently developed proton-bombardment method (explained by Shuster and Farley 2005). Because this method uses artificially produced ^3He instead of radiogenic ^4He, the method can overcome the limitations of sample size and low abundance of radiogenic ^4He. This method may provide He diffusion properties of samples from individual meteorites, a step forward which will be much more powerful than assuming the same diffusion properties for extraterrestrial phosphate samples.

He diffusion during shock metamorphism. Shock metamorphism is a very short event with an instant increase of temperature followed by rapid cooling to ambient temperature. He diffusion in phosphates is sensitive enough to constrain the temperature conditions of such rapid shock metamorphism. Figure 5 shows how the ^4He concentration in apatite would change as a function of metamorphic temperature and diffusion domain size. The metamorphic temperature variations over the shock event are simulated assuming conductive cooling after given peak shock temperatures. Based on the He diffusion modeling, Min et al. (2004) concluded that the Los Angeles Martian meteorite experienced shock metamorphism at ~3.28 Ma with a peak metamorphic temperature higher than 450 °C.

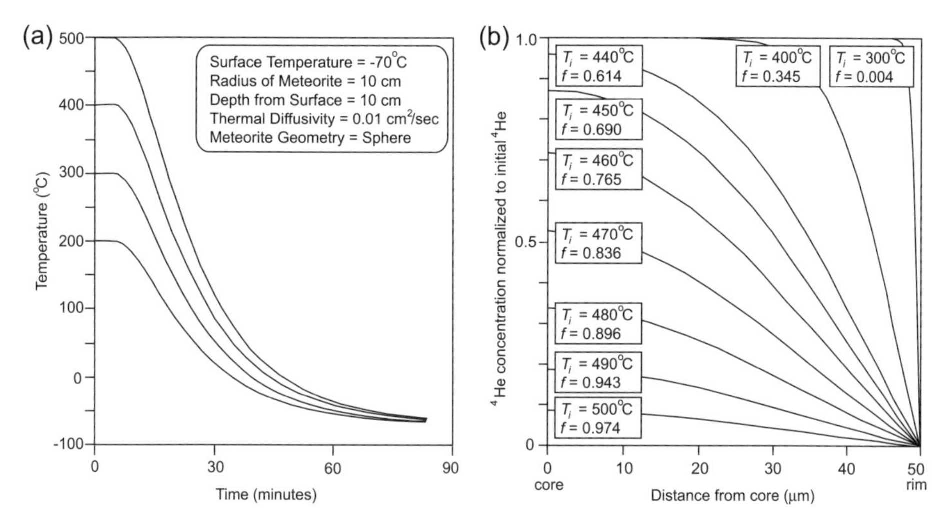

Figure 5. (a) Conductive cooling paths for samples located at the center of a 10 cm-radius meteorite. Different initial temperatures (T_i) are used for simulation, and the surface temperature of the meteorite is assumed to be −70 °C. (b) ^{4}He concentration profiles in apatite grains for different T_i. The diffusion parameters of Farley (2000: E_a = 33 kcal/mol and log D_0 = 1.5 cm^2/s) and diffusion domain radius of 50 μm were used for calculation. Total fractional ^{4}He loss (f) for each diffusion profile is also noted.

The uncertainties of this model calculation are mainly from errors associated with diffusion parameters (as discussed above) and the diffusion domain size of the samples. The shock metamorphosed mineral phases usually contain many cracks likely developed during the shock event. These cracks play an important role in He diffusion acting as fast pathways for diffusions, and making the actual diffusion domain size smaller than the crystal size. Therefore it is necessary to carefully examine the internal structures of phosphate samples to deduce the most probable diffusion domain size.

(U-Th)/He ages

Wasson and Wang (1990) calculated whole rock (U-Th)/He age for H-, L- and LL-chondrites based on previously published He isotopic data and average U and Th compositions (Fig. 6). For 16 selected L-chondrites, Heymann (1967) deduced whole rock (U-Th)/He ages ranging from <100 Ma to 4400 Ma with a peak near 500 Ma, consistent with the isochron age [~520 ± 60 Ma (1σ); Fig. 2]. These ages are interpreted as the timing of impact, followed by catastrophic breakup of the parent asteroid of the L-chondrites. The ages are, however, based on the ~20–30% ^{4}He loss postulated from ^{3}He/^{21}Ne distributions. Therefore the ages are sensitive to the degree and timing of ^{4}He loss presumably by solar heating after the impact event (Heymann 1967). The sample selection and data reduction procedures employed by Heymann (1967) were questioned by Alexeev (1998) who obtained much younger (U-Th)/He ages of ~340 ± 50 Ma (1σ) for the L-chondrites. However, ^{40}Ar/^{39}Ar ages (Bogard et al. 1976; Bogard 1979, 1995; Keil et al. 1994) and Rb/Sr dates (Fujiwara and Nakamura 1992) for the L-chondrites are concentrated near ~500 Ma supporting the Heymann's (1967) results. In addition, abundant chromite fragments with L-chondritic composition were found in mid-Ordovician (~480 Ma) sediment layers (Schmitz et al. 2003). These fragments have very short (~10^5 years) cosmic ray exposure ages implying rapid transport from asteroidal belts to the Earth (Heck et al. 2004). These findings confirm the suggestion of massive destruction of parent body of L-chondrites at ~500 Ma. Kring et al. (1996) assigned two more impact episodes at ~880 Ma and ~20 Ma for the L-chondrites.

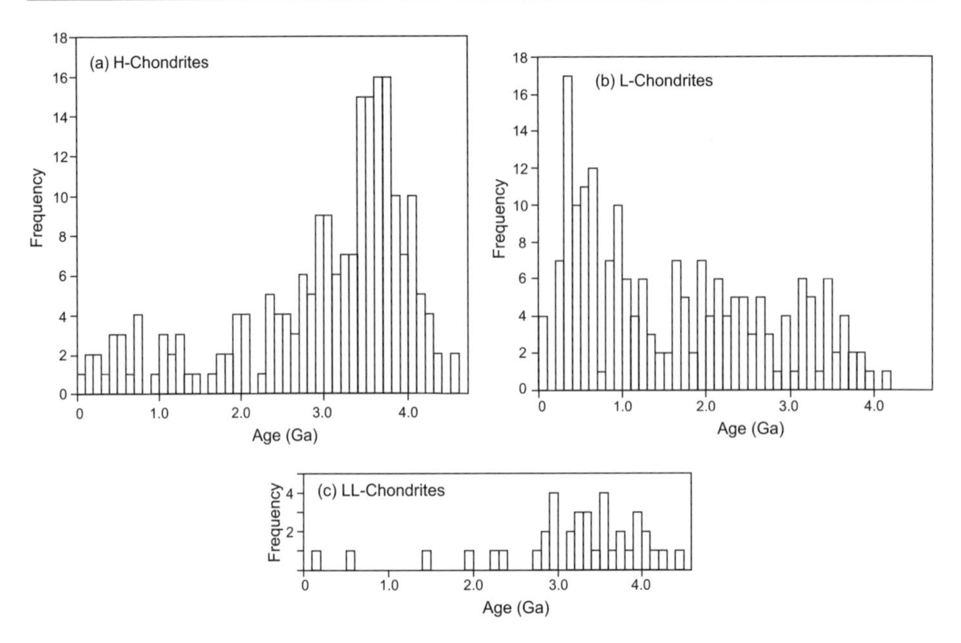

Figure 6. (U-Th)/He age distribution for (a) H-, (b) L- and (c) LL- chondrites. H- and LL- chondrites show similar patterns with broad peaks near 3-4 Ga, whereas L-chondrites have a much younger peak at ~400–500 Ma. [Used by permission of The Meteoritical Society, from Wasson and Wang (1990) *Meteoritics*, Vol. 26, p. 161-167, Fig. 2]

The H- and LL-chondrite have generally older ages than L-chondrites. The age distributions of the H- and LL-chondrites show broad scattering with peaks near 3–4 Ga (Fig. 6a,c). These ages are younger than the $^{40}Ar/^{39}Ar$ ages (> ~4.2 Ga), and the main explanations for ^4He loss is impact-induced shock metamorphism. Alexeev (1998) selected six H-chondrites having young $^{40}Ar/^{39}Ar$ ages, and deduced a significant heating at ~200 Ma for these samples. The whole rock ages of 25 achondrites of eucrites, howardites and shergottites were estimated to be 0.1–4.5 Ga, although the meanings of these estimated ages are not clear (Heymann et al. 1968).

For the Acapulco meteorite, a (U-Th)/He ages of ~3.1 Ga was reported from whole rock analyses (Palme et al. 1981). Single grain analyses applied to the phosphates from the same meteorite yielded a peak near 4.5 Ga and a few young ages (Min et al. 2003). After examining the age distribution, they suggested very heterogeneous thermal events within a few mm scale, with the primary cooling occurring after high temperature metamorphism at > 4538 ± 64 Ma (2σ). For the LA Martian meteorite, Min et al. (2004) also performed single grain analyses for phosphates, obtaining very young (U-Th)/He ages of ~3.28 ± 0.15 Ma (2σ), consistent with the exposure age (3.1 ± 0.2 Ma).

These two recent studies show that the single grain method can be used to uncover the cooling histories of the early solar system as well as to characterize young shock metamorphism. The modern (U-Th)/He method may give significantly different results from whole rock analyses, as shown for the Acapulco meteorite. Also, the large uncertainties caused by cosmogenic ^4He corrections made for whole rock ages (e.g., Heymann 1967 vs. Alexeev 1998) can be significantly reduced by performing single grain analyses. It is very likely that many of the whole rock ages in the literature (e.g., summarized by Wasson and Wang 1991) will demand revision using the improved (U-Th)/He method.

Limitations

The high sensitivity of He diffusion to temperature is one of the merits of this dating method in probing low-temperature thermal histories. However, such high sensitivity may cause significant He loss by solar heating or mild thermal disturbances. A few L-chondrites with young ^{3}He- and ^{21}Ne- exposure ages have low ^{4}He abundances best explained by loss of those noble gases (^{3}He, ^{4}He and ^{21}Ne) due to solar heating after the meteorites were exposed to cosmic rays (Crabb and Schultz 1981). The samples experienced ^{4}He loss by solar heating can be distinguished from undisturbed samples by checking the cosmic ray exposure ages. Meteorites having exceptionally young ^{3}He-exposure ages may have experienced He loss after they were exposed to cosmic rays, thus yielding erroneously young (U-Th)/He ages.

Another difficulty is from identifying and quantifying trapped ^{4}He component in certain meteorites. As discussed above, meteorites having high ^{20}Ne/^{22}Ne or trapped solar wind component should be avoided if the trapped component cannot be precisely measured. In addition, caution is required for primary meteorites which contain a trapped planetary component, although the amount of the planetary ^{4}He in target minerals (e.g., phosphates) is believed to be very small.

^{244}Pu FISSION TRACK METHOD

Fundamentals

The ^{244}Pu, a radionuclide with half life of 82 m.y., decays by spontaneous fission producing Xe and nuclear tracks (Kuroda 1960; Reynolds 1960). To calculate an absolute age, it is necessary to know the amount of ^{244}Pu at the time of track retention and the number of fission tracks produced by ^{244}Pu. However, because of the large uncertainties in estimating the extinct-^{244}Pu concentrations in specific samples during the early solar system and difficulties in discriminating ^{244}Pu fission tracks from other sources, this method is used only for relative age determinations. ^{244}Pu fission processes also register tracks in the phases adjacent to the Pu-rich phosphates. The track densities in the neighboring phases are compared with those of phosphates, and converted to cooling rates provided that the track annealing properties in each mineral are well understood. The deduced cooling rates or relative age differences can be anchored using other thermochronometers (e.g., ^{40}Ar/^{39}Ar) which yield absolute ages.

The basis of this method is similar to U fission track dating, which is commonly used for the terrestrial samples. An immediate question is why ^{244}Pu fission tracks are used instead of U fission tracks for the meteoritic samples. This choice results from much more effective ^{244}Pu (spontaneous fission decay constant = 1.058×10^{-11}/yr) production of fission tracks compared with ^{238}U (spontaneous fission decay constant = 8.46×10^{-17}/yr). Even though U was more abundant than Pu in the early solar system (^{244}Pu/^{238}U < 0.02 in bulk ordinary chondrites; Hagee et al. 1990), it was shown that Pu produces > 10 times more fission tracks than U in merrillites with relatively high Pu/U (Crozaz et al. 1989; Pellas et al. 1997).

History

After the ^{244}Pu and ^{238}U were created through nucleosynthesis, the nuclides began to spontaneously fission, generating Xe and fission tracks. The generated fission tracks were not retained until the solid body cooled down below a certain temperature. Fleischer et al. (1965) showed how the time interval (ΔT_0) between cessation of nucleosynthesis and beginning of fission track retention can be estimated from fission track data of the Moore County (achondrite) and Toluca (iron) meteorites. However, because these estimates are sensitive to initial Pu/U values, they can only provide ΔT_0 with large uncertainties.

The technique was further developed to estimate cooling rates for individual meteorites (Pellas and Strozer 1974, 1975, 1981) based on the fission track record in U- and Pu- rich phosphates of merrillite and apatite, and adjacent minerals such as feldspar, pyroxene and olivine. Because the annealing properties of tracks in these minerals are different, it should be possible to deduce cooling rates by examining fission tracks registered in the phosphates and on the surface of adjacent phases. Pellas and Storzer (1981) estimated cooling rates for individual meteorites of H- and L- chondrites over temperature range of ~300–100 °C. Crozaz and Tasker (1981) applied the ^{244}Pu fission track dating to mesosiderites and suggested more rapid cooling rate than previously estimated at low-temperature regime.

The difficulty in applying the ^{244}Pu fission track dating comes from the fact that the ^{244}Pu is extinct and estimating initial ^{244}Pu concentrations in individual phases is not a trivial issue. Because ^{244}Pu does not have any isotopes having long half life to be used as a standard, U (or Nd) is commonly used to estimate initial ^{244}Pu with an assumption that the Pu/U (or Pu/Nd) is undifferentiated (thus constant) during chemical processes in the early solar system. However, it is known to be invalid in many meteorites (Pellas and Storzer 1975). Crozaz et al. (1989) calculated the initial ^{244}Pu concentrations in single phosphate crystals in 21 ordinary chondrites by counting fission tracks and assuming the ages of the samples. By comparing the estimated ^{244}Pu contents with measured U and REE, they discovered no systematic correlation between Pu and U contents. Although they found that Nd is more likely to behave similar to Pu, they showed that none of these elements can be used to trace ^{244}Pu. Additional complexity comes from (1) possible heterogeneous distribution of Pu within phosphate crystals in a single meteorite, (2) tracks produced by interactions with cosmic rays and (3) cracks developed by shock events. Due to these and other difficulties, Crozaz et al. (1989) concluded that ^{244}Pu fission track dating could not produce reliable chronologic results. Later, the Pu/U and Pu/Nd variations for ordinary chondrites were better documented on a whole rock scale (Hagee et al. 1990) and an effective way of discriminating cosmic ray-produced tracks were developed (Pellas et al. 1989, 1997; Lavielle et al. 1992).

Pellas et al (1997) deduced cooling rates of the Acapulco meteorites in the range of ~280–90 °C by carefully correcting the observed tracks in merrillites, apatites and adjacent orthopyroxenes. They also determined ^{40}Ar/^{39}Ar ages of feldspar, and used these data to anchor the relative ages determined from ^{244}Pu fission track dating, finally obtaining a complete cooling path in the range of ~280–90 °C. Trieloff et al. (2003) recently applied the ^{244}Pu and ^{40}Ar/^{39}Ar dating methods to single grains from 14 H-chondrites, discovering correlation between the metamorphic grades and cooling rates of these samples.

Age correction

Cosmic ray tracks. Cosmic rays have energies high enough to penetrate into meteorites, producing tracks similar to fission tracks. The density of the cosmic ray tracks is dependent on the exposure age and location of the sample in the parent meteoroid. Samples from inner parts of meteorites with young exposure ages are preferred for ^{244}Pu fission track dating because these samples are better shielded against cosmic rays. The cosmic track density can be estimated by measuring track densities of Pu- and U- free phases in the same meteorite body. For the Acapulco apatite and merrillite crystals, with an exposure age of ~7 Ma, less than a 2% contribution was estimated (Pellas et al. 1997).

Cosmic rays also induce spallation reactions (disintegration of the target nuclei into protons, neutrons and light nuclei). Recoil during the spallation reactions also generates tracks in meteorites, which are generally short (< 2 μm; Fleischer et al. 1967). The spallation reaction is dependent on the chemical composition of the target materials as well as the exposure age and shielding effect. Therefore, spallation tracks in phosphates cannot be simply deduced from spallation tracks in other Pu- and U- free phases. There are two ways to estimate the production rate of spallation tracks in minerals. Because spallation tracks are more easily annealed than fission tracks, the spallation tracks can be selectively annealed by successive heating experiments. From this method, Pellas et al. (1989) deduced the production rate of ~5.3 tracks/cm^2-yr for merrillites in the Forest Vale chondrite. The production rate can then be calculated by comparing the number of annealed tracks and exposure age of the meteorite. Another approach is irradiating the samples with protons having energies similar to the cosmic rays. Using this technique, Pellas et al. (1989) estimated ~4 tracks/cm^2-yr for merrillites from the Forest Vale chondrite whereas Perron (1993) suggested ~6 tracks/cm^2-yr and ~1.6 tracks/cm^2-yr for merrillite and apatite, respectively. Direct application of these estimates to other samples may introduce uncertainties, however, as the deduced production rates are sensitive to many factors including shielding effect and orientation of the crystals (Perron 1993). Also, if some spallation tracks have been annealed by minor thermal events, simple calculation of the spallation effect using the production rate and exposure age may overestimate the true spallation track density. It is recommended to use both approaches for reliable track density estimation.

For Acapulco merrillite, ~10–15% tracks are from the spallation reactions (Pellas et al. 1997). For merrillites from the Forest Vale chondrite having very long exposure age of ~76 Ma, the contribution of cosmic ray-induced tracks (cosmic ray tracks + spallation tracks) is ~50% when the production rate of Pellas et al. (1997) was used. The contribution may significantly larger if the production rate of Perron (1993) is considered. Therefore, careful interpretation of the track data is required, particularly for apatites having relatively low Pu/U and samples with long exposure ages. The spallation track densities in orthopyroxene are ~2 orders of magnitude smaller than for merrillites (Fleischer et al. 1971).

U fission tracks and cracks. U is another source of fission tracks. This effect can be corrected if the U concentration can be estimated. One way to estimate U concentration is analyzing samples using ICP-MS or a neutron activation method, and assuming the same U concentrations for the target materials. A more direct way is by irradiating the track-measured samples and detectors (e.g., mica) attached to them, then counting the number of artificially generated tracks registered in the monitors. This process is similar to the normal U fission track method applied to terrestrial materials.

Because U is more enriched in apatite (~1–10 ppm) than in merrillite (<1 ppm), the U fission tracks are more abundant in apatite. For meteorites with ages > ~4 Ga, the contribution of U fission tracks is ~1% in merrillites, whereas up to ~50% in apatites (Pellas and Storzer 1981). For this reason, merrillites with low U content are more preferred over apatite for ^{244}Pu fission track dating.

Shock induced cracks developed in samples may be confused with fission tracks. In many cases, it is difficult to distinguish these cracks from fission tracks, thus the ^{244}Pu fission track method is limited to the meteorites having no shock features. Even unshocked meteorites may have cracks, requiring careful examination of the samples. For example, Pellas et al. (1997) argued that more than 50% of the tracks in apatites from the unshocked meteorite of Acapulco are actually cracks (see Table 2 in Pellas et al. 1997), although they did not find any crack components from merrillites and orthopyroxenes in the same meteorite. To overcome these complications, it will be necessary to develop a more direct way to discriminate cracks from fission tracks.

Track registration efficiency. The efficiency of track registration varies for different phases, thus the measured track densities need to be corrected accordingly. The registration efficiency can be estimated by irradiating polished target crystals attached on a thick source of fission fragments of known activity (e.g., ^{252}Cf). Commonly a mica standard is also irradiated with the samples, and densities of the generated fission tracks in the standard are compared with those in the samples (Pellas et al. 1997).

Annealing properties

Understanding the annealing properties of tracks in each mineral phase is essential to deduce thermochronologic information from track data. The general trend of retention temperatures is known to be in the order of feldspar > orthopyroxene > merrillite ≥ apatite (Fleisher et al. 1975; Pellas and Storzer 1975). From empirical calibrations, Pellas and Strorzer (1981) suggested approximate retention temperatures of ~327 (357) ± 30 °C for feldspar, ~277 (307) ± 20 °C for orthopyroxene and ~92 (117) ± 15 °C for phosphate for cooling rates of 1–3 °C /Ma (each number in parentheses corresponds to retention temperatures for a cooling rate of 8–20 °C/Ma). More reliable data were obtained by Mold et al. (1984) who performed annealing experiments for fission tracks in merrillite and apatite from the Estacado chondrite (Fig. 7). By extrapolating the results to natural conditions, they obtained closure temperatures (linearly extrapolated intercept from the number of tracks vs. temperature plot) of 84 ±10 °C and 66 ±10 °C for merrillite and apatite, respectively, at a cooling rate of 1 °C/Ma. They also performed annealing experiments for induced tracks in merrillites from the Bondoc mesosiderite, and obtained a higher closure temperature of 134 ± 15 °C at a cooling rate of 0.2 °C/Ma. Retention temperatures of ~120 ± 25 °C and ~ 90 ± 20 °C were suggested to be the most reliable (Pellas et al. 1997; Trieloff et al. 2003). The retention temperatures of orthopyroxene were estimated to be 267–307 °C for different cooling rates of 0.5-100 °C/Ma (unpublished data cited in Pellas et al. 1997).

^{244}Pu fission track data

Pellas et al. (1997) applied ^{244}Pu fission track and ^{40}Ar/^{39}Ar dating methods to the Acapulco meteorite which is believed to have a common parent body with Lodranites. Based on fission track data from orthopyroxene, merrillite and apatite, slow cooling (~1.7 ± 0.5 °C/Ma) was deduced in temperature range of ~280–90 °C. The ^{40}Ar/^{39}Ar experiments applied to bulk samples yielded a plateau age of 4514 ± 32 Ma (2σ) and a plagioclase closure temperature of 290 °C. Because the deduced Ar closure temperature in plagioclase is indistinguishable from the track retention temperature of orthopyroxene, they could anchor the orthopyroxene fission track ages using plagioclase ^{40}Ar/^{39}Ar data, and concluded that the ^{244}Pu fission track ages are ~4410 Ma and ~4400 Ma for merrillite and apatite, respectively.

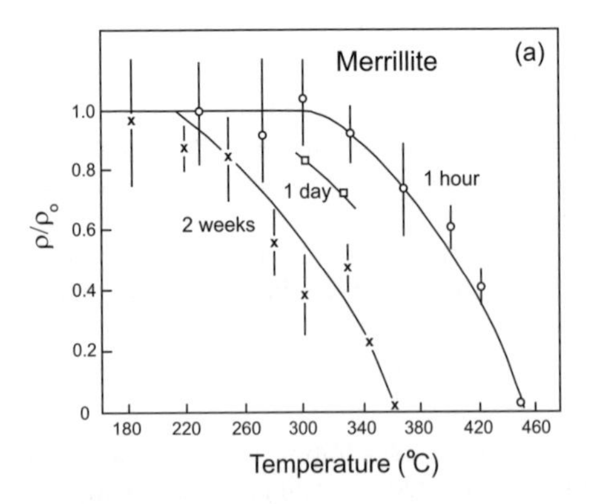

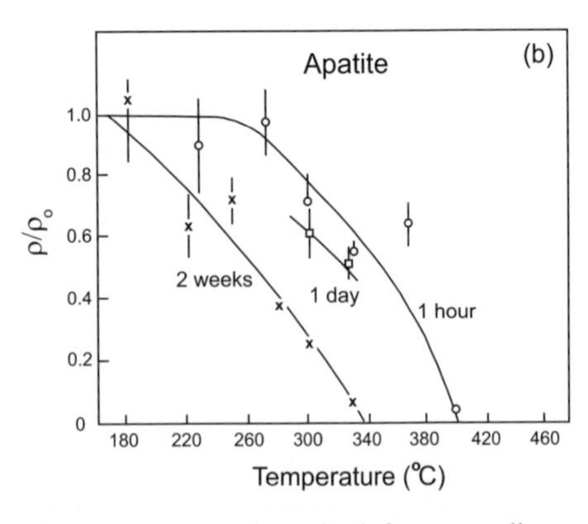

Figure 7. Track density, annealing temperature and time relationship determined from annealing experiments for (a) merrillite and (b) apatite from the Estacado chondrite. From these annealing experiments, closure temperatures of 84 ±10 °C and 66 ±10 °C were determined for merrillite and apatite, respectively. [Used by permission of Elsevier, from Mold et al. (1984) *Nuclear Tracks and Radiation Measurements*, Vol. 9, p. 119-128, Figs. 1 and 2]

The reliability of the traditional decay constant of ^{40}K (5.543×10^{-10}/yr; Steiger and Jäger 1977) was questioned (Min et al. 2000), however, and Renne (2000) suggested possibility of older $^{40}Ar/^{39}Ar$ ages (4554 Ma) for the Acapulco meteorite based on single grain analyses and updated ^{40}K decay constants. The suggestion of increasing $^{40}Ar/^{39}Ar$ ages was generally accepted, although the degree of shift is debated due to the large uncertainties of the ^{40}K decay constants and difficulties in determining precise ages for meteorite samples (Renne 2000, 2001; Trieloff et al. 2001). If the true $^{40}Ar/^{39}Ar$ age of the Acapulco is older than the values based on the traditional decay constant of Steiger and Jäger (1977) by ~47 Ma (Renne 2000, 2001) or ~30 Ma (Trieloff et al. 2001, 2003), the ^{244}Pu fission track age of the merrillite becomes 4460 ± 30 Ma (2σ) or 4443 ± 30 Ma (2σ), respectively. Min et al. (2003) determined (U-Th)/He ages [>4538 ± 64 Ma (2σ)] from single crystals of apatite suggesting higher cooling rate than deduced by Pellas et al. (1997), questioning the retentivity of the fission tracks by later stage thermal disturbances. Nevertheless, the revised ^{244}Pu fission track ages and (U-Th)/He ages [>4538 ± 64 Ma (2σ)] slightly overlap within 2σ uncertainties, and more accurate and precise age estimates on the low-temperature regime are required to better constrain ^{40}K decay constants as well as to understand cooling history of the Acapulco meteorite.

The ^{244}Pu fission track method is most intensively applied to unshocked H-chondrites which are presumed to have originated from a single parent body. The H-chondrites are divided into four groups depending on their metamorphic grades: from H3 (the least metamorphosed) to H6 (the most metamorphosed). Trieloff et al. (2003) recently inferred low-temperature (280–120 °C) cooling histories of these samples by applying $^{40}Ar/^{39}Ar$ method to feldspar and pyroxene separates and whole rock samples, and by applying ^{244}Pu fission track dating to merrillite and orthopyroxene crystals. After correcting the decay constant effect for $^{40}Ar/^{39}Ar$ age calculation, they found a systematic increase of cooling rates from <4 °C/Ma for H6 chondrites to >8 °C/Ma for H4 samples (Fig. 8). These results support a layered structure

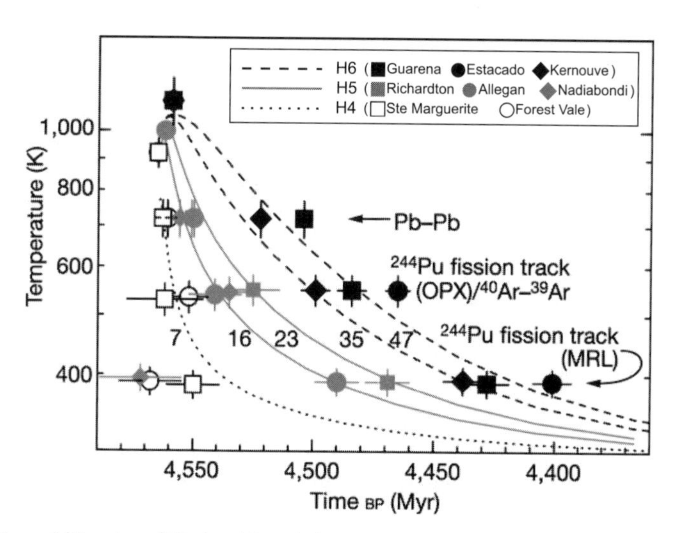

Figure 8. Thermal histories of H chondrites deduced from various thermochronometers. The cooling histories in the low-temperature regime (280–120 °C) are from ^{244}Pu fission track data linked with $^{40}Ar/^{39}Ar$ results (Trieloff et al. 2003). The $^{40}Ar/^{39}Ar$ ages were up by 30 Myr to account for the decay constant effect (Renne 2000, 2001; Trieloff 2001). Cooling curves are calculated for a chondritic 100-km diameter asteroid internally heated by ^{26}Al decay for depths of 47, 35 (dashed), 23, 16 (solid) and 7 km (dotted). [Used by permission of Nature Publishing Group (*http://www.nature.com*), from Trieloff et al. (2003) *Nature*, Vol. 422, p. 502-506, Fig. 3b]

model for H parent body where rocks in the center were heated to higher temperatures followed by slow cooling, whereas the outer shell was heated less and cooled down rapidly. Another implication is the main heat source of the early metamorphism in the parent body of H-chondrites was short-lived radionuclides, most likely ^{26}Al (Lee et al. 1976).

Limitations

Although ^{244}Pu fission track dating is a powerful tool for estimating cooling rates in many meteorites, it should be used with great care due to following limitations. (1) Because of the difficulties in accurately estimating the initial ^{244}Pu concentration in target materials, the ^{244}Pu method, if not calibrated to other chronometers, cannot yield reliable absolute ages. The method can yield age differences among phases, however, which can be converted into cooling rates. (2) Extinct-^{244}Pu has a short half life (82 m.y.) and its initial abundance in the early solar system was very small, thus the method can yield cooling rates for sample histories >~4 Ga. This would not be applicable to shocked or young meteorites (e.g., Martian meteorites). (3) The ^{244}Pu fission tracks must be distinguished from U fission tracks or other dislocations. These corrections introduce large uncertainties in estimating the density of ^{244}Pu fission tracks, most seriously for apatite having relatively low Pu/U.

CONCLUDING REMARKS

The (U-Th)/He dating can provide absolute ages for when the body passed through ~100 °C , whereas the ^{244}Pu fission track method may yield cooling rates over ~300–100 °C for old (> ~4 Ga) meteorites. The combination of these two methods provides the best constraints for low-temperature thermal histories of meteorites.

The (U-Th)/He method applied to single grains may provide useful information on the timing and temperature conditions of impact events. This information can be compared with cosmic ray exposure ages to understand physical processes (ejection or breakup) of the target meteorites. If the material was exposed to cosmic rays before its ejection, the cosmic ray exposure age will be older than the timing of ejection-related shock metamorphism. If the material was ejected from parent asteroid, but resided in a large body before it exposed to cosmic rays by later breakup processes, the exposure age will be younger than the timing of primary impact. The exposure ages will be same as the (U-Th)/He ages if the ejection, shock metamorphism and exposure happen within a brief period (e.g., LA Martian meteorite). Related questions are: (1) How much heat would be generated by impact events and breakup episodes? (2) What would be the distribution of the generated heat in the body? (3) Which physical processes can explain these heat distributions? Quantitative estimations of the temperature (and pressure) conditions as well as timing of the shock metamorphism may be deduced by studying He diffusion in extraterrestrial materials.

ACKNOWLEDGMENTS

Constructive reviews from Mario Trieloff and an anonymous reviewer greatly improved the original manuscript. Comments from Kim Knight, Jeremy Hourigan and Paul Renne are gratefully appreciated.

REFERENCES

Alexeev VA (1998) Parent bodies of L and H chondrites: times of catastrophic events. Meteoritics Planet Sci 33:145-152

Allègre CJ, Manhès G, Göpel C (1995) The age of the Earth. Earth Planet Sci Lett 59:1445-1456

Amelin Y, Krot AN, Hutcheon ID, Ulyanov AA (2002) Lead isotopic ages of chondrules and calcium-aluminum-rich inclusions. Science 297:1678-1683
Arrol WJ, Jacobi RB, Paneth FA (1942) Meteorites and the age of the solar system. Nature 149:235-238
Ash RD, Knott SF, Turner G (1996) A 4-Gyr shock age for a martian meteorite and implications for the cratering history of Mars. Nature 380:57-59
Bauer CA (1947) Production of helium in meteorites by cosmic radiation. Phys Rev 72:354-355
Bauer CA (1948a) The absorption of cosmic radiation in meteorites. Phys Rev 74:225-226
Bauer CA (1948b) Rate of production of helium in meteorites by cosmic radiation. Phys Rev 74:501-502
Begemann F, Ludwig KR, Lugmair GW, Min K, Nyquist LE, Patchett PJ, Renne PR, Shih C-Y, Villa IM, Walker RJ (2001) Call for an improved set of decay constants for geochronological use. Geochim Cosmochim Acta 65:111-121
Bennett ME III, McSween HY Jr. (1996) Revised model calculations for the thermal histories of ordinary chondrite parent bodies. Meteoritics Planet Sci 31:783-792
Bogard DD (1979) Chronology of asteroid collisions as recorded in meteorites. In: Asteroids. Gehrels T (ed) Arizona University Press, Tucson, p 558-578
Bogard DD (1995) Impact ages of meteorites: A synthesis. Meteoritics 30:244-268
Bogard DD, Garrison DH (1999) Argon-39-argon-40 'ages' and trapped argon in Martian shergottites, Chassigny, and Alan Hills 84001. Meteoritics Planet Sci 34:451-473
Bogard DD, Hirsch WC (1980) $^{40}Ar/^{39}Ar$ dating, Ar diffusion properties, and cooling rate determinations of severely schocked chondrites. Geochim Cosmochim Acta 44:1667-1682
Bogard DD, Horz F, Johnson P (1987) Shock effects and argon loss in samples of the Leedey L6 chondrite experimentally shocked to 29-70 GPa pressures. Geochim Cosmochim Acta 51:2035-2044
Bogard DD, Husein L, Wright RJ (1976) ^{40}Ar-^{39}Ar dating of collisional events in chondrite parent bodies. J Geophys Res 81:5664-5678
Chackett KF, Golden J, Mercer ER, Paneth FA, Reasbeck P (1950) The Beddgelert meteorite. Geochim Cosmochim Acta 1:3-14
Cohen BA, Swindle TD, Kring DA (2000) Support for the lunar cataclysm hypothesis from lunar meteorite impact melt ages. Science 290:1754-1756
Crabb J, Schultz L (1981) Cosmic-ray exposure ages of the ordinary chondrites and their significance for parent body stratigraphy. Geochim Cosmochim Acta 45:2151-2160
Crozaz G (1979) Uranium and thorium microdistributions in stony meteorites. Geochim Cosmochim Acta 43: 127-136
Crozaz G, Pellas P, Bourot-Denise M, de Chazal SM, Fiéni C, Lundberg LL, Zinner E (1989) Plutonium, uranium and rare earths in the phosphates of ordinary chondrites-the quest for a chronometer. Earth Planet Sci Lett 93:157-169
Crozaz G, Sibley SF, Tasker DR (1982) Uranium in the silicate inclusions of stony-iron and iron meteorites. Geochim Cosmochim Acta 46:749-754
Crozaz G, Tasker DR (1981) Thermal history of mesosiderites revisited. Geochim Cosmochim Acta 45:2037-2046
Culler TS, Becker TA, Muller RA, Renne PR (2000) Lunar impact history from $^{40}Ar/^{39}Ar$ dating of glass spherules. Science 287:1785-1788
Eugster O (1988) Cosmic-ray production rates for ^{3}He, ^{21}Ne, ^{38}Ar, ^{83}Kr, and ^{126}Xe in chondrites based on ^{81}Kr-Kr exposure ages. Geochim Cosmochim Acta 52:1649-1662
Farley KA, Wolf RA, Silver LT (1996) The effect of long alpha-stopping distances on (U-Th)/He ages. Geochim Cosmochim Acta 60:4223-4229
Farley KA (2000) Helium diffusion from apatite: General behavior as illustrated by Durango fluorapatite. J Geophys Res 105:2903-2914
Fleischer RL, Hart HR, Comstock GM, Evwaraye AO (1971) The particle track record of the Ocean of Storms. Proc Lunar Sci Conf III: 2259-2568
Fleischer RL, Price PB, Walker RM (1965) Spontaneous fission tracks from extinct Pu^{244} in meteorites and the early history of the solar system. J Geophys Res 70:2703-2707
Fleischer RL, Price PB, Walker RM (1975) Nuclear Tracks in Solids. University of California Press, Berkeley
Fleischer RL, Price PB, Walker RM, Maurette M (1967) Origins of fossil charged particle-tracks in meteorites. J Geophys Res 72:331-353
Fujiwara T, Nakamura N (1992) Additional evidence of a young impact-melting event on the L-chondrite parent body. Lunar Planet Sci XXIII: 387-388
Ganguly J, Hexiong Y, Ghose S (1994) Thermal history of mesosiderites: Quantitative constraints from compositional zoning and Fe-Mg ordering in orthopyroxenes. Geochim Cosmochim Acta 58:2711-2723
Ganguly J, Stimpfl M (2000) Cation ordering in orthopyroxenes from two stony-iron meteorites: Implications for cooling rates and metal-silicate mixing. Geochim Cosmochim Acta 64:1291-1297

Gilmour JD, Saxton JM (2001) A time-scale of formation of the first solids. Phil Trans R Soc London A 359: 2037-2048
Göpel C, Manhès G, Allegrè CJ (1994) U-Pb systematics of phosphates from equilibrated ordinary chondrites. Earth Planet Sci Lett 121:153-171
Grimm RE (1985) Penecontemporaneous metamorphism, fragmentation, and reassembly of ordinary chondrite parent bodies. J Geophys Res (B2) 90:2022-2028
Haack H, Scott ERD, Rasmussen KL (1996) Thermal and shock history of mesosiderites and their large parent asteroid. Geochim Cosmochim Acta 60:2609-2619
Hagee B, Bernatowicz TJ, Podosek FA, Johnson ML, Burnett DS, Tatsumoto M (1990) Actinide abundances in ordinary chondrites. Geochim Cosmochim Acta 54:2847-2858
Harrison TM, Zeitler PK (2005) Fundamentals of noble gas thermochronometry. Rev Mineral Geochem 58: 123-149
Heck PR, Schmitz B, Baur H, Halliday AN, Wieler R (2004) Fast delivery of meteorites to Earth after a major asteroid collision. Nature 430:323-325
Heymann D (1967) On the origin of hypersthene chondrites: Ages and shock effects of black chondrites. Icarus 6:189-221
Heymann D, Mazor E, Anders E (1968) Ages of the Ca-rich achondrites. In: Astrophysics and Space Science Library. Vol 12. Meteorite research. Millman PM (ed) D Reidel Publishing Company, Dordrecht, p 125-170
Hourigan JK, Reiners PW, Brandon MT (2005) U-Th zonation-dependent alpha-ejection in (U-Th)/He chronometry. Geochim Cosmochim Acta 69:3349-3365
Huntley HE (1948) Production of helium by cosmic rays. Nature 161:356
Keil K, Haack H, Scott ERD (1994) Catastrophic fragmentation of asteroids: evidence from meteorites. Planet Space Sci 42:1109-1122
Kring DA, Swindle TD, Britt DT, Grier JA (1996) Cat Mountain: a meteoritic sample of an impact-melted asteroid regolith. J Geophys Res 101:29353-29371
Kuroda PK (1960) Nuclear fission in the early history of the earth. Nature 187:36-38
Kwon J, Min K, Bickel PJ, Renne PR (2002) Statistical methods for jointly estimating the decay constant of ^{40}K and the age of a dating standard. Math Geol 34:457-474
Lavielle B, Marti K, Pellas P, Perron C (1992) Search for ^{248}Cm in the early solar system. Meteoritics 27: 382-386
Lee T, Papanastassiou DA, Wasserburg GW (1976) Demonstration of ^{26}Mg excess in Allende and evidence for ^{26}Al. Geophys Res Lett 3:109-112
Leya I, Lange H-J, Neumann S, Wieler R, Michel R (2000) The production of cosmogenic nuclides in stony meteoroids by galactic cosmic-ray particles. Meteoritics Planet Sci 35:259-286
Lorenzetti S, Eugster O, Busemann H, Marti K, Burbine TH, McCoy T (2003) History and origin of aubrites: Geochim Cosmochim Acta 67:557-571
Masarik J, Nishizumi K, Reedy RC (2001) Production rates of cosmogenic helium-3, neon-21, and neon-22 in ordinary chondrites and the lunar surface. Meteoritics Planet Sci 36:643-650
McCoy TJ, Keil K, Scott ERD, Haack H (1993) Genesis of the IIICD iron meteorites: Evidence from silicate-bearing inclusions. Meteoritics 28:552-560
Min K, Mundil R, Renne PR, Ludwig KR (2000) A test for systematic errors in ^{40}Ar/^{39}Ar geochronology through comparison with U/Pb analysis of a 1.1-Ga rhyolite. Geochim Cosmochim Acta 64:73-98
Min K, Farley KA, Renne PR, Marti K (2003) Single grain (U-Th)/He ages from phosphates in Acapulco meteorite and implications for thermal history. Earth Planet Sci Lett 209:323-336
Min K, Reiners PW (2005) Low-temperature thermal history of Martian meteorite ALH84001 from (U-Th)/He thermochronometry. Lunar Planet Sci XXXVI: Abstract 2214 (CD-ROM)
Min K, Reiners PW, Nicolescu S, Greenwood JP (2004) Age and temperature of shock metamorphism of Martian meteorite Los Angeles from (U-Th)/He thermochronometry. Geology 32:677-680
Minster J-F, Birck J-L, Allègre CJ (1982) Absolute age of formation of chondrites studied by the ^{87}Rb-^{87}Sr method. Nature 300:414-419
Mittlefehldt DW, Lindstrom MM, Bogard DD, Garrison DH, Field SW (1996) Acapulco- and Lodran-like achondrites: Petrology, geochemistry, chronology, and origin. Geochim Cosmochim Acta 60:867-882
Miyamoto M, Fujii N, Takeda H (1981) Ordinary chondrite parent body: An internal heating model. Proc Lunar Planet Sci Conf XII: 1145-1152
Mold P, Bull RK, Durrani SA (1984) Fission-track annealing characteristics of meteoritic phosphates. Nucl Tracks Rad Meas 9:119-128
Moren AE, Goldstein JI (1978) Cooling rate variations of group IVA iron meteorites. Earth Planet Sci Lett 40: 151-161
Moren AE, Goldstein JI (1979) Cooling rates of group IVA iron meteorites determined from a ternary Fe-Ni-P model. Earth Planet Sci Lett 43:182-196

Ott U (2002) Noble gases in meteorites - trapped components. Rev Mineral Geochem. 47:71-100

Ott U, Mack R, Chang S (1981) Noble-gas-rich separates from the Allende meteorite. Geochim Cosmochim Acta 45:1751-1788

Palme H, Schultz L, Spettel B, Weber HW, W nke H (1981) The Acapulco meteorite: chemistry, mineralogy and irradiation effects. Geochim Cosmochim Acta 45:727-752

Paneth FA, Reasbeck P, Mayne KI (1952) Helium 3 content and age of meteorites. Geochim Cosmochim Acta 2:300-303

Paneth FA, Urry Wm.D, Koeck W (1930) The age of iron meteorites. Nature 125:490-491

Patchett PJ, Tatsumoto M (1980) Lu-Hf total-rock isochron for the eucrite meteorites. Nature 288:571-574

Pellas P, Fi ni C, Perron C (1989) Spallation and fission tracks in chondritic merrillites. Meteoritics 24:314-315

Pellas P, Fi ni C, Trieloff M, Jessberger EK (1997) The cooling history of the Acapulco meteorite as recorded by the ^{244}Pu and ^{40}Ar-^{39}Ar chronometers. Geochim Cosmochim Acta 61:3477-3501

Pellas P, Storzer D (1974) The plutonium-244 fission track record in ordinary chondrites: Implications for cooling rates. Meteoritics 9:388-390

Pellas P, Storzer D (1975) Uranium and plutonium in chondritic phosphates. Meteoritics 10:471-473

Pellas P, Storzer D (1981) ^{244}Pu fission track thermometry and its application to stony meteorites. Proc Royal Soc Lond A374:253-270

Perron C (1993) Cosmic ray-induced spallation recoil tracks in meteoritic phosphates: simulation at the CERN synchrocyclotron. Nucl Tracks Rad Meas 22:739-744

Rasmussen KL (1981) The cooling rates of iron meteorites-a new approach. Icarus 45:564-576

Rasmussen KL (1982) Determination of the cooling rates and nucleation histories of eight group IVA iron meteorites using local bulk Ni and P variation. Icarus 52:444-453

Rasmussen KL (1989) Cooling rates of group IIIAB iron meteorites. Icarus 80:315-325

Rasmussen KL, Haack H, Ulff-M ller F (2001) Metallographic cooling rates of group IIF iron meteorites. Meteoritics Planet Sci 36:883-896

Rasmussen KL, Ulff-M ller F, Haack H (1995) The thermal evolution of IVA iron meteorites: Evidence from metallographic cooling rates. Geochim Cosmochim Acta 59:3049-3059

Reiners PW, Farley KA (1999) Helium diffusion and (U-Th)/He thermochronometry of titanite. Geochim Cosmochim Acta 63:3845-3859

Reiners PW, Farley KA, Hickes HJ (2002) He diffusion and (U-Th)/He thermochronometry of zircon: initial results from Fish Canyon Tuff and Gold Butte. Tectonophys 349:297-308

Reisener RJ, Goldstein JI (2003) Ordinary chondrite metallography: Part 1. Fe-Ni taenite cooling experiments. Meteoritics Planet Sci 38:1669-1678

Reisener RJ, Goldstein JI (2003) Ordinary chondrite metallography: Part 2. Formation of zoned and unzoned metal particles in relatively unshocked H, L, and LL chondrites. Meteoritics Planet Sci 38:1679-1696

Renne PR (2000) ^{40}Ar/^{39}Ar age of plagioclase from Acapulco meteorite and the problem of systematic errors in cosmochronology. Earth Planet Sci Lett 175:13-26

Renne PR (2001) Reply to comment on ^{40}Ar/^{39}Ar age of plagioclase from Acapulco meteorite and the problem of systematic errors in cosmochronology by Mario Trieloff, Elmar K. Jessberger and Christine Fiéni. Earth Planet Sci Lett 190:271-273

Reynolds JH (1960) Determination of the age of the elements. Phys Rev Lett 4:8-10

Schaeffer GA, Schaeffer OA (1977) ^{39}Ar-^{40}Ar ages of lunar rocks. Proc Lunar Sci Conf VIII: 2253-2300

Schaeffer OA, Zähringer J (1959) High-sensitivity mass spectrometric measurement of stable helium and argon isotopes produced by high-energy protons in iron. Phys Rev 113:674-678

Schmitz B, H ggstr m T, Tassinari M (2003) Sediment-dispersed extraterrestrial chromite traces a major asteroid disruption event. Science 300:961-964

Schultz L, Franke L (2000) Helium, neon, and argon in meteorite-A data collection, update 2000. Max-Planck-Institute f r Chemie, Mainz, CD-ROM (*schultz@mpch-mainz.mpg.de*)

Scott ERD, Rajan RS (1981) Metallic minerals, thermal histories and parent bodies of some xenolithic, ordinary chondrite meteorites. Geochim Cosmochim Acta 45:53-67

Shuster DL, Farley KA (2005) ^{4}He/^{3}He thermochronometry: theory, practice, and potential complications. Rev Mineral Geochem 58:181-203

Steiger RH, Jäger E (1977) Subcommission on geochronology: convention on the use of decay constants in geo- and cosmochronology. Earth Planet Sci Lett 36:359-362

Swindle TD (2002) Martian noble gases. Rev Mineral Geochem 47:171-190

Taylor GJ, Maggiore P, Scott ERD, Rubin AE, Keil K (1987) Original structures, and fragmentation and reassembly histories of asteroids: evidence from meteorites. Icarus 69:1-13

Tilton GR (1988) Age of the solar systems: In: Meteorites and the Early Solar System. Kerridge JF, Mathews MS (ed) Arizona University Press, Tucson, p 259-275

Trieloff M, Jessberger EK, Fi ni C (2001) Comments on $^{40}Ar/^{39}Ar$ age of plagioclase from Acapulco meteorite and the problem of systematic errors in cosmochronology by Paul R. Renne. Earth Planet Sci Lett 190: 267-269
Trieloff M, Jessberger EK, Herrwerth I, Hopp J, Fi ni C, Ghelis M, Bourot-Denise M, Pellas P (2003) Structure and thermal history of the H-chondrite parent asteroid revealed by thermochronometry. Nature 422:502-506
Turner G, Enright MC, Cadogan PH (1978) The early history of chondrite parent bodies inferred from ^{40}Ar-^{39}Ar ages. Proc Lunar Sci Conf IX: 989-1025
Turner G, Huneke JC, Podosek FA, Wasserburg GJ (1971) ^{40}Ar-^{39}Ar ages and cosmic ray exposure ages of Apollo 14 samples. Earth Planet Sci Lett 12:19-35
Turner G, Knott SF, Ash RD, Gilmour JD (1997) Ar-Ar chronology of the Martian meteorite ALH84001: Evidence for the timing of the early bombardment of Mars. Geochim Cosmochim Acta 61:3835-3850
Wasson JT, Wang S (1991) The histories of ordinary chondrite parent bodies: U, Th-He age distributions. Meteoritics 26:161-167
Wieler R (1998) The solar noble gas record in lunar samples and meteorites. Space Sci Rev 85:303-314
Wieler R (2002) Cosmic-ray-produced noble gases in meteorites. Rev Mineral Geochem 47:125-170
Wood JA (1967) Chondrites: Their metallic minerals, thermal histories, and parent planets. Icarus 6:1-49
Yang C-W, Williams DB, Goldstein JI (1997a) Low-temperature phase decomposition in metal from iron, stony-iron, and stony meteorites. Geochim Cosmochim Acta 61:2943-2956
Yang C-W, Williams DB, Goldstein JI (1997b) A new empirical cooling rate indicator for meteorites based on the size of the cloudy zone of the metallic phases. Meteoritics Planet Sci 32:423-429
Zeitler PK, Herczeg AL, McDougall I, Honda M (1987) U-Th-He dating of apatite: a potential thermochronometer. Geochim Cosmochim Acta 51:5865-2868
Zema M, Domeneghetti MC, Molin GM (1996) Thermal history of Acapulco and ALHA81261 acapulcoites constrained by Fe^{2+}-Mg ordering in orthopyroxene. Earth Planet Sci Lett 144:359-367
Zipfel J, Palme H, Kennedy AK, Hutcheon ID (1995) Chemical composition and origin of the Acapulco meteorite. Geochim Cosmochim Acta 59:3607-3627

APPENDIX: SAMPLE PREPARATION AND ANALYTICAL PROCEDURES

The meteorite chips are gently crushed in a mortar agate and sieved until a certain amount of desirable size fractions are available. For meteorites having large concentrations of phosphates, it is possible to apply standard mineral separation procedures using heavy liquids and a magnetic separator. Many meteorites, however, have only minor amounts of phosphates, thus there are chances of losing significant amounts of samples during processing. In these cases, it is preferred to use X-ray methods to identify phosphates. The fractions are rinsed with alcohol or acetone briefly, and spread on double-sided carbon sticky tape for secondary electron microprobe (SEM) analyses. The grains are X-ray mapped to identify any phosphate grains, and the selected phosphates are further mapped at higher resolution. Energy Dispersive Spectra (EDS) for each phosphate is obtained to distinguish merrillite, fluoapatite and chlorapatite. The individual grains are retrieved, washed and wrapped with Pt tubes for He analyses. Each sample package is degassed using Nd-YAG laser or CO_2 laser systems. The phosphates (merrillites and apatites) can be degassed (>99%) by applying intensive laser heating for ~7 minutes (Min et al. 2003) or even for ~3 minutes (Min et al. 2004). The degassed samples are then retrieved from the laser cell, dissolved with fluoric and nitric acids, and U, Th and Sm contents are measured using ICP-MS (Inductively Coupled Plasma Mass Spectrometry).

For many Martian meteorites having small phosphate grains and young ages, the U, Th and He contents in single grains are relatively low. Many of the samples yielded signals less than 5 times the level of measured blanks, requiring careful treatment of the samples.

Reviews in Mineralogy & Geochemistry
Vol. 58, pp. 589-622, 2005

Computational Tools for Low-Temperature Thermochronometer Interpretation

Todd A. Ehlers, Tehmasp Chaudhri, Santosh Kumar

Department of Geological Sciences
University of Michigan
Ann Arbor, Michigan, 48109-1005, U.S.A.
tehlers@umich.edu

Chris W. Fuller, Sean D. Willett

Department of Earth and Space Sciences
University of Washington
Seattle, Washington, 98195, U.S.A.

Richard A. Ketcham

Jackson School of Geosciences
University of Texas at Austin
Austin, Texas, 78712-0254, U.S.A.

Mark T. Brandon

Department of Geology and Geophysics
Yale University
New Haven, Connecticut, 06520-8109, U.S.A.

David X. Belton, Barry P. Kohn, Andrew J.W. Gleadow

School of Earth Sciences
The University of Melbourne
Victoria 3010, Australia

Tibor J. Dunai

Faculty of Earth and Life Sciences
Vrije Universiteit
De Boelelaan 1085, 1081 HV
Amsterdam, The Netherlands

Frank Q. Fu

School of Geosciences
University of Sydney, NSW 2006, Australia and
CSIRO Exploration and Mining, PO Box 1130, Bentley, WA, Australia

INTRODUCTION

This volume highlights several applications of thermochronology to different geologic settings, as well as modern techniques for modeling thermochronometer data. Geologic interpretations of thermochronometer data are greatly enhanced by quantitative analysis of the data (e.g., track length distributions, noble gas concentrations, etc.), and/or consideration of the thermal field samples cooled through. Unfortunately, the computational tools required for a rigorous analysis and interpretation of the data are often developed and used by individual labs, but rarely distributed to the general public. One reason for this is the lack of a good

1529-6466/05/0058-0022$05.00 DOI: 10.2138/rmg.2005.58.22

medium to communicate recent software developments, and/or incomplete documentation and development of user-friendly codes worth distributing outside individual labs. The purpose of this chapter is to highlight several user-friendly computer programs that aid in the simulation and interpretation of thermochronometer data and crustal thermal fields. Most of the software presented here is suitable for both teaching and research purposes.

Table 1 summarizes the different programs discussed in this chapter, the contributing authors behind each program[1], the computer operating system each program runs on, and the application or purpose each program is intended for simulating. The intent of this chapter is not to provide a user manual for each program, but rather an introduction to concepts and physical processes each program simulates. All software discussed here is freely available for non-profit applications and can be downloaded from *http://www.minsocam.org/MSA/RIM/* under the link for Volume 58, *Low-Temperature Thermochronology*. This web page contains executable versions of the software, copyright information, example input and output data files, and, in some cases, additional documentation and user manuals. The availability of source code for modifying each program varies, and interested persons are encouraged to contact the leading contributor. Regular users of these programs are encouraged to contact the software contributor to receive software updates and bug fixes. Software updates will occasionally be posted on the MSA web server as well. Software users should be aware of the caveats associated with use of these programs[2].

The remainder of this chapter is dedicated to highlighting the functionality of each software package. Several items are discussed for each software package. These items include: (1) the intended application of each program and motivation behind its development, (2) the general theory and equations solved in the program, (3) the input required to run the program, and (3) the output generated.

TERRA: FORWARD MODELING EXHUMATION HISTORIES AND THERMOCHRONOMETER AGES

Contributors: T. Ehlers, T. Chaudhri, S. Kumar, C. Fuller, S. Willett
(tehlers@umich.edu)

The TERRA (Thermochronometer Exhumation Record Recovery Analysis) software simulates rock thermal histories and thermochronometer ages generated during exhumation. There are two linked modules to this program. The "Thermal Calculation" module calculates cooling histories for either 1D transient heat transfer with variable erosion rates, or 2D steady-

[1] ***Suggested format for referencing software described in this chapter:*** Each computer program described in this chapter was developed by different contributors (Table 1). To assure each contributor receives appropriate credit for their contribution please include the contributor and software names in any citations used in publications. For example, programs in this volume could be cited in the following way, "We computed erosion rates using the program AGE2EDOT developed by M. Brandon and summarized in Ehlers et al. (2005)." If additional publications are available describing the software (e.g., HeFTy; Ketcham et al. 2005) please also reference those publications.

[2] ***Caveats related to software use and distribution:*** There are two caveats associated with the software provided in this volume. (1) The software is provided as a courtesy to the broader scientific community and in some cases represents years of effort by the contributors. The software can be used free of charge for non-profit research, but persons with commercial or for-profit interests and applications should contact the contributors concerning the legalities of its use. (2) Although every effort has been made by the contributors to assure the software provided functions properly and is free of errors, no guarantee of the validity of the software and results calculated can be made. Thus, application of the software is done at the users own risk and the software contributors are not accountable for the results calculated.

Table 1. Summary of software contributions.

Program Name	*Contributors*	*Application / Purpose*	*Operating Systems Supported*
TERRA	T. Ehlers, T. Chaudhri, S. Kumar, C. Fuller, S. Willett	Forward modeling thermochronometer ages, apatite and zircon fission track and (U-Th)/He	Windows®, Linux
HeFTy	R. Ketcham	Forward and inverse modeling apatite fission track, (U-Th)/He, vitrinite reflectance data	Windows®
BINOMFIT	M. Brandon	Calculate ages and uncertainties for concordant and mixed distributions of FT grain ages	Windows®
CLOSURE, AGE2EDOT RESPTIME	M. Brandon	Calculate effective closure temperatures, erosion rates, and isotherm displacement for several thermochronometer systems	Windows®
FTIndex	R. Ketcham	Calculated index temperatures and lengths of fission track annealing models	Windows®
TASC	D. Belton, B. Kohn, A. Gleadow	Fission track age spectrum calculations	Windows® (Excel®)
DECOMP	T. Dunai	Forward modeling (U-Th)/He age evolution curves	Windows®
4DTHERM	F. Fu	Thermal and exhumation history of intrusions	Windows®, Linux, Mac®

state heat transfer for user defined periodic topography and erosion rates. The "Age Prediction" module uses the thermal histories generated in the first module, or any other user defined thermal histories, to calculate cooling rate dependent apatite and zircon (U-Th)/He ages, and apatite and zircon fission track ages. The program is platform independent and runs on Windows® XP, and Linux® operating systems. Future releases of the program will run on the Macintosh® operating system as well. A graphical user interface (GUI) is provided for user input. However, the age-prediction program can also run without the graphical interface at the command line under the Linux® or DOS® operating systems if batch processing of multiple thermal histories from other, more sophisticated, thermal models is desired. The source code for these programs is available on request and written using C++ and Qt.

TERRA – 1D and 2D thermal history calculations

The TERRA thermal calculation module is well suited for users to explore different erosion/exhumation scenarios and their influence on subsurface temperatures. The influence of different rates and durations of erosion, as well as topographic geometries, can be easily evaluated so users develop an intuition for crustal thermal processes.

Figure 1 illustrates the user interface, or menu, for the calculation of 1D transient thermal fields during erosion. This program can be used to predict sample cooling histories for different user defined rates and durations of erosion. It is particularly useful for evaluating the transient evolution of subsurface temperatures and cooling histories during the early stages (e.g., first 5 m.y.) of mountain building and erosion. The program operates by solving the 1D transient advection-diffusion equation for a homogeneous medium with no heat production (for details,

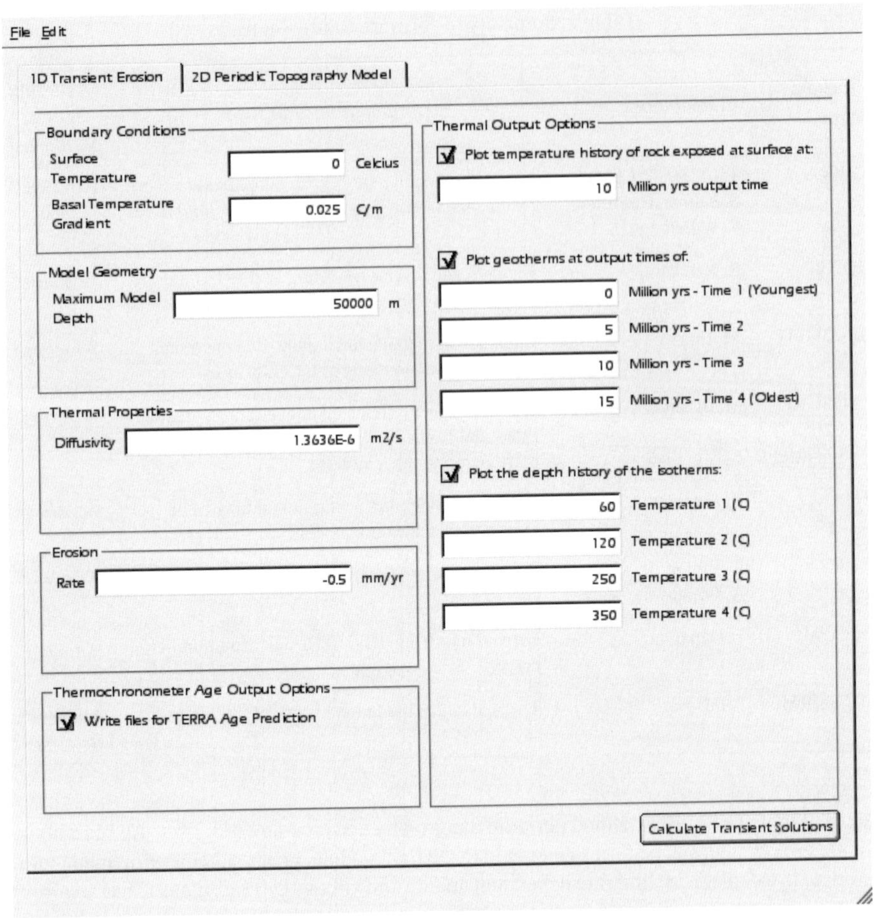

Figure 1. TERRA interface for calculation of 1D transient thermal histories for variable erosion rates.

see Eqns. 6 and 7 in Ehlers 2005). Rock thermal histories are recorded as they approach the surface. User defined inputs include thermal boundary conditions and material properties, model geometry, and rate of erosion. Several plots are generated from this program, including: (1) the thermal history of exhumed samples (Fig. 2), (2) geotherms at different time intervals (Fig. 2), and (3) the depth history of user defined temperatures (e.g., closure temperatures). The cooling history of a rock exposed at the surface is written to an output file for loading into the age prediction program.

Figure 3 illustrates the TERRA user interface for calculating the steady-state 2D thermal field beneath periodic topography. This program can be used to determine the position of isotherms and rock cooling histories under uniformly eroding topography. This program is particularly useful for evaluating the curvature of closure isotherms beneath topography of different wavelengths and amplitudes and potential topographic effects on cooling ages. The program operates by solving the steady-state 2D advection-diffusion equation for a uniformly eroding medium with depth dependent radiogenic heat production and periodic topography (Manktelow and Grasemann 1997). User defined inputs include thermal boundary conditions,

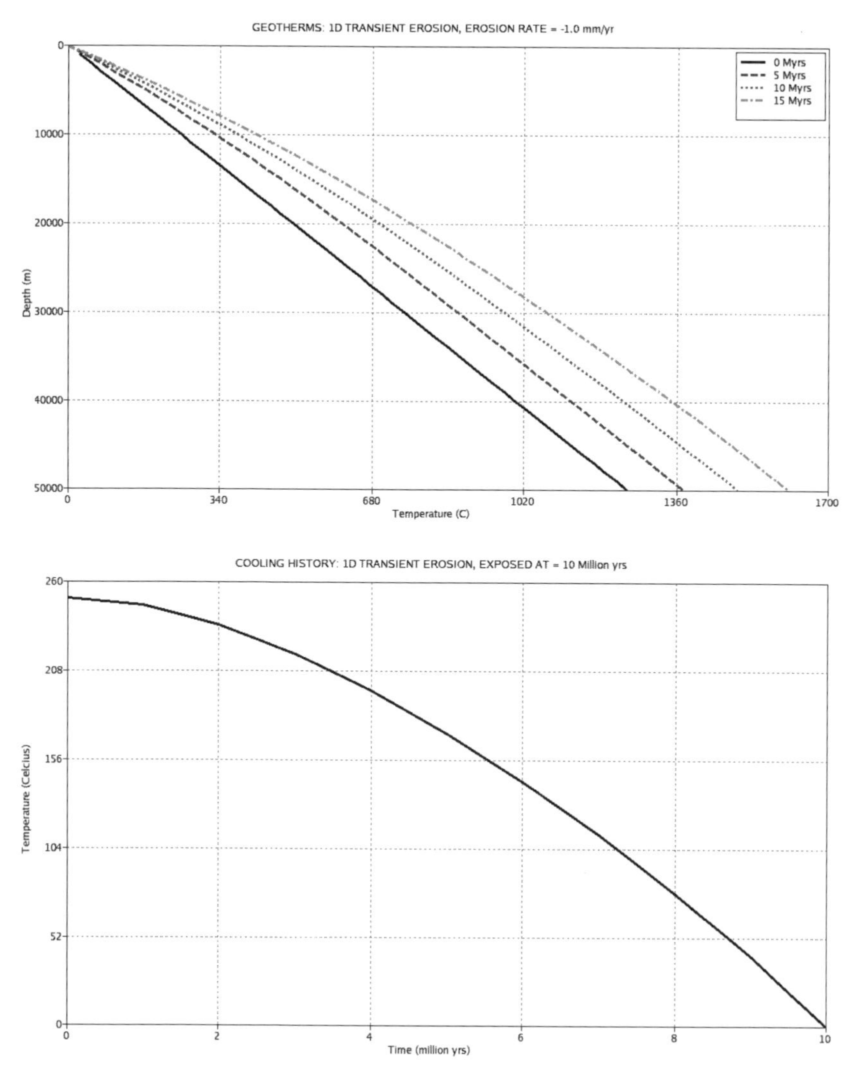

Figure 2. Example of TERRA predicted geotherms (top) and exhumed sample thermal history (bottom) plots from the 1D transient thermal calculation.

model geometry (topographic wavelength and amplitude), erosion rate, and material properties. Two plots are generated from this program: (1) The steady-state cooling history of rocks exposed between a ridge and adjacent valley bottom, and (2) the topography and steady-state position of user defined isotherms beneath the topography. Thermal histories of samples exposed between a ridge and valley are output to a file for loading into the age prediction program (see below).

All plots generated in this 2D periodic topography and the 1D transient erosion menus have options to print or save the plot. Other options available in the plot windows include the ability to zoom in and out to different parts of the plot. The menus also include default input parameters to help users unfamiliar with common thermal properties of rocks use the programs.

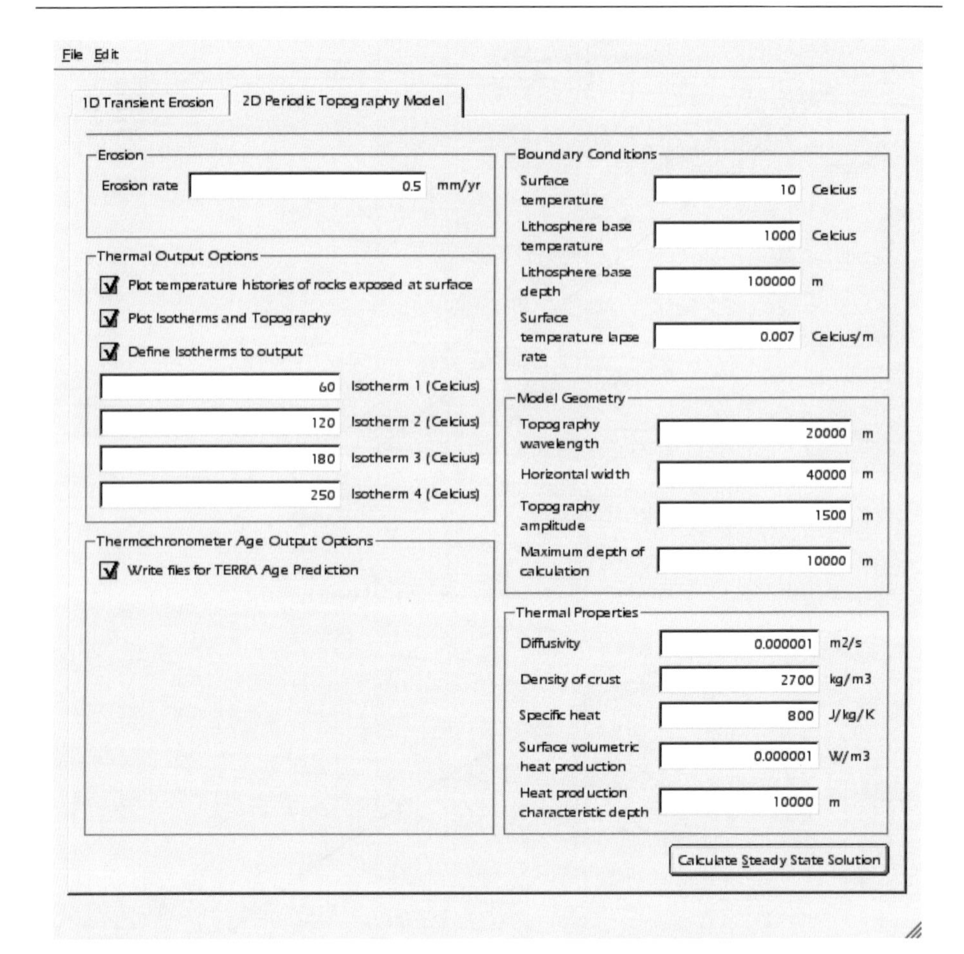

Figure 3. TERRA input window for calculating steady-state cooling histories beneath 2D periodic topography.

TERRA – thermochronometer age prediction

The TERRA program also calculates predicted low-temperature thermochronometer ages from user defined thermal histories. Predicted ages for several thermochronometer systems are possible including: apatite and zircon (U-Th)/He ages and ^{4}He concentration profiles across grains, multi-kinetic apatite fission track ages, and zircon fission track ages. Zircon fission track ages are calculated using an effective closure temperature.

Figure 4 shows two example input 'tabs' for the TERRA age prediction module. The first step in using the age prediction module is to load in files with rock thermal and coordinate position (optional) histories. The thermal and coordinate history for either a single sample, or for multiple samples, can be loaded to calculate predicted ages. The ability of this program to load multiple thermal histories for batch processing predicted ages is particularly useful for interpreting cooling histories calculated from more complex 2D or 3D thermal models (e.g., Ehlers and Farley 2003, Ehlers et al. 2003). After a thermal history is loaded, any or all of the thermochronometer ages listed at the top of the window can be calculated.

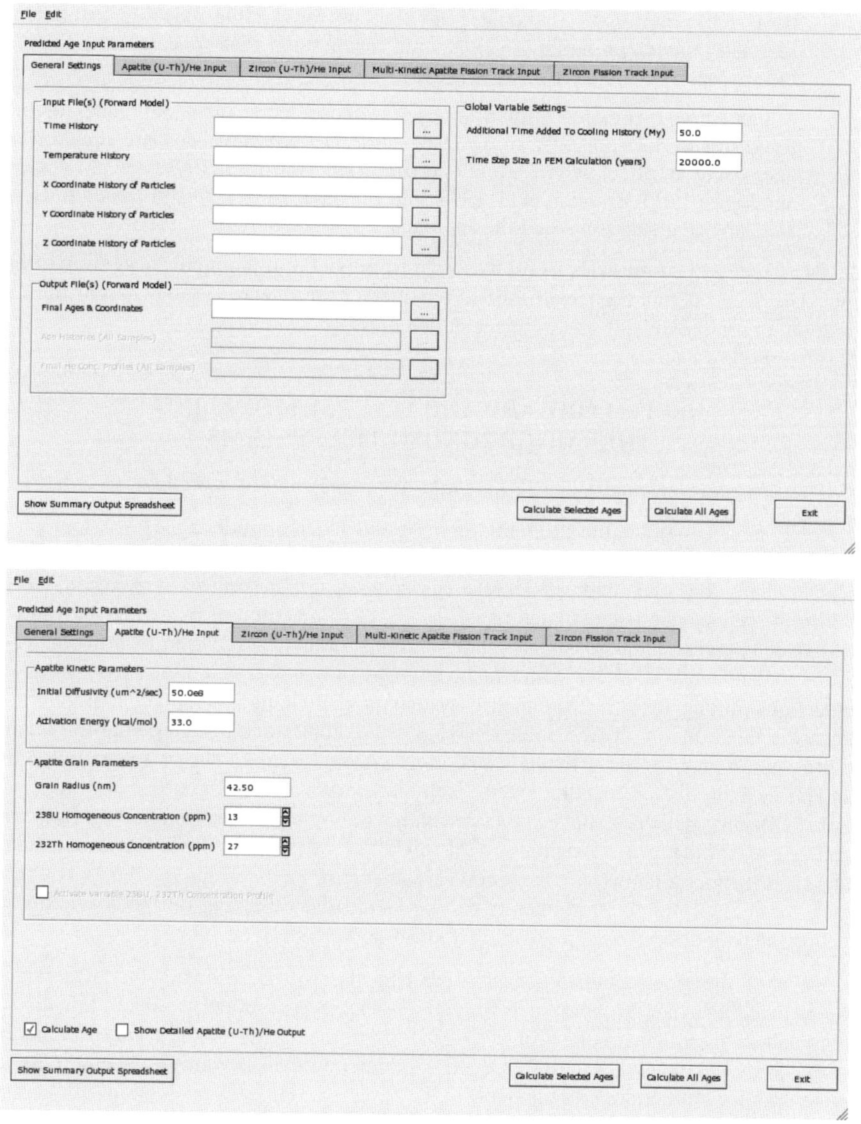

Figure 4. TERRA input windows for loading thermal history input files (top) and apatite (U-Th)/He age prediction (bottom). The window tabs for other thermochronometer systems have similar layouts.

The methods used for calculating each of the thermochronometer ages are as follows:

- *Apatite and zircon (U-Th)/He ages*: (U-Th)/He ages are calculated by solving the transient spherical ingrowth diffusion equation for homogeneous distributions of U and Th, and temperature dependent diffusivity (e.g., Farley 2000; Reiners et al. 2004). The equation is solved using a spherical finite element model as described in Ehlers et al. (2001, 2003).

- *Apatite fission track ages:* Apatite fission track ages are calculated using the mutikinetic annealing model of Ketcham et al. (1999, 2000). This model is similar to the HeFTy program discussed later and accounts for variable annealing behavior as it correlates to etch pit width (D_{par}). Apatite fission track ages are calculated following the approach developed by Fuller (2003).

- *Zircon fission track ages*: Unlike apatite, a well developed annealing model for zircon fission tracks does not exist. As a consequence, TERRA calculates zircon fission track ages using an effective closure temperature (Dodson 1973, 1979; see also Brandon and Vance 1992; Brandon et al. 1998; and Batt et al. 2001). Zircon fission track ages are also calculated following the approach developed by Fuller (2003).

Calculated ages and coordinates of each sample (if provided in the input files) are written as output. Users can define their own kinetic parameters for (U-Th)/He and fission track age prediction, or use the default values to get started quickly.

HeFTy: FORWARD AND INVERSE MODELING THERMOCHRONOMETER SYSTEMS

Contributor: R. Ketcham (ketcham@mail.utexas.edu)

HeFTy is a computer program for forward and inverse modeling of low-temperature thermochronometric systems, including apatite fission-track, (U-Th)/He, and vitrinite reflectance. Fission-track and (U-Th)/He calculations are described in Ketcham (2005), and vitrinite reflectance is calculated using the EasyRo% method of Sweeney and Burnham (1990). HeFTy can simultaneously calculate solutions for up to seven thermochronometers. In addition to data analysis, HeFTy can also be an instructional aid, as it allows easy, interactive comparisons among thermochronometric systems and their parameters. It reproduces and extends the functionality of AFTSolve (Ketcham et al. 2000), and is intended to fully replace that program. It runs on the Windows® operating system. Figure 5 shows an example screen from HeFTy, with one file window open. Each file window in HeFTy corresponds to a single sample or locality, which can have associated with it multiple thermochronometers. To create a model, use one of the 3 "new" buttons on the top left of the task bar: the first creates a "blank" model (no thermochronometers), the second creates an AFT model, and the third a (U-Th)/He model. Additional thermochronometers can be included by clicking on the "+" button and selecting the type desired. Up to seven thermochronometers can be added.

For each thermochronometer, a new tab page is created, allowing model parameters to be changed and data to be entered for fitting. Whenever a model is added, this tab page is automatically brought to the front. To generate numbers, go to the "Time-Temperature History" tab. Clicking and dragging in the time-temperature history graph causes the forward model to run for all thermochronometers.

Data to be compared to forward modeling results or fitted using inversion can be typed directly into the program interface, and tabular data (such as fission-track lengths and single-grain ages) can be transferred from a table or spreadsheet using the clipboard (i.e., cutting and pasting). The easiest and most foolproof way to enter such data is using tab-delimited text files, which can be exported from any spreadsheet program. Included with the HeFTy distribution is the Microsoft Excel® file "Import templates.XLS", which contains examples of all of the various input formats HeFTy supports.

The inverse modeling module is accessed by clicking the button to the right of the "+" button. A new window comes up for controlling the inversion processing (Fig. 6), and mouse clicks and motions in the time-temperature history graph create box constraints through which inverse thermal histories are forced to pass. Constraints can overlap, although rules are enforced to ensure that they occur in an unambiguous sequence. The amount of complexity

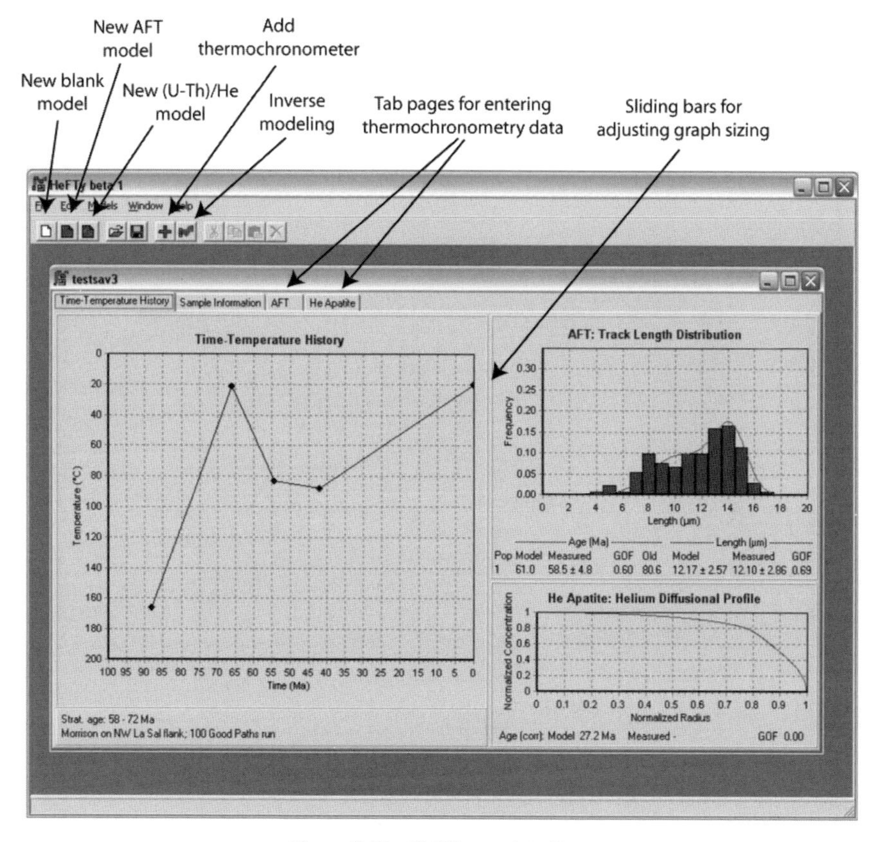

Figure 5. The HeFTy user interface.

allowed between each constraint (see Ketcham 2005) is controllable by right-clicking on the label in the middle of the line segments connecting constraint midpoints. In the example shown here, the first (leftmost) code of "1I" indicates that the time-temperature path segment is halved one time, with "intermediate" complexity; "2G/5" indicates that the segment is halved twice (into four sub segments), with the randomizer favoring a "gradual" history with a maximum heating rate of 5 °C/m.y.; and "3E/10" indicates that the segment is halved three times with the possibility for "episodic" histories (prone to sudden changes), but with a maximum cooling rate at all times of 10 °C/m.y.

FTINDEX: INDEX TEMPERATURES FROM FISSION TRACK DATA

Contributor: R. Ketcham (ketcham@mail.utexas.edu)

FTIndex is a program for calculating index temperatures and lengths that describe the geological time scale predictions of fission-track annealing models, as proposed by Ketcham et al. (1999). It is written in the C programming language. Both an executable version (for Windows®) and the source code are available for download.

Three index temperatures describe high-temperature annealing behavior. The fading temperature (T_F) is the downhole temperature required to fully anneal a fission-track population

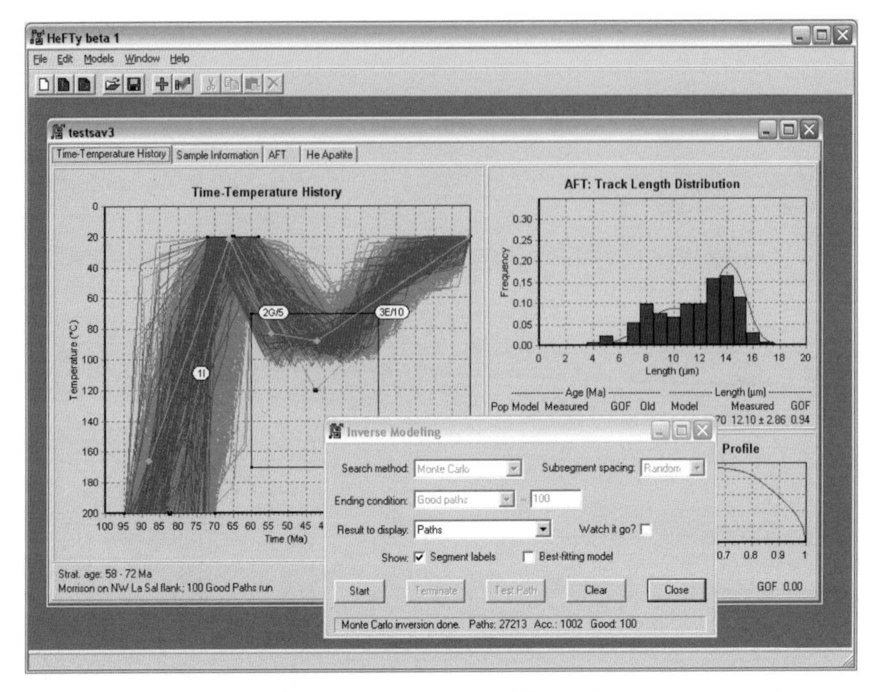

Figure 6. HeFTy inverse modeling mode.

after a certain number of millions of years, signified by the second subscript. For example, $T_{F,30}$ indicates the temperature required to fully anneal a population of fission tracks after 30 million years. The closure temperature (T_C) is the temperature experienced by a mineral at the time given by its age, assuming a constant-rate cooling history given by the second subscript—for example, $T_{C,10}$ means closure temperature given a 10 °C/m.y. cooling history. The total annealing temperature, T_A, is the temperature at which a fission-track population, fully anneals given a constant heating rate; it is equivalent to the oldest remaining population after a constant cooling episode. Again, the second subscript gives the heating/cooling rate. For both T_C and T_A, cooling is assumed to stop at 20 °C.

Low-temperature annealing behavior is modeled by estimating fission-track length reduction given a proposed time-temperature history. The best-constrained low-temperature history is from Vrolijk et al. (1992), which used apatites from deep-sea drill cores from the East Mariana Basin and utilized proposed time-temperature histories based on independent evidence. Two histories are provided by Vrolijk et al. (1992), which are end-members the bound uncertainties in rifting time in the East Mariana Basin. FTIndex calculates reduced length for both end-member cases and the average between them. To convert these to actual lengths, they must be multiplied by an estimated initial track length. The mean length of apatites measured by Vrolijk et al. (1992) was 14.6 ± 0.1 μm. Note that if an annealing model is based on c-axis projected lengths then the results will have to be converted to unprojected means to be compared to this value; this can be done with the relation:

$$r_m = 1.396 r_c - 0.4017 \tag{1}$$

In general, to match the Vrolijk et al. (1992) data to within the uncertainty of the initial track length, the reduced length value for non-projected lengths should be in the range 0.89–

0.92, and reduced c-axis projected lengths should be in the range 0.92–0.95. A second low-temperature test is for Fish Canyon tuff, for which the assumed thermal history is a constant 20 °C for 27.8 million years.

FTIndex program operation

The program works by reading a text file that contains the coefficients describing one or more annealing model. It then calculates the index temperatures and writes them to a tab-delimited text file called bench.txt.

An example input file called IndexTest.txt has been provided in Table 2 to illustrate the file format. The basic format is a tab-delimited text file, with eleven columns and the first line a series of headers as shown above. The first column is a model name that will be repeated in the output file. The name should have no spaces. The second column is a numerical code signifying the form of the modeling equation; this form defines the meaning of the numbers in columns six through eleven, marked c_0 through c_5. The third column denotes whether the model describes mean length or mean c-axis projected length; if the former it should be zero, otherwise one. The fourth and fifth columns are for r_{mr0} and κ parameters to translate lengths from one form of apatite to another (Ketcham et al. 1999), according to the equation:

$$r_{lr} = \left(\frac{r_{mr} - r_{mr0}}{1 - r_{mr0}}\right)^{\kappa} \tag{2}$$

where r_{lr} is the reduced length of a less-resistant apatite, and r_{mr} is the reduced length of the more-resistant apatite whose annealing is described by parameters c_0 through c_5. If a conversion such as this is not desired then set r_{mr0} to 0 and κ to 1.

Unless otherwise noted, for all model equations, time (t) is in seconds and temperature (T) is in Kelvin, and r denotes reduced length while l denotes mean length. The model codes and forms are:

0: Fanning linear, based on Laslett et al. (1987) and Crowley et al. (1991).

$$\frac{\left[\left(1 - r^{c_5}\right)/c_5\right]^{c_4} - 1}{c_4} = c_0 + c_1 \frac{\ln(t) - c_2}{(1/T) - c_3} \tag{3}$$

1: Fanning curvilinear, based on Crowley et al. (1991) and Ketcham et al. (1999).

$$\frac{\left[\left(1 - r^{c_5}\right)/c_5\right]^{c_4} - 1}{c_4} = c_0 + c_1 \frac{\ln(t) - c_2}{\ln(1/T) - c_3} \tag{4}$$

Table 2: Example FTIndex input file format.

Name	Model	Lc	rmr0	kappa	C0	C1	C2	C3	C4	C5
L87Dur	0	0	0	1	-4.87	0.000168	-28.12	0	0.35	2.7
C90Dur	2	0	0	1	1.81	0.206	40.6	16.21		
LG96Fap	3	0	0	1	16.713	-4.879	0.000187	-33.385	0.000295	0.3333
K99DRLm	1	0	0	1	-106.18	2.1965	-155.9	-9.7864	-0.48078	-6.3626
K99RNLc	1	1	0	1	-61.311	1.292	-100.53	-8.7225	-0.35878	-2.9633
K99MLcRN	1	1	0.846	0.179	-19.844	0.38951	-51.253	-7.6423	-0.12327	-11.988
K99MLc165	1	1	0.84	0.16	-19.844	0.38951	-51.253	-7.6423	-0.12327	-11.988

2: Model of Carlson (1990).

$$l_{as} = c_3 - c_0\left(\frac{kT}{h}\right)^{c_1} \exp\left(\frac{-c_1 c_2}{RT}\right) t^{c_1} \tag{5}$$

Where l_{as} is mean length due to axial shortening, k is Boltzmann's constant (3.2997×10^{-27} kcal·K^{-1}), h is Planck's constant (1.5836×10^{-37} kcal·s), and R is the universal gas constant (1.987×10^{-3} kcal·mol^{-1}·K^{-1}). Note that there are only four fitted parameters in this model, and thus only four should be entered in the file; columns 10 and 11 should be left blank.

3: Model of Laslett and Galbraith (1996).

$$l = c_0\left[1 - \exp\left(c_1 + c_2 \frac{\ln(t) - c_3}{(1/T) - c_4}\right)\right]^{c_5} \tag{6}$$

Following their convention, in this case the time units are hours rather than seconds. If one fits the model using seconds, as with the other equations listed here, simply add ln(3600) = 8.18869 to c_3.

BINOMFIT: A WINDOWS® PROGRAM FOR ESTIMATING FISSION-TRACK AGES FOR CONCORDANT AND MIXED GRAIN AGE DISTRIBUTIONS

Contributor: Mark T. Brandon (mark.brandon@yale.edu)

The BINOMFIT program calculates ages and uncertainties for both concordant and mixed distributions of fission-track (FT) grain ages. The current program runs on the Windows® platforms. It incorporates features from two older DOS programs, called ZETAAGE and BINOMFIT. The ZETAAGE algorithm provides an exact estimation of FT ages and uncertainties (Sneyd 1984), regardless of track density. This capability is essential for estimating ages for samples with low track densities, which is common in young FT apatite samples (< ~15 Ma). The BINOMFIT algorithm is based on the decomposition method of Galbraith and Green (1990). An automated search routine has been added to the Windows® version of BINOMFIT, which automatically determines the optimal number of components in a FT grain-age distribution. This feature makes the program much easier to use for beginners. The original DOS versions of BINOMFIT and ZETAAGE are still maintained given that they are sometimes faster to use for experts. The source code for these programs is available on request.

Introduction to BINOMFIT

A common problem in FT dating is the interpretation of discordant fission-track grain age (FTGA) distributions. Discordance refers to the situation where the variance of the grain ages in FTGA distribution is greater than expected for analytical error alone. The widely used χ^2 test provides the main method for assessing if a distribution is "over-dispersed" relative to the expectation for count statistics for radioactive decay (Galbraith 1981).

Mixed distributions are expected for samples that have "detrital" FT ages, such as a sandstone with unreset zircon FT ages. Discordance is sometimes observed in zircon FT dating of volcanic tuffs, where the cause is likely due to contamination by older zircons.

Reset FT samples are also commonly discordant. In some cases, the discordance is taken as evidence for partial resetting, but the cause is more commonly due to differences in annealing properties, as caused by variations in composition for apatite or variations in radiation damage for zircon. Heterogeneous annealing is expected for reset sandstones given that the dated apatites and zircons are derived from a variety of sources (Brandon et al. 1998), but this result is also found in some plutonic rocks as well, where one might expect that the dated mineral would be more homogeneous (O'Sullivan and Parrish 1995).

The binomial "peak-fitting" method of Galbraith and Green (1990) and Galbraith and Laslett (1993) is an excellent method for decomposing a mixed FT grain age distribution. This method was implemented by Brandon (1992) in a DOS program called BINOMFIT. The binomial peak-fitting method has been extensively used over the last 15 years and works very well for real FTGA distributions. A big advantage of the method is that it provides a one-component solution that is equivalent to the FT pooled age. Thus, the program works equally well for concordant and mixed FTGA samples.

Galbraith and Laslett (1993) introduced the term "minimum age," which can be loosely viewed as the pooled age of the largest concordant fraction of young grain ages in a FTGA distribution. Binomial peak fitting can also be used to estimate the minimum age of a distribution—equal to the youngest component in the distribution—while at the same time providing information about older components.

The FT minimum ages have been found useful for a variety of studies. For detrital zircon FT studies of volcanoclastic rocks, the minimum age is commonly a useful proxy for the depositional age of the rock (Brandon and Vance 1992; Garver et al. 1999, 2000; Stewart and Brandon 2004). Likewise, the minimum age for a tuff can remove biases due to older contaminant grains. For reset rocks, the minimum age represents the time of closure for that fraction of grains with the lowest retention for fission tracks, such as fluorapatites in apatite FT dating (e.g., Brandon et al. 1998) and radiation-damaged zircons in zircon FT dating (e.g., Brandon and Vance 1992).

Using BINOMFIT

BINOMFIT is distributed in a compressed zip file, called BinomfitInstall.zip. You need to download the file to your system, and place it in a temporary directory. Close all non-essential programs in Windows® to avoid conflicts with active processes during the setup. Double click the file to launch the decompression process, which will create three files needed for the installation process: setup.exe, setup.lst, and binomfit.cab.

The setup process is launched by double clicking setup.exe. This will install BINOMFIT into your program files area (e.g., C:/Program Files/Binomfit), and will also create a group entry labeled BINOMFIT in the Start Menu. Within this group are short cuts to the BINOMFIT program, a ReadMe file (i.e., this document in html format), and Documentation (in Adobe pdf format). The remaining support files (Mscomctl.ocx, Comdlg32.ocx, Msvcrt.dll, Scrrun.dll, Msstdfmt.dll, Msdbrptr.dll) will be placed in the application subdirectory as well, to avoid conflict with system DLLs and OCXs. After the installation is complete, you can delete the setup files. The installation is registered in Control Panel. As a result, you can use the "Add and Remove Program" option in Control Panel to uninstall the entire package.

The distribution package contains detailed instructions about how to construct a data file for input. Also included are several different types of data files and examples of typical output.

Figure 7 shows the program after completion of an automatic search for an optimal number of best-fit peaks. The buttons on the menu bar can be used to select between different types of plots, including a density plot (as shown) or a radial plot.

The Print option in the File menu will send a full report of grain ages, best-fit solution and graphical copies of all plots to a designated printer. The printed output for the Mount Tom FT sample is shown in Figure 7 is included in the documentation.

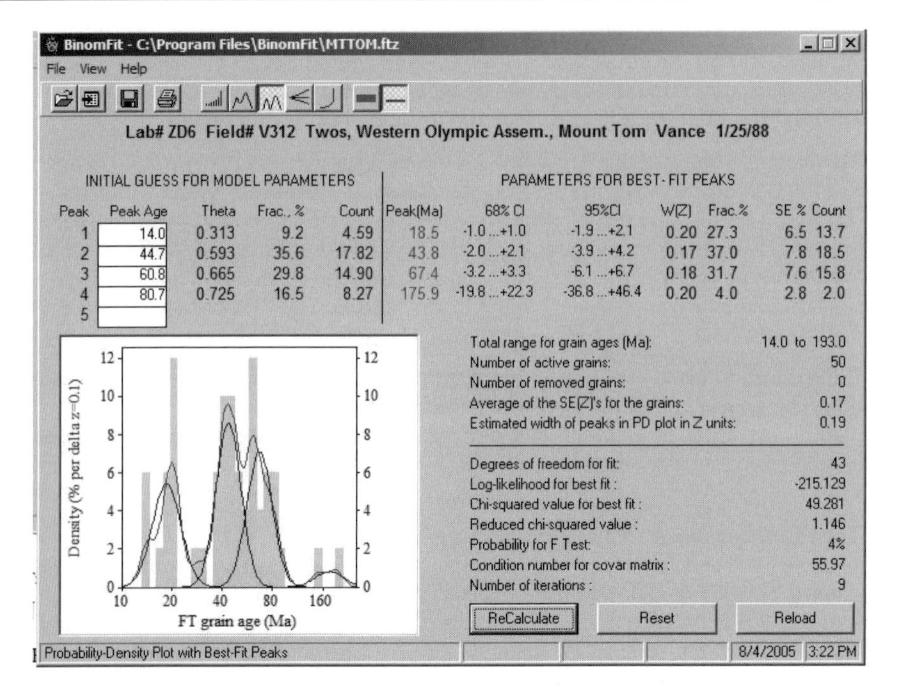

Figure 7. BINOMFIT best-fit solution for the Mount Tom zircon FT sample using the automatic peak-search mode.

The Save option in the File menu allows the user to save plot files, which can be imported into a graphics program (e.g., Excel®, SigmaPlot®) to prepare formal plots for publication. Plot files are available for probability density plots, radial plots, and P(*F*) plots. The files are annotated to add in the construction of the relevant plot. Data file can be saved. This allows a permanent record of datasets that have constructed in BINOMFIT by removal of grain ages or by merging multiple data files.

PROGRAMS FOR ILLUSTRATING CLOSURE, PARTIAL RETENTION, AND THE RESPONSE OF COOLING AGES TO EROSION: CLOSURE, AGE2EDOT, AND RESPTIME

Contributor: Mark T. Brandon (mark.brandon@yale.edu)

CLOSURE, AGE2EDOT, and RESPTIME are a set of simple programs that were first developed in Brandon et al. (1998). They were designed to help demonstrate the influence of steady and transient erosion on fission-track (FT) cooling ages. The programs have since been extended to include (U-Th)/He and $^{40}Ar/^{39}Ar$ ages, as well as FT ages. The programs now include modern diffusion data for all of the minerals commonly dated for thermochronometry, ranging from He dating of apatite to Ar dating of hornblende.

CLOSURE is a standard Windows®-style program, whereas AGE2EDOT and RESPTIME are console-style programs. All of the programs are compiled for the Windows® operating system. Each consists of a single exe file. Setup involves copying the file into a suitable directory. The programs are started by double-clicking the file name. The programs require no input files. Rather, the user is guided by a set of prompts and questions to supply the necessary input parameters for the calculation of interest. Results are output to a window for CLOSURE and to an output file for AGE2EDOT and RESPTIME. In all cases, the output is organized with tab-separated columns, so that it can be easily imported into a plotting program, such as Excel® or SigmaPlot®.

The source code for the programs is available on request.

Methods for CLOSURE

The CLOSURE program provides a compilation of the data needed to calculate effective closure temperatures and partial retention temperatures for all of the minerals commonly dated by the He, FT, and Ar methods (Figure 8). Laboratory diffusion experiments have demonstrated that, on laboratory time scales, the diffusivity of He and Ar are well fit by the following relationship:

$$D = D_0 \exp\left[\frac{-E_a - PV_a}{RT}\right] \tag{7}$$

where D_0 is the frequency factor (m^2 s^{-1}), E_a is the activation energy (J mol^{-1}), P is pressure (Pa), V_a is the activation volume (m^3 mol^{-1}), T is temperature (K), R is the gas law constant (8.3145 J mol^{-1} K^{-1}), and D is the diffusivity (m^2 s^{-1}). The PV_a term is commonly set to zero, since its contribution is generally small relative to E_a. D_0 and E_a are compiled in Tables 1 and 2 for the main minerals dated by the He and Ar method.

The *partial retention zone* is defined in two ways. This concept was first introduced to account for partial annealing of fission tracks when held at a steady temperature. Laboratory heating experiments were used to define the time-temperature conditions that caused 90% and 10% retention of the initial density of fossil tracks. The retention behavior was considered for loss only, without regard for the production of new tracks during the heating event. We use the term loss-only PRZ to refer to this kind of partial retention zone for He, FT, and Ar dating. Figures 9, 10, and 11 show examples of loss-only PRZs. They help to illustrate the time and

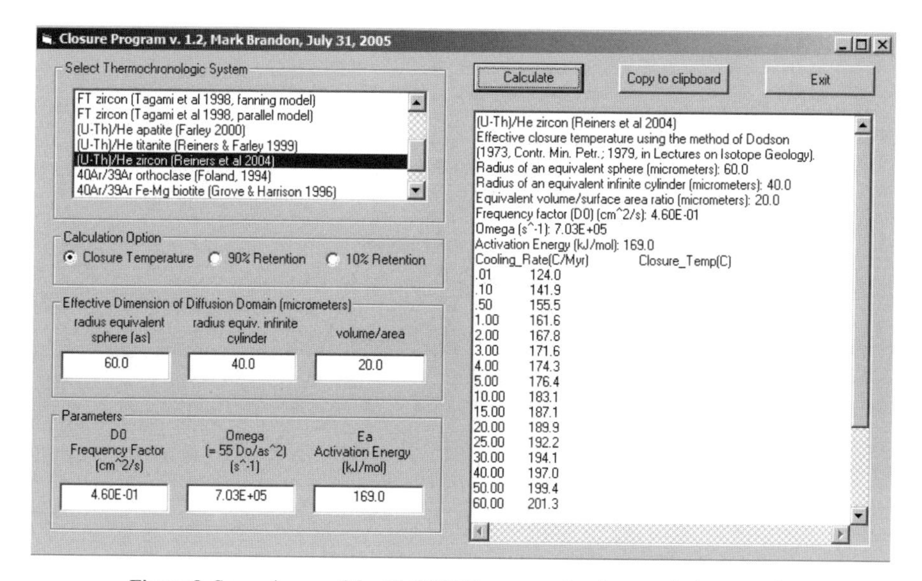

Figure 8. Screen image of the CLOSURE program showing a typical run result.

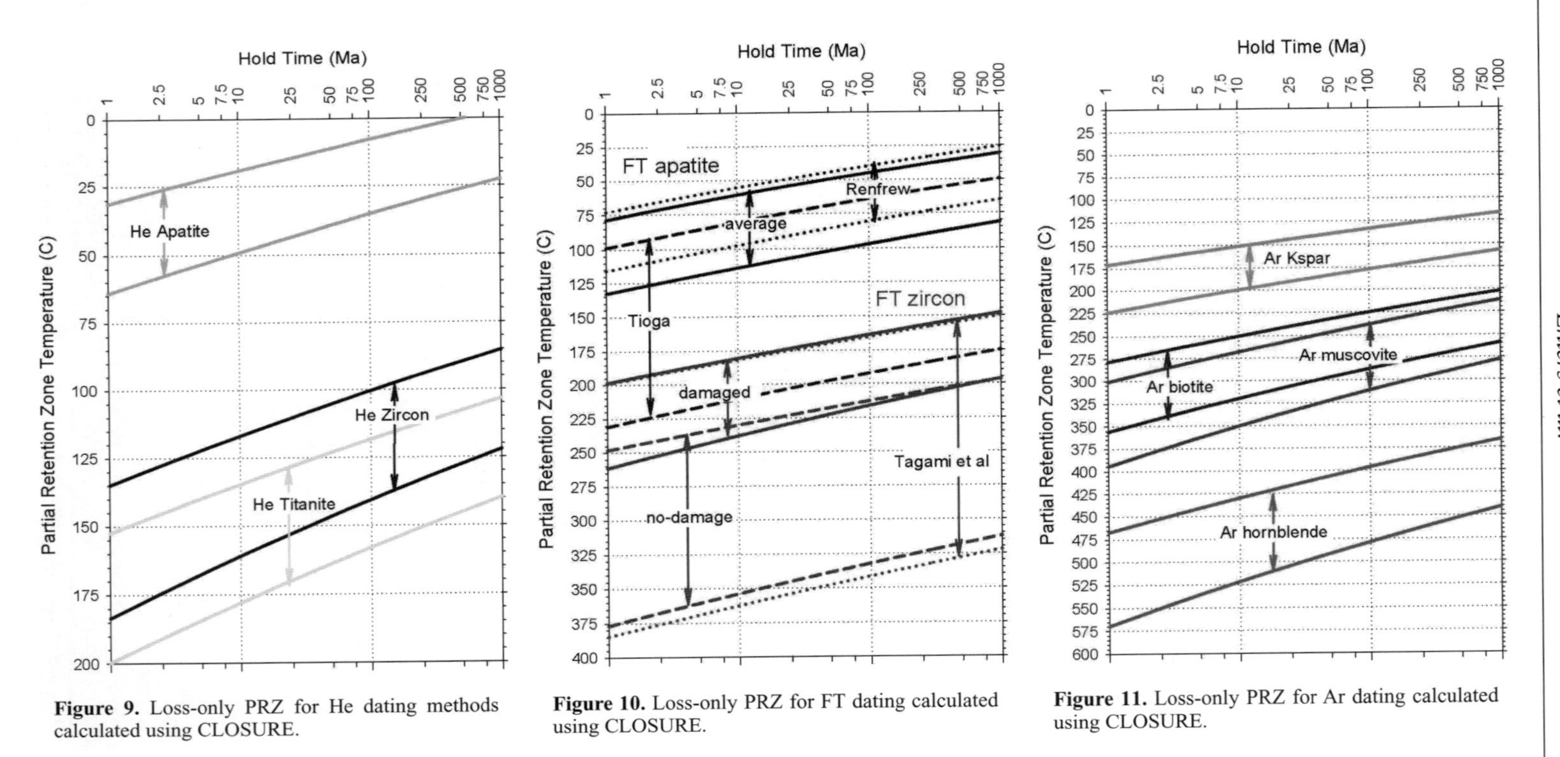

Figure 9. Loss-only PRZ for He dating methods calculated using CLOSURE.

Figure 10. Loss-only PRZ for FT dating calculated using CLOSURE.

Figure 11. Loss-only PRZ for Ar dating calculated using CLOSURE.

temperature conditions needed to fully preserve or fully reset a He, FT, or Ar cooling age. The loss-only PRZ is calculated for He and Ar dating using the exact version of the loss-only diffusion equation (spherical geometry) from McDougall and Harrison (1999). The equations require D_0 and E_a as parameters, as given in Tables 3 and 4.

Wolf and Farley (1998) defined a different kind of PRZ, one that accounted for both production and loss of the ^{4}He, FT, or ^{40}Ar. The limits of the loss-and-production PRZ are defined by the temperatures needed to maintain a He, FT, or Ar age that is 90% or 10% of the hold time. The time-temperature conditions associated with this 90% and 10% retention are determined using the spherical He loss-and-production equation. The equations require D_0 and E_a as parameters. This kind of PRZ is useful for considering how the measured age for a thermochronometer will change down a borehole, as a function of downward increasing but otherwise steady temperatures.

Table 3. Closure parameters for He and FT dating.

Method (*references*)	E_a ($kJ\ mol^{-1}$)	D_0 ($cm^2\ s^{-1}$)	a_s [4] (μm)	[5] (s^{-1})	$T_{c,10}$ [6] (°C)
(U-Th)/He apatite (Farley 2000)	138	50	60	7.64×10^7	67
(U-Th)/He zircon (Reiners et al. 2004)	169	0.46	60	7.03×10^5	183
(U-Th)/He titanite (Reiners and Farley 1999)	187	60	150	1.47×10^7	200
FT apatite[1] (average composition[2]) (Ketcham et al. 1999)	147	—	—	2.05×10^6	116
FT Renfrew apatite[3] (low retentivity) (Ketcham et al. 1999)	138	—	—	5.08×10^5	104
FT Tioga apatite[3] (high retentivity) (Ketcham et al. 1999)	187	—	—	1.57×10^8	177
FT apatite (Durango) (Laslett et al. 1987; Green 1988)	187	—	—	9.83×10^{11}	113
FT zircon[1] (natural, radiation damaged) (Brandon et al. 1998)	208	—	—	1.00×10^8	232
FT zircon (no radiation damaged) (Rahn et al. 2004; fanning model)	321	—	—	5.66×10^{13}	342
FT zircon (Tagami et al. 1998; fanning model)	324	—	—	1.64×10^{14}	338
FT zircon (Tagami et al. 1998; parallel model)	297	—	—	2.56×10^{12}	326

Footnotes:

1) Recommended values for most geologic applications.

2) Average composition was taken from Table 4 in Carlson et al. (1999). Equation 6 in Carlson et al. (1999) was used to estimate r_{mr0} = 0.810 for this composition. Closure parameters were then estimated using the HeFTy program (Ketcham 2005).

3) Closure parameters were estimated from HeFTy and r_{mr0} = 0.8464 and 0.1398 for Renfrew and Tioga apatites, respectively, as reported in Ketcham et al. (1999).

4) a_s is the effective spherical radius for the diffusion domain. Shown here are typical values.

5) is measured directly for FT thermochronometers, and is equal to $55D_0a_s^{-2}$ for He and Ar thermochronometers.

6) $T_{c,10}$ is the effective closure temperature for 10 °C/m.y. cooling rate and specified a_s value.

Table 4. Closure parameters for Ar dating.

Method (references)	E_a *(kJ mol^{-1})*	D_0 *(cm^2 s^{-1})*	a_s [1] *(μm)*	Ω [2] *(s^{-1})*	$T_{c,10}$ [3] *(°C)*
$^{40}Ar/^{39}Ar$ K-feldspar (orthoclase) (Foland 1994)	183	9.80×10^{-3}	10	5.39×10^5	223
$^{40}Ar/^{39}Ar$ Fe-Mg biotite (Grove and Harrison 1996)	197	7.50×10^{-2}	750 (500)	733	348
$^{40}Ar/^{39}Ar$ muscovite (Robbins 1972; Hames and Bowring 1994)	180	4.00×10^{-4}	750 (500)	3.91	380
$^{40}Ar/^{39}Ar$ hornblende (Harrison 1981)	268	6.00×10^{-2}	500	1320	553

Footnotes:

1) a_s is the effective spherical radius for the diffusion domain. Shown here are typical values. Muscovite and biotite have cylindrical diffusion domains, with typical cylindrical radii shown in parentheses. For these cases, a_s is approximated by multiplying the cylindrical radius by 1.5.

2) Ω is equal to $55D_0a_s^{-2}$ for Ar thermochronometers.

3) $T_{c,10}$ is the effective closure temperature for 10 °C/m.y. cooling rate and specified a_s value.

The CLOSURE program estimates both types of PRZ for He and Ar methods. It only calculates the loss-only PRZ for FT methods. Annealing models, such as HeFTy (Ketcham 2005) could be used to calculate a loss-and-production PRZ for the FT apatite system, but there is no such model yet for annealing and production for the FT zircon system. The 90% and 10% retention isopleths are determined from time-temperature results from laboratory stepwise-heating experiments. The isopleths commonly have an exponential form,

$$t = \Omega^{-1} \exp\left[-\frac{E_a}{RT}\right] \tag{8}$$

where E_a is the activation energy, Ω is the normalized frequency factor, R is the gas law constant, T is the steady temperature (K), and t is the hold time (s). This can be recast into the typical Arrhenius relation,

$$\ln[t] = -\ln[\Omega] - \frac{E_a}{RT} \tag{9}$$

This approach was used to determine E_a and Ω for 90% and 10% retention minerals dated by the fission track method (Table 5).

Dodson (1973, 1979) showed that for a steady rate of cooling, one could identify an *effective closure temperature* T_c, which corresponds to the temperature at the time indicated by the cooling age measured for the thermochronometer (Fig. 12). We emphasize that T_c is only defined for the case of steady cooling through the PRZ, but this assumption is reasonable for many eroding mountain belts, given the narrow temperature range for the PRZ and the slow response of the thermal field to external changes. In contrast, this assumption will likely fail for areas affected by local igneous intrusions, hydrothermal circulation, or depositional burial.

Dodson (1973, 1979) estimated T_c using

$$\dot{T}(T_c) = \frac{-\Omega R T_c^2}{E_a} \exp\left[-\frac{E_a}{RT_c}\right] \tag{10}$$

where $\dot{T}(T_c) = (\partial T/\partial t)_{T=T_c}$ ($\dot{T} < 0$ for cooling), R is the gas law constant, and the normalized

Table 5. Parameters for FT partial retention zones.

Method (references)	*Retention Level*	E_a *(kJ mol*$^{-1}$*)*	Ω [4] *(s*$^{-1}$*)*
FT apatite[1] (average composition[2]) (Ketcham et al. 1999)	90%	127	2.67×10^5
	10%	161	1.55×10^7
FT Renfrew apatite[3] (low retentivity) (Ketcham et al. 1999)	90%	124	1.91×10^5
	10%	150	4.39×10^6
FT Tioga apatite[3] (high retentivity) (Ketcham et al. 1999)	90%	140	1.41×10^6
	10%	232	3.38×10^{10}
FT Durango apatite (Laslett et al. 1987; Green 1988)	90%	160	1.02×10^{12}
	10%	195	2.07×10^{12}
FT zircon[1] (natural, radiation damaged) (Brandon et al. 1998)	90%	225	2.62×10^{11}
	10%	221	1.24×10^8
FT zircon (no radiation damage) (Rahn et al. 2004; fanning model)	90%	272	5.66×10^{13}
	10%	339	5.66×10^{13}
FT zircon (Tagami et al. 1998; fanning model)	90%	231	1.09×10^{12}
	10%	359	1.02×10^{15}
FT zircon (Tagami et al. 1998; parallel model)	90%	297	5.94×10^{15}
	10%	297	1.51×10^{11}

Footnotes:

1) Recommended values for most geologic applications.
2) Average composition was taken from Table 4 in Carlson et al. (1999). Equation 6 in Carlson et al. (1999) was used to estimate $r_{mr0} = 0.810$ for this composition. PRZ parameters were then estimated using the HeFTy program (Ketcham 2005).
3) PRZ parameters were estimated from HeFTy and $r_{mr0} = 0.8464$ and 0.1398 for Renfrew and Tioga apatites, respectively, as reported in Ketcham et al. (1999).
4) Ω is measured directly for FT thermochronometers

frequency factor Ω and activation energy E_a are closure parameters, as defined in Tables 3 and 4. The cooling rate at T_c is given by $\dot{T} = (\partial T/\partial z)_{Tc}\,\dot{\varepsilon}(\tau_c)$, where $(\partial T/\partial z)_{T_c}$ is the thermal gradient at the closure isotherm, and $\dot{\varepsilon}(\tau_c)$ is the erosion rate at τ_c, the time of closure.

For He and Ar dating, $\Omega = 55D_0/a_s$, where a_s is the equivalent spherical radius for the diffusion domain. Tables 3 and 4 list typical values for a_s, but the user will need to judge if a more suitable value is appropriate given the specifics about what has been dated. Most of the minerals dated by He and Ar have isotropic diffusion properties, meaning that He and Ar diffuse at equal rates in all directions (muscovite and biotite are exceptions that are discussed below). Furthermore, the diffusion domains are commonly at the scale of the full mineral grain. The dated minerals may have anisotropic shapes. As an example, zircons and apatites tend to occur as elongate prisms. We can calculate an approximate equivalent spherical radius using

$$a_s \approx 3\frac{V}{A} \tag{11}$$

(Fechtig and Kalbitzer 1966; Meesters and Dunai 2002), where V and A refer to the volume and surface area of the mineral grain.

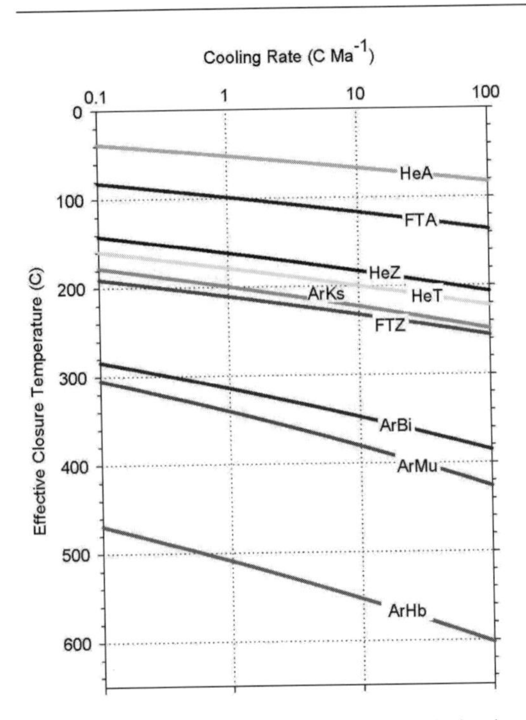

Figure 12. Effective closure temperatures calculated using CLOSURE.

Biotite and muscovite are anisotropic, with the fast direction of diffusion parallel to the basal plane, indicating cylindrical diffusion geometry. Equations are available to solve for cylindrical diffusion, but we have opted to approximate the solution by converting the cylindrical radius a_c of the mica grains into an equivalent spherical radius, where a_s = 1.5 a_c, and then using this radius with the spherical solution for the diffusion equations. These approximate scaling relationships are shown for the values under the label "Effective Dimensions of Diffusion Domain (micrometers)." These approximations are very good for calculating 90% retention and effective closure temperatures. They work less well for calculating the time-temperature conditions for 10% retention.

The Dodson equation can also be applied to the fission-track system (Dodson 1979). He recommended using E_a and Ω determined from the 50% retention isopleth from time-temperature heating experiments. Fission tracks contain a range of defects, created by the flight of the two energetic fragments created by the fission decay reaction of ^{238}U. This situation accounts for why the annealing process has a range of E_a values, which increase with increasing annealing of initial tracks. This observation is thought to mean that there is a range of activation energies needed to drive the diffusion involved in repairing this lattice damage (e.g., Ketcham et al. 1999). We do not know the size of the diffusion domain, which means that we cannot measure D_0. Nonetheless, we can measure Ω, which is all that is needed to use the Dodson closure equation. Dodson (1979) argued that the 50% retention isopleth provides the best average values for E_a and Ω, given that the cooling path for closure requires moving through the PRZ.

We have estimated these parameters from a range of FT annealing experiments (Table 5). We have compared the T_c values calculated with the Dodson equation with those estimated by more complex FT models, such HeFTy. In general, the Dodson estimates for T_c are within ~1 °C relative to those given by numerical models.

Methods for AGE2EDOT

AGE2EDOT estimates the cooling age for a thermochronometer that was exhumed by steady erosion at a constant rate (Fig. 13). The thermal field is represented by the steady-state solution for an infinite layer with a thickness L (km), a thermal diffusivity κ (km^2 Ma^{-1}), a uniform internal heat production H_T, a steady surface temperature T_s and an estimate of the near-surface thermal gradient for no erosion. These thermal parameters are usually estimated, at least in part, from heat flow studies. We use, as an example, thermal parameters for the active convergent orogen in the northern Apennines of Italy, where L ~ 30 km, κ ~ 27.4 km^2 Ma^{-1},

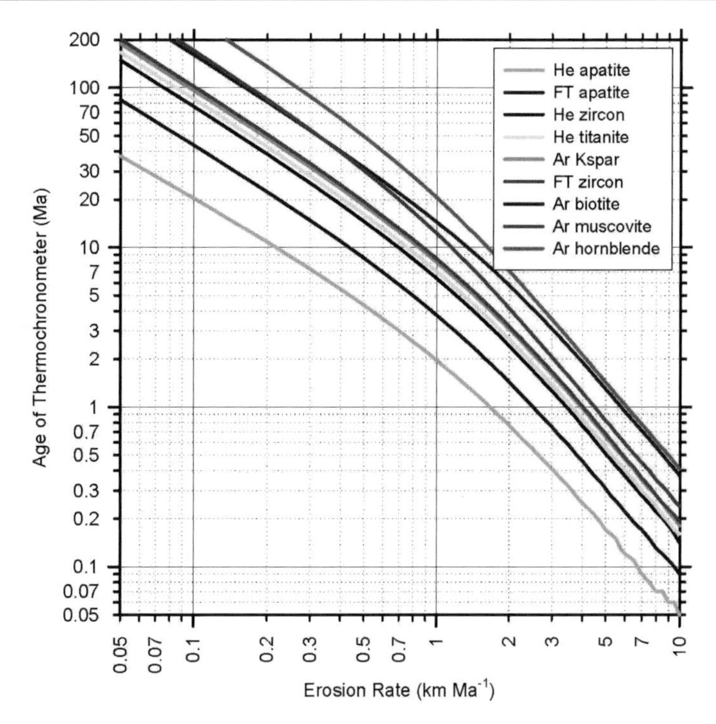

Figure 13. Age versus erosion rate for all of the major thermochronometers. Calculated using AGE2EDOT.

$H_T \sim 4.5$ °C Ma^{-1}, $T_s \sim 14$ °C, and the zero-erosion surface thermal gradient would be ~20 °C km^{-1}. The calculated basal temperature is 540 °C, which is held constant throughout the calculation. Material moves through the layer at a constant velocity u. One can envision that this situation simulates a steady-state orogen where underplating is occurring at the same rate as erosion, with $u = \dot{\varepsilon}$. The thickness of the orogen remains steady and the vertical velocity through the wedge would be approximately uniform and steady.

The thermal model provides a full description of the temperature and thermal gradient as a function of depth. The cooling rate is $\dot{T}(T_c) = \left(\partial T/\partial z\right)_{T_c} \dot{\varepsilon}$. Given the cooling rate and the temperature with respective to depth, we can use the Dodson equation to solve for T_c, and for the depth of closure z_c. The predicted cooling age is given by $z_c/\dot{\varepsilon}$.

AGE2EDOT gives a full prediction of how the cooling age for the specified thermochronometer will change as a function of erosion rate. Faster erosion causes isotherms, including the closure isotherm, to migrate closer to the surface. The steeper thermal gradient causes a faster rate of cooling and thus a greater T. The net effect is that the closure depth becomes shallower with faster erosion, but this effect is reduced by the response of the increase in T_c caused by faster cooling.

Brandon et al. (1998) provides more details about this calculation. Figure 13 is an example of the relationship between erosion rate and cooling ages for all of the major thermochronometers. The thermal parameters used are those discussed above for the northern Apennines. An example of the input data and results is given in the file: AGE2EDOT.output available online at the software download site associated with this chapter.

Methods for RESPTIME

RESPTIME calculates the migration of the closure isotherm due to an instantaneous change in erosion rate. The program is similar to AGE2EDOT but it uses a finite-difference algorithm to solve for the evolution of 1D thermal field in an infinite layer. Figure 14 shows plots of the response of all of the major thermochronometers to a instantaneous change from no erosion before 0 Ma to steady erosion at a rate of 1 km Ma^{-1}. Thermal parameters used are identical to the northern Apennines values used for the example for AGE2EDOT. The distribution package includes a sample output file from this example (see the file called RESPTME.output). Note that L was increased to 50 km for the Ar muscovite and Ar hornblende calculations, in order to ensure that T_c remained within the layer for these high-temperature thermochronometers.

The motion of the closure isotherm is represented in Figure 14 by its normalized velocity, which is defined by the vertical velocity of the isotherm divided by the erosion rate. A normalized velocity of zero means that the closure isotherm has reached a steady-state position; whereas a normalized velocity of one means that the isotherm is moving upward at the same rate as the rock. There would be no cooling at this stage.

Figure 14 shows that the normalized velocity for the closure isotherm for the He thermochronometer slows down to <10% within 2.5 Ma following the start of fast erosion. In contrast, it takes 16 Ma for the Ar muscovite system to reach the 10% level. This example highlights the importance for using low-temperature thermochronometers for measuring erosion rates.

TASC: COOLING ONSET AGES AND EVENT TIMING IN NATURAL SAMPLES FROM FISSION TRACK LENGTH DATA

Contributors: D. Belton, B. Kohn, A. Gleadow (dxbelton@unimelb.edu.au)

Track age spectrum calculator (TASC) is an Microsoft Excel® worksheet built for Microsoft Windows® XP that provides a means of gaining additional information from the raw track length and apparent age data routinely collected for fission track samples. By recalculating the track density to incorporate the length-dependent probabilities, the user can determine the age of the oldest measured track—the "cooling onset" age—which is effectively the time the sample passed through the base of the partial annealing zone (PAZ) and began to retain tracks. Quantifying this age permits each individual track in the traditional length histogram to be allocated an equivalent age, thus allowing it to be recast as a "track age spectrum". The calculations are independent of chemistry and mineralogy and the output incorporates uncertainties reflecting age error and anisotropy. While the time information is quantitative, the temperature information is largely qualitative. The "cooling onset age" and indicators of cooling style and timing extracted by this approach provide a very useful guide to the thermal history before inverse modeling of the sample is attempted. TASC enables robust histories to be rapidly developed, particularly in shield areas where stratigraphic and structural control may be minimal or absent. Other applications include: (1) rapid identification of timing in the simultaneous cooling of vertical profiles—the timing of which is frequently masked or ambiguous in those samples residing in or above the palaeo-PAZ; (2) use of the "cooling onset age" to contour maps of total exhumation (where suitably calibrated); (3) generation of "event" spectra (analogous to detrital zircon age spectra) from a regional series of samples. The method may also prove of significant value in the study of other minerals such as sphene and zircon, where annealing models are still evolving.

Background to the TASC program

The TASC approach is based on the logic of Laslett et al. (1982) who considered the probability of a track intersecting an arbitrary surface through a grain. They concluded that the

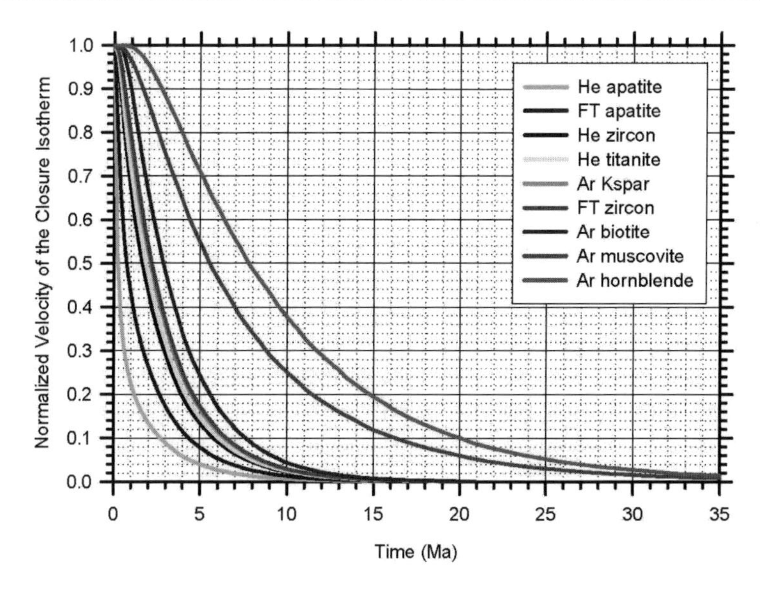

Figure 14. Response time of closure isotherms for the listed thermochronometers due to an instantaneous increase in erosion rate at time 0 Ma, from no erosion to 1 km Ma^{-1}. Calculated using RESPTIME.

sampling or intersection probability is directly proportional to the track length (i.e., a linear length-density relationship). Their argument was in essence, that a 16 μm track is twice as likely to intersect a polished surface, and thus be sampled in density measurements, as an 8 μm track and 4 times as likely to be sampled as a 4 μm track. Laslett et al. (1982) also provided a mathematically rigorous assessment of the treatment for the case of TINTS and TINCLES (Lal et al. 1969) routinely measured in samples.

If, for arguments sake, fission tracks in a given sample have an initial track length of ~16 μm, then the inverse of the Laslett et al. (1982) argument can be restated as follows: For a given "true" or debiased track-length distribution, the "true" number of 8 μm tracks that would have intersected the surface at formation length, will be double the observed amount, since because of annealing, there is only a 50% probability of measuring the shortened track. Similarly, to adjust for the observation probability of only 25%, the "true" number of 4 μm tracks will be four times the observed amount. This correction is equivalent to extending all tracks in a variable length distribution back to their initial formation length (Fig. 15). Once the true probability is determined for each measured length in the distribution (or each bin interval if only binned data are available), the recalculated total density of tracks that would be expected to intersect the surface is used to modify the traditional age equation (Fleischer et al. 1975).

Having determined the "cooling onset age", it is a simple matter to rescale the traditional length histogram with the relationship between the number of track lengths and the total number of "rescaled tracks". Each bin is allocated its appropriate number of rescaled tracks based on their intersection probability, which allows the boundary of each bin to be defined by a specific time. In order to allocate some measure of uncertainty (e.g., due to annealing anisotropy) to the result, the entire TASC procedure is repeated at one standard deviation (±) using the standard deviations recorded for Durango apatite by Green et al. (1986). This uncertainty gives a plausible, though not necessarily statistically robust, estimate of the uncertainty of the bin ages.

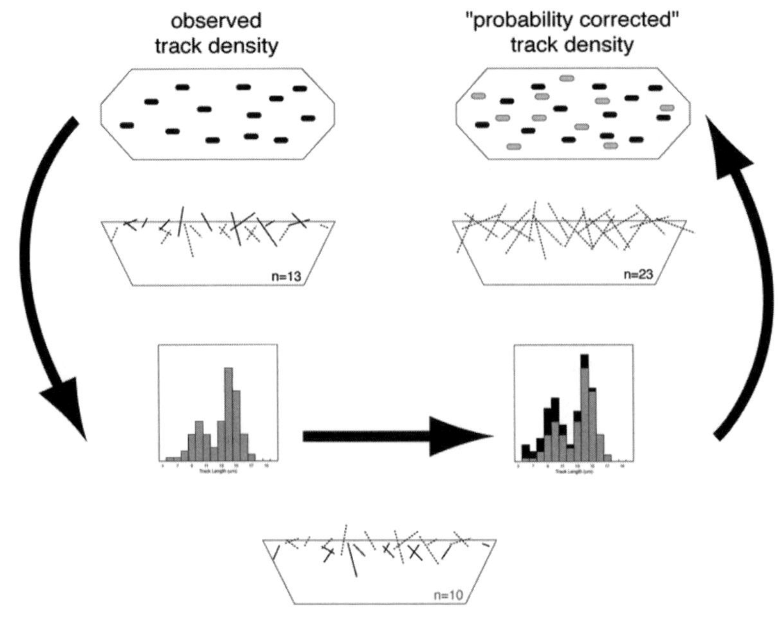

Figure 15. Schematic of the TASC calculation. Observed track density is a function of the intersection probability of the mixture of short and long tracks. When the density is adjusted for the expected intersections that would occur if all tracks retained their initial formation length, then the cooling onset age is revealed. This age reflects the point in time at which fission tracks began to be retained, rather than completely annealed, (nominally 110 °C) as the sample moved up through the crustal isotherms.

Applications of the TASC program

Recasting the traditional length histogram produces the "track age spectrum" which retains all the original thermal history information from the length histogram (Figs. 16 and 17) including, for example, the skewed distribution of an accelerated cooling or the bimodality of a reheating event (Gleadow et al. 1986). However, the track age spectrum also enables the timing of these events to be readily estimated, and in some cases gives an indication of the "severity" of a thermal perturbation.

When applied to vertical profiles, TASC analysis enables rapid identification of events recorded in samples both above and below a "break in slope." Estimates of exhumation rates, particularly above a "break in slope", can be derived from "onset cooling ages." Semi-quantitative timing information extracted from the track age spectra of grouped samples (e.g., vertical profiles or localized tectonic blocks) can be effectively and concisely summarized on the event spectrum in much the same way as detrital zircon fission track age populations are illustrated in probability plots (Brandon 1992). This enables the user to readily identify important thermal events at one site and is a powerful means of evaluating the regional thermal histories. Because the TASC calculation is independent of chemistry and mineralogy, it has potential application in the study of other minerals such as zircon, where annealing models are still evolving.

By itself, the "cooling onset" age derived from TASC can be used as the basis of extended regional contour maps giving a large-scale overview of regional cooling histories. In contrast

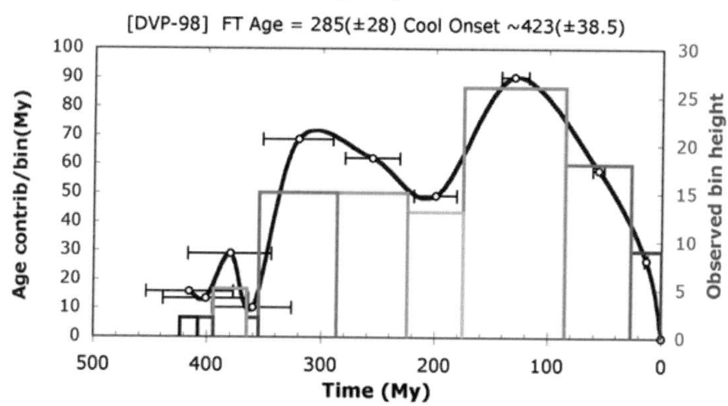

Figure 16. Track age spectrum for a cratonic sample with the complex thermal history (illustrated in Fig. 17). The spectrum retains the typical bimodal distribution seen in the traditional length histogram and indicative of a reheated sample. The TASC results are predict the style of cooling independently modeled using inverse methods (Fig. 17).

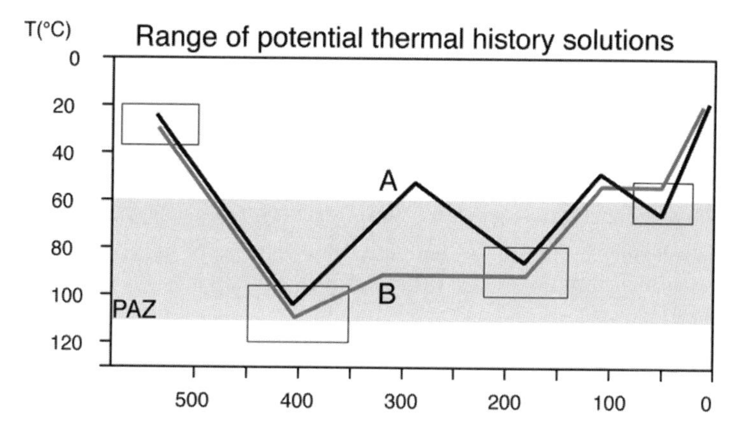

Figure 17. Range of thermal history solutions (curves A and B) for a sample from central Australia (Belton et al. 2004). The thermal histories were determined independently of the TASC solution (see Fig. 16) and are based on geological constraints and inverse modeling (MonteTrax; Gallagher 1995).

to traditional maps relying on apparent fission track ages, the cooling onset age provides an unambiguous indicator of the time that a sample either (1) entered the base of the PAZ or (2) began to retain tracks again following a thermal reset (e.g., Fig. 17). With some additional assumptions regarding composition and cooling rates, this information can be inverted to derive exhumation estimates for large areas.

The TASC approach should prove to be a powerful precursor to current well-established inverse modeling techniques such as MonteTrax (Gallagher 1995) and AFTSolve (Ketcham et al. 2000). It cannot replace these methods but will complement them, so that in many cases the additional information garnered from TASC will enable robust model histories to be rapidly developed.

Using the TASC program

The program is provided as an Microsoft Excel® worksheet to suit the Microsoft Windows® XP platform. No installation is required, simply copy the file to your hard drive, click on the file and it will open under Excel®. It is recommended that you create a permanent folder for the TASC file *before* you start it for the first time. This ensures that the macros in the work sheet remain correctly linked to the TASC toolbar and it provides a single location for the output text files as well. Users may have to alter security permissions on their computers to enable the Visual Basic® macros in this program to function.

TASC controls

TASC has its own tool bar and each of the control buttons is replicated in the worksheet. These controls permit the user to:

1. Clear previous input data prior to the next calculation.
2. Adjust spectrum *x*-axis. This may need to be done if the apparent fission track age input is altered during calculation.
3. Choose color plots for easy comparison between the length histogram and track age spectra or select black and white to print the plots directly from Excel®.
4. Export output as a text file ("sample name spectra") containing the *x* and *y* ranges for plotting the spectra in other software packages.
5. Select from a choice of five quick examples showing the spectrum for different thermal history styles. These examples are based on both synthetic and real data. The user manual provides more detailed information on a number of complex examples with case histories.

TASC inputs

The following inputs are required for TASC to perform its calculation:

1. Binned lengths – these can be entered individually or pasted. Note, only lengths greater than 5 μm are used in the calculation...OR...
2. Individual track lengths – again these can be entered one at a time however most users will prefer to cut and paste. TASC will accept up to a maximum of 130 individual tracks.
3. Apparent fission track age in Ma
4. Error on the apparent age ($\pm 1\ \sigma$) in Ma
5. Name of the sample.
6. Initial (or maximum) track length – closely approximating the formation length of fission tracks in the sample. The default is 16.3 μm, however in natural samples you may wish to use the longest individual length observed in the sample.

TASC outputs

Screen outputs from TASC are:

1. Cooling Onset Age – which marks the age of the oldest retained fission track.
2. Fission track density multiplier – which is used to adjust the traditional apparent fission track age.
3. Track age spectrum, consisting of age calibrated histogram bins (notice the bins now have differing widths) with an additional smoothed curve indicating the central value for each bin and error bars for each bin. This curve is useful for rapidly interpreting the results.

4. Traditional length histogram.
5. A text file is output on request and contains the input age, error and length bins as well as all the output data allowing for plotting in other packages.

DECOMP: FORWARD MODELING AGE EVOLUTION OF (U-TH)/HE AGES

Contributor: T. Dunai (tibor.dunai@falw.vu.nl)

The program DECOMP is designed to calculate age evolution curves for U-Th-He thermochronology (Forward modeling). It is an amendment to the papers of Meesters and Dunai (2002a,b) and is described in Dunai (2005). DECOMP computes age evolution curves for spherical symmetry. It can also be used as an accurate approximation for other geometries if spheres of identical surface to volume ratio and properly transposed zonation of parent nuclides are used. Any results obtained with this program must be properly referenced if they are used in publications and presentations. Proper referencing is referring to *both* Meesters and Dunai (2002, part II) *and* the program (DECOMP; by A. Bikker, A.G.C.A. Meesters and T.J. Dunai).

The target audiences for DECOMP are beginners and experienced practitioners in thermochronology. The functionality of the software lends itself to several applications.

1. Qualitative assessment of how changes of sample parameters (diffusion parameters, surface/volume ratios, emission distance) and model temperature histories affect ages. This aspect is particularly useful for teaching/learning what the most sensitive parameters for a certain system are.
2. Quantitative forward modeling of any time temperature history.
3. Qualitative assessment of the effect of parent nuclide zonation.
4. Although DECOMP was originally designed for forward modeling of (U-Th)/He ages, forward modeling of any thermochronological system that is governed by volume diffusion, and of which the diffusion parameters (D_0, E_a) of the radionuclide are known, is possible.

How to use DECOMP

DECOMP provides a user-friendly interface that allows repeated calculations of (U-Th)/He ages (e.g., Meesters and Dunai 2002). The results can be exported to standard spreadsheet programs. On first starting the program the user is asked to fill in, step by step, the parameters necessary for a computation. In most cases default values or exemplary temperature histories are supplied, which of course can be changed. Both the left and right buttons of the mouse can be used to have access to options (such as adding and/or editing points to the temperature histories). On finishing this introduction the actual window of DECOMP will appear. After the use quits DECOMP for the first time the introduction will not reappear on restarting DECOMP. Starting DECOMP is straightforward: double click on the icon of the executable file (DECOMP.EXE).

An example of the DECOMP program interface is shown in this volume (e.g., Fig. 3; Dunai 2005). There are several input options associated with this interface required for using the program:

1. Button panel
   - *Save* and *open* geometric parameters, temperature histories, constants and annotations.

- *Quit* the program.
- *Recalculate* the age evolution diagram.
- *Edit constants*: number of eigenvalues used for computation; activation energy (the non-SI unit [cal/mol] is used here as currently most values are still published in [cal/mol], for conversion 1 joule = 0.239 cal); pre-exponential factor D_0 [cm^2/s].

2. Input parameters describing the geometry of a sample:
   - *Sphere radius:* Radius (in microns) of a sphere of identical surface to volume ratio as the sample.
   - *Outer and inner zone radius:* Gives the outer and inner radius (in microns) of the zone containing the parent nuclides. The shape of the zone containing the parent nuclides is visualized by the red zone in the circle to the right of the input field that is changing simultaneously while changing values. Samples with a homogenous parent nuclide distribution have outer zone radius equal to the sphere radius and an inner radius of 0 microns.
   - *Emission distance:* Emission distance (in microns) of alpha particles.

Temperature history plot

The temperature and real time at the position of the tip of the cursor are indicated in the frame at the top right of the diagram. Note that the real time represents time elapsed since beginning of the model simulation. When using this plot, the following operations can be performed using the indicated mouse buttons:

- Right mouse button:

  copy the numeric values of temperature history (comma-delimited table) to be pasted (e.g., in a spreadsheet program).

  add and delete points to/from the temperature history

  table edit: edit the numeric values of existing points of the temperature history

- Left mouse button:

  drag axis to scale

  drag existing points

Age evolution plot

Initially the age-evolution diagram is blank. After pressing the recalculation button the age evolution is calculated for the current temperature history. The numeric values curve can be copied (right mouse button) and pasted into a spreadsheet program (table comma delimited). The calculated age and real time at the position of the tip of the cursor are given in the frame at the top right of the diagram. Note, in this plot the calculated age represents the age as it would be calculated from U, Th and He concentrations in a sample.

4DTHERM: THERMAL AND EXHUMATION HISTORY OF INTRUSIONS

Contributor: F. Fu (qifu6346@mail.usyd.edu.au)

4DTHERM v.1.2 is an inverse thermal modeling program. It uses a 2D explicit finite difference solution which addresses conduction cooling, latent heat of crystallization/fusion, convection within magma bodies, and hydrothermal circulation induced by magma intrusion as well as exhumation and erosion processes. 4DTHERM was implemented in Java programming language and is platform-independent. It has been tested in Windows® ME, XP and Linux/Unix OS, and should run in Mac® OS X and other operating systems. It has a interactive graphical user interface (GUI) for constructing geologic units, inputting and modifying computational parameters, editing model scenarios, and controlling modeling procedures. It provides graphic windows (Fig. 18) to observe the cooling process in a 2D cross section and to view the plotted thermal histories of pre-defined positions. Final modeling results can be saved in formatted text files whereas step-by-step running status is displayed in the status board.

4DTHERM applications

The primary application of this model is to derive thermal and exhumation histories of igneous intrusions directly from multiple thermochronometers. 4DTHERM quantifies a number of parameters related to the dynamic processes of magmatic-hydrothermal cooling, timing and duration of hydrothermal activity, and denudation histories of igneous intrusions and related mineralization (McInnes et al. 2005a,b). It also computes both cooling and exhumation histories of igneous intrusions. A brief overview of the processes simulated is as follows:

Modeling of magmatic-hydrothermal cooling. 4DTHERM simulates the conductive and convective cooling processes of igneous bodies. It computes the distributions and variations of temperature in both igneous bodies and country rocks throughout the whole cooling process. It records the time and duration of magmatic-hydrothermal activities induced by intrusions.

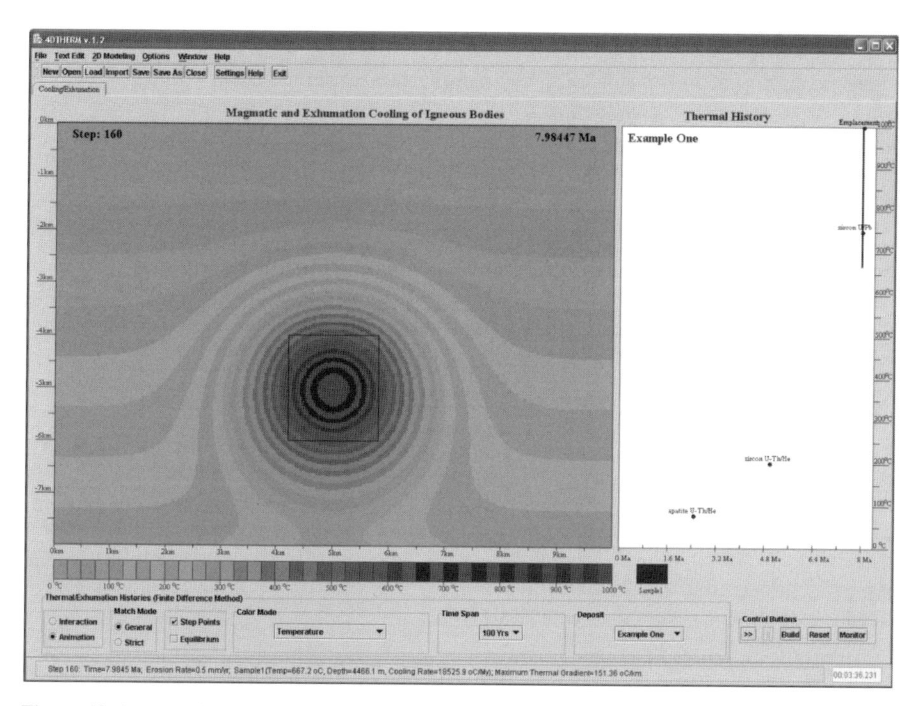

Figure 18. A screenshot of 4DTHERM v.1.2. The large window on the left displays the distribution of temperature in vertical cross section with an isotherm of 25 °C for each color band, and the black rectangle in the middle of the window indicates the position of a cooling igneous body. The white window on the right is plotting the cooling curve for a sample from the intrusion and the three small dots are the input age data obtained from sample of the intrusion or country rock used to constrain the cooling curve. Below the two windows are the temperature legend, control panel and status board, respectvily.

It also records the cooling histories and calculates the cooling rates for some pre-defined positions (e.g., sample positions in igneous bodies and/or country rocks).

The algorithm behind the model solves a system of equations of conservation of mass, momentum, and energy. The mathematic formulations for an incompressible fluid in a porous medium can be expressed as (Cathles 1977; Parmentier and Spooner 1978; Turcotte and Schubert 2002):

$$\nabla\left(\rho_f U\right)=0 \tag{12}$$

$$U=-\frac{k}{\mu_f}\left(\nabla P+\rho_f g\right) \tag{13}$$

$$\rho_b C_b \frac{\partial T}{\partial t}=\lambda_b \nabla^2 T-\left(\rho_f C_f U T\right)+Q \tag{14}$$

where T is temperature, t time, density, C specific heat, conductivity, U Darcy velocity, Q internal heat source, P pressure, k permeability, μ dynamic viscosity and g the acceleration of gravity, and subscript f denotes fluid property, and b bulk (rock + fluid) property. Equation (12) represents conservation of mass flow of the fluid, Equation (13) is conservation of momentum expressed by Darcy's law. Equation (14) represents conservation of energy where rock and fluid are assumed to be in local thermal equilibrium at temperature T. The system of equations is solved using an explicit finite difference scheme.

Reconstruction of exhumation histories of igneous bodies. The reconstruction of exhumation histories of igneous bodies is achieved by determining the emplacement states of igneous intrusions and calculating the erosion rates for both the country rocks and igneous rocks. The calculation of erosion rates is mainly based on the input thermochronometer data, whereas the emplacement depth and eroded thickness of the igneous body are determined through inverse modeling using an iterative "trial and error" strategy. For detailed algorithms and an example of the application of 4DTHERM see Appendix II of McInnes et al. (2005b).

Currently, 4DTHERM can support the modeling of regular and irregular shapes of igneous bodies and of multiple intrusion events. It is possible to produce a unique solution to the emplacement depth and eroded thickness of igneous units through the use of multiple geochronology datasets (i.e., each sample containing three or more thermochronology ages) from multiple sample locations in the same igneous unit.

4DTHERM inputs

Data inputs to the model include: (1) size and shape of igneous and country rock units; (2) residual sizes, position, and properties of igneous bodies; (3) age data and corresponding nominal closure temperatures, and (4) sample position. The construction of geologic bodies (including country rocks and igneous rocks) can be done either through the "Geobody Building" panel (for regular shapes only), or by drawing directly on the model using the mouse (for both regular and irregular shapes) (Fig. 19). Alternatively, these input data can be loaded separately from formatted text files.

4DTHERM provides a default geologic background initialized with a set of initial parameters (e.g., surface temperature, thermal gradient, basal heat flow, lithology, thermal conductivity, specific heat, and density, etc.), which can be edited using the "Parameter Settings" panel.

4DTHERM outputs

The outputs available after each successful test include: (1) digitized cooling curves with highlighted key points (Fig. 20); (2) parameters calculated during the modeling run such as

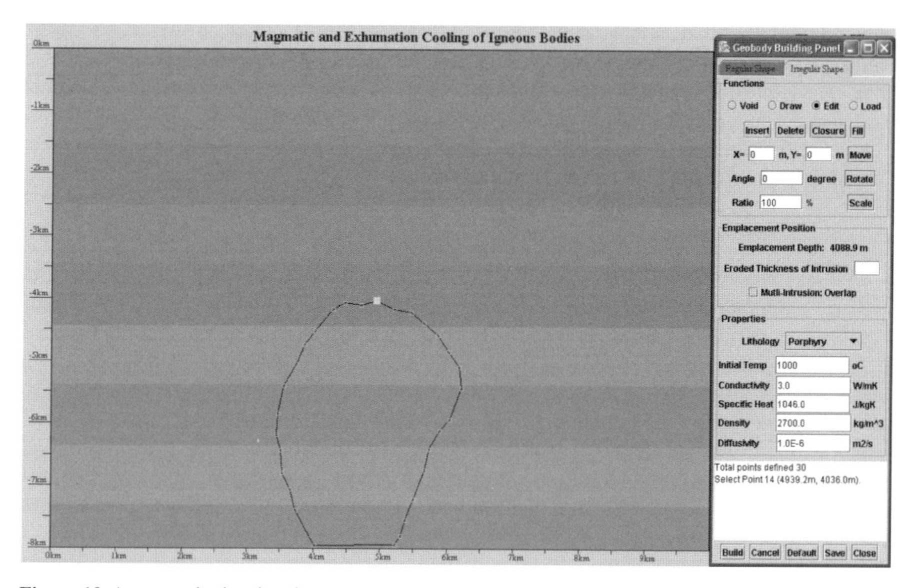

Figure 19. An example showing the construction of an igneous body by mouse drawing. The igneous body drawn on the cross section can be resized, scaled and rotated through the Geobody Building Panel on the right. The properties of the igneous body can be assigned via the input text fields in the panel.

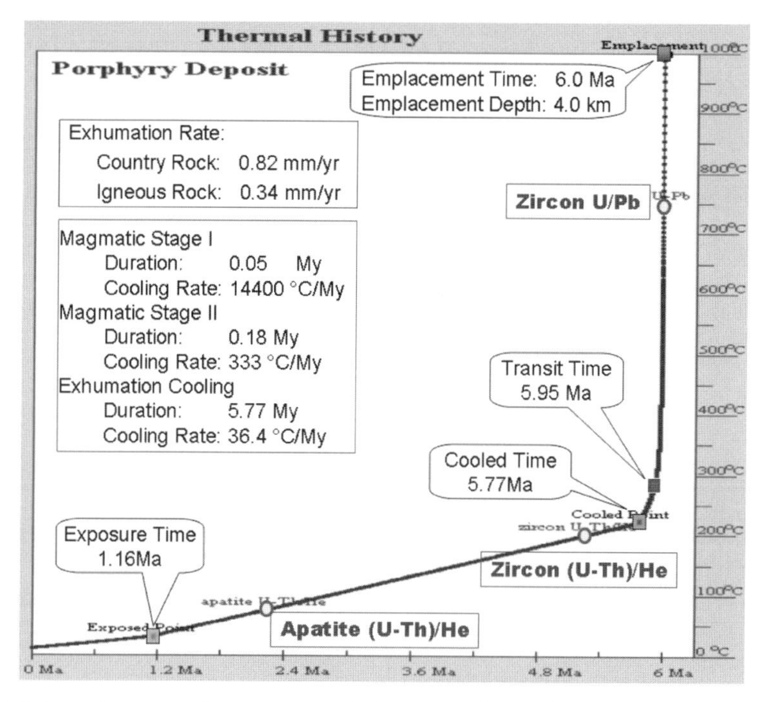

Figure 20. Example showing the output available in a successful modeling case. The modeled cooling curve for an igneous body matches all three age data (small red circles). Some critical time points are highlighted by small colored rectangles. Cooling rates, exhumation rates and many other parameters can be calculated based on the digitized cooling curve.

depth and time of emplacement, cooling rates of different stages, exhumation/erosion rates for country rocks and igneous rocks, eroded thicknesses of intrusions, and the exposure time if the sample was exhumed to the surface; (3) detailed step-by-step modeling states. These data can be saved in formatted text files for further examination. In addition, the instant visualization of temperature distribution in the 2D model at each time step is another source of output and can be saved using "Print Screen" button on the keyboard.

CONCLUDING REMARKS

The computational tools presented in this chapter (Table 1) are useful for simulating processes on a variety of temporal and spatial scales. Crustal and magmatic thermal processes can be simulated with the 4DTHERM RESPTIME, AGE2EDOT, and TERRA programs. Simulation of fission track annealing and analysis of track length distributions are possible with HeFTy, FTIndex, TASC, BINOMFIT, CLOSURE and TERRA programs. And finally, prediction of (U-Th)/He ages is possible with HeFTy, DECOMP, and TERRA programs. Future applications of thermochronometer data to different geologic problems will hopefully benefit from the continued development and free distribution of programs such as these. New directions of thermochronometer related software development will hopefully focus on not only implementing new and improved kinetic models for different thermochronometer systems, but also on relating different geologic processes (e.g., landscape evolution, shear heating on faults, etc.) to testable scenarios of predicted thermochronometer ages.

ACKNOWLEDGMENTS

Development of the TERRA program was made possible through support to TAE from the US National Science Foundation (EAR 0409289, 0309779, and 0196414) and the University of Michigan Undergraduate Research Opportunity Program. The TASC program was made possible through generous support provided by the Australian Research Council (ARC) and the Australian Institute for Nuclear Science and Engineering (AINSE). Development of BINOMFIT was supported by NSF grant OPP-9911925 to MTB. Igor Boreyko of the Institute of the Lithosphere of Marginal Seas (Russian Academy of Sciences, Moscow) did the initial conversion of BINOMFIT into Visual Basic® with the assistance of Mark Brandon and Alex Soloviev. Development of CLOSURE, AGE2EDOT, and RESPTIME were supported by grant EAR-0208652 to MTB. Jason Barnes and Greg Stock are thanked for comments that lead to the improvement of this manuscript.

REFERENCES

Batt GE, Brandon MT, Farley KA,Roden-Tice MK (2001) Tectonic synthesis of the Olympic Mountains segment of the Cascadia wedge, using two-dimensional thermal and kinematic modeling of thermochronological ages. J Geophys Res 106:731-746

Belton DX, Brown RW, Kohn BP, Fink D, Farley KA (2004) Quantitative resolution of the debate over antiquity of the central Australian landscape: implications for the tectonic and geomorphic stability of cratonic interiors. Earth Planet Sci Lett 219:21-34

Brandon MT (1992) Decomposition of fission-track grain-age distributions. Am J Sci 292:535-564.

Brandon MT (1996) Probability density plots for fission-track grain age distributions. Radiat Meas 26:663-676

Brandon MT, Roden-Tice MK, Garver JI (1998) Late Cenozoic exhumation of the Cascadia accretionary wedge in the Olympic Mountains, northwest Washington State. GSA Bulletin 110:985-1009

Brandon MT, Vance JA (1992) Fission-track ages of detrital zircon grains: implications for the tectonic evolution of the Cenozoic Olympic subduction complex. Am J Sci 292:565-636

Carlson WD (1990) Mechanisms and kinetics of apatite fission-track annealing. Am Mineral 75:1120-1139

Carlson WD, Donelick RA, Ketcham RA, (1999) Variability of apatite fission-track annealing kinetics. I. Experimental results. Am Mineral 84:1213-1223

Cathles LM (1977) An analysis of the cooling of intrusives by ground water convection which includes boiling: Econ Geol 72:804-826

Crowley KD, Cameron M, Schaefer RL (1991) Experimental studies of annealing etched fission tracks in fluorapatite. Geochim Cosmochim Acta 55:1449-1465

Dodson MH (1973) Closure temperature in cooling geochronological and petrological systems. Contrib Mineral Petrol 40:259-274

Dodson MH (1976) Kinetic processes and thermal history of slowly cooling solids. Nature 259: 551-553

Dodson MH (1979) Theory of cooling ages. *In:* Lectures in Isotope Geology. Jager E, Hunziker JC (eds), Springer-Verlag, New York, p. 194-206

Dunai TJ (2005) Forward modeling and interpretation of (U-Th)/He ages. Rev Mineral Geochem 58:259-274

Ehlers T, Farley K (2003) Apatite (U-Th)/He thermochronometry; methods and applications to problems in tectonic and surface processes. Earth Planet Sci Lett 206(1-2):1-14

Ehlers TA, Armstrong PA, Chapman D (2001) Normal fault thermal regimes and the interpretation of low temperature thermochronometers. Phys Earth Planet Interiors 126:179-194

Ehlers TA, Willett SD, Armstrong PA, Chapman DS (2003) Exhumation of the central Wasatch Mountains, Utah. 2. Thermokinematic model of exhumation, erosion, and thermochronometer interpretation. J Geophys Res 108(B3):2173, doi:10.1029/2001JB001723

Farley K (2000) Helium diffusion from apatite; general behavior as illustrated by Durango fluorapatite. J Geophys Res 105(B2):2903-2914

Fechtig H, Kalbitzer S (1966) The diffusion of argon in potassium-bearing solids. *In:* Potassium Argon Dating. Schaeffer OA, Zähringer J (eds), Springer-Verlag, New York, p. 68-107

Fleischer RL, Price PB, Walker RM (1975) Nuclear Tracks in Solids: Principles and Applications. University of California, Berkeley

Foland KA (1994) Argon diffusion in feldspars. *In:* Feldspars and Their Reactions. Parsons I (ed), Kluwer, Dordrecht, p. 415-447

Fuller CW (2002) Thermochronometry and Thermomechanical Modeling of the Taiwan Orogen. MS Thesis, University of Washington, Seattle, Washington

Galbraith RF (1981) On statistical models for fission-track counts. J Math Geology 13:471-478

Galbraith RF, Green PF (1990) Estimating the component ages in a finite mixture. Nuclear Tracks Radiat Meas 17:197-206

Galbraith RF, Laslett GM (1993) Statistical models for mixed fission track ages. Nuclear Tracks Radiat Meas 21:459-470

Gallagher K (1995) Evolving temperature histories from apatite fission-track data. Earth Planet Sci Lett 136: 421-435

Garver JI, Brandon MT, Roden TMK, Kamp PJJ (1999) Exhumation history of orogenic highlands determined by detrital fission-track thermochronology. *In:* Exhumation processes; Normal Faulting, Ductile Flow and Erosion. Geological Society Special Publications, Vol. 154. Ring U, Brandon MT, Lister GS, Willett SD (eds) Geological Society of London, London, p. 283-304

Garver JI, Soloviev AV, Bullen ME, Brandon MT (2000) Towards a more complete record of magmatism and exhumation in continental arcs using detrital fission track thermochronometry: Phys Chem Earth, Part A 25:565-570

Gleadow AJW, Duddy IR, Green PF, Lovering JF (1986) Confined fission track lengths in apatite—a diagnostic tool for thermal history analysis. Contrib Mineral Petrol 94:405-415

Green PF, Duddy IR, Gleadow AJW, Tingate PR, Laslett GM (1986) Thermal annealing of fission tracks in apatite. 1. A qualitative description. Chem Geol Isot Geosci Sect 59: 237-253

Green PF (1988) The relationship between track shortening and fission-track age reduction in apatite: combined influences of inherent instability, annealing anisotropy, and length bias and system calibration: Earth Planet Sci Lett 89:335-352

Grove M, Harrison TM (1996) $^{40}Ar^*$ diffusion in Fe-rich biotite: Am Mineral 81:940-951

Hames WE, Bowring SA (1994) An empirical evaluation of the argon diffusion geometry in muscovite: Earth Planet Sci Lett 124:161-167

Harrison TM (1981) Diffusion of ^{40}Ar in hornblende. Contrib Mineral Petrol 78:324-331

Ketcham RA, Donelick RA, Donelick MB (2000) AFTSolve: A program for multi-kinetic modeling of apatite fission-track data. Geol Mat Res 2:(electronic)

Ketcham RA, Donelick RA, Carlson WD (1999) Variability of apatite fission-track annealing kinetics III: Extrapolation to geological time scales. Am Mineral 84:1235-1255

Ketcham RA (2005) Forward and inverse modeling of low-temperature thermochronometry data. Rev Mineral Geochem 58:275-314

Lal D, Rajan RS, Tamhane AS (1969) Chemical composition of nuclei of Z > 22 in cosmic rays using meteoric minerals as detectors. Nature 221:33-37

Laslett GM, Galbraith RF (1996) Statistical modelling of thermal annealing of fission tracks in apatite. Geochim Cosmochim Acta 60:5117-5131

Laslett GM, Green PF, Duddy IR, Gleadow AJW (1987) Thermal annealing of fission tracks in apatite 2. A quantitative analysis. Chem Geol 65:1-13

Laslett GM, Kendall WS, Gleadow AJW, Duddy IR (1982) Bias in measurement of fission-track length measurements. Nucl Tracks 6:79-85

Mancktelow N, Grasemann B (1997) Time-dependent effects of heat advection and topography on cooling histories during erosion. Tectonophysics 270(3-4):167-195

McDougall I, Harrison TM (1999) Geochronology and Thermochronology by the $^{40}Ar/^{39}Ar$ Method, Second Edition. Oxford University Press, Oxford

McInnes BIA, Evans NJ, Fu FQ, Garwin S, Belousova E, Griffin WL, Bertens A, Sukarna D, Permanadewi S, Andrew RL, Deckart K (2005a) Thermal history analysis of select Chilean, Indonesian and Iranian porphyry Cu-Mo-Au deposits. *In:* Superporphyry Copper and Gold Deposits – A Global Perspective. Porter TM (ed), PGC Publishing, Adelaide, p. 27-42

McInnes BIA, Evans NJ, Fu FQ, Garwin S (2005b), Application of thermochronology to hydrothermal ore deposits. Rev Mineral Geochem 58:467-498

Meesters AGCA, Dunai TJ (2002a) Solving the production-diffusion equation for finite diffusion domains of various shapes. I. Implications for low-temperature (U-Th)/He thermochronometry. Chem Geol 186/3-4: 337-348

Meesters AGCA, Dunai TJ (2002b) Solving the production-diffusion equation for finite diffusion domains of various shapes. II. Application to cases with alpha-ejection and non-homogenous distribution of the source. Chem Geol 186/3-4:351-369

O'Sullivan PB, Parrish RR (1995) The importance of apatite composition and single-grain ages when interpreting fission track data from plutonic rocks: a case study from the Coast Ranges, British Columbia. Earth Planet Sci Lett 132:213-224

Parmentier EM, Spooner ETC (1978) A theoretical study of hydrothermal convection and the origin of the ophiolitic sulphide ore deposits of Cyprus. Earth Planet Sci Lett 40:33-44

Rahn MK, Brandon MT, Batt GE, Garver JI (2004) A zero-damage model for fission-track annealing in zircon: Am Mineral 89:473-484

Reiners PW, Farley KA (1999) Helium diffusion and (U-Th)/He thermochronometry of titanite. Geochim Cosmochim Acta 63:3845-3859

Reiners PW, Spell TL, Nicolescu S, Zanetti KA (2004) Zircon (U-Th)/He thermochronometry: He diffusion and comparisons with $^{40}Ar/^{39}Ar$ dating. Geochim Cosmochim Acta 68:1857-1887

Robbins GA (1972) Radiogenic argon diffusion in muscovite under hydrothermal conditions. MS thesis, Brown University, Providence, Rhode Island

Sneyd AD (1984) A computer program for calculating exact confidence intervals for age in fission-track dating. Computers Geosci 10:339-345

Stewart RJ, Brandon MT (2004) Detrital zircon fission-track ages for the "Hoh Formation": Implications for late Cenozoic evolution of the Cascadia subduction wedge. Geol Soc Am Bull 116:60-75

Sweeney JJ, Burnham AK (1990) Evaluation of a simple model of vitrinite reflectance based on chemical kinetics. Am Assoc Petrol Geol Bull 74:1559-1670

Tagami T, Galbraith RF, Yamada R, Laslett GM (1998) Revised annealing kinetics of fission tracks in zircon and geologic implications. *In:* Advances in Fission-Track Geochronology. Van den Haute P, De Corte F (eds) Kluwer Academic Publishers, Dordrecht, p. 99-112

Turcotte DL, Schubert G (2002) Geodynamics, 2nd ed. Cambridge, New York

Vrolijk P, Donelick RA, Queng J, Cloos M (1992) Testing models of fission track annealing in apatite in a simple thermal setting: site 800, leg 129. *In*: Proceedings of the Ocean Drilling Program, Scientific Results. Larson RL, Lancelot Y (eds)129:169-176

Wolf RA, Farley KA, Kass DM (1998) Modeling of the temperature sensitivity of the apatite (U-Th)/ He thermochronometer. Chem Geol 148:105-114